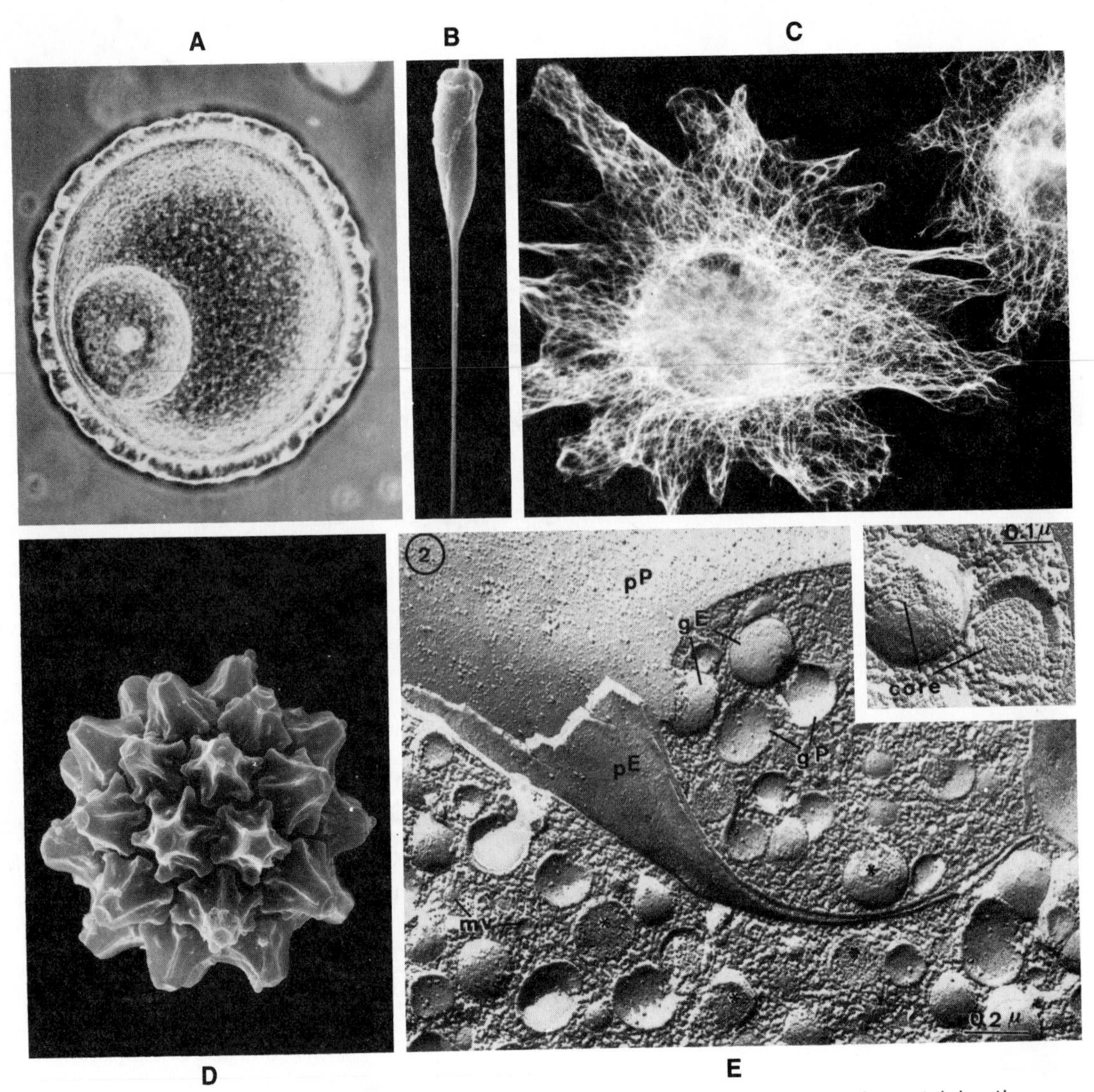

A, Egg of the West Coast seastar *Patiria miniata.* Note the large germinal vesicle containing the nuclear elements and the heavy protective membrane over the plasma membrane. (Phase optics, ×185. By courtesy of Dr. Richard Boolootian.)

B, Sperm head of the Florida chiton *Certaozona squalida,* head length from tip to flagellar attachment (top) 8 μm. (Scanning electron micrograph by courtesy of Dr. John Pearse.)

C, Cytoskeleton of a mouse fibroblast (3-T-3) cell as seen by fluorescence microscopy of a specimen containing immunofluorescent-labeled tubulin in the microtubules. Body diameter of cells about 60 μm. (By courtesy of Drs. Klaus Weber and Mary Osborn, Max Planck Institut für Biophysikalische Chemie, Göttingen, Germany. From K. Weber, 1976. Cold Spring Harbor Laboratory, Book A. pp. 403–417.)

D, Egg of the chiton *Cyanoplax hartwegii* (ca. 230 μm. diameter) to show ornate protective hull over the fragile plasma membrane. Scanning electron micrograph. (By courtesy of Dr. John Pearse.)

E, Overview of two freeze-fractured neurohypophyseal axons. The fracture has exposed the inner leaflet (pP) of one axonal plasma membrane and the outer leaflet (pE) of the other axonal membrane. Microvacuoles (mv) and large secretion vacuoles are seen in both cells (concave face at gP, convex face at gE). The core of a vacuole is seen in the enlarged insert. A vacuole extruding to the exterior of the cell membrane is seen on the left just below the middle of the figure. (Theodosis, D. T., Dreifuss, J. J., and Orsi, L.: J. Cell Biol. *78*: 542–553.)

CELL PHYSIOLOGY

Fifth Edition

ARTHUR C. GIESE

Stanford University
Stanford, California

SAUNDERS COLLEGE / *Philadelphia*

Saunders College Publishing
West Washington Square
Philadelphia, PA 19105

Cover illustration: *Ectothiorhodospira mobilis* photosynthetic purple sulfur bacterium (×123,700). (Courtesy of S. W. Watson, Woods Hole Oceanographic Institute.)

Cell Physiology ISBN 0-7216-4120-2

 Press of W. B. Saunders Company. Library of Congress catalog card number 79-4728

0123 147 98765432

To the memory of Raina
who beautified everything she touched

PREFACE TO THE FIFTH EDITION

During a hospital stay in 1975, as I read proof of my book *Living With Our Sun's Ultraviolet Rays,* I conceived of a new approach to cell physiology. I realized that life arose and developed under the influence of radiations that supplied most of the energy for chemical evolution and, once cells appeared, for mutations. The evolution of the atmosphere following the liberation of oxygen by photosynthetic organisms had a further effect on life, resulting in the attenuation of the sun's deadly short wave ultraviolet rays thereby making possible life in shallow waters and on land. Approaching cell physiology from the standpoint of the origin and development of life would integrate consideration of radiation effects on life where it was most relevant, rather than treating radiation effects separately, as was previously done. The damaging effects of radiations are now considered chiefly as they affect cell division, at the end of the book. Reorganizing the text on this basis provides greater unity but requires recasting both initial and final sections of the text. I believe the greater unity and focus merit the effort.

The sections between those at the beginning and the end of the text have been somewhat less completely modified. The major changes are twofold, first updating and second, removal of older or less relevant material. As a consequence, historical material and references have suffered, but the newer references make it possible for those interested to find the older references. By this means, and by eliminating or relegating to an appendix discussions of topics (pH, for example) that are now well treated in other courses, the book has been kept from expanding.

I am indebted to the many readers, students and teachers alike, for their suggestions, even though some are diametrically opposed. I hope my decisions in these cases are to the liking of a larger proportion of readers. I am also indebted to my editors who have taken a personal interest in the text. I acknowledge with appreciation the expert typing, often from difficult copy, by Ms Arlene Novak and the many illustrations by Ms Stephanie Williams, both of whom generously accommodated me in spite of their busy schedules. I wish to acknowledge with thanks helpful criticisms and suggestions from colleagues, especially J. Kachergis on design, D. Regnery on cytogenetics, F. Halberg and L. N. Edmunds on chronobiology, A. Bajer and R. M. Iverson on cell division, and J. M. Brown on effects of ionizing radiation. I am especially grateful to Carl May, my editorial consultant, whose meticulous, searching, and highly constructive critique of the manuscript not only improved its quality and content but also brought about better coordination between text, figures, and tables.

I now miss the comments of my late wife, who was my severest critic in previous editions of the text, but was taken from my side by a medical mistake in the use of antibiotics. When I initially dedicated the text to her, it had more meaning than my love for her. The text was conceived jointly by us when we took a course in cell biology from S. C. Brooks, at a time, however, when explanations on a molecular basis were only beginning to be made, although we were hopeful about the possibilities for these. It was unfortunate that I was teaching everything but cell physiology during the Depression and its aftermath until an opportunity arose in 1947 and I again began thinking of a text. When I felt I had the experience to begin writing, I had her cooperation and her overall sense of beauty to improve the writing, for which I am forever grateful. She was always the inspiration to keep up what is undeniably an enormous effort and her whole life was a model in the pursuit of excellence.

In the treatment of complex matters I have at times given one interpretation rather than going into all the possible explanations; I realize the danger of incompleteness, but I feel that confusing the student by all the explanations is even worse. I have often said the problem is open. The student can find other interpretations in references cited for the purpose. For no biological phenomenon is an explanation final. Some explanations appear best in the light of the evidence before us at present.

As the text is now conceived, we proceed from the origin of cells to the study of the prokaryotic cell, the evolution of photosynthesis, and the origin of the eukaryotic cell. The next block of chapters briefly surveys the nature of cell organelles. This is followed by the study of energetics—thermodynamics, kinetics, metabolic pathways for the liberation of energy to do the work of the cell, photosynthesis, and biosynthesis and its regulation. Three chapters discuss passage of substances through the cell membrane, followed by sections on the topics of excitability, contractility, cellular rhythmic activity, cell growth and division, and finally the effects of various agents on cell division. The dangers to cells of pollution and especially of radiation damage become apparent in the last section and questions are then raised as to man's future. For "Man," after all, in the words of Claude (Science 1975, p. 433), "like other organisms is so perfectly coordinated that he may easily forget that he is a colony of cells in action, and that it is the cells which achieve this through him, that he has the illusion of accomplishing himself. It is the cells which create and maintain in us, during the span of our lives, our will to live and survive, to search and experiment and to struggle." What affects cells affects man. If he is not to be eliminated, he must preserve his cell environment. Cell physiology lays the basis for an understanding of what our cells can and cannot tolerate and why.

ARTHUR C. GIESE

Stanford, California

PREFACE TO THE FIRST EDITION

"Cell Physiology" as presented here is patterned after a set of lectures as they are given in my class at Stanford University. Although several excellent textbooks in cellular physiology are available and a number of good books bordering on the field have appeared, there exists no single book which gives a brief account of the subject. Students have asked repeatedly that this need be fulfilled; that a book be written which describes in simple language and in bold outline the major problems of cellular physiology, explaining their interrelationships and the current status of each of them without confusing the beginner with details or taking him into controversies upon which even the experts cannot agree.

Introducing a topic in cellular physiology by presenting controversial issues has often left students without an anchorage of fact by which to evaluate anything at all. On the other hand, I have been impressed by the tendency of many students to accept uncritically and to memorize anything written in a textbook. It is certainly not possible to avoid conflicting information on some subjects, since an open mind must be maintained on all real issues; but controversies on less important points can be minimized so that the major achievements in cellular physiology are not completely obscured. To help the student understand the scientific approach and develop an attitude of critical evaluation, I have found it extremely useful to include fundamental laboratory work and to conduct weekly discussions focusing on more specific controversial problems. And it is expected that after this primary orientation in the subject, students may become interested enough to turn to the more complete and analytical studies cited in references at the end of each chapter.

I have, moreover, confined the subject matter in this book to that dealing primarily with the cell, since this is basic to studies on multicellular organisms. By avoiding treatment of problems which relate to the organization of cells into organisms, it is possible to develop a more closely knit body of information, such as could be presented to a class in the course of a quarter or a semester. Classes in plant physiology, comparative animal physiology, mammalian physiology, and bacterial physiology are logical sequels to the course in cellular physiology.

At Stanford a one quarter course in cellular physiology is required of biology majors, most of whom take it at the junior or senior level. The student will have taken elementary botany, zoology, physics and introductory and organic chemistry. He is, therefore, unprepared for a rigorous physical-chemical treatment of the subject. However, the experience of the department over the years seems to have justified the premise that an introduction to cellular physiology at the elementary level is more beneficial than restriction of the course to a handful of graduate students with a broader background. Consequently, derivations of equations on physical-chemical principles or more extended discussions, if included at all, are put in appendices because this makes the material available to those students who are really interested without impeding the flow of the subject for those who are not.

I am indebted to Miss Ruth Ogren, scientific illustrator, for preparing many of the illustrations, and to Professor Hadley Kirkman of Stanford University, Professor David Waugh of Massachusetts Institute of Technology, and Dr. Richard Boolootian for use of some photographs. The courtesy of numerous publishers who have permitted copying of figures from books or periodicals is also acknowledged with gratitude. I am also indebted to Dr. George Palade of the Rockefeller University for electron micrographs used in the frontispiece, and to Dr. John Bennett, for the figure of the metabolic mill.

I wish to express my appreciation for many helpful comments and criticisms made by my colleagues and assistants, and especially to Professor J. P. Baumberger of Stanford University, Professor Jack Myers of the University of Texas, Professor John Spikes of the University of Utah, and to my students, Dr. Ray Iverson, Dr. David Shepard and Dr. Raymond Sanders, for a critical reading of the manuscript or parts of it. However, I must assume responsibility for opinions expressed and for omissions or errors which may still remain after proofreading. My main hope is that students will find this account interesting and stimulating.

ARTHUR C. GIESE

Stanford, California

CONTENTS

CHAPTER 14

CHAPTER 15

CHAPTER 16

Section IV

Transport Across Cell Membranes

Section V

Excitability and Contractility

Section VI

Biological Rhythms 523

Section VII

The Cell Cycle 545

INTRODUCTION

At the time of its birth the earth could have had no life and so no cells, regardless of its mode of origin. Currently it is thought that "A great cloud of gas and dust contracted through interstellar space 4.6 billion years ago, far out along one of the arms of our spiral galaxy. The cloud collapsed and spun more rapidly, forming a disk. At some stage a body collected at the center of the disk that was so massive, dense and hot that its nuclear fuel ignited and it became a star: the sun. At some stage the surrounding dust particles accreted to form planets bound in orbit around the sun and satellites in orbit around some of the planets" (Cameron, 1975). One of the planets was our earth, which had a satellite, the moon.

If the early earth had an atmosphere, it was one like that of the sun, with a preponderance of hydrogen and helium and the presently "rare" gases argon, neon, krypton, and xenon. These gases were lost because the earth's gravitational field could not hold them. A secondary atmosphere appeared, most likely formed of outgassing from the bowels of the earth by way of volcanoes and fumaroles, from products of radioactive decay and of rocks as they disintegrated, though other hypotheses have been suggested.

There is little agreement about the exact composition of the earth's secondary atmosphere except that it contained considerable water vapor (as do present outgassings) but little, if any, molecular oxygen. The other gases were probably nitrogen, hydrogen sulfide, carbon dioxide, hydrogen, and possibly methane, ammonia, and some hydrogen cyanide. Much of the carbon dioxide, whatever its origin, was later locked in sedimentary rocks. A small amount of oxygen was produced by photolysis of water that had absorbed short ultraviolet (UV) radiation entering the atmosphere at the time. Photolysis of water was self-limiting because when oxygen had accumulated to about 1/10,000 that of the present atmospheric level, the oxygen itself absorbed the very wavelengths that photolyzed water. Therefore, oxygen could not accumulate by this mechanism. However, there is some difference in opinion as to the exact level at which water photolysis became self-limiting. The prevalent opinion is that oxygen accumulated beyond the very low ultraviolet-photolytic level only after it was liberated from water by photosynthetic organisms (see Chapter 2).

Lack of oxygen in the earth's early atmosphere was critical to the evolution of life. Without oxygen (and ozone formed from it) to absorb the short ultraviolet wavelengths, much photochemically active ultraviolet radiation could reach the earth's surface, penetrate bodies of water, and induce synthesis of organic compounds. Furthermore, without oxygen the initial organic compounds synthesized from inorganic precursors by ultraviolet rays would not have been oxidized; therefore, these organic compounds would have persisted, permitting them to be used as precursors in additional syntheses.

As water accumulated on the earth's surface, it formed pools and streams, and as these drained into basins, they formed the oceans. Into the oceans were washed soluble minerals along with a sampling of the less soluble ones, some with catalytic activity. Seawater thus became rich in sodium and magnesium with lesser amounts of potassium and calcium. Carbon dioxide probably formed a buffer system in seawater, maintaining a relatively constant acidity just above neutrality. It was perhaps the major environmental buffer system then as at present. The other gases of air also dissolved in seawater.

The stage was thus set for *biochemical evolution*—the development of a wide variety of organic compounds that characterize life. The question arises, Could the chemicals of which the cell is composed have arisen on the primeval earth? This was answered most clearly by Wald (1954, p. 49): "In a sense organisms demonstrate to us what organic reactions and products are *possible*. We can be certain that, given time, all of these must occur. Every substance that has been found in an organism displays thereby the finite probability of its occurrence. Hence, given time, it should arise spontaneously." Presumably there was ample time for such biochemical evolution in the 1.2 billion years before fossil evidence of life appeared in the earth's crust, although this might be contested. (Radioactive decay measurements in meteorites indicate that the earth is probably about 4.6 billion years old; the old-

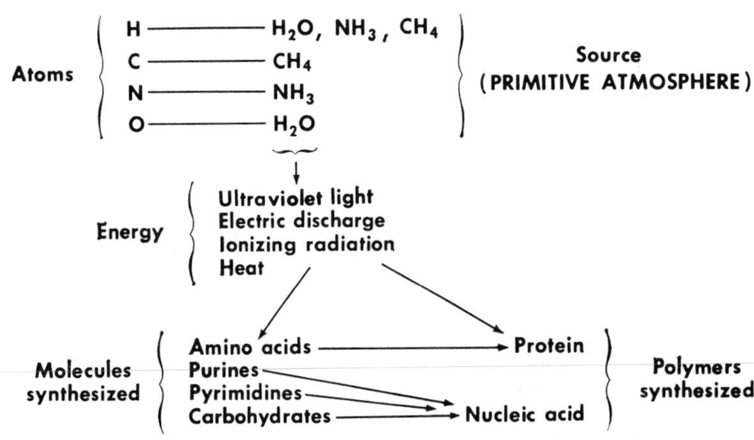

Figure A. Chemical evolution: gases in the primeval atmosphere and possible synthesis of organic compounds on the primitive earth. (From Young and Ponnamperuma, 1964: Early Evolution of Life. D. C. Heath & Co., Boston (BSCS Pamphlets #11) by permission of the Biological Sciences Curriculum Study.)

est fossils yet found are considered to be about 3.4 billion years old.)

The sources of energy for inducing biochemical syntheses were ultraviolet radiation, ionizing radiation from space and radioactive decay, lightning, impact, and heat (Ponnamperuma, 1968). Ultraviolet radiation was probably the most prominent source (Fig. A). Although the reduced compounds of the earth's primitive atmosphere absorb in the short ultraviolet portion of the spectrum—methane at wavelengths shorter than 150 nm; water, 170 nm; and ammonia, less than 220 nm—it is likely that sufficient short wavelength ultraviolet radiation still passed through the atmosphere to energize various organic syntheses.

Using such radiations, electric discharges, and heat, investigators have been able in the laboratory to synthesize various carbohydrates, amino acids, polypeptides, proteinoids, organic bases characteristic of nucleic acids, porphyrins (important as catalysts) and lipids (Gabel and Ponnamperuma, 1972; Fox and Dose, 1977; Calvin, 1975). Calvin and Calvin (1964, p. 168) have stated, "Not only can the building blocks of today's organisms be generated by abiogenic processes, but the basic 'energy currency' [adenosine triphosphate: ATP] used by all organisms can be formed in a similar abiogenic conversion of the prime energy sources, ionizing radiation and light."

In the absence of life on the primeval earth, degradation of organic compounds would have been very slow, because neither organisms nor oxidation would have affected them in an abiotic and nearly anaerobic planet. Therefore, mixtures of various types of organic chemicals might have persisted for prolonged periods. Oparin (1924) postulated that such compounds could have accumulated to form organic "soups" in the environmental waters. Once polymers of amino acids, called polypeptides, and polymers of the nucleic acid subunits (nucleotides), called polynucleotides, became available, as well as the high energy phosphates and a number of catalysts (based upon porphyrin and metal ions), a good deal of the machinery of life existed. This is true also because the three-dimensional structures of proteins and nucleic acids required in the cell are the most stable states for these molecules and therefore formed spontaneously once the primary backbone structures (polypeptides and polynucleotides) became available. Even such structures as cell membranes, in many instances, will re-form spontaneously from extracted lipids and proteins, the membrane representing the state of greatest stability of its components. Such considerations do not minimize the basic remaining problem, to explain the initial "spark" of life. Presumably in the period antedating the appearance of the organisms providing the oldest fossils, the first living cells evolved from a mass of such organic compounds synthesized on the abiotic, anaerobic earth.

Metabolism, excitability, growth, and reproduction together characterize life, although each of these properties may be found separately in lifeless physical systems. It is difficult to conceive of replication of the hereditary sub-

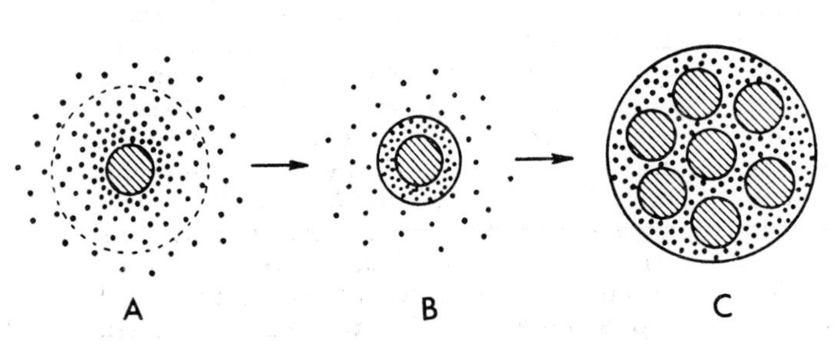

Figure B. Diagram of the process of coacervation. A, Colloidal particle with diffuse hydration layer; B, hydration layer reduced and delimited; C, beginning of the coacervation phenomenon. When partial discharge or dehydration occurs, the less fully hydrated particles of the colloid may combine to form larger units with less solvent bound to them than the sum of what they had before. (From Bungenberg de Jong, 1932: Protoplasma 15: 110.)

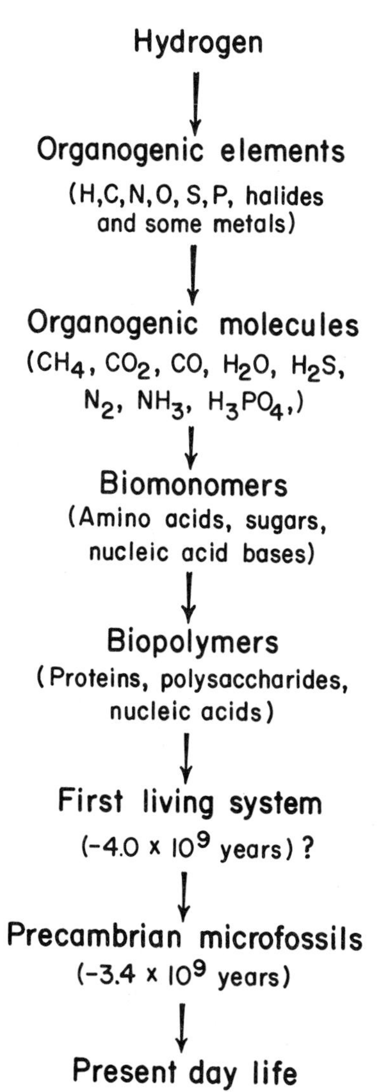

Figure C. *Time sequence of evolution from the elements to the present. The first two units represent element formation; the second three, chemical evolution; the last three, biological evolution. (After Calvin, 1975: Am. Sci.* 63: *171.)*

stance of a cell without nucleic acids, although it has been speculated that perhaps enough information is present in the protein molecule for replication of some mechanism other than the one finally adopted by cells in the course of evolution. Watson (1976), however, points out that the accuracy of a protein template in replication would be so low as to be useless. From this standpoint the interesting models of cells, such as the *microspheres* (proteinoid globules) that appear when heat-treated amino acid mixtures are dissolved in water (Fox and Dose, 1977) or the *coacervates,* systems in which highly hydrated polymers separate from the surrounding solutions on loss of some water of hydration (Fig. B), are only suggestive of cells. Coacervates with incorporated enzymes may carry out some digestive, oxidative, and even photosynthetic reactions, thereby simulating some cell functions (Oparin, 1972). However, neither microspheres nor coacervates can perform the functions that nucleic acids do in replication of the information in the cell.

It appears necessary also that the reactions in the primeval cell be confined inside a compartment, separated from the environment by a selectively permeable membrane, presumably containing lipid. Finally, it is an indication of its orderliness that the living cell generally produces only one of two possible optical isomers of organic compounds, e.g., the D or L forms of amino acids (Wald, 1957). The abiotic syntheses that are achieved give mixtures of both isomers.

The stages by which life presumably evolved from the organic soups are being investigated and, except in broad outline (Fig. C), there are many differences of opinion as to the way in which this event might have occurred. These are not reviewed here but are considered elsewhere (e.g., Oparin, 1972; Miller and Orgel, 1974; Fox and Dose, 1977). We shall assume that the compounds for sustaining life were available and that life had already appeared. The plan of this text is, then, primarily to study present-day cells and their functions and relate these to the events and forces involved in their creation.

The origin of cells on earth may serve to explain their fitness in terms of the environment into which they developed, as well as give perspective on the functions characterizing cells: metabolism, excitability, growth, and reproduction. For example, all cells are capable of temporary anaerobic existence, perhaps a relic of the anaerobic period in the earth's history when energy liberation without oxygen was the only way cells could carry out their life activities.

The approach to cell physiology in the present edition provides evolutionary perspective to cell functions, suggesting more experimentation than previous ones. However, as in the past, the interpretations of cell functions presented are those considered most probable on the basis of the facts presently available. It should be taken for granted that while the facts remain, their interpretation may change as additional data render another interpretation more probable. It is in the nature of science that no interpretation can be taken as final; at best, it is only the one most probable in view of the available information.

APPENDIX

UNITS USED IN CELL PHYSIOLOGY

The units used in this text are generally those recommended by the Systéme Internationale d'Unités.

The unit of *length* is the meter (m); of *mass*, the kilogram (kg); of *time*, the second (s) though often the minute (min) or hour (h) are also used; of *work*, the joule (J); of *power*, the watt (W); W = Js^{-1}; of *energy content*, the calorie (1 cal = 4.184 J); of *temperature*, degrees kelvin (K), though usually given in degrees Celsius (C); of *charge*, the coulomb (c); of *capacitance*, the farad (F); of *frequency*, the hertz (Hz = s^{-1}).

The unit of *wavelength* is the nanometer (nm), although the Ångstrom (Å) is used for X-rays and gamma-rays (1 nm = 10 Å). For dosimetry the *energy fluence* (F) was recommended by the International Commission of Radiation Units and Measurements although it would seem that the abbreviation (F) had been preempted as the unit of capacitance, as indicated above, and *dose* (D) is common in the literature. The dose is the amount of energy crossing unit area normal to the direction of propagation and is measured in Jm^{-2} rather than in ergs mm^{-2} as formerly used (1 Jm^{-2} = 10 ergs mm^{-2}). Energy *fluence rate* (formerly the intensity) is the energy per unit area per unit time ($Jm^{-2}s^{-1}$).

Subdivision or multiplication of the basic units is given by the following abbreviations: 10^9 = giga, 10^6 = mega, 10^3 = kilo, 10^{-2} = centi, 10^{-3} = milli, 10^{-6} = micro, 10^{-9} = nano, and 10^{-12} = pico.

Molecular weight is given in daltons, a unit equal to the weight of a hydrogen atom. I have designated the molecular weight in daltons only at the beginning of the text; to save space it is implied elsewhere. Although the custom is not yet fully established, there is a tendency to omit the word daltons in dealing with compounds of low molecular weight but to use it for compounds of high molecular weight such as proteins and nucleic acids.

REFERENCES CITED

Calvin, M., 1975: Chemical evolution. Am. Sci. *178*: 169–177.

Calvin, M. and Calvin, D. J., 1964: Atom to Adam. Am. Sci. *52*: 163–186.

Cameron, A. G. W., 1975: The origin and evolution of the solar system. Sci. Am. (Sept.) *233*: 32–41.

Dayhoff, M. O., 1972: Evolution of proteins. *In* Exobiology. Ponnamperuma, ed. North Holland Publishing Co., Amsterdam, pp. 266–300.

Dickerson, R. E., 1978: Chemical evolution and the origin of life. Sci. Am. (Sept.) *239*: 70–86.

Farley, J., 1977: The Spontaneous Generation Controversy from Descartes to Oparin. Johns Hopkins University Press, Baltimore.

Fox, S. W. and Dose, K., eds., 1977: Molecular Evolution and the Origin of Life. Rev. Ed. W. H. Freeman Co., San Francisco.

Gabel, N. W. and Ponnamperuma, C., 1972: Primordial organic chemistry. Exobiology. Ponnamperuma, ed. North Holland Publishing Co., pp. 95–135.

International Conference on the Origin of Life (5th, 1977). Kyoto, Japan. 1978: Business Center for Academic Societies.

Lahav, N., White, D. and Chang, S., 1978: Peptide formation in the Prebiotic Era: Thermal condensation of glycine in fluctuating clay environment. Science *201*: 67–69.

Miller, S. L. and Orgel, L. F., 1974: The Origin of Life on Earth. Prentice-Hall, Englewood Cliffs, N. J.

Oparin, A., 1924: The Origin of Life (in Russian). Izdatelstvo Moskovsky, Moscow.

Oparin, A. I., 1972: The appearance of life in the universe. *In* Exobiology. Ponnamperuma, ed. North Holland Publishing Co., Amsterdam, pp. 1–15.

Ponnamperuma, C., 1968: Ultraviolet radiation and the origin of life. *In* Photophysiology. Giese, ed. Academic Press, New York, Vol. 3, pp. 253–267.

Scientific American, 1975: The Solar System. (Sept.) *233*, No. 3. (Especially pp. 22–57, 82–105.)

Wald, G., 1957: The origin of optical activity. Ann. N. Y. Acad. Sci. *69*: 255–376.

Wald, G., 1954: The origin of life. Sci. Am. (Aug.) *191*: 45–53.

Walker, J. C. G., 1977: Evolution of the Atmosphere. Macmillan, New York.

Watson, J. D., 1976: Molecular Biology of the Gene, 3rd Ed. W. A. Benjamin, Menlo Park, Calif.

Section I

ORIGIN AND EVOLUTION OF CELLS

Cells are of two types: prokaryotic, without a nucleus, and eukaryotic, with a nucleus. In the fossil record prokaryotic cells precede eukaryotic cells by a considerable period of time. In prokaryotic cells the chromosome that bears the hereditary units (genes) lies naked in the cytoplasm; in eukaryotic cells the multiple chromosomes are combined with proteins and lie within a double-membraned envelope and are generally visible only during cell division. In eukaryotic cells many compartments are present within which are isolated enzymes for performing various cellular functions, such as oxidative enzymes for liberation of energy, digestive enzymes, and enzymes for synthesis of macromolecules. In prokaryotic cells these activities are carried out either on the cell membranes and its involutions or within the cytoplasm.

In this section the two types of cells are described and their metabolic activity is considered in a general way. On the primitive earth without atmospheric oxygen (anaerobic), prokaryotic cells could liberate energy from nutrients only anaerobically (Chapter 1). The more efficient aerobic metabolism could develop only after oxygen appeared in the atmosphere. This occurred with the evolution of photosynthetic cells that liberated oxygen from water using the energy of sunlight (Chapter 2). Atmospheric evolution was a consequence of the evolution of photosynthetic cells. One of the consequences of oxygen in the atmosphere was the formation of a stratospheric ozone layer that, along with oxygen, removed the deadly short ultraviolet rays from sunlight reaching the earth's surface. This made life possible in shallow water and on land. Eukaryotic cells evolved with the appearance of an aerobic atmosphere (Chapter 3). In time, cells became adapted to diverse conditions, including extreme environments on the earth's surface (Chapter 4).

CHAPTER 1

ORIGIN AND NATURE OF PROKARYOTIC CELLS

ORIGIN

The earliest fossils of 3.4 billion years ago superficially look much like the bacteria of today (Fig. 1.1). We do not know what preceded them, although there may have been a series of protoplasmic masses ("eobionts") with the capacity to divide, but without cell walls. Therefore, they presumably disintegrated, leaving no trace. Those investigators working on the origin of life may ultimately determine stages in the evolution of cells.

We assume that the fossil bacteria probably resembled the bacteria of today in most respects, perhaps including their chemistry. Even the simplest cells now living are quite complex and are made up of a myriad of chemicals. These chemicals fall into a number of categories: proteins, nucleic acids, carbohydrates, lipids, miscellaneous organic molecules of relatively low molecular weight, water, and a variety of salts. The nucleic acids are of two major types, deoxyribonucleic acid (DNA) and ribonucleic acid (RNA).

A living cell is characterized by the fact that it carries on chemical activity (metabolism) by which it liberates energy for its life activities, remains excitable and responsive to its environment, and synthesizes cell substances for growth and cell division. Abiotic systems have been created that grow, or metabolize to some extent, or are excitable, or replicate DNA, or even divide, but none has been synthesized that carries out all these life activities simultaneously.

The earliest cells to appear on earth probably had a limited capacity to synthesize compounds and therefore incorporated most of what they used from the rich organic "soups," a milieu that presumably contained almost all the compounds needed by cells. Their main distinction would perhaps have been their capacity to replicate genic DNA, which contained the information for the few syntheses that the cells carried out. Present-day viruses,* which carry out very few syntheses, have low-molecular-weight DNA, and the early prokaryotes may have had rather small DNA molecules. The few proteins coded by the DNA of such cells other than those that made up the structure of the cell, may have been the enzymes needed to drive the syntheses performed.

Organisms that use ready-made organic compounds present in an environment (for example, a culture medium) are spoken of as *heterotrophs* ("ones that feed upon others"). The early cells were heterotrophic; they did not synthesize organic material from inorganic sources as do *autotrophs* ("self feeders"), such as chlorophyll-containing bacteria and plants. The early heterotrophs obtained the compounds ready-made from the organic solutions in which they lived. The organic material may initially have been quite abundant in the sea, where life is considered to have originated, but it was finite. As the cells multiplied, certain compounds became limiting. It is likely that mutants then appeared that could synthesize some limiting substance, enabling them to survive in the nutrient-depleted environment. As each nutrient became depleted locally, mutants presumably developed that were capable of synthesizing it. Finally, organisms capable of synthesizing most of their nutrients appeared.

*Most viruses are considered to be escaped fragments from cellular genomes, coding primarily for viral replication and coating proteins. Some large viruses with a cell membrane may be degenerate cells with a highly simplified genome.

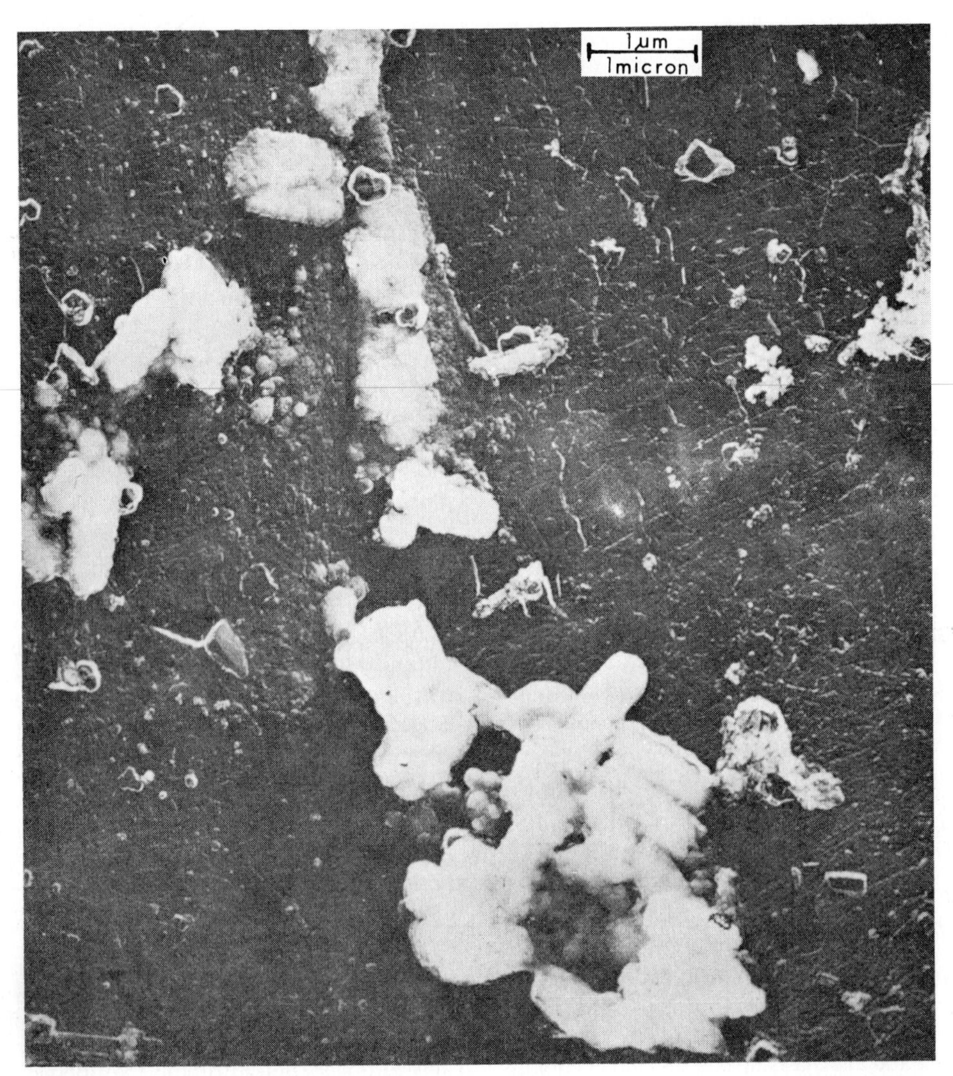

Figure 1.1. *Fossil bacteria, Precambrian. (From Schopf* et al., *1965: Science* 149: *1165.)*

This sequence of events is spoken of as the *heterotroph hypothesis* (Horowitz, 1945).

This hypothesis gains plausibility because organisms in the anaerobic phase of the earth's history were bombarded by short wavelength ultraviolet radiation such as we use in the laboratory today to produce mutants for genetic and biochemical research. This radiation kills many cells, but survivors show genic changes; most of these are genetic deficiencies, but some are adaptive. It is thought that such radiation speeded mutation on the anaerobic earth. The chemical changes that produce mutation in DNA are discussed in Chapter 15.

Many bacteria live on very simple diets; for example, *Escherichia coli* in our gut (Fig. 1.2) needs only ammonium chloride, magnesium sulfate, potassium phosphate (KH_2PO_4), sodium phosphate (Na_2HPO_4), glucose, and water for its growth. From this simple medium it can produce the multitude of nucleic acids, proteins, and smaller organic compounds that it needs, and it does so rapidly, dividing every hour on such a medium at 37°C. However, *E. coli* can also use organic nutrients, for example, amino acids, and, when supplied with such ready-made building blocks, it grows more rapidly, dividing every 20 minutes at 37°C (Watson, 1976).

STRUCTURE OF PROKARYOTIC CELLS

The structure of a prokaryotic cell as exemplified by the colon bacillus *(Escherichia coli)* (Figs. 1.2 and 1.3), a rod-shaped cell 2 μm long and 1 μm in diameter, is relatively simple. A gram-negative *cell wall* (Fig. 1.4) encloses the cytoplasm in which is found a *nucleoid* or nuclear region composed of a naked chromosome

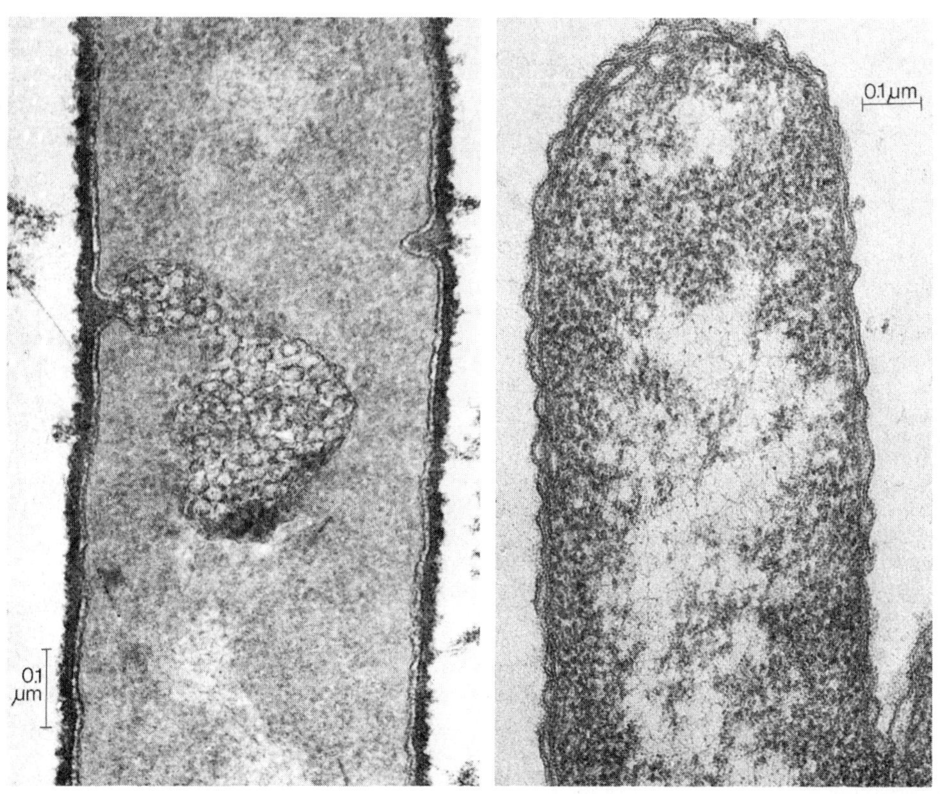

Figure 1.2. Electron micrographs showing the relatively undifferentiated structure of Bacillus subtilis (left) *and* Escherichia coli (right). *The vesiculate body (mesosome) in* B. subtilis *appears to participate in the formation of cross walls on division of the cell. Nucleoids are the lighter areas in both* B. subtilis *and* E. coli. *The cell wall (dark outer covering) is closely applied to the plasma membrane in* B. subtilis *but is thinner and separated from the plasma membrane in* E. coli. *(From Iterson, 1965: Bacteriol. Rev.* 29: *299.)*

of deoxyribonucleic acid (DNA) (see Fig. 1.17). A cell *plasma membrane* abuts against the cell wall and serves to determine the entry and exit of molecules into and from the cell. In the cytoplasm are many small structures called *ribosomes,* on the surface of which proteins are synthesized. Little more structure is discernible, even with the high resolution of the electron microscope. Many enzymes, beyond the limit of resolution of the electron microscope, are present in the cytoplasm. Some of the enzymes are involved in the synthesis of protein and other substances needed for cell growth. Other enzymes that take part in energy-liberating reactions occur in the cytoplasm; those concerned with aerobic reactions are present in the cell membrane. Several types of ribonucleic acid (RNA) that take part in protein synthesis are also present in the cytoplasm, but they are likewise beyond the resolution of the electron microscope. Other bacteria, for example, *Bacillus subtilis,* are very similar to *E. coli* in structure (Fig. 1.2). Photosynthetic prokaryotic cells differ structurally from *E. coli,* possessing either chromatophores or cytoplasmic membranes that carry the photosynthetic pigments (see Chapter 2).

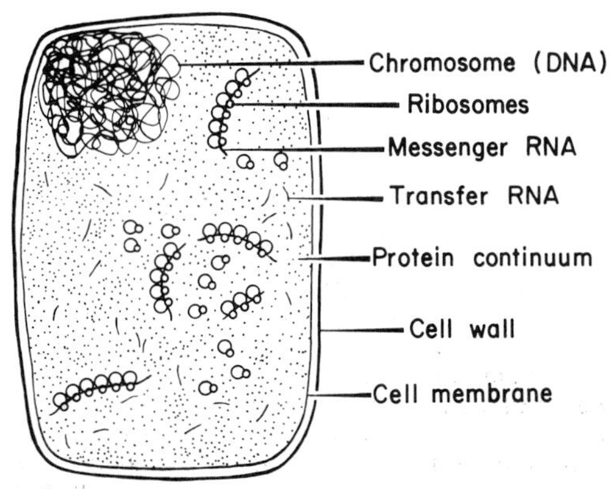

Figure 1.3. Diagram of a bacterium with internal structure.

Although simple in structure, *E. coli* and other prokaryotic cells are exceedingly complex at the molecular level, in this respect rivaling the eukaryotic cell. Many of the same chemical compounds are found in both prokaryotic and eukaryotic cells, but certain ones are unique either to one or the other type of cell. Chemicals common to both types of cells are considered in this chapter. Those unique to eukaryotic cells are considered in Chapter 3.

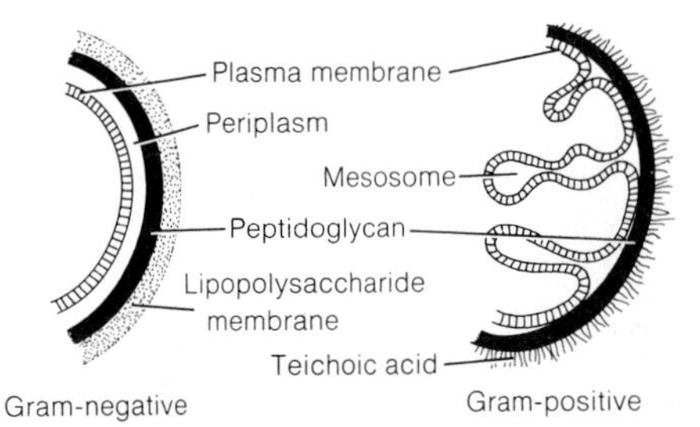

Figure 1.4. *Gram-negative and gram-positive cell walls. (From Saier and Stiles, 1975: Molecular Dynamics in Biological Membranes. Springer Verlag, New York, p. 5.)*

CHEMICAL COMPOSITION OF PROKARYOTIC CELLS

A cell such as *E. coli* has a sampling of all the classes of compounds found in cells: water, proteins, carbohydrates, lipids, nucleic acids, and salts. The 10 nm* thick cell wall that covers the surface of a bacterium (shown in Figure 1.2) is composed of proteins, lipids, and polysaccharides. Interior to it is a cell membrane, also 10 nm thick, composed of lipid and protein. The DNA is present as a long thin molecule occupying about 20 percent of the cell's interior. Surrounding it in the cytoplasm are from 20,000 to 30,000 ribosomes, each composed of 40 percent proteins and 60 percent RNA. About 70 percent of the cell mass is water. Various enzymes composed of protein and a variety of smaller organic molecules are also present. Proteins constitute 15 percent of the mass; about half of them function as enzymes, the remainder constitute structures. RNA constitutes about 6 percent of the cell mass, DNA about 1 percent, carbohydrates and precursors constitute 3 percent, lipids and precursors constitute 2 percent, inorganic ions (Na, K, Mg, Ca, Fe, Cl, PO_4, and SO_4) constitute 1 percent, and a variety of amino acids and small organic molecules make up the rest.

The proteinaceous respiratory enzymes in sequences (chains) occupy the inner surface of the cell membrane. It is estimated that between 3000 and 6000 different types of molecules are present in an *E. coli* cell (Watson, 1976).

It is necessary to consider the properties of the classes of compounds present in cells before discussing cell functions. These are considered in order.

Water

The general consensus of those interested in the early history of the earth is that water vapor outgassed from volcanoes and fumaroles condensed on the primitive earth's surface to form liquid water. Otherwise life would have been impossible.

*nanometer= 10^{-9} meter, a billionth of a meter.

Water constitutes 70 to 80 percent of prokaryotic cells; its prominence diminishes in dormant cells such as bacterial spores. Offhand, one might not expect structure in a cell made up of 70 percent water and 15 percent protein, but water forms bonds with itself and with other molecules.

The shape of a water molecule is that of an isosceles triangle; the intermolecular O—H distance is nearly 0.099 nm, and the H—O—H angle approximates 105°. The powerful attraction of the oxygen nucleus tends to draw electrons away from the protons (H nuclei), leaving the region around them with a net positive charge. Since only two pairs of electrons around the oxygen atom are shared with the protons, the two other pairs in the eight-electron shell point outward from the O—H bond, attracted to the net positive charge of other water molecules with a force of about 4500 cal/mole.* A

*Bond strengths are measured in calories of energy that would be required to break the bond in one gram mole of the compound.

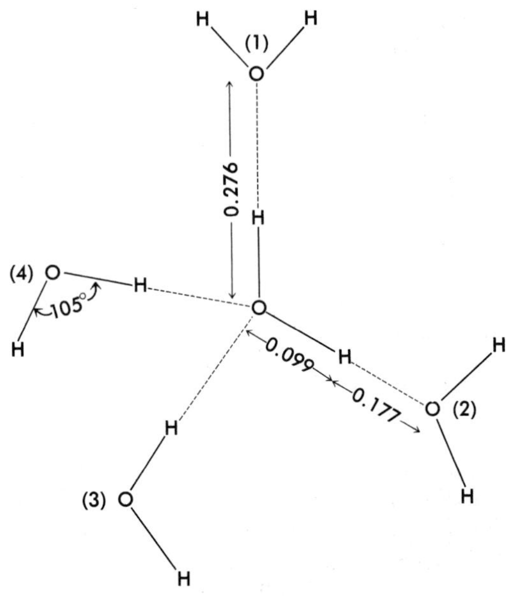

Figure 1.5. *The bipolar nature of the water molecule. The diagram shows the orientation of water molecules through tetrahedral coordination in ice. Molecules (1) and (2) lie entirely in the plane of the paper; molecule (3) lies above this plane, and molecule (4) below it, so that oxygens (1), (2), (3) and (4) lie at the corners of a regular tetrahedron. In water the orientation is not as regular, but multiples occur bound by hydrogen bonds between the molecules (shown in dashes). The distances are given in nanometers. (From Edsall and Wyman, 1958: Biophysical Chemistry. Academic Press, New York.)*

group of water molecules therefore forms a tetrahedron around the oxygen atom, the positively charged region of one molecule tending to orient itself toward a negatively charged region of one of its neighbors, as shown in Figure 1.5 (see Ben-Naim, 1974).

Water forms hydrogen bonds of like strength with electronegative atoms on other molecules, such as the oxygen atom covalently bound to hydrogen in the —OH of a carboxyl group and nitrogen covalently bound to hydrogen in —NH of amino groups in proteins. Hydrogen bonding stabilizes the structure of proteins in this manner, not only at amino and carboxyl groups but also at any other electronegative atoms in protein molecules. Hydrogen bonding is also important in stabilizing nucleic acids in their characteristic structure, as is discussed later in this chapter. Water also forms hydrogen bonds with small organic compounds.

Hydrogen bonds, with about one-tenth the strength of covalent bonds, are broken by small changes in temperature. This makes for considerable flexibility in hydrogen-bonded structures, enabling them to change with internal or external environmental changes (Joesten and Schaud, 1974).

Proteins

Proteins are responsible for the characteristic structure of the cell. They are molecules of large molecular weight, ranging from about 6000 to many million daltons. Their large size probably accounts for many of their peculiar

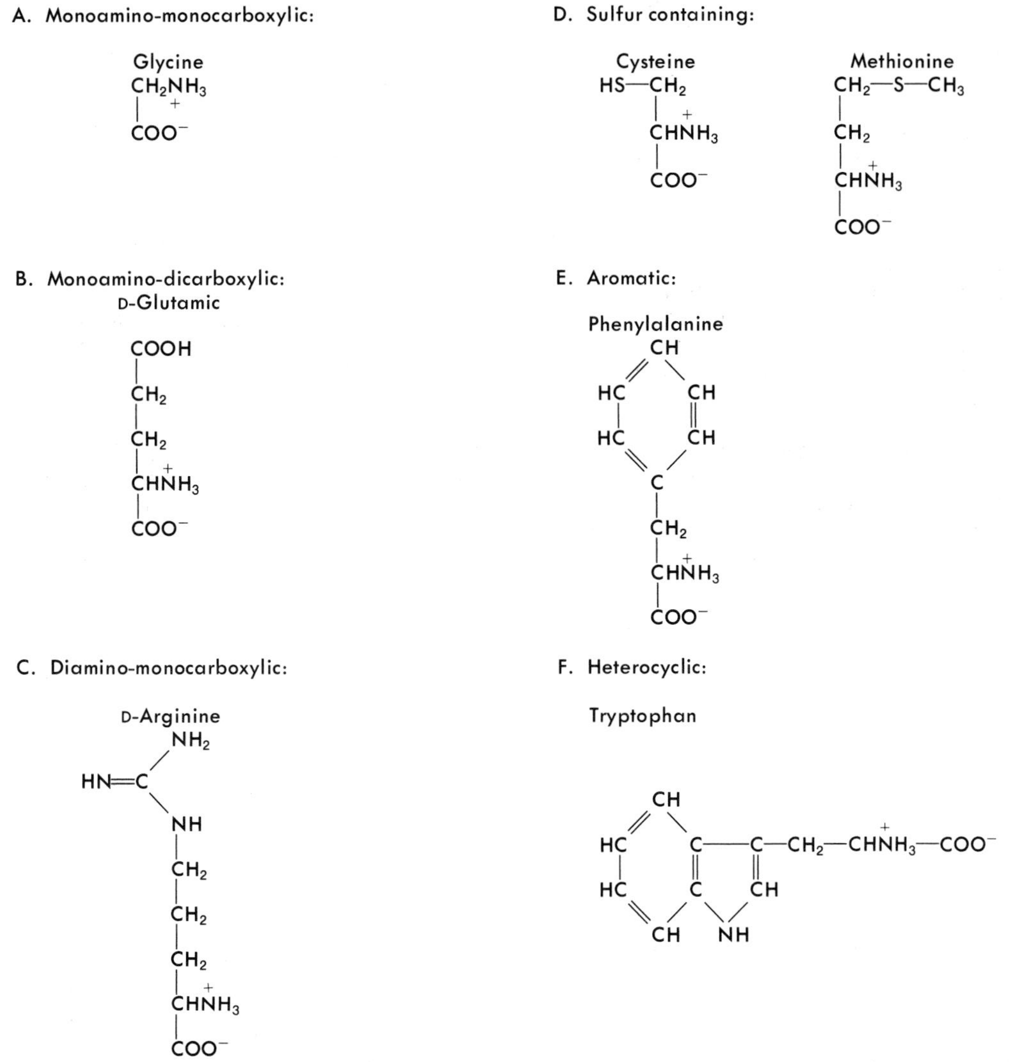

A. Monoamino-monocarboxylic:

Glycine
$CH_2\overset{+}{N}H_3$
|
COO^-

B. Monoamino-dicarboxylic:
D-Glutamic

COOH
|
CH_2
|
CH_2
|
$CH\overset{+}{N}H_3$
|
COO^-

C. Diamino-monocarboxylic:

D-Arginine
NH_2
/
HN=C
\
NH
|
CH_2
|
CH_2
|
CH_2
|
$CH\overset{+}{N}H_3$
|
COO^-

D. Sulfur containing:

Cysteine
HS—CH_2
|
$CH\overset{+}{N}H_3$
|
COO^-

Methionine
CH_2—S—CH_3
|
CH_2
|
$CH\overset{+}{N}H_3$
|
COO^-

E. Aromatic:

Phenylalanine
CH
HC CH
HC CH
C
|
CH_2
|
$CH\overset{+}{N}H_3$
|
COO^-

F. Heterocyclic:

Tryptophan

CH
HC C—C—CH_2—$CH\overset{+}{N}H_3$—COO^-
HC C CH
CH NH

__Figure 1.6.__ Representative amino acids. The amino acids probably occur in the cell in ionic forms, the ionization depending upon the pH or hydrogen ion concentration of the cell. They are here shown in zwitterion (dipolar ion) form, the proton being attached to the amino group.

TABLE 1.1. TYPES OF NATURAL AMINO ACIDS

A. Monoamino-monocarboxylic	D. Hydroxyl-containing
Glycine (Gly)*	Threonine (Thr)
Alanine (Ala)	Serine (Ser)
Valine (Val)	E. Sulfur-containing
Leucine (Leu)	Cystine (Cys or Cy-S)
Isoleucine (Ileu)	Methionine (Met)
B. Monoamino-dicarboxylic	F. Aromatic
Glutamic (Glu)	Phenylalanine (Phe)
Aspartic (Asp)	Tyrosine (Tyr)
C. Diamino-monocarboxylic	G. Heterocyclic
Arginine (Arg)	Tryptophan (Tryp)
Lysine (Lys)	Proline (Pro)
Hydroxylysine (Hlys)	Hydroxyproline (Hpro)
	Histidine (His)

*Abbreviations in parentheses.

properties (Haschemeyer and Haschemeyer, 1973). They also act as enzymes or parts of enzymes, combining with molecules to catalyze reactions involving those molecules.

All proteins contain carbon, hydrogen, oxygen, nitrogen, and usually sulfur, and some have phosphorus. Proteins are made up of chains of amino acids (Fig. 1.6 and Table 1.1). The amino acids of a protein are united to one another by their respective amino and carboxyl groups, forming *peptide bonds* (Fig. 1.7). In the cell, only phosphorylated amino acids catalyzed by appropriate enzymes react to form peptide bonds (see Chapter 15). The chain of peptide bonds between successive amino acid residues constitutes the backbone of the protein molecule and its primary structure (Fig. 1.8). However, other linkages are necessary to explain the native structure of a protein. Some 20 kinds of amino acids are present in proteins (Table 1.1), but all the amino acids are not present in every protein and those present occur in proportions characteristic for each protein. The number of proteins that can be formed from only 20 types of amino acids is very large. This is not difficult to comprehend if we make an analogy between the combinations possible with amino acids and with letters of the alphabet, since the entire English language is derived from 26 letters combined in diverse ways.

When both the molecular weight of a protein and the relative proportions of its constituent amino acids are known, an empirical formula in terms of its amino acid content may be written for it (Sanger, 1959). Thus, the empirical formula for ribonuclease, an enzyme catalyzing digestion of ribonucleic acid, is:

$$Asp_4, (AspNH_2)_{11}, Glu_6, (GluNH_2)_6, Gly_3, Ala_{12}, Val_9, Leu_2, Ileu_3, Ser_{15}, Thr_{10}, Cys_8, Met_4, Pro_4, Phe_3, Tyr_6, His_4, Lys_{10}, Arg_{14}$$

$AspNH_2$ and $GluNH_2$ refer to the amines of

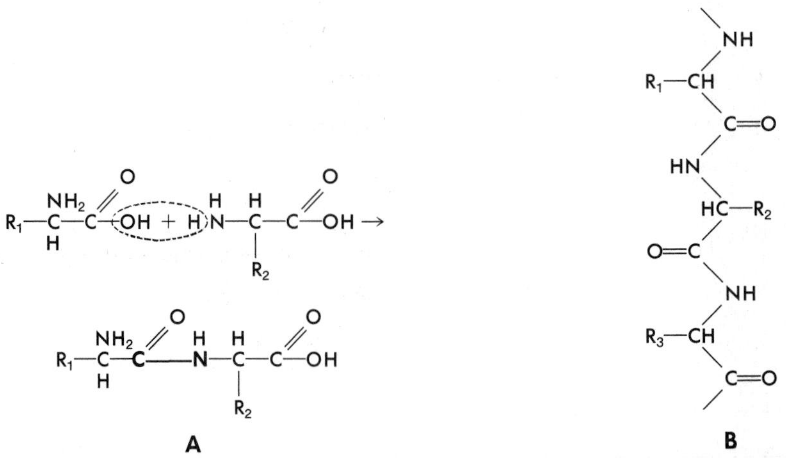

A B

Figure 1.7. A, *Simplified representation of the formation of a peptide bond between two amino acids. The N and C of the peptide bond are shown in bold type. (See Chapter 15 for the actual mode of peptide bond formation.)* B, *The peptide backbone of a protein molecule (schematic).*

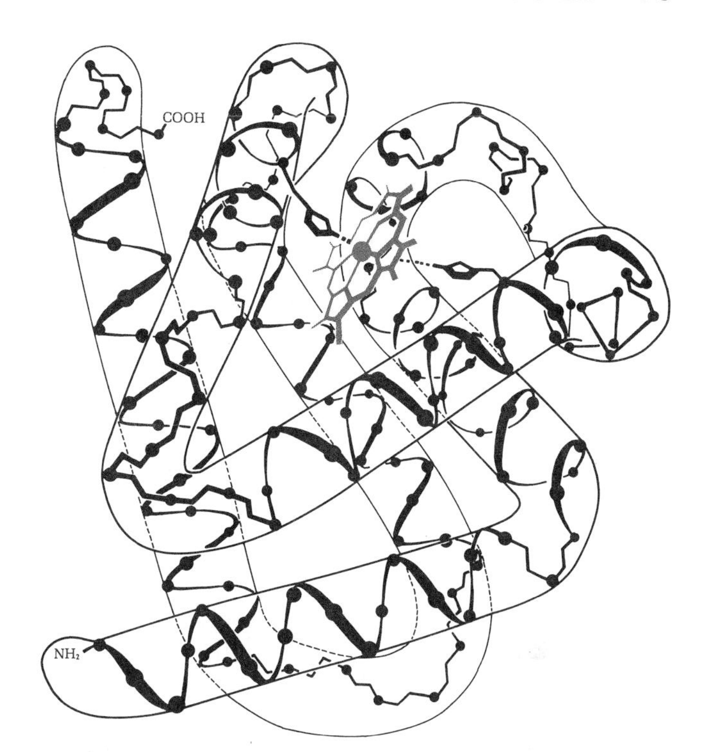

Figure 1.8. *The three-dimensional structure of a protein, myoglobin, as derived from x-ray diffraction analysis. The polypeptide backbone is shown in black, the heme group containing the iron (shown as a grey ball) is shown in grey. (From Dickerson, 1964: In H. Neurath (ed.) The Proteins, Ed. 2, Vol. 2. Academic Press, New York, p. 634.)*

aspartic and glutamic acids, respectively (see Table 1.1 for meaning of the abbreviations for amino acids). Ribonuclease consists of 124 amino acids of 19 different kinds. Quite a few other proteins have now been similarly analyzed.

In some cases the analysis of protein structures has been pursued to the point at which not only the empirical but also the structural formulas have been determined. For example, this was done for insulin (molecular weight 6000), illustrated in Figure 1.9. The procedure is long and difficult. It is necessary to determine first, by end-group analysis, how many polypeptide chains are present in the molecule. It is obvious that a polypeptide chain can have only one terminal α-amino group, and such an amino group can be made to react with dinitrofluorobenzene to form a bright yellow dinitrophenol derivative. When a known molar concentration of the protein is hydrolyzed by acid or enzymatically, the marked peptide from which the number of polypeptide chains per protein molecule can be determined can be isolated and determined quantitatively. This is checked by determining the number of terminal carboxyl groups per protein molecule, using an enzyme specific to the terminal carboxyl groups (carboxypeptidase, see Chapter 10).

Next, the disulfide bonds linking the polypeptide chains are broken by performic acid. Then each of the polypeptides, separated from one another by polyacrylamide gel electrophoresis, must be hydrolyzed by methods that cleave them in certain places only. Fortunately, selective enzymes are available, and these, combined with acid or alkaline hydrolysis, made it possible to shatter the chain in many places. The overlapping sequences in the chromatographed products can then be noted (see appendix to Chapter 15). It thus proved possible to identify all amino acids in the two polypeptide chains making up the insulin molecule. Furthermore, it was possible to determine the order of the various amino acids in the two polypeptide chains. Determination of the positions of cross linkages between the two chains by disulfide groups proved difficult, but successful methods were devised. This intensive analytical study (Sanger, 1959), continuing for 10 years, won the Nobel Prize for Sanger in 1958.

At present amino acid sequences of increasing numbers of proteins are being determined. Among the ones analyzed earlier were insulin (from several animal species), adrenocorticotropic hormone, bovine glucagon, bovine pigment cell-stimulating hormone, the protein constituent of tobacco mosaic virus, and ribonuclease (Lehninger, 1975).

The three-dimensional arrangement of lysozyme was the first to be determined (Phillips, 1966), and the arrangement in ribonuclease, myoglobin, hemoglobin, and others has also been worked out. The three-dimensional arrangement of protein molecules is being actively studied by x-ray diffraction methods (Westofer and Ristand, 1973); an example is shown in Figure 1.8. It is interesting that when hemoglobin combines with oxygen,

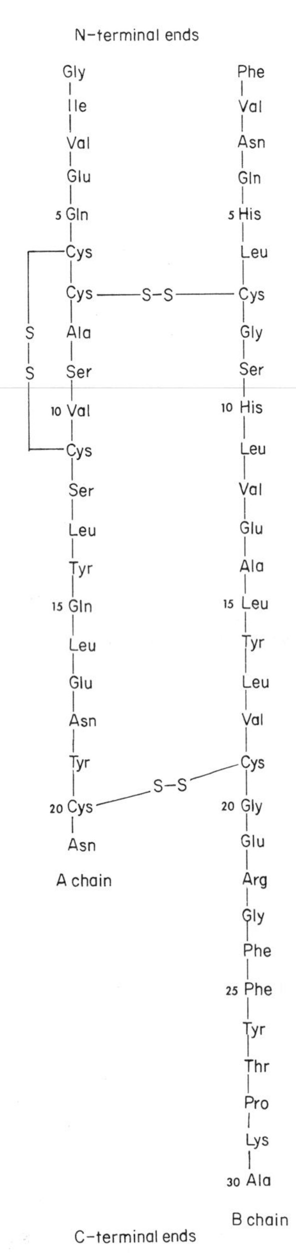

Figure 1.9. *The sequence of amino acids in a molecule of the protein insulin. The amino acid abbreviations are explained in Table 1.1. (From Lehninger, 1975: Biochemistry, 2nd Ed. Worth Publishers, New York.)*

its structure undergoes a change detectable by x-ray crystallography. Such a change may also accompany combination of an enzyme with its substrate (see Chapter 16).

SOLUBILITY

Proteins differ from one another in solubility. In the past, when little was known of their structure, solubility was one of the main criteria for classification of proteins. At present such a classification is interesting primarily from an operational point of view. For example, some proteins (albumins) are soluble in neutral distilled water, some (globulins) in dilute salt solutions, some (glutelins) in weak acid or alkaline solutions, some (gliadins) in alcohol, and some (keratins in fingernails) are relatively insoluble.

PROTEINS AS ZWITTERIONS

All proteins show certain characteristics in common. Like amino acids, proteins are *zwitterions;* that is, they are charged either positively or negatively. At the isoelectric pH they are electrically neutral and will migrate toward neither the anode nor the cathode of an electric field. As with amino acids, the isoelectric pH of a protein depends upon the relative number of basic and acidic groups present in the molecules. An excess of acid groups on a protein molecule—for example, a molecule with many polycarboxy amino acids—shifts the isoelectric point to the acid side of neutrality, as in serum albumin with an isoelectric pH of 4.7. An excess of basic groups on a protein—for example, one with many polyamino acids —shifts the isoelectric pH to the basic side, as in gliadin, with an isoelectric pH of 9.0. On the acid side of the isoelectric pH, the protein dissociates as a base and is positively charged. On the alkaline side, the protein dissociates as an acid and is negatively charged.

When the charges such as those on the amino and carboxyl groups of a protein molecule are neutralized by addition of salts (e.g., ammonium sulfate), proteins may be precipitated or *salted out.* Salting out can be used in purification, since each protein is precipitated by a specific concentration of salt and can be separated in this manner from other proteins. However, at present polyacrylamide gel electrophoresis is most frequently used (see Appendix 5.5).

GLOBULAR AND FIBROUS PROTEINS

Proteins may be globular or fibrous. Globular proteins such as albumin and globulin consist of one or more peptide chains held in configuration by hydrogen bonds and cohesive forces of like-to-like radicals (e.g., methyl groups of neighboring amino acids). Globular proteins unfold when the cohesive and hydrogen bonds are broken by heat or other agents that denature proteins (Johnson *et al.,* 1974).

Two kinds of fibrous proteins have been found in some prokaryotic cells: *flagellin* of which bacterial flagella are made and *pilin* of which the pili (extremely fine filamentous appendages by which some bacteria agglutinate)

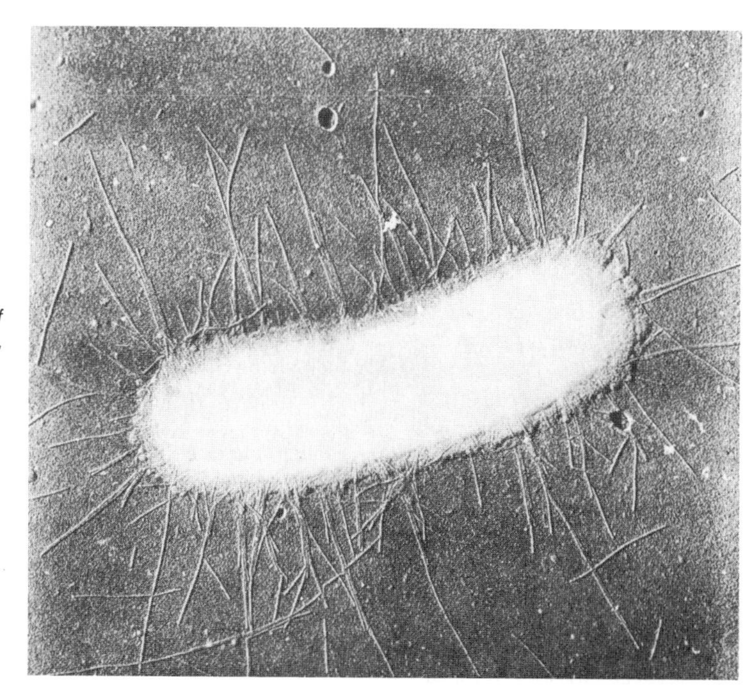

Figure 1.10. *Electron micrograph of* E. coli *showing pili. (From Brinton, Trans. N.Y. Acad. Sci. 27: 1005.)*

are made. Both proteins are unique to prokaryotes.

Flagellin is a protein of molecular weight 40,000. The individual molecules aggregate into three intertwined strands. The flagellar activity arises not in the flagellum but in the basal body, to which each flagellum is attached. A detached flagellum does not rotate. As will be seen later, the bacterial flagellum is distinctive to prokaryotes and quite different in organization and action from the flagellum of a flagellated (or ciliated) eukaryotic cell. Bacterial flagella may be quite long—as much as 10 to 15 μm, many times the length of a bacterium—and 0.01 to 0.035 μm in diameter.

Flagellin contains many residues of the amino acid ϵ-N-methyllysine, which is also found in muscle actin. It is interesting that at the appropriate pH and salt concentration the disaggregated flagellin molecules reaggregate to form a flagellum, indicating spontaneous bonding between the molecules of flagellin.

Pili are of several kinds, some longer than others, all very fine and filamentous (Fig. 1.10). They may be 0.5 to 20 μm in length and 0.005 to 0.020 μm in diameter. Rigid and immobile, they are visible only under the electron microscope, but their agglutinating action results in films of bacteria that are visible on the surface of broth cultures. Pili also agglutinate other cells (e.g., plant cells, yeast, red blood corpuscles, and other animal cells). Pili, in conjugating species of certain bacteria, form the tube for exchange of chromosomal material. Pili are composed of a protein subunit with a molecular weight of about 17,000 arranged helically to form a single rigid filament with a central hollow core.

Both flagella and pili can be removed from a bacterium by mechanical agitation without affecting viability. Some pili act as receptors for phage particles, and others may be associated with the disease-causing activity of some pathogenic bacteria.

NATURE OF STRUCTURAL BONDS BETWEEN PROTEINS

Four main types of interactions are postulated between protein molecules in the cell and shown in Figure 1.11: (1) *Homopolar cohesive "bonds"* are van der Waals forces of the type that hold a paraffin crystal together with a bond strength between 1000 and 2000 calories per mole (I in Fig. 1.11). In proteins, this type of bond is formed by interlaced methyl groups or other hydrophobic groups of adjacent molecules (Table 1.2). Such bonds in proteins are as easily broken in an environment of heat as they are in paraffin. (2) *Heteropolar cohesive bonds,* such as dipole to dipole attraction and hydrogen bonds, result from attraction of neighboring proteins by residual valences (II in Fig. 1.11). These bonds are broken by mild heat, the bond strength being 5000 calories per mole or less. (3) *Heteropolar valency bonds* (Coulomb forces), such as those that form a salt or ester linkage, are stronger than the first two kinds and are not broken by mild heat (III in Fig. 1.11). (4) *Homopolar valency bonds* involve formation of a bridge (as exemplified in the elimination of hydrogen between two sulfhydryl radicals, IV

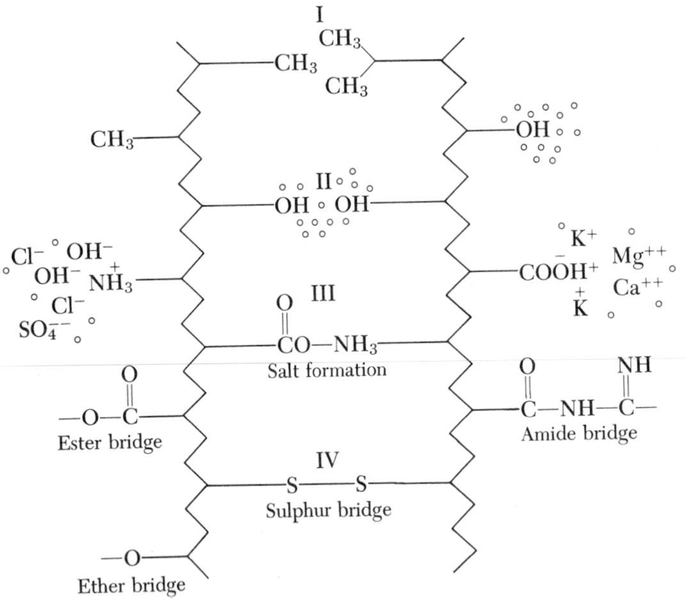

Figure 1.11. *Schematic representation of types of bonds possible between neighboring polypeptide chains of cytoplasmic proteins. The small circles represent water molecules. (From Frey-Wyssling, 1953: Submicroscopic Morphology of Protoplasm. 2nd Ed. Elsevier Publishing Co., Amsterdam.)*

in Fig. 1.11). These, like heteropolar valency bonds, are also relatively strong bonds of the order of 50,000 calories per mole. All these types of bonds are illustrated in Figure 1.11.

The bonds between phosphatides (heteropolar valency bonds) should be added to the list, since they are likely to play a role of importance in explanations of such phenomena as the structure maintained in highly hydrated proteins in which bonds of the first three classes are likely to have been broken. It should also be realized that such ions as magnesium and calcium help in linking carboxyl groups. Water plays an important part in hydrophilic bonding as well, becoming part of the cellular structure via hydrogen bonding.

Weak bonds such as homopolar or heteropolar cohesive bonds are important in many biological interactions. For example, they give shape to polypeptides and polynucleotides. Hydrogen bonds (heteropolar cohesive bonds) are important in maintaining the DNA double helix. Weak bonds are easily broken when they occur singly, but in ordered arrays as in DNA, they exist for a long time. The multiple hydrogen bonds can be broken when the DNA is heated (melting temperature). The atoms are farther apart from one another in a weak bond than in a strong (covalent) bond. For example, hydrogen atoms in a hydrogen molecule (H:H) are 0.0074 nm apart, whereas when held together by a non-polar cohesive bond, the hydrogen atoms are 0.12 nm apart. Strong bonds are unlikely to fall apart under physiological conditions. Therefore, they do not give the structural flexibility to the cytoplasm under changing environmental conditions that weak bonds do. As might be expected, enzymes are not required to break weak bonds.

Primary secondary and tertiary structures occur within the protein molecule. The *primary structure* is that expressed by the structural

TABLE 1.2. HYDROPHOBIC AND HYDROPHILIC RADICALS OF PROTEINS*

Hydrophilic (Lipophobic)		***Hydrophobic (Lipophilic)***	
Carboxyl	—COOH	Methyl	$—CH_3$
Hydroxyl	—OH	Methylene	$—CH_2—$ or $=CH_2$
Aldehyde	—CHO	Ethyl	$—C_2H_5$
Carbonyl	—CO	Propyl	$—C_3H_7$
Amino	$—NH_2$	Alkyl	$—C_nH_{2n-1}$
Imino	=NH	Isoprene	$—C_5H_8—$
Amido	$—CONH_2$	Phenyl	$—C_6H_5$
Imido	—CORNH		
Sulfhydryl	—SH		

*After Frey-Wyssling, 1948: Submicroscopic Morphology of Protoplasm and Its Derivatives. Elsevier, Amsterdam. The solubility of the hydrophilic groups in water decreases progressively from top to bottom, whereas the solubility of the hydrophobic groups in lipids increases from top to bottom. Hydrophilic bonds are attracted to other hydrophilic bonds and hydrophobic to hydrophobic.

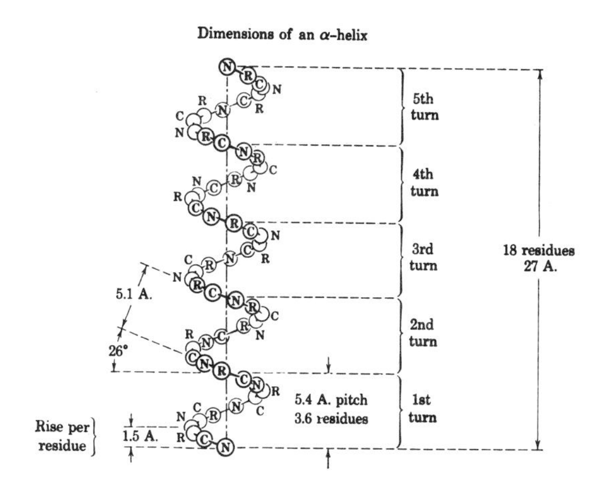

Figure 1.12. The Pauling-Corey 3.6 residue α helix. R refers to a radical (methyl, phenyl, etc.), C stands for the carbon atom and N for the nitrogen atom. (From Anfinsen, 1959: The Molecular Basis of Evolution. John Wiley and Sons, New York.)

chemical formula and depends entirely upon chemical valence bonds: the peptide bonds unite amino acid residues in the peptide backbone of a protein and covalent bonds forming fixed sites of cross linkage, such as disulfide bonds between half-cystine residues. *Secondary structural bonds* are the result of hydrogen bond formation, as, for example, the amide (CONH) linkages between C=O and NH^- groups on amino acid residues of the polypeptide chain. The α helix shown in Figure 1.12 is the result of secondary bonds. The α helix has the maximum number of amide bonds for spirals of various pitch values considered; therefore, it is probably the most stable. *Tertiary structure* and disulfide linkages are considered necessary to stabilize the α helix in solution and probably only those parts of proteins properly anchored by such bonds can maintain the helical configuration. The tertiary structure is the result of (1) van der Waals interactions between hydrophobic groups, agglomerated by mutual repulsion of solvent, and (2) those special hydrogen bonds that exist between hydroxyl groups of tyrosine and the ε amino groups of lysine and various electronegative groups along the peptide chain.

A protein molecule may consist of several polypeptide chains associated either end-to-end or laterally to form a single unit indicating *quaternary structure*. For example, hemoglobin consists of four polypeptide chains, two designated α and two β, each type with a characteristic amino acid sequence. These polypeptide chains take a configuration or characteristic shape within the hemoglobin molecule (Lehninger, 1975).

Membrane-bound enzymes lose most of their activity when removed from the membrane. Addition of membrane lipids often reactivates the isolated enzymes. Some membrane-bound enzymes require association with other enzymes for activity. The enzyme conformation requiring association with a membrane or with other enzymes has been called *quintinary structure* (Hochachka and Somero, 1973).

Lipids

Lipids are a diverse chemical assembly that includes fats, waxes, phospholipids, carotenoids and sterols. Lipids serve not only as food reserves but also in cell structures, notably cell membranes. They have one property in common: they are all soluble in fat solvents.

Fat is a triglyceride ester of glycerol and three fatty acid molecules. For instance, if three butyric acid molecules are linked to glycerol, the product is tributyrin; and if three palmitic acid molecules are linked to glycerol, the product is tripalmitin (Fig. 1.13*A*). Triglycerides are often spoken of as neutral fats, since they have no free acid or basic groups. The three fatty acid molecules in neutral fat may be of two or three kinds. Neutral fats occur in cells primarily as food reserves (Gurr and James, 1971).

Phospholipids, containing phosphorus, are of great importance in the cell. The simplest phospholipid is phosphatidic acid (Fig. 1.13*B*), in which two fatty acids and one phosphoric acid are attached to the glycerol residue. In most phospholipids the phosphoric acid, in turn, is combined with an organic base by another ester linkage, as shown in Figure 1.13*C*.

Since phospholipids possess both acid and basic groups, they behave as zwitterions. Because of their ionic water-attracting (hydrophilic) groups and their fatty acid fat-attracting and water-repelling (lipophilic or hydrophobic, respectively) groups, phospholipids are somewhat soluble in both water and fats (or fat solvents). Therefore, they serve an important role in the cell in binding water soluble and fat soluble compounds together. For example, lecithin, one of the phospholipids (Fig. 1.13*C*), is especially important in the cell membrane; by its hydrophilic groups it maintains the continuity between the aqueous outside and the aqueous inside of the cell, yet fat-soluble materials dissolve in it and enter the cell because of its hydrophobic groups.

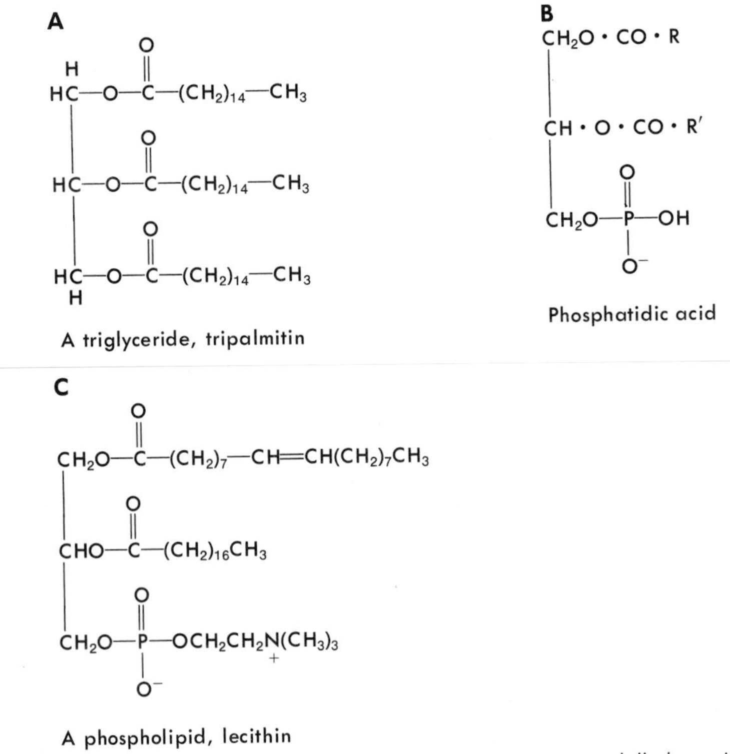

A

A triglyceride, tripalmitin

B

Phosphatidic acid

C

A phospholipid, lecithin

Figure 1.13. *Representative lipids found in cells. Phospholipids are especially important in membranes; neutral lipids (triglycerides) are primarily nutrient reserves.*

Waxes are esters of higher aliphatic alcohols and higher fatty acids of longer chain length than the fatty acids of other lipid classes (generally C_{24} to C_{30} or even to C_{36}).

Carotenoids (carotenes and xanthophylls), an important group of red and yellow pigments present in photosynthetic organisms, for example, blue-green algae, are included among lipids because of their solubility in fat solvents. Chemically, they are tetraterpine hydrocarbons (a terpene has 10 carbon atoms) with alternate double bonds and either acyclic (without a ring)—for example, spirilloxanthin, a carotenoid present in purple bacteria, or cyclic (with a ring)—for example, carotene, a dicyclic tetraterpene present in both prokaryotic and eukaryotic cells.

Carotenes consist of carbon and hydrogen only (Fig. 1.14), whereas xanthophylls contain, in addition, oxygen (in the rings as hydroxy or carbonyl groups, or both).

Carbohydrates

Carbohydrates are composed of carbon, hydrogen, and oxygen present in the proportions of one carbon to two hydrogens to one oxygen

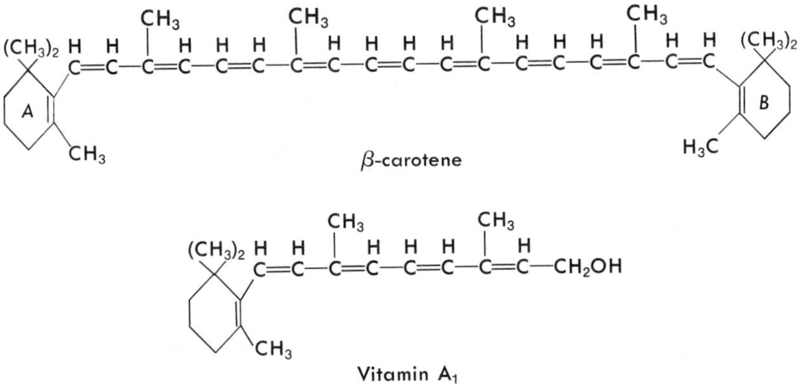

β-carotene

Vitamin A_1

Figure 1.14. *β-carotene and vitamin A_1. Note that vitamin A_1 is essentially half a β-carotene molecule, in which units A and B are mirror images of one another.*

(CH_2O). Representatives of the main groups of carbohydrates are monosaccharides, oligosaccharides (composed of two or more monosaccharides), polysaccharides, and mucopolysaccharides.

The *monosaccharides* are known as pentoses if they have five carbons and hexoses if they have six carbons (Fig. 1.15). Ribose present in RNA and deoxyribose present in DNA are examples of pentoses. Glucose (dextrose, or

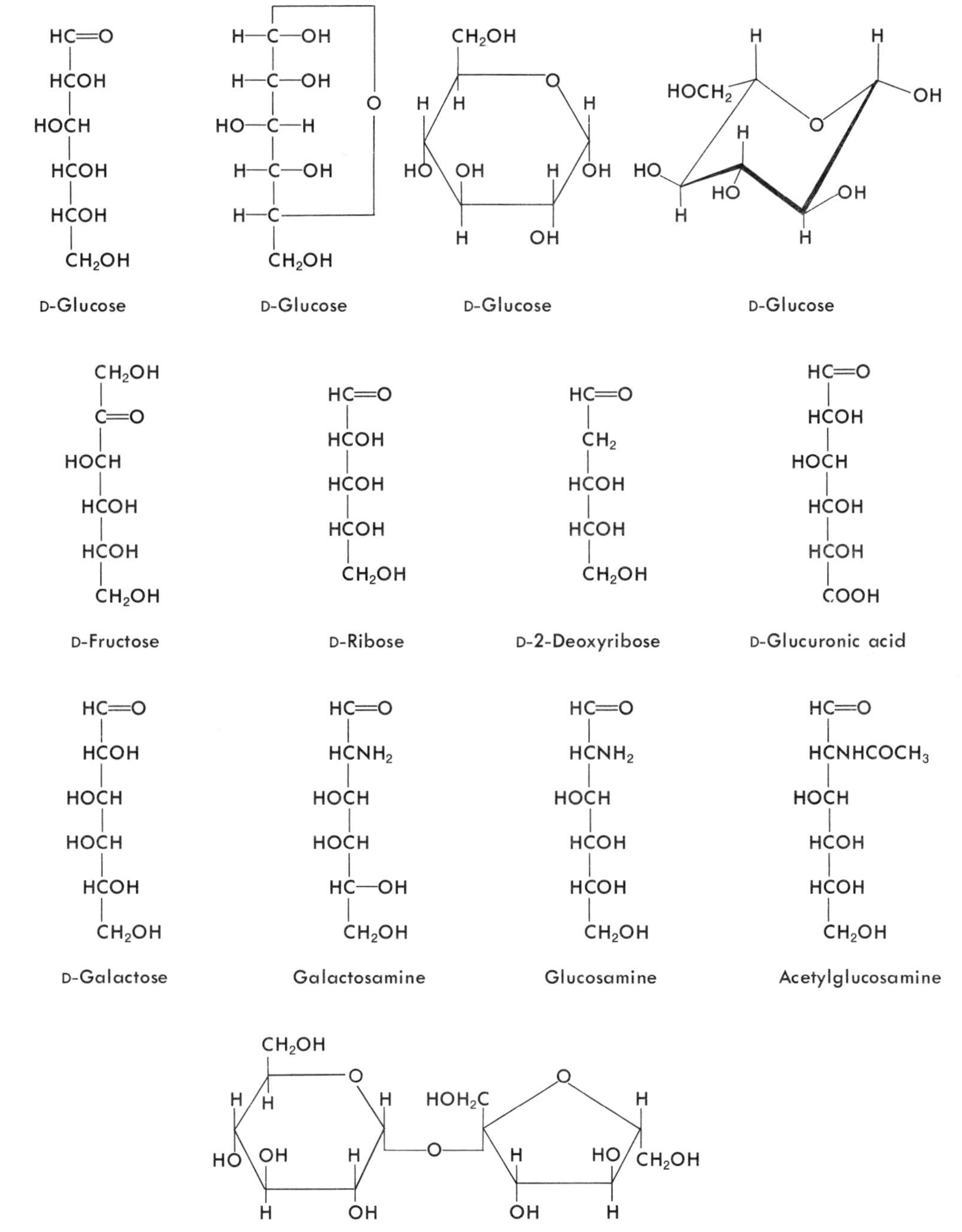

D-Glucose D-Glucose D-Glucose D-Glucose

D-Fructose D-Ribose D-2-Deoxyribose D-Glucuronic acid

D-Galactose Galactosamine Glucosamine Acetylglucosamine

Sucrose

Figure 1.15. *Representative carbohydrates and their derivatives. Each substance may be represented in the four forms shown for glucose. The fourth form of glucose is only one of the three-dimensional conformations.*

grape sugar) and fructose (levulose, or fruit sugar) are examples of hexoses. Glucose, which is of great importance in cellular metabolism, is discussed at length in Chapters 10 and 12. Although it may be stored as glucose, it is usually polymerized to an insoluble form such as glycogen (animal starch) in animal cells and starch in plant cells. There are enzymes in cells that are able to digest these insoluble food reserves, mobilizing them for use when need arises. These monosaccharides are produced by photosynthetic cells, including the prokaryotic blue-green algae. However produced, they are of great importance in nutrition of bacteria.

The commonest *oligosaccharides* are the disaccharides of which the best known is sucrose or table sugar, stored in cells of sugar cane and sugar beets (Fig. 1.15). Another important disaccharide is lactose, present in milk secreted by mammary gland cells. These sugars are also sources of bacterial nutrition.

The *polysaccharides* of prokaryotic cell walls are heteropolymers, chiefly *lipopolysaccharides,* complexes of pentoses and lipids. Bacteria also produce a wide variety of distinctive *mucopolysaccharides* in the gummy capsules outside their cell walls. Mucopolysaccharides are polymers of various monosaccharides and their substituted derivatives (Davis *et al.*, 1973). They are specific to species and even vary within one species. For example, *Streptococcus pneumoniae* has 75 immunologically distinct types. Unique among bacteria is *Streptococcus pyogenes* (of blood poisoning infamy), which produces hyaluronic acid, a mucopolysaccharide composed of glucose derivatives (glucuronic acid and acetylglycosamine). Hyaluronic acid, which is the animal intercellular cement, is also prominent in vertebrate connective tissues. Pathogens that produce hyaluronidase penetrate host tissues effectively (Frobisher *et al.*, 1974). Pectin, a galacturonic acid polymer, is found in the capsules of blue-green algae, as are gummy polymers of glucose (glucans). Cellulose, the main polysaccharide found in higher plant cells, has been described among prokaryotes only in the vinegar bacteria *(Acetobacter xylinum),* which produce a fine surface mat of cellulose fibers in cultures.

Nucleic Acids

Over 80 years ago Miescher extracted nucleic acids and nucleoproteins from pus cells and fish sperm in high concentrations of salt (1 to 2 molar NaCl). He found that about 60 percent of the solids in sperm were nucleic acids, 35 percent were proteins and 5 percent were lipids, salts and carbohydrates. Since then extensive information has accumulated on the chemistry of the nucleic acids. Testes have been widely used in such studies, but in bacteria 1 percent of the net weight consists of DNA and 6 percent of RNA. Therefore, bacteria are also excellent sources of nucleic acids.

In 1944 Avery and co-workers discovered that it was the DNA and not any other substance in an extract of a smooth (virulent) strain of *Pneumococcus* applied to a rough (nonvirulent) strain, which transformed the latter into a virulent strain. Decisive evidence for DNA as the genetic substance in chromosomes was thus provided. Other lines of evidence supported this finding; for example, viral DNA and not protein was the genetic substance of DNA viruses.

The discovery by Chargaff of the molar equivalence of certain bases in DNA (for example, adenine equal to thymine, guanine to cytosine), was later crucial to interpreting the three-dimensional structure of DNA. Meanwhile, the x-ray diffraction studies of DNA by Franklin and Wilkins provided the data for the development of a model of DNA by others. Many of these exciting developments are covered in treatises on nucleic acids (Davidson, 1976).

Two types of nucleic acids have been identified in all cells: deoxyribonucleic acid (DNA) and ribonucleic acid (RNA). The prokaryotic cell chromosome consists of a long molecule of DNA, many times the length of the bacterium, packed in the "nucleoid." Ribosomes consist of ribosomal RNA (rRNA) and protein. Transfer RNA (tRNA) and messenger RNA (mRNA) are two other types of RNA found in all cells.

Each nucleic acid is considered to be composed of nucleotides (Fig. 1.16). Differences between the two nucleic acids include differences in pentoses, pyrimidine bases, and size of molecule. The pentose called deoxyribose and the pyrimidine thymine are characteristic of DNA, whereas the pentose D-ribose and the pyrimidine uracil are characteristic of RNA (Fig. 1.17). The molecular weight of DNA is many millions, whereas that of RNA is usually considered to be smaller. The cellular molecular weight of neither is actually known, since values depend upon the method of preparation. Thus, freshly isolated tobacco mosaic virus RNA has a molecular weight of 300,000 but decomposes spontaneously to units of molecular weight of 61,000. Alkaline hydrolysis decomposes these into units of molecular weight of 15,000 (Spencer, 1972).

DEOXYRIBONUCLEIC ACID (DNA)

The structure of DNA has been the center of research attention given to this molecule, especially studies employing x-rays. X-ray diffraction patterns indicate that dried fibers of pure

Purine, the parent compound

Pyrimidine, the parent compound

Adenine
(6-aminopurine)

Guanine
(2-amino-6-oxopurine)

Thymine
(5-methyl-2,4-dioxopyrimidine)

Cytosine
(4-amino-2-oxopyrimidine)

Uracil
(2,4-dioxopyrimidine)

5-Methylcytosine

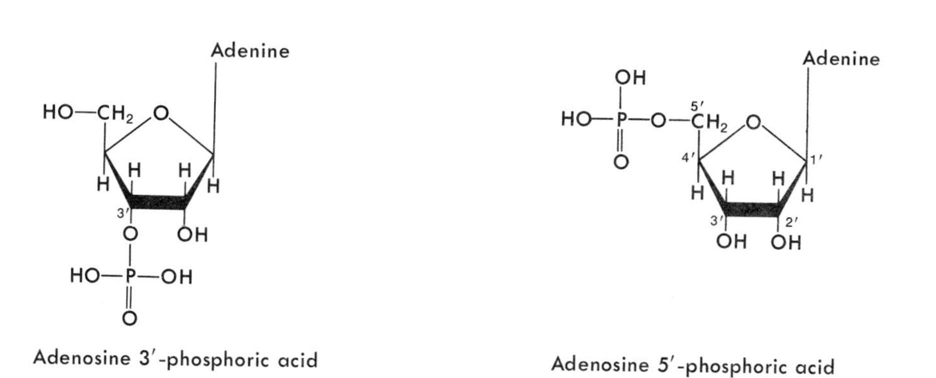

Adenosine 3′-phosphoric acid

Adenosine 5′-phosphoric acid

Figure 1.16. *Purines and pyrimidines found in nucleic acids, nucleotide units. In nucleotides the phosphate group may be in any one of several locations. In deoxyribonucleotides there are only two positions in deoxyribose that can be esterified with phosphoric acid, namely the 3′ and 5′; in ribonucleotides, the phosphate group may be at the 2′, 3′ or 5′ position. Nucleotides that occur in free form in cells are mainly those with the phosphate group in the 5′ position.*

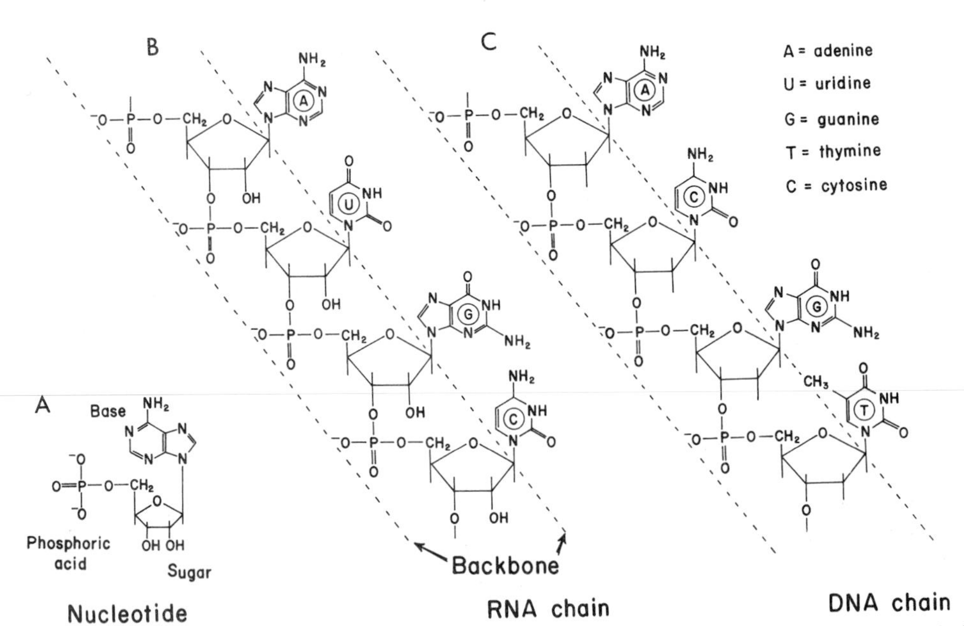

Figure 1.17. *A, Nucleotides present in DNA and RNA strands. A shows a nucleotide, B the RNA chain of nucleotides, A standing for adenine, U for uridine, G for guanidine and C for cytosine bases; C is a DNA chain of nucleotides, T stands for thymine. (After Giese, 1976: Living With Our Sun's Ultraviolet Rays. Plenum Press, New York.)*

DNA have a crystalline structure, although moist fibers form a less perfect pattern (paracrystalline). The x-ray refracting units are much farther apart than would have been anticipated on the basis of a simple linear arrangement of known constituents in nucleic acids (phosphoric acid, pentose, and base). In 1953–54, Watson and Crick found that they could interpret the x-ray-diffraction pattern in the data of Franklin and Wilkins by proposing a double helix model as the structure of a nucleic acid fiber instead of a simple arrangement. When such a linear fiber was coiled into a helix, refraction occurred periodically only from those units that were in line at any one time, their periodicity depending upon the pitch of the helix. From the calculated distances between atoms of the constituents of nucleic acid, these investigators decided that the backbone of the nucleic acid consists of phosphoric acid residues and pentose sugar, and the inward pointing side chains consist of the purine and pyrimidine bases. When the purine base of one helix in the model is juxtaposed to the pyrimidine base of the other helix, the fit of the two helices in the model is excellent (Fig. 1.18).

Calculation of the angles between units in the molecule predicted from x-ray diffraction data showed that this model fits the requirements for the structure of DNA best. Watson and Crick therefore proposed that the two helices of DNA are linked through these bases by way of hydrogen bonds, adenine being linked to thymine, and cytosine to guanine (Fig. 1.19). In extracts of DNA the amount of

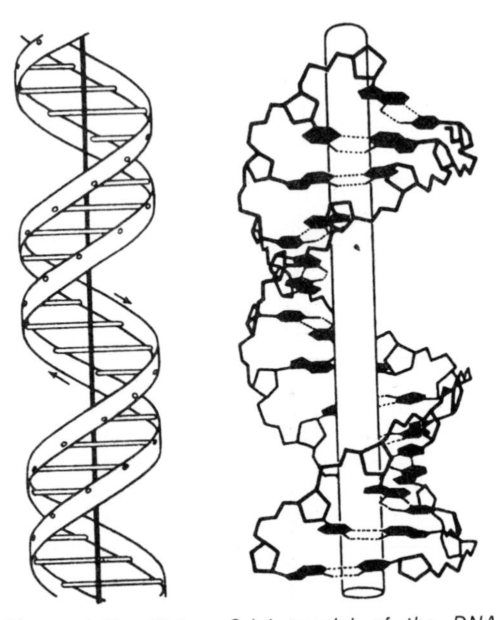

Figure 1.18. *Watson-Crick model of the DNA molecule, on the left shown as consisting of two DNA strands (illustrated in Figure 1.17C) running spirally in antiparallel directions as indicated by arrows; on the right the purine bases are represented by the black hexagonal and pentagonal units, the pyrimidine bases by the black hexagonal units, the hydrogen bonds connecting purines of one chain with pyrimidines of the other shown by dotted lines. The rod in the center of the DNA molecule is a support for the model, not a real structure. (From Davidson, 1977: The Biochemistry of the Nucleic Acids. 8th Ed. Methuen, London.)*

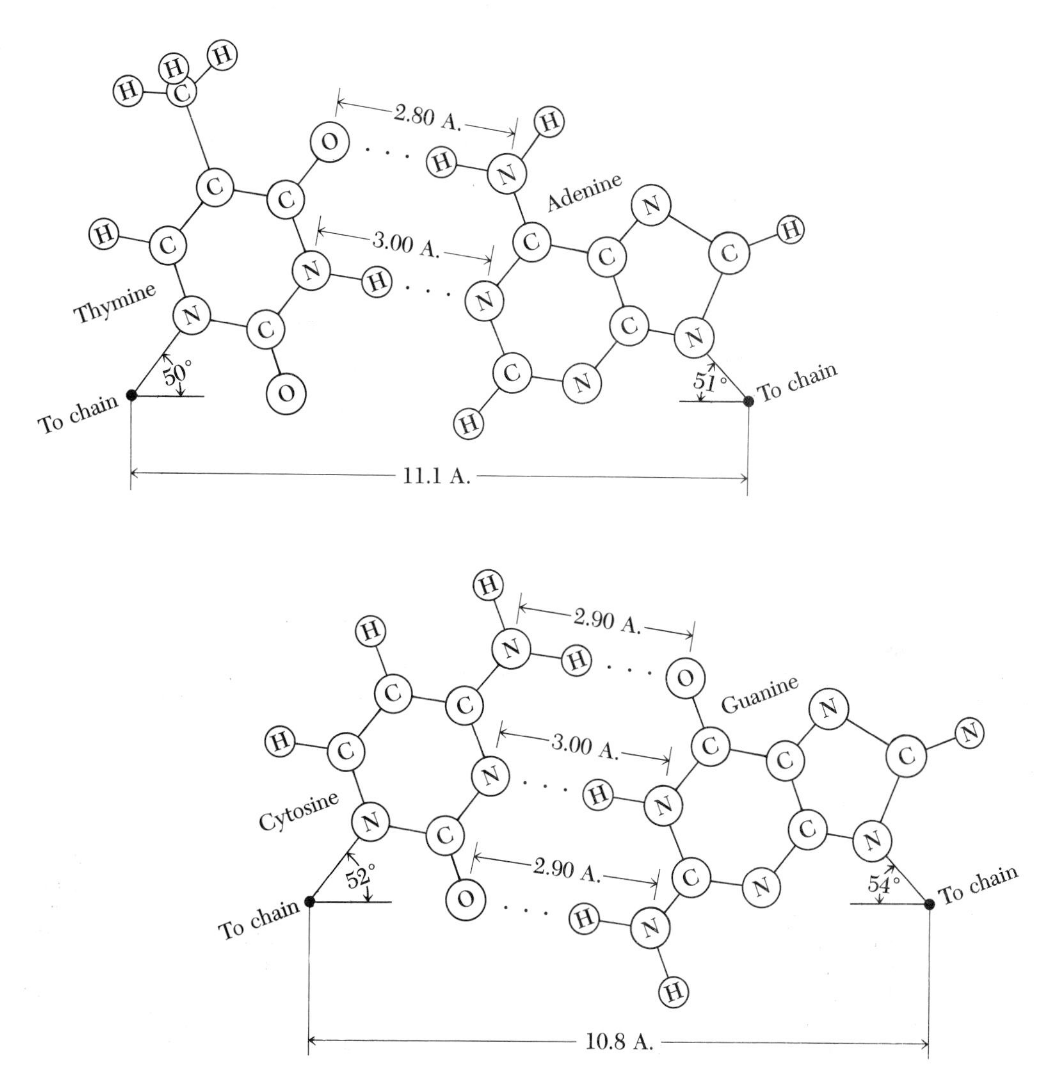

Figure 1.19. *The pairing of adenine and thymine (top) and cytosine and guanine (bottom) by means of hydrogen bonding. (From Anfinsen, 1959: The Molecular Basis of Evolution. John Wiley & Sons, New York.)*

adenine is always equal to that of thymine, and the amount of cytosine is equal to that of guanine, as shown by the earlier work of Chargaff. The only way in which such hydrogen bonding can occur is to have the nucleotide chains antiparallel; in this position the bases are complementary to one another. The sequence reading from the position of the phosphate group (5′ to 3′) is opposite in the two strands. The model therefore satisfactorily accounts for the content of bases in deoxyribonucleic acid.

The puzzling change in the x-ray diffraction pattern between moist and dry DNA is also interpretable with the helical model as being the result of slippage between the two helices in the moist state, causing displacement of the perfect crystalline pattern of the dry fiber. The model also explained how replication occurred: the fiber unwinds at one end. Synthesis of the complementary strand then occurs on each separate strand by DNA polymerase, as described in Chapter 15.

RIBONUCLEIC ACID (RNA)

Every cell contains three kinds of RNA: transfer (tRNA), messenger (mRNA), and ribosomal (rRNA).

Transfer RNA, which transports the activated amino acids during protein synthesis, has a low molecular weight (about 25,000). There are more kinds of transfer RNA than there are amino acids, that is, more than 20; sometimes two of them handle the same amino acid. *Messenger* RNA, which, as its name implies, carries information from DNA to the ribosome where protein synthesis occurs under its direction, has a variable molecular weight, of the order of 500,000. *Ribosomal* RNA is

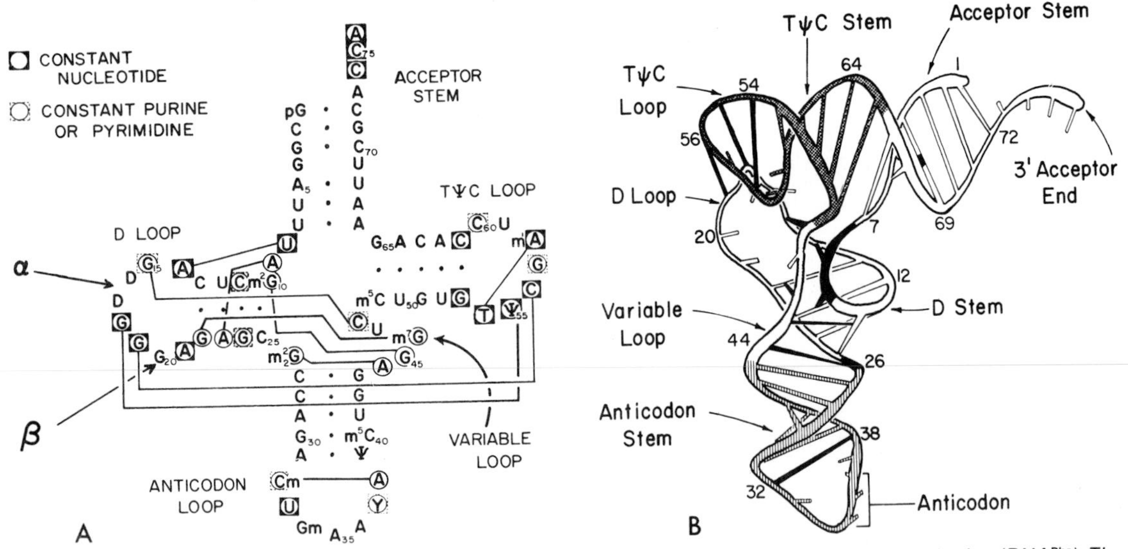

Figure 1.20. A, *Cloverleaf diagram of the nucleotide sequence of yeast transfer RNA for phenylalanine (RNA^Phe). The solid lines connecting circled nucleotides indicate tertiary hydrogen bonding between bases. Solid squares around nucleotides indicate that they are constant; dashed squares indicate that they are always purines or pyrimidines. The regions α and β in the D loop contain one to three nucleotides in different tRNA sequences.* B, *The folding of the ribose phosphate backbone of yeast tRNA^Phe is shown as a coiled tube; the numbers refer to nucleotide residues in the sequence. Hydrogen-bonding interactions between bases are shown as cross rungs. Tertiary interactions between bases are solid black. Bases that are not involved in hydrogen bonding to other bases are shown as shortened rods attached to the backbone. (From Quigley and Rich, 1976: Science* 194: *796–806. Copyright 1976 by The American Association for the Advancement of Science.)*

thought to have a molecular weight of several million, depending upon the method of preparation.

RNA is single stranded. However, the strand may twist back upon itself, so that portions of it may come in contact, at which points connections (by hydrogen bonds) may be formed, giving RNA a pseudohelical structure. Evidence from x-ray diffraction studies and from changes in ultraviolet absorption after breakage of hydrogen bonds by heat indicates that the RNA helix is imperfect, with hydrogen bonds between neighboring guanine and cytosine and between neighboring adenine and uracil residues. The defects in the helix in the form of unpaired bases are reflected in the somewhat unequal proportions of adenine to uracil and guanine to cytosine. The probable structure of a tRNA molecule (phenylalanine tRNA) is shown in Figure 1.20.

It is known that ribosomal RNA is always associated with structural protein in the ribosome. The ribosome contains about 35 to 60 percent protein, which is rich in arginine and lysine residues. The unpaired bases of the RNA helix perhaps serve as attachment sites for the amino acid residues of the protein. Pure ribosomes probably contain no lipid (Nomura *et al.*, 1974).

Salts

Salts are required in nutrient media for all prokaryotes. While considerable information on these requirements is available, salt requirements are discussed in Chapter 3 in connection with eukaryotic cells because the examples have more relevance to our personal experience. However, in addition to the salt requirements listed for *E. coli,* prokaryotes require many micronutrients, as do eukaryotic cells (Chapter 3). Many of these micronutrients are needed for enzyme function (see Chapters 10 and 12).

VARIATION IN ORGANIZATION OF PROKARYOTIC CELLS

Because bacteria have been extensively investigated, they alone will be considered here, but much of what is said is applicable to blue-green algae, as well. Indeed, blue-green algae are sometimes classified as the Cyanobacteria. Blue-green algae are considered in Chapter 2.

Bacterial cells are covered with a protective cell wall from which the contents can be made to retract if the bacterium is placed in a concentrated solution of sugar or salt. In such circumstances, the cell membrane (plasma membrane) forms the outer boundary of the living material of the cell, now called the *protoplast* (Fig. 1.21). The cell wall may be digested with the enzyme lysozyme, exposing the protoplast free of the cell wall; the protoplast will swell or shrink on decrease or increase of the salt concentration dissolved in the medium. The cell

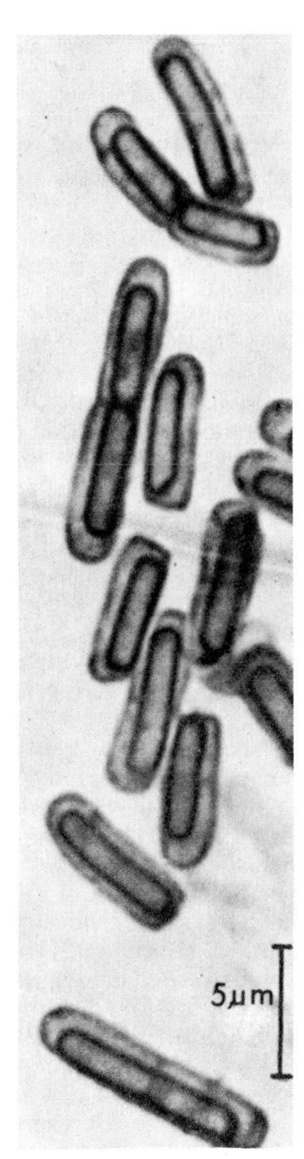

Figure 1.21. Retraction of the cell contents from the cell wall induced by exposure to ether vapor and drying on glass before fixation. Stained with Victoria blue. (From Robinow, 1960: In The Cell. Vol. 4. Brachet and Mirsky, eds. Academic Press, New York.)

wall apparently plays a protective function but is not essential to initiate growth, since protoplasts grow. But bacteria with cell walls resist osmotic imbalance and other unfavorable conditions better than protoplasts (Frobisher *et al.*, 1974).

In some bacteria, for example, *Bacillus megatherium,* RNA is present in the cell surface. It takes up the Gram stain consisting of crystal violet and iodine and is therefore said to be gram positive. On the other hand some bacteria, which lack RNA in the surface and have a lipid coating, for example, *E. coli,* do not take up the Gram stain and are called gram negative. Thick-walled resting stages called *spores,* which resist high temperature and other unfavorable conditions, are formed by some bacteria (for example, the genera *Bacillus* and *Clostridium*).

Bacteria are filled with ribosomes. As in eukaryotic cells the ribosomes cluster to form polyribosomes during the synthesis of proteins (see Chapters 7 and 15). Prokaryotic ribosomes are smaller than those in the cytoplasm of eukaryotes. The plasma membrane, which is rich in oxidative enzymes, probably performs the function usually carried out by the mitochondria of plant and animal cells. It is of interest that the plasma membrane of the large bacterium *Thiovulvum majus* sinks deep into the cytoplasm, forming a multilayered structure called the *mesosome.* This is true of many bacteria, although the size of the mesosome varies. The tubercle bacillus has concentric cell membranes, suggesting incipient mitochondria.

Some bacteria have even more elaborate membrane systems, comparable to photosynthetic chlorophyll membranes of blue-green algae, for example, *Nitrosocystis oceanus,* a chemosynthetic bacterium (Fig. 1.22) that oxidizes ammonium salt to nitrate. Elaborate membrane systems appear when there is considerable enzymatic activity (Watson, 1976).

Many bacteria have *nucleoids* that grow and divide and stain with the dyes that affect eukaryotic chromosomes. These bodies condense before division and become more diffuse thereafter. In some bacteria the mesosomes are attached to the nucleoids and act almost as centrioles of eukaryotic cells in guiding the nucleoids to the ends of a dividing cell (Frobisher *et al.*, 1974).

Cell walls of prokaryotes, both bacteria and blue-green algae, are composed of heteropolymers (polymers made up of several types of molecules) called peptidoglycans or mucopeptides. The molecules in the polymers are acetylglucosamine and acetylmuramic acid. Linked to the muramic acid is a short peptide that varies in composition with the species. The polymers are cross-linked with short peptides to make a grill. Muramic acid, diaminopimelic acid, and D-amino acids are unique to prokaryotes.

The peptidoglycans alone do not constitute the bacterial cell wall, which is very complex and consists of a number of layers. The cell walls of gram-negative bacteria are thinner (about 10 to 15 nm) and contain 5 to 15 percent peptidoglycans, 35 percent phospholipids, about 15 percent protein, and up to 50 percent lipopolysaccharide. Cell walls of gram-positive bacteria are about twice as thick as those of the gram-negative kind and are made up of about 20 to 80 percent peptidoglycan, 1 to 3 percent

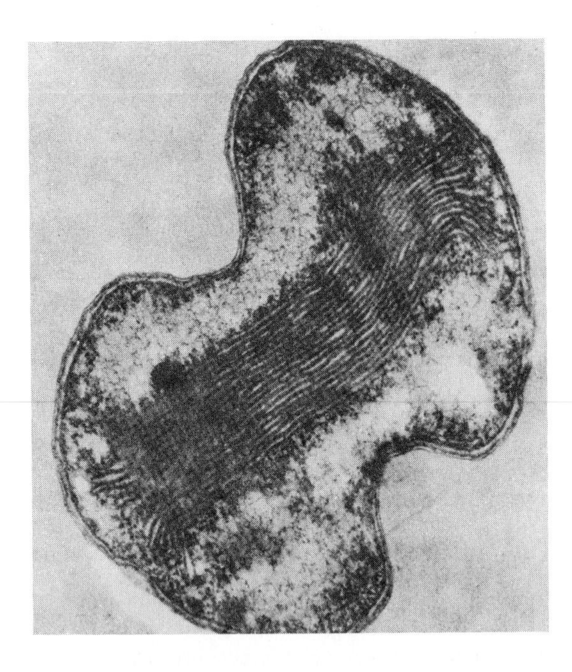

Figure 1.22. *Thin section of* Nitrosocystis oceanus *showing the beginning of constrictive division and a membranous organelle consisting of 20 vesicles so flattened that the lumen is only 10 nm thick; these lamellae almost completely traverse the organism, cutting across both cytoplasm and nucleoplasm. Such membrane systems may be specialized mechanisms for acquiring energy from conversion of* NH_4^- *to* NO_2^-*. (From Murray and Watson, 1965: Structure of* Nitrosocystis oceanus *and comparison with* Nitrosomonas *and* Nitrobacter*. Courtesy R. G. E. Murray and permission of the American Society for Microbiology.)*

lipid and some teichoic acids, as well as other substances. (Teichoic acid is a polymer containing chains of glycerol or ribityl residues joined to each other by phosphate groups.) The peptidoglycans permeate the cell wall in a more homogeneous way than in gram-negative bacteria.

Penicillin stops cell division in gram-positive bacteria because it inhibits development of the cell wall in growing bacteria. Streptomycin is bacteriostatic to many gram-negative forms.

Outside the cell wall of most, if not all, bacteria is found an encapsulating layer of slime (capsule) or firmer matter made up of polysaccharides, lipid, and sometimes other materials. These encapsulating materials are important to bacteriologists as antigens, but to bacteria they are primarily additional mechanical protection.

At present investigators are fractionating cells and purifying bacterial cell walls to study their constituents biochemically as well as by electron microscopy and x-ray analysis. It has become evident, however, that the cell walls of bacteria may serve as a permeability barrier to a greater extent than was previously recognized, acting as a molecular sieve by excluding some molecules, such as the antibiotic actinomycin D in *E. coli.* When the cell wall is removed by lysozyme, actinomycin D enters, exerting its characteristic inhibition of DNA-dependent RNA synthesis. Similar results have been found for other antibiotics that fail to affect normal cells. Cell walls may also act as ionic exchangers and as adsorbants for proteins (Chrispeels, 1976).

Rickettsias, which cause Rocky Mountain spotted fever, endemic typhus, and murine typhus, and *Chlamydia,* which cause psittacosis, are of about the size of small bacteria and in general their organization clearly places them in the same category as bacteria, where they are now classified. However, to some workers they appear to be escaped mitochondria, and to others, a stage between viruses and bacteria. Like bacteria, they possess a number of respiratory enzymes. With one exception, they have been cultured only in host cells.

THE PLEUROPNEUMONIA-LIKE ORGANISMS (PPLO)

The smallest cells known are found in the group of organisms formerly known as pleuropneumonia-like organisms (PPLO; now called Mycoplasmataceae or just mycoplasmas) (Fig. 1.23). *Mycoplasma laidlawii,* a free-living strain growing in sewage, is the smallest cell so far studied. It has a definite life cycle. In culture it exists in bodies of three sizes, minute elemental bodies 0.1 μm in diameter, intermediate-size bodies and full-size cells about 1 μm in diameter. Studies indicate that the small bodies develop into large cells, which divide to form more of their kind. *Mycoplasma gallisepticum,* the cause of chronic respiratory disease in poultry, is somewhat larger (about 0.25 μm) than *M. laidlawii* and has been especially useful in biochemical studies.

Chemical analysis of *M. gallisepticum* indicates that the nonaqueous portion of the cell contains 4 percent DNA and 8 percent RNA, each in greater proportion to other compounds

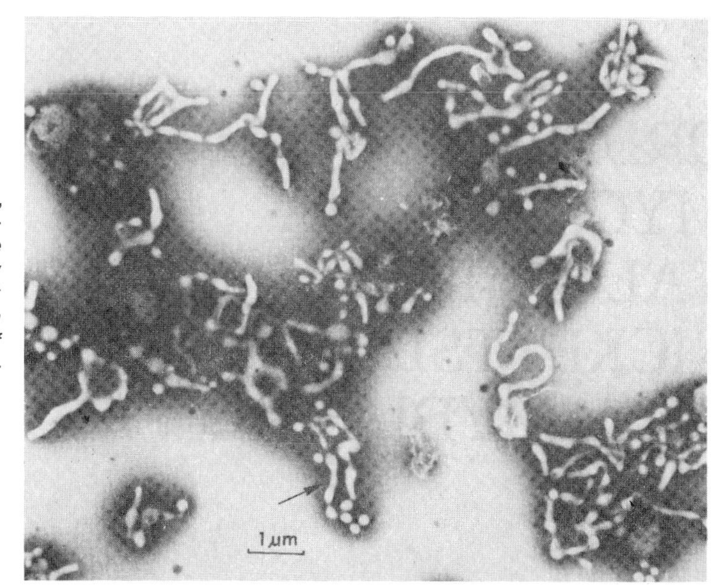

Figure 1.23. Mycoplasma pneumoniae, *negatively stained. The electron micrograph shows numerous forms of variable morphology, some of which are clearly ring-shaped with lobes and some that have beaded filaments (arrow). (From Frobisher* et al., *1974: Fundamentals of Microbiology. W. B. Saunders Co., Philadelphia.)*

than is found in eukaryotic cells but in keeping with, although greater than, that found in bacteria. The DNA is a two-stranded circular filament. The RNA, combined with protein, forms granules resembling ribosomes; tRNA and mRNA are also present. The proteins are similar to those of larger cells; over 40 different kinds of enzymes have been identified. Lipids are also present, among them cholesterol and cholesterol esters, compounds that are characteristic of animal cells and not found in bacteria or blue-green algae (Hayflick, 1969; Smith, 1971).

It has been estimated that at least 100 kinds of enzymes are necessary for cell metabolism and that more than one molecule of each must be present because thermal agitation in the cell leads to degradation of some of them. Furthermore, these enzymes must be encoded, and it presumably takes a DNA molecular weight of about 1,000,000 daltons to encode one of them. Even a *M. gallisepticum* cell of diameter 0.25 μm has only enough DNA to make one molecule of molecular weight 45,000,000 daltons or 45 DNA molecules of molecular weight 1,000,000 daltons. Thus, although the biochemistry of the smallest cells is evidently not unique, these cells pose problems that are difficult to answer (Razin, 1973).

PPLO pose problems because, although they resemble bacteria in many respects and are prokaryotes, they have some resemblances to eukaryotes: the sterols mentioned earlier and the lack of cell walls. The plasma membrane is flexible and electron micrographs disclose it to be a "unit membrane" similar in dimensions to the unit membrane of cells in general.

Free-living PPLO might perhaps serve as a model of the minimum size that the cells that evolved on the primitive earth had to attain to be able to carry out all their functions. They live in a complex medium from which they draw many essential nutrients, much as we think primitive cells did. However, present-day PPLO could be bacteria that have lost the capacity to manufacture many of their constituents because of life in an environment providing these substances and have secondarily diminished in size.

ORGANIZATION OF VIRUSES

Viruses are parasitic on living cells but are not themselves cells. They replicate using the biochemical apparatus of bacteria or eukaryotic cells for their own benefit.

Viruses are much smaller than common bacteria. For example, the large vaccinia virus measures 0.21 μm in diameter and a small virus such as that causing hoof and mouth disease is only 0.01 μm in diameter. The bacterium *E. coli* measures about 1 μm in diameter. However, some PPLO, which have many of the characteristics of bacteria, have "small bodies" in the life cycle that measure only 0.125 to 0.150 μm in diameter. Viruses depend for their existence upon host cells; they lack enzymes for energy-yielding reactions. Some viruses possess enzymes enabling them to penetrate cells, presumably digesting the cell wall in bacteria and mucopolysaccharides in animal cells (Fenner *et al.*, 1974). Viruses may be of various shapes. Some have cubic symmetry, the most common form being the icosahedron (a solid with 20 faces, each of which is an equilateral triangle), as seen in adenovirus, herpesvirus, and polyomas. Some have helical

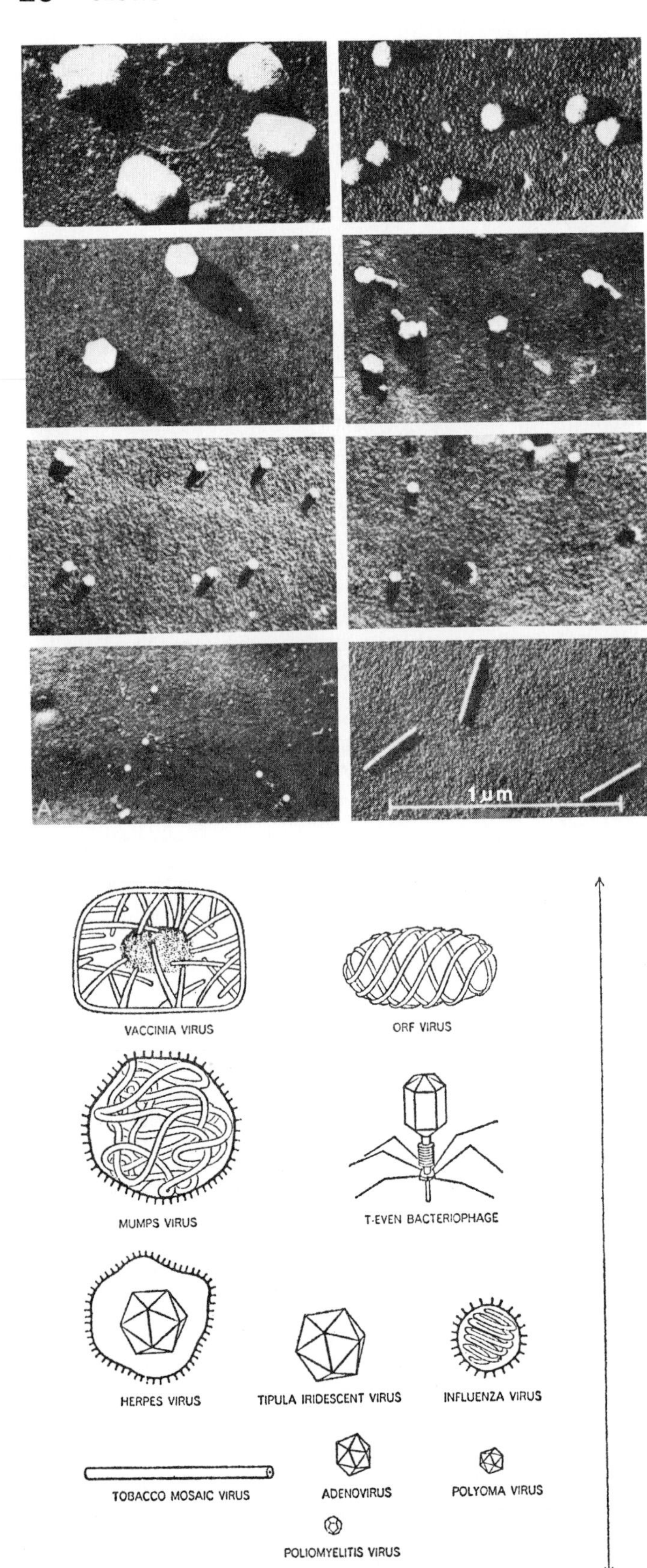

Figure 1.24. *Electron micrographs showing the variation in size and shape of animal, plant, and bacterial viruses. From top to bottom: left, vaccinia,* Tipula iridescens, *T3 bacteriophage, poliomyelitis. Right. influenza, T2 bacteriophage, rabbit papilloma, tobacco mosaic. (From Fraenkel-Conrat, 1962: Design and Function at the Threshold of Life: The Viruses. Academic Press, New York.) See also Figure 7.12.* B, *Diagrams of various viruses. The arrow represents 1 µm (× 100,000). (From Smith, 1965: The Biology of Viruses. Oxford University Press, London.)*

symmetry, as seen in tobacco mosaic virus, the myxomas of influenza, and others have complex symmetry, as in the T-even phages of *E. coli,* which resemble tadpoles, having a body and a tail (Fenner *et al.*, 1974) (Fig. 1.24).

All viruses consist of a protein coat over a nucleic acid core. Unlike cells, viruses contain only one kind of nucleic acid; RNA is present in plant viruses, while either DNA or RNA is present in animal and bacterial viruses.

Viruses may contain double-stranded DNA, as in the T-even phages of *E. coli,* single-stranded DNA, as in ϕX174 and S13 phages of *E. coli,* or RNA, as in some bacteriophages (MS2), some small animal viruses, and all plant viruses. RNA, too, may be single-stranded as in poliovirus, or double-stranded, as in many other viruses. One virus, satellite tobacco necrosis virus, even is parasitic on another virus; it has an RNA genome only sufficient to code for its protein coat and is dependent on its associated virus for replication (Matthews, 1971). Viruses, like cells, can mutate and change characteristics, for example, virulence.

In general, animal viruses appear to be different from bacterial viruses (bacteriophages or phages), but less is known of them because, like plant viruses, they present so many experimental difficulties. All viruses lack the unit membranes characteristic of cells, but they are delimited from the medium by a protein envelope called a *capsid.* In addition, some viruses in mammalian cells have an envelope formed of lipids and lipoproteins (Fenner *et al.*, 1974). Viruses lack organelles characteristic of bacteria, and represent a different category of organization from cells.

Phages, the viruses attacking bacteria, have been most extensively studied and form the basis for the following discussion. Marker techniques demonstrate that, once the phage enters the bacterium, the phage nucleic acid takes command, induces hydrolysis of bacterial nucleic acids, and commands synthesis of virus nucleic acid and protein. The virus protein coat is synthesized from molecules absorbed by the bacteria. In *E. coli* phage, packaging of virus nucleic acid in protein coats begins about 10 minutes after infection and continues for another 20 minutes, when the process is complete. After a delay, which varies with the virus and conditions, the bacterium is ruptured (lysed), liberating between 20 and 100 complete phage particles. When small viruses are involved, many thousands of particles may be produced (Watson, 1976).

Phages may behave in one of three ways. First, they may replicate and lyse the bacteria as just described (phages that cause lysis are called virulent or *lytic phages*). Second, some small phages may replicate and be extruded individually without lysing bacterial hosts. Third, phages may become incorporated into the chromosome of the bacterium and remain in nonreplicating (repressed) state, except that phage nucleic acid incorporated into the host chromosome divides when the host cell genome replicates. That the phage has altered the hereditary characteristics of the bacteria is shown by the resistance (immunity) of the bacteria so infected to attacks by homologous (similar) phages. This is achieved by synthesis of a repressor that prevents multiplication of homologous phages (Frobisher *et al.*, 1974). Phages that "hide" in this manner, dividing only when the bacterium divides, are temperate or *lysogenic* phages (i.e., giving rise to lytic phages). Occasionally the lysogenic phage is derepressed and undergoes rapid replication, usually when its host cell has been injured. Nothing comparable to phage lysogeny has been demonstrated in plant or animal viruses. However, some tumor-inducing animal viruses transform host cells, changing their growth characteristics, a process similar in nature to lysogeny (Fenner *et al.*, 1974).

The process of damaging bacteria infected with lysogenic phages by radiation or other unfavorable agents leads to rapid development of the phage into the lytic form. The particles are then released in infectious form. Lysis of an occasional bacterium in an undamaged population can spread the infection to many previously uninfected members of the population, in this way maintaining a mild infection. Lysogeny indicates the intimate relationship between phage and bacterium, a concept of great importance in microbial genetics (see Chapter 7).

ENERGY RELEASE IN PROKARYOTES

Energy by which living machines operate ultimately comes from oxidation-reductions. *Oxidation* is any reaction in which electrons are passed from one substance (electron donor) to another (electron acceptor); this passage proceeds down a potential gradient, in the course of which energy is released, equivalent to the number of electrons passed multiplied by the potential difference between donor and acceptor (see Chapter 11). *Fermentation* is an oxidation occurring in the absence of oxygen. In fermentations, hydrogen ions are usually transferred along with the electron; therefore, the oxidation-reduction may be represented in the following manner:

$$DH_2 \rightarrow D + 2H^+ + 2e$$
$$2H^+ + 2e + A \rightarrow AH_2 + \text{energy} \qquad (1.1)$$

DH_2 is the hydrogen donor, A is the hydrogen acceptor. To yield energy, AH_2 must have less chemical potential energy than DH_2. The difference in chemical potential energy (roughly the difference in energy content) is the energy available to the organism. Usually the energy is stored as a high-energy phosphate bond (see Chapter 12), the commonest of which is the nucleotide adenosine triphosphate, ATP. On breakdown of the high-energy phosphate, inorganic phosphate and adenosine are given off and energy is liberated to do the work of the cell, such as movement and synthesis. The details of these reactions are discussed in Chapter 12.

Some steps in fermentation of organic compounds also involve decarboxylations, that is, removal of CO_2 from a carboxyl (COOH) group in a molecule. Such reactions are catalyzed by enzymes known as decarboxylases. In the anaerobic period of life on earth, fermentations were the only means for liberating energy for life processes. Presumably, the kinds of fermentations that occur today are a relic of those times; for example, decarboxylation of pyruvic acid yields acetaldehyde. In yeast, acetaldehyde acting as a hydrogen acceptor is reduced to ethyl alcohol. In the presence of oxy-

TABLE 1.3. PRINCIPAL CLASSES OF CELLULAR SUGAR FERMENTATIONS*

Class of Fermentations	***Principal Products from Hexose Sugars***	***Responsible Organisms***
Alcoholic	Ethanol (CH_3—CH_2OH) CO_2	Yeasts Rarely bacteria (e.g., *Zymomonas*)
Lactic (homo-lactic)	Lactic acid (CH_3—CHOH—COOH)	Certain lactic acid bacteria (e.g., *Streptococcus lactis*) (Some protozoa and fungi) (Mammalian tissue cells)
Mixed lactic (hetero-lactic)	Lactic acid Ethanol CO_2	Certain lactic acid bacteria (e.g., *Leuconostoc*)
Mixed acid	Lactic acid Ethanol Acetic acid (CH_3—COOH) H_2 and CO_2 or formic acid (HCOOH)	The colon-dysentery-typhoid group of bacteria (e.g., *E. coli*) Some pseudomonads
Butylene glycol	Similar to mixed acid but including 2,3 butylene glycol (CH_3—CHOH—CHOH—CH_3)	The genus *Aerobacter* and related organisms *Bacillus polymyxa* Some pseudomonads
Butyric-butanol-acetone	Butyric acid (CH_3—CH_2—CH_2—COOH) Acetic acid H_2 CO_2 Sometimes butanol (CH_3—CH_2—CH_2—CH_2OH) Ethanol Acetone (CH_3—CO—CH_3) Isopropyl alcohol (CH_3—CHOH—CH_3)	Certain anaerobic bacteria (e.g., *Clostridium*) *Bacillus macerans* (Some anaerobic protozoa)
Propionic	Propionic acid (CH_3—CH_2—COOH) Acetic acid CO_2	*Propionibacterium* and related anaerobic bacteria

*From Stanier, Doudoroff and Adelberg, 1965: The Microbial World, 2nd Ed. Prentice-Hall, Englewood Cliffs, N.J., p. 263.

TABLE 1.4. HEAT OF COMBUSTION OF GLUCOSE FERMENTATION PRODUCTS*

Product	Carbons	Heat of Combustion kcal/mole
Ethanol	2	327.6
Lactic acid	3	326.0
Acetic acid	2	209.4
Butyric acid	4	524.3
Butanol	4	638.6
Acetone	3	426.8
Isopropyl alcohol	3	474.8
Propionic acid	3	367.2
Formic acid	1	62.8

*Data from Handbook of Chemistry and Physics, Chemical Rubber Publishing Co., Cleveland, Ohio. The heat of combustion of pyruvic acid is 279.1 kcal per mole and that of acetaldehyde, 279.0 kcal per mole.

gen yeast can oxidize alcohol completely to carbon dioxide and water, but in the absence of oxygen it produces alcohol until its growth is inhibited by the accumulation of alcohol.

A wide variety of fermentative products are produced by various organisms.* Some of these are shown in Table 1.3. No one organism produces all of them. For example, a lactic acid bacterium (e.g., *Lactobacillus bulgaricus*) produces lactic acid, while another bacterium, *Clostridium acetobutylicum,* produces butyl alcohol, acetone, and ethyl alcohol, as well as H_2 and CO_2. Propionic acid bacteria produce propionic acid and CO_2. All of these and others are used in industrial processes. As an example, propionic acid bacteria are used in fermenting milk in cheese manufacturing. The holes in Swiss cheese are made by carbon dioxide, while the flavor comes largely from propionic acid.

During the anaerobic phase of the earth's history, fermentation was the only mode of obtaining energy. However, it is quite inefficient. A considerable mass of substrate is needed, because the products still retain much of the chemical potential energy of the organic compounds (Broda, 1975). A measure of the remaining energy is obtained by determining the heats of combustion of the compounds. Some values given in Table 1.4 illustrate the amount of energy remaining in some of these products. Reduced compounds required for syntheses were obtained by an anaerobic oxidation of glucose resulting in the formation of pentoses (pentose shunt), as discussed in Chapter 12.

*Some workers consider methanogenic (methane-producing) anaerobic bacteria as perhaps the most primitive of cells. These bacteria not only reduce CO_2 to CH_4 but fix CO_2 in a manner unique to them. The sequence of bases in their RNA is different from that in other cells. It is still too early to evaluate their position in the scheme of classification (Maugh, 1977: Science *198*: 812).

ANAEROBIC PHOTOSYNTHETIC BACTERIA

Some bacteria synthesize their own cell substance from carbon dioxide and other inorganic compounds using the energy of sunlight. Unlike green plants, they use hydrogen sulfide or other hydrogen donors instead of water:

$$CO_2 + 2H_2S \rightarrow CH_2O + H_2O + 2S \quad (1.2)$$

Photosynthesis is an oxidation reduction, the hydrogen donor in this case being H_2S, the hydrogen acceptor CO_2. The oxygen from carbon dioxide appears in the water. Other photosynthetic bacteria use organic compounds as hydrogen donors to convert CO_2 into cell substance. Since hydrogen sulfide is limited in quantity, except as cycled from decomposing organisms, and organic compounds are hardly an efficient source of hydrogen donors when their supply in the environment is limited, photosynthetic bacteria probably represent a blind alley along which bacterial evolution travelled.

It is interesting that some photosynthetic bacteria (e.g., *Ectothiorhodospira mobilis*), carry their pigmentary complexes of chlorophyll and carotenoids on assemblies of tubular membranes (Fig. 1.25) superficially like those in the chemosynthetic bacterium *Nitrocystis oceanus* described earlier, and suggestive of thylakoids, or in chromatophores (Fig. 1.26), small spherical bodies that in aggregate expose a considerable area of pigmented membrane to light. However, bacteriochlorophyll differs in detail from chlorophyll found in blue-green algae and higher plants, and absorbs at longer and shorter wavelengths than the chlorophyll of other photosynthetic organisms (Fig. 2.4). The photosynthetic bacteria also lack the photochemical system that evolves oxygen from water in other photosynthetic organisms (see Chapter 14).

For photosynthesis to have impact on biological evolution a major step was still needed: the use of water as the hydrogen donor with the evolution of oxygen as a byproduct. This was accomplished by the blue-green algae as recounted in the next chapter.

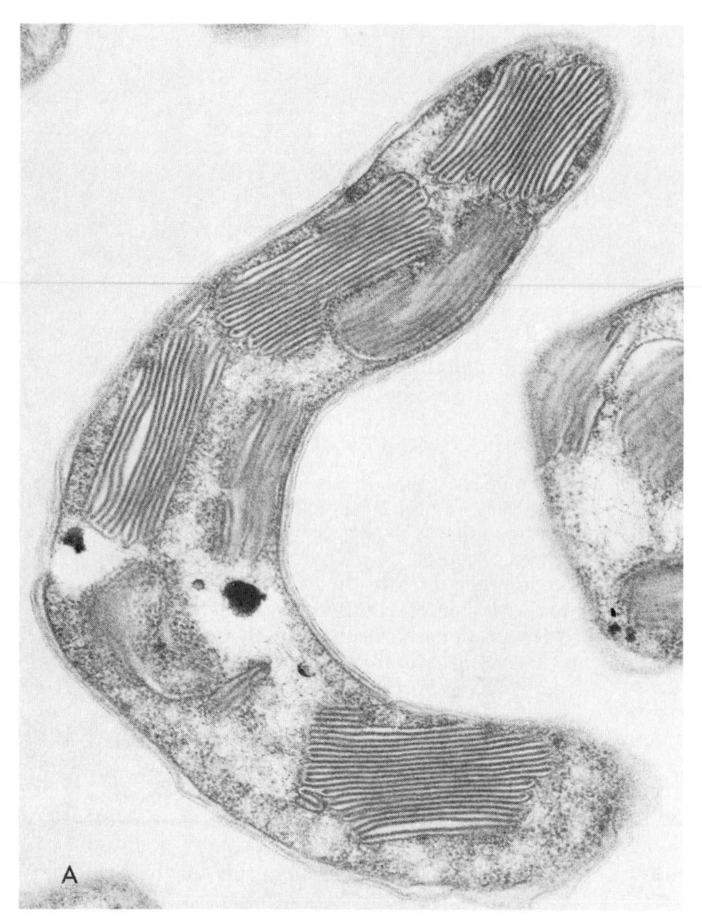

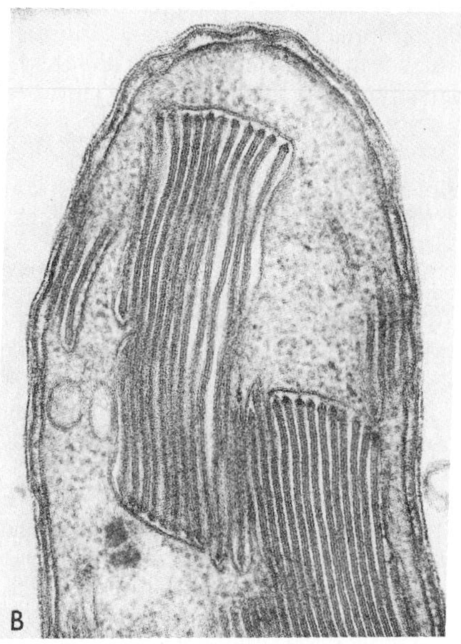

Figure 1.25. *Membranes containing bacteriochlorophyll and carotenoids found in the bacterium* Ectothiorhodospira mobilis, *a photosynthetic sulfur bacterium.* A, *The entire organism showing how extensively the pigment-bearing membranes have developed (× 86,200).* B, *A small section of the bacterium at higher magnification to show details of the membrane assemblies (× 123,000). (By courtesy of S. W. Watson.)*

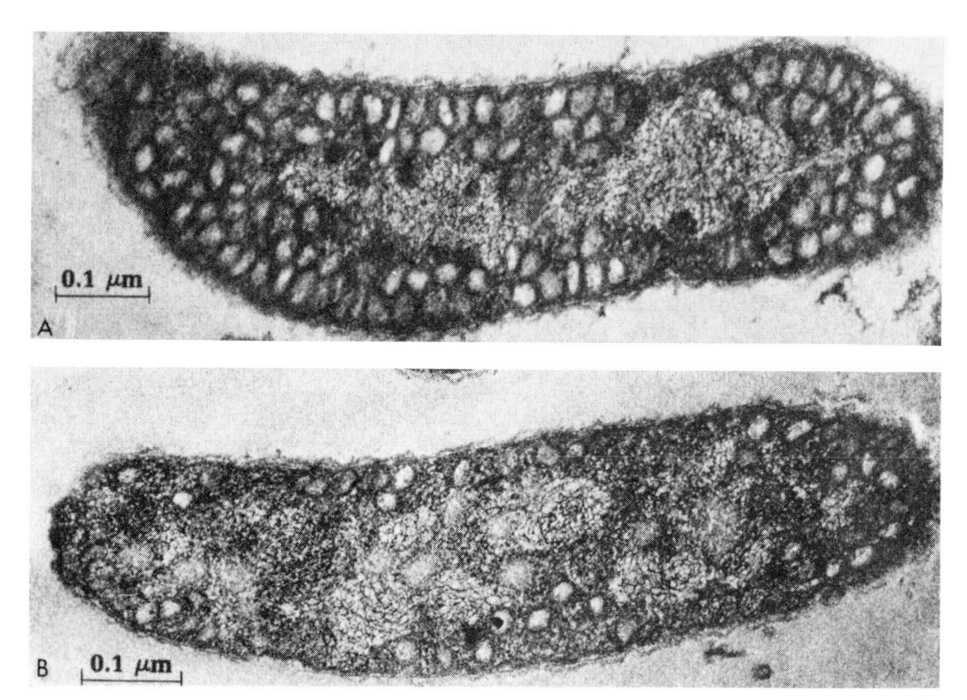

Figure 1.26. *Electron micrographs of thin sections of the purple nonsulfur bacterium,* Rhodospirillum rubrum, *illustrating changes in the quantity of internal membranes as a function of growth conditions (× 46,200).* A, *Cells grown anaerobically in dim light.* B, *Cells grown anaerobically in bright light. (Courtesy of Germaine Cohen-Bazire. From Stanier* et al., *1976: The Microbial World. Prentice-Hall, Englewood Cliffs, N.J.)*

LITERATURE CITED AND GENERAL REFERENCES

Abercrombie, M., 1973: A Dictionary of Biology. Penguin Books, Baltimore.

Altman, P. L., 1972: Biology Data Book. 2nd Ed. Federation of Am. Soc. for Exp. Biol., Bethesda.

Avery, O. T., MacCleod, C. M. and McCarty, M., 1944: Studies on the chemical nature of the substance inducing transformation of pneumococcal types. Induction of a DNA fraction isolated from *Pneumococcus* III. J. Exp. Med. *79*: 137–158.

Ben-Naim, A., 1974: Water and Aqueous Solutions. Plenum Press, New York.

Broda, E., 1975: The Evolution of the Bioenergetic Processes. Pergamon, New York. (Of very great relevance to this chapter.)

Carpenter, P. L., 1977: Microbiology. 4th Ed. W. B. Saunders Co., Philadelphia. (Elementary but useful reference.)

Casjens, C. and King, J., 1975: Virus assembly. Ann. Rev. Biochem. *44*: 555–611.

Chargaff, E., 1950: Chemical specificity of nucleic acids and mechanism of their enzymatic degradation. Experimentia *6*: 201–210.

Chargaff, E., 1972: Preface to a grammar of biology. Science *172*: 637–642.

Chrispeels, M. J., 1976: Biosynthesis, intracellular transport and secretion of extracellular macromolecules. Ann. Rev. Plant Physiol. *27*: 19–38.

Davis, B. P., Dulbecco, R., Eisen, H. N., Ginsberg, H. S. and Wood, W. B., Jr., 1973: Principles of Microbiology and Immunology. 2nd Ed. Harper and Row, New York.

Davidson, J. N., 1976: The Biochemistry of the Nucleic Acids. 8th Ed. Methuen, London.

Fenner, F., McAusland, B. R., Nims, C. A., Sambrook, J. and White, D. O., 1974: The Biology of Animal Viruses. Academic Press, New York.

Fox, S. W. and Dose, K., 1972: Molecular Evolution and Origin of Life. W. H. Freeman Co., San Francisco.

Frobisher, M., Hinsdill, R. D., Crabtree, K. T. and Goodheart, C. R., 1974: Fundamentals of Microbiology. 9th Ed. W. B. Saunders Co., Philadelphia.

Gibor, A., 1976: Compiler. Conditions for Life; Readings from Scientific American. W. H. Freeman Co., San Francisco.

Gurr, M. I. and James, A. T., 1971: Lipid Biochemistry: an Introduction. Cornell University Press, Ithaca, New York.

Hagler, A. T. and Scheraga, H. A., 1973: Current status of the water-structure problem: application to proteins. Ann. N. Y Acad. Sci. *204*: 51–78.

Hanawalt, P. C. and Haynes, R. H., eds., 1973: The Chemical Basis of Life. An Introduction to Molecular and Cell Biology: Readings from the Scientific American. W. H. Freeman Co., San Francisco. (Some articles are dated but are still the best introduction to the subjects at an elementary level.)

Haschemeyer, R. H. and Haschemeyer, A. E. V.,

1973: Proteins, A Guide to Study by Physical and Chemical Methods. John Wiley & Sons, New York.
Hayflick, L., 1969: The Mycoplasmatales and the L-Phase of Bacteria. Academic Press, New York.
Hazlewood, C. F., ed., 1973: Physicochemical status of ions and water in living tissues and model systems. Ann. N. Y. Acad. Sci. *204*: 1–631.
Hochachka, P. and Somero, G., 1973: The Strategy of Biochemical Adaptation. W. B. Saunders Co., Philadelphia.
Horowitz, N., 1945: On the origin of biochemical syntheses. Proc. Natl. Acad. Sci. U.S. *31*: 153–157.
Joesten, M. D. and Schaad, L. J., 1974: Hydrogen Bonding. Marcel Dekker, Inc., New York.
Johnson, F. H., Eyring, H. and Stover, B. J., 1974: The Theory of Rate Processes in Biology and Medicine. John Wiley & Sons, New York.
Knoll, A. H. and Barghoorn, F. S., 1975: Precambrian eukaryotic organisms: a reassessment of the evidence. Science. *190*: 52–54.
Lehninger, A. L., 1975: Biochemistry. 2nd Ed. Worth Publishers, New York. (The best overall reference for this chapter.)
Matthews, C. K., 1971: Bacteriophage Biochemistry. Van Nostrand Reinhold Co., New York.
Meynell, G. G., 1973: Bacterial Plasmids: Conjugation, Colicinogeny, and Transmissible Drug Resistance. Massachusetts Institute of Technology Press, Cambridge.
Mortlock, D. P., 1976: Catabolism of unusual carbohydrates by microorganisms. Adv. Microbiol. Physiol. *13*: 1–53.
Nomura, M., Tissieres, A. and Lenyel, P., eds., 1974: Ribosomes. Cold Spring Harbor Laboratory, Cold Spring Harbor, New York.
Phillips, D. C., 1966: The three dimensional structure of an enzyme molecule. Sci. Am. (Nov.) *215*: 78–90.
Phillips, D. M., 1971: Histones and Nucleohistones. Plenum Press, New York.
Prescott, D. M., ed., 1976: Methods in Cell Biology. A Multivolume Treatise. Vol. 12. Yeast Cells. Academic Press, New York. (Treatment of various topics by specialists.)
Preston, R. D., 1974: The Physical Chemistry of Plant Cell Walls. Chapman and Hall, London.
Quigley, G. J. and Rich, A., 1976: Structural domains of transfer RNA molecules. Science *194*: 796–806.
Razin, S., 1973: Physiology of Mycoplasmids. Adv. Microbiol. Physiol. *10*: 1–80.
Richmond, M. H. and Wiederman, B., 1974: Plasmids and bacterial evolution. *In* Evolution in the Microbial World. Carlile and Shekel, eds. Cambridge University Press, New York, pp. 59–85.
Sanger, F., 1959: Chemistry of insulin. Science *129*: 1340–1344. (Nobel lecture.)
Schopf, J. W., 1975: The age of microscopic life. Endeavour *34*: 51–58.
Schopf, J. W., 1978: The evolution of the earliest cells. Sci. Am. (Sept.) *239*: 110–138.
Spencer, J. H., 1972: The Physics and Chemistry of DNA and RNA. W. B. Saunders Co., Philadelphia.
Smith, P. F., 1971: The Biology of Mycoplasmas. Academic Press, New York.
Stanier, R. Y., Adelberg, E. A. and Ingraham, J. R., 1976: The Microbial World. 4th Ed. Prentice-Hall, Englewood Cliffs, New Jersey.
Stanier, R. Y., Rogers, H. J. and Ward, B. J., eds., 1978: Relations Between the Structure and Function in the Prokaryotic Cell. Symp. Soc. Gen. Microbiol. Vol. 28.
Valentine, J. W., 1978: The evolution of multicellular plants and animals. Sci. Am. (Sept.) *239*: 141–158.
Watson, J. D., 1976: Molecular Biology of the Gene. 3rd Ed. W. A. Benjamin, Menlo Park, Ca. (Excellent.)
Watson, J. D. and Crick, F. H. C., 1953: A structure for deoxyribose nucleic acid. Nature *171*: 737–738. (A classic.)
Westofer, D. B. and Ristand, S., 1972: Acquisition of three dimensional structure of proteins. Ann. Rev. Biochem. *42*: 135–158.
Wilson, A. C., Carlson, S. S. and White, T. J., 1977: Biochemical evolution. Ann. Rev. Biochem. *46*: 573–639.

CHAPTER 2

EVOLUTION OF PHOTOSYNTHESIS, NITROGEN FIXATION, AND THE ATMOSPHERE

Heterotrophs could multiply and evolve on the anaerobic earth only as long as the accumulated supply of abiotically synthesized organic nutrients and their recycling permitted. The anaerobic prokaryotic cells required a relatively large amount of organic nutrient, inasmuch as much potential chemical energy remained in the waste products of fermentative metabolism, as shown in Table 1.3 and Table 1.4. Because depletion of organic sources was inevitable, further development of the heterotrophs depended upon addition to the organic stores from another source.

The most profound step in evolution of life occurred when photosynthetic blue-green algae appeared on earth. In comparison with the limited organic nutrients in the oceans, carbon dioxide and water are virtually boundless, as is light energy from the sun. The blue-green algae make their cell substance (indicated below as CH_2O) from inorganic molecules, fixing inorganic carbon of CO_2 into organic carbon of body structure:

$$DH_2 + A \rightarrow AH_2 + D \quad (2.1)$$

where DH_2 is the hydrogen donor and A is the hydrogen acceptor. Or, more specifically:

$$2H_2{}^{18}O + CO_2 \rightarrow (CH_2O) + {}^{18}O_2 + H_2O \quad (2.2)$$

Water provided the hydrogen to reduce carbon dioxide and the oxygen released came from the oxygen in water, as was demonstrated experimentally by using tracer-labeled oxygen (^{18}O) in the water molecules (Ruben *et al.*, 1941).

Oxygen liberated from water by photosynthesis was to have a profound effect on the future of life in at least two important ways: first, it made possible aerobic metabolism that is twenty times as efficient as anaerobic metabolism; and second, it served for photoproduction of an ozone layer in the stratosphere high in the atmosphere. The ozone layer provided a shield against the photochemically potent ultraviolet radiation (210 to 310 nm) (see Fig. 2.11) that had previously reached the anaerobic earth's surface and prevented the growth of organisms.

THE BLUE-GREEN ALGAE

How many mutations it took to produce a blue-green algal cell from an anaerobic bacterium is not known. On the basis of products of radioactive decay (the radioactive clock), the oldest blue-green algal fossils are dated as 2.5 billion years old (Mahler and Raff, 1975).* Although bacteria and blue-green algae are in different taxonomic groups, both are prokaryotes. (Even though Cyanobacteria is the name used by some taxonomists, the more common designation of blue-green algae will be used here.) In both bacteria and blue-green algae, the chromosome lies naked in the cytoplasm as a tightly coiled DNA helix in contact with the cell membrane. Like some photo-

*In 1977 Knoll and Barghoorn reported what appear to be blue-green algae possibly 3.5 billion years old seen in very thin sections of rocks of that age. The finding awaits verification.

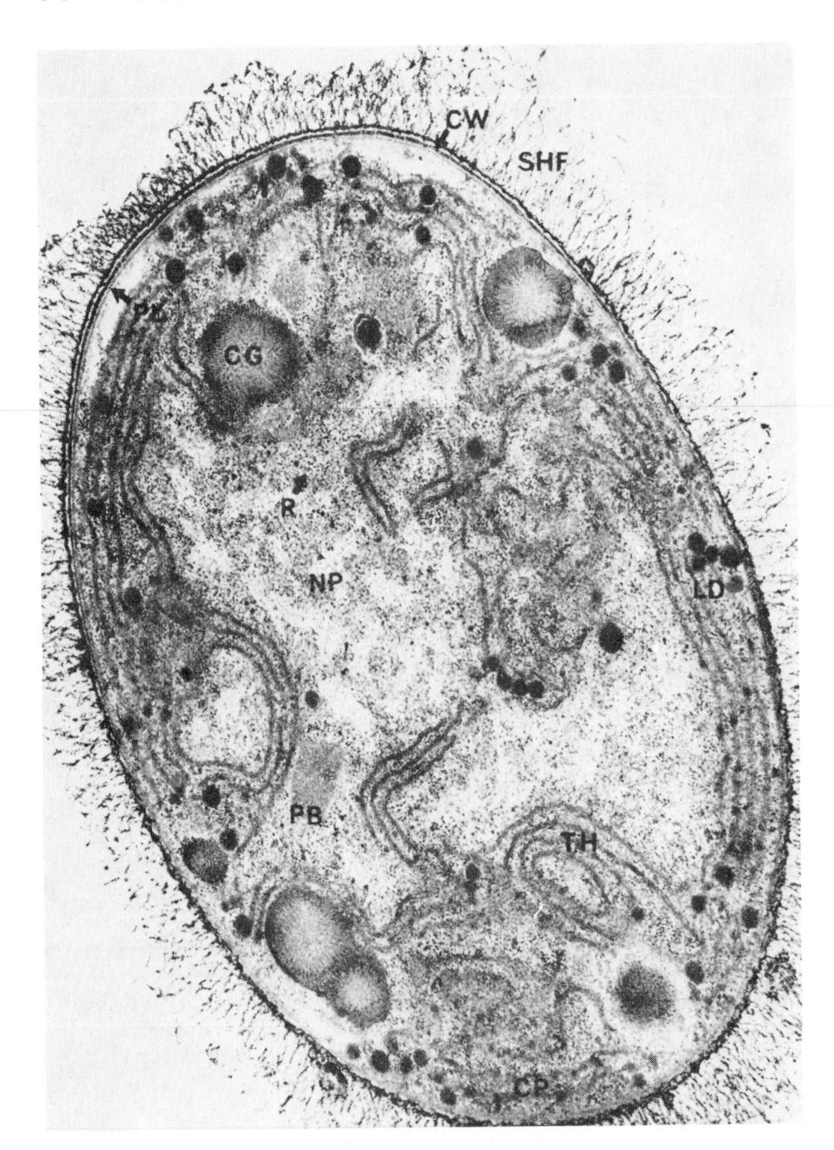

Figure 2.1. *Electron micrograph of a section of* Anabaena variabilis, *illustrating the characteristic ultrastructure of the cell of a blue-green alga. CG, cyanophycin granule of unknown function; CW, cell wall; LD, lipid droplets; NP, nucleoplasmic region; PB, polyhedral body; PL, plasmalemma; R, ribosomes; SHF, sheath fiber; TH, thylakoids (×33,000). (From Leak, 1967: J. Ultrastruct. Res. 20: 190.)*

synthetic bacteria, the blue-green algae have aggregations of membranes lined with assemblies of chlorophyll and carotenoid molecules (Fig. 2.1). The chlorophyll of the blue-green algae is "true" chlorophyll (Fig. 2.2), differing chemically from bacteriochlorophyll of photosynthetic bacteria. The membranes (thylakoids) that contain chlorophyll, a bluish pigment, and carotenoids lie in the cell matrix, not in chloroplasts as in eukaryotic plant cells (Fogg, 1973).

The most dramatically important photosynthetic innovation distinguishing blue-green algae from photosynthetic bacteria is their ability to use water as a hydrogen donor in place of hydrogen sulfide or organic compounds. It is considered unlikely that the blue-green algae evolved from membrane-pigmented photosynthetic bacteria; instead, they probably evolved independently from anaerobic heterotrophic bacteria (Mahler and Raff, 1975), although a difference of opinion exists on this subject (Broda, 1975). Blue-green algal cells are larger than true bacteria and some species form colonies in masses or filaments, much as do some species of bacteria. Some large filamentous bacteria, for example, species of the genus *Beggiatoa,* are thought to have evolved from blue-green algae, similar to the genus *Oscillatoria,* by loss of photosynthetic pigments (Broda, 1975).

The blue-green algae have accessory pigments known as *phycobilins:* blue ones called *phycocyanin* and *allophycocyanin* and in lesser amounts a related red one called *phycoerythrin* (Fig. 2.3). These are not present in photosynthetic bacteria. The accessory pigments harvest light of wavelengths not absorbed by

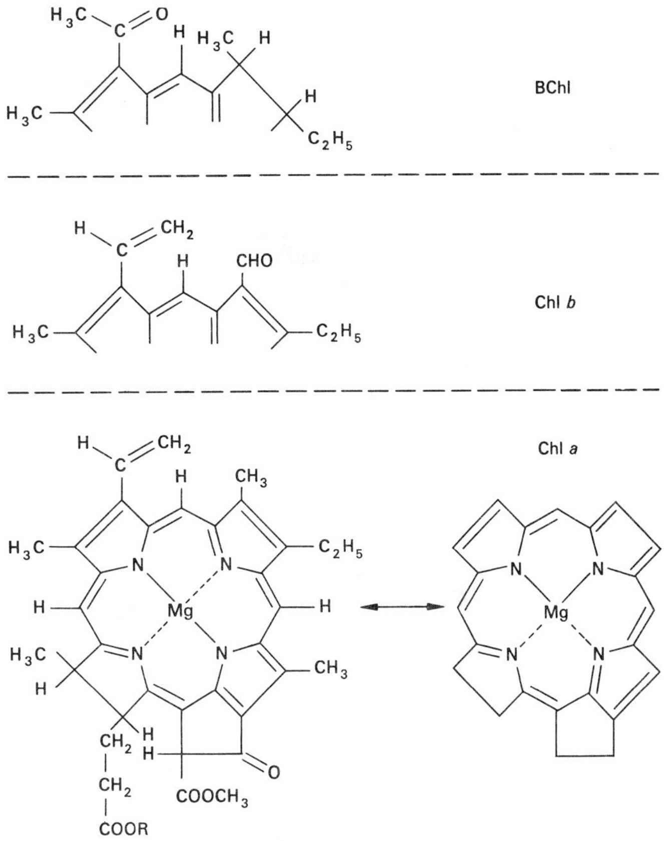

Figure 2.2. *The chemical structures of chlorophylls* a *and* b *and bacteriochlorophyll (BChl). The presence of a carbon atom is implied at each unlabeled junction of bonds. In Chl* a *and* b, *but not in BChl, the pattern of alternating single and double bonds is in resonance with the one sketched at the right, The residue R is a long-chain hydrocarbon,* $C_{20}H_{39}$, *or phytyl, in Chl* a *and* b *and something similar in BChl. (From Clayton, 1971: Light and Living Matter. Vol. 2. McGraw-Hill Book Co., New York.)*

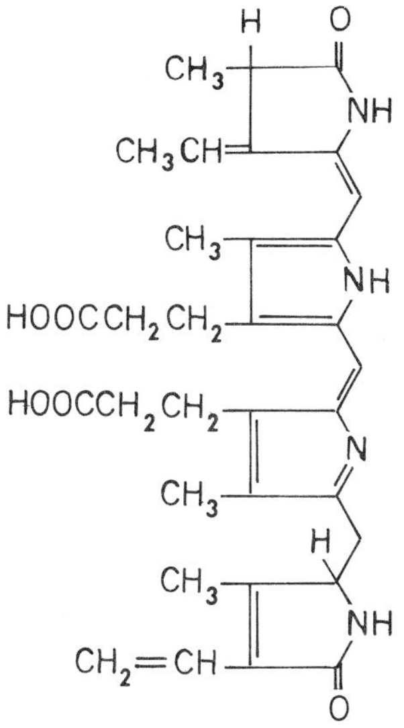

Figure 2.3. *Phycoerythrobilin, the open porphyrin that attaches to a protein to form the red pigment* phycoerythrin *in blue-green and red algae. In the blue pigment* phycocyanin, *the molecule attached to the protein called phycocyanobilin is also an open porphyrin with the* $CH_2{=}CH_2$ *on the fourth pyrrole replaced by* CH_3CH_3. *(From Lehninger, 1975: Biochemistry. Worth Publishers, New York, p. 597.)*

chlorophyll and relay the energy to chlorophyll *a* for use in photosynthesis. The chlorophylls of blue-green algae are chemically like those of higher plants (Fig. 2.2), and their absorption spectra resemble those of green plants rather than those of photosynthetic bacteria (Fig. 2.4). This reinforces the suggestion that blue-green algae originated from prokaryote ancestors other than photosynthetic bacteria (Stanier, 1975), although this point is controversial. Photosynthesis in blue-green algae is very similar to that in eukaryotic photosynthetic plants, as discussed in Chapter 14.

Using the energy of sunlight, the blue-green algae synthesize all their cell substance from carbon dioxide, water, and salts. From the immediate products of photosynthesis they then synthesize all the other molecules needed for growth and cell division.

When blue-green algae are exposed to blue light, they accumulate more red phycobilin pigment than the closely related blue phycobilin; the latter is present in higher concentration in cells exposed to white light. Conversely, when exposed to red light, they accumulate more blue pigment than red (Bogorod, 1975).

Initially, the blue-green algae inhabited nooks and crannies of the early oceans shaded from sunlight, or perhaps sunlit water contain-

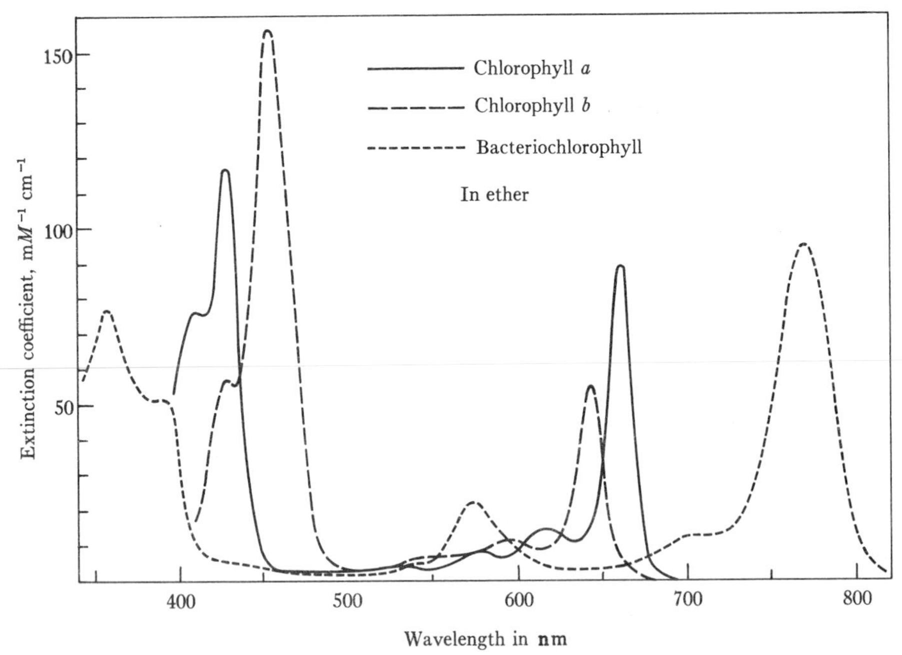

Figure 2.4. *Absorption spectra of various chlorophylls. The absorption spectra in the red and infrared portions of the spectrum from 600 to 800 nm are of special interest. Note that chlorophyll* a *absorbs at longer red wavelengths than chlorophyll* b *and that bacteriochlorophyll absorbs in the infrared portion, just beyond the limit of vision in the red range. (From Clayton, 1965: Modern Physics in Photosynthesis. Blaisdell Publishing Co., Waltham, Mass.)*

ing organic matter that absorbed the lethal short wavelength ultraviolet radiation in sunlight. Possibly, dead layers of algae shielded layers of live ones growing under them, as suggested by the matted remains of fossil blue-green algae. Only after the blue-green algae became widespread could they make significant changes in the oxygen partial pressure of the atmosphere. How long it took to make enough oxygen to produce an ozone shield and whether it was the work of blue-green algae alone is not known, but ultimately an ozone shield developed. The ozone shield permitted blue-green algae to invade large bodies of water and to rapidly increase atmospheric oxygen. Oscillations in oxygen partial pressure in the atmosphere are recorded in ancient rocks, with alternate layers of oxidized and reduced strata known as the banded iron formations (1.8 to 2.2 billion years ago). Thick red ferric iron deposits that formed later indicate the continuous presence of an aerobic atmosphere (Cloud, 1974).

The development of blue-green algae made organic matter available for growth of anaerobic bacteria and for the evolution of aerobic heterotrophic bacteria. Aerobic bacteria developed enzyme chains on the cell membrane, and in some of them, mesosomes provided an increased surface to accommodate more respiratory enzymes. The more effective use of nutrients permitted larger bacterial populations to develop. Genetic variety in large populations fostered more rapid evolution of aerobic bacteria, which invaded a wide variety of habitats, including extreme environments (see Chapter 4).

OXYGEN AS A POISON

During the anaerobic phase of life on earth, fermentation served as the source of energy and the hydrogen transferred during cellular oxidoreductions was accepted by organic compounds, which were thereby reduced. Such reduced compounds were excreted as the end products of metabolism. After the evolution of photosynthetic organisms that used water as the hydrogen donor and liberated oxygen into the atmosphere, it was possible for oxygen to have become the hydrogen acceptor, combining with hydrogen to form an innocuous end product, water. However, the first step in the partial reduction of oxygen is the poisonous oxidant, superoxide $O_2^{\dot{-}}$. Obligate anaerobes without enzymes to dispose of superoxide are killed by it, and aerobic cells tolerate oxygen only because they have the enzymes to dispose of superoxide. At least two enzymes are necessary: superoxide dismutase, which combines two molecules of superoxide to form hydrogen peroxide and oxygen:

$$O_2^{\dot{-}} + O_2^{\dot{-}} + 2H^+ \xrightarrow{\text{superoxide dismutase}} H_2O_2 + O_2 \quad (2.3)$$

and catalase, which disposes of hydrogen peroxide, itself an oxidant:

$$H_2O_2 + H_2O_2 \xrightarrow{\text{catalase}} 2H_2O + O_2 \quad (2.4)$$

Energy is liberated in these reactions, but because it is not coupled to the formation of a high energy-yielding compound, such as adenosine triphosphate (ATP) that can be stored for later use by the cell, it is dissipated as heat (Fridovich, 1975).

When aerobic oxidations in cells pass through a pathway that includes cell pigments known as cytochromes, hydrogen combines with oxygen to form water, accompanied by the conservation of much of the bond energy as ATP (see Chapter 12). On the other hand, oxidations from dehydrogenases directly to oxygen, for example, by way of "yellow enzymes" (flavoprotein dehydrogenases) and a number of "oxidases" for organic compounds (e.g., amino acid oxidases), produce hydrogen peroxide. With these reactions cellular catalase is necessary to prevent oxidation of essential cell components by H_2O_2. Hydrogen peroxide so formed can also be disposed of by using it to oxidize organic molecules in the presence of peroxidases. Catalase may also function as a peroxidase when the concentration of peroxide is low:

$$DH_2 + A \rightarrow D + AH_2 \quad (2.1)$$

$$DH_2 + H_2O_2 \xrightarrow{\text{catalase}} D + 2H_2O \quad (2.5)$$

where D is the oxidized hydrogen donor, DH_2. The energy liberated in peroxidations is not coupled to an energy-storing system in the cell. Even in eukaryotes, in which both peroxidases and catalase are present in small, intracellular, membrane-bound bodies called peroxisomes, oxidations are not coupled to conserve the energy liberated. It is interesting that a special cell organelle should have been developed just to dispose of undesirable oxidants, indicating how important it was for cell viability to dispose of undesirable oxidants (Fridovich, 1974). Peroxidase is also present in the cell wall, perinuclear space, endoplasmic reticulum, and Golgi bodies in the brown alga *Ectocarpus* (Oliveira and Bisalputra, 1976).

Bioluminescence has been suggested as another way that cells may have avoided oxy-

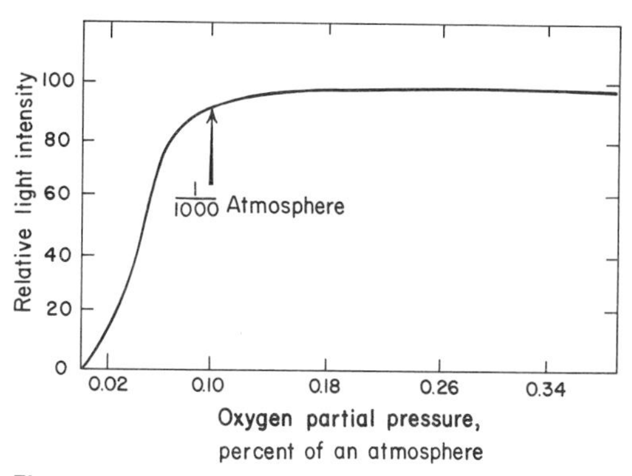

Figure 2.5. *Relation of the luminescence intensity of a freshwater bacterium* (Vibrio phosphorescens) *to oxygen partial pressure expressed in percent of an atmosphere. (After Shapiro, 1934: J. Cell Comp. Physiol. 4: 313.)*

gen's poisonous action (Seliger, 1975). Luminous organisms often respond to extremely low partial pressures of oxygen. For example, *Vibrio phosphorescens* (Fig. 2.5) begins to glow at an oxygen partial pressure of only 0.0005 mm mercury. Hydrogen peroxide formed in this reaction combines with an aldehyde to form the corresponding acid and water. Since bioluminescence occurs in unrelated bacteria, it may once have been more widespread. The importance of bioluminescence as oxygen-detoxifier has been seriously questioned (Broda, 1975).

PROKARYOTE METABOLISM

Bacteria are life's most versatile biochemists—they oxidize almost all natural organic compounds as well as most organic compounds synthesized by chemists, although some man-made plastics and detergents appear to be nonbiodegradable. When bacteria lack an enzyme (a "constitutive" enzyme) for the job on hand, they often synthesize one to fit the available substrate (an "adaptive" enzyme).

A few species of oxidative bacteria fix CO_2 into cell matter by chemosynthesis, using the energy released by oxidation of inorganic substances, such as hydrogen sulfide, carbon monoxide, ammonia, hydrogen, and ferrous iron (Table 2.1). Inasmuch as such substances are limited in quantity, these organisms represent an interesting but minor branch in the evolution of organisms.

Blue-green algae, like bacteria, have invaded extreme environments: high temperatures, low temperatures, high salinities, unbalanced salt solutions, extremes of pH, and so on (see Chapter 4). Because many of them can reduce atmospheric nitrogen to ammonia, blue-green algae can grow where "fixed" nitrogen (nitrate or ammonia) is not available.

TABLE 2.1. EXAMPLES OF OXIDATIONS BY CHEMOAUTOTROPHIC BACTERIA*

Organism	*Reaction*
Pseudomonas carboxydovorans†	$6H_2 + 2O_2 + CO_2 \rightarrow 5H_2O + (CH_2O)$ (cell matter)
Thiobacillus ferro-oxidans†	$4FeCO_3 + O_2 + 6H_2O \rightarrow 4Fe(OH)_3 + 4CO_2$ + cell matter
Thiobacillus thio-oxidans	$H_2S_2O_3 + 2O_2 \rightarrow 2H_2SO_4$ + cell matter
Nitrosomonas europea	$2NH_3 + 3O_2 \rightarrow 2HNO_2 + 2H_2O$ + cell matter
Nitrobacter winogradskyi	$2HNO_2 + O_2 \rightarrow 2HNO_3$ + cell matter

*Many substrate atoms or molecules may be oxidized by chemoautotrophic bacteria to accomplish reduction of one molecule of CO_2, despite the fact that the energy released per mole is high. For example, 74 kcal are released per hydrogen molecule oxidized. Oxidation of HNO_2 to HNO_3, on the other hand, yields only 17 kcal per mole.

†*Pseudomonas carboxydovorans* was formerly *Hydrogenomonas, Thiobacillus ferro-oxidans* was formerly *Ferrobacillus.*

In spite of the success of prokaryotes in terms of sheer numbers and the variety of habitats that they occupy, they were unable to progress beyond colony formation. The eukaryotes alone evolved multicellularity. Perhaps eukaryotes also experimented with colonial forms, and intermediate stages, now lost, appeared before multicellular species evolved. Although the anaerobic period during which life first appeared (Pre-Cambrian) is sometimes called the Age of Prokaryotes, the prokaryotes are still very much in evidence as they attack both live and dead eukaryotes, and they play a major role in the cycles of matter on earth (see Chapter 12). The subsequent aerobic phase (Phanerozoic) is called the Age of Eukaryotes because they became prominent during this period.

Many modern prokaryotes are facultative anaerobes; they can ferment substrates, albeit with less efficiency than by oxidation, but they continue to live and multiply so long as enough substrate is available (Fig. 1.26 and Table 1.4). Energy liberated in fermentation is stored in high-energy phosphate bands (e.g., ATP). Alternatively, when oxygen is available, the same species oxidize the fermentation products completely to water and oxygen, making much more effective use of the substrate.

GLYCOLYSIS AND THE EVOLUTION OF OXIDASES

Anaerobic metabolism, named *glycolysis* in animal cells (because stored glycogen is used) and *fermentation* in microorganisms, is the most ancient type of metabolism and the only kind available for generation of metabolic energy from organic nutrients on the anaerobic earth. The initial steps up to the formation of pyruvic acid are the same (Fig. 2.6); the pathways then diverge. In animal cells reduction of pyruvic acid in absence of oxygen results in the formation of lactic acid. In bacteria a wide variety of other compounds are formed, depending upon the type of metabolism adopted (see Table 1.3).

The development of a sequence of oxidases (see Chapter 12), using oxygen as the hydrogen acceptor and producing water as the end product, was of great evolutionary moment and enabled prokaryotes to invade niches not previously available to them. These enzymes, present on membranes, are synthesized by all aerobic cells, prokaryotic and eukaryotic. Oxidative metabolism, employing the oxidases, is a somewhat advanced evolutionary acquisition. Today, bacteria other than strict anaerobes decompose oxidatively the very "waste" products produced by their anaerobic ancestors (Broda, 1975). This permits much larger populations to develop and speeds evolution. However, in all aerobic cells, prokaryotic and eukaryotic, fermentative pathways are retained to form the initial steps in metabolic breakdown of glucose, although eukaryotes have chosen only a few of the many fermentations developed by prokaryotes (Fig. 2.6). In all cells the enzymes for fermentative reactions are soluble and present in the aqueous medium of the cell, not on membranes.

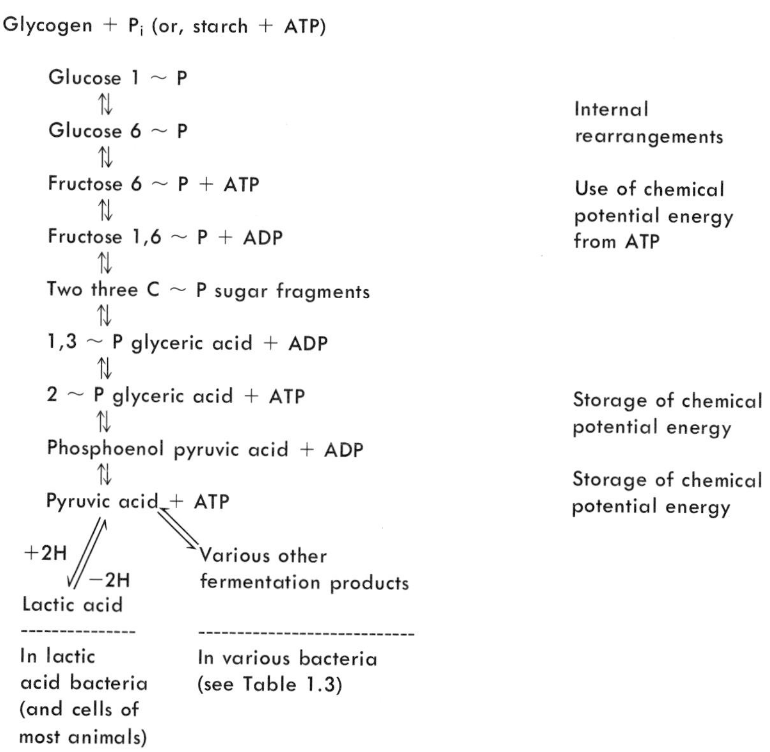

Figure 2.6. *Generalized glycolytic scheme. P_i stands for inorganic phosphate; ~ P stands for high-energy phosphate from ATP; ADP, for adenosine diphosphate with no readily available potential chemical energy; ATP for adenosine triphosphate, with a readily transferable high-energy phosphate bond capable of performing biological work.*

EVOLUTION OF PHOTOSYNTHETIC EUKARYOTIC CELLS

The first fossil cells identifiable as eukaryotes are considered to be 1.3 billion years old (Mahler and Raff, 1975). No fossil records of stages intermediate between prokaryotes and eukaryotes have been found. The fossil cells show evidence of true nuclei and other eukaryotic organelles. Some hypotheses of the origin of eukaryotes are considered in Chapter 3.

Eukaryotic photosynthetic cells are characterized both by the presence of a nucleus in a double-membraned envelope and various membrane-enclosed cytoplasmic organelles (compartments), including one or more chloroplasts containing chlorophyll (Fig. 2.7). Much ATP is produced in chloroplasts during photosynthesis (see Chapter 14). The mitochondrion, present in both eukaryotic plant and animal cells, is the seat of the chains of oxidative enzymes that oxidize nutrients completely to carbon dioxide and water, forming many high-energy phosphate compounds (ATP) in the course of the reactions. It seems likely that eukaryotic cells did not evolve until sufficient oxygen was present in both atmosphere and water to permit functioning of the mitochondrial enzymes. All eukaryotes are capable of oxidative metabolism, although some of them are facultatively anaerobic, briefly tolerating lack of oxygen while obtaining energy by fermentative metabolism. Eukaryotic organisms that live in this manner for prolonged periods occur where oxygen supply is limited, as in mud flats between tides (tubeworms) and in the intestines of animals (*Ascaris*).

Aerobic existence permitting oxidative metabolism had a distinct selective advantage. As Nobel laureate George Wald has put it, "Fermentation was so profligate a way of life that photosynthesis could do little more than keep up with it. To use an economic analogy, photosynthesis brought organisms to the subsistence level, respiration provided it with capital. It is mainly this capital that they invested in the great enterprise of organic evolution." That prokaryotes did not produce multicellular species, despite their chemical virtuosity which included such activity as aerobic metabolism, suggests that prokaryotic cell organization was inadequate for the purpose.

Both heterotrophic and photosynthetic eukaryotic cells probably evolved nearly the same time. The unicellular eukaryotic algae that dominate the sea seasonally are the major crop supplying organic matter to the sea. Diatoms are probably most important in this respect, but flagellates, especially dinoflagellates, are also important. The total amount of

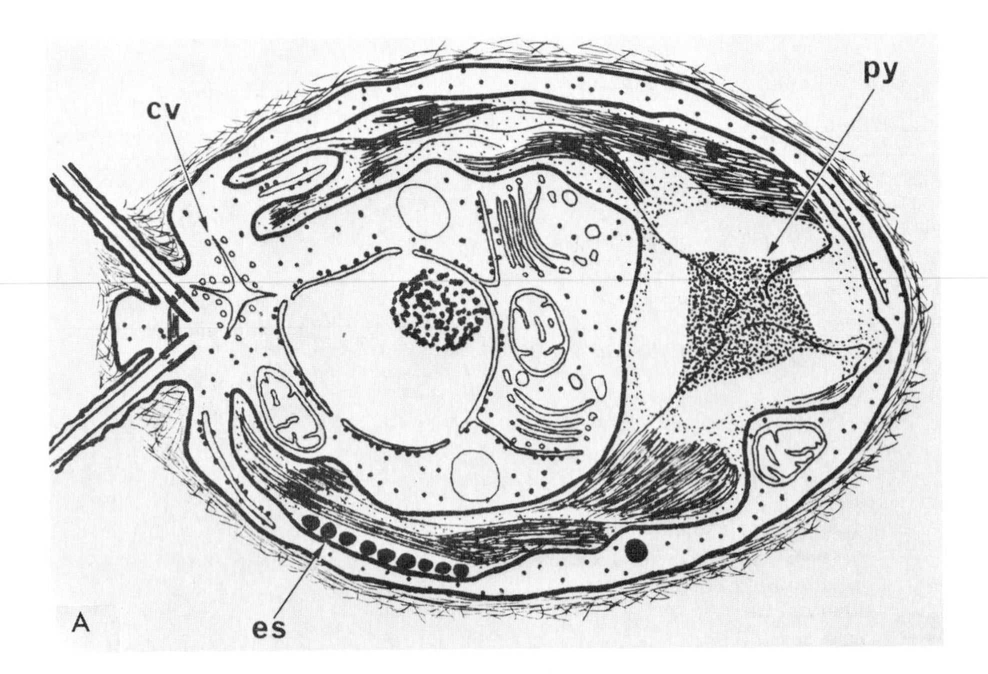

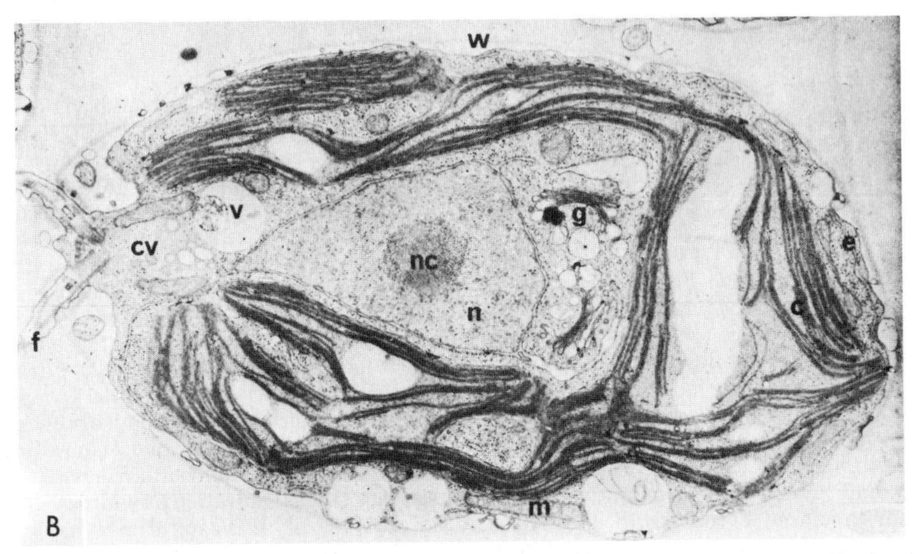

Figure 2.7. A, *Diagrammatic representation of* Chlamydomonas. *Some organelles labeled in the electron micrograph are recognizable here. Others out of the plane of sectioning in the electron micrograph include the eyespot (es), pyrenoid (py) surrounded by starch plates, and contractile vacuole (cv), part of which is visible in the electron micrograph. (From Pickett-Heaps, 1975: Green Algae. © 1975 by Sinauer Associates, Sunderland, Mass.)* B, Chlamydomonas reinhardii. *Electron micrograph of an interphase cell, showing typical distribution of some organelles, including the nucleus (n), single large chloroplast (c), nucleolus (nc), Golgi bodies (g), flagella (f), and flagellar apparatus near components of the contractile vacuole (cv), cell wall (w), mitochondria (m), endoplasmic reticulum (e), and assorted vacuoles (v) (×13,000). (From Goodenough and Porter, 1968: J. Cell Biol.* 38: *403.)*

carbon dioxide fixed in the sea is perhaps about equal to that on land. Three distinct algal types —green, red, and brown—all occur as unicellular as well as multicellular species. All the algae have chlorophylls and carotenoids housed in chloroplasts, as in green algae; the red algae in addition have phycobilins like those in blue-green algae, except that the red pigment predominates, hence the name (Gantt, 1975). The brown algae, besides chlorophyll and other carotenoids, have the brown-colored carotenoid, fucoxanthin, which gives the characteristic color to the group. The accessory pigments in these algae and others not considered collect wavelengths of light not absorbed by chlorophyll and relay the energy to chlorophyll *a*, where it can be used to reduce carbon dioxide (see Chapter 14).

How eukaryotic algae may have arisen from the blue-green algae has not been determined and remains in the realm of speculation (Stanier, 1975). It is part and parcel of the problem of the origin of eukaryotic cells from prokaryotic cells (see Chapter 3).

ATMOSPHERIC OXYGEN FROM PHOTOSYNTHESIS

To a biogeologist the sudden appearance of fossils of all the invertebrate phyla in the Cambrian strata at the very beginning of the Phanerozoic Eon (Fig. 2.8 and Table 2.2) suggests that an important change had occurred at that time. Some workers correlate the evolution of the major taxa with the time when a critical level of oxygen, one-hundredth of the present atmospheric level, had been achieved. This is the partial pressure of oxygen at which cellular respiration (oxidative metabolism) supersedes fermentation (anaerobic metabolism), providing the advantages of aerobic respiration over fermentation.

As the atmospheric oxygen partial pressure continued to increase, it became possible for organisms to invade the land. At least land invasion was possible to the extent that the cells were protected by the ozone layer from the deadly shortwave ultraviolet rays that would have made life on land impossible in the anaerobic stage. Plants gradually covered the land masses, ultimately with the lush vegetation of the Carboniferous period (the coal and oil "measures"), when the atmospheric oxygen partial pressure may even have exceeded that of today.

Life first came upon land about 400 million years ago (Fig. 2.8 and Table 2.2). The timing that has been established for various events may change as more evidence is obtained, but it is not likely that what we know as the sequence of the events will change, since it is based on positions of fossils in the strata (Schopf, 1975).

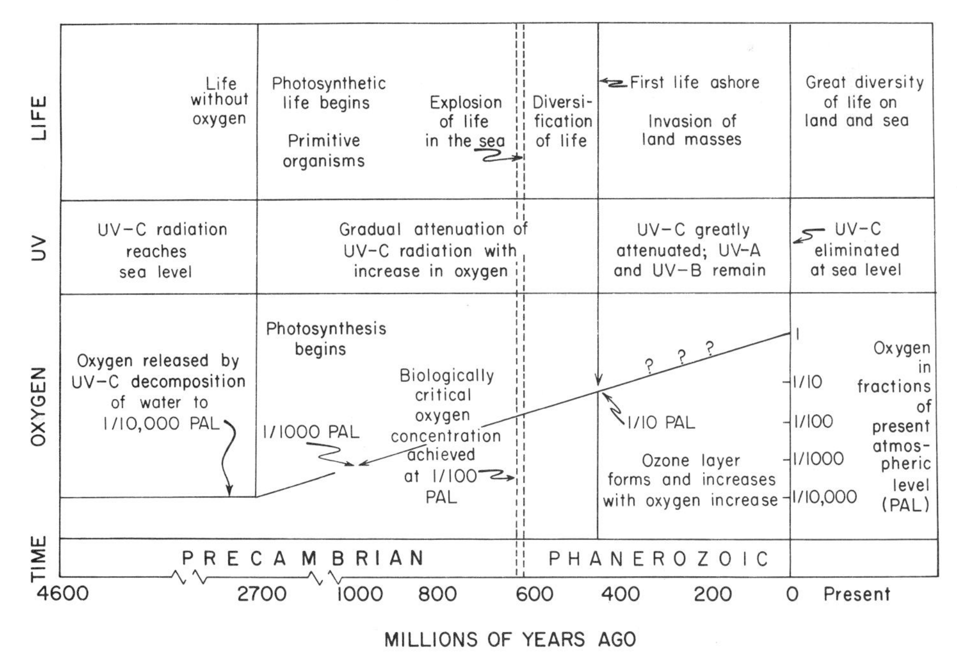

Figure 2.8. *The relation between availability of oxygen and the evolution of life in the sea and on land, as suggested by fossils and correlations with deposits in minerals. The approximate timing is based on radioactive decomposition products in the rocks and fossils. Not all workers in the field agree with the timing and the details presented here from data in Berkner and Marshall, 1965:* Journal of Atmospheric Science, *p. 257. The oldest fossil cells are now dated at 3.4 billion years, the oldest photosynthetic cells at 2 billion years ago. (From Giese, 1976: Living With Our Sun's Ultraviolet Rays. Plenum Press, New York.)*

TABLE 2.2. A TIMETABLE OF PRECAMBRIAN EVENTS AND FOSSIL OCCURRENCES*

Age (Years × 10⁹)	*Events or Occurrence*
3.7	Oldest dated terrestrial rocks; Greenland
3.4	Oldest dated sedimentary rocks containing globular microstructures of possible biogenic origin; Onverwacht series, South Africa
2.7 to 3.1	Oldest known stromatolite; Bulawaya, Rhodesia
3.0	Oldest fossil rod-shaped bacteria; Fig Tree series, South Africa
2.2 to 2.3	Oldest fossil blue-green algae showing cell diversification; Transvaal sequence, South Africa
~2.0	Diversified prokaryotic microbiota; Gunflint iron formation, Canada
~1.8	First free oxygen in atmosphere
~1.2 to 1.5	Oldest fossil cells resembling eukaryotes;† Beck Springs formation, California; Bungle-Bungle dolomite, Australia
0.9	Oldest large assemblage of eukaryotic cells and evidence for meiosis; Bitter Springs formation, Australia
0.6 to 0.7	Oldest fossil multicellular animals
0.6	End of Precambrian

*From Mahler and Raff, 1975: Int. Rev. Cytol. *43*: 5. All dating is approximate and tentative.

†Knoll and Barghoorn (1975: Science *190*: 52–54) claimed that no convincing evidence has been found for the existence of eukaryotes earlier than 900 million years ago, but in a lecture April 17, 1978, Schopf gave evidence for the existence of eukaryotes 1.2 billion years ago. See Schopf, 1978.

As plants invaded the land they were followed by animals, whose cells were now protected by the ozone layer from damaging ultraviolet radiations.

NITROGEN FIXATION

One of the limiting factors for growth of plant crops on land or sea is *"fixed" nitrogen,* that is, nitrogen in the form of ammonia or nitrate. Therefore, organisms capable of fixing nitrogen were important to development of large plant populations, and large plant populations were important for the evolution of an aerobic atmosphere by the release of oxygen in photosynthesis. Animals depend upon plants for their nitrogenous compounds, usually ingested as proteins containing amino nitrogen.

Most blue-green algae fix nitrogen by reducing atmospheric gaseous nitrogen with hydrogen available from metabolic reactions, including the reduced compounds formed during photosynthesis. Reduction of nitrogen occurs in the presence of the enzyme nitrogenase found in all nitrogen fixers. Nitrogen on land and sea is also fixed by heterotrophic prokaryotes, for example, species of the aerobic bacterium *Azotobacter* in loose soil and species of the anaerobic bacterium *Clostridium* in mucky soil. Similar prokaryotes are present in the sea. Nitrogen-fixing bacteria may also live symbiotically with some plants, for example, *Rhizobium,* in legume root nodules. Some of the main steps in the nitrogen cycle are shown in Figure 2.9.

The origin and evolution of nitrogen-fixing bacteria are highly controversial subjects. Because some workers consider nitrogenase an ancient enzyme, nitrogen fixation is said to be a primitive characteristic (Broda, 1975). However, the early anaerobic earth probably had adequate fixed nitrogen in the form of ammonia (Postgate, 1978). Nitrogen fixation would then have had little survival value. Furthermore, anaerobic nitrogen fixation is energetically very expensive: 15 ATP molecules are required to reduce one nitrogen molecule to ammonia. Fermentations do not supply a cell with ATP to nearly the extent that aerobic oxidations do (see Chapter 12).

If indeed nitrogenase is an ancient enzyme, it may have had an important function other than nitrogen fixation. A suggestion is detoxication by reduction of such cell poisons as cyanide and cyanogen, both of which are thought to have been in significant concentration in the anaerobic phase of the earth's history (Postgate, 1978).

Nitrogen fixation may have begun when fixed nitrogen became a limiting factor for life, probably after the appearance of oxygen in the atmosphere. More rapid growth of populations with the more efficient aerobic metabolism would have led to quicker exhaustion of available fixed nitrogen. Capacity for nitrogen fixation would then have had great survival value.

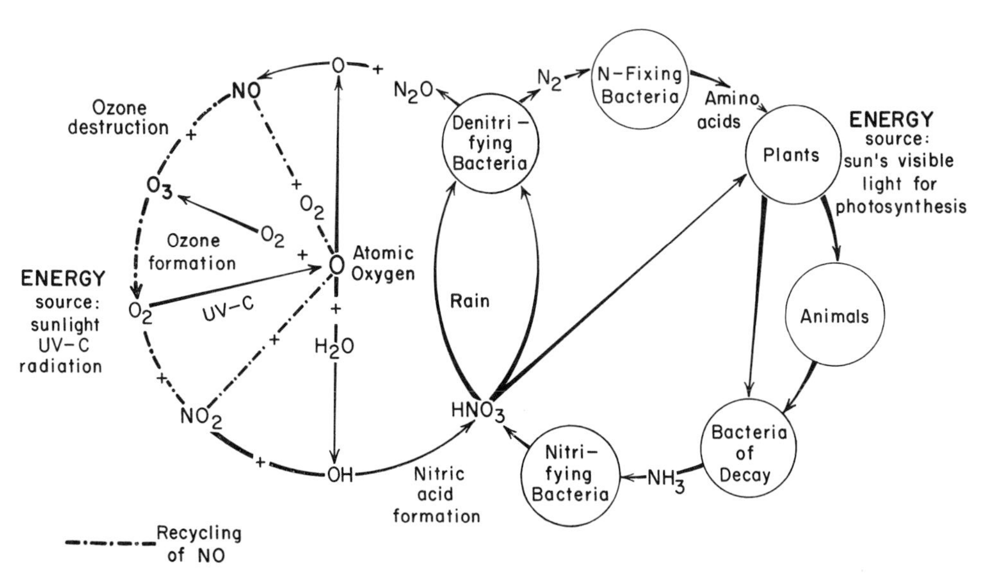

Figure 2.9. *The nitrogen cycle (right) with the newly recognized secondary cycle (left) involving stratospheric nitrogen oxides (N_2O, nitrous oxide; NO, nitric oxide; NO_2, nitrogen dioxide; HNO_3, nitric acid), nitrogen (N_2), ammonia (NH_3), ozone (O_3), molecular oxygen (O_2), atomic oxygen (O), water (H_2O), and hydroxyl radical (OH). (Data from Johnston, 1972: Proc. Nat. Acad. Sci., U.S.A. 69: 2371, and Giese, 1976: Living With Our Sun's Ultraviolet Rays. Plenum Press, New York.)*

Nitrogen fixation is widely, although seemingly haphazardly, spread throughout the prokaryotes. No relationship can be traced between various nitrogen-fixing bacteria. Some are anaerobes *(Clostridium pasteurianum),* some facultative anaerobes *(Bacillus polymyxa),* and some obligate aerobes (the genus *Azotobacter*). Nitrogen fixation is widespread in many genera of four families of blue-green algae, although not universal. Furthermore, nitrogen fixation is found in many species of purple photosynthetic bacteria, and one species of green sulfur bacteria. In photosynthetic prokaryotes, light energy is used to fix nitrogen. Nitrogen fixation confers selective value on many blue-green algae because they need only carbon dioxide, water, and a few salts to synthesize all their growth requirements (Postgate, 1978).

Since nonphotosynthetic nitrogen fixation is expensive (on the average 15 ATP molecules are required to reduce one molecule of N_2 to NH_3), little occurs in soil or water in the presence of minimal supplies of oxidizable substrate. But photosynthetic nitrogen fixation is not so limited; nutrient supplies are abundantly provided by photosynthesis. Therefore, it is of great significance in the economy of nature (Postgate, 1978).

Since oxygen reduces or even stops nitrogenase activity in cells, the preceding discussion may seem paradoxical. *In vitro* nitrogenase of some aerobes is irreversibly inactivated by exposure to oxygen, only emphasizing the paradox. The issue is not fully resolved, although some plausible suggestions have been made. One is that nitrogenase is compartmented in cells, giving protection from ready access to oxygen. For example, nitrogenase from *Azotobacter* is attached to particles that can be separated from a supernatant by centrifugation. The aerobic enzymes of the cell are on the cell membrane. The aerobic processes take up most of the oxygen reaching the cell. Another suggestion is that the gums secreted by nitrogen fixers may decrease entry of oxygen.

The cost of respiratory protection of nitrogenase in nitrogen-fixing organisms is best illustrated by some examples. A laboratory culture of the aerobic soil microbe *Azotobacter chroococcum* uses 1 gram of glucose to fix 10 to 15 mg of nitrogen. If the oxygen partial pressure is decreased to a low value, 1 gram of glucose suffices to fix 45 mg of nitrogen. Conversely, if the oxygen partial pressure is much increased, 1 gm of glucose will fix only 1 mg of nitrogen. At high oxygen partial pressures production of ATP is also decreased to about one third that in low oxygen partial pressure, oxygen consumption being largely to protect the nitrogenase in the cell from oxygen (Postgate, 1978).

Nitrogenase of some species of nitrogen-fixing organisms appears to undergo a conformational change rendering it inactive but resistant to oxygen at high oxygen partial pressures. The process is reversible.

In nitrogen-fixing *Rhizobium* species that associate with leguminous plants to form root nodules, another protective device against excess oxygen has been developed. The cells of the nodule contain *leghemoglobin,* reddish in color like our blood pigment. Leghemoglobin serves as a store of oxygen such that oxygen is made available and released to the nodule cells only when the oxygen partial pressure has fallen to a level that is not toxic to the bacteria (Postgate, 1978).

In many blue-green algae, particulate compartmentation may be similar to that in bacteria, but the problem of separation of nitrogenase and oxygen is even more important because oxygen is produced within the cell by photosynthesis. In filamentous blue-green algae with heterocysts, nitrogen fixation is probably confined to these special cells, which have three cell walls and a minimum of thylakoid membranes and chlorophyll for photosynthesis (Tyagi, 1975). However, it has not been demonstrated that nitrogen fixation occurs only there.

A more likely explanation of the paradox for both nitrogen-fixing bacteria and blue-green algae is that nitrogen fixation may occur only when the oxygen partial pressure in the cell is low. When the oxygen partial pressure is high, nitrogenase becomes reversibly inactivated. In blue-green algae the oxygen partial pressure is low at low light intensities (morning, evening, and stormy days). Sufficient fixed nitrogen may then be supplied only by such "part-time labor" (Postgate, 1978).

A few eukaryotes are said to fix nitrogen, for instance, some yeasts and fungal mycorrhiza on tree rootlets (Frobisher *et al.,* 1974). Evidence for nitrogen fixation in these cases is not convincing (Postgate, 1978), so nitrogen fixation appears to be limited to prokaryotes. It may be possible to transfer the ability to fix nitrogen from the genome of a nitrogen-fixing prokaryote to a eukaryote (Venkalaraman, 1975), and this is often mentioned in connection with genetic engineering using plasmids.

If no way were available to relase nitrogen from the fixed into the gaseous form, accumulation of fixed nitrogen would interfere with the circulation of matter that characterizes ecosystems. In nature, nitrifying bacteria oxidize amino nitrogen to nitrates and denitrifying bacteria reduce nitrates to nitrogen, using nitrates as receptors for metabolic hydrogen. This whole set of relations is called the *nitrogen cycle* (Delwiche, 1970) (see also Fig. 2.9).

ULTRAVIOLET RADIATION BEFORE AND AFTER PHOTOSYNTHESIS

Although we are uncertain of the atmosphere's composition during the anaerobic phase of the earth's history, the types of gaseous atmospheres postulated by various workers would not completely absorb the short wavelength end of the sun's spectrum. Therefore we assume that considerable short wavelength ultraviolet radiation reached the earth's anaerobic surface.

The sun is a radiating body with a surface temperature of about 5600 K,* emitting a characteristic spectrum of radiations from short wavelength ultraviolet through the visible and into the far infrared portion of the spectrum (Table 2.3 and Fig. 2.10). X-rays are given off at the short end of the spectrum (beyond the ultraviolet) and radio waves at the long end, both at very low intensity (Parker, 1975). A considerable portion of sunlight is scattered and reflected back into space (Fig. 4.3).

The sun's short ultraviolet wavelength radiation reaching the earth's surface in anaerobic times probably provided most of the energy for prebiotic synthesis of organic compounds. It served to increase the rate of mutation, thereby speeding the evolution of primitive organisms. But it also limited the distribution of life to habitats such as crevices, shade, and deep or murky water where they were protected from direct sunlight exposure.

For biological purposes we may divide the ultraviolet spectrum into three regions: UV-A, UV-B, and UV-C. The UV-A extends from 390 to 320 nm (a nanometer is a billionth of a meter, 10^{-9} m, and is the presently accepted unit of wavelength), the UV-B from 320 to 286 nm, the latter the shortest wavelength presently reaching the earth, and the UV-C from 286 nm to the shortest wavelengths in sunlight, presumably about 150 nm.

In very large doses UV-A radiation is harmful (Parrish *et al.,* 1978). UV-B is important in synthesis of vitamin D and is the sunburning portion of present-day sunlight reaching the earth's surface. UV-C radiation is both photobiologically damaging (biocidal), killing all types of exposed cells, and photochemically active, causing deterioration of plastics. UV-C radiation is used in the laboratory to induce hereditary changes in organisms, to study damage and repair in nucleic acids, and to sterilize surfaces in laboratories and hospitals (Giese, 1976).

Comparison of the sun's spectrum striking the outside of our atmosphere (studied by rockets and satellites) and the spectrum transmitted to the earth's surface shows that much short ultraviolet radiation is absorbed by our atmosphere. High in the atmosphere, oxygen absorbs very short-wave UV-C rays with a peak at 150 nm (Fig. 2.11), which splits the oxygen molecule (O_2) into oxygen atoms (O).

---

*Absolute temperature: K is −273.16°C.

TABLE 2.3. TYPES OF RADIATION AFFECTING LIFE

Type of Radiation	Natural Source	Subdivision	Wavelength in nm*
Gamma	Radioactive minerals Cosmic rays		0.0001–0.14
X-rays	Sun—low fluence		0.0005–20
		Hard	0.0005–0.1
		Soft	0.1–20
Ultraviolet†	Sun		40–390
		UV-C	40–286
		UV-B	286–320
		UV-A	320–390
Visible	Sun		390–780
		Violet	390–430
		Blue	430–470
		Blue-green	470–500
		Green	500–530
		Yellow-Green	530–560
		Yellow	560–590
		Orange	590–620
		Red	620–780
Infrared	Sun		$780–4 \times 10^5$
		Near	$780–2 \times 10^3$
		Far	$2 \times 10^3–4 \times 10^5$
Hertzian waves	Sun—low fluence		$10^5–3 \times 10^{13}$
		Space heating	$10^5–10^6$
		Radio	$10^6–10^{12}$
		Radar	$10^6–10^9$
		Television	$10^9–10^{11}$
Power A.C.	?		10^{15}

*1 nm (nanometer) = 10^{-9} meter, 1 nm = 1 mμ = 10 Å (angstrom). Although various units of length are used to measure wavelength of light, the nanometer is recommended.

†These subdivisions of the ultraviolet spectrum are arbitrary, but they facilitate a simpler presentation of the subject matter of this book. Even these subdivisions are not always given the same limits; thus, the originator of these subdivisions, Coblentz (*Journal of the American Medical Association 99* [1943]: 125) gave UV-A as 400–315 nm; UV-B as 315–280 nm, and UV-C as less than 280 nm. Photobiologists prefer the subdivisions: near-UV radiation, 390–310 nm; far-UV radiation, 310–200 nm, and vacuum-UV radiation, 200–40 nm; their use would be cumbersome here. From Giese (1976).

Oxygen atoms unite with oxygen molecules to form ozone (O_3). Ozone absorbs UV-C radiation with a peak at 260 nm, splitting the ozone to oxygen molecules and oxygen atoms, most of which combine (equation 2.9) with oxygen again to form ozone:

$$O_2 \xrightarrow[\text{peak 150 nm}]{\text{UV-C radiation}} 2O \quad (2.6)$$

$$O + O_2 \rightarrow O_3 \quad (2.7)$$

$$O_3 \xrightarrow[\text{peak 260 nm}]{\text{UV-C radiation}} O_2 + O \quad (2.8)$$

$$O + O_2 \rightarrow O_3 \quad (2.9)$$

The UV-C radiation absorbed is degraded to heat, warming the ozone layer. This makes for an inversion between the troposphere next to the earth and the stratosphere above it (Fig. 2.12). As a consequence, there is little turbulent exchange between the two layers, but continuous slow exchange takes place in both directions by diffusion.

The ozone layer lies in the lower part of the stratosphere about 15 to 35 kilometers above sea level (Fig. 2.12). Its height varies at different latitudes, and with season, time of day, and solar flares. The total amount of ozone in the stratosphere is quite small. If the ozone layer were placed under standard conditions of one atmosphere pressure, equivalent to 760 mm mercury, and at 0°C, it would have a partial pressure of 2.4 to 2.6 mm at the equator and 3.1 to 4.6 mm at 70° N latitude (Smith, 1973). Under the same conditions oxygen would have a partial pressure of 152 mm mercury.

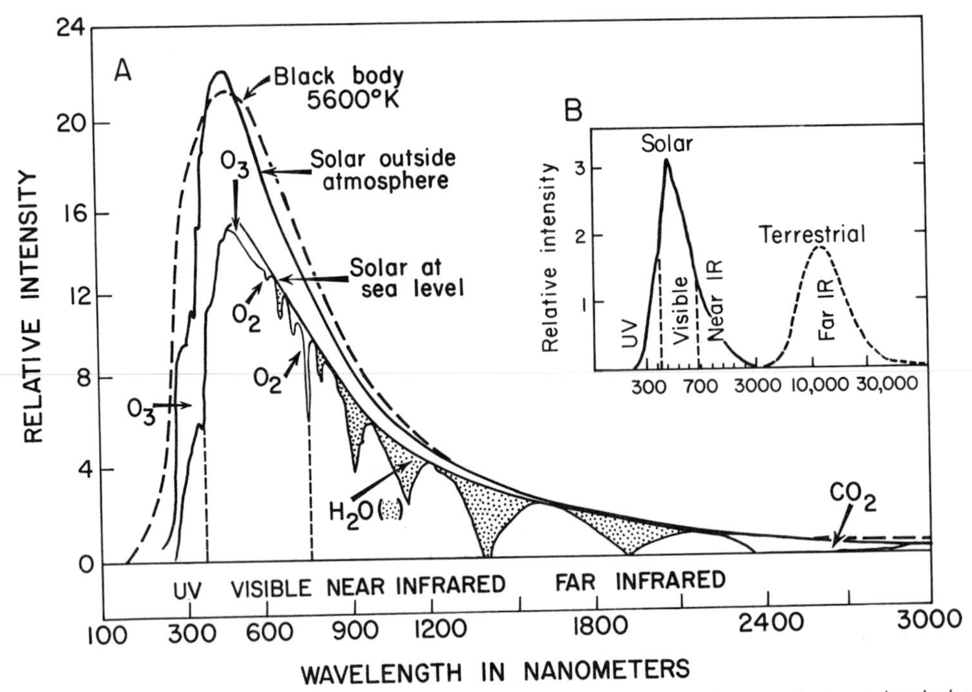

Figure 2.10. A, *The sun's spectrum at the outer surface of our atmosphere and at sea level showing absorption of various wavelengths of radiation by atmospheric gases (O_3 = ozone, O_2 = oxygen, H_2O = water, CO_2 = carbon dioxide). The intensity of radiation is given along the vertical scale and the wavelength in nanometers on the horizontal scale. Note absorption of light: by ozone in the ultraviolet (UV) and visible part of the spectrum, by water vapor over most of the infrared (IR) part of the spectrum, and by carbon dioxide in the very long wavelength infrared spectrum. Intensity of sunlight given in 10 million ergs per square meter per 10 nm bandwidth. (After Hynek, 1951: Astrophysics. McGraw-Hill, New York.)* B, *Comparison of extraterrestrial solar radiation with normalized terrestrial radiation in terms of relative intensity on the vertical axis (different scales for sun and earth) plotted against wavelength of radiation on the horizontal axis. Note that the earth's radiation is entirely in the infrared spectrum. The sun is a "black body" radiator at 5600 K (where K stands for Kelvin, or absolute temperature), while the earth is a black body radiator at 250 K. For comparison to something familiar, the ordinary tungsten filament lamp is a radiator at 2760 K. The radiation from the earth is only a small fraction of that from the sun and would not show on the same vertical axis. (From Giese, 1976: Living With Our Sun's Ultraviolet Rays. Plenum Press, New York.)*

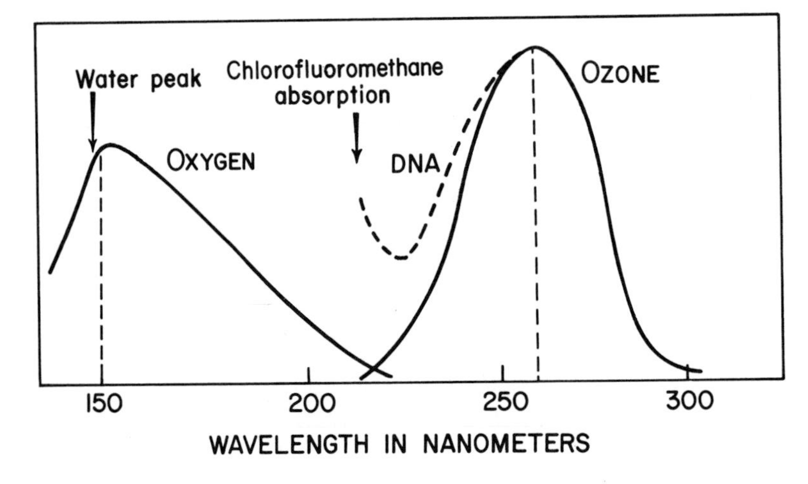

Figure 2.11. *Schematic representation of UV absorption bands of oxygen, ozone, DNA, water, and chlorofluoromethanes. Note that ozone absorbs over much the same ultraviolet range as DNA, and chlorofluoromethanes absorb in the window between oxygen and ozone absorption. Because water absorbs in the same range as oxygen, photolysis of water is a self-limiting reaction. The oxygen and ozone peaks are approximately normalized to permit adding other data. (From Giese, 1976: Living With Our Sun's Ultraviolet Rays. Plenum Press, New York.)*

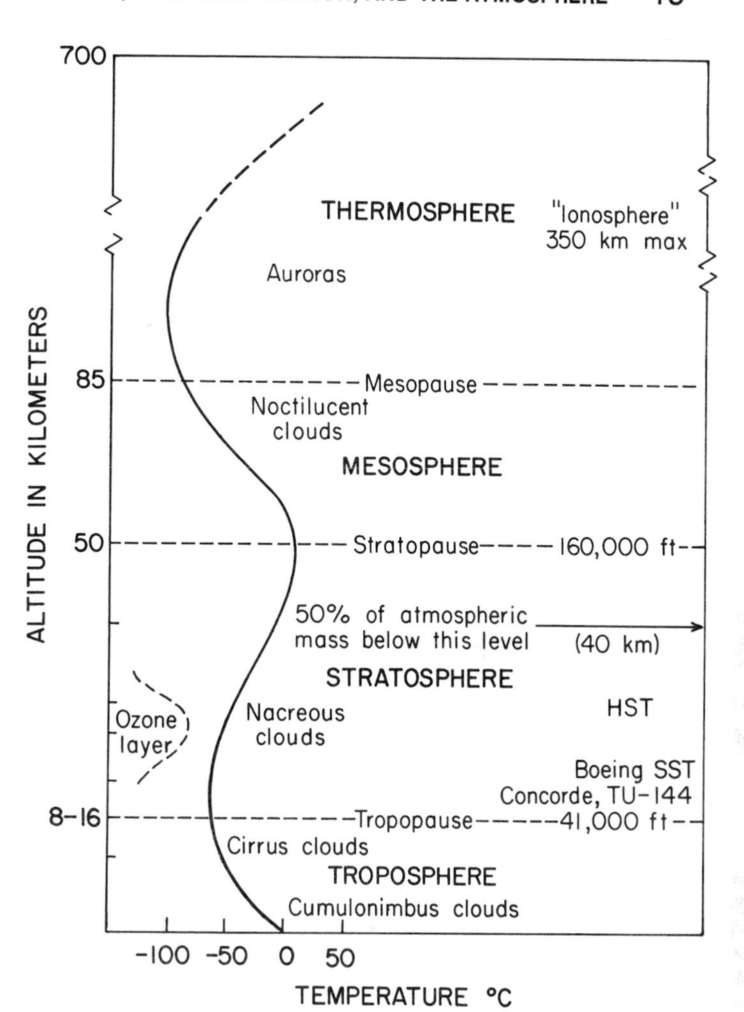

Figure 2.12. *The stratification of the earth's atmosphere. The height of the tropopause varies with latitude (from 8 km at high latitude to 16 km at low latitude), seasons, and various atmospheric conditions. The vertical sinuous solid curve is the temperature curve. The ozone layer at mid-latitude indicated begins about 15 kilometers above sea level and reaches a peak 25–30 kilometers above sea level. Abbreviations: SST, supersonic transport; HST, hypersonic transport; Tu-144, Soviet Tupolev SST. Atmospheric temperature at the earth's surface, shown arbitrarily as 0°C, varies with location and season. (From Giese, 1976: Living With Our Sun's Ultraviolet Rays. Plenum Press, New York.)*

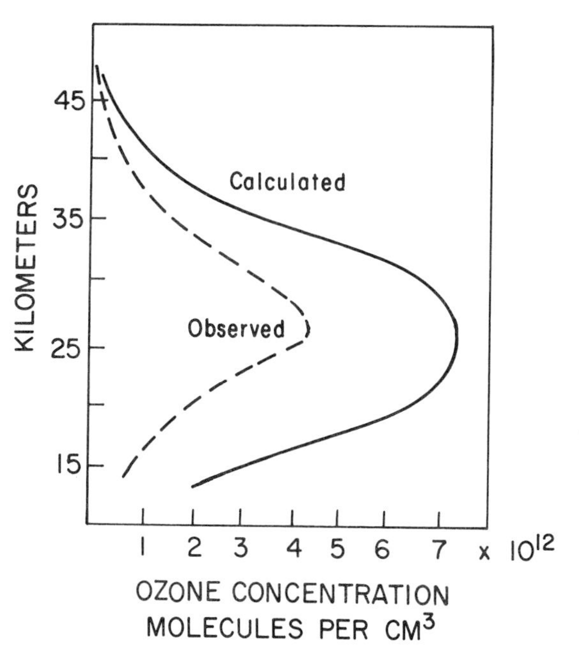

Figure 2.13. *Calculated ozone profile (solid curve), determined on the basis of photochemical reactions occurring in the stratosphere in the absence of nitrogen oxides (at the equator) and observed ozone profile (dashed curve), resulting from decomposition of some of the ozone by nitrogen oxides. To simplify calculations, a uniform distribution of NO and NO_2 was assumed in the stratosphere as well as a static atmosphere. (After Johnston, 1972: Proc. Natl. Acad. Sci. U.S.A. 69: 2370.)*

Clearly the ozone layer depends upon the presence of oxygen in the atmosphere; an atmosphere without oxygen cannot form an ozone layer. When the earth was anaerobic it had no ozone layer. Even on the anaerobic earth a very small amount of oxygen was formed from the photolysis of water, which has a strong ultraviolet absorption band near the ozone absorption peak (Fig. 2.11). Photolysis results in production of hydrogen and oxygen, and oxygen so formed could form ozone. However, the amount of water decomposition is small because the very wavelengths that cause its decomposition are also absorbed by oxygen. Therefore, the maximal amount of oxygen produced in this manner has been calculated to equilibrate at about 0.0001 of the present atmospheric level. The small amount of ozone formed from this partial pressure of oxygen could hardly protect life on earth from sunlight.

REGULATION OF THE OZONE LAYER BY THE NITROGEN CYCLE

Calculations extrapolated from irradiated samples of oxygen and nitrogen mixtures approximating air indicate that more ozone should be present in the ozone layer than is actually found in samples from the stratospheric ozone layer. These experiments assumed that air contained no ozone-decomposing substances. However, unpolluted air at the surface of the earth, in addition to 78.1 percent nitrogen (N_2) and 20.9 per cent oxygen (O_2), also contains 0.033 percent carbon dioxide (CO_2), and small amounts of helium (He), krypton (KR), xenon (Xe), hydrogen (H_2), methane (CH_4), and nitrous oxide (N_2O). Samples of air taken with balloons and rockets at progressively higher atmospheric levels contain the same relative proportions of various gases in air but the concentration of each gas per unit volume decreases with height. Most of the gases, other than oxygen, are photochemically inert but nitrous oxide (N_2O) is not. When it was added to air samples in appropriate concentration, a considerably lower ozone equilibrium concentration was observed following ultraviolet irradiation (Johnston, 1972). It turns out that N_2O reacts with atomic oxygen to form nitric oxide (NO). Nitric oxide reacts with ozone to form oxygen and nitrogen dioxide (NO_2). NO_2 may react with atomic oxygen to form oxygen molecules and NO. If it reacts with the OH radical, it forms nitric acid and diffuses downward into the troposphere, where it is present in lower concentration and is washed out by rain. Under natural conditions, therefore, the equilibrium concentration of ozone in the stratosphere is regulated by the amount of nitrous oxide reaching the stratosphere from the earth's surface (see Fig. 2.9) (Johnston, 1972).

EFFECT OF OZONE ON ULTRAVIOLET RADIATION

The ozone layer eliminates UV-C radiation from sunlight entering the atmosphere and attenuates the UV-B radiation reaching the earth's surface, as shown in Figure 2.14. Between the period when the earth was anaerobic

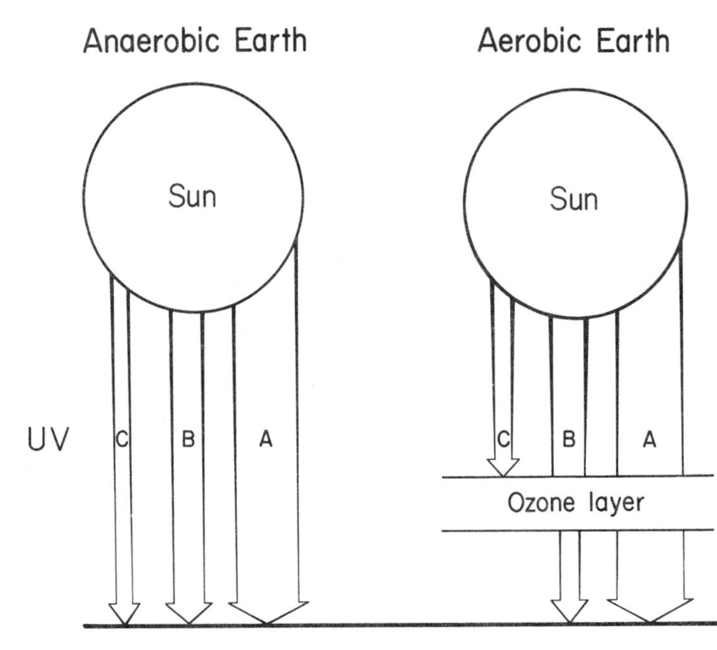

Figure 2.14. Ultraviolet radiation reaching the earth's surface from the sun during the anaerobic and present ozone-modified aerobic conditions. (From Giese, 1976: Living With Our Sun's Ultraviolet Rays. Plenum Press, New York.)

and its present aerobic state, wide variations may have been present in the ultraviolet radiation reaching the earth's surface as the ozone partial pressure in the stratosphere gradually increased with increasing atmospheric partial pressures of oxygen. If, during the great coal ages, the oxygen partial pressure was higher than today, the ozone layer may have been richer than it is today.

At other times, it is possible that the ozone layer may have been attenuated during solar flares that produced much nitric oxide in the stratosphere. Some organisms may have been exterminated during periods of attenuated ozone partial pressure, permitting potent ultraviolet rays to reach the earth's surface. Great solar flares probably occurred when the earth's magnetic field was reversing, at the critical time when, before the reversal process was completed, it practically disappeared. Solar particles normally deflected away from it were allowed to strike the earth and produce high stratospheric nitric oxide concentrations. Nitric oxide destroys ozone catalytically.

Ultraviolet radiation injures cells primarily by altering their DNA, which is extremely vulnerable because any section is generally present only in duplicate. Note that the DNA absorption spectrum parallels that of ozone except at very short ultraviolet wavelengths, where the DNA absorbs more than ozone (Fig. 2.11). This shows how vital the ozone layer is for protection of life on earth. Without an ozone layer, or with a greatly attenuated one, cellular DNA would be rapidly damaged by sunlight. Unless the cell's DNA is repaired, it stops replicating, resulting in the cell's reproductive death. Analysis of the mechanism of ultraviolet radiation action on cells can be more meaningfully taken up after considering the mechanism of DNA replication and the synthesis of proteins (see Chapter 15) and with cell division (see Chapter 26).

HOW CELLS IN MULTICELLULAR ORGANISMS ARE SUPPLIED WITH OXYGEN

Oxidative processes made possible much more efficient use of cell nutrients; therefore, a ready oxygen supply benefits cells that would otherwise have to depend upon glycolysis (animal fermentation) as a source of energy. However, oxygen is not very soluble in water (Table 2.4), especially at higher temperatures and salinities. Even at room temperature only about 0.5 ml of oxygen is dissolved in a liter of tap water. This poses a problem that multicellular organisms have solved in a number of ways.

TABLE 2.4. SOLUBILITY COEFFICIENTS OF OXYGEN, SHOWING VARIATION WITH TEMPERATURE AND CHLORINITY*

Temperature °C	*Chlorinity in gm/kg* 0	15	17	20
	Solubility Coefficients of O_2			
0	0.0489	0.0406	0.0395	0.0378
5	0.0429	0.0359	0.0350	0.0336
10	0.0380	0.0321	0.0313	0.0301
15	0.0342	0.0292	0.0284	0.0274
20	0.0310	0.0267	0.0262	0.0253
25	0.0283	0.0246	0.0240	0.0231

*The solubility coefficient of a gas is the volume of the gas (reduced to 0°C and 760 mm Hg) dissolved per volume of distilled water when the partial pressure of the gas is one atmosphere. To convert this to ml oxygen per liter at 20°C, multiply the coefficient 0.31 or 31 ml per liter by the partial pressure of oxygen in air, 20.99, to get 6.5 ml. (Data from Sverdrup et al., 1942 and Handbook of Chemistry and Physics.)

By day, chlorophyll-containing plant cells are supplied with oxygen by photosynthesis and excess oxygen escapes through surfaces of cells or through leaf stomata. For respiration in cells without chlorophyll and in all cells at night, terrestrial plants take oxygen from air and aquatic plants from water. Since the rate of plant metabolism is low, plant cells are sufficiently supplied by diffusion through the surfaces. Owing to oxygen's low solubility in water, many land plants are easily killed by submerging their roots in water; roots cannot get sufficient oxygen when so "drowned." True aquatic plants generally have large surface areas through which the oxygen may be taken up at night. Like bacteria and animals, plants have a limited capacity for anaerobic respiration (fermentation), but their characteristic metabolism is oxidative much as in animals.

The first multicellular animals were probably so small that sufficient oxygen could diffuse through the few layers of cells to satisfy their cellular oxygen requirements. As multicellular animals became bigger, the problem of supplying their cells became critical. The answer was to develop extensive ventilated surface areas such as gills and lungs, where blood could pick up the oxygen, and a circulatory system to carry the oxygen to the cells and remove the carbon dioxide produced in respiration. However, the solubility of oxygen in blood is no greater than in water. The answer was to develop blood pigments with a high binding affinity for oxygen. The affinity of blood pigment for oxygen must be sufficient to permit loading the gas, yet not so strong as to prevent unloading of oxygen needed by cells. The de-

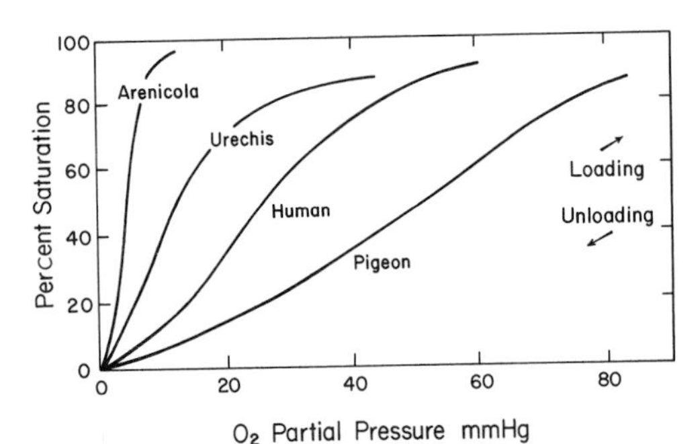

Figure 2.15. Oxygen dissociation curves for several animals, in percent saturation of hemoglobin as a function of oxygen partial pressure in mm Hg. Arenicola *is a marine annelid worm and* Urechis *is a marine echiuroid worm, both living in burrows. (After Prosser, ed., 1973: Comparative Animal Physiology. 3rd Ed. W. B. Saunders Co., Philadelphia.)*

tails of the mechanisms for gaseous exchange and supply to cells, while interesting, are the subjects of comparative physiology. What primarily interests us here is the way blood pigments function.

Our red blood cell hemoglobin is a good example of loose binding between oxygen and carrier because it is oxygenated by binding, not oxidized. A milliliter of our blood carries 40 times as much oxygen as would be carried in an equal volume of water. Many invertebrates have also developed oxygen carriers (blood pigments) to supply their cells. Some of these pigments are hemoglobins, some resemble hemoglobin in containing iron, and still others are hemocyanins, copper-containing pigments that serve the same purpose.

When oxygen bound by blood pigments is carried to tissue spaces, contact with the carbon dioxide that has been released in cellular respiration loosens the binding of the pigment for oxygen, and the oxygen is readily released. A characteristic "loading" and "unloading" partial pressure for a blood pigment varies with circumstances and with species (Fig. 2.15). Our blood carries as much oxygen as if air were piped directly to the tissues; its advantage over the tracheal system of insects is that it circulates the supply of oxygen rapidly and continuously.

Oxygen release in photosynthesis had a profound effect on all life, prokaryotic and eukaryotic. It also determined the course of atmospheric evolution from the anaerobic to the aerobic stage that, in turn, determined the type of ultraviolet radiation impinging on life. Removal of short-wave ultraviolet radiation by ozone, produced from oxygen's absorption of even shorter ultraviolet wavelengths, made life possible at the surface of land and sea. In subsequent chapters the importance of these events in the earth's early history will become even clearer.

LITERATURE CITED AND GENERAL REFERENCES

Brin, W. J., 1977: Biological nitrogen fixation. Sci. Am. (Mar.) *216*: 68–81.

Bogorod, L., 1975: Phycobilins and complementary chromatic adaptation. Ann. Rev. Plant Physiol. *26*: 369–401.

Broda, E., 1975: The Evolution of the Bioenergetic Process. Pergamon Press, Oxford.

Burris, R. H., arranger, 1978: Future of Biological N_2 Fixation. Bioscience *28*: 563–592.

Carlile, M. J. and Shekel, J. J., eds., 1974: Evolution in the Microbial World. Cambridge University Press, Cambridge.

Carr, N. G. and Whitton, B. A., eds., 1973: The Biology of Blue-Green Algae. University of California Press, Berkeley.

Cloud, P. E., 1974: Evolution of ecosystems. Am. Sci. *62*: 54–66.

Cloud, P. E. and Gibor, A., 1970: The oxygen cycle. Sci. Am. (Sept.) *223*: 110–123.

Delwiche, C. C., 1970: The nitrogen cycle. Sci. Am. (Sept.) *223*: 137–146.

Fogg, G. E., Stewart, W. D. P., Fay, P. and Walsby, A. L., 1973: The Blue-Green Algae. Academic Press, New York.

Fridovich, I., 1975: Superoxide dismutase. Ann. Rev. Biochem. *44*: 147–159.

Fridovich, I., 1977: Oxygen is toxic! Bioscience *22*: 462–466.

Frobisher, M., Hinsdill, R. D., Crabtree, K. T. and Goodheart, C. R., 1974: Fundamentals of Microbiology. W. B. Saunders Co., Philadelphia.

Gantt, F., 1975: Phycobilisomes: light harvesting pigment complex. Bioscience *25*: 781–788.

Gibson, E. G., 1973: The Quiet Sun. NASA, Washington, D.C.

Giese, A. C., 1976: Living With Our Sun's Ultraviolet Rays. Plenum Press, New York.

Johnston, H., 1972: Newly recognized vital nitrogen cycle. Proc. Natl. Acad. Sci. U.S.A. *69*: 2369–2372.

Knoll, A. H. and Barghoorn, E. S., 1975: Precambrian eukaryotic organisms: a reassessment of the evidence. Science *190*: 52–54.

Mahler, H. R. and Raff, R. A., 1975: The evolutionary origin of the mitochondria, a non-symbiotic model. Int. Rev. Cytol. *43*: 1–124.

Michelson, A. M., McCord, J. M. and Fridovich, I., 1977: Superoxide and Superoxide Dismutase. Academic Press, New York.

Miller, S. L. and Orgel, L. E., 1974: The Origins of Life on Earth. Prentice-Hall, Englewood Cliffs, N.J.

Oliveira, L. and Bisalputra, T., 1976: Studies on the brown alga *Ectocarpus* in culture: structural localization of enzyme activities. Can. J. Bot. *54*: 913–922.

Parker, E. N., 1975: The sun. Sci. Am. (Sept.) *233*: 43–50.

Parrish, J. A., Anderson, R. R., Urbach, F. and Pitts, D., 1978: UV-A: Human Biologic Responses to Ultraviolet Radiation With Special Emphasis on Longwave Ultraviolet. Plenum Press, New York.
Pickett-Heaps, J. D., 1975: Green Algae. Sinauer Assoc., Sunderland, Massachusetts.
Pfenig, N., 1977: Phototrophic green and purple bacteria: a comparative systematic study. Ann. Rev. Microbiol. *31*: 275–290.
Postgate, J., 1974: Evolution within nitrogen fixing systems. *In*: Evolution in the Microbial World. Carlile and Shekel, eds. Cambridge University Press, Cambridge, pp. 263–292.
Postgate, J., 1978: Nitrogen Fixation, Edward Arnold, London.
Ruben, S., Randall, M., Kamen, M. and Hyde, J. L., 1941: Heavy oxygen (^{18}O) as a tracer in the study of photosynthesis. J. Am. Chem. Soc. *63*: 877–879.
Schopf, J. W., 1975: The age of microscopic life. Endeavour *34*: 51–58.
Schopf, J. W. and Oehler, D. Z., 1976: How old are the eukaryotes? Science *193*: 47–49.
Schwartz, H. M., 1973: Toxic oxygen effects. Int. Rev. Cytol. *35*: 321–343.
Seliger, H. H., 1975: The origin of bioluminescence. Photochem. Photobiol. *21*: 355–361.
Silver, I. A., Evecinska, M. and Bicher, H. I., 1977: Oxygen Transport to Tissue. Plenum Press, New York.
Smith, K. C., ed., 1973: Biological Impacts of Increased Intensities of Solar Ultraviolet Radiation. Nat. Acad. Sci. Nat. Acad. Engineer. Washington, D.C.
Stewart, W. D. P., ed., 1976: Nitrogen Fixation by Free-living Microorganisms. Cambridge University Press, Cambridge.
Stanier, R. Y., 1974: The origins of photosynthesis in eukaryotes. *In* Evolution of the Microbial World. Carlile and Shekel, eds. Cambridge University Press, Cambridge, pp. 219–240.
Tyagi, V. V. S., 1975: The heterocysts of blue green algae (myxophyceae). Biol. Rev. *50*: 247–284.
Venkalaraman, G. C., 1975: Progress in biological nitrogen-fixation. Acta Biol. Indica *3*: 23–29.
Walker, J. C. G., 1977: Evolution of the Atmosphere. Macmillan, New York.
Winter, H. C. and Burris, R. H., 1976: Nitrogenase. Ann. Rev. Biochem. *45*: 409–426.

CHAPTER 3

ORIGIN AND NATURE OF EUKARYOTIC CELLS

Eukaryotic cells of animals and plants are generally larger than prokaryotic cells and more highly organized (Hughes *et al.*, 1974). As the name implies, cells of eukaryotes have a "true nucleus" enclosed in a double-membraned envelope and a cytoplasm organized into membrane-bound organelles suspended in the cytoplasmic matrix. Some of the basic similarities and differences between prokaryotic and eukaryotic cells are summarized in Table 3.1.

On the basis of computer studies of amino acid sequences in a number of proteins common to both prokaryotes and eukaryotes, it would appear that the eukaryotes diverged from the prokaryotes 2.7 times farther back than the four major groups of eukaryotes diverged from one another (Broda, 1975). The divergence between prokaryotes and eukaryotes is inferred to be more profound than the divergence between plant and animal forms. The classification of cells into prokaryotes and eukaryotes by Roger Stanier and Cornelis van Niel in 1962 was a profound and seminal concept.

The chromosomes of the eukaryotic cell are more complex than the circular naked DNA molecule of the prokaryotic cell, with the DNA bound to basic (histone) and acidic proteins. Furthermore, in diploid cells, two chromosomes of a kind are present, and there may be many kinds of chromosomes. The chromosomes evolve into chromatin during the interphase in the cell cycle and condense during mitosis and meiosis; these events do not happen in the cell cycle of prokaryotes. The enclosure of a nucleus in an envelope provides an intracellular barrier within which the complex nuclear activities can be isolated from those in the cytoplasm, for example, division of chromosomes, synthesis of ribosomes, coordination of gene activity with cytoplasmic activity, provision of sites for nuclear metabolic reactions (Wischnitzer, 1974). The details of nuclear organization and function are dealt with in Chapters 7 and 15.

The membrane-bound organelles in eukaryotic cells are of several kinds: mitochondria, endoplasmic reticulum, Golgi bodies, lysosomes, peroxisomes, ribosomes, vacuoles, centrioles (usually lacking in plant cells), and plastids (including photosynthetic chloroplasts present in plant cells). Representative eukaryotic cells are depicted in Figures 3.1 and 3.2. The organelles segregate the varied activities of the cell. For example, mitochondria are the site of oxidative energy metabolism, separated from the glycolytic processes in the cell matrix. The endoplasmic reticulum-Golgi complex (ERG) is involved in carbohydrate and lipid synthesis; it also carries out the assembly of products of protein synthesis, which occurs on ribosomes covering the endoplasmic reticulum. Lysosomes and peroxisomes are enzyme-containing products of the ERG complex. Vacuoles, especially prominent in plant cells, serve a number of functions, for example, storage and transfer of substances, as well as serving as a hydrostatic skeleton in the case of central sap vacuoles.

No one has yet found fossil or biochemical evidence for transitions between prokaryotic and eukaryotic cells. Both prokaryotes and eukaryotes have some basic features in common, such as the genetic code and the synthesis of proteins on ribosomes with the aid of messenger RNA and transfer RNA. Both prokaryotic and eukaryotic cells may be either photosynthetic or heterotrophic, and many of the metabolic pathways are similar (see Chapter 12). Both prokaryotes and eukaryotes store high-energy phosphates to perform cell work.

TABLE 3.1. GENERAL ORGANIZATION OF PROKARYOTE AND EUKARYOTE CELLS

Component	*Prokaryote*	*Eukaryote*
Genome	Naked DNA molecule in cytoplasmic matrix	"True nucleus" in nuclear envelope.
Chromosome	Circular DNA molecule	Paired chromosomes in nucleus, diffuse in interphase, compact during mitosis and meiosis. Circular DNA molecule in mitochondria and chloroplasts.
Chromosome composition	DNA	DNA plus basic histone and acidic proteins. DNA in mitochondria and chloroplasts.
Chromosome number*	Single	Two of a kind, often numerous kinds; single DNA molecule in mitochondria and chloroplasts.
Chromosome division	DNA replication	Mitosis and meiosis after DNA replication; DNA replication in mitochondria and chloroplasts.
Glycolytic enzymes	Glycolytic (fermentative) enzymes in cell matrix	Glycolytic enzymes in cytoplasmic matrix.
Oxidative enzymes	On cell membrane and its extension (mesosome)	On membranes, including cristae of mitochondria and in peroxisomes.
Hydrolytic enzymes	On cell membranes	In lysosomes.
DNA and RNA synthesis	On DNA molecule	On chromosomal DNA, in mitochondria and chloroplasts on DNA molecule.
Protein synthesis	On ribosomes in cell matrix	On ribosomes in cytoplasmic matrix and on rough endoplasmic reticulum; packaged by endoplasmic reticulum-Golgi body. In nucleus, mitochondria and chloroplasts.
Lipid and carbohydrate synthesis	Matrix or membranes?	Endoplasmic reticulum, Golgi body.
Microfilaments microtubules	Absent	Probably universal.
Ribosomes	Small (70S)	Large 80S (small, 50 to 60S, in mitochondria and chloroplasts).
Cell wall	Generally present	Generally present on plant cells but absent in animal cells.
Centriole	Absent	Present in animal cells, rare in plant cells.
Glycocalyx	Generally present	Generally present, especially on algae, protozoans, and animal cells.
Flagella	Composed of flagellin. Unknown energy source for movement.	Composed of tubulin. ATP energy source for movement.

*Plasmids, small pieces of genic DNA, are present in both prokaryotic and eukaryotic cells.

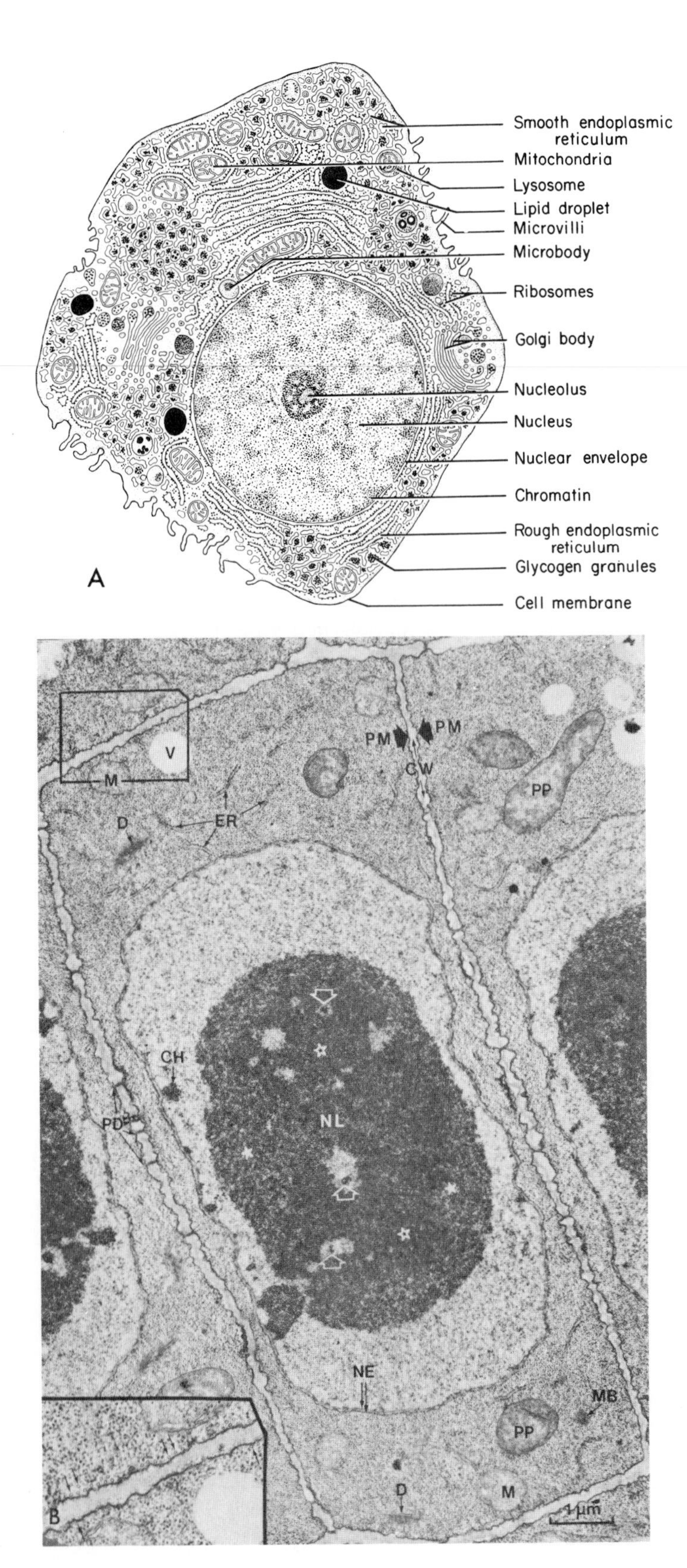

Figure 3.1. A, *Diagram of a eukaryotic cell as exemplified by a cell from the liver (hepatocyte). Between cell divisions the nucleus, enclosed in a double-membrane envelope, has diffuse genetic DNA called chromatin. The continuum of the cell between the organelles is proteinaceous as in a bacterium. Chromosomes generally appear during cell division (see Chapter 26). (From Lentz, 1971: Cell Fine Structure. W. B. Saunders Co., Philadelphia.)* B, *Electron micrograph of a plant cell (root tip of cress-*Lepidium staivum*) (×20,000). CW, cell wall; PM, plasma membrane; PD, plasmodesmata connections between cell walls; V, vacuoles; ER, endoplasmic reticulum to which ribosomes are attached (rough ER), some ribosomes are free in the matrix; M, mitochondria; PP, proplastids; NE, nuclear envelope; NL, nucleolus; CH, chromatin; inset, lower left (×40,000), shows cross sections of microtubules. (From Gunning and Steer, 1975: Ultrastructure and the Biology of Plant Cells. Edward Arnold, London.)*

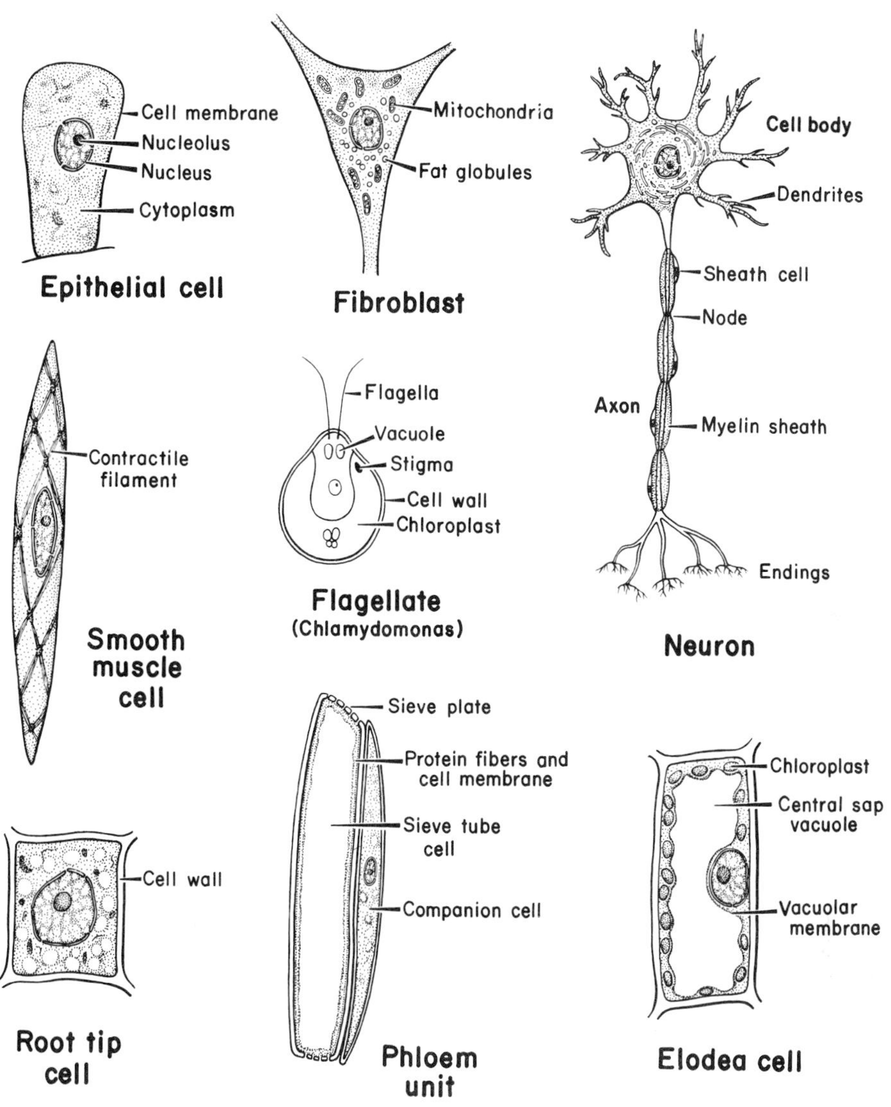

Figure 3.2. *Some examples of animal and plant cells: epithelial cell; fibroblast; neuron; smooth muscle cell;* Chlamydomonas, *a flagellate; root tip cell; phloem unit;* Elodea *cell (freshwater plant).*

But profound differences are seen in their biochemistry, some of which are described in this chapter.

ORGANIZATION

The cytoplasm has two major compartments: a lipoidal hydrophobic ("water-hating") membrane system and a hydrophilic ("water-loving") cytoplasmic matrix between the membrane system (Fig. 3.1). A lipoidal membrane system covers the surface of the cell, encloses the nucleus, the mitochondria, the endoplasmic reticulum-Golgi body complex and its products, including the lysosomes, peroxisomes, and vacuoles, and in plant cells, the plastids. The cytoplasmic matrix, in contrast, is rich in water-soluble enzymatic and structural proteins in various states of aggregation.

Before the advent of electron microscopy, it had been observed that during centrifugation, natural or introduced particles often moved in discontinuous sudden jumps as if they were breaking through a brush heap of linear molecules in a gel. Furthermore, the cells showed viscosity changes with temperature and other conditions, as if structural changes were occurring in the cell matrix. These were interpreted as gel-to-sol or sol-to-gel changes as protein bonds were either broken or made. Therefore, it was no surprise that when good electron microscopic methods were used, microfilaments and microtubules were discovered in the cytoplasmic matrix between the membranous compartments (Bryan, 1974; Spooner, 1975).

Microtubules and microfilaments are now known to be universal constituents of eukaryotic cells. They provide a cytoskeleton in the cytoplasmic matrix between such organelles as the Golgi complex and the endoplasmic reticulum. It was probably the microtubules and microfilaments that interfered with movements of particles in the cytoplasmic matrix during centrifugation in the experiments referred to (Snell *et al.*, 1974; Warfield and Bouck, 1974) (see frontispiece).

It is likely that as cell size increased during the course of evolution, microfilaments and microtubules became necessary to support the extensive membranellar systems in the otherwise relatively fluid cytoplasmic matrix. Segregation of the "powerhouse of the cell," a term given mitochondria by Philip Siekevitz, and of "cell factories" for syntheses became necessary to avoid interference with each other's activities. The "ancient" enzymes for anaerobic respiration (called glycolysis in animal cells and fermentation in microbes) remained in the cell matrix, presumably much as in the prokaryotic ancestors of eukaryotic cells.

In the early twentieth century biochemical analyses of both enzymes and their substrates in entire cells disclosed disappointingly low levels of both enzymes and their substrates. This led to postulation of possible concentration of both enzymes and substrates to levels sufficient for reaction onto the surfaces of colloidal particles in the matrix. It is now apparent that within the compartments, bounded by membranes that have selective permeability to molecules, and enzymes concentrated on surfaces, both enzymes and substrates are locally present in effective concentrations.

Glycolysis yields a net increase of only four high-energy phosphates, but oxidation of products of glycolysis in the mitochondria produces 32 more. It is therefore important that products of glycolysis enter the mitochondria. This appears to be the case, and even more compartmentation takes place within the mitochondria. It is also important to the metabolism of the cell that the energy-liberating processes be enzymatically coupled to phosphate carriers that can store this energy for future use. The mitochondrion is admirably suited for this purpose, as will be clear when the details are considered in Chapter 12. It is evident that compartmentation isolates enzymatic activities but permits exchanges of compounds between them.

Compartmentation also isolates enzymes that are potentially dangerous to the cell, for example, the hydrolases in lysosomes. Lysosomes combine with phagosomes that contain worn-out cell organelles and supply the enzymes needed to digest the contents. If the hydrolases were liberated in the cytoplasmic matrix, they would digest the cell. This happens to cells in the prickle cell layer of our skin that have been "sunburned" by ultraviolet radiation.

It is of interest to note that although a eukaryotic cell normally contains but one nucleus, it may have about 500 mitochondria, 5×10^5 ribosomes, and probably 5×10^8 enzyme molecules. A green plant cell will commonly have about 50 chloroplasts. This multiplicity of units is a characteristic feature of eukaryotic cell organization. Even the nucleus of a diploid cell has two chromosomes of a kind and duplicate genes, one in each member of a homologous chromosome pair.

ORIGIN OF EUKARYOTIC CELLS

The lack of fossil records of cells intermediate between prokaryotic and eukaryotic has puzzled cell physiologists. Some have dismissed the lack of intermediates as the result of selection; intermediate forms may have been ill adapted and so eliminated without trace.

On the other hand, it is also possible that eukaryotes arose from prokaryotes as a result of symbiotic combinations of several prokaryotes, the *endosymbiosis hypothesis* (Margulis, 1970). For example, chloroplasts initially may have been nothing more than blue-green algae symbiotic with another prokaryote. A suggestive observation in this respect is an organism called *Cyanophora paradoxa* that was thought to have blue-green chloroplasts. It turned out to be a cryptomonad protozoan with an internal blue-green symbiont that, inside the cell, resembles a chloroplast. Another example of symbiotic association is *Paramecium bursaria,* which has green bodies that turn out to be a species of *Chlorella* living symbiotically with the protozoan. Each of the symbionts can be grown separately, the alga as a photosynthetic autotroph and the *Paramecium,* without symbionts, as a heterotroph (Margulis, 1970).

If chloroplasts in photosynthetic plant cells arose from endosymbiotic blue-green algae, they have lost independence in the long association between the two kinds of cells. The strongest argument for the symbiotic origin of chloroplasts is the presence within the chloroplast of a distinctive chloroplast DNA genome that codes for some of the chloroplast characters. The chloroplast chromosome is circular like that in a prokaryote. Chloroplasts arise from minute bodies, and multiplication of chloroplasts, while regulated by the nucleus, is governed by the chloroplast genome. Chloroplasts also have ribosomes resembling those in prokaryotes rather than eukaryotes.

Arguments for the symbiotic origin of mitochondria in eukaryotic cells similar to those for chloroplasts can now be made; mitochondria have a separate genome that codes for their characters but is also under control of the nucleus (Margulis, 1970). The mitochondrial chromosome is a circular molecule of DNA and the ribosomes are small, like those in a prokaryote.

Margulis also makes a case for eukaryotic flagella and cilia taking origin from spirochetes that attached themselves to a prokaryotic cell and then became part of the cell. The granule at the base of cilia and eukaryotic flagella is considered homologous to the original prokaryote (spirochete) genome. It is thus interesting and suggestive that the flagellate *Mixotricha paradoxa* present in the termite gut moves not by its flagella but by action of attached spirochetes very regularly arranged on the surface of the cell.

Perhaps the most telling evidence against an endosymbiotic origin of cilia in eukaryotic cells is the lack of DNA in basal bodies as shown by studies with enzymes that selectively remove DNA or RNA (Hartman *et al.*, 1974; Hartman, 1975). The enzyme that hydrolyzes DNA (DNase) has no effect on the fine structure of the basal body of the cilium while the enzyme that removes RNA (RNase) does (Dippel, 1976). DNase has no effect on the tubulin-assembling property of basal bodies (tested by injection into eggs of the clawed toad *Xenopus*) while RNase does. The RNA in basal bodies (and in centrioles presumably derived from them and active in mitosis) (see Chapter 26) appears to be different in composition from transfer and ribosomal RNA although biochemical details are lacking (see Chapters 7 and 15). The function of the RNA in basal body and centriolar regions is considered to be the organization of microtubules—in basal bodies to form functional cilia, in centrioles to organize the division spindle (Heidemann *et al.*, 1977).

Stanier (1970) has proposed that perhaps the major cleavage in the evolution of prokaryotic and eukaryotic cells occurred as a result of external surface differences. Almost uniformly, prokaryotes are bound by a cell wall. It may be that in early evolution, cells that were without such walls but were still in an anaerobic environment developed a capacity for endocytosis, that is, eating other cells and particles, much as ameboid cells do. With an adequate source of food, they could afford the wasteful glycolytic metabolism. Natural selection of cells would then have proceeded along two lines. One of these was the development of a variety of energy-liberating mechanisms including, finally, the highly efficient oxidative metabolism using oxygen as the terminal hydrogen acceptor, and the other was selection for the development of more effective endocytotic mechanisms. Some present-day amoebas lack mitochondria, although whether this is a primitive trait or a secondary loss because of the way of life of the organism is not known (Jeon, 1973). Natural selection of the walled prokaryotes has led to the unrivaled diversity and virtuosity of prokaryotic, as compared to eukaryotic, metabolism. On the other hand, selection for endocytotic mechanisms led to development of larger and more highly organized cells with structures lacking in prokaryotes, for example, the endoplasmic reticulum-Golgi complex for synthesis and packaging of cell products, and the microtubules and microfilaments for movement. Such protoeukaryotic cells could later have acquired mitochondria by endosymbiosis. The similarity of mitochondrial systems and aerobic metabolic pathways in prokaryotes argues for the acquisition of respiratory metabolism over a relatively short evolutionary period.

Not everyone has accepted the symbiotic origin of mitochondria and chloroplasts and other organelles, although the hypothesis is appealing. Mahler and Raff (1975) point out that mitochondrial DNA could have arisen by insertion of a nuclear-derived plasmid (piece of DNA) into the organelle. Amassing evidence from cytochrome evolution, they make a plausible case for their hypothesis. They allow that the evidence for symbiotic origin of chloroplasts is more convincing than that for mitochondria but ask for an open mind on the problem, a view seconded by Bogorad (1975).

There is no evident way to account for the origin of the endoplasmic reticulum-Golgi body complex by symbiosis. It may be the consequence of the cell's enlargement and the need for organization of its compartments to separate and facilitate the many cell syntheses. The nuclear envelope may have arisen from endoplasmic reticulum membranes abutting upon it, or, in turn, the membranes of the nuclear envelope may have given rise to the endoplasmic reticulum. Mesosomal membranes appear in some bacteria as extensions of the cell plasma membrane, suggesting another possible origin for the endoplasmic reticulum-Golgi body complex. However, stages intermediate between the mesosomal enlargements and the endoplasmic reticulum-Golgi complex have not been found, although the large mesosome of the nitrifying bacterium *Nitrocystis oceanus* (see Fig. 1.24) is suggestive.

Symbiosis also fails to explain the much more highly organized nucleus of eukaryotes. Furthermore, the genome is present on a number of chromosomes, two of each kind, and each chromosome is composed not only of DNA as in a prokaryote but also of proteins. Especially prominent is the low-molecular-weight basic protein histone that associates with the DNA. Mitotic cell division, meiotic formation of gametes, and the centriolar polar apparatus for eukaryotic cell division have no counterpart in the prokaryote.

Nass (1969) points out that dinoflagellate chromosomes, like those of prokaryotes, lack histones, and each chromosome is enclosed in a membrane. They also contain a peculiar DNA base not found in other eukaryotes (Rae, 1976). A true mitotic spindle is lacking; the chromosomes line up in a plane but the spindle fibers pass between them. The dinoflagellate nucleus could have arisen by fusion of a group of prokaryotic cells (in a colony?), each one at first maintaining its individuality. The nuclear membrane could have arisen by fusion of the individual chromosomal membranes. Admittedly tenuous, the speculation is nonetheless interesting.

Regardless of how eukaryotes may have originated, they appeared in the fossil record about 1.2 billion years ago and their appearance is thought to correspond with that of atmospheric oxygen in sufficient quantity to permit aerobic metabolism (Table 2.2). The mitochondrion with its assemblies of enzymes allows the eukaryotic cell to extract every bit of energy from substrates being metabolized, using oxygen as the hydrogen acceptor.

Discovery of the fact that organelles existed in the eukaryote cell awaited the development of the electron microscope because some organelles cannot be seen with the light microscope and others are just on the borderline of resolution. The term cell, however, was used to describe compartments in cork by Robert Hooke in 1665; Anton van Leeuwenhoek described bacteria and protozoans in 1675, and Robert Brown described the nucleus in 1835 and recognized its possible importance to the cell. The cell doctrine formulated by Matthias Schleiden and Theodor Schwann in 1839 led to the acceptance of the cell as a unit of structure and function.

CYTOPLASMIC MATRIX, MICROFILAMENTS, AND MICROTUBULES

The glycolytic enzymes present in the aqueous cytoplasmic matrix are discussed in Chapter 12. Microfilaments and microtubules in flagella and muscle cells are considered in Chapters 22 and 23. But we shall here consider structural microtubules and microfilaments in cells.

Microfilaments are rods of indefinite length and about 4 to 5 nm in thickness (Fig. 3.3). They are most evident in muscle but probably occur in most eukaryotic cells, including even amoebas and undifferentiated plant cells. Microfilaments provide a cytoskeleton especially noticeable in cells of odd shape, such as red blood cells, in which microfilaments are attached to the inner surface of the membrane and provide stability to the disc-shaped membrane (Hepler and Palevitz, 1974).

Microtubules function as part of the cytoskeleton, especially in cells with long motile processes, for example, in the long axopodia of sun-ray animalcules (heliozoan protozoans). They participate in cell differentiation and determination of polarity. They also may serve as channels for transport of macromolecules in the cell's interior.

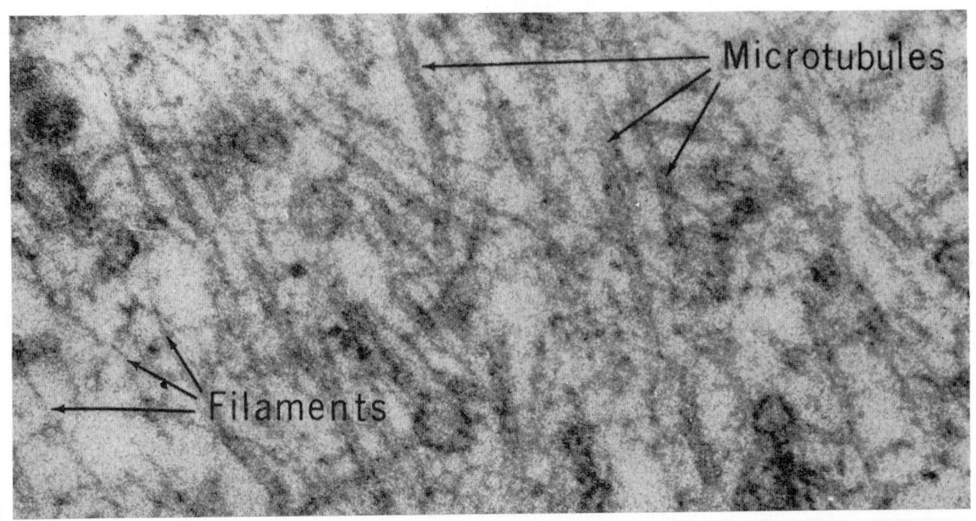

Figure 3.3. A, *Microtubules and microfilaments from cells of rat kidney glomeruli (×120,000). (From Fawcett, 1966: The Cell. W. B. Saunders Co., by courtesy of Dr. Arnold Schechter.)*

Illustration continued on the following page

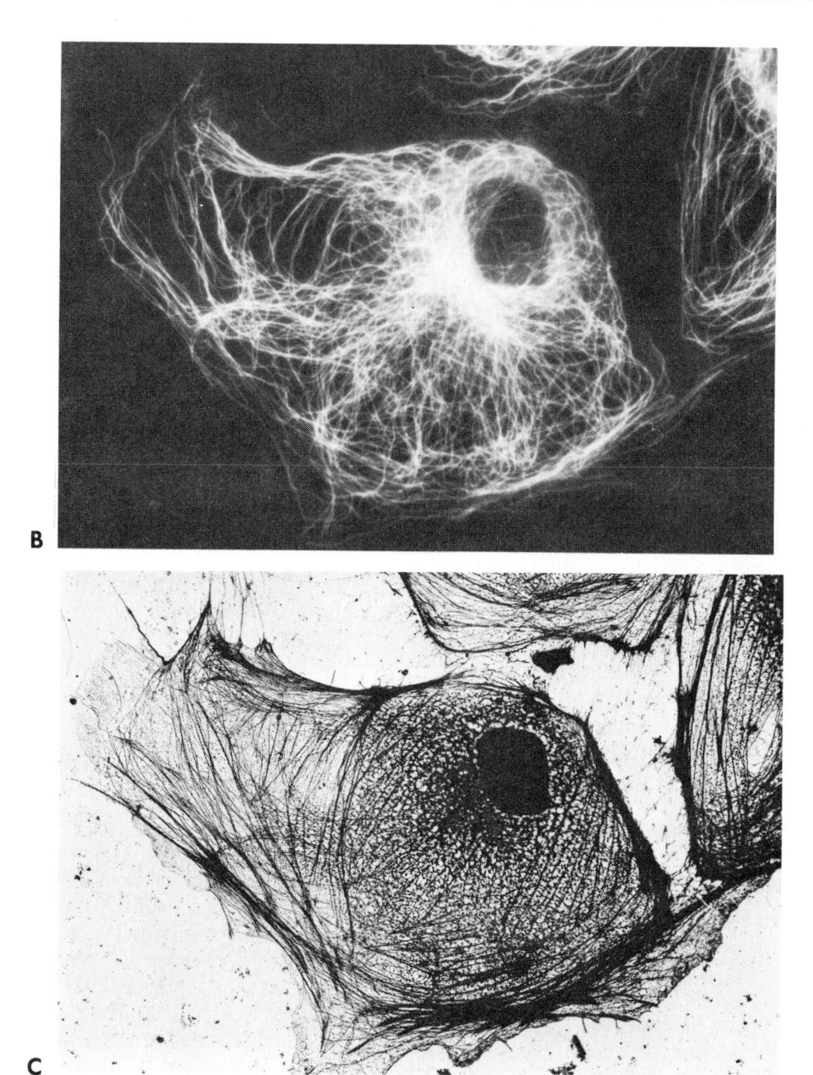

Figure 3.3 Continued. *B, Fluorescent micrograph and C, electron micrograph of the cytoskeleton of the same rat kangaroo cell. In B the cell was treated with fluorescent tubulin antibodies and photographed with a fluorescence microscope (×1,140); in C the cell was treated to stain the microtubules with an electron opaque substance (uranyl oxalate) and photographed with the electron microscope (×1,270). There are regions of correspondence between electron microscopic and fluorescence microscopic studies of fibrous structures. Lack of correspondence in other regions suggests the presence of fibrous structures other than microtubules; in another study these structures were found to stain with fluorescent antibody for actin. (From Osborn, Webster and Weber, 1978: J. Cell Biol. 77: R27–R34.)*

Microtubules are found in the cytoplasmic matrix of all eukaryotic cells (Fig. 3.3), although they also form the motile elements of cilia and flagella (see Chapter 22). Microtubules are generally 20 to 27 nm thick and several micrometers long, and are composed of 13 microfilaments, with a cross-sectional spacing of 4.5 nm, enclosing a central lumen. Each filament is composed of subunits.

Microtubules are labile, disappearing in many cells at 0°C. Those in the mitotic spindle disappear on treatment with the drug colchicine. Those in cilia and flagella are resistant to both treatments.

Microtubules are composed of the protein tubulin, which exists in two forms, A and B, both of similar molecular weight (110,000 to 120,000). Tubulin appears to be similar in most species tested. Microtubules assemble spontaneously from a concentrated solution of tubulin containing magnesium and a high-energy phosphate (guanosine triphosphate, GTP).

Tests with fluorescent antibodies to tubulin and gel electrophoresis on extracts of gliding bacteria and spirochetes suggest the presence in prokaryotes of microtubules, although these are smaller than in eukaryotes. If chemical analysis of the protein of which the presumptive microtubules are composed should show them to be tubulin, the findings will support the hypothesis that cilia in eukaryotes arose from association of a eukaryotic cell with microtubule-containing prokaryotes (Margulis *et al.,* 1978).

MEMBRANE-BOUND ORGANELLES

The endoplasmic reticulum-Golgi body complex consists of tubules or sacs, the latter sometimes in stacks, all interconnected by tubules in a system through which molecules are distributed. The structure, composition and function of these organelles are considered in Chapter 5. Mitochondria are membranous organelles with an outer enclosing membrane and an inner membrane invaginated to form internal shelves or cristae, which provide a large surface area to hold organized enzyme systems of oxidative metabolism. The structure of a chloroplast resembles somewhat that of a mitochondrion; both energy transducing systems are considered in Chapter 6. Cell membranes are considered in Chapter 8.

SIZE RANGE AND DIVERSITY OF EUKARYOTIC CELLS

All who have studied plant and animal cells know that they are diverse in size and form. Size variation of eukaryotic cells is summarized in Table 3.2. The dimensions of most eukaryotic cells range from 2.0 μm to about 40 μm; the smallest eukaryotic cells, *Leishmania* bodies, relatives of trypanosomes, are about 2 μm in diameter—about the size of large bacteria—while the largest ones, such as cycad ovules and ostrich eggs, are 20–50 mm in diameter. In both these large cells the nucleus and cytoplasm occupy a very small part of the total mass; both cells contain large stores of nutrient for development of the embryo. In *Valonia,* another relatively large cell listed in Table 3.2, a very large central sap vacuole and a massive cell wall make up most of the cell mass.

It is of interest to point out that cells are examples of a remarkable degree of miniaturization. Thus, in its minute mass, a human sperm carries half the genetic determinants that will form the mature individual. For that matter, any cell has within its minute confines the genetic determinants to synthesize all the compounds needed for its functioning.

The eukaryotic cell nucleus also varies in size, but less than the entire cell. Even the largest cells have a nucleus comparable in size to that in small cells.

TABLE 3.2. RANGE IN SIZE OF EUKARYOTIC CELLS

Mass in Grams	*Organism or Cell*
10^3 to 10^2	Dinosaur eggs, ostrich eggs, cycad ovules (diameter, 20 cm)
10^1	*Valonia macrophysa* (mature)
10^0	*Valonia ventricosa* (mature)
10^{-1}	*Nitella* (large internode) (diameter, 1 mm)
10^{-2} to 10^{-3}	Frog eggs
10^{-4}	Striated muscle of man
10^{-5}	Human ovum
10^{-6}	Large *Paramecium* (diameter, 60 μm)
	Large sensory neuron of dog
10^{-7}	Average *Vorticella*
	Human smooth muscle fiber
	Human liver cell (diameter, 20 μm)
10^{-8}	Dysentery ameba
10^{-9}	Frog erythrocyte
	Malarial parasite
	Human sperm
	Smallest Protozoa *(Monas)*

One need only list the types of animal cells to visualize their diversity in form for performance of specialized functions. Consider the differences between epithelial, muscular (smooth and skeletal), nervous, and connective tissue; and cartilage, bone, blood, eggs, sperm, and protozoans. Some cells, such as the basal cells in skin epidermis, are relatively unspecialized in organization; others, such as nerve cells, are highly specialized for particular functions. A few types of animal cells are included in Figure 3.2.

In plants, consider cells in epidermal, vascular (xylem, phloem), supporting (bast fibers), parenchyma, cambial tissues, and unicellular and filamentous algae and fungi to picture their extreme diversity. A few types are shown in Figure 3.2.

VARIATION IN ORGANIZATION OF ANIMAL CELLS

A *tissue* is a mass of similar cells, usually contiguous, held together in a supporting matrix, secreted by the cells that perform a common function, and usually forming a part of an *organ.* Intercellular fluid is almost always found in tissues (Fig. 3.4). Every cell must obtain supplies and pour wastes into the internal fluid.

Epithelial cells are generally closely packed, like bricks in a pavement; they form

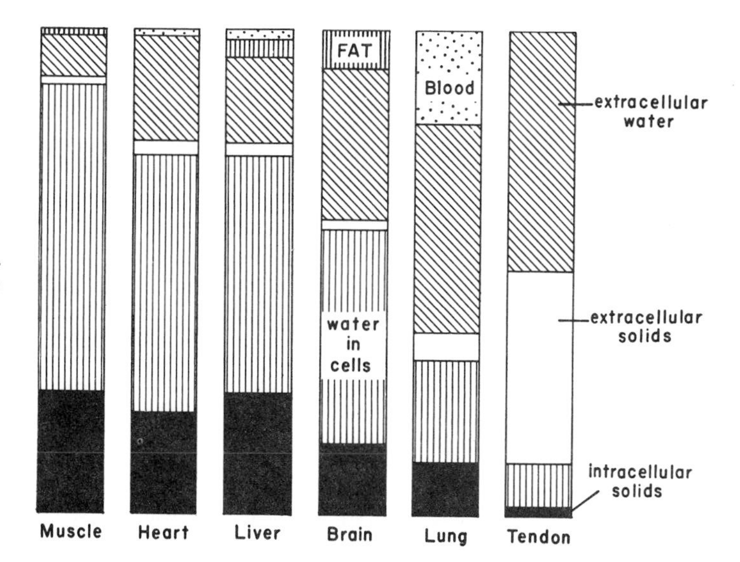

Figure 3.4. Extracellular material in animal tissues. (Adapted from Lowry, 1943: Biol. Symp. 10: 239.)

solid protective outside layers on animals and line their internal cavities and tubules. Epithelial cells may have equal diameters in all directions or they may be flattened, elongated, or columnar. Columnar epithelial cells in glands produce secretions; epithelial cells in gonads give rise to gametes. Epithelial cells have typical cell organelles, but in glands their mitochondria and Golgi complexes are particularly well developed. External epithelium may secrete a cuticle, as in the earthworm, or a skeleton, as in arthropods.

Connective tissue contains few cells and much extracellular material, for example, cartilage, tendon, bone, and adipose (fatty) tissue. During early embryonic development connective tissue cells lie close together, but they secrete a gelatinous matrix that separates the cells (Tracey, 1968). In this gelatinous matrix are laid the fibers of connective tissue and the salts of bone.

In *muscle tissue* (smooth, cardiac, and skeletal) the cells are considerably modified for contractile function. They are elongated and contain contractile fibrils. Vertebrate visceral muscle cells are spindle-shaped with a single central nucleus and fine elongate contractile fibrillae. The vertebrate skeletal muscle fiber consists of the confluent bodies of numerous cells, each indicated by a spindle-shaped nucleus (see Fig. 3.2). Elongated fibrils with striations in register pack the cytoplasm, and large mitochondria are present; mitochondria are especially conspicuous in the highly active flight muscles of vertebrates and insects. In vertebrate cardiac muscle, the fibers are striated but different in organization from the skeletal muscle fiber. Invertebrate muscle cells are diverse, but they all have contractile fibrils, sometimes striated (e.g., insects, crustaceans).

Nerve cells *(neurons)*, highly specialized for conduction of nerve impulses, have a nucleus-containing cell body from which extend long fibers called *axons* and short fibers called *dendrites* (see Fig. 3.2). A well-developed Golgi complex is found in the neuron cell body; mitochondria occur chiefly at the synapses. The fibers contain granular deposits of nucleic acids. A vertebrate neuron covered with an insulating envelope consisting of many layers of apposed cell membranes and enclosed by sheath cells is said to be a *myelinated fiber* (see Chapter 20). The myelin sheath adds bulk to the vertebrate nervous system; because of it, the brain, for example, contains considerable extracellular material (see Fig. 3.4). However, outgoing fibers from autonomic nerve ganglia located outside the cord are not myelinated. Invertebrate nerve fibers have little if any myelin.

Blood cells are also highly specialized. For instance, mature red blood cells in mammals *(erythrocytes)* are enucleate (without a nucleus) and lack cell organelles; they contain mainly the oxygen-carrying pigment hemoglobin. The white cells of vertebrates and invertebrates—both the granular, amoeboid *leukocytes* and the more or less spherical, clear *lymphocytes*—are typical cells with nuclei and other cell organelles.

VARIATION IN ORGANIZATION OF PLANT CELLS

Plant cells characteristically possess plastids and vacuoles and are covered by a cell wall. They have mitochondria, ribosomes, and a nucleus as in animal cells, and structures similar to the Golgi complex, usually called *dictyosomes* (Gunning and Steer, 1975).

Botanists do not agree on classification of plant tissues, but following are lists of a variety of cells found: epidermal, undifferentiated (parenchyma), vascular (xylem and phloem), sclerenchyma and supporting, or bast, fibers. A few of these cell types are shown in Figure 3.2. Epidermal cells line external surfaces of plant structures. Embryonic plant tissue (e.g., the cambium), except for cell walls, is made up of cells similar in appearance and function to animal epithelial cells. Numerous undifferentiated plant cells, called *parenchyma,* occur in the stem cortex, filling spaces between conducting or supporting tissues. They also make up the leaf photosynthetic mesophyll and palisade cell layer. The elongated living *phloem* cells, abutting upon one another, distribute the manufactured food to cells throughout the plant. The *xylem,* consisting of tubular remains of cells, distributes the water and salts. In adult plants xylem ducts are entirely extracellular, the cellular material having disappeared. Cells with thickened wall, such as the gritty stone cells in the pear, are representatives of *sclerenchyma* tissue. Supporting "cells" or bast fibers, consisting largely of secretions of cellulose and a lignin, occur along the vascular tissues and contribute to the plant skeleton. Supporting fibers in the adult plant are entirely extracellular. As a result, an organ such as a mature stem, which contains an abundance of ducts and supporting fibers, has much extracellular material. In contrast, a bud or a leaf is made up largely of relatively undifferentiated living cells.

CHEMICAL COMPOSITION OF EUKARYOTIC CELLS

The protoplasm of a plant or animal cell is likely to contain approximately 75 to 85 percent water, 10 to 20 percent protein, 2 to 3 percent lipid, 1 percent carbohydrate and about 1 percent salt and miscellaneous substances. Muscle, made up largely of cells, contains somewhat less than 80 percent water, close to 20 percent protein, about 1 percent carbohydrate, and about 1 percent lipid and salt. In its chemical constituents the eukaryotic cell thus resembles the prokaryotic cell. However, the relative amounts of DNA and RNA in eukaryotic cells are conspicuously less than in prokaryotic cells; for example, about 0.4 percent of a rat liver cell is DNA and 0.7 percent RNA, while about 1 percent of an *E. coli* cell is DNA and 6 percent RNA (Watson, 1976).

Although the most prominent constituent of a cell is water, the substance that gives the cell its characteristic structure is protein. Lipids are important in cell membranes; carbohydrates generally serve as nutrient reserves, but in plant cells they also make up the cell walls. It is instructive to compare the relative numbers of molecules of cell constituents, as is done in Table 3.3. For purposes of calculation deoxyribonucleic acid (DNA) is arbitrarily assigned a molecular weight of 10^7, ribonucleic acid (RNA) 5×10^5, and protein 3.6×10^4. Calculation shows that more than 10 million water molecules are present for every DNA molecule and about 18,000 for every protein molecule. Nucleic acids (DNA and RNA together) constitute about 1 percent of the net weight of a cell (1.2 percent of the liver cell, for example), yet they have a profound effect on the functioning of the cell. With gentle extraction procedures DNA may have a molecular weight several times 10^7, and ribosomal RNA may have a molecular weight ten times larger than 10^5. The disparity between the number of nucleic acid molecules and other molecules in the cell is therefore much greater than that given in Table 3.3.

Water

Active tissues have more water than inactive ones (see Fig. 3.4); for example, white matter of the brain with the fibrous connections of

TABLE 3.3. RELATIVE NUMBER OF MOLECULES OF VARIOUS TYPES OF CELLULAR MATERIALS

Substance	Percent	Average Molecular Weight	Molarity in Cell	Number of Molecules Relative to Protein	Number of Molecules Relative to DNA
Water	85	18	47.2	1.7×10^4	1.2×10^8
Protein	10	36,000	0.0028	1.0	7.0×10^3
DNA	0.4	10^7	0.0000004	—	1.0
RNA	0.7	4.0×10^5	0.0000175	—	4.4×10^1
Lipid	2	700	0.028	10^1	7.0×10^4
Other organic materials	0.4	250	0.016	6	4.0×10^4
Inorganic	1.5	55	0.272	10^2	6.9×10^5

neurons contains about 68 percent water, whereas the gray matter, made up of neuron cell bodies, contains about 85 percent.

Water occurs in the cell in two forms, free and bound, although partition between the two is indefinite. *Free water* is available for metabolic processes; *bound water,* loosely attached to protein molecules by dipole attraction, presumably is not. Imbibition is an expression of adsorption of water to protein. Water bound to protein molecules by hydrogen bonds forms part of the structure of protoplasm. Each amino group of a gelatin molecule is capable of binding 2.6 molecules of water. It is likely that only a small fraction of the water available in the cell is bound, the remainder being free to act as solvent for metabolic reactants.

Salts

Salts, present in all cells, are necessary for life in both prokaryotic and eukaryotic cells, although salt function has perhaps been more thoroughly studied in eukaryotic cells. Cells immersed in distilled water lose salt and die. The cell's requirement for salts is illustrated in man by muscle cramps induced by loss of sodium chloride following excessive perspiration. Loss of sodium from muscle cells is accompanied by leakage of potassium into the extracellular fluid, inducing the cramps. Since potassium reenters the cells after intake of sodium chloride, cramps are quickly relieved by ingestion of a salted food or drink. "Miners always salt their beer" is an old saying.

The salt content in cells of representative tissues and blood serum is given in Figure 3.5. It will be noted that inside cells potassium is the dominant cation, magnesium next in concentration, while sodium and calcium are present in relatively small amounts. The dominant anion inside cells is phosphate, part of it soluble in acid, part insoluble. A second important anion is bicarbonate. In some cells (erythrocytes and liver cells) chloride is also present. Besides the major salts, small and varying quantities of many elements are also found in cells. For example, considerable iron is present, whereas only microconcentrations of manganese and copper (less than 0.05 percent) are found. Vanadium, zinc, nickel, molybdenum, and even tin are found in variable but always minute quantities (Florkin, 1971). Some of these elements are necessary for enzyme function (see Chapters 10 and 12). Tin, vanadium, fluorine, and silicon also appear to be needed (Table 3.4) (Florkin, 1971). Some kelps concentrate phenomenal amounts of iodine and some tunicates concentrate considerable vanadium, although the reason for this is not clear in either case.

It is interesting that in body fluids such as serum, sodium is present in highest concentrations (Fig. 3.5), whereas only a small amount of potassium is found, the reverse of the situation in cells. Such a difference in concentration of sodium and potassium inside and outside cells is vital to their functioning, as is most readily seen in studies on highly excitable tissues such as muscle and nerve (see Chapters 20 and 21).

Proteins

Some classes of proteins are similar in prokaryotes and eukaryotes, and in both some proteins are globular and others are fibrous. However, some distinctive differences also occur. For example, *histones,* which are low-molecular-weight basic proteins that complex with DNA, are lacking in prokaryotes; instead, prokaryotes have polyamines. Two types of fibrous proteins, *collagen* and the *elastin-keratin-myosin-fibrinogen* type are conspicuous cell products in eukaryotic tissues but are lacking in prokaryotes. Collagen is found in skin, connective tissue and tendons. *Elastin* is present in elastic skin fibers, connective tissue and tendons; *keratin* occurs in the horny vertebrate epidermal cells and constitutes hair and feathers, horny epidermal cells, fingernails, and horns; *myosin* is one of the muscle fibril compo-

TABLE 3.4. ELEMENTS FOUND IN CELLS OF ORGANISMS AND THEIR ATOMIC NUMBERS*

Most Abundant		***Next Most Abundant***		***Trace Elements***			
H	1	Na	11	F	9	Ni	28
C	6	Mg	12	Si	14	Cu	29
N	7	P	15	V	23	Zn	30
O	8	S	16	Cr	24	Se	34
		Cl	17	Mn	25	Mo	42
		K	19	Fe	26	Sn	50
		Ca	20	Co	27	I	53

*Data from Frieden (1972) and Miller and Neathery (1977).

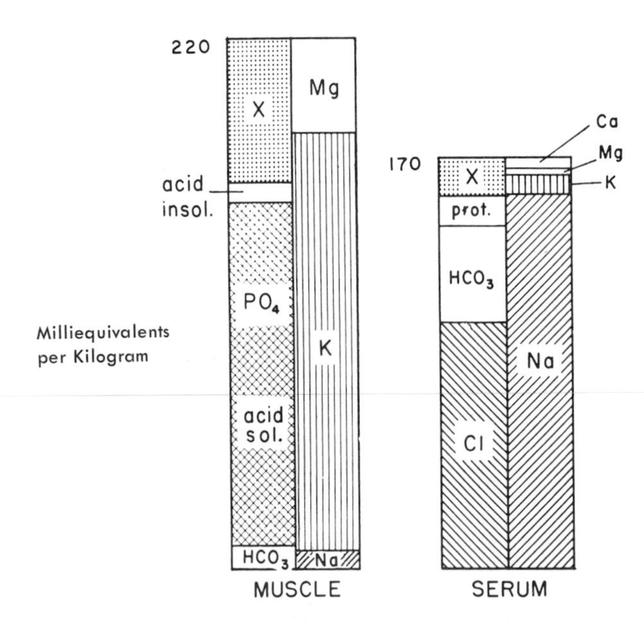

Figure 3.5. *Comparison of ion content of muscle tissue and blood serum. Anions are shown on the left, cations on the right. (After Lowry, 1943: Biol. Symp. 10: 241.)*

nents, while *fibrinogen* makes up the fibers of a blood clot. Prokaryotes lack these proteins but have two other fibrous proteins lacking in eukaryotes: *flagellin* in bacterial flagella and *pilin*. Pilin is present in pili, the fine filamentous appendages on the surface of many bacteria (see Chapter 1).

The collagen molecule, an extremely long thin rodlet and the most asymmetrical molecule yet isolated, consists of three peptide chains, each an α helix (see Fig. 1.12), coiled into a larger helix. Each collagen molecule is made up of subunits called tropocollagen. The complete sequence of amino acids in collagen has not yet been determined (Lehninger, 1975). Each chain incorporates considerable proline, hydroxyproline, and glycine (Bornstein, 1974). Glycine, the smallest of the three amino acid residues, occurs every fourth position on each chain. This makes it possible for the bulkier amino acid residues to fit into the links of the triple strand. The adjacent tropocollagen units are cross-linked by covalent bonding between lysine residues, forming dehydrolysinonorleusine. The banding seen in an electron micrograph (Fig. 3.6) is the result of regular orientation of amino acids in collagen fibers. As a consequence of the hydrogen bonding, tissue collagen is more resistant to heat than dissolved, molecularly dispersed collagen. When a collagen solution is warmed it forms a gelatin gel that has one third the molecular weight of collagen. Collagen fibers give the skin its suppleness and a tendon its strength. It is interesting that collagen in all members of the animal kingdom, from sponges to humans (Matsumura, 1976), is similar. Seemingly, a good "invention," once acquired, is retained in the repertoire of organisms.

Elastic connective tissue fibers (Fig. 3.7) are much shorter and thinner than collagen fibers. They stretch, reversibly, to double their length. All these proteins give x-ray diffraction patterns related to that of natural unstretched mammalian hair (α keratin) (Anfinsen and Scheraga, 1975). Elastic fibers stretched in hot water or alkali assume a second form called β keratin. The latter gives a simpler x-ray diffraction pattern than the former. Fibers released while still immersed contract to two-thirds of their original length, giving a supercontracted form. Electron micrographs of elastic fibers reveal bundles of microfibrils about 11 nm in diameter embedded in a more abundant amorphous matrix called elastin (Bloom and Fawcett, 1975).

Analyses of the α form of the elastin-keratin-myosin-fibrinogen proteins by monochromatic x-ray bombardment indicate a fundamental crystallite structure showing regular periodic distances between refracting units (Liljas and Rosmann, 1974). In fibrous proteins the only arrangement of amino acids that could be in keeping with the x-ray diffraction pictures is thought to be a helix. Knowing the stereochemical requirements of the peptide backbone, that is, of the distances between atoms and of the possible angles they make with one another, scientists can make models that summarize most of the data on protein structure. X-ray analysis of the β or stretched form of the elastin-keratin-myosin-fibrinogen proteins shows that the atomic arrangements are no longer spiral as in the common α helix of globular proteins and the collagen fibrous protein, but may be best represented as a pleated ribbon polypeptide chain (Fig. 3.8) (Baldwin, 1975).

The structure that a protein assumes is the

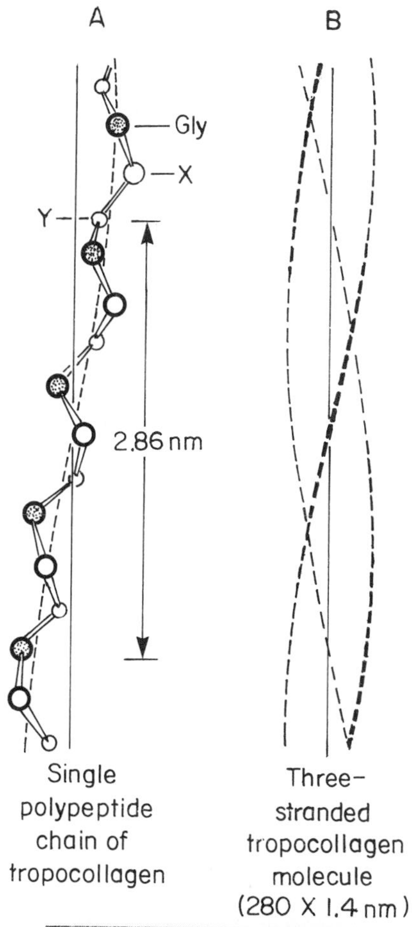

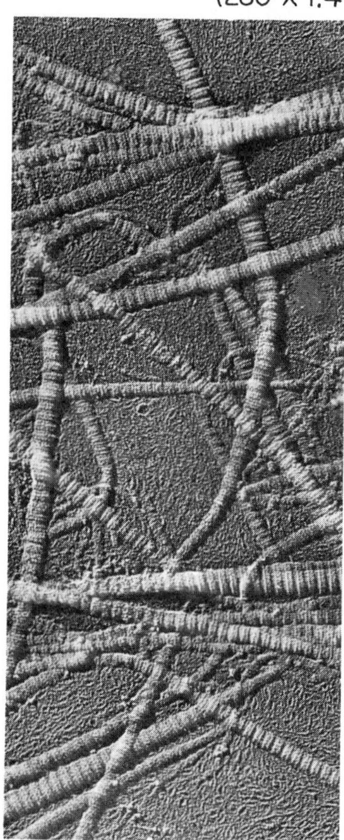

most thermodynamically probable configuration and is determined by its content of amino acids. Thus, proline and hydroxyproline are not compatible with an α helix (see Fig. 1.12), which is found only where these residues are absent. The sequence (glycine-serine-glycine-alanine-glycine-alanine)$_n$ induces a β sheet structure (Fig. 3.8) while a sequence such as (glycine-X-proline)$_n$ induces a collagen triple helix (see Fig. 3.6*B*). It is more difficult to indicate what sequence of amino acids produces an α helix than to list some of the conditions under which it will not form, for example, formation will not occur in the presence of large proline concentrations or of amino acid residues with bulky side chains branching from the β carbon of the amino acid. The α helix is a stable configuration lacking steric strain and the most stable form of protein in the absence of factors which would prevent its formation (Lehninger, 1975).

These examples of protein structure are presented to illustrate the possibility of providing a molecular explanation of the x-ray data and many of the facts known about the chemistry of proteins. The models at hand already help to interpret chemical and biological data with more consistency than was previously possible.

Colloidal Properties of Proteins

In the colloidal state of matter the physics and chemistry of surfaces is dominant. Colloids display characteristic behavior. Because of their large molecular size and each particle's large surface area, proteins show all the characteristics of the colloidal state, even though they may be dispersed as individual molecules in water. As expected, proteins do not pass through natural membranes; therefore, solutions of proteins are placed in cellophane sausage tubing and immersed in dis-

Figure 3.6. *A, A single polypeptide chain of collagen. Gly is glycine, X and Y are other amino acids. The sequences glycine-X-proline, glycine-proline-X, and glycine-X-hydroxyproline are common in collagen. X may be any one of a number of amino acid residues. The sequence of amino acids varies with species. (From Lehninger, 1975: Biochemistry. 2nd Ed. Worth Publishers, New York.) B, A triple-stranded tropocollagen molecule, the unit of which collagen is composed. (From Lehninger, 1975: Biochemistry. 2nd Ed. Worth Publishers, New York.) C, Electron micrograph of collagen. Fibrils with typical collagen period (649A), precipitated from collagen solution by dialysis against 1 per cent NaCl. Chromium shadowed. (Courtesy of J. Gross, F. O. Schmitt, and J. H. Highberger; from Bloom and Fawcett, 1975: A Textbook of Histology. W. B. Saunders Co., Philadelphia.)*

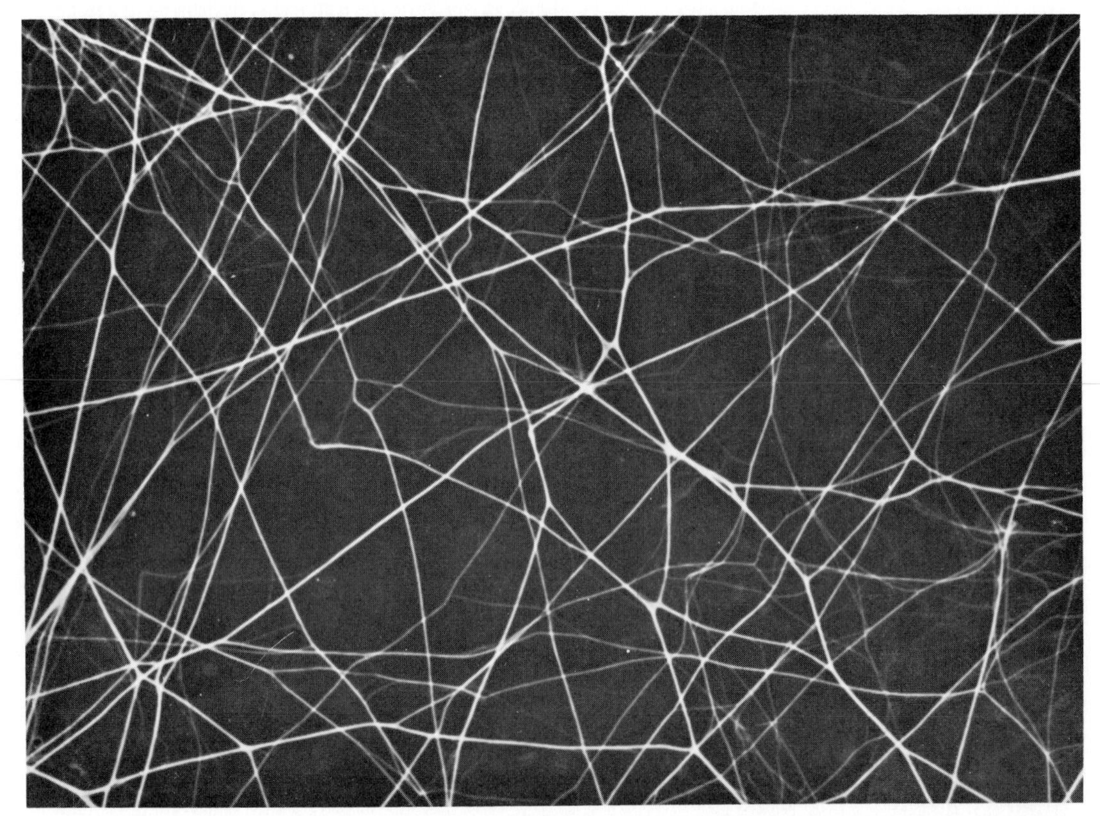

Figure 3.7. *Elastic fibers in a spread of rat mesentery stained with resorcin-fuchsin; the photomicrograph is printed as a negative. Notice that the fibers are smaller and less variable in size than collagen bundles and they branch and anastomose to form a network (×550). (After D. W. Fawcett.* In *Greep, R. O., ed., 1953: Histology. The Blakiston Co., Philadelphia. Used by permission of McGraw-Hill Book Co.)*

tilled water or buffer to free them of crystalloids (dialysis). Only crystalloids, such as salts or glucose, pass from the enclosed solution into the distilled water or buffer.

Protein molecules form a phase separate from the aqueous suspending medium. Consequently, proteins in water show a typical Tyndall cone in a beam of light, like the light scattered from dust particles in a sunbeam passing through a chink in a shade, although the opalescence is less marked than in a metallic colloid such as gold sol, because the refractive index of proteins is only slightly greater than that of water.

The interplay of surface-active forces at the molecular interfaces in a protein solution is manifested by the high viscosity that such a solution quite often shows. The viscosity is a

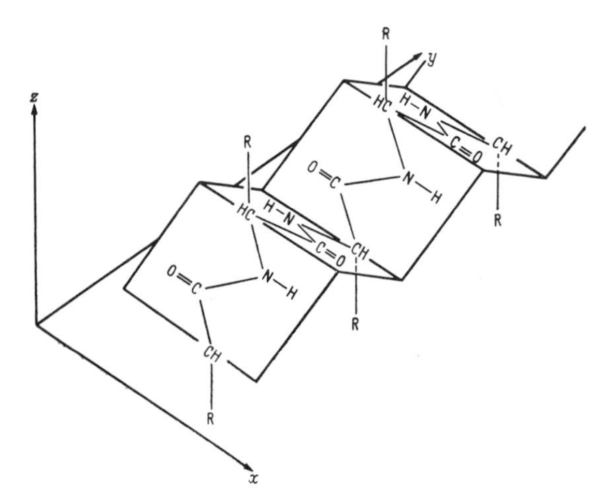

Figure 3.8. *Pleated ribbon polypeptide chain to represent the arrangement of the amino acids in the β or stretched fibers of the keratin-myosin-elastin-fibrinogen group of proteins, for example, in silk. (From Springall, 1954: The Structural Chemistry of Proteins. Academic Press, New York.)*

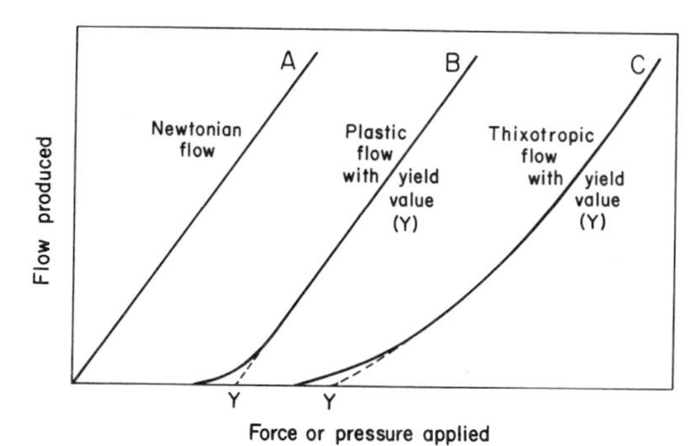

Figure 3.9. *Flow-force diagrams:* A, *for gases and liquids;* B, *for butter and clay;* C, *for proteins. (From McBain, 1950: Colloid Science. D. C. Heath & Co., Boston.)*

measure of the resistance of the protein molecules to flow past one another when force is applied to the solution in a confined space. Resistance to flow is thought to result from the attraction of the molecules for one another as well as from asymmetry in their structure.

A protein molecule will both attract and repel another molecule of the same kind, depending upon the distance separating them. Repulsion occurs because the overall charge (positive or negative) is similar. But if two molecules approach each other so closely that local residual valency forces can act, they may be sufficiently attracted to each other to form a hydrogen bond. In dynamic state, hydrogen bonds are continually made and broken. As a consequence of such bonds, protein molecules initially resist flow past each other when a force is first applied to a protein solution in a tube. The pressure required to initiate flow is called the *yield value* and represents the force required to break these temporary bonds. Even after it begins, the flow may not be directly proportional to the pressure applied and gradually increases as more bonds are broken (Fig. 3.9). Such a pattern of flow is said to be thixotropic. In absence of such bonds, flow is proportional to the force applied, showing a constant increment with each increment of force past the yield value, as in Newtonian flow that begins as soon as force is applied, or in plastic flow (e.g., in butter).

Some colloids exist in either of two states, *sol* or *gel,* depending upon the conditions. The sol state is fluid because the particles are not able to form more than very temporary bonds with one another. Gelation of proteins probably results from the formation of numerous stronger linkages between protein molecules. Electron microscopic study of a gel reveals a structure resembling a brush-heap meshwork of intertwining fibers.

Proteins require no stabilizing agents to disperse in water because protein ions repel one another, and each protein ion attracts and orients the water molecules surrounding it,

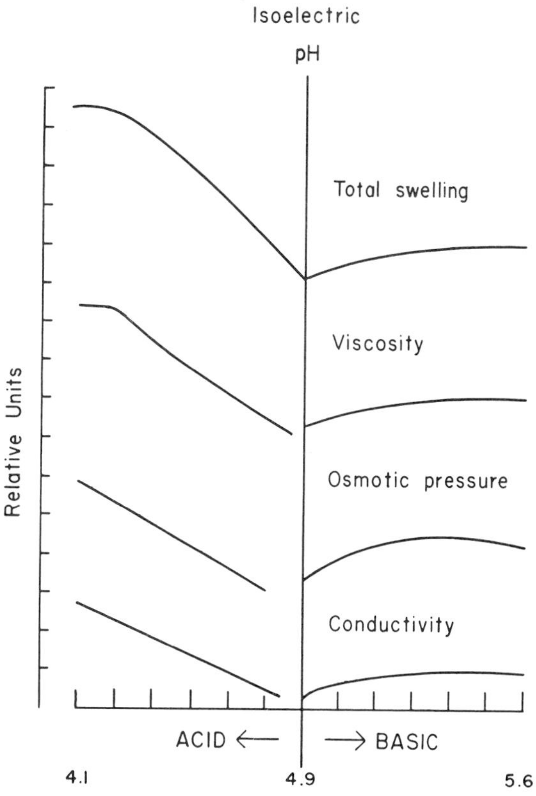

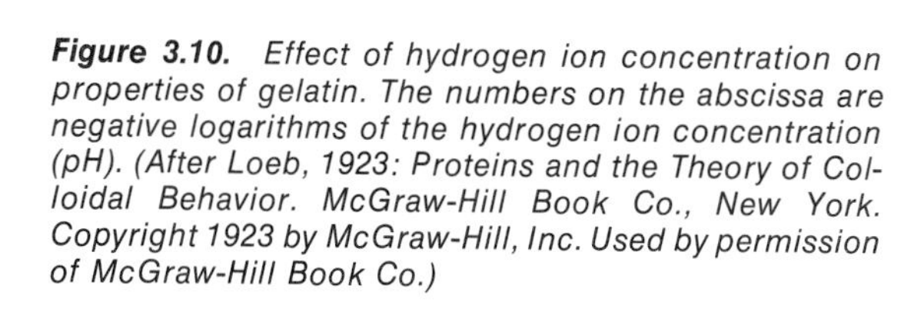

Figure 3.10. *Effect of hydrogen ion concentration on properties of gelatin. The numbers on the abscissa are negative logarithms of the hydrogen ion concentration (pH). (After Loeb, 1923: Proteins and the Theory of Colloidal Behavior. McGraw-Hill Book Co., New York. Copyright 1923 by McGraw-Hill, Inc. Used by permission of McGraw-Hill Book Co.)*

thereby binding itself to the solvent. Each water molecule acts as a dipole with positive and negative ends.

Imbibition depends upon the *hydration* of proteins. Dry proteins adsorb (imbibe) water with great avidity, even against the osmotic pressure of saturated lithium chloride (about 1000 atmospheres). When gelatin has adsorbed four times its weight in water, it no longer takes up water against a high osmotic pressure. Seeds take up water against an osmotic pressure of more than 1000 atmospheres because of the dehydrated state of their proteins, but the force of uptake falls very rapidly as their proteins become hydrated.

Many properties of proteins such as viscosity, imbibition, and osmotic pressure are affected by the ionic environment and the pH of the solution (Fig. 3.10). This is to be expected because proteins are ionized; the greater the pH difference from the isoelectric point the more they are ionized. Their behavior in response to salt ions and pH can usually be explained by how these factors affect protein ionization (for proteins as zwitterions see Chapter 4).

Lipids

Many lipids occur commonly in both prokaryotes and eukaryotes (see Chapter 1). One group of lipids present in eukaryotes and not found in the prokaryotes are the steroids, aromatic alcohols soluble in fat solvents. The most common sterol is cholesterol (Fig. 3.11), of ill fame for its role in blockage of human blood vessels, leading to heart attacks. In vertebrates sterols are of importance as the D vitamins and as sex hormones. In eukaryotes sterols play a major role in structure of cell membranes; this distinguishes them from cell membranes of prokaryotes, which lack sterols (see Chapter 8). Lipids combine with proteins to form lipoproteins of importance in cell functions (Morrisett *et al.*, 1975).

Carbohydrates

Many of the carbohydrates found in prokaryotes are also present in eukaryotes (see Chapter 1).

Cholesterol

Vitamin D_3

Fucosterol

Corticosterone

Estradiol

Testosterone

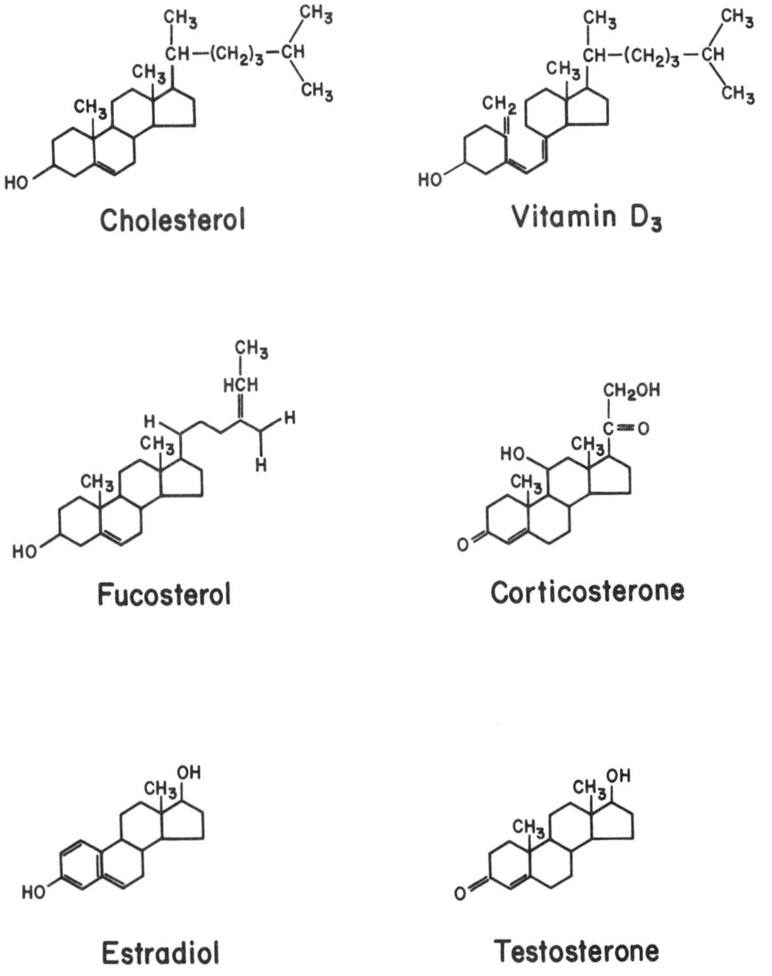

Figure 3.11. Examples of steroids in plants and animals. Cholesterol is important in animal cell membranes; vitamin D_3 (cholecalciferol) is formed by irradiation of 7-dehydrocholesterol; fucosterol is a plant sterol; corticosterone is an important vertebrate metabolic regulatory hormone produced by the adrenal cortex; estradiol is a vertebrate female sex hormone, and testosterone, a vertebrate male hormone.

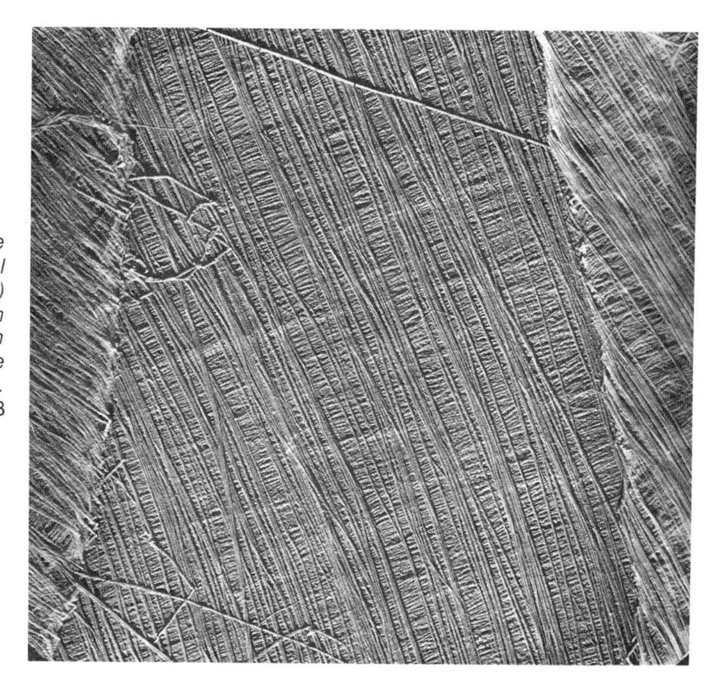

Figure 3.12. Electron micrograph of the arrangements of cellulose fibers in the cell walls of Cladophora rupestris *(green alga) (×24,000). Note the three directions in which the fibers run. Smaller units, of which the fibers are composed, may be seen in the isolated fibers at the top and the bottom. (From Preston, 1961: Proc. R. Soc.* B *154: 70–94.)*

It will be recalled that cellulose is found in prokaryotes in only one recorded instance, the fibers formed by acetic acid bacteria. However, cellulose is very common in eukaryotes, especially as the major constituent of plant cell walls. The plant cell wall cellulose bundles appear to be linked together by small units of the polysaccharides arabinogalactan, xyloglucan and rhamnogalacturan (Alberscheim, 1975). A *cell wall* is always present in plant cells, protecting the cell membrane. In most higher plants the cell wall is composed chiefly of cellulose secreted by the cell. The adjacent walls of abutting cells are cemented together by the D-galacturonic acid polymer *pectin* (see p. 20). Old cell walls may be thickened by a *lignin,* a group of ring carbohydrates, and surfaces of epidermal cells may be covered with a waxy coating that reduces water loss.

The architecture of the plant cell wall has been extensively studied by x-ray diffraction, polarized microscopy, and electron microscopy. It appears that the cytoplasm lays down an interwoven network of fibers (primary wall) to which lamellar deposits (secondary wall) are later added. Spaces remain between the fibers, and the network design gives maximal strength for a minimum of material (Preston, 1974).

The first rudiment of a future cell wall in a cell about to divide is seen as a series of droplets on the midline between the two daughter cells and at right angles to the spindle (see Chapter 26). Half an hour later, birefringence (doubly refracting) is observed, probably at the moment of formation of the cellulose fibrils on either side of the pectic materials forming the middle lamella (Preston, 1974).

In *Valonia,* the unit fibers are composed of α cellulose and may be quite long, about 10 μm, and astonishingly constant in diameter: 15 to 30 nm. The fibers are ribbon-like and lie with the broader face in the plane of the cell wall. X-ray evidence indicates "periods" of 500 glucose residues, each one 250 nm long (Preston, 1974). X-ray mapping before and after successive stripping of these fibers revealed two sets of fibers: one meridional and the other following a shallow, spiral path; the two kinds of fibers were laid down in layers following a definite pattern of succession, a total of 200 to 300 layers being present.

The primary cell wall of many other plant cells, however, has fibers that are distinctly interwoven, not in plane sheets as in *Valonia.* This interweaving indicates that the secreting cytoplasm penetrates the fibril meshwork. The secondary cell wall has parallel fibers of constant thickness laid down in close texture and forming lamellae (Fig. 3.12). The intimacy of contact between cytoplasm and cell wall is shown also upon plasmolysis by the adhesion of the cytoplasm to the growing tips of cell wall fibers (Albersheim, 1975).

Cellulose is found in the cell walls of a few animal groups, for example, in the tunicates, which are primitive relatives of the chordates. Cellulose is also produced by cellular amoeboid slime molds and some fungi.

A concentrated preparation of the enzyme cellulase removes the walls from the cells in root, leaf, and other parts of higher plants and liberates naked protoplasts bounded by the plasma membrane. Protoplasts of plant cells act like any other naked cells; they become round and swell or shrink depending upon the concentration of solutes in the surrounding medium.

The cell wall may have a degree of plasticity. For example, stamen hair cells of *Tradescantia,* placed in hypotonic solutions, may extend 100 percent and again return to their original size on replacement into isotonic solutions. In the growing tip of a plant, cell walls must be plasticized before they can elongate, to accommodate a manifold increase in length of the cell wall during maximal extension (Preston, 1974).

Cell wall composition is not always the same. For example, although the cotton fiber is 90 percent cellulose by dry weight, only 2 percent of a red algal cell wall is cellulose, other polysaccharides of complex nature making up the remainder. The cell walls of some algae in warm seas are made up entirely of polysaccharides other than cellulose; for example, xylan, a polymer of the pentose sugar xylose, makes up the cell wall of the green algae *Bryopsis,* and mannan, a polymer of the hexose sugar mannose, is present in the wall of the large single-celled green alga *Acetabularia.* While xylan fibers similar to cellulose fibers of land plants are found in the cell walls of *Bryopsis,* granules or short rods (shown by x-ray analysis to be regularly oriented) are found in the cell wall of *Acetabularia* (Preston, 1974). Cell walls of fungi, including yeast, are made up largely of chitin, a glucosamine polymer (see Fig. 1.15). In the chitinous skeletons of insects and crustaceans, proteins are also present.

APPENDIX

3.1 STAINING TECHNIQUE FOR LIVING AND FIXED CELLS

Vital dyes penetrate living cells and stain structures without serious injury. For example, Janus green B stains mitochondria selectively and methylene blue stains the Golgi complex. Vital dyes are not entirely harmless, but they kill only after long exposure. Although helpful, the vital dye technique has limited use because many cell organelles are not selectively stained or are too small to see with light microscopy.

Most early work on cell organelles was done with fixed and stained preparations. *Fixing agents,* such as formalin, alcohol, acids, salts of heavy metals or mixtures of these render proteins insoluble. Next, water is removed from the fixed tissues by *dehydrating agents,* such as alcohol, and the tissues are embedded in paraffin and sectioned with a microtome. The sections are affixed to slides, the paraffin removed with xylol and the slides washed in xylol-alcohol. By washing in decreasing concentrations of alcohol the sections are partially hydrated and the proteinaceous matter of the cell is then differentially stained to distinguish the structures present. Natural dyes (e.g., hematoxylin) or basic anilin dyes (e.g., safranin and basic fuchsin) stain the nucleus selectively, and acid dyes (e.g., orange G, eosin, and fast green) stain the cytoplasm. The sections are then dehydrated with alcohol. To reduce scattering of light, it is necessary to replace the alcoholic medium with a substance having the same refractive index as that of the protein particles. This is accomplished with a *clearing agent,* such as xylol, which infiltrates among the protein particles. The preparation is then mounted in balsam, which has a refractive index about equal to that of the cell proteins, making it possible to see the stained structures clearly (Galligher and Kozloff, 1971).

3.2 AUTORADIOGRAPHY

If a cell is suspended in a solution of radioactively labeled substance, it is possible to determine where in the cell incorporation occurs and to what extent. For example, ^{3}H-uridine may be presented to cells for a brief time and the cells are mounted on a slide and fixed. A layer of photographic emulsion is deposited over the preparation. The radioactive emission from the ^{3}H-uridine will reduce the silver halide to silver in the emulsion. On development, the grains affected by the radiation (beta particles of low energy value) appear as black grains or traces over the structure into which the ^{3}H-uridine had been incorporated. If the grains appear over the nucleus we know that the nucleus incorporates uridine into ribonucleic acid (RNA). Similar experiments can be performed with ^{3}H-thymidine (Fig. 3.13), which is selectively incorporated into deoxyribonucleic acid (DNA), and with ^{3}H-amino acids, which are incorporated into proteins. Autoradiography has been of inestimable value in localizing the sites of syntheses in the cell. Recurrent reference is made to the use of this technique in subsequent chapters.

3.3 MICROSCOPICAL TECHNIQUES

The *light microscope* makes it possible to see structures larger than the limit of resolution for the naked eye (Fig. 3.14). The smallest ob-

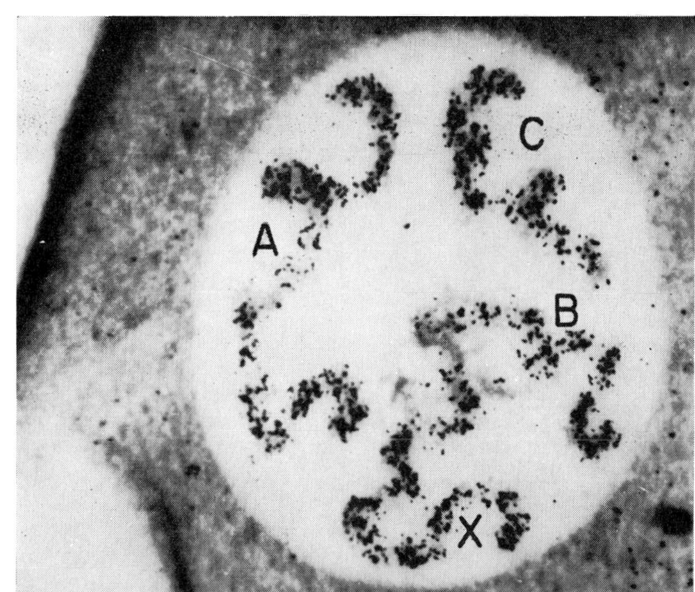

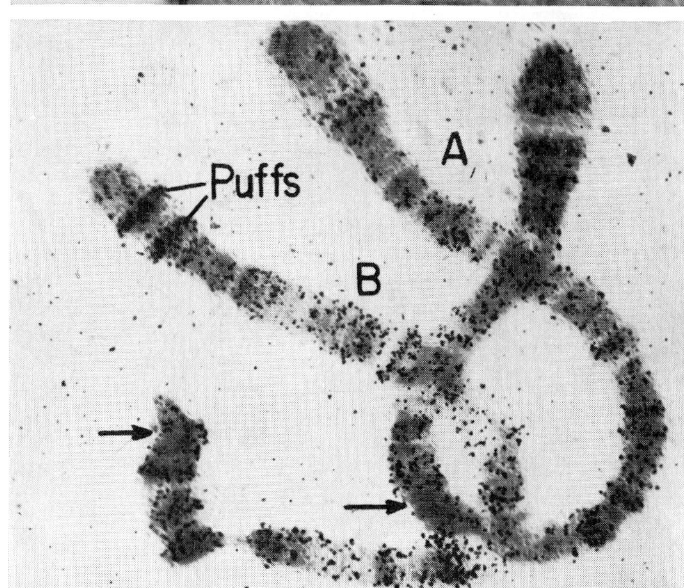

Figure 3.13. *The autoradiographic method. The nucleus and polytenic (multiple-stranded) chromosomes of the larvae of the fly* Rhynchosciara angelae *injected 24 hours earlier with tritiated thymidine. (Tritium emits electrons of 10,000 electron volts.) In the upper part are shown the four chromosomes of a cell. In the stained chromosomes below it can be seen that the localization of the thymidine ^3H is in the bands of the chromosomes, indicating that incorporation occurs only here. ^{32}P may also be used for labeling. (By courtesy of A. Ficq and C. Pavan.)*

ject, called d, resolved (seen clearly and distinctly from similar sized objects separated by like small distances) may be determined from the equation:

$$d = \frac{\lambda}{NA + na} \quad (3.1)$$

where λ is the wavelength of light, NA the objective numerical aperture, and na the condenser numerical aperture. Since the numerical aperture of the condenser is best when equal to that of the objective, the equation may be written:

$$d = \lambda/2NA \quad (3.2)$$

$$\text{where } NA = n \sin \alpha \quad (3.3)$$

n is the refractive index (ratio of the sine of the angle of incidence to the angle of refraction) of the microscope light path and α is half the angular aperture of the beam entering the condenser. When immersion oil is used between condenser and slide, and slide and objective, the numerical aperture may be as large as 1.6, although in practice it is seldom over 1.4. The factor that limits resolution with the light microscope is therefore the wavelength of light. Under optimal conditions the smallest object seen is theoretically about one-third the wavelength of visible light, or about 0.13 μm in blue light. Usually the limit is about 0.2 μm because the eye is most sensitive to the yellow-green part of the spectrum (0.55 μm) and a numerical aperture of 1.4. An object of this size is visible only if the structure shows maximum

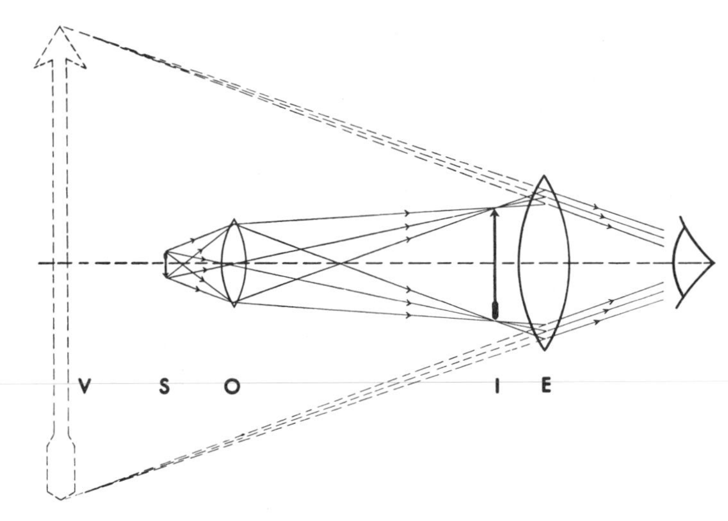

Figure 3.14. Principle of the compound microscope. The intermediate image (I) of the subject (S) formed by the objective (O) is enlarged by the eyepiece (E). The virtual image (V) is seen by the eye. A real image of V is recorded with a camera. (From Photography through the Microscope, Kodak Scientific & Technical Data Book P-2. 4th Ed. Eastman Kodak Company, Rochester, N.Y., 1966.)

contrast. In order to increase the contrast between unstained structures and to obtain greater resolution of details of cell organelles, various types of microscopes have been developed. Each of these is considered in turn (Bradbury, 1967; Slayter, 1970; Rochow and Rochow, 1978).

The *darkfield microscope* has a special condenser with a dark stop to remove light from the center of the field. The object is illuminated only by an oblique beam of light coming through the margin of the dark stop, as seen in Figure 3.15. The objects, seen by light reflected and scattered from interfaces, are outlines of the nucleus, mitochondria, oil droplets, vacuoles, and various inclusions. In cells undergoing division, the centrosomes, asters, spindles, and chromosomes may also be observed. Objects smaller than those seen with the light microscope may be detected but not resolved with dark field microscopy.

The *polarizing microscope* is a light microscope with polarizing optics in the ocular and condensers, each transmitting only plane-polarized light. The condenser is then spoken of as the *polarizer* and the ocular as the *analyzer*. If the optics of the condenser and ocular are crossed (at 90° to one another) no light is transmitted unless an interposed object rotates the plane of the light.

Light can be polarized, that is, separated into two rays that vibrate at right angles to one another, by passing it through a doubly refracting crystal such as calcite. The *Nicol prism,* one of the best known polarizers, is made by cutting a rhomb of Iceland spar (a pure, transparent form of calcite) in halves along the line of one of the optical faces, polishing the cut faces and cementing them together with Canada balsam. As seen in Figure 3.16, unpolarized light entering one end is separated by the prism into two beams, one vibrating in the plane of the figure, the other at right angles to it. One of these beams is totally reflected by the layer of Canada balsam and is usually absorbed by the opaque case covering part of the rhomb. The

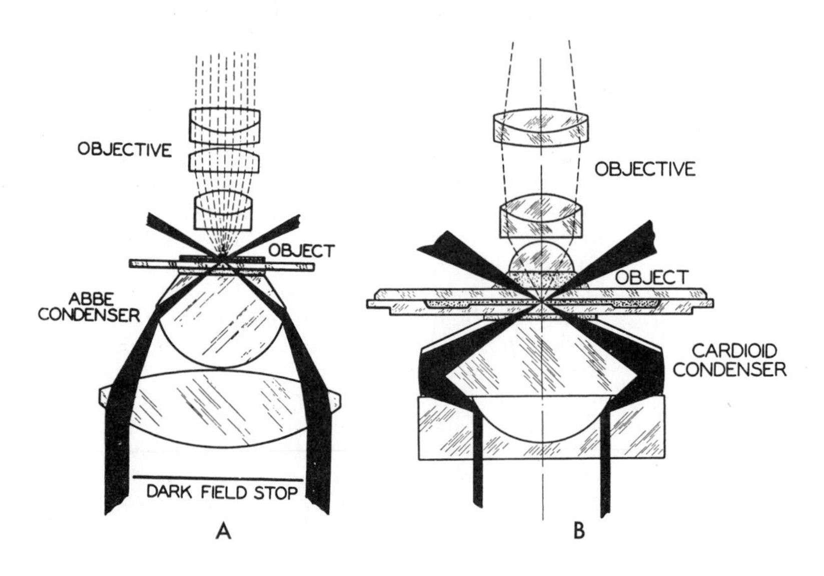

Figure 3.15. Darkfield microscopy. The dark field stop shown in A is much less effective than the dark field condenser shown in B. (By permission of Bausch and Lomb, Inc., Rochester, N.Y.)

other beam is transmitted. Hence, a Nicol prism transmits plane-polarized light. A Nicol prism may be used in the microscope condenser, another in the ocular. If the second Nicol prism is used in series with the first, and in the same position, the polarized light is transmitted. If, however, the second prism is rotated through 90° in relation to the first prism, the light fails to pass and the microscope field is black. But if an object like quartz, which polarizes light, is placed in the beam of light between the two prisms, and if the two Nicols are at right angles to each other (crossed), a brilliantly lit object is seen against a dark background, indicating that the object has polarized the light passing through it. If the second prism is rotated away from 90°, the field becomes dark; the angle at which the light is extinguished is a measure of the degree to which the plane of polarized light has been rotated by the interposed object.

Various natural substances are observed to be *isotropic;* that is, light passes through them equally well in all directions. Others, which affect the passage of light unequally in different directions, are *anisotropic* or *birefringent.* The ray passing directly through a crystal is called the *ordinary ray;* the ray that is deviated is called the *extraordinary ray.* If the velocity of the extraordinary ray passing through such a material is less than the velocity of the ordinary ray, the crystal is said to be negative (e.g., quartz). If the extraordinary ray has a velocity greater than the ordinary, the crystal is said to be positive. *Crystalline* or *intrinsic birefringence* is found where the bonds between molecules (or ions) have a regular asymmetrical arrangement. The birefringence in this case is independent of the refractive index of the medium.

Dichroism occurs when the absorption of a given wavelength of polarized light changes with the orientation of the object. For example, the degree of absorption at wavelength 260 nm by nucleic acids in sperm cells depends upon whether the light is parallel to, or at right angles to, the oriented purine and pyrimidine residues. Dichroism is a sensitive indicator of orientation of molecules in a structure, seldom seen in biological objects.

Birefringence of inanimate objects, then, is the result of their crystalline nature. Birefringence of biological materials indicates that their molecules are oriented with respect to one another much as in a crystal. Some particles—for example, nucleoprotein molecules—line up only if the solution is flowing; what they show is therefore called *flow* or *streaming birefringence.* This indicates that the molecules are elongate and line up in the path of least resistance to a current, just as sticks line up in a brook. Organisms are more likely to show *form birefringence,* because of regular orientation of tiny particles (rods or planes) differing in refractive index from the surrounding medium. Birefringence in such cases disappears when the material is immersed in a medium of the same refractive index. Oriented rods have different refractive indices for directions parallel to and at right angles to the major axis. Therefore, by making observations with polarized light one may determine in which direction the rods lie in a given structure.

Polarizing optics have been especially useful in study of biological membranes (see Chapter 8) and of the dividing cells (see Chapter 26). They are also useful in studying fibrillar structures.

The *phase contrast microscope,* which has the same resolving power as the ordinary light microscope, enables one to see more clearly structures in the living cell by taking advantage of the slight differences in refractive index between any two structures to improve their visibility. The phase contrast microscope does this by the use of a special diaphragm in the condenser and a phase plate in the objective. The annular diaphragm used in the condenser of the phase microscope permits light to pass through the condenser as a hollow cone, the remaining light being absorbed (Fig. 3.17). This cone is focused on the object. The phase plate placed at the back focal plane of the objective is a transparent disk containing a groove (or elevation) of such a size and shape as to coincide with the direct image of the substage annular diaphragm that is formed at the back focal plane when no object is viewed. If an object is placed between the condenser and the objective, in addition to the direct image a number of overlapping diffracted images of the diaphragm then appear at the back focal plane of the objective. The depth of the groove (or the elevation) in the phase plate of the objective is so made that the two sets of rays forming the direct image and the diffracted image differ in optical path by a quarter wavelength of the illuminating beam of light (Fig. 3.17*B*). Under these conditions the phase difference, which is not seen by the eye, is converted to an intensity difference that we see. In the bright contrast phase optical system the two sets of rays are added to make a brighter image as in Figure 3.17 at *D,* whereas in dark contrast phase they partially cancel one another, making a more contrasting darker image, as in Figure 3.17 at *C* (Ross, 1967).

The phase contrast microscope (bright or dark contrast may be used), in which a small difference of refractive index is exaggerated, enabling one to distinguish adjacent structures, has made it possible to observe structures previously very difficult or impossible to see. The behavior of chromosomes of living cells during mitosis or meiosis can be followed with

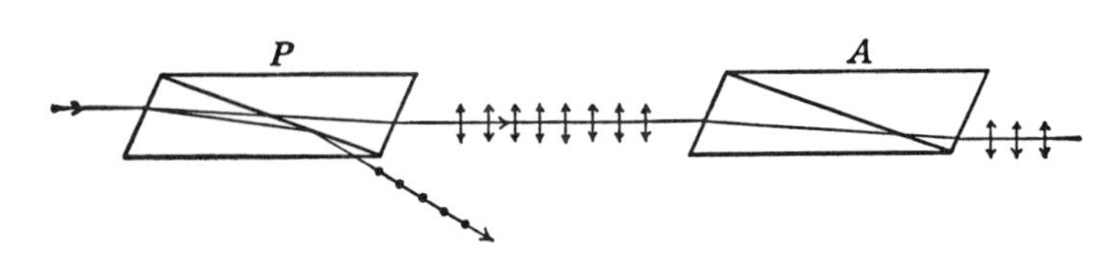

Figure 3.16. *Polarization of light by a Nicol prism. P stands for polarizer, A for analyzer. Only the beam in the plane of the sheet has been permitted to pass; that at right angles is deflected. The second Nicol prism, if oriented like the first, has no effect on the polarized light. However, if it is rotated through 90 degrees, the beam will be extinguished. (From Stewart and Gingrich, 1959: Physics. 6th Ed. Ginn & Co., Boston.)*

ease, and many of the other organelles may be studied. The phase contrast microscope has furnished the most convincing evidence that many of the organelles identified in fixed and stained preparations of the cell are real and not artifacts. However, the images are often surrounded by a halo and the difference in intensity between two areas of unequal refractive index diminishes with the distance from their boundary, thus causing some distortion.

The *interference microscope* avoids the residual halo that accompanies the image seen under a phase microscope because the aperture of the objective of such a microscope is not limited by a phase plate and the same area of the objective is used for both interfering beams. The relative phase of the two beams is varied continuously so that any part of an object can be given maximum or minimum contrast at will (Ross, 1967).

The *Nomarski interference microscope* developed by Nomarski in 1955 uses polarized light for illumination. Entering light is divided and recombined by use of Wollaston prisms (much like the Nicol prisms in Figure 3.16 but cemented by a substance other than Canada balsam). One of Nomarski's major innovations was to tilt the prism; by this maneuver he was able to obtain different degrees of "shear" between the refracted images. This exaggerates the difference in refractive index between object and background. As a result even such structures in a dividing cell as spindle fibers, which are not seen clearly with interference microscopes previously in use, can be observed clearly with the Nomarski interference microscope (see Fig. 26.2). Consequently the microscope is in widespread use for the study of structures in living cells (Ross, 1967). Originally manufactured by the Polish Optical Works in Warsaw, it is now also being manufactured by Zeiss in Jena and Zeiss in Oberchochen and perhaps elsewhere.

The *ultraviolet microscope* looks like a light microscope but its lenses are made of quartz to permit transmission of wavelengths of invisible ultraviolet as short as 220 nm (visible light covers the range from violet at 390 nm to the far red at 780 nm). The invisible ultraviolet light is then projected upon a screen which fluoresces in the visible spectrum. Usually, however, it is recorded photographically. Because chromosomes with considerable DNA absorb more short ultraviolet radiation than cytoplasm, they are readily seen in photographs taken with ultraviolet light at 260 nm, the peak of absorption by DNA (Fig. 3.18). With the ultraviolet microscope, resolution of fine structure is about twice that obtained with the ordinary light microscope, as may be judged from application of equations 3.1 and 3.2 (the wavelength of short ultraviolet light being about half that of the visible). Unfortunately, ultraviolet radiation damages living cells, so that prolonged observation of cells so illuminated is not possible.

An instrument that may serve as a valuable adjunct to all the microscopes considered here is the *television camera attachment.* Since the electron tube used in it "sees" at low light intensities and can be made sensitive to a restricted part of the spectrum (a narrow band in the green, blue or ultraviolet ranges), much clearer pictures can be obtained than with ordinary microscopy. Furthermore, a prolonged illumination is avoided since the image may be viewed on a phosphorescent screen that retains the image for a minute or two after exposure. Momentary illumination with light of low intensity suffices. This attachment, therefore, is especially useful for viewing living cells under ultraviolet light.

Fluorescence microscopy is used especially to localize structures containing molecules to which dye-containing fluorescent antibodies have been developed. The microscope resembles a dark field microscope that transmits long-wavelength (ca. 360 nm) ultraviolet light (Wehry, 1976). The technique now has widespread use in clinical diagnosis, wherein fluorescent antibodies specific for a suspected pathogen are used.

The *electron microscope,* which was developed on the eve of World War II, has proved a powerful tool for analysis of large chemical molecules, viruses, and cellular structures (Meek, 1976). In the electron microscope the beams of electrons are focused by magnets that serve as lenses (see Fig. 3.19). The object must be viewed on a fluorescent screen or pho-

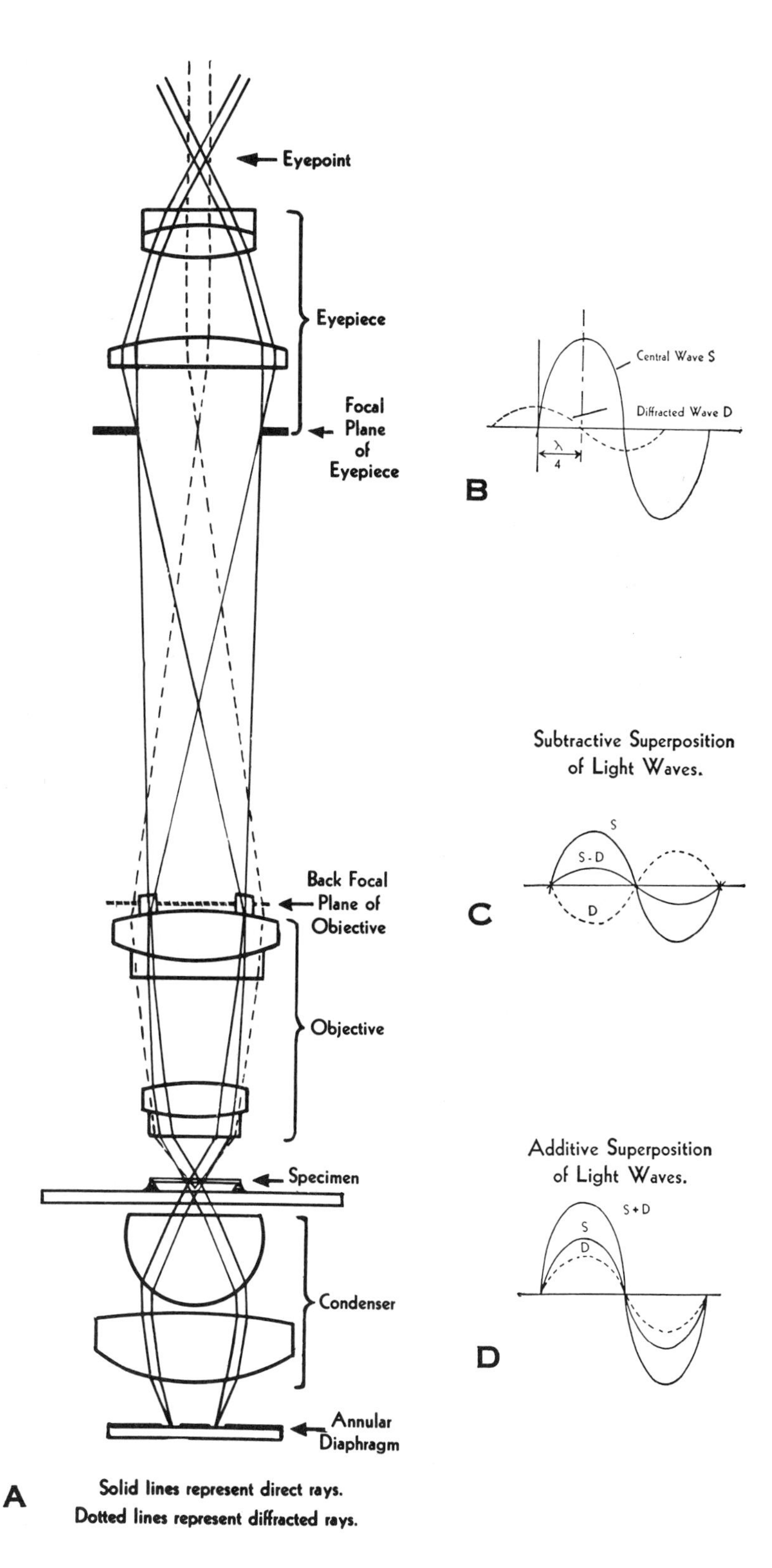

Figure 3.17. A *shows the light path in a phase microscope.* B *shows the normal retardation by ¼ wavelength of light diffracted by an object and its difference in phase from the light passing through the surrounding medium. By phase optics the two waves are superimposed to subtract from each other as in dark contrast phase shown in* C, *or to reinforce each other in bright contrast phase as shown in* D. *(By permission of the American Optical Company, Buffalo, N.Y.)*

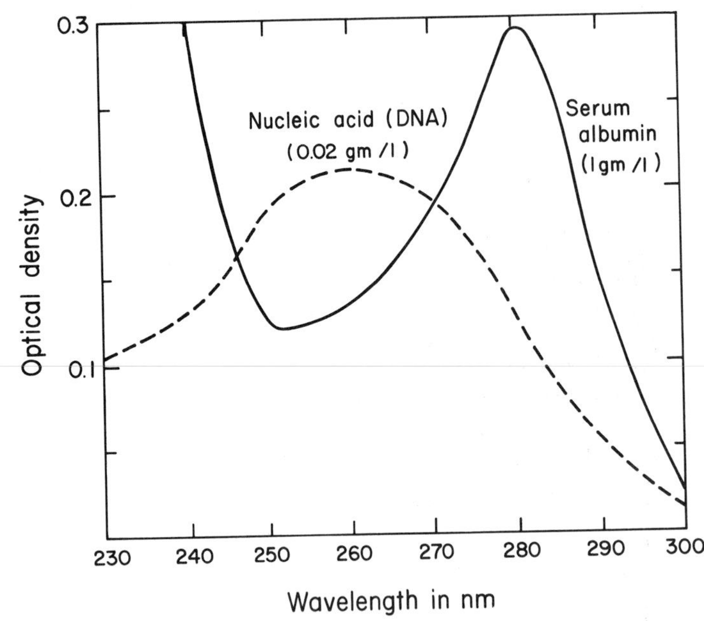

Figure 3.18. *Absorption spectra of nucleic acid and serum albumin. The optical density is highest where absorption is greatest; therefore, the peaks in the figure indicate the regions of greatest absorption. Note that albumin has a peak at 280 nm whereas nucleic acid has a peak at 260 nm. Also of interest is the relative absorption of ultraviolet radiation by the two compounds; for instance, a solution of only 0.02 gm per liter of nucleic acid absorbs at its region of maximum absorption almost as much of the radiation as does a solution of 1 gm per liter of albumin at its maximum.*

tographed. The electron microscope has a resolving power 400 (theoretically 2000) times that of the light microscope. Although the wavelengths accompanying the electron beams are short (e.g., a 50,000 volt electron beam is accompanied by a wavelength of 0.005 nm), the numerical aperture of the electron microscope is only about 0.0005.

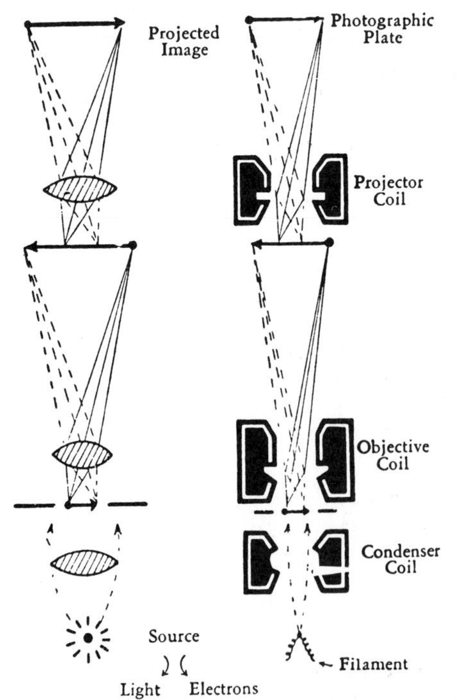

Figure 3.19. *Light and electron microscope equivalent. (Courtesy of Carl Zeiss, Inc., Oberkochon, Würtemberg.)*

The image results from differential scattering of electrons from molecular constituents of the cell. Scattering is greater the denser the material, regardless of its chemical composition. Even a gas scatters electrons; therefore, biological objects must be studied in a vacuum, since electron scattering from the gas molecules would obscure the differential scattering of the structures in the biological object. The preparations must be dry because the water vapor in the evacuated preparation would have the same effect as a gas. A thin slice of the object must be used; otherwise the electrons would all be scattered and the electron micrograph would reveal no detail. Since biological material varies little in density, it must be stained differentially, i.e., with some structures taking up more stain than others. *Positive staining* is obtained by fixing the cell in salts of heavy metals, e.g., osmium tetroxide or potassium permanganate. In this case the electron-dense (i.e., opaque to electrons) metal adheres differentially to structures of the cell, causing them to scatter the electrons more than the surroundings. In other cases *negative staining* is used; that is, the object is infiltrated with a heavy metal compound such as phosphotungstic acid or uranyl acetate. These materials do not adhere to the biological structures but fill the interstices between them, making these areas electron-dense in contrast to the biological structures, which are then seen as relatively electron-transparent structures.

The main disadvantages attendant upon use of the electron microscope are the required thinness of sections and the necessity of studying them dry (in a vacuum). The first difficulty has been overcome largely by the development of microtomes that can make sections 0.2 to 0.02 μm in thickness (Meek, 1976). However, removal of water and salts dissolved in it as well as the use of stains is likely to alter structure; therefore, membranes, cytoplasm, or formed structural components seen by electron microscopy must be interpreted with some caution. Notwithstanding these facts, it will become apparent from studies presented later that much has been learned about the cell by electron microscopy.

Because of the small contrast between some biological objects and the background even after staining, *shadowcasting* with heavy metals is often used, especially to bring out three-dimensional topography. Heavy metals, especially gold, platinum, and chromium, are evaporated in a vacuum at an oblique angle to the object. The metal piles up like blowing sand or snow and outlines the object. Shadowing may be performed from several directions by moving the object during the process of evaporation (Meek, 1976).

When the object is too thick for shadowcasting a *replica* can be made. In this case the object is subjected *in vacuo* to evaporation of platinum and carbon and becomes coated to the depth of many nanometers. The preparation is then treated with hydroxide to remove the

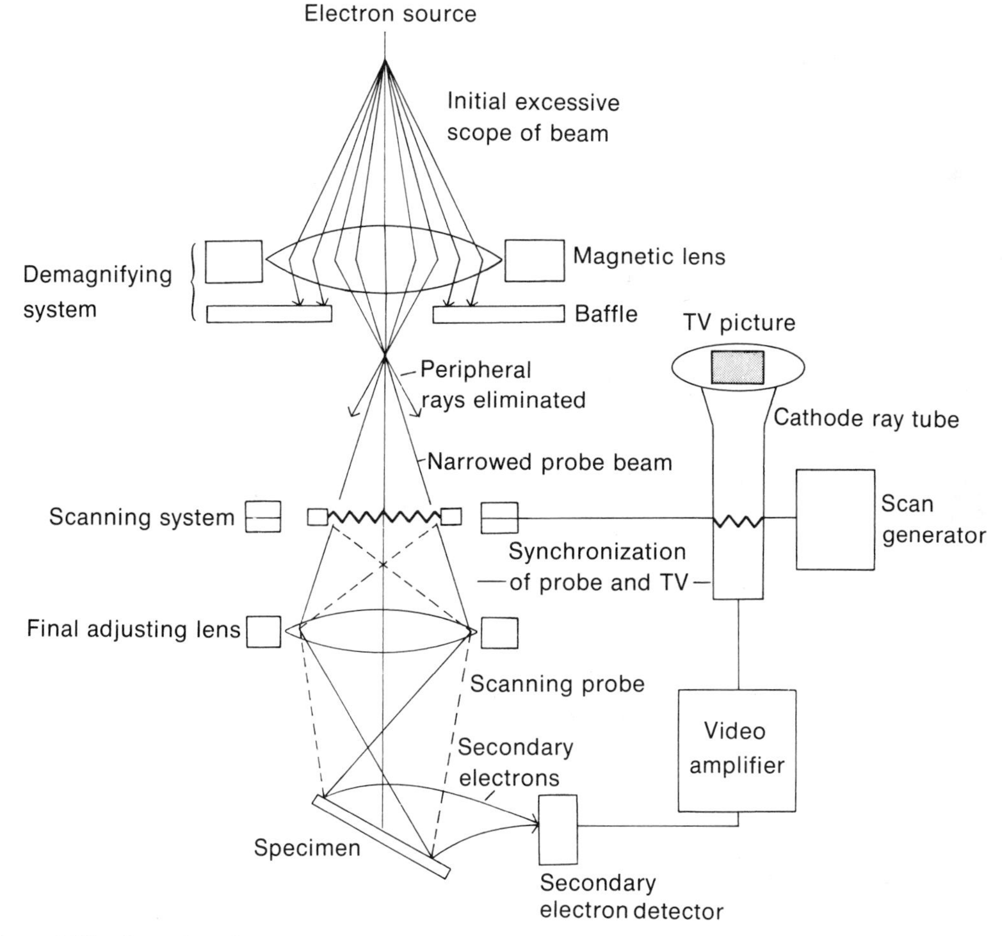

Figure 3.20. *Scanning electron microscope. Electrons from the cathode source are accelerated through a grid and grounded anode, the whole assembly constituting the "electron gun." The electron beam, too broad at first, is reduced, or demagnified, by magnetic lens systems. Passing through the deflector yoke of a scanning coil system, and a final lens system, the beam, now much narrowed and adjusted, scans the sample. Secondary electrons scattered from the sample are collected by the detector. They are transmitted to a video amplifier and into the TV cathode ray tube, the scanning rate of which is synchronized with that of the probe of the microscope. Thus, any given point on the TV screen, at any moment, represents a point on the specimen. The final image is really a series of pictures of different points on the sample, each about 10 nm in diameter, or the diameter of the probe. These are seen in such rapid succession as to provide, for the eye, a unified view of the entire surface of the sample. (From Frobisher* et al., *1974: Fundamentals of Microbiology. W. B. Saunders Co., Philadelphia, p. 56.)*

specimen and the replica is floated onto another grid and prepared for electron microscopic study.

Surface spreading has also been a useful technique for study of cell fine structure. The spread layer is mounted on a grid, dried, and shadowed for electron microscopic study. In this manner a bacterial protoplast (bacterium from which the cell wall has been removed with the enzyme lysozyme) that is ruptured osmotically on the surface of a vessel filled with distilled water is found to liberate its chromosome as a single long fiber. The same technique has been used for a study of interphase chromosomes in cells of plants and animals and for a study of microtubules present in some cells.

The images of various cell organelles seen with the electron microscope have been criticized because they are structures seen in cells that have been fixed and stained. Are they real or are they artifacts introduced by the chemical procedures and drying? It is disquieting also that different fixatives sometimes give different images; for example, osmium tetroxide fixation shows a plasma membrane consisting of a single line, whereas permanganate fixation shows the plasma membrane consisting of two outer electron-dense layers and an inner less electron-dense layer. Using a method called *freeze etching,* it is possible to study the cell organelles with the electron microscope. The object to be studied is placed in 20 percent glycerol (antifreeze, see Chapter 4) and is frozen at −100°C. It is mounted on a chilled holder and splintered with a knife along natural cleavage planes, usually in the middle of membranes. Occasionally the structure is cross-fractured, giving cross sections of organelles. The splintered preparation is freeze-dried and covered with a platinum and carbon coating in

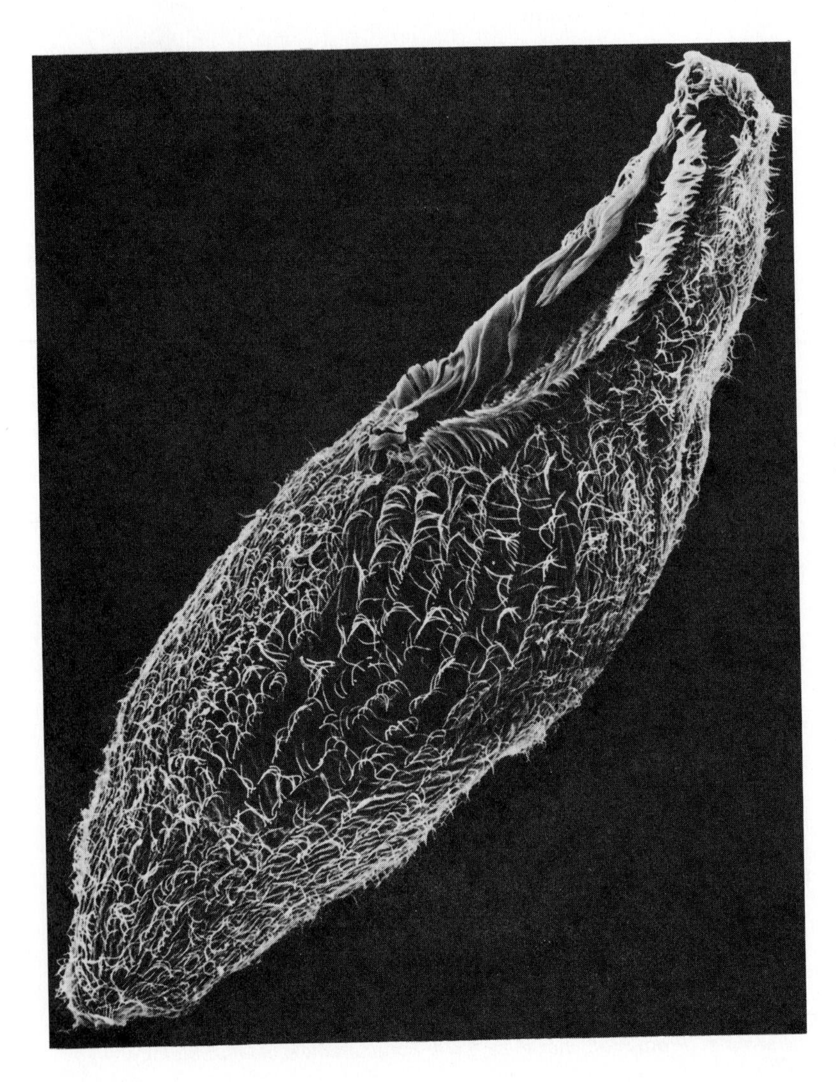

Figure 3.21. *Scanning electron micrograph of the ciliate* Blepharisma americanum. *The ciliary rows and fused ciliary "paddles" (membranelles of the mouth) are especially prominent. (Courtesy of D. Shamlian and H. Wessenberg, San Francisco State University.)*

the high vacuum of a freezing ultramicrotome. When the preparation is dry the vacuum is broken and the preparation is placed in water to float the replica off the carrier. The replica is then washed in basic solution to remove the cellular material and the replica is mounted on a grid and dried for electron microscopy.

Such replicas give the outlines of various cell structures and verify the structures seen in thin fixed and stained sections of the mitochondria, smooth and rough endoplasmic reticulum, double nuclear envelope with pores and plugs, unit plasma membrane of the cell, chloroplasts, and so forth. The views of structures are somewhat three-dimensional. Since some cells (e.g., yeast) subjected to these freezing and drying procedures revive and show no evident change in properties, it is assumed that electron micrographs of cells so prepared give images of cell structures in as nearly normal condition as is at present possible. The fact that the structures compare favorably in size and appearance to those seen in fixed and stained cells is, in some measure, confirmation of the validity of the conventional method using fixatives.

Another electron microscope that is coming into wider use is the *scanning electron microscope* (Figs. 3.20 and 3.21). While in most respects it resembles the transmission electron microscope, its light source is a beam of electrons which scan the specimen, producing a television-like image (Kimoto and Russ, 1969; Everhart and Hayes, 1972). At present the scanning electron microscope is most effective in observing topography of structural features such as cilia, surface pits, and contours, which it does with great clarity, but future models may permit resolution as high as with the transmission electron microscope.

Scanning electron microscopy gives three-dimensional effects that can only be surmised by reconstruction from transmission electron microscopy. Furthermore, the cells need not be fixed; therefore, they are observed in a more natural condition.

Ion, x-ray and ultrasound microscopes are currently being applied to a variety of problems. Their value to the study of cell structure remains to be determined (Rochow and Rochow, 1978).

3.4 MICROCINEMATOGRAPHY

In many cases considerable information has been obtained by time-lapse photomicrography. Details that escape the observer on a single viewing become apparent after viewing the record repeatedly. Familiar examples are mitosis and meiosis in cells and irradiation of cells with a microbeam. The components used in such studies are now familiar; the details must be sought elsewhere (Rose, 1963).

3.5 MICROMANIPULATORS

Since the end of the last century micromanipulators of one kind or another have been in use, first for handling single cells and later for microsurgery upon them (see Fig. 3.22). The pneumatic micromanipulator of De Fonbrunne (Fig. 3.23) has had wide use for microsurgery because of its ease of operation, speed and sensitivity of response. Essentially it consists of three pumps, two horizontal and one vertical, which connect by way of rubber tubes to the pneumatic capsules of the instrument holder (for needle or pipette). The movements in the horizontal plane of a single lever or joy stick are reflected at a ratio of anywhere from 1:50 to 1:2500, according to the setting decided upon. The vertical plane level is set by the screw on the lever. Production of microforges for the controlled manufacture of microtools has also facilitated microdissection.

Micromanipulation makes the microscope more than an observational tool, and properties of many of the cell organelles have been explored

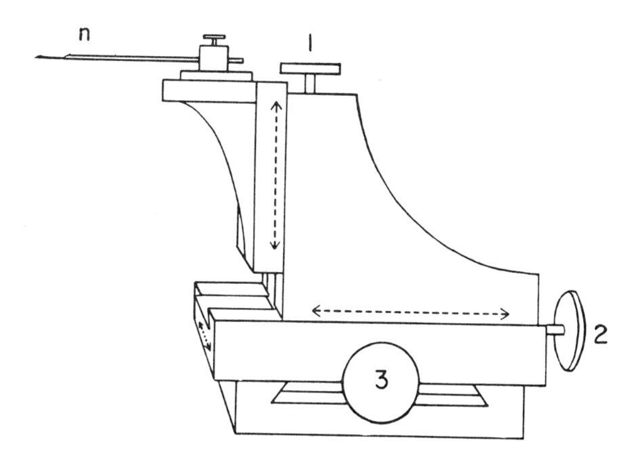

Figure 3.22. *One type of micromanipulator. It consists essentially of a microlathe enabling movements of the needle (n) in the three coordinates of space by manipulation of screws 1, 2, and 3. Another instrument (a mirror image of the one shown) stands on the other side of the microscope, all mounted on a heavy stand. (After Taylor, 1925: Zool., University of California. Publ. 26: 443.)*

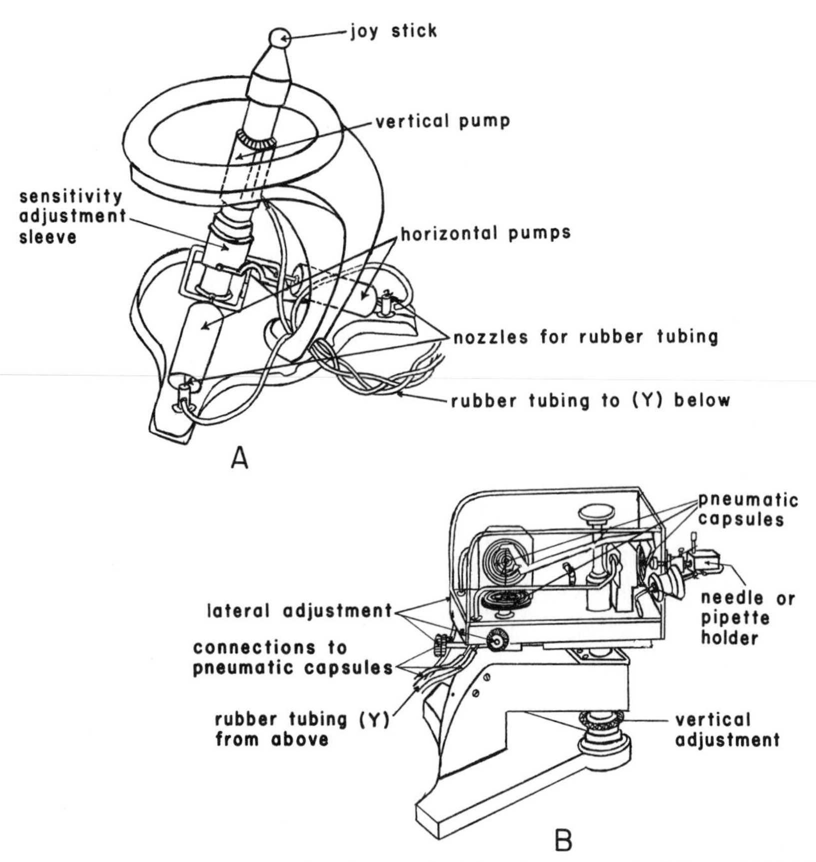

Figure 3.23. *The De Fonbrunne pneumatic micromanipulator. A movement of the control or joy stick is duplicated by the needle (or other instrument) with a reduction in magnitude of $^1/_{50}$ to $^1/_{2500}$, set at the sensitivity adjustment sleeve. This is achieved through passage of the pressure change through three pumps, two horizontal and one vertical, attached to three pneumatic capsules with rubber tubing, the capsules in turn controlling the needle or instrument holder. Two such micromanipulators are often used together, one to hold, the other to cut or manipulate a structure viewed under the microscope. (By courtesy of A. S. Aloe Company, St. Louis, Mo.)*

by its use. Important information on the function of the nucleus of cells has been gained by transplantation of nuclei between cells (see Chapter 7). Even bacteria have been dissected under high power (El-Badry, 1964). Micromanipulators are essential for inserting microelectrodes and micropipettes into cells (see Chapters 20 and 21).

LITERATURE CITED AND GENERAL REFERENCES

Adams, E., 1978: Invertebrate collagens. Science *202*: 591–598.

Albershiem, P., 1975: The walls of growing plant cells. Sci. Am. (Apr.) *232*: 81–95.

Allsopp, A., 1969: Phylogenetic relationships of prokaryotes and the origin of the eukaryote cell. New Phytologist *68*: 591–612.

Anfinsen, C. B. and Scheraga, H. A., 1975: Experimental and theoretical aspects of protein folding. Adv. Protein Chem. *29*: 205–300.

Bailey, A. J., Robins, S. P. and Belian, G., 1974: Biological significance of intermolecular crosslinks of collagen. Nature *25*: 105–109.

Baldwin, R. G., 1975: Intermediates in protein folding reactions and the mechanism of protein folding. Ann. Rev. Biochem. *44*: 453–475.

Baumeister, W., 1978: Biological horizons in molecular microscopy. Cytobiologie-European J. Cell Biol. *17*: 246–297. (Review.)

Bessis, M., 1965: Introductory lecture: microirradiation of cells. *In* Recent Progress in Photobiology. Bowen, ed. Blackwell, Oxford,

Bloom, W. and Fawcett, D. W., 1975: A Textbook of Histology. W. B. Saunders Co., Philadelphia.

Blundell, T., Dodson, G., Hoagland, D. and Merola, D., 1972: Insulin: the structure in the crystals and its reflection in chemistry and biology. Adv. Protein Res. *26*: 279–402.

Bogorad, L., 1975: Evolution of organelles and eukaryotic genomes. Science *188*: 891–898.
Bornstein, P., 1974: The biosynthesis of collagen. Ann. Rev. Biochem. *44*: 567–603.
Bradbury, S., 1967: The Evolution of the Microscope. Pergamon Press, New York.
Broda, E., 1975: The Evolution of the Bioenergetic Processes. Pergamon Press, New York.
Bryan, J., 1974: Microtubules. Bioscience *24*: 701–711.
Burrells, W., 1977: Microscope Technique, Rev. Ed. John Wiley & Sons, New York.
Bush, V., Duryee, W. R. and Hastings, J. A., 1953: An electric micromanipulator. Rev. Sci. Instr. *24*: 487–489.
Dippell, R. V., 1976: Effects of nuclease and protease digestion on the ultrastructure of *Paramecium* basal bodies. J. Cell Biol. *69*: 622–637.
Dodge, J. D., 1973: The Fine Structure of Algal Cells. Academic Press, New York.
Dupouy, G., 1973: Three-megavolt electron microscopy. Endeavour *32*: 66–70.
El-Bawdry, H., 1964: Micromanipulators and Micromanipulation. Academic Press, New York.
Elinov, N. P. and Kurlova, N. A., 1976: Chemical composition of the cells of *Aureabasidium* (Pullularia) *pullulans*. Mikrobiologiya (Rus.) *45*: 111–115.
Euler, H. V. and Hahn, L., 1948: Concentrations of RNA and DNA in animal tissues. Archiv. Biochem. *17*: 285–291.
Everhart, T. E. and Hayes, T. L., 1972: The scanning electron microscope. Sci. Am. (Jan.) *226*: 55–69.
Florkin, M., ed., 1971: Comprehensive Biochemistry. Vol. 21. Vitamins and Trace Elements. Elsevier, Amsterdam.
Frieden, E., 1972: The chemical elements of life. Sci. Am. (Jul.) *227*: 52–60.
Galligher, A. E. and Kozloff, E. N., 1971: Essentials for Practical Microtechnique. 2nd Ed. Lea and Febiger, Philadelphia.
Goss, J. A., 1973: Physiology of Plants and their Cells. Pergamon Press, New York.
Gunning, B. E. S. and Steer, M. W., 1975: Ultrastructure and the Biology of Plant Cells. Crane Russak and Co., New York.
Hartman, H., 1975: The centriole and the cell. J. Theor. Biol. *51*: 501–509.
Hartman, H., Puma, J. P. and Gurney, T., Jr., 1974: Evidence for the association of RNA with the ciliary basal bodies of *Tetrahymena*. J. Cell Sci. *16*: 241–259.
Hayat, M. A., 1972: Basic Electron Microscopic Techniques. Van Nostrand Reinhold Co., New York.
Hayat, M. A., 1970–1977: Principles and Techniques of Electron Microscopy: Biological Applications. Vols. 1–7. Van Nostrand Reinhold Co., New York.
Hayat, M. A., 1977: Introduction to Biological Scanning Electron Microscopy. Pergamon Press, New York.
Heidemann, S. R., Sanders, G. and Kirschner, M. W., 1977: Evidence for a functional role of RNA in centrioles. Cell *10*: 337–350.
Hepler, P. K. and Palevitz, B. A., 1974: Microtubules and microfilaments. Ann. Rev. Plant Physiol. *25*: 309–362.
Hughes, D. E., Lloyd, D. and Brightwell, R., 1970: Structure, function and distribution of organelles in prokaryotic and eukaryotic microbes. Soc. Exp. Biol. Symp. *20*: 295–322.
Jeon, K., ed., 1973: The Biology of Ameba. Academic Press, New York.
Kimoto, S. and Russ, J. C., 1969: The characteristics and applications of the scanning electron microscope. Am. Sci. *57*: 112–133.
Klotz, I. M., Langerman, M. and Darnall, D., 1970: Quaternary structure of proteins. Ann. Rev. Biochem. *39*: 25–62.
Konarev, V. G., 1972: Cytochemistry and Histochemistry of Plants. Israel Program for Scientific Translation, Jerusalem.
Krogman, D., 1973: The Biochemistry of Green Plants. Prentice-Hall, Englewood Cliffs, N.J.
Lehninger, A. L., 1975: Biochemistry, 2nd Ed. Worth Publishers, New York.
Lemon, R. A. and Quate, C. F., 1975: Acoustic microscopy: biomedical applications. Science *188*: 905–912.
Liljas, A. and Rosmann, M. G., 1974: X-ray studies of protein interactions. Ann. Rev. Biochem. *43*: 475–507.
Mahler, H. R. and Raff, R. A., 1975: The evolutionary origin of the mitochondria, a non-symbiotic model. Int. Rev. Cytol. *43*: 1–124.
Margulis, L., 1970: Origin of Eukaryotic Cells. Yale University Press, New Haven.
Margulis, L., To, L. and Chase, D., 1978: Microtubules in prokaryotes. Science *200*: 1118–1124.
Marx, J. L., 1975: Actin and myosin: role in non-muscle cells. Science *189*: 34–37.
Matsumura, T., 1972: Relationship between amino acid composition and differentiation of collagen. Int. J. Biochem. *3*: 265–274.
McQuade, A. B., 1977: Origin of nucleate organisms. Quart. Rev. Biol. *52*: 249–262.
Meek, G. A., 1976: Practical Electron Microscopy for Biologists. John Wiley & Sons, New York.
Mercer, E. N. and Birbeck, H. S., 1972: Electron Microscopy: A Handbook for Biologists. Blackwell, Oxford.
Miller, W. J. and Neathery, M. W., 1977: Newly recognized trace mineral elements and their role in animal nutrition. Bioscience *27*: 674–679.
Morrisett, J. D., Jackson, R. L. and Gotts, A. M. Jr., 1975: Lipoproteins: structure and function. Ann. Rev. Biochem. *44*: 183–207.
Muscatine, L., 1972: Chloroplasts and algae as symbionts in mollusks. Int. Rev. Cytol. *36*: 137–169.
Nass, S., 1969: The significance of the structural and functional similarities of bacteria and metazoa. Int. Rev. Cytol. *25*: 55–129.
Parsons, D. F., 1974: Structure of wet specimens in electron microscopy. Science *186*: 407–414.
Preston, R. D., 1974: The Physical Biology of Plant Cell Walls. Chapman and Hall, London.
Rae, P. M. N., 1976: Hydroxymethyluracil in eukaryote DNA: a natural feature of the Pyrophyta (dinoflagellates). Science *194*: 1062–1064.
Rochow, T. G. and Rochow, E. G., 1978: An Introduction to Microscopy by Means of Light, Electrons, Ions, X-rays and Ultrasound. Plenum Press, New York.
Rose, G. G., 1963: Cinematography in Cell Biology. Academic Press, New York.
Ross, K. F. A., 1967: Phase Contrast and Interference Microscopy for Cell Biologists. St. Martin's, New York.

Sager, R., 1972: Cytoplasmic Genes and Organelles. Academic Press, New York.

Schwartz, R. M. and Dayhoff, M. O., 1978: Origins of prokaryotes, eukaryotes, mitochondria, and chloroplasts. Science *199*: 395–403.

Simpson, L., 1972: The kinetoplast of the hemoflagellates. Int. Rev. Cytol. *32*: 139–207.

Slayter, E. M., 1970: Optical Methods in Biology. John Wiley & Sons, New York.

Smith, H., ed., 1978: Molecular Biology of Plant Cells. Blackwell Scientific, Oxford.

Snell, W. J., Dentler, W. L., Haimo, L. T., Binder, L. I. and Rosenbaum, J. L., 1974: Assembly of chick brain tubulin onto isolated basal bodies of *Chlamydomonas reinhardi*. Science *185*: 357–360.

Spooner, B. S., 1975: Microfilaments, microtubules, and extracellular materials in morphogenesis. Bioscience *25*: 440–450.

Stanier, R. Y., 1970: Some aspects of the biology of cells and their possible evolutionary significance. Soc. Gen. Microbiol. Symp. *20*: 1–38.

Tracey, M. V., 1968: The biochemistry of supporting materials in organisms. Adv. Comp. Physiol. Biochem. *3*: 233–270.

Traub, W. A. and Piez, K. A., 1971: The chemistry and structure of collagen. Adv. Protein Chem. *25*: 243–352.

Warfield, R. K. N. and Bouck, G. B., 1974: Microtubule-macrotubule transitions: intermediates after exposure to the mitotic inhibitor vinblastine. Science *186*: 1219–1221.

Watson, J. D., 1976: The Molecular Biology of the Gene, 3rd Ed. W. A. Benjamin, Menlo Park, Calif.

Wehry, E. L., ed. 1976: Modern Fluorescence Microscopy. Vol. 2. Plenum Press, New York.

Williams, R. J. P., 1970: Biochemistry of sodium, potassium, magnesium and calcium. Quart. Rev. Chem. Soc. *24*: 331–365.

Wischnitzer, S., 1974: The nuclear envelope: its structural and functional significance. Endeavour *33*: 137–142.

Woodhead-Galloway, J. and Hukins, D. W. L., 1976: Molecular biology of cartilage. Endeavour *35*: 73–78.

Yamanaka, T., 1972: Evolution of the cytochrome molecule. Adv. Biophys. *33*: 227–276.

CHAPTER 4

ADJUSTMENT OF CELLS TO DIVERSE ENVIRONMENTS

Paleontological evidence suggests that life originated in the sea. In terms of major taxonomic groups much of the animal kingdom is still in the sea. However, invaders in fresh waters and on land are probably more familiar to most of us than marine organisms are, particularly the major terrestrial plant groups. Furthermore, life has spread to the most diverse habitats, so that cells are subjected to extremes of salt concentrations, pH, temperature, lack of oxygen, and hydrostatic pressure. The resistance of microorganisms to extreme environments has recently been reviewed (Kushner, 1978). Some examples of such environmental adjustments are presented in this chapter.

WATER IN THE CELL ENVIRONMENT

A cell must always have access to water, inasmuch as all cell reactions occur in aqueous solutions. In unicellular organisms this is achieved by direct contact with environmental water. The complex plant or animal has an outer coating that is largely impervious to water, but each of its cells has contact with the aqueous environment of the internal medium.

Water makes up 80 percent of the cell. When the cell loses its free water, life is suspended or extinguished, although a small amount of water remains bound to the proteins. Activity reappears in desiccated, dormant cells only when water again becomes available in quantity to permit reactions in solution. The cells of dried mosses, lichens, rotifers, tardigrades, and nematodes last for years and resume activity quickly when moistened, provided other conditions are favorable. Many protozoa encyst and some bacteria sporulate when subjected to seasonal drying. Dried resistant stages are especially characteristic of fresh-water and terrestrial organisms.

Water as Solvent

Water is the best solvent known; more substances dissolve in it than in any other solvent. Electrolytes dissolved in water readily ionize because water molecules, by virtue of their own charged nature (Fig. 1.5), act as small dipoles* attracting the ions of the solute, thereby weakening the attraction between ions. Ionization and ionic reactions are of great importance to life. For all its importance as a solvent, water is remarkably inert; most substances are unchanged chemically when dissolved in it. Therefore, they remain in solution in the cell, unaltered until such time as they are utilized.

Water Required for Hydrolysis and Oxidation

Nutrients are generally stored as insoluble polymers in cells. Before use they must be hydrolyzed, that is, split by addition of water (see Chapter 10). Solid food taken into cells, for example, into amebas or leukocytes, must also be hydrolyzed (digested). A dry mixture of urease and urea allowed to equilibrate at various relative humidities begins hydrolysis at about 50 percent relative humidity. This was

*Conceptualization of the structure of water continues to be controversial. The complex subject is covered elsewhere for those interested (Drost-Hansen, 1972).

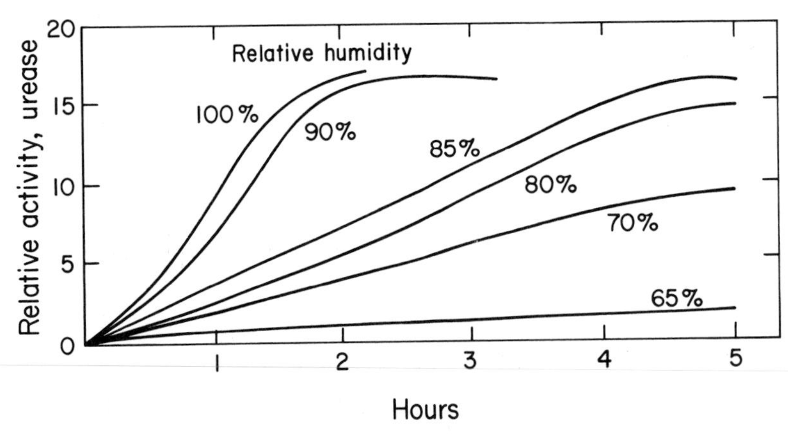

Figure 4.1. Effect of increasing relative atmospheric humidities on urease activity. Lyophilized mixture of urease and urea, 20°C. (From Skujins and McLaren, 1967: Science 158: 1569–1570. Copyright 1967 by the American Association for the Advancement of Science.)

determined by measuring the rate of $^{14}CO_2$ formation from ^{14}C urea. Relative hydrolytic activity increased with increasing humidity, being highest at 100 percent (Fig. 4.1). The rate of hydrolysis of urea by urease paralleled the adsorption of water by the enzyme. The experimenters conclude:

"Evidently the limiting factor in this reaction is the availability of water molecules to the active sites of the enzyme: at 60% relative humidity the amount of water sorbed by urease is only 1.3 moles per mole of side-chain polar groups, corrected for side-chain amide. Thus only loosely bound water can be utilized in hydrolysis, that is water sorbed in excess of the stoichiometric minimum of one molecule of water per polar site." (Skujins and McLaren, 1967, p. 1570.)

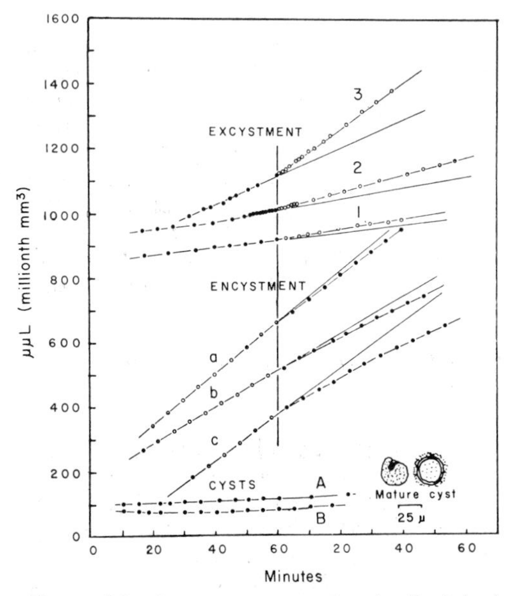

Figure 4.2. Oxygen consumption (ordinate) of a single ciliate protozoan (Bresslaua) *in a microrespirometer during encystment when hydration decreases and excystment when hydration increases. Note that during encystment respiration falls progressively (a, b, c). Cysts (A, B) show little respiration. During excystment respiration gradually rises (1, 2, 3). (After Scholander* et al., *1952: Biol. Bull.* 102: *182.)*

Oxidations, the means by which a cell liberates energy from nutrients for performing its work, occur only in aqueous solution. A dry mixture of the appropriate enzymes and substrates does not react. Oxygen consumption of cells has been measured both while they were dehydrating for encystment and rehydrating during excystment. Oxygen consumption decreases during dehydration, falling to an unmeasurable rate in fully formed cysts; conversely, oxygen consumption increases rapidly with rehydration (Fig. 4.2). Water present in dormant organisms is largely unavailable for chemical reaction. Thus, water is essential for *hydrolyses* and for biological *oxidations,* the two chemical reactions most fundamental to the living cell.

Thermal Stabilization

Among water's most important temperature stabilizing properties are its high specific heat, its high heat of vaporization and its high heat of fusion during melting of ice (Henderson, 1927; Ben-Naim, 1974). How these properties provide a better control on the earth's surface as well as in the cell's external and internal environment can be best illustrated by an example. About 10^{15} calories of the sun's radiant energy per square kilometer per year fall upon the earth's surface at the equator, the equivalent of that released by burning 700,000 tons of coal. If a dry land mass of rock and soil with low specific heat were to absorb this much energy, it would rise to a lethal temperature by day and fall below the freezing point of water by night, as it apparently does on the moon. However, water on earth and water vapor and carbon dioxide in the earth's atmosphere absorb some of the incoming sunlight, preventing overheating by day, and infrared rays reradiated from the land are absorbed by atmospheric water vapor, thereby slowing the earth's cooling by night.

This so-called "greenhouse effect" of atmospheric water vapor, moderating the temperature of land masses, is of special importance to life (Fig. 4.3).

Because of water's high specific heat (one cal to raise one gm of water 1°C), the temperature of the earth's great bodies of water is increased much less than that of land masses. The *heat capacity* is the number of calories required to raise the temperature of a body by 1°C. The larger a body of water the greater is its heat capacity. The vastness of the oceans therefore makes them resist temperature change, tempering the climate of nearby land masses.

The oceans moderate the temperature on earth in yet another way. Because of the high heat of vaporization of water (529 cal to evaporate one gm of water), a tremendous amount of the sun's heat is absorbed during evaporation from the ocean surface. The vapor retains the heat until the water-laden air is chilled in cooler regions of the earth. Evaporation of water from surface cells in organisms cools them, for example, in human sweating and plant transpiration.

Water's high heat of fusion (the 80 cal required to melt one gm of ice to one gm of water) also serves to reduce radical changes in temperature, since water forms ice only after considerable loss of heat, and it melts only when this much heat is replaced. The system:

$$[1 \text{ gm water}]_{0°C} \underset{+80 \text{ cal heat}}{\overset{-80 \text{ cal heat}}{\rightleftharpoons}} [1 \text{ gm ice}]_{0°C} \quad (4.1)$$

remains at 0°C during withdrawal of heat, while ice forms, or during addition of heat, while ice melts.

An additional thermostabilizing property of water is its peak specific gravity (weight per unit volume) at 4°C. Water cooled below 4°C rises and freezes only at the surface. Were it not for this, lakes would freeze solid and become uninhabitable. Finally, water is a very much better heat conductor than other common liquids. Consequently, heat absorbed is readily distributed. This equalizes the temperature, not only in the environment but also inside cells.

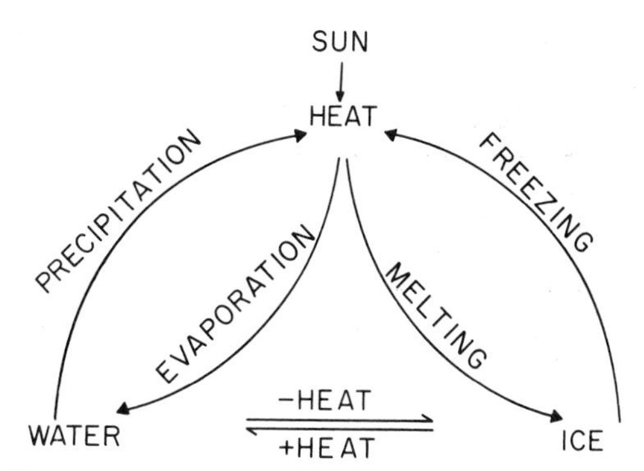

Figure 4.4. Heat exchanges between different states of water.

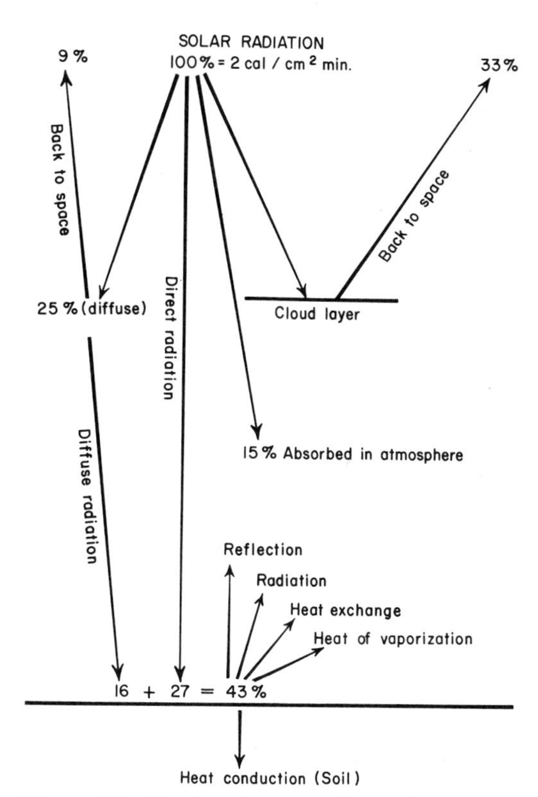

Figure 4.3. Heat exchange at the earth's surface and in the atmosphere at midday. The numbers represent average values. (Modified from Allen, 1960: Comparative Biochemistry. Vol. 1. Florkin and Mason, eds. Academic Press, New York, p. 487.)

Special Properties

Water has several properties that make it particularly useful as part of an organism's external and internal environment. It is *transparent to light,* permitting photosynthesis in chlorophyll-bearing plants below the water surface. The *high surface tension* of water enables it to move closer to the soil surface than would otherwise be possible, making it available to life on land.

Most cell nutrients are water soluble and probably enter cells by channels permeable to aqueous solutions. However, it was also important that the primitive cell isolate itself from the environment by a membrane containing lipid in which only some channels are penetrated by water-soluble molecules.

TABLE 4.1. COMPOSITION OF SOME TYPICAL NATURAL WATERS IN GRAMS PER LITER*

Water	Na	K	Ca	Mg	Cl	SO_4	CO_3
Sea water	10.7	0.39	0.42	1.34	19.3	2.69	0.073
Hard fresh water	0.021	0.016	0.065	0.014	0.041	0.025	0.119
Soft fresh water	0.016	—	0.010	0.00053	0.019	0.007	0.012

*Data from Baldwin, 1948: An Introduction to Comparative Biochemistry. Cambridge University Press, Cambridge.

SALTS IN THE CELL ENVIRONMENT

Although water is necessary for life most cells do not tolerate distilled water and live only in a medium containing some salts. Melted snow and freshly fallen rain water seldom contain more than 0.003 percent salt, yet some organisms live in this water. Hard fresh water contains about 0.1 percent salt, calcium being the predominant cation and carbonate the predominant anion. Sea water contains about 3 percent salt (30 gm per liter) and brine in which salt is beginning to crystallize, about 30 percent (Table 4.1), which some halophilic organisms can tolerate (Larsen, 1967). The principal salts in sea water are the chlorides, sulfates, and carbonates of sodium, potassium, calcium, and magnesium. Sodium is the predominant cation and chloride the predominant anion (Table 4.1). Biogeologists think that the ancient seas probably resembled present-day sea water in composition, although the ancient seas may have been somewhat more dilute in salt content (Rubey, 1951).

Every cell contains salts (see Chapters 1 and 3), and salt is present in the cell's internal environment. The blood salt concentration in most marine invertebrates is equal to sea water (3 percent), while in all other animals it is less than that of sea water, being 0.68 to 0.9 percent in vertebrates and 0.3 to 0.7 percent in fresh water and terrestrial invertebrates (Prosser, 1973).

The relative proportions of various ions most favorable for biological functions are approximately those of sea water (Table 4.2). As an example, take the heart of a frog that stops beating when placed in pure sodium chloride solution equal in salt concentration to blood. If calcium ion is added, the heart resumes beat-

TABLE 4.2. RELATIVE IONIC COMPOSITION OF SEA WATER AS COMPARED WITH THE RELATIVE IONIC COMPOSITION OF THE BLOOD OR BODY FLUID OF VARIOUS ANIMALS*

(The Na ion in all cases is taken as 100.)

A

Location of Sea Water	Ion					
	Na	K	Ca	Mg	Cl	SO_4
Woods Hole, Mass., USA	100	2.74	2.79	13.94	136.8	7.10
Japan	100	2.14	2.28	11.95	119.0	5.95

B

Animal	Classification	Na	K	Ca	Mg	Cl	SO_4
Aurelia	Coelenterate	100	2.90	2.15	10.18	113.05	5.15
Strongylocentrotus	Echinoderm	100	2.30	2.28	11.21	116.1	5.71
Phascolosoma	Sipunculid	100	10.07	2.78	—	114.06	—
Venus	Mollusk	100	1.66	2.17	5.70	117.3	5.84
Carcinus	Crustacean	100	2.32	2.51	3.70	105.2	3.90
Cambarus	Crustacean	100	3.09	2.60	6.70	30.9	—
Hydrophilus	Insect	100	11.1	0.92	16.8	33.6	0.12
Lophius	Fish	100	2.85	1.01	1.61	71.9	—
Frog	Amphibian	100	2.40	1.92	1.15	71.4	—
Man	Mammal	100	3.99	1.78	0.66	83.97	1.73

*Calculated from data in Prosser, ed., 1973, Comparative Animal Physiology, 3rd Ed. W. B. Saunders Co., Philadelphia.

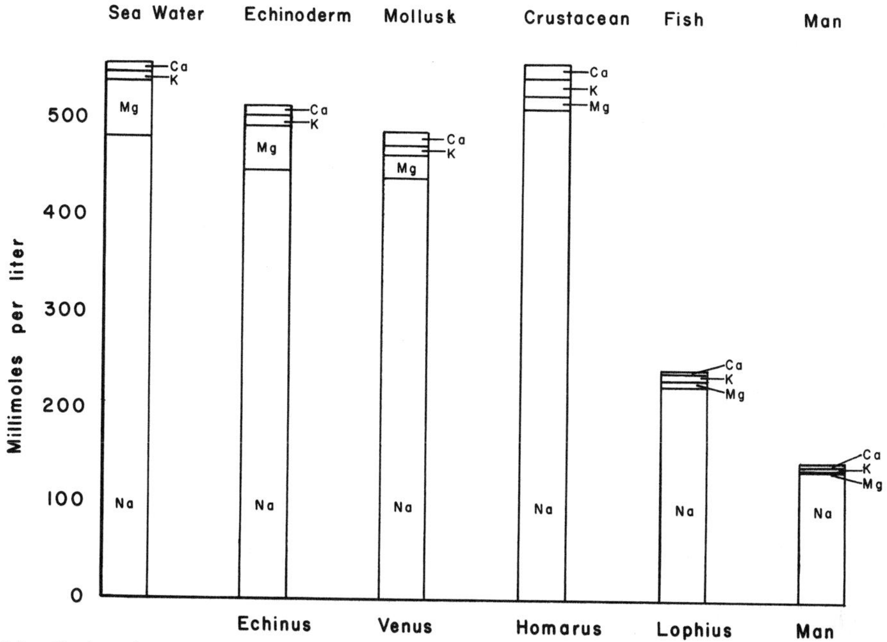

Figure 4.5. Cations in sea water, body fluid, and blood. Note the decreased amount of magnesium in the three forms on the right and the relative increase of potassium. The much lower concentration of total salt in the vertebrate is also readily noticeable. Echinus *is a sea urchin,* Venus *a clam,* Homarus *a lobster and* Lophius *a fish.*

ing, but the beat is normal only when potassium is also added, with the three salts present in the relative proportions characteristic of blood and sea water. The cations are said to act antagonistically to one another (see Chapter 17). Therefore, not only the total concentration of salt (osmotic factor) but also the concentration of each relative to the others (antagonism) is important in a *balanced medium* such as blood or sea water. The specific functions of each of the ions in cell functions are discussed in other chapters. Magnesium and calcium have been of special interest as enzyme activators as well as in other respects, but sodium and potassium are also important in this regard (Williams, 1970).

Although neither the relative nor the absolute concentrations of salts are the same in blood and sea water, as seen in Figure 4.5, the striking similarity in the relative concentrations of the ions suggests a relationship between sea water and blood. Paleontologic evidence indicates that life originated in the sea, and the blood as an internal environment appears to have simulated sea water (Prosser, 1973).

However, some differences are found between blood and sea water. Pantin (1931) suggested that in the course of evolution it was advantageous for an active species to eliminate from the blood an ion with narcotic action such as magnesium. Differences in the proportions of ions between blood and sea water could be maintained by marine animals only as kidneys developed, because the differences are the result of a balance between the input and outgo of salts as determined by kidney function. A steady state is achieved, characteristic of the blood of each species. Organisms (e.g., echinoderms) whose body fluids have about the same relative concentration of the various ions as sea water, have no kidneys whereas those with good kidneys (e.g., mollusks, crustaceans, and vertebrates) have blood ions most unlike sea water.

In hard fresh water the concentration of calcium relative to sodium is higher than in sea water. This is important to life in fresh waters since calcium retards the transfer of molecules and ions across the cell membrane, protecting cells from losing useful salts to the dilute external medium (see Chapter 17).

Halophilic Organisms

Quite a few unicellular and multicellular organisms tolerate sea water (about 3 percent salt equivalent to 0.55 M NaCl), but the number that can live in solutions that have several times marine salinity is not large. Salt-tolerant organisms are found in salt lakes, salt muds, salt evaporating pools, and brine used to preserve food. Among the organisms that occur in highly saline habitats are bacteria, blue-green algae, yeasts (e.g., *Torula wehmeri,* some species of *Debaromyces*), molds (e.g., *Sporendonema epizoum),* phytoflagellates (e.g., *Dunaliella salina*), ciliates (e.g., *Fabrea salina*), green algae, and some crustaceans, such as the brine shrimp *Artemia salina*. Many of the microorganisms accumulate reddish carotenoid pigment. Even a halophilic eukaryotic photosynthetic alga, *Chlamydomonas* sp. (Yamada and Okamoto, 1961), a halophilic photosynthetic bacterium (Raymond and Sistrom, 1969), and a halophilic bacterium that grows in saturated lithium chloride (Siegel and Roberts, 1966) have been described.

Bacteria may be facultative or obligate halophiles. The facultative halophiles, among which are *Sarcina morrhuae* and *Micrococcus morrhuae,* tolerate salinities equivalent to between 0.9 and 3.7 M NaCl but do not require high salt concentrations for growth. They grow quite well at much lower salt concentrations and are found in sea water, intestines of various animals, soil, and even fresh water. Obligate halophiles, members of the genus *Halobacterium,* grow at salinities ranging from about 3.7 to 5.5 M NaCl; they grow very slowly at half this concentration and swell and lyse when placed at one-fifth to one-tenth optimal concentration. Obligate halophiles divide only about every 8 hours even in solutions of optimal salt concentration (Forsythe and Kushner, 1970).

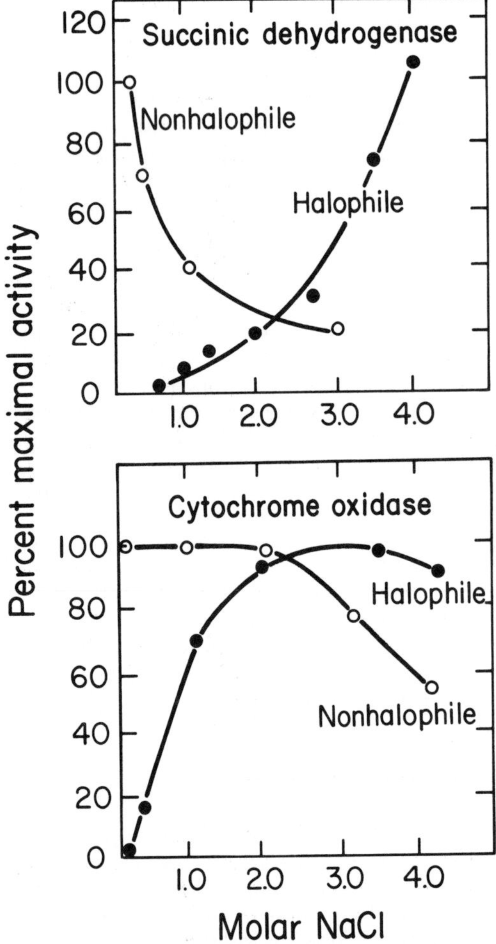

Figure 4.6. Effect of sodium chloride on enzymes extracted from representative halophilic and nonhalophilic bacteria. In the upper figure, the nonhalophile is Micrococcus denitrificans, *the halophile is* Halobacterium salinarium. *In the lower figure the nonhalophile is a species of* Pseudomonas, *the halophile is* Halobacterium salinarium. *(After Larsen, 1962:* In *The Bacteria. Gunsalus and Stanier, eds. Academic Press, New York.)*

Experiments show that the salt concentration inside halophilic bacteria is generally less than that outside, but it increases as the external concentration increases. Thus in *Micrococcus halodenitrificans* grown in 2 to 4 percent salt, the internal concentration of salt was 1.5 percent; in 8 percent the internal concentration was 5 percent; but in 22 percent salt, it was also 5 percent. In *Sarcina littoralis* when the external concentration of salt was 22 percent, the internal concentration was 18 percent. Interestingly, internal potassium increases more rapidly than internal sodium (Larsen, 1967).

Since a halophile's internal salt concentration is elevated above that in nonhalophiles, its enzymes must be able to operate at higher salt concentrations than enzymes in ordinary cells. To test this, the activity of some 20 enzymes extracted from halophiles (mostly *Halobacterium*) was determined at various salinities. Enzymes of nonhalophiles proved more sensitive to increasing salt concentration than those of halophiles (Fig. 4.6). Removal of salt by dialysis against water inactivates halophilic enzymes, usually irreversibly; therefore, salts stabilize these enzymes against denaturation or dissociation or both (Larsen, 1967).

Organisms in Bitterns

In evaporating pools of sea water, calcium carbonate precipitates at a specific gravity of 1.057 to 1.068 and calcium sulfate at a specific gravity of 1.095 to 1.190. Sodium chloride crystallizes between a specific gravity of 1.2014 and 1.2109, at which time some magnesium chloride and magnesium sulfate also crystallize. The extremely bitter remaining solution consists mainly of chlorides, bromides, and sulfates of potassium and magnesium, appropriately named the *bittern.* At a still higher specific gravity, magnesium chloride and magnesium sulfate crystallize. It is evident that the ratios of various ions characteristic of sea water are altered as soon as any salt precipitates or crystallizes and that the ionic imbalance of the remaining solution becomes more extreme with each crystallization. Such ionic imbalance has marked effects on cellular permeability (see Chapters 17 and 18). *Artemia salina* tolerates the loss of calcium carbonate; salt manufacturers call it the "clearer worm" because as *Artemia* feeds upon microscopic organisms, the initial fine precipitate of calcium carbonate is incidentally trapped in the slime and formed into larger granules that settle. *Artemia* and the flagellate *Dunaliella* endure the solution until sodium chloride begins to crystallize, swimming slowly through the viscous fluid but no longer growing or reproducing. When sodium chloride crystallizes, both *Artemia* and *Dunaliella* die, presumably unable to tolerate the ionic imbalance of the bittern. Some bacteria survive but little is known about them. Cells of bittern organisms must have adaptations enabling them to tolerate salt imbalance not acceptable to cells of other organisms (see Chapter 17).

RESISTANCE OF CELLS TO ACIDITY AND ALKALINITY

Acidity and alkalinity are measured by a function called pH, the negative logarithm of the hydrogen ion concentration (see Appendix 4.1). An amoeba tolerates an environmental pH

TABLE 4.3. pH VALUES OF ENVIRONMENTAL MEDIA AND SOME ORGANISMS FOUND THEREIN*

Type of Habitat	*pH Range*	*Medium*	*Characteristic Organisms*
Very acid	0.0	Sulfur and salts	*Thiobacillus thiooxidans.*
	1.0–3.0	Organic	The fungus, *Merulius lacrymans.*
	2.0	Organic	Acetate flagellates after adaptation.
	3.2–4.6	Peat bogs	Mosses: *Sphagnum* and *Drepanocladus.*
	3.4	Vinegar	Vinegar eels (nematodes), vinegar bacteria.
	3.2	Soil water	High moor vegetation.
	3.3–3.7	Organic	The flagellate, *Astasia,* in tryptone (lower limit).
Somewhat acid	5.2	Sour milk	Lactic acid bacteria.
	5.7–5.8	Distilled water Fresh rain	
Somewhat acid to alkaline	5.2–9.0	Soil water	Peat pit mosses.
	6.2–9.2	Neutro-alkaline lakes	Mosses at edge of lakes.
	4.5–8.5	Soil water in most soils	Various forest and field plants.
Nearly neutral	6.8–7.0	Freshly boiled (cooled) distilled water	
	6.5–8.0	Drinking water	
	6.8–8.6	River water	River plants and animals.
	6.2–8.2	Springs	Algae around springs.
	7.8–8.6	Sea water	Marine plants and animals.
Alkaline	7.5–9.1	Soil water	Greasewood, rabbit bush, salt grass and brome grass growing on alkaline flats.
	9.0–9.6	Organic medium	*Astasia,* extreme alkaline range.
	10.0	Alkaline desert pool	Cyprinid fishes (Denio, Nevada).
	10.5	Organic medium	Dermatophytes.
	10.5	Inorganic	Larvae of *Aedes matronius.*
	9.3–11.1	Organic medium	Several species of *Penicillium,* one of which tolerates pH 11.1; also some species of *Aspergillus.*

*Data for *Thiobacillus* from Stanier *et al.,* 1963: The Microbial World; for fungi from Wolf, 1947: The Fungi, II, and Leise and James, 1946: Arch. Dermat. & Syph. *53*:481; for mosses from Sørensen, 1948: Dansk Botanisk Arkiv., *12*:5, and Richards and Verdoorn, 1932: Manual of Bryology; for Salt Lake from Flowers, 1933: The Bryologist, 34; for acetate flagellates, Hutner and Provasoli, 1951, *In* Lwoff: Biochem. & Physiol. of the Protozoa, I; for *Astasia* from Schoenborn, 1949; J. Exp. Zool., *111;* for desert pool near Denio, personal communication from Eminger Stewart; for *Aedes* from Beadle, 1932: J. Linn. Soc. London, Zool. *38;* others from the Handbook of Chemistry and Physics, 1964, 45th Ed., and various other sources.

Note: More examples are given for the extremes because endurance of these conditions is so unusual. Most organisms occur in an environment slightly to the acid or alkaline side of neutrality.

range from 4.2 to 8.2, and yeast from 4.5 to 8.5, a 10,000-fold change in hydrogen ion concentration. As seen from the data in Table 4.3, vinegar bacteria and cells in vinegar eels (nematodes), certain mosses, and some fungi tolerate a more acid range; cells of dermatophyte fungi and plants in alkali flats tolerate a more alkaline range. Cells of most organisms tolerate a pH range from 6.0 to 8.0. It is interesting that blue-green algae do not tolerate acidity and grow only in near neutral or slightly alkaline habitats. Brock (1973) has suggested that perhaps such a limitation occurred in the early history of the earth, leaving the acid niche to green algae and thereby greatly enhancing their opportunities of developing without competition with blue-green algae.

The hydrogen ion does not penetrate healthy cells readily, but it may influence the penetration of other molecules. For example, carbonic acid (like other weak acids), which is largely molecular at low pH values, readily penetrates cells, whereas in ionized state at higher pH values it does not (and, in fact, may be withdrawn from the cell). The converse occurs with a weak base, such as ammonia, that is largely molecular at high pH values.

When cells are damaged, however, hydrogen ions are able to penetrate the cell membrane. For example, healthy stomach lining cells tolerate a high hydrogen ion concentration at pH 1.0 produced by secretion of 0.1 N HCl. But when these cells are injured by detergents (for example, bile salts) hydrogen ions enter, damaging the cells (Davenport, 1972). This leads to ulceration.

Although the pH of environmental water is highly variable (Table 4.3), the pH of body fluids and blood that bathe the cells of multicellular organisms is held within narrow limits, as seen in Table 4.4. For example, human blood is very closely regulated at a pH of 7.4. If the blood pH falls to 7.0, acidotic coma and death ensue; at pH 7.8, life terminates in tetany.

TABLE 4.4. pH VALUES OF BODY FLUIDS, BLOOD AND PLANT FLUIDS

pH Range	*Fluids*
7.3–7.5	Blood, human
7.3–7.5	Spinal fluid, human
6.9–7.2	Whole blood, dog
6.7–7.9	Body fluids or blood of various invertebrates
6.6–7.6	Milk, human
6.9–7.5	Saliva, human
1.0–3.0	Gastric contents, human
4.8–8.2	Duodenal contents, human
2.9–3.3	Apple juice freshly extracted
2.2–2.4	Lemon juice
1.8–2.0	Lime juice
5.0–5.2	Expressed juice of *Sedum* leaf

The body fluid pH of many invertebrates lies in the same range as that in vertebrates; however, that of a few is not neutral. For example, the body fluid of ascidians is alkaline (pH 7.8) and the body fluid pH inside a coelenterate may equal that of sea water, approximately 8.2. On the other hand, the body fluid of metamorphosing insects may be as acid as pH 6.8.

The pH inside cells usually varies in a narrow range to either side of neutrality.

BUFFERS

A buffer is a mixture of a weak acid and its salt, for example, carbonic acid and bicarbonate. When acid (or base) is added to a buffer, the pH, which depends upon the ratio of salt to undissociated acid, changes slowly because the salt suppresses the dissociation of the acid (or base) by the common ion effect (Fig. 4.7).

Weak acids (or bases) with several dissociation constants are often used as buffers in biological experiments (Table 4.5). Each dissociation constant of a multivalent acid (or base) represents the midpoint of one of its buffer ranges. The effective buffer range of each buffer system is about one pH unit to either side of the dissociation constant, for example, in phosphoric acid, to either side of pH 2.12, 7.20, or 12.66 (Fig. 4.8).

When a strong acid and one of its salts are used to make a buffer system, buffering action

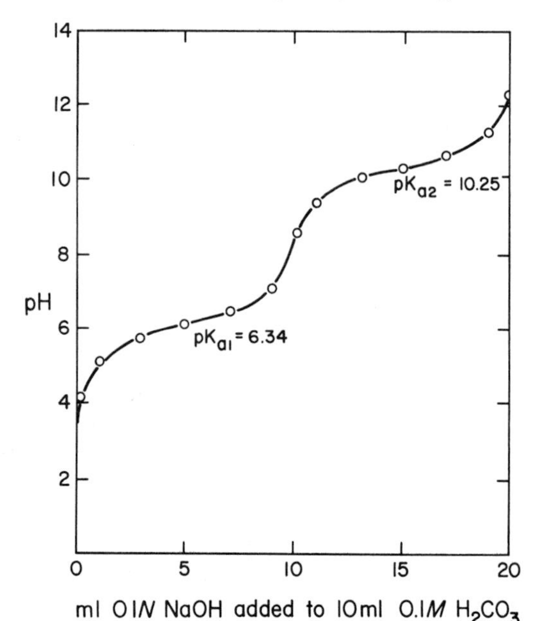

Figure 4.7. *Change in hydrogen ion concentration (measured by pH) when carbonic acid is titrated with a base. Note the two dissociation constants of carbonic acid (25°C).*

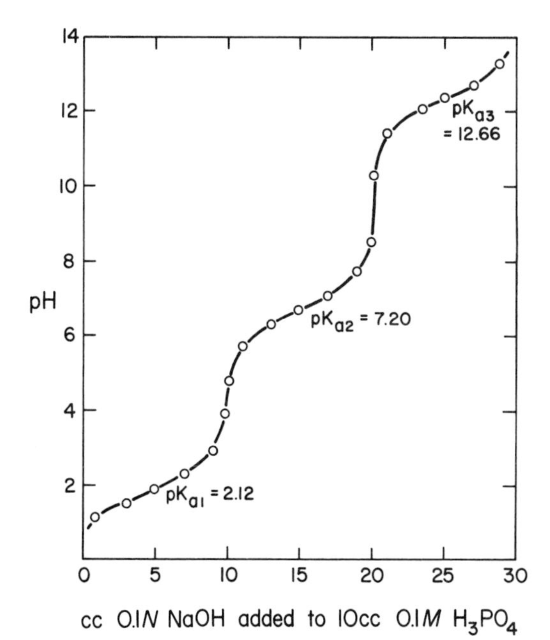

Figure 4.8. *Change in hydrogen ion concentration (measured by pH) when phosphoric acid is titrated with a base. Note the three dissociation constants (25°C).*

is restricted to the low pH range, because no matter how much salt is present, the acid remains completely dissociated and the pH is kept low. A strong acid and its salt also resist pH change on addition of base, changing no more than one pH unit until about 80 percent of the acid is neutralized.

Buffer Capacity

Buffer capacity is measured by the pH change following addition of acid or base. Buffer capacity is greatest at a pH (the pK, see appendix 4.1) near the midpoint of its range. The higher the concentration of the buffer the greater its capacity to absorb acid or base without change in pH. Dilution of a buffer therefore reduces its capacity.

Buffers in the Cell and Cell Environment

Carbonic acid and phosphoric acid are the main buffers in the cell. Proteins, amino acids, fatty acids, and various organic acids (for example, citric acid) may also serve as cellular buffers, although to a lesser extent.

Carbonic acid plays a major role as buffer in the cell environment because of its omnipresence. Carbon dioxide is present in the air and soil and in natural waters as carbonic acid. It is difficult to displace carbon dioxide from water; even after water is boiled and freed of gases, it rapidly becomes acidic from reentry of carbon dioxide. Carbon dioxide is highly soluble in water. At room temperature the volume of gaseous carbon dioxide dissolved in water is equal to the volume of water. An equilibrium always exists between carbon dioxide in air and water; water can never remove it completely from the air, nor air from water.

TABLE 4.5. STANDARD BUFFER MIXTURES*

Name of Buffer	*Constituents*	*pH Range*
Clark and Lubs	Phthalate-HCl	2.2–3.8 (20°C)
Clark and Lubs	Phthalate-NaOH	4.0–6.2 (20°C)
Clark and Lubs	KH_2PO_4-NaOH	5.8–8.0 (20°C)
Clark and Lubs	Boric acid, KCl-NaOH	7.8–10.0 (20°C)
Clark and Lubs	HCl-KCl	1.0–2.2 (20°C)
Sørensen	Glycine-NaCl-NaOH	8.58–12.97 (18°C)
Sørensen	Borate-NaOH	9.24–12.38 (18°C)
Sørensen	Borate-HCl	7.61–9.23 (20°C)
Sørensen	Citrate-HCl	1.04–4.96 (18°C)
Sørensen	Glycine-HCl	1.04–3.68 (18°C)
Sørensen	Na_2HPO_4-KH_2PO_4	5.29–8.04 (18°C)
Sørensen	Citrate-NaOH	4.96–6.69 (20°C)
Palitzsch	Borax-borate	8.69–6.77 (18°C)
McIlvaine	Citric-NaH_2PO_4	2.2–8.0
Tris(hydroxymethyl) aminomethane	Tris + HCl	7.0–9.0 (20°C)
Kolthoff and Vleeschhouwer	Na_2CO_3-borax	9.2–11.0 (18°C)
Kolthoff and Vleeschhouwer	Na_2HPO_4-NaOH	11.0–12.0 (18°C)

*Data from Clark, 1928. Data for Tris(hydroxymethyl)aminomethane from Whitehead, 1959: J. Chem. Educ. *36*:297; and Bulletin 106, Sigma Chemical Co., St. Louis.

HEAT RESISTANCE OF ACTIVE CELLS

The major source of the earth's heat is the sun. A local source of heat in the earth's crust is radioactive decomposition; still another source of heat is the compression of the earth's crust. Both these local heat sources may be the underlying cause of hot springs, fumaroles and volcanism. A lesser but growing source is man's industry.

Biokinetic Zone

All cells are exposed to thermal variation. Because most cells tolerate only a small range of thermal changes, temperature affects distribution and activity of organisms. A temperature optimal for one cell function, such as respiration, may not be optimal for another, such as growth. Furthermore, a temperature endured during brief exposure may prove harmful on longer exposure. It is good experimental procedure to determine the effect of a temperature range on any process under study.

The narrow range of temperatures called the *biokinetic zone,* within which most cells are active, lies approximately between 10° and 45°C. Some desert plants function best at the upper limit of the biokinetic zone (Bjorkman *et al.*, 1972), while marine organisms in cool waters function best near the lower limits. Some organisms subcultured at progressively higher (or lower) temperatures, adapt (by selection?) to higher (or lower) temperatures.

Thermophiles

Thermophilic (heat-loving) algae, fungi, and bacteria normally live at higher temperatures than their *mesophilic* relatives. For example, some algae and bacteria live in hot springs at temperatures of 60° and 70°C, and some strains of the blue-green alga *Synechrococcus* grow best between 53° and 67°C. The blue-green alga *Phormidium* has been reported to exist even at 89°C, a temperature that would kill most active cells. Some thermophilic bacteria in hot springs of Yellowstone National Park can grow in super-heated pools at 93.5° to 95.5°C, which is above the boiling point of water (92°C) at this altitude (Brock, 1967). This suggests that if water remains in the liquid state, some form of life is possible regardless of temperature (Mitchell, 1974).

It is common knowledge that the activities of organisms are primarily catalyzed by proteinaceous enzymes and that many proteins are altered by heat; for instance, many enzymes are inactivated between 60° and 100°C. Some thermophilic bacteria will grow at high temperatures because they appear to synthesize enzymes faster than heat destroys them. They grow only when supplied with amino acids and vitamins that they make at lower temperatures. This suggests breakdown of the corresponding synthetic mechanisms at high temperatures (Crabb *et al.,* 1975).

However, thermophilic bacteria are generally more resistant to heat because their enzymes continue to function at high temperatures. Such thermophiles have enzymes with stronger hydrogen bonding than do enzymes of mesophiles and are unable to live at lower temperatures, presumably because at low temperatures their enzyme bonding becomes too strong for normal catalytic action. Since adaptation to temperature is not "all or none" and some species of blue-green algae (e.g., *Synechrococcus lividus*), show different optimal temperature ranges (Fig. 4.9) (Castenholz, 1973), different isozymes are suggested, permitting adaptation to these ranges (Hochachka and Somero, 1973; Somero, 1975a).

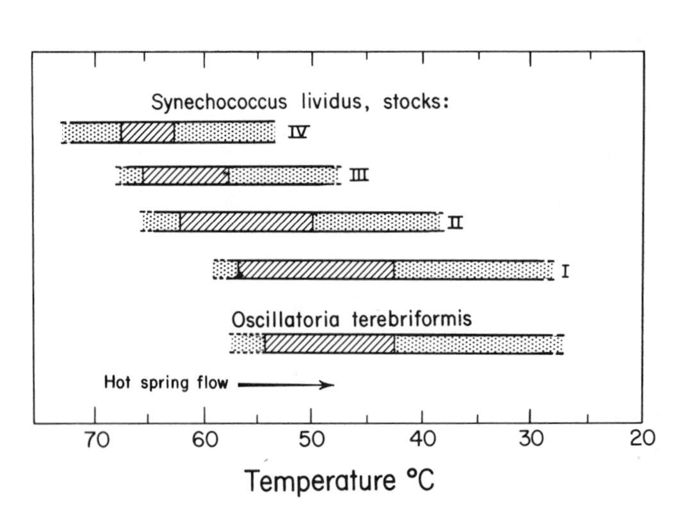

Figure 4.9. *Schematic plan of a distribution of blue-green algal species in a thermal stream at Hunter's Hot Springs, Oregon. The arrow indicates the direction of flow. The horizontal bars represent growth ranges of several stocks of* Synechococcus *and of one stock of* Oscillatoria. *(After Castenholz, 1973:* In *The Biology of Blue-Green Algae. Carr and Whitten, eds. University of California Press, Berkeley, pp. 379–414.)*

Protein Denaturation

In denaturation of proteins such as enzymes by heat, the spatial arrangement of the polypeptide chains within the protein molecule is changed from that of the native protein to a more disordered arrangement. Denaturation may alter quaternary, tertiary, and secondary structure but *not* the primary structure of the molecule. Loss of the unique structure of a protein changes its chemical, physical, and biological properties. For example, denatured protein binds water less readily and is less soluble, and denatured enzyme is no longer active catalytically. Both fibrous and globular proteins are subject to denaturation (Johnson *et al.*, 1974a).

As hydrogen bonds are broken by heat, proteins appear to unfold during denaturation. Since unfolding exposes hydrophobic groups that repel water, the protein becomes less soluble in water and likely to precipitate on denaturation. Denaturation in some proteins is accompanied by an increase in total volume of solution, as measured by a dilatometer flask that has a finely calibrated capillary tip. The increase in protein volume during denaturation may amount to 0.2 ml per 100 gm of protein. Volume increase may be accounted for in the following manner: when the protein is denatured and unfolded, the hydrophobic radicals exposed repel water; the capacity of the protein to bind water is thereby reduced and the volume of solution is increased (Johnson *et al.*, 1974a).

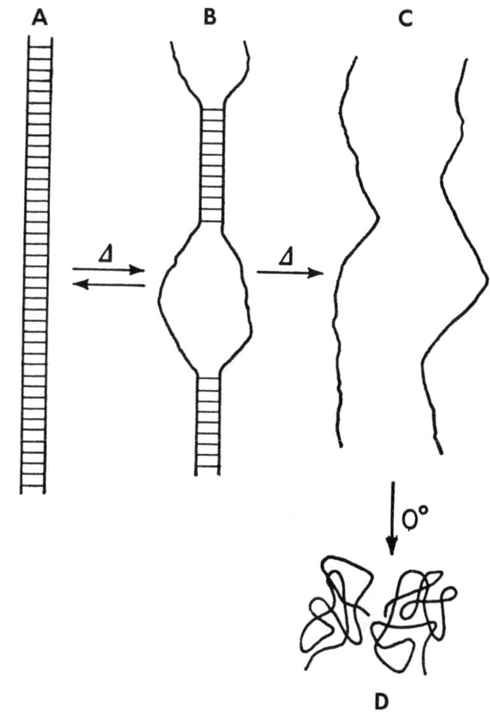

Figure 4.10. *Thermal denaturation of* A, *double-stranded, native DNA proceeding through state* B *of partial strand dissociation to state* C, *complete separation at temperatures above the melting point. Upon rapid cooling, completely denatured DNA* C *collapses into structure* D. *Renaturation thus occurs from* B *but not from* C. *(From Szybalski, 1967:* In *Thermobiology. Rose, ed. Academic Press, New York.)*

Nucleic Acid Denaturation

While a lesser heat sensitivity of thermophile enzymes may account for their greater heat resistance, a possible difference in heat sensitivity of other cell constituents (e.g., DNA, RNA, and cell membranes), should also be considered.

The hydrogen bonds between the two DNA strands are broken by heat. When the strands separate from one another, they may fail to recombine in the same manner (renature) when the temperature is lowered (Fig. 4.10). Because it is accompanied by progressively increased absorption at 260 nm, breaking the hydrogen bonds of DNA may be followed quantitatively by spectrophotometry (Fig. 4.11). It has been established that the DNA denaturation temperature is elevated by increasing the ratio of guanine–cytosine to adenine–thymine (GC:AT) base pairs in the DNA molecule (Szybalski, 1967), but it has not been demonstrated that the DNA of thermophiles is richer in G-C base pairs than the DNA of mesophiles, or that the DNA of thermophiles is completely denatured at temperatures that kill the cells. Therefore, the higher heat resistance of thermophiles cannot be attributed to a heat-resistant DNA.

Although heat injury and damage to the cellular DNA have not been correlated, interestingly the DNA of a thermophilic strain of *Bacillus subtilis* growing at 55°C transforms (1 in 10^4) a mesophilic strain. This indicates that information for thermophily rests in the DNA. The cell's gene for thermophily is linked to the streptomycin-resistant locus (Farrell and Rose, 1967).

The ribosomes of the thermophile *Bacillus stearothermophilus* grown at 68°C resist exposure to heat better than those from a mesophile, *E. coli* (Fig. 4.12). Because the thermophile ribosome protein moiety is not more resistant to heat than the mesophile ribosome protein, and thermophile polyamines (bound to RNA) do not protect mesophile ribosomes from high temperatures, it is suggested that the greater resistance of thermophile ribosomes might reside in their RNA components. While the G-C to A-U base pair ratio is higher in thermophile than in mesophile ribosomal RNA, the difference is not considered significant. Little differ-

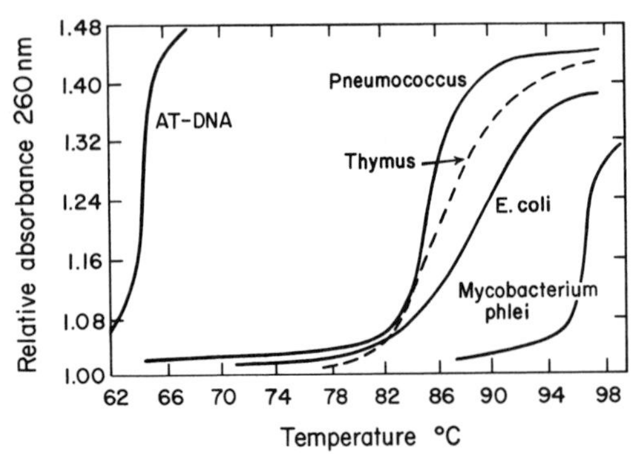

Figure 4.11. Absorbance (at 260 nm) versus temperature curves for various samples of DNA, showing development of hyperchromaticity with rise in temperature. (Modified from Doty, 1961: Harvey Lect. 55: 103.)

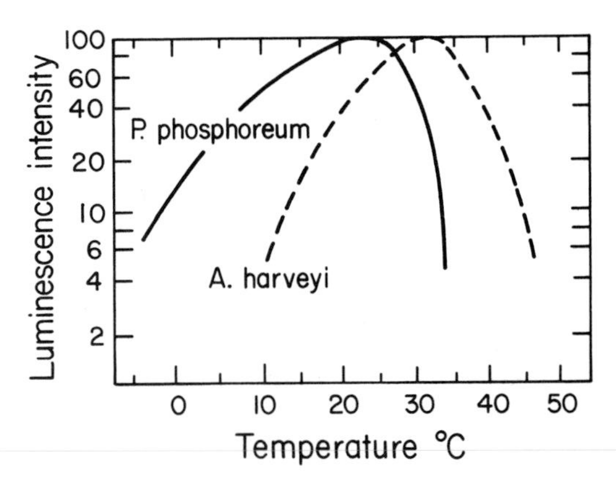

Figure 4.13. Relation between temperature and luminescence intensity for two species of bacteria (Photobacterium phosphoreum *and* Achromobacter harveyi) *growing at widely different temperature ranges. The maximum intensity was arbitrarily taken as 100 and the data were recalculated to this scale. (After Johnson* et al.*, 1942: J. Cell Comp. Physiol. 20: 253.)*

ence also exists in the properties of the thermophile and mesophile transfer RNA's (Mangiantini *et al.*, 1965).

A cell's thermal tolerance depends upon the temperature range to which it has been previously adapted, and for how long. For example, some polar fish die when exposed to 6°C; others, at 10° or 15°C. Temperate zone marine animals may be killed at 30° or 35°C, whereas fresh water and terrestrial animals resist temperatures as high as 40° to 45°C.

Enzyme Inactivation

It is well known that many enzymes, reversibly inactivated by mild heat, are irreversibly inactivated by high temperatures. The action of heat on an enzyme can be visibly illustrated with bioluminescence. In bioluminescence the heat-stable substrate luciferin is oxidized in the presence of the enzyme luciferase with the emission of light, the intensity of which measures the reaction rate. Luminescence increases to a maximum intensity with rise in temperature until at still higher temperatures it declines and is finally quenched. If the temperature is lowered, luminescence reappears, provided neither temperature nor exposure has been excessive (Fig. 4.13).

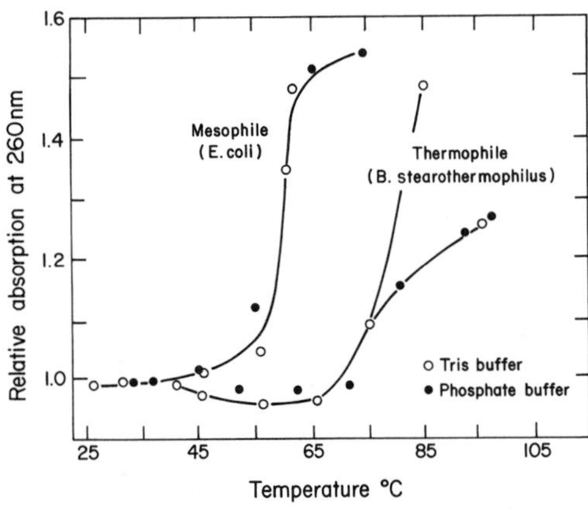

Figure 4.12. Difference in melting temperatures of RNA from ribosomes of a mesophile (E. coli) *and a thermophile* (B. stearothermophilus). *The higher melting temperature of the RNA from the thermophile indicates stronger bonding than in the mesophile. Magnesium acetate added to buffers. (From Mangiantini* et al., *1965: Biochem. Biophys. Acta* 103: *252–274.)*

Inactivation of the luminous system in the bacteria by heat is probably concomitant with inactivation of other cell enzymes. High temperature shifts the equilibrium of each thermolabile enzyme in favor of the denatured, inactive form.

$$\underset{\text{(active)}}{\text{Native enzyme}} \overset{\text{heat}}{\rightleftharpoons} \underset{\text{(inactive)}}{\text{denatured enzyme}} \quad (4.2)$$

Enzyme denaturation is reversible after brief exposure but often irreversible after prolonged exposure. If some enzymes in a cell are so damaged, a process essential to life may cease, leading to death.

It is possible that derangement in cell lipids is the cause of heat injury (McElhaney, 1974; Welker, 1976). This topic will be considered in Chapter 17.

COLD RESISTANCE OF ACTIVE CELLS

Most cells, although they can survive near-freezing temperatures, become relatively inactive as the temperature falls to the freezing point. A drop in body temperature (hypothermia), both natural (e.g., in hibernators) and artificial, as in experiments, results in decreased metabolism. However, *cryophilic* (cold-resistant) organisms such as polar fishes and polar invertebrates are active at 0°C or below. Some molds grow slowly even at −4° to −6.7°C and bacteria grow slowly in ice cream at −10°C.

Because the Antarctic polar region is a large land mass of low heat capacity compared to water, it is colder than the Arctic, a small land mass surrounded by water. Sea water cannot fall below −1.9°C without freezing and therefore it acts as a reservoir of heat. As water freezes, it gives off heat. The Arctic is therefore richer in land organisms than its opposite polar region. Both Arctic and Antarctic seas have a large and diversified marine fauna, although many of the marine forms must live at −1.9°C essentially all their lives. At McMurdo in the Antarctic, for example, the temperature of sea water varies only about 0.1°C annually (Somero, 1975a).

Organisms that live and grow at low temperatures are called *cryophiles* (cold-loving). Cryophily is accompanied by special adaptations that prevent freezing and death. The freezing point of marine invertebrates is lowered by raising the osmotic pressure of the body fluids. Some antarctic fishes (e.g., *Trematomus*) have developed a glycoprotein antifreeze, composed of two amino acid residues, alanine and threonine, and two sugar residues, N-acetyl-galactosamine and galactose. These components occur in repeating four residue units of alanyl-alanyl-threonyl-*o*-disaccharide. The glycoprotein in the serum is present in different molecular weights—10,500, 17,000, and 21,500—probably structured as random coils. The glycoprotein in very low concentration lowers the freezing point (DeVries, 1974, 1975; Feeney and Osuga, 1976). In the presence of sodium borate, the glycoprotein loses its antifreeze action, which returns with removal of the salt by dialysis. These data "suggest that the glycoproteins function either by structuring water in such a way that it prevents water molecules from forming an ice lattice or by coating the surfaces of ice crystals, preventing water molecules from settling on these surfaces and furthering crystal growth" (De Vries and Somero, 1971).

Even in environments just above the freezing point of water an organism must acclimate, otherwise its enzymes cannot react at a rate sufficient to maintain life. In organisms that acclimate to seasonal winter and summer temperatures, metabolic reaction rate after acclimation is about the same for both seasons. Acclimation probably occurs by synthesis of isozymes with higher substrate affinity in winter and lower substrate affinity in summer.

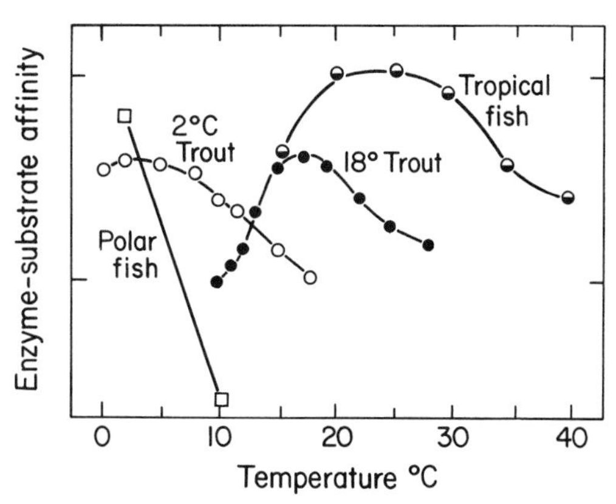

Figure 4.14. *Relative enzyme-substrate affinity for brain acetylcholinesterase from polar fish (Antarctic* Trematomus*), trout acclimated to 2°C and 18°C, and a tropical fish (Amazon* Electrophorus*). At low concentrations of substrate in biological systems, the enzyme-substrate affinity is a major determinant in reaction velocity (see Chapter 10). (Reciprocals of data for Michaelis constants presented in Baldwin, J., 1970: Ph.D. Dissertation, University of British Columbia.)*

The greater the enzyme's substrate affinity, the greater its activity. This shows strikingly in a comparison between an antarctic fish, a temperate fish, and a subtropical fish (eel) shown in Figure 4.14 (Hochachka and Somero, 1973).

FREEZING AND THAWING OF CELLS

When an organism is slowly exposed to a freezing temperature, water freezes outside the cells, thereby concentrating the solution surrounding the cells. Consequently, water diffuses out of the cells into the bathing fluid, and the solutes become more concentrated inside the cell. The freezing point inside the cell is thereby lowered. However, the increased solute concentration is injurious to the cell, as shown by direct application of equally concentrated solutions to the cells. In some cases as much as 60 to 90 percent of the free cell water may be withdrawn. Many cells endure being nearly frozen at −5°C, reimbibing water on thawing without apparent injury (Mazur, 1975).

Even when cells are slowly cooled below −40°C, ice crystals may form inside because external ice crystals serve as nucleators through cell membrane channels. However, if the water inside the cell has already been reduced by 90 percent, crystals do not form (Mazur, 1975).

The time factor in freezing is also of great consideration. Rapid freezing results in ice formation (as small crystals) both inside and outside cells. This may not only concentrate solutes to the point of injury but also the internal ice crystals may disrupt cell organelles. The effect is augmented when cells are stored at low temperature, allowing for growth of large ice crystals.

Freezing and thawing alter cell ultrastructure; membranes of the endoplasmic reticulum, mitochondria, and the plasma membrane are deranged, although they often recover when warmed (Trump *et al.*, 1965). Freezing may dislodge lipid from cell membranes, making them so fragile that they disintegrate on warming. The increased solute concentration and pH change during freezing may split enzymes into their composite polypeptide chains. Warming may result in abnormal reassociation; the altered quaternary structure that forms has little or no catalytic activity (Mazur, 1975).

After both slow and rapid freezing the thawing rate is especially important. For example, if the ice melts slowly, the cell is exposed to a high electrolyte concentration at a higher temperature than during freezing. Also, ice crystals may grow inside cells. A cell that survived freezing might still be damaged on thawing. Rapid thawing is therefore especially important to minimize injury.

Glycerol in the suspending medium penetrates cells, lowering the freezing point of both bathing medium and cells, acting as an "antifreeze." When freezing occurs, less ice forms than in absence of glycerol, and the crystals are smaller because of the strong hydrogen bonding of glycerol for water. Concentration of electrolytes by freezing is thereby reduced; for this reason glycerol is sometimes spoken of as an "electrolyte buffer." Bacteria and phages that are readily killed by freezing in broth have been shown to survive eight cycles of freezing and thawing (Fig. 4.15) in 15 percent glycerol. However, as glycerol is toxic, it must be removed by dialysis after thawing. Glycerol is especially useful in slow freezing procedures. Treated in this manner, cells can be stored at low temperatures for long periods. Even eggs in mammalian ovaries have been kept for months

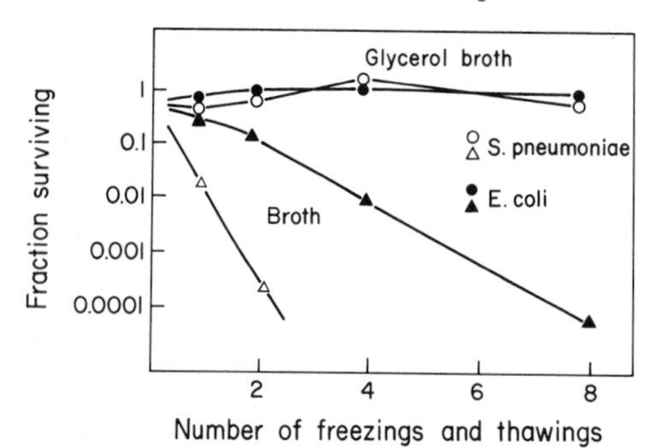

Figure 4.15. *A, Survival of bacteria after repeated freezing (5 minutes at −80°C) and thawing, with and without glycerol (15 percent). (After Hollander and Nell, 1954. From Wood, 1956: Adv. Biol. Med. Phys. 4: 119.)*

at temperatures well below freezing and then successfully transplanted. When cells are frozen with glycerol, the rate of thawing matters little.

Some arctic organisms dehydrate before winter, retaining between 25 and 90 percent of summer moisture. When subjected to very low temperatures, arctic organisms may freeze, yet survive. Respiration continues at a much reduced rate. Ice crystals can be seen (e.g., in tissue spaces of midge larvae) or demonstrated in frozen organisms by a decrease in the specific gravity (Scholander *et al.*, 1953). However, a fish completely frozen in ice has never revived, despite stories to the contrary. Therefore, cells of polar organisms are only more frost resistant than cells of temperate forms.

Cells in some organisms are injured even by a temperature above the freezing point of water. Cells so injured show decreased respiration, inadequate for the exclusion of certain ions (see Chapter 18).

THERMAL RESISTANCE OF DORMANT CELLS

Many cells suspend life activities and become dormant during winter and dry spells. Bacteria and lower plants form spores, protozoans, and some invertebrates form cysts, and higher plants form seeds in which the cells are highly resistant to temperature change. Activity is renewed only when water becomes available and other conditions are favorable.

Dehydrated dormant stages of organisms are more resistant to both heat and cold. Cysts and spores, for example, resist exposure to dry heat of over 100°C, although even they may be chemically sensitized to heat (Alderton and Snell, 1969). Therefore, as a means of killing dormant organisms dry sterilization requires a temperature of 150° to 170°C for half an hour or so. Moist sterilization, on the other hand, requires only 15 minutes at 115°C (15 pounds of pressure in an autoclave or pressure cooker). Boiling (100°C) kills some cysts and spores, but it actually induces germination of others, and a second boiling is necessary to kill the bacteria that have emerged from the spores. Pasteurization (heating at 60°C for half an hour) kills vegetative cells only, not dormant spores in milk.

The greater thermal resistance of dormant organisms is not yet fully explained. It is known that the lower the water content of proteins the higher the temperature required to denature them (Fig. 4.16). Similarly, seeds of desert plants, highly resistant to heat, contain only 5 percent water, whereas the active organism, highly sensitive to heat, contains 80 to 90 percent water (Precht *et al.*, 1973).

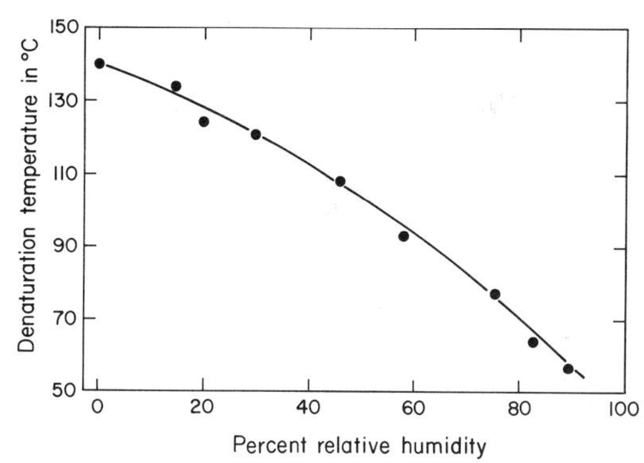

Figure 4.16. Relation between relative humidity and temperature for denaturation of egg albumin. (The data for denaturation of half the egg albumin in 60 minutes are from Barker, 1933: J. Gen. Physiol. 17: 27.)

Some spores and cysts germinate after exposure to temperatures between −250°C and −272°C (close to absolute zero, −273.18°C), yet active cells do not survive −183°C. Dehydrated cells have little water free to form ice. However, the fact that hibernating insects and cocoons of insects resist low temperatures even though not dehydrated suggests that factors other than dehydration may be involved (e.g., antifreeze, known to occur in some insects).

Because life on earth exists within a rather limited range of temperature, exobiologists consider determination of the temperatures of other planets (by means of probes with spacecrafts such as Mariner II to Venus) of utmost importance. The existence of life on Venus seems unlikely because of the planet's high temperature, and questionable on Mars and all distant planets because of their low temperatures. But until adequate experiments have been performed, one must keep an open mind.

LACK OF OXYGEN AS A LIMITING FACTOR

All cells are capable of temporary anaerobic existence. Many bacteria and protozoans tolerate anaerobiosis for prolonged periods, some indefinitely. Even animal and plant cells tolerate anoxia briefly. However, cells of vertebrate central nervous systems suffer quickly from oxygen deprivation.

Violent muscular exertion results in an oxygen debt, because energy is liberated largely by glycolysis that proceeds in the cells without oxygen. The oxygen debt is paid off when exertion eases and the products of glycolysis are either completely oxidized or rebuilt into glycogen for future use (Guyton, 1976).

It is doubtful that many animals are obligate anaerobes, although a few can live under relatively anaerobic conditions. Under favorable conditions cells of animal parasites lacking oxygen depend entirely upon glycolysis for release of energy. Sea animals in the intertidal mud flats are periodically subjected to low oxygen in the stagnant water of pools or burrows at low tide (Steen, 1971; Hochachka *et al.*, 1973). Some symbiotic protozoans such as the flagellates constituting the cellulose-digesting termite fauna live under relatively anaerobic conditions and are killed by exposure to oxygen. Ciliate protozoans in the rumen of a cow are probably under completely anaerobic conditions (Hungate, 1966). However, some obligate anaerobic bacteria exist; perhaps they are relics of the anaerobic era of the earth's history.

Plant cells tolerate anaerobiosis better than animal cells, although a young maize seedling dies after 24 hours in nitrogen at 18°C; the actively growing cells of seedlings are more sensitive than those of older plants. The nature of the anaerobic processes in plant cells, termed fermentation, has not been fully elucidated, although in many cases alcohol is formed and carbon dioxide is evolved (Devlin, 1975).

PRESSURE AS AN ENVIRONMENTAL FACTOR

Because dredgings from deep Mediterranean waters disclosed no life, it was once thought that no life existed in the ocean depths. Early in the last century, however, several peculiar animals were brought up on cables laid in the Atlantic, and later some peculiar fishes were caught where water wells up near Madeira off the northwest coast of Africa. Since that time, towing, dredging, and trawling have disclosed life at various ocean levels, including the greatest depths (Fig. 4.17).

Life is not only present in deep waters but it is diverse, with bacteria and representatives of all the animal phyla being found. Plants, however, are largely confined to the upper 100 meters of the ocean; some microscopic forms occur in the unlighted deep waters where they presumably live on nutrients released by decomposition of dead organisms. The initial failure to find life in the Mediterranean Sea has been attributed to lack of oxygen in the particular regions tested, because later explorations revealed a rich fauna.

The cells of deep-water organisms are under great hydrostatic pressures. For every 10 meters below sea level (slightly over 5 fathoms or 33.4 feet) the pressure increases by about 1 atmosphere (0.987 atmosphere). At the bottom of the Marianas Trench near the Philippines, which is over 10,000 meters deep, the pressure

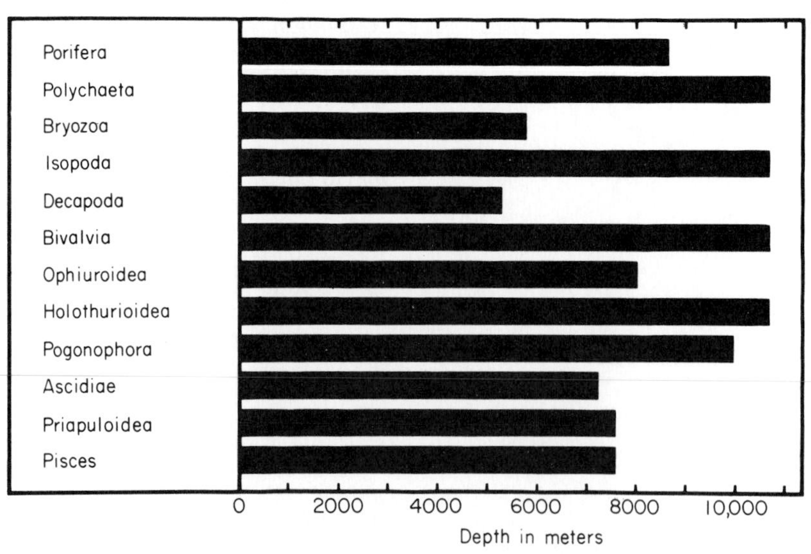

Figure 4.17. *Greatest depths of distribution of various groups of bottom-dwelling animals. Porifera are sponges; Polychaeta, segmented marine worms; Bryozoa, moss animals; Isopoda, small crustaceans; Decapoda, crustaceans such as crabs, lobsters, etc.; Bivalvia, clams, mussels, etc.; Ophiuroidea, brittle stars; Holothuroidea, sea cucumbers; Pogonophora, wormlike creatures with chordate affinities; Ascidiae, tunicates; Priapuloidea, wormlike deep sea creatures; Pisces, fish. (After Zenkevitch; from Flügel and Schlieper, 1970:* In *High Pressure Effects on Cellular Processes. Zimmerman, ed. Academic Press, New York.)*

is over 1000 atmospheres. It would seem that organisms of the deep seas have developed special adaptations enabling them to live under great hydrostatic pressure (Fig. 4.17) (Kitching, 1972; Johnson *et al.*, 1974b; Macdonald, 1975). One of these adaptations is the synthesis of isozymes that function well under high hydrostatic pressure (Siebenaller and Somero, 1978).

The nature of cell adaptations to tolerate hydrostatic pressure has been investigated to a limited extent only, primarily with marine eggs and microorganisms. Sea urchin eggs *(Arbacia punctulata)*, beginning to cleave, revert to the one-celled stage under pressures of more than 400 atmospheres. Contraction and gelation required for cleavage are suppressed. If the pressure is released after brief exposure, cleavage is resumed. About the same amount of pressure suppresses formation of pseudopodia in *Amoeba proteus*; the viscosity of the cytoplasm falls to a value so low that both food vacuoles and nucleus sink to the bottom of the cell. It is thus apparent that pressure has a striking effect on sol-gel relationships in the cell. The "sol-gel" relationship has been investigated at the fine structural level and found to depend upon microtubules (see Chapter 3). These are normally present in structures such as the furrows of a cleaving egg. Hydrostatic pressure causes dissolution of microtubules and the loss of shape and rigidity of the structures they support. When after brief application pressure is released, these structures usually reform (Johnson *et al.*, 1974b). Growth also is suppressed by pressure, and cultures of such bacteria as *Bacillus subtilis* and *Pseudomonas fluorescens* fail to grow at pressures of 300 atmospheres. Replication of phage in *Escherichia coli* is suppressed by 300 atmospheres but growth resumes as soon as the pressure is released. Yet some bacteria that grow readily even at 600 atmospheres, such as *Bacillus submarinus,* have been isolated from the ocean depths. In general, however, bacteria under pressure, even those from the depths, grow slowly (Zobell and Kim, 1972). These laboratory findings have been verified by sinking to 5000 meters labeled nutrient media inoculated with bacteria from sea water that was collected 200 meters down. The rate of nutrient utilization and presumably growth of the bacteria in the depths was one tenth to one hundredth that of surface controls (Jannasch *et al.*, 1971).

Enzymatic reactions are affected by pressure. For example, luminescence, a result of the oxidation of a substrate luciferin by the enzyme luciferase is gradually quenched by hydrostatic pressure. If at this point the pressure is released, luminescence reappears. This has been interpreted as a result of renaturation of luciferase (luciferin is not pressure sensitive) denatured by pressure (Johnson *et al.*, 1974b).

Pressure denaturation of enzymes (and some other proteins) presumably occurs by tightening of the enzyme configuration as the molecular volume is reduced. In a simplified sense the catalytically active groups could be considered to move inward in the molecule, where they cannot bind to substrates. Such binding is necessary for enzyme action.

Since most cell reactions are catalyzed by enzymes, it is thought that suppression of cell activities by pressure is a result of action upon critical cell enzymes. For example, high pressure (666 atmospheres) reduces protein and nucleic acid synthesis. It is found to reduce or prevent the incorporation of labeled amino acids into proteins (Fig. 4.18) and of labeled pyrimidine bases into nucleic acids in *E. coli*, although the effects depend partly on the temperature and other conditions. Low pressure (about 270 atmospheres) may even stimulate incorporation of labeled precursors, while moderate pressure (about 400 atmospheres) generally has little effect. Similar effects were found with animal cells (Zimmerman, 1970; Albright, 1975).

Induction of an enzyme to handle maltose in *E. coli* is blocked immediately by high pressure. This suggests action of pressure by preventing production of messenger RNA needed to give information for synthesis of the enzymatic protein (Landau, 1970).

Since thermal denaturation of enzymes occurs with an increase in volume, while pressure denaturation occurs with a decrease in volume,

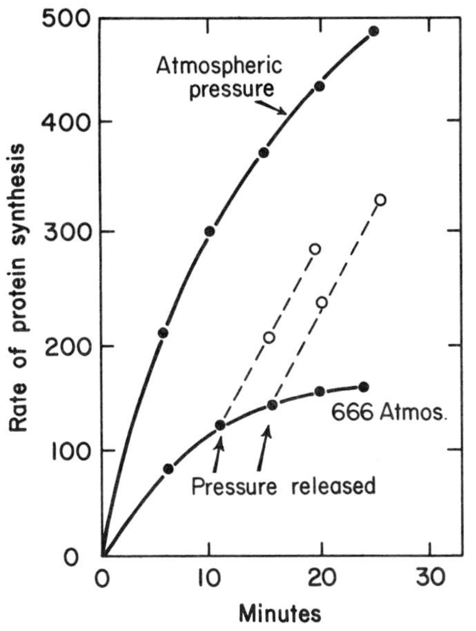

Figure 4.18. *The rate of protein synthesis at atmospheric pressure and at 666 atmospheres hydrostatic pressure. Protein synthesis was measured in counts of radioactive emissions per minute of* ^{14}C*-glycine taken up by* E. coli. *Rate of incorporation is given by the slope of the lines plotting counts against time. Note that the rate of incorporation falls to nearly zero under 666 atmospheres but rapid incorporation is resumed almost immediately upon release of pressure. (After Zimmerman, A. M., 1966: Science* 153: *1273–1274. Copyright 1966 by the American Association for the Advancement of Science.)*

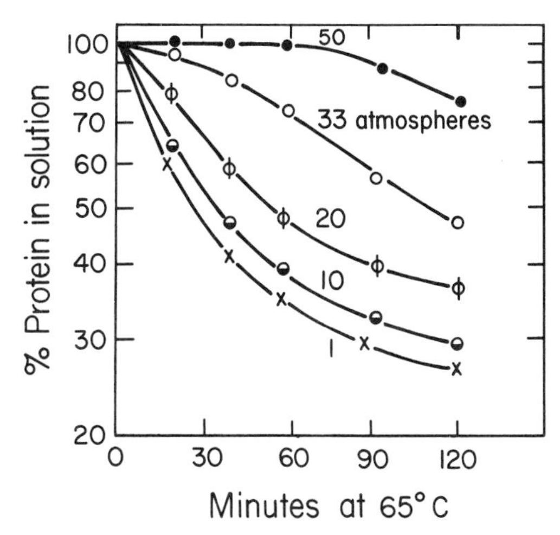

Figure 4.19. *Degree of precipitation of protein at 65°C when subjected to various pressures. (After Johnson* et al., *1974: The Theory of Rate Processes in Biology and Medicine. Wiley-Interscience, New York.)*

pressure and temperature should counteract one another. This has been found to be the case. For example, whereas a protein at atmospheric pressure is quickly denatured by exposure to 65°C, it is denatured at this temperature slowly if subjected to hydrostatic pressure. For example, when subjected to 50 atmospheres pressure, only a small amount of protein is denatured by a two-hour exposure to 65°C (Fig. 4.19). Suppression of bacterial luminescence by heat can also be reversed by hydrostatic pressure, as can thermal inhibition of cleavage in sea urchin eggs and thermal loss of virus infectivity in *E. coli*. These experiments all suggest that the key enzyme(s) involved in each process is inactivated by changing its steric configuration. Enzymes must have the right configuration to be catalytically active (Johnson *et al.*, 1974b).

In accordance with these suppositions, surface bacteria that cannot tolerate the high pressures of deep waters at low temperatures have been shown to tolerate them and grow if the temperature is increased. Perhaps the truly barophilic (literally, pressure-loving) bacteria and other organisms are those whose enzymes have the proper configuration even under great pressures at relatively low temperatures of deep water (around 4°C). However, few species of bacteria have been found that require high pressure for growth (Zobell and Kim, 1972; Johnson *et al.*, 1974b). Even the bacteria that grow at the high pressure of the ocean depths respire, decompose nutrients, and incorporate tracers very slowly (Jannasch *et al.*, 1977).

It is not known by what mechanism pressure causes sex reversal in a copepod (Vacquier and Belser, 1965) or mutation in *Euglena* (Gross, 1965).

RADIATION IN THE CELL ENVIRONMENT

Cells are exposed to a wide range of radiation, chiefly from the sun and from radioactive elements in the earth's crust (Bridges, 1976). Unicellular organisms and organisms composed of a small number of cells are particularly vulnerable. In larger organisms only outer skin cells and those in the eyes are vulnerable to the sun's radiation, since adsorption by superficial cells protects those below them. A variety of protective screens have evolved. Some unicellular organisms are pigmented, as are the superficial layers of cells in larger organisms. Protective pigments may develop after exposure to sunlight, as in tanning of human skin. Large animals may cover the outer body surface with screens opaque to sunlight, such as exoskeletons, shells, scales, feathers, and hair. Sunlight has little ionizing radiation at present, although during the times when the earth's magnetic field was reversing, the solar wind may have come to earth instead of being deflected. At such times intense ionizing radiation may have enveloped parts of the earth. Some researchers have suggested such episodes to explain the extinction of species in the past. Ionizing radiation penetrates through pigments and other light shields and may reach deep into the body of a larger organism; x-ray penetration depends upon wavelength, penetration being deeper for short than for long wavelengths.

In addition to the defensive devices listed, as a further accommodation cells can repair radiation damage. Without such repair they could not survive natural radiations. Some cells are highly radioresistant and photoresistant because of their very effective repair of radiation damage. Since most radiation damage interferes with DNA replication, discussion of repair mechanisms is deferred until DNA replication has been considered (see Chapters 15 and 27).

ADAPTABILITY OF PROKARYOTES AND EUKARYOTES

Cells have invaded a wide variety of environments, displaying a capacity to withstand drying, wide ranges of salinities, pH, high and low temperatures, lack of oxygen, and hydrostatic pressure. In each case cells adapted to one environment are at a disadvantage in another, because their chemistry is attuned to the special environment in which they live.

In comparing prokaryotic and eukaryotic cells, it becomes evident that in general prokaryotes are more adaptable than eukaryotes and occupy the most extreme environmental niches. It was previously noted that prokaryotes are also biochemically more versatile, utilizing even synthetics not found in nature as a source of energy. The greater adaptability of prokaryotes may rest on their high mutation rates and their ability to meet a new situation by adaptive enzymes. Eukaryotes acclimate to seasonal change in temperature by isozymes, but in general they are less adaptive to change than prokaryotes.

Eukaryotes lost some of their plasticity as they developed more specialized structures. While in some respects less flexible than prokaryotes, eukaryotes have achieved success by developing highly specialized cells and have developed organs and organ systems of great complexity and effective function. No prokaryote can fly.

APPENDIX

4.1 DISSOCIATION OF WEAK ACIDS, WEAK BASES, AND WATER

An *acid* forms hydrogen ions (positively charged hydrogen atoms or protons) in solution. A *base* yields hydroxyl ions in solution and reacts with an acid to form a salt and water.*

Strong acids and strong bases are completely ionized and dissociated in water; the degree of dissociation of weak acids and weak bases depends upon the strength of the acid or base.

When a weak acid, HA, is dissolved in water, it dissociates, and if the velocity of the reaction in the direction of dissociation is given by V_1,

*According to Brönsted (1928), an *acid* is a substance capable of giving off a proton (hydrogen ion), and a *base* is one capable of combining with a proton. By this definition the chloride ion is a base because it is capable of combining with a hydrogen ion. Since water combines with hydrogen ions to form *hydronium* ions, H_3O^+, it is a base, but since H_3O^+ may yield a proton, the hydronium ion is an acid. This method of defining acids and bases has special use for nonaqueous systems.

The *hydronium ion* is always formed by combination of a proton with water whenever free protons are dissociated from acids. Proof of the presence of H_3O^+ is obtained from x-ray diffraction study of the crystalline hydrates, for example, of perchloric acid. The distances between the ions identified indicate H_3O^+ and ClO_4^-. The presence of hydronium ions rather than hydrogen ions has been assumed throughout the discussion here and in subsequent chapters, but following custom, all formulations (since they apply equally well to hydronium or hydrogen ions) are given in terms of hydrogen ions.

and the velocity of association of the ions to re-form molecules of acid is given by V_2, at equilibrium we have the following relationship:

$$HA \underset{V_2}{\overset{V_1}{\rightleftharpoons}} H^+ + A^- \quad (4.3)$$

The value of V_1 increases as the acid dissociates more completely, and the equilibrium is then said to lie to the far right. Under a given set of conditions the velocity of the reaction to the right depends primarily on the concentration of HA:

$$V_1 = k_1 [HA] \quad (4.4)$$

where k_1 represents a constant.

The velocity of the reaction to the left depends upon the concentration of H^+ and A^-.

$$V_2 = k_2 [H^+][A^-] \quad (4.5)$$

where k_2 represents another constant.

When equilibrium is reached the quantity of acid dissociating is equal to that reassociating; therefore, V_2 equals V_1, and

$$k_2 [H^+][A^-] = k_1 [HA]. \quad (4.6)$$

By dividing each side by k_2 and by [HA], expression of the law of mass action is obtained:

$$\frac{[H^+][A^-]}{[HA]} = \frac{k_1}{k_2} = K_a \quad (4.7)$$

where K_a is the dissociation constant of the weak acid or the ratio of dissociation to association. At room temperature for acetic acid, K_a is 1.85×10^{-5} (or 43 molecules in 10,000); for formic acid, it is 2.4×10^{-4}; and for boric acid, 6.4×10^{-10}. The smaller the dissociation constant the weaker the acid. A parallel formulation might be developed for dissociation of weak bases.

Water also dissociates very slightly, and gives rise to both hydrogen ions (like an acid) and hydroxyl ions (like a base). For the dissociation of water the following equations may be written:

$$H_2O \rightleftharpoons H^+ + OH^- \quad (4.8)$$

$$\frac{[H^+][OH^-]}{[H_2O]} = K \quad (4.9)$$

where K is the dissociation constant of water.

Then,

$$[H^+][OH^-] = K [H_2O] \quad (4.10)$$

Since the degree of dissociation is so slight, $[H_2O]$ may be considered as approximately constant. Therefore it may be combined with K, giving K_w:

$$[H^+][OH^-] = K_w: \quad (4.11)$$

The experimentally determined electrical conductivity of pure water at room temperature indicates that the concentration of hydrogen (or hydroxyl) ions is approximately 10^{-7}N.

Therefore,

$$K_w = 10^{-7} \times 10^{-7} = 10^{-14} \quad (4.12)$$

The dissociation of water, as is true of dissociation in general, depends upon the temperature. For example, K_w is 0.115×10^{-14} at 0°C., 10^{-14} at room temperature, and 9.614×10^{-14} at 60°C. Since the hydrogen ion is one of the important biological catalysts, the increase in hydrogen ion concentration with rise in temperature may contribute to the increased activity of reactions in organisms when the temperature is raised.

4.2 THE pH SCALE

Sørensen chose to express the concentration of hydrogen ions by negative logarithms of the hydrogen ion concentration. This function he called the pH:

$$pH = -\log_{10} [H^+] \quad (4.13)$$

The pH scale ranges from 0 to 14 or from 1 N $[H^+]$ to 1 N $[OH^-]$ by units of 0.1 N.

Because, in water, the concentration of hydrogen ions equals the concentration of hydroxyl ions, water is considered neutral. Since the hydrogen ion concentration of water is 10^{-7}, at room temperature the pH of water is 7.0. Therefore, a pH lower than 7.0 indicates a concentration of hydrogen ions greater than that found in water and the solution is considered acid, a pH higher than 7.0, basic.

4.3 MEASUREMENT OF THE STRENGTH OF A WEAK ACID BY THE pK_a

By applying the law of mass action to dissociation of a weak acid, HA, the following expressions may be written:

$$HA \underset{V_2}{\overset{V_1}{\rightleftharpoons}} H^+ + A^- \quad (4.3)$$

$\frac{[H^+][A^-]}{[HA]} = K_a$, the acid dissociation constant (equation 4.7).

Solving for $[H^+]$:

$$[H^+] = \frac{K_a[HA]}{[A^-]} \quad (4.14)$$

Taking the negative logarithm of each side,

$$-\log_{10}[H^+] = -\log_{10} K_a - \log_{10}\frac{[HA]}{[A^-]} \quad (4.15)$$

By definition, $-\log_{10}[H^+] = pH$, and $-\log_{10} K_a = pK_a$; and since

$$-\log_{10}\frac{[HA]}{[A^-]} = +\log_{10}\frac{[A^-]}{[HA]}$$

substitution of these values in equation 4.15 gives:

$$pH = pK_a + \log_{10}\frac{[A^-]}{[HA]} \quad (4.16)$$

When the salt of a weak acid is added to a solution of that acid, the dissociation of the acid is suppressed by the common ion effect (law of mass action). Consequently, most of the acid is present in the undissociated form and the concentration of anion in solution becomes approximately equal to the concentration of dissolved salt. By applying equation 4.16 to this case it will be seen that the term for anion concentration may be replaced by a term for salt concentration, and the acid concentration may be considered approximately equal to the total acid. The equation may now be written:

$$pH = pK_a + \log_{10}\frac{[salt]}{[acid]} \quad (4.17)$$

This equation is known as the Henderson-Hasselbalch equation for buffer action.

When the concentrations of salt and acid are equal, the second term drops out of the Henderson-Hasselbalch equation, since the $\log_{10}$ of 1 is zero. The pH then equals the pK_a. By making the salt concentration equal the acid concentration it is relatively easy to determine the pK_a values for acids.

It should be noted that a large pK_a indicates a weak acid, a small pK_a a strong acid. Multivalent acids possess several dissociation constants, for each of which a pK_a may be determined.

Once the pK_a of an acid is known, the ratio of acid to salt required to attain a desired pH may be readily calculated. Conversely, if the concentrations of salt and acid are known, the pH of the mixture may be readily calculated. However, it must be remembered that the Henderson-Hasselbalch equation is an approximation only and that it holds best when salt and acid are in nearly equal concentration. It is most in error when applied to widely divergent ratios of acid and salt. As the pK_a is also a function of the total salt concentration, an experimentally determined pK_a may differ significantly from a handbook value that is cited for the weak acid and its salt at a given concentration in the absence of extraneous salts.

4.4 AMPHOLYTES OR ZWITTERIONS

Ampholytes are inorganic and organic substances that can act either as acids or as bases; that is, they can either give off or take up hydrogen ions. If an acid is added, an ampholyte will dissociate as a base. If a base is added, the ampholyte will dissociate as an acid.

An amino acid has no net charge at its *isoelectric pH*. The isoelectric pH is defined as the pH at which there is no net accumulation of amino acid at either anode or cathode of an electric field, and the ions are therefore considered to be equally charged negatively and positively.

Evidence indicates that amino acids exist as zwitterions, or bipolar ions, rather than as uncharged molecules. For instance, they have a high solubility in water; they have a highly polar nature, shown by their behavior in an electromagnetic field; and they have relatively high boiling points—properties which cannot be explained if uncharged molecules are assumed. The isoelectric pH is different for different amino acids (6.1 for glycine, 5.7 for serine, 3.0 for aspartic acid), and it depends upon the number and relative strength of acidic and basic groups in the amino acid, e.g., their tendency to ionize. For a monocarboxylic, monoamino amino acid the isoelectric pH (pI) is the sum of its acidic pK and its basic pK divided by 2. On the acid side of the isoelectric pH an amino acid is positively charged, and on the alkaline side it is negatively charged:

$$\underset{\text{zwitterion}}{H_3\overset{+}{N}—R—COO^-} + H^+ \rightleftharpoons H_3\overset{+}{N}—R—COOH \quad (4.18)$$

$$HO^- + \underset{\text{zwitterion}}{H_3\overset{+}{N}—R—COO^-} \rightleftharpoons H_2N—R—COO^- + H_2O \quad (4.19)$$

It must not be construed that proteins dissociate as if they were a multitude of amino acids. The peptide backbone of the protein binds most of the amino and carboxyl groups of the constituent amino acids. Only the end groups of the peptide chains and the extra amino and carboxyl groups of the polyamino and polycarboxy amino acids and special groups such as the hydroxy group of tyrosine are free to dissociate as the pH of the solution containing the protein is changed.

4.5 METHODS OF MEASURING pH

In the pH meter, a glass electrode and a calomel electrode are dipped into the unknown solution, as shown in Figure 4.20. The glass electrode consists of a bubble of glass containing a solution of known concentration of hydrochloric acid, usually 0.1 normal with respect to hydrogen ions, into which extends a silver-silver chloride wire connected to the potentiometer. The calomel electrode contains calomel (Hg_2Cl_2) in contact on one side with saturated potassium chloride and on the other with mercury from which a platinum wire connects to the potentiometer.

A potential, proportional to the difference in concentration between the two solutions arises because of a difference in concentration between the hydrogen ions inside the glass electrode and in the unknown solution, the two being separated by the glass membrane through which only the hydrogen ions can pass. While hydrogen ions, in sufficient numbers to set up a potential, pass from the solution in which they are more concentrated through the glass electrode toward the other solution, actual leakage of acid does not occur, because anions cannot accompany the hydrogen ions.

In a potentiometer the potential developed between the two solutions inside and outside the glass electrode is balanced by that from a battery, the point of balance indicated by a null point on the dial of the galvanometer. In the pH meter the calomel electrode is used as a reference electrode against which the potential between the two solutions inside and outside the glass electrode is measured (see Chapter 11). In practice, the potential is amplified, and the dial of the potentiometer slide wire, which then measures the potential, is calibrated to read directly in pH units.

4.6 RELATION BETWEEN CONCENTRATION AND ACTIVITY OF ELECTROLYTES

The pH of a weak acid is less than would be expected on the basis of the normality of the acid, because dissociation is incomplete. However, the concentration of hydrogen ions of even a strong acid like hydrochloric, which is considered to be completely dissociated, is also somewhat less than would be expected on the basis of its normality. This is because the electrostatic attraction between oppositely charged ions in solutions prevents free activity of the ions. Some hydrogen ions are therefore not free to act at any given moment. If strong acid is sufficiently diluted, however, the concentration of hydrogen ions as measured by pH approaches the normality of the acid, because as interionic distance increases, ionic interference decreases to the vanishing point. This is strictly true only at infinite dilution.

Since in solutions of electrolytes ionic interference reduces the effectiveness of ions so that their concentration seems lower than it is, the concept of activity has been introduced. *Activity* is the effective concentration of an ion in solution. The *activity coefficient* is defined as the ratio of the activity, as measured by some property such as the depression of the freezing point of a solution, to the true concentration (molality):

$$\text{activity coefficient } (\gamma) = \frac{\text{activity}}{\text{concentration}} \quad (4.20)$$

The activity coefficient, which is usually less than one, increases as the solution becomes more dilute, and approaches unity at infinite dilution when the attractive forces between oppositely charged ions become negligible.

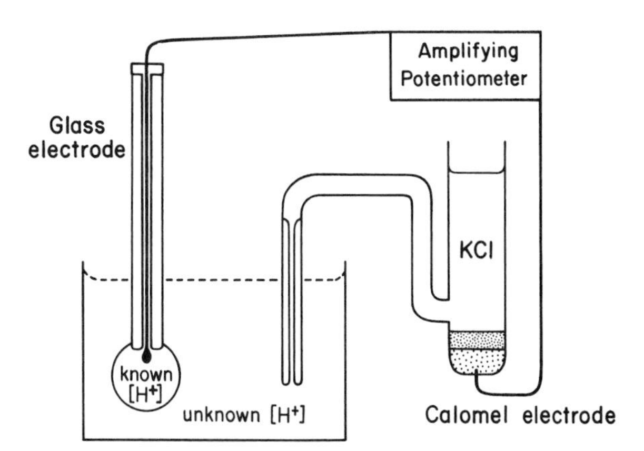

Figure 4.20. *Simplified diagram of the electrometric determination of pH with a glass electrode. The sample whose pH is to be determined is placed in the vessel marked "unknown [H+]" into which dip the glass electrode (containing 0.1N H+) and the calomel electrode. A potential is developed between the solution in the vessel and the glass electrode because of the difference in hydrogen ion concentrations between the two. This potential is amplified and measured by a potentiometer. The dial on the instrument reads directly in pH units. For details of the calomel electrode see Figure 11.1.*

Since pH measures the hydrogen ions free to act, it is a measure of the hydrogen ion activity, and pH should therefore be more strictly defined as:

$$pH = -\log_{10} [\text{hydrogen ion activity}] \quad (4.21)$$

When this is done the pK_a obtained for a weak acid is designated pK'_a.

4.7 BUFFER MIXTURES USING A WEAK BASE AND ITS SALT

For making buffers it is possible to calculate the ratio of the salt of a weak base to the base (BOH) in much the same way as is done for a weak acid. From the relationship,

$$BOH \rightleftharpoons B^+ + OH^- \quad (4.22)$$

the Henderson-Hasselbalch equation is derived

$$pOH = pK_b + \log_{10} \frac{[\text{salt}]}{[\text{base}]} \quad (4.23)$$

Since

$$[H^+] \cdot [OH^-] = 10^{-14} \quad (4.24)$$

and

$$pH + pOH = 14 \quad (4.25)$$

or

$$pOH = 14 - pH \quad (4.26)$$

Then substitution of equation 4.26 for pOH in equation 4.23 gives:

$$14 - pH = pK_b + \log_{10} \frac{[\text{salt}]}{[\text{base}]} \quad (4.27)$$

or

$$-pH = -14 + pK_b + \log_{10} \frac{[\text{salt}]}{[\text{base}]} \quad (4.28)$$

or

$$pH = 14 - pK_b - \log_{10} \frac{[\text{salt}]}{[\text{base}]} \quad (4.29)$$

LITERATURE CITED AND GENERAL REFERENCES

Albright, L. F., 1975: The influence of hydrostatic pressure upon biochemical activities of heterotrophic bacteria. Can. J. Microbiol. *21*: 1406–1412.

Alderton, G. and Snell, N., 1969: Bacterial spores: chemical sensitization to heat. Science *163*: 1212–1213.

Ben-Naim, A., 1974: Water and Aqueous Solutions. Plenum Press, New York.

Björkman, O., Pearcy, R. W., Harrison, A. T. and Mooney, M., 1972: Photosynthetic adaptation to high temperature: a field study in Death Valley, California. Science *175*: 786–789.

Bridges, B. A., 1976: Survival of bacteria following exposure to ultraviolet and ionizing radiations. Soc. Gen. Microbiol. Symp. *26*: 183–208.

Brock, T. D., 1967: Life at high temperatures. Science *158*: 1012–1019.

Brock, T. D., 1973: Lower pH limit for the existence of blue-green algae: evolutionary and ecological implications. Science *179*: 480–483.

Brönsted, J. N., 1928: Acid and basic catalysis. Chem. Rev. *5*: 231–238.

Castenholz, R. W., 1973: Ecology of blue-green algae in hot springs. *In* The Biology of Blue Green Algae. Carr and Whitton, eds. Blackwell, London, pp. 379–414.

Clayton, R. K., 1971: Light and Living Matter. Vol. 2, The Biological Part. McGraw-Hill Book Co., New York.

Crabb, J. W., Murdock, A. L. and Amelunxen, R. E., 1975: A proposed mechanism of thermophily in facultative thermophiles. Biochem. Biophys. Res. Communs. *62*: 627–633.

Davenport, H. W., 1972: Why does the stomach not digest itself? Sci. Am. (June) *226*: 87–93.

Devlin, R. M., 1975: Plant Physiology. Van Nostrand Reinhold Co., New York.

DeVries, A. L. 1974: Survival at freezing temperatures. *In* Biochemical and Biophysical Perspectives in Marine Biology. Malins and Sargent, eds. Academic Press, New York, pp. 290–330.

DeVries, A. L., 1975: The role of macromolecular antifreezes in cold water fishes. Comp. Biochem. Physiol. *52A*: 193–199.

DeVries, A. L. and Somero, G. N., 1971: The physiology and biochemistry of low temperature adaptation in Antarctic marine animals. Antarctic Symposium on Ice and Water Masses, Tokyo, pp. 101–113.

Drost-Hansen, W., 1972: Effects of pressure on the structure of water in various aqueous systems. Soc. Exp. Biol. Symp. *26*: 61–101.

Dugan, P. R., 1972: Biochemical Ecology of Water Pollution. Plenum Press, New York.

Farrell, J. and Rose, A. H., 1967: Temperature effects on microorganisms. *In* Thermobiology. Rose, ed. Academic Press, New York, pp. 147–218.

Feeney, R. F. and Osuga, D. T., 1976: Comparative biochemistry of Antarctic proteins. Comp. Biochem. Physiol. (A) *54*: 281–286.

Flowers, T. J., Troke, P. F. and Yeo, A. R., 1977: The mechanism of salt tolerance in halophytes. Ann. Rev. Plant Physiol. *28*: 89–121.

Forsythe, M. P. and Kushner, D. J., 1970: Nutrition and distribution of salt responses in populations of moderately halophilic bacteria. Can. J. Microbiol. *16*: 253–261.

Gould, G. W. and Dring, G. J., 1974: Mechanism of spore heat resistance. Adv. Microb. Physiol. *11*: 137–164.

Gross, J. A., 1965: Pressure-induced color mutation of *Euglena gracilis*. Science *147*: 741–742.

Guyton, A. C., 1976: Textbook of Medical Physiology. 5th Ed. W. B. Saunders Co., Philadelphia.

Heinen, W., 1971: Growth conditions and temperature dependent substrate specificity of two extremely thermophilic bacteria. Arch. Mikrobiol. *76*: 2–17.

Heinrich, M. R., ed., 1975: Extreme Environments. Academic Press, New York. (See especially chapters by Reid [temperature and tRNA], Stenesh [thermophiles], Livingdahl and Sherod [proteins from thermophiles], and Singleton [amino acids from thermophiles].)

Henderson, L. J., 1927: Fitness of the Environment. Macmillan Co., New York. (A classic.)

Hochachka, P., Fields, J. and Mustafa, T., 1973: Animal life without oxygen: basic biochemical mechanisms. Am. Zool. *13*: 543–555.

Hochachka, P. and Somero, G. N., 1973: Strategies of Biochemical Adaptation. W. B. Saunders Co., Philadelphia.

Hungate, R. F., 1966: The Rumen and Its Microbes. Academic Press, New York.

Jannasch, H. W., Eimhjellen, K., Wilson, C. O. and Farmanfarmaian, A., 1971: Microbial degradation of organic matter in the deep sea. Science *171*: 652–675.

Jannasch, H. W. and Wirsen, C. O., 1977: Microbial life in the deep sea. Sci. Am. (Jun.) *236*: 42–52.

Johnson, F. H., Eyring, H. and Stover, B. J., 1974a: Temperature. *In* The Theory of Rate Processes in Biology and Medicine. Wiley-Interscience, New York, pp. 155–272.

Johnson, F. H., Eyring, H. and Stover, B. J., 1974b: Hydrostatic pressure and molecular volume changes. *In* The Theory of Rate Processes in Biology and Medicine. Wiley-Interscience, New York, pp. 273–369.

Kitching, J. A., 1972: The effects of pressure on organisms. A survey of progress. Soc. Exp. Biol. Symp. *25*: 473–482.

Kushner, D. J., 1968: Halophilic bacteria. Adv. Appl. Microbiol. *10*: 73–99.

Kushner, D. J., ed., 1978: Microbial Life in Extreme Environments. Academic Press, New York. (Temperature, pressure, pH, salinity, irradiation, and heavy metals.)

Landau, J. V., 1970: Hydrostatic pressure on the biosynthesis of macromolecules. *In* High Pressure Effects on Cellular Processes. Zimmerman, ed. Academic Press, New York, pp. 45–70.

Larsen, H., 1967: Biochemical aspects of extreme halophilism. *In* Advances in Microbial Physiology. Vol. 1. Rose and Wilkinson, eds. Academic Press, New York, pp. 97–132.

Macdonald, A. C., 1975: Physiological Aspects of Deep Sea Biology. Cambridge University Press, New York.

McElhaney, R. N., 1974: The effect of alterations in the physical state of the membrane lipids on the ability of *Acholeplasma laidlawii* to grow at various temperatures. J. Mol. Biol. *84*: 145–157.

MacLeod, R. A. and Calcott, P. H., 1976: Cold shock and freezing damage to microbes. Soc. Gen. Microbiol. Symp. *26*: 81–109.

Mangiantini, M. T., Tecce, G., Tochi, G. and Trentalance, A., 1965: A study of ribosomes and of RNA from a thermophilic organism. Biochim. Biophys. Acta *103*: 252–274.

Marquis, R. E., 1976: High pressure microbial physiology. Adv. Microbiol. Physiol. *14*: 159–241.

Maxwell, K. E., 1976: Environment of Life. 2nd Ed. Dickenson Publishing Co., Inc., Encino, Calif.

Mazur, P., 1975: Fundamental aspects of freezing injury in cells. *In* Proc. First Intersectional Congress of IAMS, Hasegawa, ed. Vol. 5. Microbial Products, Freezing and Freeze-drying. Science Council of Japan, Tokyo, pp. 577–586.

Meeks, J. C. and Castenholz, R. W., 1971: Growth and photosynthesis in an extreme thermophile *Synechrococcus lividus* (Cyanophyta). Arch. Mikrobiol. *78*: 25–41.

Meryman, H. T., ed., 1966: Cryobiology. Academic Press, New York.

Mitchell, R., 1974: The evolution of thermophily in hot springs. Quart. Rev. Biol. *49*: 229–242.

Morita, R. Y., 1976: Survival of bacteria in cold and moderate hydrostatic pressure environment with special reference to phychrophilic and barophilic bacteria. Soc. Gen. Microbiol. Symp. *26*: 279–298.

Morowitz, H. J., 1971: Energy Flow in Biology. Academic Press, New York.

Novitsky, J. A. and Morita, R. Y., 1977: Survival of a psychrophilic marine *Vibrio* under long term nutrient starvation. Appl. Environ. Microbiol. *33*: 635–641.

Pantin, C. F. A., 1931: The origin and composition of the body fluids in animals. Biol. Rev. *6*: 459–482.

Pollard, E. C. and Weller, P. K., 1966: The effect of hydrostatic pressure on the synthetic processes in bacteria. Biochim. Biophys. Acta *112*: 573–580.

Precht, H., Christophersen, J. and Hensel, H., 1973: Temperature and Life. Springer-Verlag, New York.

Prosser, C. L., ed., 1973: Comparative Animal Physiology. 3rd Ed. W. B. Saunders Co., Philadelphia.

Racker, E. and Stoeckenius, W., 1974: Reconstitution of purple membrane vesicles catalyzing light-driven proton uptake and adenosine triphosphate formation. J. Biol. Chem. *249*: 662–664.

Raymond, J. C. and Sistrom, W. R., 1969: The isolation and preliminary characterization of a halophilic photosynthetic bacterium. Archiv. Mikrobiol. *59*: 255–268.

Rose, P. H., ed., 1967: Thermobiology. Academic Press, New York.

Rubey, W. W., 1951: Geologic history of sea water. Bull. Geol. Soc. Am. *62*: 1111–1147.

Scientific American issues on the environment: Sept. 1955, on The Planet Earth; Sept. 1956, The Universe; Sept. 1962, The Antarctic; Sept. 1969, The Ocean; Sept. 1970, The Biosphere.

Schlieper, C., 1972: Comparative investigations on the pressure tolerance of marine invertebrates and fish. Soc. Exp. Biol. Symp. *26*: 197–207.

Scholander, P. F., Claff, C. L. and Sveinsson, S. L., 1952: Respiratory studies on single cells. II. Observations on oxygen consumption in single protozoans. Biol. Bull. *102*: 157–177.

Scholander, P. F., Flagg, W., Hock, R. J. and Irving, L., 1953: Studies on physiology of frozen plants and animals in Arctic. J. Cell Comp. Physiol. *42*(Suppl. 1): 1–56. (Dated, but still a very informative article.)

Siebenaller, J. and Somero, G., 1978: Adaptive differences in lactic dehydrogenase in congeneric fishes living at different depths. Science *201*: 255–257.

Siegel, S. M. and Roberts, K., 1966: The microbiology of saturated salt solutions and other harsh environments. I. Growth of a salt-dependent bacterial form in LiCl saturated nutrient broth. Proc. Natl. Acad. Sci. U.S.A. *56*: 1505–1508.

Skujins, J. J. and McLaren, A. D., 1967: Enzyme reaction rates at limited water activities. Science *158*: 1569–1570.
Sleigh, M. A. and MacDonald, A. G., eds., 1972: The Effects of Pressure on Organisms. Cambridge University Press, Cambridge.
Somero, G. N., 1975a: Temperature as a selective factor in protein evolution: the adaptational strategy of "Compromise." J. Exp. Zool. *194*: 175–188.
Somero, G. N., 1975b: The roles of isozymes in adaptation to varying temperatures (Markert, ed.). *In* Isozymes II, Physiological Functions. Academic Press, New York, pp. 221–239.
Steen, J. B., 1971: Comparative Physiology of Respiratory Mechanisms. Academic Press, New York.
Stein, D. B. and Searcy, D. G., 1978: Physiologically important stabilization of DNA by a prokaryotic histone-like protein. Science *202*: 219–221.
Szybalski, W., 1967: Effects of elevated temperatures on DNA and some polynucleotides. *In* Thermobiology. Rose, ed. Academic Press, New York, pp. 73–122.
Tansey, M. and Brock, T., 1972: The upper temperature limit for eukaryotic organisms. Proc. Nat. Acad. Sci. U.S.A. *69*: 2426–2428.
Theyer, D. W., ed., 1975: Microbial Interaction with the Physical Environment. John Wiley & Sons, New York.
Trump, B. F., Young, D. E., Arnold, E. A. and Stowell, R. E., 1965: Effects of freezing and thawing on the structure, chemical composition and function of cytoplasmic structure. Fed. Proc. *24*: S144–S168.
Vacquier, V. D., and Belser, W. L., 1965: Sex conversion induced by hydrostatic pressure in the marine copepod *Tigriopus californicus*. Science *150*: 1619–1621.
Wald, G., 1959: Life and light. Sci. Am. (Oct.), *201*: 92–108.
Welker, N. E., 1976: Microbial endurance and resistance to heat stress. Soc. Gen. Microbiol. Symp. *26*: 241–277.
Wickstrom, C. E. and Castenholz, R. W., 1973: Thermophilic ostracod; aquatic metazoan with the highest known temperature tolerance. Science *181*: 1063–1064.
Williams, R. J. P., 1970: Biochemistry of sodium, potassium, magnesium and calcium. Quart. Rev. Chem. Soc. *24*: 331–365.
Yamada, H. and Okamoto, H., 1961: Isolation and pure culture of a halophilic strain of *Chlamydomonas*. Z. Allg. Microbiol. *1*: 245–250.
Zimmerman, A. M., ed., 1970: Pressure Effects on Cellular Processes. Academic Press, New York. (Especially chapters on biosynthesis of macromolecules [J. V. Landau], selected biological systems [R. Y. Morita and R. R. Becker], morphology and life processes in bacteria [G. Zobell], on marine invertebrates and fishes [H. Flügel and C. Schlieper], synthesis in marine eggs [A. M. Zimmerman], and mechanisms of cell division [D. Marsland]).
Zobell, G. F. and Kim, J., 1972: Effects of deep sea pressures on microbial enzyme systems. Soc. Exp. Biol. Symp. *26*: 135–146.

Section II

CELL ORGANELLES

Perhaps the greatest contribution to our knowledge of the structure and function of cell organelles came from the application of electron microscopy to cells concomitantly with the separation of cell organelles by fractional centrifugation and chemical studies of their composition. These studies, along with determination of the chemical and structural changes during cycles of biochemical activity in cells and cell division, opened up a whole new world of information even now not fully exploited. The Nobel Prize in physiology and medicine in 1974 was awarded to three pioneers in these studies: Albert Claude, Christian de Duve, and George Palade.

Electron microscopic studies on prokaryotes failed to disclose structures in the cell other than a cell wall, a membrane, a chromosome consisting of a tightly wound double-stranded DNA molecule, ribosomes, mesosomes, and in photosynthetic cells, chlorophyll-bearing chromatophores or thylakoid-like membranes—all in a cytoplasmic matrix. Much has been learned of prokaryotic membranous components that are the seat of oxidative metabolism in chemoautotrophic and oxidative heterotrophic bacteria and of photosynthesis in autotrophic species. The chromosomes have also been extensively investigated.

Electron microscopic studies on eukaryotes have verified or disclosed the existence of a number of distinctive cell organelles, including the doubly-membraned nucleus containing the chromosomes, the membranous vacuolar system consisting of the endoplasmic reticulum, the Golgi complex, and their products (lysosomes, microbodies, etc.), and the double-membraned mitochondria and chloroplasts. In addition, the cell matrix, which appears essentially similar in both prokaryotic and eukaryotic cells under the light microscope, has been shown to contain microfilaments and microtubules in eukaryotic cells. The cytoplasmic matrix, considered to be a "cytoskeleton," contains 20 to 25 percent of the cellular protein, including enzymes catalyzing glycolysis and activation of amino acids for protein synthesis on the ribosomes in both types of cells. The cytoplasmic matrix is thought by some to represent an ancient component of cells, harking back to the anaerobic period in the earth's history, when glycolysis was the only mode of energy liberation.

One set of eukaryotic organelles—the cytoplasmic vacuolar system, consisting of the endoplasmic reticulum, Golgi bodies and their products—forms a convenient and related unit for study in Chapter 5. The energy transducers, mitochondria and chloroplasts, are treated in Chapter 6. The nucleus and chromosomes are taken up in Chapter 7, and the molecular organization and properties of cell membranes in Chapter 8.

CHAPTER 5

THE ENDOPLASMIC RETICULUM–GOLGI BODY COMPLEX

In embryonic eukaryotic plant and animal cells the canals and cavities of the endoplasmic reticulum-Golgi complex are little developed, but with cell differentiation, they develop extensively. These membrane systems serve to separate each synthetic system in the endoplasmic reticulum from synthetic and assembly areas in the Golgi complex and from their packaged products (lysosomes, peroxisomes, secretory granules, and droplets).

THE ENDOPLASMIC RETICULUM

The *endoplasmic reticulum* is organized into a network of thin membrane-bound cavities that vary considerably in size and shape under different physiological conditions. In some cells the network appears to consist of fine tubules 50 to 100 nm in diameter (Figs. 5.1, 5.2, 5.3) and in some places enlargements form flattened sacs and vesicles called *cisternae*. In initial studies the network appeared to be confined to the endoplasm of the cell, where it is most prominent, but thin sections of cells demonstrated that the reticulum extended into the peripheral part of the cell (the ectoplasm), even connecting with the cell membrane. Serial sections of cells that permit reconstruction of structure in three dimensions showed connections between the plasma membrane, the endoplasmic reticulum, and the Golgi complex, as well as the nuclear envelope. The endoplasmic reticulum is poorly developed in spermatocytes and entirely lacking in red blood cells (erythrocytes).

The endoplasmic reticulum is differentiated into *smooth* and *rough* portions, the latter studded with ribosomes. In adipose tissue cells, cells that store brown fat, cells of the adrenal cortex, and in liver cells after injection of some drugs, the smooth endoplasmic reticulum is well developed (Fig. 5.2). From the smooth sac many tubules of small diameter pervade the cytoplasmic matrix. In cells producing proteins such as enzymes and gland secretions for export, the rough endoplasmic reticulum is especially well developed (Figs. 5.1, 5.3). Some conception of the magnitude of the tubular system may be obtained from knowing the fact that 1 ml of liver has been calculated to have a surface area of 11 square meters (Palade, 1956).

When labeled amino acid precursors to protein are used to study protein synthesis in a gland cell, the label appears first over the polysomes (aggregations of ribosomes on messenger RNA) of the rough endoplasmic reticulum, then in the cisternae of the endoplasmic reticulum, next in the sacs of the Golgi assembly, and finally in the zymogen granules (enzyme precursor) that are extruded from the cells bordering the gland's lumen. Just how the protein gets from the ribosome surface to the cavity of the tubules and sacs is not known, although it could pass directly into the lumen as it is synthesized (Satir, 1975).

The membranes of the endoplasmic reticulum separate the cytoplasm from the lumen content of the canals and vesicles. The membranes possess selective permeability properties, restricting entry or exit of materials into the lumen (see Chapter 18). It is possible that molecules are assembled inside the membrane for storage, and that the canalicular system allows intercommunication between the outside and the inside of the cell, because connections have been seen to the plasma membrane.

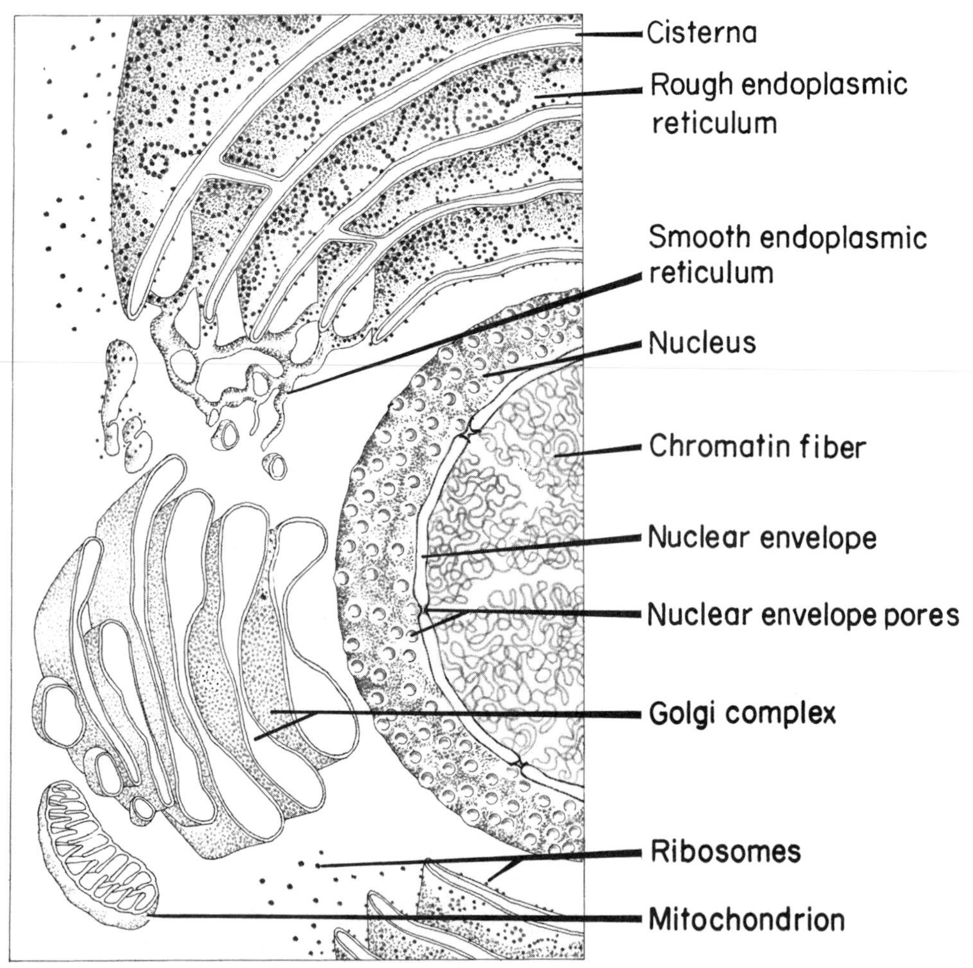

Figure 5.1. *Three-dimensional diagram of the eukaryotic cell's vacuolar system—endoplasmic reticulum and Golgi complex and the nucleus and a mitochondrion. Note that the cisternae of the rough endoplasmic reticulum are interconnected and studded with ribosomes. Some cisternae are extended as tubules of the smooth endoplasmic reticulum. Some ribosomes may be seen free in the cytoplasmic matrix. The unit membranes on the nucleus and mitochondria were left as single lines because of size limitation. (Relabeled from DeRobertis* et al., *1975: Cell Biology. 6th Ed. W. B. Saunders Co., Philadelphia, p. 178.)*

In muscle, the endoplasmic ("sarcoplasmic") reticulum serves to conduct excitation from the cell surface to contractile myofibrils deep inside. In this process calcium ions are secreted. Immediately afterward the calcium is actively reabsorbed (see Chapter 23).

The membranes of the smooth endoplasmic reticulum contain enzymes that are involved in the synthesis of sterols, phosphatides, triglycerides, and other lipids and therefore probably play an active part in lipid metabolism.

Autoradiographic evidence from incorporation of ^{3}H-glucose suggests that enzymes catalyzing glycogen synthesis occur in the membranes of the smooth endoplasmic reticulum (Cardell, 1977). Injection of phenobarbitol into rats leads to proliferation of the smooth endoplasmic reticulum, suggesting synthesis of enzymes active in detoxifying drugs (Allison, 1967; Kupfer, 1970). The endoplasmic reticulum membranes in liver cells contain enzymes that deaminate, hydroxylate, form aromatic ring compounds, and catalyze other chemical reactions. The endoplasmic reticulum therefore has numerous and diverse functions in the cell.

Origin

The origin of the endoplasmic reticulum is not as yet known. Invagination of the plasma membrane of the cell might give rise to a canalicular system, somewhat like the mesosome of a bacterium, but no fully convincing evidence has been produced that such a process occurs on a large scale. The endoplasmic reticulum could also originate from the nuclear

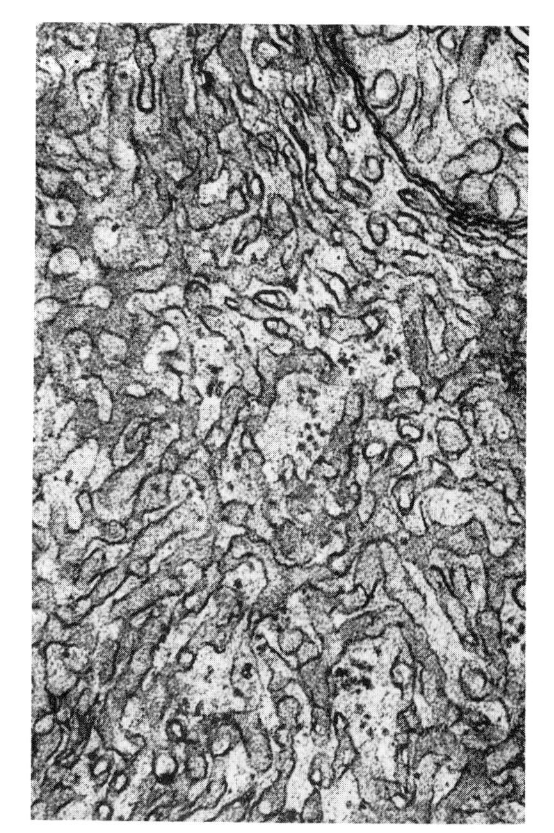

Figure 5.2. *Smooth endoplasmic reticulum of the human adrenal cortex cell (× 50,000). (Courtesy J. Long from Bloom and Fawcett, 1975: A Textbook of Histology. 10th Ed. W. B. Saunders Co., Philadelphia, p. 44.)*

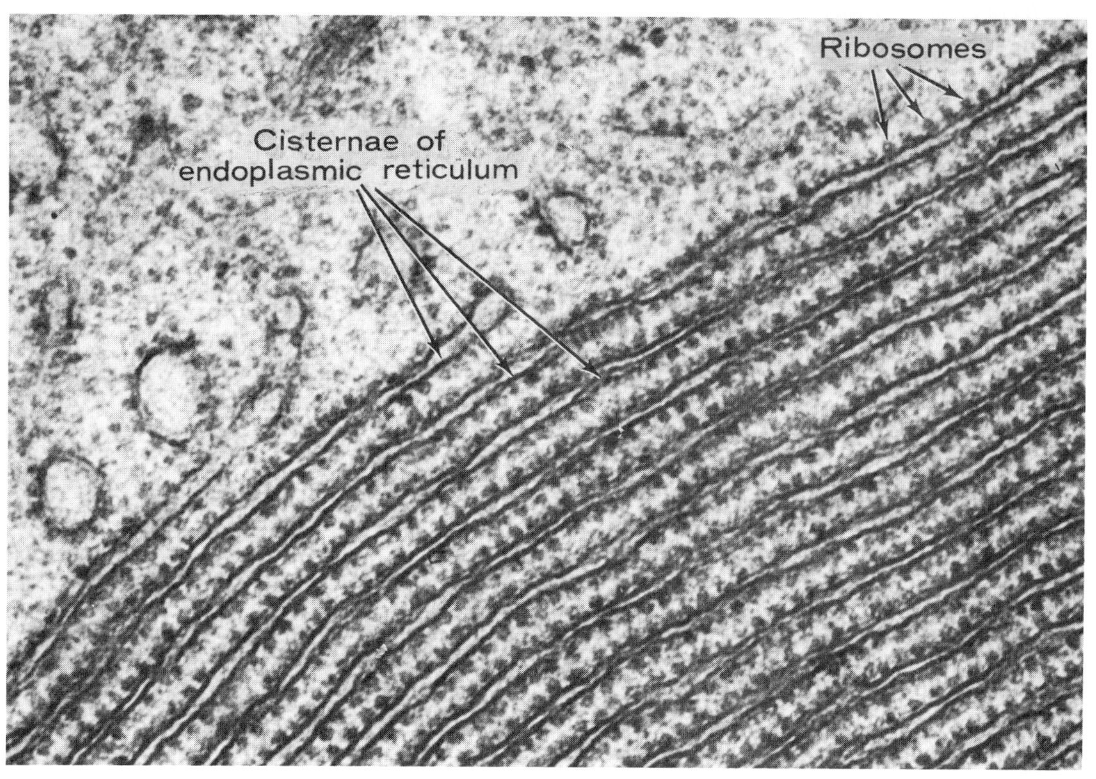

Figure 5.3. *Rough endoplasmic reticulum from a glandular cell. Many cisternae of the endoplasmic reticulum are arranged in parallel. On the outer surface of their limiting membranes are great numbers of ribosomes about 15 nm in diameter. (From Bloom and Fawcett, 1975: A Textbook of Histology. 10th Ed. W. B. Saunders Co., Philadelphia, p. 41.)*

envelope, which it resembles physically and chemically; however, because at the end of mitosis the nuclear envelope re-forms from the endoplasmic reticulum, this is questionable. The endoplasmic reticulum grows by expansion of preexisting membranes, there being no *de novo* origin. Tracer-labeled components are incorporated into preexisting membranes (Dallner *et al.*, 1966; DeRobertis *et al.*, 1975).

Chemical Composition

The endoplasmic reticulum membranes, which are about 5 to 6 nm in thickness, consist largely of lipid (30 to 50 percent) and protein (40 to 60 percent), with some RNA. They show the typical sandwich structure seen in electron micrographs of "unit" membranes (see Chapter 8). The lipids are mostly phospholipids (50 to 66 percent) of which lecithin (Fig. 1.13*C*) predominates, but considerable cephalin may be present. Cholesterol (Fig. 3.11*A*) and cholesterol esters are also found. More lipid in relation to protein is present in smooth than in rough membranes. The protein is partly structural and partly enzymatic, the endoplasmic reticulum being a region where active synthesis of many compounds occurs (Depicrre and Dallner, 1975). It would seem that the RNA found in the endoplasmic reticulum might be an experimental contaminant, but after the ribosomes were removed by fractionation, RNA still constituted about 10 percent of the total dry weight of the smooth membranes, an amount seemingly beyond contaminant quantity (Siekevitz, 1963). The large number of protein types present in the membranes of the rough endoplasmic reticulum in pancreatic cells is indicated by 30 polypeptide bands obtained when the proteins are fractionated on acrylamide gel. The molecular weight of the proteins varies from 5,000 to 150,000.

Membrane Turnover and Function

The turnover of the endoplasmic reticulum membranes can be judged by the half-lives of some of the enzymes; some are as short as 40 to 60 hours, others 16 days. This indicates that not all portions of the membrane are replaced at equal rates, the process occurring piecemeal (Dallner *et al.*, 1966).

Some evidence exists for flow of molecules in the tubules and cisternae (cavities of the sacs of the endoplasmic reticulum). Bulk transport of particles or very large molecules may also occur by pseudopod-like enclosure of the particle by the membrane, a process described in Chapter 19.

Like all the membranes in the cell, those surrounding the endoplasmic reticulum are selective in permitting passage of molecules, some of which may be actively absorbed at the expense of metabolic energy. Such transport is considered in Chapters 17 and 18.

RIBOSOMES

In 1953 George Palade focused attention on electron-opaque basophilic particles of macromolecular dimensions (10 to 25 nm) usually attached to the outer surfaces of the endoplasmic reticulum and its vesicles (Figs. 5.1 and 5.3). In actively growing cells such particles may also be free within the ground substance (Fig. 5.1). Such particles are especially numerous in synthetically active cells and fewer in less active and starved cells. As expected, they are numerous in cancer cells. The particles, composed of ribonucleic acid and proteins, were named *ribosomes*.

Chemical Composition

When a suspension of cells or a tissue is centrifuged following mechanical disintegration in a sucrose medium or in a non-polar solvent and followed by removal of the largest bodies, the particles form sediment at different levels according to their size and specific gravity (Fig. 5.4). The submicroscopic particles that settle out after prolonged centrifugation at high speeds vary in dimensions from 10 to 150 nm. The pellet obtained by such centrifugation contains 50 to 60 percent of the ribonucleic acid of the cell and constitutes 15 to 20 percent of the cell mass. Because of their small size, the particles making up the pellet have been called *microsomes* (Fig. 5.5). Electron microscopic studies of ultrathin sections of the centrifugal microsome pellet from rat liver shows them to be fragments of the endoplasmic reticulum, appearing as isolated vesicles, tubules, and cisternae with attached ribosomal granules. A similar preparation of homogenized pancreas cells shows round vesicles with numerous attached ribosomal granules. The range in size of microsomes obtained by centrifugation is probably the result of different degrees of fragmentation of the endoplasmic reticulum during homogenization.

That ribosomes are engaged in protein synthesis was demonstrated by tracer studies. For example, ^{14}C-leucine is incorporated into protein by the microsomes with attached ribosomes. If the microsomes are subjected to the action of ribonuclease, many of the ribosomes disappear and synthesis of protein is reduced or

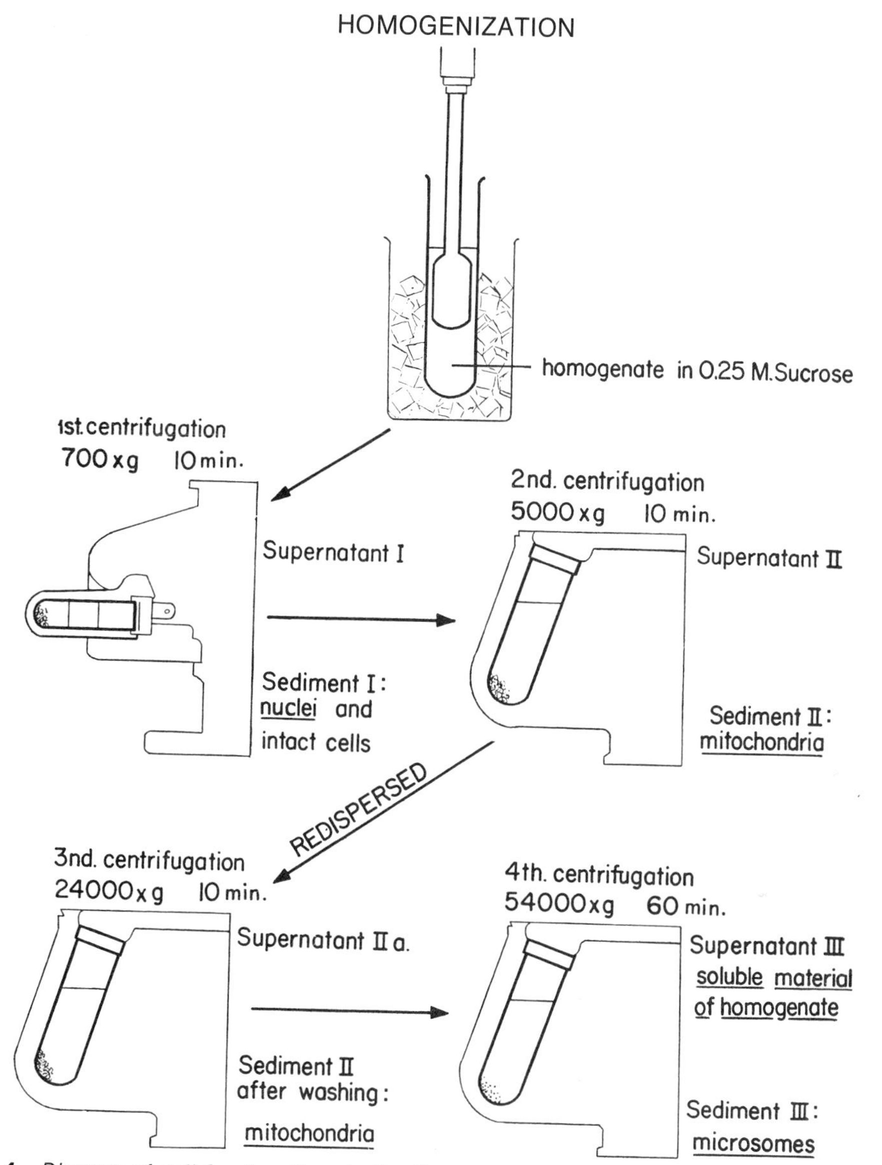

Figure 5.4. *Diagram of cell fractionation. In the first step, an ordinary table centrifuge is used. In the succeeding steps, a preparative centrifuge is used; 10′ means ten minutes (see Appendix 5.2). (From DeRobertis* et al., *1975: Cell Biology. 6th Ed. W. B. Saunders Co., Philadelphia, p. 110.)*

ceases, depending on how long the ribonuclease has acted.

The ribosomes may not all be alike. Several kinds of RNA have been identified from ribosomes by analysis of their purine and pyrimidine bases (Nomura, 1969, 1976).

Synthesis of proteins occurs on ribosomes whether they are attached to the outer surface of the endoplasmic reticulum or scattered in the cell matrix. Synthesized proteins are transported through the endoplasmic reticulum membranes and assembled into globules within the cisternae and canals in cells that produce "proteins for transport" (Palade, 1956). Protein may later appear in the form of granules (250 to 350 nm in diameter) just outside the Golgi complex, as in the case of zymogen granules synthesized by pancreatic cells.

Prokaryotic and Eukaryotic Ribosomes Compared

Ribosomes are found in every type of prokaryotic and eukaryotic cell, even the smallest pleuropneumonia-like organisms

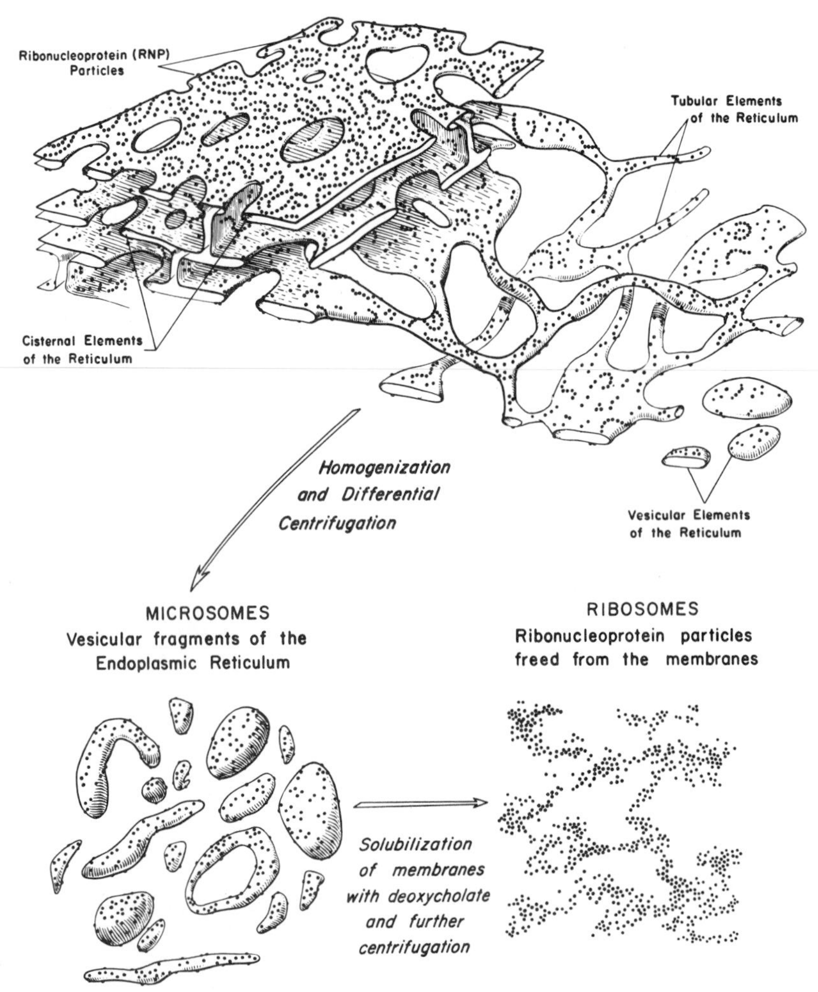

Figure 5.5. *Schematic representation of the relations between the rough endoplasmic reticulum and the microsomes and ribosome fractions isolated by differential centrifugation. The ribosomes are always on the cytoplasmic surface of the membrane defining the tubules and cisternae. (From Bloom and Fawcett, 1975: A Textbook of Histology. 10th Ed. W. B. Saunders Co., Philadelphia, p. 43.)*

(PPLO) (mycoplasmas) examined. Ribosomes consist almost entirely of protein and RNA, the RNA constituting 40 to 60 percent of the dry weight.

In electron micrographs of tissue sections, ribosomes, such as in Figure 5.3, appear as electron-opaque bodies about 15 nm in diameter. Isolated ribosomes have been studied in the electron microscope by shadow casting and negative staining. In *E. coli* ribosomes the 30S particle (Fig. 5.6) is asymmetrical and is represented by a prolate ellipsoid 14 by 17 nm. With negative staining the 50S particle is dome-shaped, and the 30S particle is flatter and fits into the 50S subunit (Watson, 1976).

The RNA of the ribosome is associated with structural proteins of average molecular weight 25,000. The end groups appear to be the same for proteins attached to the 30S and the 50S subunits (and the 40S and 60S units in the ribosomes of eukaryotic cells) in cells studied (Fig. 5.6). It will be recalled from Chapter 1 that ribosomal RNA forms an imperfect helix, and it has been shown that the amino acid residues of the protein attach to the unpaired bases of the RNA (Garrett and Wittmann, 1973).

The ribosomes of eukaryotic cells are more spherical than those of bacteria and more highly hydrated. Hydrated particles are much larger than dry ones visualized in electron micrographs. Much less is known about eukaryotic than about prokaryotic ribosomes.

The apparent identity in construction of ribosomes in all prokaryotes and a similar identity in the larger ribosomes in all eukaryotes emphasizes the profound difference between the two types of cells. Yet despite these differences, protein synthesis occurs in essentially the same manner in both. The fact that eukaryotic genes can be translated by bacteria into protein

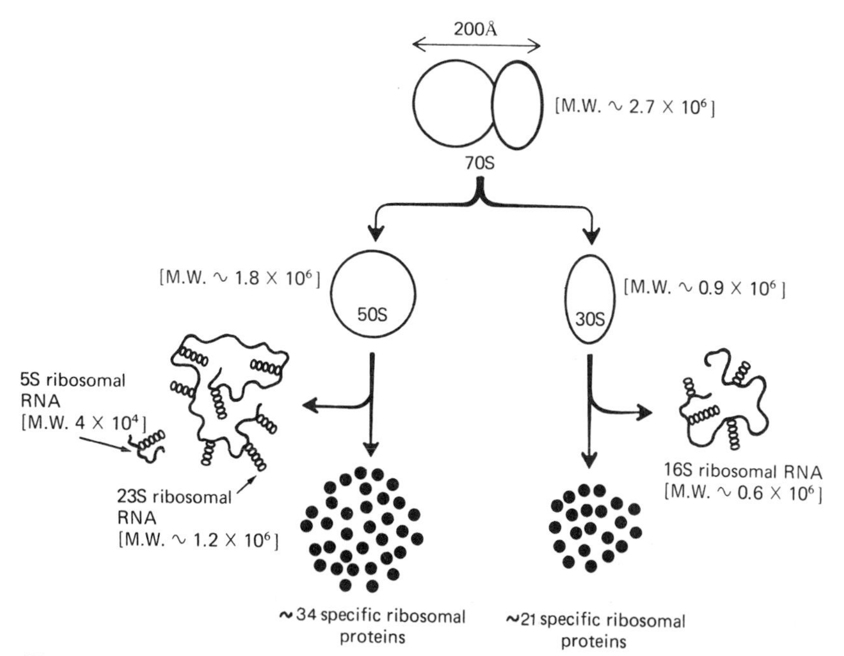

Figure 5.6. *The structure of the* E. coli *ribosome. It is usually called the 70S ribosome since 70S (S-Svedberg, the sedimentation constant) is a measure of how fast this ribosome sediments in the centrifuge. Likewise, the designations 30S and 50S are the sedimentation constants of the smaller and larger ribosomal subunits; 16S and 23S are the sedimentation constants of the smaller and larger ribosomal RNA molecules. All bacterial ribosomes have sizes similar to those of* E. coli, *possessing 30S and 50S subunits. In eukaryotes (including yeast) ribosomes are somewhat larger (80S), with 40S and 60S subunits. (From Watson, 1976: Molecular Biology of the Gene, 3rd Ed. W. A. Benjamin, Menlo Park, Calif., p. 317.)*

reemphasizes this similarity (see Chapters 7 and 15). Seemingly, once the genetic code had become established, it was kept unmodified during evolution and directs protein synthesis in the same manner in both types of cells. The small size of chloroplast and mitochondrial ribosomes of eukaryotes is intriguingly suggestive of the origin of these organelles from symbiotic prokaryotes (see Chapter 3).

Polysome Size

It is known that the number of ribosomes associated with messenger RNA to form a *polysome* in both prokaryotic and eukaryotic cells varies with the size of protein molecule being synthesized (Fig. 5.7). When only one protein (hemoglobin) is produced, as in mammalian reticulocytes (immature red blood cells that have lost the nucleus), only four to six (average five) ribosomes are attached to the messenger RNA, each presumably forming the same protein. When many proteins are being synthesized in a cell, for example, in a HeLa tumor cell, the number of ribosomes per messenger RNA varies from a few to 30 or even more (Rich, 1963; Novikoff and Holtzman, 1976).

Protein synthesis occurs mainly on the surface of the endoplasmic reticulum where the ribosomes are attached, but the protein gets into the lumen of the canals and cisternae. Palade (1956) described protein granules 250 to 350 nm in diameter in canals of the endoplasmic reticulum. The protein moves through these canals to the Golgi complex, as, for example, in pancreatic cells that synthesize enzymic proteins. The extruded vesicles containing the enzymes coalesce with the plasma membrane at specific loci, resulting in extrusion of their contents into the pancreatic tubules (Satir, 1975).

THE GOLGI COMPLEX

The *Golgi complex,* like the endoplasmic reticulum, of which it is probably an extension, is a canalicular system with stacked sacs, but it is always smooth and devoid of ribosomes. The fact that it may be selectively stained by osmium tetroxide and silver salt indicates its chemical uniqueness. It was by these staining reactions (Fig. 5.8) that the Italian physician Camillo Golgi in 1898 first recognized the complex in nerve cells of the barn owl and the cat, long before any evidence was available for an endoplasmic reticulum.

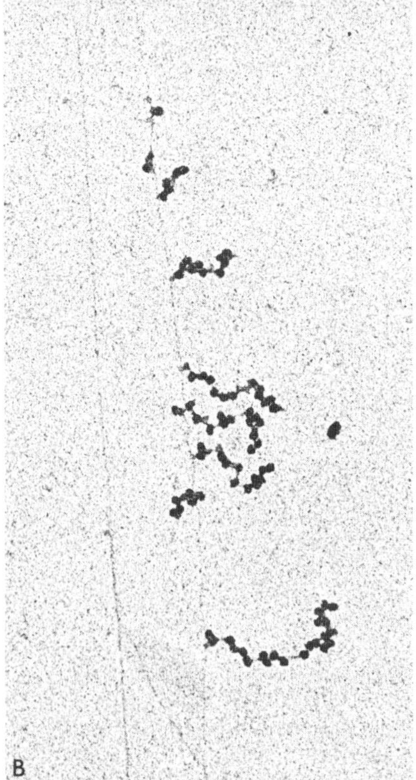

Figure 5.7. A, *Electron micrograph of polysomes in a preparation from rabbit reticulocytes synthesizing hemoglobin. Most of the ribosomes appear in clusters that change from compact to relatively open. (From Warner et al., 1962: Science 183, 1400. Copyright 1962 by the American Association for the Advancement of Science.)* B, *Polysomes on an extruded chromosome of* E. coli *(the vertical thin thread-like fibers). The thin fibers issuing from the chromosome are mRNA and the black bodies attached to the mRNA are the ribosomes. Translation into protein is probably also occurring but is not visible. The figure therefore shows a bacterial gene in action. (By courtesy of Dr. Barbara Hamkalo, University of California, Irvine. Published in Science* 169: *392–395, 1970).*

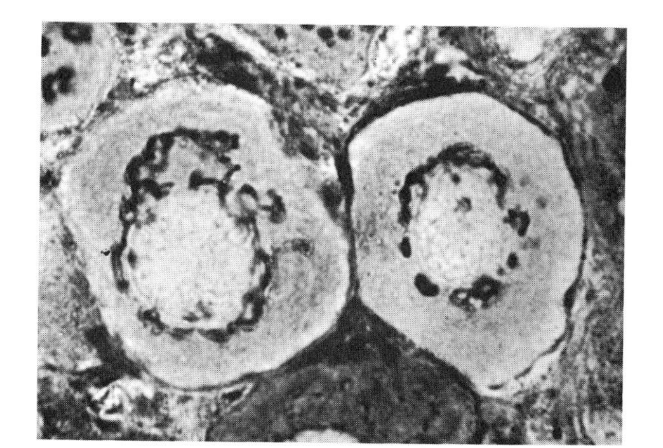

Figure 5.8. *The Golgi complex (black structures) in cervical spinal ganglion cells of a female guinea pig. Kolatchev-Nassanov fixation (× 2000). (By courtesy of Professor Hadley Kirkman.)*

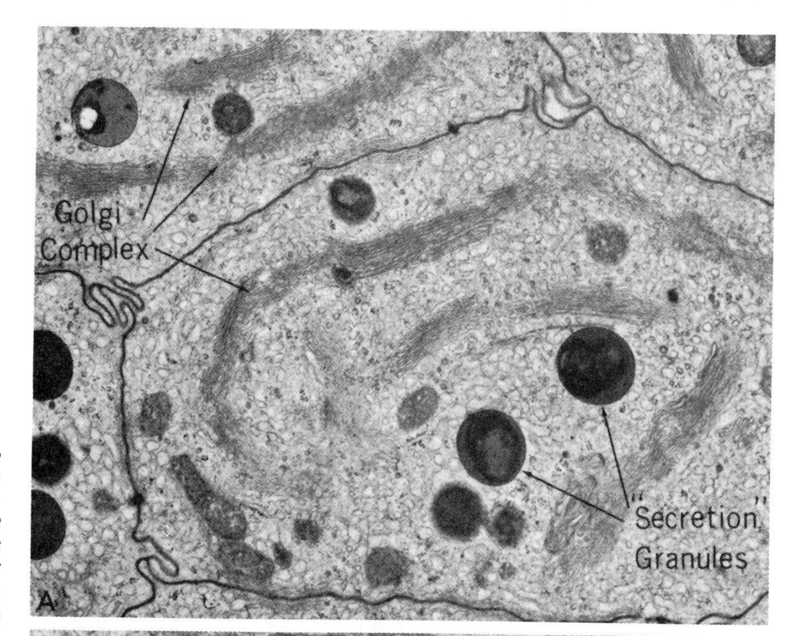

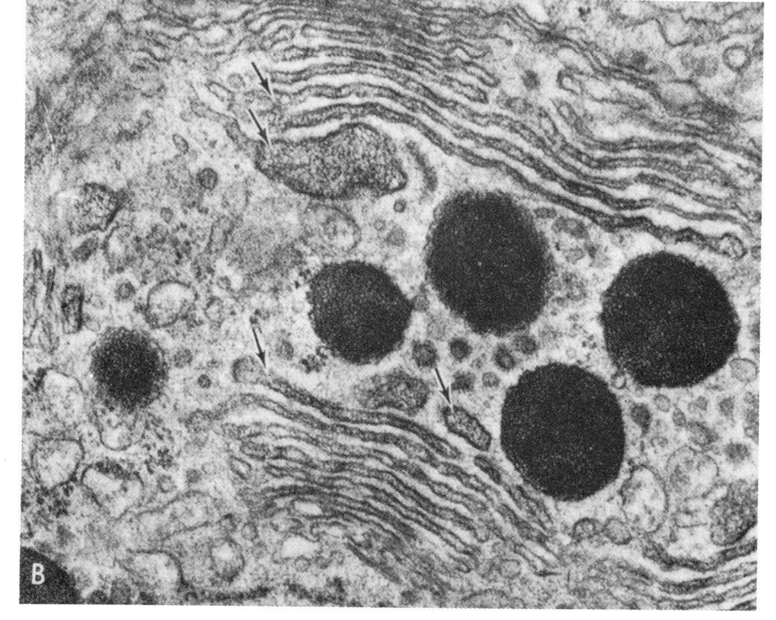

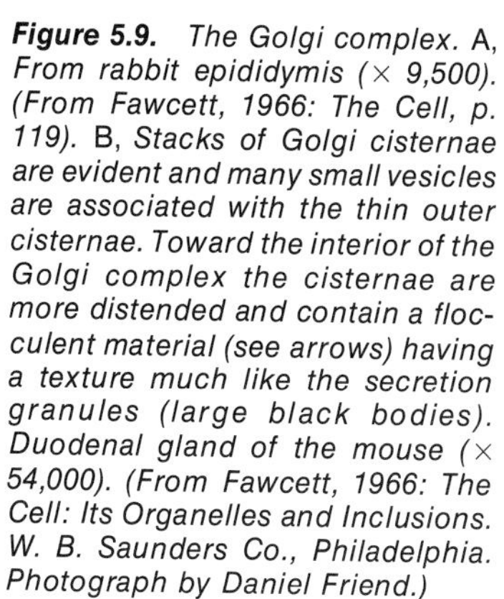

Figure 5.9. *The Golgi complex.* A, *From rabbit epididymis (× 9,500). (From Fawcett, 1966: The Cell, p. 119).* B, *Stacks of Golgi cisternae are evident and many small vesicles are associated with the thin outer cisternae. Toward the interior of the Golgi complex the cisternae are more distended and contain a flocculent material (see arrows) having a texture much like the secretion granules (large black bodies). Duodenal gland of the mouse (× 54,000). (From Fawcett, 1966: The Cell: Its Organelles and Inclusions. W. B. Saunders Co., Philadelphia. Photograph by Daniel Friend.)*

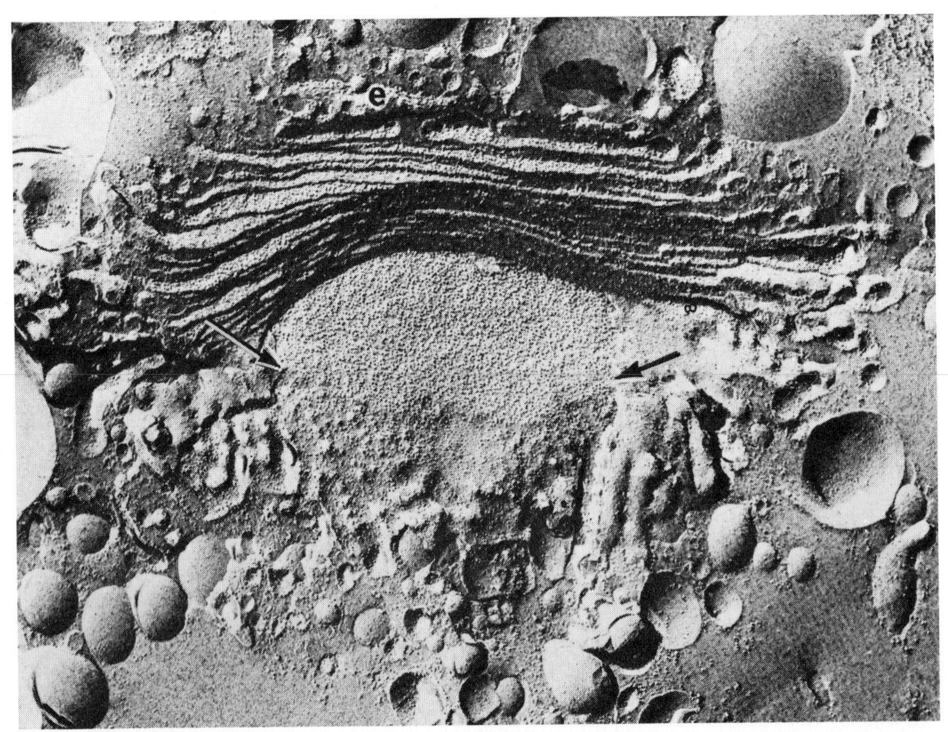

Figure 5.10. *Golgi complex (dictyosome) of the alga* Micrasterias denticulata *as revealed in a freeze etching. The fracture plane has passed almost vertically through most of the cisternae except the last one. The association of the endoplasmic reticulum (e) with the proximal or forming face of the Golgi complex is clear. At the other face numerous granules can be seen on the surface of the membrane and these particles decrease sharply in density as vesicle formation begins (between arrows) (× 60,000). (From Staehelin and Kiermayer, 1970: J. Cell Sci. 7: 787).*

Under favorable conditions the Golgi complex may be stained with methylene blue and observed for a limited time in live cells. It is also visible without staining under the phase microscope. The presence of phosphatases, especially thiamine pyrophosphatase detected by cytochemical tests at the electron microscope level, verifies the reality of the organelle and shows it to correspond to the structure formerly detected only by staining methods (Novikoff and Holtzman, 1976). Electron micrographs of thin sections of cells indicate that the canals of the Golgi complex are of various shapes, usually consisting of stacks of flattened sacs (cisternae) associated with small vesicles and vacuoles of various sizes (Figs. 5.9 and 5.10). The surface of the canal and sacs is always smooth, like the smooth endoplasmic reticulum, in contrast to the ribosome-encrusted outer surface of the rough endoplasmic reticulum (Whaley, 1975).

The Golgi complex is present in almost all eukaryotic cells, but is perhaps more prominent in vertebrate than in invertebrate and plant cells. In plant cells it has been collectively called the *dictyosome* and may consist of many flattened sacs (Fig. 5.10). It is absent from red blood cells. The Golgi complex is organized in basically different ways in different types of cells, but its organization for any one kind of cell is usually the same. For example, it is small in muscle fibers, but it is large and well developed in secretory cells and neurons, where it is clearly a reticulate structure. It occupies different positions in different kinds of cells, being polar, between nucleus and periphery, in cells of ectodermal origin, circumnuclear in neurons, and located elsewhere in other cells (Bloom and Fawcett, 1975). When many units are present in plant and invertebrate cells, they are likely to be scattered throughout the cytoplasm.

Maintaining a healthy Golgi complex depends upon the presence of a cell nucleus, in the absence of which the complex decreases in size and may disappear. Yet within half an hour after renucleation of an amoeba, small curved cisternae of the Golgi complex reappear. Starvation reversibly attenuates the Golgi complex, which redevelops soon after the cells are resupplied with nutrient. This occurs only after the rough endoplasmic reticulum has redeveloped; therefore, proteins needed for reconstituting the Golgi complex must come from the endoplasmic reticulum (Northcote, 1971).

The surface of the Golgi complex is covered

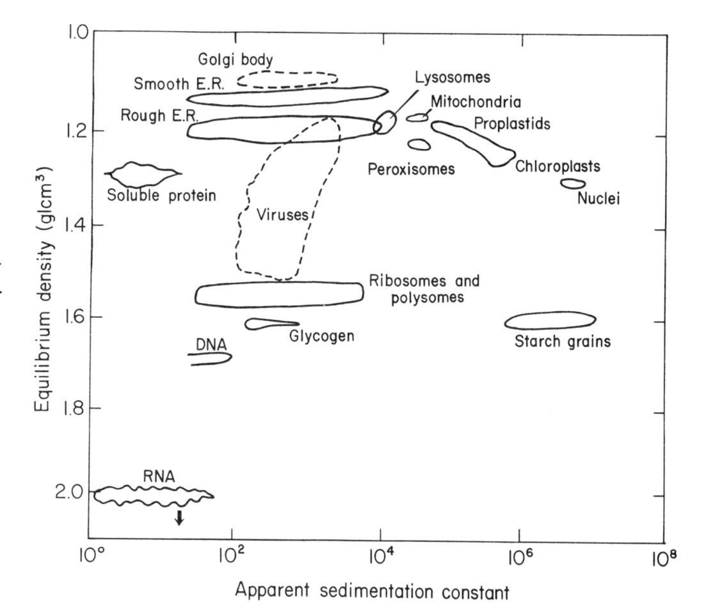

Figure 5.11. Sedimentation of subcellular particles. (After Umbreit et al., *1972: Manometric and Biochemical Methods, 4th Ed. Burgess Publishing Co., Minneapolis.)*

by a unit membrane. The two faces of the membrane have different staining properties; only the outer part reacts with silver salts and osmic acid, indicating a difference in chemical composition of the inner and outer surfaces. The Golgi sacs nearest the endoplasmic reticulum resemble it in structure and chemistry, while those farthest away resemble the plasma membrane, indicating a polarity for the organelle. In the export of secretions, small vesicles from the Golgi complex coalesce with the plasma membrane, extruding the secretion onto the surface of the cell. The Golgi membrane then becomes part of the plasma membrane (Northcote, 1971).

The inside of the Golgi complex appears to be the consistency of a fluid, since needles inserted into the canals can be moved with no indication of resistance. However, this might be the result of experimental injury. In centrifugation experiments the entire Golgi complex moves as a unit and is displaced toward the centripetal pole, indicating a specific gravity less than that of the surrounding cytoplasm. The Golgi membranes are found in the interface between fluids of specific gravity 1.09 and 1.13 (Fig. 5.11).

Chemical Nature

Membranes of the Golgi complex, isolated by fragmenting cells and fractionally centrifuging them in a sucrose density gradient, have

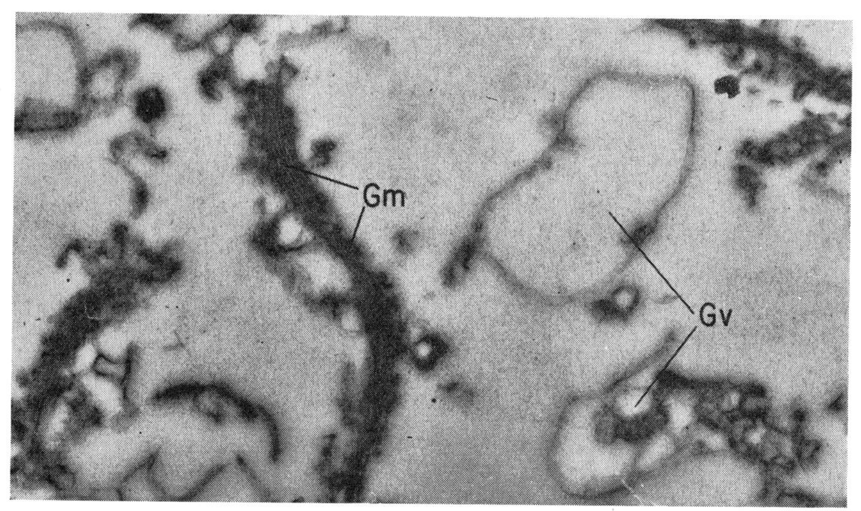

Figure 5.12. *Fragments of the Golgi complex after homogenization of cells of the epididymis. Gm, Golgi membranes; Gv, Golgi vacuoles. (By courtesy of A. J. Dalton.)*

been processed by scientists for chemical studies (Fig. 5.12). The membranes contain more lipid than the endoplasmic reticulum. The kinds of lipids are considered elsewhere (for example, De Robertis *et al.,* 1975). A number of enzymes but only slight amounts of nucleic acid are present, at least when the membranes are washed free of contaminating ribosomes. Acid phosphatase, an enzyme characteristic of the Golgi complex, is found in several times the amount present in general cell homogenates.

Function

Membranes of the Golgi complex absorb tracer-labeled monosaccharides and sulfur-containing precursors of sulfated polysaccharides, as shown by appearance in autoradiographs of tracer labels over the Golgi complex before they appear elsewhere. The membranes appear to be the locus of synthesis of mucopolysaccharides and glycoproteins. The proteins synthesized in the endoplasmic reticulum pass to the Golgi complex where the locally synthesized polysaccharides are added to form glycoproteins.

Mucus (consisting of mucins, or mucoproteins) is formed in the Golgi complex. Some mucus is later incorporated into the outer surface of the cell membrane, some exported. The Golgi membranes proliferate along with those of the rough endoplasmic reticulum in early stages of mucus formation in cells of many species (Satir, 1975). Dictyosomes (Golgi complexes) of plant cells lay the polysaccharide cell wall matrix in a dividing cell (Northcote, 1971).

The Golgi complex also packages secretions—for example, enzymes of the pancreatic cells synthesized in the rough endoplasmic reticulum, and in many cells the lipid globules synthesized in the smooth endoplasmic reticulum. It also gives rise to the acrosome in sperms, the nematocysts in *Hydra* (and probably other coelenterates), the trichocysts in ciliates, for example, *Paramecium,* and comparable mucus bodies in *Tetrahymena* (Satir, 1975). Primary lysosomes may bud off from the Golgi membranes, although they may also originate from the endoplasmic reticulum.

LYSOSOMES

Lysosomes, membrane-bound organelles containing hydrolytic enzymes, are usually of smaller dimensions than mitochondria; in a liver cell they are about 0.4 μm in diameter (Fig. 5.13). The cell interior is protected from autolysis by lysosomal enzymes by the membrane packaging (Pitt, 1975).

Lysosomes have centrifugal properties between those of mitochondria and ribosomes and

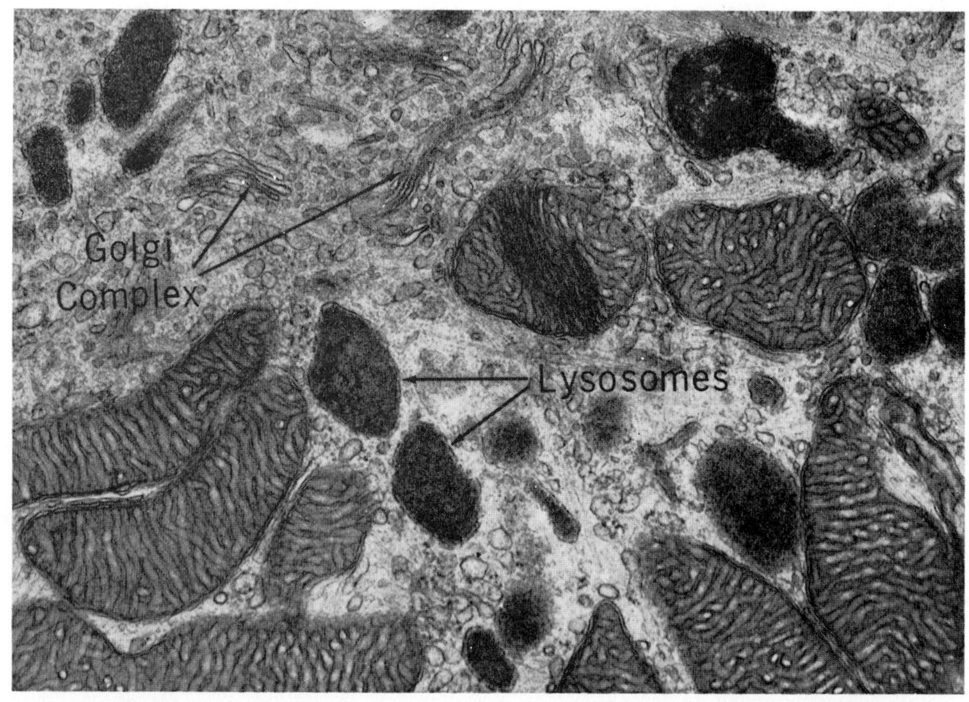

Figure 5.13. Lysosomes in cells of the hamster adrenal cortex (× 25,000). Note also the Golgi complex and numerous mitochondria. (From Fawcett, 1966: The Cell: Its Organelles and Inclusions. W. B. Saunders Co., Philadelphia.)

have been studied in a restricted number of cell types (Fig. 5.11). They have, however, been demonstrated in protozoans, insects, amphibians, and mammals, and lysosome-like structures are also found in plants (Gunning and Steer, 1975; Matile, 1975). Lysosomes range in size from 0.25 to 0.8 μm and under the electron microscope they appear dense and finely granular, individual granules having a diameter of about 5.5 to 8.0 nm (Fig. 5.13). They are distinguished from other cytoplasmic particles by their hydrolytic enzymes and acid phosphatase. About 36 hydrolytic enzymes have been identified in lysosomes, including those that digest proteins (5), nucleic acids (4), polysaccharides (15), lipids (6), organic-linked sulfate (2), and organic-linked phosphate (4), although all do not occur in any one lysosome. There are many kinds of lysosomes containing different spectra of enzymes (Novikoff and Holtzman, 1976).

Lysosomes are covered by a single unit membrane. The enzymes may be released from lysosomes by lysosome labilizers: action of a blender, freezing and thawing, ultraviolet radiations, vitamin K, and detergents, all of which disrupt the membrane. Cortisone (a hormone), chloroquine (an antimalarial), and cholesterol, all of which strengthen the membrane, are called lysosome stabilizers (DeDuve, 1969).

Although lysosomal-type enzymes have been found in the supernatant of disrupted cells it is possible that they have escaped from lysosomes ruptured during fragmentation. Lysosomes of injured or dying cells appear to rupture spontaneously, thereby lysing decrepit cells. They also enable a developing organism to degrade outgrown structures, for example, cells in the tail of a metamorphosing tadpole.

Lysosomes facilitate turnover of organelles in normal cells. Fragments of cell organelles (e.g., mitochondria) in vacuoles are lysed after union of vacuoles with lysosomes. Undigestible residues remaining after digestion of particulate matter are extruded when the lysosome membrane combines with the plasma membrane (DeDuve, 1969) (Fig. 5.14).

Lysosome malfunction may lead to disease, for example, when glycogen taken up by lysosomes is not digested (Pompe's disease). Conversely, rupture of lysosomes in skin cells exposed to direct sunlight leads to the pathologic changes following sunburn. The enzymes liberated kill cells in the epidermis, leading to blistering and later to "peeling" of a layer of epidermis.

Lysosomes may originate either from the endoplasmic reticulum or the Golgi complex, or both. In studies with radioactive amino acids, autoradiographic labeling in the cell occurs over the lysosomes only after labeling in both endoplasmic reticulum and Golgi complex. Protein granules are often seen in enlargements of the endoplasmic reticulum, which, if pinched off, would produce bodies of the size and character of lysosomes (Pitt, 1975). The endoplasmic reticulum is well known for synthesis of protein for export. Lysosomes may represent another agent of export.

MICROBODIES

By differential centrifugation *microbodies* are separated from lysosomes in fragmented liver cells (DeDuve, 1969). Being smaller, microbodies have higher S values than lysosomes (see Fig. 5.11). And except for their smaller size, microbodies resemble lysosomes in electron micrographs (Fig. 5.15*A, B*). They are ovoid and bound in a single membrane; they measure about 0.3 to 1.5 μm in diameter and have now been well characterized. Microbodies are widespread and have been found in a variety of animal cells including protozoans, yeast, and many plant cells. They are probably characteristic of eukaryotic cells (Gunning and Steer, 1975), and in some cells they are more numer-

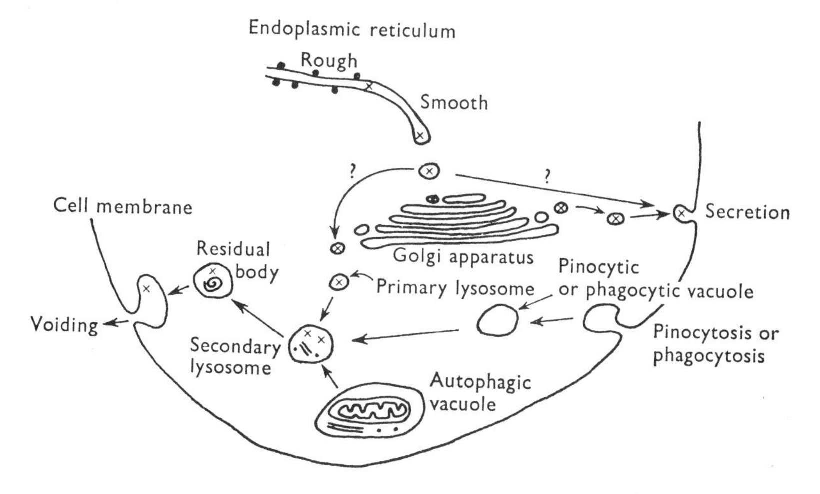

Figure 5.14. *Relationship between various membranes on animal cells. Hydrolytic enzymes (denoted by crosses) are given as an example of a product of protein synthesis at the rough endoplasmic reticulum. Secondary lysosomes are formed on fusion of primary lysosomes with autophagic and/or pinocytic or phagocytic vacuoles. (From Hughes, Lloyd, and Brightwell, 1970: Soc. Exp. Biol. Symp. 20: 295–322.)*

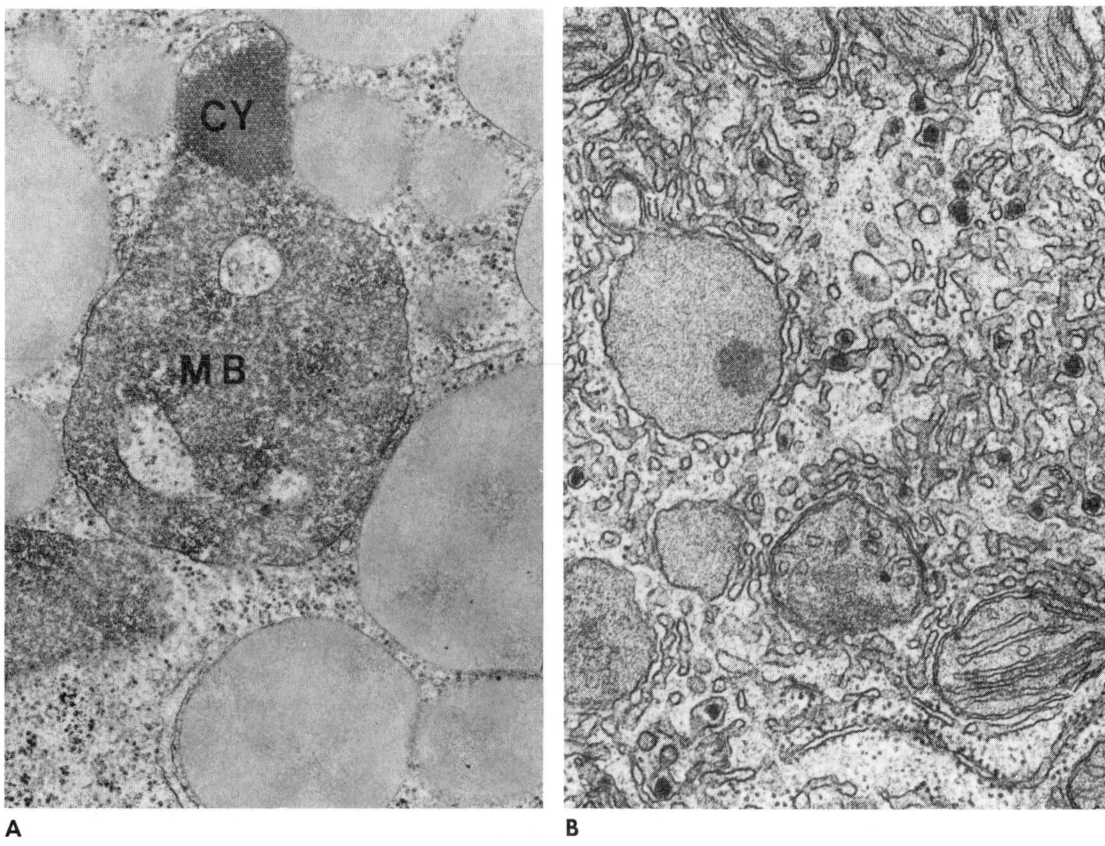

Figure 5.15. A, *Microbody (MB, glyoxysome) in sunflower seed cotyledon cell lying against lipid globules. CY is a crystal, probably of catalase (× 31,000). (From Gruber and Newcomb, 1970: Planta 93: 269.)* B, *Microbodies (larger structures with grey interior) in liver cell. These are peroxisomes scattered in smooth endoplasmic reticulum. The spherical dense bodies are lipoprotein bodies newly synthesized by the endoplasmic reticulum. (Courtesy of R. Bolender, from Bloom and Fawcett, 1975: Textbook of Histology. 10th Ed. W. B. Saunders Co., Philadelphia, p. 707.)*

ous than lysosomes. Microbodies are of at least two types: glyoxysomes and peroxisomes.

Glyoxysomes are found only in plant cells (Fig. 5.15*A*). They contain enzymes making possible the conversion of lipid into carbohydrate in the course of which hydrogen peroxide is produced. They also contain catalase, which decomposes hydrogen peroxide into water and oxygen (Lehninger, 1975). The chemical details of the glyoxylate cycle are considered in Chapter 12. It is of greater interest here to point out that hydrogen peroxide is a cell poison and that in the course of evolution confinement of the enzymes that produce it and that decompose it within membranes has become an effective mechanism for preventing damage to cell structures.

Peroxisomes are found in both animal and plant cells. In animal cells both L and D amino acid oxidases and urate oxidase are confined to peroxisomes and the substrate oxidation results in the formation of hydrogen peroxide. The catalase present decomposes the hydrogen peroxide into water and oxygen, as in glyoxysomes (Lehninger, 1975). In plant cells photorespiration occurs during photosynthesis; in the process some of the immediate photosynthetic products are used, resulting in the formation of hydrogen peroxide. The peroxisomes make possible removal of the peroxide as in animal peroxisomes (Gunning and Steer, 1975). The significance of peroxisomes to plant cells in photosynthesis is considered in Chapter 14. As in the case of glyoxysomes, confinement of catalase and the enzymes producing hydrogen peroxide in membrane-bound peroxisomes prevents cell damage.

It is likely that sequestering catalase and enzymes that produce hydrogen peroxide in microbodies was an early evolutionary eukaryotic cell development, judging from the prevalence of microbodies and their similarity in plant and animal cells. Oxygen, on its appearance in the earth's early atmosphere, was a cell poison, partly because of its role in the production of more highly oxidizing substances, including hydrogen peroxide (see Chapter 2). Among the consequences of adaptation of eukaryotic cell life to oxygen was the development of microbodies to confine the reac-

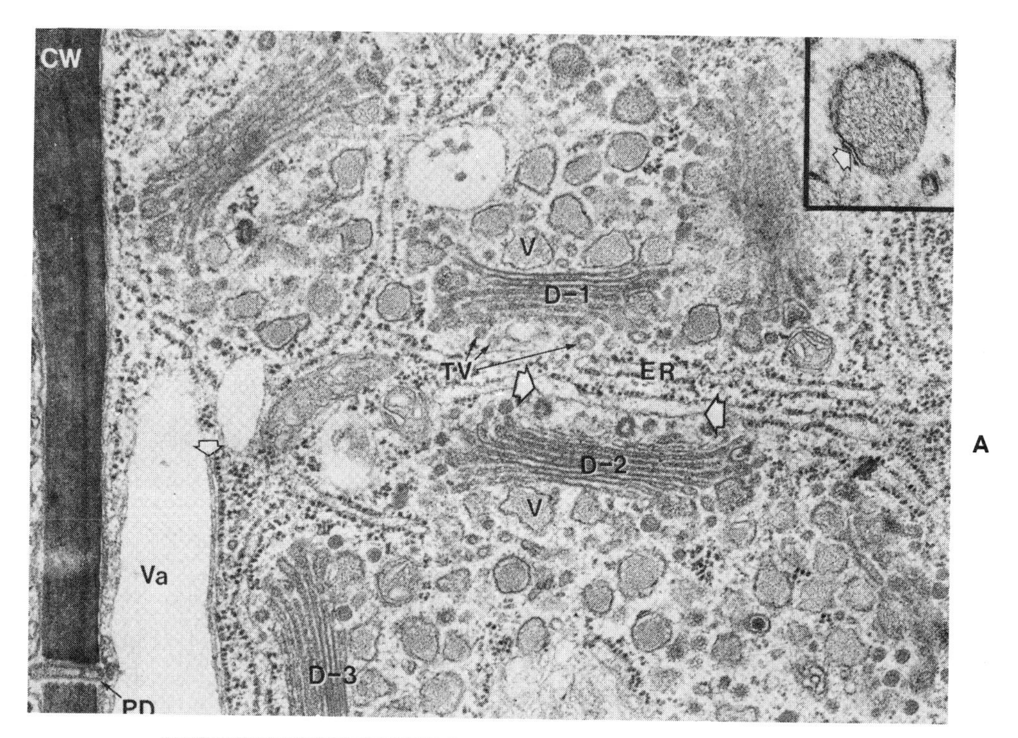

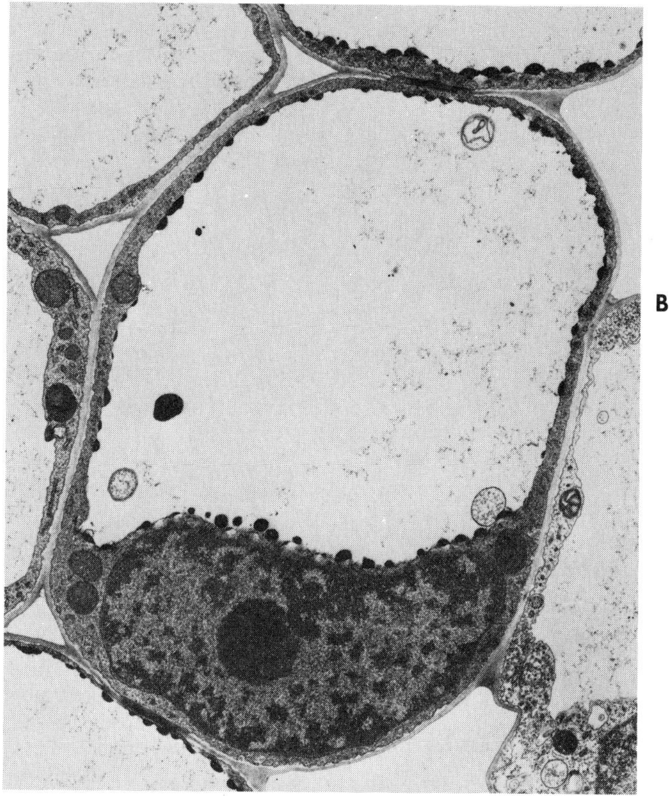

Figure 5.16. A, *Vacuole and other structures in the green alga* Bulbochaete. *VA, vacuole; CW, cell wall; PD, plasdesmata (intercellular protoplasmic connections); dictyosomes (D-1, D-2, D-3); TV, transitional vesicles; V, vesicles with granular and fibrillar contents, insert at upper right shows one enlarged. Note that the vacuole (VA) has a tripartite membrane (arrow), as does the vesicle at upper right. (From Gunning and Steer, 1975: Plant Cell Biology: An Ultrastructural Approach. Edward Arnold, London, plate 27.)* B, *Vacuole in the marsh plant* Limonium. *In a mature plant the vacuole occupies much of the cell volume. The nucleus and mitochondria are found in the thin layer of cytoplasm surrounding the vacuole. The cell is separated from its neighbors by cell walls. (From Ledbetter and Porter, 1970: An Atlas of Plant Structure. Springer-Verlag, New York.)*

tions that produce hydrogen peroxide, as well as the inclusion of an enzyme to decompose it. Catalase has one of the highest turnover numbers known among enzymes; that is, one molecule of catalase can decompose more molecules of the substrate (hydrogen peroxide, in this case) per second than perhaps any other enzyme.

It might be pointed out that many bacteria, like plants, can decompose fats and synthesize carbohydrates from the resulting two-carbon fragments. As in plants, hydrogen peroxide is produced in these reactions and is decomposed by catalase. Perhaps a degree of compartmentation of catalase and these enzymes occurs on the cell membrane of prokaryotes, but possibly the small size of the cell and the proximity of catalase to hydrogen peroxide permits sufficiently rapid decomposition to prevent cell damage.

Little is known of the detailed structure of the membranes of the endoplasmic reticulum, Golgi complex, lysosomes, and peroxisomes. In general they appear to be similar to the plasma membrane (see Chapter 8). Permeability of organellar membranes has been studied mainly with those of the endoplasmic reticulum, especially in muscle cells (see Chapter 23).

VACUOLES

A vacuolar system of very small dimensions is found in the endoplasmic reticulum-Golgi complex of cells and in all cells that perform pinocytosis (see Chapter 19). However, both plant and animal cells may also have much larger *vacuoles* (Fig. 5.16*A*, *B*). For example, in many plant cells the salts and organic molecules in the large central sap vacuoles maintain the turgor of rigidity of the cells by uptake of water. A variety of vacuoles is present in protozoans; some *(pinosomes)* are minute, formed by "cell drinking" (the engulfment of some of the surrounding fluid by invagination of the cell membrane). Protozoan food vacuoles, or *phagosomes,* are large and are formed by invagination of the mouth membrane, which surrounds food particles or organisms and encloses some of the environmental fluid. Protozoan *contractile vacuoles* may be formed in either of two ways: by coalescence of many spherical vacuoles or by filling with fluid from channels in the cytoplasm. Contractile vacoules are of special interest because of their dynamic nature and cyclic appearance and disappearance. Contractile vacuoles are also present in some freshwater unicellular algae.

The vacuole of a mature plant cell generally occupies about 50 percent of the cell volume (in extreme cases, 90 percent). Generally lacking in cells of motile organisms, the central vacuole is prominent in cells of sessile plants. A sessile plant depends upon diffusion (and convection) to supply it with raw materials for photosynthesis by day and oxygen for respiration at night. The central vacuole permits maximal surface exposure of the cytoplasm and its organelles for this purpose and in addition stores ions essential for syntheses and growth. The sessile plant's need for maximal surface exposure for photosynthesis (whence its dendritic form) is thus met by the central vacuole with its high water content and the cellulose skeleton—both "cheap" items thermodynamically (Wiebe, 1978).

MISCELLANEOUS CELL INCLUSIONS

Occasionally oil droplets and oil globules are seen in the cytoplasm of cells in tissue culture, in marine eggs, and in marine protozoans known as radiolarians. These oil droplets are probably stores of nutrient or sometimes devices for reducing the specific gravity of the cell, making flotation easier. These lipids are synthesized in the smooth endoplasmic reticulum. Oil droplets lack membranes. Many cells also contain glycogen granules as nutrient reserves (Fig. 5.17). These also lack membranes. In many cells crystalline inclusions may be seen and in aging cells, pigment deposits.

SUMMARY

Although to some extent the membranes of the endoplasmic reticulum, the Golgi complex, lysosomes, peroxisomes, and the nuclear envelope differ from one another, they all show general similarity to the plasma membrane of the cell. The interconnections from the surface of the cell to the endoplasmic reticulum, and from the latter to the Golgi complex and then to the outer nuclear membrane, point to a common source with ensuing modifications. However, no agreement seems to exist on whether the plasma membrane or the nuclear membrane, if either, is the point of origin of the internal membranes. After cell division the nuclear envelope is formed anew from the membranes of the endoplasmic reticulum. Transitions in size and composition are seen from the endoplasmic reticulum membranes to the Golgi complex and from the latter to the plasma membrane. In any case, whatever their origin, the vacuolar system has membranes that permit intercommunication between all organelles in the system. Protein formed in the rough endoplasmic reticulum and lipids formed in the smooth endoplasmic reticulum membranes pass to the Golgi complex. Thence by way of membrane-enclosed vesicles they move to the exterior by exocytosis.

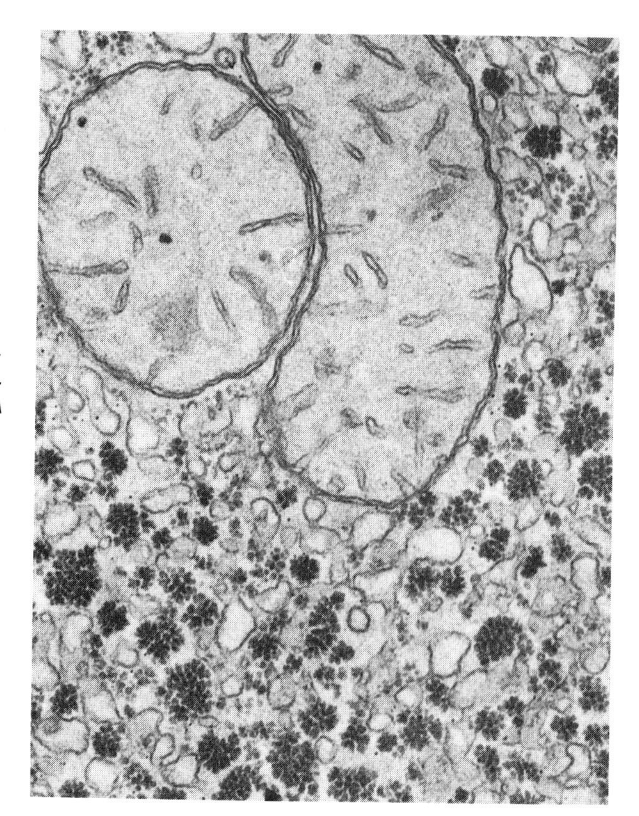

Figure 5.17. Glycogen aggregations in a hamster liver cell associated with the outlines of the endoplasmic reticulum. (From Bloom and Fawcett, 1975: A Textbook of Histology. 10th Ed. W. B. Saunders Co., Philadelphia, p. 710.)

APPENDIX

5.1 SEPARATION OF CELL ORGANELLES

When a centrifuged cell is observed through the centrifuge microscope (Fig. 5.18), the organelles, differing in specific gravity and size, are seen to move as discrete units. For example, in centrifuged eggs of the sea urchin *Arbacia*, the red pigment granules are thrown to the bottom of the cell; above them the yolk is deposited, followed by a layer of mitochondria; finally the oil droplets are collected at the top of the egg, forming an oil cap. Between the zone of the mitochondria and the oil cap is a clear zone of fine granular ground substance within which the nucleus of the egg comes to lie (Fig. 5.19).

If cells are first fragmented and then centrifuged, the fragments of higher specific gravity settle more rapidly than those of lower specific gravity, other factors being equal, and larger particles settle more rapidly than smaller ones of equal specific gravity. On this basis various particulates within the cell can then be segregated from one another. Large structures such as nuclei are readily separated from smaller structures, as are also mitochondria and large granules (Fig. 5.19). After all these larger structures have been removed, minute submicroscopic bodies remain that can

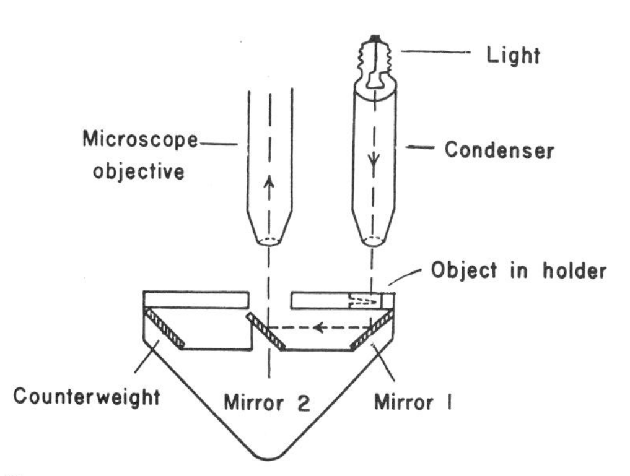

Figure 5.18. Optical system for centrifuge microscope. The condenser throws an image of the incandescent lamp on the object. The object is under the light for a fraction of a second during each revolution of the disk, but the image on mirror 1 reflected to mirror 2 is seen as a continuous moving picture because the eye, over the microscope in the axis of rotation of the centrifuge, fuses the many images coming to it per second. (After Harvey, 1938: J. Appl. Physics 9: 73.)

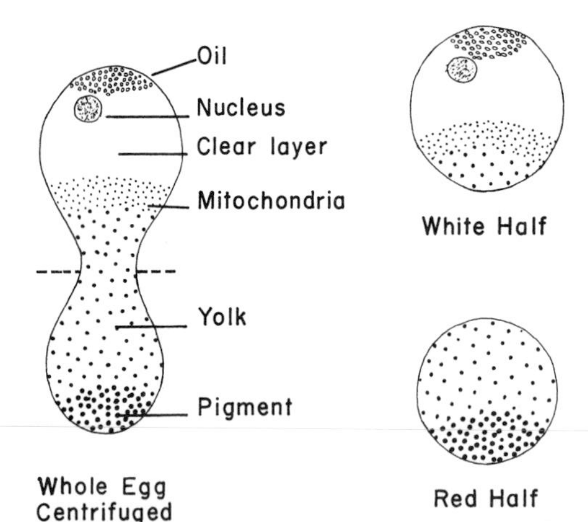

Figure 5.19. *Stratification in a centrifuged egg of the sea urchin* Arbacia punctulata. *(After Harvey, 1941: Biol. Bull.* 81: *114.)*

be centrifuged down at very high speeds. These, collectively called the microsomes, include ribosomes torn from the endoplasmic reticulum and spheroid fragments of the endoplasmic reticulum that had been disrupted in the procedure (see Fig. 5.5).

With special care even fragile bodies like Golgi structures can be isolated for biochemical studies. Through these procedures, coupled with tracer-labeled precursors, much has been discovered about the function of many of the organelles.

5.2 CENTRIFUGATION FOR STUDY OF SMALL PARTICLES AND MACROMOLECULES

Differential centrifugation for separating cell organelles after cell disruption may be accomplished with a conventional centrifuge. Centrifuges for separation of macromolecules and for determination of their buoyant density, molecular weight, and other properties are of two types: preparative and analytical. Both use high centrifugal forces, mostly from 50,000 to 150,000 times gravity (g); in the analytical centrifuge, forces up to 300,000g may be used. In the *preparative centrifuge,* which is a refrigerated conventional type of centrifuge running at high speeds with either bucket or angle rotors for holding the samples (see Fig. 5.4), the sample to be centrifuged is placed in a plastic tube. After centrifugation, samples may be withdrawn by puncturing the bottom of the centrifuge tube and removing banded fractions (aliquots) drop by drop, applying suction to the top of the centrifuge tube to slow release of the fluid. The *analytical centrifuge* (Fig. 5.20) is modified in design to accommodate a cuvette (usually of quartz to transmit far ultraviolet light), through which the separation of constituents into layers can be photographed during centrifugation.

The suspension medium onto which macromolecules are placed is of a density such as to permit layering of molecules along a density gradient. This makes possible their physical separation and determination of their physical constants. To separate larger particles, brief periods of centrifugation (a few minutes to several hours) suffice; for separation of macromolecules 36 hours or longer may be required. Gradients are produced by layering solutions of decreasing density (e.g., sucrose) or they may be formed as a consequence of protracted centrifugation of some solutions, e.g., cesium chloride, which because of its high density (1.790 at 20°C, for a 60 percent solution) becomes stratified. Cesium chloride is required in studies on nucleic acids because of their high densities.

While for steep gradients bucket rotors are desirable, for many procedures angle-type rotors are used (Fig. 5.20). Although the macromolecules are layered at angles in the latter rotors, the layers reorient when, after centrifugation, the tubes are set upright. The larger volume present in an oblique layer increases the separation of the bands when set upright, since the volume of fluid between bands is accommodated into a narrower diameter.

High-speed centrifuges are run under conditions of minimum friction during centrifugation. They are refrigerated and at very high speeds they are evacuated to reduce air friction. In this way, even after long periods of centrifugation the temperature does not rise to alter the substances under study.

In both types of centrifuge one can layer constituents of extracts adequately for separation and purification without achieving equilibrium in the density gradient, thus requiring shorter periods of centrifugation. The banded material can be withdrawn fractionally for further study. In both types of centrifuge the rate at which banding is accomplished may be measured and certain physical properties may be determined from such information. Because the rate of banding can best be determined by photography of substances during centrifugation, the analytical centrifuge is much more useful for such determinations.

The buoyant density may be determined with the preparative centrifuge by measuring the concentration of tracer in the fractions from various levels of the centrifuge tube (see Fig. 5.21). For example, if in place of thymine its analogue, 5-bromouracil, is supplied to *E. coli,* it is incorporated into *E. coli* DNA. Since bromouracil has a higher molecular weight than thymine, the DNA formed from it has a higher density than normal DNA containing

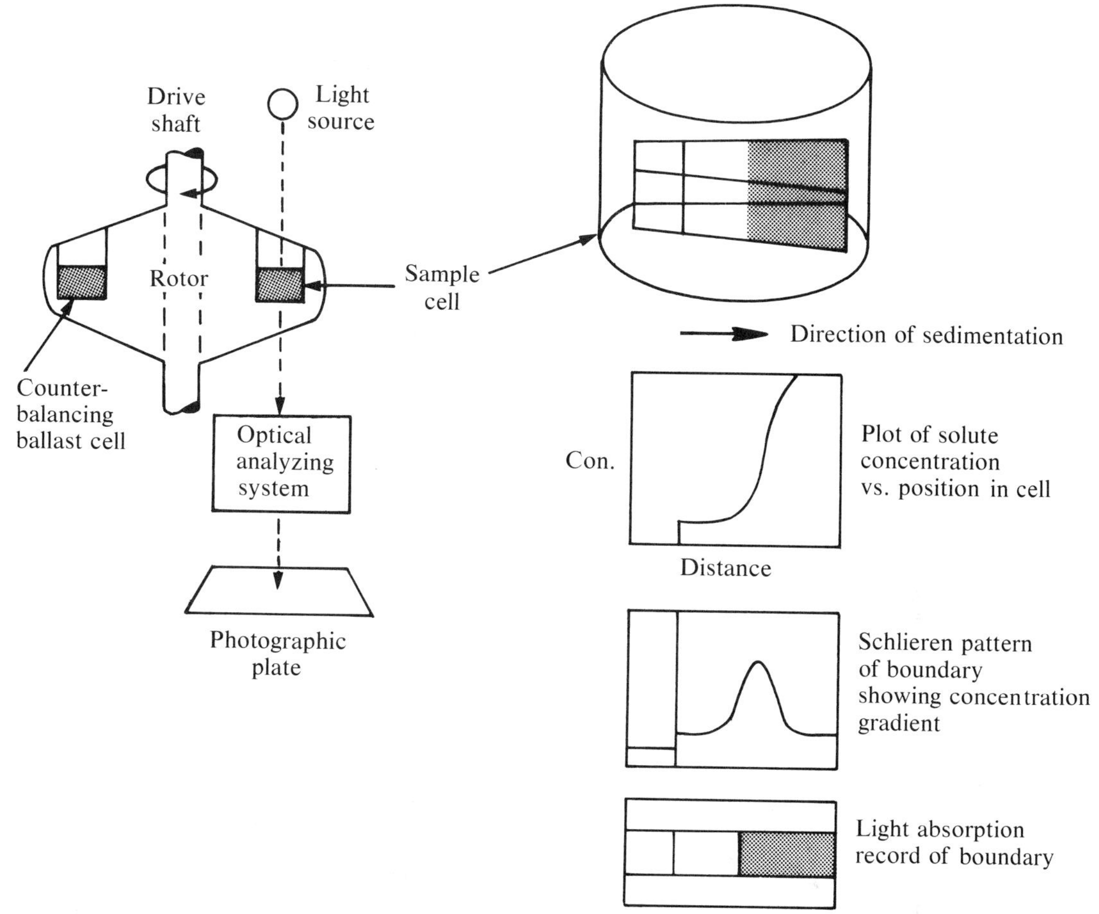

Figure 5.20. *The analytical ultracentrifuge for determination of properties of macromolecules in cells, such as buoyant density, molecular weight, etc. At right is shown how optical measurements are made while a sample is being centrifuged. (Illustration courtesy of Beckman Instruments, Inc.)*

thymine. Such DNA can be separated from normal DNA by isopycnic (same density) centrifugation in a cesium chloride gradient, because each type of DNA will take equilibrium positions in the gradient corresponding to its density (Fig. 5.21). Long centrifugation (about 36 hours) is required to achieve such equilibrium.

For determination of molecular weights of enzymes (or other proteins), nucleic acids, and viruses, the analytical centrifuge is used. Since these molecules absorb far ultraviolet rays but not visible light, the bands in the centrifuge cuvette are generally photographed with far ultraviolet light. However, accumulation of any of these molecules in a band leads to a localized change in refractive index which, by appropriate optical systems (Schlieren optics), can be measured by visible light. A detailed description of the optical arrangements in an analytical centrifuge may be found in treatises on physical methods in biochemistry.

To determine the molecular weight, M, of some substance from centrifugation experiments it is essential to know the sedimentation constant, s, and the diffusion constant, D, for the molecule in question:

$$M = \frac{RT}{D(1 - \widetilde{V}d_m)} \cdot (s) \qquad (5.1)$$

where R is the gas constant, T the absolute temperature, D is the diffusion constant, which can be determined by the rate at which the band of protein moves in the absence of centrifugal force, $\widetilde{V}$ is the partial specific volume of the protein in question, which can be determined from the density of a solution of that protein as a function of its anhydrous concentration, d_m is the density of the suspending medium, and s is the rate at which the molecule travels at distance x from the center of rotation.

$$s = \frac{dx/dt}{x\omega^2} \qquad (5.2)$$

where dx/dt gives the rate of travel of x and ω is the angular velocity or velocity of revolution in radians per second.

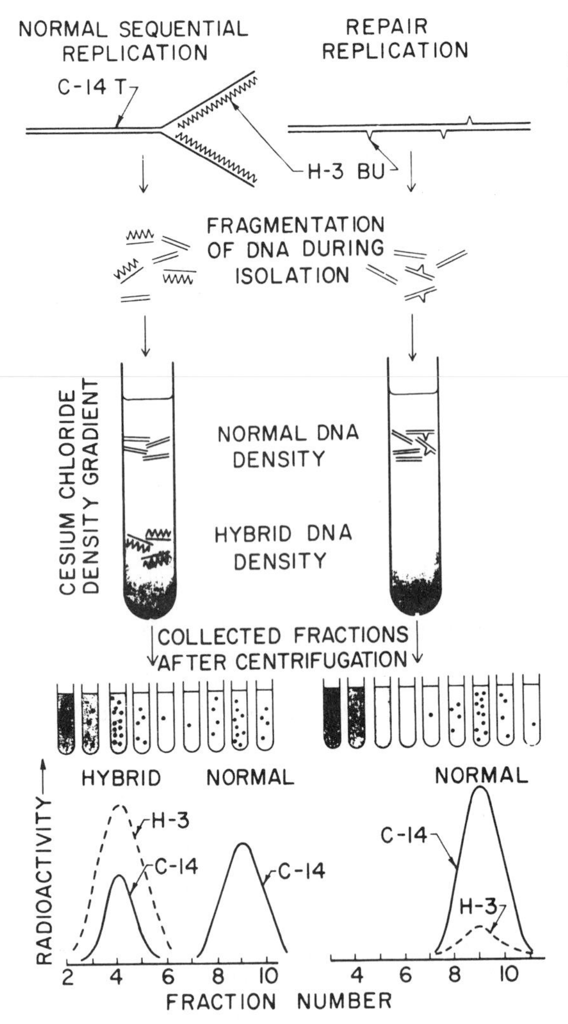

Figure 5.21. *Separation of molecules of different buoyant density by isopycnic equilibrium gradient centrifugation. On the left is shown the labeling of replicating* E. coli *DNA in which the thymine analogue, bromouracil, is substituted for thymine in the nutrient supplied. Because bromouracil has a higher molecular weight than thymine, the DNA replicated in its presence has a higher density; in a cesium chloride gradient it will therefore band nearer the bottom of the tube than the DNA containing thymine only. Because only a small quantity of DNA is available, it is detected by tracer methods. For this purpose the DNA is first radioactively labeled by growth of the* E. coli *in ^{14}C-labeled thymine; subsequently the cells are density-labeled at the same time as they are radioactively labeled by using ^{3}H-bromouracil. When the denser band is withdrawn from near the bottom of the centrifuge tube, the concentration of DNA in the fraction so withdrawn can be determined by tracer counting, since the denser DNA molecules have both ^{14}C and ^{3}H labels. As expected, peaks for both tracer labels are in the lower fraction numbers (high density) and both reach a peak at fraction 4. Since not all the DNA has replicated and is therefore not labeled with bromouracil, the normal, less dense DNA, with thymine only, will appear in higher fraction number (higher layers of the centrifuge tube) and contain only the ^{14}C label of the thymine.*

On the right of the figure is shown repair replication (e.g., after damage to the DNA by ultraviolet light). In this case exposure is such as to label only during repair, not replication of the DNA of the ^{14}C-labeled cells. Since the amount of ^{3}H-bromouracil incorporated during such repair is small compared to that incorporated during replication, the DNA which contains repair patches with bromouracil is not changed sufficiently in density to band differently from the unrepaired DNA strands. The ^{3}H-labeled DNA therefore occurs in the higher fraction numbers only. (From "The Repair of DNA," by P. C. Hanawalt and R. H. Haynes. Copyright © 1967 by Scientific American, Inc. All rights reserved.)

The sedimentation constant, s, varies from 1 to 200×10^{-13} sec for proteins. To avoid negative exponents the sedimentation constant is converted by multiplying by 10^{13}, and the unit of sedimentation so obtained is called the Svedberg, S, named in honor of a pioneer investigator in the field. The Svedberg unit, or more simply, the Svedberg, describes the velocity achieved per unit of force by a particle (molecule) moving through a liquid. While for proteins the S ranges from 1 to 200, for larger molecules or particles it ranges over a much wider spectrum.

The analytical centrifuge is also used to study sedimentation equilibrium. In this case, the distribution of protein molecules is followed in the cell periodically until it no longer moves—that is, it has achieved an equilibrium position. The gravitational field required is less than in studies on molecular weight, but the centrifugation time required is longer.

Appropriate centrifugal procedures may be used to determine the size and shape of macromolecules, as well as some of their other characteristics. Much of the information on nucleic acids, proteins, viruses, and some particulates in the cell has been obtained by centrifugal measurements.

For many biochemical determinations large quantities of cellular particulates or molecules are required. For this purpose the *zonal centrifuge* is useful. The principles for the operation of the zonal centrifuge are the same as for the centrifuges already described, but the marginal cavity of the rotor takes the place of tubes so that much larger volumes of solution can be centrifuged at a time. The apparatus is also so designed that solutions can be added or removed during spinning. Such centrifuges and their operation are described in detail by Umbreit and his colleagues (1972).

5.3 LOCALIZATION OF CHEMICALS IN CELL ORGANELLES: CYTOCHEMISTRY

The aims of cytochemistry are to determine what chemicals are present in each cell organelle, how much of each is present and where they are located. To accomplish these objectives the chemicals must be identified by microchemical reactions and quantitatively determined. This is an ambitious and inclusive program (Komarov, 1972). It is possible here to give only a sampling of the methods used in such studies.

In one of the methods of choice, tissues are freeze-dried in vacuum at low temperatures, a method that minimizes shrinkage. The proteins and other cell constituents are thus precipitated *in situ*. A specimen dried in this manner may be directly embedded in paraffin or some polymer for sectioning, and the sections stained or otherwise chemically treated to identify cell constituents.

If tissue sections are stained or otherwise treated in aqueous solutions, the dehydrated proteins must first be rendered insoluble by fixing agents that do not alter the activity of enzymes; glutaraldehyde, or other suitable fixing agents may be used for this purpose. Specific staining of each chemical of a cell organelle is the object of cytochemistry. For example, the identification and distribution of some enzymes can be determined by their color reactions. A number of specific color reactions have been developed for enzymes, nucleic acids, proteins, lipids and carbohydrates.

Nucleic acids can also be determined by ultraviolet spectrophotometry, because of the differential absorption of short wavelength ultraviolet rays by nucleic acids and proteins. Nucleic acids absorb maximally at 260 nm because of their high content of purines and pyrimidines that absorb in this range. Proteins absorb maximally at about 280 nm because of the presence of aromatic and heterocyclic amino acids, as shown in Fig. 3.18.

In the autoradiographic method (discussed in Appendix 3.2 of Chapter 3) cells or tissue sections fed labeled nutrients (e.g., with ^{32}P) are coated with a photographic emulsion. After exposure, the film is developed, fixed, and examined microscopically. The density of the labels in the developed film measures the relative concentrations of the radioactive nutrient in various cell organelles (see Fig. 3.13).

Sometimes enzymes are used to remove one protoplasmic constituent, making possible identification of the remaining constituents. For example, pepsin and trypsin remove protein from a *Drosophila* chromosome but leave the nucleic acids fixed with lanthanum salts (see also Chapter 7).

Still another type of cytochemical technique, used for determining the localization of salts in the cell, is *microincineration*. Entire cells or tissue slices are placed in an electric furnace at a temperature sufficiently high to oxidize organic materials. Only the ash of cell content remains. From the ash *in situ* it is possible to identify the kind and amount of salts present in each organelle (Fig. 5.22). Calcium is concentrated around chromosomes, and more iron is present in the nucleus than in the cytoplasm. Iron can be recognized by the red color of its oxide. Not all elements are retained in the ash,

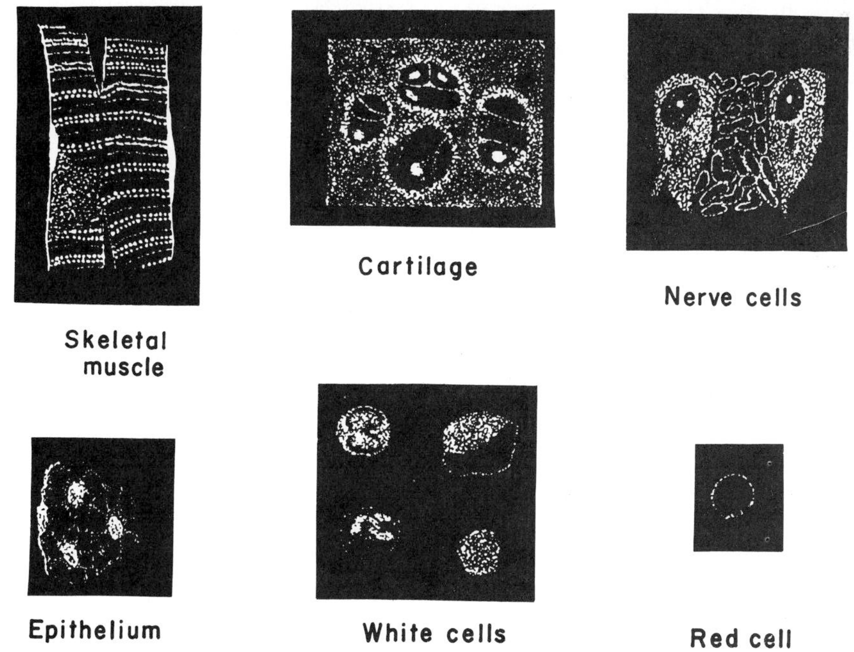

Figure 5.22. Microincinerated cells. Muscle from rectus abdominis of cat, cartilage from femur of cat embryo, nerve cells from cat brain, epithelium of corpus luteum of cat, blood cells from human. (Adapted from Scott, 1933: Am. J. Anat. 53: 243.)

however; for example, sodium and potassium disappear because they volatilize at red heat. Their distribution may be studied by first converting them into sulfates that do not volatilize at incineration temperatures. Some materials may be identified by the use of x-ray (Engstrom, 1966).

In some cases the quantity of a cell constituent may be determined by the amount of light transmitted through a section of an organelle stained with a specific dye and comparing it with the light transmitted by an equal thickness of a standard solution of the substance. Ultraviolet spectrophotometry of nucleic acid has been made quantitative by comparing transmission of ultraviolet light through cell organelles and the same thickness of nucleic acid solutions.

Electron microscopy makes possible localization of enzymatic reactions within cell organelles. For this purpose cells are fixed in an appropriate aldehyde that does not inactivate the enzyme. From the products of their action on added substrates, the enzymes are localized in relation to fine structure in the cells.

Electron microscopy has also been combined with autoradiography for localization of labeled nutrients and salts in constituents of cell organelles (Caro, 1966). Both ^{3}H-labeled and ^{14}C-labeled nutrients have been used for this purpose, the first giving much better resolution than the second because of the short path traveled by the emitted radiation before absorption in the emulsion. By this means it has been possible to show, for example, that mitochondria incorporate labeled amino acids.

Cytochemistry, like many other approaches to biological problems, has its pitfalls. Purity of reagents is often of paramount importance, since the tests are performed with minute quantities at the cell level. Yet some biological reagents (e.g., enzyme preparations) are difficult to purify and control. Also, diffusion of a water-soluble constituent of a cell may occur during application of the very agent used for its identification. In addition, the dynamic state of the cell, the continuous relocation of material in it, just prior to fixation, must be kept in mind. Nonetheless, cytochemistry has added a wealth of material to our knowledge of cellular constituents and structure.

5.4 PAPER CHROMATOGRAPHY

In the general chromatographic method as introduced into biology by Tswett in 1906, various adsorbants (e.g., starch, aluminum oxide) are poured as a slurry into a tube (see Fig. 5.23*A*). An unknown is poured on the widened top of the tube and then "developed" by the slow trickle of an appropriate solvent through the column. The rate at which any compound migrates through the column depends upon the balance between its affinity for the solvent and its adsorption on the particles of the column. This method of separating compounds is spoken of as *adsorption chromatography*.

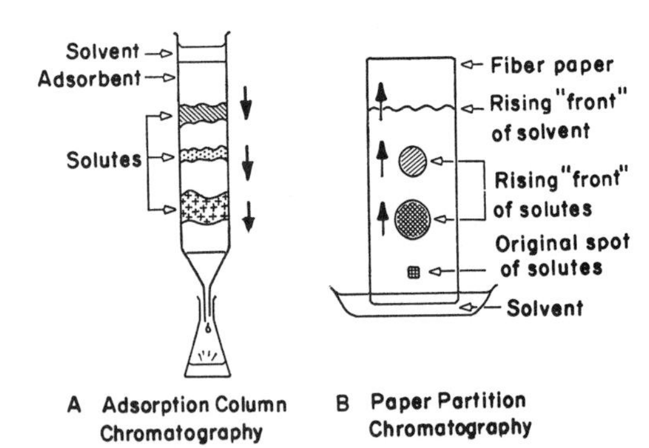

Figure 5.23. *Column and paper chromatography. (From Cantarow and Schepartz, 1967: Biochemistry. 4th Ed. W. B. Saunders Co., Philadelphia.)*

Another chromatographic method depends upon partition of a solute between two solvents in which the solute has differential solubility. If a solute has a greater solubility in one solvent than in another, the two solvents being immiscible, it is possible to remove and concentrate the solute in the solvent in which it has greater solubility.

Something like this separation can be achieved on a column of some powdered substance if only one of the solvents wets the powder. The solvent wetting the powder will form a stationary solvent phase, and the powder in this case holds the solvent that wets it. In practice, the solute in the appropriate organic solvent is added to the top of the wet column and developed by washing with additional aliquots of the solvent. As the second solvent, immiscible with the first, moves through the column, the solute is partitioned between the two solvents, as if an infinite number of separatory funnels were being used in series to extract the solute. The solute moves as a band on the column. If instead of one solute several have been dissolved in the solution poured on the top of the column, they will separate as several bands on the column because of their differences in partition between the two solvents (Strain, 1958).

Paper chromatography, as developed in 1941 by Martin and Synge, is a variation of Tswett's partition chromatographic method. It is a method based also on the differential solubility of compounds in solvents, and it achieves separation by partition of the solute between the two solvents. The paper acts as a support for one of the solvents (in theory, at least).

The solution to be tested is placed as a spot on a large sheet of filter paper. When the spot is dry the paper is placed with one edge dipping into a tray containing the mixture of solvents, the choice of solvents depending upon the solubility of the solutes to be chromatographed (Fig. 5.23*B*). After the solvent front has moved the desired distance beyond the sample spot, the solvent front is marked, and the paper is dried and sprayed with some agent to "develop" the spots of materials that have migrated from the original sample spot.

For example, a mixture of amino acids is spotted on filter paper, and the paper is dipped into a butyl alcohol-water mixture. Diverse amino acids are differentially soluble in these two solvents, some being more soluble in water, others more soluble in butyl alcohol. The alcohol-water mixture migrates along the filter paper fibers upward through the sample spot, and the various substances with differential solubility migrate to different distances from the original spot, those most soluble in butyl alcohol moving farthest. After the solvent front has advanced past the spot to a distance considered from experience to be adequate, the line of the solvent front is marked and the paper is dried. For detection of amino acids the paper is then sprayed with ninhydrin, which forms a purplish color on reacting with an amino acid. Control chromatograms, made from known mixtures of amino acids, are run in parallel with the unknown. From the positions of known amino acids the unknown acids can be determined. From the depth of the color reaction quantitative data may be gathered (Zweig and Whitaker, 1971).

The ratio of the distance traveled by any particular substance to the distance covered by the solvent is known as the *ratio of the fronts* or R_f. This ratio is usually constant for a given substance in a particular solvent and under given conditions (e.g., temperature). Thus the R_f values determined from controls of known amino acids under similar conditions may be used as standards for identification of unknown amino acids.

When a "one-dimensional" chromatogram does not adequately separate the components of a mixture, the once-developed paper is placed, after it has dried, in a pan of another mixture of solvents in such a way that the movement of the second mixture of solvents occurs at right angles to the movement of the first, producing a "two-dimensional" chromatogram (Zweig and Whitaker, 1971).

When the substances separated by paper chromatography are radioactive, they can be visualized by photography. The filter paper is pressed against a sensitive photographic film. The ionizing radiations from the radioactive material will so affect the film that on development the film will show darkening in proportion to the radiation, the amount of radiation depending upon the concentration of the radioactive substance. From the R_f values of controls and from the density of the darkened areas on film exposed to known concentrations of radioactive material, identification and quantitative analysis of an unknown material are possible, even when the unknown is present only in traces, as, for example, after a few seconds of photosynthesis.

5.5 POLYACRYLAMIDE GEL ELECTROPHORESIS

Polyacrylamide gel electrophoresis has been extensively used in recent years for fractionation of proteins into distinct bands, especially enzymes from a mixture of proteins as well as high molecular weight RNA. Acrylamide (CH_2:$CHCONH_2$) is polymerized with appropriate reagents and quickly poured into open-ended tubes temporarily stoppered at the lower end. The polymer can be made to provide an effective median pore radius of 0.5 to 3 nm by adjusting the acrylamide concentration, the concentration of the cross-linking reagent, and the pH (3 to 11) to optimize charge separation. When the gel sets, the substance to be investigated is poured on top of a tube and an electrode is inserted at the top and another in a container in which the lower end of the tube (with the stopper now removed) dips. Usually rows of

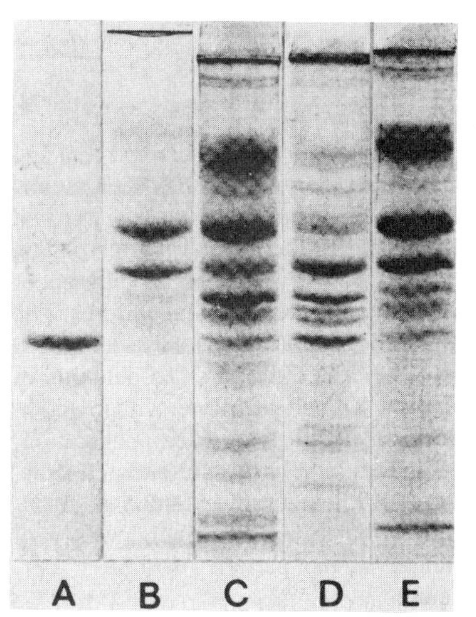

Figure 5.24. *Polyacrylamide gel electrophoresis (4 hours at 2 mA) of sodium dodecyl sulfate (detergent) treated urea-soluble mouse proteins.* A, *Mouse muscle actin;* B, *mouse stratum corneum (outer layer) of skin;* C, *mouse epidermal cells before culture;* D, *attached cells from mouse epidermal culture;* E, *unattached cells from mouse epidermal cultures four hours after subculture on glass dish. The bands of high molecular weight in* B *to* E *are noncovalently bound aggregates of the two keratin proteins. (From Steinert and Yuspa, 1978: Science* 200: *1491–1493. Copyright 1978 by the American Association for the Advancement of Science.)*

tubes in special racks are used (Chrambach and Rodbard, 1971), but the principle can be illustrated with a single tube. In a 5 to 15 percent acrylamide gel with appropriate low voltage as many as 15 to 30 bands in a mixture of proteins may be separated from one another (Fig. 5.24). In this way, for example, isozymes, which are enzymes of the same function but differing in amino acid constitution and with different affinities for a substrate, may be resolved into distinct bands. The gel acts essentially like a sieve through which small molecules migrate more rapidly than large ones of the same charge, and those of higher charge migrate faster than those of lower charge, but of the same size. Many variants of the method are in use, some with a pH gradient in the tube that focuses the bands more clearly (Korte, 1974). From data obtained by acrylamide electrophoresis one may estimate molecular size of the substance in question. For details of the equipment and of the polymerization process the reader is referred to a report by Chrambach and Rodbard, 1971).

LITERATURE CITED AND GENERAL REFERENCES

Allison, A., 1967: Lysosomes and disease. Sci. Am. (Nov.) *217*: 62–72.

Animal Ribosomes: Experimental Studies of the Last Five Years. MSS Information Corp., New York, 1973.

Bloom, W. and Fawcett, D. W., 1975: A Textbook of Histology. 10th Ed. W. B. Saunders Co., Philadelphia.

Branton, D., 1964: Fine structures in freeze-etched *Allium cepa*. J. Ultrastruct. Res. *11*: 401–411.

Cardell, R. R., Jr., 1977: Smooth endoplasmic reticulum in rat hepatocytes during glycogen deposition and depletion. Int. Rev. Cytol. *48*: 221–279.

Caro, L., 1966: Progress in high-resolution autoradiography. Prog. Biophys. Mol. Biol. *16*: 171–190.

Chrambach, A. and Rodbard, D., 1971: Polyacrylamide gel electrophoresis. Science *172*: 440–451.

Coffey, R. L., Legwinska, L., Oliver, M. and Martonosi, A., 1975: Mechanism of ATP hydrolysis by sarcoplasmic reticulum. Arch. Biochem. Biophys. *170*: 37–48.

Cunningham, W. P., Morre, D. J. and Mollenhauer, H. H., 1966: Structure of isolated plant Golgi apparatus revealed by negative staining. J. Cell Biol. *28*: 169–179.

Dallner, G., Siekevitz, P. and Palade, G. E., 1966: Biogenesis of endoplasmic reticulum membranes. J. Cell Biol. *30*: 73–118.

Dean, R. T., 1975: Concerning a possible mechanism for selective capture of cytoplasmic proteins. Biochem. Biophys. Res. Commun. *67*: 604–609.

DeDuve, C., 1969: Lysosomes in retrospect. *In* Lysosomes in Biology and Pathology. Vol. 1. Dingle, ed. North Holland Publishing Co., Amsterdam, pp. 3–40.

Depicrre, J. W. and Dallner, G., 1975: Structural aspects of the membrane of the endoplasmic reticulum. Biochim. Biophys. Acta *415*: 411–472.

DeRobertis, E. D. P., Saez, F. A. and DeRobertis, E. N. F., 1975: Cell Biology. 6th Ed. W. B. Saunders Co., Philadelphia.

Dingle, J. T., compiler, 1969: Lysosomes in Biology and Pathology. North Holland Publishing Co., Wiley Interscience, New York.

Dingle, J. T., ed., 1972: Lysosomes: A Laboratory Handbook. North Holland Publishing Co., Amsterdam.

Dutta, G. P., 1974: Recent advances in cytochemistry and ultrastructure of cytoplasmic inclusions in Ciliophora (Protozoa). Int. Rev. Cytol. *39*: 285–343.

Engelman, D. M. and Moore, P. B., 1976: Neutron-scattering studies of the ribosome. Sci. Am. (Oct.) *235*: 44–54.

Engstrom, A., 1966: X-ray microscopy and X-ray absorption analysis. *In* Physical Techniques in Biological Research. Vol. 3A. Pollister, ed. Academic Press, New York, pp. 87–117.

Garrett, R. A. and Wittmann, H. G., 1973: Structure and function of the ribosome. Endeavour *32*: 8–14.

Gunning, B. E. S. and Steer, M. W., 1975: Ultrastructure and the Biology of Plant Cells. Edward Arnold, London.

Ham, A. W., 1974: Histology. 7th Ed. J. B. Lippincott Co., Philadelphia.

Hogg, J. F., ed., 1969: The nature and function of peroxisomes (microbodies, glyoxysomes). Ann. N. Y. Acad. Sci. *168*: 211–381.

Holtzman, E., 1976: Lysosomes: A Survey. Springer-Verlag, New York.

Hruban, Z. and Rechcigl, M., Jr., 1969: Microbodies and Related Particles. Academic Press, New York (Int. Rev. Cytol. Supplement).

Hughes, D. E., Lloyd, D. and Brightwell, R., 1970: Structure, function and distribution of organelles in prokaryotic and eukaryotic microbes. Symp. Soc. Gen. Microbiol. *20*: 295–322.

Kelley, W. S. and Schaechter, M., 1968: The life cycle of bacterial ribosomes. *In* Advances in Microbial Physiology. Vol. 2. Rose and Wilkinson, eds. Academic Press, New York, pp. 89–142.

Kolata, G. B., 1975: Ribosomes (I): genetic studies with viruses. Science *109*: 136 and 183–184.

Korte, F., ed., 1974: Methodicum Chimicum. Vol. I. Analytical Methods. Part B. Academic Press, New York.

Kupfer, D., 1970: Enzyme induction by drugs. Bioscience *20*: 705–714.

Lehninger, A. L., 1975: Biochemistry. 2nd Ed. Worth Publishers, New York.

Mahler, H. R. and Raff, R. A., 1975: The evolutionary origin of the mitochondria, a non-symbiotic model. Int. Rev. Cytol. *43*: 1–124. Ribosomal function 92–96.

Matile, P., 1975: The Lytic Compartment of Plant Cells. Cell Biology Monographs, Vol. 1. Springer-Verlag, New York.

Matile, P., 1978: Biochemistry and function of vacuoles. Ann. Rev. Plant Physiol. *29*:193–213.

Maugh, T. M., 1976: Ribosomes (II): a complicated structure begins to emerge. Science *190*: 258–260.

Nanninga, N., 1973: Structural aspects of ribosomes. Int. Rev. Cytol. *35*: 125–188.

Neutra, M. and LeBlond, C. P., 1969: The Golgi apparatus. Sci. Am. (Feb.) *220*: 100–107.

Nomura, M., 1969: Ribosomes. Sci. Am. (Oct.) *221*: 28–35.

Nomura, M., ed., 1976: Ribosomes. Cold Spring Harbor Laboratory, Cold Spring Harbor, New York.

Northcote, D. H., 1971: The Golgi apparatus. Endeavour *30*: 26–33.

Novikoff, A. B. and Holtzman, E., 1976: Cells and Organelles. 2nd Ed. Holt, Rinehart & Winston, New York.

Palade, G. E., 1956: Intracisternal granules in the exocrine cells of the pancreas. J. Biophys. Biochem. Cytol. *2*: 417–422.

Pitt, D., 1975: Lysosomes and Cell Functions. Longman Inc., New York.

Rich, A., 1963: Polyribosomes. Sci. Am. (Dec.) *209*: 44–53.

Satir, B., 1975: The final steps in secretion. Sci. Am. (Oct.) *233*: 29–37.

Siekevitz, F., 1963: Protoplasm: endoplasmic reticulum and microsomes and their properties. Ann. Rev. Physiol. *25*: 15–40.

Society for Experimental Biology, 1970. Control of Organelle Development. Cambridge University Press, Cambridge.

Strain, H. H., 1958: Chloroplast Pigments and Chromatographic Analysis. Pennsylvania State University, University Park, PA.

Umbreit, W. W., Burris, R. H. and Stauffer, J. F., 1972: Manometric and Biochemical Techniques: A Manual Describing Methods Applicable to the Study of Tissue Metabolism. 5th Ed. Burgess Publishing Co., Minneapolis, Minn.

Watson, J. D., 1976: Molecular Biology of the Gene. 3rd Ed. W. A. Benjamin, Menlo Park, Calif.

Whaley, W. G., 1975: The Golgi Apparatus. Cell Biology Monographs Vol. 2, Springer-Verlag, New York.

Wiebe, H. H., 1978: The significance of plant vacuoles. Bioscience *28*: 327–331.

Wittmann, H. G., 1970: Comparison of ribosomes from prokaryotes and eukaryotes. Symp. Soc. Gen. Microbiol. *20*: 55–76.

Zweig, G. and Whitaker, J., 1971: Paper Chromatography and Electrophoresis. Academic Press, New York.

CHAPTER 6

ENERGY TRANSDUCERS: MITOCHONDRIA AND CHLOROPLASTS

MITOCHONDRIA

Eukaryotic cells, both plant and animal, are typically aerobic—an evolutionary offshoot of the liberation of oxygen into the atmosphere after the advent of photosynthesis (see Chapter 2). The mitochondrion, which is a distinctly aerobic organelle, is a symbol of eukaryotic emphasis on an aerobic existence. Mitochondria have been called the "powerhouses" of the eukaryotic cell (Siekevitz, 1957); oxygen serves as a hydrogen acceptor with formation of water, permitting reactions that make it possible to extract energy still available in the metabolic products of glycolysis (see Fig. 2.6). Water and carbon dioxide formed as the end products of oxidative metabolism are innocuous, not harmful, as are some products of fermentation.

With use of coupled reactions, the energy liberated by mitochondrial systems is incorporated mainly into high-energy phosphate bonds of ATP. ATP then diffuses to all centers where work is performed in the cell (see Chapter 12). High-energy phosphates power the synthesis of compounds needed for cellular repair, maintenance, growth, secretion, transport of molecules through membranes, cellular motility, maintenance of excitability, and so on. Mitochondria are structurally well suited to accommodate the extensive enzyme systems required to liberate energy efficiently.

Mitochondria were first seen in 1880 by Kölliker, who teased them out of muscle cells. Subsequently, they were extensively studied in a variety of cells, mostly in fixed and stained preparations. Mitochondria have also been studied in living cells stained with the vital dye Janus green B. Although not entirely harmless, the dye enables one to observe how, in a living cell, the mitochondria change in size and shape as conditions vary. It also permits one to see how mitochondria move about, especially during mitosis and cell division. Mitochondria may also be seen with the phase microscope. However, most has been learned about mitochondria by studies with the electron microscope coupled with biochemical determinations (Green and Bonem, 1970; Tadeschi, 1976).

Isolation

Mitochondria were first isolated by fractional centrifugation of ruptured cells by Bensley and Hoerr in 1934 and recognized as sites of cellular respiration by Hoogeboom and co-workers in 1948. Separated from other cell particulates, mitochondria maintain their identity. Biochemical analyses of mitochondria demonstrated the presence of many enzymes concerned with aerobic respiration and energy transfer (see Chapter 12). Mitochondria are often concentrated in regions of great cellular activity, such as the secretory surfaces (glandular cells), absorptive surfaces (intestinal lining cells), near the impulse-propagating nodes of a nerve cell, and in active muscle fibers. These locations permit mitochondria to deliver most directly the high-energy compounds needed for the corresponding cell activities.

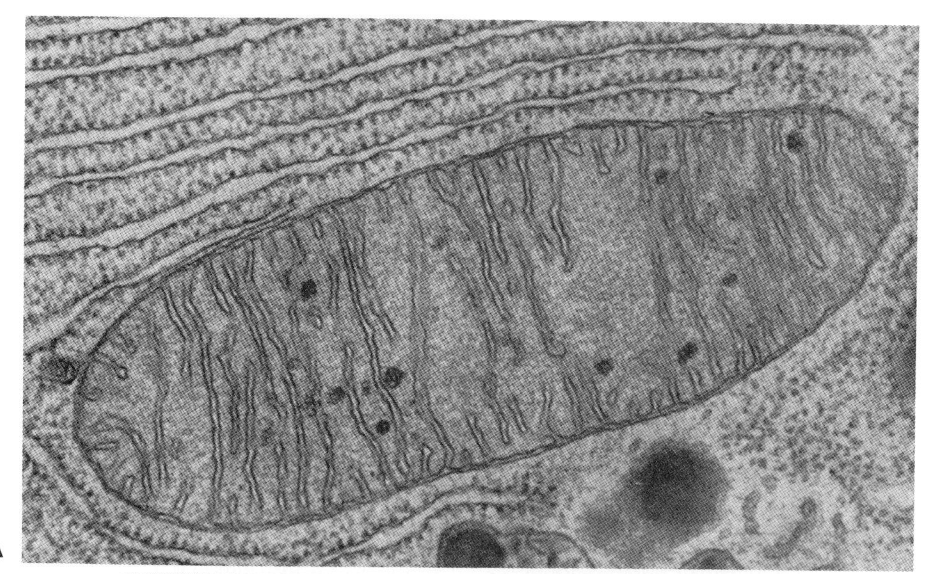

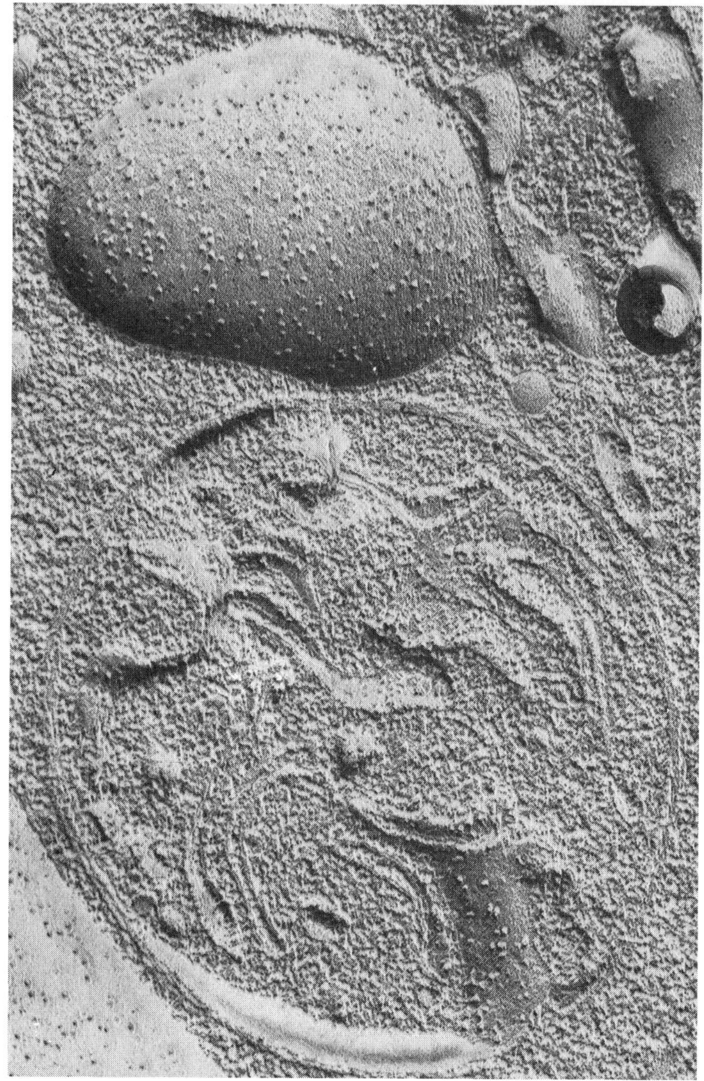

Figure 6.1. A, *Electron micrograph of a typical mitochondrion from a bat pancreas cell showing the cristae, matrix and matrix granules. The endoplasmic reticulum studded with ribosomes shows in the upper left, and free ribosomes and lysosomes in the lower right (× 79,000). (From K. R. Porter in Bloom and Fawcett, 1975: A Textbook of Histology. 10th Ed. W. B. Saunders Co., Philadelphia, p. 46.)* B, *Freeze etching of a mitochondrion from onion root tip. Note the cristae with the granular protein particles on the inner membranes, the larger bodies on the outer membrane at the bottom, and the double nature of the membranes. (By courtesy of D. Branton.)*

Size and Shape

Mitochondria (Figs. 6.1 to 6.5) are usually 0.5 to 1.0 μm in cross sectional diameter and vary in length up to a maximum of 7 μm. They may be filamentous or granular and may change from one form to another depending upon the physiological conditions of the cells and the way the cells are treated. They may also attach to one another and then dissociate at a later time. Occasionally they are enlarged at one end, having the shape of a tennis racket. In mammals the shape and form of the mitochondria varies with hormonal state; for example, after removal of the hypophysis the form and structure of the mitochondria in the adrenal cortex is greatly altered. It returns to normal on injection of adrenocorticotropic hormone. Under unfavorable conditions mitochondria become abnormal, sometimes fusing to form large bodies, sometimes vesiculating. Mitochondria in germinal tissues of some species transform into egg yolk granules (Bereiter-Hahn, 1976).

Organization

The structural organization of mitochondria is not resolvable by light microscopy; its detailing awaited development of the electron microscope. Electron micrographs revealed that mitochondria have a complex organization. Perhaps the most common overall structure is a double-walled rod or cylinder with rounded ends (Fig. 6.1). The inner wall or membrane of the mitochondrion is 6 to 8 nm thick and in most mitochondria is extended as a projection called the *crista* (or crest) inward almost across the internal cavity of the mitochondrion. Many cristae may be present. The cristae greatly extend the surface area exposed to the inside cavity of the mitochondrion, providing ample space for accommodation of enzyme assemblies. The cristae usually run at right angles to the long axis of the rod-shaped mitochondrion. The space (matrix) between these infoldings is filled with a relatively dense fluid material in which no structure is normally seen with the electron microscope, although fibrils and granules have been reported in some cases. The outer membrane of the mitochondrion is about 6 nm thick and is separated from the inner membrane by a clear space, the outer chamber (Fig. 6.2).

Many variations on the fundamental mitochondrial architecture are found in different types of cells. In active kidney cells the cristae may be very numerous transverse sheets, almost like a stack of coins, whereas in less active cells they may be few in number. In some cells the cristae are longitudinal folds paralleling the long axis of the mitochondrion (Fig. 6.3). In many protozoan cells, in insect flight muscle cells, and in adrenal cells the infolding may take the form of tubules rather

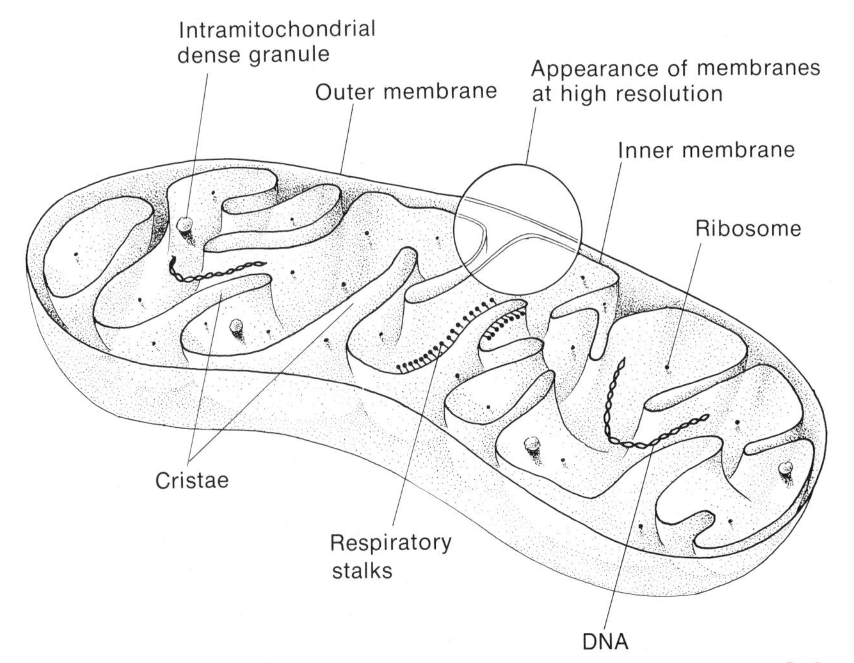

Figure 6.2. *Diagram of a mitochondrion. (From DeWitt, 1977: Biology of the Cell. W. B. Saunders Co., Philadelphia.)*

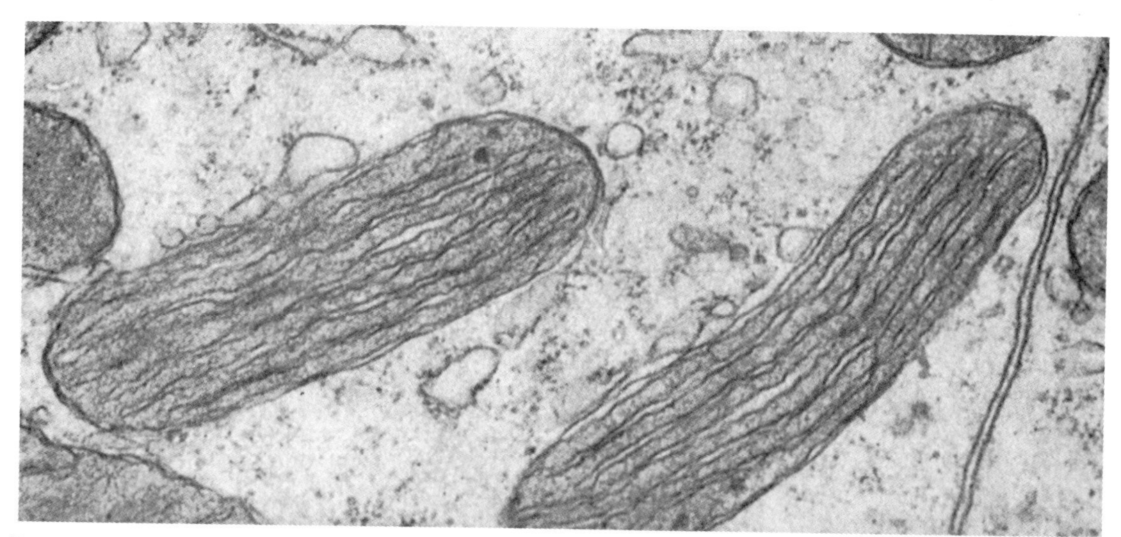

Figure 6.3. Longitudinal mitochondrial cristae in a proximal kidney tubule cell from a starved summer frog with low enzyme activity (× 65,000). (From Fawcett, 1966: The Cell. W. B. Saunders Co., Philadelphia, p. 93.)

than lamellar infoldings (Fig. 6.4). Mitochondria of different cells of the same species may have different types of infoldings; for example, tubules occur in insect flight muscle cells, while lamellar cristae occur in leg muscle cells. In heart muscle and in the giant multinucleate amoeba (*Chaos chaos*), the mitochondrial tubules may have an interesting zigzag pattern (Fig. 6.5). Mitochondria in cells of some species even branch.

Location and Number

Mitochondria show active movement, especially in cells undergoing division. In some cells, however, they appear to be stationary (Frederic, 1958). They may aggregate as rings around the sperm tail, be present in rows between contractile units in muscle cells, and aggregate at nodes of neurons where impulse transmission occurs. In secretory cells they accumulate at the base of the cells, in the retina at one end of the rod cells, and in kidney tubule cells they accumulate along infoldings of the cell membrane.

In a rat liver cell as many as 1600 mitochondria have been recorded; in some oocytes a hundred times as many may be present, and in the giant amoeba (*Chaos chaos*) half a million have been estimated. Mitochondria are quite numerous in brown fat cells to which they im-

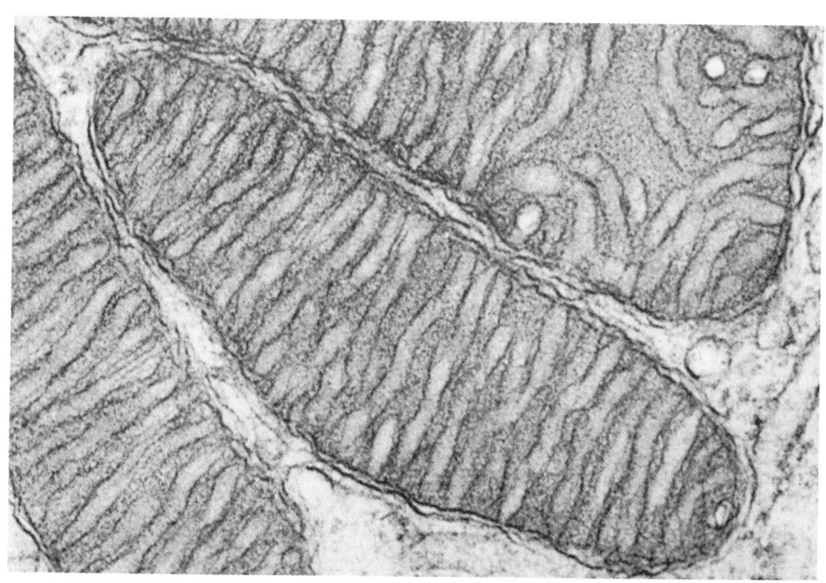

Figure 6.4. Tubular mitochondrial cristae in a cell from a hamster adrenal cortex cell (× 54,000). (From Fawcett, 1966: The Cell. W. B. Saunders Co., Philadelphia, p. 75.)

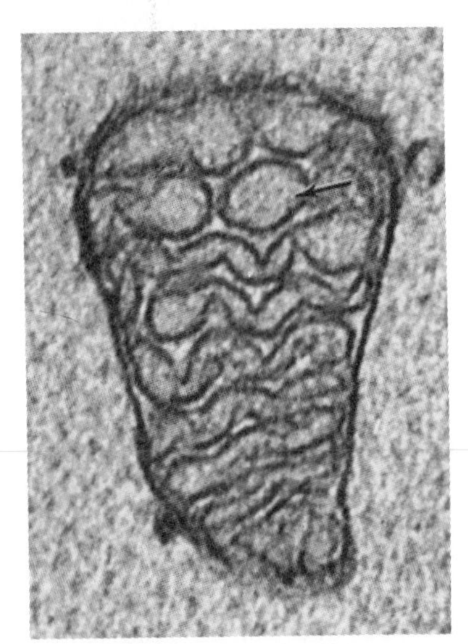

Figure 6.5. *Zigzag mitochondrial cristae in cells of the cat right ventricle papillary muscle (× 123,000). (From Fawcett, 1966: The Cell. W. B. Saunders Co., Philadelphia, p. 91.)*

part their color. Plant cells generally have fewer mitochondria than animal cells. The number of mitochondria decreases in cancerous cells, perhaps because of the tumor's lowered oxidative and increased glycolytic activity. The single-celled eukaryotic algae *Micromonas* and *Chlamydomonas* have only one large mitochondrion (Grobe and Arnold, 1975) and one amoeba, *Pelomyxa palustris,* has none. A claim has been made that for yeast, reconstruction of cell sections indicates that the mitochondrion is single, with connections between the enlargements that are separated on mechanical disruption of the cell (Hoffman and Avers, 1973). This finding does not seem to have been generally substantiated.

It is interesting that a relationship has been found between total body weight, liver mitochondrial mass and oxygen utilization in a series of animals, emphasizing the mitochondrion as the prime oxygen consumer in the eukaryote cell (Lehninger, 1975).

Chemical Composition

Because the specific gravity of mitochondria is greater than that of the cytoplasmic matrix (the larger fragments having first been removed), ultracentrifugation of mechanically disintegrated cells at 20,000 times gravity deposits the mitochondria at the bottom of a centrifuge tube (see Figs. 5.11 and 5.19). The mitochondria so separated, washed free of other particles, contain 65 to 70 percent protein, 25 to 30 percent lipids, 0.5 percent RNA, and a small amount of DNA. The sulfur content is also relatively high because considerable sulfhydryl (SH) is present on active groups of some enzymes (Lehninger, 1975).

Of the proteins, the largest portion probably consists of enzymes, but structural proteins constitute about 30 percent of the total protein. Structural protein added to some purified mitochondrial enzymes in the right proportion increases their activity. An actomyosin-like protein that splits adenosine triphosphate (ATP) and contracts on addition of ATP has been reported from liver mitochondria. The membranous component of the centrifugate is most effective in restoring enzymatic activity to the reconstructed mitochondria; this indicates the importance of the appropriate steric configuration for mitochondrial enzyme function. Phospholipids are also important for enzyme function, since their addition to purified enzymes also increases activity.

Mitochondrial lipids are largely phospholipids, lecithin being prominent, but triglycerides, phosphatidic acid (see Fig. 1.13*A*–*C*) and cholesterol (see Fig. 3.11) are present as well. Also found is a lipid characteristic of the mitochondria, called cardiolipin, which consists of three phosphatidyl and one diglyceryl radicals. Mitochondrial enzymes involved in

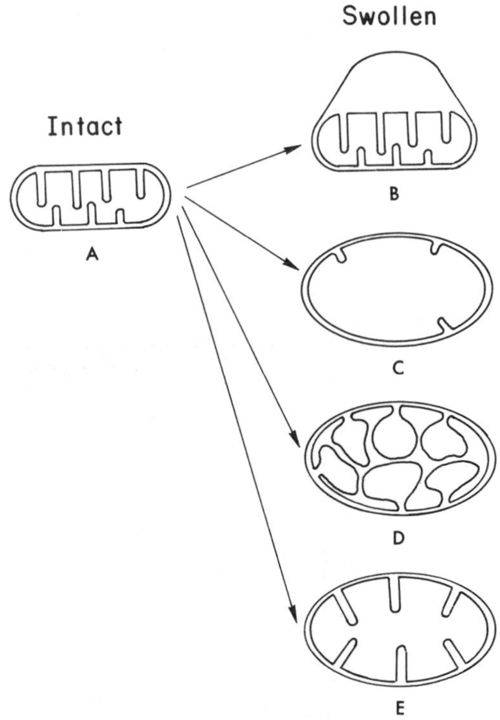

Figure 6.6. *Structure of a normal intact mitochondrion (A) and of different types of swelling. Entrance of solutes in the outer chamber produces dilation of the intermembranal space (B), or of the intracristal spaces (D). Penetration in the inner chamber produces dilution of the matrix with (C) or without (E) unfolding of the crests. (From Lehninger, 1962: Physiol. Rev. 42: 467–517.)*

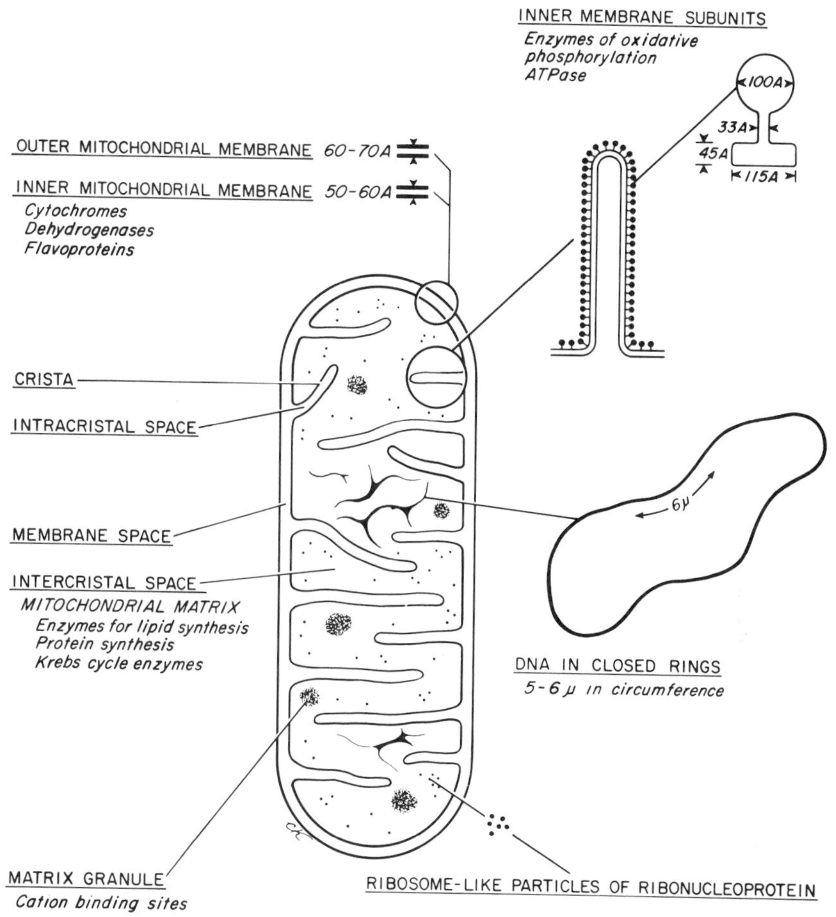

Figure 6.7. *Schematic representation of the structural components of the mitochondrion and some of the functions of the components. (From Bloom and Fawcett, 1975: A Textbook of Histology. 10th Ed. W. B. Saunders Co., Philadelphia, p. 47.)*

energy liberation from products of glycolysis are discussed in Chapters 10 and 12.

Mitochondrial structural protein is insoluble in water but may be solubilized by anionic detergents. Removal of detergent by dialysis permits reprecipitation of the protein (Lehninger, 1975).

Figure 6.7 is a diagram of the structural components of a mitochondrion. The general function of each component is indicated. Attention is focused on the enzymes for liberation of energy and for synthesis.

Permeability

The mitochondrion is capable of accumulating calcium, magnesium, and phosphate against concentration gradients; this requires expenditure of energy (Carafoli and Crompton, 1976) (see Chapter 18). Potassium uptake into mitochondria is also energy-linked. In this respect the mitochondrial membrane resembles the plasma membrane of the cell (see Chapters 8 and 18). Accumulation of ions in the mitochondrion affects its structure and results in swelling. Accumulation of ions is regulated by the inner membrane, the outer membrane being freely permeable to inorganic ions, water, sucrose, and other molecules of molecular weights up to 10,000 (Novikoff and Holtzman, 1976).

The inner mitochondrial membrane uses carriers to translocate ions and molecules across the membrane. Proteins present in the inner membrane serve as the carriers. Transport of this type also occurs across the plasma membrane of the cell and is discussed in Chapter 18.

The inner membrane of the mitochondrion, like the cell membrane, is selectively permeable, and the "rules" governing the relative rates of entry of various non-polar and polar materials into mitochondria resemble those observed for the plasma membrane (see Chapter 17). Mitochondria, like entire cells, swell when placed in an osmotic concentration lower than

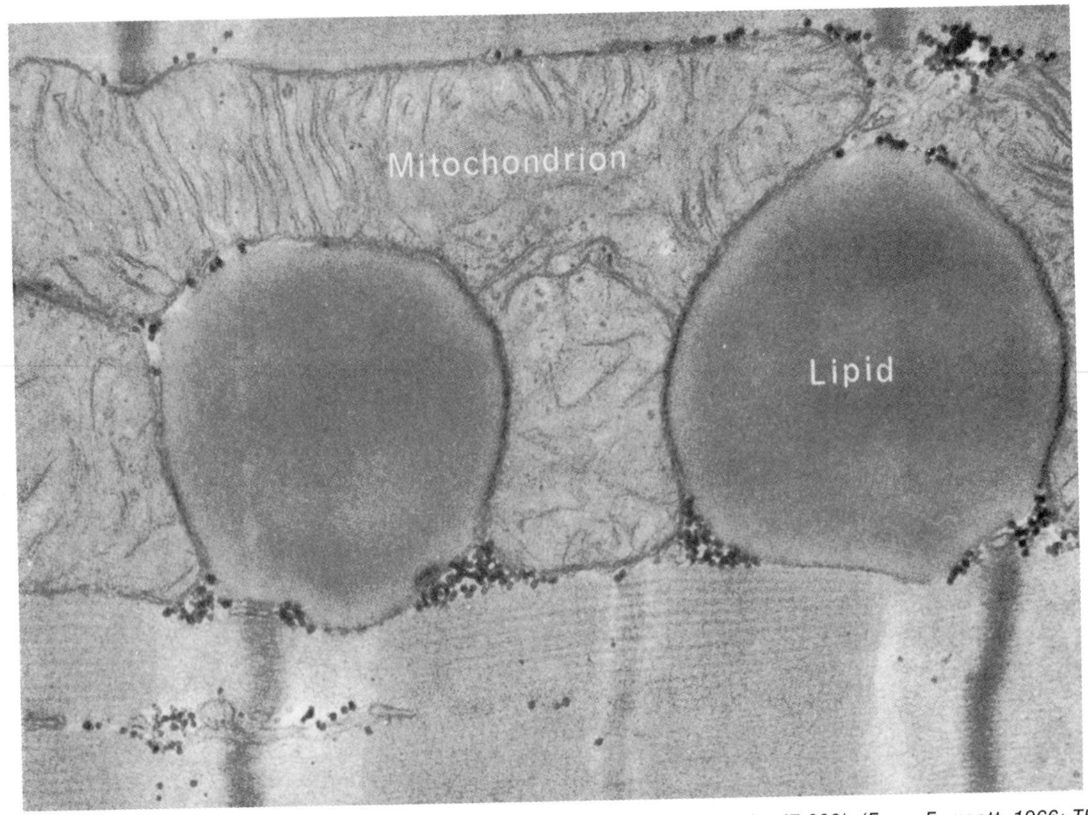

Figure 6.8. *Mitochondria in direct contact with lipid globules; heart muscle (× 47,000). (From Fawcett, 1966: The Cell. W. B. Saunders Co., Philadelphia, p. 103.)*

that characteristic of the cell and shrink in higher osmotic concentrations (Fig. 6.6). Isolated mitochondria may swell to five times their normal volume yet remain intact. Therefore, the membranes are sturdy structures (Tedeschi, 1976).

Enzymes

Many enzymes have been isolated from mitochondria fragmented by the surfactant deoxycholate (a mild detergent). Some of these enzymes are firmly bound to the membranes and resist purification. However, it is established that all the cellular enzymes that effect the complete aerobic breakdown of nutrients to carbon dioxide and water occur in the mitochondria, as do enzymes enabling the cell to transfer the released energy to stable high-energy compounds (high-energy phosphates such as ATP) (see Chapter 12). These compounds of high chemical potential energy are the agents by which the work of the cell is performed, only a small amount (about 5 percent) being formed outside the mitochondria. Therefore, the condition and activity of the mitochondria are likely to influence practically all cell functions. The involvement of mitochondria in such functions is considered in later chapters.

When an animal is starved, the mitochondria in many of its cells come in contact with the stored lipid (Fig. 6.8). The contact between mitochondrion and lipid droplet is so tight that the inner membrane makes contact with the lipid droplet, permitting direct action on the lipid of fatty acid oxidases known to be present in the inner mitochondrial membrane.

To promote enzyme-catalyzed reactions characteristic of mitochondria in the test tube, numerous co-factors (e.g., salts and vitamin B components) must be added in a concentration higher than when the enzymes are inside the cell. This and the difficulty of removing active enzymes from the mitochondrial membranes make it likely that their activity depends upon their specific steric spatial relationships in the membranes (Racker, 1970). The lamellar (made of thin sheets) structure of the mitochondrion provides a large surface area to accommodate the enzymes and a three-dimensional matrix for juxtaposing enzymes involved in sequences of reactions (see Chapter 12). Although under higher resolving power of the electron microscope the inner membrane of the mitochondrion appears to have tennis racket-like or F_1 bodies 7 nm in diameter on

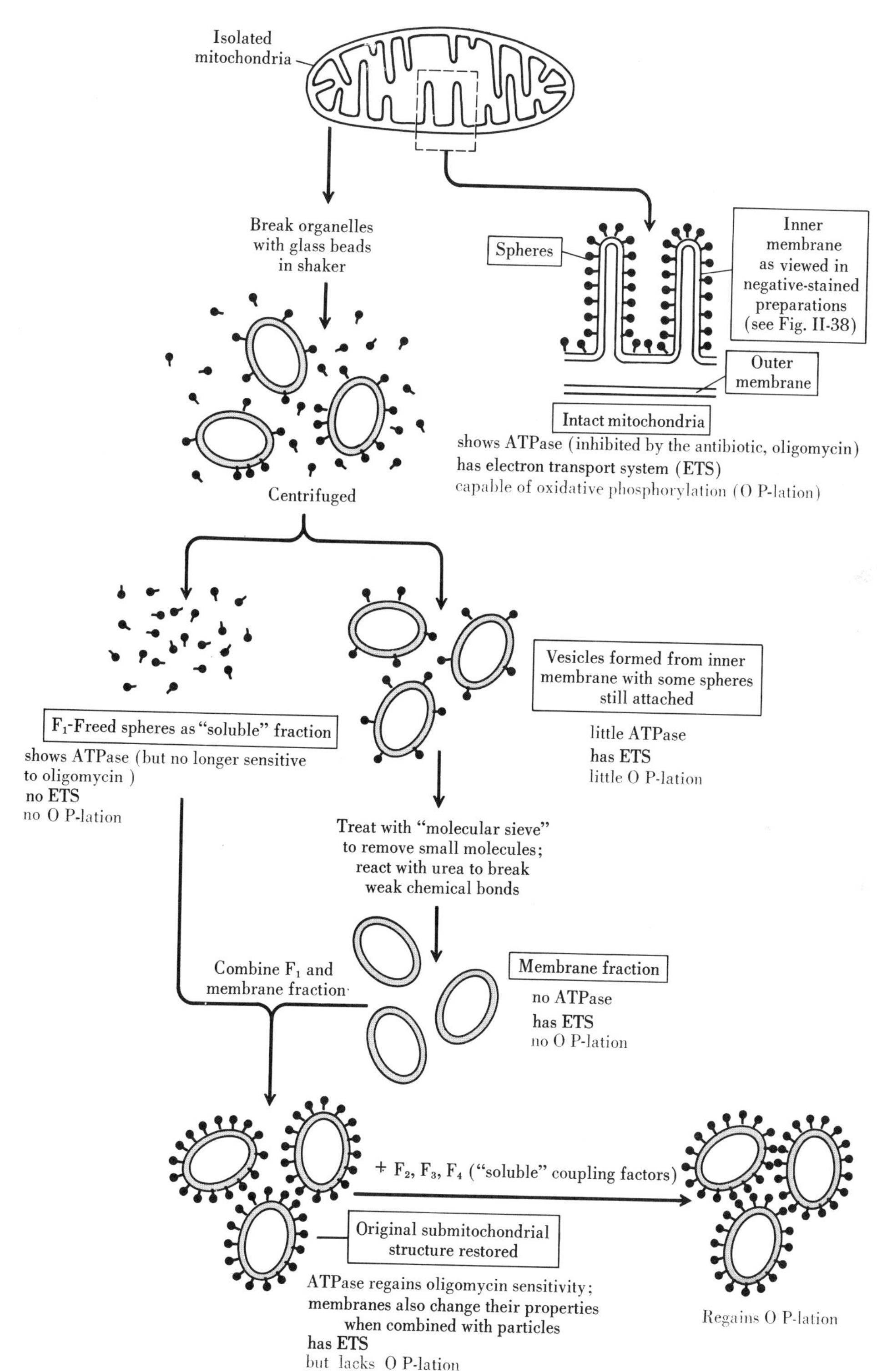

Figure 6.9. *Experimental degradation and reconstitution of the particles of the mitochondrial inner membrane. When the F_1 (racket-like) bodies are removed with urea, the particles are nonphosphorylating, although they still carry out respiration. When the coupling factors (F_2, F_3, F_4 and F_1) are reassociated, they perform both respiration and phosphorylation. (From Novikoff and Holtzman, 1970: Cells and Organelles. Copyright 1970 by Holt, Rinehart and Winston, New York. Used by permission of Holt, Rinehart and Winston, Inc.)*

pedicels 3.5 nm long (see Fig. 6.9), these proved to be solely ATPases (see below).

When mitochondria, separated from other organelles of ruptured cells, were fragmented by shaking with glass beads, small vesicles were formed that still carried out oxidation of products of glycolysis and breakdown of ATP (ATPase activity), although some of the F_1 bodies had been removed. Treatment with oligomycin (an antibiotic) inhibits ATPase activity but not other enzyme activities. The vesicles can be freed of the remaining F_1 bodies by treatment with urea. The vesicles, now consisting only of mitochondrial membranes, lack ATPase activity but carry on oxidation of products of glycolysis. Oligomycin has no effect on their enzymatic activity. The F_1 bodies attached to the inner mitochondrial membrane that are removed by fragmentation can be separated from the vesicles by fractional centrifugation because they are so much smaller. If a suspension of the F_1 bodies is combined with a suspension of vesicles freed of such bodies, the two recombine and all three mitochondrial enzymatic activities are once again evident: oxidative phosphorylation, electron transport, and ATPase activity (Fig. 6.9). The preparation now becomes sensitive to oligomycin as in the original fragments of the mitochondria (Racker, 1970). The F_1 bodies are therefore the site of the ATPase.

It has become possible to separate the two mitochondrial membranes, inner and outer, by selective detergent action inducing swelling, followed by fractional centrifugation. The outer membrane is lighter (40 percent lipid) than the inner membrane (20 percent lipid). Isolated inner membranes contain much of the mitochondrial matrix and the respiratory enzymes. The inner membrane has the enzymes for oxidative processes and for phosphorylations producing ATP. Details of these reactions are given in Chapter 12.

Since very small mitochondrial fragments are capable of all three types of enzymatic activity described, it is evident that the enzymes concerned are present in repetitive assemblies on the inner mitochondrial membrane. The repetitiveness is visualized only in the case of the ATPase, which apparently is localized in the F_1 bodies. Calculations suggest 650 respiratory assemblies in each square micrometer of membrane surface. The characteristics of the outer mitochondrial membrane are less well known than those of the inner membrane, but evidence is present for some oxidative enzymes on the inner surface (Novikoff and Holtzman, 1976).

It has already been pointed out that mitochondria make intimate contact with some portions of the cell involved in active work, being present, for example, among the fibrils in muscle cells (see Fig. 23.12). They have also been described in contact with the ribosomes on the endoplasmic reticulum, and it has been suggested that they might supply energy for protein synthesis directly. If this were true, all dividing cells and secretory cells actively producing proteins should show more frequent contacts than they actually do. Ultimately, however, mitochondria supply energy for most eukaryotic cell work.

Biogenesis: Origin and Development

When cells divide, mitochondria orient themselves relatively symmetrically on either side of a dividing cell and are partitioned fairly equally between the two daughter cells (Kroon and Saccone, 1974). They then divide to reconstitute the number of mitochondria characteristic of the cell. Dumbbell-shaped mitochondria have been seen in fixed preparations, and Frederic (1958) has observed by cinematography the entire process of fission (as well as fusion in some cases). When only one mitochondrion is present, it divides *before* the nucleus does, and the spatial organization in the cell is such as to insure passage of one mitochondrion into each cell (Gunning and Steer, 1975). A mitochondrion always arises from a mitochondrion; no evidence exists that they arise anew. In early embryonic stages the mitochondria may be small and undifferentiated, but they can be identified even in gametes (eggs and sperm).

Luck (1963) attacked the problem of mitochondrial origin by 10-minute pulse labeling with ^{14}C-choline for a choline-requiring mutant of the mold *Neurospora*. The choline was found incorporated into the mitochondrial phospholipid lecithin. At various times after the "washout with nonradioactive choline" samples were removed and tested. It was found that the label, measured by autoradiography, persisted in each mitochondrion through three mass-doubling cycles of growth; the radioactivity was distributed among all the mitochondria observed and decreased with each cell division. The experiment indicates that the mitochondria grow by addition of lecithin to the existing mitochondrial framework, which in turn suggests that mitochondria arise by division. Had they arisen anew, newly formed, unlabeled mitochondria should exist side by side with labeled mitochondria, the latter decreasing in proportion with each mass doubling, but with the radioactive grain count per labeled mitochondrion remaining the same. A similar conclusion was reached from experiments with ^{14}C-valine incorporated into proteins of tumor cell mitochondria (Halder *et al.*, 1966; Packer *et al.*, 1976).

On the basis of decay of label it appears that mitochondria are quite labile. Methionine labeled with ^{35}S and acetate labeled with ^{14}C

disappear at rates to give a half-life of about 10 days for cytochrome *c*, (found only in mitochondria of these cells), soluble protein, insoluble protein, and lipid in mitochondria of rat liver cells. A half-life of a similar order of magnitude was found for mitochondria of cells from other organs of the rat, as tested with tritiated water. In other words, all the protein and all the lipid in the mitochondria are replaced about every 20 days (Schatz and Mason, 1974).

Yeast grown anaerobically lacks typical mitochondria, showing only some membranes containing a few enzymes. When the yeast is returned to air, these membranes soon form true mitochondria with all the respiratory enzymes.

DNA Genome and Mutations

Mitochondrial DNA, which constitutes only 0.2 percent of the DNA in an animal cell, is circular (Fig. 6.10) like that of prokaryotic cells (Birky, 1976). Mitochondrial DNA occurs in multiples, from two to six molecules being present per mitochondrion in most cells tested, but as many as 50 are present in a yeast mitochondrion. DNA base ratios in mitochondrial DNA resemble those of prokaryotic cells more than those in nuclei of the eukaryotic cells in which they occur (Saccone and Quagliariello, 1975).

Mitochondrial DNA forms a loop from which replication begins and passes in one direction. A large proportion of the DNA may occur in the looped form. However, only a small proportion of the DNA replicates at any one time; replication is thus not synchronized and continues throughout the cell cycle not in coordination with cyclic nuclear DNA replication. DNA synthesis is slower in mitochondrial than in nuclear DNA. That mitochondrial DNA is independent of nuclear DNA is shown by selective drug inhibition; nuclear DNA replication can be reduced by 90 percent without an observable effect on mitochondrial DNA replication.

The molecular weight of mitochondrial ribosomes is low like that of prokaryotic ribosomes rather than like that of cytoplasmic ribosomes of the same cell.

Mitochondrial DNA codes for intrinsic ribosomal RNA, transfer RNA, ribosomal proteins and a few other proteins, those regulating enzymes and structure (Schatz and Mason, 1974; Tsagoloff, 1977). This limited array might be expected considering the small size of the DNA molecule, in most mammalian cells only about 5 μm, although in yeast it is much larger (26 μm). Many of the mitochondrial proteins, especially enzymatic ones, are coded by nuclear DNA (Fig. 6.11). All soluble enzymes of the mitochondrial matrix and many of the respiratory enzymes on the membranes are formed outside and enter the organelle to be

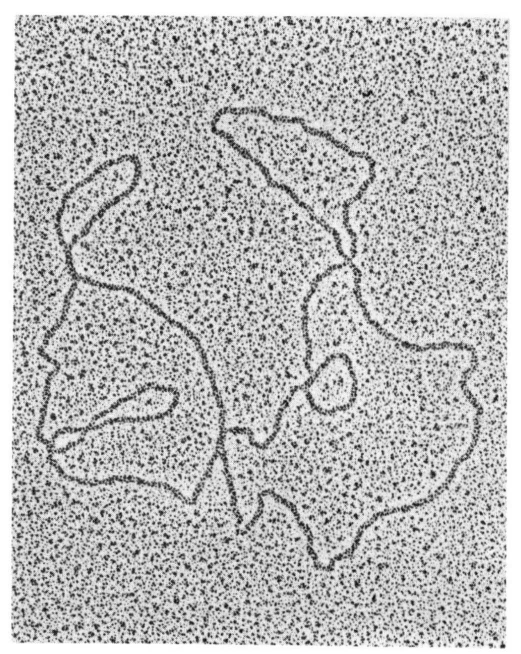

Figure 6.10. *DNA from mitochondria of the toad* Xenopus *forming a circle 5.6 to 7.0 μm long. (By courtesy of I. B. Dawid and D. R. Wolstenholme. From Bloom and Fawcett, 1975: A Textbook of Histology. 10th Ed. W. B. Saunders Co., Philadelphia, p. 49.)*

integrated into the membranes. Whatever the evolutionary origin of the mitochondrion, this indicates that a close relationship has developed between the mitochondrion and the nucleus. The mitochondrion, in spite of possessing its own DNA, is not as independent of the nucleus as it first appears to be.

Mitochondrial DNA is capable of mutation; in *Neurospora* mutants of the types called "pokey," the structural protein is altered at the site of one amino acid. Furthermore, injection of mutated mitochondria into wild type *Neurospora,* but not injection of nuclei of such mutants, induces changes characteristic of the particular mitochondrial mutant. Mutations not only originate in mitochondria but also can be transmitted by them in "cytoplasmic inheritance" (inheritance not determined by nuclear genes). That mitochondrial gene changes may influence structures in the cell other than mitochondria is shown by the "pokey" *Neurospora* mutant, in which a mutation in the mitochondrial DNA induces changes also in the structural proteins of microsomal and nuclear membranes.

Inasmuch as a mutation in the mitochondria is an independent event transmitted to descendants of that cell line, an organism may have cell mosaics (partly of one genotype, partly of another) that are dependent upon the mutations which have occurred in mitochondria, even when the nuclei of the cells have maintained nuclear identity. Mitochondrial change

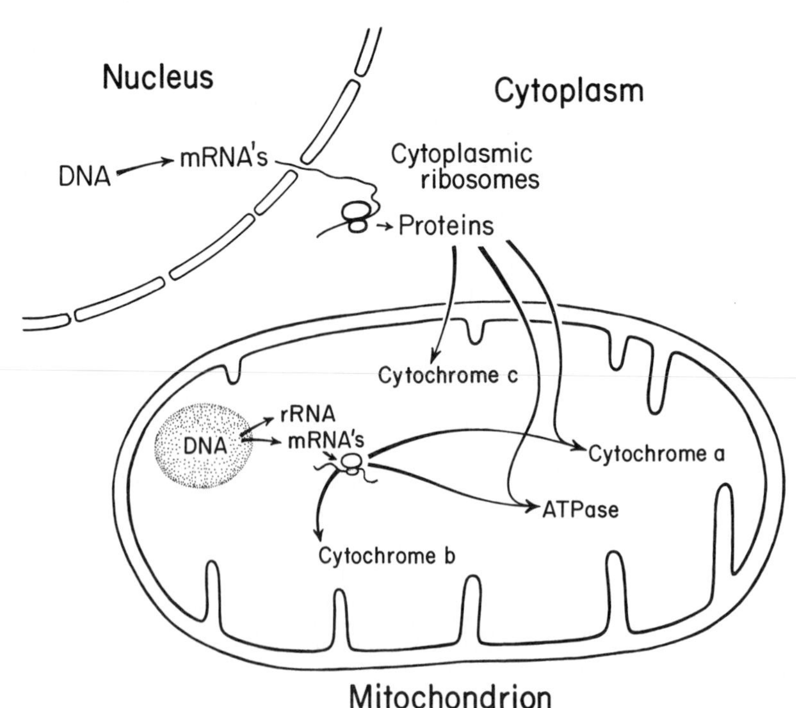

Figure 6.11. Diagramatic representation of transcription of messenger RNA on DNA and translation of the message into synthesis of proteins (protein part of cytochromes, ATPase, and ribosomes) functioning inside the mitochondrion. (After Birky, 1976: Bioscience 26: 26–33.)

is therefore an added source of heterogeneity in living things (Birky, 1976).

CHLOROPLASTS

Chloroplasts are one type of plastid found in higher plants, the major other types being *amyloplasts* (starch-storing) and *chromoplasts* (storing other than photosynthetic pigments). All plastids have a common origin as protoplastids and share many common characteristics (Gunning and Steer, 1975).

Chloroplasts, first seen by Leeuwenhoek, are photosynthetic organelles of eukaryotic plants. Like mitochondria, they are enclosed in double membranes. Being larger than mitochondria and colored, most are easily seen with the light microscope. Chloroplasts may be single as in some algae, *Spirogyra* for example, or many, generally 20 to 40 per cell, occasionally 500 or even more.

Chloroplasts may be spherical, discoidal or ovoid, except in algae where the chloroplast may be cup-shaped *(Chlamydomonas),* spiral *(Spirogyra)* or star-shaped *(Zygnema).* Single chloroplasts are often quite massive (see Fig. 2.7).

Chloroplasts may be isolated by centrifuging macerated leaves. Some chloroplasts are injured during maceration. If green plant cells are enzymatically liberated from their cell walls as protoplasts before the chloroplasts are isolated, the chloroplasts obtained are more normal and active than those from ground leaves. They maintain a level of photosynthesis comparable to that in the leaf (Ratham and Edwards, 1976).

Chloroplasts in cells of shade plants are larger than those in sun plants, the commonest ones being 4 to 6 μm in major axis, although some may be as large as 10 μm. Single chloroplasts of algal cells are usually much larger.

Chloroplasts, like mitochondria and nuclei, are enveloped by two membranes separated by a gap 10 to 20 nm thick. The membranes resemble the plasma membrane and are of about the same thickness (60 nm). The outer membrane is smooth, and the inner membrane is much folded into the chloroplast *stroma;* this provides considerably more surface area than would a smooth membrane. It is the inner membrane that has highly selective permeability to molecules, reminiscent of the inner mitochondrial membrane.

Organization

The chloroplast somewhat resembles the mitochondrion: both organelles are enveloped by two membranes and both have enzyme-bearing internal membranes. Both are also concerned with energy transductions in the cell. However, although chlorophyll-bearing membranes originate from the chloroplast inner membrane, they later lie free in the stroma (the nonmembranous chloroplast matrix), whereas the mitochondrial cristae remain as invaginations of the inner membrane even

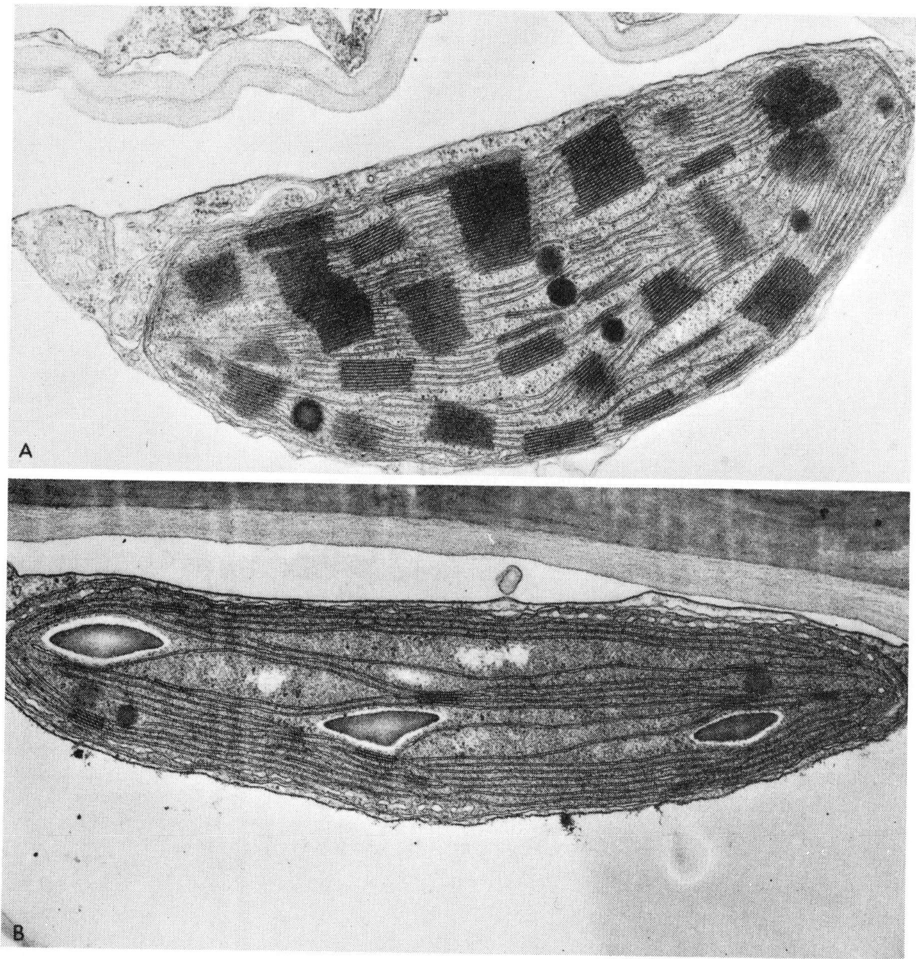

Figure 6.12. A, *Fine structure of a chloroplast from the mesophyll cells of maize (× 20,000). Note the massed thylakoids forming the grana and the dark globules of uncertain function.* B, *Fine structure of a chloroplast from a cell forming a sheath around a vascular bundle in maize (× 20,000). Note the lack of grana and the three large spindle-shaped starch granules. A peripheral reticulum is seen around the borders of the chloroplast. The significance of these two types of organization of chloroplasts is considered in Chapter 14 on photosynthesis. (By courtesy of D. G. Bishop. From Bishop* et al., *1971:* In *Photosynthesis and Photorespiration. Hatch, Osmond and Slatyer, eds. John Wiley and Sons, New York.)*

in mature cells. In section, under the electron microscope, the internal chloroplast membranes, called *thylakoids,* appear to be a series of lamellae. Here and there the lamellae are punctuated with ovoid, more densely lamellated units called *grana.* However, the thylakoid is actually a single closed, flattened continuous sac in the chloroplast (Figs. 6.12 and 6.13) (Gunning and Steer, 1975).

Under an electron microscope a granum appears to consist of a series of membrane discs packed back to back, seemingly like a stack of coins. However, each disc is interconnected at an angle to all the other discs in a granum by tubules called *frets.* By branching, a fret connects a disc to each of the other discs in turn (Fig. 6.14).

Thylakoids provide a large membrane area to hold the photosynthetic pigments and enzymes. For example, 10 granum discs, each 0.25 μm in diameter, provide an area of 200 μm^2. Thylakoids, containing the photic apparatus for photosynthesis, permit separation of the light reactions that occur there from the dark reactions in the chloroplast stroma that fix carbon dioxide into carbohydrates (or organic acids). ATP and reduced coenzymes diffuse from the thylakoids where they are formed into the stroma, where they are used as energy source and reductant, respectively, for the fixation of carbon dioxide.

Maintenance of grana and frets in a chloroplast requires the presence of inorganic salts. If the salts are replaced by organic buffer, the granum opens up to form an extensive membrane system. Replacement of the ions leads to spontaneous reformation of grana and frets, this conformation apparently representing a

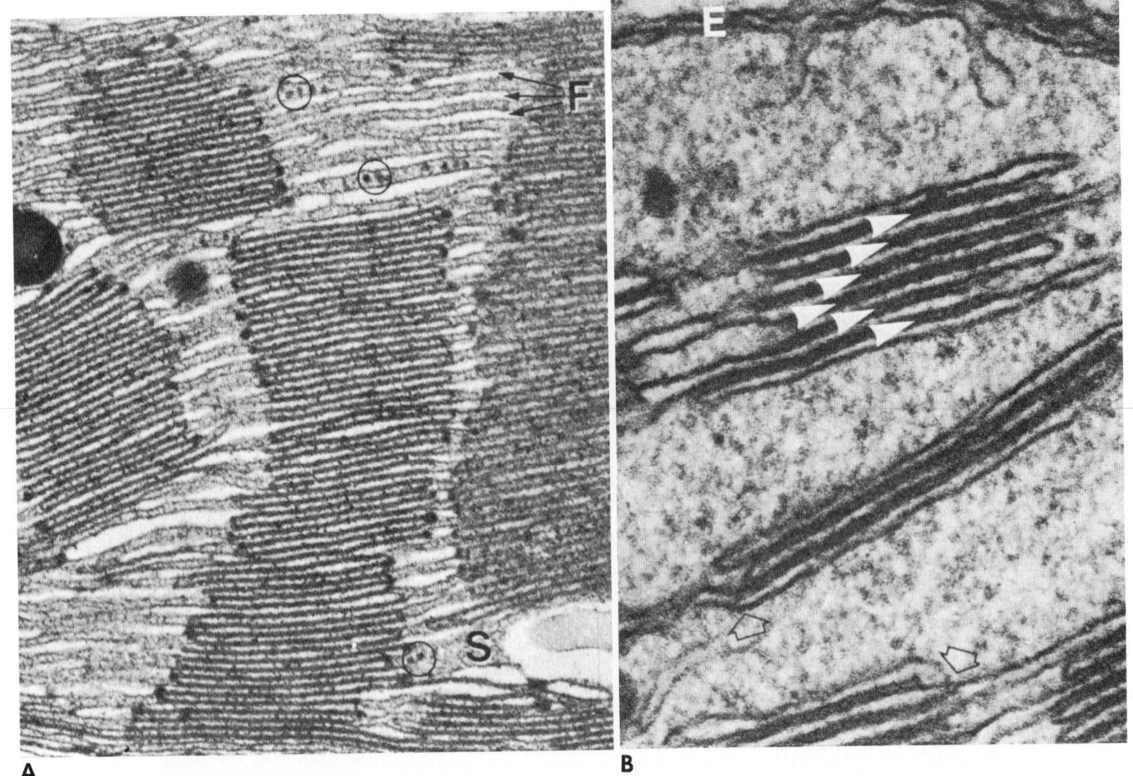

Figure 6.13. A, *Grana of a corn cell* (Zea mays) *(× 75,000). Note that the stroma (S) is much divided by the frets (F). A few ribosomes (circled) can be seen in the stroma. The frets are somewhat swollen. (From Gunning and Steer, 1975: Biology of Plant Cells. An Ultrastructural Approach. Edward Arnold, London, p. 247.)* B, *Ultrathin section of lupine leaf chloroplast showing the chloroplast envelope (E) and the internal membrane system. Note invaginated envelope at the top and right. A space between the granum discs is visible (white arrow). Continuities between fret channels and discs are shown by open arrows (× 140,000). (From Gunning and Steer, 1975: Biology of Plant Cells, An Ultrastructural Approach. Edward Arnold, London, p. 249.)*

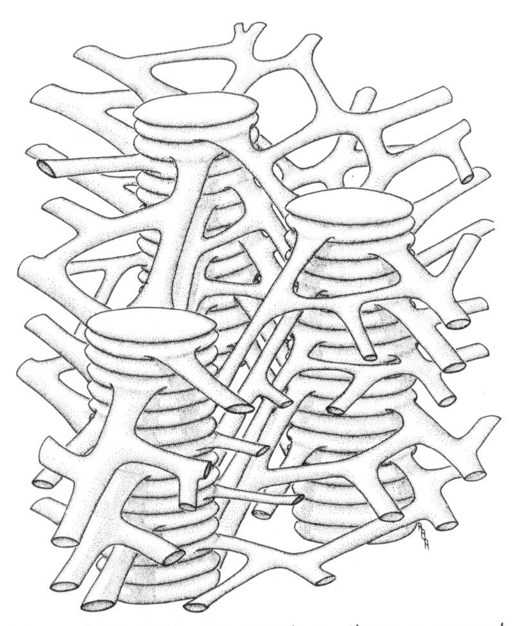

Figure 6.14. This diagram shows three grana and their associated frets. The fret system is highly dissected and interconnects not only grana but also different discs within a granum. (From Weier, Stocking, Thomson and Drever, 1963: J. Ultrastruct. Res. 8: 122–145.)

lower energy state than the open membrane.

Grana may be separated as individual bodies from the rest of the organelles either mechanically or by detergents. Grana vary in size from 0.25 to 1.7 μm in various species; in spinach chloroplasts they are about 0.6 μm in diameter.

The structure of lamellae and grana has been determined from studies with the electron microscope. The layers making up a single membrane of a granum differ in electron opacity. Chemical analysis of isolated grana indicates that, per unit weight, they contain more lipids of various types and more chlorophyll than the other membranous portions of the chloroplast. Using inferences from all available data, the structure of lamellae and grana has been reconstructed as shown in Figure 6.15.

A freeze-fractured chloroplast is shown in Figure 6.16*A*. Thylakoids are evident in the preparation (arrows). Grana are also visible in the upper left hand corner.

Still smaller bodies are present inside the grana (Fig. 6.16*B*); these are called the *quantasomes*. Each quantasome contains 250 molecules of chlorophyll, the minimum amount found necessary for photosynthesis.

Grana may be present in chloroplasts of a plant in some locations and absent from those

Figure 6.15. *Diagram of probable macromolecular structure of grana of chlorophyll. "Its" is the intrathylakoid space. (By courtesy of A. J. Hodge.)*

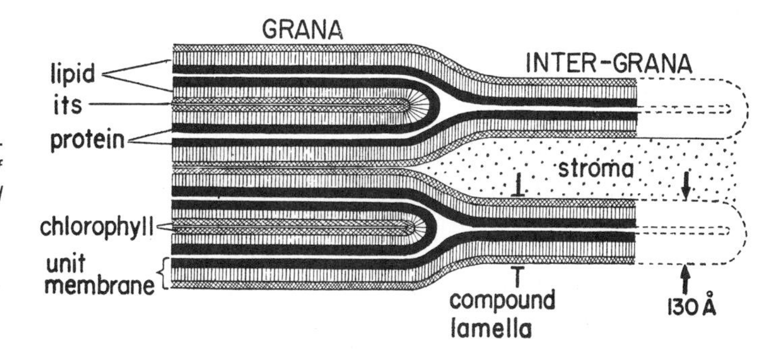

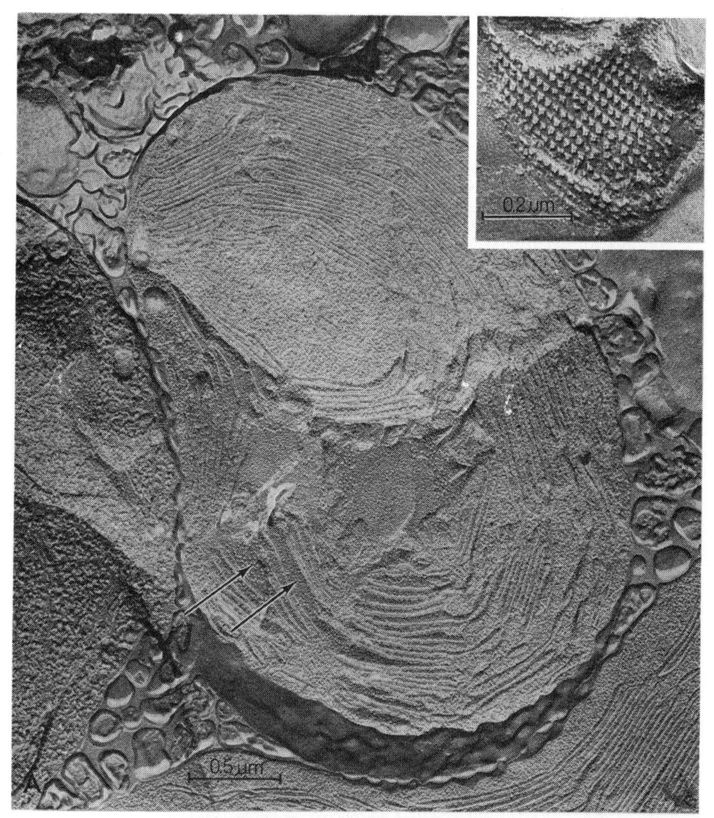

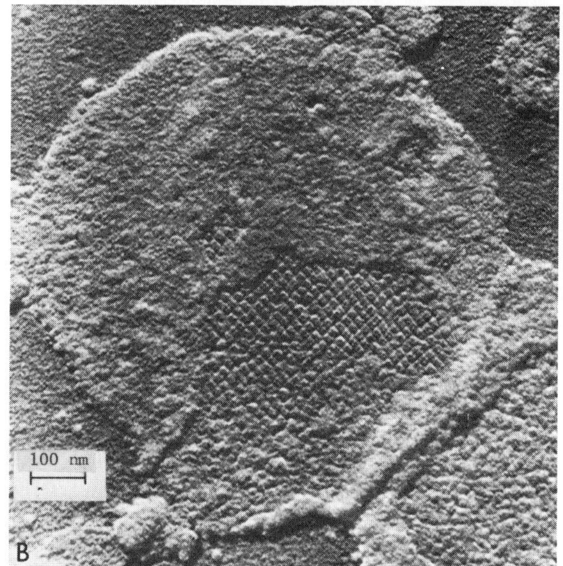

Figure 6.16. A, *Freeze etching of a spinach chloroplast. Arrows point to face views of thylakoid membranes; insert shows quantasome array. (From Branton, 1968: Photophysiology* 3: *212.)* B, *A single lamella of a granum in a spinach leaf cell, showing a quantasome. Osmium tetroxide fixed, heavy metal-shadowed preparation. Each quantasome is about 15.5 nm long, 18 nm wide and 10 nm thick; a quantasome has a molecular weight of about* 2×10^6 *and contains 230 chlorophyll molecules. (By courtesy of Dr. R. Park. From Park and Biggins, 1964: Science* 144: *1010.)*

in others; for example, in maize the mesophyll chloroplasts have grana, but sheath parenchyma chloroplasts do not (Fig. 6.12) (Bishop *et al.*, 1971). Grana appear to be absent from algae.

Chemical Composition and Molecular Organization of Constituents

Chloroplasts are unable to live outside the cell, although under favorable conditions they carry out photosynthesis when provided with the appropriate chemicals. The more delicate the method of chloroplast extrusion the higher the rate of the reactions, although they probably never quite equal the rate in the leaf. However, it can be assumed that findings on isolated chloroplasts bear a fairly close resemblance to what occurs in normal leaf chloroplasts.

Chloroplasts, isolated by centrifugation of mechanically disrupted spinach leaf cells, contain about 56 percent protein, 32 percent lipids and 8 percent chlorophyll. Among the lipids are carotenoids and xanthophylls, as well as triglycerides, steroids, phospholipids, sulfolipids, and galactolipids. The chlorophyll exists in combination with some of the protein and with carotenoids. Combination of chlorophyll (see Fig. 2.2) with β-carotene (see Fig. 14.6) makes separation of the two pigments more difficult. In mutants lacking carotenoids that have many (unsaturated) double bonds, illumination in presence of oxygen leads to damage from chlorophyll-catalyzed photooxidation. It is therefore believed that by serving as a reductant, β-carotene (with many unsaturated bonds) protects chlorophyll from photooxidation in wild-type cells.

A thylakoid sac, whether in a granum or intergranal, is always double; the space between the two layers is called the *intrathylakoid space* (Fig. 6.15). To study the membrane components it is necessary to use more delicate methods than for study of their gross chemistry. When such methods are used, it becomes evident from electron micrographs that, much as in any membrane, each thylakoid membrane is composed of lipid bilayers, with hydrophobic groups facing inward, and hydrophilic groups in contact with proteins inserted through the membrane (see Chapter 8). The proteins are of two types: extrinsic and intrinsic. The extrinsic proteins are extractable with ease because they are not firmly bound and are removed by mild treatment with hydrophilic reagents. The intrinsic proteins, to which the extrinsic proteins are weakly bound, are embedded deep in the lipid layer, perhaps penetrating through it, and they have hydrophobic groups that interact with the lipids. Intrinsic proteins are removed with more difficulty and require hydrophobic reagents. The surface of the thylakoid facing the chloroplast stroma has more extrinsic protein than the surface bordering another thylakoid, one a protein called the *coupling factor*. The coupling factor may be involved in stacking for formation of grana. Some of the proteins in the thylakoids are enzymes concerned with the photosynthetic light reactions (ATP production and formation of reductant) and are discussed in Chapter 14. Enzymes for the dark reactions occur in the stroma (Anderson, 1975). The chloroplast is known to be a complete photosynthetic unit (Arnon, 1955).

Lipids form half the mass of a spinach chloroplast membrane and some, sulfolipids and galactolipids, are unique to chloroplasts. Galactolipids form 40 percent of the total membrane lipid. Present also are polyunsaturated fatty acids, the main one linolenic (C_{18}) acid (Anderson, 1975). Chlorophyll, making up 40 percent of the lipid mass, is probably attached to proteins with the ring structure (see Fig. 2.2) toward the membrane surface, although too little information is presently available for more than suppositions (Anderson, 1975). The thylakoid membranes are most likely in a mobile liquid state so that the molecules shown in diagrams should not be construed to represent more than temporary structures, much as in the plasma membrane (see Chapter 8).

Biogenesis

During development chloroplasts originate from minute submicroscopic amoeboid *proplastids*. The proplastids have double membranes like the mitochondria. When light is available and the proplastid reaches a diameter of 1 μm, its inner membrane invaginates to form lamellae and differentiates to the mature condition found in the chloroplast (Fig. 6.17). The sensor that absorbs the light and mediates chloroplast development has been found to be a bluish pigment, phytochrome, present in minute amounts (see Fig. 24.14*B*) (Mohr, 1978). Chloroplasts have also been seen to elongate and divide in some plants (Kamiya, 1972).

When a culture of *Euglena* is kept in the dark, chloroplast lamellae and chlorophyll disappear. Restoring the cells to the light is followed by the simultaneous recovery of lamellation and chlorophyll content. Chloroplasts remain sensitive to light and their structure continues to be controlled by exposure to light (Schiff, 1974). Chlorophyll development may also be prevented by some antibiotics.

When leaves receive little light, the precursor for development of thylakoid lamellae may be present but the lamellae fail to develop, although a prolamellar body may be present (Fig. 6.18). In dim light a few lamellae develop; this results in an *etioplast* rather than a chloroplast, as seen in etiolated (elongate yellow-green) plants, such as sprouts on potatoes kept in darkness.

DNA Genome

Chloroplasts contain DNA and are capable of self-duplication. Chloroplast DNA is distinct from nuclear DNA, as shown by attempts to conjugate the two DNA's after they are opened by heat treatment (see Chapter 7). The DNA may appear in a chloroplast as a number of nucleoids resembling those of bacteria. Each nucleoid may be polyploid, consisting of a number of circular DNA molecules.

The larger the chloroplast the larger the nucleoid(s). During growth the DNA in chloroplasts of a species increases. For example, in a chloroplast of a young beet plant only 0.9 μm^3 DNA may be present; in a chloroplast of a mature plant it increases to 3.7 μm^3 (Gunning and Steer, 1975; Birky, 1976).

Although the chloroplast DNA is of higher molecular weight than mitochondrial DNA of the same species, only a fraction of the chloroplast protein is synthesized in chloroplasts. Nuclear control of chloroplast characters is evident, as is nuclear control in some mitochondrial characters (Birky, 1976).

Because some chloroplast genes affect characters of the organelle, uniparental heredity (formerly called maternal heredity) is observed. Such heredity has been extensively studied in the green alga *Chlamydomonas* (Sager, 1972).

Chloroplasts have the three types of RNA required for synthesis of protein: ribosomal, transfer (to carry the activated amino acids), and messenger (giving the ribosome information for the kind of protein to be synthesized). As in mitochondria, the ribosomes in chloroplasts are small, like those in prokaryotes. Some proteins, lipids, and carbohydrates needed for chloroplast function enter the organelle from

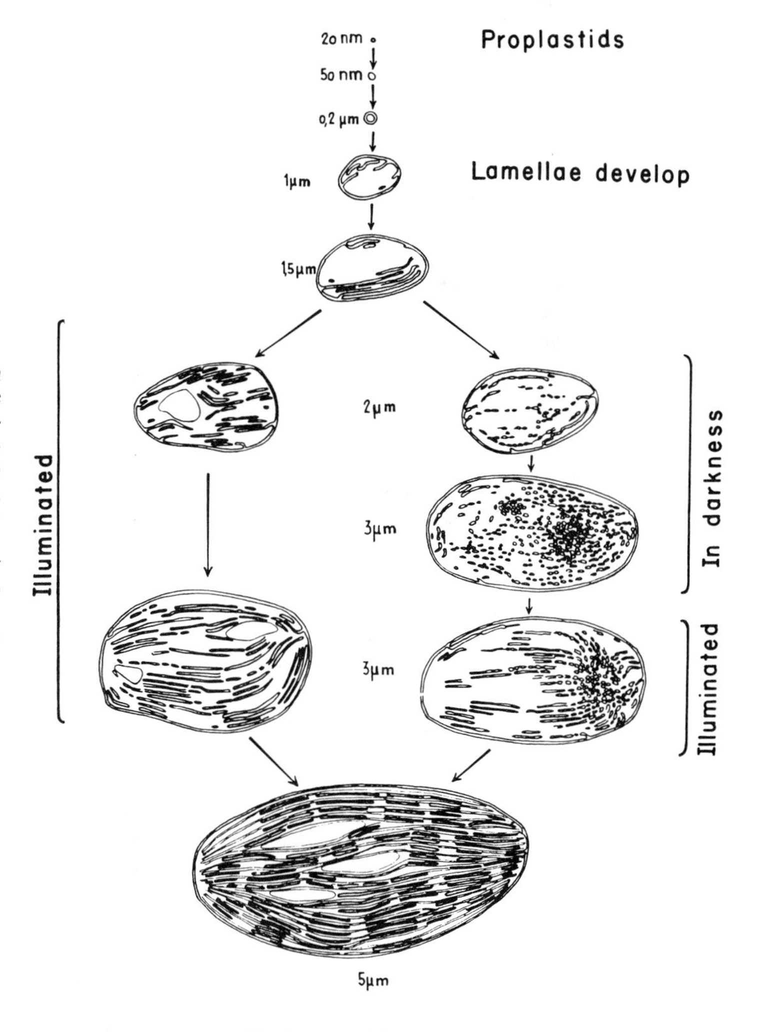

Figure 6.17. *The development of a chloroplast from a submicroscopic proplastid. When the illuminated proplastid is 1 nm in diameter, its inner surface membrane begins to invaginate to form lamellae. The proplastids grow and, in light, progressively develop lamellae to form the mature chloroplast* (left). *In darkness, however, the lamellae break up into vesicles* (right), *but on reillumination the vesicles unite to reform the lamellae. (From Muhlethaler and Frey-Wyssling, 1959: J. Biophys. Biochem. Cytol. 6: 509.)*

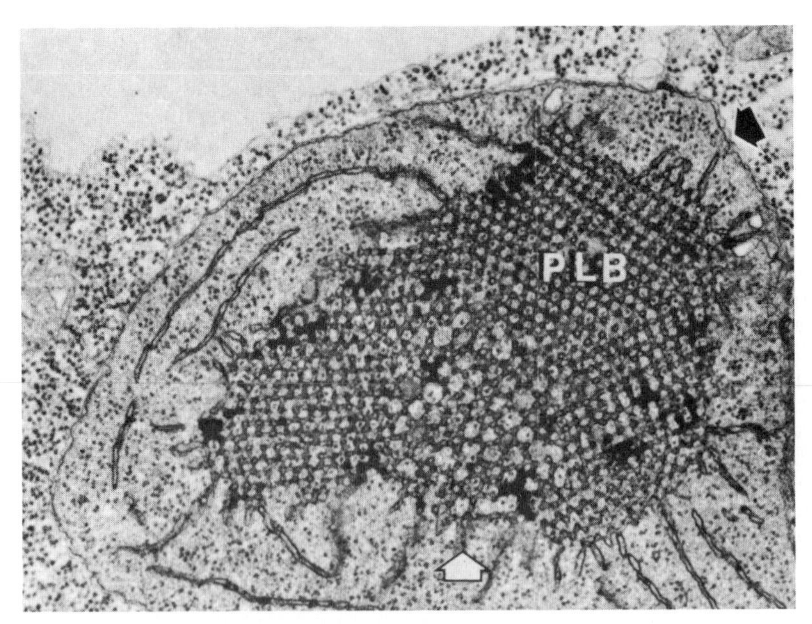

Figure 6.18. *Prolamellar body from an etiolated oat plant. Note the two surrounding membranes of the chloroplast (dark arrow) and numerous plastid ribosomes. The prolamellar body (PLB) has semicrystalline lattices that develop into lamellae on illumination (× 47,000). (From Gunning and Steer, 1975: Ultrastructure and the Biology of Plant Cells. Edward Arnold, London, p. 255.)*

the cytoplasmic matrix. Chloroplast ribosomes associate to form polysomes for synthesis of proteins, just as they do in the cytoplasm (Gunning and Steer, 1975).

Chloroplast DNA in *Euglena* is much more sensitive to short wavelength ultraviolet radiation than nuclear DNA. Chloroplast precursors (proplastids) containing DNA are inactivated by exposure to such radiation at doses that have no detectable effect on cell division. Colorless colonies are formed as a consequence (in darkness). However, colorless colonies can be maintained only if nutrients normally obtained by photosynthesis are supplied (Diamond *et al.*,

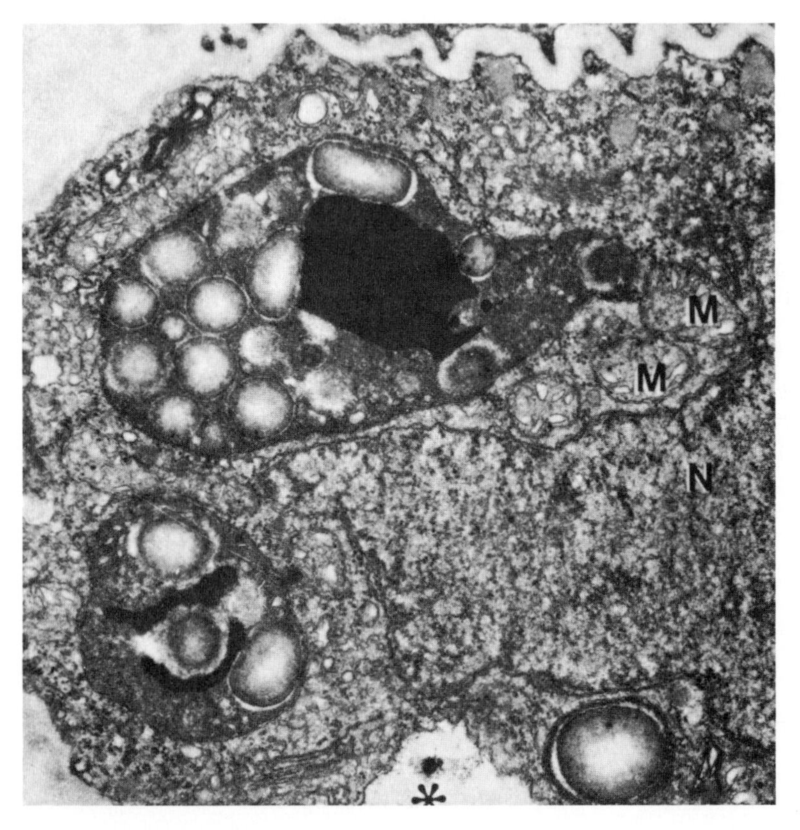

Figure 6.19. *Amyloplast from young root cap cell of* Cosmea. *Note the roundish starch grains and electron dense (black) inclusion, probably a lipoprotein. Mitochondria (M) and part of the nucleus (N) are evident, and the asterisk shows electron dense particles in a vacuole (× 26,000). (From Gunning and Steer, 1975: Ultrastructure and the Biology of Plant Cells. Edward Arnold, London, p. 261.)*

1975). The high sensitivity of chloroplast DNA is reminiscent of like sensitivity of bacterial DNA as compared to DNA in eukaryotic cells.

AMYLOPLASTS

All plant plastids begin development as proplastids in much the same manner. *Amyloplasts,* starch-storing organelles, are characteristically found in regions of little or no illumination, for example, in the potato tuber. Initially they may have a few thylakoid membranes, but these are ultimately obscured as starch grains accumulate. Amyloplasts have nucleoids and ribosomes; however, except in early stages of development, they are overshadowed by stored starch (Fig. 6.19). Long-storage amyloplasts represent the extreme form of the organelle.

As implied, amyloplasts of all types, from chloroplasts to long-term storage amyloplasts, are found in higher plants. By day chloroplasts temporarily accumulate starch grains in the pyrenoid (starch-forming organelle). At night the starch is translocated to regions of growth or storage. Starch is generally stored in amyloplast stroma. However, in red algae the starch (called floridean starch) is stored in the cytoplasm. Enzymes for polymerizing glucose to starch are present in the amyloplast, but the ATP to energize the synthesis must come from outside the amyloplast—either from chloroplasts in light or from mitochondria in darkness (Gunning and Steer, 1975).

CHROMOPLASTS

Strictly speaking, chromoplasts are nonphotosynthetic plastids storing red or brown pigments (see Fig. 6.20). Chromoplasts may develop from chloroplasts by accumulation of nonphotosynthetic pigments to the exclusion of chlorophyll function, for example, the red carotenoid, lycopin, in tomatoes. Chromoplasts account for the bright red, orange, and yellow colors of flowers. Chromoplast pigments occur as droplets, filaments, or crystals and the crystals may become so large as to disrupt the chromoplast.

Although chromoplasts have DNA, the genes that control pigment synthesis are located in the nucleus. Genic control of carotenoid synthesis is interestingly shown in corn *(Zea mays),* in which a mutant nuclear gene codes for accumulation of the tomato pigment lycopin instead of β-carotene. It is thought that carotenoid accumulation in carrots is the result of a natural mutation selected by man (Gunning and Steer, 1975).

SUMMARY

Mitochondria and chloroplasts have a number of features in common: envelopment in double membranes, enzyme-encrusted membranes developing from ingrowth of the inner enveloping membrane, and energy-transducing action. However, differences also exist: the cristae of mitochondria continue their attachment to the inner membrane, whereas

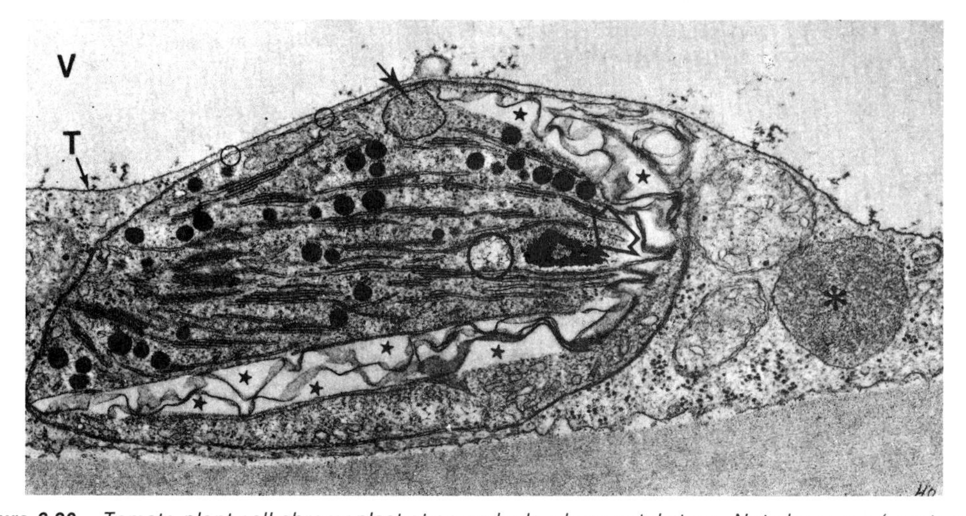

Figure 6.20. Tomato plant cell chromoplast at an early developmental stage. Note lycopene (carotenoid) crystals (stars), the thylakoids, the double envelope (small circles), the membrane-bound inclusion (small arrow), electron dense mass (large arrow), nucleoid area (large area). Outside the chromoplast note the large vacuole (V) bounded by a membrane (T) and to the right the two mitochondria and a microbody (asterisk). (From Gunning and Steer, 1975: Biology of Plant Cells. An Ultrastructural Approach. Edward Arnold, London, p. 263.)

during development all trace of such attachment disappears from chloroplast thylakoids. Mitochondria liberate ATP and reductants from breakdown of glucose and other compounds, thylakoids produce both ATP and reductants by action of light. Photosynthetic plant cells with chloroplasts also contain mitochondria, but mitochondrial activity accounts for less than 1/20 of the ATP and reductant found in the cell. Both mitochondria and chloroplasts divide and distribute about equally between the two daughter cells following division.

In both mitochondria and chloroplasts the outer enclosing membrane is freely permeable to small molecules while the inner one is highly selective, retaining compounds of importance to the organelle while permitting others to pass inward when they are needed in the cell. Both organelles swell in solutions of osmotic strength lower than the surrounding cytoplasm and shrink in solutions of higher osmotic strength. The cristae and chloroplast thylakoids are similar in containing an assortment of structured enzyme assemblies important for their respective functions, but different in that pigmentary systems play a prominent part in thylakoid structure. Mitochondria are present in differentiated gametes; proplastids are present in plant gametes.

In both mitochondria and chloroplasts the genome consists of nucleoids that themselves consist of one or more circular DNA molecules. The genome of mitochondria is smaller than that of chloroplasts. Both organelles have ribosomes and synthesize some of their organellar proteins, but many of their proteins are synthesized in the cytoplasm under nuclear control and later enter the organelles. Amyloplasts and chromoplasts, like chloroplasts, originate from proplastids and share many of their properties with chloroplasts.

LITERATURE CITED AND GENERAL REFERENCES

Anderson, J., 1975: The molecular organization of chloroplast thylakoids. Biochem. Biophys. Acta *416*: 191–235.

Arnon, D. I., 1955: The chloroplast as a complete photosynthetic system. Science *122*: 9–16.

Bereiter-Hahn, J., 1976: Fine structure and dynamic shape relations of mitochondria. Cytolobiologie *12*: 429–439.

Birky, C. W. J., 1976: The inheritance of genes in mitochondria and chloroplasts. Bioscience *26*: 26–33.

Bishop, D. G., Anderson, K. S. and Smellie, R. M., 1971: Lamellar structure and function in relation to photochemical activity. *In* Photosynthesis and Photorespiration. Hatch, Osmond and Slatyer, eds. John Wiley & Sons, New York, pp. 372–381.

Borst, P., 1969: Biochemistry and function of mitochondria. *In* Handbook of Molecular Cytology. Lima-de-Faria, ed. North Holland Publishing Co., Amsterdam, pp. 914–942.

Brunner, G. and Neupert, W., 1968: Turnover of outer and inner membrane proteins of rat liver mitochondria. Fed. Eur. Biochem. Soc. Lett. *1*: 153–155.

Calvayrac, R., Umlente, R. and Butow, R. C., 1971: *Euglena gracilis:* formation of giant mitochondria. Science *173*: 252–254.

Carafoli, E. and Crompton, M., 1976: Calcium ions and mitochondria. Soc. Exp. Biol. Symp. *30*: 81–115.

Cohen, S. S., 1973: Mitochondria and chloroplasts revisited. Am. Sci. *61*: 437–445.

Crotty, W. J. and Ledbetter, M. C., 1973: Membrane continuities involving chloroplasts and other organelles in plant cells. Science *182*: 839–841.

Diamond, J., Schiff, J. A. and Kelner, A., 1975: Photoreactivating enzyme from *Euglena* and control of its intracellular level. Arch. Biochem. Biophys. *167*: 603–614.

Fawcett, D. W., 1966: The Cell: Its Organelles and Inclusions. W. B. Saunders Co., Philadelphia.

Frederic, S., 1958: Recherches cytologiques sur le chondriome normal ou soumis à l'expérimentation dans des cellules vivantes cultivées *in vitro*. Arch. Biol. *69*: 167–349.

Gibbs, M., ed., 1971: Structure and Function of Chloroplasts. Springer-Verlag, New York.

Granick, S. and Gibor, A., 1967: The DNA of chloroplast, mitochondria and centrioles. Advances in Nucleic Acid Research pp. 143–186.

Green, D. E., 1964: The mitochondrion. Sci. Am. (Jan.) *210*: 63–74.

Green, D. E. and Bonem, H., 1970: Energy and the Mitochondrion. Academic Press, New York.

Grobe, B. and Arnold, D. G., 1975: Evidence of large ramified mitochondria in *Chlamydomonas reinhardtii*. Protoplasma *86*: 291–294.

Gunning, B. E. S. and Steer, M. W., 1975: Plant Cell Biology. An Ultrastructural Approach. Edward Arnold, London.

Halder, D., Freeman, K. and Work, T. S., 1966: Biogenesis of mitochondria. Nature *211*: 9–12.

Hoffmann, H. P. and Avers, C. J., 1973: Mitochondrion of yeast: ultrastructural evidence for one giant, branched organelle per cell. Science *181*: 749–751.

Kamiya, T., 1972: Cell elongation and division of chloroplasts. J. Exp. Bot. *23*: 62–64.

Kroon, A. M. and Saccone, C., eds., 1974: The Biogenesis of Mitochondria. Academic Press, New York.

Laetsch, W. M., 1971: Chloroplast structural relationships in leaves of C4 plants. *In* Photosynthesis and Photorespiration. Hatch, Osmond and Slatyer, eds. Wiley-Interscience, New York, pp. 323–349.

Lehninger, A. L., 1965: The Mitochondrion: Molecular Basis of Structure and Function. W. A. Benjamin, New York.

Lehninger, A. L., 1975: Biochemistry. 2nd Ed. Worth Publishers, New York.

Luck, D. J. L., 1963: Formation of mitochondria in *Neurospora crassa.* A quantitative autoradiographic study. J. Cell Biol. *16*: 483–499.

Mahler, H. R. and Raff, R. A., 1975: The evolutionary origin of the mitochondria, a non-symbiotic model. Int. Rev. Cytol. *43*: 1–124.

Margulis, L., 1971: Symbiosis and evolution. Sci. Am. (Aug.) *225*: 48–57.

Mohr, H., 1978: Phytochrome and chloroplast development. Endeavour N. S. *1*: 107–114.

Novikoff, A. and Holtzman, E., 1976: Cells and Organelles. 2nd Ed. Holt, Rinehart and Winston, New York.

Packer, L. and Gomez-Puyou, A., eds., 1976: Mitochondria: Bioenergetics, Biogenesis and Membrane Structure. Academic Press, New York.

Park, R. B., 1966: Subunits of chloroplast structure and quantum conversion in photosynthesis. Int. Rev. Cytol. *20*: 67–95.

Racker, E., 1968: The membrane of the mitochondrion. Sci. Am. (Feb.) *218*: 32–39.

Racker, E., 1970: Membranes of Mitochondria and Chloroplasts. Van Nostrand Reinhold Co., New York.

Ratham, C. K. M. and Edwards, G. E., 1976: Protoplasts as a tool for isolating functional chloroplasts from leaves. Plant Cell Physiol. *17*: 177–186.

Saccone, C. and Quagliariello, E., 1975: Biochemical studies of mitochondrial transcription and translation. Int. Rev. Cytol. *43*: 125–165.

Sager, R., 1972: Cytoplasmic Genes and Organelles. Academic Press, New York.

Sato, S., 1972: Mitochondria. University Park Press, Baltimore.

Schatz, G. and Mason, T. L., 1974: Biosynthesis of mitochondrial proteins. Ann. Rev. Biochem. *43*: 51–87.

Schiff, J. A., 1974: The control of chloroplast differentiation in *Euglena. In* Proc. Third Intern. Congress Photosynthesis. Avron, ed. Elsevier, Amsterdam, pp. 1691–1717.

Schwartz, R. M. and Dayhoff, M. O., 1978: Origins of prokaryotes, eukaryotes, mitochondria, and chloroplasts. Science *199*: 395–403.

Siekevitz, P., 1957: Powerhouse of the cell. Sci. Am. (Jul.) *197*: 131–140.

Taylor, D. L., 1970: Chloroplasts as symbiotic organelles. Int. Rev. Cytol. *27*: 29–64.

Tedeschi, H., 1976: Mitochondria. Structure, Biogenesis and Transducing Functions. Springer-Verlag, New York.

Tsagoloff, A., 1977: Genetic and translational capabilities of the mitochondrion. Bioscience *27*: 18–23.

Uzzell, T. and Spolsky, C., 1974: Mitochondria and plastids as endosymbionts: a revival of special creation? Am. Sci. *62*: 334–343.

Wilkie, D., 1970: Reproduction of mitochondria and chloroplasts. Symp. Soc. Gen. Microbiol. *20*: 381–399.

CHAPTER 7

THE NUCLEUS, CHROMOSOMES, AND GENES

In prokaryotic cells under optimal conditions growth and chromosome (DNA) replication are continuous; by contrast, in eukaryotic cells, even under optimal conditions, DNA synthesis in the chromosome is limited to the S (for synthetic) phase of the cell cycle. The entire eukaryotic cell cycle may be abbreviated as $M\text{-}G_1\text{-}S\text{-}G_2)^n$, where M stands for mitosis and cell division, G_1 for gap 1 preceding S, S for DNA synthesis (replication), G_2 for gap 2 succeeding S, and n for the number of cell cycles. In eukaryotes S lasts for about half the cell cycle, and the entire cell cycle is likely to occupy 12 to 24 hours, as compared to 20 to 60 minutes for many bacteria (see Chapters 25 to 27).

The more prolonged cell cycles of eukaryotic cells are partly a result of the greater complexity and larger size of eukaryotic chromosomes, their enclosure in a nuclear envelope, and the mitotic distribution of replicated chromosomes to each daughter cell. However, the main reason for the prolonged cycles is perhaps the slowness of RNA transcription from DNA (see Chapter 15). In both prokaryotes and eukaryotes the genes are composed of DNA and function in a similar manner. Much of the present chapter is devoted to the special organization of the genic substance in the eukaryotic cell.

How, in the course of evolution, the eukaryotic nucleus arose from the naked DNA chromosome in the prokaryote cytoplasm is not known. However, the separation of nucleus and cytoplasm may have become essential for the effective function of an increasingly complex genome, a genome that coded for the numerous organelles and structures characteristic of the eukaryotic cell.

Ever since its discovery by Robert Brown in 1835 the nucleus has been recognized as an important organelle in the eukaryotic cell. Although the behavior of chromosomes during the cell cycle was determined by cytological studies long ago, much of our knowledge of nuclear chemistry and molecular biology is of fairly recent date.

RELATIONS BETWEEN NUCLEUS AND CYTOPLASM

A well-knit and delicately balanced relationship exists between the nucleus and the cytoplasm of a eukaryotic cell, as seen when a cell is enucleated or nuclei of different species are interchanged. For example, when an *Amoeba proteus* is cut in two, one piece with and one without a nucleus (Fig. 7.1), the enucleated half continues to digest food previously ingested but will take in no more, nor will it respond to stimuli. It continues to absorb and bind phosphates (as shown by tracer tests) but at a lesser and continuously declining rate. It will maintain a contractile vacuole already present but will not replace an excised one. Its cytoplasmic membranes (including Golgi complex and endoplasmic reticulum) degenerate (Flickinger, 1968). It soon rounds up and becomes inactive, and eventually dies. The nuclear portion, by contrast, takes in food, reacts normally to stimuli, and replaces an excised contractile vacuole; it grows and in time divides (Jeon and Danielli, 1971).

An amoeba from which the nucleus has been removed by a hooked needle behaves like an

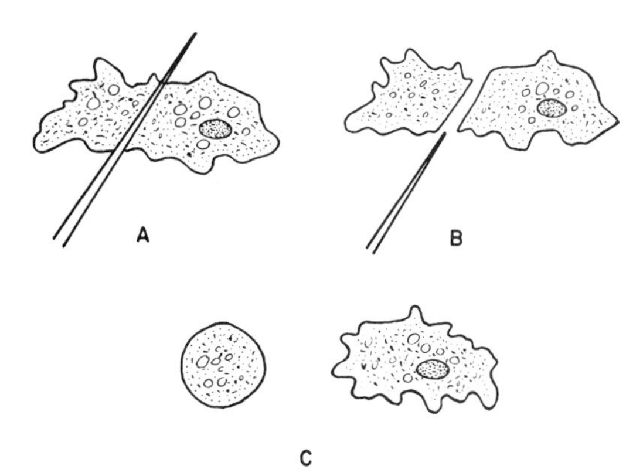

Figure 7.1. Fate of nucleate and enucleate pieces of an amoeba.

enucleated half. Its activities are restored when it is renucleated. Since the nucleus deteriorates even when kept outside the cytoplasm only briefly, renucleation is usually accomplished by placing donor and recipient amoebas of the same species close together and, with a hooked needle, quickly pulling the nucleus out of one and putting it into the other.

Although the amoeba cytoplasm deteriorates after the nucleus is removed, it can be revived when a new nucleus is inserted, even after a lapse of six days. The percentage of individuals surviving thereafter declines gradually, until after 12 days only a few renucleated cells recover.

Enucleation and renucleation of amphibian eggs have also been successfully accomplished in a similar manner. In fact, a nucleus taken from a cell of a blastula or gastrula inserted into an enucleated egg of the same species gives rise to a normal embryo.

Similar nuclear transfers have been made between species of *Acetabularia* (Fig. 7.2), a sizable single-celled alga that has a foot, a stalk, and a cap (Gibor, 1966; Brachet and Bonotto, 1970). The cap has a characteristic shape for each species and if removed is easily replaced. When the nucleus from an individual of one species of *Acetabularia* is transferred to an enucleated individual of another species from which the cap has also been removed, the regenerated cap is intermediate in form between the two species. If more than one donor nucleus is introduced, the regenerated cap has more resemblance to that of the donor than that of the recipient species. The nucleus therefore exerts a quantitative morphogenetic influence; that is, the influence is in proportion to "dose" of nuclei.

The nucleus of *Acetabularia* does not divide until just before gamete formation, at the end of its life cycle (Fig. 7.2). Division of a nucleus from a "young" cell can be induced, much before its time, if the nucleus of such a cell is introduced into an enucleated "old" cell of the same species about to form gametes. Apparently the "old" cytoplasm develops substances that induce nuclear division. A two-way influence seems to exist—nucleus upon cytoplasm and cytoplasm upon nucleus.

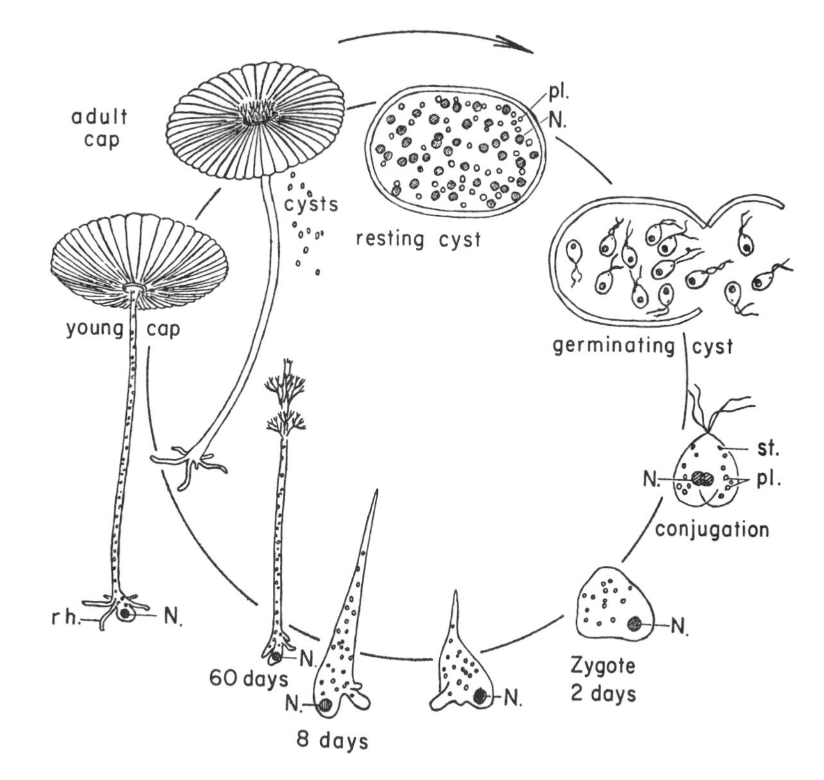

Figure 7.2. Life cycle of Acetabularia mediterranea. N., nucleus; rh., rhizoid; st., stigma on the isogametes; pl., plastids. The nucleus shown in the rhizoid of a cell with a young cap later migrates upwards into the cap and divides many times to form, in the mature cell, the cysts, one of which is shown in section as a resting cyst. Note, too, that the plastids present in each gamete in the germinating cyst multiply in the developing zygote and later fill the stalk (dots) and cap. (From Brachet, 1957: Biochemical Cytology. Academic Press, New York.)

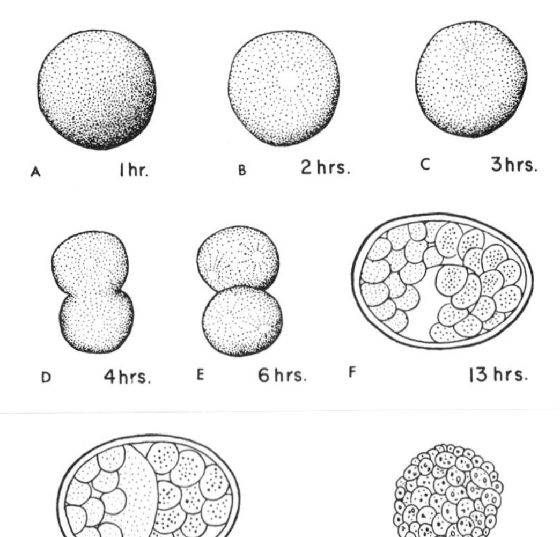

Figure 7.3. *Development of an enucleated* Arbacia *(sea urchin) egg artificially stimulated to divide. Note aster formation in* B *and* C*, division in* D *and* E*, and the many cells that develop later. No further division occurs after the stage shown in* H*, although the cells vacuolate and live for a month. (After Harvey, 1936: Biol. Bull. 71: 101.)*

An enucleated unfertilized sea urchin egg may be stimulated to divide by brief immersion in solutions of salt content higher than sea water. In the total absence of a nucleus it undergoes divisions to develop a multicellular ball of cells that later degenerates (Fig. 7.3). A nucleus is clearly necessary for continued viability and differentiation of embryonic cells.

A nucleus without its cytoplasm is also relatively helpless. An amoeba nucleus kept in physiological salt solution dies within a few hours. The nucleus of the alga *Acetabularia* lives a little longer (Berger *et al.*, 1976). Nuclei in a culture of rapidly proliferating mouse fibroblast L cells synthesize RNA; therefore, under most favorable conditions nuclei without cytoplasm can persist for many hours and carry on some synthetic functions (Schaeffer, 1976). However, the cytoplasm's importance becomes apparent with the observation that even the sperm's thin cover of cytoplasm enables sperm from choice stock animals to survive for long periods at low temperatures, permitting storage in nutrient solutions for artificial insemination. Cell nucleus and cytoplasm therefore constitute a mutualistic association (Goldstein, 1973).

The nuclear cytology of eukaryotic cells and the sequence of events which almost miraculously takes place, step by step, in the nucleus during cell division have been the subject of thorough study. The nuclear cycle includes the events by which haploid gametes (containing only one of each kind of chromosome) form the diploid zygote (containing two of each kind of chromosome). The zygote divides by mitosis to form the cells of the adult and, during gametogenesis in the adult, the haploid gametes. This and the correlations between chromosomal behavior and the distribution of genes are outlined in introductory textbooks of biology and in more detail in textbooks of cytology and genetics (e.g., Wolfe, 1972; Suzuki and Griffiths, 1976). Further consideration of these topics is omitted here.

Cell Fusion

Revealing information on nuclear-cytoplasmic relations has been obtained by *cell fusion* (Harris, 1970; Gurdon, 1975; Bernhardt, 1976). Cell fusion, the conjoining of two cells, was observed to occur spontaneously at a very low frequency in tissue cultures of mammalian cells. Cell fusion has been induced with much greater frequency either by the use of a virus or of chemicals. The Sendai virus, an RNA virus that carries a bit of lipoprotein cell membrane as it buds out of cells, is widely used. The virus is first inactivated by ultraviolet radiation or chemicals (e.g., β-propiolactone) to prevent its multiplication in the cells. Then the virus coating apparently fuses with the cell surface receptors of both cell lines and leads to attachment and fusion of the cells from different origin in the mixed tissue culture. The chemicals used for cell fusion are primarily lysolecithin for animal cells and polyethylene glycol for plant protoplasts (cells without cell walls). These chemicals probably labilize the cell membranes of both cell lines.

Cells of the same tissue at different stages of the cell cycle, cells of different tissues of the same species, cells of closely related species, cells of distantly related species, and even cells of plant and animal kingdoms have been fused. Tobacco plant protoplasts and human HeLa tumor cells were recently fused after treatment with polyethylene glycol (Jones *et al.*, 1976). However, the nuclei remain separate, forming *heterokaryons,* not true hybrids.

Cell fusion occurs only when appropriate conditions are provided: among other conditions, a pH between 7.6 and 8.6, presence of calcium and manganese ions, and an available source of energy, either for glycolysis or oxidative metabolism. Anesthetics stop cell fusion (Gurdon, 1975).

Heterokaryons, that is, cells with nuclei of the two strains, are first produced as a result of cell fusion. Sometimes more than two cells fuse and more than two nuclei are present in the resulting hybrid cell. When more than two nuclei are present, the cell line is likely to be abnormal, although not always so, depending upon the cells fused. Heterokaryons between mutant strains of the same species have been known for a long time and have been used for analyses of biochemical mutants, each one lacking a gene coding for an enzyme in a different part of a metabolic pathway for the production of a growth requirement (Beadle, 1959). Heterokaryons from fusion of cells of different tissues in animals and from fusion of cells of different species are of more recent date. Proof of the separate origin of nuclei can be obtained by treating one of the cell lines with a labeled DNA precursor before fusion and then demonstrating that the label occurs in only one type of nucleus in a heterokaryon.

Nuclei in heterokaryons may divide at the same time but remain separate. However, in some cases a single spindle may form and the resulting single metaphase figure contains all the chromosomes from both cell lines *(synkaryon).* The daughter nuclei enlarge to accommodate the polyploid (having three or more times the haploid number) chromosome number. Synkaryons, true hybrids, tend to lose the chromosomes of one or the other original cell lines over the course of numerous divisions. For example, in early studies on somatic hybrids between human and mouse fibroblasts, mouse chromosomes were retained in full but most of the human chromosomes were lost after a number of mitoses. The situation can be reversed under some conditions. However, since different human chromosomes are retained in different cells of the mouse-human somatic hybrids, the cultures are very useful in assigning characters to specific chromosomes and in mapping characters on the respective human chromosomes retained (Ephrussi, 1972; Ruddle and Kucherlapati, 1974) (Fig. 7.4). True hybrids between cells of different classes of animals (e.g., chick and mammals, frog and mammals) and between different phyla probably do not exist (Puck, 1974).

The percentage of successful somatic hybrids is relatively small, about 0.1 to 10 percent of the fused cells. By biochemically selective media, which discourage both pure lines and encourage the hybrid, the few somatic hybrids that occur can be isolated much as in the case of the biochemical heterokaryons of *Neurospora* mentioned previously.

By the use of somatic hybridization much has been learned about many aspects of cell physiology, some of which is discussed in later chapters. Of special interest to nuclear-cytoplasmic interactions is the reactivation of a nucleus that no longer promotes DNA synthesis, for example, the nucleus of a mature chick red blood cell (erythrocyte). Red blood cells of birds contain nuclei. If the chick erythrocyte is treated with Sendai virus, the cytoplasm is hemolyzed (disrupted and emptied) leaving a nucleus in a "ghost" cytoplasm. The ghost can be fused with any one of a number of cells from mouse to human. The erythrocyte nucleus in

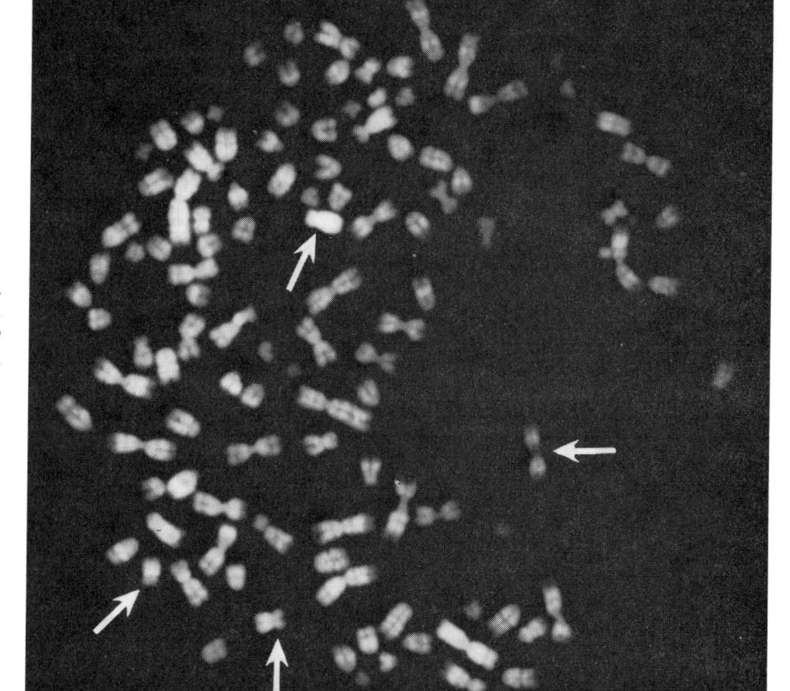

Figure 7.4. *Arrows point out several remaining human chromosomes in a fluorescently stained human-mouse hybrid cell. (By courtesy of R. S. Kucherlapati.)*

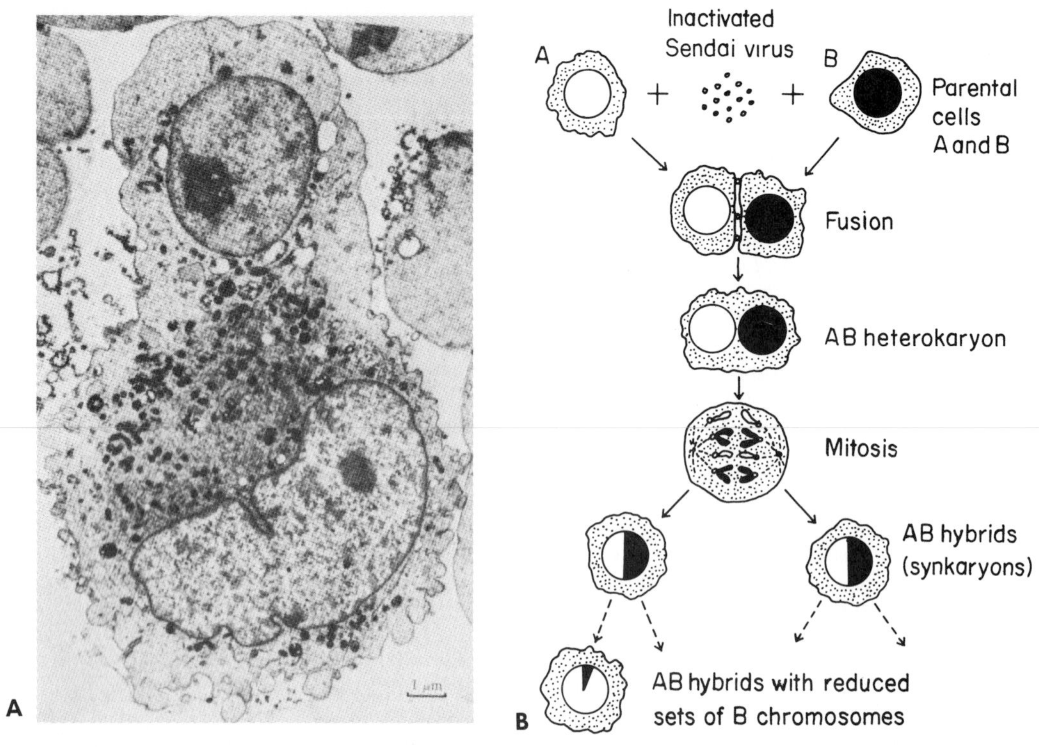

Figure 7.5. A, *An immature erythrocyte (above) from a 5-day-old chick embryo is in the process of fusing with a mouse tissue culture cell. Erythrocyte cytoplasm engaged in hemoglobin synthesis is thereby introduced into the hybrid cell. (From Harris, 1970: Cell Fusion. Harvard University Press, Cambridge, p. 70.)* B, *Schematic representation of Sendai virus-induced fusion of two mononucleate cells A and B from different species to form a binucleate heterokaryon AB, in which the nuclei fuse to form a synkaryon followed by mitosis. The AB hybrids continue to divide while gradually eliminating most of the chromosomes originating from the parental cell B. (From Ringertz and Savage, 1976: Cell Hybrids. Academic Press, New York, p. 2.)*

the heterokaryon enlarges by uptake of protein from the cytoplasm of the host cell and begins DNA synthesis. It is evident that an influence from the cytoplasm of the host cell has activated the chick erythrocyte nucleus. The same experiment can be done with a frog erythrocyte. In neither case is the cytoplasm of the fused cell necessary for activation, so activation of DNA synthesis is general, not species-specific, coming from the foreign host cell cytoplasm.

Neither a mature erythrocyte nor a hybrid of an erythrocyte nucleus with a host cell (e.g., a fibroblast) from another species synthesizes hemoglobin. If instead of a mature erythrocyte an immature one from a 3-day-old embryo chick is used, the Sendai virus does not hemolyze the cytoplasm and the whole cell can be fused with a cell of another species (Fig. 7.5). Such a heterokaryon actively synthesizes hemoglobin. If the nucleus of the young erythrocyte alone is introduced (after hemolysis by other than Sendai virus), hemoglobin synthesis is not induced. Apparently, it is not the nucleus but the cytoplasm of the young cell still synthesizing hemoglobin that has this activating action in the heterokaryon. The story is actually much more complicated than described, but the details do not alter the conclusion that the influence between nucleus and cytoplasm is a two-way affair (Harris, 1970).

NUCLEAR ORGANIZATION

The internal organization of the living resting nucleus is not revealed by light microscopy. During mitosis and meiosis the chromosomes may be seen with phase microscopy (see Appendix 3.3), which exaggerates slight differences in refractive index between cell constituents. Time-lapse photography permits continual observation of a single chromosome preparation throughout a cycle of meiosis or mitosis without the distortion that usually attends fixation and staining. Time-lapse photography makes evident not only the sequence of events but also the dynamic nature of mitosis, including chromosome distribution, the movements of cell organelles (especially mitochondria), and the changes in cell shape (see Chapter 26). Chromosomes are also discernible in ultraviolet photomicrographs of cells in mitosis or meiosis, because they absorb short wavelengths more strongly than the cytoplasm (Fig. 7.6).

When visible, chromosomes may be reached by micromanipulator needles and moved about in the nucleus or in the cell (see Appendix 3.5), indicating that they are discrete units. Chromosomes have also been released from a suspension of nuclei (obtained by fractional

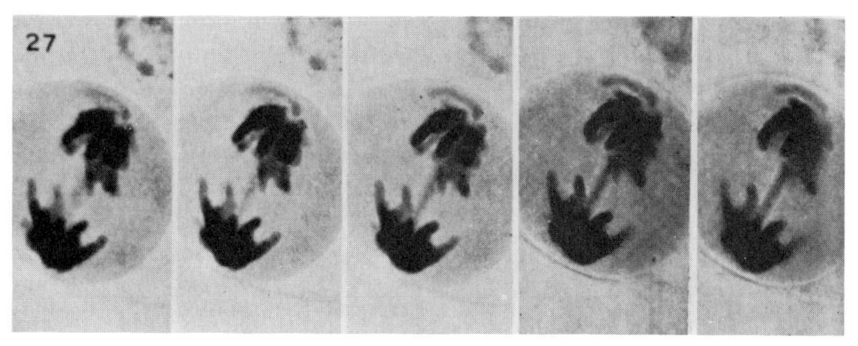

Figure 7.6. Chromosomes of a dividing cell appear grey or black in a photomicrograph taken with UV-C radiation because the DNA in the chromosomes absorbs so much of the incident radiation; these are living, unstained sperm-forming cells of a grasshopper (Melanoplus femur rubrum) *under quartz optics. The five figures are for the same cell at different focal levels (× 1200). In this phase of division (anaphase, secondary spermatocyte) the chromosomes condense at opposite poles of the cell. (From Lucas and Stark, 1931: J. Morphol.* 52: *91.)*

centrifugation), and they maintain their identity through a variety of manipulations.

Nuclear mass varies with cell type and species. For example, the nucleus of a mammalian liver cell constitutes 10 to 18 percent of the cell mass, whereas that of a thymus cell of the same species represents 60 percent of the cell mass. In cells that store much nutrient, for example, yolk or fat, and in plant cells with a large central sap vacuole, the nuclear mass may be small but still proportional to the small amount of active cytoplasm enclosing the food reserves or the vacuole.

Under the microscope the resting eukaryotic cell nucleus usually shows a nucleolus and chromatin (consisting of "threads" with attached granules) suspended in a proteinaceous continuum, the entire mass being surrounded by a nuclear envelope. The two-membraned nuclear envelope appears to develop as an extension of the endoplasmic reticulum applied to the nucleus and subsequently modified. Each membrane of the nuclear envelope is essentially a unit membrane (see Chapter 8) and the two membranes are separated by a space about 25 nm wide. The nuclear envelope contains *annuli* or pores that control the passage of some molecules and particles, even some ribosome components, between nucleus and cytoplasm. At the annulus the inner and outer membranes of the nuclear envelope coalesce. In some cells a single membrane covers the pore of the annulus. It has also been suggested that the annulus may serve as a sphincter, alternately decreasing and increasing the size of the pore with varying conditions, and some evidence suggests the presence of myosin in the annulus area (DuPraw, 1970). A segment of a nuclear envelope and a number of annuli are shown in Figure 7.7 and its permeability properties are discussed in Chapter 8.

The electron micrograph of an intermitotic

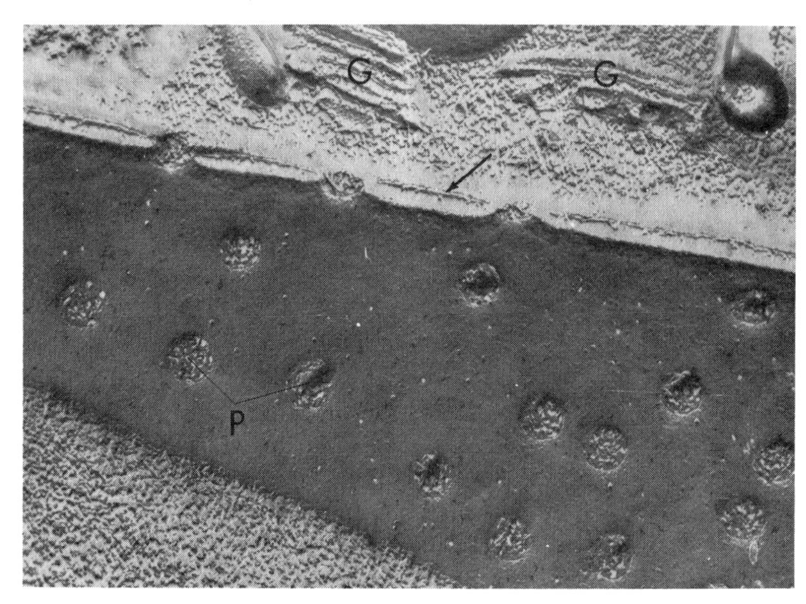

Figure 7.7. Surface view of the inner nuclear membrane of an onion root tip cell, freeze-etched. Note the numerous annuli (P) through which the cytoplasm may be seen. In the lower left corner may be seen the nuclear contents. The outer unit membrane of the nuclear envelope is indicated by the arrow above. Also seen in the upper part of the illustration are two cross-fractured Golgi complexes (G) (× 84,000). (By courtesy of D. Branton.)

nucleus in Figure 6.19 shows a rather irregular accumulation of granular material, probably representing DNA and protein. Perhaps resolution is still inadequate to reveal the organization of the DNA in such nuclei. Chromosomes condense out of the chromatin during mitosis (somatic cell division) and meiosis (formation of gametes). Both processes are characteristic of eukaryotic cells and absent in prokaryotic cells. At the close of both mitosis and meiosis most of the chromosomal substance disperses into chromatin but presumably retains linear structure. The process by which this occurs is not understood.

In electron micrographs eukaryote chromosomes appear to be accumulations of stringy dense granules, without limiting membranes (Fig. 7.8). In some spermatocytes thread-like elements are reported, with delicate laterally extending side chains the size of nucleic acid helices. Much more has been learned by osmotic extrusion of chromosomes on a water surface to enable them to spread on the surface film (see section on Molecular Structure of Chromosomes).

Euchromatin and Heterochromatin

At telophase much of the chromosomal substance disperses into chromatin. That substance fully dispersed is called *euchromatin* ("true" chromatin) while the portion that remains compacted during interphase is called *heterochromatin* (the "other" chromatin). Heterochromatin occurs in several places: adjacent to centromeres where mitotic spindle "fibers" attach to chromosomes, at ends of chromosomes (telomeres), in the region of a chromosome acting as an organizer of a nucleolus, and in other portions between euchromatin. Heterochromatin remains visible during interphase (in stained preparations). It is now known that heterochromatin is not transcribed into RNA during interphase whereas euchromatin is and that heterochromatin is the last to be replicated. Some

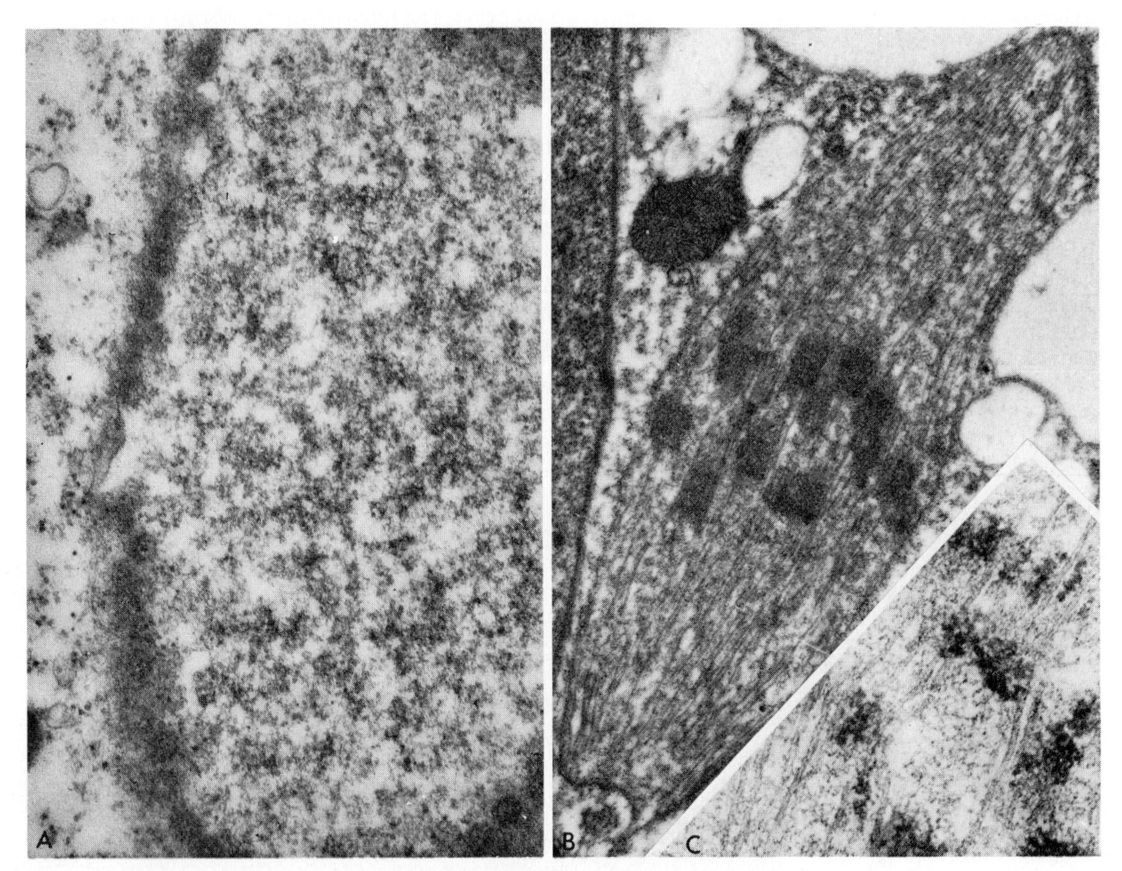

Figure 7.8. *Electron micrographs of nuclei of the ciliate* Blepharisma. A, *Grazing section of the macronucleus of* B. japonicum *showing somewhat fibrous internal structure and nuclear pores lining the left side and the extreme right corner of the photograph, each pore with a central opening (× 45,000).* B, *Mitotic spindle of the micronucleus of* B. sp. *showing spindle fibers and chromosomes halfway between the poles (× 37,000).* C, *Somewhat higher magnification (× 46,000) of a micronucleus similar to that in* B, *showing spindle fibers and chromosomes, including a pair of separated sister chromosomes in the center of the field. Note that some spindle fibers attach to chromosomes while others pass directly through the equator of the spindle. (By courtesy of R. A. Jenkins, University of Wyoming.)*

chromosomal regions that are euchromatic in the cells of the embryo become heterochromatic in cells of the organism in later life. No function is known for the chromatin adjacent to the centromeres that becomes permanently heterochromatic after embryonic cleavages cease.

The chromosome that serves as nucleolus organizer has considerable heterochromatin, but scattered in it are uncompacted regions with repetitive genes for rRNA and tRNA. Considerable portions of the heterochromatic chromosomes have redundant nucleotides, seemingly coding for nothing. These simple redundant sequences of nucleotides perhaps serve as spacers between active genes, others as transcriptional stops. The detailed function of heterochromatin is yet to be revealed (Watson, 1976).

CHEMISTRY OF THE NUCLEUS

Their chemistry can be studied readily because nuclei can be separated from other cell constituents by fractional centrifugation. The cells are first disrupted with distilled water, by freezing and thawing, or by applying lytic agents such as saponin and detergents. The fragmented cells are placed in solutions of sucrose and calcium chloride (or citric acid) and differentially centrifuged and washed with the solution. Unfortunately, some substances are leached from the nucleus, but adding appropriate salts and adjusting the pH of the medium minimizes losses. Isolated nuclei in buffered sucrose retain some metabolic ability and can incorporate radioactive precursors into their nucleic acids and proteins.

Nuclei have also been isolated from frozen dried tissues in nonaqueous solvents of graded density. Such nuclei retain all their water and acid-soluble constituents and nuclear proteins. With carefully designed procedures this is also possible with aqueous solvents (e.g., buffered sucrose). The two types of analysis complement each other.

Some constituents are always present in nuclei: DNA; RNA; proteins of two kinds, histone and non-histone; some lipids; various organic phosphorus compounds; and various inorganic compounds, mostly salts (Davidson, 1976). DNA in salmon sperm heads, which are mostly occupied by the nucleus, may constitute 48.5 percent of the dry, fat-free substance. Generally less than half the dry weight is DNA but the amount varies from species to species. Proteins in nuclei vary with nutrition, falling with starvation; DNA is said to remain constant (Mirsky, 1950–51). However, in a marine bacterium, starvation for six weeks halved the DNA concentration in the cells (Novitsky and Morita, 1977).

As mentioned in Chapter 1, three kinds of RNA are present in cells: ribosomal (rRNA), transfer (tRNA), and messenger (mRNA). RNA contains nucleotides of the purines adenine and guanine and the pyrimidines cytosine and uracil. In any one of these RNAs the ratio of the bases to one another is constant for a species. However, rRNA, tRNA, and mRNA separated from one another disclose different base ratios, indicating differences in their chemical composition. The tRNAs have a low molecular weight of about 25,000, the rRNAs have molecular weights running to several million, and the mRNAs are variable but probably from half a million to several million. Differences also exist in the affinities of the various RNAs for other molecules (Lehninger, 1975). The rRNA is synthesized and assembled in the nucleolus, the tRNAs and the mRNAs are synthesized on the chromosomes, and all the RNAs enter the cytoplasm through the pores in the nuclear envelope.

The nucleus contains a number of enzymes and performs metabolism, including synthesis of DNA and various RNAs. Investigators now seem to agree in general that the nucleus lacks the enzymes for aerobic metabolism that are found in mitochondria (see Chapter 12), but that it contains enzymes for anaerobic metabolism and for formation of high-energy phosphates. It also contains enzymes for coenzyme synthesis (nicotinamide adenine dinucleotide, or NAD) (see Chapter 10).

Studies of nuclear contents other than chromatin (nuclear "sap"), removed by micropipetting from amphibian oocytes, showed it to contain proteins and salts in solution but no detectable nucleic acid (Davidson, 1976).

A variety of proteins is present in the nucleus: nucleoproteins, enzymes, and structural proteins. The nucleoproteins form two classes, deoxyribonucleoproteins and ribonucleoproteins.

Deoxyribonucleoproteins, which largely form the chromosomes, consist primarily of histones and DNA in about equal amounts. However, chromosomes also contain non-histone proteins in smaller amounts. *Histones* are of five main kinds: now designated H1, H2a, H2b, H3 and H4 (Isenberg and Spiker, 1977). They vary in molecular weight from 21,000 for H1 to 11,000 for H4 and in amino acid composition. The histones are separated from one another on acrylamide gel by electrophoresis (see Appendix 5.5) and differ in amino acid composition (Stein and Stein, 1976). They are differentiated from one another largely on the basis of their content of the basic amino acids lysine and arginine. They all lack the aromatic amino acid tryptophan (Lehninger, 1975). Histones make up about 54 percent of the proteins in a liver cell nucleus. The *non-histone proteins,* sometimes called *acidic proteins,* contain the aromatic amino acid tryptophan and are a diverse lot. A considerable amount of the contractile proteins actin, myosin, tropomyosin, and tubulin are said to be present (Dauvas *et al.,* 1975). They constitute about 10 percent of the protein in a liver cell. Non-histone proteins have a more rapid turnover than histones do.

Both histones and non-histone proteins are synthesized in the cytoplasm and enter the nucleus through the nuclear envelope. Histones are synthesized only when DNA is replicated, whereas non-histone proteins are synthesized continuously. Histones induce a compact structure in the chromosome. Some investigators also consider histones as stabilizers against heat damage (Tashiro and Kurakawa, 1976) and against nucleases (Toczko *et al.,* 1975). Activation and repression of genic expression are thought to be carried out by non-histone proteins. However, the mechanism by which this is accomplished in eukaryotic cells is less clear than it is in prokaryotic cells (see Chapter 15).

THE NUCLEOLUS

A *nucleolus* is present in the nucleus of most cells, but it is inconspicuous or absent in sperm and muscle fibers. It is conspicuous in active cells such as neurons and in glandular secretory cells. The nucleolus becomes enlarged during periods of synthetic activity and atrophies during quiescent stages. Some cells possess multiple nucleoli (Davidson, 1976; Ghosh, 1976; Jordan, 1978).

Nucleoli are not delimited from the nucleus by a membrane visible with the electron microscope (Fig. 3.1). A dense, granular peripheral portion and a filamentous portion may be seen in the nucleolus with the electron microscope. The nucleolus forms around the nucleolus-organizing portion of a chromosome at what under light microscopy appears to be a constriction. The constriction appears opaque under electron microscopy and contains the ribosomal genes in a *nucleonema.* The chromosome follows a tortuous path through the granular and fibrillar components of the nucleolus (Jordan, 1978). Nucleoli exposed to labeled RNA precursors (e.g., ^{3}H-uridine) show labeled RNA first along the nucleonema. After a lapse of time the label appears over all regions of the nucleolus and still later appears in the cytoplasm. The filaments and granules seen with electron microscopy are stages in the synthesis of ribosomes.

Nucleoli are formed from loops of one or more chromosomes combining with specific proteins. These loops represent multiple, tandemly arranged genes for rRNA. The genes transcribe rRNAs of 45S value that are later degraded to form 28S and 18S rRNA. The 28S rRNA combines with newly formed ribosomal proteins to form the 60S ribosome subunit, which escapes into the cytoplasm through a nuclear pore. The 18S rRNA leaves the nucleolus and combines with proteins in the cytoplasm to form the 40S subunit of the ribosome. The ribosomal proteins are synthesized in the cytoplasm (Watson, 1976). The enzymes in the nucleolus are primarily for RNA synthesis and for synthesis of the coenzyme nicotinamide adenine dinucleotide (see Chapter 15).

When the nucleolus is absent, as in some mutants of the toad *Xenopus,* the cells of the embryo make only about 5 percent of the ribosomal RNA formed in the cells with a nucleolus. Mutant cells make normal amounts of tRNA and mRNA, but the embryos die because ribosomes are required to synthesize proteins for embryonic growth. When nucleoli are removed from cells experimentally, the cells die. Irradiation of the nucleolus with an ultraviolet microbeam stops passage of tRNA, mRNA, and rRNA from nucleus to cytoplasm. Therefore, the nucleolus has a controlling role in addition to its synthetic role (Sidebottom and Deak, 1976).

LOCALIZATION OF DNA AND RNA IN NUCLEI

Nucleic acids can be located in nuclei by ultraviolet photomicrography. This method is based on the fact that no other protoplasmic constituents absorb ultraviolet light significantly at the 260 nm absorption peak of nucleic acid (see Fig. 3.13). Therefore, the density of the image in photomicrographs taken at this wavelength measures the relative nucleic acid content in the cell.

It is even possible to estimate the amount of nucleic acid present in a nucleus, for example, by comparing the optical density of the photograph through a nucleus with known nucleic acid concentrations. The measuring equipment is shown in Figure 7.9, and a graph relating the absorption spectra for nucleus and cytoplasm in Figure 7.10. Because both DNA and RNA absorb ultraviolet radiation about equally well, it is necessary to use specific staining to localize each of the nucleic acids separately.

The Feulgen reaction detects DNA in cells. After fixation, tissue sections or cells are hydrolyzed with hydrochloric acid to liberate the aldehyde group in the deoxyribose sugar of DNA and then treated with leucofuchsin (basic fuchsin in sulfurous acid). The aldehyde reacts with leucofuchsin to give a purplish color, the depth of which can be estimated by spectrophotometry. Deoxyribonuclease re-

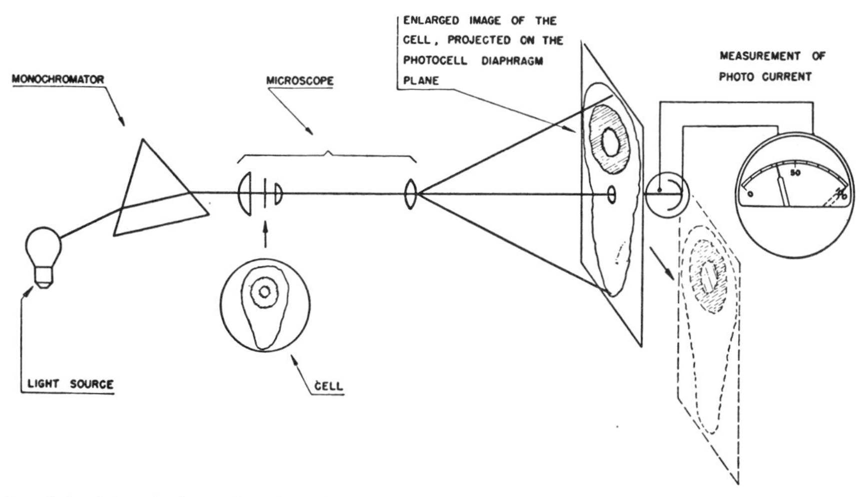

Figure 7.9. *Principle of photoelectric microspectrophotometry, showing equipment of Caspersson (diagrammatic). (From Caspersson, 1950: Cell Growth and Cell Function. W. W. Norton & Co., New York.)*

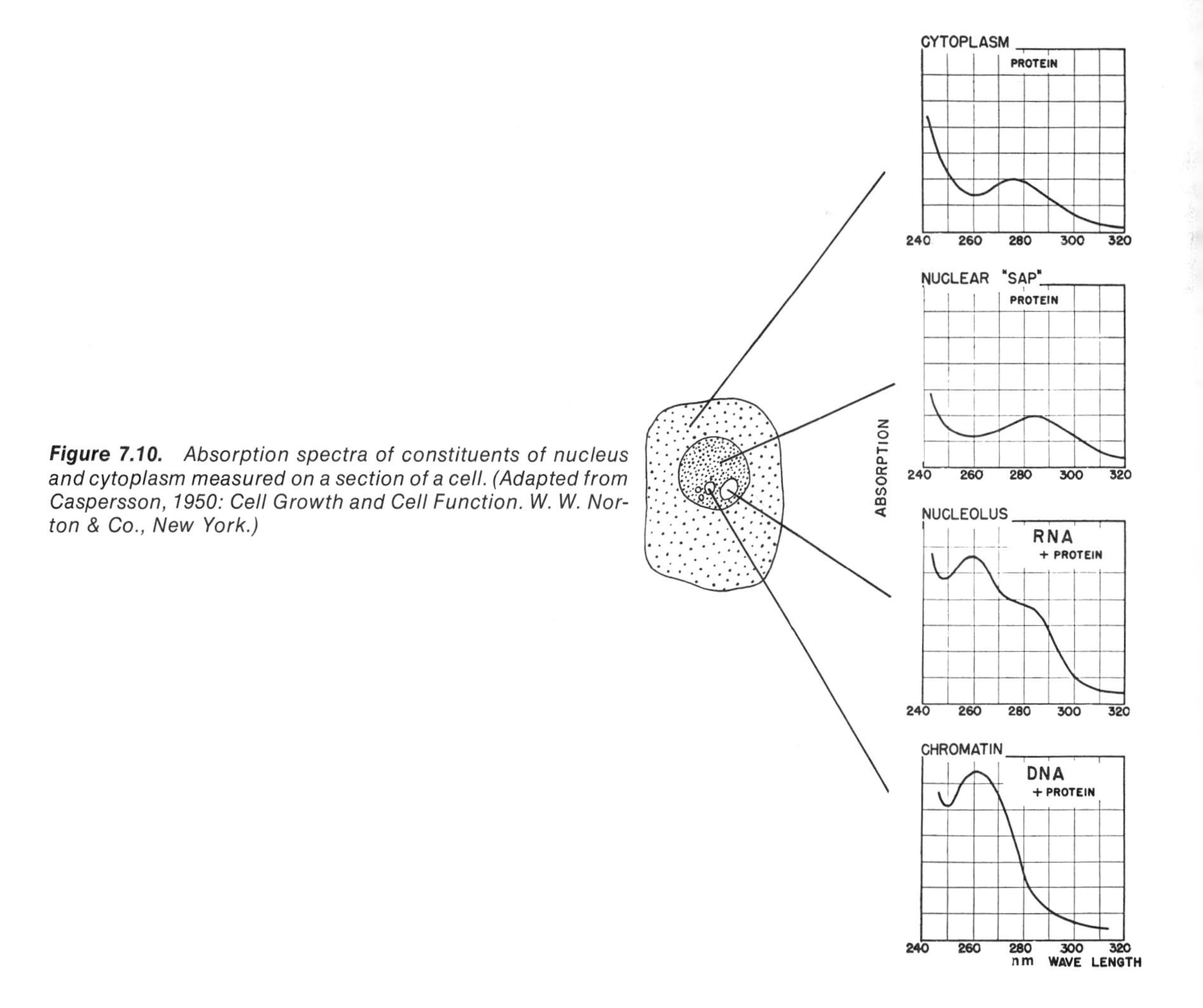

Figure 7.10. *Absorption spectra of constituents of nucleus and cytoplasm measured on a section of a cell. (Adapted from Caspersson, 1950: Cell Growth and Cell Function. W. W. Norton & Co., New York.)*

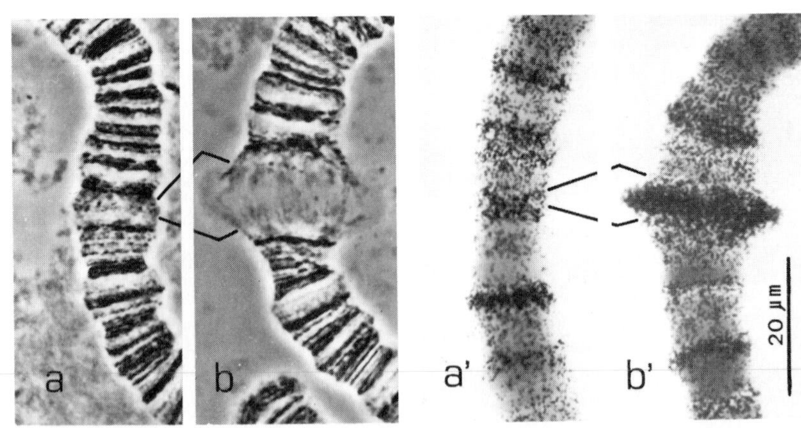

Figure 7.11. *Portion of a polytene chromosome in a salivary gland cell of the fly* Trichosa pubescens. a, *Band across the polytene chromosome revealed by selectively staining DNA with aceto-orcein and acidic fast green.* b, *Puff induced 100 min after treatment with diethyl ether in a cell similar to the one in* a. a′, *^{3}H-uridine incorporation into chromosome X of the salivary gland cell of a control* Trichosa larva, *indicating RNA synthesis upon the DNA.* b′, *^{3}H-uridine incorporation into a similar chromosome as in* a′ *100 min after treatment with diethyl ether. The dots in both* a′ *and* b′ *are produced by radiation from the incorporated tracer molecule. Note that there is more RNA synthesis in the puff than in any of the other bands, indicating excitation of RNA synthesis in this region by the treatment with diethyl ether. (From Amabis and Janczur, 1978: J. Cell Biol.* 78: *1–7.)*

moves DNA selectively; after such treatment the Feulgen test should be negative.

Ribonucleic acid may be selectively stained with the dye pyronin; RNA content may then be estimated spectrophotometrically. When RNA is selectively removed by ribonuclease, the characteristic test should be lacking.

DNA is visible in bands on metaphase chromosomes, perhaps most clearly seen in the multiply replicated (polytene) chromosomes in salivary gland cells of flies (Fig. 7.11). In interphase nuclei DNA is visible primarily in chromatin granules that give a positive reaction with the Feulgen test.

RNA is present in considerable amounts in the nucleolus and is also found in small amounts on chromosomes. However, it will be recalled (see Chapter 3) that the major part of the cell's RNA is in the cytoplasmic ribosomes. Transfer RNA and mRNA are present in solution in the cytoplasmic matrix unless affixed to the ribosomes. A small amount of RNA is also present in mitochondria and chloroplasts. Cytochemical methods cannot be used to differentiate among the three kinds of RNA, so fractional precipitation and quantitative analysis are necessary for this.

The RNA content of nucleus and cytoplasm varies with activity cycles of the cell. During prophase in a cell about to divide, the nucleolar RNA gradually passes into the cytoplasm when the nucleolus disappears and the chromosomes re-form. The mitotic and meiotic nucleolus is re-formed from chromatin of the nucleolus organizer chromosome, which remains directly associated with the nucleolus during the interphase. If the nucleolus organizer is damaged by radiation, multiple small nucleoli may form. The cytoplasmic RNA increases in quantity during cell growth preceding mitosis and is partitioned equally between the daughter cells.

RNA accumulates in both nucleus (especially in the nucleolus) and the cytoplasm during high metabolic activity or growth, as in regenerating nerve cells, active neurons, gland cells, cells infected with virus, and tumor cells. Actively metabolizing yeast cells contain a large amount of RNA, but starved ones have little. In fact, starved cells in general show RNA depletion. RNA also varies with other physiological conditions such as lack of oxygen and presence of metabolic poisons.

Since RNA varies so markedly with conditions and the metabolic state of cells, one might expect it to be quite labile, degraded and rebuilt readily. This is indeed indicated by the quick incorporation of radioactive tracers such as ^{32}P in phosphate and ^{14}C in precursors of the purine and pyrimidine bases into RNA. It is interesting that the RNA is labile not just in dividing cells but also in active cells that are not dividing. Consequently, the RNA turnover must accompany various cell activities. Nuclear RNA (chiefly nucleolar) appears to be more readily turned over than cytoplasmic, but this may be an artifact of the methods used (Davidson, 1976).

REPEATED DNA

Most genes in eukaryotes are present singly per haploid set of chromosomes, but some genes of a chromosome are present in multiples, for example, the genes coding for rRNA. Some sequences of nucleotides appear to code for nothing and may represent spacers or terminators of transcription (punctuation). Sometimes these sequences are extraordinarily numerous (Goodenough, 1978). Sometimes the entire genome is multiple, as in the polytene chromosomes of salivary glands in fly larvae, although usually it is diploid. Sometimes the genes present singly are called the *genetic* DNA and the multiple genome the *metabolic* DNA, although this distinction is not generally emphasized; the multiple DNA is spoken of as redundant DNA (Britten and Kohne, 1970; Flamm, 1972; Pelc, 1972).

In ciliates the separation of genetic and metabolic DNA is complete, the genetic DNA being in the micronucleus, which is essentially inactive in RNA synthesis. The metabolic DNA is present in the macronucleus, and this structure is the center of active RNA synthesis in the cell. Thus ciliate stocks lacking micronuclei but possessing macronuclei grow for many generations, but they do not survive conjugation, at which time the metabolic DNA in the macronucleus normally disperses into the cytoplasm and a new macronucleus arises from the new zygotic micronucleus. On the other hand, individuals from which the macronucleus has been excised survive only briefly, although the injury suffered in a mock operation with removal of an equivalent amount of cell content is quickly healed. It is perhaps significant that the macronucleus of ciliates has nucleoli and annuli in keeping with its metabolic functions, whereas the micronucleus does not (e.g., in *Blepharisma*). Macronuclear DNA is necessary for regeneration of a *Blepharisma* after injury or transection (DuPraw, 1970).

As indicated previously, metabolic DNA serves for transcription of ribosomal RNA, as a spacer between genes in the chromosomes, and as a span for attachment and movement of chromosomes (Rees and Jones, 1972; Nagl, 1976). In ciliates the metabolic DNA obviously plays a much wider role. It is possible that more specific roles will be assigned to metabolic DNA in other cells when more is learned about their function.

CONSTANCY IN NUCLEAR DNA OF A SPECIES

Regardless of whether the nucleus tested is that of a liver cell, an erythrocyte, a nerve cell, or any other cell of a given species, the same amount of DNA is found per haploid set of chromosomes in a resting cell nucleus. Gametes, with only one haploid set of chromosomes, possess half the DNA of a diploid cell. Polyploid nuclei have the haploid quantity of DNA multiplied by the degree of the polyploidy.

Although nuclear DNA per set of chromosomes is constant, cytoplasmic DNA may vary. Usually so little is present in the cytoplasm (mostly in mitochondria and chloroplasts) that it is difficult to measure. However, nurse cells of the ovaries of animals and ovules of many plants inject DNA into the egg cytoplasm where it accumulates. In egg cells of some echinoderm species the quantity of DNA may be many times that for a haploid set of chromosomes. This is believed to provide a ready supply of DNA for the rapid divisions following fertilization.

The quantity of DNA in a species does not necessarily indicate its evolutionary position. Even though in invertebrates DNA content per cell progressively increases with increasing degree of complexity (Rees and Jones, 1972), little relation is found between DNA content and phylogenetic position in vertebrates (Szarski, 1976).

The amount of DNA in a dividing cell is duplicated in late interphase (or in very early prophase in some cases) long before the cell is ready to divide. The DNA therefore varies in discrete amounts, the unit being the amount per haploid set of chromosomes.

It is interesting that chromosomal DNA does not decrease when cells are starved, although RNA and protein decrease markedly. It would appear that the genes, necessary for instituting growth on return of favorable conditions, are protected. This seeming "adaptation" results primarily perhaps from the general metabolic inertness of DNA. Studies with ^{32}P indicate that the phosphorus of DNA is not readily exchanged or incorporated, except when DNA is being replicated before cell division. However, some labels are exchanged to a limited extent, even in DNA.

MOLECULAR STRUCTURE OF THE CHROMOSOME*

Chromosomes may be visualized by releasing them from bacteriophages or cells after osmotic shock on the surface of distilled water. A chromosome so released spreads out on the surface of the water and can be picked up on a grid, fixed, stained, and seen under the electron microscope. Since a phage particle consists of little other than a DNA double helix molecule and a coat of protein (Fig. 7.12) and under the conditions of the experiment protein is not released on the surface of the water, one sees the DNA chromosome alone; what is observed is a circular molecule digestible by DNase† (Fig. 7.13). In the head of the phage the chromosome is tightly

*The subject of a symposium volume at Cold Spring Harbor, New York, 1974. Summarized in Watson, 1976.

†Abbreviation for deoxyribonuclease.

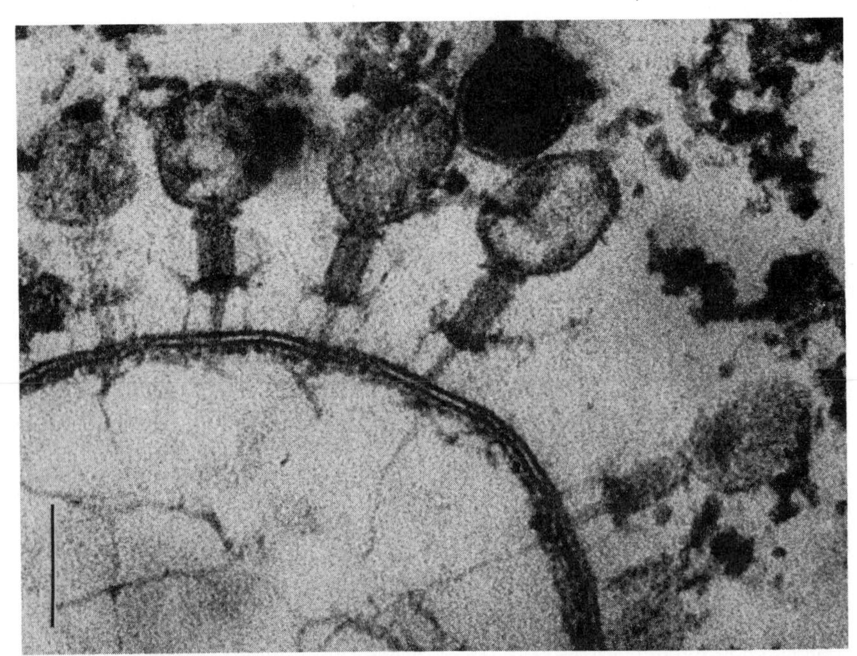

Figure 7.12. *Several T4 phages injecting their thread-like DNA through the* E. coli *cell wall. The cell wall is the circular double membrane. The head (large ovoid body), tail sheath below the head, the tail core within the sheath, and the base plate of the phage are visible. (From Simon and Anderson, 1967: Virology.* 32: *279–297.)*

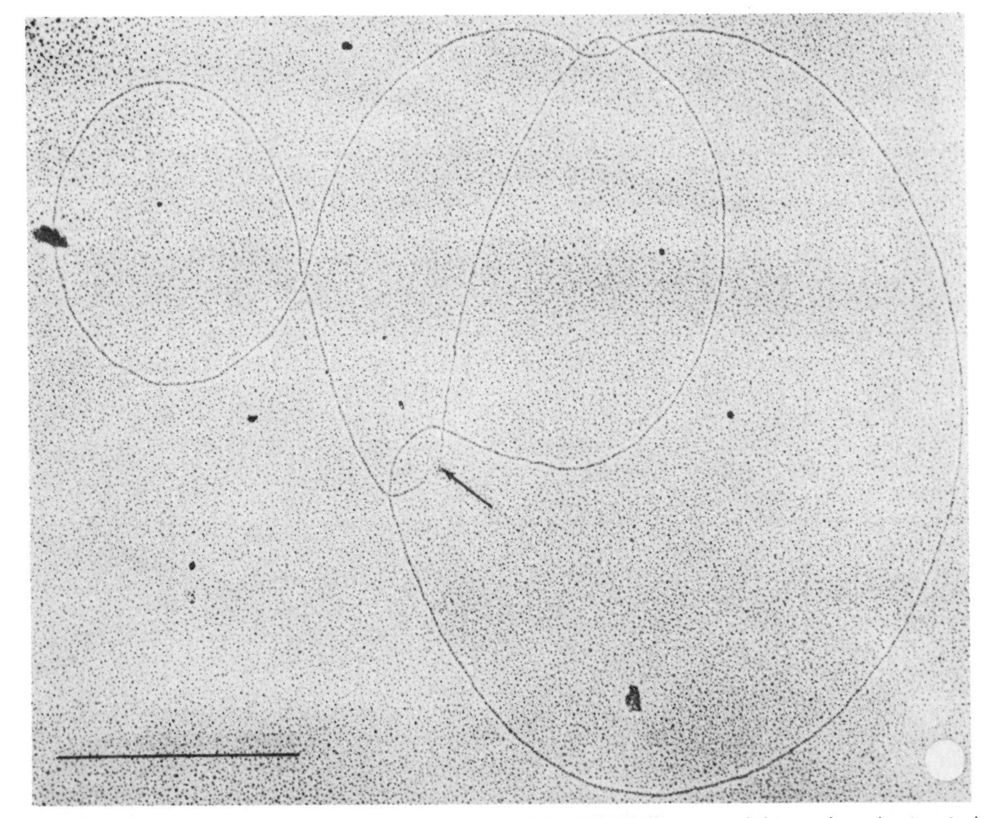

Figure 7.13. *Electron micrograph of lambda phage DNA ($\times$ 44,000). The material was phenol extracted and spread as a monolayer on a surface of water from a solution of DNA and the basic protein cytochrome* c. *The arrow indicates a discontinuity, perhaps the place where replication begins. (From Ris and Chandler, 1963: Cold Spring Harbor Symp. Quant. Biol.* 28: *1.)*

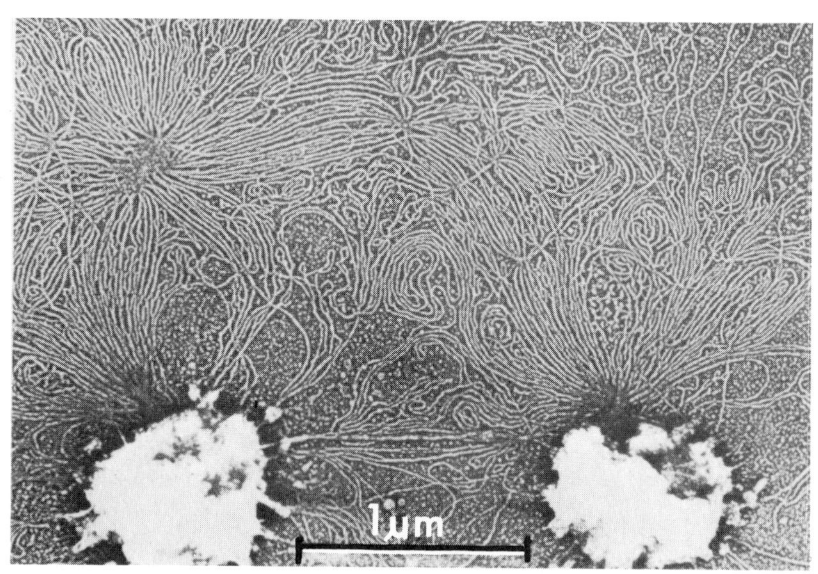

Figure 7.14. *Portion of two spread bacterial protoplasts with liberated DNA fibers that run together in the middle of the figure. Platinum shadow cast (× 30,000). (From Kleinschmidt and Lang, 1962: Proc. 5th Intern. Congress Electron Microscopy 2: 1–8.)*

packed. When it is spread out from a T phage it measures 52 μm, yet it must fit into a head only 95 nm in diameter, indicating a reduction in length (or packing ratio) of 520 to 1 (DuPraw, 1970). The chromosomes of bacteria (Fig. 7.14), chloroplasts, and mitochondria are rather similar to those of phage although they are longer, but those from the nucleus of eukaryotes are more complex and different in appearance. Nonetheless, nuclear chromosomes also contain a double-stranded DNA molecule as the basic hereditary substance.

When a eukaryotic chromosome becomes visible in prophase, it consists of two chromatids (daughter chromosomes, still adherent) resulting from replication of DNA during the S phase of the cell cycle (Whitehouse, 1973). Each chromatid is in the molecular sense a chromosome, and the duplex nature of the visible chromosome is a result of the mitotic process. In the following account, when the molecular nature of the eukaryote chromosome is considered, what is meant is the single unit and not the duplex unit seen in the condensed "chromosome" partaking of mitosis. The paired "chromosomes" of a first division meiotic cell are tetrads of single units (chromatids).

Phages

The chromosome of a ϕX174 phage is a single DNA filament, but during replication a complementary DNA strand is made. The active chromosome is thus a double-stranded DNA molecule (Sinsheimer, 1962).

While most phages have a limited number of genes, they code for coat proteins, enzymes for penetrating cells, and enzymes for some phases of replication. However, the satellite tobacco necrosis virus has only enough genome (RNA in its case) to code for its protein coat, all the remaining proteins being coded by another virus on which the satellite virus is parasitic (Goodheart, 1975; Fenner *et al.,* 1974).

Prokaryotes

The prokaryotic chromosome can be visualized by applying bacteria, whose cell walls have been removed by lysozyme, to the surface of a trough containing distilled water. The bacteria rupture as a result of water uptake and the contents are spread by surface forces. The cell contents picked up on a grid and studied by electron microscopy disclose a filament of DNA (digestible by deoxyribonuclease) many times the length of the bacterium. It consists of a single two-stranded DNA molecule forming a circle (Fig. 7.15). The chromosome may also be seen in highly packed condition in electron micrographs of bacterial nucleoids. In this case also the diameter of the fiber is 2 nm. An *E. coli* chromosome about 1.35 mm long is packed into a nucleoid about 1 μm long, a packing ratio of about 1000 to 1. The bacterial chromosome is looped and packed into the nucleoid held together by some RNA that appears to form a core. Polyamines are also present. The genome in such a bacterium is capable of coding for the hundreds of proteins, enzymic and structural, required for its life activities (see Chapter 15).

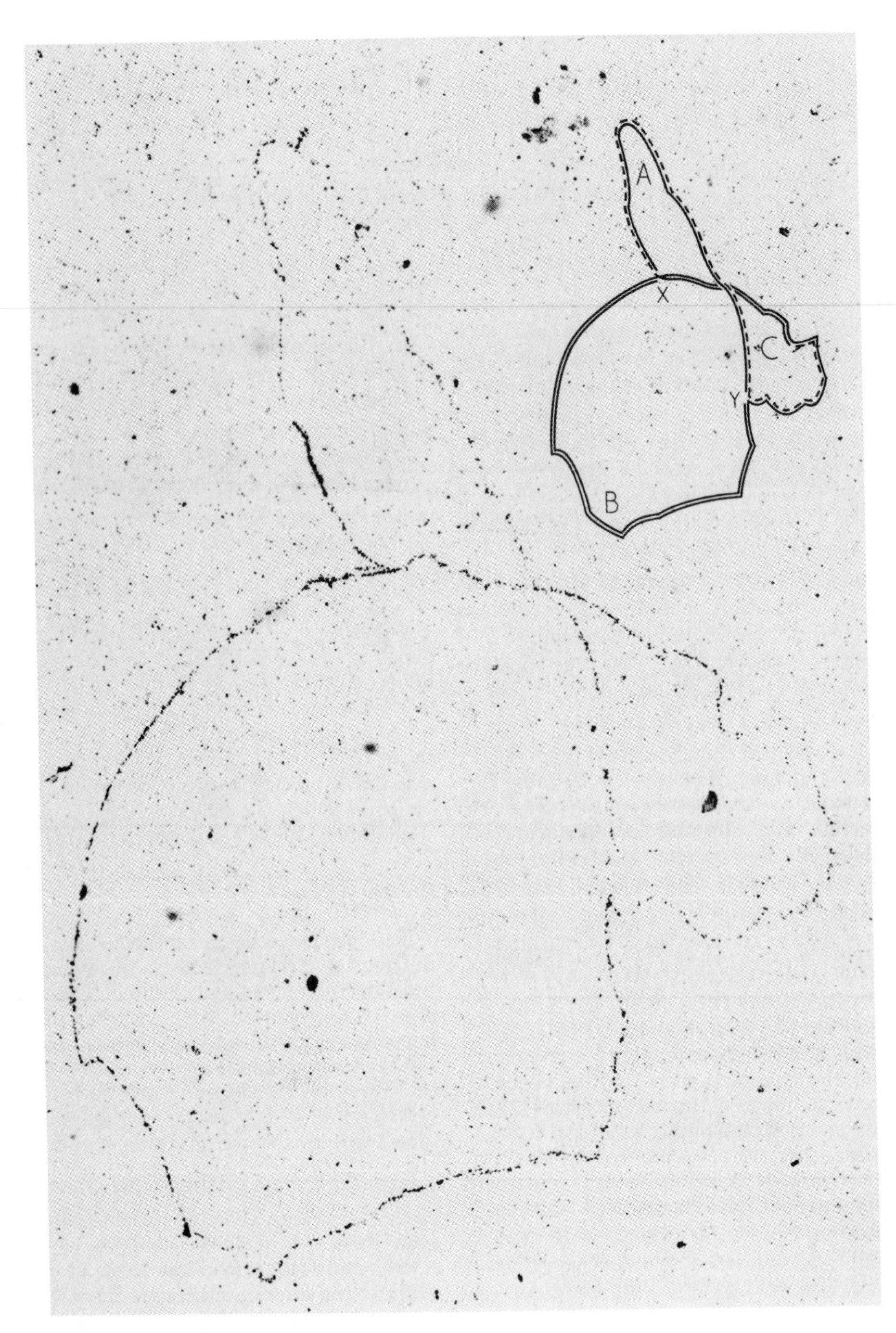

Figure 7.15. *Autoradiograph of the chromosome of* E. coli *(strain K12 Hfr) labeled with tritiated thymidine for two generations and extracted by treatment of the cells with lysozyme. The inset shows the same chromosome in diagram. The chromosome has completed two-thirds of its second round of duplication. Part of the still unduplicated section is marked with tracer (Y to C). (From Cairns, 1963: Cold Spring Harbor Symp. Quant. Biol.* 28: *44.)*

During mating in *E. coli,* a chromosome passes from the donor to the recipient through a tube generally constructed of a filament (pilus connecting the two). Because replication of DNA is required for chromosome transfer, it is thought that transfer takes place as replication of the chromosome proceeds, part of one of the replicates passing into the recipient cell at a rate of about 11 μm per minute at 37°C (DuPraw, 1970).

A linear model for the chromosome of *E. coli* is in keeping with the distribution of label during division of bacteria. Bacteria grown on a medium containing labeled nitrogen (^{15}N) incorporate the nitrogen into the DNA. Subsequent growth on ordinary nitrogen shows that in samples tested after one generation only one half the DNA was labeled; after two generations, only one fourth was. This demonstrates that the original DNA strand remains intact, the new DNA strand being synthesized from unlabeled nitrogen (the elegant experiment of Meselson and Stahl, 1958). Such replication is spoken of as *semiconservative* replication.

Equally convincing are experiments with *E. coli* grown on ^3H-thymidine and then lysed with a mild detergent on a membrane to spread the chromosome. On the coating of photographic emulsion the radiation from the thymidine reduces the silver salt to silver, which on development discloses a circular chromosome, often with a replicating point (Fig. 7.15). The growing point of the chromosome is attached to the membrane (or mesosome if present) where oxidative phosphorylation provides energy for DNA replication. Similar experiments have been performed with chromosomes from mitochondria and chloroplasts, with like results.

In addition to the single large circular chromosome, most bacteria contain up to 20 smaller circular DNA molecules called *plasmids* or *episomes;* they replicate along with the large chromosome. Some episomes become incorporated into the host cell chromosome. Plasmids may be transferred during bacterial conjugation and carry phenotypic characters into the recipient cell (Campbell, 1969). Plasmids move about widely in the microbial world, sometimes picking up genes from host cells and transferring them to another cell. In a way, plasmid interchange is a precursor of sexual reproduction. Gene transfers do not show species boundaries but they are less efficient the greater the evolutionary distance between donor and recipient (Davis, 1977). Viruses may also be incorporated into the bacterial genome acting in much the same way (Campbell, 1976). These transfers of genome have been named "jumping genes" (Kolata, 1976). Transfer of DNA into eukaryotic cells can also occur via plasmids that have incorporated a bit of host DNA and is the basis of genetic engineering (see Chapter 15).

Eukaryotes

Evidence has accumulated to indicate that a linear double-stranded DNA molecule is the central core of the eukaryotic chromosome, which is otherwise quite different in organization from that in prokaryotes (Hood *et al.,* 1975). For example, treatment with DNase causes chromosome disintegration whereas treatment with trypsin does not. This indicates that the continuum in a chromosome is its DNA.

The apparent lack of continuity between the chromatin granules in interphase chromatin treated with the Feulgen reagent is an artifact. Decisive evidence for the presence of DNA in the regions between the chromatin granules has been obtained by staining with acridine orange. The molecules of acridine orange align themselves parallel to the purine and pyrimidine bases in nucleic acids, and exposure to long wavelength ultraviolet radiation evokes fluorescence: orange in RNA and yellow-green in DNA. The fluorescence measurements indicate that about 5 percent as much DNA is present between granules as in them. It therefore seems likely that the thin segments between granules are not observable in Feulgen-stained preparations because the color is just too faint to be seen with the light microscope.

Convincing evidence for linear arrangement of a DNA molecule in a eukaryotic chromosome comes from experiments using ^3H-thymidine to label the DNA in chromosomes, for example, in bean seedlings. When broad bean *(Vicia faba)* seedlings were incubated in the labeled thymidine, the 12 chromosomes became labeled, as was evident from correspondence between the location of silver grains in the photographic emulsion and the microscopically visible chromosomes in control root cells (Taylor *et al.,* 1957; Taylor, 1959).

Eukaryotic chromosomes may also be spread on the surface of distilled water after the cells are ruptured by osmotic shock (Fig. 7.16), in the manner already described for bacteria. Nucleated erythrocytes, cells of ascites tumors, and other cells so treated disclose thick lumpy fibers (20 nm in average diameter, 30 nm at the lumps) with a core about 15 nm in diameter (Cairns, 1966). Whether the fibers isolated from eukaryotic cells in this manner are made up of a single highly packed double-stranded DNA molecule covered by a protein coat that adds to its diameter is not detectable by electron microscopy. However, when the DNA is labeled with tracer-marked DNA precursors, a

Figure 7.16. Fibrous fine structure of a pair of honey bee metaphase chromatids after surface spreading and critical point drying (× 18,600). (By courtesy of E. DuPraw; from DuPraw and Rae, 1966: Nature 212: 598–600.)

single string of incorporated radioactive atoms is revealed. Because a eukaryotic chromosome shows incorporation of labeled thymidine in many places simultaneously, replication must begin simultaneously in each of the *DNA units* (replicons). Each unit is about 50–60 μm long in a hamster cell, and the entire genome consists of about 50,000 such units (Edenberg and Huberman, 1975). When a labeled eukaryotic chromosome is replicated, the labeled DNA is passed intact to only one of the progeny (semiconservative replication). In contrast, tracer labels in both RNA and protein are lost by continued dilution in the course of four generations.

Electron microscopic studies of condensing chromosomes show impacting of the fibers into compact chromosomes. Digestion with trypsin permits the impacted strands of DNA to "spring out" as if they had been bound by the protein. The point of attachment of the chromosome to a spindle fiber is thinned by digestion with trypsin but still shows the presence of DNA (DuPraw, 1970).

The model of chromosome molecular organization as a linear DNA double helix covered by protein has appeal because it is easier to see how it permits partition of hereditary substance between daughter chromatids during cell division. In a cell about to undergo division the two strands of the DNA double helix split lengthwise and separate at the replication fork; each half helix may then re-form its complementary half as required by the Watson-Crick model of DNA. The chromosomes of eukaryotes are much longer than those of prokaryotes and contain more DNA; a single human chromosome may contain 7 cm of linear DNA molecule. The total length of the human diploid genome is about 174 cm of DNA. It is about 37 meters in the plant *Trillium* and 97 meters in polytenic chromosomes of *Drosophila* (DeRobertis *et al.*, 1975).

Chromosomes undergo condensation and decondensation during mitosis and meiosis, except for heterochromatic regions, which remain condensed even during interphase. Histones, combined with the phosphoric acid groups of the nucleic acids on a one-to-one basis, serve for chromosome configuration and perhaps in other ways as well. A number of intermediate stages in chromosome organization have been observed in some primitive eukaryotes such as dinoflagellates and unicellular red algae, indicating a possible evolution from the prokaryotic state (DuPraw, 1970; Hood *et al.*, 1975).

In the giant chromosomes of the salivary gland cells of *Drosophila* larvae (Fig. 7.11), the individual deoxyribonucleohistone units believed to be the fundamental units of the chromosomes have multiplied without cell division to the point that the polyteny (multiplicity) may reach 1000 strands of DNA (as determined by DNA content) and the total length 2000 μm. This total length may be compared to a length of 7.5 μm in normal tissue cells of *Drosophila*. Because they are lined up in phase with one another, the Feulgen-positive DNA-containing regions of the units form bands in register, which are readily seen with the light microscope.

LAMPBRUSH CHROMOSOMES

In vertebrate oocytes (egg-forming cells) just before the first meiotic division, the chromo-

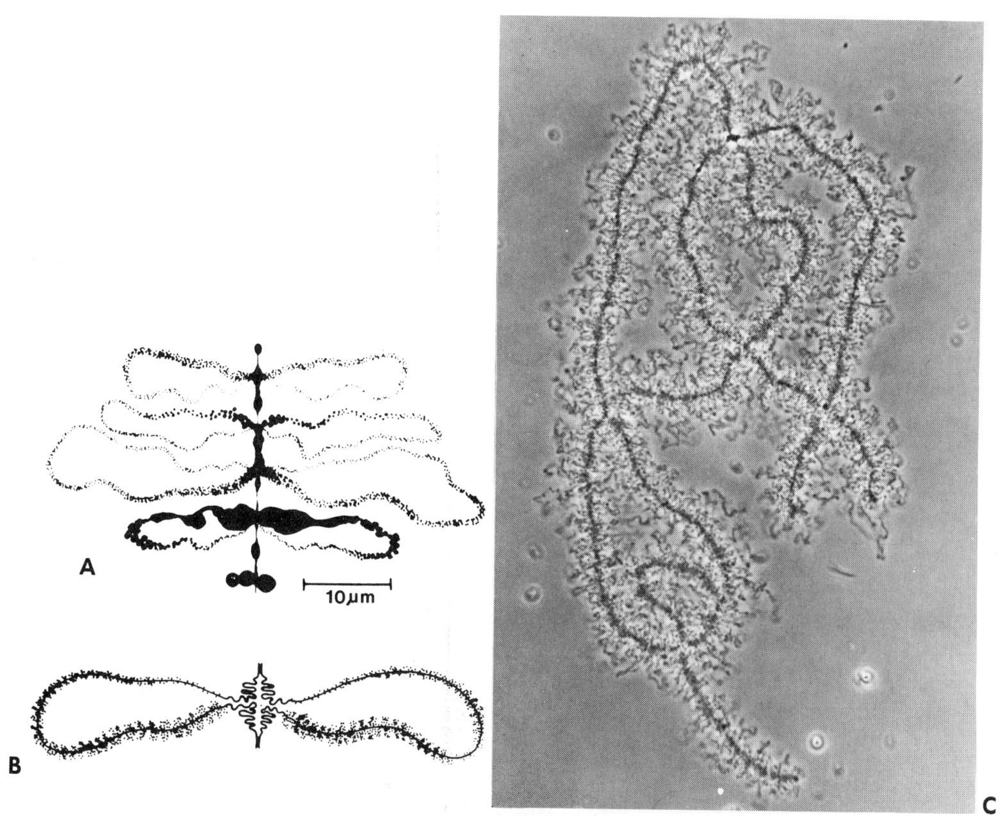

Figure 7.17. *A, Segment of a lampbrush chromosome of a newt* (Triturus), *showing characteristic paired loops projecting from the main axis of Feulgen-positive granules. B, Interpretation of the chromosome structure of a pair of loops in terms of two continuous chromatids. It is thought that the loop consists of a very thin DNA strand surrounded by ribonucleoprotein. (From Gall, 1963: Nature* 198: *36.) C, Lampbrush chromosome (bivalent) from oocyte of the newt* Triturus viridescens *(phase contrast, × 680). (By courtesy of Joseph Gall.)*

some shows many elongated loops of uncompacted DNA, although the remainder of the chromosome is compacted in the form of chromomeres (Fig. 7.17*A*). The DNA loops (genes?) are sites of very active RNA synthesis. The RNA molecules (rRNA) develop from the loop making it look like tinsel (Fig. 7.17*B*). Such chromosomes are called *lampbrush chromosomes*. Under very high resolution electron microscopy RNA polymerase is visible while nascent (newly transcribed) RNA is in the form of strands. Labeled precursors of RNA (^{3}H-uridine, for example) are actively incorporated into the developing RNA strands.

Akin to development on the lampbrush chromosomes are the puffs that appear on polytene chromosomes of salivary glands in fly larvae. At times the polytene chromosomes show RNA enlargements spoken of as puffs (Fig. 7.18). Puffs are regions of active RNA synthesis, as shown by rapid incorporation of ^{3}H-uridine into the RNA. An inhibitor such as actinomycin D, which stops DNA-dependent RNA synthesis, prevents the development of puffs. At different times during development, and at the same time in different tissues, different regions of chromosomes develop puffs. Analysis of RNA base ratios in the puffs in different chromosome regions shows that each has a unique chemical composition (Beermann and Clever, 1964).

It is thought that puffing results from intensive RNA transcription in response to the temporary, pressing developmental needs of the larva. Specific regions of the chromosome may be excited to puffing activity by injecting a larva with the insect-molting hormone ecdysone. It is likely that the hormone activates specific genes and that considerable messenger RNA is formed in response (Fig. 7.11) (Beermann and Clever, 1964; Davidson, 1976).

BEADED STRUCTURE OF CHROMATIN

Considerable excitement was occasioned by the finding that when chromatin extracted from the resting nucleus was treated in a particular manner, the ultimate structure observable under the electron microscope appeared to be a series of beads, called *nucleosomes* (Kornberg, 1974). Removal of the lysine-rich H1 histone causes the very compact chromatin structure to

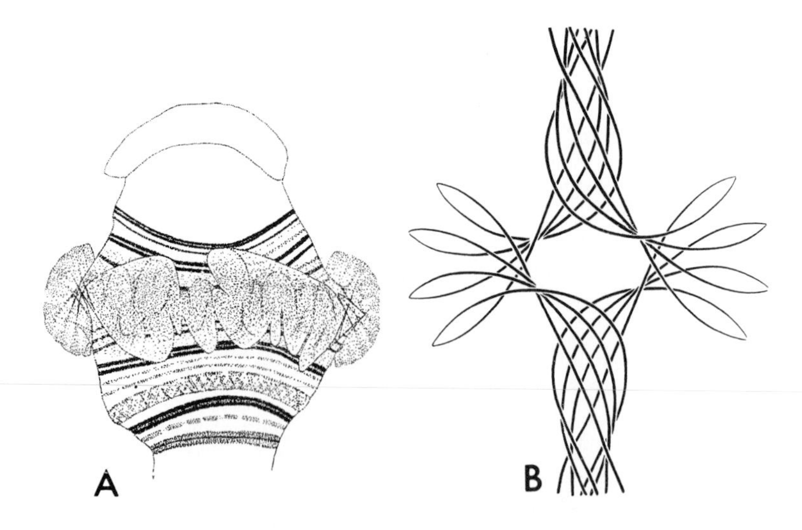

Figure 7.18. A, *Puffing in a segment of a giant chromosome of a midge* (Chironomus) *salivary gland cell.* B, *Interpretive diagram of the chromosome chromatids in the region of a puff. There would actually be a thousand or more chromatids in a giant chromosome rather than the few shown here. (From Beermann, 1952: Chromosoma* 5: *139.)*

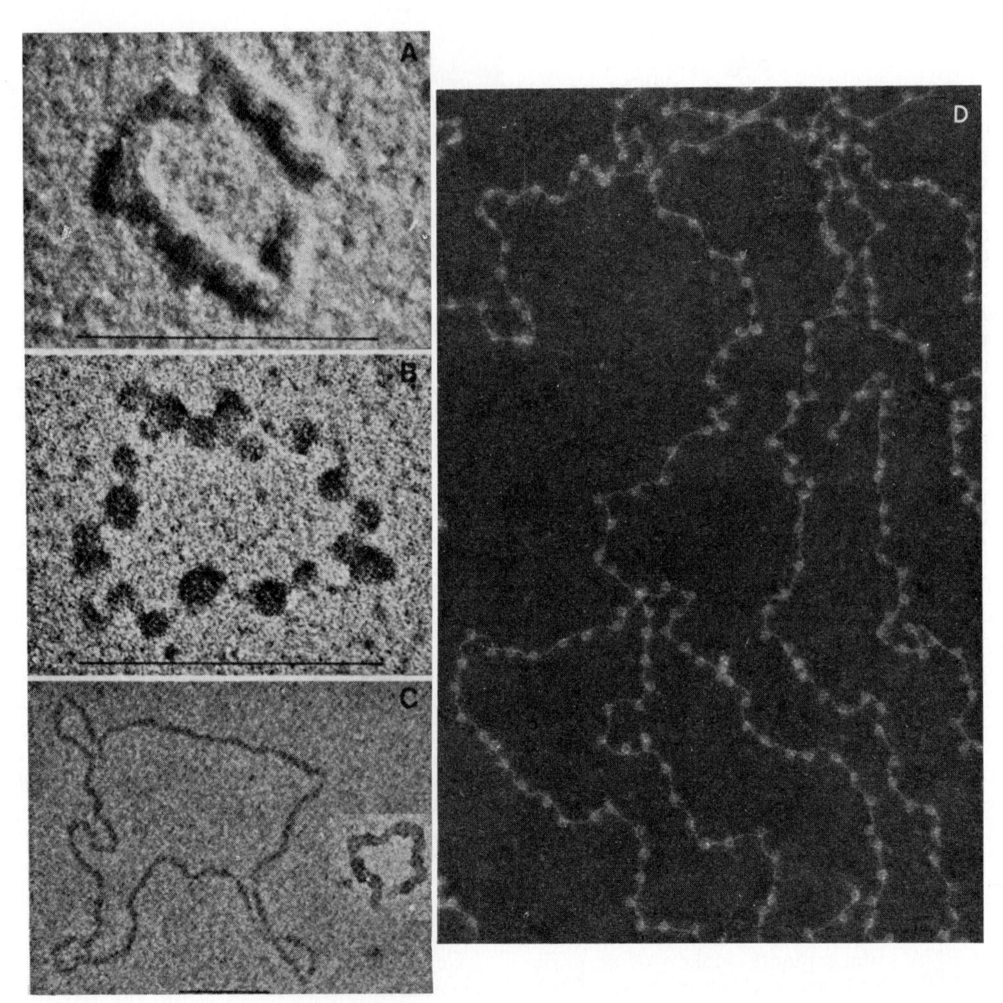

Figure 7.19. *Simian virus (SV40) minichromosomes. High resolution electron micrographs of the SV40 minichromosome in its native* (A), *beaded* (B), *and deproteinized* (C) *state. The bars represent 100 nm; the inset in* (C) *compares a native minichromosome at an equal magnification with the deproteinized DNA. (From Griffith, 1975: Science* 187: *1207. Copyright 1975 by the American Association for the Advancement of Science.)* D, *Darkfield electron micrograph of the nucleosomes (minichromosomes from a chicken erythrocyte nucleus (× 260,000). (By courtesy of A. L. Olins and D. E. Olins.)*

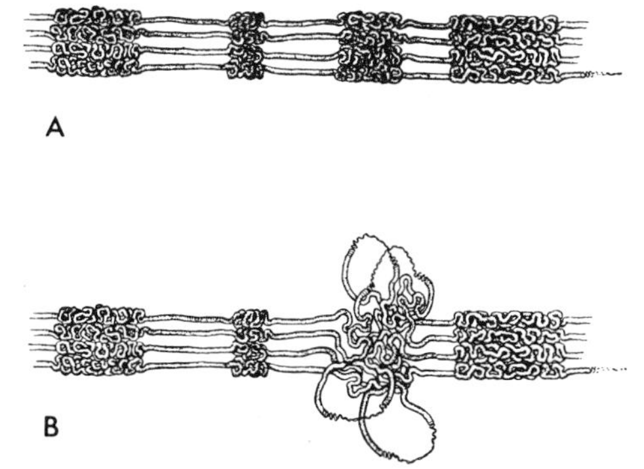

Figure 7.20. A, *A model of band and interband structure in a giant salivary gland chromosome according to DuPraw and Rae.* B, *The same model showing a possible method of "puffing" in one of the bands. (From DuPraw and Rae, 1966: Nature 212: 598–600.)*

open up and reveal the nucleosomes. Each nucleosome appears to consist of a DNA section composed of about 200 base pairs and an equal weight of histones (H2a, H2b, H3, and H4). The histones and DNA can be separated, and when again mixed they form the "string of beads." Addition of H1 histone leads to a more compacted structure (Oudet *et al.*, 1975). These findings for animal cell chromatin also hold for plant cell chromatin (Nagl, 1976). Interestingly, a virus DNA mixed with the appropriate histones leads to the formation of a similar chain of beads (Fig. 7.19) indicating that such structure results from interaction between DNA and histone regardless of their source (Griffiths, 1975).

The full significance of these findings cannot be assessed until more information is available. Nonetheless, formation of nucleohistone beads appears to be of importance in the packing of DNA in eukaryotic cells. It is likely that other types of packing are also involved in keeping such a long eukaryotic chromosome fiber within the confines in which it occurs in the chromosome. One model is shown in Figure 7.20. A model of replication in a eukaryotic chromosome is shown in Figure 7.21.

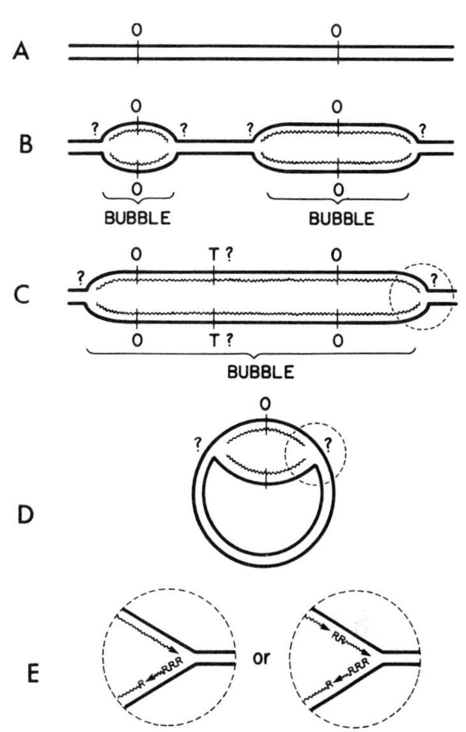

Figure 7.21. Bidirectional replication in eukaryotic chromosomes. A–C, *Diagrams showing three stages in the replication of a stretch of chromosomal DNA containing two replicating units (replicons).* A, *Prior to replication;* B, *after initiation in both replicons;* C, *after the replicons have fused to form a single, still growing, bubble;* D, *a replicating papova virus (SV40 or polyoma) DNA molecule;* E, *enlargements of the replication fork: two models, with semidiscontinuous synthesis shown on the left and totally discontinuous synthesis on the right. O stands for origin; T stands for terminus; RR indicates RNA stretch serving as a primer;* arrowhead *indicates 3′ end of growing chains.* Question marks *indicate areas of uncertainty. (From Edenberg and Huberman, 1975: Ann. Rev. Genetics* 9: *246.)*

CHEMICAL NATURE OF THE GENE

No cell survives enucleation because synthesis of enzymes and other needed molecules soon stops and the cell deteriorates. Some cells die much sooner than others; mammalian red blood cells last for several days. The information as to what kind of protein is to be synthesized ultimately comes from the genes, and enzyme proteins are needed for all cellular molecules that are synthesized.

It is now evident that the gene is a segment of the double-stranded DNA molecule that consists of the four nucleotides: adenine, thymine, cytosine, and guanine arranged in a specific sequence with the nucleotides on one strand paired with complementary nucleotides on the opposite strand. These nucleotides, making up the double-stranded helical DNA molecule, are now considered to serve as a language that gives all the information to the cytoplasm to account for inheritance; groups of three adjacent nucleotides on one strand form a "word" (codon) that specifies a particular amino acid (Watson, 1976). Some of the evidence supporting this view is summarized here. Historical articles of a semipopular nature but written by outstanding scientists are collected in a useful volume (Hanawalt and Haynes, 1973).

Viral DNA Genome

About 45 percent of the dry weight of a phage

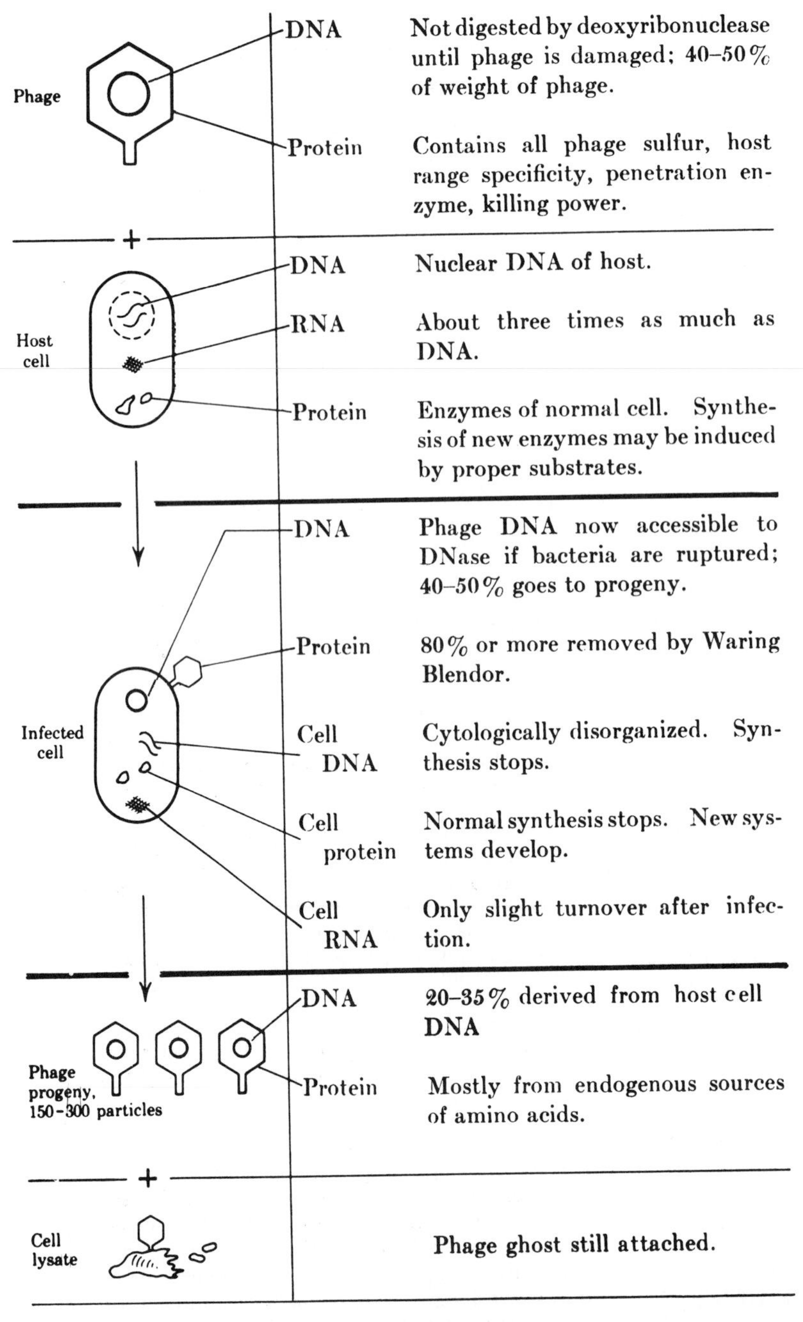

Figure 7.22. The events in multiplication of phage in a bacterial cell. Specifically, the diagram shows what happens in the course of infection of the colon bacillus with bacteriophages of the T series. (Modified from Hotchkiss, R. D., 1954: The Nucleic Acids. Vol. 2. Academic Press, New York; from Anfinsen, 1959: The Molecular Basis of Evolution. John Wiley & Sons, New York.)

(T-2 phage of *E. coli*) consists of DNA. Shortly after adsorption of an active phage particle to the surface of a bacterium, its DNA enters the cell and replicates (Fig. 7.22), forming many phage particles. This occurs even if the phage protein coat originally adhering to the surface of the bacterium is washed off, leaving not more than 1 percent of ^{35}S-labeled phage protein with the bacterium. On the other hand, phage "ghosts," with the DNA removed by osmotic shock, still adsorb to the host cells but do not reproduce. Since a phage particle carries a variety of genes arranged in linear order in a chromosome, it is clear that its genetic information is carried by the DNA alone.

RNA replaces DNA as the genome in plant viruses and in some animal viruses. In some viruses, RNA induces DNA synthesis by *reverse transcription* in the host cell, and the DNA, complementary to the original RNA, now induces transcription of RNA for the new virus particles (Temin, 1972).

In one experiment virus protein and virus RNA were separated from wild-type and mutant strains of a virus. When the protein of the mutant virus was mixed with the wild-type virus RNA, a new "hybrid" virus of the wild-type RNA with the mutant protein coat was formed. Such a virus infected plant cells and reproduced, but when the protein formed by the hybrid virus was isolated, it proved to be like that induced by the wild-type RNA.

Transformation

Equally direct evidence for the importance of DNA in heredity is the *transformation of Streptococcus pneumoniae*. Addition of DNA from a virulent strain makes a nonvirulent strain virulent; addition of protein or RNA from the virulent strain does not. It would appear that DNA of the virulent strain is incorporated into the chromosome of the nonvirulent strain and induces synthesis of DNA like itself, since the change is hereditary (Avery *et al.,* 1944). Similar changes in genome have been induced in other microorganisms.

Conjugation

During *conjugation* in *E. coli* part of a chromosome of the donor is transferred to the recipient (Fig. 7.23). When the donor is a mutant strain with an observable character, for example, resistance to an antibiotic, the offspring of the recipient will show that character. Since only DNA is transferred during conjugation the conclusion is inescapable that DNA determines the hereditary characteristics of a cell.

Lysogeny

Lysogenic phages are those that can be incorporated into the genome of infected bacteria.

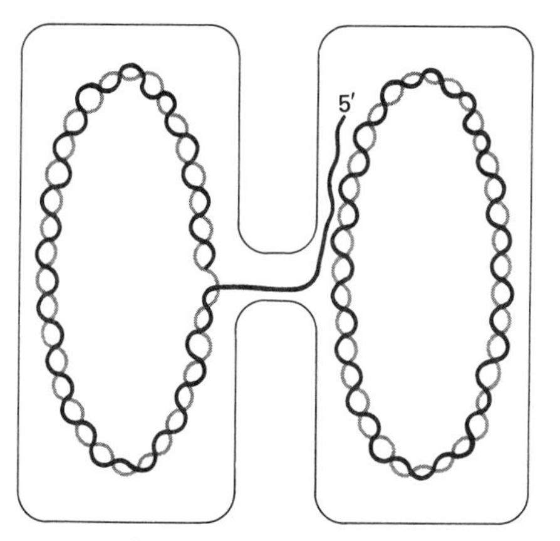

Figure 7.23. Conjugation in E. coli *with transfer of single-stranded DNA from a male bacterium into a female via a rolling circle mechanism. The 5' refers to the free end of the phosphate residue in the nucleic acid (see Chapter 15). From Watson, 1970: The Molecular Biology of the Gene. 2nd ed. W. A. Benjamin, New York.)*

The bacterium thereby gains a characteristic: resistance from further infection by that type of phage. The lysogenic phage in the bacterial genome may be replicated for many generations without harming the bacterium (Fig. 7.24). Should the bacterium be damaged, as, for example, by ultraviolet radiation, the phage becomes activated and separates from the bacterial chromosome, multiplies, and lyses the bacterium (Campbell, 1976). Since only the DNA of the phage enters the bacterium, the inherited character depends only on DNA.

It is likely that in lysogeny the phage genome is accepted into the bacterial chromosome by crossing over, as shown in Figure 7.25. In conjugation (Fig. 7.23) the piece of male chromosome entering the female cell is probably accepted in a similar way.

Transduction

Akin to lysogeny is *transduction* in bacteria, in which genetic material from one strain of bacteria is transferred to another strain of the same species by phages. For example, each of two strains of *Salmonella typhimurium* lacked the ability to synthesize one of the amino acids, but each one had a deficiency for a different amino acid. Even when a membrane through which bacteria could not pass was interposed between suspensions of the two strains to prevent mating, a new strain of bacteria appeared that produced both amino acids for which the two original strains of bacteria were deficient. The new strain of bacteria was able to grow in the medium deficient for the two amino acids. The only carrier of a gene for production of a missing amino acid that could pass through the membrane barrier is a phage (Zinder, 1958).

It has now been shown that the phage, which transmits the information from one strain of bacterium to another, is often incapable of independent multiplication and cannot lyse the infected bacteria. Presumably the phage genetic material has become defective. Perhaps this has happened because by a mistake during its replication in the preceding bacterium, a section of its chromosome has been replaced by a section of the bacterial chromosome containing the gene for the synthesis of the missing amino acid. The phage's very effectiveness in transduction can thus prevent its further effectiveness as a phage.

THE ROLE OF THE GENE

It can now be stated unequivocally that the gene is a subdivision of the DNA molecule that codes for a polypeptide (except in RNA viruses, in which RNA has this role). Since a polypeptide usually consists of several hundred amino acids, with each amino acid coded by a set of

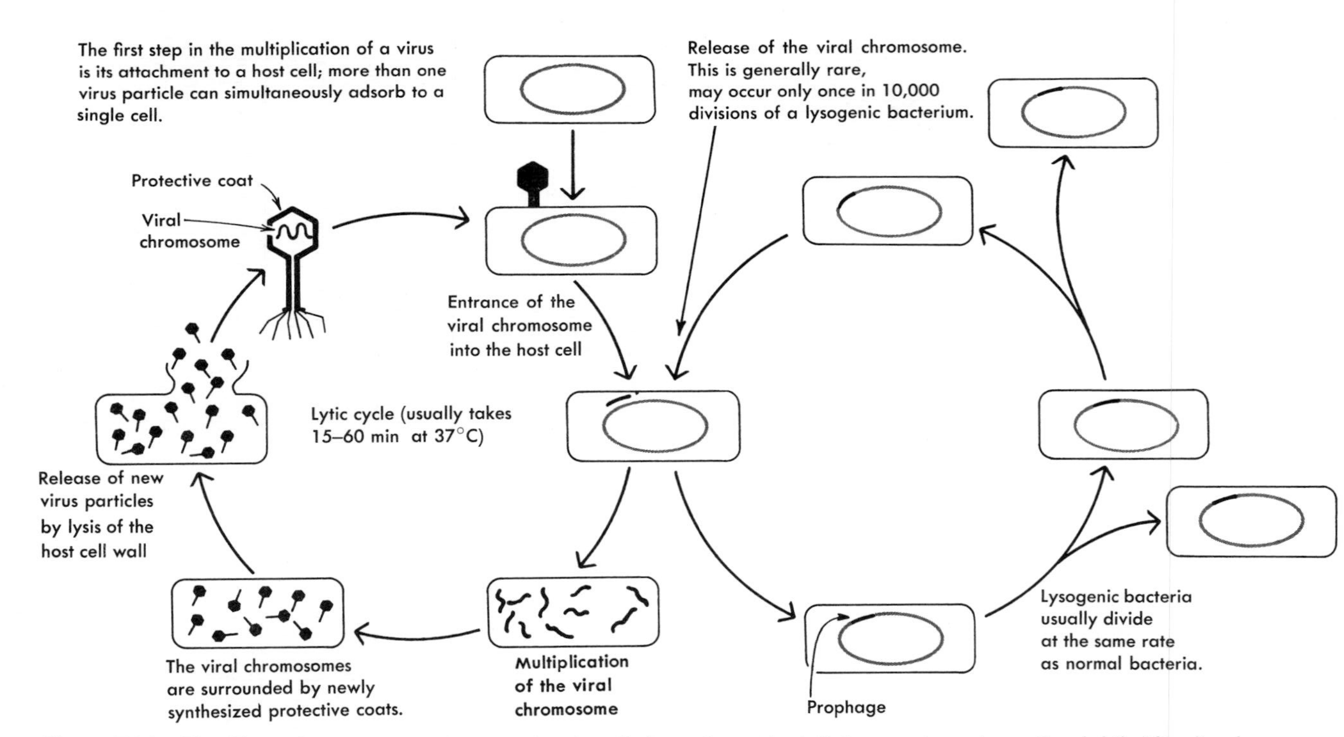

Figure 7.24. *The life cycle of a lysogenic bacterial virus (lytic cycle on the left, lysogenic cycle on the right). After its chromosome enters a host cell, it sometimes multiplies immediately like a lytic virus and at other times becomes transformed into prophage. The lytic phase of its life cycle is identical to the complete life cycle of a lytic (nonlysogenic) virus. Lytic bacterial viruses are so called because their multiplication results in the rupture (lysis) of the bacteria. (From Watson, 1976: The Molecular Biology of the Gene. 3rd Ed. W. A. Benjamin, New York.)*

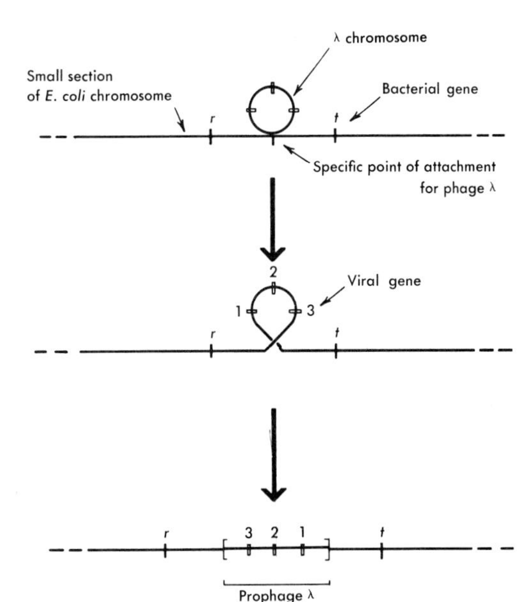

Figure 7.25. *Insertion of the chromosome of phage λ into the* E. coli *chromosome by crossing over. (From Watson, 1970: The Molecular Biology of the Gene. 2nd Ed. W. A. Benjamin, New York.)*

three nucleotide bases, an average gene has been determined to consist of about 1500 nucleotide pairs (molecular weight 10^6) (Watson, 1976).

It is of interest that some proteins consist of several polypeptide chains. For example, the hemoglobin molecule consists of two α and two β polypeptide chains. Sometimes the respective polypeptides that form a given molecule are encoded on successive portions of the genome, but this is not the case with hemoglobin (Ingram, 1972).

Since a typical mutation represents a substitution, deletion, or addition of a nucleotide in one of the sets of three bases coding for an amino acid, and since about 1500 nucleotides containing such bases are present in a gene, one might expect that crossover and recombination might occur within a gene. Data proving that this does indeed occur have been obtained for the so-called rII gene of the T-4 virus of *E. coli*. From crossover data obtained on about 1000 mutants it is possible to construct a gene map, and it is found that the arrangement of mutations in a gene is linear, resembling in this respect the arrangement of genes in a chromosome (Benzer, 1961; Watson, 1976).

COMPARISON OF THE GENOME IN BACTERIA, VIRUSES, AND EUKARYOTIC CELLS

The genome of the bacterium *E. coli* is better known and more completely studied than any other (Watson, 1976). About 2000 to 3000 genes coding for the various types of proteins necessary to the life of *E. coli* (or any other bacterial cell) are thought to exist, about a third of which have been described. The *E. coli* DNA molecule makes up only 1 percent of the cell mass and is present as a single double-stranded molecule about 1000 times the cell length, presumably coiled up like a tangled skein of yarn. Each gene has a molecular weight of about 10^6 and contains about 1500 base pairs. During replication, it is thought that the growing point of the DNA at which the nucleotides are being added is accompanied by a separation of the two strands in the DNA molecule (see Chapter 15). Since replication in *E. coli* under most favorable conditions takes only 20 minutes, the growing point must move at a rate of about 40 to 50 μm per minute.

E. coli (wild type) can live in the absence of other organisms on a diet of glucose and a variety of salts, including a source of nitrogen and sulfur. Therefore, it is capable of synthesizing all of its other organic cellular requirements. The pleuropneumonia-like organisms (PPLO) or mycoplasmas (see Chapter 1) are minute organisms that usually live as parasites on other organisms. The genome of the free-living PPLO *Mycoplasma laidlawii* is estimated to contain about 700 genes coding for as many proteins. As pointed out before, the PPLO organisms, when free-living, must be supplied with many vitamins, amino acids, purine and pyrimidine bases, ribose and deoxyribose, and other complex molecules that they cannot synthesize. The genome of this species, which is much smaller than in *E. coli,* thus suffices to code for the remaining proteins, presumably enzymes involved in growth and reproduction of their kind (Watson, 1976).

Even more reduced in size is the genome of viruses. Viruses, as has been mentioned, are parasites that depend upon the host cell for most of their synthetic machinery. A DNA virus is shown to consist of the DNA molecule (chromosome), a coat of one or more proteins, and sometimes one or more enzymes. However, it codes for mRNA synthesis but depends upon the host's tRNA and rRNA. The T-2 phage, for example, probably possesses several hundred genes. It can therefore code for the enzymes necessary for its replication and packaging. The genome of the phage may contribute to that of the bacterium. For example, the T-2 phage infecting a thymineless (thymine-minus) mutant of *E. coli* incapable of thymine synthesis will induce in the thymineless host cell the synthesis of thymine, the code for which the phage contains in its genome (Watson, 1976).

Some phages have much smaller genomes than the T-2 phage, for example, lambda (λ) phage, which has only about 15 to 20 genes; none of its genic information complements that of the bacterium. Although in many cases phage infection represses host DNA synthesis, lambda phage infection does not. It often even joins the genome of the host and becomes dormant as a lysogenic phage.

The smallest viruses contain only three to five genes. The F2 RNA virus of *E. coli* has a nucleic acid chain that contains only 3000 nucleotides and so can code for approximately 1000 amino acids; of these, 100 are used to construct proteins of the virus coat. Probably about 100 of the remaining 900 amino acids are present in the F2-specific RNA synthetase. The remainder of the amino acids probably form several proteins (enzymes?) involved in virus reproduction.

A virus even smaller than F2 is found in tobacco cells infected with tobacco necrosis virus. Its nucleic acid contains only about 1000 nucleotides, which would be sufficient to code for one average size protein. Necrosis virus never successfully invades a host cell alone and multiplies only when the host cell is also infected with a larger (6000-nucleotide) RNA virus. Presumably the small virus uses the enzymes of the larger virus (e.g., the RNA synthetase) for its own replication (Watson, 1976).

The smallest DNA virus is polyoma, a virus that causes tumors in the mouse. The polyoma virus inhabits the cell nucleus. It has about 5000 nucleotide pairs in its DNA molecule, sufficient to code for several proteins. These are probably enzymes and virus coat proteins (Watson, 1976). The finding of overlapping genes in phage ϕX174 makes possible coding for more proteins than previously estimated (Barrell and Hutchinson, 1976).

In contrast, the genome of eukaryotic cells is vastly more complex than that of bacteria or viruses (Wischnitzer, 1973). This is apparent from the number of genes present in eukaryotic cells. For instance, the nucleus of a *Drosophila* cell is estimated to possess 5,000 to 10,000 genes. The human haploid genome, as found in a sperm, is estimated to contain at least 23,000 to 100,000 genes (Hartl, 1977).

SUMMARY

The cell nucleus and cytoplasm constitute a reciprocal system of delicately balanced interrelationships. Experiments have shown that the cell cytoplasm cannot survive for long without a nucleus because many of its activities are initiated and regulated by the nucleus. The nucleus also influences cytoplasmic characteristics and exerts morphogenetic influence on differentiation of the cell. On the other hand, a nucleus without its cytoplasm is helpless and soon dies.

The nucleus of ruptured cells can be separated from other cell constituents and studied by cytochemical, biochemical, and biophysical methods. The nucleus is the center of most of the cellular DNA, only a small amount being present in mitochondria, and in green plant cells, in chloroplasts as well. The proteins in the chromosome are of two types, histone and nonhistone. Histones are low-molecular-weight proteins with a high proportion of the basic amino acids, lysine and arginine, and are lacking the aromatic amino acid tryptophan. Histones are combined with DNA in the formation of chromatin in the resting nucleus and the central chromosome fiber in condensed chromosomes. The histones are probably primarily structural packing proteins for chromosomal DNA. Non-histone proteins are of higher molecular weight than histones. They have little lysine and arginine but contain tryptophan. They are perhaps partly structural but may also be enzymatic and probably serve as repressors and activators of genes.

A nucleolus is present in the nucleus of most cells and contains most of the nuclear RNA. The nucleolus can be isolated from fragmented nuclei by fractional centrifugation. A small amount of DNA of chromosomal origin, the nucleolar organizer, is present at the periphery of the nucleolus. The primary function of the nucleolus appears to be the synthesis of ribosomal RNA.

The chromosome in viruses and bacteria usually consists of a single long double-stranded DNA molecule, most often joined at the ends to form a circle. In RNA viruses the genome consists of an RNA molecule that directs DNA synthesis of a template for its reproduction in the host cell. In some viruses the chromosome is single stranded, for example, in the ϕX174 (DNA) and tobacco mosaic (RNA) viruses. In mitochondria and chloroplasts the chromosome(s) is a double-stranded DNA molecule in circular form also.

The nuclear chromosome in eukaryotic cells

is probably a long DNA molecule (or a series of such molecules bound to histone, or protamine in some sperm), much compacted, but without a membrane separating it from the rest of the nucleus. Investigators have not yet agreed completely on the fundamental structural plan of the eukaryotic chromosome. Replication of DNA occurs sequentially along the length of the chromosome in viruses and prokaryotic cells, but it may begin at many places in a chromosome of a eukaryotic cell, presumably on each subunit of the long DNA molecule in each chromosome.

A gene is a subdivision of a DNA molecule that codes for a polypeptide, each triplet of polynucleotide bases within the gene coding for an amino acid (see Chapter 15). Some genes coding for important cell syntheses, for example, of rRNA, are present in multiple copies. Some series of nucleotides appear to serve as spacers between functional genes and in viruses, within genes as well. A typical mutation consists of a substitution for, deletion of, or addition to, one of the three nucleotides coding for an amino acid. Since crossovers and recombinations can occur in genes, it is possible to construct gene maps, as has been done in a few instances. The genome of some viruses may consist of only a few to several hundred genes, that of bacteria several thousand, and that of eukaryotic organisms, several to many orders of magnitude greater.

LITERATURE CITED AND GENERAL REFERENCES

Avery, O. T., MacLeod, C. M. and McCarty, M., 1944: Studies on the chemical nature of the substance inducing transformation of pneumococcal types. Induction of transformation of a DNA section isolated from *Pneumococcus,* Type III. J. Exp. Med. *79*: 137–158.

Barget, S. M., Moore, C. and Sharp, P. A., 1976: Spliced segments at the 5′-terminus of adenovirus 2, late in RNA. Proc. Nat. Acad. Sci. U.S.A. *74*: 3171.

Barrell, B. and Hutchinson, G. M., 1976: Overlapping genes in bacteriophage φX174. Nature *264*: 64.

Beadle, G. W., 1959: Genes and biological enigmas. *In* Science in Progress. Baitsell, ed. Vol. *6*: 184–265. Yale University Press, New Haven.

Beermann, W., and Clever, J., 1964: Chromosome puffs. Sci. Am. (Apr.) *210*: 50–58.

Benzer, S., 1961: Genetic fine structure. Harvey Lectures *56*: 1–21.

Berger, S., Niemann, R. and Schweiger, H. G., 1976: *Acetabularia* nucleus after 24 hours in an artificial medium. Protoplasma *85*: 115–118.

Bernhardt, H. P., 1976: The control of gene expression in somatic cell hybrids. Int. Rev. Cytol. *47*: 289–326.

Brachet, J. and Bonotto, S., eds., 1970: International Symposium on *Acetabularia*. Academic Press, New York.

Britten, R. J. and Kohne, D. E., 1970: Repeated segments of DNA. Sci. Am. (Apr.) *222*: 24–31.

Bustin, M., Goldblatt, D. and Sperling, R., 1976: Chromosome structure visualized by immunoelectron microscopy. Cell *7*: 297–304.

Cairns, J., 1966: The bacterial chromosome. Sci. Am. (Jan.) *214*: 37–44.

Callan, H. G., 1963: The nature of lampbrush chromosomes. Int. Rev. Cytol. *15*: 1–34.

Campbell, A. M., 1969: Episomes. Harper & Row, New York.

Campbell, A. M., 1976: How viruses insert their DNA into the DNA of the host cell. Sci. Am. (Dec.) *235*: 103–122.

Cantell, K., 1978: Towards the clinical use of interferon. Endeavour *2*NS: 27–30.

Chromosome Structure and Function 1974. Cold Spring Harbor Symp. Quant. Biol. Vol. 38.

Crick, F. H. C., 1954: The structure of the hereditary material. Sci. Am. (Oct.) *194*: 54–66.

Darnell, J. E., Jr., 1978: Implications of RNA-RNA splicing in evolution of eukaryotic cells. Science *202*: 1257–1260.

Dauvas, A. S., Harrington, C. A. and Bonner, J., 1975: Major non-histone proteins of rat liver chromatin: Identification of myosin, actin, tubulin and tropomyosin. Proc. Nat. Acad. Sci. U.S.A. *72*: 3902–3906.

Davidson, J. N., 1976: The Biochemistry of the Nucleic Acids. 8th Ed. Academic Press, New York.

Davis, B. D., 1977: The recombinant DNA scenarios: Andromeda strain, chimera and golem. Am. Sci. *65*: 547–555.

DeRobertis, E. D. P., Saez, F. A. and DeRobertis, E. M. F., 1975: Cell Biology. 5th Ed. W. B. Saunders Co., Philadelphia.

DuPraw, E. J., 1970: DNA and Chromosomes. Holt, Rinehart & Winston, New York.

Edenberg, H. J. and Huberman, J. A., 1975: Eukaryotic chromosome replication. Ann. Rev. Genetics *9*: 245–284.

Elgin, S. C. and Weintraub, H., 1975: Chromosome proteins and chromatin structure. Ann. Rev. Biochem. *44*: 725–774.

Ephrussi, B., 1972: Hybridization of Somatic Cells. Princeton University Press, N.J.

Fenner, F., McAuslin, B. R., Mims, C. D., Sambrook, J. and White, D. V., 1974: The Biology of Animal Viruses. Academic Press, New York.

Flamm, W. G., 1972: Highly repetitive sequences of DNA in chromosomes. Int. Rev. Cytol. *32*: 1–51.

Flickinger, C. J., 1968: The effects of enucleation on the cytoplasmic membranes of *Amoeba proteus*. J. Cell Biol. *37*: 300–315.

Fraenkel-Conrat, H. and Wagner, R. R., 1974: Comprehensive Virology. (Multivolume treatise.) Plenum Press, New York.

Ghosh, S., 1976: The nucleolar structure. Int. Rev. Cytol. *44*: 1–28.

Gibor, A., 1966: *Acetabularia:* a useful giant cell. Sci. Am. (Nov.) *25*: 118–124.

Goldstein, L., 1973: Nucleocytoplasmic interactions in amoebae. *In* The Biology of Amoeba. Kwang, J., ed. Academic Press, New York, pp. 479–504.

Goodenough, V., 1978: Genetics. 2nd Ed. Holt, Rinehart & Winston, New York.
Goodheart, C. R., 1975: An Introduction to Virology. 2nd Ed. W. B. Saunders Co., Philadelphia.
Griffiths, J. D., 1975: Chromatin structure deduced from a minichromosome. Science *187*: 1202–1203.
Gurdon, S., 1975: Cell fusion of some subcellular particles of heterokaryons and hybrids. J. Cell Biol. *61*: 257–280.
Hanawalt, P. C. and Haynes, R. H., eds., 1973: The Chemical Basis of Life. W. H. Freeman Co., San Francisco.
Harris, H., 1970: Cell Fusion. Harvard University Press, Cambridge.
Hartl, D. L., 1977: Our Uncertain Heritage: Genetics and Human Diversity. J. B. Lippincott Co., Philadelphia.
Hood, L. E., Wilson, J. H. and Wood, W. B., 1975: Molecular Biology of Eukaryotic Cells. W. A. Benjamin, Menlo Park, Calif.
Hotchkiss, R. D. and Weiss, E., 1956: Transformed bacteria. Sci. Am. (Nov.) *195*: 48–53.
Ingram, V., 1972: Biosynthesis of Macromolecules. 2nd Ed. W. A. Benjamin, Menlo Park, Calif.
Isenberg, I. and Spiker, S., 1977: An approach to a functional definition of histones; a current stock taken. *In* Molecular Human Genetics. Sparkes, Corning and Fox, eds. Academic Press, New York.
Jenkins, R. A., 1977: The role of microtubules in macronuclear division of *Blepharisma*. J. Protozool. *24*: 264–275.
Jeon, K. W. and Danielli, J. F., 1971: Micrurgical studies with large, free-living amebae. Int. Rev. Cytol. *30*: 49–89.
Jones, C. W., Mastrangelo, I. R., Smith, H. H., Liu, H. Z. and Meck, R. A., 1976: Interkingdom fusion between human (HeLa) cells and tobacco hybrid (GGLL) protoplasts. Science *193*: 401–403.
Jordan, E. G., 1978: The Nucleolus. 2nd Ed. Carolina Biol. Supply Co., Publication Div., Burlington, N.C. (Printed by Oxford Press.)
Kolata, G. B., 1976: Jumping genes: a common occurrence in cells. Science *193*: 392–394.
Kornberg, R. D., 1974: Chromatin structure: repeating units of histones and DNA. Science *184*: 868–871.
Kornberg, R. D., 1977: Structure of chromatin. Ann. Rev. Biochem. *46*: 931–954.
Kornberg, R. D. and Thomas, J. O., 1974: Chromatin structure: oligomeres of the histones. Science *184*: 865–868.
Lacy, E. and Axel, R., 1975: Analysis of DNA of isolated chromatin units. Proc. Nat. Acad. Sci. U.S.A. *72*: 3978–3982.
Lehninger, A. L., 1975: Biochemistry, 2nd Ed. Worth Publishers, New York.
Luria, S. E., Darrell, J. E., Jr., Baltimore, D. and Campbell, A., 1978: General Virology. John Wiley & Sons, New York.
Maugh, T. H. II, 1976: The artificial gene. It's synthesized and it works in cells. Science *194*: 44.
Meselson, M. and Stahl, F. W., 1958: The replication of DNA in *E. coli*. Proc. Natl. Acad. Sci. U.S.A. *44*: 671–682.
Mirsky, A. E., 1950–51: The chemical composition of chromosomes. Harvey Lectures *46*: 98–115.
Nagl, W., 1976: Nuclear organization. Ann. Rev. Plant Physiol. *27*: 39–69.
Novitsky, J. A. and Morita, R. Y., 1977: Survival of a psychrophilic marine vibrio under long-term nutrient starvation. Appl. Environ. Microbiol. *33*: 635–641.
Olins, A. E. and Olins, D. E., 1974: Spheroid chromatin units (mu bodies). Science *183*: 330.
Oudet, P., Gross-Bellard, M. and Chambon, P., 1975: Electron submicroscopic biochemical evidence that chromatin is a repeating unit. Cell *4*: 281–300.
Pelc, S. R., 1972: Metabolic DNA in ciliated Protozoa, salivary gland chromosomes and mammalian cells. Int. Rev. Cytol. *32*: 327–353.
Puck, T., 1974: Hybridization and somatic cell genetics. *In* Somatic Cell Hybridization. Davidson and de la Cruz, eds. Raven Press, New York.
Recombinant DNA. Science *196*: 159–221. (Series of articles.)
Rees, H. and Jones, R. N., 1972: The origin of the wide species variation in nuclear DNA content. Int. Rev. Cytol. *32*: 53–92.
Rinch, T. T., Noll, M. and Kornberg, A. R., 1975: Electron micrographs of defined lengths of chromatin. Proc. Nat. Acad. Sci. U.S.A. *72*: 3320–3322.
Ringertz, N. R. and Savage, R. E., 1976: Cell Hybrids. Academic Press, New York.
Ruddle, F. and Kucherlapati, R. S., 1974: Hybrid cells and human genes. Sci. Am. (Jul.) *231*: 36–44.
Schaeffer, K., 1976: The RNA synthetic capacity of nuclei isolated from cultured mouse cells. Biochem. Biophys. Res. Commun. *68*: 219–226.
Shaw, D. C., Walker, J. C., Northrup, F. D., Barrell, B. G., Godson, G. N. and Fiddes, J. C., 1978: Gene K, a new overlapping gene in bacteriophage G4. Nature *272*: 510–515.
Sidebottom, E. and Deak, I. I., 1976: The function of nucleus in expression of genetic information: studies with hybrid animal cells. Int. Rev. Cytol. *44*: 29–53.
Sinsheimer, R. L., 1962: Single stranded DNA. Sci. Am. (Jul.) *207*: 109–116.
Sparrow, A. H. and Naumann, A. F., 1976: Evolution of genome size by DNA doublings. Science *192*: 524–528.
Stein, G. and Stein, J., 1976: Chromosomal proteins: their role in the regulation of gene expression. Bioscience *26*: 488–498.
Suzuki, P. T. and Griffiths, A. J. F., 1976: An Introduction to Genetic Analysis. W. H. Freeman Co., San Francisco.
Szarski, H., 1976: Cell size and nuclear DNA control in vertebrates. Int. Rev. Cytol. *44*: 93–111.
Tashiro, T. and Kurokawa, M., 1976: A contribution of non-histone proteins to the formation of chromatin. Eur. J. Biochem. *68*: 569–577.
Taylor, J. H., 1959: Autoradiographic studies of nucleic acids and proteins during meiosis in *Lillium longiflorum*. Am. J. Bot. *46*: 477–484.
Taylor, J. H., Woods, P. S. and Hughes, W. G., 1957: The organization and duplication of chromosomes as revealed by autoradiographic studies using tritium labelled thymidine. Proc. Nat. Acad. Sci. U.S.A. *42*: 122–128.
Temin, H. M., 1972: RNA-directed DNA synthesis. Sci. Am. (Jan.) *226*: 25–33.
Toczko, K., Kalinski, A. and Zukowska, A., 1975: On the arrangement of histones in chromatin. Acta Biochim. Pol. *22*: 251–256.
Watson, J. D., 1976: The Molecular Biology of the Gene. 3rd Ed. W. A. Benjamin, Menlo Park, Calif.
Watson, J. D. and Crick, F. H. C., 1953: Molecular structure of nucleic acids. A structure for deoxyribose nucleic acid. Nature *171*: 737–738.
Whitehouse, H. L. K., 1973: Towards an Understand-

ing of the Mechanism of Heredity. 2nd Ed. St. Martin's Press, New York.

Wischnitzer, S., 1973: The submicroscopic morphology of the interphase nucleus. Int. Rev. Cytol. *34*: 1–48.

Wolfe, S., 1972: Biology of the Cell. Wadsworth Publishing Co., Inc., Belmont, Calif.

Wollman, E. L., and Jacob, F., 1956: Sexuality in bacteria. Sci. Am. (Jul.) *195*: 109–118.

Woodcock, C. L. F., ed., 1977: Progress in *Acetabularia* Research. Academic Press, New York.

Zinder, N. D., 1958: Transduction in bacteria. Sci. Am. (Nov.) *199*: 38–43.

CHAPTER 8

CELL MEMBRANES

THE CELL SURFACE

The cytoplasm of every cell is bounded by a *plasma membrane* so thin and fragile that it was not visualized until the advent of the electron microscope. It is therefore no surprise that the plasma membrane is covered by protective coatings: in bacteria and plants by a cell wall (see Chapters 1 [Fig. 1.4] and 3 [Figs. 3.2, 3.12]), in animal cells by a galactoprotein *glycocalyx* (Luft, 1976) (see Chapter 5). The glycocalyx is helpful in cell adhesion and migration (Oppenheimer, 1978). In many bacteria and plant cells a glycocalyx is present in addition to the cell wall. Bacteria stick to surfaces by strands of polysaccharides of the glycocalyx (Costerton *et al.,* 1978).

In eggs of marine animals the protective coats are visible even with the light microscope. For example, surrounding a sea urchin egg a *jelly layer* covers the *vitelline membrane,* within which is the plasma membrane, as shown in Figure 8.1. The jelly layer can be most readily visualized by the addition of India ink; the colorless eggs and membranes stand out against the contrasting grey background. Both layers are secreted by the ovary and may be removed without damaging the plasma membrane; the jelly can be removed by acidified sea water, the tough, elastic vitelline membrane by microdissection or by washing in isotonic potassium chloride solution. An egg, denuded of the jelly and the vitelline membrane, becomes sticky. Such a "naked" egg, although more susceptible to injury, nevertheless develops normally when fertilized. The presence of a plasma membrane on such a naked sea urchin egg can be demonstrated when it visibly crinkles as it is touched by a microdissection needle.

When a normal egg is fertilized, the vitelline membrane is raised as a *fertilization membrane,* and a space then separates it from the surface of the egg. This space is filled with fluid except for some strands connecting the membrane to the surface of the egg. Around the surface of the egg is then secreted a *hyaline plasma layer.* This becomes important as the cement that binds the blastomeres together when the egg undergoes cleavage.

If an egg, freed of its extraneous membranes, is sawed in two with a needle, two spheres are obtained, indicating that the cut surface has developed a membrane capable of retaining the cell substance. Even when an egg is torn with a needle in sea water, a new membrane forms (Fig. 8.2*A*). However, if the membrane is torn in calcium-free sea water, the contents of the egg disperse in the solution (Fig. 8.2*B*).

As cells differentiate into tissues they are cemented together by a substance much like the hyaline plasma layer that binds together the blastomeres of an embryo. In blood capillaries, *cellular cement* acts like a filter through which blood plasma passes into the surrounding tissue spaces. The intercellular cement is a mucopolysaccharide, the prevalent one being hyaluronic acid. Hyaluronic acid contains D-glucuronic acid and N-acetyl-D-glucosamine in equivalent quantities, each alternating with the other in a linear chain (Fig. 8.3). Even in low concentration hyaluronic acid forms a firm gel.

In ciliate protozoans the plasma membrane is protected by a tough proteinaceous *pellicle* that superficially resembles the cell wall of a plant cell. However, when a ciliate is placed in hypertonic solution, the elastic pellicle decreases in size to fit the shrunken protozoan.

In lieu of a pellicle, an amoeba is coated with a polysaccharide slime, as are even some cells with a pellicle or a cell wall (e.g., desmids and diatoms). An amoeba pulled with a microneedle and released may snap back to its orig-

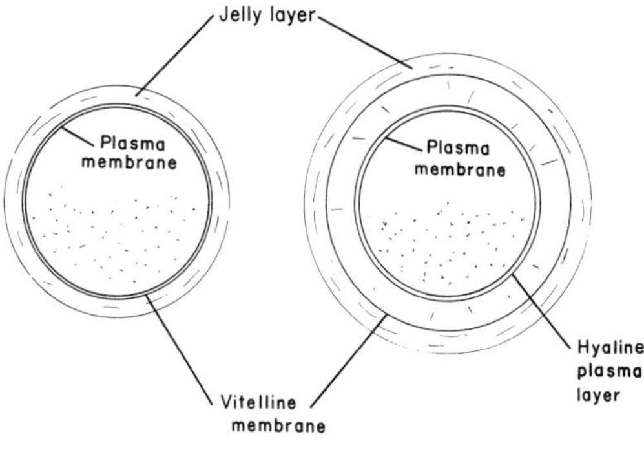

Figure 8.1. *Diagrams of unfertilized and fertilized sea urchin eggs to show extraneous layers. The plasma membrane is indicated, although it is not actually possible to see it under the light microscope. The vitelline membrane is raised during fertilization, becoming the fertilization membrane.*

inal position as a result of contraction of the elastic slime. The slime serves as a glycocalyx protecting the plasma membrane (Fig. 8.4).

It is curious that while the plasma membrane is considered delicate, it remains intact when cell walls are removed, for example, in halophilic bacteria, by treatment with 2 molar NaCl, and in other bacteria and plant cells by enzymes. However, if supplied with nutrient, such cells soon synthesize protective walls about them (Blaurock *et al.*, 1976; Salema and Brandao, 1976).

FINE STRUCTURE OF CELL MEMBRANES

Electron microscopic and other studies of the nerve cell myelin sheath laid the foundation for the fine structure concept of the plasma membrane of the cell. As can be seen diagrammatically in Figure 8.5, the axon of the nerve cell depresses the plasma membrane of the sheath cell and, rotating in the cell (or being spirally wrapped in the cell), it gradually accumulates around it a number of double membranes in which it comes to lie. Therefore, the myelin sheath consists essentially of numerous plasma membranes derived from the sheath cell and wound around the axon of the nerve cell (Robertson, 1964).

Analysis of the myelin sheath by x-ray diffraction had already demonstrated repeating units of great regularity. Electron micrographs confirmed and extended these studies. High resolution of ultrathin myelin cross sections indicate a *unit membrane* of 7.5 nm consisting of two layers staining with osmium tetroxide,

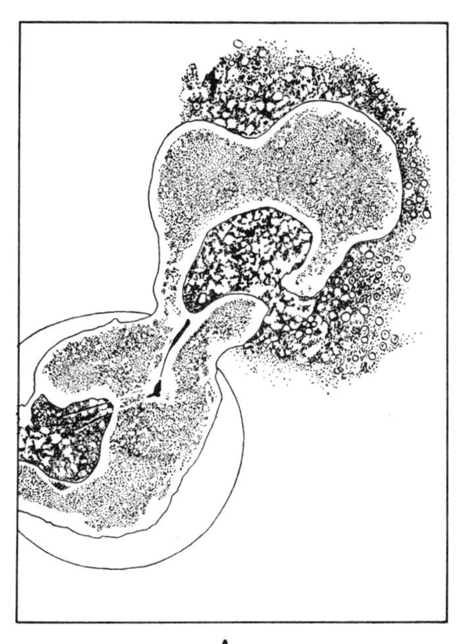
A

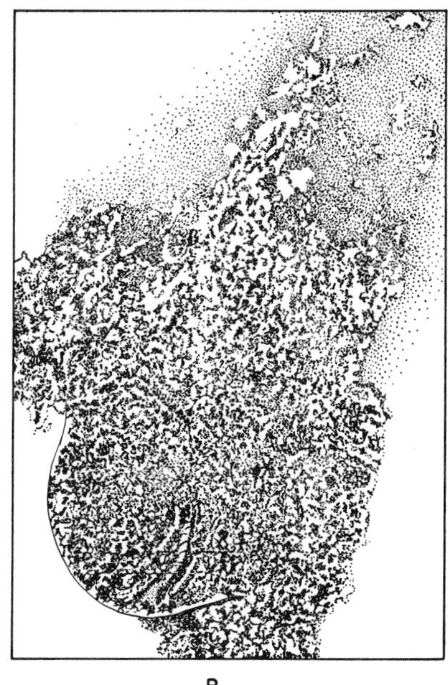
B

Figure 8.2. A, *Surface precipitation reaction forming a membrane around an egg of a starfish* (Patiria miniata) *crushed in sea water.* B, *The membrane fails to develop when the egg is crushed in calcium-free sea water. (After a photograph by Richard Boolootian.)*

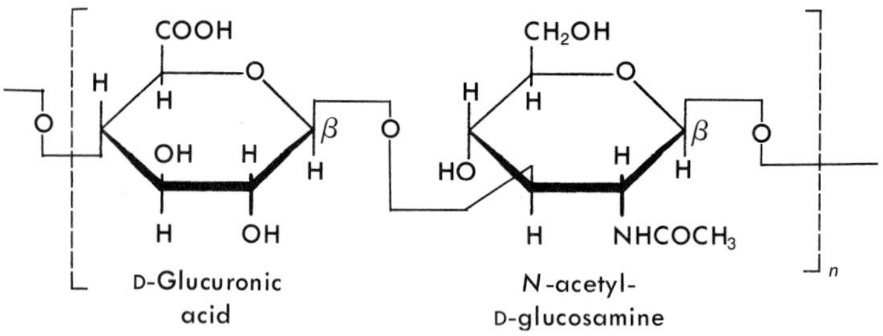

Figure 8.3. *Hyaluronic acid.*

each about 2.5 to 4.0 nm wide and a less dense band between them about 2.0 to 2.5 nm wide (Fig. 8.6). Because of the way the myelin sheath is formed during development (Fig. 8.5), such unit membranes occur in pairs, in apposition to each other within the sheath. Many such pairs with some material between them (shown between membranes in Fig. 8.5*B*) make up the myelin sheath.

Such a basic unit membrane as seen in the sheath was found to be general for a wide variety of cells (e.g., cells of intestinal epithelium, nerves, skin, liver, pancreas, kidney, endothelium, blood, and muscle of a variety of species of vertebrates and invertebrates, including ciliates and amoebas). Even more remarkable was the finding of a similar unit in plant cell membranes as well (Gunning and Steer, 1975). Since membranes of the endoplasmic reticulum, mitochondria, chloroplasts, and Golgi complex show the same unit structure, it would appear to be common to all cell membranes (see Figures 5.2, 5.3, 5.9, 5.10, 6.1, 6.7).

Electron microscopic studies also reveal extensions of the plasma membranes by folds. An intestinal epithelium cell may have as many as 3000 *microvilli*, and similar structures are seen in brush border cells of kidney tubules. These microvilli (Fig. 8.7) are believed to increase the cell surface for uptake of nutrients from the gut. In other cells, invaginations and irregularities also occur, but not to the spectacular degree they do in cells whose major activity is exchange of substances with the environment. All these membrane extensions have approximately the thickness of a unit membrane (Robertson, 1964).

It has been suggested that possibly all membrane systems in the cell have their origins as invaginations of the plasma membrane or evaginations of the nuclear envelope. Alternately, their basic similarity could originate from action of similar forces acting upon a pool of common substances. Because the membranous systems in the cytoplasm vary with conditions, it is presumed that they are in a state of dynamic equilibrium, breaking down and reforming continually. The plasma membrane is also considered to be a dynamic structure (Saier and Stiles, 1975).

CHEMICAL CONSTITUENTS OF CELL MEMBRANES

Studies on the plasma membrane of bacterial

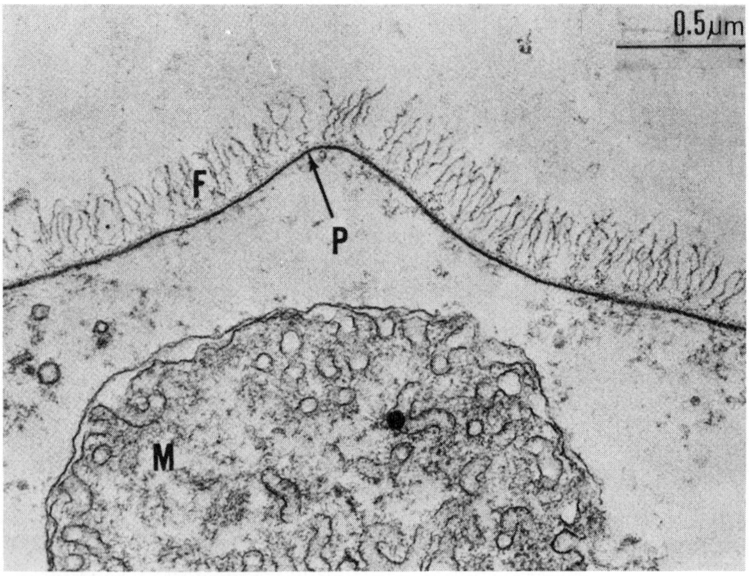

Figure 8.4. *Electron micrograph showing fringe (F) on the exterior surface of the plasma membrane (P) of the giant amoeba* Pelomyxa carolinensis. *Note that the plasma membrane is coated with a thin layer from which the filaments protrude. The mitochondrion (M) is within 0.5 μm of the cell surface. Fixed in 2 percent OsO_4 in 20 mM $CaCl_2$. Stained in uranyl acetate saturated in methanol. (From Daniels, 1973: In* The Biology of Amoeba. *Kwang, J., ed. Academic Press, New York, p. 137.)*

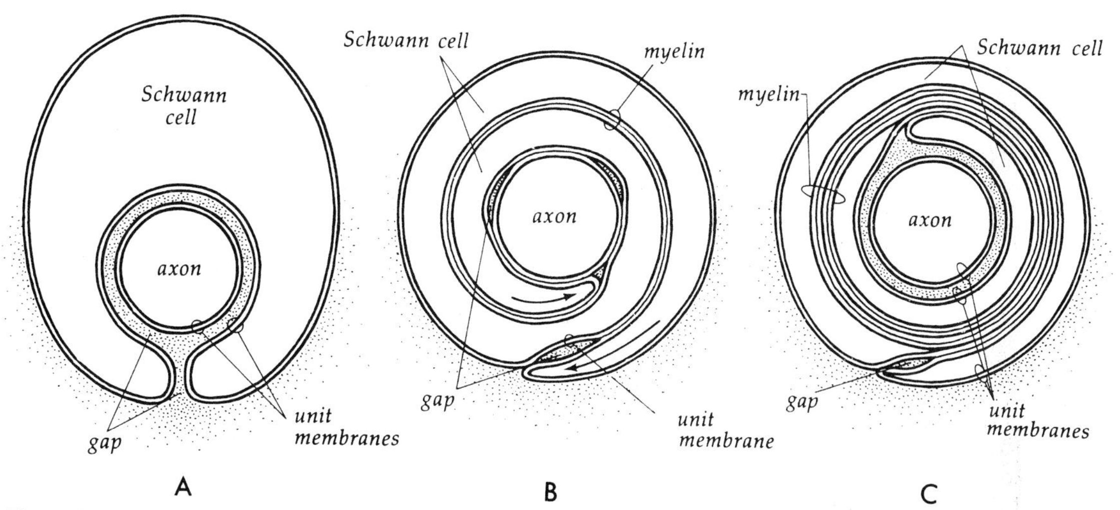

Figure 8.5. Development of the myelin sheath. In A, the earliest stage of development, a single axon is shown embedded in a Schwann cell. In B, the Schwann cell membrane has begun to overlap in a spiral around the axon. Note that the intercellular gap between apposed membranes is eliminated. At a later stage, C, the spiral is more extensive. The membranes of the Schwann cell become closely packed, in some regions eliminating both the intercellular and cytoplasmic spaces. (Redrawn, original courtesy of J. D. Robertson. From Wolfe, 1972: Biology of the Cell. Wadsworth Publishing Co., Belmont, Calif.)

and plant protoplasts,* animal cells, and membranes of cell organelles (mitochondria, chloroplasts, endoplasmic reticulum, microsomes, Golgi complexes, and lysosomes) biochemically as well as by electron microscopy and x-ray diffraction, indicate not only the common denominators among all types of membranes but also some important differences between them. Membranes are being taken apart and reassembled, and by omission or addition the effect of each chemical component on structure and function of the membrane is assessed (Razin, 1972). It is difficult yet to draw conclusions because much of the factual information has not been assimilated into guiding principles. Conflicting and even contradictory statements are often found in books, reviews, and symposia that have appeared in rapid succession (Singer, 1974; Saier and Stiles, 1975; Harrison and Lunt, 1976; Cook, 1976; Fox and Keith, 1976; Quinn, 1976). All this attests to an extraordinary interest in membrane biology, the membrane being considered by some as vital for cell function as DNA.

*Cells from which the walls have been removed by enzymes.

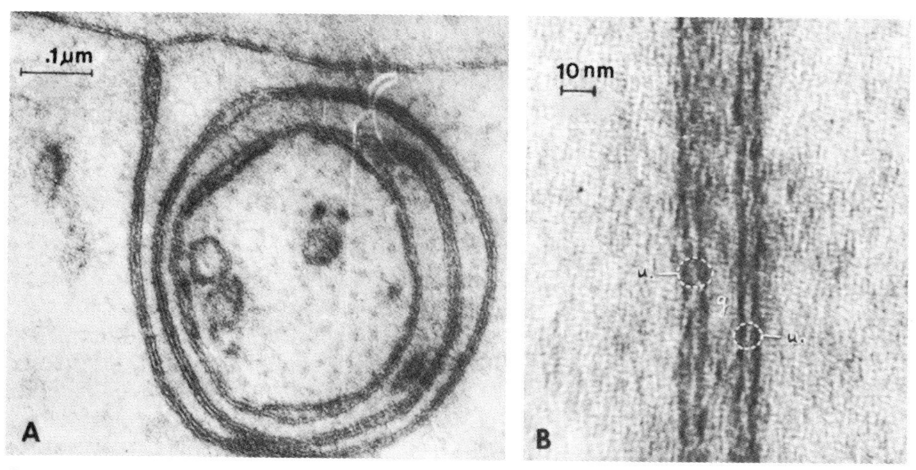

Figure 8.6. Electron micrographs of unit membranes in the myelin sheath. A, Unit membranes are clearly seen at the top of the figure and pairs of such membranes are seen beyond the point at which the unit membranes meet. Each unit membrane is made up of three layers. B, Higher magnification of a pair of unit membranes (approximately 7.5 nm wide) each member of the pair being indicated by u; g shows the material between the pair of unit membranes. (From Robertson, 1959: Biochem. Soc. Symp. 16: 3.)

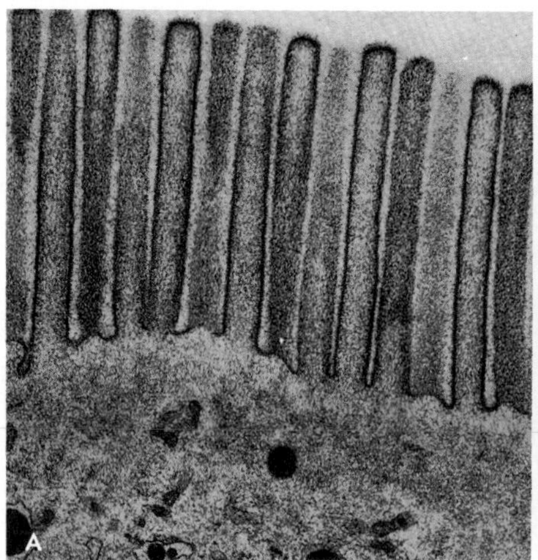

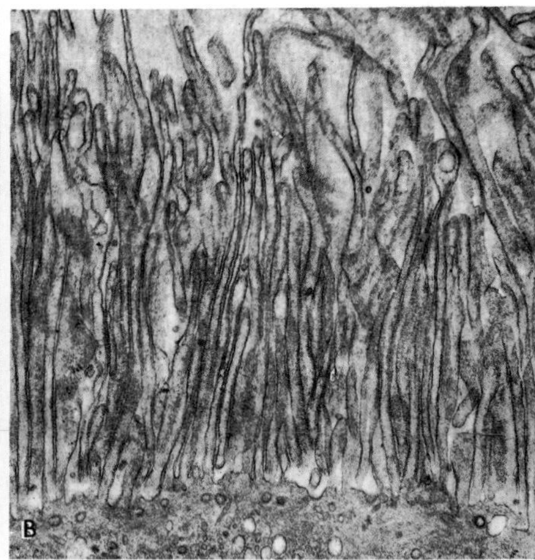

Figure 8.7. *Microvilli.* A, *Intestine of the rat (phosphate buffered osmium fixation, uranyl acetate and Karnovsky A lead staining, × 22,000).* B, *Epithelium of rabbit epididymis (collidine buffered glutaraldehyde and osmium fixation, lead citrate staining, ×16,000). Unit membranes are not clearly visible in* A *but show on some microvilli in the center of* B. *(From Fawcett, 1966: The Cell: Its Organelles and Inclusions. W. B. Saunders Co., Philadelphia.)*

Most of what is said in this chapter relates primarily to the plasma membrane, but the general principles probably apply to all cell membranes. The special attributes of the mitochondrial membranes are considered in connection with energy liberation in the cell (see Chapter 12), while those of the endoplasmic reticulum are considered in connection with synthesis of protein (see Chapter 15) and maintenance of the ionic environment in muscle fibers (see Chapter 22).

A single plasma membrane is too small for microchemical analysis; an analysis requires a mass of cell membranes. Membranes of erythrocyte ghosts, bacterial and plant protoplasts, liver cells, amoebas, microsomes, and mitochondria have been subjected to chemical analysis. Only a few examples are considered here. Other examples are considered elsewhere (Saier and Stiles, 1975; Quinn, 1976).

The plasma membranes of erythrocytes, myelin sheath cells, and bacteria have been extensively analyzed and are given special attention here, although other membranes will be considered in some instances.

Upon hemolysis an erythrocyte liberates hemoglobin. The ghost that is left behind appears to consist of nothing but the plasma membrane; it may continue to swell and shrink in response to concentration of substances in the medium. Analysis of massed ghosts indicates a ratio of protein to lipid of 1.1. Myelin consisting of multiple membranes of sheath cells wrapped about axons (see Figs. 8.5 and 8.6) has a protein-to-lipid ratio of only 0.25 to 0.5; *Bacillus megaterium,* 5.4; and *Mycoplasma laidlawii,* 0.4. Protein-to-lipid ratios for various organelles of rat tissues are given in Table 8.1; especially noticeable is the abundance of protein in membranes of the endoplasmic reticulum, mitochondria (note especially the inner membrane), nuclear envelope, and the Golgi body.

The membrane lipids are mainly phospholipids and cholesterol (Table 8.2). The *phospholipids*, which make up between 55 and 75 percent of the total lipid content, consist chiefly of *lecithin* and *cephalin* (see Fig. 1.13). The remainder consist of sphingolipids (with an amino group) and glycolipids (conjugates with

TABLE 8.1. PROTEIN AND LIPID RATIOS OF RAT TISSUE MEMBRANES*

Membrane	*Protein/ Lipid*	*Cholesterol/Polar Lipid*
Plasma membranes		
Myelin	0.25†	0.95
Erythrocyte	1.1	1.0
Rat liver cells	1.5	0.5
Nuclear membranes	2.0	0.11
Endoplasmic reticulum		
Rough-surface	2.5	0.10
Smooth-surface	2.1	0.11
Mitochondrial membranes		
Inner membrane	3.6	0.02
Outer membrane	1.2	0.04
Golgi membranes	2.4	—

*From Quinn, 1976: Molecular Biology of Cell Membranes. University Park Press, Baltimore, Md., p. 31.

†Myelin from central nervous system; higher ratios (ca. 0.5) are found in peripheral nerve myelin.

TABLE 8.2. LIPID CONSTITUENTS OF MEMBRANES

A. Phospholipids

Phosphatidic acid (P.A.):

$$
\begin{array}{lll}
 & & \quad\quad\;\; O \\
 & & \quad\quad\;\; \| \\
\quad\; O & CH_2{-}O{-}C{-}R & \\
\quad\; \| & | & \\
R'{-}C{-}O{-} & CH & \quad O \\
 & | & \quad \| \\
 & CH_2{-}O{-}P{-}OH^{*} & \\
 & \quad\quad\quad\;\; \backslash & \\
 & \quad\quad\quad\quad OH &
\end{array}
$$

1. Phosphatidyl serine/P.A.-serine: $-CH_2-CH(NH_3^{+})(CO_2^{-})$

$$
\begin{array}{l}
\quad\quad\quad\; NH_3^{+} \\
\quad\quad\quad / \\
-CH_2-CH \\
\quad\quad\quad \backslash \\
\quad\quad\quad\; CO_2^{-}
\end{array}
$$

2. Phosphatidyl ethanolamine/P.A.-ethanolamine: $-CH_2-CH_2-NH_3^{+}$
3. Phosphatidyl choline/P.A.-choline: $-CH_2-CH_2-N^{+}(CH_3)_3$
4. Phosphatidyl glycerol/P.A.-glycerol
5. Cardiolipin/P.A.-glycerol-P.A.

B. Sphingolipids

Ceramide:

$$
\begin{array}{ll}
\quad\;\; O & CHOH-CH{=}CH-R \\
\quad\;\; \| & | \\
R'-C-NH- & CH \\
 & | \\
 & CH_2-OH^{*}
\end{array}
$$

1. Sphingomyelin: ceramide-P-choline
2. Cerebrosides: ceramide-sugar
3. Gangliosides: ceramide-sugars

C. Sterols

1. Cholesterol* (see Fig. 3.11, top left).

From M. H. Saier, Jr., and C. D. Stiles, 1975: Molecular Dynamics in Biological Membranes. Springer Verlag, New York, p. 14.

*Site of derivitization in the complete lipid.

carbohydrates). Nor are the phospholipids in membranes of different kinds of cells the same. Details of the membrane lipids may be found elsewhere (Quinn, 1976). The proportion of cholesterol to phospholipid varies in membranes of various vertebrate cells, and cholesterol is completely lacking in bacteria and plant cells, although it is found in mycoplasmas and in viral envelopes derived from animal cells. Cholesterol reduces the permeability of lipid bilayers in the liquid-crystalline state and reduces movement of lipid molecules. It also strengthens plasma membranes in animal cells, which lack cell walls for protection.

The protein in erythrocyte plasma membranes can be fractionated by gel electrophoresis into six major proteins and nine minor ones. The cytoplasmic surface of the plasma membrane has a loosely attached, high-molecular-weight fibrous protein *spectrin* and another not yet characterized. The large amount of protein that can be extracted from red blood cell membranes under mild conditions is not characteristic of the subcellular membranes of organelles in which a greater proportion of the proteins are firmly bound to the membranes. The enzyme D-glyceraldehyde-3-phosphate dehydrogenase is combined to the inner surface of the erythrocyte plasma membrane but is extractable in 3M NaCl. The amount of easily removed protein is variable in plant and bacterial cells.

Bacterial plasma membranes with a large proportion of protein to lipid have about 100 different proteins, probably the result of the large number of enzymes, respiratory and hydrolytic, present there (Quinn, 1976). The plasma membranes of other cells have enzymes, among which are ATPases. Mitochondrial inner membranes, with a very large proportion of protein to lipid (Table 8.1), have a large number of enzymes participating in respiration and production of ATP, much like the bacterial plasma membranes. Enzymes are also prominent in the membranes of the endoplasmic reticulum, the Golgi bodies, and chloroplast membranes. Lipids in cell membranes may also influence the catalytic activity of membrane-bound enzymes (Quinn, 1976). These enzymes will be considered with the functions of these structures (see Chapters 12, 15, 18 and 23).

Glycoproteins are mucoproteins, conjugates of carbohydrates and proteins, present over the animal cell plasma membrane, and may form a considerable component of the membrane protein. While largely protective, they are also important as antigens (Bengelsdorf, 1976). An example is the agglutination of red blood cells by influenza virus, resulting from reaction between enzymatically active groups on the surface of the virus particles with the surface mucopolysaccharides on the red blood cells. The widespread nature of immunological reactions to all types of cells indicates the fairly general presence of such mucoids in their surfaces. (In gram-negative bacteria, glycolipids serve as antigens.) The mucoids of the foreign cells serve as antigens against which the host organism develops antibodies. Glycoproteins may also be produced by some plant and bacterial cells. With the exception of the Golgi bodies (where carbohydrates are conjugated to proteins to form glycoproteins) intracellular membranes have little, if any, carbohydrate.

Salts are also present in cell membranes, and microincineration tests (see Fig. 5.22) suggest that some of these are present in higher concentration in the membranes than elsewhere in cells. Water present in cell membranes forms part of the membrane structure as it does in all cell constituents.

THICKNESS OF CELL MEMBRANES

The dimensions of the cell membrane are within the limits of resolution by the electron microscope. Studies of flattened entire cell membranes and of thin sections of cell membranes indicate a thickness of about 6.0 to 7.5 nm for the plasma membrane (see Fig. 8.5*A* and 8.6*B*). Unit membranes from different parts of a cell are not necessarily of exactly the same dimensions. By measuring the peak-to-peak electron density on unit membranes of electron micrographs of bullfrog cells it was found that the membranes of the Golgi vesicles and of synaptic vesicles of neurons have about the same dimensions as the plasma membrane, whereas the unit membranes of the nuclear envelope, the endoplasmic reticulum, and the Golgi lamellae are only 85 to 90 percent as thick (Yamamoto, 1963) (Fig. 8.8). Sometimes the plasma membrane is thicker than 7.5 nm and may be as large as 10 nm (Korn, 1966). The cross sectional thickness of the cell membranes depends in part on the method of fixation; they are thicker when fixed in permanganate than when fixed in osmium tetroxide.

STRUCTURE OF CELL MEMBRANES

The cell membrane must act as a permeabil-

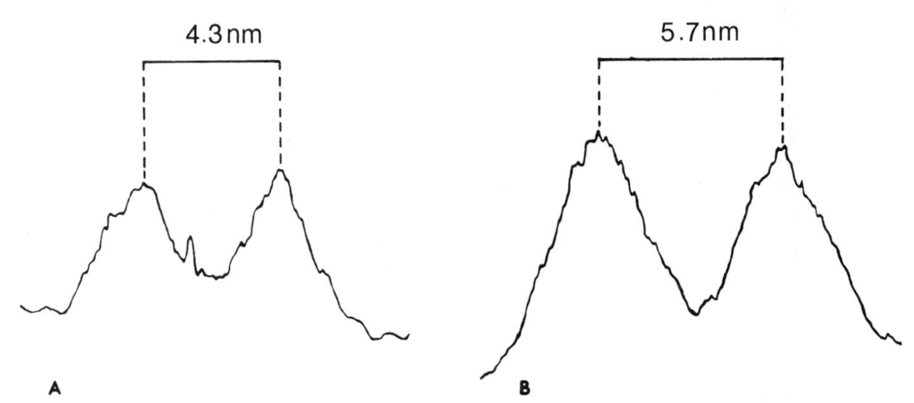

Figure 8.8. *Thickness of the unit membrane as determined by densimetric recording on electron micrographs of various membranes.* A, *Endoplasmic reticulum membrane of bullfrog ganglion cell.* B, *Plasma membrane of the same cell. Magnification of the electron micrographs is* × 225,000. *The value in nanometers is the peak-to-peak value; the thickness of the membrane should be measured from the initial increase in density over the background on either side. Peak-to-peak measurement is more accurate. (From Yamamoto, 1963: J. Cell Biol.* 17: *413.)*

ity barrier, it must be an electrical insulator, and it must have considerable mechanical strength to remain intact throughout the lifetime of the cell. At the same time, however, many compounds have to be able to pass through the membrane: metal ions, amino acids, and sugars that often must be accumulated against a concentration gradient. The structure of the cell membrane must therefore be a compromise: it must be fluid enough to allow passage of required molecules but not so fluid that the contents of the cell are continually leaking out. It is this fine balance that has to be understood (Lee, 1975).

No technique yet exists that permits direct observation of the molecular arrangements of proteins and lipids in the cell membranes by which they perform their demanding tasks. All such information on membrane structure has been inferred from indirect evidence. It is therefore to be expected that many membrane models have been suggested.

Whether the carbohydrate listed in biochemical analyses of the membrane fraction of cells is a structural component or a contaminant from extraneous coats on cells is not clear. The carbohydrate membrane fraction is only a small proportion of the total: in red cell ghosts 5 to 7 percent of dry weight, and in mouse liver cells about 1 percent of dry weight. In human red cells sialic acids (e.g., N-acetyl neuraminic acid) comprise more than a third of the carbohydrate, the remainder being made up of neutral sugars and hexosamines. A layer of carbohydrate rich in acidic groups has been reported from the surface of most types of cells in the rat and other animals; this substance is outside the unit membrane, as are the extraneous coats described at the beginning of the chapter (Malhotra, 1970).

The striking correlation between lipid solubility and permeability led Overton, more than 50 years ago, to postulate a lipid coating on the cell. Langmuir had shown that fatty acid molecules at an interface stand at right angles to the surface of the water, and the same was thought to be likely in the plasma membrane. When, in 1925, Gorter and Grendel reported that all the extracted lipids from a red blood cell occupy an area about twice that for a single molecular film, a membrane was conceived to have a double layer of lipid molecules, each with its hydrophobic end to the other, at right angles to the surface. While this view rested upon errors that cancelled each other out, the lipid-bilayer became a popular model for the cell membranes. Other data from various sources have verified the presence of lipid-bilayers in cell membranes (Lee, 1975).

The plasma membrane as visualized in 1952 by Danielli and Davson consists of two layers of lipid molecules arranged radially with their hydrophobic hydrocarbon chains toward each

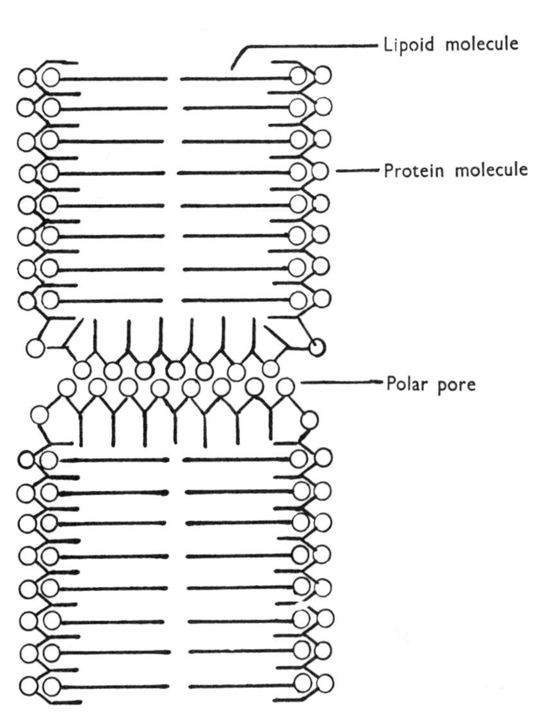

Figure 8.9. *Hypothetical molecular structure of the plasma membrane of the cell. (From Danielli, 1954: Colston Papers 7: 1.)*

other, and with their respective polar groups arranged outwardly and inwardly, the entire double layer of lipid molecules being sandwiched between two layers of protein (Fig. 8.9). Electron micrographs of cell membranes fixed in permanganate, which binds protein, show this tripartite structure and would thus appear to support the Davson-Danielli "sandwich" model. The outer layers of protein of the membrane were thought to become electron opaque on fixation with permanganate, whereas the inner lipid "filling" did not.

However, the ready passage of substances through the plasma membrane in proportion to their lipid solubility and the ready action of lipases (e.g., lecithinase) and lipid solvents on cell membranes made it seem likely that lipid molecules were also present on the surface of the membranes.

More convincing proof of the inadequacy of the sandwich model of cell membranes came from studies with freeze-etching electron microscopy. Preparations in this case are frozen (unfixed) and then cracked with a microtome knife along natural cleavage planes (see Chapter 3). Examination of membranes by this technique discloses no layers of proteins but rather spheroids or globules, some embedded in the lipid mass (Fig. 8.10), some passing completely through from the cytoplasm to the cell surface. Enzymatic tests indicate the continuum to be lipid whereas the globular particles are pro-

Figure 8.10. The lipid-globular protein mosaic model of membrane structure: schematic cross-sectional view. The phospholipids are arranged as a discontinuous bilayer with their ionic and polar heads in contact with water. Some lipid may be structurally differentiated from the bulk of the lipid, but this is not explicitly shown in the figure. The integral proteins, with the heavy lines representing the folded polypeptide chains, are shown as globular molecules partially embedded in, and partially protruding from, the membrane. The protruding parts have on their surfaces the ionic residues (− and +) of the protein, while the nonpolar residues are largely in the embedded parts; accordingly, the protein molecules have affinity for both water-soluble and oil-soluble molecules. The degree to which the integral proteins shown are embedded and, in particular, whether they span the entire membrane thickness depend on the size and structure of the molecules. The arrow marks the plane of cleavage to be expected in freeze-etching experiments. (From Lenard and Singer, 1970: Proc. Nat. Acad. Sci. U.S.A.: 65: 720 and Singer, 1971: In Structure and Function of Biological Membranes. Rothfield, Ed. Academic Press, New York, p. 145.)

teins. Most workers have therefore abandoned the sandwich concept of the cell membranes in favor of this *fluid mosaic model* (Figs. 8.10 and 8.11). The membrane is presently conceived to consist of a lipid bilayer in which, and sometimes through which, integral protein molecules are present. The lipid continuum is then the gate for lipid soluble molecules, and the proteins spanning the entire membrane are the gates for water soluble molecules (Quinn, 1976).

The layers in the plasma membrane are considered to be dynamic, not static. Pores of small size, about 0.7 nm in diameter, have been postulated from experiments on the penetration of water and salts into cells (Solomon, 1960). Pores 1.0–1.4 nm have been estimated from data on midge fly salivary gland cell junctions (Simpson *et al.,* 1977; Orci *et al.,* 1977). Such pores, however, if they occur, are below the limit of resolution of the electron microscope. Larger molecules probably must dissolve into the structure of the membrane to gain entry.

An alternate older view of a cell membrane is that it is made up of lipid-protein subunits (Stoeckenius, 1970). However, studies of thin cross sections of plasma membranes by electron microscopy do not show a homogeneous membrane with radial heterogeneity such as might be expected of this structure, although the radii of protein might conceivably be below the limit of resolution. Still another view of the plasma membrane is a series of cylinders of lipid between layers of protein (Korn, 1966).

Many investigators using a variety of chemical and physical techniques have presented data, some of it seemingly in conflict with the lipid bilayer model. Consequently, additional cell membrane models have been suggested, each one influenced by some feature of membrane structure or permeability which the investigator thought was better explained by a new model. Some are modifications of the lipid bilayer model, others are ingenious new concepts. Even as early as 1971 Hendler examined the evidence for and against the lipid bilayer model and the alternate ones suggested. He concluded that "the bulk of the new information tends to strengthen the evidence for the lipid bilayer."

The data of recent years have increased our understanding of the state of the lipids and proteins in the membrane. Thus it is known that the cell membranes change with changed environment. For example, at certain temperatures, which depend on the culture temperature and species of organism, a cell membrane (in a calorimeter) undergoes a phase transition in which its thickness decreases and its area increases (Quinn, 1976). X-ray diffraction data show that the lipid molecules, which below the phase transition temperature are oriented perpendicular to the surface of the membrane in what is sometimes called the crystalline or "gel" state, pass into a less well-oriented state, called the liquid crystalline state, involving at least 75 percent of the molecules. Presumably temperature breaks many of the weak bonds

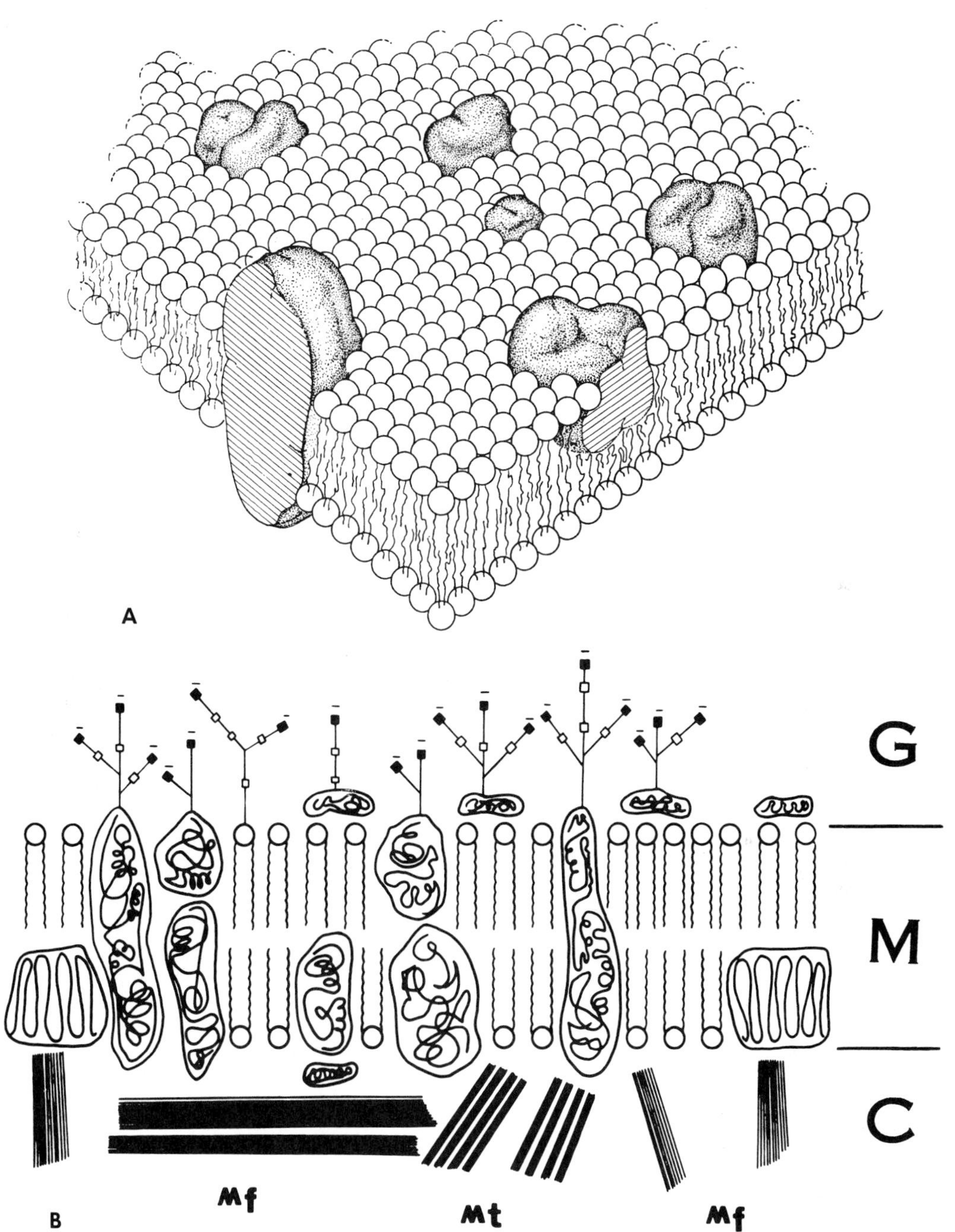

Figure 8.11. A, *The lipid-globular protein mosaic model with a lipid matrix (the fluid mosaic model); schematic three-dimensional and cross-sectional views. The solid bodies with stippled surfaces represent the globular integral proteins, which at long range are randomly distributed in the plane of the membrane. At short range, some may form specific aggregates, as shown. In cross section and in other details, the legend of Fig. 8.10 applies. Peripheral proteins, easily washed from the membrane, are not shown, nor are the carbohydrates attached to the integral proteins at the surface indicated. (From S. J. Singer and G. L. Nicolson, 1972: Science* 175: *720–731. Copyright 1972 by the American Association for the Advancement of Science.)* B, *Diagram of cell membrane organization showing three major components:* M, *the membrane matrix consisting of a lipid bilayer with hydrophilic ends directed outward and hydrophobic ends directed inward, and peripheral proteins and integral proteins embedded within the bilayer;* G, *the glycocalyx consisting of charged glycolipid and glycoprotein residues protruding above the surface; and* C, *cytofibrillar components consisting of structural microtubules (Mt) and contractile microfilaments (Mf) associated with the underside of the membrane. Both the lipids and the proteins are capable of lateral mobility in the membrane plane. All three components of the cell periphery play a role in transmembrane mechanisms which affect topographic modulation and other cellular functions. (From Pollock, 1978: Am. Zool.* 18: *25–69.)*

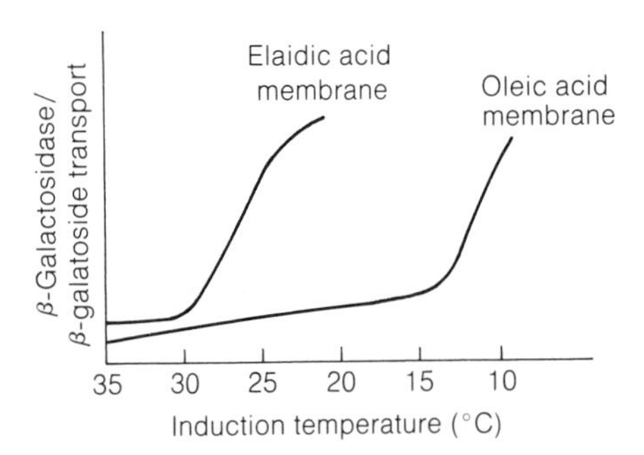

Figure 8.12. *Elaidic acid (trans-9-octadecenoic acid) has a melting point of 51.5°C; oleic acid (the cis-form) has a melting point of 14°C. (From Saier and Stiles, 1975: Molecular Dynamics in Biological Membranes. Springer Verlag, New York, p. 39.)*

between the hydrophobic tails of the lipid molecules; reduction in the number of bonds leads to a decreased orientation with respect to the membrane surface, although the overall direction of the lipid molecules is still more or less perpendicular to the membrane surface. The membrane in this state is said to be fluid and its permeability to ions and molecules is increased (Fig. 8.12). Presumably the remaining 25 percent of the hydrophobic lipid tails form weak bonds with hydrophobic groups in proteins (Engelman, 1970). Furthermore, by using mutants that fail to make certain lipids but that will incorporate lipids supplied to them, it has been possible to make membranes that have phase transition temperatures in accordance with the lipids supplied (Saier and Stiles, 1975; Silbert, 1975). The cell membranes are thus dynamic structures, responding in structural detail and permeability to changes in the environment. Just what changes occur in response to particular environmental influences and how these affect the permeability of the membranes to various substances remains to be investigated.

A case illustrating the importance of membrane fluidity to enzyme function is that of the calcium transport enzyme of the sarcoplasmic reticulum membrane vesicles. When 99 percent of the lipid in the membrane has been replaced by dioleoyl lecithin, with a phase transition at −22°C, the enzyme is active over the entire temperature range tested. When, however, the substituent is dimyristoyl lecithin with a transition at 24°C, the ATPase is inactive below this temperature.

Perhaps the most convincing evidence for fluidity of cell membranes comes from experiments with fused cell lines. When each type of cell has a distinctive antigen identifiable by a characteristic fluorescently colored antibody, it is possible to show that when the cells first fuse each of the two antibodies is present only on the antigen at one end of the combined cell, for example, one red end and one green. However, after a lapse of time the antibodies become completely mixed on the fused cell surface. The most reasonable explanation is that the antigens move about laterally in the fluid membrane (Quinn, 1976). A curious confirmation of membrane fluidity comes from the observation of continuous rotation of a rod (axostyle) in flagellates of a termite gut that pass through the membrane to the exterior. Only a fluid membrane could permit such rotation without damage to the cell (Tamm, 1976).

The fluidity of cell membranes permits molecular movements of several types: along the long axis of the membrane, sidewise, and in rotation along the perpendicular to the membrane. The configurations of both membrane lipids and proteins are naturally (thermodynamically) most stable: hydrophilic groups are in contact with other hydrophilic groups, and hydrophobic groups are in contact with other hydrophobic groups.

It is interesting that the myelin sheath, once considered the model for cell membrane structure, is perhaps the least typical of cell membranes. Not only does it have a very high lipid to protein ratio, unlike plasma membranes and internal membranes (Table 8.1), but it contains much cholesterol and is largely in the crystalline state, thus it is much more rigid than most cell membranes. However, it is well adapted for its purpose as a protective as well as an electrically insulating layer for nerve fibers.

MEMBRANE ASYMMETRY

The two faces of a cell membrane are not alike. For example, although the same lipids are present in both faces of the plasma membrane, they are present in different proportions. Although the polypeptides may be similar, they are oriented in a specific manner with one end always upward or downward. Plasma membrane carbohydrates, whether bound to lipids or proteins, are always external (Rothman and Lenard, 1977).

A variety of methods has been used to demonstrate membrane asymmetry: external proteins by studies of the action of enzymes on intact cells, internal proteins by work with inside-out vesicles from erythrocyte ghosts. All but two of the major proteins in the erythrocyte membrane are exposed only at the cytoplasmic (inside) surface. The other two major proteins (actually glycoproteins) are exposed at both surfaces: the amino terminal outward, the carboxy terminals inward. No protein has been found symmetrically distributed, or unexposed, at either surface. Protein asymmetry has been demonstrated in enveloped animal viruses, platelets, mitochondria, endoplasmic reticulum, sarcoplasmic reticulum, lymphocyte plasma membrane, intestinal brush border membrane, and bacterial membranes.

For studies of lipid asymmetry, enzymatic techniques have also been used. Membrane lipids arise from the cytoplasmic surface but are present in different amounts in the inner and outer monolayers forming the lipid bilayer. Transmembrane movement of lipids (e.g., from outer to inner monolayer) occurs in erythrocytes and bacteria not by flip-flop movement but rather by a protein-catalyzed mechanism. Although each cell membrane studied appears to have such differences between the inner and outer lipid monolayers, the details vary considerably. The two layers may or may not differ in fluidity, as shown by spin label measurements (Rothman and Lenard, 1977).

CELL SURFACE MEMBRANE RECEPTORS

Plasma membranes of cells have a variety of receptors for viruses, lectins (glycoproteins), and other proteins (Sharon, 1972). Combination of the molecule with a receptor initiates a cellular response. For example, when a polypeptide hormone such as the growth hormone reacts with a cell surface receptor, a lipoprotein enzyme adenyl cyclase is activated. This enzyme catalyzes the synthesis of cyclic 3′-5′-AMP from ATP. The cyclic AMP in turn serves as an intracellular messenger, exciting synthesis of proteins and growth. Discovery of this mechanism solved the riddle of how a polypeptide hormone could enter a cell—it does not have to! It need only affect the cell surface. Proteins and polypeptides may enter by pinocytosis also (see Chapter 19).

Activation of adenyl cyclase by polypeptides, probably a general phenomenon, has now been observed in a variety of animal cells and in yeasts and fungi, evoking cell activity in coordination with the demands of the environment. The structure of adenyl cyclase is not known; therefore, its change in configuration on activation is still under study.

Excess cyclic AMP is disposed of by an intracellular enzyme, phosphodiesterase, which converts cyclic AMP to AMP. Inactivation of excess cyclic AMP is part of cell regulation.

Among membrane receptors are antigen group substances for blood groups, recognition sites of cell (tissue) types (Greaves, 1975), sites for membrane responses, and sites for cell transformation by viruses and foreign substances (lectins, the glycoprotein phytohemagglutinins of kidney beans, for example) (Quinn, 1976; Bengelsdorf, 1976). Cell transformation is discussed in Chapter 25.

THE NUCLEAR ENVELOPE

The *nuclear envelope* is of profound importance. During interphase the nucleus can pass information to the cytoplasm only through the nuclear envelope, controlling or directing the synthetic activities occurring in the cytoplasm of the cell (Wischnitzer, 1974).

Isolated envelopes from nuclei of amphibian cells show an outer membrane of protein and lipid and an inner one containing a relatively insoluble protein in addition to other constituents. The protein in both appears to be elastin-like and capable of being stretched reversibly. A small amount of carbohydrate (chiefly hydrolyzable to glucosamine) is found in both membranes, and on the outer membrane of the envelope RNA is found, presumably being derived largely or entirely from attached ribosomes.

The nuclear envelope has a characteristic phospholipid complement that distinguishes it from other subcellular membranes. About 35 percent of the lipid consists of phosphatidyl choline, phosphatidyl ethanolamine, and phosphatidyl inositol. Nuclear envelopes from different tissues and from different species have similar constitutions. Some 20 different proteins are present in liver cell nuclear envelopes, the total protein representing about 60 to 75 percent of the envelope mass (Wischnitzer, 1974).

The nuclear envelope gives rise to cytoplasmic lamellae that become studded with ribosomes during very active synthesis in a few cell types, such as salamander oocytes, and tumor cells. The general significance of such lamellae has not yet been appraised (Wischnitzer, 1974). Perhaps they give rise to the rough endoplasmic reticulum.

Early studies suggested that the nuclear envelope was selectively permeable to solutes and water, much like the plasma membrane of the cell. When marine eggs are placed in hypertonic solution, the nucleus shrinks along with the general shrinkage of the cell and in a hypotonic solution the nucleus swells like the rest of the cell. The nucleus of some cells has been shown to accumulate labeled sodium (^{22}Na), which suggests selective uptake of salts. However, a highly porous boundary would also permit the nucleus to shrink or swell osmotically when the cytoplasm does. Furthermore, such behavior of the nucleus could also result from shrinking or swelling of the proteins of the nucleus (Davson, 1970).

Some experiments suggest that the nuclear envelope is more porous than the plasma membrane. For example, heparin and sulfated polymanuronic acid, added to a suspension of isolated nuclei from rat liver cells, displace highly polymerized DNA, a very large molecule, which then passes out of the nucleus through the nuclear membrane. Furthermore, nucleases are able to penetrate isolated nuclei of rat liver cells and to digest the nucleic acids in very short periods, indicating that these large molecules have entered the nuclei rapidly. Still, it could be argued that such nuclei removed from the cytoplasm of cells are abnormal.

However, experiments performed with nuclei still inside the cytoplasm avoid the criticism leveled against those performed on isolated nuclei. Such experiments have demonstrated that nuclear envelopes of some cells are more porous than plasma membranes and that they permit macromolecules to enter and leave the nuclei. Thus, in cases where ribonuclease and deoxyribonuclease are able to penetrate an intact cell, they also enter the nucleus and hydrolyze the nucleic acids in the nucleus. The appearance of fluorescent antibodies (proteins) in nuclei of embryos also indicates that the nuclear membrane has a greater porosity than the cell membrane.

The most direct evidence on the nature of the nuclear envelope is from electron micrographs (see Fig. 7.7). These clearly show two unit membranes similar to the unit plasma membranes. According to some investigators, the *nuclear membranes* might be considered modified extensions of the endoplasmic reticulum or the cisternae surrounding the nucleus, but others suggest that nuclear membranes give rise to the endoplasmic reticulum (Wischnitzer, 1974).

The outer nuclear membrane has annuli 40 to 70 nm in diameter, and these have been seen to extend through the inner membrane as well in cells of a variety of tissues. On the other hand, in some electron micrographs such annuli are not seen as clearly in the inner membrane, and, considering the differences in techniques used by different researchers, the status of these annuli in some cells must be considered uncertain until resolved by further study. The annuli in the outer nuclear membrane often appear to be plugged, but electron micrographs do not permit one to tell whether the plug remains tight at all times. Colloidal gold particles from 2.5 to 8.5 nm in diameter were found to penetrate the amoeba nucleus rapidly, those from 8.6 to 10.6 nm more slowly, while larger particles (10.7 to 17 nm) tested did not penetrate at all.

Ferretin (MW 450,000) does not pass through the nuclear pores, although bovine albumin (MW 62,000) does very slowly. Yet RNA labeled with ^{32}P passed from the nucleus to the cytoplasm of an unlabeled amoeba into which it had been inserted. It seems likely that RNA in general passes through the center of the annular pore. Most of the RNA in the cytoplasm is synthesized in the nucleus (Wischnitzer, 1974). Pinocytosis or evidence for it from cytoplasm to nucleus has not been described. It has also been suggested that the annuli are sphincter-like in nature, permitting closure of the pore in the center of an annulus. The annuli are well visualized in freeze-etched preparations (see Fig. 7.7) (Mühlethaler, 1971). However, the exact nature of the nuclear pores is little understood, as shown by the variety of interpretive models.

The possibility of another kind of communication between nucleus and cytoplasm has been suggested, namely, outpouching from chromosomes, forming blebs on the nuclear membrane that separate into the cytoplasm as vesicles or laminae. These structures are covered by membranes derived from the nuclear envelope. The overall importance of blebbing is still to be assessed (Wischnitzer, 1974).

A potential of 13 mV between nucleoplasm (negative) and cytoplasm (probably the ground substance) has been observed in gland cells of *Drosophila,* as well as a high resistance to passage of an electric current across the nuclear envelope. Similar results have been obtained with *Acetabularia* (Yazukov *et al.,* 1974). This could be a formidable barrier to ions even as small as K^+, Na^+, and Cl^-. The pores seen in electron micrographs of these cells are perhaps artifacts or they have become plugged with a substance equally resistant to passage of electricity as the plasma membrane. On the other hand, no potential and low resistance to passage of electric current was reported across the nuclear envelope of amphibian oocytes (Loewenstein and Kanno, 1963). The potential across the nuclear envelope of the midge *Chironomus* salivary gland cell showed a change in response to the injection of insect molting hormone, ecdysone, suggesting changes in its permeability (Loewenstein, 1965).

It is possible that some of the controversy concerning nuclear envelope behavior stems from the fact that the envelope behaves differently at different stages of the cell activity cycle. Perhaps correlation of nuclear envelope behavior with activity will reveal significant relations to cell requirements at different stages of the cell cycle.

CELL-TO-CELL CONTACTS

Cells in a tissue, although in contact with one another, most often are separated from one another by a distance equal to one or more unit membranes (10 to 30 nm), the space between them being filled with some amorphous and fibrillar substances. At most contacts between nerve cells (synapses) a similar spacing is also found. However, at some synapses (e.g., in the

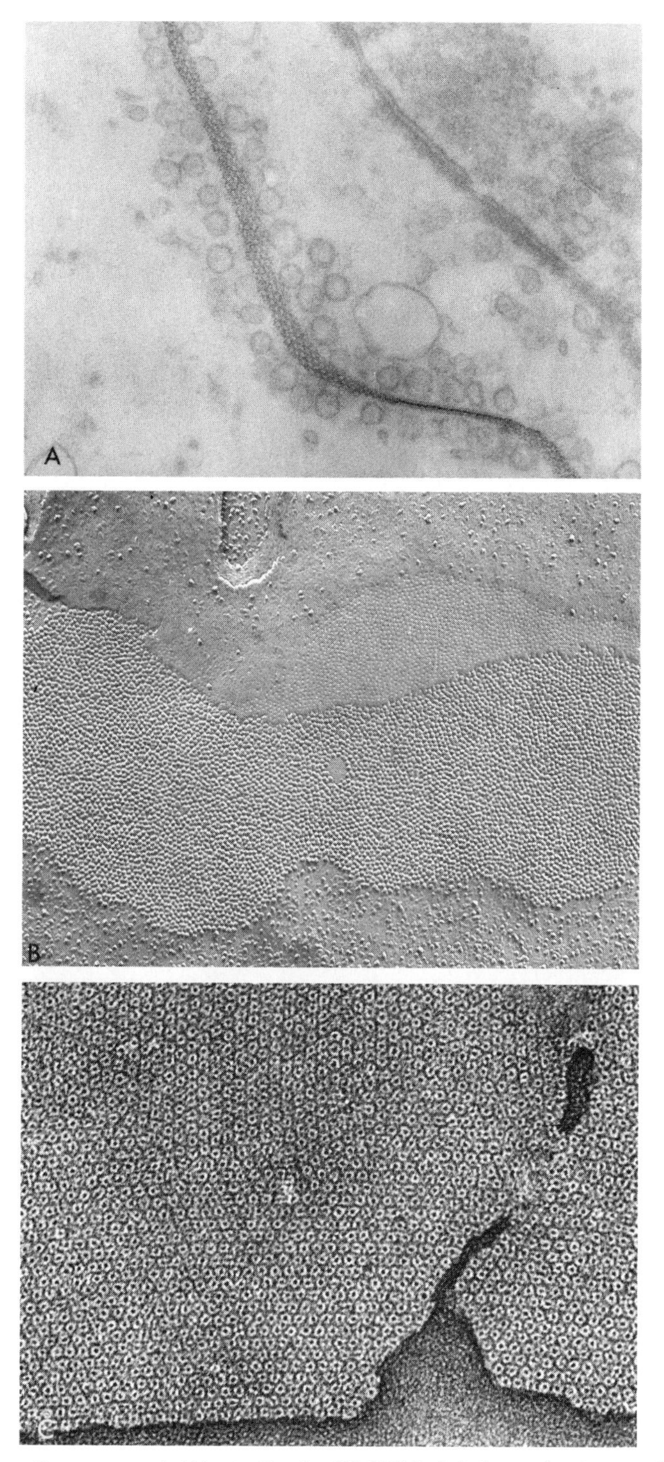

Figure 8.13. *Gap junctions as seen in thin section (× 100,000) in* A, *in freeze fracture replica (× 100,000) in* B, *and in an isolated unit (× 410,000) in* C. *The crayfish gap junction in* A *has been soaked in electron-opaque lanthanum hydroxide that penetrated the narrow intercellular space as seen in the vertical section (lower right). In tangential section a hexagonal array of particles is outlined. The freeze fracture replica in* B *shows a cytoplasmic half-membrane face of the underlying cell and a corresponding array of pits on the external half membrane face of the overlying cell. The stained isolated gap junction in* C *shows the cylindrical particles with stain-filled pores through which in life the cells have direct communication with each other. (*A *and* B *by courtesy of Camillo Peracchia, University of Rochester School of Medicine,* C *by courtesy of N. Bernard Gilula of Rockefeller University. Published in Staehelin and Hull, 1978: Sci. Am. (May)* 238: *141–152.)*

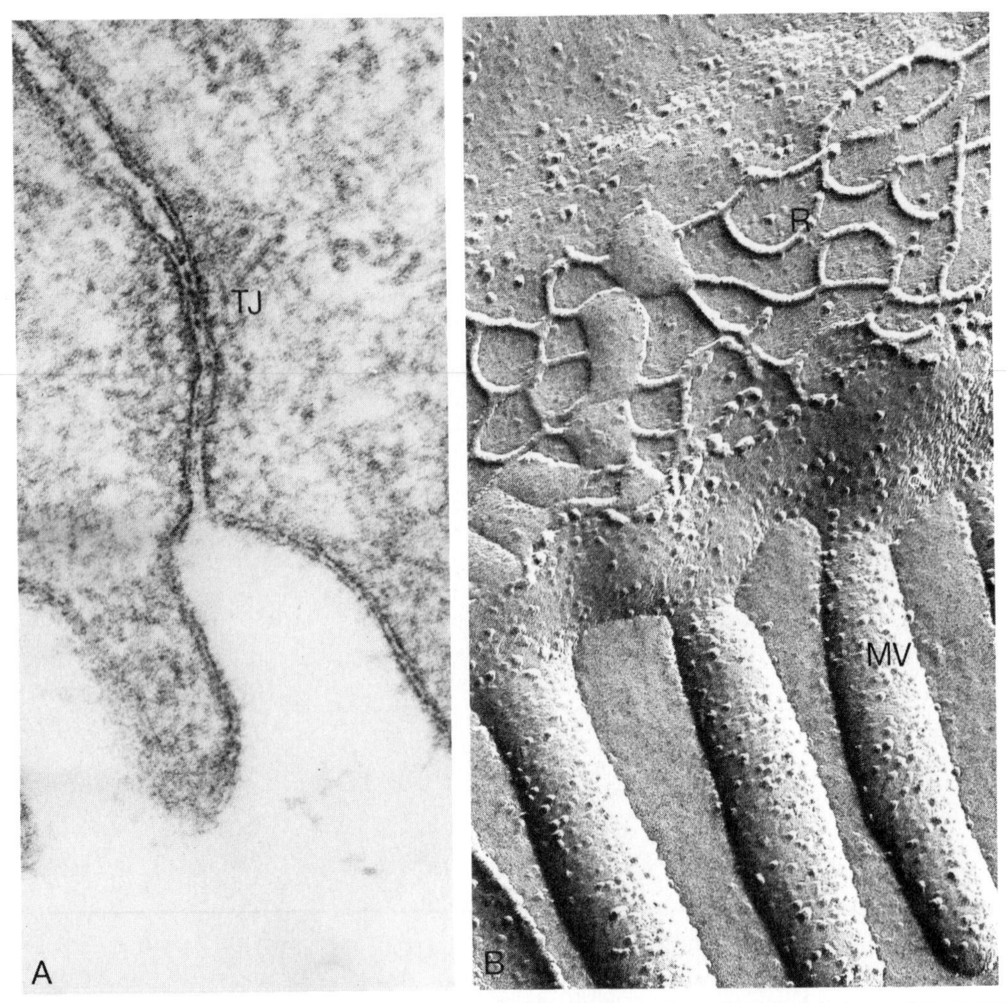

Figure 8.14. *Tight junction (TJ) as seen by transmission electron microscopy in* A *between two mucus-secreting cells of chick epidermis (× 200,000) and in* B *by freeze fracture between cells in the epithelium of the mouse large intestine (× 140,000). Note the ridges (R) between cells. The finger-like structure below (MV) are the microvilli. (By courtesy of D. S. Friend, M.D., University of California Medical Center, San Francisco.* A *from Elias and Friend, 1976: J. Cell Biol.* 68: *173–188.)*

giant fibers of the crayfish that mediate a rapid tail flip, in the Mauthner cells of fishes that mediate a rapid escape response, in some embryonic cells, and in parts of the mammalian brain) *gap junctions* are found. In these, the cell membranes of two cells are not more than 2 to 3 nm apart but are connected by hexagonally arranged cylindrical units through the central hole of which ions and molecules of less than 1000 molecular weight can pass directly from one cell to the other (Fig. 8.13). Larger molecules may sift between the cells but do not enter the cells (Staehelin and Hull, 1978). When a current is injected into one cell the voltage shift of the cell causes a similar voltage shift in the adjacent cell. A dye injected into one cell also appears in the electrically coupled cell. Synapses of this type are called electrotonic synapses; the permeability of such synapses is regulated by the level of calcium ions within the cell (Staehelin and Hull, 1978) (see Chapter 21).

By electron microscopic and x-ray diffraction studies the cylinders, 7 nm in diameter, have been shown to be made up of aggregated dumbbell-shaped subunits with a hole in the center. The dumbbell-shaped particles appear to be composed of one protein, connexin, with a molecular weight of 18,000 and consisting of two chains of amino acid units. The pores in the cylindrical units of gap junctions appear to close when the cell is injured, preventing leakage from the healthy cells.

Still another type of cell contact is the *tight junction,* especially characteristic of epithelia. However, they are also found in a variety of cells, for example, in some kidney tubules. Electron microscopy has demonstrated that in the region of a tight junction the two adjacent cells are fused at a number of points providing intimate contact that prevents the passage of molecules across the epithelial sheet except through the cells. Freeze fracture electron microscopy has revealed that the tight junction is achieved by a network of ridges on the cytoplasmic half face of the plasma membrane (the split in freeze-etching occurs between the lipid bilayers) and complementary grooves on the external half membrane face. The ridges are in turn composed of particles (the integral membrane proteins) forming a kind of zipper holding the membranes together so that no intercellular space is observable (Fig. 8.14) (Staehelin and Hull, 1978).

Tight junctions of different tissues vary in number of ridges and corresponding grooves of the sealing strands. The greater the number of such strands the greater the transepithelial electrical resistance indicating more effective sealing. The sealing network between cells in a tight junction is remarkably resistant to stretch.

Cells also may have specialized contacts, *desmosomes,* where the unit membranes of two cells are separated by the usual distance (10 to 30 nm) but where the innermost layer of each of the apposed plasma membranes is thickened by amorphous material, subjacent to which is a layer of fibrils (Fig. 8.15). The fibrils seem to anchor the desmosome to the cell. Desmosomes occurring on adjacent cells are always present in pairs. However, when a cell borders on a basement membrane secreted by the cells of a tissue, a single unit of desmosome (half desmosome) may be present anchoring the cell to the basement membrane (Kelly, 1966). Desmosomes are common in epithelial cells but they may be present in cells of the nervous system, and perhaps elsewhere as well. Desmosomes provide mechanical resilience to the groups of cells they bind together.

At least two types of desmosomes have been recognized: the belt desmosome and the spot desmosome. The *belt desmosome* forms a band linking adjacent epithelial cells just below the tight junction (Fig. 8.13). The filaments of the belt desmosome contain actin, suggesting that they contract—indeed, when ATP and calcium ions are added to a preparation, contraction occurs. Perhaps contraction serves to close spaces caused by killing or injury to some epithelial cells preventing loss of soluble constituents of the cells. The *spot desmosome* is a button-like contact between plasma mem-

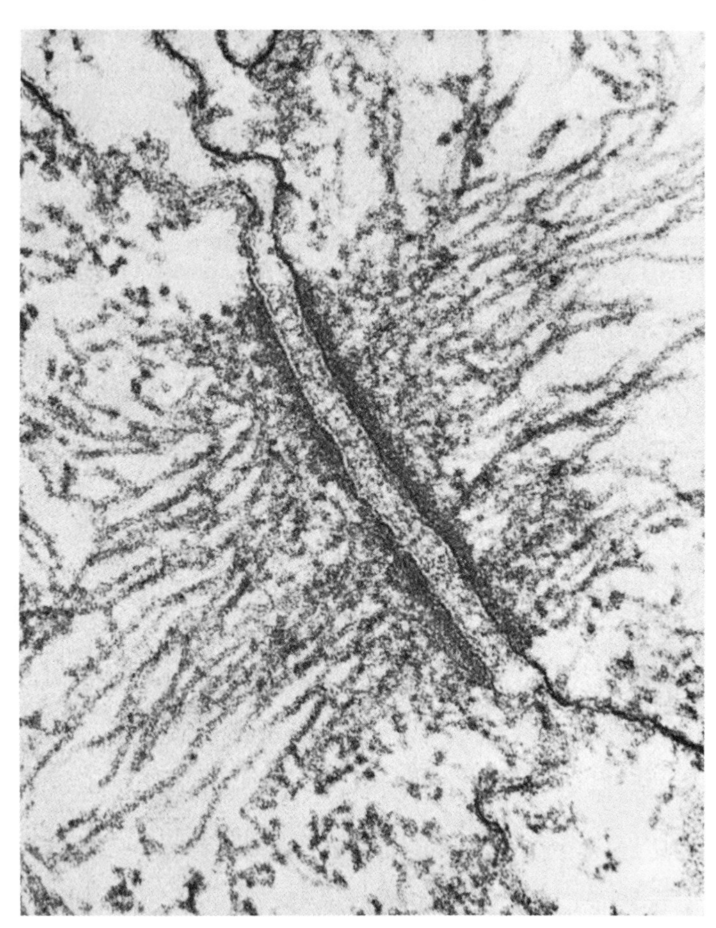

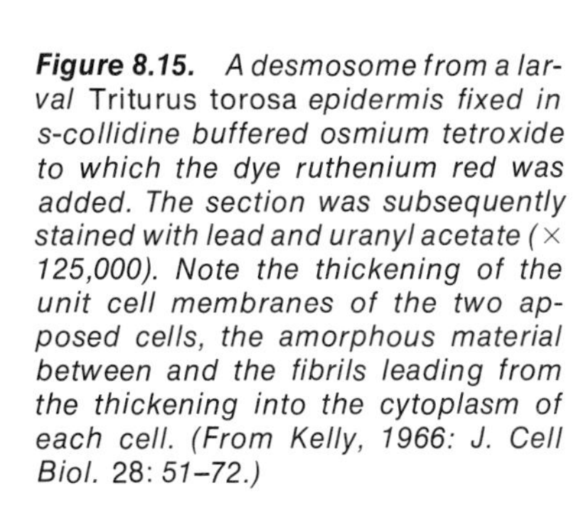

Figure 8.15. *A desmosome from a larval* Triturus torosa *epidermis fixed in s-collidine buffered osmium tetroxide to which the dye ruthenium red was added. The section was subsequently stained with lead and uranyl acetate (× 125,000). Note the thickening of the unit cell membranes of the two apposed cells, the amorphous material between and the fibrils leading from the thickening into the cytoplasm of each cell. (From Kelly, 1966: J. Cell Biol. 28: 51–72.)*

branes of adjacent cells, analogous to a spot weld or rivet. In the 30 nm space between the two cells is a mass of filamentous material with a denser central stratum (Fig. 8.15). In the cytoplasm of each of the adjacent cells is a disk-shaped plaque with filaments (tonofilaments) passing through it and into the plaques of the spot desmosome (Staehelin and Hull, 1978).

COMPARTMENTATION OF THE CELL BY MEMBRANES

As already discussed, membranes are present on many of the cell organelles: double-unit membranes on the mitochondria, single-unit membranes on the endoplasmic reticulum, Golgi complex, and lysosomes, and a nuclear envelope consisting of two unit membranes. The membranes serve to compartmentalize the cell, each compartment presumably being able to regulate its contents by virtue of the selective permeability of its membrane.

Cell membranes are not all alike in composition (see Table 8.1). The smooth and rough endoplasmic reticulum is similar in composition to the plasma membrane, while the Golgi body membrane is between that of the plasma and endoplasmic reticulum membranes in composition. Mitochondrial membranes are characterized by cardiolipin, present predominantly on the inner membrane and confined almost completely to this organelle. Cholesterol is present mainly on the outer membrane. The protein-to-lipid ratio is quite characteristic of the various organelles as seen in Table 8.1.

Mitochondria are able to concentrate magnesium, potassium, and phosphate against a concentration gradient, also at the expense of metabolic energy. The endoplasmic (sarcoplasmic) reticulum of muscle cells concentrates calcium ions against a concentration gradient at the expense of metabolic energy. The Golgi complex likewise displays a capacity for such selective activity. The lysosomes contain a ready supply of hydrolytic enzymes but the membrane confines them and prevents their action on the cytoplasm until released. The membranes around all these organelles thus serve as "buffers," protecting the reactions going on within them from the changes that may be occurring in the ground substance of the cytoplasm. The cytoplasm is, in turn, itself buffered from the changes occurring in the bathing fluid by the plasma membrane.

The nucleus within a nuclear envelope may be similarly protected from changes occurring in the cytoplasm. However, it has been argued that even large molecules have access to the nucleus because of the large pores seen in the nuclear envelope (see Fig. 7.7). The maintenance of a potential difference between nucleus and cytoplasm has already been mentioned. When found, such a potential indicates that ionic selectivity is present; therefore, selectivity for large molecules should also exist. Since plugs have been seen in pores of the nuclear envelope (Branton, 1964), it is possible that they serve to stopper the pores, at least at times. This subject is of great interest and needs further study (Severs, 1976).

SUMMARY

Cell membranes are now considered to be fluid mosaics, primarily of lipid and protein rather than the formerly accepted tripartite sandwich of the unit membrane. Cell membranes are thought to be similar, whether at the surface of the cell or on organelles inside the cell. As expected, some differences have been shown to exist between the various membranes; even on an organelle, the outer membrane may be quite different in chemical composition and function from the inner membrane. On such double-membraned structures as mitochondria and chloroplasts, the plasma membrane may have considerable cholesterol, as in the myelin sheath cells, or little, as in some animal cells, or none, as in plant cells and bacteria.

Cell membranes are dynamic structures, changing in composition and structure with changes in conditions. However, they always consist of a basic lipid bilayer into which, and through which, pass proteins, the latter serving as gates for water-soluble substances. Some of the proteins are enzymes serving a variety of functions.

Extraneous to the plasma membrane are protective layers such as the jelly coat of marine eggs and the glycocalyx of many animal cells, some plant cells and some bacteria. The extraneous coatings may vary considerably in chemical composition. Some of them serve as antigens.

LITERATURE CITED AND GENERAL REFERENCES

Bengelsdorf, I. S., 1976: Cell membrane receptors. Bioscience *26*: 400–402.

Blaurock, A. E., Stoeckenius, W., Oesterhelt, D. and Sauerhof, G. L., 1976: Structure of the cell envelope of *Halobacterium halobium*. J. Cell Biol. *71*: 1–22.

Branton, D., 1964: Fine structure in freeze-etched *Allium cepa*. J. Ultrastruct. Res. *11*: 401–411.

Branton, D. and Deamer, D. W., 1972: Membrane Structure. Protoplasmatologia. Vol. 2, Part E1. Springer Verlag, New York.

Branton, D. and Park, R. B., eds., 1968: Papers on Biological Membrane Structure. Little, Brown & Co., Boston.

Brown, J. C. and Hunt, R. C., 1978: Lectins. Int. Rev. Cytol. *52*: 277–345.

Capra, J. D. and Edmundson, A. B., 1977: The antibody combining site. Sci. Am. (Jan.) *236*: 50–59.

Cook, G. M. W., 1970: Role of glycoproteins in cell membranes. Haidemenakis, ed. Molecular Biology 7th Internat. Congr., pp. 179–207.

Cook, J. S., ed., 1976: Biogenesis and Turnover of Membrane Macromolecules. Soc. Gen. Physiol. Series Vol. 31, Raven Press, New York.

Costerton, J. W., Geesey, G. G. and Cheng, K-J., 1978: How bacteria stick. Sci. Am. (Jun.) *238*: 86–95.

Davson, H., 1970: A Textbook of General Physiology, 4th Ed. Vol. 1, Williams & Wilkins, Baltimore.

Davson, H. and Danielli, J. F., 1952: The Permeability of Natural Membranes. 2nd Ed. Macmillan Co., New York.

Eisenberg, M. and McLaughlin, S., 1976: Lipid bilayers as models of biological membranes. Bioscience *26*: 436–443.

Ellar, D. J., 1965: Biosynthesis of protective surface structures of prokaryotic and eukaryotic cells. Symp. Soc. Gen. Microbiol. *20*: 167–202.

Engelman, D. M., 1970: X-ray diffraction studies of phase transitions in the membrane of *Mycoplasma*. J. Mol. Biol. *47*: 115–117.

Farquhar, M. G. and Palade, G. E., 1965: Cell junctions in amphibian skin. J. Cell Biol. *25*: 263–291.

Fox, C. F., 1972: The structure of cell membranes. Sci. Am. (Feb.) *226*: 31–38.

Fox, C. F. and Keith, A., 1976: Membrane Molecular Biology. Sinauer Associates, Stamford, Conn.

Gray, E. G., 1964: Electron microscopy of the cell surface. Endeavour *23*: 61–65.

Greaves, M. F., 1975: Outline Studies in Biological Recognition. Halsted Press (Wiley), New York.

Green, D. E. and Brucker, R. F., 1972: The molecular principles of biological membrane construction and function. Bioscience *22*: 13–19.

Gunning, B. E. S. and Steer, N. W., 1975: Ultrastructure and the Biology of Plant Cells. Crane, Russak & Co., Inc., New York.

Harrison, R. and Lunt, G., 1976: Biological Membranes: Their Structure and Function. Halsted Press (Wiley), New York.

Hendler, W. W., 1971: Biological membrane ultrastructure. Physiol. Rev. *51*: 66–97.

Hokin, L. E. and Hokin, M. R., 1965: The chemistry of cell membranes. Sci. Am. (Oct.) *213*: 78–86.

Kanno, Y. and Loewenstein, W. R., 1964: Low resistance coupling between gland cells. Some observations on intercellular contact membranes and intercellular space. Nature *201*: 194–195.

Kelly, D. E., 1966: Fine structure of desmosomes, hemidesmosomes and an adepidermal globular layer in developing newt epidermis. J. Cell Biol. *28*: 51–72.

Korn, E. D., 1966: Structure of biological membranes. Science *153*: 1491–1498.

Lee, A. G., 1975: Interactions within biological membranes. Endeavour *34*: 67–71.

Lodish, H. F. and Rothman, J. E., 1979: The assembly of cell membranes. Sci. Am. (Jan.) *240*: 48–63.

Loewenstein, W. R., 1965: Permeability of a nuclear membrane: changes during normal development and changes induced by growth hormone. Science *150*: 909–910.

Loewenstein, W. R. and Kanno, V., 1963: The electrical conductance and potential across the membrane of some cell nuclei. J. Cell Biol. *16*: 421–425.

Lucy, J. A., 1969: Lysosomal membranes. *In* Lysosomes, Biology, Pathology. Dingle, ed. Vol. 2. North Holland Publishing Co., Amsterdam, pp. 313–341.

Luft, J. H., 1976: The structure and properties of the cell surface coat. Int. Rev. Cytol. *45*: 291–382.

Maddy, A. H., 1976: Biochemical Analysis of Membranes. Halsted Press (Wiley), New York.

Malhotra, S. K., 1970: Organization of the cellular membrane. Prog. Biophys. Mol. Biol. *20*: 27–131.

Martinez-Palomo, A., 1970: The surface coat of cells. Int. Rev. Cytol. *29*: 29–75.

Membrane Biology (journal), Vol. 27, 1977.

Morré, D. J., 1975: Membrane biogenesis. Ann. Rev. Plant Physiol. *16*: 441–481.

Muhlethaler, K., 1971: Studies on freeze etching of cell nuclei. Int. Rev. Cytol. *31*: 1–19.

Nicolson, G. L., 1974: The interactions of lectins with animal cell surfaces. Int. Rev. Cytol. *39*: 89–190.

O'Neill, C. H., 1964: Isolation and properties of the cell surface membrane of *Amoeba*. Exp. Cell Res. *35*: 477–496.

Oppenheimer, S. B., 1978: Cell surface carbohydrates in adhesion and migration. Am. Zool. *18*: 13–23.

Orci, A., Perrelet, A., Malaise-Legre, F. and Vassalli, P., 1977: Pore-like structures in biological membranes. J. Cell. Sci. *25*: 157–162.

Pollock, E. G., 1978: Fine structural analysis of animal cell surfaces: membrane and cell surface topography. Am. Zool. *18*: 25–69.

Quinn, O. J., 1976: The Molecular Biology of Cell Membranes. University Park Press, Baltimore.

Razin, S., 1972: Reconstitution of cell membranes. Biochim. Biophys. Acta *265*: 241–296.

Revel, J. P., 1974: Contact junctions between cells. Transport at the cell level. Soc. Exp. Biol. Symp. *28*: 447–462.

Robertson, J. D., 1964: Unit membranes: a review with recent new studies of experimental alterations and a new subunit structure in synaptic membranes. *In* Cellular Membranes in Development. Locke, ed. Academic Press, New York, pp. 1–81.

Robertson, J. D., 1969: Molecular structure of biological membranes. *In* Handbook of Molecular Cytology. Lima-de-Faria, ed. North Holland Publishing Co., Amsterdam, pp. 1403–1443.

Rothman, J. F. and Lenard, J., 1977: Membrane asymmetry. Science *195*: 743–758.

Saier, M., Jr. and Stiles, C. D., 1975: Molecular Dynamics in Biological Membranes. Springer Verlag, New York.

Salema, R. and Brandao, I., 1976: Higher plant protoplasts isolated with pectinase. J. Submicros. Cytol. *8*: 89–94.

Salton, M. J. and Owen, P., 1976: Bacterial membrane structure. Ann. Rev. Microbiol. *30*: 351–382.

Severs, N. J., 1976: Nuclear envelope inclusions demonstrated by freeze-fracture. Cytobios. *16*: 125–132.

Sharon, N., 1972: Lectins. Sci. Am. (Jun.) *236*: 108–119.

Silbert, D. F., 1975: Genetic modification of membrane lipid. Ann. Rev. Biochem. *44*: 315–339.

Simpson, I., Rose, B. and Loewenstein, W. R., 1977:

Size limit of molecules permeating the junctional membrane channels. Science *195*: 294–296.
Singer, S. J., 1974: The molecular organization of membranes. Ann. Rev. Biochem. *43*: 805–833.
Singer, S. J. and Nicolson, G. L., 1972: The fluid mosaic model of the structure of the cell membrane. Science *175*: 720–731.
Solomon, A. K., 1960: Pores in the cell membrane. Sci. Am. (Dec.) *203*: 146–156.
Solomon, A. K. and Karnovsky, M., 1978: Molecular Specialization and Symmetry in Membrane Function. Harvard University Press, Cambridge, Mass.
Staehelin, L. A. and Hull, B. E., 1978: Junctions between cells. Sci. Am. (May) *238*: 141–152.
Stein, J. M., Tourtelotte, M. E., Reinert, J. C., McElhaney, R. N. and Rader, R. L., 1969: Calorimetric evidence for the liquid crystalline state of lipids in a biomembrane. Proc. Nat. Acad. Sci. U.S.A. *63*: 104.
Stoeckenius, W., 1969: Current models for the structure of biological membranes. J. Cell Biol. *42*: 613–646.
Stoeckenius, W., 1970: Electron microscopy of mitochondrial and model membranes. *In* Membranes of Mitochondria and Chloroplasts. Racker, ed. Reinhold, New York, pp. 53–90.
Tamm, S. L., 1976: Properties of a rotor in eukaryotic cells. *In* Cell Motility. Goldman, Pollard and Rosenbaum, eds. Cold Spring Harbor, New York, pp. 949–967.
Tribe, M., 1976: Cell Membranes, Cambridge University Press, New York.
Vanderkooi, G. and Green, D. E., 1971: New insights into biological membrane structure. Bioscience *21*: 409–415.
Wallach, D. F. H., ed., 1972: The Dynamic Structure of Cell Membranes. Springer Verlag, New York.
Wischnitzer, S., 1974: The nuclear envelope: its ultrastructure and functional significance. Endeavour *33*: 137–142.
Yamamoto, T., 1963: On the thickness of the unit membrane. J. Cell Biol., *17*: 413–421.
Yazukov, A. A., Rogatvich, N. P., Yasinovskii, V. G. and Zubarev, T. N., 1974: Bioelectric potential of isolated nuclei of *Acetabularia*. Fiziol. Rast. *21*: 714–720.

Section III

CONVERSIONS OF ENERGY AND MATTER IN THE CELL

Cells live at the expense of chemical energy obtained from nutrients such as glucose. In addition to nutrients cells require salts and vitamins (coenzymes) and aerobic cells require oxygen. By day photosynthetic cells make their nutrients, by night they use up some of these nutrients but they also continue to convert nutrients of one type to another and to synthesize vitamins. Nonphotosynthetic cells depend upon photosynthetic cells as the ultimate source of their nutrients and vitamins.

Cells in multicellular organisms, removed from immediate contact with an aqueous environment, depend upon supplies reaching them from the blood in higher animals, and from the vascular bundle in higher plants. The services of supply and the organ systems by which these ends are achieved are considered in textbooks on comparative physiology of animals and plants. The focus in this section is on cell nutrition and on the principles governing the direction of cellular reactions (Chapter 9), the velocity of cellular reactions (Chapter 10), redox potentials (Chapter 11), metabolic pathways in the cell (Chapter 12), oxygen consumption and bioluminescence (Chapter 13), photosynthesis (Chapter 14), biosynthesis (Chapter 15), and regulation of biosynthesis in cells (Chapter 16).

CHAPTER 9

THE DIRECTION OF CELLULAR REACTIONS: THERMODYNAMICS

Cells are living physical-chemical systems that absorb, transduce, and utilize energy in maintaining the highly improbable state that is life. Energy is the capacity to do work. Life cannot be sustained without the continuous expenditure of energy. The principles that govern energy transfers and determine the direction of reactions in physical-chemical systems are also applicable to cells and these principles are the subject of this chapter.

Energy may be classified as of two types: kinetic and potential. For cellular reactions the kinetic energy of molecules is of primary interest; *kinetic energy* of molecules is the energy of motion. The amount of kinetic energy available to molecules is proportional to the absolute temperature. At absolute zero (−273.18°C) kinetic energy ceases to exist.

The kinetic energy of molecules may be inferred from Brownian movement of very small but microscopically visible particles that move about more actively as the temperature rises. *Brownian movement* is attributed to random motion of molecules colliding with visible particles to which they impart sufficient energy for movement (Fig. 9.1).

Potential energy, either bound or positional, is energy with a high potential for doing work. Potential energy may be chemical, electrical, photic, positional (as, for example, a stone on the top of a hill), or atomic.

The occurrence of interconversions between different types of potential energy and the conversion of potential energy to kinetic energy should be obvious to anyone who has turned on a light or toasted some bread. Conversions of chemical potential energy to other forms of potential energy (other chemical configurations in syntheses, production of electric potentials, and movement) and to kinetic energy (heat) occur continuously in cells. The principles governing the relations between potential and kinetic energy are embodied in the laws of thermodynamics.

Measurements of conversions from potential to kinetic energy must be made in an arbitrary portion of matter* known as the *system,* for example, a cell and its immediate surroundings or an organism and its immediate surroundings. A system will absorb energy from, or lose energy to, its surroundings (environment) unless an attempt is made to insulate it from its environment. This can be done, for example, by having a reaction occur in a thermos bottle where it becomes an *isolated system.* The size of the system to be isolated depends upon the need of the problem at hand—it may be a single cell or a single organism, each in its immediate environment, or it may be a larger fraction of the universe, the limit being the universe itself.

The laws of thermodynamics are consequently independent of our conception of the nature and structure of matter. Classical thermodynamics deals with reactions at equilibrium and is therefore called *equilibrium thermodynamics*. Equilibrium thermodynamics is concerned with the change in form of energy in a system from the *initial state,* before the process has taken place, to the *final state,* after the process has occurred.

Ascertaining the total energy of a system at

*Matter might be defined as that which occupies space and of which objects are made.

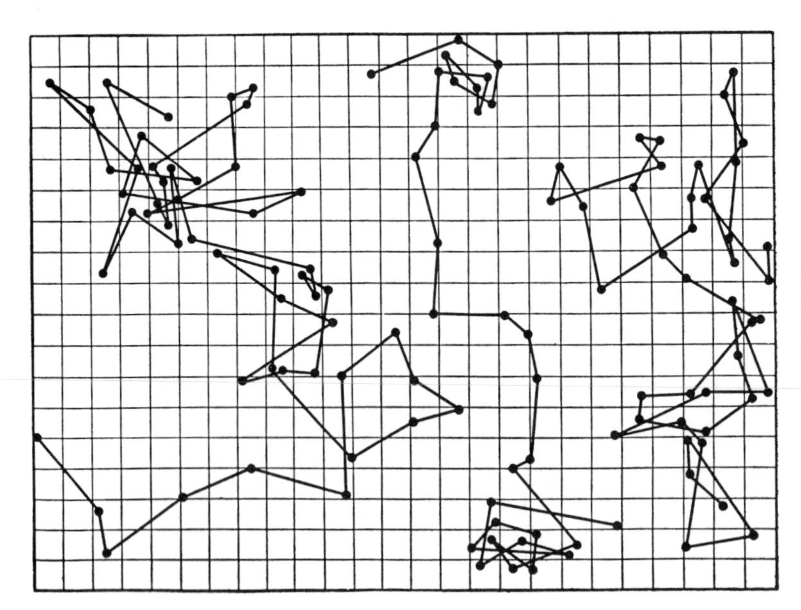

Figure 9.1. *Brownian movement of three particles observed under a microscope. (From Perrin, 1913: Les Atomes. Alcan, Paris.)*

equilibrium either in the initial or the final state requires specification of properties of the system that are difficult to ascertain (e.g., heat content, temperature, pressure, volume, composition). Therefore, experimenters have concerned themselves only with what is readily amenable to measurement, namely the *change* in energy content between the initial and the final states of the system, both in physical and biological systems.

The number of steps taken between the initial state and the final state of the system is also of no consequence, provided the system arrives at the same final state. Since reactions in cells occur by pathways usually quite different from those in physical systems, this finding is of crucial importance for application of thermodynamics to the cell.

CHANGE IN FORM OF ENERGY: CONSERVATION OF ENERGY

In this day of energy shortages everyone is aware that the chemical potential energy of gas, oil, and coal is converted into movement in engines or heat (kinetic energy), by which we warm our homes and cook. We are also aware that energy can neither be created nor destroyed, it merely changes form. This principle, known as the *conservation of energy,* can be tested in an isolated system in which total conversion of a given amount of chemical potential energy, for example, can be measured by the amount of heat energy produced.

It is very difficult to completely isolate a system because it requires a container across the boundaries of which heat neither enters nor leaves. A thermos bottle approximates such a container. Hot liquid placed in it will remain hot because the vacuum between the two walls of the flask isolates the liquid from the environment. The greater the vacuum the better the isolation of the system from the environment because heat is conducted only very slowly by the few molecules of air that remain in the vacuum. The degree of imperfection in the isolation is indicated by the time required for the liquid to cool. Heat is lost to the environment chiefly by conduction through the plug, although some is also lost by radiation.

To measure energy exchanges in a chemical system, a piece of equipment called a *calorimeter,* achieving isolation by insulation, is used (Fig. 9.2). The outermost jacket of a calorimeter is made of a good insulator and surrounds a chamber filled with water, which absorbs the heat released by the sample in the innermost chamber. The temperature of the actively stirred water is determined with a thermometer. Since perfect insulation of a calorimeter has not been attained, corrections must be made for thermal leaks by measuring the amount of electric energy required to produce an equal change in temperature in a comparable time lapse.

When, in such an isolated system, a given amount of chemical, electrical, or photic energy disappears, it is transformed into an equivalent amount of heat. It follows that the total energy in an isolated system will neither increase nor decrease. These statements of the *first law of thermodynamics,* the law of conservation of energy, are part of our everyday experience.

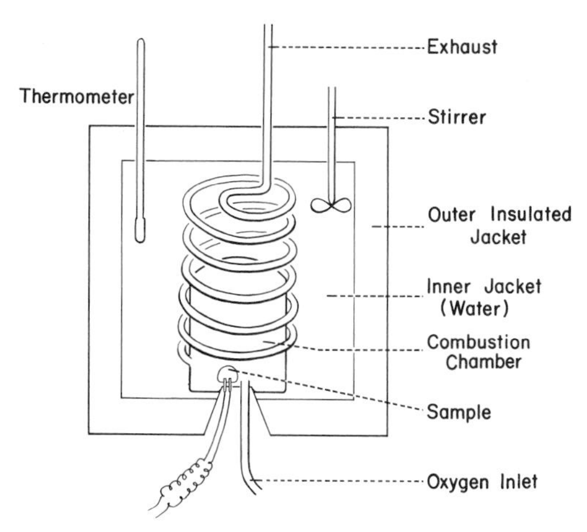

Figure 9.2. *A bomb calorimeter for determining heat of combustion of compounds.*

THE DIRECTION OF REACTIONS: THE SECOND LAW OF THERMODYNAMICS

Everyone knows that the potential energy of sunlight is converted to heat on absorption by a dark surface but that production of light from heat requires incandescence, indicating the relatively small probability of heat being converted to light. Everyone knows that water runs downhill, not uphill, its potential energy and capacity to do work decreasing as it moves. Everyone knows that heat energy flows from a warm body to a cold one. It is also widely known that only a fraction of sunlight striking a plant is converted into the potential energy of photosynthetic products, the rest being dissipated as heat. In all these cases energy is transformed in such a manner that its capacity to do useful work decreases. Potential energy ultimately becomes converted to kinetic energy capable of doing less work. The fundamental principle of the *second law of thermodynamics* both for physical systems and life is thus part of the daily experience of mankind, namely, that spontaneous reactions have direction. For this reason Sir Arthur Eddington called the second law of thermodynamics "time's arrow" (Layzer, 1975). The second law of thermodynamics can be stated in many ways, some of which will appear in subsequent sections of this chapter.

Molecular potential energy (e.g., chemical, photic) has a high degree of orderliness, whereas molecular kinetic energy is random. As potential energy changes to kinetic energy, its randomness or disorder increases. *Entropy** is a function used to measure a system's randomness. Random molecular energy is less capable of doing work than orderly potential energy. Entropy in a natural system tends to increase because an increase in randomness in the molecular world is more likely than the reverse. In other words, reactions proceed in such a way as to reach maximum entropy at equilibrium (another way of stating the second law of thermodynamics).

*At absolute zero the entropy of a crystalline system is zero since the probability of randomness is nil. Entropy can therefore be measured using absolute zero as a baseline (see Appendix 9.2).

The second law of thermodynamics stipulates that when one type of potential energy is converted into another, some of the potential energy is transformed into kinetic energy. A reaction always proceeds in such a direction that the work capacity of its products decreases. The maximum work capacity of the potential energy in a loose sense may be called its *free energy*. Natural reactions always produce a decrease in the free energy of the system. This is still another statement of the second law and shows why natural reactions have direction.

Also in a loose sense, one may speak of potential energy as high-grade energy, recognizing that it has high work capacity, while kinetic energy may be considered as low-grade energy because it has much lower work capacity. Ultimately all the potential energy available in a system may be converted to kinetic energy or heat, and, considering our universe as an isolated system, it is possible that ultimately the storehouse of potential energy may become exhausted. This is sometimes referred to as entropy doom.

If in a system isolated in a calorimeter (Fig. 9.2) all the potential energy is converted into heat, and no work is done by the system, the total potential energy conversion from the initial to the final state can be measured by the amount of heat evolved. For example, the chemical potential energy of a substance can be transformed completely to heat by combustion of the substance held in a platinum container and ignited electrically. Oxygen, usually under pressure, is supplied as needed and the carbon

TABLE 9.1. HEATS OF COMBUSTION OF VARIOUS ORGANIC COMPOUNDS IN CALORIES PER MOLE*

Compound	Calories per mole
Stearic acid	2,711,800
Sucrose	1,349,600
Glucose	673,000
Glycerol	397,000
Alanine	387,700
Ethyl alcohol	327,600
Lactic acid	326,000
Pyruvic acid	279,100
Acetaldehyde	279,000
Acetic acid	209,400

*Data from Handbook of Chemistry and Physics, 1968. 49th Ed. Chemical Rubber Co., Cleveland; data for pyruvic acid from the International Critical Tables V: 165, 1929.

dioxide released in the reaction is measured in the exhaust. If the volume of water in the calorimeter jacket is known and its temperature increase is measured, the kinetic energy produced in the reaction appearing as heat can be calculated. The *calorie* (cal) is the amount of heat required to raise 1 gram of water 1 degree centigrade (from 14.5 to 15.5°C). For example, complete combustion of glucose ($C_6H_{12}O_6$) releases 673,000 cal/mole. The calorie content of a large number of compounds has been measured calorimetrically and some values are cited in Table 9.1. The calorimeter can also be used to measure other potential energy conversions, for example, electricity to heat, light to heat, and so on.

ENERGY STORAGE

Potential energy is not easily stored. For example, no means exists for storing light energy except on a transient basis. It is, of course, converted to, and stored as, chemical energy in green plants (see Chapter 14), which conserve less than 1 percent of the light energy striking the earth's surface. Electrical energy also may be stored only briefly and in small amounts in condensers. Chemical compounds, although serving as good stores of potential chemical energy if kept dry and anaerobic, release their energy on oxidation. Having a high work capacity, light, electrical, and chemical potential energy transform to molecular kinetic energy capable of performing less work. Ultimately, all the potential energy in a system is converted to kinetic energy (heat).

Chemical compounds are the best stores of potential energy. This makes life possible. Plant cells transform the sun's light energy into chemical potential energy, which animals and microorganisms tap to obtain the energy necessary for their life activities. We use such energy reserves from our crops, and we exploit the energy reserves of coal and oil, plant products of ancient geological eras. At present, to a small extent, chemical and photic energy are converted directly to electrical energy to power space satellites, radio stations in remote areas, television repetition stations in alpine regions and the like (Gregory, 1969). Because such direct conversions occur with higher efficiency than conventional conversions, such studies may have vast future practical applications (Fickett, 1978). Fuel cells that convert waste organic matter to electricity are also under extensive study (Vielstich, 1970). We also are learning to use atomic energy effectively, thus opening an entirely new energy reserve (Belitsky, 1976) and tapping underground geothermal energy (Barnes, 1972).

ENERGY EXCHANGE AND LIFE

When a gram molecular weight of glucose is burned in a calorimeter, 6 moles of oxygen are used, 6 moles of carbon dioxide are evolved and 673,000 calories of heat are liberated. If the living cell obeys the laws of thermodynamics, it should also use 6 moles of oxygen, evolve 6 moles of carbon dioxide and liberate 673,000 calories of heat in a calorimeter for each gram molecular weight of glucose "burned." This assumes that only carbohydrate is being used by the living cells during the experiment. (This subject is further discussed in Chapter 13.)

At about the time of the American Revolution, and even before the law of conservation of energy was fully appreciated, the great French scientists Antoine Lavoisier and Pierre de Laplace devised a primitive calorimeter to study the nature of energy conversions in living organisms (Fig. 9.3). Surrounding a cage containing a guinea pig they placed an ice jacket to absorb heat produced during the experiment, and the heat was measured by the amount of water obtained from the melted ice. They isolated this system by surrounding it with another ice jacket, which absorbed all heat entering from the environment. So long as ice was present, both ice jackets remained at 0°C; therefore, no heat was transferred from one jacket to the other although some ice melted in each, as shown by control tests. Thus the inner chamber was effectively isolated.

Lavoisier and Laplace collected the water from the ice melted in the inner chamber, and from its volume (79.7 calories are required to melt one gram of ice) they determined the total heat produced by the guinea pig in 10 hours (enough to melt 341 grams of ice). In a parallel experiment they determined the carbon dioxide produced in the same period. Comparison of the energy liberated by the oxidation of the same amount of carbon, on the one hand, in chemical combustion (enough to melt 326.7 grams of ice), and, on the other, in the animal, indicated that the values were of the same order of magnitude (Gabriel and Vogel, 1955). The difference in the values probably stems from the fact that in the organism carbohydrate is burned rather than carbon alone. When this is taken into consideration, as in modern calorimetry, the agreement is much closer (Guyton, 1976).

Calorimeters in use today are more elaborate than the one used by Lavoiser and Laplace, but the essential plan is the same: to isolate a given system by good insulation and to measure the heat evolved, the oxygen consumed and the carbon dioxide produced (Fig. 9.4). Heat is caught in water passed through the chamber at a constant rate, and the temperature of the water is determined by thermocouples at entry and exit points. Oxygen is determined by flow

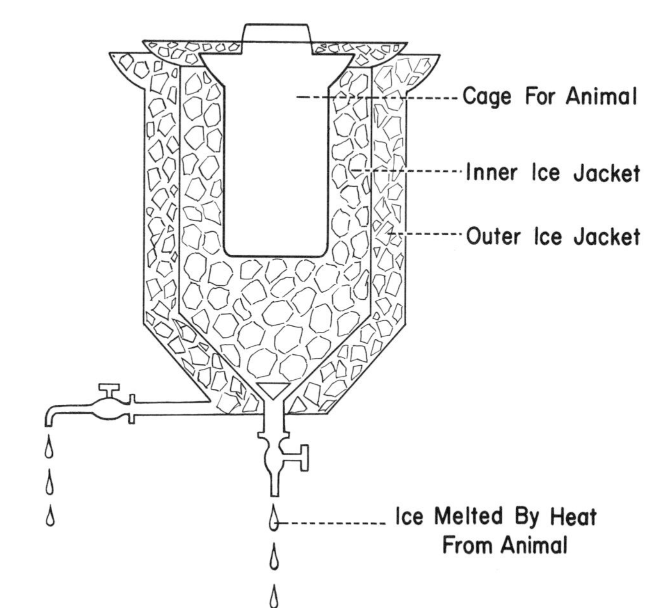

Figure 9.3. Diagram of type of calorimeter used by Lavoisier and Laplace.

meters, carbon dioxide by absorption in soda lime (a mixture of sodium hydroxide and calcium hydroxide), and water vapor by absorption in sulfuric acid.

The oxidation of a compound in the cell occurs by pathways different from those in the calorimeter. However, chemists have demonstrated that regardless of the pathway by which the combustion occurs, if a compound is oxidized to the same degree (in the case of the carbohydrate cited, completely to carbon dioxide and water), the amount of energy released is the same. Since according to the first law of thermodynamics energy is neither created nor destroyed, this could have been inferred.

The calorimetric findings are of profound significance since they indicate that in living cells, as in the inanimate world, energy transformations are described by the same physical-chemical laws. Since energy transformations underlie all work done by a cell, they are basic to all studies in cell physiology.

The cell oxidizes only a fraction of the food it obtains; much chemical potential energy is stored in compounds as reserve potential energy for future use. These reserves accumulate in cells as oil globules and starch or glycogen granules. Occasionally protein is also stored. In multicellular plants reserves are found in cells of bulbs, tubers, roots, seeds, and fruit as starches, sugars, fats, or even proteins. In multicellular animals glycogen is stored in the liver and muscles, and fat in fat bodies. At times when the diet provides insufficient food the reserves are utilized first, and when these are gone the cell substance itself is attacked.

Heat is produced by metabolism in all living cells and sometimes accumulates, as evidenced by the warmth of human tissues, by the heat of

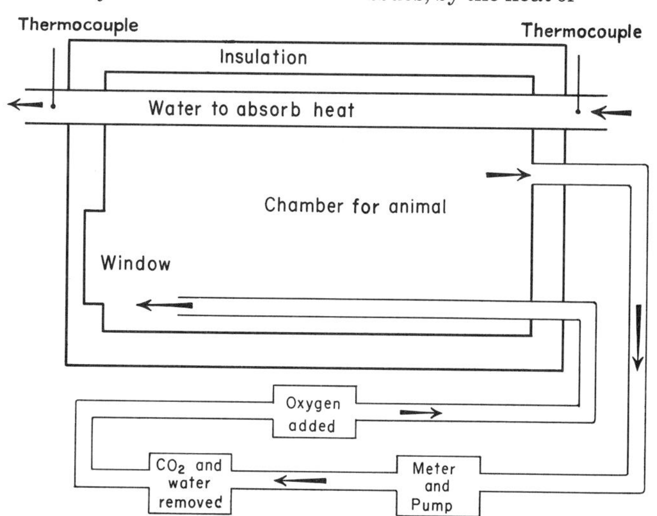

Figure 9.4. Diagram of a modern type of calorimeter.

the manure pile, by the heat of germinating seeds, and by the heat produced in a beehive. Even cold-blooded fish produce heat, but this is rapidly dissipated into the environment.

COUPLED REACTIONS IN THE CELL

A reaction that occurs with a decrease in free energy ($-\Delta G$) is said to be *exergonic,* as in oxidation of glucose. A reaction that occurs with an increase in free energy ($+\Delta G$) is said to be *endergonic,* as in photosynthesis. In cellular biochemical reactions free energy is not truly liberated and dissipated as heat, but much of it is conserved in high-energy bonds. In this form it is passed on to other molecules to be used by them to form new bonds. Cell work is accomplished at the expense of such bond energy, as illustrated in Figure 9.5, for a whole series of biological reactions. Each of these types of reactions is dealt with in more detail elsewhere (see Chapters 12, 15, 18, 21, and 23).

The free energy (G) of a reaction, best defined mathematically (see Appendix 9.1), is the net work obtainable (at constant temperature and pressure) when the reaction occurs under reversible conditions. It is not the same as the energy (heat) liberated by a reaction, because in a reaction energy may be absorbed from the surroundings so that the free energy may be greater than the heat of reaction.

The entropy in a system tends to increase and the free energy to decrease (second law). Yet in cells, reactions occur in which the free energy is known to increase, for example, in the synthesis of proteins and nucleic acids. This is not contrary to the second law of thermodynamics because while the free energy of the compounds synthesized increases, the free energy of the system as a whole decreases. The cell supplies energy for such biological syntheses in the form of labile high-energy phosphate bonds (see Chapter 12) from exergonic reactions coupled with endergonic reactions. By this means a complex compound may be synthesized, in seeming contradiction to the second law of thermodynamics. However, only a fraction of the free energy of the exergonic reaction is utilized in the synthesis so that the net effect is the conversion of some of the free energy in the system to kinetic energy, with a decrease in free energy and an increase in entropy.

As an example of such coupling consider the reaction between coenzyme A (CoA) and acetic acid to form the compound acetyl-CoA. The reaction is endergonic and proceeds with an increase in free energy. However, it is coupled to the exergonic breakdown of adenosine triphosphate (ATP) to adenosine diphosphate (ADP) and inorganic phosphate, a reaction that proceeds with a decrease in free energy. Because the decrease is greater than the increase, the coupled reactions proceed with a net de-

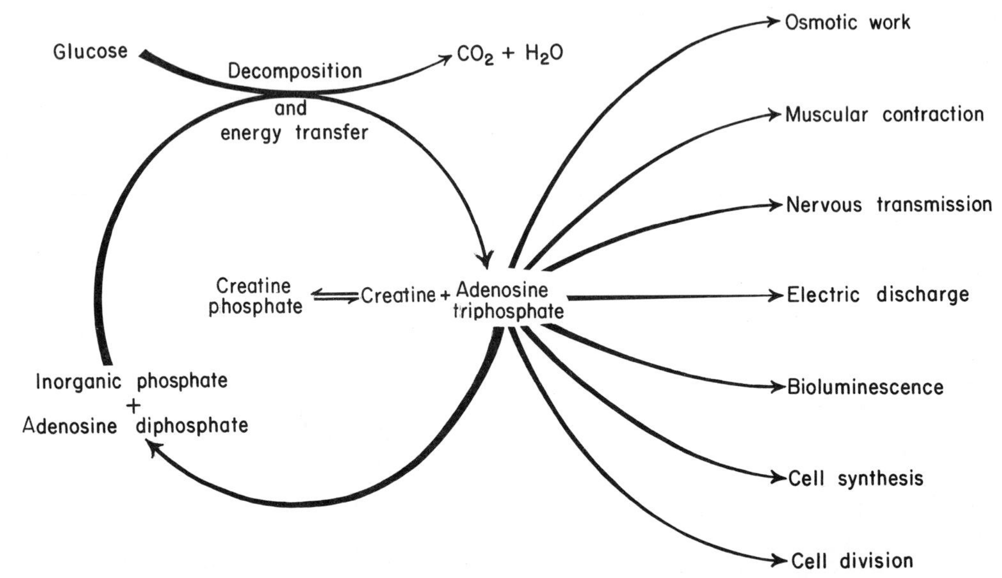

Figure 9.5. *High-energy phosphate bonds and cell work. Energy from glucose (or other cell foods) made available during its anaerobic and aerobic decomposition in the cells is transferred to an inorganic phosphate, which is thus linked by a resonant high-energy phosphate bond to adenosine diphosphate (ADP), making it into adenosine triphosphate (ATP). The ATP may transfer its high-energy phosphate (ΔG = about 7300 calories per mole) to creatine, forming creatine phosphate as a storage depot of high-energy bonds, or it may transfer the energy during coupled reactions performing all kinds of cell work. Various types of cell work are discussed in subsequent chapters. High-energy phosphate bonds are probably stored in metabolizing parts of all cells. (For other high-energy phosphate compounds, see Chapter 12.)*

crease in free energy. Thus although *negentropy,* or higher order, comes about among some parts during a synthesis, the system as a whole shows a net increase in entropy.

In cellular coupled reactions, the participants normally share atoms. The major problem, however, is to determine, at the molecular level, the nature of the constraint that prevents the reaction from merely going downhill without doing work, making possible coupling with transference of the energy. In some cases the constraint is the stability of the high-energy molecule, which "forbids" the uncoupled breakdown. For example, in the presence of the enzyme creatine kinase, creatine phosphate readily transfers its high-energy phosphate to adenosine diphosphate, forming, as a result, adenosine triphosphate. But it does not do this without the enzyme. Enzymes thus facilitate reactions along pathways useful to the cell (see Chapter 10) (Wilkie, 1960).

CHEMICAL POTENTIAL

It is evident that a free energy measurement applies to the entire system under consideration and is therefore an extensive property of the system referring to the total amount. On the other hand, such quantities as density, concentration, temperature, and pressure are intensive properties measuring degree for a fraction of the system (e.g., per unit volume). In biological studies it is often necessary to know not only the free energy change but also such intensive properties, especially the concentration of reactants. The value of such information can be illustrated by an example. A solution of a substance is capable of doing more work when in high than in low concentration, just as a gas can do more work when at high than at low pressure. The *chemical potential* measures the system's capacity to do useful work as it undergoes chemical or physical change.

An example of chemical potential is the phosphate transfer potential of ATP, which often drives a coupled endergonic chemical reaction (discussed in Chapters 12, 14, and 18). An example of physical change is the chemical potential of a solute of high concentration of a substance that moves spontaneously into a cell containing a lower concentration.

When the chemical potential is negative, for example, when a solute must move against a concentration gradient, movement can occur only if the reaction is coupled with an exergonic reaction of sufficiently high chemical potential. When the movement of ions must occur against an electrical gradient, it will be accomplished only if coupled with a reaction of sufficiently high chemical potential to supply energy greater than the work required to move the ions "upstairs" against the electrical potential gradient (see Chapter 18).

THE CELL AND THERMODYNAMICS

Biological reactions all take place in solution (e.g., gases are dissolved before being used in the cell); therefore, with few exceptions, such reactions may be considered as occurring at constant pressure and constant volume (changes in volume being negligible in cellular reactions). Even more important, cells can function when they are at uniform temperature. Thermal gradients may occur temporarily in cells when heat is produced and thermal conductivity is limited, but the thermal gradients are incidental, not essential to cellular functions. Therefore, the second law of thermodynamics can be applied in a simplified form because, in a system at constant temperature, no heat gradient exists for conversion of heat energy into potential energy or work, as in a steam engine.

Some chemical potential energy capable of doing work is continually converted into heat during cellular reactions; this kinetic energy is used to maintain the temperature of the cell and is incapable of being transformed again into potential energy. As discussed previously, the different kinds of potential free energy (e.g., mechanical energy, electrical energy, chemical energy, light) can be interconverted, but the conversion is never complete. In the cell, that part of the free energy not converted into another form of potential energy must appear as heat (in accordance with the first law of thermodynamics). The degree to which free energy is transformed to heat is therefore a measure of the inefficiency of the conversion process. Consequently, free energy must be continually supplied to the cell from outside to maintain life.

It must be pointed out, however, that a cell's kinetic energy is not useless. The cell requires an adequate supply of thermal energy to maintain its constituents in a physical state suitable for metabolism and to supply the activation energy for many vital reactions (Allen, 1960) (see Chapter 10).

When reactions are not coupled, no work is done and the free energy appears as heat. When coupling between two reactions is perfect and no free energy is transformed to kinetic energy, as in a reversible process, the work done equals the free energy change (ΔG). In the cell, only a fraction of the free energy, whose size (ϵ) depends upon the effectiveness of the coupling mechanism, is made available for work:

$$\text{Work done} = -\epsilon\, \Delta G \qquad (9.1)$$

The remainder of the free energy is then given off as heat.

It should also be obvious from thermodynamics that a cell must be supplied with simple molecules of high chemical potential energy that can be built into the complex molecules characteristic of cells. Examples of such complex molecules are the nucleic acids and proteins that are denatured very rapidly at high temperatures, indicating that their structure quickly becomes increasingly random during denaturation.

The living cell is never in equilibrium with its environment. Rather, it maintains a steady state and constitutes an *open system* in continual energy exchange with its environment. Since classical thermodynamic relationships were developed for a system in equilibrium, the question arises whether such reasoning is applicable to cellular reactions.

The accumulation, storage, and expenditure of free energy by cells is independent of whether the cells are in a steady state or in equilibrium. A given reaction occurring in the cell cannot liberate more energy or less energy than it liberates in the test tube. Thermodynamic properties of cellular reactions cannot be determined by classical thermodynamic methods because the cell is not in equilibrium with its environment. For this purpose, the subject of thermodynamics of irreversible processes or open systems is being intensively studied at the present time. This is one instance in which a biological problem prompted study of a neglected field of physical chemistry (Wisniewski *et al.*, 1976). For his development of thermodynamics of irreversible processes, Prigogine was awarded the Nobel Prize in chemistry in 1977 (see Prigogine, 1978).

Classic thermodynamics is limited in another respect. Although it predicts the direction for a reaction of known ΔG, it cannot predict the reaction rate. A physical-chemical reaction-rate theory, which relates the free energy change to the reaction rate, has been used for such studies. The relation between reaction rate and free energy change is linear only when a reaction proceeds near equilibrium. From reaction rate studies it can be shown that an open system, in which several reactions are proceeding, does not show sustained oscillations. Indeed, because each individual reaction continues to proceed at a constant rate, the overall reactions settle down to a steady state in which the amounts and concentrations of the various reactants remain constant even though the reactions are proceeding. This steady state is also one in which occurs the lowest possible rate of free energy dissipation. When extraneous changes are introduced into the system, the system likewise alters spontaneously in such a way as to reduce to a minimum its free energy dissipation. The steady state thus has somewhat the appearance of an equilibrium and is similar to what goes on in the living cell (Wilkie, 1960). Therefore, to this extent thermodynamics can be applied to cellular reactions (McClare, 1971).

APPENDIX

The quantitative relations between energy liberated in a reaction (ΔH), free energy (ΔG), entropy (ΔS) and the equilibrium constant (K) are given below. They will take on more meaning after consideration of cellular metabolism (see Chapter 12).

9.1 FREE ENERGY

ΔH, the heat of reaction (energy content of the products minus the energy content of the reactants), is written with a negative sign (−ΔH) if heat is lost to the environment, and with a plus sign (+ΔH) if heat is gained from the environment during the reaction. In calorimetric combustion of glucose when no external work is done:

$$C_6H_{12}O_6 + 6O_2 \rightarrow 6CO_2 + 6H_2O \qquad (9.2)$$

$$\Delta H = \text{energy content of products} - \text{energy content of reactants} = -673{,}000 \text{ cal/mole (at 25°C)}$$

Of greater theoretical interest than the heat of reaction (ΔH) is the *change in free energy* (ΔG),* which represents the maximum work that can be obtained from a reaction at constant temperature. The change in free energy (free

*The free energy is given as ΔG° if determined at standard state, that is, 25°C., and for gases, 1 atmosphere pressure. For liquids ΔG° implies the change in free energy when one mole of reactant forms one mole of product, each being maintained at one molal concentration. For biological reactions, which in the cell occur at about pH 7.0 and at very low concentrations, the designation ΔG′ is often used. Free energy is here represented by G in honor of J. Willard Gibbs, who established thermodynamics as a science, and is often called *Gibbs free energy*.

energy of products minus the free energy of reactants) of a reaction occurring reversibly at constant temperature (isothermally) and constant pressure may be defined in terms of changes in entropy (ΔS) and changes in heat of reaction (ΔH):

$$\Delta G = \text{free energy of products} - \text{free energy of reactants, or}$$
$$\Delta G = \Delta H - T\,\Delta S \qquad (9.3)$$

where T is the absolute temperature, and ΔS the change in entropy.

The free energy change (ΔG) of a reaction bears no simple relation to the change in energy content (ΔH). Although the two quantities are often similar in size and sign, they are sometimes very different.

The change in free energy (ΔG) may be greater than the ΔH, if heat is absorbed from the surroundings during the reaction and more useful work is accomplished than possible from only the energy liberated in the reaction. For instance, in the combustion of glucose, ΔG (at 25°C) is −686,000 calories per mole while ΔH is −673,000 calories per mole. The difference between ΔG and ΔH is the entropy term:

$$\Delta G = \Delta H - T\,\Delta S$$
$$-686{,}000 = -673{,}000 - 13{,}000$$

Dividing by 298 to obtain the entropy term at standard state (25°C, or 298° absolute, 1 atmosphere pressure) gives ΔS = +43.6 cal/mole. On the other hand, when the work performed during a reaction is less than the energy liberated in the reaction, the ΔG of the reaction is less than the ΔH.

Free energy changes occurring during a variety of types of reactions encountered in biological processes are given in Table 9.2. When the free energy change in a reaction is small, the kinetic energy of activated molecules may be sufficient to react (Chapter 10).

It is possible to relate the change in free energy to the equilibrium constant for a reaction. In the reaction:

$$A + B \rightleftharpoons C + D \qquad (9.4)$$

at equilibrium, application of the law of mass action gives the following expression:

$$\frac{[C][D]}{[A][B]} = K \qquad (9.5)$$

where K is the equilibrium constant.

The change in free energy (ΔG) for the reaction is related to its equilibrium constant (K) in the following way:

$$\Delta G = -RT\,ln\,K \qquad (9.6)$$

where R is the gas constant (about 2.0 cal/mole/degree), T is the absolute temperature, and *ln* is the natural logarithm (2.3 times the logarithm to the base 10). ΔG can thus be determined by the ratio of product(s) to reactant(s). For example, in the conversion of glucose-1-phosphate to glucose-6-phosphate (in the presence of the enzyme phosphoglucomutase) the proportion of glucose-6-phosphate to glucose-1-phosphate at equilibrium was found to be 0.019 to 0.001. The equilibrium constant K_{eq} is therefore 19. Substitution of this value in equation 9.4 enables one to de-

TABLE 9.2. STANDARD FREE ENERGY CHANGES AT pH 7 AND 25°C OF SOME CHEMICAL REACTIONS*

	ΔG′†
Oxidation	
glucose + $6O_2 \rightarrow 6CO_2 + 6H_2O$	−686,000
lactic acid + $3O_2 \rightarrow 3CO_2 + 3H_2O$	−326,000
palmitic acid + $23O_2 \rightarrow 16CO_2 + 16H_2O$	−2,338,000
Hydrolysis	
sucrose + H_2O → glucose + fructose	−5,500
glucose 6-phosphate + H_2O → glucose + H_3PO_4	−3,300
glycylglycine + H_2O → 2 glycine	−4,600
Rearrangement	
glucose-1-phosphate → glucose-6-phosphate	−1,745
fructose-6-phosphate → glucose-6-phosphate	−400
Ionization	
$CH_3COOH + H_2O \rightarrow H_3O^+ + CH_3COO^-$	+6,310
Elimination	
malate → fumarate + H_2O	+750

*From Lehninger, 1971: Bioenergetics. 2nd Ed. W. A. Benjamin, Menlo Park, Calif.
†ΔG′ in cal/mole.

TABLE 9.3. THE NUMERICAL RELATIONSHIP BETWEEN EQUILIBRIUM CONSTANT AND $\Delta G'$ AT 25°C*

$K_{eq.}$	$\Delta G'$
0.001	4089
0.01	2726
0.1	1363
1.0	0.0
10.0	−1363
100.0	−2726
1000.0	−4089

*From Lehninger, 1971: Bioenergetics. 2nd Ed. W. A. Benjamin, Menlo Park, Calif.

termine the ΔG value at standard state (1 molal concentration, 25°C):

$$\Delta G = -RT\, ln\, K = -1.987 \times 298 \times ln\, 19 = -1745 \text{ cal/mole}$$

(Lehninger, 1971).

If K is 1, no reaction will occur when equivalent concentrations of the reactants are mixed. If K is greater than 1 (and ΔG is minus) reaction 9.4 will proceed to the right with a decrease in free energy. If K is less than 1 (and ΔG is plus), reaction 9.4 will proceed to the right only if energy is supplied, as can be seen by the relation between K and ΔG in Table 9.3.

9.2 ENTROPY AND THE SECOND LAW OF THERMODYNAMICS

The total energy change in a reaction consists of two components—the utilizable free energy change (ΔG) and the nonutilizable energy change ($T\,\Delta S$), or change in entropy times the absolute temperature:

$$\Delta H = \Delta G = T\,\Delta S \quad (9.7)$$

Equation 9.7 rearranged, is the same as equation 9.3.

For a specific reaction,

$$A \rightleftharpoons B \quad (9.8)$$

the ΔH can be determined if the heat of combustion of each of the reactants, A and B, is determined. This can be done in a calorimeter (see Fig. 9.2). If, for this hypothetical example, the heat of combustion of A is taken to be 5000 cal/mole and that of B, 2000 cal/mole, then for the reaction at the arbitrary temperature of 27°C (300 degrees absolute):

$$(A - B) = (5000 - 2000)$$

or 3000 calories of heat would be given up; therefore $\Delta H = -3000$ cal/mole.

If we assume that the entropy change on going from A to B was 10, then substituting this value into equation 9.7,

$$\Delta G = \Delta H - T\,\Delta S, \text{ or } (-3000) - (300 \times 10) = -6000 \text{ cal/mole.}$$

In this case the entropy term contributes significantly to the free energy change. The entropy term may be large or small (Lehninger, 1971).

Although some reactions can be performed in such a way that they are practically reversible, all spontaneous reactions are to a considerable degree irreversible; that is, the equilibrium is shifted to favor the reactants, the reverse reaction indicates its degree of irreversibility. If the symbol S designates entropy, and ΔS, the difference in entropy between the beginning (S_a) and the end (S_b) of the reaction, then:

$$\Delta S = S_a - S_b = q/T \quad (9.9)$$

where q is the heat absorbed from the surroundings and T is the absolute temperature.

The second law of thermodynamics may now be stated in terms of entropy: the entropy of a system may not decrease during spontaneous reactions. Part of a system may show a decrease of entropy while another part shows an increase, for example, in coupled reactions that occur in many biological syntheses. However, the entropy of the entire system cannot decrease.

9.3 CONVERSION OF MASS TO ENERGY

According to Einstein's theory of relativity, the equivalence of mass and energy can be represented by the equation

$$E = mc^2 \quad (9.10)$$

where E is the energy, m the mass converted in the atomic shrinkage and c the velocity of light. If m is given in grams and c is taken as 3×10^{10} cm per sec, then E will be in ergs: $E = m(9 \times 10^{20})$. To convert to calories, divide the value for E (in ergs) by 4.18×10^7.

The total mass of a uranium-235 nucleus and a neutron is known to be greater than the mass of the immediate products resulting from atomic fission. For fission of a kilogram of uranium-235, the decrease in mass is just about 1 gram. This amounts to about 2.1×10^{13} calories and is equivalent to 20,000 tons of TNT (trinitrotoluene), or to the daily output of the Hoover Dam (2.3×10^7 kilowatt hours of electricity) (Shilling, 1964). Put in another way, fission of one uranium-235 atom liberates a million times as much energy as the explosion of one molecule of TNT. For the mechanism of fission the reader is referred to textbooks of atomic physics.

Fission of uranium-235 is what is now used to produce the steam heat to run the motor-generators that produce electricity. The environmentalists' objection to the dangerous fission products could be obviated if instead of fission we could master production of heat by fusion of hydrogen to helium, as in the sun, a process still beyond our reach. Fewer people would probably oppose the widespread use of nuclear power then, because no harmful fission products result from the reaction.

The conversion of matter to energy indicates that perhaps the laws of conservation of matter and energy should be stated in even broader terms than was customary in the past; e.g., matter and energy can be neither created nor destroyed although under certain conditions they can be converted from one to another. This implies that matter may also be converted from energy, a view for which some evidence exists.

Whereas matter circulates between the environment and living cells, coming in to form parts of cells, returning to the environment upon death, and again finding its way into cells of another organism at a future time, movement of energy is unidirectional. Potential energy comes into the green plant cell as light from the sun; some of it is stored as chemical energy in carbohydrates and other foods, and some of it is converted to kinetic energy (heat), in which form it leaves the living cell. Thus, in nature the flow of matter is cyclic (see Chapter 12), whereas, essentially, the flow of energy is not.

LITERATURE CITED AND GENERAL REFERENCES

Allen, M. B., 1960: Utilization of thermal energy by living organisms. Comp. Biochem. *1*: 487–514.

Barnes, J., 1972: Geothermal power. Sci. Am. (Jan.) *226*: 70–77.

Becker, W. M., 1977: Energy and the Living Cell. J. B. Lippincott Co., Philadelphia.

Belitsky, B. 1976: Fusion power—a step in the right direction. New Scientist *71*: 447–448.

Calvin, M., 1976: Photosynthesis as a resource for energy and materials. Am. Sci. *64*: 266–278.

Cox, K. E. and Williamson, K. D., eds., 1977: Hydrogen: its Technology and Implications. Vol. 5. Implications of Hydrogen Energy. Cole Press, Inc., Cleveland.

Fickett, A. P., 1978: Fuel-cell power plants. Sci. Am. (Dec.) *239*: 70–76.

Gabriel, M., and Fogel, S., eds., 1955: Great Experiments in Biology. Prentice-Hall, Englewood Cliffs, N. J.

Gates, D. M., 1971: The flow of energy in the biosphere. Sci. Am. (Sept.) *224*: 88–100.

Gregory, A. L., 1969: Fuel cells. Endeavour *28*: 8–12.

Guyton, A., 1976: A Textbook of Medical Physiology, 5th Ed. W. B. Saunders Co., Philadelphia.

Hill, T. L., 1977: Energy Transduction in Biology. Academic Press, New York.

Hubbert, M. K., 1971: The energy resources of the earth. Sci. Am. (Sept.) *225*: 61–70.

Johnson, W. D., Jr., 1977: The prospects for photovoltaic conversion. Am. Sci. *65*: 729–736.

Layzer, D., 1975: The arrow of time. Sci. Am. (Dec.) *233*: 56–69.

Lehninger, A. L., 1971: Bioenergetics, 2nd Ed. W. A. Benjamin, Menlo Park, Calif.

Lewis, K., 1966: Biochemical fuel cells. Bacteriol. Rev. *30*: 101–113.

McClare, C. W. F., 1971: Chemical machines, Maxwell's demon and living organisms. J. Theor. Biol. *30*: 1–34.

Mitchell, W., Jr., ed., 1963: Fuel Cells. Academic Press, New York.

Mitsui, A., Miyachi, S., San Pietro, A. and Tamura, S., eds., 1977: Biological Solar Energy Conversion. Academic Press, New York.

Morowitz, H. J., 1970: Entropy for Biologists. An Introduction to Thermodynamics. Academic Press, New York.

Morowitz, H. J., 1978: Foundations of Bioenergetics. Academic Press, New York.

Prigogine, E., 1978: Time, structure, and fluctuations. Science *201*: 777–785. (Nobel lecture.)

Proctor, W. G., 1978: Negative absolute temperature. Sci. Am. (Aug.) *239*: 90–99.

Science issue of April 19, 1974, on Energy. Vol. *184*: 245–386.

Scientific American, Sept. 1971: Energy and Power especially chapters on The energy resources of the earth (Hubbert), The flow of energy in the biosphere (Gates), and The conversion of energy (Summers), Vol. *224,* No. 3.

Shilling, C. W., 1964: Atomic Energy Encyclopedia in the Life Sciences. W. B. Saunders Co., Philadelphia.

Smith, C. J., 1975: Problems with entropy in biology. Biosystems *7*: 259–265.

Vielstich, W., 1970: Fuel Cells (tr. Ives). Wiley-Interscience, New York.

Summers, C. M., 1971: The conversion of energy. Sci. Am. (Sept.) *224*: 148–160.

Wilkie, D. R., 1960: Thermodynamics and the interpretation of biological heat measurements. Prog. Biophys. *10*: 260–298.

Wisniewski, S., Staniszewski, B. and Szymanik, R., 1976: Thermodynamics of Nonequilibrium Processes. D. Reidel Publishing Co., Hingham, Mass.

Woodwell, G. M., 1970: The energy cycle of the biosphere. Sci. Am. (Sept.) *223*: 64–74.

CHAPTER 10

VELOCITY OF CELL REACTIONS: KINETICS

In a cell metabolic reactions and syntheses occur at environmental or body temperature and at atmospheric pressure. The chemist applies heat and high pressure to achieve most industrial reactions and syntheses. The purpose of this chapter is to determine how the cell can perform its operations under what to a chemist might appear to be impossible limitations.

The chemist applies heat to increase the kinetic energy of the reactants. Kinetic energy is energy of motion (see Chapter 9). Molecular kinetic energy increases in proportion to rise in absolute temperature. Room temperature on the absolute scale is about 293°C, and a 10-degree rise in temperature therefore amounts to 10/293 or about 1/30 increase in kinetic energy. Yet experimental evidence indicates that with a 10° rise in temperature, thermochemical reactions double in rate, as do many cellular reactions. The increase in reaction rate with increase in temperature is called the *temperature coefficient* or *temperature quotient,* abbreviated Q_{10}. In Table 10.1 are given a few temperature coefficients for the increase in reaction rates of known chemical reactions and biological processes.

ENERGY OF ACTIVATION

It is evident that an increase in reaction rate with an increase in temperature depends on something other than the increase in kinetic energy of the system. The Swedish chemist Svante Arrhenius in 1889 had a plausible explanation for a twofold or greater rise in thermochemical reaction rate with a temperature increase of 10°C. He pointed out that at a given temperature not all molecules have the same amount of kinetic energy. Following random collisions, some molecules accumulate more energy than others. Some become energy-rich, some remain energy-poor, and most are somewhere in between (Fig. 10.1). Arrhenius (1915) suggested that the energy-rich molecules are most likely to react on collision and that the reaction rate depends upon the concentration of such activated molecules. In sum, to undergo thermochemical reaction a molecule must acquire a critical amount of energy called the *energy of activation.*

Calculation shows that several times as many activated molecules are present at the higher temperature than at 10° lower, as shown in Figure 10.1. Therefore, it seems most likely that the increase in reaction rate with rise in temperature is attributable to the disproportionate increase in the fraction of molecules with the required activation energy for a particular reaction (Lehninger, 1971).

Consequently, the velocity of a thermochemical reaction appears to be proportional to the frequency with which the state of activation is achieved by the molecules in a given system. Since a temperature rise increases the chances for molecules to acquire the required energy of activation, it increases the velocity of thermochemical reactions. As an example, consider a spontaneous reaction that proceeds with a decrease in free energy. It will go rapidly if little activation energy is required; on the other hand, it will proceed slowly if much activation energy is required. Heat that enables more molecules to acquire the requisite activation energy will increase the reaction rate (Becker, 1977).

When in a reaction a molecule moves from one energy level to another, it must cross over

TABLE 10.1. INCREASE IN RATE OF THERMOCHEMICAL REACTIONS WITH A 10° RISE IN TEMPERATURE*

Reactions in Inanimate System	Q_{10}	***Temp. Range °C***	***Reactions in Organisms***	Q_{10}	***Temp. Range °C***
Reaction between peroxide and hydriotic acid	2.080	0–10	Photosynthesis at high light intensity	1.6–4.3	4–30
	2.078	10–20			
	1.940	20–30			
	1.932	30–40			
Activity of malt amylase	2.2	10–20	Rate of cleavage of Arbacia eggs	3.9	7–17
				3.3	8–19
				2.6	15–25
				1.7	20–30
Most thermochemical reactions (van't Hoff rule)	2.3		Contractility of duodenum	2.42	28–38
			Respiration of pea seedlings	2.4	10–20
Protein coagulation:			Heat killing:		
egg albumin	635	69–76.3	spores	2–10	40–140
hemoglobin	13.8	60–70.4	bacteria	12–136	48–59
			protozoans	891–1000	36–43
			rabbit leukocytes	28.8	38–60

*Most of the data are from Belehradek, 1935: Temperature and Living Matter. Protoplasma-Monographien. Vol. 8. Borntraeger Verlag, Berlin. If data are available for an interval other than 10°C, the temperature coefficient may be obtained from the van't Hoff equation: $\log Q_{10} = \frac{10}{t_2 - t_1} \log \frac{k_2}{k_1}$, where k_2 is the velocity at the higher temperature, t_2, and k_1 is the velocity at the lower temperature, t_1.

an energy "barrier," even though the overall effect is to decrease the free energy, and the reaction becomes thermodynamically possible. The barrier is the activation energy that a molecule must acquire. Anything that increases the chances of the molecule passing over this barrier increases its rate of reaction (Johnson *et al.*, 1974). These concepts are summarized in Figure 10.2. Activation energy can be calculated from experimental data relating reaction rate to temperature (see Appendix 10.1).

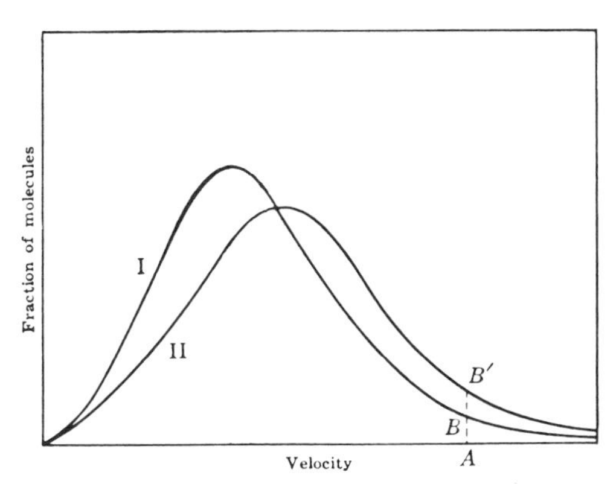

Figure 10.1. *Increase in kinetic energy of molecules as measured by velocity when the temperature is raised by 10°C from that in I to that in II. The kinetic energy required for activation, as given at ABB′, is approximately doubled. (After Glasstone, 1946: The Elements of Physical Chemistry, D. van Nostrand Co., New York.)*

You may recall that in an introductory chemistry course you usually heated a test tube with the Bunsen burner to induce a reaction between the reagents. You were supplying the activation energy to enable the molecules to cross this energy barrier.

LOWERING THE ACTIVATION ENERGY BY ENZYMES

Cells cannot supply a large amount of heat to activate molecules as we do in a laboratory to facilitate many chemical reactions. In the cells of most organisms reactions occur at essentially constant temperature, since the temperature varies over only a few degrees, even in fever or after intense muscular exercise. Organisms have "solved" the problem of supplying the activation energy through the evolution of enzyme catalysts to lower the energy of activation. Even at the body temperature of cold- and warm-blooded organisms, adequate numbers of molecules are activated. Chemists similarly make wide use of catalysts in industrial processes for the same reason; namely, less en-

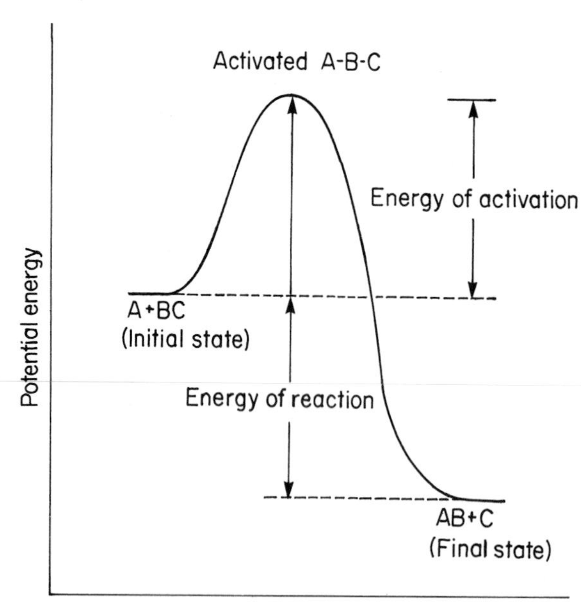

Figure 10.2. Potential energy diagram for a thermochemical reaction. Before reacting the molecules A and BC must accumulate sufficient energy of activation to pass over the energy barrier represented by the rising curve, even though the total reaction to AB plus C occurs with a loss in free energy (approximately indicated by the energy of reaction).

ergy is required to promote the catalyzed chemical reactions.

An *enzyme* is a *biocatalyst.* It is made up of protein, and like other catalysts* it accelerates the reaction rate without itself being used up (except incidentally in side reactions). It is safe to assume that most cellular reactions are catalyzed by enzymes. Only a few cellular reactions, some examples of which will be cited later, happen rapidly enough without them. The main function of a catalyst is to lower the energy of activation required of a molecule before it can react. In other words, the catalyst lowers the energy barrier over which a molecule must pass to react (Fig. 10.3). It thus increases the reaction rate. For example, the activation energy for the decomposition of hydrogen peroxide to water and oxygen is 18,000 calories per mole. In the presence of the enzyme catalase, which occurs in all aerobic cells, the activation energy is only 5500 calories per mole. The advantage of enzymes to the cell, which must operate at relatively low temperature and atmospheric pressure, is self-evident. Some enzymes increase reaction velocity by a factor of 10^8 to 10^{20} (Lehninger, 1975).

Another important aspect of enzyme action is

*The term catalyst is used by the physical chemist to denote a substance that alters the rate of a reaction. Consequently, negative catalysis (the slowing down of a reaction) is possible. The word is used here in the popular sense, a substance that accelerates a reaction (Bender and Brubaker, 1973).

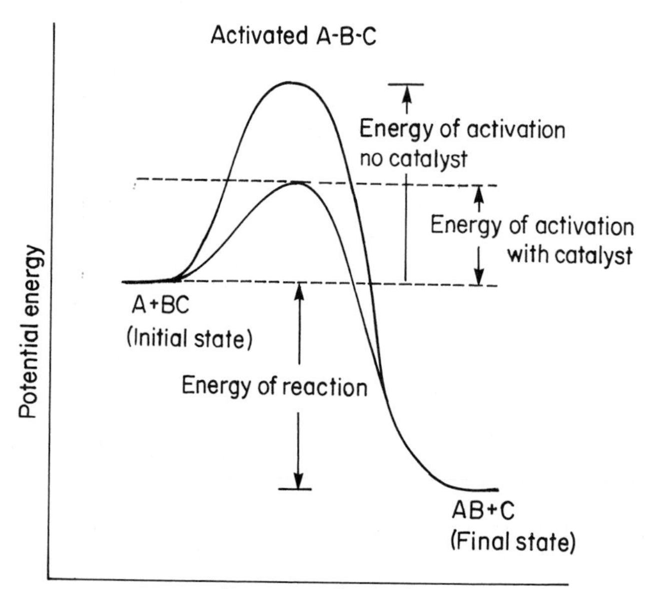

Figure 10.3. Energy diagram for a chemical reaction uncatalyzed and catalyzed. The dashed line indicates the much reduced energy barrier in the presence of catalyst.

the channeling of cellular reactions along certain pathways. Many reactions may be thermodynamically possible (spontaneous) in a cell but they do not occur at a finite rate. Selected reactions may be promoted or detained by supplying or not supplying the appropriate enzymes. In a sense, by supplying the required enzymes, the cell achieves selective control over a large series of possible reactions (Wilkie, 1960).

It is not known whether it is possible for the cell to make enzymes to catalyze any chosen reaction, or whether enzymes can be made only for certain types of reactions. If enzymes can be made only for certain reactions, then the metabolic chain of reactions not only must satisfy the energetic requirements but also it must be channeled by means of the enzymes that the cell is capable of making.

Although an enzyme cannot increase the amount of energy obtainable from a given chemical reaction, it can enormously increase the rate at which energy is obtained and direct it into useful channels. The effect of the enzyme is thus to raise what might be called chemical conductance, thereby permitting a given rate of energy production with a lower loss through dissipation of free energy.

CLASSIFICATION OF ENZYMES

Enzymes are classified into categories depending on the type of reaction they catalyze

TABLE 10.2. CLASSIFICATION OF ENZYMES

*1.** ***Oxidoreductases***
Oxidation reductions on:

1.1 >CH—OH
1.2 >C=O
1.3 >CH=CH—
1.4 >CH—NH_2
1.5 >CH—NH—
1.6 NADH; NADPH

2. ***Transferases***
Transfer of:
2.1 One-carbon groups
2.2 Aldehydic or ketonic groups
2.3 Acyl †groups
2.4 Glycosyl groups‡
2.7 Phosphate groups
2.8 S-containing groups

3. ***Hydrolases***
Hydrolysis of:
3.1 Esters
3.2 Glycosidic bonds
3.4 Peptide bonds
3.5 Other C—N bonds
3.6 Acid anhydrides (R—C(=O)—O—C(=O)—R)

4. ***Lyases***‖
Formation of or addition to the following double bonds:

4.1 >C=C<
4.2 >C=O
4.3 >C=N—

5. ***Isomerases***
5.1 Racemases§

6. ***Ligases (Synthetases)***
Joining of two molecules, with ATP cleavage by bonding:

6.1 C to O: C—O
6.2 C to S: C—S
6.3 C to N: C—N
6.4 C to C: C—C

*These code numbers for enzymes were agreed upon by the Commission on Enzymes of the International Union of Biochemistry.
†Organic radical derived from an organic acid by removal of an OH.
‡Sugar radical in a glycoside.
‖Formation of double bonds by removal of radicals or addition of radicals to double bonds.
§Transformation of one isomer to another, as glucose to fructose.
From Lehninger, 1975: Biochemistry, 2nd Ed. Worth Publishers, New York, p. 184.

(Table 10.2). To illustrate properties of enzymes we shall mainly use hydrolases and oxidoreductases but examples of all other types of enzymes appear in other chapters. While enzymes catalyze diverse chemical reactions, they also have many properties in common: they are all proteins and they require cofactors, which may be metals or ions or even large organic molecules. Enzymes usually have a high degree of specificity, they are inhibited by a variety of poisons often specific to particular enzymes, they are sensitive to the environment in which they act, and they show similar reaction kinetics. These properties are illustrated in one or another of the examples discussed.

HYDROLASES

In hydrolysis by the addition of water, a compound is split into fragments. The manner in which proteins, fats, and polysaccharides are hydrolyzed is of special importance because these compounds are necessary for maintenance of life, yet they are too large to penetrate through cell membranes. Several hydrolyses are illustrated in Figure 10.4.

Hydrolyses are spontaneous reactions accompanied by liberation of a small amount of energy; that is, they occur with a net decrease in free energy. However, in the absence of a catalyst, they often proceed too slowly to be measured. For instance, it is an everyday experience that starch does not hydrolyze to sugar when suspended in water (provided that growth of microorganisms is prevented). This means that the number of molecules having the requisite energy of activation is too small to give a perceptible change. However, addition of a small amount of acid catalyzes hydrolysis. If heat is also applied, the reaction is accelerated because more molecules acquire the necessary activation energy. In cells, hydrolyses are catalyzed by enzymes called *hydrolases*. Hydrolysis is also spoken of as digestion. Hydrolases are present in all cells (in eukaryotic cells in lysosomes) and make possible both intercel-

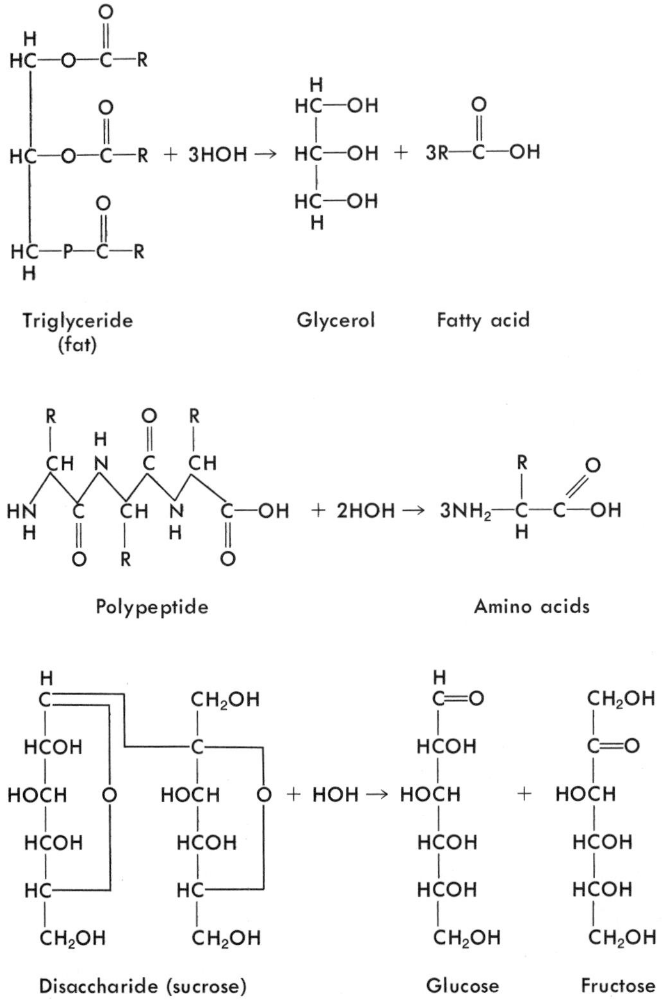

Triglyceride (fat) + 3HOH → Glycerol + 3R—C(=O)—OH (Fatty acid)

Polypeptide + 2HOH → $3NH_2$—CHR—C(=O)—OH (Amino acids)

Disaccharide (sucrose) + HOH → Glucose + Fructose

Figure 10.4. Diagrams of the digestion of food by hydrolysis.

lular and intracellular movements of proteins, lipids, and carbohydrates.

In unicellular organisms, digestion is often intracellular. For example, *Amoeba* and *Paramecium* form vacuoles enclosing food that is hydrolyzed therein after union of the vacuoles with lysosomes containing hydrolases. Cell foods resulting from hydrolysis pass through the vacuolar membrane into the cell, and the undigestible residues are voided from the cell. Bacteria, however, secrete enzymes onto the substrate, which is then liquefied by hydrolysis, and cell foods in the hydrolysate are absorbed by the bacteria.

In multicellular animals with differentiated tissues, digestion of food occurs in steps, each localized in a compartment of the digestive system. For example, protein digestion in vertebrates begins in the stomach (acid phase) and then continues in the small intestine (alkaline phase), assisted by the action of pancreatic enzymes and intestinal wall juices, respectively.

Digested cell foods in multicellular animals pass into the blood and are distributed to all body cells. The excess is stored in insoluble form, e.g., glycogen (in liver and muscle), fat (in adipose tissue), and protein (in muscle). By hydrolase activity, foods in storage depots can be mobilized when the need arises. They then move into the blood for relocation to places where required.

In multicellular green plants, digestion is usually intracellular, and digested cell foods are moved into the sap and carried to each cell. Here, too, excess of manufactured food is stored, e.g., sugar (in fruit), starch (in potatoes), fat (in nuts), and protein (in beans). Digestion is concerned mainly with hydrolysis of foods into a form suitable for transport from storage depots to growing parts. An exception is found in carnivorous plants, in which digestion, resembling the process in animals, may occur at the tips of glandular hairs, as in the sundew *Drosera*, or in a cavity, as in a modified leaf of the pitcher plant *Darlingtonia*.

SPECIFICITY OF HYDROLASES

Proteins are polymers of amino acids; glycogen and starch are polymers of monosaccharides; fats are conjugates of glycerol and fatty acids. Large polymeric molecules are usually hydrolyzed in several steps, each step being catalyzed by a specific enzyme. For example, internally acting *endopeptidases* split a protein molecule into smaller fragments, from the ends of which amino acids are hydrolyzed by *exopeptidases.* The exopeptidases perform their work more effectively after the action of endopeptidases, since more protein fragment ends become available.

Three familiar endopeptidases are *pepsin,* which is secreted by vertebrate stomach cells, and *chymotrypsin* and *trypsin,* both present in vertebrate pancreatic juice. Each one attacks a particular linkage between amino acids: pepsin attacks on the amino side of the peptide bond between tyrosine (or phenylalanine, tryptophan, and methionine) and other amino acids (Fig. 10.5). Chymotrypsin acts on the carboxyl end of the bond between tyrosine (or phenylalanine, tryptophan, and methionine) and other amino acids. Trypsin acts at the carboxyl end of lysine or arginine residues (Lehninger, 1975).

Exopeptidases (polypeptidases) are also specific in action. Aminopolypeptidases attack a polypeptide on the amino end group; carboxypolypeptidases attack a polypeptide on the carboxyl end (Fig. 10.5). The result of each reaction is a free amino acid and a polypeptide

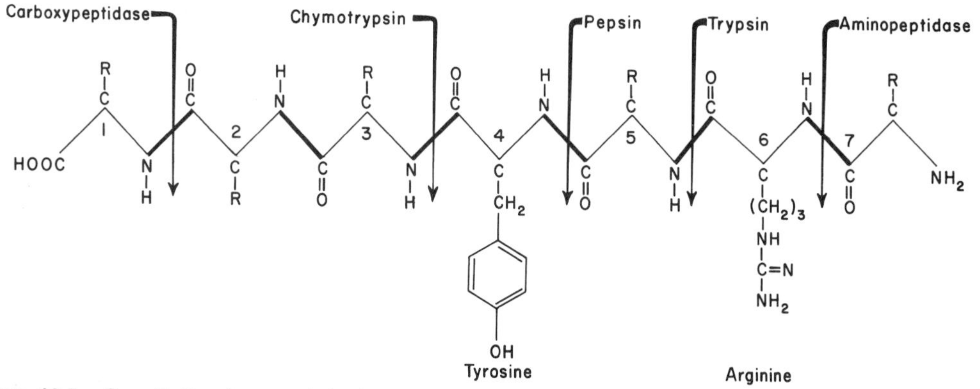

Figure 10.5. *Specificity of proteolytic (protein-splitting) enzymes. The numbers refer to the amino acid residues, only two of which, tyrosine and arginine, are labeled. The polypeptidases are specific to the free ends of a protein molecule: one (carboxypolypeptidase) to the carboxyl end (left) of a protein molecule or peptide, the other (aminopolypeptidase) to the amino end (right) of such molecules. Pepsin is specific to the amino side of tyrosine, phenylalanine, tryptophan, and methionine residues inside a protein molecule; chymotrypsin is specific to the carboxyl side of such residues; and trypsin is specific to the carboxyl side of arginine or lysine residues.*

molecule reduced in molecular weight. Dipeptidases split dipeptides into two amino acids. Dipeptidases found in the small intestine are specific to particular dipeptides, such as glycylalanine, for example.

The *carbohydrases* are also specific: α-amylases attack bonds between glucose residues in starch randomly; β-amylases attack only the second link from the nonreducing end of the chain and liberate maltose. Digestion of various hexosans (polymerized hexoses) (see Fig. 1.15) yields monosaccharides, primarily glucose. Plant cells also produce and store pentosans (polymerized pentoses) that their enzymes split into pentoses.

Certain animal and plant cells have special carbohydrases enabling them to split unusual hexosans or pentosans. For example, a cellulase in the flagellates that inhabit the termite gut enables them to digest wood cellulose to glucose, and a cellulase in cecum bacteria in cattle and other herbivores digests cellulose in grass and hay (Hungate, 1966). Inulase secreted by tuber cells in the Jerusalem artichoke digests stores of the fructose polymer inulin to fructose when need arises.

Lipases hydrolyze fats by attacking the ester linkages between alcohols such as glycerol and fatty acids in nonspecific action. Provided the acid is organic, its chemical identity and that of the alcohol radical with which it is combined do not appear to matter. Thus one lipase usually acts catalytically upon them all, presumably because of a similar steric configuration of the bonds. However, lipases from different organisms are not identical, since their pH optima vary. For example, the pancreatic lipase of humans acts best at pH 7.0, whereas that of the castor bean has an optimum at pH 5.0. But in either case the products of complete digestion of fats are fatty acids and glycerol. The products of digestion of phospholipids, on the other hand, include, in addition to fatty acids and glycerol, phosphoric acid, and an organic base. For example, hydrolysis of lecithin (see Fig. 1.13*C*) yields fatty acids, glycerol, phosphoric acid, and choline.

Since an enzyme only accelerates a reaction, it cannot make a thermodynamically impossible reaction proceed. The enzyme accelerates a reversible reaction in both directions; at a given time its direction depends upon the law of mass action. In other words, the reaction's equilibrium constant is not changed by an enzyme. For example, the hydrolysis of fat by lipase proceeds to equilibrium, at which time fat, fatty acid, and glycerol are mixed. If more fatty acids and glycerol are added to this mixture, fat will be synthesized until a new equilibrium is reached (Fig. 10.6). Hydrolyses of carbohydrates and proteins, however, are not similarly reversible; disaccharides and polysaccharides can be synthesized from monosaccharides possessing a high-energy phosphate bond, as a source of energy as well as a specific synthetase rather than a hydrolase (Lehninger, 1975). Polymerization thus occurs by a pathway entirely different from hydrolysis. Similarly, synthesis of proteins requires phosphorylated amino acids and enzymes other than hydrolases (see Chapter 15).

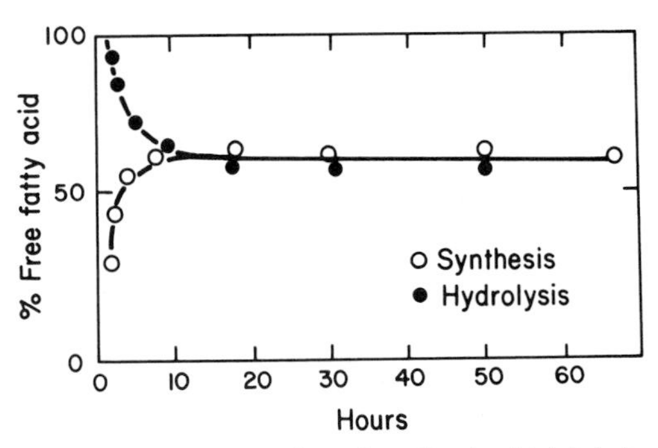

Figure 10.6. *Hydrolysis and synthesis of triolein by* Ricinus *(castor bean plant) lipase. (After Parsons, and data of Armstrong and Gosney, as cited in Baldwin, 1967: Dynamic Aspects of Biochemistry, 5th Ed. Cambridge University Press, Cambridge.)*

CHEMICAL NATURE OF ENZYMES

Sumner, in 1926, first succeeded in crystallizing urease from jack bean seeds. Urease hydrolyzes urea to carbon dioxide and ammonia. Over 2000 enzymes of all classes are now identified and at least 200 have been crystallized (Lehninger, 1975).

Several methods have been used in purification of enzymes. For example, hydrolytic enzymes were originally purified by fractional salting out of proteins from digestive juices or from a mash of digestive tissue and subsequent testing of the activities of both the supernatant and the precipitated material. Since some proteins in a mixture are precipitated by a low concentration of a salt such as ammonium sulfate and others by a higher concentration, separation of the proteins in a mixture is possible only by successive fractional salting out operations. At present a variety of chromatographic procedures are being used for enzyme purification (see Appendix 5.4). Ionic resins that selectively remove molecules of particular sizes permit even quicker separation of enzymes from other materials than classic chromatography.

Even crystalline enzymes often contain impurities, but they may be further freed of minute quantities of contaminating protein through purification by Tiselius electrophoresis. This method takes advantage of the fact that a protein molecule is positively charged on the acid side of its isoelectric pH and negatively charged on the alkaline side; therefore, it migrates in an electric field except at its isoelectric pH. Three factors affect the electrophoretic mobility of a protein: its size, symmetry, and charge density. Movement is hindered by large size, by asymmetry, and by decrease in charge density. As a result, proteins in a mixture moving in response to an electric field are "sorted out" by these three variables, and the concentration of each one in a gradient may be measured optically (e.g., with ultraviolet light or, more commonly, by the change in refractive index). Each protein may be removed separately from the electrophoresis tube or from the appropriate region of the filter paper when paper electrophoresis is used. Each fraction can be tested for enzyme activity. Even electrophoretic separation is not final proof of purity, since two kinds of protein may form an aggregate and move together.

Proof of enzyme purity can be obtained, however, by the method called *end-group analysis*. This is based on the knowledge that any single peptide chain has only one terminal amino and one terminal carboxyl group. The terminal amino group can be labeled, for example, with the dye 2,4-dinitrofluorobenzene, and comparable methods are available for identifying the carboxyl group. If the protein consists of a single polypeptide strand, only one labeled amino acid should appear per molecule when the protein is hydrolyzed and chromatographed or separated on an ion exchange column. If the protein consists of two (or more) chains, two (or more) terminal amino acids should appear per molecule, but always in the same stoichiometric ratios in each sample tested. From a mixture of proteins an exact stoichiometric ratio is unlikely (Lehninger, 1975).

All hydrolytic enzymes purified by these procedures show properties characteristic of proteins: color reactions, ultraviolet absorption spectra, inactivation by heat and radiations, sensitivity to pH and salts, and so on. Therefore, the specific properties of an enzyme probably depend upon the chemical nature of the protein of which it is composed. However, in some enzymes proteins are complexed with other ions or molecules.

ACTIVATION OF HYDROLASES

Many hydrolases appear to associate with salts that are important to their catalytic activity. For example, when ptyalin (salivary amylase) is freed of sodium chloride by dialysis, neither protein nor sodium chloride alone can effect the hydrolysis of starch. However, a mixture of the two is catalytically active. It has been shown that the chloride ion in sodium chloride is essential for enzyme activation. Many of the elements needed for enzymatic reactions are present in minute quantities in plant and animal cells and may be associated in a similar manner with such enzymes (Table 10.3). However, not all hydrolases are activated by salts. Some, like pepsin, trypsin, and chymotrypsin, are not known to require salt ions for their catalytic action (Gutfreund, 1965).

Hydrolases often leave cells in which they are manufactured as enzyme precursors in an inactive state. For example, *pepsinogen*, the inactive forerunner of the pepsin secreted from stomach lining cells (chief cells) into the lumen, is activated by hydrochloric acid (H^+ ions) coming from other secretory cells (parietal cells) nearby and also by the autocatalytic action of pepsin itself. Activation of pepsinogen involves removal of a polypeptide, since the molecular weight of pepsinogen differs from that of pepsin by the equivalent of the molecular weight (4000) of a small polypeptide, as shown in Figure 10.7 (Lehninger, 1975). The suggestion has been made that the small polypeptide masks the active spots of the enzyme before removal.

Enzymes that require free sulfhydryl groups (—SH) for catalytic activity must be reduced after purification to restore their catalytic activity since during purification the sulfhydryl groups are oxidized. For example, the protease *papain*, as extracted from the papaya, is inactive until it is treated with glutathione, hydrogen cyanide, or hydrogen sulfide, which reduce the enzyme's disulfide bonds (S—S) to sulfhydryl radicals (—SH). The enzyme is active only

TABLE 10.3. SPECIFIC METAL ACTIVATION OF SOME HYDROLASES*

Enzyme	*Hydrolyses Catalyzed*	*Metal Required*
Prolidase	Glycylproline + HOH → proline + glycine	Mn
Carboxypeptidase	Carbobenzoxyglycyl-l-leucine + HOH → carbobenzoxyglycine + leucine	Mg
Glycylglycine dipeptidase	Glycylglycine + HOH → 2 glycine	Zn

*These hydrolyses will proceed only if the metals are present in small quantities, indicating that the metals are catalytic in action. The peculiar substrate of the second reaction is used because analytic procedures are easier with such synthetic compounds. (Data from McElroy and Nason, 1954: Ann. Rev. Plant Physiol. *5*: 1–30, and from Smith and Hanson, 1949: J. Biol. Chem. *179*.)

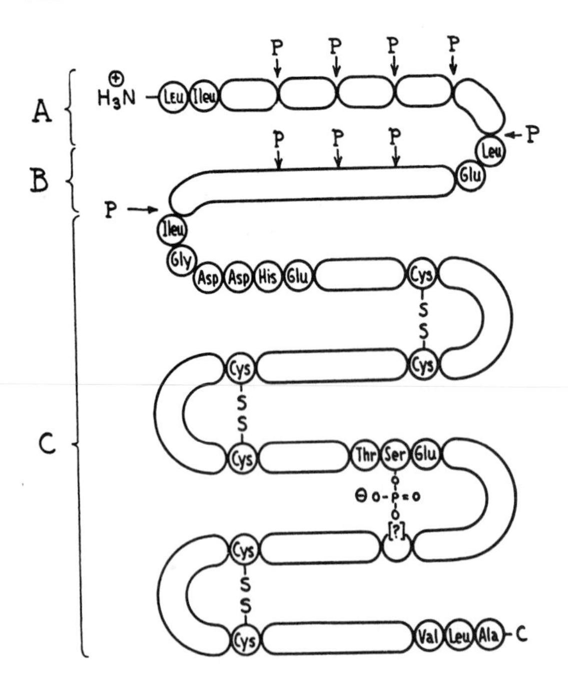

Figure 10.7. The structure of pepsinogen. The principal attack by hydrogen ions (of HCl) and pepsin is at points marked P, releasing miscellaneous peptides (A), pepsin inhibitor (B), and pepsin (C). The entire sequence of amino acids has now been determined, but the details are not pertinent to the issue here. (From Boyer et al., 1960: The Enzymes. Vol. 4. Academic Press, New York.)

when one of these radicals is present in the reduced state. Cathepsin, a protein-digesting enzyme in the liver, is also active only in its reduced state.

OTHER ENZYMES INVOLVED IN CELLULAR METABOLISM

In addition to the hydrolases, a number of other kinds of enzymes (isomerases, oxidoreductases, transferases, lysases, and ligases) take part in the degradation and synthesis of nutrients in the cell (Table 10.2). Although some of these enzymes are known only in impure state in cell fragments, others have been crystallized and purified (Lehninger, 1975; Boyer *et al.*, 1970).

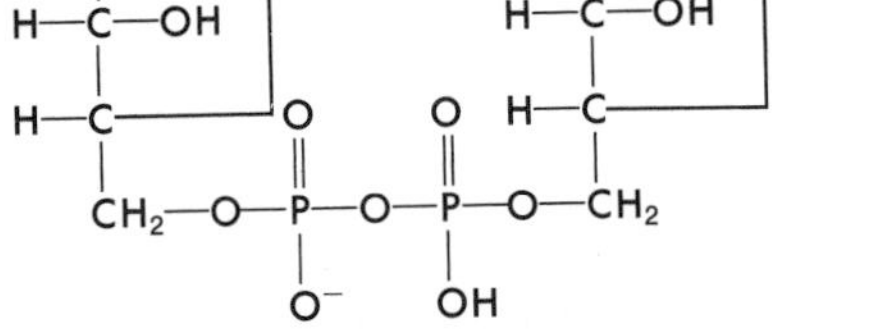

Figure 10.8. Nicotinamide adenine dinucleotide (NAD^+), the coenzyme or prosthetic group of some of the dehydrogenases. Nicotinamide adenine dinucleotide phosphate (NADP) is similar in structure except that it has an extra phosphoric acid residue attached to the pentose of adenosine (upper right side of the molecule). The reduced form of NAD^+ may be designated NADH. The same conventions are to be used with $NADP^+$.

Transferases catalyze transference of various radicals, such as the high-energy phosphates, methyl groups, acetyl groups, and amino groups. For instance, the enzyme hexokinase, a phosphorylase, catalyzes the phosphorylation of glucose, in the course of which a high-energy phosphate bond is transferred from an ATP to a glucose molecule, forming glucose-6-phosphate.

Lyases are enzymes that remove a group of atoms from a substrate without hydrolysis and with the formation of a double bond. Examples are decarboxylases, which remove carbon dioxide, and deaminases, which remove the amino group from an amino acid with the formation of a corresponding keto acid, for example, glutamic acid yields α-ketoglutaric acid on deamination.

Isomerases catalyze internal change in molecules such as the shift from glucose-1-phosphate to fructose-6-phosphate in both anaerobic and aerobic phases of metabolism.

Ligases are enzymes that bond two substrates using adenosine triphosphate (ATP) as the source of energy. Examples of ligases are amino acid-RNA ligase and acyl-CoA synthetase. The various types of covalent bonds listed in Table 10.2 are formed by ligase activity.

Oxidoreductases are concerned with electron and hydrogen ion transfers. Examples are cytochrome oxidase, which transfers electrons, and coenzyme I (nicotinamide adenine dinucleotide or NAD) (see Fig. 10.8), which transfers hydrogen ions and electrons. Oxidoreductases are

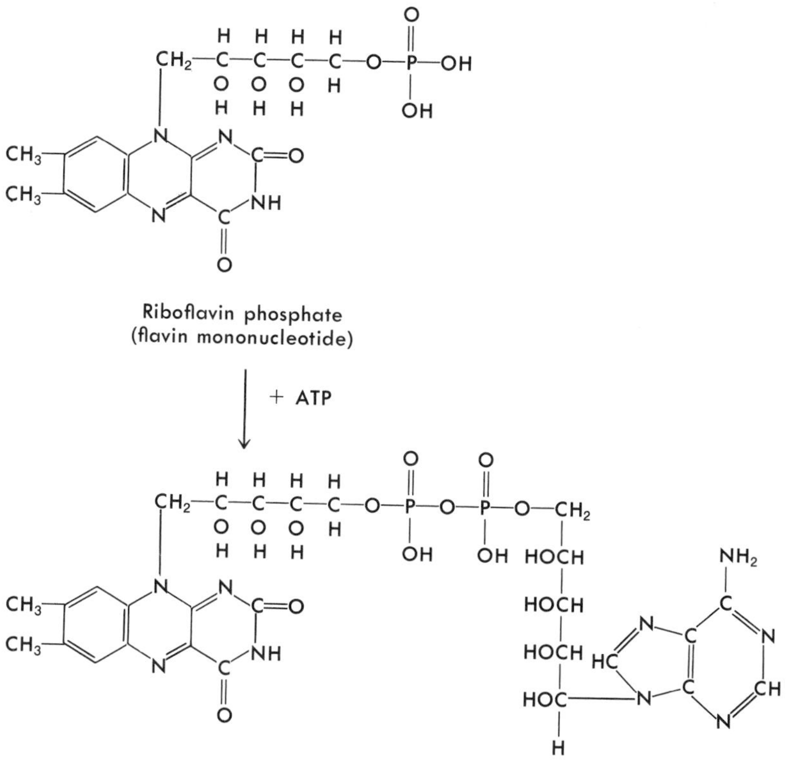

Riboflavin phosphate
(flavin mononucleotide)

+ ATP

Flavin adenine dinucleotide (FAD)

Figure 10.9. *Riboflavin, a part of the prosthetic group of "yellow" enzymes. Both yellow enzymes, namely flavin adenine mononucleotide (FMN) and flavin adenine dinucleotide (FAD), participate primarily in transport of hydrogen ions and electrons from prosthetic groups of one of the dehydrogenases (e.g., nicotinamide adenine dinucleotide) to the cytochromes. Each prosthetic group has specific functions, e.g., FMN is the prosthetic group of lactic dehydrogenase and FAD is the prosthetic group of aldehyde oxidase, D-amino oxidase, and fatty acyl-CoA dehydrogenases.*

specific to various types of bonds listed in Table 10.2.

All oxidoreductases consist of a protein carrier and a prosthetic group and have some properties in common. The *prosthetic group* is an organic radical, sometimes of considerable complexity, that is attached to the protein. Neither protein nor prosthetic group alone is enzymatically active. For example, when yellow flavoprotein enzyme (flavin adenine dinucleotide + protein) is dialyzed against distilled water, the yellow prosthetic group containing riboflavin (vitamin B_2) diffuses out, leaving the protein in the cellophane bag (see Fig. 10.9). Neither the protein inside nor the prosthetic group outside the cellophane bag shows enzymatic activity. When mixed, the two form the active enzyme. In the cell the two are probably present mostly in the associated form, since the dissociation constant for this enzyme is small.

In some oxidoreductases, such as the *dehydrogenases,* the prosthetic group is usually dissociated from the protein and associates with it only when acting on the substrate. When the dissociation constant is large, the prosthetic group is called a *coenzyme.* Two coenzymes (Fig. 10.8) are known for dehydrogenases—coenzyme I, or nicotinamide adenine dinucleotide (NAD^+),* and coenzyme II, or nicotinamide adenine dinucleotide phosphate ($NADP^+$).† $NADP^+$ is most frequently the coenzyme in synthetic reactions and as NADPH + H^+, represents a "reduction reservoir", that is, a source of H for reduction. It also functions in the pentose oxidative shunt (see Fig. 12.12), whereas NAD^+ is the coenzyme for most degradative dehydrogenations. All dehydrogenases formed by union of these prosthetic groups (coenzymes) with various proteins are highly

*Coenzyme I was previously designated diphosphopyridine nucleotide (DPN).

†Coenzyme II was previously designated triphosphopyridine nucleotide (TPN).

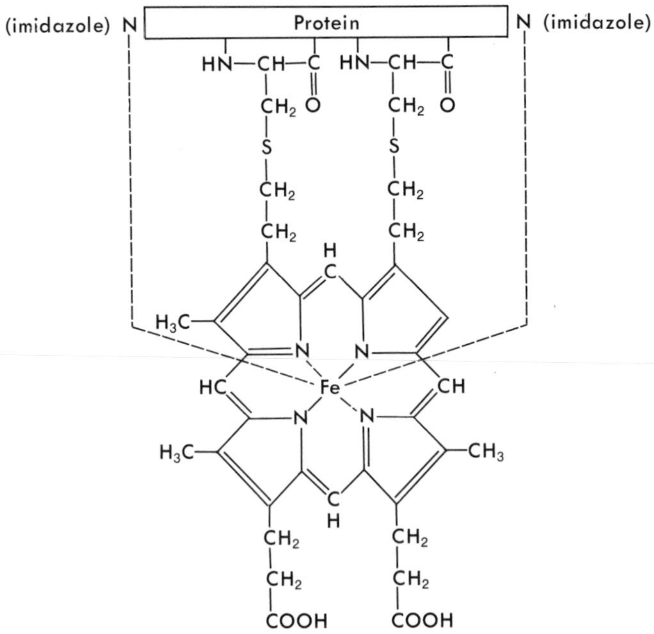

Figure 10.10. Reduced cytochrome c.

specific as to substrate. Dissociation of enzyme into protein carrier and prosthetic group appears advantageous, since the same prosthetic group can associate with various proteins, each combination being specific to some substrate. The specificity of a dehydrogenase for its substrate, therefore, must be due largely to the protein moiety of the enzyme rather than the coenzyme. Dehydrogenases are important in aerobic metabolism, as will become apparent in Chapter 12.

Cytochromes are enzymes that transfer electrons from flavoproteins or other carrier enzymes to cytochrome oxidase. Cytochrome oxidase in turn passes the electrons to oxygen, activating it and enabling it to combine with the hydrogen ions to form water. Therefore, cytochromes are vital to aerobic metabolism of cells. Cytochromes (Fig. 10.10), a number of which exist in all aerobic cells, are cell pigments with a prosthetic group resembling heme of hemoglobin. Cytochrome oxidase also has a similar prosthetic group. In cytochromes and cytochrome oxidase the iron changes from the oxidized (Fe^{+++}) to the reduced (Fe^{++}) state as an electron is received, and from the reduced to the oxidized state as an electron is passed on to oxygen. In hemoglobin the iron is always in the reduced state and adds oxygen (becomes oxygenated) only loosely, without changing valence.

The cytochromes and cytochrome oxidase are tetrapyrrole compounds, four pyrroles surrounding the iron. Tetrapyrroles without the

TABLE 10.4. ELEMENTS NECESSARY FOR SOME OXIDOREDUCTASES*

Enzyme	*Reaction*	*Metal*
Carbonic anhydrase	$CO_2 + H_2O \rightleftarrows H_2CO_3$	Zn
Inorganic pyrophosphatase	Pyrophosphate + $H_2O \rightarrow PO_4$	Mg†
Catalase	$2H_2O_2 \rightarrow 2H_2O + O_2$	Fe
Cytochromes	Electron transport	Fe
Tyrosinase	Tyrosine + $\frac{1}{2}O_2 \rightarrow$ hallachrome	Cu
Ascorbic acid oxidase	Ascorbic acid $\rightarrow$ dehydroascorbic acid	Cu

*From McElroy and Swanson, 1953: Sci. Am. (Jan.) *188*.
†Magnesium is required for activity of many enzymes, for example, transferases.

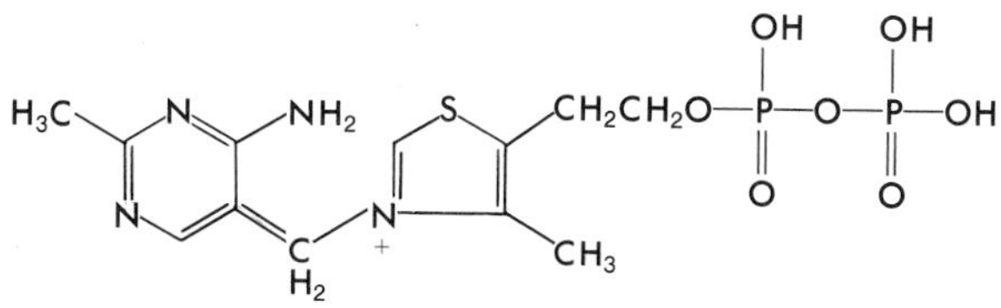

$$H_3C$$ … NH₂ … S … CH_2CH_2O—P(OH)(=O)—O—P(OH)(=O)—OH … N … CH_3 … $C H_2$

Figure 10.11. *Thiamine pyrophosphate, or cocarboxylase.*

metal are called *porphyrins*. When a metal is present they are called metalloporphyrins. Cytochrome and heme are examples of metalloporphyrins in which iron is present. It is interesting to note that the plant pigment chlorophyll (see Fig. 2.2) is a metalloporphyrin in which magnesium is present. Porphyrins, which can be produced by electrical discharges in gases simulating the primitive earth's atmosphere, were present when life began (see Chapter 1).

Many of the enzymes catalyzing oxidoreductions are quite specific in action. Thus, succinic acid dehydrogenase catalyzes dehydrogenation of succinic acid only, and malic acid dehydrogenase catalyzes dehydrogenation of malic acid only. However, enzymes exist that catalyze a number of reactions, being specific not to an individual molecule but to a particular steric configuration of atoms that might occur in a number of different molecules. For instance, isocitric dehydrogenase catalyzes dehydrogenation of both isocitric and oxalosuccinic acids in the Krebs cycle (see Chapter 12). Some enzymes participating in oxidoreductions are activated by elements, several of which are listed in Table 10.4.

Members of the vitamin B complex serve as components of prosthetic groups of oxidoreductases and synthetases. This vital role accounts for the presence of certain B vitamins in cells of all organisms examined. One example is vita-

TABLE 10.5. VITAMIN FUNCTION*

Class	*Chemical Nature*	*Cellular Function*	*Effect of Lack on Vertebrates*
A	Carotenoid	Visual purple; growth	Deficient growth; night blindness
D	Sterol	Mobilizes bone salts	Rickets, osteomalacia
E	Tocopherol	Unknown	In rat, defects of reproduction
B_1	Thiamine	Coenzyme for pyruvate metabolism	Beriberi
B_2	Riboflavin	Flavoprotein coenzyme	Cataract
Niacin	Nicotinic acid amide	Dehydrogenase coenzyme	Pellagra
Pantothen	Pantothenic acid	Part of coenzyme A	Dermatitis
B_6	Pyridoxine	Coenzyme for amino acid conversions	Dermatitis, nervous disorders
H	Biotin	Coenzyme in CO_2 fixation in C_4 acids	Dermatitis
Folic acid	Pteroylglutamic acid	Coenzyme for "one" carbon metabolism	Anemia
B_{12}	Cyanocobalamin	Coenzyme for methyl transfer; nucleic acid and lipid metabolism	Anemia
C	Ascorbic acid	Maintain optimal redox potential?	Scurvy
K	1,4-naphthoquinone acetate	Prothrombin	Hemorrhage

*Data from Kutsky, R., 1973: Handbook of Vitamins and Hormones. Van Nostrand Reinhold Co., New York. Reprinted by permission of Van Nostrand Reinhold Co.

The amino acids tryptophan, phenylalanine, lysine, histidine, leucine, isoleucine, threonine, methionine, valine and arginine are required by the rat. The requirement is not always the same for all types of animals. Most vertebrates also require essential fatty acids (linoleic, linolenic and arachidonic), methyl compounds (choline, methionine), and sulfhydryl compounds (cysteine and glutathione) for growth.

min B_1, or thiamine, which forms part of the prosthetic group of cocarboxylase (Fig. 10.11), an enzyme important in the metabolism of pyruvic acid. Some of the pertinent information on the metabolic role of these and other vitamins is summarized in Table 10.5, along with a summary of superficial symptoms that result from lack of dietary vitamins. Pathological symptoms here are the cumulative effects of prolonged metabolic disturbances.

EFFECT OF ENVIRONMENT ON ENZYME ACTIVITY

Provided the substrate is present in excess, the rate of an enzyme-catalyzed reaction increases proportionally with increasing concentration of the enzyme. For a given concentration of enzyme the relation between the rate of the reaction and substrate concentration has been formulated in most general terms as:

$$\text{Rate} = k[S]^n \qquad (10.1)$$

In this equation k and n are constants and [S] is the substrate concentration. In some cases increasing substrate concentration (in the lower part of the concentration range) proportionally increases the reaction rate. In that case, exponent n in equation 10.1 has the value of unity (first order). When a high concentration of substrate proves inhibitory, the exponent n is said to have a negative value. In some other cases, much more complex relations than these exist between concentration of substrate and rate of the enzyme-catalyzed reaction.*

*When n = 2, the reaction is second order. This type of reaction and the more complex types are considered in biochemistry textbooks (e.g., Lehninger, 1975).

TABLE 10.6. DENATURATION OF TRYPSIN BY HEAT*

T°C	Percent Denatured	T°C	Percent Denatured
42	32.8	45	57.4
43	39.2	48	80.4
44	50.0	50	87.8

*Data from Northrop, 1948: Crystalline Enzymes. Columbia University Press, New York.

If substrate concentration is varied, but the enzyme concentration is kept constant, a limiting rate is reached that is the maximal turnover rate of the enzyme at the particular temperature and pH of the experiment. Examples of the effect of enzyme concentration and substrate concentration on reaction rate are given in Figures 10.12 and 10.13, respectively. These data suggest that formation of an enzyme-substrate complex is the first step in any enzyme-catalyzed reactions (Lehninger, 1975).

When the concentration of enzyme is so low as to be rate limiting and the concentration of substrate is varied, the reaction rate may prove independent of substrate concentration. In that case n in equation 10.1 is zero, and the substrate term drops out of the equation. Qualitatively speaking, the enzyme becomes saturated with substrate. This becomes the "bottleneck" determining the reaction rate, and further increase in substrate concentration has a negligible effect. A reaction is *zero order* if it is independent of the reactant concentrations.

Temperature also affects enzyme activity. Experiments with hydrolytic enzymes indicate that hydrolysis is maximal at the temperature optimal for the hydrolase. The optimal temperature is usually low for enzymes in cells of polar plants and polar cold-blooded animals, very

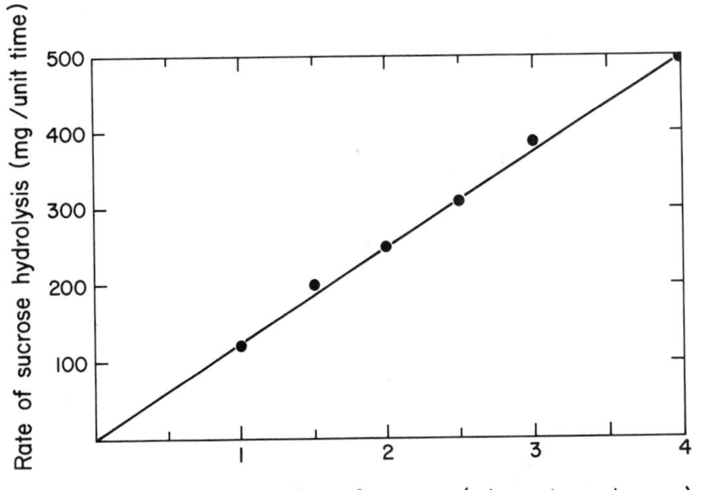

Figure 10.12. Effect of concentration of enzyme on reaction rate when an excess of substrate (sucrose in this instance) is present. (After Baldwin, 1963: Dynamic Aspects of Biochemistry. 4th Ed. Cambridge University Press, Cambridge.)

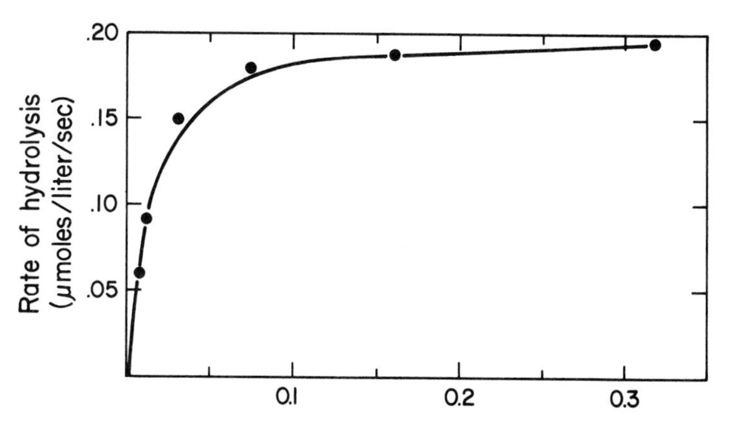

Figure 10.13. *Effect of substrate concentration on the reaction rate when a limited amount of enzyme is present. Note that the enzyme quickly becomes saturated with substrate so that further increase in substrate concentration has no effect. (After Queilet, 1952: Arch. Biochem. Biophys.* 39: *41.)*

high for enzymes from some thermophilic bacteria, and intermediate for enzymes from cells of temperate species. Seemingly, the optimal temperature for the catalytic activities of hydrolytic enzymes relates to the temperature natural for the particular cell.

A temperature increase to the optimum is accompanied by a corresponding increase in the rate of hydrolysis (Fig. 10.14) because of the increased number of molecules with the required activation energy.

Enzyme activity declines if the temperature is increased above the optimal level (Fig. 10.14). This has been explained in the following manner. Only the native or unaltered form of an enzyme is catalytically active; therefore, other things being equal, the rate of a hydrolytic reaction will depend upon the concentration of the native enzyme. As temperature rises, more of the enzyme becomes denatured, as shown in Table 10.6. As the quantity of native enzyme decreases, the reaction rate declines. Enzyme structure required for catalytic action is maintained primarily by hydrogen bonds and some sulfur-to-sulfur linkages. Many of these bonds are readily broken by heat leading to a more random arrangement of higher entropy (Lehninger, 1971).

Most enzymes are denatured irreversibly at a temperature between 50° and 80°C. However, some are resistant to high temperature; for example, crystalline trypsin is denatured reversibly even at 100°C and so is Taka-Diastase. The latter is denatured irreversibly only above 140°C. Each enzyme is catalytically inactive while at the limit of its high temperature tolerance but becomes active when cooled (Johnson *et al.,* 1974).

In cold-blooded animals a seasonal change in temperature induces production of isozymes (multiple forms of an enzyme differing in amino acid composition and properties) that differ

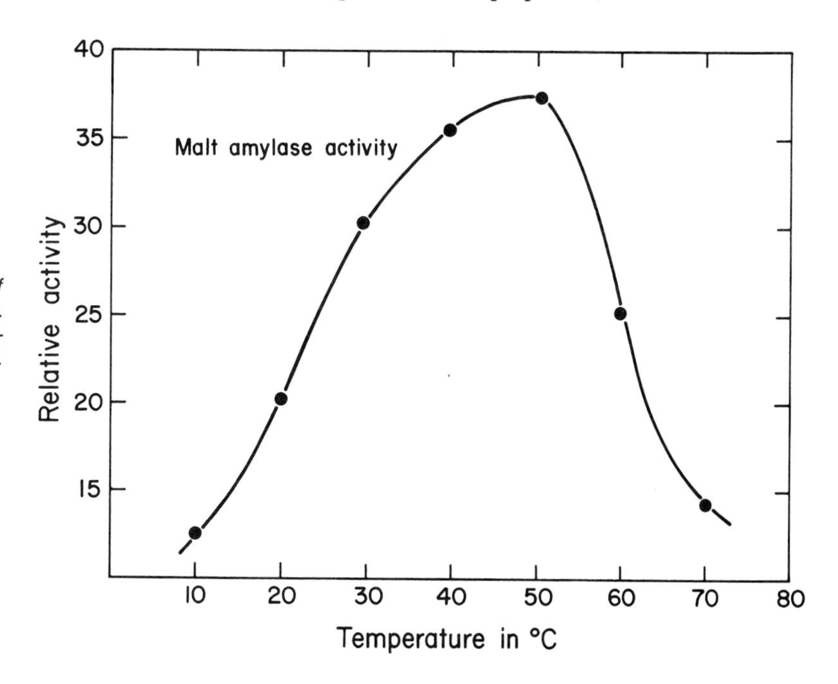

Figure 10.14. *The effect of temperature on enzyme activity. (Data from Gortner, 1949: Outlines of Biochemistry. 3rd Ed. John Wiley & Sons, New York.)*

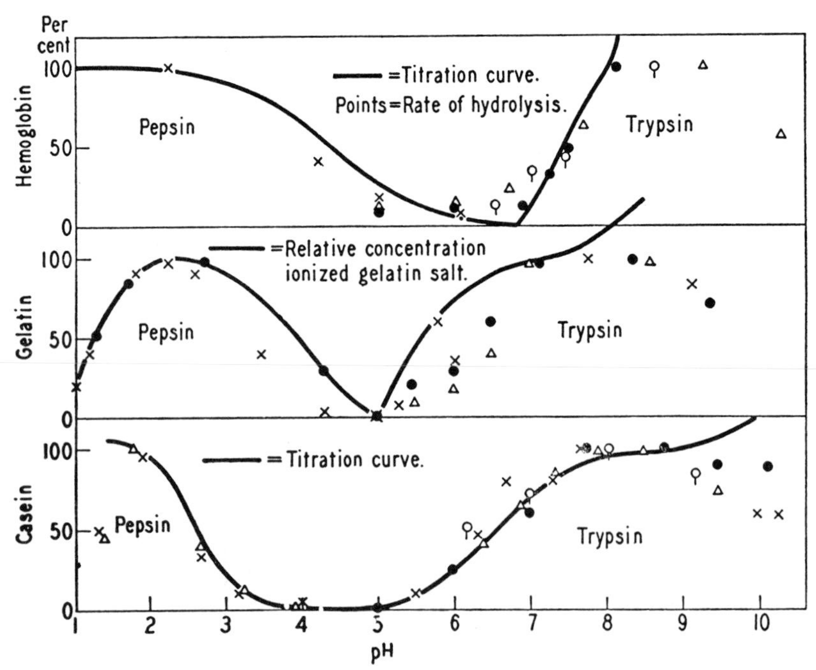

Figure 10.15. *The effect of pH upon the activity of pepsin and trypsin. The ordinate measures the relative rates of hydrolysis. Note that the pH effect varies with each of the three proteins. (From Northrup, 1939: Crystalline Enzymes, Columbia University Press, New York.)*

from the original enzymes in the strength of their substrate affinities. At low temperature the isozyme affinity for substrate is increased, while at higher temperature it is decreased, making possible the achievement of equal enzyme activity over the entire environmental range of the organism (Hochachka and Somero, 1973).

Enzymes are also inactivated by heavy metal salts that act as denaturants and protein precipitants. Alkaloidal reagents such as phosphotungstic acid inactivate enzymes because they precipitate proteins.

One of the most interesting influences on enzymatic activity is pH. Each enzyme is catalytically active within a limited pH range (Fig. 10.15). For example, the optimal range for the activity of pepsin is between pH 1.5 and 2.5, while that for trypsin is between 8 and 11. Activity declines gradually on either side of the optimum. The pH sensitivity of enzymes is thought to result from the necessity of a particular charge distribution around the active site in the enzyme for optimal activity. Since enzymes are composed of amino acids in different proportions, the optimal pH for an enzyme is apt to depend upon its amino acid composition.

Enzymes can also be induced where none were present to handle new substrates different from the one on which the organisms had been growing, an occurrence quite common in bacteria and also found in plants and animals. Thus *E. coli* when growing on glucose has no galactosidase that would enable it to hydrolyze the disaccharide lactose, but by substituting lactose for glucose in the culture medium (see Chapter 18) a galactosidase can be induced. In sum, the enzyme constitution of an organism is dynamic and adaptable to changes in environmental conditions, enabling the organism to survive changes in environment and subject only to limitations in its genome.

MECHANISM OF ENZYME ACTION

It is recognized that an enzyme forms an intimate union with the substrate upon which it acts, as shown schematically for pepsin in Figure 10.16. In this case apparently the electrical charges on the enzyme protein attract the oppositely charged radicals on the substrate. The precise nature of the union is not known and may vary with different enzymes (Mildvan, 1974).

The hydrolysis of a polypeptide by a peptidase, in which a metal ion is often associated with a protein (see Table 10.3) has been envisaged in the following manner. The metal ion attaches itself by residual valency to a polypeptide molecule on one side of the peptide bond, while the protein portion of the enzyme attaches itself on the other side of the same peptide bond. The stress this places on the peptide

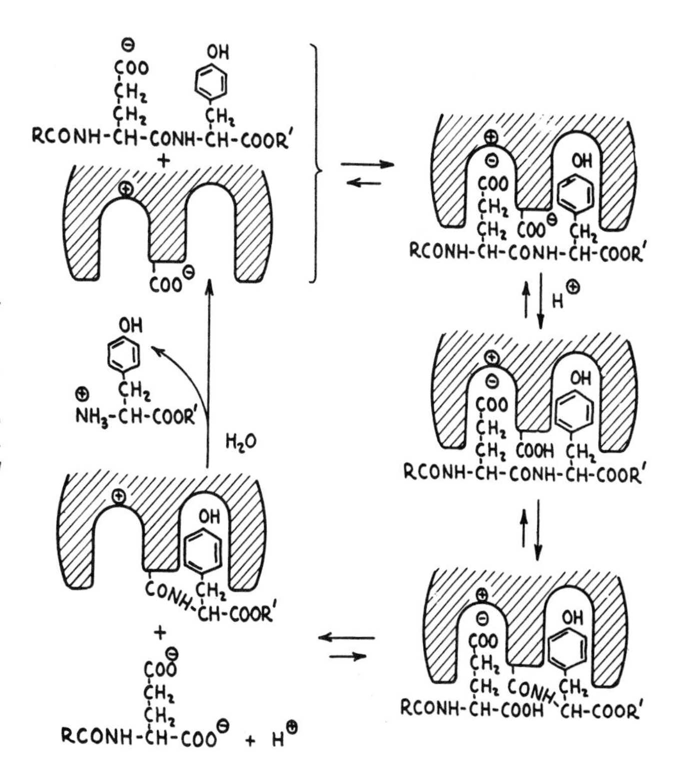

Figure 10.16. *A proposed hypothetical mechanism of action of pepsin; N-acyl-L-glutamyl-L-tyrosine ester represented as substrate. (From Boyer et al., 1960: The Enzymes, Vol. 4. Academic Press, New York.) The sequence of amino acids, their three-dimensional arrangement, and the nature of the active spots have now been determined for pepsin, trypsin, chymotrypsin, and elastase. (D. M. Chatton, unpublished.)*

bond weakens the bond between the amino and carboxyl groups of two adjacent amino acid residues of the polypeptide, thereby reducing the energy required for breaking the peptide bond. Hydrogen ions, always present, then cause its hydrolysis, even at room temperature. In absence of an enzyme, a high temperature is required for this hydrolysis.

It has been possible to resolve the three-dimensional detail of an enzyme crystal at the 0.2 nm level with x-ray diffraction methods, which permit analysis of the positions of amino acid residues by joining lines (contours) of equal electron density in x-ray diffraction pictures. Such a study has been made for the hydrolytic enzyme lysozyme, which digests the cell walls of bacteria, by computer analysis of some 10,000 contours.

Lysozyme appears to be a single polypeptide chain of 129 residues of 20 different kinds of amino acids (Fig. 10.17*A*). The sequence of amino acids in lysozyme has been identified by the methods already discussed for insulin in Chapter 1. From x-ray diffraction studies it is evident that amino acid residues that are numbered 5 through 15, 24 through 34, and 88 through 96 represent three lengths of alpha helix structure, although the helical sections are somewhat distorted. The amino acid residues with polar hydrophilic acid and basic groups are on the surface of the molecule; those with hydrophobic nonpolar groups (e.g., the methyl groups in leucine and isoleucine) are on the inside of the molecule. The crystal of lysozyme contains water of crystallization to the extent of about 35 percent by weight. Parts of the polypeptide chain, especially near the amino terminal end, are folded into stable conformations and appear to act as centers around which the remainder of the molecule is in turn folded. The end result is a lysozyme molecule which consists of a helix in a gap between two "wings" formed by other residues with their hydrophobic groups turned deep into the molecule. The gap is not filled, and a deep cleft running up on the side of the molecule appears to be the *active site* (place at which binding occurs) on the enzyme molecule (Phillips, 1966; Sigman and Mooser, 1975). The remaining amino acid residues are wound around the globular unit formed by the terminal amino acid end of the polypeptide.

By using a polymer of an amino sugar (e.g., N-acetyl muramic acid), which lysozyme binds but does not hydrolyze, and studying the x-ray diffraction pattern of the enzyme-sugar complex, it was found that the sugar binds in several places within the cleft of the lysozyme molecule (Fig. 10.17*B*). Activation presumably occurs by distortion resulting from binding. Although x-ray determinations do not permit study of enzyme in action, they supply three-dimensional data that give stronger evidence for the probable mechanism of enzyme action than had previously been available (Koshland and Neet, 1968). Such a cleft also appears in

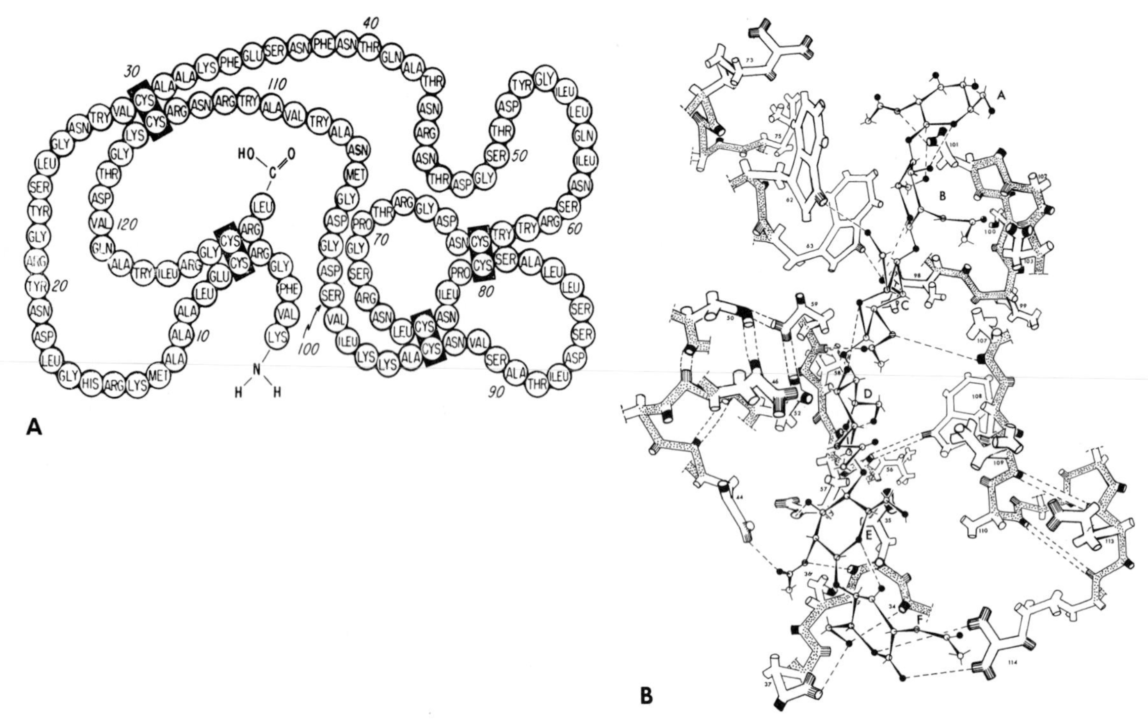

Figure 10.17. *Structure of lysozyme.* A, *Two dimensional structure showing the sequence of amino acids; the abbreviations are explained in Table 1.1. S-S represents a disulfide linkage; the numbers, the amino acid sequence from the initial amino acid lysine at (1) the amino end.* B *shows the three dimensional atomic arrangement of the lysozyme molecule in the neighborhood of the cleft, with a molecule of hexa-N-acetylglucosamine bound to the enzyme. Sugar residues A, B, C, D, E and F are shown bound. The linkage hydrolyzed is suggested to be between residues D and E. (From Phillips, 1967: Proc. Nat. Acad. Sci. U.S.A.* 57: *493.)*

other enzymes that have been under intensive study (e.g., chymotrypsin, trypsin, and elastase).

Not all portions of an enzyme molecule are necessary for its function. Thus, papain may lose 120 of its 180 amino acid residues without losing its enzyme activity. On the other hand, a small change in a critical part of an enzyme molecule may alter or destroy its activity. Removal of one histidine residue destroys the hydrolytic activity of chymotrypsin, and blocking the sulfhydryl groups of the protein of phosphoglyceraldehyde dehydrogenase destroys its dehydrogenase activity.

Experimental evidence indicates that oxidoreductases like hydrolases combine firmly with the substrate. For example, peroxidase, which absorbs light maximally in the visible spectrum at wavelengths 498, 538, 548, and 645 nm on addition of peroxide, shows maximal absorption without peroxide at 561 and 630.5 nm. Such a change in absorption spectrum indicates that the union between substrate and enzyme must be rather intimate, since the absorption spectrum of a compound is not so altered except by a chemical change (Baldwin, 1967).

Evidence of union between enzyme and substrate is also indicated by experiments on bioluminescence. An enzyme, luciferase, catalyzes the oxidation of a substrate, luciferin, with the emission of light. If luciferin and luciferase are mixed before oxygen is admitted, the oxidation takes less time than if oxygen is first added to each separately and the two are then mixed (Fig. 10.18). This indicates that first a complex of luciferin and luciferase develops, followed by oxidation of the complex by atmospheric oxygen:

$$\text{luciferin} + \text{luciferase} \longrightarrow \text{luciferin-luciferase complex} \quad (10.2)$$

$$\text{luciferin-luciferase complex} + O_2 \longrightarrow \text{oxyluciferin} + \text{luciferase} + \text{light} \quad (10.3)$$

Because formation of the complex luciferin-luciferase is slower than its oxidation, a mixture of luciferin and luciferase in which the complex has already formed gives off light more rapidly on admission of oxygen than will the reaction upon addition of luciferin and oxygen to luciferase and oxygen. The difference in time in the two cases measures the time required for association of luciferin and luciferase (Chance *et al.,* 1940).

From such evidence as this it is thought that the substrates involved in oxidoreductions are

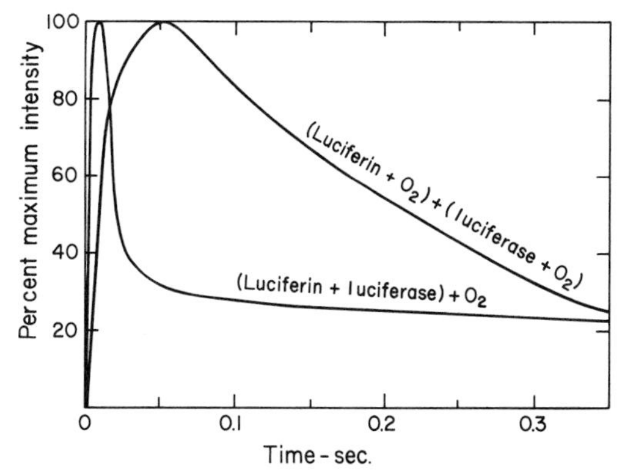

Figure 10.18. *The rate of light emission when luciferin (substrate) and luciferase of the Japanese crustacean* Cypridina *are mixed, in one case (1), before admission of oxygen and in the other (2) when each has been mixed with oxygen before being put together. (After Chance* et al.*, 1940: J. Cell. Comp. Physiol.* 15: *195.)*

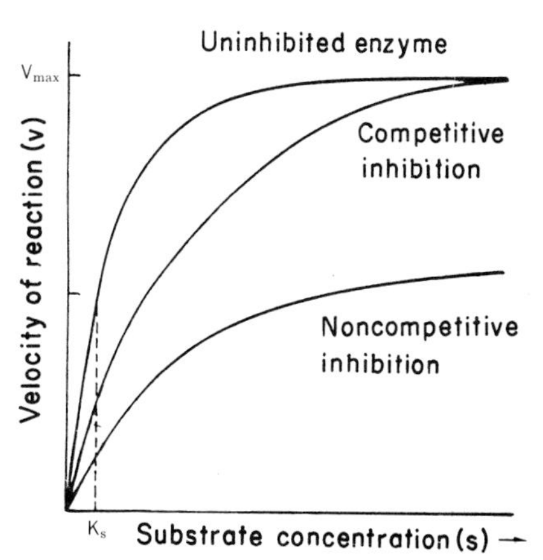

Figure 10.20. *The rate of a catalyzed reaction as a function of substrate concentration is plotted for an active enzyme and for an enzyme inhibited by either a competitive or a noncompetitive inhibitor. K_S is equivalent to K_M in Figure 10.22 and represents substrate concentration at one-half maximum reaction velocity. (From Sizer, 1957: Science* 125: *34. Copyright 1957 by the American Association for the Advancement of Science.)*

bound selectively to the surfaces of the enzymes concerned. As a result, the energy of activation is lowered (see Fig. 10.3).

INHIBITION OF ENZYMES BY POISONS

Since enzymes have one or more sites at which they combine with a substrate, inactivation of these sites by an enzyme poison prevents the enzyme from acting on its substrate. Poisons are effective to different degrees and act upon enzymes in different ways.

In *competitive inhibition* a nonmetabolizable molecule A resembles a metabolizable molecule B sufficiently to be bound to the enzyme. A, competing with B for space on the enzyme surface, reduces the attachment of B molecules. For example, malonic acid resembles succinic acid (Fig. 10.19) because each has two carboxyl groups. Malonic acid inhibits the action of succinic dehydrogenase on succinic acid by "clogging the enzyme's active spots,"

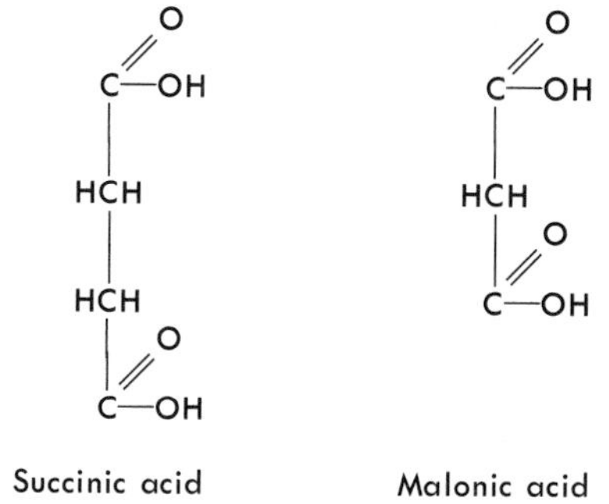

O=C—OH | HCH | HCH | O=C—OH

Succinic acid

O=C—OH | HCH | O=C—OH

Malonic acid

Figure 10.19. *Structural formulas of succinic and malonic acids.*

and since the malonic acid is not metabolized, it remains attached to the enzyme. Because the complex of malonic acid and succinic dehydrogenase dissociates at a finite rate given by the dissociation constant, an excess of succinic acid will reverse the action of malonic acid (law of mass action). In general, when competitive inhibition occurs, an excess of substrate reverses the action of the poison, as shown in Figure 10.20.

Some molecules that resemble the metabolizable substance, but are apparently attached to the enzyme at some point other than the one that binds the substrate, exhibit *noncompetitive inhibition* of enzymes. In such cases inhibition is not relieved by high substrate concentration even at a low inhibitor concentration (Fig. 10.20). In other cases, however, when the inhibitor concentration is low, a high concentration of substrate reverses the inhibition. For instance, inhibition of bacterial metabolism by a small concentration of sulfonamide is relieved by a high concentration of the vitamin B factor folic acid, the prosthetic group of the "one-carbon metabolism" (see Table 10.5) enzyme (Sizer, 1957). Poisoning usually acts as if a fraction of the enzyme had been removed, reducing the enzyme concentration.

Enzymes that depend upon the presence of sulfhydryl (—SH) groups for their activity are inactivated by salts of heavy metals or alkylating agents. Mercury and copper salts inactivate some hydrolases by combining with the sulfhydryls. Iodoacetic acid inhibits some oxidative enzymes by reacting with their sulfhydryl groups (alkylation). Some enzymes active in the

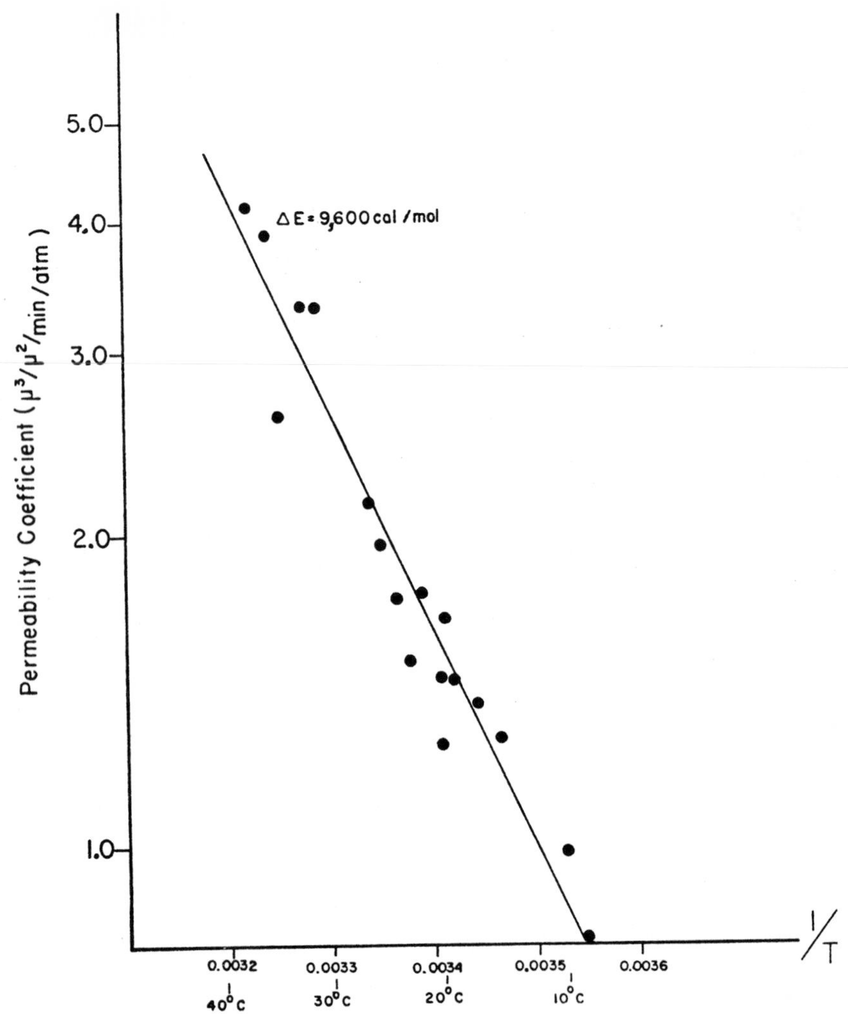

Figure 10.21. *Relationship between the logarithm of the permeability coefficient of Ehrlich ascites tumor cells to water and the reciprocal of the absolute temperature; the energy of activation determined from the plot is 9600 cal per mole. (From Heampling, 1960: J. Gen. Physiol. 44: 374.)*

anaerobic phase of sugar decomposition require sulfhydryl groups; therefore, they are readily inhibited by iodoacetic acid. This type of inhibition, depending upon the concentration of the poison and not on the ratio of poison to substrate, is also a noncompetitive inhibition.

Members of a class of enzymes requiring metal ions such as Mg^{++}, Mn^{++}, Co^{++}, Zn^{++}, or Fe^{++} for their activity are inactivated by substances that combine with these metals. For example, cyanide binds iron in cytochrome oxidase, an enzyme in the main pathway of oxidation-reduction in the cell (see Chapter 12). In this case, also, inhibition depends upon the concentration of the poison, not on the ratio of poison to substrate.

Some enzyme poisons produce irreversible effects by chemical modification of the enzyme's structure, thereby altering its affinity for a substrate. For example, iodoacetamide irreversibly reacts with —SH groups forming covalent derivatives of the enzyme. The organophosphorus compound diisopropylphosphate (DFP) reacts with the serine residue of an enzyme's active site, forming a covalent linkage with an OH group. This may occur, for example, in the enzyme acetylcholine esterase, an important enzyme in the nervous system (see Chapters 20 and 21), and for this reason DFP is used as a nerve poison and an insecticide.

When a poison inactivates an enzyme gradually, its irreversible effect may not be obvious at once. Furthermore, catalysis of the reaction may continue at a very low level.

By varying the substrate concentration one can generally distinguish between competitive and noncompetitive inhibition of enzymes by poisons (Sizer, 1957). The effects of competitive and noncompetitive inhibitors may also be compared on the basis of the Michaelis-Menten formulation (see Fig. 10.22).

Oxygen poisons some enzymes, especially in anaerobes. This is generally thought to be a relic of the anaerobic period in the earth's history when enzymes evolved in the absence of oxygen. In the case of the enzyme nitrogenase required for nitrogen fixation and active only in the absence of oxygen, the interpretation is controversial. If life evolved in an atmosphere with an abundance of ammonia as some have postulated (see Chapter 2), such an enzyme would have had no survival value. On the other hand, if the earth's atmosphere consisted of only carbon dioxide and nitrogen, as others postulate to be the case after the intense bombardment of the earth by meteorites, then nitrogenase would have had survival value and would represent an ancient acquisition.

Many other types of enzyme poisons exist besides the ones considered here. For example, oxidoreductions and formation of high energy phosphates are uncoupled by dinitrophenol (DNP), as a result of which the energy liberated is dissipated as heat. Some of these other enzyme poisons are cited in later chapters where discussion of them is pertinent.

APPENDIX

10.1 THE ARRHENIUS EQUATION

Arrhenius found a way to determine experimentally the energy of activation. He measured the rate of a thermochemical reaction at different temperatures and plotted the natural logarithm (ln) of the velocity (k), at each absolute temperature (T). He obtained a straight line. An example of a similar plot for a biological reaction is shown in Figure 10.21. The equation for this relation is:

$$\frac{d\,ln\,k}{dT} = \frac{a}{T^2} \qquad (10.4)$$

or

$$ln\,k = c - \frac{a}{T} \qquad (10.5)$$

where a is the slope and c is the intercept on the temperature axis, 1/T. He reasoned that the reaction rate depends upon the number of molecules possessing the necessary activation energy, E. The slope, a, measures the reaction rate. On the basis of theoretical deductions, the relation of a to E was given by Arrhenius as: a = E/R, where R is the gas constant.

Substitution into equation 10.4 gives:

$$\frac{d\,ln\,k}{dT} = \frac{E}{RT^2} \qquad (10.6)$$

Integrated, this becomes:

$$\frac{k_2}{k_1} = e^{-\frac{E}{R}\left[\frac{1}{T_1} - \frac{1}{T_2}\right]} \qquad (10.7)$$

or,

$$ln\,\frac{k_2}{k_1} = \frac{E}{R}\left[\frac{1}{T_1} - \frac{1}{T_2}\right] \qquad (10.8)$$

where E is the activation energy; k_1 and k_2 are the velocity constants for the reaction rates at two different absolute temperatures, T_1 and T_2; *ln* is the natural logarithm; and R is the gas constant. E may be determined by plotting the logarithm of the reaction rate against 1/T (see Fig. 10.21).

To calculate the number of molecules with the energy of activation before and after a 10-degree rise in temperature, Arrhenius used the Maxwell-Boltzmann distribution law. Calculation shows that several times as many activated molecules are present at the higher temperature than at 10° lower, as shown in Figure 10.1. Therefore, it seems most likely that the increase in reaction rate with rise in temperature is attributable to the disproportionate increase in the fraction of molecules with the required activation energy for a particular reaction (Lehninger, 1971).*

10.2 THE MICHAELIS-MENTEN EQUATION FOR ENZYME KINETICS

Assuming that enzyme (E) and substrate (S) combine reversibly to form a complex (ES) in enzyme-catalyzed reactions: enzyme + substrate → complex, or

$$E + S \underset{k_2}{\overset{k_1}{\rightleftharpoons}} ES \qquad (10.9)$$

where k_1 is the velocity constant of the formation of complex (ES, and k_2 is the velocity constant for dissociation of (ES). If complex → enzyme + products, or

$$ES \underset{k_4}{\overset{k_3}{\rightleftharpoons}} E + P \qquad (10.10)$$

then k_3 is the velocity constant for decomposition of the complex (ES) with the formation of reaction products (P), and k_4 is the velocity constant for the recombination of E and P,

*Strictly speaking, the E value determined in this manner is actually a much more complicated entity according to the transition state theory of Eyring and is preferably spoken of as the *temperature characteristic, μ*, rather than E, the energy of activation.

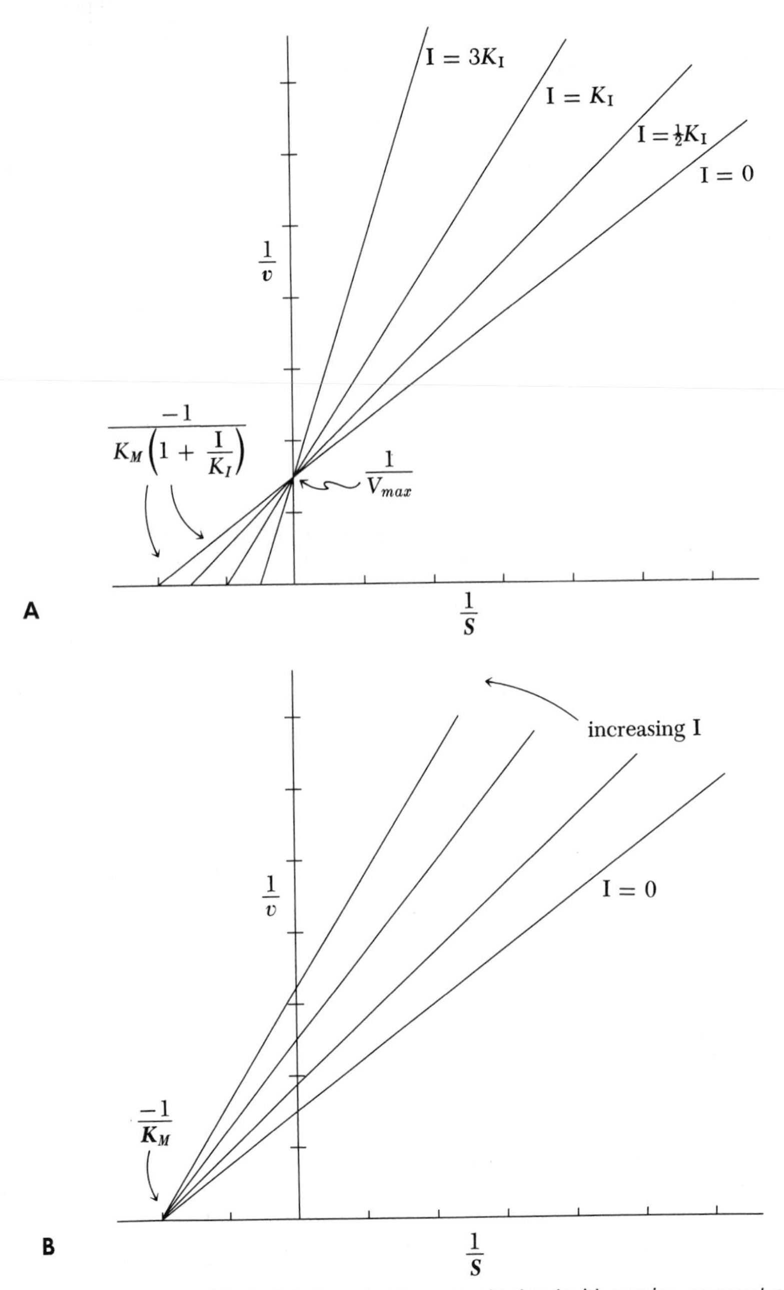

Figure 10.22. A, *Lineweaver-Burk plots for reaction rate obtained with varying concentrations, I, of a purely competitive inhibitor. There is a single intercept on the ordinate because* V_{max} *is the same in all cases, but the intercept on the abscissa changes.* B, *Lineweaver-Burk plots obtained for reaction rates with varying concentrations of a purely noncompetitive inhibitor.* K_M *is constant in all cases, so there is a single intercept on the abscissa. The inhibitor has the same effect as removing enzyme, so* V_{max} *and the intercept on the ordinate change. (From McGilvery, 1970: Biochemistry. W. B. Saunders Co., Philadelphia.)*

a reaction so slow in biological situations that it may be neglected. Reaction k_3 is considered to be the slowest of the other three reactions (k_1, k_2, k_3); therefore, it sets the pace for the overall reaction since E cannot be recovered until complex ES is decomposed.

Under certain conditions, the rate of the enzymatic reaction will also depend upon the concentration of the substrate. In the present formulation, concentration of substrate may be considered so large that the amount of S bound by E at any time is negligible compared to total concentration of S. This relation was formulated quantitatively by Michaelis and Menten in 1913 on the basis of the law of mass action. The derivation below follows Briggs and Haldane, as outlined by Lehninger (1975). The derivation of the Michaelis-Menten equation is given in detail as a key to a better understanding of the symbols in plots of the variables, but of itself is not important to remember.

The rate of formation of ES from E and S may be given as:

$$\frac{d[ES]}{dt} = k_1([E] - [ES])[S] \qquad (10.11)$$

and the rate of breakdown of ES may be written as:

$$\frac{-d[ES]}{dt} = k_2[ES] + k_3[ES] \qquad (10.12)$$

At steady state when the rate of formation of ES equals its rate of decomposition, and [ES] may be considered constant,

$$k_1([E] - [ES])[S] = k_2[ES] + k_3[ES] \qquad (10.13)$$

Or, rearranging:

$$\frac{[S]([E] - [ES])}{[ES]} = \frac{k_2 + k_3}{k_1} = K_M \qquad (10.14)$$

where K_M is the *Michaelis constant,* which replaces, for convenience, the expression $\frac{k_2 + k_3}{k_1}$.

Solving for the steady state concentration of [ES]:

$$[ES] = \frac{[E][S]}{K_M + [S]} \qquad (10.15)$$

The initial rate, v, of an enzymatic reaction is proportional to the concentration of the complex ES:

$$v = k_3[ES] \qquad (10.16)$$

And, when substrate concentration is so high that essentially all the enzyme is combined with substrate as ES, maximal velocity is achieved:

$$V_{max} = k_3[E] \qquad (10.17)$$

where [E] represents the total enzyme concentration.

Substituting in equation 10.16 the value of [ES] from equation 10.15, we have:

$$v = k_3 \frac{[E][S]}{K_M + [S]} \qquad (10.18)$$

Dividing equation 10.18 by equation 10.17:

$$\frac{v}{V_{max}} = \frac{k_3 \frac{[E][S]}{K_M + [S]}}{k_3[E]} \qquad (10.19)$$

and, solving for v,

$$v = \frac{V_{max}[S]}{K_M + [S]} \qquad (10.20)$$

This (10.20) relationship is called the *Michaelis-Menten equation.* When $v = \frac{1}{2} V_{max}$, then:

$$\frac{V_{max}}{2} = \frac{V_{max}[S]}{K_M + [S]} \qquad (10.21)$$

and dividing each side by V_{max}:

$$\frac{1}{2} = \frac{[S]}{K_M + [S]}$$

Rearranging: $K_M + [S] = 2[S]$

$$K_M = [S] \qquad (10.22)$$

Thus the Michaelis constant K_M is equal to the concentration of substrate (in moles per liter) at which the velocity of the reaction is half maximal.

While the K_M could be determined from a plot of the velocity of the reaction against the substrate concentration (Fig. 10.13), it is much easier to determine it from a relationship giving a straight line. This can be done by taking the reciprocal of equation 10.20:

$$\frac{1}{v} = \frac{1}{V_{max}[S]/(K_M + [S])} = \frac{K_M + [S]}{V_{max}[S]} \qquad (10.23)$$

TABLE 10.7. K_M FOR SOME ENZYMES*

Enzyme and Substrate	*K_M (mM)*
Catalase	
H_2O_2	25
Hexokinase	
Glucose	0.15
Fructose	1.5
Chymotrypsin	
N-Benzoyltyrosinamide	2.5
N-Formyltyrosinamide	12.0
N-Acetyltyrosinamide	32
Glycyltyrosinamide	122
Carbonic anhydrase	
HCO_3^-	9.0
Glutamate dehydrogenase	
Glutamate	0.12
α-Ketoglutarate	2.0
NH_4^+	57
NAD_{ox}	0.025
NAD_{red}	0.018

*From Lehninger, 1975: Biochemistry. 2nd Ed. Worth Publishers, New York, p. 193.

Rearranging, it gives:

$$\frac{1}{v} = \frac{K_M}{V_{max}[S]} + \frac{[S]}{V_{max}[S]}$$

Simplifying, we have:

$$\frac{1}{v} = \frac{K_M}{V_{max}} \cdot \frac{1}{[S]} + \frac{1}{V_{max}} \qquad (10.24)$$

Plotting $\frac{1}{v}$ against $\frac{1}{[S]}$, as suggested by Lineweaver and Burk, gives a straight line (Fig. 10.22).

The Michaelis-Menten equation has been of great use in enzyme studies, and determination of K_M values at which the velocity of the enzyme reaction is half maximal is standard laboratory procedure in enzymology and related subjects. A few K_M values for several enzymes are given in Table 10.7. As is seen in Chapter 18, the Michaelis-Menten equation applies to the substrate-carrier complex as well, and a rate constant, equivalent to K_M for enzymes, is a valuable quantitative measure of the affinity of substrate for carrier in permeability studies.

The Michaelis-Menten substrate-enzyme complex is not the activated complex, the latter being present in much lower concentration. The substrate-enzyme complex must pass into an activated form before it can decompose into the final products of the reaction.

Tests of this expression of the velocity on many biological reactions, as affected by substrate in the lower concentration range, indicate good quantitative agreement with prediction. This is usually done by a double reciprocal plot 1/v, where v stands for the velocity at any time, against 1/S, where S stands for the substrate concentration (Fig. 10.22). This gives assurance that the kinetics of enzymatic reactions are subject to the law of mass action and indicates the incisiveness of the formulation of the relationships between enzyme and substrate by Michaelis and Menten.

LITERATURE CITED AND GENERAL REFERENCES

Arrhenius, S., 1915: Quantitative Laws in Biological Chemistry. G. Bell & Sons Ltd., London.

Baldwin, E., 1967: Dynamic Aspects of Biochemistry. 5th Ed. Cambridge University Press, London.

Becker, W. M., 1977: Energy and the Living Cell. J. B. Lippincott Co., Philadelphia.

Bender, M. L. and Brubacher, L. J., 1973: Catalysis and Enzyme Action. McGraw-Hill, New York.

Boyer, P. D., Lardy, H., and Myrback, K., 1970: The Enzymes. Vol. 4. Hydrolysis. Multivolume Treatise, 3rd Ed. Academic Press, New York.

Brewer, G. J., 1970: An Introduction to Isozyme Techniques. Academic Press, New York.

Chance, B., Harvey, E. N., Johnson, F. H. and Millikan, G., 1940: The kinetics of bioluminescent flashes, a study in consecutive reactions. J. Cell Comp. Physiol. *15*: 195–215.

Colowick, S. P. and Kaplan, N. D., 1955–1978: Methods in Enzymology. 48 Vols. Academic Press, New York.

Frieden, E., 1959: The enzyme-substrate complex. Sci. Am. (Aug.) *201*: 119–125.

Gutfreund, H., 1965: An Introduction to the Study of Enzymes. John Wiley and Sons, New York.

Hochachka, P. and Somero, G. N., 1973: Strategies of Biochemical Adaptation. W. B. Saunders Co., Philadelphia.

Hungate, R. E., 1966: The Rumen and Its Microbes. Academic Press, New York.

Johnson, F. H., Eyring, H. and Storer, D. S., 1974: The Theory of Rate Processes in Biology and Medicine. Wiley-Interscience, New York, pp. 155–272.

Koshland, D. E., Jr. and Neet, K. E., 1968: The catalysis and regulatory properties of enzymes. Ann. Rev. Biochem. *37*: 159–210.

Lehninger, A. L., 1971: Bioenergetics. 2nd Ed. W. A. Benjamin, Menlo Park, Calif.

Lehninger, A. L., 1975: Biochemistry. 2nd Ed. Worth Publishers, New York.

Low, P. S. and Somero, G. N., 1975: Activation volumes in enzymic catalysis: their source and modification by low molecular weight solutes. Proc. Nat. Acad. Sci. U.S.A. *72*: 3014–3018.

McGilvery, R. W., 1975: Biochemical Concepts. W. B. Saunders Co., Philadelphia.

Markert, C. L., ed., 1975: Isozymes. 4 Vols (Molecular Structure, Physiological Function, Developmental Biology, Genetics and Evolution). Academic Press, New York.

Masters, C. J. and Holmes, R. S., 1974: Isozymes, multiple enzyme forms and phylogeny. Adv. Comp. Physiol. Biochem. *5*: 110–195.

Mildvan, A. C., 1974: Mechanism of enzyme action. Ann. Rev. Biochem. *43*: 357–399.

Phillips, D. C., 1966: The three dimensional structure of an enzyme molecule. Sci. Am. (Nov.) *215*: 78–90.

Porcellati, G. and di Jeso, F., 1971: Membrane-Bound Enzymes. Plenum Press, New York.

Sigman, D. S. and Mooser, G., 1975: Chemical studies of enzyme active sites. Ann. Rev. Biochem. *44*: 889–931.

Sizer, I. W., 1957: Chemical aspects of enzyme inhibition. Science *125*: 34–59.

Wilkie, D. R., 1960: Thermodynamics and the interpretation of biological heat measurements. Prog. Biophys. Mol. Biol. *10*: 260–298.

CHAPTER 11

REDOX POTENTIALS

In the electron transport chain of cellular oxidation, the hydrogen and electrons removed from a substrate during glycolysis and the tricarboxylic acid cycle are passed to oxygen by a series of enzymes (see Fig. 11.6). The particular sequence of transfers from one enzyme to another is determined by the *oxidation-reduction (redox) potential* of each component, namely, its tendency to give up or take on electrons. Transfer of electrons is down an electrical potential gradient. Much as a ball rolls from one step of a staircase to another down a gravitational gradient, so an electron in a biological oxidation-reduction moves from one enzyme to another down an electrical potential gradient. As discussed in Chapter 12, the energy liberated in the process is conserved in the high-energy phosphate bond of ATP *(oxidative phosphorylation)*. In photosynthesis, light excites an electron in chlorophyll to a reducing potential; as the electron drops down the potential gradient it liberates energy to enzymes and carriers along the way, and this energy is then conserved in the high-energy phosphate bond of ATP *(photophosphorylation)*. The light energy is thus transduced to chemical bond energy (see Chapter 14).

The study of redox potentials of biological systems is of great significance because these potentials determine the particular sequence of reactions occurring in a cell. While the biochemical reactions of the cell are the center of interest, it is first necessary to explain the electrode techniques in simpler inorganic systems.

REDOX POTENTIALS AT ELECTRODES IMMERSED IN ELECTROLYTES

The redox potential of an element—its tendency to give up or take on electrons—is a characteristic property of the element and depends upon its nuclear and electronic constitution. Members of one class of elements such as hydrogen give up electrons readily, forming ions in solution, whereas those of another class such as oxygen take up electrons instead.

An electrode is a device for conducting electrons. It may conduct them to or from an electrolyte solution, an electric arc, or a vacuum tube. It is usually made of metal or of a metal and one of its salts, but a gas such as hydrogen may serve as an electrode when adsorbed on the surface of an inert metal like "spongy" platinum. The calomel electrode (Fig. 11.1) consisting of mercury and mercurous chloride is widely used in electrode potential studies because of its pH independence and its stability. The platinum electrode is used in redox studies of organic substances because it does not react with a solution in which it is immersed.

The electrode potential (or voltage) of an element is measured by immersing it in a solution of one of its salts and, by use of a potentiometer, determining the potential difference between it and some standard electrode as a baseline. For example, zinc, which like hydrogen gives up its electrons readily, dissolves when immersed as an electrode in a solution of one of its salts. As it dissolves it forms ions in the solution, leaving the electrons on the electrode:

$$Zn \underset{\text{reduction}}{\overset{\text{oxidation}}{\rightleftharpoons}} Zn^{++} + 2e \qquad (11.1)$$

or, in more general terms:

$$\text{Metal} \underset{\text{reduction}}{\overset{\text{oxidation}}{\rightleftharpoons}} \text{ion} + \text{electron} \qquad (11.2)$$

An electrode made of zinc in contact with one of

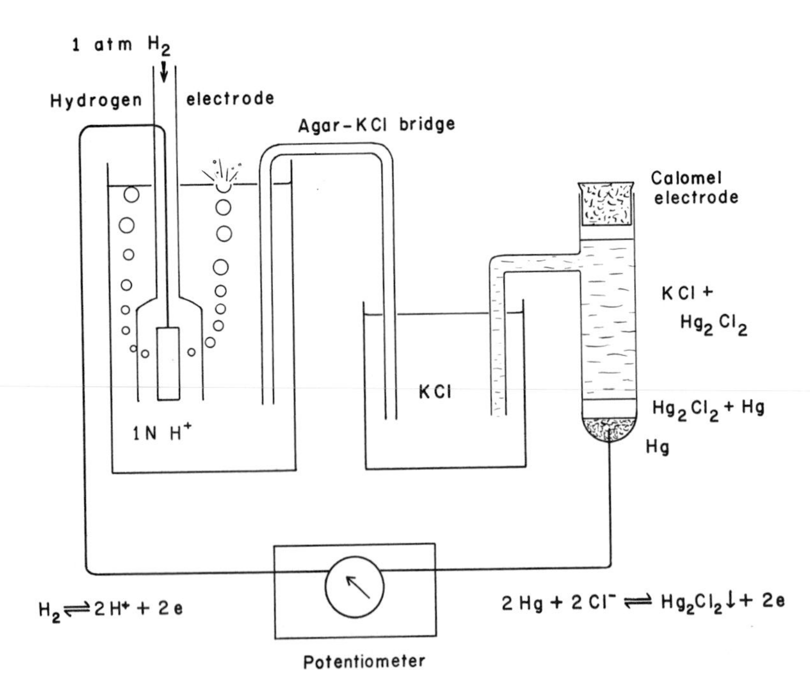

Figure 11.1. *The hydrogen electrode as used to determine the potential of another electrode, in this case the calomel electrode. The normal hydrogen electrode is taken as 0.0, and against it the calomel electrode has a potential of +0.2458 volt. The potentials of other electrodes can be determined by substituting the electrodes, one at a time, for the calomel electrode. The potential difference between the two electrodes is measured by a potentiometer in which an unknown emf (electromotive force) is matched against a known emf (divided as desired by passing it through a slide wire resistance) from a battery. The matching is registered by a null point on a galvanometer. In this way the data in Table 11.1 and in Figure 11.2 were determined.*

its salts is therefore negatively charged due to the electrons that remain on the electrode. The greater the tendency of an element (electrode) to go into solution, other things being equal, the greater will be the potential developed between the solution and the electrode. On the other hand, copper, like oxygen, does not dissolve when immersed as an electrode in one of its salts. Instead, it delivers electrons to the cupric ions in the salt solution, and the cupric ions plate out as copper on the metal. An electrode made up of such a metal in contact with one of

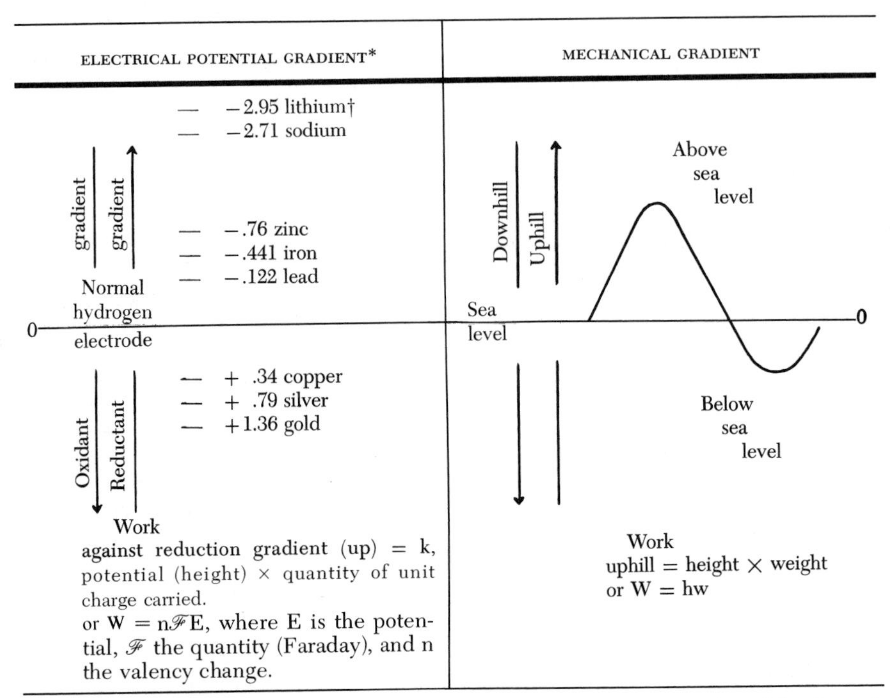

* Potentials at pH 0.0. Note that the sign of the charge is not that of the ion but of the electrode in contact with a solution of ions.

† The sign of the potential represents the reduction potential (see Table 17.1). Oxidant means oxidizing agent and reductant, reducing agent.

Figure 11.2. *Electromotive potentials and a mechanical analogue.*

its salts is therefore always positive due to its loss of electrons in discharging copper ions. All metals, on the basis of similar experimental data, can be arranged in an *electromotive series,* which measures their relative potentials (Table 11.1 and Fig. 11.2).

The potential of the "normal" hydrogen electrode (Fig. 11.1) taken at standard state is arbitrarily taken as zero and serves as a reference point for potentials of all other electrodes taken relative to it (Fig. 11.2). The *standard state* for the normal hydrogen electrode means that gaseous hydrogen at 1 atmosphere pressure is continuously bubbled past a spongy platinum electrode kept in a solution 1N with respect to hydrogen ions.

To measure the potential, at a given temperature, of a metal electrode immersed in a solution of one of its own salts that is 1N in activity, one must connect the electrode through a potentiometer with a hydrogen electrode. For example, it is found by this means that at 25°C the normal magnesium electrode, measured against the normal hydrogen electrode, registers a potential of −1.55 volts. A saturated calomel electrode containing mercury in contact with mercurous chloride, which in turn is in contact with potassium chloride (Fig. 11.1), registers a potential of +0.2458 volt against the normal hydrogen electrode.

It should be emphasized that biological convention places the strongest reducing agents at the top of tables and graphs (Table 11.1 and Fig. 11.2). When electrons are passed in response to an oxidation gradient during oxidation-reductions, that is, from reducing potential to oxidizing potential (downward in Fig. 11.2), the free energy change is the product of the change in potential through which the electrons fall, multiplied by the number of electrons:

$$\Delta G_0 = -n\mathcal{F}\ \Delta E \qquad (11.3)$$

where ΔG_0 is the standard free energy change; n is the number of equivalents in one stoichiometric occurrence; $\mathcal{F}$ is the Faraday constant, 96,494 coulombs. This constant is a product of the charge on an electron (1.592×10^{-19} coulombs) and the gram mole number of electrons (Avogadro's number, 6.024×10^{23}, the number of molecules in a gram molecular weight). ΔE is the change in potential in volts.

The product is in volt-coulombs, or joules. To convert to calories, a more familiar unit, it is necessary to divide by 4.185 because a calorie is equal to 4.185 joules. It will be recalled that the free energy for a reaction represents the maximal work that can be performed by that reaction, i.e., its chemical potential.

Conversely, when a gram mole of electrons moves up a reduction gradient from oxidizing potential to reducing potential, the same amount of free energy (work) must be expended on the system (upward in Fig. 11.2).

As will become evident in Chapter 12, it is instructive to calculate the free energy available from moving a gram mole of electrons from $NADPH_2$ (−0.324 volt) to the level of oxygen (the oxygen electrode, +0.815 volt at pH 7.0):

$$\Delta G = -n\mathcal{F}\ \Delta E = -(1)\ (96{,}500)\ (1.14)/4.185 = -26{,}300 \text{ cal}$$

The energy in calories for any voltage change in the series of enzymes and carriers in the cell may be calculated in the same way.

In order to emphasize the *reducing capacity* of a substance, that is, its capacity to donate electrons, the biological convention followed here assigns a negative sign to electrode potentials on the reducing side of the normal hydrogen electrode (Clark *et al.*, 1928; Clark, 1960).

An atom that gives up electrons and becomes a positively charged ion is a *reducing agent (reductant),* and the greater its tendency to give

TABLE 11.1. NORMAL ELECTRODE POTENTIALS AT 25°C*

Electrode	*Electrode Reaction*	*Normal† Electrode Potential (Volts) Against Hydrogen Electrode*
Li^+, Li	$Li = Li^+ + e$	−2.9595
Na^+, Na	$Na = Na^+ + e$	−2.7146
Zn^{++}, Zn	$Zn = Zn^{++} + 2e$	−0.7618
Fe^{++}, Fe	$Fe = Fe^{++} + 2e$	−0.441
Pb^{++}, Pb	$Pb = Pb^{++} + 2e$	−0.122
H^+, H_2 (1 atm.)	$½H_2 = H^+ + e$	0.0000‡
Cu^{++}, Cu	$Cu = Cu^{++} + 2e$	+0.3441
Ag^+, Ag	$Ag = Ag^+ + e$	+0.7978
Au^{+++}, Au	$Au = Au^{+++} + 3e$	+1.36

*Data from West, 1956: Textbook of Biophysical Chemistry. 2nd Ed. Macmillan Co., New York. The electrode potential is the reduction potential and measures the free energy of reduction; its sign is arbitrary. This convention is now followed by biologists the world over and by European physical chemists.

†All salts 1 N in activity (standard state).

‡At pH 0.0 (1 N H^+).

up electrons, the greater is its reducing potential. The greater the negative electrode potential, the greater is the reducing tendency of the element. Elements (or ions) with highly negative electrode potentials reduce those less negative, and, on the other side of the hydrogen electrode potential, the elements with less positive electrode potentials reduce the more positive ones.

Conversely, anything that takes on electrons, like an oxygen molecule or a copper ion, is an *oxidizing agent (oxidant)*. The greater the tendency of a substance to take on electrons, the greater is its oxidizing potential. Therefore, the more positive the electrode potential, the greater is the oxidizing potential of the element. Elements or ions with highly positive potentials oxidize those less positive, and on the other side of the hydrogen electrode potential the less negative elements oxidize the more negative ones.

To summarize, then, in the electromotive series as listed in Table 11.1, any element is capable of oxidizing the one above it and of reducing the one below it. In slightly different terms, each element is an oxidizing agent to those above it in the electromotive series and a reducing agent to those below it. Therefore, the electromotive series represents an *oxidation-reduction,* or *redox series* and the electrode potential of each element is a measure of the oxidation and reduction capacity of the element.

MEASUREMENT OF REDOX POTENTIALS OF ORGANIC COMPOUNDS

Organic materials cannot be made to serve as electrodes for measurement of oxidation-reduction potentials in the laboratory, yet in the cell there are oxidation-reduction systems similar to those discussed. For example, in biological oxidation-reduction reactions one often encounters changes in the valencies of iron in iron-porphyrin enzymes, such as cytochrome and cytochrome oxidase:

$$\underset{\text{Reductant}}{Fe^{++}} \underset{\text{reduction}}{\overset{\text{oxidation}}{\rightleftharpoons}} \underset{\text{Oxidant}}{Fe^{+++}} + e \qquad (11.4)$$

One also encounters organic systems in which hydrogen and electrons are transferred (in pairs) in oxidation-reduction reactions:

$$\underset{\text{Hydroquinone}}{H_2Q} \underset{\text{reduction}}{\overset{\text{oxidation}}{\rightleftharpoons}} \underset{\text{Quinone}}{Q} + 2H^+ + 2e \qquad (11.5)$$

Fortunately, however, such redox potentials can be measured indirectly by the use of the platinum electrode. Platinum, an inert metal, does not dissolve when placed in contact with a

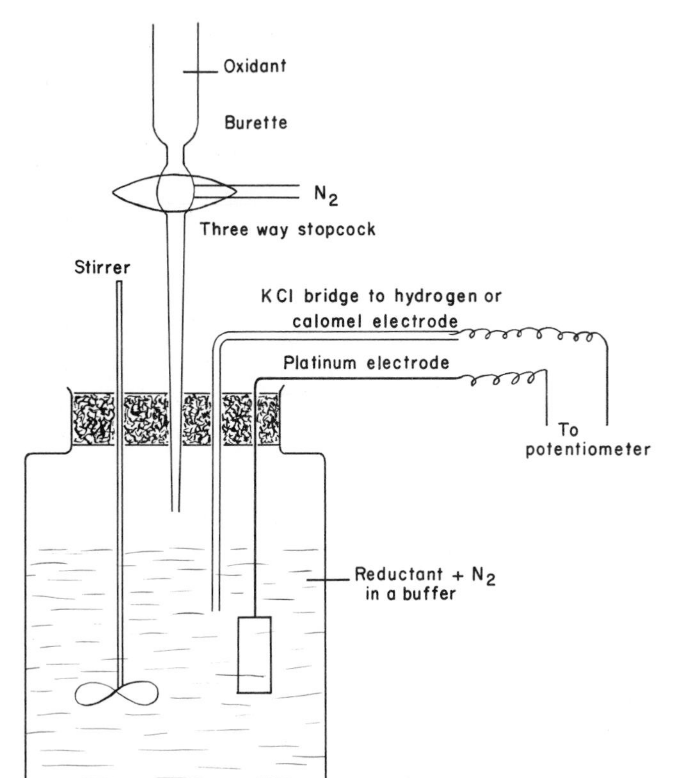

Figure 11.3. *Apparatus for measuring oxidation-reduction potentials. Although a redox potential is always given as between a platinum and a hydrogen electrode (the latter being the standard of reference), a calomel electrode is found in practice to be more convenient than the hydrogen electrode for a laboratory determination. The potential of the calomel electrode against the hydrogen electrode must be taken into account.*

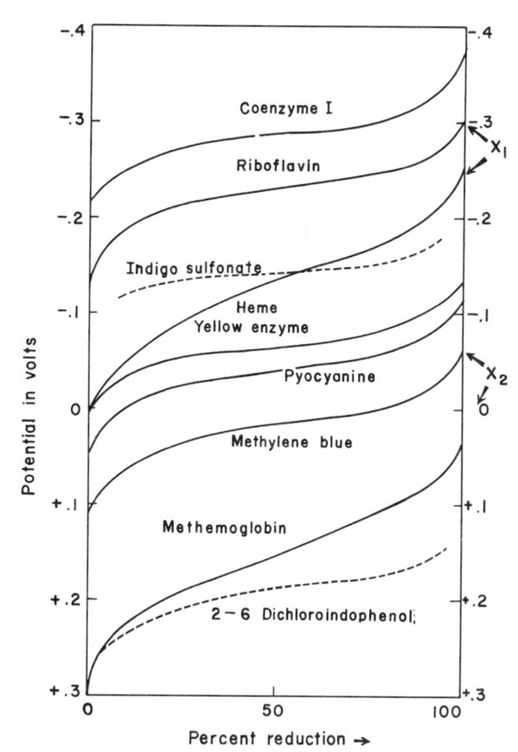

Figure 11.4. *Oxidation-reduction titration curves for dyes and cellular constituents.* X_1 *indicates the state of an amoeba in nitrogen;* X_2 *represents its state in air.* Note: *The convention of placing the potential of greater reduction capacity at the top is followed here and in succeeding graphs. (Data at pH 7.0 from Clark, 1938: J. Appl. Physics* 9: *102. Data on indigo sulfonate and indophenol at pH 7.4 from Clark, 1928: U. S. Hygienic Lab. Bul.* 151: *308.)*

solution and does not form an oxidation-reduction system with it. Although chemically inert, platinum is a good conductor of electrons, which it transfers from the solution to the potentiometer. The redox potential of an organic system, like that of the inorganic system, is always measured against the normal hydrogen electrode at 0.0 as a reference standard (Clark, 1960).

The apparatus for such a determination is illustrated in Figure 11.3. The reductant must be kept reduced by gassing with nitrogen free of oxygen. Aliquots of oxidant are added (titrated) to the reductant, the mixture is stirred after each addition and the potential of the mixture is measured with the potentiometer. Redox potentials for different ratios of oxidant to reductant are plotted in Figure 11.4, giving a curve for each redox system. The curves resemble titration curves for weak acids. It is interesting to note that near the middle of the curves, addition of either oxidant or reductant has little effect on the redox potential, a phenomenon resembling the buffer action of a weak acid (Clark, 1960).

Although such measurements are made without too much difficulty, it is more convenient to determine the redox potential (E) of a particular mixture indirectly, through the use of Peter's equation (for derivation see Appendix 11.1). For a temperature of 30°C and a valency change of 2, and converting from natural logarithms *(ln)* to $\log_{10}$, the equation is:

$$E = E_0 + 0.030 \log_{10} \frac{[\text{oxidant}]}{[\text{reductant}]} \qquad (11.6)$$

Here [oxidant] and [reductant] are the given concentrations of the particular oxidant and reductant studied, and E_0, which is experimentally determined, is the redox potential of a mixture containing an equimolal concentration of the oxidant and reductant. The factor 0.030 includes a term for the absolute temperature; therefore, it is always necessary to specify the temperature when any data on redox potentials are presented.

It is clear that once the E_0 for any given redox system (e.g., ferrous to ferric, or hydroquinone to quinone) is found, the redox potential at any other ratio of concentrations for the same redox system can be quickly calculated by substituting in equation 11.6. For example, for a system with an E_0' of -0.48 volt, a solution with 75 percent oxidant to 25 percent reductant is $-0.48 + 0.03 \log 3$, or -0.496 volt.

THE EFFECT OF pH ON REDOX POTENTIALS

When an oxidation-reduction reaction involves a hydrogen atom of an organic molecule, the hydrogen ion concentration of the solution becomes of great importance. For example, in the oxidation of hydroquinone mentioned before (see equation 11.5), a change in hydrogen ion concentration is found to influence the equilibrium between hydroquinone and quinone (law of mass action). This influences the redox potential by a definite factor (see Appendix 11.2). This fact is of great importance to the cell since many oxidation-reduction reactions in the cell involve not only electron transfer but also hydrogen transfer. It should be noted that decreasing the hydrogen ion concentration increases the reducing powers of most of the redox systems, as shown in Figure 11.5. This is to be expected since the hydrogen ion itself is an oxidant:

$$\underset{\text{Reductant}}{\tfrac{1}{2}H_2} \xrightleftharpoons[\text{reduction}]{\text{oxidation}} \underset{\text{Oxidant}}{H^+ + e} \qquad (11.7)$$

Furthermore, it is important to know the pH

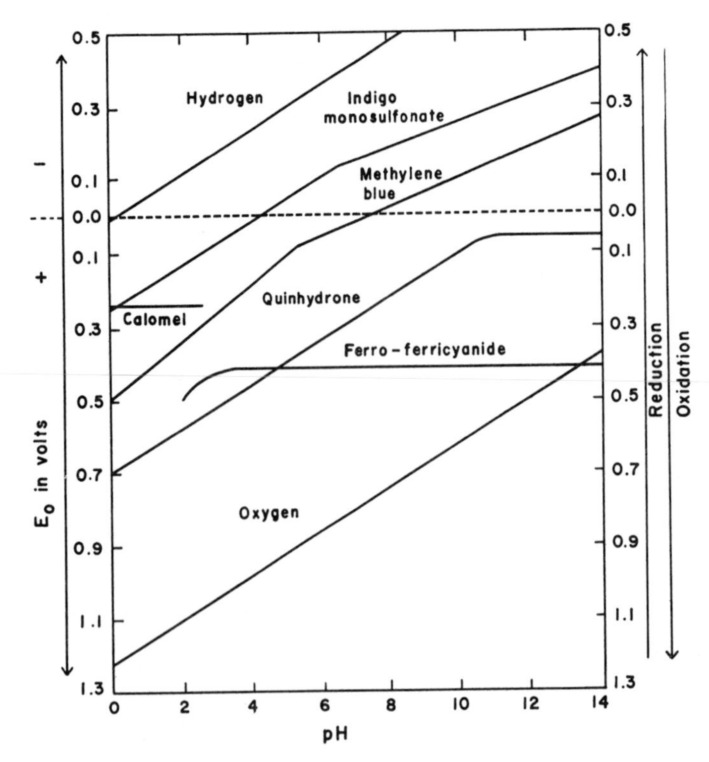

Figure 11.5. The effect of pH on oxidation-reduction potential when concentration of oxidant equals concentration of reductant. (Data from Gortner, 1949: Outlines of Biochemistry. 3rd Ed. John Wiley & Sons, New York.)

of a redox system because the redox potential of enzyme systems cannot be measured at the pH of the normal hydrogen electrode (pH 0.0 or 1N in respect to hydrogen ions). At such a pH enzymes would be denatured and the results would not have biological significance. Cellular reactions are carried on at pH 7.0 (or near it), as is shown in Chapter 12. The redox potentials of biological systems are therefore referred to the hydrogen electrode at pH 7.0 as a standard. A hydrogen electrode at pH 7.0 and at 30°C has a potential of −0.42 volt against the normal standard hydrogen electrode.

Therefore, when comparisons of redox potentials are made at a pH other than 0.0, the pH and the temperature should be specified. The redox potential of a substance under such conditions is designated E′ rather than E, and the potential of an equimolal mixture of reductant and oxidant under such conditions is designated E'_0 rather than E_0.

POTENTIALS DEVELOPED BY BIOLOGICAL REDOX SYSTEMS

Why does a dehydrogenase pass the electrons it has removed from a substrate to flavoproteins in glycolysis and the tricarboxylic cycle? Why, in turn, do the flavoproteins pass the electrons to each of a number of cytochromes in the electron chain, eventually reaching cytochrome oxidase, from which they are transferred to oxygen? On the basis of the principles developed in this chapter, and on the basis of the redox potential measurements, it is now possible to give a reasonable explanation for such sequences, and the ideas developed here are basic to understanding the pathways of cellular energy metabolism in Chapter 12. Values (at the indicated temperatures) of redox potentials for sample extracted systems are given in Table 11.2, along with reference values for the hydrogen and the oxygen electrode potentials. These values are E'_0 and were determined at or near pH 7.0. It will be seen from this table that each enzyme system develops a characteristic redox potential, the entire range of values lying between the values of the hydrogen electrode at one end and the oxygen electrode at the other. The enzyme systems and substrates in a cell, therefore, form a "biological electromotive series," or what is called the "electron transport chain," in which each system is theoretically capable of reducing the one below it and oxidizing the one above it.

Some bacteria have the enzyme hydrogenase that enables them to liberate hydrogen from nutrient substrate under anaerobic conditions. This indicates a reducing potential greater than the hydrogen electrode. Some nitrogen-fixing bacteria also liberate hydrogen under anaerobic conditions, either by way of hydrogenase or by way of nitrogenase when nitrogen is lacking, or liberate hydrogen simultaneously with nitrogen fixation (Postgate, 1978).

It has been found that the most rapid pathway for the oxidation-reduction reactions in the cell is that which makes use of the entire sequence of steps from dehydrogenases to cytochrome oxidase (Fig. 11.6). When this pathway is abridged—for example, when flavoprotein passes electrons to oxygen directly, instead of by way of cytochrome and cytochrome oxidase—the pathway is slow. This is demonstrated when a respiring cell is poisoned with a low concentration of cyanide, which inactivates the cytochrome oxidase. Respiration continues but only at a small fraction of its former rate.

Such a series of reactions, in which each reaction involves a small amount of energy exchange and a small difference in redox potential, has great thermodynamic advantage. A reaction occurring in a large number of steps, in each of which the energy exchange is small, is closer to being reversible and thermodynamically more efficient than one occurring in a small number of steps, each with a large energy exchange. It is to be noted that the steps in oxidative decomposition of cell nutrients more or less correspond to the free energy required to make 1 ATP; hence, the steps are correlated with the number of ATP molecules formed from ADP and inorganic phosphate. Consider, for example, the decomposition of one molecule of glucose by way of glycolysis and the Krebs cycle. This process may yield a total of 36 high-energy phosphate bonds (in ATP and CP, see Chapter 12). If for purposes of calculation we take the standard free energy change for one gram mole of ATP being converted to ADP and inorganic phosphate as 7300 cal/mole, the yield per gram mole of glucose decomposed will be 277,400 cal. The ΔG for degradation of glucose to CO_2 and H_2O is $-686{,}000$ cal/mole. The efficiency of the process on this basis is therefore about 40 percent, a testimonial to the success of a cell in solving its thermodynamic problem, as compared with a heat engine, in which the maximal possible but seldom achieved efficiency is 34 percent. It is probable that the efficiency in the cell is even greater than 40 percent (see Chapter 12).

The presence of at least five cytochromes, b, c_1, c, a, and a_3, the redox potential of each being successively closer to the redox potential of oxygen in the order given, is probably interpretable on these grounds. Since the cytochromes have the same prosthetic group, an iron-porphyrin compound, the specificity of each cytochrome probably depends upon the protein to which the prosthetic group is attached. It is the nature of the protein and its union to the prosthetic group that gives each of these protein-porphyrin complexes a characteristic E_0' value.

Altogether about 15 biological redox systems with metabolic function have been studied. However, few if any of them are amenable to exact theoretical treatment because the derivation of the oxidation-reduction equation (see Appendix 11.1) depends upon a system in equilibrium, and living systems develop steady states approaching and approximating but never reaching equilibrium (see Chapter 9).

COUPLING OF UNIVALENT AND DIVALENT ENZYMATIC REDOX SYSTEMS

During an oxidation-reduction reaction at the beginning of the electron transport chain (Fig. 11.6) the reduced flavoprotein molecule transfers two electrons. Originally it was thought necessary for the flavoprotein molecule to transfer both electrons at once, one to each of two ferricytochrome *b* molecules, since the iron in the ferricytochrome *b* is capable of receiving only one electron. But the reaction of one flavoprotein molecule with two cytochrome *b* molecules at one time requires a triple collision. Such an occurrence is much less probable than the reaction of one flavoprotein molecule with the two cytochrome *b* molecules in sequence, one at a time. As a two-step reaction, each step would require a collision between only two molecules.*

This problem, whether oxidation-reduction reactions involving the exchange of two electrons occur in one step or in two steps, formed the basis of a series of brilliant studies by Michaelis (1935, 1940). His solution of the problem can best be illustrated by his study of the oxidation of hydroquinone, as a model system. Michaelis demonstrated that the curves (Fig. 11.7) obtained for the redox potentials when increments of oxidizing material are added to hydroquinone indicate first the loss of one electron, then another, in successive steps. The loss of the first electron is thought to correspond to the formation of a semiquinone, which is a type of *free radical.* (The free radical formed by loss of the first electron from hydroquinone is highly stable, existing as an insoluble substance called quinhydrone, which separates from the solution of hydroquinone in the form of dark green crystals.) The second electron is lost from the semiquinone, and the semiquinone becomes quinone, the fully oxidized form (Fig. 11.8). Perhaps semiquinones, or free radicals, are always formed whenever two hydrogens and two electrons are transferred in oxidation-reduction reactions (Michaelis, 1930) but are too short-lived to be demonstrable by these methods (see paramagnetic methods in the next section).

An example of biological interest is the formation of a semiquinone during the oxidation-reduction of riboflavin (vitamin B_2). In this

*Since some flavoproteins shuttle between fully reduced and oxidized forms, simultaneous passage of two electrons is inferred. Possibly the single steps occur so rapidly that intermediates cannot be observed (Lehninger, 1975).

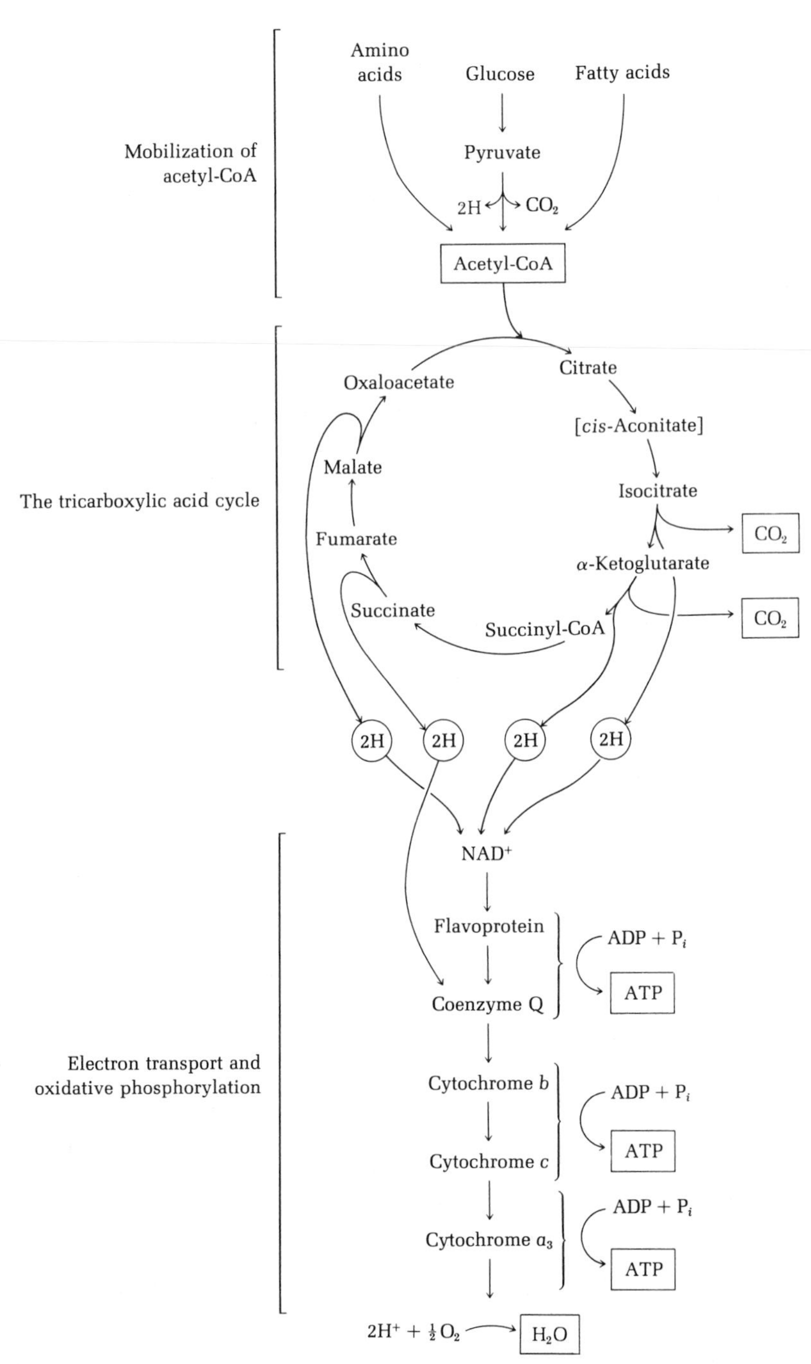

Figure 11.6. *Flow chart for cellular oxidative metabolism following formation of pyruvate in glycolysis (top). Acetyl CoA formed from pyruvate, as well as from amino acid and fatty acid metabolism, is the activated form of acetate "fed" into the tricarboxylic acid cycle. Hydrogen atoms (hydrogen ions and electrons) are transferred through NAD, flavoprotein, and Coenzyme Q, at which point hydrogen ions are liberated into the cell continuum, where they combine with oxygen to form water (bottom of scheme). The electrons are transported down the voltage gradient from reducing to oxidizing potential, liberating their energy for the production of ATP from ADP and P_i (right of figure). A total of 36 moles of ATP is produced per mole of glucose decomposed to carbon dioxide and water (see Table 12.2). (From Lehninger, 1975: Biochemistry. 2nd Ed. Worth Publishers, New York, p. 445.)*

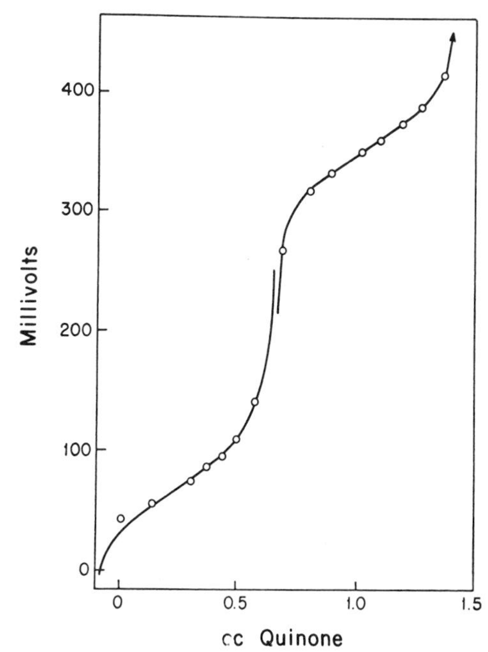

Figure 11.7. *Potentiometric demonstration of single electron transfer in an oxidation-reduction. Potential values in millivolts are plotted for increasing amounts of oxidant (quinone) added to the reductant (α-oxyphenazine in 1N HCl, pH 0.08). The temperature is 30°C. Note that for each of the two electrons transferred from reductant to oxidant a separate curve (inflection) is obtained, indicating the transfer of a single electron at a time. (From Michaelis, 1931: J. Biol. Chem. 92: 221.)*

reaction two hydrogen ions and two electrons are transferred stepwise. Riboflavin is colorless when reduced in acid solution by a reducing agent, but when it is exposed to air by shaking, it turns a bright orange color until it is fully oxidized, at which stage it becomes yellow. In this example, loss of a single electron results in the formation of an orange colored semiquinone; loss of the second electron results in the yellow color of the quinone form of riboflavin (Fig. 11.9). It is evident in this case

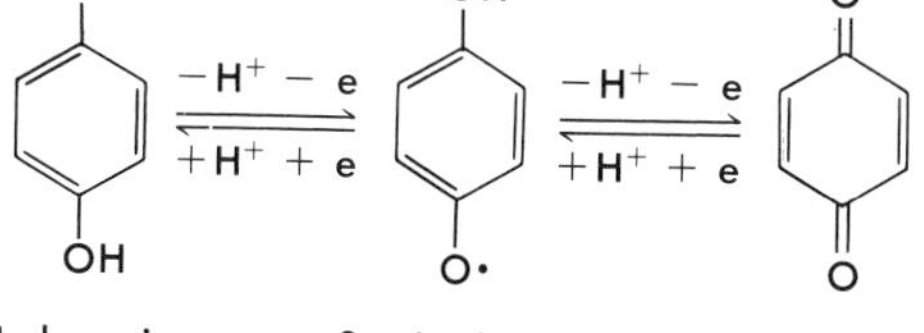

Figure 11.8. *Free radical formation (semiquinone) in the oxidation of hydroquinone to quinone. The oxidation occurs in two steps, one hydrogen ion and electron coming off at a time. The semiquinone has an unpaired electron that is paired after the second hydrogen ion and electron are removed.*

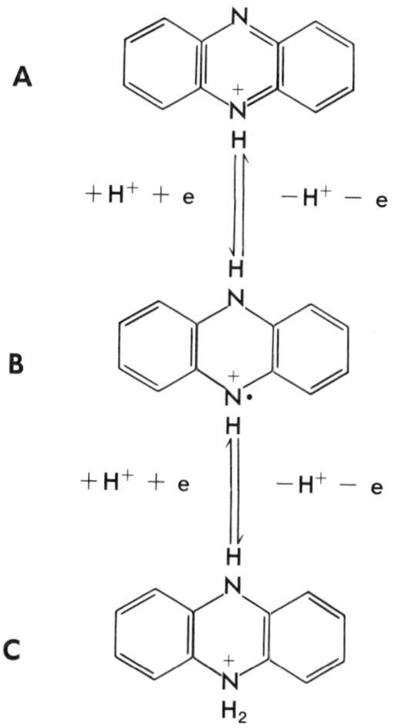

Figure 11.9. *Free radical formation with single electron and hydrogen ion transfer in riboflavin. Note the change in double bonds in the middle ring as electrons and hydrogen ions are transferred. A, The quinone; B, the semiquinone; and C, the hydroquinone. (After Leach, 1954: Adv. Enzymol. 15: 1.)*

that the electrons are transferred one at a time from a riboflavin molecule to an electron acceptor. The color changes can be followed spectrophotometrically in the visible spectrum (Fig. 11.10) and the concentration of the free radical and the rate of reaction ascertained by this means (Blois, 1961).

A similar oxidation-reduction reaction occurs in the prosthetic group of the flavoprotein enzyme. This enzyme is colorless in the reduced form and yellow when oxidized. Here, too, it is suggested that electrons need to be transferred one at a time from flavoprotein enzymes to an electron acceptor (cytochrome *b*). Since proteins stabilize free radicals of compounds attached to them, flavoprotein is particularly favorable as an electron transporting and coupling mechanism (Beinert, 1961). It has been suggested that possibly two electron transfers may occur (Lehninger, 1975).

A marked color change is seen in flavodoxin present as an electron carrier in some nitrogen-fixing bacteria, for example, the soil aerobe *Azotobacter chroococcum*. Flavodoxin is yellow in the oxidized state, a striking blue as the semiquinone, and colorless in the reduced state. The colorless form reacts with the enzyme nitrogenase enabling it to carry out the reduction of N_2 to NH_3 (Postgate, 1978).

In nitrogen fixation, electrons appear to be transferred singly. The nitrogen-fixing enzyme nitrogenase consists of two iron-containing

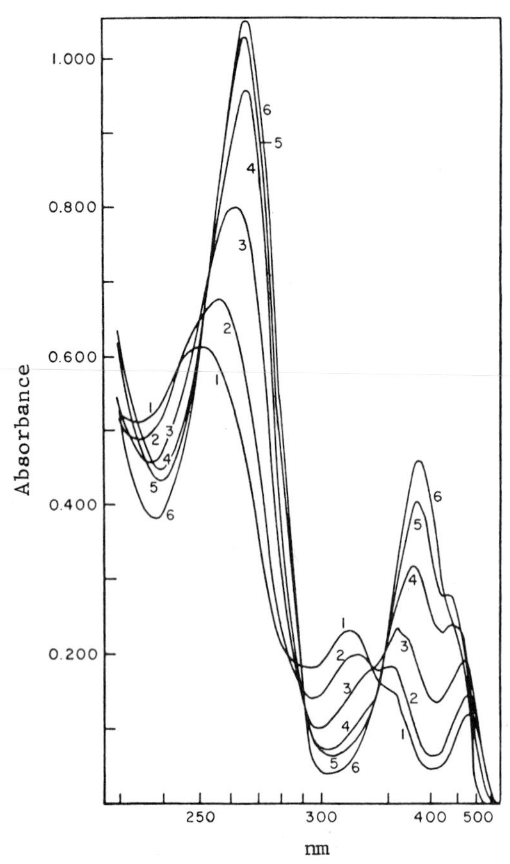

Figure 11.10. *Absorption spectra of flavin mononucleotide (6.4 × 10⁻⁵M in 1N HCl) at successive oxidation states from about 75 percent reduction (curve 1) to full oxidation (curve 6); light path 0.5 cm, temperature 28°C; reduced with metallic Zn. (From Beinert, 1956: J. Am. Chem. Soc. 78: 5323.)*

proteins, P_2 and P_1; P_1 also contains molybdenum in addition to iron. The proteins exist dissociated from one another until P_2 is reduced by ferredoxin (or flavodoxin in some nitrogen-fixing bacteria). The reduced P_2, activated by magnesium ion and ATP, attaches to P_1, transferring an electron to form the active enzyme. This permits attachment of N_2 to the molybdenum in P_1. P_2 then dissociates from P_1 and performs several more rounds of reduction, the transferred electrons permitting reduction stepwise of N_2 to NH_3. ATP is decomposed to ADP and P_2 in the course of these reactions (Postgate, 1978).

No problem is posed by transfers among cytochromes *b*, c_1, *c*, *a*, and a_3 (cytochrome oxidase), because in each case a single electron is passed and only two molecules need collide. The activation of oxygen, however, involves two electrons, but at a given time only one oxygen is coupled with cytochrome oxidase (see Chapter 12 and Fig. 12.9). If the organized systems of enzymes on the cristae of the mitochondria were to serve as semiconductors (imperfect conductors but capable of passing electrons), it would be unnecessary to postulate sequential collisions between enzymes and substrate molecules (Lehninger, 1975).

DETECTION OF FREE RADICALS BY THEIR MAGNETIC PROPERTIES

Although in some cases free radicals can be followed spectrophotometrically in visible and ultraviolet light, some reactions may be so rapid that other methods with quicker response would be advantageous. Furthermore, if such methods not only permit studies with extracts, or mixtures of enzymes and substrate but also are applicable to living cells, they would provide great advantages over conventional spectrophotometry. One could then follow reactions in living cells under a variety of conditions. This is the case with electron spin resonance and, especially, with nuclear magnetic resonance. Although the theory may seem complex and the equipment difficult to comprehend at first, essentially in both cases we are dealing with spectrophotometry in the radiowave region. The radiation penetrates the cell and causes little if any damage (possibly a slight heating effect). The records obtained are somewhat more complex to analyze than conventional spectrographic results but essentially similar in genre. It is this aspect that the reader should remember rather than the equipment and details of theory.

A spinning electron generates a magnetic field. In atoms or molecules the electrons are paired and spinning in opposite directions; therefore, their magnetic fields cancel out. A free radical, having an unpaired electron, is *paramagnetic;* that is, it has a magnetic moment because of the unpaired electron spin. Application of a magnetic field of appropriate strength to such a substance aligns its magnetic axes in the direction of the field at the expense of energy from the magnetic field. By measuring the absorption of energy from the magnetic field, information can be obtained about the kinds of free radicals present and their relative concentrations.

Michaelis brought a reaction mixture suspected of having free radicals between the poles of an electromagnet (Fig. 11.11), and after bringing it to a balance point on a sensitive balance, he turned on the current to the electromagnet. The balance point was displaced. The degree of displacement of the balance point measures the change in weight of the reaction mixture before and after application of the magnetic field and is proportional to the concentration of free radicals present. This method is suitable only when free radicals are produced

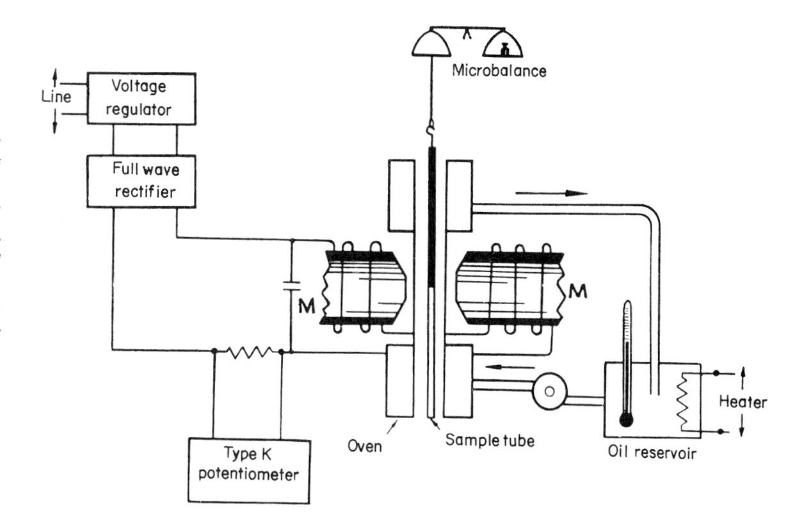

Figure 11.11. *Balance method for measurement of paramagnetism of free radicals. The Gouy apparatus has a thermostat for the sample and means for control and measurement of the magnetizing current. M refers to the electromagnet. (From Blois, 1956:* In Physical Techniques in Biological Research. *Vol. 2. Oster and Pollister, eds. Academic Press, New York.)*

in slow reactions and in relatively high concentrations (Chance, 1961). However, it is possible to refine the method for use in rapid reactions and the method, so modified, is still in limited use (Brill, 1961; Chance *et al.,* 1962).

A much more sensitive and rapid method for detecting the magnetic moments of unpaired electrons of free radicals is the *electron paramagnetic resonance* (EPR) or *electron spin resonance* (ESR) method (Fig. 11.12). Here detection of free radicals depends upon the fact that in a magnetic field the orientation of a free electron with the field differs by a significant and measurable amount of energy from its orientation against the field. The electron, spinning as it does continually, generates a magnetic field. Application of an external magnet causes the electron to precess, just as tipping from the vertical causes a gyroscope (or a top) to precess rather than tip over. The precession of the electron is in the microwave frequency range (about 3 cm or 10,000 megacycles per second). When microwaves travel down a rectangular waveguide (the tube used to guide such waves) they produce a rotating magnetic field at any fixed point. The material to be tested is placed in a side wall of the waveguide (Fig. 11.12), the radiowaves are turned on, and the external magnetic field is applied to make the electrons precess. When the precession rate reaches resonance with the radiowave frequency and the electrons flip to align their fields with the applied magnetic field, they extract energy from the radiowaves and the reading on a receiver at the end of the tube dips accordingly.

The energy involved in causing the flip is given by the following equation:

$$h\nu = g\beta H \tag{11.8}$$

where $h\nu$ is the quantum of electromagnetic energy; h is Planck's constant (6.624×10^{-27} erg seconds); ν is the radiowave frequency sec^{-1}; g is the spectroscopic splitting factor which is 2.0023 for an electron exhibiting pure "spin only" paramagnetism; β is the Bohr magneton $= 0.927 \times 10^{-20}$ erg/gauss; and H is the magnetic field strength in gauss.

In most organic free radicals a small interaction between the electron spin and the electron orbit causes the g-value to deviate from the spin-only value of 2.0023. The g-value varies in the narrow range of 2.000 to 2.0070 for the materials of greatest importance to the cell. Each type of free radical shows a characteristic g-value within this range.

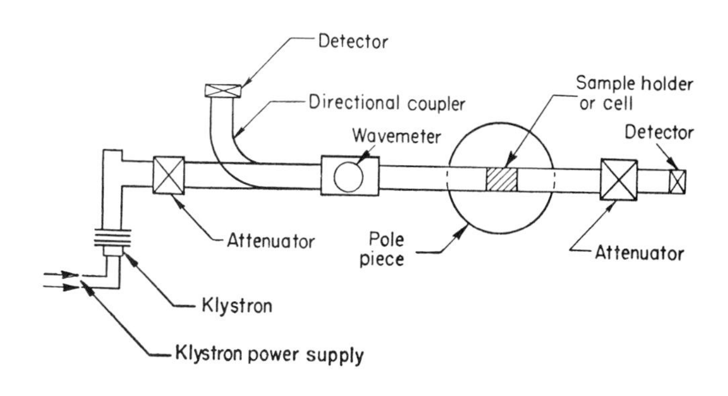

Figure 11.12. *Typical experimental arrangement used to observe electron paramagnetic resonance at microwave frequencies. The klystron is the source of the waves. The sample holder is in the side wall of the waveguide. (From Blois, 1956:* In Physical Techniques in Biological Research. *Vol. 2. Oster and Pollister, eds. Academic Press, New York.)*

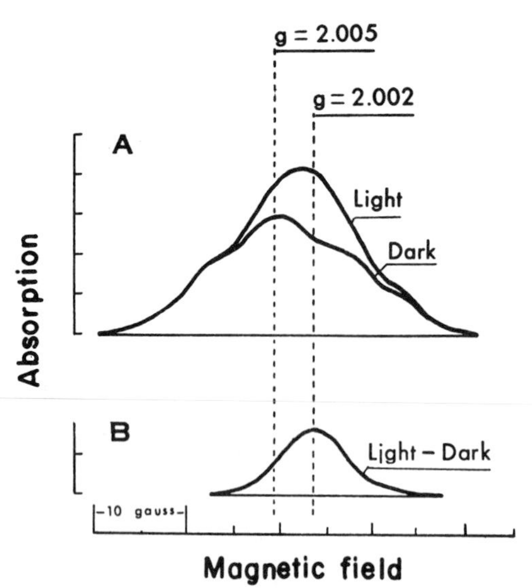

Figure 11.13. A, *Absorption of radiowaves in electron spin resonance of spinach chloroplasts illuminated in one case and in the dark in the other.* B, *Difference between the light and dark absorptions. The meaning of g is defined in the text. The data are integrals of the areas under the deflections in the magnetic field. (From Commoner, 1957: Science 126: 4. Copyright 1957 by the American Association for the Advancement of Science.)*

Free radicals have been found much more widespread in chemical and biological systems than had previously been realized. Free radicals have been detected not only in oxidation-reductions but also in a large number of other reactions of biological interest such as photochemical reactions, thermal cleavage of molecules in chemical systems, photosynthesis, bioluminescence and enzymatic reactions in cells, and as a result of action of radiations upon cells (Beinert and Sands, 1961; Blois and Weaver, 1964; Weaver and Weaver, 1972).

Biological studies of free radicals initially had to be made with lyophilized (low temperature, vacuum-dried) cells because water absorbs so strongly the microwaves used for detection of free radicals. However, it has proved possible to make the measurements with frozen biological materials, thus enabling determination of the free radicals present within the materials at the moment of freezing. Free radicals are usually short lived, but in a frozen system they last for a longer time and concentrations of free radicals of the order of 10^{-6} to 10^{-7} M have been found in intact frozen cells (Herzfeld and Bass, 1957; Blois *et al.*, 1961). Recent improvements in physical methods have made possible the detection of free radicals in unfrozen, undried biological systems, opening the possibility of gathering data during the course of biological reactions, rather than only at the moment of freezing (Blois and Weaver, 1964).

An example of results obtained with ESR analysis is shown in Figure 11.13 for chloroplasts (from spinach leaves) carrying on photosynthesis. It seems that excitation of the photosynthetic system in chloroplasts by absorption of light results in the appearance of free radicals that interact with the magnetic field to give the resultant curve (light deflections minus dark deflections). By this means it has been possible to study the two photochemical systems in photosynthesis (see Chapter 14), identifying signals for a particular system by use of a mutant lacking the other system. Then the rapid exchanges between the two systems can be studied on wild-type cells (Weaver and Weaver, 1972).

Recently the free radical, superoxide ($O_2^{\dot{-}}$), has been found to occur in a wide range of biological reactions, including bioluminescence, photosynthesis, nitrogen fixation, and many other biological oxidations (especially those involving nicotinamide dinucleotide, riboflavin, etc.). The activity of superoxide and its breakdown by the enzyme superoxide dismutase can be followed by electron spin resonance, renewing interest in such studies (Michelson *et al.*, 1977). In some strains of "strictly anaerobic bacteria," such as the sulfate-reducing *Desulfovibrio* and *Clostridium,* electron spin resonance studies indicate the presence of superoxide free radicals when oxygen is admitted to a culture. Presumably mutants have appeared that synthesize superoxide dismutase and catalase and thereby are protected from occasional entry of oxygen from the environment (Hachikian *et al.*, 1977).

The main pattern traced on the record of ESR on the receiver is from the unpaired electron in the free radical, but the fine structure is the result of the interaction of its field with the fields of certain surrounding atomic nuclei. Local fields produced by the nuclei may be strong enough to "split" the electron resonance. It was hoped that this "hyperfine" structure would "fingerprint" free radicals in biological systems and permit their identification, but really well resolved patterns have not thus far characterized biological systems, although consistent improvement has been noted (Weaver and Weaver, 1972; Schwartz, 1972). The method continues to be useful, especially in photosynthetic studies, because of the distinctive signals for the two photosystems involved in photosynthesis (Brown, 1977) (see Chapter 14). ESR is also very useful in studies on nitrogen fixation because the enzyme nitrogenase contains iron in several valency states, an ideal condition for ESR studies (Postgate, 1978). More excitement has been caused by related studies using nuclear magnetic resonance discussed in the next section.

NUCLEAR MAGNETIC RESONANCE

Although nuclear magnetic resonance in bulk materials was observed in 1945, it was not applied to biological substances until relatively recently. At present there is intense interest and activity among scientists about nuclear magnetic measurements of biological materials (Dwek, 1973; Knowles *et al.*, 1976; Opella, 1977).

Because the nucleus of a molecule or an atom has a positive charge, associated with its spin is a circulation of electric charge that gives rise to a magnetic moment. When a nucleus (e.g., a proton in a hydrogen atom) is placed in a uniform magnetic field, the torque upon it orients it so that its north-seeking pole points toward the north of the magnetic field. The nucleus does not assume a stable orientation but precesses around the applied field, like a gyroscope in a gravitational field. As the nucleus orients itself in the magnetic field, it absorbs a quantum of energy from the radiofrequency radiation imposed upon it.

For nuclear magnetic studies a permanent magnet or an electromagnet may be used. Because the signal (useful measurement) to noise (static) ratio is greater in a stronger magnetic field, strong magnets are desirable, and much progress has been made in the production of powerful magnetic fields in superconducting materials. For details of instrumentation, both with regard to magnetic fields and radiofrequency radiation, the reader is referred to textbooks on the subject (e.g., Dwek, 1973; James, 1975; Knowles *et al.*, 1976). In general, radiowaves are provided much as in electron spin resonance and recording of spectra is achieved in a somewhat similar manner (Fig. 11.14). The radiofrequency used must have the

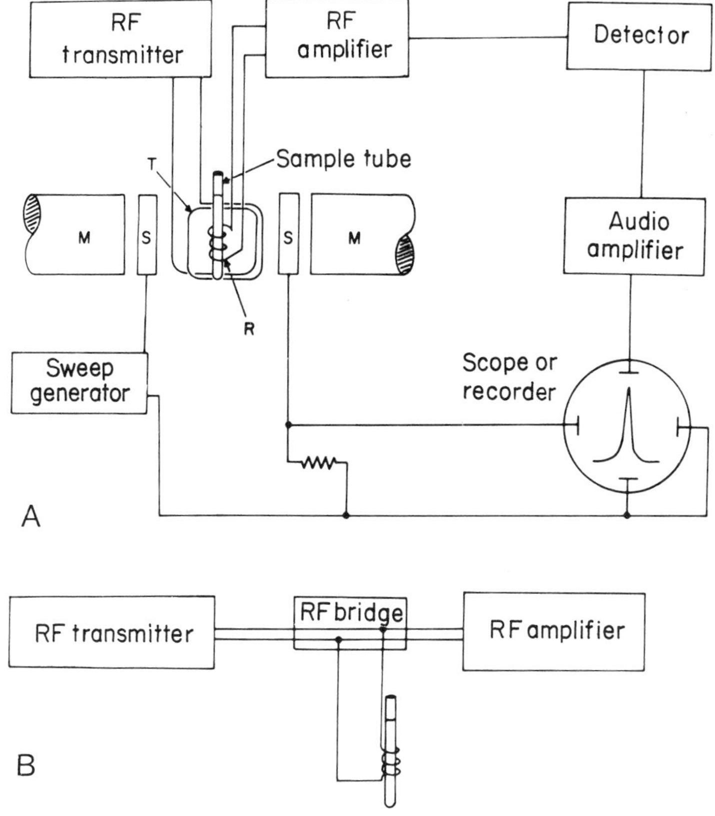

Figure 11.14. *Equipment for study of nuclear magnetic resonance.* A, *Block diagram of a double-coil nuclear magnetic resonance spectrometer.* B, *Modification of the equipment for a single coil spectrometer. RF = radio frequency; M = magnet; S = sweep coils; T = transmitter; R = receiver coil. The sample is placed in the cylindrical glass tube between the pole pieces of the magnet M and the tube is placed inside the receiver coil R, which is so oriented that the receiver, the coil axis, the transmitter coil axis, and the magnetic field are at right angles to one another. The RF transmitter applies an oscillating radiofrequency field to the sample via the transmitter coil at a frequency appropriate to achieve resonance at the field strength of the magnet. The magnet is swept through the resonance condition by varying the strength of the magnet's sweep coils, S, with a sweep generator. The magnetic field strength and the frequency are very close to the resonance condition before and after the sweep current is applied. (From James, 1975: Nuclear Magnetic Resonance in Biochemistry. Academic Press, New York, p. 126.)*

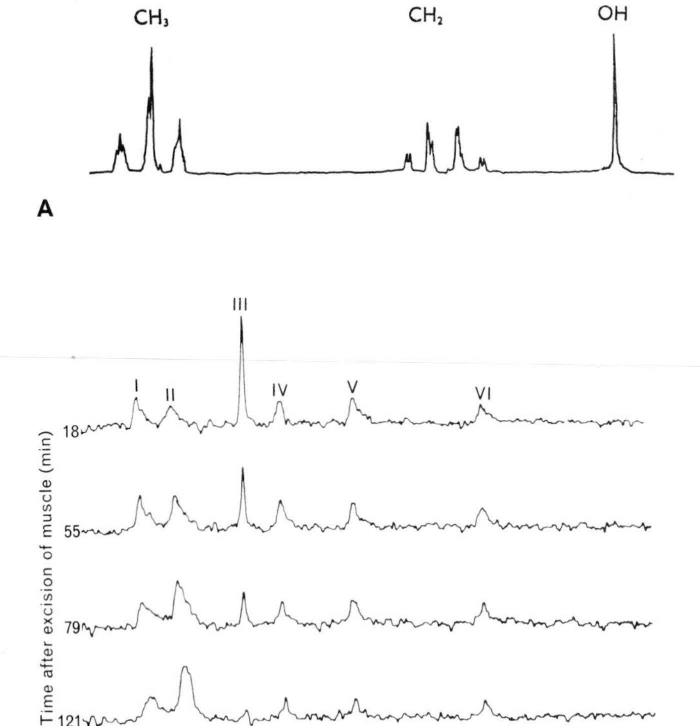

Figure 11.15. *Examples of NMR spectra.* A, *Proton resonance spectrum of ethyl alcohol showing fine structure arising from second order interaction. (From Richards, 1957: Endeavour* 16: *192.)* B, *^{31}P NMR spectrum of an intact muscle from the hind leg of a rat, recorded at 129 megacycles/sec. Temperature 20°C. Labels to peaks: I, sugar phosphate and phospholipid; II, inorganic phosphate; III, creatine phosphate; IV, V, and VI, ATP (γ, α, βATP's, respectively). The times indicated are the midpoints of the 50 scan spectral accumulations (referred to excision time as zero). Note the rapid decline in creatine phosphate, the rise in inorganic phosphate, and the gradual decline in ATP, as the high-energy phosphates are used up. (From Richards, 1975: Endeavour* 34: *122.)*

resonance frequency of the same magnitude as the nucleus of interest (e.g., proton, phosphorus, etc.).

Of greater biological interest are the kinds of nuclei that can be investigated, the type of problems elucidated by nuclear magnetic resonance, and the type of spectra obtained. Proton nuclei are generally of great biological interest and give interesting spectra. For example, the proton nuclear magnetic spectrum of ethanol has three peaks (Fig. 11.15*A*) arising from the methyl, methylene, and hydroxyl radicals, each of which has protons; the first has three, the second two, and the third only one. The height of a peak is proportional to the number of protons in each radical. It is also possible to distinguish between structural differences, for example, cis- and trans- isomers. A marker such as ^{13}C, which distributes itself over all the carbons in a compound, has revolutionized studies of the structure of carbon compounds because the position of a carbon in an organic compound determines its characteristic absorption spectrum.

Sometimes it is possible to get information from nuclear magnetic spectra not otherwise available. In simple molecules, for example, diketene, it was possible to get decisive evidence as to which of two possible structures postulated was the correct one. Biological molecules have also been investigated by this technique. However, in complex molecules such as proteins, with many protons in the sequence of amino acids of which they are composed, the proton spectrum "is imperfectly resolved and of alarming complexity" (Richards, 1975). Therefore, it is necessary to use paramagnetic ligands (binders) to separate off particular portions of the molecule. By this means simpler and more readily interpretable absorption spectra may be obtained. Another approach is to alter molecules in a small way and to analyze the difference in spectra from the unaltered and altered molecules, thereby excluding the common denominators.

Considerable success has been obtained in studies on biological molecules containing phosphorus ^{31}P in phosphates. For example, by using appropriate radiofrequencies for the phosphorus nucleus in phosphate and varying the angle of the applied magnetic field, it was possible to show that the phosphorus-containing polar groups of phospholipids in a bilayer undergo limited internal movement and that their average orientation is perpendicular to the plane of the bilayer membrane. It has also been possible to study the absolute concentration of phosphorus in muscular tissue and to follow the changes in concentration, state of ionization, and the molecular movement of the phosphorus compounds (Fig. 11.15*B*) (Richards, 1975). Deuterium, sodium, and potassium are among the other atomic nuclei available for nuclear magnetic resonance studies. Imaging of tumors and diagnosing cancer are some newer applications of the method (Damadian *et al.*, 1976).

There are many other applications of nuclear

magnetic resonance, some of which have already been mentioned in other chapters. However, the details in this field and the interpretations are quite complex and beyond the scope of this discussion. Reviews and textbooks on the subject may be consulted by those interested (Dwek, 1973; James, 1975; Opella, 1977).

DETERMINATION OF REDOX POTENTIALS IN THE LIVING CELL

Cell activities generally influence the redox potentials of the environment by reducing the oxygen tension and shifting the potential toward a more negative value. Changes in redox potentials, whether naturally or artificially produced, profoundly influence growth and differentiation of some cells (Cater, 1960).

It is interesting to determine whether the redox potential of a whole cell, or some part of it, is changed when the cell is subjected to various environmental conditions. A number of approaches have been used to measure intracellular redox potentials. Some investigators, for instance, have inserted microelectrodes into a cell, but satisfactory electrodes are not easy to make, and it is not possible to be certain that the cell remains in healthy condition (Cater, 1960).

Another method used for measuring redox potentials of cells is to inject into a living cell some dye which might act as an oxidation-reduction indicator, determining from the color change whether the dye is reduced or oxidized. By injecting, in succession, dyes covering a range of E_0' values, the likely E_0' value of the cell may be determined. Of course, the E_0, E_0', and titration curves must first be determined

TABLE 11.2. NORMAL OXIDATION-REDUCTION POTENTIALS OF SOME BIOLOGICALLY IMPORTANT SYSTEMS AT pH 7.0

System	E_0'	*T in °C*
Ketoglutarate ⇌ succinate + CO_2 + $2H^+$ + 2e	−0.68	—‡
Ferredoxin reducing substance	−0.60	—‖
Isocitrate ⇌ α-ketoglutarate + CO_2	−0.48	—
Ferredoxin	−0.432	—§
Formate ⇌ CO_2 + H_2	−0.420	38
H_2 ⇌ $2H^+$ + 2e	−0.414	25
NADH + H^+ ⇌ NAD^+ + $2H^+$ + 2e	−0.317	30†
NADPH + H^+ ⇌ $NADP^+$ + $2H^+$ + 2e	−0.316	30†
Horseradish oxidase	−0.27	—†
$FADH_2$ ⇌ FAD + $2H^+$ + 2e	−0.219	30†
$FMNH_2$ ⇌ FMN + $2H^+$ + 2e	−0.219	30†
Lactate ⇌ pyruvate + $2H^+$ + 2e	−0.180	35
Malate ⇌ oxaloacetate + $2H^+$ + 2e	−0.102	37
Reduced flavin enzyme ⇌ flavin enzyme + $2H^+$ + 2e	−0.063	38
Vitamin K	−0.060	—¶
Luciferin* ⇌ oxyluciferin + $2H^+$ + 2e	−0.050	?*
Ferrocytochrome *b* ⇌ ferricytochrome *b* + e	−0.04	25
Succinate ⇌ fumarate + $2H^+$ + 2e	−0.015	30
Plastoquinone	0.00	—‖
Decarboxylase	+0.19	—†
Ferrocytochrome *c* ⇌ ferricytochrome *c* + e	+0.26	25
Ferrocytochrome *a* ⇌ ferricytochrome *a* + e	+0.29	25
Ferrocytochrome a_3 ⇌ ferricytochrome a_3 + e	?	—‡
Plastocyanin	+0.36	—‖
P_{700}	+0.40	—‖
Coenzyme Q (in ethanol)	+0.54	—
H_2O ⇌ $\frac{1}{2}O_2$ + $2H^+$ + 2e	+0.815	25

Data from Goddard, 1945: Potentials in all cases are at or near neutrality, except:
*From McElroy and Strehler, 1954: Bact. Rev. *18*.
†From Clark, 1960: Oxidation Reduction Potentials of Organic Systems. Williams & Wilkins, Baltimore.
‡From Goddard and Bonner, 1960: *In* Plant Physiology, a Treatise. Steward, ed. Academic Press, New York. Goddard and Bonner give the NADPH/$NADP^+$ system as −0.324, and NADH/NAD^+ as −0.320.
§From Tagawa and Arnon, 1962: Nature *195*:537–543. The value cited is for spinach ferredoxin.
‖Rabinowitch and Govindjee, 1971: Photosynthesis. Wiley, New York. The relative positions of Cyt *f* (preceding plastocyanin) and Cyt b_3 (preceding plastoquinone) in green plant cells have been determined but the exact values have not.
¶Lehninger, 1965: The Mitochondrion. W. A. Benjamin, Menlo Park, Calif.

separately for the various oxidizable dyes used. For example, at pH 7.0 methylene blue has an E_0' of +0.011 volt, pyocyanin of −0.034 volt and indophenol just below the E'_0 of cytochrome *c* (−0.26 volt). As might be expected, in the absence of oxygen the overall redox potential of a cell was found to be more negative than in the presence of oxygen (Hewitt, 1950).

Since the cytoplasm of the cell is a polyphasic colloid and since it has many particles, organelles, and surfaces upon or within which many enzymatic reactions may occur independently at the same time, it would be of interest to study the potentials of such isolated fragments. A few such studies have been made, particularly with mitochondria, which are active centers of aerobic processes (Cater, 1960).

In summary, oxidation-reduction studies on biological systems indicate that the sequence of reactions in the cell occurs along a most probable biological electromotive series. Not only enzyme kinetics but also enzyme thermodynamics appear to be in accord with physical-chemical principles. The fundamental basis of enzymatic action is well established, although much remains to be learned about the manner in which enzymes carry out their functions (Lehninger, 1975).

APPENDIX

11.1 DERIVATION OF PETER'S EQUATION FOR REDOX POTENTIALS

To derive Peter's equation (11.6), it is necessary to equate the mechanical work to the electrical work done in reducing 1 gram mole of metal:

$$\text{Metal} \rightleftharpoons e_{\text{solution}} + \text{ion} \qquad (11.2)$$

If for purposes of argument the electron e is considered to act as a gas particle, the formulation of laws applying to gases will apply to electrons which are at pressure e_{metal} in the metal and e_{solution} in the solution.

If a gas is allowed to expand slowly in a cylinder by pushing a piston over the distance l (an isothermal, reversible, ideal expansion) (Fig. 11.16), an equilibrium existing all the time, the work done (W) is the force (F) times the distance (l):

$$W = Fl \qquad (11.9)$$

If the piston moves only the distance (dl) the work done (dW) equals:

$$dW = Fdl \qquad (11.10)$$

Since
$$P = F/\text{area} \qquad (11.11)$$

then
$$F = P \cdot \text{area} \qquad (11.12)$$

Substituting this value of F into equation 11.10:

$$dW = P \cdot \text{area} \cdot dl = PdV \qquad (11.13)$$

since area × distance = volume.

According to the gas law,

$$PV = RT \text{ or } P = RT/V \qquad (11.14)$$

Substituting this value of P into equation 11.13 gives:

$$dW = \frac{RT}{V}\,dV, \text{ or } RT\,\frac{dV}{V} \qquad (11.15)$$

Integrating the equation between V_1 and V_2 gives:

$$W = RT\,ln\,\frac{V_2}{V_1} \qquad (11.16)$$

Here R is the gas constant (8.315 joules per degree), T the absolute temperature, and *ln* the natural logarithm (2.302 × $\log_{10}$).

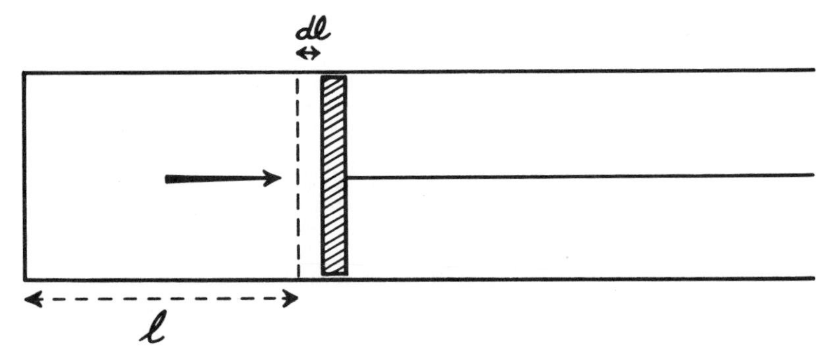

***Figure 11.16.** Scheme of the mechanical work equated to the electrical work in Peter's equation.*

But $$V_2 = \frac{RT}{P_2}$$

and $$V_1 = \frac{RT}{P_1} \quad (11.17)$$

Substituting these values of V_2 and V_1 in equation 11.16,

$$W = RT\ ln \frac{RT/P_2}{RT/P_1} = RT\ ln \frac{P_1}{P_2} \quad (11.18)$$

Since for purposes of the derivation the electrons are considered to act as gas particles, they should exert pressure. If the pressure of the electrons in the metal e_{metal} is taken as equal to P_1 and $e_{solution}$ as equal to P_2, substitution of these values in equation 11.18 gives:

$$W = RT\ ln \frac{e_{metal}}{e_{solution}} \quad (11.19)$$

The electrical work (W) done in the same process is a product of the valency of the metal, the Faraday or quantity of electricity transferred, and the potential against which the work is done:

$$W = n\mathscr{F}E \quad (11.20)$$

Here n is the valency of the metal, $\mathscr{F}$ the Faraday (96,500 coulombs), and E the potential in volts. Since the work for reduction of a gram mole of metal is the same whichever way it is calculated, then:

$$n\mathscr{F}E = RT\ ln \frac{e_{metal}}{e_{solution}} \quad (11.21)$$

$$E = \frac{RT}{n\mathscr{F}} ln \frac{e_{metal}}{e_{solution}} \quad (11.22)$$

$$E = \frac{RT}{n\mathscr{F}} ln\ e_{metal} - \frac{RT}{n\mathscr{F}} ln\ e_{solution} \quad (11.23)$$

If $RT/n\mathscr{F}\ ln\ e_{metal}$ is set at C′, since it is a property characteristic of each metal, then:

$$E = C' - \frac{RT}{n\mathscr{F}} ln\ e_{solution} \quad (11.24)$$

However, since metal = $e_{solution}$ + ion, or, in more general terms:

$$\text{Reductant} = e_{solution} + \text{oxidant} \quad (11.25)$$

then by the law of mass action:

$$\frac{[e_{sol.}][ox.]}{[red.]} = K \quad (11.26)$$

$$[e_{sol.}] = K\frac{[red.]}{[ox.]} \quad (11.27)$$

Substituting 11.27 into 11.24,

$$E = C' - \frac{RT}{n\mathscr{F}} ln \frac{[red.]}{[ox.]} - \frac{RT}{n\mathscr{F}} ln\ K \quad (11.28)$$

Since C′ − $(RT/n\mathscr{F})\ ln$ K is a constant for any metal (or oxidation-reduction system) it may be set equal to E_0:

$$E_0 = C' - \frac{RT}{n\mathscr{F}} ln\ K \quad (11.29)$$

Then substituting this value of E_0 in equation 11.28 gives:

$$E = E_0 - \frac{RT}{n\mathscr{F}} ln \frac{[red.]}{[ox.]} \quad (11.30)$$

or

$$E = E_0 + \frac{RT}{n\mathscr{F}} ln \frac{[ox.]}{[red.]} \quad (11.31)$$

Peter's equation may be simplified if it is applied under a special set of conditions, for example, at 30°C, or 303° absolute, and with n equal to 2, as was done in equation 11.6,

$$E = E_0 + 0.030 \log_{10} \frac{[oxidant]}{[reductant]} \quad (11.6)$$

Also, by setting,

$$[oxidant] = [reductant],$$

in which case,

$$\log_{10} \frac{[oxidant]}{[reductant]} = 0,$$

then: $$E = E_0$$

E_0 values for various oxidation systems are given in Table 11.2. Comparisons of the redox potentials of compounds are most frequently made on this basis.

11.2 EQUATION FOR THE EFFECT OF pH ON REDOX POTENTIALS

If hydrogen ions are produced in the course of an oxidation-reduction reaction, a second term must be added to Peter's equation. Thus:

$$H_2Q \rightleftharpoons Q + 2H^+ + 2e \quad (11.32)$$

Reductant in solution	Oxidant in solution	On electrode

The equation corresponding to this reaction is:

$$E = E_0 + \frac{RT}{n\mathscr{F}} ln \frac{[ox.]}{[red.]} + \frac{RT}{\mathscr{F}} ln [H^+] \quad (11.33)$$

If we substitute numerical values for R, n (the valency change, taken as 2 for organic redox systems since two electrons are generally involved, and as 1 for hydrogen since only one electron is involved), $\mathscr{F}$, and T (take T as 30°C, or 303° absolute) and give the equation in terms of $\log_{10}$, the equation becomes:

$$E = E_0 + 0.03 \log_{10} \frac{[\text{oxidant}]}{[\text{reductant}]} - 0.06\ (\text{pH}) \tag{11.34}$$

LITERATURE CITED AND GENERAL REFERENCES

Beinert, H. and Sands, R. H., 1961: Semiquinone formation of flavins and flavoproteins. *In* Free Radicals in Biological Systems. Blois, ed. Academic Press, New York, pp. 17–52.

Blois, M. S., 1956: Magnetic methods. *In* Physical Techniques in Biological Research. Oster and Pollister, eds. Academic Press, New York, Vol. 2, pp. 393–440.

Blois, M. S., ed. 1961: Free Radicals in Biological Systems. Academic Press, New York.

Blois, M. S. and Weaver, E. C., 1964: Electron spin resonance and its application to photophysiology. *In* Photophysiology. Giese, ed. Academic Press, New York, Vol. 1, pp. 35–63.

Bray, R. C., 1969: Electron paramagnetic resonance in biochemistry. Fed. Eur. Biochem. Soc. Lett, *5*: 1–6.

Brill, A., 1961: The detection of free-radical intermediates in biochemical reactions by their magnetic susceptibility. *In* Free Radicals in Biological Systems. Blois, ed. Academic Press, New York, pp. 53–74.

Brown, J. S., 1977: Spectroscopy of chlorophyll in biological and synthetic systems. Photochem. Photobiol. *26*: 319–326.

Cater, D. B., 1960: Oxygen tension and oxidation-reduction potential in living tissues. Prog. Biophys. *10*: 153–194.

Chance, B., 1961: Free radicals and enzyme-substrate compounds. A tribute to Lenor Michaelis. *In* Free Radicals in Biological Systems. Blois, ed. Academic Press, New York, pp. 1–16.

Chance, B., Cohen, P., Jobsis, F. and Schoener, B., 1962. Intracellular oxidation reduction states in vivo. Science *137*: 499–506.

Clark, W. M., 1960: Oxidation Reduction Potentials of Organic Systems. Williams & Wilkins, Baltimore.

Clark, W. M. and collaborators, 1928: Studies on Oxidation-Reduction. Hygienic Lab. Bull. No. 151. A collection of 15 papers.

Damadian, R., Minkoff, L., Goldsmith, M., Stanford, M. and Koutcher, J., 1976: Field focusing nuclear magnetic resonance: visualization of a tumor in a live animal. Science *194*: 1430–1431.

Dwek, R. A., 1973: Nuclear Magnetic Resonance. Applications to Enzyme Systems. Oxford University Press, London.

Dyer, J. R., 1965: Applications of Absorption Spectroscopy of Organic Compounds. Prentice-Hall, Englewood Cliffs. N.J. (Chapter 4 on nuclear magnetic resonance.)

Eyring, H., Henderson, D. and Just, W., eds., 1970: Physical Chemistry, An Advanced Treatise. Vol. 4. Molecular Properties. Academic Press, New York.

Hachikian, C. E., Le Gall, J., and Bell, G. R., 1977: Significance of superoxide dismutase and catalase activities in the strict anaerobes, sulfate-reducing bacteria. *In* Superoxide and Superoxide Dismutase. Michelson, McCord, and Fridovich, eds. Academic Press, New York, pp. 159–172.

Herzfeld, C. M. and Bass, A. M., 1957: Frozen free radicals. Sci. Am. (Mar.) *196*: 90–104.

Hewitt, L. F., 1950: Oxidation-Reduction Potentials in Bacteriology and Biochemistry. 2nd Ed. Williams & Wilkins, Baltimore.

Ingram, D. J. E., 1969: Biological and Biochemical Applications of Electron Spin Resonance. Adam Hilger, London.

James, T., 1975: Nuclear Magnetic Resonance in Biochemistry. Principles and Applications. Academic Press, New York.

Jardetzky, O., 1962: Introduction to magnetic resonance spectroscopy: methods and biochemical applications. *In* Methods of Biochemical Analysis. Glick, ed. *9*: 235–410.

Johannsen, C. B. and Coyle, T. D., 1972: Nuclear double magnetic resonance. Endeavour *31*: 11–15.

Knowles, P. F., Marsh, D. and Rattle, W. E., 1976: An Introduction to the Theory and Practice of NMR and ESR in Biological Systems. John Wiley & Sons, New York.

Lasker, S. F. and Milvy, P., eds., 1973: Electron spin resonance and nuclear magnetic resonance in biology and medicine. Ann. N. Y. Acad. Sci. *222*: 1–1124.

Lehninger, A. L., 1975: Biochemistry. 2nd Ed. Worth Publishers, New York.

Michaelis, L., 1930: Oxidation-Reduction Potentials. J. B. Lippincott Co., Philadelphia.

Michaelis, L., 1935: Semiquinones, the intermediate steps of reversible organic oxidation-reduction. Chem. Rev. *16*: 243–286.

Michaelis, L., 1940: Occurrence and significance of semiquinone radicals. Ann. N. Y. Acad. Sci. *40*: 39–76.

Opella, S. J., 1977: Biological nuclear magnetic spectroscopy. Science *198*: 158–165.

Postgate, J., 1978: Nitrogen Fixation. Edward Arnold, London.

Pryor, W. A., 1970: Free radicals in biological systems. Sci. Am. (Aug.) *223*: 70–83.

Pryor, W. A., ed., 1977: Free Radicals in Biology. Vol. 3. Academic Press, New York.

Richards, R. E., 1975: Nuclear magnetic resonance spectroscopy of biochemical materials. Endeavour *34*: 118–122.

Schwartz, H. M., 1972: Biological Applications of Electron Spin Resonance. John Wiley & Sons, New York.

Weaver, E. C. and Weaver, H., 1972: Electron resonance studies on photosynthetic systems. *In* Photophysiology. Giese, ed. Academic Press, New York, Vol. 7, pp. 1–32.

CHAPTER 12

THE RELEASE OF ENERGY IN CELLS

The life activities of the cell demand both compounds of high-energy potential, which can release energy quickly for performing cell work, and a reduction pool for synthesis of the compounds necessary for life, in addition to "building block" molecules for syntheses. As shown in this chapter, cellular oxidoreductions, anaerobic and aerobic, supply both these needs using cell nutrients. Considered first, glycolysis, which precedes the aerobic reactions and is probably a relic of the anaerobic period of life on earth, releases only about 5 percent of the energy in cell nutrients. The aerobic reactions that follow glycolysis release the remaining 95 percent of the energy remaining in the products of glycolysis and supply most of the high-energy compounds (in phosphorus and sulfur bonds) along with a reduction pool (NADH). The reduction pool serves to carry the hydrogen released in the oxidative dehydrogenations to oxygen with the formation of water, accompanied by a considerable release of energy. The reduction pool for syntheses comes largely from another pathway of glucose degradation, the pentose shunt, which liberates high-energy bonds along with production of a pool of NADPH. The pentose shunt is considered last.

In essence, the energy possessed by the highly reduced compounds serving as cell nutrients is liberated in the oxidoreductions that continue from a potential near the hydrogen electrode at pH 7 to the level of the oxygen electrode at pH 7 (see Chapter 11). In the process NAD is reduced to NADH + H^+. The hydrogen ions, being soluble in water, pass into the continuum of the cell, later combining with oxygen to form water. The electrons, insoluble in water, are transferred by a series of electron transport carriers down the potential gradient. In the process of electron transport some of the energy released is conserved in high-energy phosphate bonds of ATP, as discussed later in this chapter.

THE NATURE OF OXIDOREDUCTIONS

The very word *oxidation* implies a process in which a substance combines with oxygen—for example, the union of hydrogen and oxygen in the formation of water—and early workers so defined it. Hydrogen gives up electrons to oxygen and is thereby oxidized; oxygen, taking up the electrons, is thereby reduced:

$$2H_2 + O_2 \rightarrow 2H_2O \qquad (12.1)$$

In many cases, however, electrons are given up even when oxygen is not present; for example, in the formation of ferric from ferrous ion:

$$Fe^{++} \underset{\text{reduction}}{\overset{\text{oxidation}}{\rightleftharpoons}} Fe^{+++} + e \qquad (12.2)$$

Fe^{++} is the electron donor that is oxidized to Fe^{+++}. In the reverse reaction Fe^{+++} is the electron acceptor that is reduced to Fe^{++}. *Oxidation,* in its most generalized sense, is at present defined as the donation of electrons from one molecule (ion, atom, radical) to another, and *reduction* is defined as the acceptance of electrons by one molecule (ion, atom, radical) from another. The two processes always occur simultaneously.

In many oxidoreductions that occur in the cell, hydrogen transfer accompanies electron transfer. For example, when succinic acid is oxidized to fumaric acid, two hydrogen ions and two electrons are removed. In this case, oxidation consists of the transfer of hydrogen ions and electrons from succinic acid to the enzyme catalyzing the reaction:

$$\underset{\text{Succinic acid}}{\begin{array}{c} COOH \\ | \\ CH_2 \\ | \\ CH_2 \\ | \\ COOH \end{array}} \xrightarrow[\text{Succinic dehydrogenase}]{-(2H^+ + 2e)} \underset{\text{Fumaric acid}}{\begin{array}{c} COOH \\ | \\ CH \\ \| \\ CH \\ | \\ COOH \end{array}} \qquad (12.3)$$

The donor of hydrogen ions and electrons is a reductant called the *hydrogen donor*. On giving up hydrogen ions and electrons it is oxidized. The enzyme catalyzing the reaction and accepting hydrogen ions and electrons is an oxidant called the *hydrogen acceptor*. It is reduced in the process. A generalized equation for such an oxidation-reduction reaction may be written as follows:

$$\underset{\text{H-donor}}{H_2D} + \underset{\text{H-acceptor}}{A} \xrightarrow{\text{oxidation}} \underset{\text{Oxidized H-donor}}{D} + \underset{\text{Reduced H-acceptor}}{H_2A} \qquad (12.4)$$

When oxygen is present in a cell, it generally acts as the terminal hydrogen and electron acceptor, becoming reduced to water in the process. Oxygen readily accepts electrons from molecules (or from ions, radicals, and so forth) and it is because of this property that oxidations in which oxygen participates were studied first. In the major pathway or "main line" of oxidation in the cell, oxygen serves as the final hydrogen acceptor, and water (H_2O) is the end product. When flavoproteins pass hydrogen ions and electrons to oxygen in other than the main line of oxidation, H_2O_2 is usually formed.

STEPWISE RELEASE OF ENERGY DURING OXIDOREDUCTIONS

Whereas many cell hydrolyses proceed with small changes in free energy, cell oxidoreductions usually are accompanied by a considerable change in free energy. Cell work is directly or indirectly accomplished by this free energy. Chemical potential energy is usually stored in cells in compounds containing so-called high-energy phosphate bonds (e.g., adenosine triphosphate), from which the energy can be liberated and used for synthesis of other compounds in the cell.

All the energy available in a nutrient molecule is never liberated by the cell in one step. The complete oxidation of glucose into carbon dioxide and water occurs with a ΔG of $-686{,}000$ cal/mole. The cell has no way of using this much free energy at once; it could efficiently convert only a very small fraction of it. Rather, the cell degrades the glucose stepwise, with a small free energy change in each step, each step usually catalyzed by a specific enzyme. The merit of a series of stepwise reactions is that they can be coupled energetically by a common reactant and the energy liberated is transferred to high-energy phosphate bonds through the coupled reactions. A sequence of reactions involving small free energy changes is therefore advantageous to the organism. The overall efficiency, as indicated by the formation of 36 high-energy phosphate bonds during the decomposition of 1 mole of glucose, is about 40 percent (Lehninger, 1975), a very high value compared to that of machines.

PATHWAYS OF OXIDOREDUCTIONS IN CELLS

Figure 12.1 shows a general scheme for oxidation-reduction of glucose in cells, during which glucose is gradually decomposed, with the evolution of energy, into carbon dioxide and water as the end products. Glucose is first phosphorylated, and after atomic rearrangements in the molecule, it is split into two triose phosphates. Each triose phosphate, in the presence of dehydrogenases and other enzymes, is oxidized to pyruvic acid. This phase of glucose decomposition can occur in the presence or absence of oxygen and is discussed in more detail later as the anaerobic decomposition of glucose (glycolysis).

In the second phase of glucose decomposition, which occurs only in the presence of oxygen, the pyruvic acid formed as the end product of the first series of reactions described in the preceding paragraphs is first decarboxylated (carbon dioxide being given off) and the 2-carbon fragment, called acetyl, is combined with a coenzyme called *coenzyme A*, or CoA (see Fig. 12.2). The acetyl-CoA combines with a 4-carbon acid to form a 6-carbon acid. The 6-carbon acid is subsequently decarboxylated and oxidized again to form a 5-carbon acid, carbon dioxide and water being given off in the process. A series of decarboxylations follows, each one resulting in the liberation of carbon dioxide and water. As a result of this entire series of reactions glucose is completely decomposed to car-

Process	O_2 Need	Reactions	High Energy Phosphates*	Classes of Enzymes Involved
Glycolysis	−	Glucose (hexose) + ↓ ATP Hexose-phosphate ↓ 2 triose phosphates ↓ 2 pyruvic acid + energy†	~P + ADP → ATP ATP + C → ADP + CP	Phosphate-transferring enzymes Phosphate-splitting enzymes Isomerases Dehydrogenases
Krebs Tricarboxylic Acid (TCA) Cycle and Electron Transport Chain	+	↓ CO_2 + H_2O + energy‡	~P + ADP → ATP ATP + C → ADP + CP	Dehydrogenases Isomerases Flavoproteins Coenzyme Q Cytochromes Cytochrome oxidase Decarboxylase Phosphate-transferring enzymes

Figure 12.1. *General scheme of the decomposition of glucose. *The phosphates are identified as follows:*
ATP = adenosine triphosphate (adenine-pentose-phosphate ~ phosphate ~ phosphate)
ADP = adenosine diphosphate (adenine-pentose-phosphate ~ phosphate)
CP = creatine phosphate (creatine ~ phosphate)
C = creatine
In the cells of some organisms arginine takes the place of creatine.

†About 5 percent of the energy in glucose is released in glycolysis
‡The remaining 95 percent of the energy is released in the TCA cycle.

bon dioxide and water. These reactions of the aerobic decomposition of glucose are discussed in more detail later.

Part of the free energy change in the oxidation of glucose in the cell is conserved in *phosphorylated organic compounds* or the *thioester bonds* of CoA derivatives. The high-energy phosphate bond (usually written ~P) in such compounds is easily transferred from one organic compound to another (e.g., from creatine phosphate to adenosine diphosphate to form adenosine triphosphate) coupled by an enzyme. Thus the phosphate transfer potential of such a compound is high. Therefore, it serves as a convenient means of storing and distributing energy in cells (Boyer *et al.*, 1977). The free

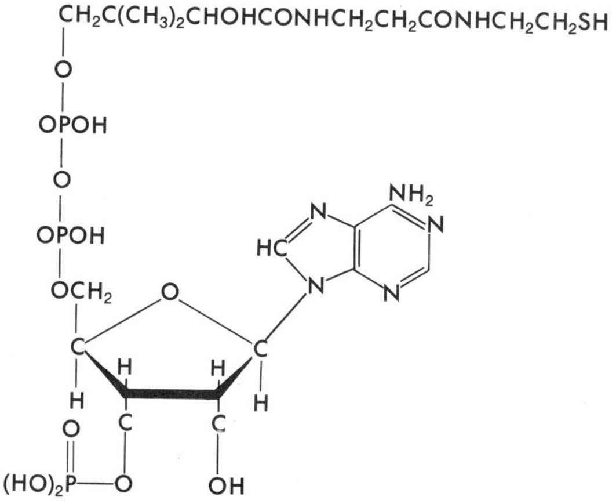

$CH_2C(CH_3)_2CHOHCONHCH_2CH_2CONHCH_2CH_2SH$ — O — OPOH — O — OPOH — OCH_2 (ribose–adenine, with $(HO)_2P=O$ on the ribose)

Figure 12.2. *Coenzyme A.*

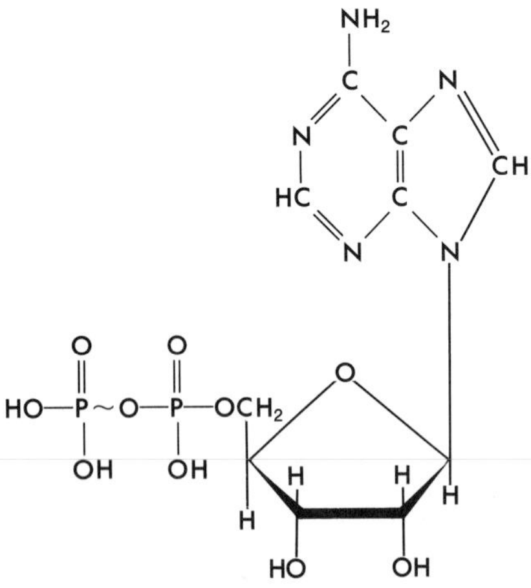

Adenosine diphosphate (ADP)

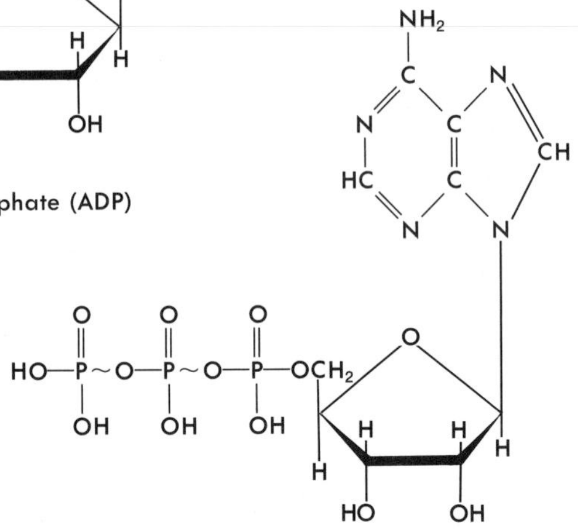

Adenosine triphosphate (ATP)

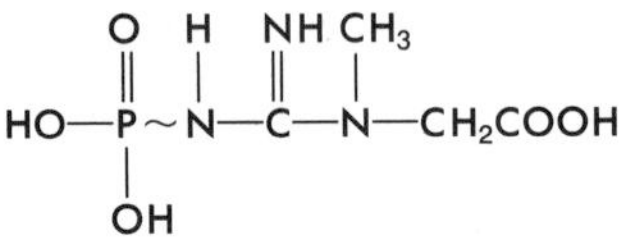

Creatine phosphate (CP)

Figure 12.3. *Compounds of high phosphate transfer potential. In ADP and ATP the nitrogen-containing moiety is adenine; it is attached to the pentose, which in turn is attached to the phosphate.*

energy of such a phosphate bond is about 7300 cal/mole at standard conditions (at 25°C and pH 7.0, when each reactant and product is at thermodynamic activity, i.e., at 1.0 molal), the exact amount depending upon the conditions of the reaction, such as pH and salt environment. When the energy of a phosphate bond is utilized in a reaction (e.g., synthesis of a compound), inorganic phosphate (low-energy content) appears as the end product (Lehninger, 1975; Racker, 1977).

Adenosine triphosphate (ATP), a nucleotide, is composed of the nitrogenous base adenine, a pentose sugar, and three phosphoric acid residues (Fig. 12.3). The bond between the pentose and the first phosphoric acid residue is a low-energy bond, but the bonds between the other two phosphoric acid residues are high-energy phosphate bonds; that is, they have a high phosphate transfer potential (Pauling, 1970).

In *adenosine diphosphate* (ADP), which has two phosphoric acid residues, only one high-energy phosphate bond is present. Free energy release from ATP usually results in the loss of one high-energy bond and formation of one ADP.* The two phosphates are therefore always present together (see Fig. 11.15*B*). In muscle cells a relatively small amount of adenosine triphosphate is present, but a larger supply of another phosphate, known as creatine phosphate (CP), exists as the high-energy

*Splitting of pyrophosphate (PP) from ATP with formation of AMP, with a greater free energy charge than in the formation of ADP, occurs in a number of synthetic reactions discussed in Chapter 15.

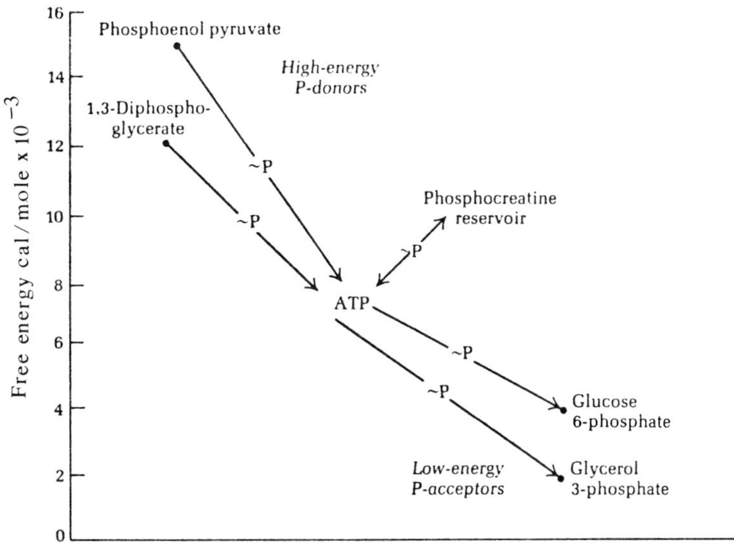

Figure 12.4. *Flow of phosphate groups from high-energy phosphate donors to low-energy accpetors via ATP-ADP system. (After Lehninger, 1975: Biochemistry. 2nd Ed. Worth Publishers, New York.)*

phosphate reservoir (Fig. 12.3). Muscle cells may contain up to 0.5 percent, wet weight, of creatine phosphate. Creatine phosphate in the presence of the appropriate enzyme may transfer its high-energy bond to ADP, forming creatine and ATP. Conversely, creatine may accept one ~P from ATP, forming CP and ADP; ATP and ADP appear to be distributors of high-energy phosphate bonds. In cells of many invertebrates *arginine phosphate* takes the place of creatine phosphate. Creatine phosphate and arginine phosphate are called *phosphagens* (Slater, 1977).

Lehninger (1975) calls attention to the fact that "high-energy phosphate bond" is a misnomer. The fact is that a large difference in free energy exists between reactants and the products, but the energy does not really reside in the phosphate bond. However, the terminology has become so prevalent that it is used here to indicate compounds of high phosphate transfer potential.

A whole series of bonds of different phosphate transfer potential occurs; for example, that of phosphoglyceryl phosphate is almost 12,000 cal/mole and ATP 7300 cal/mole, while glucose-6-phosphate is only 4000 cal/mole, all at pH 7.0 and unit molality (Fig. 12.4). The efficiency of the cell in producing high-energy phosphates from nutrients therefore should be calculated on the basis of 7300 cal/mole for ATP (Lehninger, 1971). Actually, in the cell the concentration of ATP and ADP is much less than 1 molal; furthermore, Mg^{++} complexes with ATP, ADP, and phosphate, shifting the equilibrium toward ATP hydrolysis. In the cell the free energy change is therefore much greater than 7300 cal/mole (Lehninger, 1975). The hydrolysis of ATP has such a large ΔG because the equilibrium constant for the reaction is so large ($\Delta G = -RT\,ln\,K_{eq}$) (see Chapter 9).

ANAEROBIC DECOMPOSITION OF GLUCOSE

The series of steps in glucose breakdown in the cell up to the formation of pyruvic acid, which do not require oxygen, are called *glycolysis* in animals and *fermentation* in microorganisms and plants. The various steps in the direct anaerobic decomposition of glucose are given in simplified form in Figure 12.5 and in more detail in Figure 12.6*A;* these steps are known as the Embden-Meyerhof pathway after two pioneer investigators. Glucose must first be phosphorylated at the expense of a high-energy phosphate from ATP. Glycogenolysis (which is phosphorolysis catalyzed by phosphorylase), on the other hand, liberates a phosphorylated glucose, which already has one high-energy phosphate bond (glucose-1-phosphate). In either case, a second ATP is needed before the hexose phosphate shown in the outline is split into two triose phosphates (3-phosphoglyceraldehyde and dihydroxyacetone phosphate). It should be noted that in the breakdown of triose phosphates, hydrogen is removed by dehydrogenases, and after a series of reactions pyruvic acid appears. Some energy liberated in the anaerobic process is stored in compounds of high phosphate transfer potential, converting ADP to ATP. Since some ATP is used in phosphorylating glucose at the outset, the net gain of ATP is the total stored minus that used. In making a calcu-

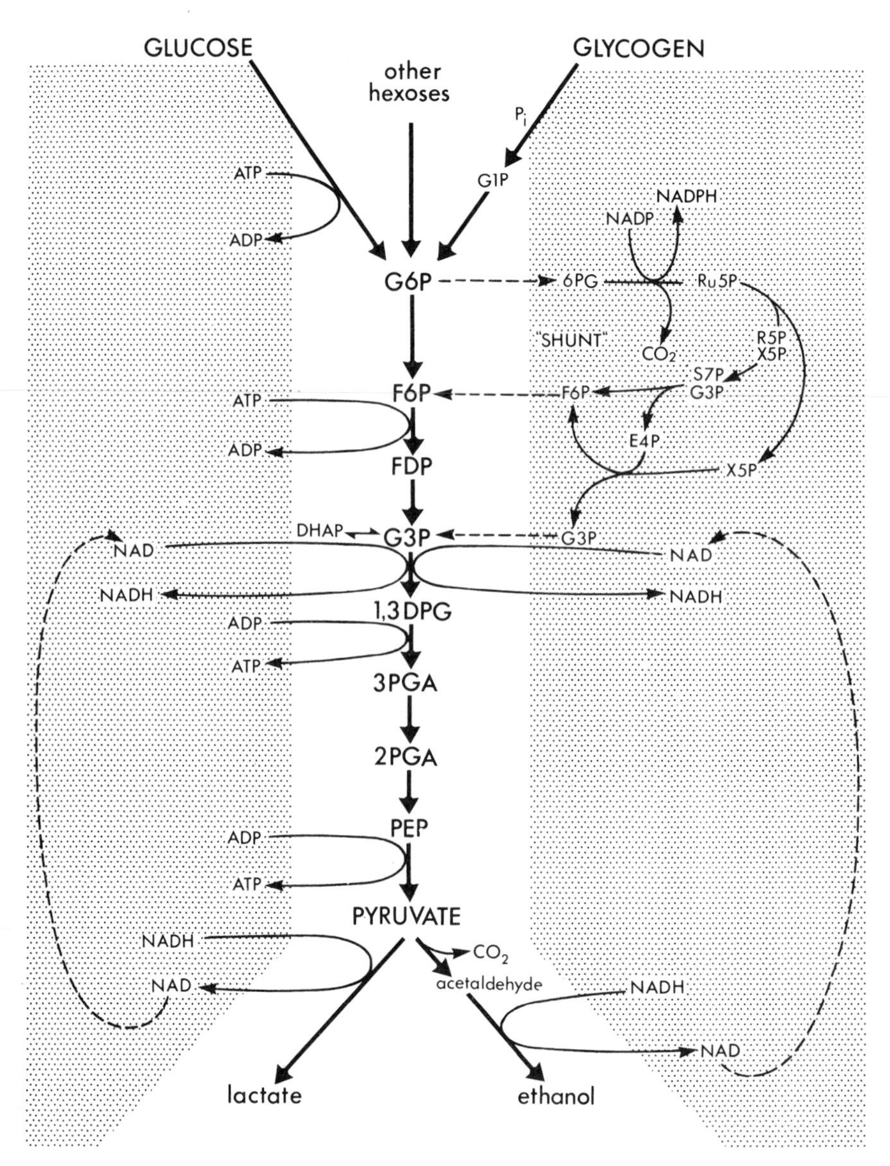

Figure 12.5. *The two basic glycolytic schemes, one leading to lactate and the other leading to ethanol as the major end product. Note that, per mole of glucose mobilized, a net gain of 2 moles of ATP is realized. (From Hochachka and Somero, 1973: Strategies of Biochemical Adaptation. W. B. Saunders Co., Philadelphia, p. 25.)*

Metabolite abbreviations:

G1P	*glucose-1-phosphate*	*DHAP*	*dihydroxyacetone phosphate*
G6P	*glucose-6-phosphate*	*1,3 DPG*	*1,3 diphosphoglycerate*
F6P	*fructose-6-phosphate*	*3 PGA*	*3 phosphoglycerate*
FDP	*fructose-1,6-diphosphate*	*2 PGA*	*2 phosphoglycerate*
G3P	*glyceraldehyde-3-phosphate*	*PEP*	*phosphoenolpyruvate*

lation of such an ATP gain, it should be remembered that each glucose gives rise to two trioses (step 5b → 6 in Fig. 12.6*A*) and that for each triose decomposed to pyruvic acid, two ADP molecules are converted to two ATP molecules. In terms of energy this means that anaerobic oxidation of glucose nets about 14,600 cal/mole starting with glucose, or about 21,900 cal/mole starting with glycogen. (Remember that at the outset one less ATP is needed for phosphorylation when glycogen is used; see top right of Figure 12.5.) The enzymes listed in Table 12.1 are involved in the reactions shown, one enzyme usually catalyzing each step.

Under anaerobic conditions, pyruvic acid acts as a hydrogen acceptor, forming lactic acid,

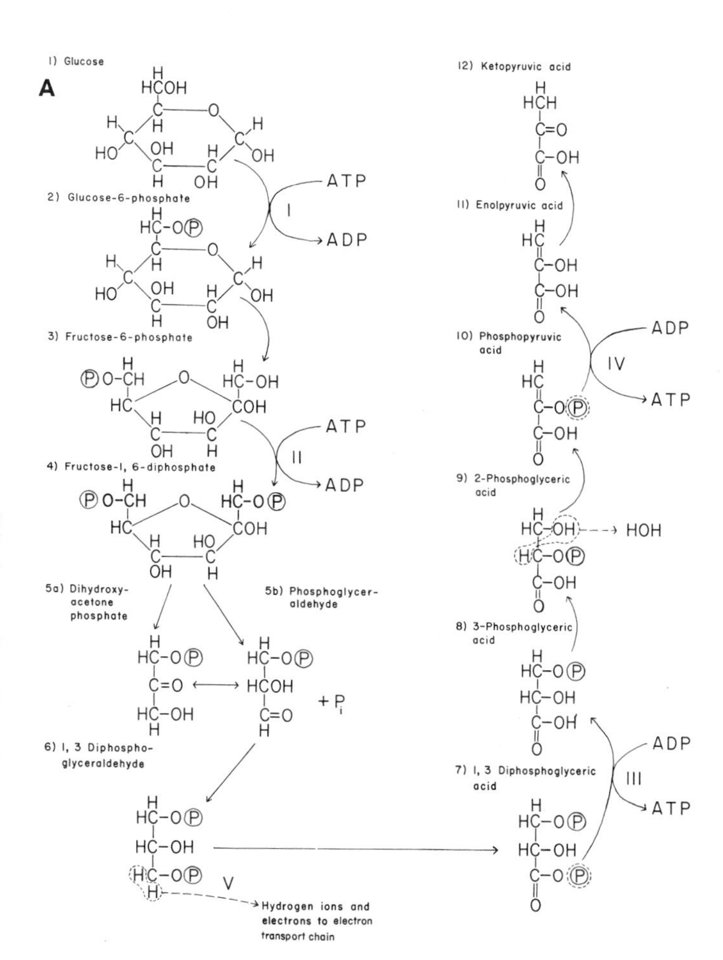

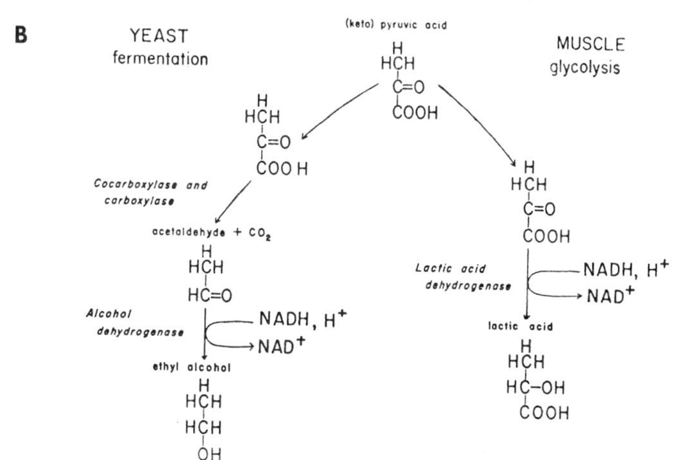

Figure 12.6. *A, Glycolysis or anaerobic decomposition of glucose. The terminal reactions are given in Figure 12.6B. The enzymes concerned in the reactions given here are shown in Table 12.1. Ⓟ refers to organically bound phosphate, P_i to inorganic phosphate. Although the reactions from the bottom of the left column in A and the whole of the right column are shown for one triose, remember that each glucose gives rise to two trioses. Therefore, in calculating yields, multiply the ATP formation in III and IV by two. The net gain of ATP in glycolysis of one glucose molecule is two (four produced minus two used). B, Terminal reactions in yeast fermentation and muscle glycolysis. These reactions follow directly after steps 11 to 12 in Figure 12.6A.*

TABLE 12.1. ENZYMES OF GLYCOLYSIS*

Reaction†	*Enzyme*	*Cofactor*	*Inhibited By*
1–2	Hexokinase and glucokinase	ATP, Mg^{2+} (Mn^{2+})	Dialysis
2–3	Glucose phosphate isomerase	—	—
3–4	6-Phosphofructokinase	ATP, Mg^{2+}	Dialysis
4–5	Fructose diphosphate aldolase	Zn^{2+}, Fe^{2+}, Ca^{2+}, K^{+}	—
5a–5b	Triose phosphate isomerase	—	—
5b–6	Spontaneous?	—	—
6–7	Glyceraldehydephosphate dehydrogenase	NAD^{+}, $HOPO_3^{=}$	Iodoacetate, dialysis
7–8	Phosphoglycerate kinase	ADP, Mg^{2+}	Dialysis
8–9	Phosphoglyceromutase	Mg^{2+}	—
9–10	Enolase	Mg^{2+}	NaF
10–11	Pyruvic kinase	ADP, Mg^{2+}	Dialysis
11–12	Spontaneous?	—	—

*Information largely from Lehninger, 1975: Biochemistry. 2nd Ed. Worth Publishers, New York.
†Refer to Figure 12.6*A* for numbers of steps listed in this column.

as shown in Figure 12.6*B*. For example, muscle cells in oxygen debt may accumulate considerable lactic acid. Part of the lactic acid is oxidized to carbon dioxide and water, and part of it is reduced to triose, which forms hexose phosphate in liver cells. The latter is stored as glycogen. Liver cells store as much as 20 percent glycogen per unit dry weight, whereas muscle cells store only about 2 percent. When glucose is needed, liver glycogen undergoes phosphorolysis and glucose is released into the blood (Guyton, 1976).

In yeast still another path is followed. Pyruvic acid is decarboxylated and the acetaldehyde thus formed is reduced to alcohol, as shown in Figure 12.6*B* in which the enzymes required for the reactions are also indicated. Alcohol dehydrogenase is sensitive to iodoacetate, the addition of which interferes with fermentation.

Neither reduction of pyruvic acid to lactic acid nor decarboxylation of pyruvic acid to acetaldehyde liberates energy; this can be inferred from the heat of combustion of each of these compounds given in Table 9.1. The heat of combustion for lactic acid is greater than for pyruvic acid, as one might expect of a more highly reduced compound. The heat of combustion of acetaldehyde is the same as that of pyruvic acid.

If the efficiency of glycolysis is calculated on the basis of the chemical potential of the ATP produced at the expense of free energy released in production of lactate from glucose (47,000 cal/mole glucose), then the efficiency is 14,600 divided by 47,000 cal/mole (for two moles of ATP produced with glycolysis of one mole of glucose to 2 moles of lactate), or 31 percent (Lehninger, 1971).

AEROBIC DECOMPOSITION OF GLUCOSE

Since only a small fraction (about 5 percent) of the chemical potential energy in a glucose

TABLE 12.2. YIELD OF ATP FROM ADP AND PHOSPHATE IN AEROBIC DEGRADATION OF A MOLE OF GLUCOSE

Reaction	*Text Figure No.*	*Reaction on Text Figure*	*Net Gain*
Initial phosphorylations	12.6*A*	I, II	−2
Substrate phosphorylation (x2)	12.6*A*	III, IV	+4
Hydrogen ion and electron transport chain			
from oxidation of 1,3-diphosphoglyceraldehyde (x2)	12.6*A*	V	+4
from pyruvic to citric acid (x2)	12.8	VI	+6
from isocitric to oxalosuccinic acid (x2)	12.8	VII	+6
from ketoglutaric to succinic acid (x2)	12.8	VIII	+6
from GTP to ATP	12.8	VIIIa	+2
from succinic to fumaric acid (x2)	12.8	IX	+4
from malic to oxaloacetic acid (x2)	12.8	X	+6
			Total +36

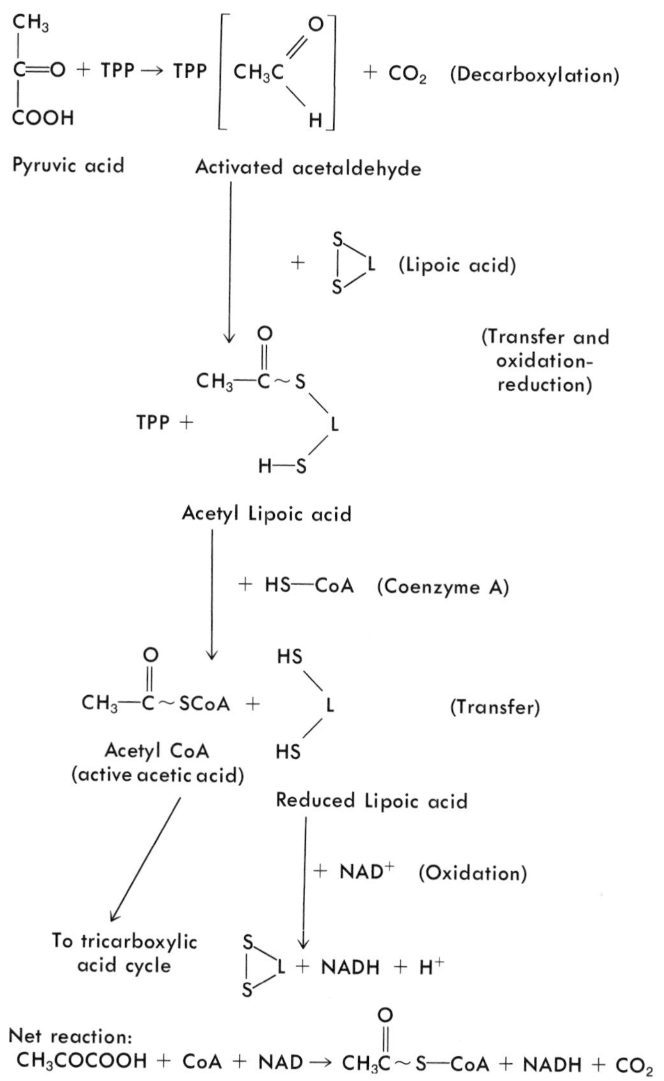

CH_3–C(=O)–COOH + TPP → TPP [CH_3C(=O)(H)] + CO_2 (Decarboxylation)

Pyruvic acid — Activated acetaldehyde

+ S–S–L (Lipoic acid)

↓

TPP + CH_3—C(=O)~S–L–S—H (Transfer and oxidation-reduction)

Acetyl Lipoic acid

↓ + HS—CoA (Coenzyme A)

CH_3—C(=O)~SCoA + HS–L–HS (Transfer)

Acetyl CoA (active acetic acid) — Reduced Lipoic acid

↓ To tricarboxylic acid cycle

↓ + NAD^+ (Oxidation)

S–S–L + NADH + H^+

Net reaction:

$CH_3COCOOH$ + CoA + NAD → CH_3C(=O)~S—CoA + NADH + CO_2

Figure 12.7. The mechanism of oxidative decarboxylation of pyruvic acid. TPP refers to thiamine pyrophosphate. (After McElroy, 1964: Cellular Physiology and Biochemistry. 2nd Ed. Prentice-Hall, Englewood Cliffs, N.J.)

molecule has been liberated by the time pyruvic acid is formed in glycolysis, much is still available for release in aerobic oxidation of pyruvate to carbon dioxide and water. This is the most important source of energy in all aerobic cells (see Table 12.2), and since these reactions occur in the mitochondria, the latter have been called the "powerhouses" of the cell (Siekevitz, 1957).

Pyruvic acid is decomposed in a series of stepwise reactions. Pyruvate combines with coenzyme A (CoA) and is decarboxylated; that is, carbon dioxide is split out of the carboxyl group of the acid, resulting in the formation of acetyl CoA as shown in Figure 12.7. Hydrogen is passed to coenzyme I (nicotinamide adenine dinucleotide, NAD). The 2-carbon fragment (acetyl) is now added to a 4-carbon acid (oxaloacetic) to form a 6-carbon acid (citric), the beginning of a series of reactions called the *tricarboxylic acid cycle* or sometimes the *citric acid cycle.* Citric acid first undergoes internal rearrangements and then is dehydrogenated. After this it is decarboxylated, giving rise to a 5-carbon acid (α-ketoglutaric). The 5-carbon acid in turn is decarboxylated in the presence of CoA (forming succinyl CoA), giving rise to a 4-carbon acid (succinic). The 4-carbon acid undergoes a series of dehydrogenations ending with the formation of oxaloacetic acid, which again becomes a receptor for an acetyl fragment. A considerable amount of energy, amounting to 40,000 or 50,000 cal/mole, is liberated in each oxidative decarboxylation. Part of this energy is stored in high-energy phosphate bonds (Table 12.2).

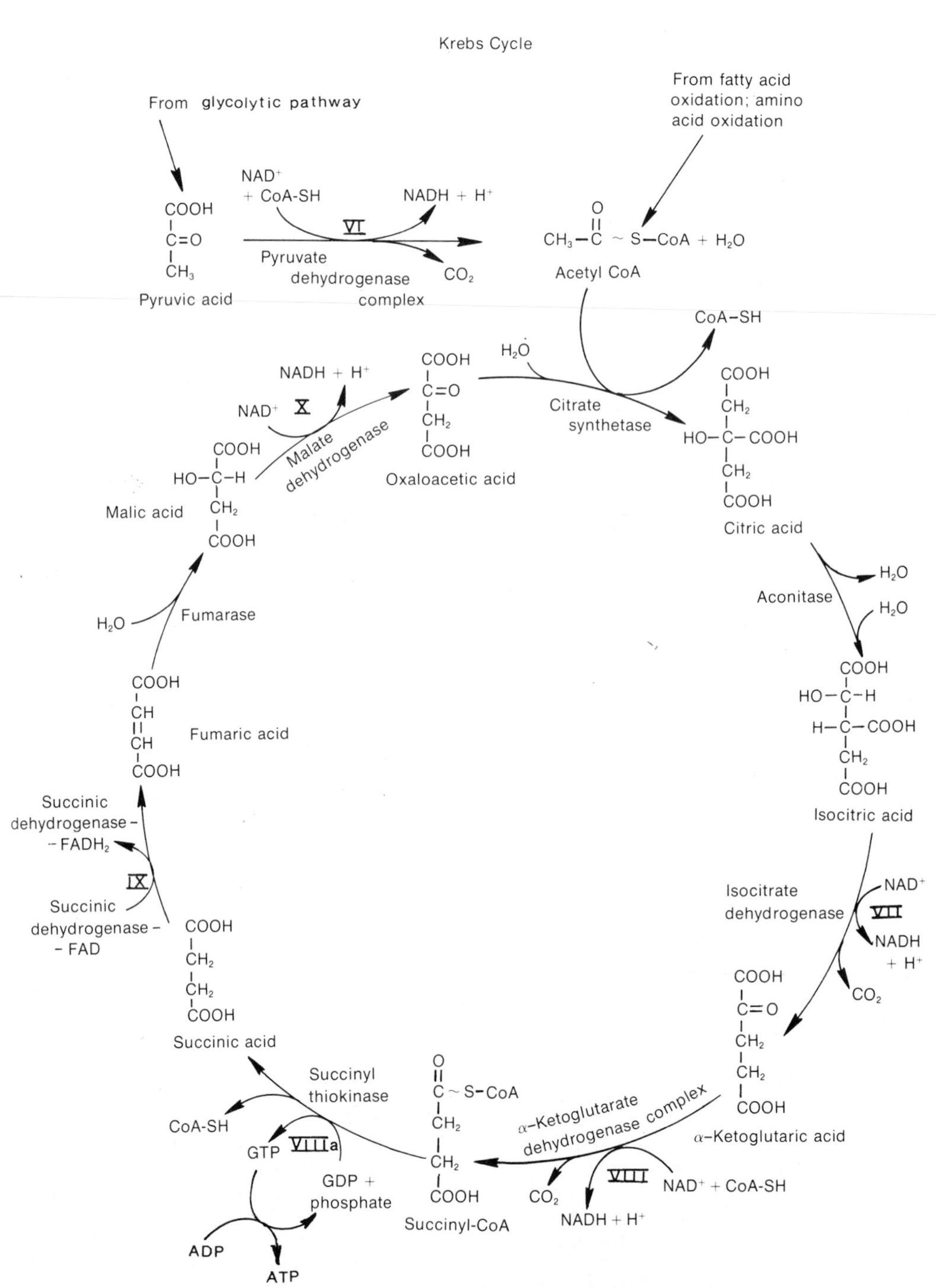

Figure 12.8. *The tricarboxylic acid cycle. Aerobic breakdown of pyruvic acid. (For anaerobic reactions see Figure 12.6.) Decarboxylation of an acid, with release of CO_2 (emphasized in this diagram), can be* reversed *under appropriate conditions in all types of cells, animal as well as green plant or microorganism, thus permitting incorporation ("fixing") of free CO_2, although in a different way than in photosynthesis (see Chapter 14). This diagram shows the origin of the hydrogen (H^+ and electrons) from various substrates. (Adapted from Bronk, 1973: Chemical Biology: An Introduction to Biochemistry. New York, Macmillan. Modified from DeWitt, 1977: Biology of the Cell. W. B. Saunders Co., Philadelphia, p. 387.)*

These events, first associated by Krebs in 1948–1949, form a cycle that has been called the *Krebs tricarboxylic acid* (TCA) *cycle* because several acids appearing in it are tricarboxylic. The function of coenzyme A in the transformation was demonstrated by Lipmann, who also determined its chemical nature. The prosthetic group of coenzyme A contains pantothenic acid, one of the B vitamins. For their elucidation of these reactions Krebs and Lipmann received a Nobel Prize in 1953. These reactions of the Krebs cycle are presented in a simplified way in Figure 11.6 and in detail in Figure 12.8. The yield of high-energy phosphates from the reactions of the Krebs cycle are listed in Table 12.2.

THE METABOLIC MILL

The Krebs tricarboxylic acid cycle has much wider significance in the animal or plant cell than is shown by the degradation of pyruvic acid. It is the means whereby fragments of organic compounds available from various metabolic reactions involving degradation of proteins, fats and carbohydrates can be utilized effectively. For example, when plant proteins in the food of an animal are digested into their constituent amino acids, various amino acids appear in proportions different from the proportions needed for synthesis of animal protein. The amino acids present in excess are deaminated, and the residues conveniently enter the Krebs cycle where, upon being oxidized, they transfer energy to compounds of high phosphate transfer potential. To illustrate, glutamic acid on deamination forms α-ketoglutaric acid; aspartic acid forms oxaloacetic acid; and alanine forms pyruvic acid. Fatty acid fragments also enter the Krebs cycle. Fatty acids are oxidized by what is called β-oxidation, in the course of which a 2-carbon acid fragment (acetyl) is split from the long chain of the fatty acid. The acetyl fragment, as described earlier, enters the Krebs cycle. Some of these relations are indicated in Figure 12.8. Details are given in textbooks of biochemistry.

Cyclic conversion into compounds of high phosphate transfer potential such as creatine phosphate or into formation of glycogen is an effective and efficient method of storing chemical potential energy. It would be infinitely more difficult for the organism to store and use the odd assortment of compounds that enter the Krebs cycle than it is to store glycogen, ATP, and CP.

The metabolic mill permits synthesis in the cell of more complex compounds from fragments, for example, of amino acids from ammonia and the appropriate keto acids (e.g., glutamic acid from ammonia and α-ketoglutaric acid). Protein synthesis from phosphorylated amino acids is dealt with elsewhere (see Chapter 15). Carbohydrate in excess of that stored in glycogen depots is converted into fat by way of synthesis of fragments from the Krebs cycle. Synthesis of fatty acids is more indirect than reversal of fatty acid degradation. It involves incorporation of carbon dioxide into acetyl CoA at the expense of energy from ATP to form a 3-carbon compound. The latter compound is combined with another acetyl CoA to form a 5-carbon compound from which, after reduction and decarboxylation, a 4-carbon fatty acid is produced. Each addition of two carbons to the fatty acid involves a similar round of reactions, details of which may be obtained from a textbook of biochemistry. In animals fat is stored in fat bodies or adipose tissue. In plants fat is often found in fruit (avocado) or in seeds (nuts). It is this overall set of interrelations in the degradation and synthesis of proteins, fat, and carbohydrates that is called the *metabolic mill.*

HYDROGEN AND ELECTRON TRANSFER; OXIDATIVE PHOSPHORYLATIONS

In the major pathway of the electron transport chain, hydrogen is removed from a substrate and transferred through a series of enzymes until it is oxidized to water, with an accompanying release of energy in the process (Fig. 12.9). As stated previously, most dehydrogenases are formed by association of coenzyme I (NAD^+) with a protein molecule specific to a particular substrate. Hydrogen is usually picked up from the substrate by the coenzyme portion of the dehydrogenase and carried to the next enzyme in the chain of events, the flavoproteins, which act as hydrogen carriers (e.g., flavin adenine dinucleotide, FAD). From the flavoprotein each hydrogen is passed to coenzyme Q (Figs. 12.9 and 12.10), from which it is subsequently discharged into the cell fluid as a hydrogen ion, and the electrons are passed on to a series of cell pigments known as cytochromes.

From the cytochromes electrons are passed to the enzyme *cytochrome oxidase,* which ultimately discharges the electrons to oxygen. Oxygen, thus excited, unites with hydrogen ions, forming water. The whole series of reactions is shown in Figure 12.9 with transfer of hydrogens and electrons indicated in each of the steps.

The *cytochromes* are cell pigments universally present in aerobic cells as shown by Keilin in 1925. They receive electrons from the flavoproteins or other carrier enzymes and pass them to cytochrome oxidase, which activates oxygen.

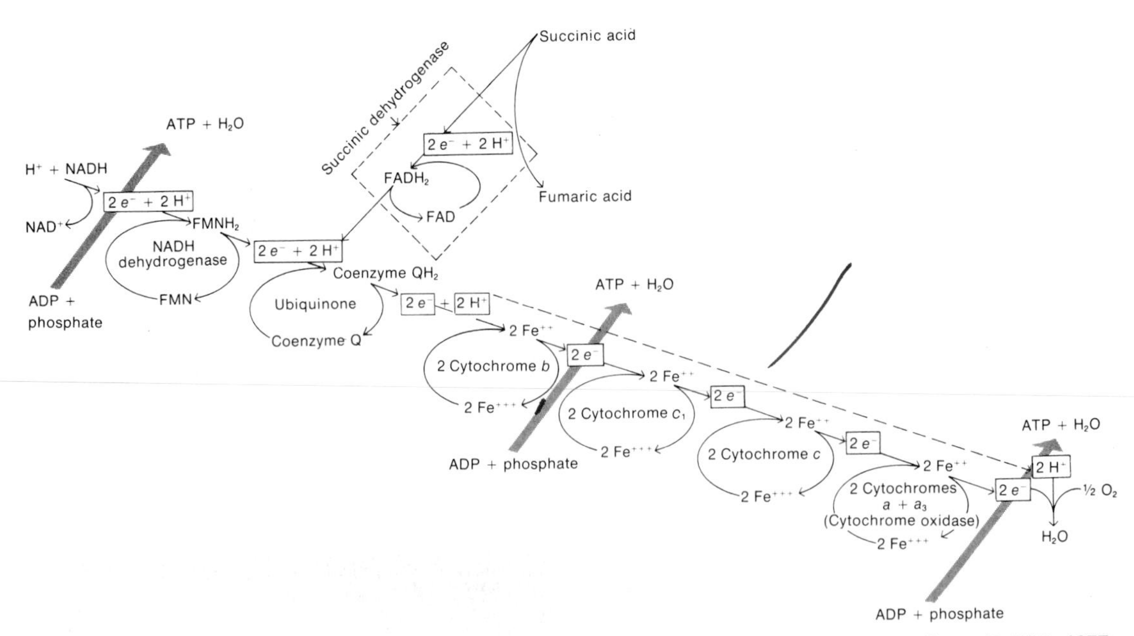

Figure 12.9. *Electron transport and ATP formation in the mitochondrial respiratory chain. (From DeWitt, 1977: Biology of the Cell. W. B. Saunders Co., Philadelphia, p. 391.)*

At least five cytochromes, a, a_3, b, c_1, and c, are usually found in cells. In bacteria several other cytochromes have also been described. In plant cells nine cytochromes have been identified. A cytochrome has a prosthetic group like the heme portion of hemoglobin (Fig. 12.11). The cytochromes are distinguishable by their absorption spectra. A peculiar cytochrome, b_5, is present in the microsomes along with a unique carbon monoxide-binding heme protein.

The enzyme *cytochrome oxidase* has cytochrome a_3 as its porphyrin prosthetic group. On receiving an electron from cytochrome a, iron in the cytochrome oxidase molecule becomes reduced, and when the electron is passed to oxygen the iron is again oxidized.

Iron-sulfur proteins exist at various points along the line of electron transfer from highly reducing to oxidized states. Seven such compounds have been identified by their electron spin resonances. Some are associated with NADH dehydrogenase, succinic acid dehydrogenase, and cytochrome c in mitochondria of eukaryotic cells, where they occur on the inner membrane. The iron-sulfur compounds known as ferredoxins found in photosynthetic bacteria

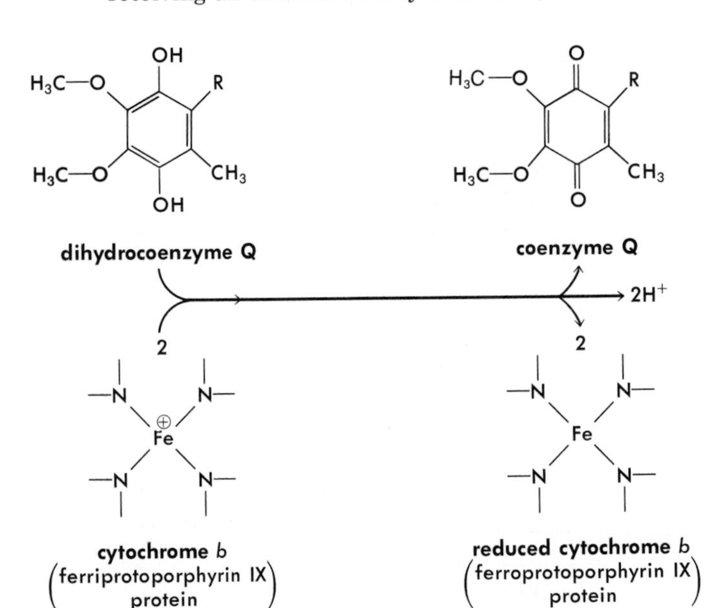

Figure 12.10. *Oxidation of reduced coenzyme Q (dihydrocoenzyme Q) to coenzyme Q, with correlated reduction of cytochrome* b. *The H^+ ions are discharged into the continuum of the cell while the electrons are passed to the cytochromes as shown in Figure 12.9. (From McGilvery, 1970: Biochemistry. W. B. Saunders Co., Philadelphia.)*

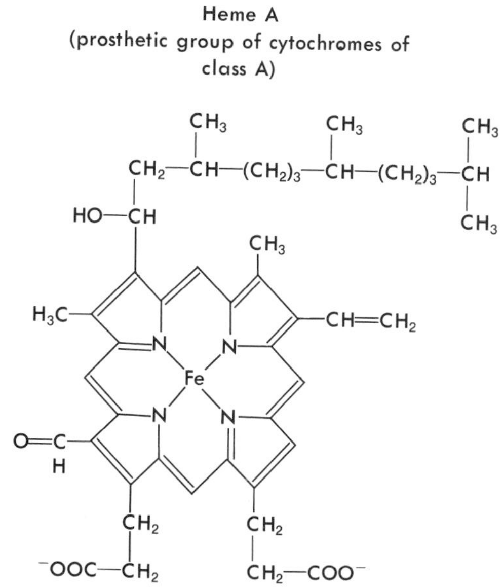

Heme A
(prosthetic group of cytochromes of
class A)

Figure 12.11. *Structure of heme A. (From Lehninger, 1975: Biochemistry. 2nd Ed. Worth Publishers, New York.)*

and photosynthetic plants are electron carriers of high reduction potential (see Chapter 14). Similar compounds are found in nonphotosynthetic bacteria; in fact, ferredoxin was first identified in anaerobic bacteria of the genus *Clostridium.* Except for cases in which they play a role in photosynthesis their exact function is not understood, although they probably serve as electron transporters. Therefore, it is difficult to include them in figures of electron transport compounds.

GENESIS OF NADPH FOR SYNTHETIC REACTIONS: THE PENTOSE SHUNT

About 90 percent of glucose degradation in aerobic cells follows the series of reactions of the Embden-Meyerhof pathway and Krebs TCA cycle just described. However, alternate pathways exist, one of the most important of these being the *pentose* (or *oxidative*) *shunt,* which may normally account for as much as 10 percent of the cell's glucose metabolism, and more under some circumstances (Lehninger, 1975).

The first reaction in the pentose shunt, as shown in Figure 12.12, is the oxidation of glucose-6-phosphate, in the presence of the enzyme glucose-6-phosphate dehydrogenase, to form 6-phosphogluconic acid by way of 6-phosphogluconolactone. The 6-phosphogluconic acid is then oxidized in the presence of coenzyme $NADP^+$ to form NADPH (the reduced form of the coenzyme), H^+, and 3-keto-6-phosphogluconic acid, which, in the presence of enzyme 3-keto-6-phosphogluconic acid decarboxylase, is decarboxylated to form ribulose-5-phosphate and carbon dioxide. The NADPH and H^+ can then reduce some compound or be oxidized through the flavin-cytochrome system in the electron transport chain, in which case water is ultimately formed. However, most of the NADPH serves as a reductant in synthetic reactions. This is an example of the almost clear-cut separation between the functions of the two coenzymes NADH and NADPH. NADH must be oxidized in the electron transport chain (Fig. 12.9) to continue its catalytic activity. NADPH, on the other hand, serves for reductions in biosynthetic activities (see Chapter 15).

Ribulose-5-phosphate can be converted into either of two pentoses that can enter into a series of reactions not yet fully identified. However, triose phosphate is formed, possibly by a pathway that is the reverse of the one in photosynthesis (see Chapter 14). A 7-carbon sugar and a 4-carbon sugar are also intermediates. The reactions resemble those in photosynthetic plants concerned with the regeneration of pentose (see Table 14.4). The net effect is as if one molecule of glucose had been completely decomposed, the other five being reconstituted.

The enzymes necessary for these reactions occur in all cells tested. The main function of the reactions is the synthesis of a "reduction pool" of NADPH and five carbon sugars when required.

Since most of the compounds in the pentose shunt are participants in photosynthetic reactions, one of them (ribulose diphosphate) "fixing" CO_2, it is not surprising that the pentose shunt pathway exists in plant cells. It is also found in some microorganisms. Cells of some tumors place a disproportionate emphasis on the pentose shunt as an oxidative pathway, compared to normal cells. The pentose shunt is additionally important as a source of pentoses needed as constituents of nucleotides in all cells.

ALTERNATE PATHWAYS

Microorganisms, perhaps the most versatile "chemists" among cells, have explored many pathways for decomposition of organic compounds. Some of these have already been considered in Chapter 1 (see Table 1.3).

Pathways in Poisoned Cells

Cyanide binds iron in the prosthetic group of cytochrome oxidase and is incapable of

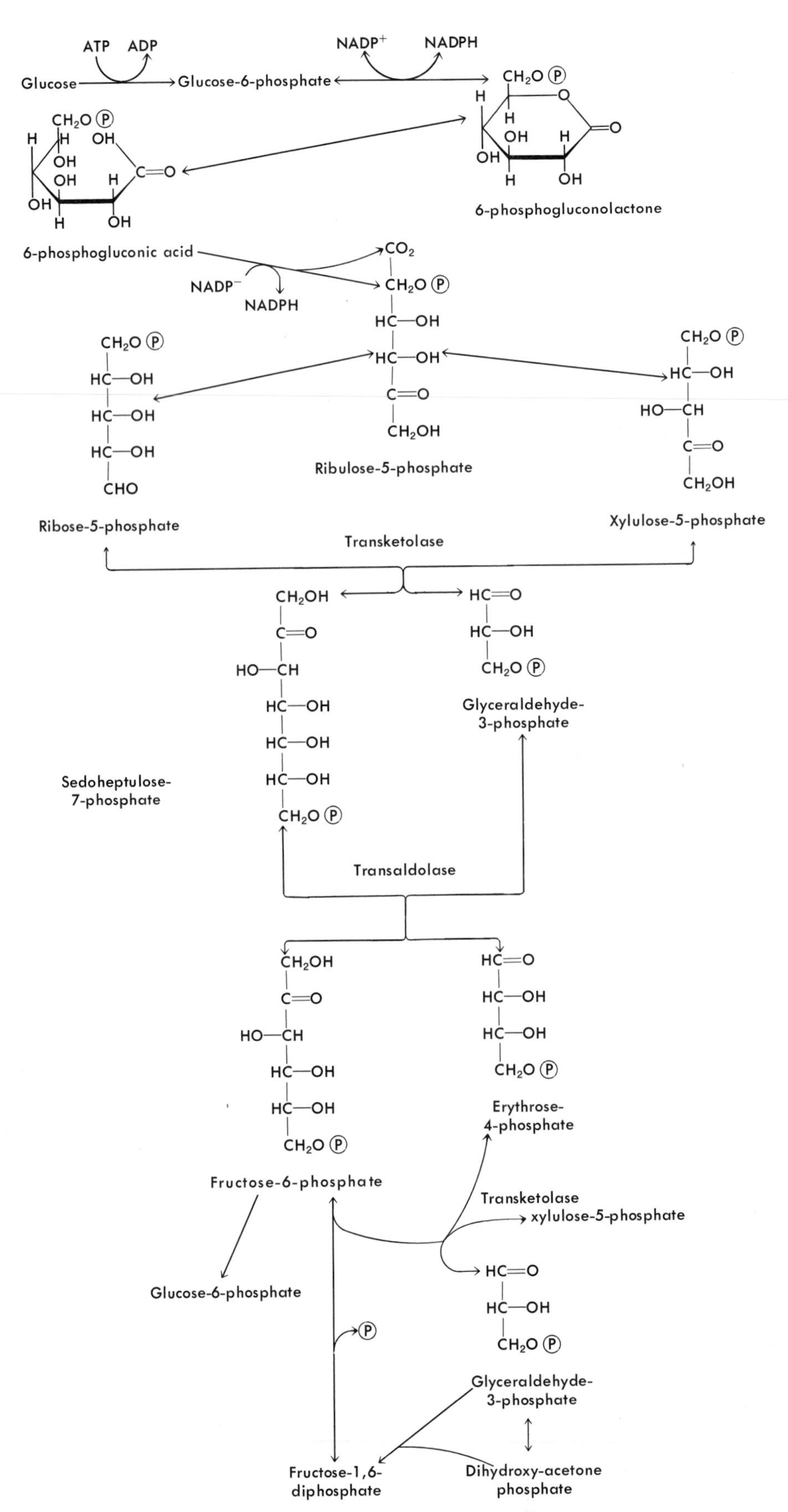

Figure 12–12. *See legend on the opposite page.*

transferring electrons. The iron remains in the oxidized state and cytochromes are unable to discharge their electrons to the enzyme. Because the absorption spectra of reduced and oxidized cytochromes are different, the effect of cyanide on the cell's respiratory condition can be detected spectroscopically, cyanide causing the absorption spectrum of reduced cytochromes to appear.

Although 0.001 molar cyanide blocks the main line of oxidation, a residual respiration, constituting about 5 to 15 percent of the total and called the cyanide-stable respiration, continues—presumably by pathways other than through cytochrome oxidase. For example, flavoproteins have been found to pass electrons slowly to oxygen, and a few oxidases specific to amino acids remove hydrogen from corresponding amino acids and pass it directly to oxygen, even in the presence of cyanide. The experiments with cyanide, therefore, demonstrate the presence of oxidative pathways other than the main line pathway of oxidation through cytochrome oxidase.

A poison such as urethan, which affects many enzymes in a rather unspecific manner, is not as useful in demonstrating alternate pathways. Urethan has a certain degree of selectivity for dehydrogenases, and in its presence the oxidized absorption spectrum of the cytochromes appears.

Poisons have been useful in many cases because when they are used to block a specific pathway, alternate pathways show up more clearly. In this way it has been possible to identify and characterize some subsidiary pathways that play a role in normal metabolism as well (Buehler and Sies, 1969).

Pathways in Tumor Cells

In studying the subject, Warburg has emphasized the glycolytic nature of tumor metabolism. Regardless of whether this is a cause or an effect of the derangement of growth that is cancer, it is interesting to note that greater emphasis occurs in many cases, although not all, upon anaerobic metabolism. Since tumors are often rapidly dividing cells without much organization, glycolysis could be favored because the supply of oxygen is probably inadequate. No special kind of metabolism unique to tumor cells has been found, the difference in reactions and enzymes being quantitative.

MECHANISM OF ATP FORMATION

In our discussions it has generally been assumed that the energy released in electron transport is transferred by coupled reactions to ADP (and P_i) with the formation of ATP, but other mechanisms are possible. Two hypotheses have been suggested: conformational change in a protein and chemiosmotic coupling.

Boyer proposed that the energy liberated in electron transport is conserved not in one covalent bond but rather in a number of weak bonds. Such a series of changes would result in a conformational change in an electron transport protein or in the coupling factor F_1-ATPase. This implies an energy-dependent shift in such weak bonds as hydrogen bonds and hydrophobic interactions. As evidence it is pointed out that the inner mitochondrial membrane undergoes rapid physical changes as electrons pass along the transport chain. Ultrastructural changes also accompany addition of ADP to a respiring mitochondrion. The data have not convinced workers generally.

Mitchell in 1961 proposed the chemiosmotic hypothesis to explain ATP formation (Fig. 12.13). The energy liberated in electron transport is said to be conserved by creating an osmotic gradient of H^+ between the inside and outside of the inner mitochondrial membrane. It has been demonstrated that protons are extruded by the inner mitochondrial membrane, rendering the inside of the membrane alkaline with respect to the outside. Return of the H^+ to the outside liberates the energy to power the addition of P_i to ADP forming ATP.

Considerable evidence supports the chemiosmotic hypothesis. Oxidative phosphorylation occurs only when the mitochondrion is intact. Hydrogen ions are known to be pumped outward by the inner mitochondrial membrane. Uncoupling agents like dinitrophenol allow escape of hydrogen ions from inside the inner mitochondrial membrane without concomitant ATP formation. When the inner mitochondrial membrane is turned inside out (after disruption), pumping of hydrogen ions now occurs inwardly; in other words, the inner mitochondrial membrane has a vectorial property (Mitchell, 1977).

One objection to the chemiosmotic hypothesis is that to provide the energy, on thermodynamic grounds, to power the formation of ATP from ADP and P_i would require a pH difference of 3.5 units. This objection has been met by the calculation that if one adds the energy available from the potential difference between the inside and the outside of the inner

Figure 12.12. *The pentose shunt pathway in heterotrophic cells. The oxidation of six molecules of glucose with the formation of six molecules of pentose is equivalent to complete oxidation of a glucose molecule to carbon dioxide and water. Note the generation of NADPH + H^+ in the first reaction above—NADPH is important as a reductant in synthetic reactions. (After Stanier,* et al., *1976: The Microbial World. Prentice-Hall, Englewood Cliffs, N.J.)*

See illustration on the opposite page

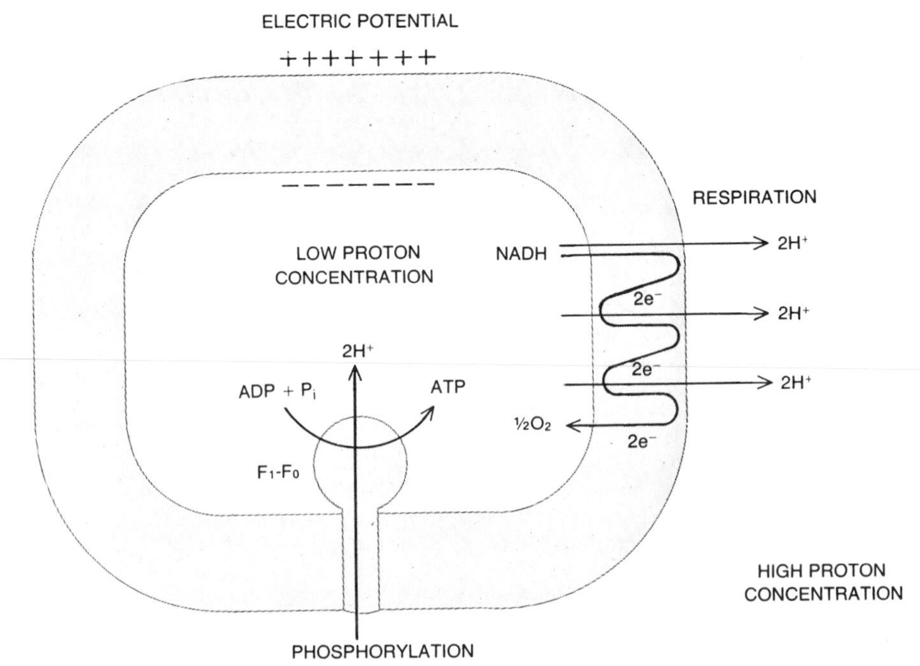

Figure 12.13. *Oxidative phosphorylation couples the release of energy by the oxidation of molecules derived from carbohydrates and fats to the synthesis of ATP, which takes place on the F_1 bodies protruding inward from the mitochondrial inner membrane. The F_1 bodies are made up of several protein subunits known collectively as F_1; they are attached to a membrane protein known as F_0. In cells with nuclei, oxidative phosphorylation takes place only in mitochondria. The basic events of the mitochondrial process are depicted here as they are interpreted in the chemiosmotic theory. Electrons and protons (or hydrogen atoms) from carbohydrates and fats are conveyed by molecules of the hydrogen carrier NADH to a system of enzymes in the mitochondrial membrane. In respiration pairs of electrons cross the membrane three times, each time transporting two protons from inside the mitochondrion to outside. The result is a gradient in proton concentration and electric potential that tends to force the protons back through the membrane. The energy of the gradient drives the process of ATP synthesis. For each two protons traversing the F_1-F_0 complex one molecule of ATP is formed from adenosine diphosphate (ADP) and inorganic phosphate (P_i). (From Hinkle and McCarty, 1978: Sci. Am. (Mar.) 238: 104–123.*

mitochondrial membrane, equal to 0.50 V, to the pH difference actually found, 1.0 pH units, sufficient energy becomes available for the formation of ATP from ADP and P_i.

While it is attractive to consider a covalently linked reaction as the mechanism of ATP formation, the facts point strongly in favor of the chemiosmotic hypothesis, for which Peter Mitchell received the Nobel Prize in 1978. The fact that the chemiosmotic hypothesis also is favored as the explanation for photophosphorylation in photosynthesis (see Chapter 14) adds support for the hypothesis as the explanation of oxidative phosphorylation. It is unlikely that the mechanism would be different in the two cases.

For our present purpose, the mechanism of phosphorylation is perhaps less important than the role of phosphorylation in biological reactions. Those interested in the details of the chemiosmotic process will find these treated in considerably more detail elsewhere (Lehninger, 1975; Mitchell, 1977; Hinkle and McCarty, 1978). Still another hypothesis for the mechanism of phosphorylation has been developed by Green and Blondin (1978), to which the interested reader is referred.

PHOTIC ATP FORMATION IN A HALOPHILIC BACTERIUM

A unique method for producing ATP, perhaps comparable to photosynthetic ATP formation in chlorophyll-bearing plants (see Chapter 14), is used by *Halobacterium halobium,* which grows in a medium equivalent to 4.3 molar in sodium chloride. The plasma membrane contains a purple pigment, bacteriorhodopsin, similar to the visual pigment (rhodopsin) in our eyes. Fully half the membrane surface is covered by patches of this pigment. The molecule consists of a protein (opsin) and a carotenoid. On absorption of light (maximal in the yellow 570 nm), the pigment causes extrusion of protons, inducing not only a greater potential difference between the inside and the outside of the cell but also a change in pH inside. The protons return through an ATPase system providing the energy to synthesize ATP from ADP and P_i. Only light absorbed by bacteriorhodopsin is effective for ATP synthesis. The bacteria can also produce ATP by oxidative phosphorylation, but not anaerobically. At low oxygen partial pressure or anaerobic conditions, the photic method of ATP synthesis is highly adaptive (Stoeckenius, 1976).

HYDROGEN PEROXIDE FORMATION AS AN END PRODUCT IN CELLULAR OXIDATION

When enzymes transport hydrogen that is removed from a substrate directly to oxygen, hydrogen peroxide rather than water is formed. An example of such a reaction is oxidative deamination of amino acids to the corresponding keto acids by D-amino oxidase. Hydrogen peroxide, a poison for many enzymes, is decomposed almost immediately through action of enzymes specific to it. In animal cells the most generally distributed enzyme of this type is *catalase,* which decomposes hydrogen peroxide to water and oxygen. Catalase, which contains iron in a porphyrin nucleus, has some peroxidase-like action and is inhibited by cyanide. In many plant cells peroxidases dispose of hydrogen peroxide by using it as a hydrogen receptor in oxidation of substrates.

METHODS FOR INVESTIGATING INTERMEDIARY METABOLISM

The concepts summarized in the preceding pages were developed only after prolonged research by diverse techniques. The history of their development is involved and cannot be considered here. However, discussion of an interesting controversy illustrates the devious routes taken in making some discoveries.

In analyzing cell respiration, O. Warburg considered activation of oxygen to be the important step. The activated oxygen, he said, would combine with the substrate. At the time, many investigators sought to analyze cell problems by making simple analogues, or models, and Warburg found that activated blood charcoal catalyzes oxidation of many metabolites present in cells by activating oxygen. The active agent in blood charcoal is iron, which is subject to poisons like cyanide and carbon monoxide. Warburg therefore searched for an iron-containing agent in the cell that activates oxygen. He obtained evidence for such an agent by the effective use of respiratory inhibitors. He found that this cellular catalyst is inhibited by cyanide and carbon monoxide. The carbon monoxide poisoning proved reversible with visible light, and from the action spectrum (relative efficiency of different wavelengths) of this reversal Warburg determined that the respiratory enzyme was a porphyrin. He called this enzyme the "atmungsferment" or respiratory enzyme. Later, isolation of the enzyme verified his deductions.

A. Wieland, Warburg's contemporary, emphasized the importance of activation of hydrogen, and as a model of cell enzymes he used palladium black, which is known to adsorb hydrogen. He found that palladium black catalyzes oxidation of some cellular organic compounds, and he postulated the presence of dehydrogenase action in the cell, with activation and removal of hydrogen. Thunberg, using different substrates with tissue preparations, was able to identify a large number of dehydrogenases and to measure their relative concentrations.

It was finally realized that Wieland and Warburg were both correct, that many enzymes are involved in cell respiration, and that both hydrogen and oxygen have to be activated (Needham, 1970). Many respiratory enzymes have since been obtained in relatively pure form, and their properties have been investigated. Some enzymes seem to be inseparable from cell membranes.

At present the few purified oxidoreductases require cofactors (metals and coenzymes) in concentrations higher than present in living cells. This might imply that the structural and functional relations between the enzymes and cofactors are optimal in the cell or perhaps that they have been altered in some way during purification (Lehninger, 1975).

For a study of steps in the breakdown or synthesis of an organic compound in cells of an organism (a field of continuing interest), the organism or a perfused organ is supplied with various diets or given certain nutrients by injection. Its urine is then examined for metabolic intermediates (that is, intermediate stages in the breakdown of a nutrient). Many nutrients can now be labeled with "tagged" nitrogen, carbon, phosphorus, sulfur, and other elements; therefore, they can be readily traced in metabolism. In addition, it is possible to determine metabolic pathways by studying cells of organisms possessing genetic metabolic abnormalities, which would make them incapable of completely decomposing some compounds. As a result, intermediates accumulate in their urine in quantities sufficient for identification. Mutants lacking some enzymatic constituent have also been produced in the bread mold *Neurospora* and in cells of other organisms (see Chapter 15). Such mutants accumulate some metabolic intermediates in large quantities, facilitating identification (Hommes and Van den Berg, 1973).

In vitro studies of intermediary metabolism have been made with cells in tissue cultures, cells of perfused organs (e.g., liver and kidney), and cells in slices of tissues. Sometimes the cells are ground to a brei for biochemical studies. Manometric studies of gaseous exchanges in cells in tissue slices have added to the knowledge of intermediary metabolism (see Chapter 13), because comparisons can be made between respiration of normal and of pathological cells or of tissue cells from animals grown on media omitting certain essential requirements.

It is also important that, as each enzyme is separated and purified, the reaction it catalyzes be studied *in vitro*. Only when the various thermodynamic and kinetic properties of the isolated system have been studied under controlled conditions can the part it plays in the whole sequence of reactions in the cell be established with any certainty.

As mentioned previously, enzymes can be separated from extraneous proteins by fractional salting out or by adsorption on resins, and they may be purified by electrophoretic procedures. Even when purification is not possible, considerable information may be obtained by the use of a partially purified enzyme, provided the extraneous material does not interfere with the reaction under study. Much has been learned even by use of heterogeneous suspensions containing many enzymes, such as entire mitochondria, fragmented mitochondria, or partially purified enzyme systems from mitochondria.

The rate of an enzymatic reaction is often measured spectrophotometrically because accurate measurements can be made quickly and continuously during the course of the reaction. Fortunately, many of the enzymes important in the reactions discussed in this chapter have markedly different absorption spectra in their oxidized and reduced states. Consequently, the relative amounts of the enzyme system in each state can be readily determined and the course of the reaction can be followed by the change in proportion of each. An example for visible light absorption is given in Figure 11.10. It is also possible to study radio-wave absorption using nuclear magnetic resonance techniques discussed in Chapter 11. Methods for studying enzymatic reactions are considered in treatises on enzymes (Colowick and Kaplan, 1955–1972; Boyer, 1976).

ENZYMES IN ORGANELLES

The enzymes involved in the anaerobic phase of sugar decomposition are apparently membrane bound, but whether to the endoplasmic reticulum or the plasma membrane is not known. On the other hand, all the enzymes of aerobic oxidation of glucose (e.g., of the Krebs tricarboxylic acid cycle) are localized in the mitochondria. Most of the energy for cell metabolism, therefore, appears to be released and transferred in the mitochondria (Siekevitz, 1957). Since the various enzymes that participate in the formation and transfer of high-energy phosphate bonds are also localized there, the role of the mitochondria appears to be the formation and liberation of compounds of high phosphate transfer potential (e.g., ATP).

As already described, the mitochondrion has a complex structure, usually with lamellate cristae protruding into the central lumen, providing a large surface area (Fig. 6.2). The problem is to localize and isolate each enzyme, to identify the individual reactions catalyzed by the enzyme and, finally, to reconstruct the enzyme system *in vitro*, especially the enzymes of the Krebs cycle, electron transport, and phosphorylation (De Pierre and Ernster, 1977). Attempts to separate many of the enzymes have failed, although some success has been achieved in submitochondrial systems obtained by fragmenting the mitochondria, the whole series of reactions now being obtained in pieces equal to 1/40 of a mitochondrion. The difficulties suggest that a biological necessity may exist for structural organization of some of these catalysts in a moderately rigid, geometrically organized constellation in the membrane in order to minimize the path distance between slowly diffusing large molecules and to increase the probability of interaction (Wainio, 1970).

Now it is apparent that enzymes that catalyze some reactions of the Krebs tricarboxylic cycle are probably present on membranes, whereas others are in the matrix (Table 12.3). By use of negative staining, fine structures called F_1 bodies have been detected on the cristae (Fig. 12.14). The dimensions of

TABLE 12.3. LOCATION OF SOME MITOCHONDRIAL ENZYMES*

Outer membrane
Monoamine oxidase
Fatty acid thiokinases
Kynurenine hydroxylase
Rotenone-insensitive cytochrome *c* reductase
Space between the membranes
Adenylate kinase
Nucleoside diphosphokinase
Inner membrane
Respiratory chain enzymes
ATP-synthesizing enzymes
α-Keto acid dehydrogenases
Succinate dehydrogenase
D-β-Hydroxybutyrate dehydrogenase
Carnitine fatty acyl transferase
Interior matrix
Citrate synthase
Isocitrate dehydrogenase
Fumarase
Malate dehydrogenase
Aconitase
Glutamate dehydrogenase
Fatty acid oxidation enzymes

*From Lehninger, 1975: Biochemistry. 2nd Ed. Worth Publishers, New York.

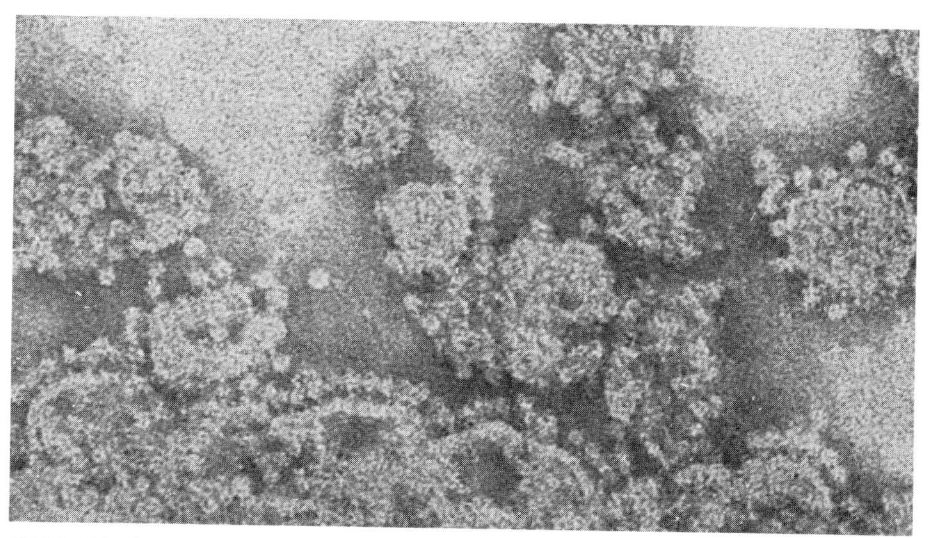

Figure 12.14. Electron transport particles on cristae of mitochondria (× 300,000). (By courtesy of Ephraim Packer, Cornell University.)

these structures are of the right order of magnitude to serve as electron-transport particles, but the reactions occur equally well when the bodies are removed (Racker, 1970). However, phosphorylation of ADP to ATP does not occur. The F_1 bodies are the seat of these reactions. When the F_1 bodies are restored to ruptured mitochondria phosphorylation again occurs. F_1 bodies are also present on the cell membrane of bacteria (e.g., *E. coli*) and in chloroplasts (see Chapter 14).

Mitochondria undergo changes in volume, indicating changes in permeability (see Chapter 18). These changes in permeability accompany oxidative reactions on membranes of the mitochondria. The mitochondrial membrane is not very permeable to ATP or ADP; therefore, it is thought that the phosphate transfer and transfer of energy occur at the membrane surface itself, ADP just outside the membrane being phosphorylated to ATP. The ratio of ATP to ADP at the surface determines the swelling of mitochondria (Lehninger, 1971; 1975).

Some evidence indicates that each mitochondrion has about 5000 to 10,000 complete sets of such enzymes. The assemblies are probably distributed equally over the cristae. It will be recalled that a cell has many mitochondria—about 500 being present in many kinds of cells (a liver cell has about 2500); therefore, ample room exists to accommodate numerous sets of enzymes (Lehninger, 1975).

Corresponding to their role in energy transfer, mitochondria are numerous in active cells and, in general, concentrate in regions of greatest activity. During cell division they may be seen actively moving about the spindle. Mitochondria appear at synapses between neurons. They are numerous in secretory cells and in muscle cells, especially in very active flight muscle cells of insects and birds.

Glycolytic reactions occur mainly in the cell's ground substance (cytosol), as do some of the syntheses.

Enzymes participating in synthesis of proteins are present in the ribosomes that adhere to the outer surfaces of the cell's endoplasmic reticulum. When cells are mechanically crushed, the endoplasmic reticulum breaks up into vacuoles, on the outer surface of which are the ribosomes. Treatment with a detergent separates the ribosomes from the lipoprotein membranes of the vacuoles. The ribosomes thus separated are concentrated by centrifugation and are found to incorporate labeled amino acids into proteins very readily when provided with mRNA, tRNA, the appropriate enzymes, and high-energy phosphates, under the proper conditions. It is on the ribosomes then that the major part of protein synthesis of the cell is performed (see Chapter 15). Lipid and steroid synthesis occur within the endoplasmic reticulum.

The nucleus contains enzymes for anaerobic metabolism and for some oxidative phosphorylation as well. The nucleus also contains enzymes for synthesis of DNA, RNA (in the nucleolus) and some proteins.

The compartmentation of functions and reactions in the cell is extensive, the compartments (organelles) being separated from one another by cell membranes. Because cell membranes are selectively permeable, each set of reactants is held in its own compartment, making possible many diverse yet simultaneous reactions required for maintenance of life in the cell. Table 12.4 reviews some of the major functions compartmented in cells.

DYNAMIC NATURE OF CELL CONSTITUENTS

Through the use of cell foods containing

TABLE 12.4. COMPARTMENTATION OF FUNCTIONS AND REACTIONS IN THE CELL*

Compartment	Functions and Reactions
Nucleus	Replication of DNA, synthesis of some nuclear proteins, transcription of mRNA and tRNA
Nucleolus	Transcription of rRNA
Mitochondria	Site of oxidative reactions: tricarboxylic acid cycle, electron transport, phosphorylation, fatty acid oxidation, amino acid oxidation
Cytosol (ground substance)	Site of glycolytic reactions, site of many reactions in the pentose shunt (phosphogluconate pathway), activation of amino acids, fatty acid synthesis, mononucleotide synthesis, synthesis of some amino acids
Endoplasmic reticulum	Lipid and steroid synthesis, hydroxylation, channeling of biosynthetic products
Ribosomes	Translation of mRNA message—protein synthesis
Golgi complex	Formation of secretory vesicles and granules (packaging), formation of plasma membrane, possibly source of lysosomes and peroxisomes
Lysosomes	Segregation of hydrolytic enzymes
Peroxisomes (microbodies)	Site of amino acid oxidase, catalase and, in plant cells, of glyoxylate reactions
Glycogen granules	Enzymes of glycogen synthesis and degradation
Chloroplasts (in chlorophyllous cells of plants)	Photosynthesis, photophosphorylation and some protein and other syntheses

*Data mainly from Lehninger, 1975: Biochemistry. 2nd Ed. Worth Publishers, New York, p. 381.

isotopes of nitrogen (^{15}N) and carbon (^{14}C) it is possible to follow their incorporation into the cell and thus to "label" cell proteins. The rate of decrease of the labeled nitrogen and carbon may then be measured. By this means it has been demonstrated that cells of all tissues lose some labeled material and replace it anew periodically. Even in the protein in connective tissue, amino acids are replaced, although less rapidly than in more active tissues. Replacement is very rapid in some cells, such as those in the liver (Schoenheimer, 1942). The proteins of rat liver have a half-life of five to six days, while the liver cell has a life of several months. The proteins of muscle turnover move slowly (30 days), while the phospholipids of brain turnover move much more slowly (200 days) (Lehninger, 1975).

The results have generally been taken to indicate that cellular protein is continually being replaced. On the other hand, it may be that individual cells in the tissues studied are dying and are being replaced (for example, in the intestine, blood, and epidermis), rather than that part of the protein in all cells is being replaced. Studies with bacteria show that nitrogen is incorporated at a steady rate so long as the cell is growing; no protein breakdown appears to occur, except when the cells are starved for nitrogen. The *dynamic state* hypothesis should be tested with a variety of animal and plant cells in tissue culture. It may be that the increases occur only with growth (or storage) and that losses of protein accompany nitrogen starvation. The histones of chromosomes are exceedingly stable, although individual atoms in molecules may be exchanged. Tests on tissue cultures suggest that cells that have stopped dividing do replace their cellular substance slowly (Mazia, 1961).

APPENDIX

12.1 CYCLES OF MATTER

In the course of cellular reactions elements enter into various combinations; for example, photosynthetic carbohydrates may be converted by plant cells into a wide variety of or-

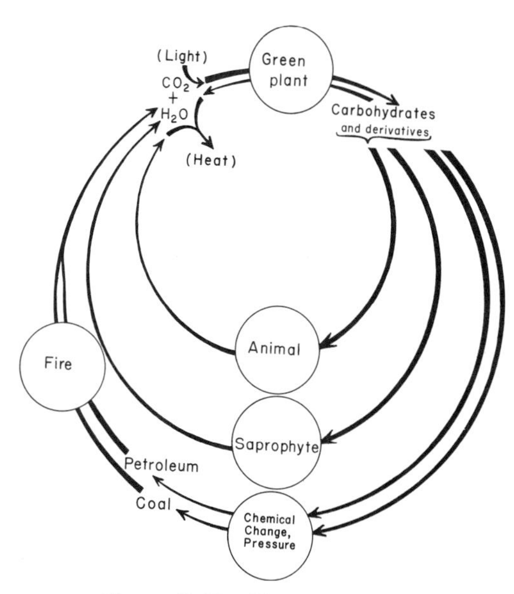

Figure 12.15. *The carbon cycle.*

ganic compounds. An animal cell converts the plant's organic compounds into compounds necessary for its own cellular activities. Dead animal and plant cells serve as nutrient for microorganisms of decay. This series of interconversions of carbon compounds, one of the cycles of matter, is known as the *carbon cycle* (Bolin, 1970), and is illustrated in Figure 12.15. The nitrogen cycle is considered in Chapter 3.

Phosphorus, a constituent of many biological compounds, including high-energy phosphates, accumulates in animals in such body parts as bones, teeth, and shells, but is recycled on death. Insoluble phosphate in rocks is liberated in soluble form by action of microorganisms and is taken up by plants, from which it reaches animals (Brock, 1966). No organisms are known that oxidize or reduce phosphorus. Therefore, the phosphorus cycle consists mainly of an alternation between organic and inorganic forms of phosphorus (Fig. 12.16). Phosphorus is supplied to the soil as insoluble inorganic phosphate in fertilizers and is liberated in soluble form by acids produced in the nitrogen and sulfur cycles.

In soil and water, phosphates occur in relatively small amounts compared to nitrogen compounds. Phosphorus therefore often limits growth of organisms. Upwellings in the sea that supply phosphates are often followed by the blooming of photosynthetic organisms.

Sulfur, a component of many cellular compounds, is liberated from dead cells either as H_2S or as SO_3, depending on the availability of oxygen and the bacteria of decay. Some bacteria oxidize H_2S as a source of energy; others use SO_3 as a hydrogen acceptor, forming H_2S. H_2S is released into the atmosphere from tidal flats of the ocean and swampy areas on land. H_2S may be oxidized to form sulfate in the atmosphere by some of the processes shown in Figure 12.17. Sulfates in droplets of sea water are carried over the land and washed down by rain. Volcanoes also contribute H_2S, SO_2, and SO_3. Under natural conditions a balance favorable for life is maintained by the operation of the sulfur cycle.

Some concern has arisen over the addition of

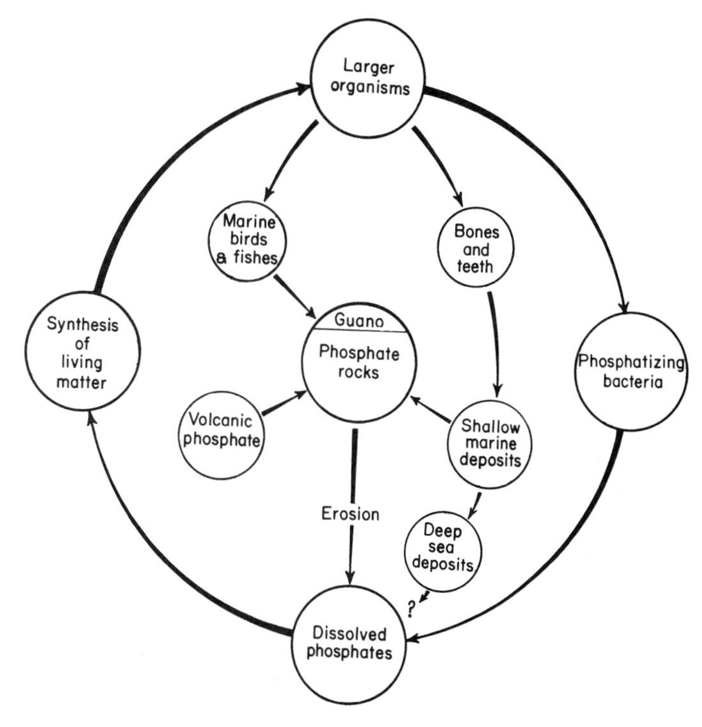

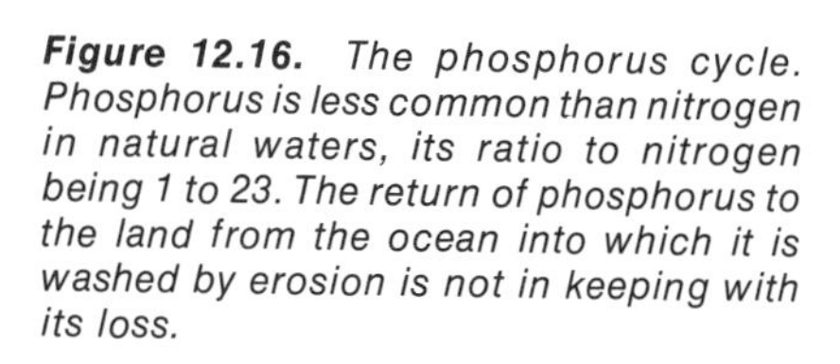

Figure 12.16. *The phosphorus cycle. Phosphorus is less common than nitrogen in natural waters, its ratio to nitrogen being 1 to 23. The return of phosphorus to the land from the ocean into which it is washed by erosion is not in keeping with its loss.*

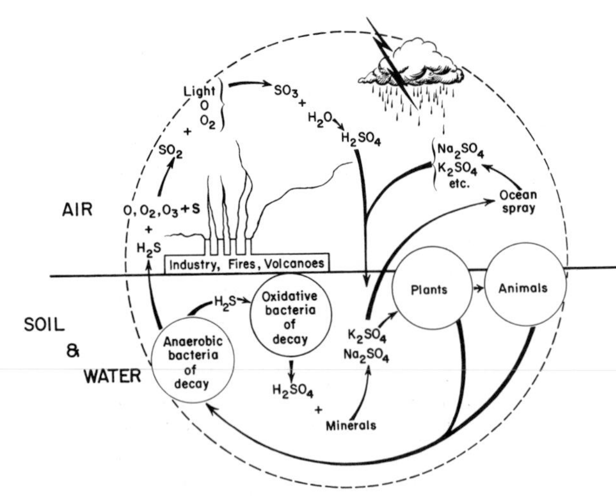

Figure 12.17. The sulfur cycle. Sulfur is accumulating in the atmosphere at a rate beyond the capacity of the rains to wash it back to the land and sea. It is especially prevalent in its compounds near industrial centers. See Chapter 27 for details.

considerable quantities of sulfur to the atmosphere as a result of industrial processes. It is estimated that we add to the atmosphere half again as much of sulfur compounds as nature does and that by the year 2000 the artificial addition will be equal to nature's. About two-thirds of the sulfur we contribute comes from burning coal, about a sixth from combustion of petroleum products, and the remainder from petroleum refining and nonferrous smelting. Therefore, we are likely to overwhelm natural processes, inasmuch as removal of sulfur compounds from the atmosphere is rather slow; contamination has developed for hundreds and even thousands of kilometers outward from industrial centers. It perhaps comes as something of a shock to most of us that the earth's atmosphere is finite and so easily disrupted.

Other elements used by the cell such as carbon, nitrogen, phosphorus and sulfur, recycle from compound to compound, cell to cell, and organism to organism, all the while forming various combinations with other elements. Only the details differ.

Since life has existed for billions of years, it may be inferred that conservation of matter is necessary for continuity of life; the same elements must be used over and over again, and, in fact, turnover rates have been calculated for some of them; see Chapter 14 for carbon dioxide and oxygen turnover (Cloud and Gibor, 1970). Atoms that today form part of one organism have probably formed part of another organism years ago and some day may form part of still another. Therefore, material transformations in the living cell are governed by the same physical-chemical laws that apply to the inanimate world.

LITERATURE CITED AND GENERAL REFERENCES

Adler, A. D., ed., 1975: The biological role of the porphyrins and related structure. Ann. N. Y. Acad. Sci. *244*: 1–694.

Bagley, S. and Nicholson, D. E., 1970: An Introduction to Metabolic Pathways. Blackwell Scientific Publications, Oxford.

Bartley, W., Kornberg, H. L. and Quayle, J. R., eds., 1970: Essays in Cell Metabolism. Hans Krebs Dedicatory Volume. Wiley-Interscience, New York.

Becker, W. M., 1977: Energy and the Living Cell. J. B. Lippincott Co., Philadelphia.

Bolin, B., 1970: The carbon cycle. Sci. Am. (Sept). *223*: 124–132.

Boyer, P. D., ed., 1976: The Enzymes. 3rd Ed. Vol. 13. Part C, Oxidation Reduction. Dehydrogenases, Oxidases, H_2O_2 Cleavage. Academic Press, New York.

Boyer, P. D., Chance, B., Ernster, L., Mitchell, P., Racker, E. and Slater, E. C., 1977: Oxidative phosphorylation and photophosphorylation. Ann. Rev. Biochem. *46*: 955–1026.

Brock, T. D., 1966: Principles of Microbial Ecology. Prentice-Hall, Englewood Cliffs, N. J.

Broda, E., 1975: The Evolution of Bioenergetic Processes. Pergamon Press, London.

Buehler, T. and Sies, H., 1969: Inhibitors, Tools in Cell Research. Springer Verlag, New York.

Chance, B., 1977: Electron transport; pathways, mechanisms and controls. Ann. Rev. Biochem. *46*: 967–980.

Cloud, P. and Gibor, A., 1970: The oxygen cycle. Sci. Am. (Sept.) *223*: 111–123.

Colowick, S. P. and Kaplan, M. D., 1955–1972: Methods in Enzymology. A continuing multivolume treatise. Academic Press, New York.

De Pierre, J. W. and Ernster, L., 1977: Enzyme topology of intracellular membranes. Ann. Rev. Biochem. *46*: 201–262.

Dickerson, R. E., 1972: The structure and history of an ancient protein. Sci. Am. (Apr.) *226*: 58–72.

Ernster, L., 1977: Chemical and chemiosmotic aspects of electron transport-linked phosphorylation. Ann. Rev. Biochem. *46*: 981–995.

Fogh, J., 1975: Human Tumor Cells *in vitro*. Plenum Press, New York.
Green, D. E., 1958: Biological oxidation. Sci. Am. (Jul.) *199*: 56–62.
Green, D. E., 1960: The synthesis of fat. Sci. Am. (Feb.) *202*: 46–51.
Green, D. E., 1964: The mitochondrion. Sci. Am. (Jan.) *210*: 63–74.
Green, D. E. and Blondin, G. A., 1978: Molecular mechanism of mitochondrial energy coupling. Bioscience *28*: 18–24.
Guyton, A. C., 1975: Textbook of Medical Physiology. 5th Ed. W. B. Saunders Co., Philadelphia.
Hayashi, O., ed., 1974: Molecular Mechanisms of Oxygen Activation. Academic Press, New York.
Hinkle, P. C. and McCarty, R. E., 1978: How cells make ATP. Sci. Am. (Mar.) *238*: 104–123.
Hommes, F. A. and Van den Berg, C. J., 1973: Inborn Errors of Metabolism. Academic Press, New York.
Kamen, M. D., 1958: A universal molecule of living matter. Sci. Am. (Aug.) *197*: 77–82.
Krebs, H. A., 1970: History of the tricarboxylic acid cycle. Perspect. Biol. Med. *14*: 154–170.
Lehninger, A. L., 1960: Energy transformation in the cell. Sci. Am. (May) *202*: 102–114.
Lehninger, A. L., 1965: The Mitochondrion. W. A. Benjamin, Menlo Park, Calif.
Lehninger, A. L., 1971: Bioenergetics. 2nd Ed. W. A. Benjamin, Menlo Park, Calif.
Lehninger, A. L., 1975: Biochemistry, 2nd Ed. Worth Publishers, New York.
Lemberg, R. and Barrett, J., 1973: Cytochromes. Academic Press, New York.
Lipmann, F., 1955: Coenzyme A and biosynthesis. Am. Sci. *43*: 37–47.
Mazia, D., 1961: Mitosis and the physiology of cell division. *In* The Cell. Vol. 3. Brachet and Mirsky, eds. Academic Press, New York, pp. 77–412.
Mitchell, P., 1977: Vectorial chemiosmotic processes. Ann. Rev. Biochem. *46*: 996–1005.
Needham, J., ed., 1970: The Chemistry of Life, Lectures on the History of Biochemistry. Cambridge University Press, Cambridge.
Pauling, L., 1970: Structure of high-energy molecules. Chem. Br. *6*: 468–472.
Penman, H. L., 1970: The water cycle. Sci. Am. (Sept.) *222*: 99–108.
Postgate, J., 1978: Nitrogen Fixation. Edward Arnold, London.
Safrany, D. R., 1974: Nitrogen fixation. Sci. Am. (Oct.) *231*: 69–80.
Racker, E., 1970: Membranes of Mitochondria and Chloroplasts. Van Nostrand Reinhold, New York.
Racker, E., 1976: A New Look at Mechanisms in Bioenergetics. Academic Press, New York.
Racker, E., 1977: Mechanism of energy transformations. Ann. Rev. Biochem. *46*: 1006–1014.
Richards, R. E., 1975: Nuclear magnetic resonance spectroscopy of biochemical materials. Endeavour *34*: 118–122.
Schoenheimer, R., 1942: The Dynamic State of Body Constituents. Cambridge University Press, Cambridge.
Scientific American issue on the Biosphere. 1970: Sept. Vol. *223*, No. 3.
Siekevitz, P., 1957: Powerhouse of the cell. Sci. Am. (Jul.) *197*: 131–140.
Slater, E. C., 1977: Mechanism of oxidative phosphorylation. Ann. Rev. Biochem. *46*: 1015–1026.
Stoeckenius, W., 1976: The purple membrane of salt-loving bacteria. Sci. Am. (Jun.) *234*: 38–46.
Stumm, W., ed., 1977: Global Chemical Cycles and Their Alteration by Man. Physical and Chemical Sci. Res. Rept. Dahlem Konferenzen.
Wainio, W., 1970: The Mammalian Mitochondrial Respiratory Chain. Academic Press, New York.
Wiskich, J. T., 1977: Mitochondrial metabolite transport. Ann. Rev. Plant Physiol. *28*: 45–69.

CHAPTER 13

OXYGEN CONSUMPTION AND BIOLUMINESCENCE OF CELLS

During aerobic oxidation of nutrients in the cell atmospheric oxygen is consumed, energy is liberated, and carbon dioxide is given off. In glycolysis and fermentation, nutrients are oxidized in the absence of atmospheric oxygen, that is, anaerobically. Glycolysis and fermentation have been considered in Chapters 1 and 12. Organisms that live in the presence of oxygen are aerobes; those that live in its absence are anaerobes. Many aerobes can, however, tolerate limited periods of anaerobiosis (Hochachka, 1975), especially parasites (Hochachka and Mustafa, 1972).

While respiration and oxygen consumption are related, the term oxygen consumption is a more limited one. Oxygen consumption is easier to measure than energy liberation. Energy liberated can be calculated by indirect calorimetry, provided other data are available, as will be discussed later. Not all energy liberated is useful—that liberated in peroxidations is not stored in high-energy phosphate bonds (see Chapter 12).

OXYGEN CONSUMPTION OF CELLS

The oxygen consumption of cells is measured by the volume of oxygen consumed (at standard temperature and pressure, STP) per unit weight of organism per unit of time. This is called the Q_{O_2} or rate of oxygen consumption. The units of measurement used are microliters (μl) per milligram per hour for small samples or milliliters (ml) per gram per hour for larger samples. Data could be cited for either wet or dry weight; the latter weight is approximately one-fourth of the former (Table 13.1).

In the manometric method for measurement of oxygen consumption a suspension of cells is placed in a vessel such as A of Figure 13.1, the center well of which contains potassium hydroxide. Vessel B contains no cells and serves as a compensator. The two vessels are connected to each other through a manometer, which is set at the zero position (level) at the start of an experiment. As oxygen is consumed, the pressure in the closed system falls and the level on the right side of the manometer rises because carbon dioxide, given off during the respiration, is absorbed by the potassium hydroxide. The volume of oxygen consumed can be determined by adding enough air from the calibrated syringe to just compensate for the change in level of the manometer. To insure equilibrium, the vessels should be shaken, preferably in a constant temperature bath.

Unicellular organisms, such as bacteria, yeast, some molds, unicellular algae, small colonial algae, protozoans, marine eggs, sperm, and red blood cells, lend themselves especially well to these studies, since suspensions of them are readily prepared and handled. Compact tissues, however, must be cut into thin slices before the Q_{O_2} of the cells can be studied on respirometric flasks. In this manner liver and brain slices have been extensively studied, and to a lesser extent sections of leaf, stem, and root.

From the data so collected it is apparent that the Q_{O_2} of diverse organisms and tissues (Table 13.1) varies widely. Some microbes, like *Azotobacter* and *Bacillus fluorescens,* respire most actively, and yeast and mold respire more actively than do protozoans and cells from animal tissues. An extensive list of Q_{O_2} values is given by Prosser (1973).

TABLE 13.1. OXYGEN CONSUMPTION (Q_{O_2}) OF CELLS, TISSUES AND ORGANISMS IN ML/GM WET WEIGHT/HOUR

Group	Organism	°C	Q_{O_2}	Remarks
Microorganisms	*Bacillus mesentericus vulgatus**	16	12.1	
	Azotobacter chroococcum†	28	500–1000	
	Bacillus fluorescens non liquefaciens‡	22	4100	
	Neurospora crassa§ (breadmold)	26	6.4	
	Saccharomyces cereviseae (yeast)	26	8–14.5	
	Paramecium¶	20	0.5	
Plants	*Verbascum thapsus** (mullein)	23	0.093	leaf
			0.204	pistil
			0.190	stamen
	*Papaver rhoeas** (poppy)	22	0.803	leaf
			0.172	pistil
			0.280	stamen
Invertebrates	*Anemonia sulcata*‡ (anemone)	18	0.0134	
	Asterias rubens‡ (seastar)	15	0.03	
	Nereis virens‡ (annelid worm)	15	0.026	
	Mytilus¶ (mussel)	20	0.02	
	Astacus¶ (crayfish)	20	0.047	
	Vanessa¶ (butterfly)	20	0.6	at rest
			100.0	in flight
Vertebrate	Carp¶	20	0.1	
	Mouse¶	37	2.5	at rest
			20.0	running
	Human¶	37	0.2	at rest
			4.0	maximal work
Animal tissues	Rat liver†	37.5	2.2–3.3	
	Rat kidney cortex†	37.5	5.2–9.0	
	Rat brain cortex†	37.5	2.7	
	Rat voluntary muscle#	37	1.5	at rest
		37	10.0	active
	Frog nerve#	15	0.02	at rest
		15	0.75	active
	Rabbit nerve#	37	0.29	at rest

All Q_{O_2} values cited in references as volume of oxygen per gram of dry weight per hour have been divided by 4 to give the value for wet weight.
*Stiles and Leach, 1952: Respiration in Plants. 3rd Ed. John Wiley, New York.
†Tabulae Biologicae, 1934:*9*.
‡Heilbrunn, 1952: A Textbook of General Physiology. 3rd Ed. W. B. Saunders, Philadelphia.
§Giese and Tatum, 1946: Arch. Biochem. *9*.
‖Giese and Swanson, 1947: J. Cell. Comp. Physiol. *30*.
¶Krogh, 1941: The Comparative Physiology of Respiratory Mechanisms. University of Pennsylvania Press, Philadelphia.
#Holmes, 1937: The Metabolism of Living Tissues. Cambridge University Press, London.

It is of interest to find out what relation exists between Q_{O_2} and size of cells. In Table 13.1, we will see that the rate of respiration in unicellular organisms varies with the ratio of cell surface to volume. Microbes, being of small linear dimensions, have a greater surface-to-volume ratio than do larger cells. When cell size increases, the Q_{O_2} declines (Zeuthen, 1970). The range in size among protozoans is ample to illustrate this point clearly (Fig. 13.2); the respiratory rate of the large multinucleate ameba *Pelomyxa carolinensis (Chaos chaos)* is only about half its increase in bulk. Probably *Pelomyxa* represents the limit of effective increase in size of a single cell. Presumably, during evolution an increase in mass brought about by aggregation of small cells proved more effective in maintaining a sufficiently high Q_{O_2} than a corresponding increase in mass of a single cell (Zeuthen, 1970).

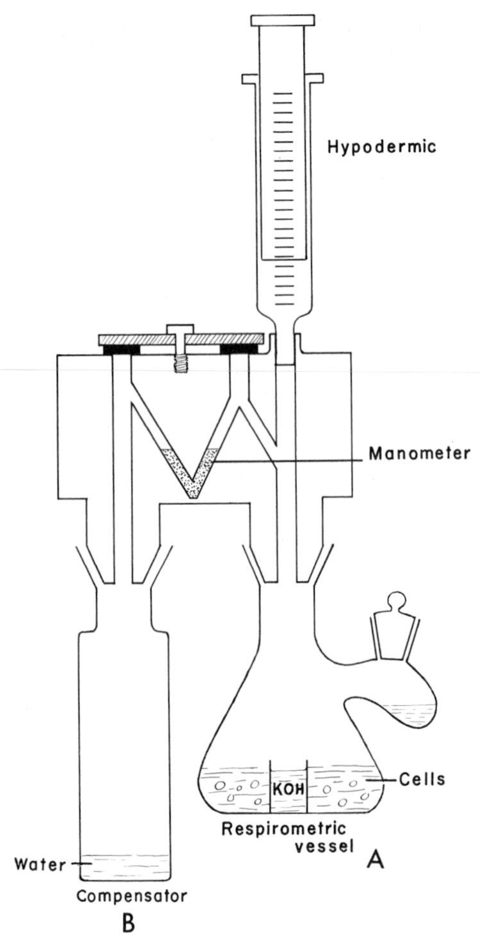

Figure 13.1. The Scholander-Wennesland modification of the Fenn-Winterstein respirometer. Usually a calibrated screw plunger is used instead of a hypodermic syringe. (After Wennesland, 1951: Science 114: 100.)

When the Q_{O_2} of larger animals and plants is compared with that of microbes, the contrast is even more striking (Table 13.1). Yet the Q_{O_2} of the entire organism or tissue is the sum of the oxygen consumption of its component cells. One probable reason for this difference is that oxygen consumption of microorganisms is generally measured in the presence of nutrient favorable for maximal activity, while the oxygen consumption of cells in larger organisms is usually measured during relatively inactive states. When measured during maximal activity, their metabolic rates are 10-fold to 100-fold greater (Table 13.1), at which time their oxygen consumption comes within the range of the oxygen consumption shown by active microorganisms.

The rate of oxygen consumption falls with increase in size of an organism. Even when an individual of a species increases in size during its development, its Q_{O_2} falls (Fig. 13.3). In warm-blooded animals the greater the ratio of surface area to volume, the greater is the heat loss and the greater is the rate of oxygen consumption. For example, the shrew loses so much heat because of its large surface-to-volume ratio that it must eat continuously to get enough food to keep warm. (The hummingbird solved this problem by becoming torpid during sleep, its temperature approximating the environment.) Consequently, a higher metabolic rate is expected in small mammals or birds than in larger ones. A limit is thus set to the minimal size of a warm-blooded form (Pearson, 1953, 1954). Whether loss of heat is also the cause of the greater Q_{O_2} found in small invertebrates is not resolved at present (Mill, 1972).

The oxygen consumed by an animal per unit time can be related to its surface and volume by means of an empirical equation:

$$O_v = KW^b \qquad (13.1)$$

where O_v is the volume of oxygen consumed per unit time, K is a constant, W, the weight of the organism and b, an exponent. For most animals tested, b = 0.73. Since O_v is the Q_{O_2} times the weight of the animal, equation 13.1 indicates that the Q_{O_2} is related to the 0.73 power of W (Steen, 1971).

One reason for decrease in Q_{O_2} with increase in size is perhaps the corresponding increase in the larger organism's content of inert materials. According to the principle of similitude (Bonner, 1952), the volume of an animal increases as the cube of its linear dimensions, whereas the strength of bones and connective tissue increases only as the square of the animal's linear dimensions. Consequently, it is the bone, cartilage, and fibrous connective tissue that increase disproportionally with increase in size. However, they are relatively inert metabolically and contribute little to the Q_{O_2}. Similarly, in plants, cell walls and supporting tissue and vascular tubes ("dead wood") add bulk without contributing metabolizing tissue.

Cells removed from animals (or plants) regardless of size, and grown in tissue culture, respire at much more nearly the same rate (Umbreit *et al.*, 1972).

THE EFFECT OF TEMPERATURE ON Q_{O_2}

The Q_{O_2} of a cell suspension increases with a 10°C rise in environmental temperature by twofold to fourfold (Fig. 13.4). However, exposure to temperatures above the optimum always results in injury, after which oxygen consumption declines.

The rate of oxygen consumption of multicellular plants, cold-blooded multicellular animals, and young mammals in which the "ther-

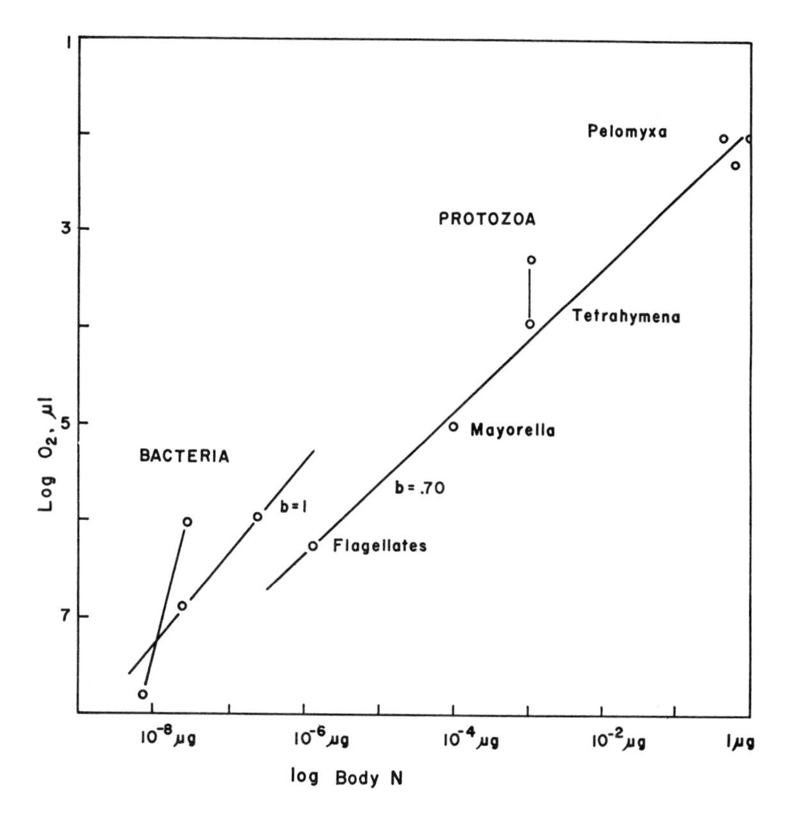

Figure 13.2. The relation between cell size, as measured by the logarithm of the cell nitrogen content (N), and the logarithm of the volume of oxygen consumed by single-celled organisms. Note that the points of protozoans fall along a straight line, the slope of which (b, the exponent in equation 13.1) is 0.70. This slope tells us that the rate of respiration increases much more slowly than the mass (size) of the protozoan. For bacteria, on the other hand, the slope (b) of the line is 1.0, indicating that over the small size range covered by the experiments the respiration increases proportionally with cell size. The higher value for b indicated in far left line has questionable significance. (After Zeuthen, 1953: Quart. Rev. Biol. 28: 3.)

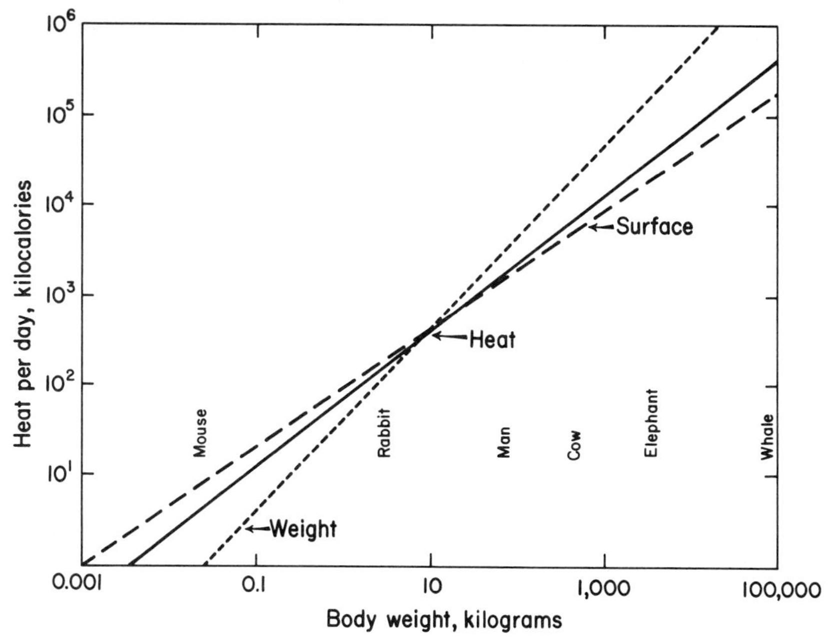

Figure 13.3. Relation between respiration (measured as heat per day) and size. Heat is shown by the solid line; surface is shown by the broken line with longer dashes, weight is shown by the broken line with shorter dashes. Note that respiration is more closely correlated with surface than with weight. (After Kleiber, 1947: Physiol. Rev. 27: 530.)

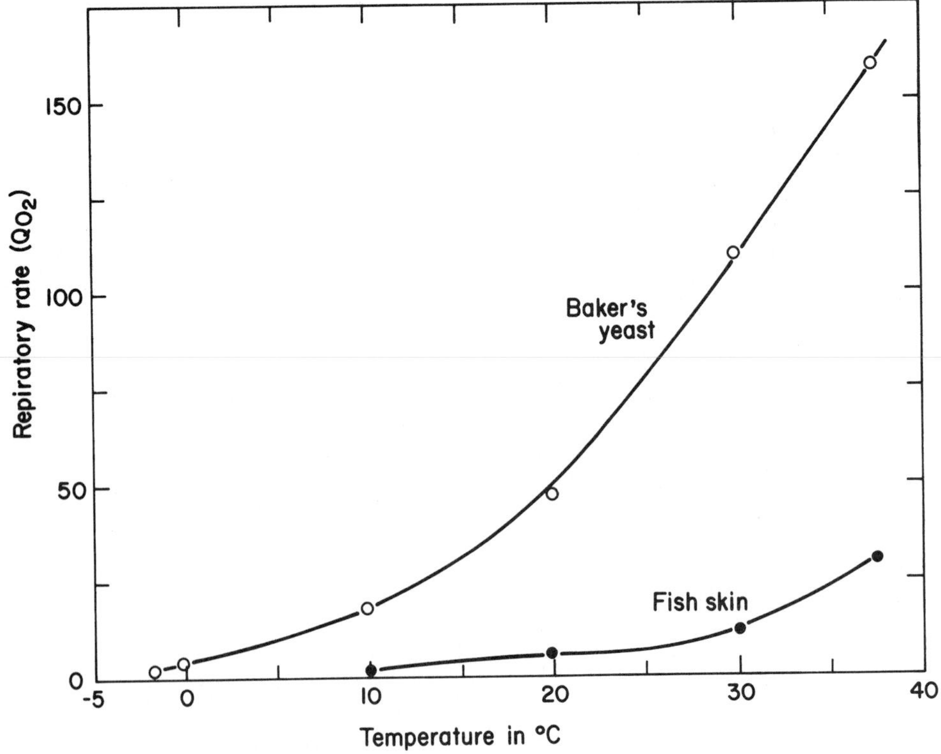

Figure 13.4. *Relation between temperature and oxygen consumption in fish skin and yeast. (Data from Tabulae Biologicae, 1934, 9: 223.)*

mostatic" mechanism is not yet in operation increases in a similar manner when the ambient temperature rises. Therefore, tissue cells respond to temperature much as do suspensions of cells. However, the Q_{O_2} of mature warm-blooded animals does not vary with temperature in the same manner because their body temperature remains relatively constant, and a fall in environmental temperature may even lead to increased oxygen consumption to release the extra heat needed for maintaining a constant temperature.

THE EFFECT OF OXYGEN PARTIAL PRESSURE ON Q_{O_2}

The relation between oxygen partial pres-

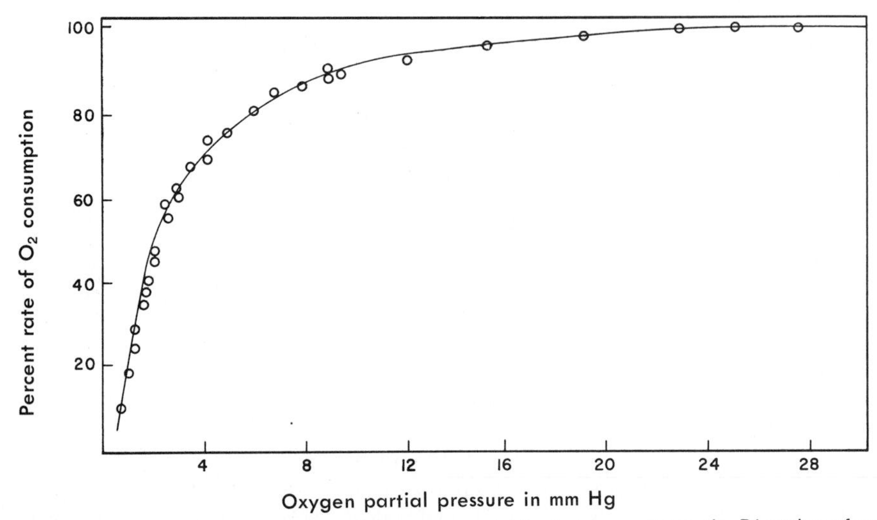

Figure 13.5. *Relation between oxygen partial pressure and the luminous bacteria. Dimming of emitted light occurs when the rate of oxygen consumption falls to 50 percent. (From Shoup, 1929–1930: J. Gen. Physiol. 13: 41.)*

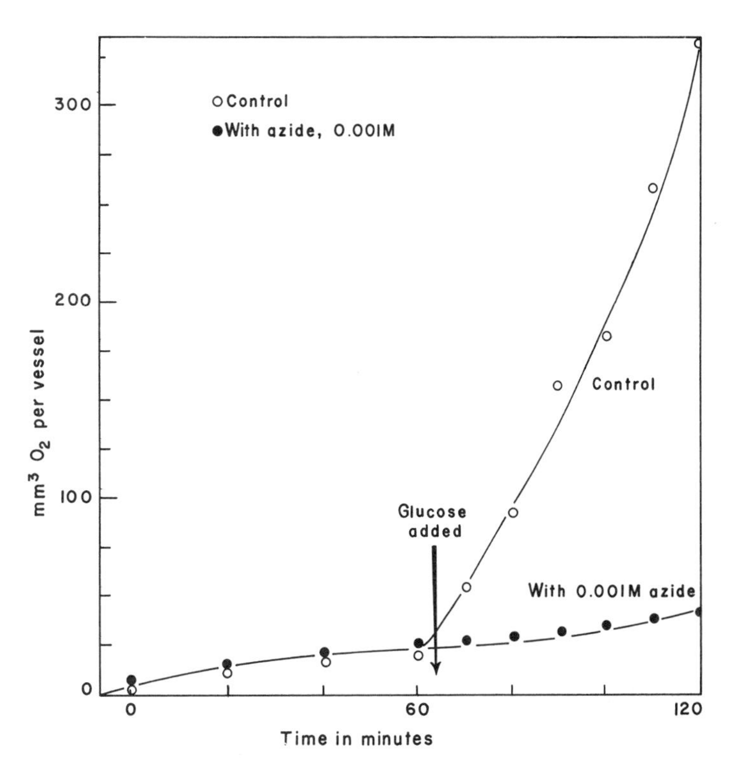

Figure 13.6. *The effect of a respiratory poison on the respiration of yeast. The azide, which resembles cyanide in its action on respiration, was added to one group of cultures (black circles) but not to the control (open circles). Note that the endogenous respiration (before the addition of glucose) is little affected by azide, since experimental and control points fall side by side, but note also that exogenous respiration (after the addition of glucose) is virtually abolished. In the same experiment 0.3M urethan had little effect on endogenous or exogenous respiration, although it strongly inhibited fermentation.*

sure (the fraction of the total pressure due to the gas in question) and oxygen consumption is of considerable interest, since oxygen must enter all living cells in dissolved state. For example, when the oxygen partial pressure of a liquid containing luminous bacteria in suspension falls to 3 percent of that in the atmosphere, the Q_{O_2} of the bacteria begins to fall. It declines rapidly to one half the normal rate when the oxygen partial pressure reaches 0.25 percent of atmospheric oxygen, as shown by the data in Figure 13.5. The curve plotting the relation between oxygen consumption and oxygen partial pressure is similar to the curve relating the amount of oxygen adsorbed by charcoal at different gas pressures. This suggests that adsorption of oxygen to the surface of the respiratory enzyme (cytochrome oxidase) perhaps underlies the relation between cell respiration and oxygen partial pressure.

RESPIRATORY INHIBITORS

Many substances inhibit respiration (Fig. 13.6). Cyanide binds the iron group in cytochrome oxidase, preventing transfer of electrons to oxygen, thereby preventing activation of oxygen. Consequently, it depresses the Q_{O_2} even when present in relatively low concentrations (about 0.001 molar). Azide resembles cyanide in its action. Ethyl urethan is much less specific and interferes with a large number of enzymes in the cell; consequently, a relatively high concentration (from 0.1 to 0.5 molar) is necessary to reduce the Q_{O_2}. Some enzymes (e.g., dehydrogenases and luciferases) are more sensitive to urethan than are other enzymes.

HYDRATION AND RATE OF OXYGEN CONSUMPTION

It is difficult to demonstrate oxygen consumption in dormant dehydrated stages of organisms such as cysts, spores, or seeds. The progressive decline in oxygen consumption of a single protozoan undergoing encystment and dehydration has been demonstrated in a microrespirometer (see Fig. 4.2). Conversely, the Q_{O_2} increases as water is absorbed during excystment. Similar experiments with wheat seeds show that as the percentage of water content in the wheat kernels increases above 14.75 percent, the respiratory rate rapidly increases. This suggests that perhaps below this value all the water is bound, whereas above it sufficient free water is present for resumption of cell activities. Experiments with cells of lichens demonstrate that the metabolically inert dried cells can be activated to a high rate of respiration (and photosynthesis in the light) about 30 minutes after wetting (Fig. 13.7). These and other experiments indicate the importance of adequate free water for maintenance of cellular reactions. The water present in dormant stages of organisms is largely *bound water,* not free to participate in chemical reactions. As discussed in Chapter 4, neither hydrolyses nor oxida-

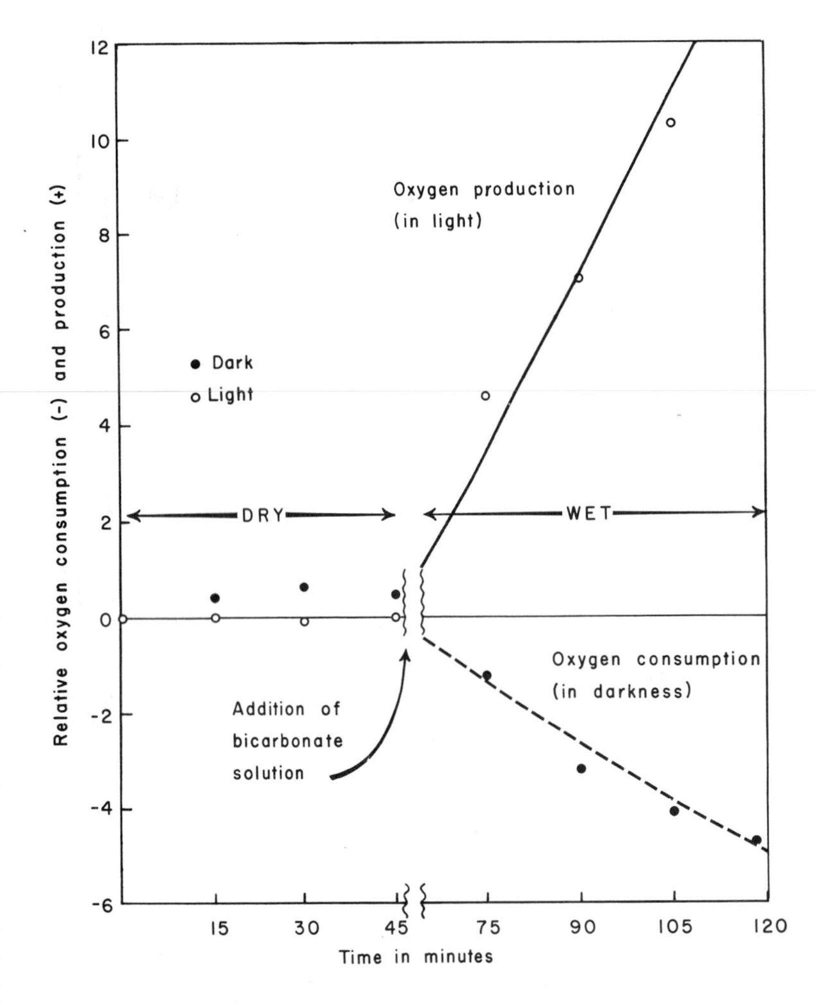

Figure 13.7. *Effect of hydration on oxygen consumption and photosynthesis in a lichen. About 0.5 gm of foliose lichen freshly collected was added to each of two flasks, one of which was tightly sealed with tinfoil to exclude light, the other illuminated. The dark flask had KOH in the side well, the other none. After addition of bicarbonate buffer as a source of CO_2 for photosynthesis, the vessels were kept open to the atmosphere for about 20 minutes. No readings were taken during this time, because desorption of gases attached to the dry lichen occurs in both cases. Readings were resumed when desorption was about complete.*

tions, both of which are required for maintenance of active existence, occur in the absence of water.

RESPIRATORY QUOTIENT (R.Q.)

The *respiratory quotient* (R.Q.) is the ratio of the carbon dioxide produced to the oxygen consumed:

$$\text{R.Q.} = \frac{CO_2}{O_2} \tag{13.2}$$

From this ratio considerable information can be gathered about the nutrients being metabolized.

The respiratory quotient for cells metabolizing carbohydrates with a general formula $(CH_2O)_n$ is 1, because a molecule of carbon dioxide is produced for every molecule of oxygen consumed during complete combustion of a carbohydrate:

$$CH_2O + O_2 \rightarrow CO_2 + H_2O; \tag{13.3}$$

$$\text{R.Q.} = \frac{CO_2}{O_2} = \frac{1}{1} = 1$$

The respiratory quotient of cells metabolizing fats is less than 1, because the amount of oxygen consumed is greater than the amount of carbon dioxide produced. Calculation of the oxygen required for complete combustion of a fat molecule such as tripalmitin, $(C_{15}H_{31}COO)_3C_3H_5$, makes this clear:

$$(C_{15}H_{31}COO)_3C_3H_5 + 72.5\ O_2 \rightarrow 51\ CO_2 + 49\ H_2O \tag{13.4}$$

$$\text{R.Q.} = \frac{CO_2}{O_2} = \frac{51}{72.5} = 0.70$$

The respiratory quotient varies with the fat metabolized. For instance, it is 0.8 for tributyrin, a fat with short-chain fatty acid.

The respiratory quotient for cells metabolizing protein is also less than 1, because proteins are made up of amino acids, which require more oxygen for complete combustion than do carbohydrates. Since the amino acids have a short carbon chain, an R.Q. of 0.8 is generally assigned to proteins.

If an organism is fed a mixed diet, the R.Q. depends upon the relative proportion of the types of nutrients used. For example, the R.Q. for humans on an average diet of carbohydrates, fats, and proteins is 0.825 (Guyton, 1976).

Occasionally the R.Q. of respiring cells is greater than 1. For example, when some of the oxygen released by the carbohydrate during its synthesis into fat is used as a hydrogen acceptor instead of atmospheric oxygen, less atmospheric oxygen is consumed. For quite a different reason, the R.Q. of cells going into oxygen debt may be temporarily greater than 1 (until the debt is paid), because more carbon dioxide is being produced by fermentative processes than can be accounted for by the cellular uptake of oxygen. The R.Q. of plant cells in the dark depends upon the nutrient being consumed (Bonner and Varner, 1977).

Because fats have a higher reduction level than carbohydrates or proteins they serve as valuable storage materials in animal and plant cells. A gram of fat has almost double the energy available in a gram of protein or carbohydrate.

HEAT EVOLVED DURING RESPIRATION

Considerable heat is developed as a byproduct of cellular respiration, aerobic and anaerobic. This is readily measured with a thermometer during the activities of microorganisms as, for example, in a manure pile. Heat produced by plant cells is more difficult to detect because it is so readily lost, but a temperature rise is demonstrable under controlled conditions. For example, considerable heat is produced by germinating seeds, and the cells of a spadix of an arum lily are said to become warm enough to attract insects.

Muscular activity of animals is accompanied by a marked increase in temperature. Even the temperature of the muscle cells of a fish may be several degrees higher than the temperature at the surface of the body. Maximal exertion in humans may be accompanied by the development of a fever temperature as high as 103°F which, after exertion has ended, falls rapidly by the cooling action of the evaporation of water from the skin. An increase in temperature, however achieved (e.g., fever), increases cellular respiration; cooling reduces it.

INDIRECT CALORIMETRY

Energy exchanges in organisms may be determined directly with a calorimeter (see Chapter 9) or indirectly by a measurement of the gaseous exchanges, since energy spent by the cells of an organism is ultimately derived from oxidation of foodstuffs. Because of its convenience, the latter method is now used almost exclusively in medicine for tests of basal and of activity metabolism.

When 1 gram molecular weight of glucose is oxidized, 6 × 22.4 liters, or a total of 134.4 liters of oxygen is consumed, and 673,000 calories of heat are released:

$$C_6H_{12}O_6 + 6\ O_2 \rightarrow 6\ CO_2 + 6\ H_2O + 673{,}000 \text{ calories} \quad (13.5)$$

Per liter of oxygen used, then, 673,000/134.4 or 5007 calories of heat are liberated. Per liter of oxygen used in combustion of fat, 4686 calories are liberated; and per liter of oxygen used for combustion of proteins, 4500 calories are liberated. When these foodstuffs are oxidized in the proportions present in the average mixed diet of humans, 1 liter of oxygen releases 4825 calories. Therefore, if in an experiment a man consumes 5 liters of oxygen in 20 minutes, he

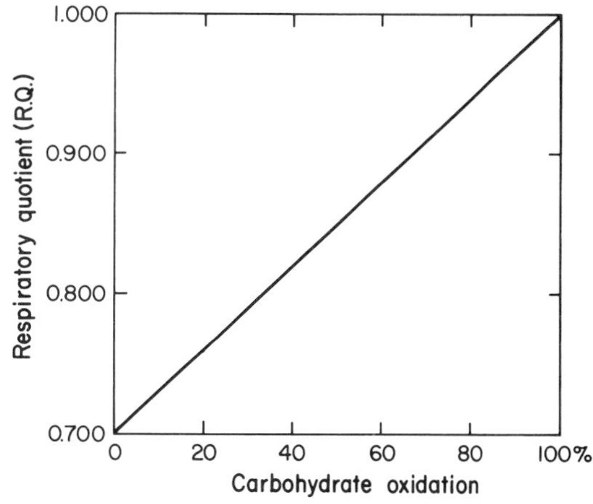

Figure 13.8. Relationship between non-protein R.Q. and percentage of the nonprotein oxygen consumption that is used in oxidation of carbohydrate. (After Ruch and Patton, 1965: Physiology and Biophysics. 19th Ed. W. B. Saunders Co., Philadelphia.)

will liberate approximately 25,125 calories of heat. This rate of respiration continued for a day liberates approximately 2×10^6 calories or 2000 kilocalories (kcal) per day. This is equivalent to the heat required to raise 20 liters of water from freezing to boiling. On the basis of the principles outlined and from the Q_{O_2}, it is possible to calculate the food requirements of cells or organisms.

From a measure of the carbon dioxide produced, the R.Q. could be calculated. From the R.Q. and the nitrogenous waste of a suspension of cells or an animal, the energy contribution of carbohydrate, fat, and protein can be determined. For example, the R.Q. of a human on a mixed diet is 0.825. One gram of urea in the urine represents the utilization of 3 grams of protein. For each gram of protein 0.95 liter of oxygen has been consumed and 0.76 liter of carbon dioxide produced. The fraction of oxygen used in combustion of the protein portion may be determined and subtracted from the total oxygen used.

The fraction of carbon dioxide produced by protein may similarly be subtracted from the total carbon dioxide produced. The remainder represents that produced by a combination of fat and carbohydrate. From these values for oxygen and carbon dioxide it is possible to obtain an R.Q. for the nonprotein component of the food utilized. By use of Figure 13.8, relating the nonprotein R.Q. to oxygen used in carbohydrate consumption, it is possible to determine the fraction of carbohydrate in the diet, the remainder constituting the fat component. Such studies show that generally the main source of energy in cells is carbohydrate (Guyton, 1976).

However, cells in some animals (butterflies, hibernators, and migrators) can utilize fat directly, as indicated by an R.Q. of about 0.7. For example, when a butterfly is fed sugar, the R.Q. rises to 1.5 or more, indicating that fat is being formed and deposited in the cells (Hochachka, 1973). Cells of starving animals use fats (and some protein) when reserves of carbohydrates are gone and, as expected, the R.Q. is then about 0.7.

The R.Q. of respiring plant cells varies. A plant cell using carbohydrate as its main nutrient source has an R.Q. near unity. When dedependent upon fat (e.g., in a germinating seedling with large fat stores), the R.Q. may be about 0.7. In some cases the R.Q. may be as low as 0.4 (*Linum,* fourth day after germination). This is believed to result from oxygen uptake during formation of carbohydrates from fats, the converse of what happens when plant and animal cells make fats from sugar. The carbohydrates are then presumably oxidized, since most plant cells show typical anaerobic and aerobic pathways like those described in Chapter 12. The R.Q. of succulent plant cells, which accumulate organic acids, may also be quite low.

CELLULAR OXIDATIONS EMITTING LIGHT

Occurrence

The flash of the firefly at evening and the luminous wake of a boat have always attracted attention, even in ancient times. Bioluminescence has been described in bacteria, fungi, protozoa, and species belonging to 40 different orders in half the 25 phyla of animals (see Figs. 13.9 and 13.10). "Flashlight" fishes have unusual luminous organs, some with symbiotic bacteria, others with innate luminous systems (McCosker, 1977). *Bioluminescence* is emission of light as a consequence of cellular oxidation of a substrate in the presence of an enzyme (Hastings and Wilson, 1976). Some "non-luminous" redox systems are found to emit enough light to be sensed by extremely sensitive light detectors (Stauff and Nimmerfall, 1969).

Robert Boyle, in 1667, showed that luminescence of fungi on rotten wood, like cellular respiration, ceases when air (later shown to be the oxygen in air) is withdrawn and that it recurs when air is readmitted. Reamur, in 1723, demonstrated that shellfish cease luminescing

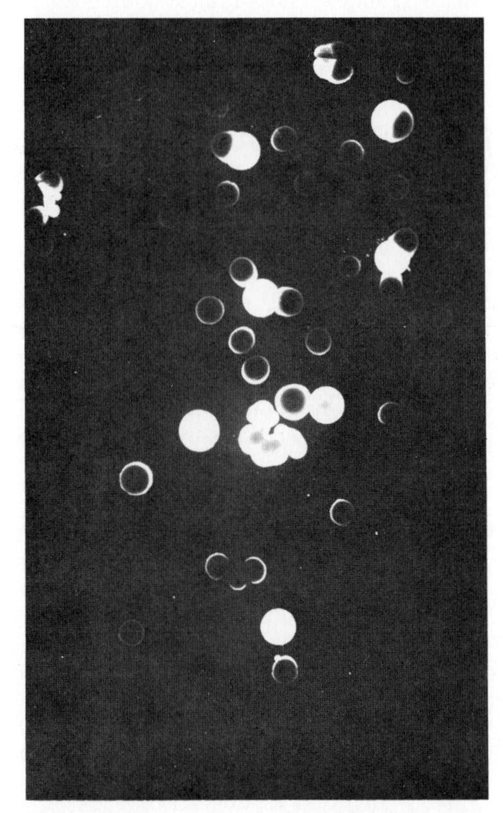

Figure 13.9. *Colonies of the luminous bacteria,* Achromobacter fischeri, *photographed using only their own light. The brilliant strain is a variant that tolerates the acidity developed in the medium and remains bright while the other strain turns dim.*

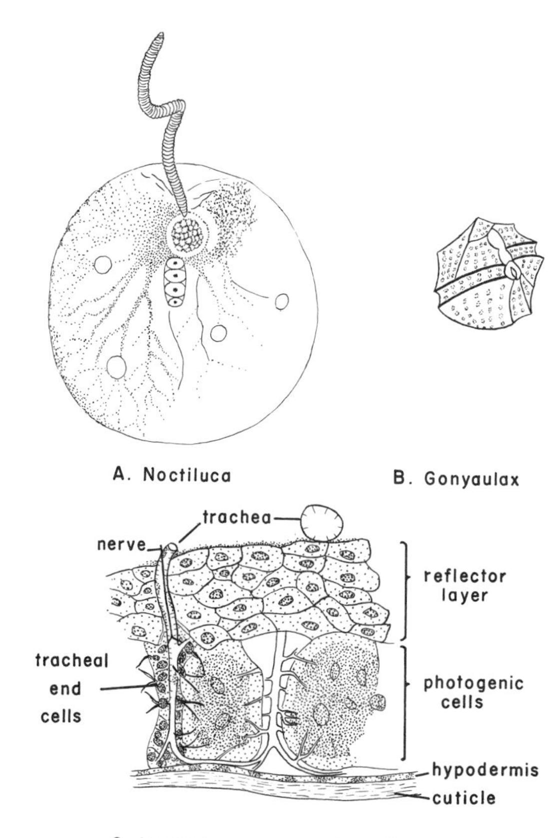

Figure 13.10. A, Noctiluca *(about 700 μm in diameter) and* B, Gonyaulax *(about 200 μm in diameter), two luminous dinoflagellates found along the Atlantic and Pacific Coasts. The wake of a boat passing through a dense population of such dinoflagellates is luminous because the movement of the water excites the protozoans to give off light. C, The luminous organ of a firefly,* Photinus. *(After Williams, 1916: J. Morphol.* 28: *145.)*

when dried, but resume luminescing when water is added. Dubois in 1885 demonstrated that luminescence results from oxidation of a heat-stable substrate, which he called *luciferin,* in the presence of a heat-sensitive enzyme, which he called *luciferase.* He found that luciferin is present in the luminous extract from a firefly in limited quantities and is soon exhausted, but that the enzyme luciferase remains essentially unchanged (Johnson, 1974).

Chemical Nature of Bioluminescence

Investigations involving extracts from luminous bacteria have demonstrated intimate relationships between luminescent systems and energy-liberating enzymatic reactions of the cell. To obtain the extract, the bacteria (grown on a synthetic medium) are harvested by centrifugation and dried in acetone cooled to −15°C. The dried powder is homogenized in water at room temperature and centrifuged to remove debris. The supernatant thus obtained is luminous, but the luminescence can be inhibited at −20°C, a temperature at which the molecules are unable to attain the energy necessary for activation. Most of the characteristics of luminescence in this extract are similar to those of the luminous system in bacteria. For example, for both the temperature optimum is about 25°C, and the emission spectrum is between 475 nm and 500 nm.

Evidence points to the probable identity of luciferin from bacteria with reduced flavin mononucleotide ($FMNH_2$), which has been shown to react with an aldehyde (RCHO) present in the cell to form a complex that is oxidized to an acid (RCOOH) with the emission of light in the presence of luciferase (Hastings and Balny, 1975):

$$FMNH_2 + RCHO + O_2 \xrightarrow{\text{luciferase}} FMN + RCOOH + H_2O + \text{light} \quad (13.6)$$

The FMN is then once again reduced, probably via NADH, by hydrogen from various dehydrogenations occurring in the bacterial cells (see Chapter 12), for example, the oxidation of malate to oxaloacetate. Various aldehydes from C_6 to C_{16} may complex *in vitro* with $FMNH_2$ to form bacterial luciferin. It is interesting to note that mutant strains of luminous bacteria *(Achromobacter fischeri),* which lack the ability to form aldehydes, and so do not luminesce, are made luminous by adding a solution of one of these aldehydes (Hastings and Wilson, 1976).

Luciferase extracted from bacteria and purified appears to require a free sulfhydryl group for its activity; therefore, it is extremely sensitive to heavy metals. Its only function appears to be the luminous reaction, although it was first believed to be a respiratory enzyme (Nicoli *et al.,* 1974). The luciferase consists of two units designated α and β, both required for activity. The α unit is highly sensitive to proteases, which fragment it. The light emitter (chromophore) appears to be a complex of luciferase and reduced flavin in a 1:1 proportion.

Bacterial luciferase is not constitutive (normally present) and must be autoinduced by a substance that accumulates in dense suspensions. Therefore, luminescence is not characteristic of a dilute suspension such as occurs in sea water. The luciferase is autoinduced when the bacteria grow on a dead fish or other organism in the sea (Hastings and Wilson, 1976).

The luciferin and luciferase systems of the East Coast fireflies have also been extracted. Firefly luciferin occurs in the lantern probably in aqueous solution and, being quite stable, can be purified in air by filter-paper chromatography or by use of appropriate resins. At a pH greater than 7.0, pure luciferin, when exposed to ultraviolet light, gives off a yellow-green fluorescence.

The luciferin luminesces only in the presence of ATP and some divalent ion, such as Mn^{++}, Mg^{++}, or Co^{++}, in addition to firefly luciferase and oxygen. The need for ATP links the luciferin-luciferase system to carbohydrate metabolism as the source of energy. Since inorganic phosphate appears as a byproduct of firefly luminescence, a degradation of compounds with high-energy phosphate bonds is indicated.

The sequence of reactions in firefly luminescence appears to be as follows:

$$LH_2 \text{ (luciferin)} + ATP \rightarrow LH_2\text{-AMP (adenyl-luciferin)} + POP \text{ (pyrophosphate)} \quad (13.7)$$

Adenyl-luciferin is regarded as the active luciferin molecule that is peroxidated by reaction with atmospheric oxygen and subsequently breaks down to adenyloxyluciferin, water, and adenylic acid:

$$LH_2\text{-AMP} + \tfrac{1}{2}O_2 \rightarrow L\text{-AMP}^* + H_2O \quad (13.8)$$

$$L\text{-AMP}^* \rightarrow L + AMP + h\nu \quad (13.9)$$

When enzymes (luciferases) of the two species of firefly were purified and tested with the luciferin of one of them, the color of the emitted light depended upon the enzyme used. However, since changes in ionic environment change the color of the light when purified luciferase from one species is used, it is probable that the way in which the luciferin is bound to the luciferase determines the color of light emitted. The binding of the luciferin to luciferase is in turn determined by the tertiary structure of the luciferase, which is probably different in various luciferases. For each luciferin molecule oxidized, 1 (0.73 to 1.03) quantum of light is emitted. This extraordinary efficiency of the process suggests an energy-stabilizing role for luciferase and great specificity of the enzyme. However, adenyloxyluciferin readily combines with luciferase, thereby acting as a potent inhibitor of the enzyme. Since coenzyme A combines with adenyloxyluciferin to form oxyluciferyl CoA and adenylic acid, it has a stimulating effect on luminescence. The pyrophosphate formed in reaction 13.7 is decomposed by the pyrophosphatase available in the cells; therefore, it does not affect the reaction.

Some of the compounds postulated by McElroy and his co-workers for these reactions have been isolated. For example, adenyl-luciferin can be formed *in vitro* from ATP and crystalline luciferin, and this compound can be hydrolyzed anaerobically to form luciferin and adenylic acid (AMP). The oxyluciferin and various phosphate compounds have also been isolated from firefly extracts. Several intermediate compounds have been postulated and although some appear more probable than others, a final explanation awaits further experimentation (Lee, 1977).

The molecular weight of crystalline firefly luciferin is 280, and its empirical formula is $C_{11}H_8N_2S_2O_3$. Oxyluciferin, which usually occurs with luciferin as a contaminant, has a molecular weight just two units less, 278. Luciferin has two pK values, one at pH 3.0 (COOH), another at pH 8.4. The structural formula of firefly luciferin has been determined and the luciferin has been synthesized.

The luciferase of fireflies, which has also been purified and crystallized, shows all the general properties of its class of enzymes. Its minimal molecular weight is about 50,000, its isoelectric pH is 6.1 to 6.2, its pH optimum is 7.8 and its temperature optimum is 23°C (Johnson, 1974).

Bioluminescence has also been studied in the extracts of a number of other organisms, e.g., *Cypridina* ("water firefly"), a Japanese ostracod crustacean, the marine dinoflagellate *Gonyaulax,* and others (Hastings and Wilson, 1976). By use of pure preparations of luciferin and luciferase of *Cypridina* it has been shown that the dark-adapted eye will perceive a flash from a solution containing 10^{-11} M luciferin and 10^{-15} M luciferase (Johnson, 1974).

It must be emphasized that the word luciferin refers not to a chemical entity, but rather to a class of compounds that may differ chemically along themselves, but all of which are oxidized in the presence of an enzyme and atmospheric oxygen (or peroxide) with the emission of light unless precharged by oxidation as in the protein aequorin from the jellyfish *Aequorea,* which emits light on addition of calcium ions (Shimomura and Johnson, 1976). Luciferase is the name given to the class of enzymes that catalyze these bioluminescent reactions. They have all the common properties of enzymes (see Chapter 10).

Luciferases do not appear to contain metal ions and are not sensitive to cyanide, but some of them are markedly sensitive to urethan (Johnson and Shimomura, 1972). Enzymatic bioluminescence has been analyzed in bacteria, fungi, protozoa, annelid worms, mollusks, crustaceans, insects, hemichordates, and fishes.

It was once thought that all enzymatic bioluminescent systems required oxygen but some have been found to use peroxide instead, and the enzyme in these cases appears to be a nonspecific peroxidase instead of a luciferase (e.g., in the acorn worm, *Balanoglossus*) (Cormier *et al.,* 1975).

While it was once thought that all bioluminescent systems were enzymatic, what appear to be nonenzymatic luminescent systems have been found in several coelenterates and have been studied in some of these and in a ctenophore *(Mnemiopsis)*, an annelid worm *(Chaetopterus)*, a mollusk *(Pholas)*, and a euphausid shrimp *(Meganyctophanes)*. (Johnson and Shinomura, 1972). In the first "nonenzymatic" bioluminescent system isolated, that of the jellyfish *Aequorea*, the luminescent substance turned out to be a protein, aequorin, to which only calcium need be added to evoke light emission, even in absence of oxygen. However, not all photoproteins are alike. Some require oxygen or peroxide, some require other salts in addition to calcium. The mechanism by which light emission is evoked is obscure, although attempts have been made to explain it on the basis that the system was charged by previous oxidation and was ready to emit light.

At present aequorin appears to be a luciferase intermediate "precharged" with stored energy, ready for an instant flash on admission of Ca^{2+}. Spent (inactive) aequorin treated with synthetic luciferin can be recharged in an oxygen-dependent reaction (Shimomura and Johnson, 1976; Cormier *et al.*, 1977).

Evidence points to the involvement of superoxide in luminescence. This is but one step in oxidation and has been treated elsewhere (Cormier *et al.*, 1977; Hatchikian *et al.*, 1977; Puget *et al.*, 1977).

Physical Nature of Bioluminescence

A molecule gives off light when the outer shell electrons in virtual orbits of the excited state return once more to their normal unexcited positions. When heat furnishes the energy for excitation, as in incandescent bodies, the low efficiency of light production (e.g., 0.42 percent in the carbon filament lamp and 2.5 percent in the tungsten filament lamp) indicates the low probability of this type of excitation. In fluorescent lamps, electrons, falling through a voltage gradient, collide with gas molecules, transferring kinetic energy to the gas molecules and thereby exciting their electrons to virtual orbits. The efficiency of conversion in this case is about five times that in the best incandescent lamps. Light may also be obtained by exciting electrons mechanically, as is done by grinding crystals.

Light is also produced by chemical reactions called chemiluminescences, in which sufficient energy is released to raise electrons of the substrate or its products to an excited state. The chemiluminescent oxidation of luminol (3-aminophthalcyclohydrazide) has been studied more extensively than most of the others (Hastings and Wilson, 1976). Bioluminescence, then, is chemiluminescence in living cells.

The intensity of light emitted in bioluminescence is very low. Lode has estimated that if the entire dome of St. Peter's in Rome were covered with a layer of luminous bacteria, the intensity of light would be about equal to that from one standard candle (Harvey, 1940).

Although the intensity of the luminescence is low, the visual efficiency of the light is high compared with artificial sources, being 95 percent for the light from the firefly *Photurus pennsylvanica*, 45 percent for the luminous bacteria, 20 percent for *Cypridina*, and 12 to 14

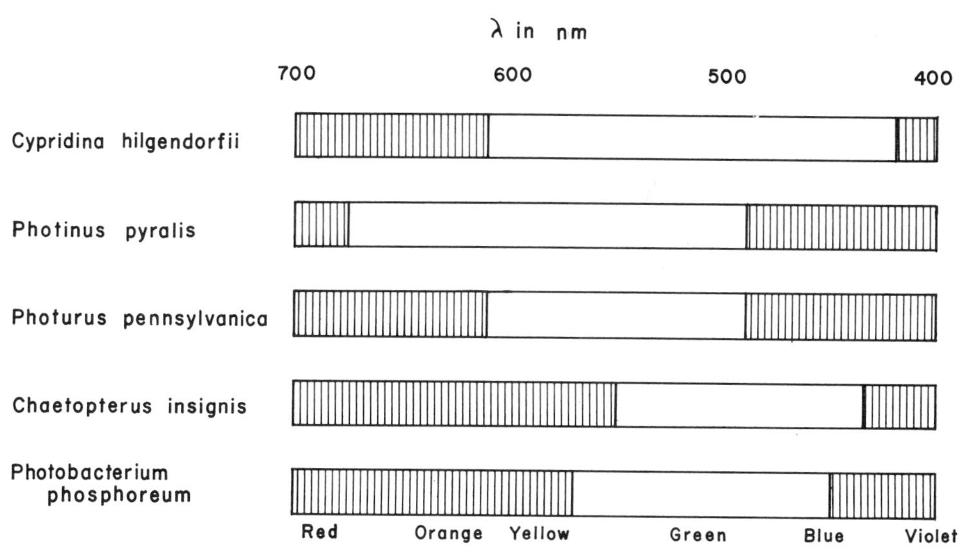

Figure 13.11. Luminescent spectra from luminous secretions or cells in various species of animals. The uppermost is from the "water firefly" of Japan, a crustacean; the next two are from American fireflies; the fourth is from a luminous tube-dwelling worm; and the last is from a luminous bacterium. To the eye the light of the first appears bluish, the light of the second yellowish, the light of the last three greenish. (Data from Harvey, 1920: The Nature of Animal Light. J. B. Lippincott Co., Philadelphia.)

percent for fluorescent lamps. In these comparisons, account is taken of the visual efficiency of the human eye at the different wavelengths in the emission spectra of the luminous organisms (Fig. 13.11). Neither infrared nor ultraviolet light has been detected in the emission spectra of luminous organisms (Harvey, 1940).

APPENDIX

13.1 METHODS FOR MEASURING RESPIRATION

Chemical Methods. The oxygen content of a sample of water may be measured by the Winkler method. This procedure involves addition of a solution of manganous chloride and of a mixture of potassium iodide and potassium hydroxide to the sample of water. The oxygen present in the sample oxidizes some of the manganous ion to manganic ion. Strong acid (hydrochloric or sulfuric) is added when the precipitate is settled. The iodide is then oxidized to iodine in proportion to the manganic ion present, and the iodine is titrated with thiosulfate, with starch used as an indicator. This method may be used to calculate oxygen consumption (or oxygen production) of cells (Giese, 1975).

Carbon dioxide produced during cell respiration may be measured if it is trapped in calcium hydroxide or barium hydroxide, which is then titrated with oxalic acid. The expired carbon dioxide may also be absorbed by a weighed sample of soda lime that is weighed again after exposure. This method is used in calorimetry (see Fig. 9.4). An improved method is discussed by Aldrich (1975).

Volumetric Methods. In the *Fenn-Winterstein apparatus* (Fig. 13.12), two flasks are attached to a capillary containing a droplet of kerosene. Into one flask, containing a well filled with hydroxide, are placed respiring cells. The corresponding flask on the other side serves as a compensator or control. As the oxygen is used, carbon dioxide is formed and is absorbed by the hydroxide. The net effect is a decrease in pressure in the vessel containing respiring cells, thus causing the kerosene drop to move toward the vessel. If the volume of the capillary is known, the oxygen consumption may be measured directly. Since a temperature change affects both flasks, its effect is canceled out, and since the system is closed to the atmosphere, it is also unaffected by barometric changes. Temperature, however, affects the rate of cell respiration (Umbreit *et al.,* 1972).

Another direct and convenient volumetric method for determination of oxygen consumption (see Fig. 13.1) is the Scholander-Wennesland modification of the Fenn-Winterstein method. A multiple unit differential respirometer has been made available by simplifying the system by using a manifold and only one reference flask (Fig. 13.13).

Manometric Methods. A method developed by Barcroft for blood-gas analysis was later modified by Warburg for measurement of gaseous exchanges such as those which occur in respiration, photosynthesis, or various chemical reactions. The essential pieces of equipment, a *manometer* and an attached vessel, are shown in Figure 13.14. The cells or tissues are placed in the flask and hydroxide is added to the center well. After thermal equilibrium has been achieved, the vessel is closed off from the atmosphere by the stopcock at the top of the manometer. As oxygen is consumed and carbon dioxide is given off, the carbon dioxide is absorbed by the hydroxide in the well, and the pressure in the system falls, causing the manometric fluid to rise in the arm toward the respiratory flask.

In making readings, the manometric fluid is first set at a fixed point on the right side of the manometer to keep the volume constant. The difference between the heights of the fluid in the two arms of the manometer is then determined after a period of respiration. By use of the gas laws, the volume of oxygen consumed can be determined from the change in pressure (Umbreit *et al.,* 1972).

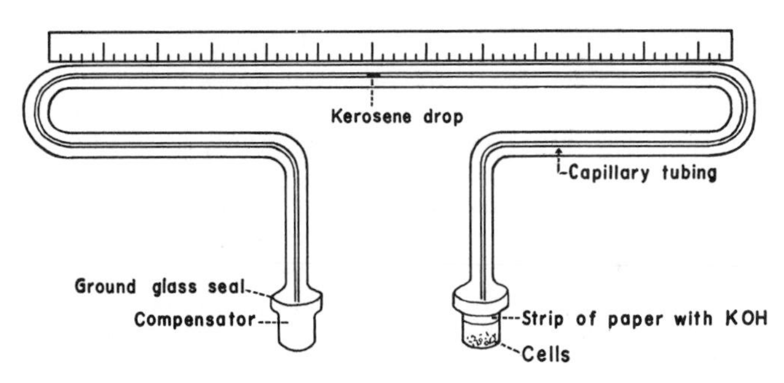

Figure 13.12. Microadaptation of the Fenn-Winterstein manometer.

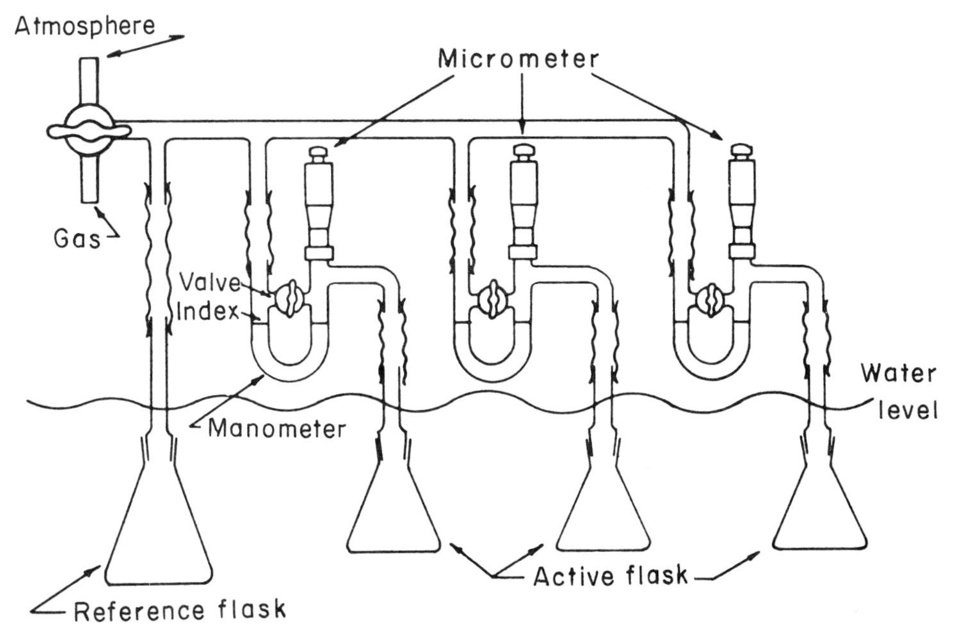

Figure 13.13. *Gilson multiple respirometer for direct reading of oxygen consumption. Because a single reference flask serves for all the respirometric flasks, instead of a reference flask for each respirometric flask as in Figure 13.1, the unit is compact. The vessels are attached to the manifold by flexible plastic tubing. (By courtesy of Gilson Electronics, Inc., 3000 West Beltline, Middleton, Wisconsin.)*

Because changes in temperature and barometric pressure affect the manometers, which are open to the atmosphere on the outer side, it is necessary to have a control manometer (called a *thermobarometer),* similar to the other manometers with respiring cells, but without cells.

The main advantage of manometric methods is their versatility. By proper management, exchanges of oxygen and carbon dioxide, glycolysis, or other acid production and any system yielding gas exchanges may be studied. The need to calibrate the vessels and manometers for quantitative measurements is a minor disadvantage.

In either volumetric or manometric techniques, it is possible to use vessels with side arms containing nutrients, poisons, or other substances of interest. These substances may be added independently of one another and at any time desired, and their effects on respiration may thus be determined. Vessels of various designs are available for special purposes (Umbreit *et al.,* 1972).

Cartesian Diver Methods. For measuring the respiratory rate of a small number of cells or of a single egg or a single protozoan, the *Cartesian diver* method (Figs. 13.15 and 13.16) has been used (Holter and Zeuthen, 1966). (Other microrespirometers have also been studied and improved—see Halprin and Gilbert, 1965; Gregg, 1966.)

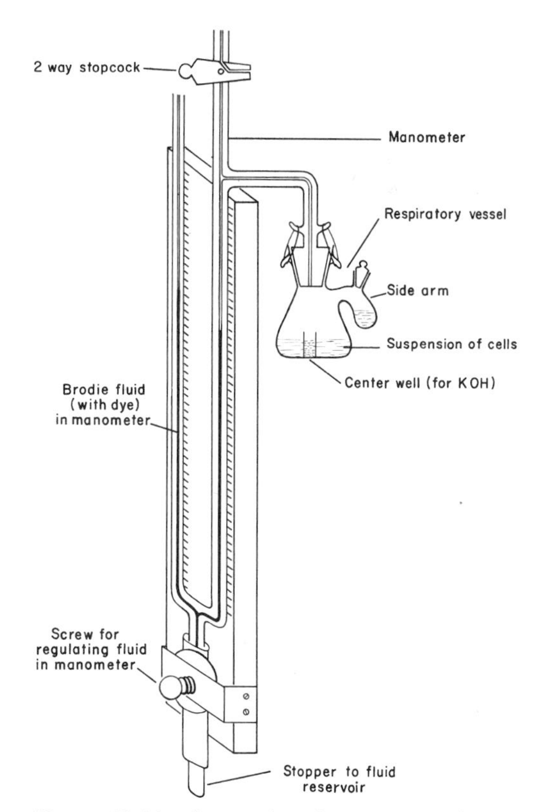

Figure 13.14. *A vessel and manometer for the Barcroft-Warburg respirometer. The stopcock is set in position for a measurement.*

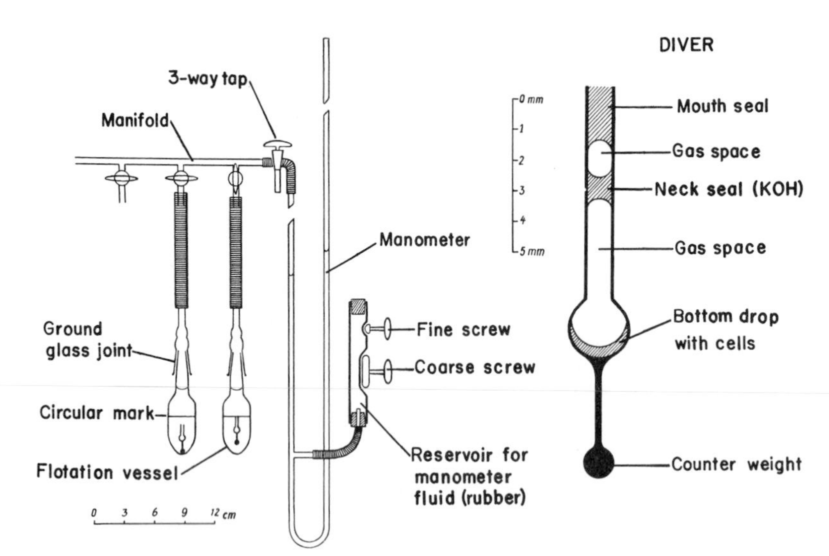

Figure 13.15. *Cartesian diver assembly. A whole series of diver chambers may be attached through a manifold to a single manometer. The flotation vessels containing the divers are attached by way of rubber tubing to the manifold as is the manometer. The reservoir is attached to the manometer by rubber tubing. (After Holter, 1943: Compt. Rend. Lab. Carlsberg Series Chimique 24: 403.)*

The Cartesian diver is a very small tube in which are placed, in succession, a drop of fluid containing the cells, a drop of hydroxide to absorb the carbon dioxide released during respiration, and an oil seal or a little plug. Each of these drops is separated by a bubble of air (see Fig. 13.15). As oxygen is consumed by the cells, and as the carbon dioxide released is absorbed by the hydroxide, the diver sinks because its buoyancy decreases. The diver may be refloated if the pressure on the "diving chamber" is lowered. In practice a definite level in the diving chamber is taken as the zero position, and the diver is kept at this level. In this way errors resulting from solution of gas in the diver at different pressures are avoided.

The Cartesian diver is used most frequently for measurement of the *relative* rates of oxygen consumption during the course of the experiments rather than for absolute measurements, but the latter can be obtained by calculation from the pressure changes measured. The disadvantage of the Cartesian diver method is that the techniques are sufficiently delicate to require extensive training for successful operation.

The Oxygen Electrode. The *oxygen (platinum) electrode* was originally developed for the determination of the oxygen content of water-containing electrolytes, a condition always met in biological experiments since organisms require salts.

In practice, a platinum electrode (Fig. 13.17) is attached to the negative terminal of a battery. The positive terminal of the battery is connected through a potentiometer or other voltage regulator to the experimental solution by a reference electrode. The potentiometer is generally set to deliver −0.5 (−0.3 to −0.9) volt and the amplified current is measured by an appropriate meter. Hydrogen ions combine with oxygen to form hydrogen peroxide at the

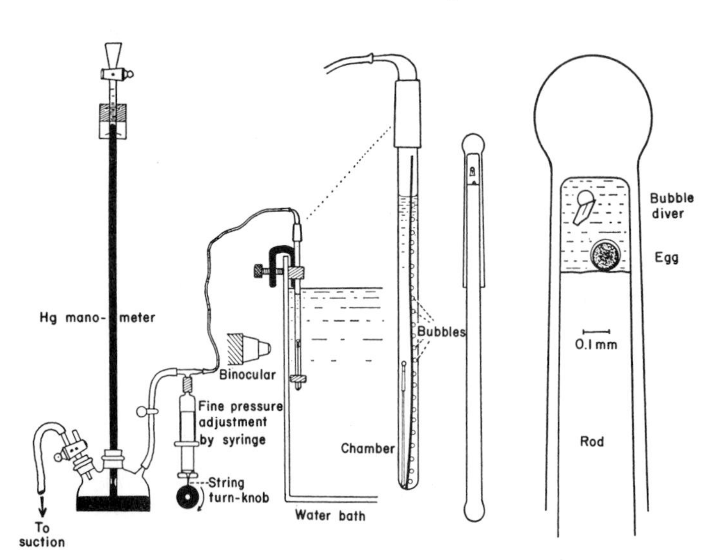

Figure 13.16. *The ultramicro or reference Cartesian diver. In this case the oxygen partial pressure of the sample containing the egg is measured by the buoyancy of the minute bubble diver, itself smaller than the egg. (After Scholander et al., 1952: Biol. Bull.* 102: *159.)*

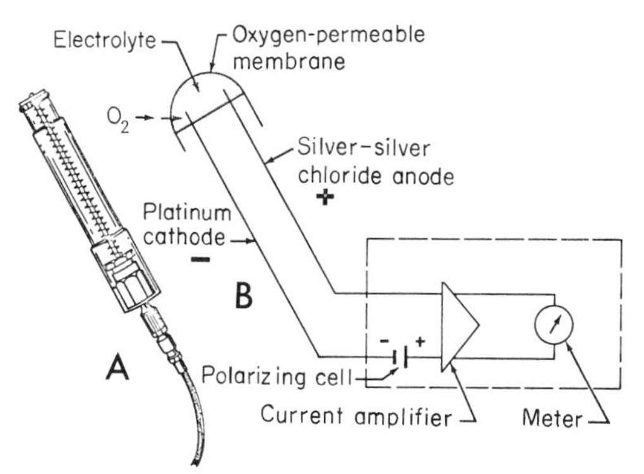

Figure 13.17. A, *One type of oxygen electrode (platinum electrode) for measuring oxygen partial pressure.* B, *Diagram of an electrode and its connections. (By courtesy of Beckman Instruments Inc., Fullerton, Calif.)*

platinum electrode surface. At a fixed voltage applied between the two electrodes the rate of discharge of hydrogen ions and the rate of formation of hydrogen peroxide is proportional to the oxygen content of the water (Fig. 13.18). The study is conducted at a low voltage because electrolysis of water will occur above 2 volts (Foster, 1966).

The apparatus is calibrated for zero oxygen content after the water is gassed with nitrogen and then the water is saturated with air or oxygen. The current of the solution is then measured to give a starting point for the experiment. The organism is placed in the solution, and the subsequent decrease in oxygen content is measured by the decrease in current. Each decrement in current represents a proportional decrease in oxygen content. Knowing the oxygen content at the start, one can calculate the respiratory rate from the rate of change in current.

The main advantage of the platinum electrode is its rapidity and accuracy. One disadvantage is the necessity of adequate stirring within the reaction vessel to make certain of equalization of conditions throughout its volume. However, shaking is necessary in most respirometric methods to insure equilibrium between the water and air. Shaking is unnecessary when the volume is small, as in a diver, or when the entire water volume is sampled by the Winkler method. The oxygen electrode also presents diffusion and electrode contamination problems, which must be taken into account (LeFevre, 1970).

The platinum electrode is highly adaptable to a wide range in size of respiring organisms. It has the advantage over all the other methods in that it records the oxygen partial pressure, which is in some cases more important to know than the oxygen content of the water or the oxygen consumed by the organism. Because it records continuously and instantaneously, it is especially valuable in following quick changes in respiration (or oxygen production) and for determining the critical oxygen partial pressure, below which the respiration of cells (or organisms) is proportional to the oxygen partial pressure. The output of the platinum electrode may be fed into a recorder for a continuous record of oxygen partial pressure over prolonged periods. An oscillating platinum electrode decreases the possibility of local exhaustion of oxygen. The platinum electrode is especially valuable when it is desirable or necessary in respirometric studies to avoid contact with

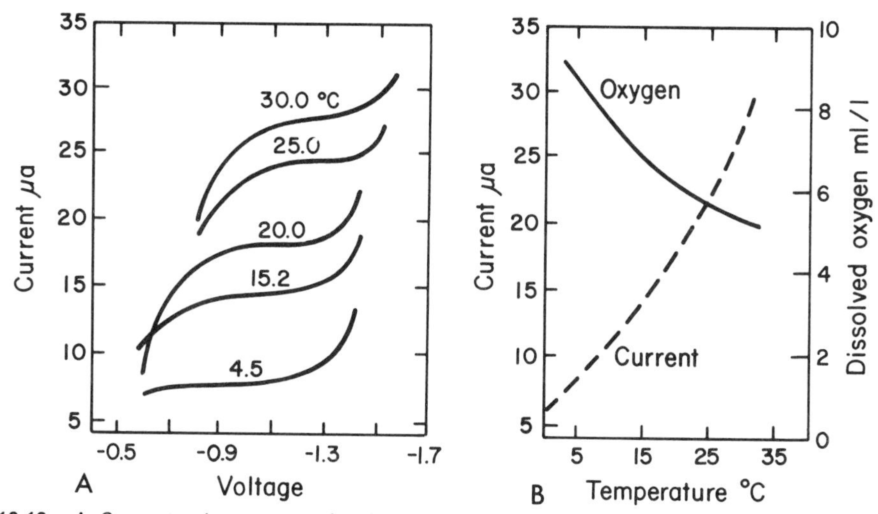

Figure 13.18. A, *Current-voltage curves for the oxygen electrode at various temperatures. In this case the reference electrode is silver–silver oxide, instead of silver–silver chloride shown in Figure 13.17*B. B, *Current and dissolved oxygen at various temperatures, measured by the oxygen electrode. The electrode response is very sensitive to temperature. (From Carritt and Kanwisher, 1959: Anal. Chem.* 31: 6.)

the gas phase, a necessary factor in manometric studies.

The platinum electrode has been used extensively in photosynthetic research where layers of algal fronds one cell thick can be placed directly on the surface of the platinum electrode. In closed vessels the dissolved oxygen in solution is continuously consumed by the electrode reaction itself, so that the current drifts slowly downward. If respiring cells are present, the downward drift is faster. Conversely, if oxygen is produced photosynthetically, the current will rise.

In the "steady state" modification of the method, the fluid in the vessel is caused to flow, is stirred, or is aerated. The oxygen electrode has also been useful in research on marine invertebrates in instances for which manometric methods were not suitable (Webster and Giese, 1975).

13.2 RELATIVE SENSITIVITIES OF VARIOUS METHODS

The Warburg method, in which accurate manometer readings can be made to 1 mm, has a sensitivity of about 1 μl. Its sensitivity can be increased by the hydraulic lever principle (Burk and Hobby, 1954). The Fenn-Winterstein manometer is more sensitive than the Warburg manometer but less sensitive than Cartesian divers. A variety of Cartesian divers has been developed. That of Zeuthen (1950) has an air volume of 1 μl and a sensitivity of 2.5 to 5.0 × 10^{-5} μl. Larger divers have a correspondingly lower sensitivity, whereas the minute divers used for single eggs (Scholander *et al.*, 1952; Scholander and Iversen, 1958) are a hundred times more sensitive. The magnetic diver is so sensitive it can be used for single mammalian cells (Brzin and Zeuthen, 1964).

LITERATURE CITED AND GENERAL REFERENCES

Aldrich, J. C., 1975: Improved method for carbon dioxide measurement in respiration. Marine Biology *29*: 277–282.

Bliss, D. and Skinner, D., 1963: Tissue Respiration in Invertebrates. American Museum of Natural History, New York.

Bonner, J., 1952: Morphogenesis. Princeton University Press, Princeton, N. J.

Bonner, J. and Varner, J. E., 1977: Plant Biochemistry. 3rd Ed. Academic Press, New York.

Brown, F. A., Jr., 1973: Bioluminescence. *In* Comparative Animal Physiology. 3rd Ed. Prosser, ed. W. B. Saunders Co., Philadelphia, pp. 951–966.

Brzin, M. and Zeuthen, E., 1964: Notes on the possible use of the magnetic diver for respiratory measurements (error 10^{-7} μl/hr). Comp. Rend. Trav. Lab. Carlsberg *34*: 427–431.

Burk, D. and Hobby, G., 1954: Hydraulic-leverage principle for magnification of sensitivity of gas change in free and fixed-volume manometry. Science *120*: 640–648.

Carlson, A. D. and Copland, J., 1978: Behavioral plasticity in the flash communication systems of fireflies. Am. Sci. *66*: 340–346.

Collicutt, J. M. and Hochachka, P. W., 1977: The anaerobic oyster heart. J. Comp. Physiol. *115*: 147–157.

Corbier, L., ed., 1972: Chemiluminescence and Bioluminescence. Proceedings, International Conference, Plenum Press, New York.

Cormier, M., Horti, K. and Anderson, J. M., 1974: Bioluminescence in coelenterates (Review). Biochim. Biophys. Acta *346*: 137–164.

Cormier, M. J., Lee, J. and Wampler, J. E., 1975: Bioluminescence: recent advances. Ann. Rev. Biochem. *44*: 225–272.

Cormier, M. J., Ward, W. W. and Charbonneau, H., 1977: Role of oxygen in coelenterate bioluminescence: evidence for enzyme-substrate, oxygen-containing intermediates. *In* Superoxide and Superoxide Dismutase. Michelson, McCord and Fridovich, eds. Academic Press, New York, pp. 451–458.

Davies, P. W., 1962: The oxygen electrode. *In* Physical Techniques in Biological Research. Vol. *4*, Nastuk, ed. Academic Press, New York, pp. 137–179.

Fee, J. A. and Valentine, J. S., 1977: Chemical and physical properties of superoxide. *In* Superoxide and Superoxide Dismutase. Michelson, McCord and Fridovich, eds. Academic Press, New York, pp. 19–60.

Foster, J. M., 1966: An oxygen electrode respirometer for measuring very low respiration rates. Anal. Biochem. *14*: 22–27.

Galt, C., 1978: Bioluminescence: dual mechanism in a planktonic tunicate produces brilliant surface display. Science *200*: 70–78.

Giese, A. C., 1975: Laboratory Manual in Cell Physiology. Rev. Ed. Boxwood Press, Pacific Grove, Calif.

Gilson, W. E., 1963: Differential respirometer of simplified and improved design. Science *141*: 531–532.

Gregg, J. A., 1966: A microrespirometer capable of quantitative substrate mixing. Exp. Cell Res. *42*: 260–264.

Guyton, A., 1976: Textbook of Medical Physiology. 5th Ed. W. B. Saunders Co., Philadelphia.

Halprin, K. M. and Gilbert, D., 1965: An improved microrespirometer for tissue slices. Anal. Biochem. *12*: 542–546.

Harvey, E. N., 1940: Living Light. Princeton University Press, Princeton, N.J.

Harvey, E. N., 1952: Bioluminescence. Academic Press, New York.

Harvey, E. N., 1957: A History of Bioluminescence. American Philosophical Society, Philadelphia.

Harvey, E. N., 1960: Bioluminescence. Comp. Biochem. *2*: 545–591.

Hastings, J. W. and Balny, C., 1975: The oxygenated bacterial luciferase-flavin intermediate. J. Biol. Chem. *250*: 7288–7293.

Hastings, J. W. and Wilson, T., 1976: Bioluminescence and chemiluminescence (Review). Photochem. Photobiol. *23*: 461–473. (Especially useful for chemical aspects.)

Hatchikian, C. E., LeGall, J. and Bell, G. R., 1977: Significance of superoxide dismutase and catalase activities in strict anaerobes, sulfate-reducing bacteria. *In* Superoxide and Superoxide Dismutase. Michelson, McCord and Fridovich, eds. Academic Press, New York, pp. 159–172.

Hochachka, P., 1973: Comparative intermediary metabolism. *In* Comparative Animal Physiology. 3rd Ed. Prosser, ed. W. B. Saunders Co., Philadelphia, pp. 212–278.

Hochachka, P. and Mustafa, T., 1972: Invertebrate facultative anaerobiosis. Science *178*: 1056–1060.

Holter, H. and Zeuthen, E., 1966: Manometric techniques for single chain cells. *In* Physical Techniques in Biological Research. Pollister, ed. Academic Press, New York, pp. 251–317.

Huntress, E. H., Stanley, N. and Parker, A. S., 1934: The oxidation of 3-aminophthal hydrazine as a lecture demonstration of chemiluminescence. J. Chem. Ed. *II*: 142–145.

Johnson, F. H., 1974: Bioluminescence and Chemiluminescence. *In* The Theory of Rate Processes in Biology and Medicine. Wiley-Interscience, New York, pp. 53–153.

Johnson, F. H. and Shimomura, O., 1972: Enzymatic and nonenzymatic bioluminescence. *In* Photophysiology. Giese, ed. Vol. 7. Academic Press, New York, pp. 275–334.

Johnson, F. H. and Shimomura, O., 1975: Bacterial and other "luciferins." Bioscience *25*: 718–722.

Johnson, F. H., Eyring, H. and Stover, B. J., 1974: Action of inhibitors in relation to concentration, temperature and hydrostatic pressure. *In* The Theory of Rate Processes in Biology. Wiley-Interscience, New York, pp. 371–548.

Lee, J., 1977: Bioluminescence. *In* The Science of Photobiology. K. C. Smith, ed. Plenum Press, New York, pp. 371–395.

LeFevre, M. F., 1970: Problems in the measurement of tissue respiration with the oxygen electrode. Bioscience *20*: 761–764.

McCosker, J. F., 1977: Flashlight fishes. Sci. Am. (Mar.) *236*: 106–114.

McElroy, W. D., 1960: Bioluminescence. Fed. Proc. *19*: 941–948.

Mill, J., 1972: Respiration in Invertebrates. St. Martin's Press, London.

Nealson, K. H., 1977: Autoinduction of bacterial luciferase. Occurrence, mechanism and significance. Arch. Microbiol. *112*: 73–79.

Nicoli, M. Z., Meighen, F. S. and Hastings, J. W., 1974: Bacterial luciferase. J. Biol. Chem. *249*: 2385–2392.

Pearson, O. P., 1953: The metabolism of humming birds. Sci. Am. (Jan.) *188*: 69–72.

Pearson, O. P., 1954: Shrews. Sci. Am. (Aug.) *191*: 66–70.

Prosser, C. L., 1973: Oxygen: respiration and metabolism. *In* Comparative Animal Physiology. 3rd Ed. Prosser, ed. W. B. Saunders Co., Philadelphia, pp. 165–211.

Puget, K., Lavelle, F. and Michelson, A. M., 1977: Superoxide dismutase from procaryote and eucaryote bioluminescent organisms. *In* Superoxide and Superoxide Dismutase. Michelson, McCord and Fridovich, eds. Academic Press, New York, pp. 139–150.

Scholander, P. F., Claff, C. L. and Sveinsson, S. L., 1952: Respiratory studies on single cells. II. Observations on the oxygen consumption in single protozoans. Biol. Bull. *102*: 178–184.

Scholander, P. F. and Iversen, O., 1958: New design of volumetric respirometer. Scand. J. Clin. Lab. Invest. *10*: 429–431.

Shimomura, O. and Johnson, F. H., 1976: Calcium-triggered luminescence of the protein aequorin. *In* Calcium in Biological Systems. Soc. Exp. Biol. Symp. *30*: 41–54.

Stauff, J. and Nimmerfall, F., 1969: Chemiluminescence of redox systems with molecular oxygen. Z. Naturforsch. *24*: 852–862.

Steen, J. B., 1971: Comparative Physiology of Respiratory Mechanisms. Academic Press, New York.

Umbreit, W. W., Burris, R. H. and Stauffer, J. F., 1972: Manometric and Biochemical Techniques: A Manual Describing Methods Applicable to the Study of Tissue Metabolism. 5th Ed. Burgess Publishing Co., Minneapolis.

Webster, S. and Giese, A. C., 1975: Oxygen consumption in the purple sea urchin with special reference to the reproductive cycle. Biol. Bull. *148*: 165–180.

Wennesland, R., 1951: A volumetric respirometer for studies of tissue metabolism. Science *114*: 100–103.

Zeuthen, E., 1950: Cartesian diver respirometer. Biol. Bull. *98*: 139–143.

Zeuthen, E., 1970: Rate of living as related to body size in organisms. Pol. Arch. Hydrobiol. *17*: 21–30.

CHAPTER 14

PHOTOSYNTHESIS

Of all the processes in nature, photosynthesis is perhaps the most fundamental. The chlorophyll-bearing plant transduces the free energy of sunlight, storing it as chemical potential energy by combining carbon dioxide and water, the end products of metabolism in organisms, to build carbohydrates. These directly or indirectly serve as the source of energy for all living beings, except chemosynthetic bacteria.

It has been calculated that each carbon dioxide molecule in the atmosphere is incorporated into a plant structure once every 300 years and that all the oxygen in the air is renewed by plants every 2000 years. Yet it is only recently that a thorough, systematic study of photosynthesis has been attempted, and botanists, chemists, physicists, biophysicists, and physiologists, working together on the photosynthetic process, have made some of the most dramatic discoveries in cell physiology (Fork, 1977).

More than 2000 years ago Aristotle observed that sunlight was necessary for greening of plants. The requirement of light for photosynthesis was next indicated by the experiments of Stephen Hales (1677–1761) and Jean Senebier (1742–1809). Then, in 1786, Jan Ingenhousz, inspired by John Priestley's work on oxygen, provided proof that in the light carbon dioxide is absorbed and oxygen is given off, whereas in the dark carbon dioxide is given off and oxygen is absorbed by green plants. In 1804, Nicholas deSaussure showed that during photosynthesis, equal volumes of carbon dioxide and oxygen are exchanged and that often equal volumes are exchanged during respiration. His was the first truly quantitative study of photosynthesis.

Otto Warburg was one of the first to use *Chlorella vulgaris,* a small, simple, unicellular green alga (Fig. 14.1), for studies of photosynthesis. Cultures of pedigreed *Chlorella* grown under light of known quality and intensity, kept at a constant temperature, and maintained in a nutrient of standard composition, are found to be more alike than the cells of a leaf. The population of *Chlorella* can be washed and suspended in a replenishable medium of known constitution for experimentation. The temperature, the concentration of carbon dioxide in equilibrium with a bicarbonate buffer, the intensity and quality of light, and the rate of diffusion of substances from the cells can all be controlled during the course of experimentation.

Such cultures of *Chlorella* and other unicellular algae are used in most studies employing manometric techniques. Thin thalli of marine algae are more convenient when the platinum oxygen electrode is used to measure oxygen production because the tissue can be held directly over the electrode. For still other studies photosynthetic bacteria are of special interest. For investigation of photophosphorylation, photoreductions, spectrophotometric changes, and dark reactions occurring during photosynthesis, chloroplasts isolated from spinach leaves and other plants and fractions of disrupted chloroplasts are most effective and are extensively used today.

PHOTOSYNTHESIS AS AN OXIDATION–REDUCTION REACTION

The overall reaction that takes place in photosynthesis might be written as follows:

$$nCO_2 + nH_2O \xrightarrow{\text{light}} (CH_2O)_n + nO_2 \qquad (14.1)$$

Here $(CH_2O)_n$ represents a carbohydrate. That the ratio of oxygen evolved to carbon dioxide consumed is 1 has been verified many times.

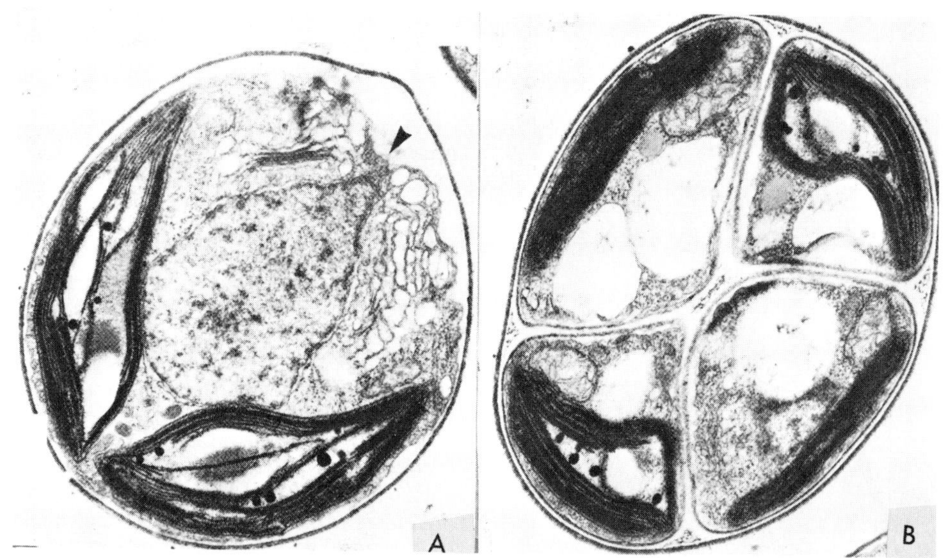

Figure 14.1. A, Chlorella, *an alga used extensively in studies on photosynthesis. Note the cup-shaped chloroplast.* B, *Four autospores are shown, but in multiple fission 8, or 16 may be formed. Individual* Chlorella *cells are small (2 to 8 μm in diameter) and lacking in definitive structures, making identification of a species difficult. This may create confusion in experimental work unless subcultures of the same stock are used. Both* A *and* B, × *30,000. (From Pickett-Heaps, 1975: Green Algae. © 1975 by Sinauer Associates, Sunderland, Mass.)*

In 1931 C. B. van Niel suggested that water is the hydrogen donor in the oxidation-reduction that occurs in photosynthesis. He based his argument on the reactions in photosynthetic purple sulfur bacteria:

$$2H_2S + CO_2 \xrightarrow{\text{light}} H_2O + 2S + CH_2O \quad (14.2)$$

If in the green plant cell H_2O served as the hydrogen donor, playing a role similar to that of H_2S in the sulfur bacteria, then by analogy, equation 14.2 may be written as follows:

$$2H_2O + CO_2 \xrightarrow{\text{light}} H_2O + O_2 + CH_2O \quad (14.3)$$

Convincing proof of the validity of van Niel's argument came from work using water labeled with the oxygen isotope ^{18}O. When a culture of *Chlorella* in a medium with labeled water is exposed to light, the gaseous oxygen that is liberated—and only the gaseous oxygen—contains the isotopic oxygen (Ruben *et al.*, 1941).

Besides H_2S, some bacteria capable of photosynthesis use a variety of hydrogen donors, including various organic compounds. Because of this it has been proposed that organisms using organic hydrogen donors are primitive ones. Presumably, as organic donors were exhausted in some early geological era, photosynthetic organisms evolved that could use water as the hydrogen donor (see Chapter 2).

LIGHT (PHOTOCHEMICAL) REACTIONS IN PHOTOSYNTHESIS

That the primary reaction in photosynthesis is photochemical is apparent from the fact that light is necessary. However, several lines of evidence indicate that secondary thermochemical reactions immediately follow the photochemical reaction.

At low light intensity the photochemical reaction is the limiting reaction in photosynthesis. In a *Chlorella* suspension illuminated with low intensities of light, the temperature

TABLE 14.1. EFFECT OF TEMPERATURE AND LIGHT INTENSITY ON THE RATE OF PHOTOSYNTHESIS*

Temp. range, °C.	4–10	10–20	20–30	5–10	16–25	15–25	25–32
Q_{10}	4.3	2.1	1.6	4.7	2.0	1.06	1.0
Rel. intensity	45	45	45	16	16	1.8	1.0
of light		High		Medium		Low	

*From Warburg, 1919: Biochem. Zeit. *100*.

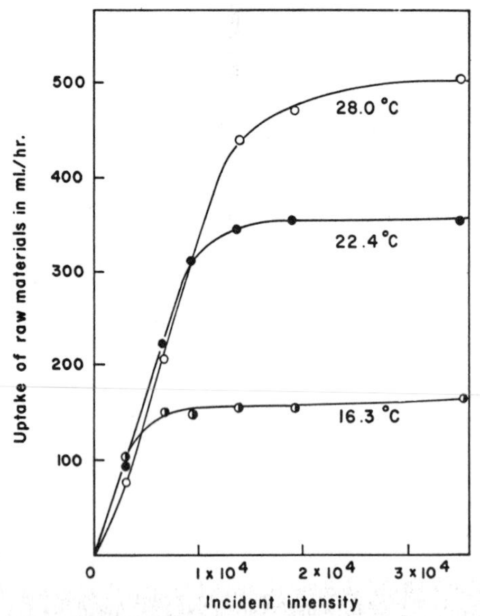

Figure 14.2. Influence of temperature on rate of photosynthesis of purple sulfur bacteria (Thiorhodaceae) at different intensities of light. Note that at low light intensity, temperature has little effect on the rate of photosynthesis, as shown by the relatively straight lines. (From Wassink et al., *1942: Enzymologia* 10: *304.)*

may be varied over fairly wide limits (15° to 32°C) without greatly altering the rate of the reaction (Table 14.1). The temperature coefficient (Q_{10} increase in rate with a 10-degree rise in temperature) of photosynthesis over this range is about 1.06. This indicates that the rate at which the chlorophyll molecules absorb and transmit the limited number of light quanta is not increased by raising the temperature. The initial portions of the curves in Figure 14.2 bring this out clearly.

When the light intensity is increased until each chlorophyll molecule is supplied with all the light quanta it can absorb (saturation), a rise in temperature increases the rate of photosynthesis. At high intensity, then, the thermochemical reactions following the photochemical reactions are rate limiting. In this case a temperature coefficient as high as 4.3 is found over the range of 10° to 32°C (Table 14.1 and Fig. 14.2).

A preparation of *Chlorella* exposed to flashing light performs more photosynthesis per unit dose of light than the same preparation does when exposed to continuous light. In other words, a given amount of light is used more efficiently if a dark period follows a light period. At low temperature, when the periods of darkness following the flashes are varied, the efficiency of a given dose of light increases with the length of the dark period, up to a maximum that may be considered as the time required for completion of the dark reactions. Some of the data are shown in Figure 14.3. The dark period is longer at lower temperatures because the rate of carbon dioxide fixation is determined by the kinetic energy of the molecules (Emerson and Arnold, 1932).

Robin Hill (1939) discovered that fresh leaves ground in water containing suitable hydrogen acceptors give off oxygen when exposed to light, even though the cells are crushed and production of carbohydrate has ceased. A variety of dyes and other compounds, such as ferricyanide, when added to the crushed cells are photochemically reduced. For example, quinone is reduced to hydroquinone (see Figure 11.8). These experiments indicate that a reduction pool of hydrogen is produced even when light is absorbed by fragmented chloroplasts.

It is interesting to note that the number of quanta of light that must be absorbed by chlorophyll in disrupted chloroplasts to liberate one molecule of oxygen is of the same order as the number required in intact green plant cells. This indicates that the process in disrupted chloroplasts is probably the same as in the intact green cells. The only difference is that under these conditions the hydrogen produced cannot be utilized to do work useful to the cell because the organized systems required for this purpose have been destroyed.

According to van Niel (1949), the decomposition of water during photosynthesis may be expressed as follows:

$$HOH + h\nu \rightarrow (H) + (OH) \qquad (14.4)$$

In this equation (H) represents a reductant, and (OH) is an oxidant of unspecified nature, possibly each attached to unknown substances X and Y, respectively, and $h\nu$ represents a quantum of light. By accumulation of (H) a reduction pool is produced that reduces the carbon dioxide

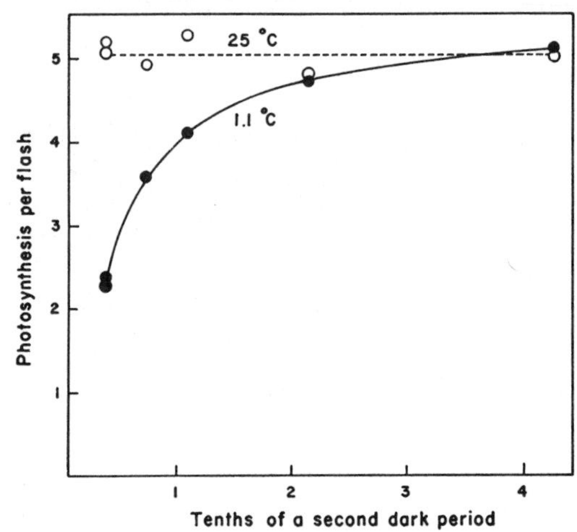

Figure 14.3. Increase in efficiency of use of light at low temperature with long dark periods between flashes. (After Emerson and Arnold, 1932: J. Gen. Physiol. 15: *403.)*

in the plant cells. Recombination of (OH) presumably forms water and oxygen:

$$(OH) + (OH) \rightarrow H_2O + \frac{1}{2}O_2 \qquad (14.5)$$

As already pointed out, the origin of oxygen from the water was demonstrated by providing the plant cells with $H_2{}^{18}O$. All the label appeared in the oxygen given off.

Van Niel's hypothesis of water decomposition by light is one of the landmarks in photosynthetic research. It postulates that light by way of chlorophyll causes a separation of oxidizing and reducing entities. "What are these entities, how (in physical detail) are they formed and what (in chemical detail) does the plant do with them?" (Clayton, 1965, p. 10). These are questions still not answered fully (Fork, 1977).

THE PHOTOSYNTHETIC UNIT

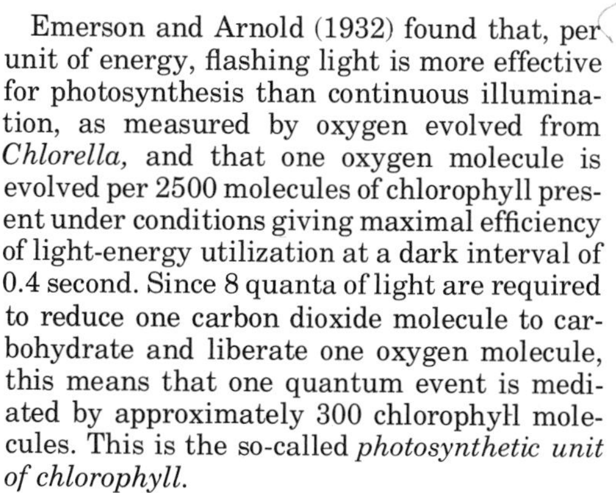

Emerson and Arnold (1932) found that, per unit of energy, flashing light is more effective for photosynthesis than continuous illumination, as measured by oxygen evolved from *Chlorella*, and that one oxygen molecule is evolved per 2500 molecules of chlorophyll present under conditions giving maximal efficiency of light-energy utilization at a dark interval of 0.4 second. Since 8 quanta of light are required to reduce one carbon dioxide molecule to carbohydrate and liberate one oxygen molecule, this means that one quantum event is mediated by approximately 300 chlorophyll molecules. This is the so-called *photosynthetic unit of chlorophyll.*

Since a plant cell achieves maximal rate of carbon dioxide assimilation and oxygen evolution almost immediately upon illumination, it is thought likely that the energy absorbed by the chlorophyll is collected at a reaction center in the photosynthetic unit. This implies that although all the 300 chlorophyll molecules serving as an antenna may collect and harvest light quanta, they are not equivalent, and only certain ones presumably participate in the photochemical reaction.

It is of interest that evidence other than flashing light data points to the likelihood of such a photosynthetic unit. Thus, the cytochromes unique to chloroplasts (e.g., cytochromes f and b_6) are present in the proportion of approximately 1 to 300 chlorophyll molecules. Also, the herbicide dichlorophenyl dimethylurea (DCMU) inhibits photosynthesis almost completely when present in a concentration of one molecule for every 300 chlorophyll molecules (Fork, 1977). Evidence has also been obtained by electron microscopic studies made with the freeze-etching technique on preparations split through chloroplast lamellae. Attached to such lamellae are structures about 10 nm in diameter (see Fig. 6.16). It is estimated that a piece of the chloroplast lamella accommodating about 300 chlorophyll molecules would be approximately this size. These units have been called *quantasomes* and have tentatively been identified as the possible photosynthetic units. Experiments indicate that the photosynthetic unit of bacteria is smaller than in green plants and algae, containing about 50 chlorophyll molecules (Rabinowitch and Govindjee, 1969).

PHOTOPHOSPHORYLATION AND PHOTOREDUCTION

Evidence has accumulated that electrons in long-lived excited states (triplet states), as well as free radicals, are formed during photosynthesis and that there is a loss of electrons from chlorophyll during photic excitation (Weaver and Weaver, 1972).

Arnon and his collaborators have obtained direct evidence of conversion in chloroplasts of the energy of photic excitation into ATP (from ADP and inorganic phosphate) and of reduction of NADP to NADPH and H^+ (Arnon, 1963). In oxidative phosphorylation, electrons pass from a large negative (reducing) potential through a series of cytochromes (the "electron cascade") to a positive (oxidizing) potential. The change in free energy during this process is conserved in production of ATP from ADP and inorganic phosphate (see Fig. 14.14).

In the chlorophyll of chloroplasts electrons are excited to high energy states by light absorption, much as water is elevated by a pump. If the "electron cascade" were coupled with electron transport carriers such as the cytochromes present in the cell, then light energy could be conserved as chemical energy of ATP and other triphosphates in a manner similar to that in oxidative phosphorylation as the electron moves from a higher to a lower reducing potential. Investigations showed that photosynthetic phosphorylation does indeed occur when a suspension of isolated chloroplasts is illuminated, provided appropriate factors and co-factors, removed during isolation procedures, are once again added. Such photophosphorylation, requiring no oxygen, is independent of the oxidative phosphorylation that occurs in mitochondria but not in chloroplasts (see Chapter 12). The green plant cell forms 30 times as much ATP by photophosphorylation as by oxidative phosphorylation.

It is possible to separate the photochemical reaction completely from the thermochemical (dark) reactions. When chloroplasts are disrupted and centrifuged, three components are obtained: a green layer, a supernatant, and a lipid layer. The green portion, consisting of *grana,* can carry out the entire photochemical reaction (e.g., the anaerobic phosphorylation of ADP to form ATP and the reduction of coenzyme NADP to NADPH* and H^+) provided catalytic amounts of several factors are supplied to the mixture: among these are vitamin K, FMN (flavin mononucleotide, one of the flavin coenzymes), and chloride ion. The other two layered components alone cannot carry on photochemical reactions, but when the products of photochemical phosphorylation and coenzyme reduction (reduction pool) are added to a mixture of these two chloroplast fractions, the result is a reduction of carbon dioxide to carbohydrate. This indicates the presence of enzymes for the dark reactions of photosynthesis in the colorless fractions.

The absorption spectra of the cytochromes in the mixture change as these reactions proceed, suggesting that the cytochromes are indeed serving as intermediates in electron transport. Furthermore, it was found that photosynthetic green plant cells have unique cytochromes (b_{563}, b_{559}, and *f*) not found in animal cells or microorganisms.

In photosynthesis two types of photophosphorylation, cyclic and noncyclic, have been described. In *cyclic photophosphorylation* ATP formation is coupled with liberation of the energy arising from electronic excitation following the absorption of light by chlorophyll; no net reducing power is stored as carbohydrate nor is oxygen liberated. Presumably the electron in chlorophyll, which was raised to an excited state by light absorption, returns to the ground state with the free energy of the return journey stored in ATP; no external source of electrons is required to return chlorophyll to its original state. Thus the electron in chlorophyll completes a cycle.

In *noncyclic photophosphorylation,* both the electrons of chlorophyll excited by light absorption and the H^+ ions from water are used to reduce NADP to NADPH, which serves as the reductant for the dark reactions of photosynthesis (discussed later). In this case, the electron contributed from the chlorophyll molecule to the reduction of NADP is not returned (hence the designation noncyclic) and must be obtained from some other source, namely the hydroxyl ion of water. The two hydroxyl ions might, in the presence of the appropriate enzyme, combine to form water and oxygen, $OH^- + OH^- \rightarrow H_2O$ and $\frac{1}{2}O_2$ (see Fig. 14.14). The reaction of the hydroxyl ions is hypothetical but plausible (Clayton, 1971).

The explanation of the mechanism of oxygen formation and liberation remains elusive. It is theoretically necessary to remove four electrons from two molecules of water to release one molecule of oxygen (O_2). Since only one electron is released when chlorophyll absorbs one quantum of light, four photoreactions are required to release one oxygen molecule. It was found by Kok in 1969 that one flash of light, intense enough to excite every reaction center, given to dark-adapted algae released no oxygen, nor did a second flash. The third flash released oxygen, as did the fourth. Kok postulated an enzyme S that successively accumulated four plus charges. This fully oxidized enzyme oxidized a water molecule, releasing a molecule of oxygen. The system thus returned to the discharged state. Superoxide ($O_2^{\dot{-}}$) appears to be present in chloroplasts during illumination and superoxide dismutase and catalase protect the cell from damage by this molecule (Allen, 1977; Lumsden *et al.,* 1977).

The oxygen-emitting system is quite labile; for example, it is destroyed by a few minutes at 50°C, by washing with Tris buffer, and by short wavelength ultraviolet radiation. There are six manganese atoms per 300 chlorophyll molecules (photosynthetic unit), of which four are released by washing. This results in a decrease in oxygen evolution on illumination. The four manganese atoms are thought to be bound to the enzyme catalyzing oxygen evolution.

It is interesting to note that photosynthetic bacteria use NAD and NADH in place of NADP and NADPH. However, a potential more reducing than -0.35 volts is required to reduce NAD to NADH and no redox potential more reducing than -0.1 volt has been recorded from bacteria. NADH must therefore be formed by reductive dephosphorylation. It is thus likely that photosynthetic bacteria use light energy only for cyclic phosphorylation.

The general phenomena discussed here are explored in greater detail later in this chapter.

THE FUNCTION OF PIGMENTS IN PHOTOSYNTHESIS*

Chlorophyll is present in all photosynthetic cells, although it is sometimes masked by the presence of other pigments. Chlorophyll exists in the cell, combined in a complex with protein. Although several varieties of chlorophyll have been identified, all have the same essential

*Sometimes the reduced form of NADP is given as $NADPH_2$ rather than NADPH + H^+, depending upon the emphasis and preference.

*By the use of x-rays, green mutant strains of *Chlorella* have been produced that absorb light and evolve oxygen but are incapable of performing the photosynthesis of carbohydrates. Such mutants can be grown in a solution of glucose and salts, indicating that they are capable of manufacturing their other requirements. In recent years a wide variety of mutants have been used to "dissect" photosynthetic processes (Levine, 1969; Bishop, 1973).

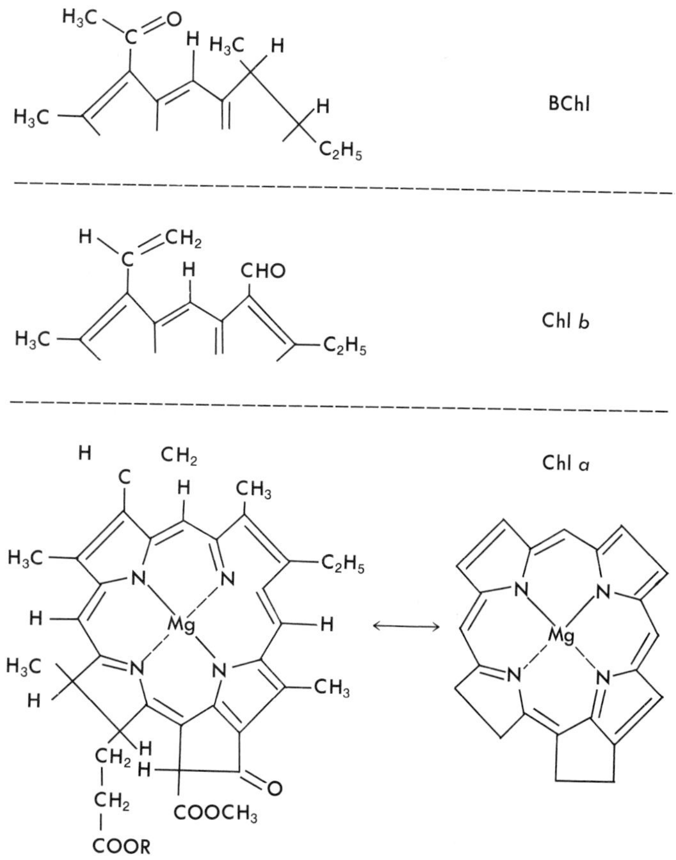

Figure 14.4. *The chemical structures of chlorophylls* a *and* b *and bacteriochlorophyll (BChl). The presence of a carbon atom is implied at each unlabeled junction of bonds. In CHl* a *and* b, *but not in BChl, the pattern of alternating single and double bonds is in resonance with the one sketched at the right. The residue R is a long-chain hydrocarbon,* $C_{20}H_{39}$, *or phytyl, in Chl* a *and* b *and something similar in BChl. (From Clayton, 1971: Light and Living Matter. Vol. 2. McGraw-Hill Book Co., New York.)*

porphyrin structure, or tetrapyrrole nucleus, shown in Figure 14.4. Magnesium is present in this nucleus and is attached to the ends of the pyrroles, forming a very stable, almost planar structure. In chemical terms, chlorophylls are methyl phytol esters of the parent dicarboxylic acids, the chlorophyllins. Phytol is a long-chain alcohol containing one double bond. The porphyrin structure is hydrophilic, whereas the phytol chain is hydrophobic. Porphyrins can be synthesized abiotically by irradiating or passing electric discharges through a mixture of gases supposedly resembling the gases on the earth's primitive atmosphere. They are also found in ancient fossils and are presumably among the most ancient molecules available for the origin of life.

The various chlorophylls differ from one another only in the side chains attached to the outer ends of the tetrapyrrole nucleus. Green plant cells have chlorophylls *a* and *b*, brown algal cells and diatoms have chlorophylls *a* and *c*, and red algal cells have chlorophylls *a* and *d*. The purple and green photosynthetic bacteria have bacteriochlorophyll, which resembles chlorophyll *a*, as shown in Figure 14.4 and Table 14.2. Cells of etiolated green plants (sprouted in the dark) contain protochlorophyll, which is quickly reduced to chlorophyll when the leaf is illuminated.

Chlorophyll a and the corresponding pigment in photosynthetic bacteria play a central role in photosynthesis. Therefore, all the other pigments have been called accessory pigments; they funnel their energy to chlorophyll *a*.

The chlorophylls absorb in the red and blue-violet parts of the spectrum (Fig. 14.5). Bacteriochlorophylls absorb in the infrared and the near ultraviolet regions. When illuminated with various wavelengths of light (including ultraviolet), chlorophyll shows a characteristic brilliant red fluorescence. It also shows a weaker band in the far red.

After illumination, chlorophyll in the plant emits light in the dark for as long as 2 minutes, a property that is inhibited when agents such as cyanide or azide are present. Since these agents also inhibit photosynthesis, light emission by chlorophyll is thought to result from reversal of some step in photosynthesis.

In all plant cells and in the photosynthetic bacteria, *carotenoids* (yellowish pigments) are present in addition to chlorophyll. The carotenoids are essentially hydrocarbons (Fig. 14.6) and fall into two main groups, the carotenes and the xanthophylls, the latter

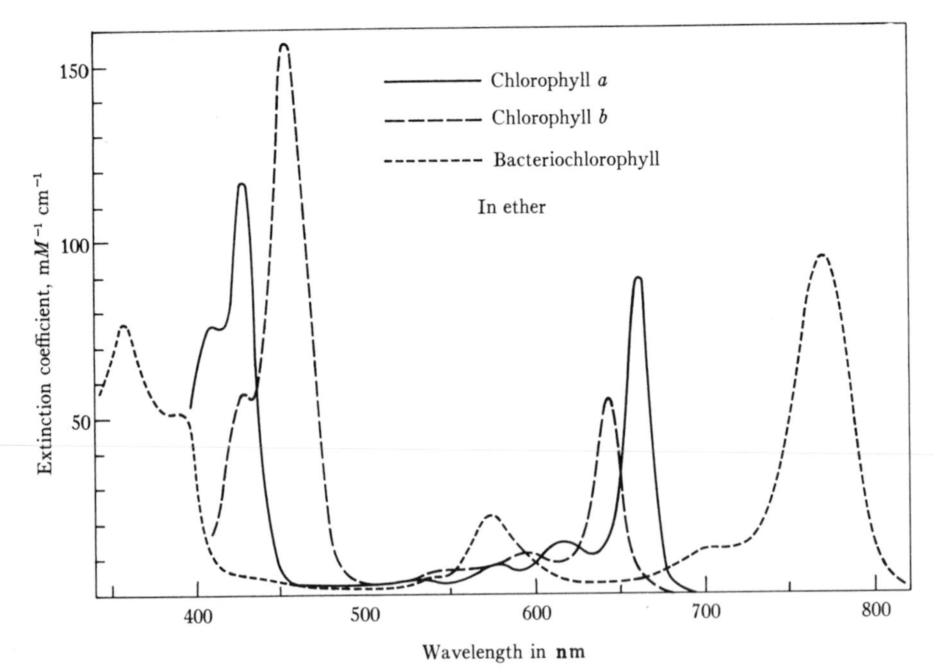

Figure 14.5. *Absorption spectra of various chlorophylls. The absorption spectra in the red and infrared portions of the spectrum from 600 to 800 nm are of special interest. Note that chlorophyll* a *absorbs at longer red wavelengths than chlorophyll* b *and that bacteriochlorophyll absorbs in the infrared portion just beyond the limit of vision in the red range. (From Clayton, 1965: Modern Physics in Photosynthesis. Blaisdell Publishing Co., Waltham, Mass.)*

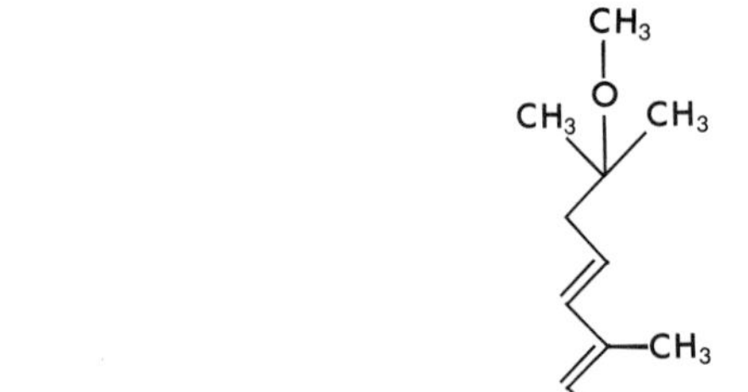

β-Carotene

Spirilloxanthin

Figure 14.6. *Structure of β-carotene and a xanthophyll, spirilloxanthin, a methylated hydroxy derivative of a carotenoid found in purple sulfur bacteria. (From Lehninger, 1975: Biochemistry, 2nd Edition, Worth Publishers, New York.)*

being hydroxy compounds otherwise similar to carotenes. Many kinds of carotenes and xanthophylls exist. The carotenoids absorb strongly in the blue-violet and ultraviolet ends of the spectrum. In the cells of each plant group, characteristic carotenoids are found; a few of the main ones are listed in Table 14.2. The carotenoids, like the chlorophylls, seem to be attached to proteins in the chloroplasts of cells.

Except in green plants and photosynthetic bacteria, carotenoids act as antennas to gather light for photosynthesis; for example, fucoxanthin in brown algae and peridinin in dinoflagellates. In all plants they probably protect the chlorophylls from photo-oxidation by their more readily oxidizable double bonds. This is most readily observed in mutants of anaerobic photosynthetic bacteria possessing carotenoids with fewer double bonds (Krinsky, 1968).

Chlorophyll absorbs strongly in red light, less effectively in blue light, and almost not at all in green light (see Fig. 14.5). Correspondingly, green plants show maximal photosynthesis in red light, lower photosynthesis in blue light, and none in green light (Fig. 14.7). Carotenoids absorb in the blue end of the spectrum, but they appear to have little light-gathering power for photosynthesis in green plants, although they do in brown algae and dinoflagellates.

The phycoerythrin (red pigment), phycocyanin, and allophycocyanin* (blue pigments) found in the red and blue-green algae are protein-linked pigments, called *phycobili-*

*Allophycocyanin has the same chromophore as phycocyanin but is combined to a different protein (Gantt, 1975).

TABLE 14.2. MAIN PIGMENTS IN PHOTOSYNTHETIC PLANTS AND BACTERIA

Groups	*Chlorophylls**	*Carotenoids*†	*Other Pigments*
Green plants	*a* and *b*	β-carotene, lutein	Anthocyans (non-photosynthetic)
Brown algae and diatoms	*a* and *c*	β-carotene, fucoxanthin	—
Dinoflagellates	*a* and *c*	β-carotene, peridinin	—
Red algae	*a* and *d*	β-carotene, α-carotene, lutein	Phycobilins‡
Blue-green algae	*a*	β-carotene, myxoxanthins	Phycobilins‡
Yellow-green algae	*a* and *e*	β-carotene, violaxanthin, vaucheriaxanthin	—
Green sulfur bacteria	Bacteriochlorophyll *a* and *c* or *d* or *e*	Chlorobactene and OH-chlorobactene§	—
Purple bacteria	Bacteriochlorophyll *a* or *b*	Aliphatic carotenoids,§ spirilloxanthin	—

*In the plant, chlorophyll *a* exists in several forms, among them those with absorbance at 662, 670, 677 and 684 nm, P700 at the center of Photosystem I. In solution no such distinctions can be found. The absorption of these various forms must depend upon their unions with proteins. Some of these varieties of chlorophyll are considered by Govindjee and Govindjee, 1974.

†All photosynthetic plants have β-carotene; α-carotene is found only in red algae and one group of green algae (Siphonales). The names listed after carotene are xanthophylls.

‡Phycobilins are phycocyanins (blue), allophycocyanins (blue), and phycoerythrins (red). Blue-green algae have a predominance of the blue pigments, red algae of the red pigment; the proportions of pigments vary depending on lighting during growth (chromatic adaptation).

§The data for green bacteria are for *Chlorobium*. Brown photosynthetic bacteria and gliding filamentous bacteria (e.g., *Chloroflexus*) have other carotenoids. See Pfenig, 1977, for details. Data for purple bacteria from Jensen, 1963. See also Stanier *et al.*, 1976 for details.

proteins (Glazer, 1976). They are destroyed by heat. The prosthetic group of phycobiliproteins is related to the bile pigments, being in each case an open tetrapyrrole lacking magnesium (Fig. 14.8). These pigments absorb light mainly in the region where chlorophyll has least absorption.

The pigments in red and blue-green algae take an active part in photosynthesis, and the highest efficiency for photosynthesis in the red

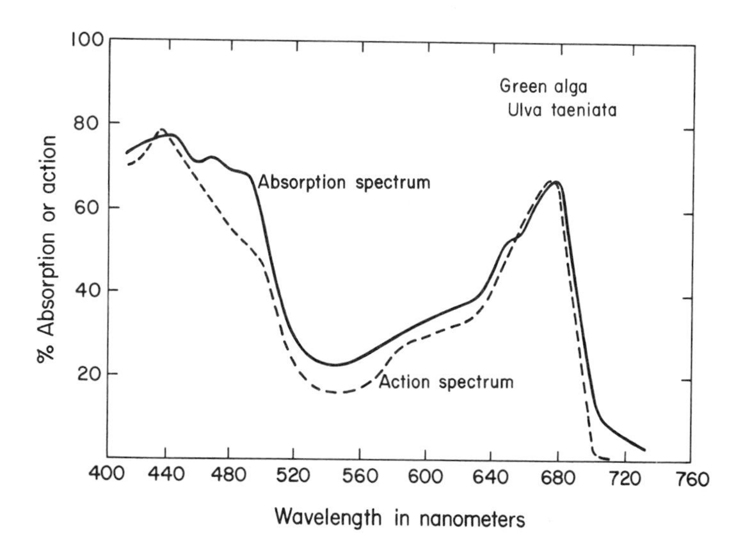

Figure 14.7. *Correspondence between absorption spectrum of the thallus (the flat blade of the plant) and action spectrum for a green alga,* Ulva taeniata. *Note that in this plant the efficiency in blue light is very high. In this case photosynthetic efficiency is high in the blue-violet, unlike that in many green plants. Presumably energy absorbed in the short wavelengths is being used in photosynthesis. (From Haxo and Blinks, 1950: J. Gen. Physiol. 33: 408.)*

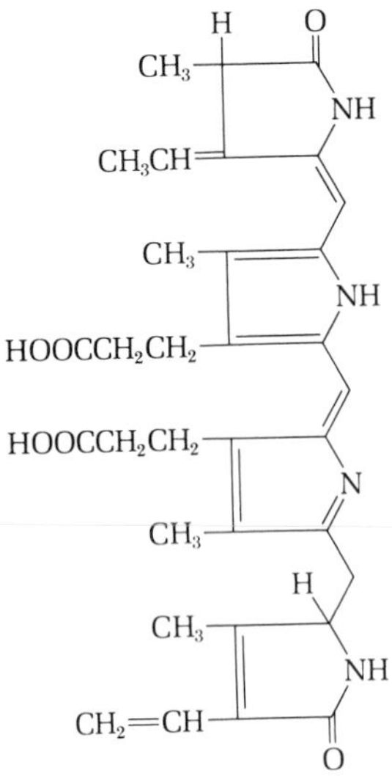

Figure 14.8. *Phycoerythrobilin, a red photosynthetic pigment. In phycocyanobilin, a blue pigment, the* $CH_2{=}CH{-}$ *group on the fourth pyrrole ring is replaced by* $CH_3CH_2{-}$.

algae, for example, is in the green part of the spectrum (Fig. 14.9), where the red pigment absorbs strongly (Fig. 14.10) and where chlorophyll absorbs least. Excitation, on absorption of light, progresses from phycoerythrin to phycocyanin (and allophycocyanin) to chlorophyll *a*. The phycobilins are arranged in rows on the thylakoids in distinct structures *(phycobilisomes)* about twice the size of ribosomes (Gantt, 1975). It is interesting to note that phytochrome, a pigment of great importance in photomorphogenetic and photoperiodic responses of plants, is similar to phycobiliproteins in chemical nature (Fig. 24.13*B*).

Although in the photosynthesizing cell accessory pigments absorb light used in photosynthesis, the fluorescence spectrum obtained on illumination is always that of chlorophyll *a*, even when plants are illuminated with wavelengths that are little absorbed by chlorophyll. Absorption of light by chlorophylls *b*, *c*, and *d* also excites chlorophyll *a* to fluorescence. This suggests that light absorbed by all the other pigments, serving as light collectors (or antennas), is transferred probably by resonance to chlorophyll *a*, which uses the energy to excite electrons to higher energetic states.

EVIDENCE FOR TWO SYSTEMS OF PHOTOCHEMICAL REACTIONS

Emerson observed a decline in efficiency of photosynthesis in monochromatic, long-wavelength red light, a long, red wavelength "drop-off" (Fig. 14.11*A*). In 1957 he showed that the photosynthetic efficiency of a long wavelength of light is enhanced by simultaneous exposure of photosynthetic plant cells to shorter wavelengths of light. In essence, if x is the rate of photosynthesis when one wavelength (e.g., long red) alone is applied, y is the rate when a shorter wavelength alone is applied and z is the rate when both are superimposed, then when enhancement occurs, z is greater than the sum, $x + y$. The fraction $z/(x + y)$ is a measure of the enhancement (Clayton, 1965). This effect has been called the *Emerson effect* (Fig. 14.11*B*). The discovery of enhancement was a milestone in photosynthetic research. Prior to it, researchers assumed that a single photochemical reaction occurred with accessory pigments serving only as light collectors, the energy so collected being transferred to chlorophyll *a*. It became evident that two photic systems were involved and that enhancement occurs only when an accessory pigment is excited along with chlorophyll *a*.

The action spectrum for enhancement obtained by constant illumination of green plant cells with a long wavelength of light and simultaneous exposure also to shorter wavelengths,

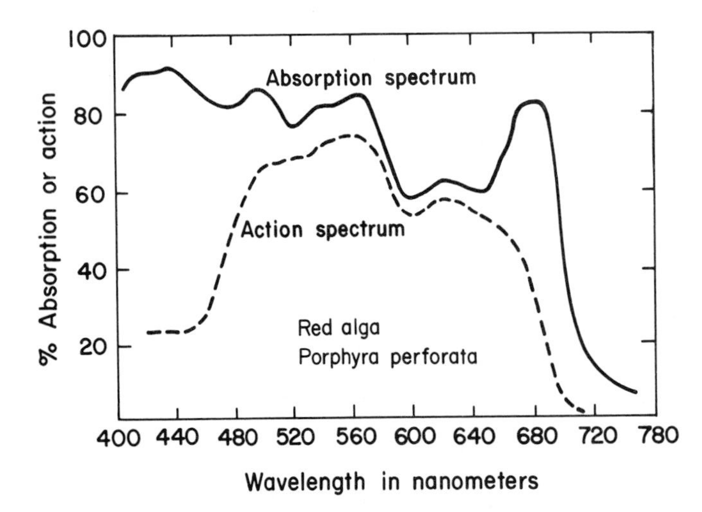

Figure 14.9. *Action and absorption spectra of a red alga,* Porphyra. *Note that the region of the spectrum absorbed most by the phycoerythrin (compare with Fig. 19.9) is most effective in photosynthesis. (From Haxo and Blinks, 1950: J. Gen. Physiol. 33: 414.)*

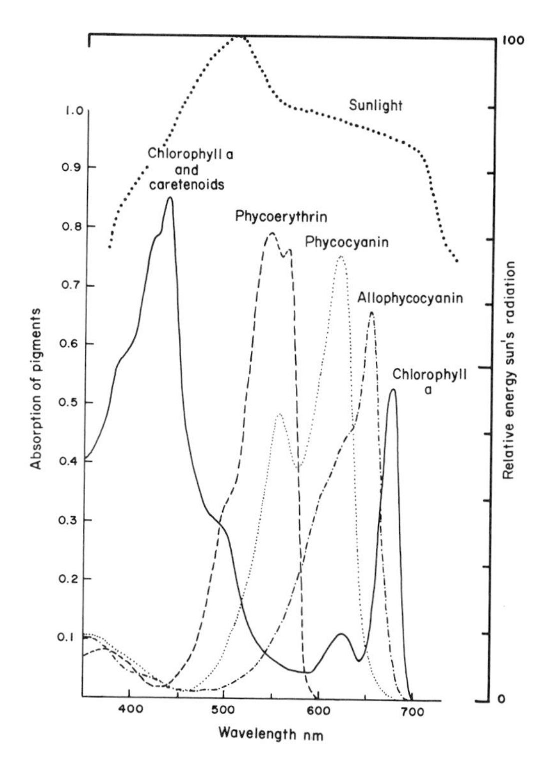

Figure 14.10. *The major peaks of chlorophyll are at about 435 and 675 nm. Phycobiliproteins (phycoerythrin, phycocyanin, and allophycocyanin), purified from phycobilisomes, effectively close the gap where chlorophyll absorbs little. The fluorescence peaks of these pigments closely overlap, forming an integrated transfer system. The relative energy of the visible part of the solar spectrum reaching the earth is shown on the top. (From Gantt, E., 1975: Bioscience 25: 781.)*

one wavelength at a time, resembles the absorption spectrum of chlorophyll *b* (Fig. 14.11*B*). Conversely, if exposure to various long wavelengths, one at a time, is superimposed over a constant illumination with a short wavelength, the action spectrum obtained is that of chlorophyll *a*. Chlorophyll *b* absorbs more strongly in the short red and chlorophyll *a* absorbs more strongly in the longer red wavelengths (see Fig. 14.5); simultaneous use of illumination with long and short wavelengths brought not only chlorophyll *b* but also chlorophyll *a* into photosynthetic activity, thus increasing the effectiveness of the photosynthetic process. These two reactions are spoken of as the two *photic reaction systems* in photosynthesis. The action spectra for enhancement in blue-green, red, and brown algae suggest that, in addition to chlorophyll, other accessory pigments are also involved in photosynthetic enhancement.

If only one of the pigments in a photosynthetic cell is excited by simultaneous exposure to two wavelengths of light absorbed by the pigment, no enhancement is observed. That is, the amount of photosynthesis induced is equal to the sum of the photosynthesis induced by each alone. Thus, 650 nm is the peak of absorption by chlorophyll *b* in *Chlorella* (see Fig. 14.5). There is no significant enhancement with superimposition of exposure to any other wavelength in the region 600 to 680 nm (the long wavelength limit of absorption by chlorophyll *b*) (Myers, 1971).

Many of the compounds involved in photosynthesis show characteristic redox potentials. Many also show absorption spectra that are characteristic of their oxidized and reduced states, making it possible to determine spectrophotometrically whenever illumination or other procedures bring about their oxidation or reduction. Measurements made with the spectrophotometer are rapid (oscillographic recording) so that an immediate change in absorbance at any given wavelength may be detected and its decline followed (Fig. 14.12). Often the absorbance changes are of a rather small order of magnitude (e.g., 1 to 2.5 percent) and for this purpose a special technique of differential spectrophotometry has been used. The differential, or rate of change of absorption (da/dt), is plotted as ΔO.D. (difference in optical density) instead of the absorption (O.D.). By this means small differences are greatly magnified, and the decay constants can be measured with much greater accuracy (Fig. 14.13; Fork and Amesz, 1970).

Electron excitation resulting from absorption of light by chlorophyll is clearly indicated by electron spin resonance studies. Such studies also provide evidence for two photic reactions in photosynthesis, a characteristic signal being associated with each one (Weaver and Weaver, 1972). At present, we can anticipate physical separation of the components of the two photic systems (Brown, 1973).

On the basis of these types of experiments the scheme shown in Figure 14.14 and Table 14.3

A. Red drop

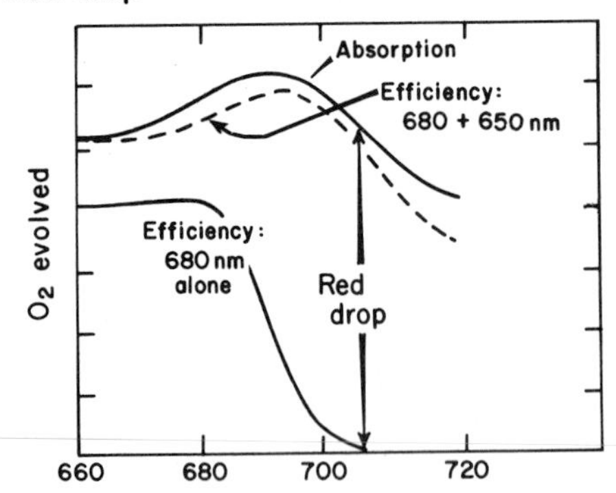

B. Emerson effect

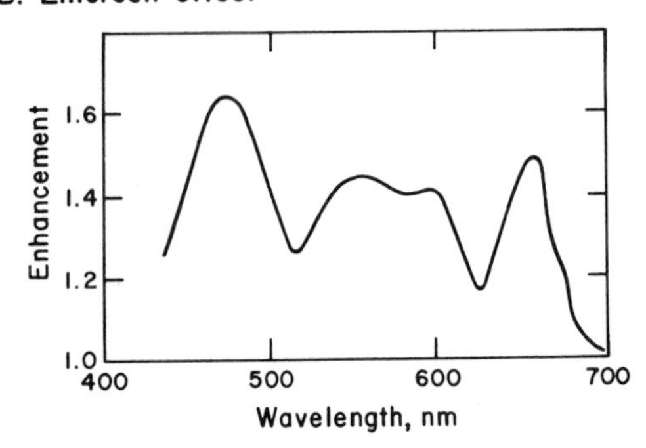

Figure 14.11. *A, Red drop in* Chlorella: *the decrease in efficiency of monochromatic light above 680 nm (wavelength on the abscissa). If supplementary light of shorter wavelengths (even 650 nm) is added, the efficiency greatly increases. This puzzling finding is now known to be the result of the adsorption of light by the second photochemical system and the consequent better use of the total light absorbed (see Figure 14.14). (After Lehninger, 1975: Biochemistry. Worth Publishers, New York, p. 601.)* B, *The Emerson effect. The action spectrum for the Emerson enhancement effect was obtained by illuminating a plant exposed to 700 nm with each of a number of shorter wavelengths. The intensity of each wavelength was adjusted to give a rate of photosynthesis equal to that at 700 nm alone. Then the two wavelengths were used simultaneously and the enhancement was determined as described in the text. (After French, 1961:* In *Life and Light. McElroy and Glass, eds. Johns Hopkins Press, Baltimore.)*

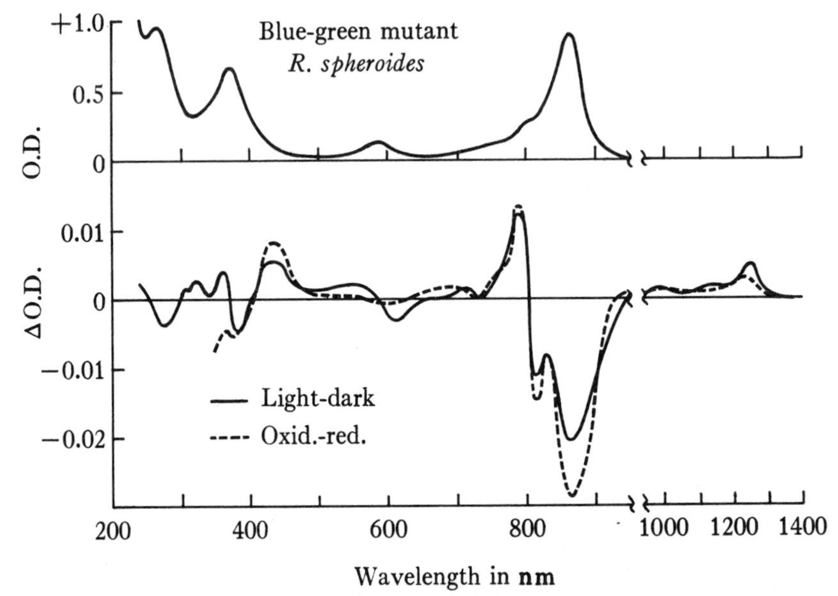

Figure 14.12. *Light-induced changes in absorption spectra of chromatophores (pigment-containing bodies) of the blue-green mutant of the purple photosynthetic bacterium* Rhodopseudomonas spheroides. *Changes reflecting cytochrome oxidation were eliminated in this measurement by saturating the chromatophore suspension with the cytochrome poison* p-*chloromercuribenzoate. The solid line shows the effects of illumination, the dashed line the effect of chemical oxidation (ferri/ferrocyanide, potential 470 mv, or* KIO_4*, potential 500 mv). (From Clayton, 1965: Modern Physics in Photosynthesis. Blaisdell Publishing Co., Waltham, Mass.)*

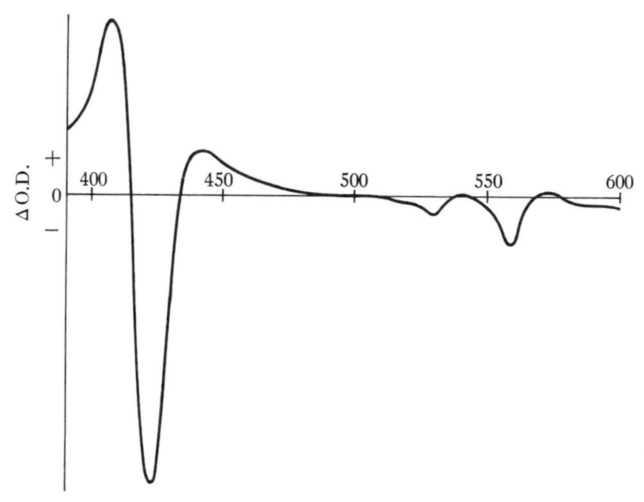

Figure 14.13. *Difference spectra showing the light-induced oxidation of cytochrome in the green photosynthetic bacterium* Chromatium *chromatophores. (From Clayton, 1965: Modern Physics in Photosynthesis. Blaisdell Publishing Co., Waltham, Mass.)*

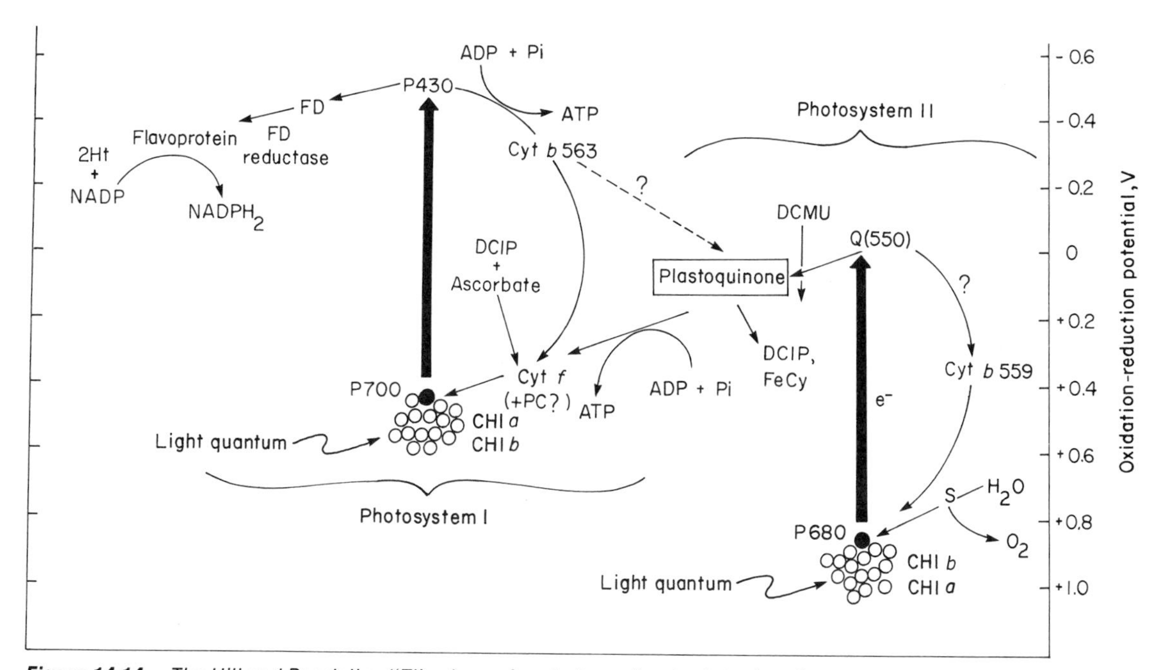

Figure 14.14. *The Hill and Bendall or "Z" scheme for photosynthesis plotted to show redox potentials of the reacting components. The upward arrow signifies the raising of an electron to a higher energy level against the thermodynamic gradient upon absorption of a photon by a reaction center. The solid circles represent P700 and P680, specialized chlorophyll molecules functioning as the reaction centers for photosystems I and II, respectively. The circles around the reaction centers represent other chlorophyll forms and accessory pigments that absorb short wavelength light and transfer the energy to the reaction centers. The different forms of chlorophyll* a *and the accessory pigments are both present in the pigment assemblages of photosystems I and II, but more pigments absorbing at short wavelengths are associated with photosystem II than with photosystem I. P430 and C550 (Q) are the primary acceptors for photosystems I and II, respectively. Cyt represents cytochrome; PC, plastocyanin; FD, ferredoxin; $NADP^+$, nicotinamide adenine dinucleotide phosphate; S, the accumulator of charge in photosystem II (manganese enzyme?); ASC, ascorbate, FeCy, ferricyanide; DCIP, dichlorophenol indophenol. Two phosphorylation sites may be coupled to electron flow. One site, for noncyclic photophosphorylation, probably occurs somewhere between plastoquinone and P700. A site for cyclic phosphorylation may be coupled to the electron pathway that includes P430 and cytochrome. For a key to abbreviations see Table 14.3. The points at which the inhibitors DCMU and DCIP act are indicated by arrows, as are the points at which FeCy and ascorbate act. (After Fork, 1977.* In *The Science of Photobiology. Smith, ed. Plenum Press, New York, p. 346.)*

TABLE 14.3. ABBREVIATIONS FOR COMPOUNDS IN PHOTOSYNTHETIC RESEARCH*

Abbreviation	*Full Name*	*Function Where Known*
Chl	Chlorophyll	Primary photosynthetic pigment; chlorophyll molecule
SI (PSI)	Photosystem I	Photosystem receiving activated electron from system II and exciting it to a high reducing potential
P700	Pigment 700	Pigment absorbing at 700 nm (red) and the reaction center for SI. Redox pot. +0.43V. Present in proportion 1:400 chlorophyll molecules
P430	Pigment 430	Probably primary acceptor of SI; may be part of bound ferredoxin complex; with more reducing redox potential than soluble ferredoxin; shows strong changes in absorbance at 430 nm with illumination
FD	Ferredoxin	Electron transfer agent of reducing potential; present in all organisms; non-heme sulfur protein
NADP	Nicotine adenine dinucleotide phosphate	Coenzyme hydrogen carrier (see Chapter 12)
FD reductase	Ferredoxin NADP reductase	Flavoprotein enzyme that reduces NADP to $NADPH_2$, the last step in photosynthetic electron transport
Cyt *f*	Cytochrome *f*	Electron carrier intermediate between SI and SII (oxidized by SI and reduced by SII); a primary electron donor to P700; present 1:400 Chl molecules
PC	Plastocyanin	An acidic protein containing 2 atoms of Cu per molecule; present in almost all plants. Many properties like those of Cyt *f* and probably a primary reaction center partner to P700
Cyt b_{563}	Cytochrome b_{563}	A cytochrome showing absorption changes at 563 nm during photosynthesis; low redox potential of −0.06V. Reduced by PC; mediates cyclic flow of electrons around SI
SII (PSII)	Photosystem II	Photosystem exciting electrons to first stage; connected to SI by way of Cyt *f*
P680	Pigment 680	Chlorophyll molecule at reaction center of SII
Q (C550)	Quencher	Possibly primary reaction center of SII, shows absorbance change at 550 nm with illumination. When Q is oxidized fluorescence is quenched. 1 Q: 400 Chl molecules
PQ	Plastoquinone	A major electron carrier, 1 PQ: 400 Chl molecules. Reduced by SII, slowly oxidized by SI
S	Enzyme S	An enzyme postulated to oxidize water and release oxygen. Contains manganese atoms; heat labile and sensitive to UV radiation.
Cyt b_{559}	Cytochrome absorbing at 559 nm	Probably mediates cyclic electron flow around SII; acts as safety valve preventing accumulation of strong oxidizing substances
DCMU	Dichlorophenyl dimethyl urea	Herbicide used to block photosynthesis between oxidation of Q and reduction of PQ
DCIP	Dichlorophenol indophenol	An electron donor used in photosynthetic studies
FeCy	Ferricyanide	An oxidizing agent used in photosynthetic studies
Asc	Ascorbate	A reducing agent used in photosynthetic studies

*A guide to photosynthetic jargon; mainly from Fork, 1977. For explanation of ATP, ADP, Pi, NADP, $NADPH_2$, etc., see Chapter 12. Photosystems I and II are also often designated photosystems 1 and 2 (PS1, PS2).

was proposed by Hill and Bendall in 1960. This is called the *series formulation* or *Z scheme* of photosynthesis because it pictures the two photic reactions in *photosystem I and photosystem II* as succeeding each other. The details are being filled in as information becomes available.

According to this formulation, shorter wavelengths of light absorbed by chlorophyll *b* in green plants (by phycobilins in red and blue-green algae, and by fucoxanthin in diatoms) enter photosystem II and excite the chlorophyll *a* in the reaction center (P680) from its ground state of +900 mv to a reduced state of about 0.0 mv. The excited electron cascades down the potential gradient through a number of steps, i.e., a quinone (plastoquinone), cytochrome *f* (at +360 mv), and finally to the pigment (chlorophyll *a*) of the longer wavelength photosystem I, usually designated as P700.

Absorption of light by photosystem I (P700) results in excitation of the electron from the level it had reached in chlorophyll *a* to about −600 mv for the ferredoxin-reducing substance to −420 mv for ferredoxin, from which it passes to NADP, reducing it to NADPH in the presence of H^+ and the enzyme ferredoxin reductase.

It will be noted that just as in metabolic oxidation in the mitochondria the free energy of an electron at a reducing potential is utilized to phosphorylate ADP, so in photosynthesis the light energy of excited electrons results in photophosphorylation. The consequence of the operation of this scheme is the production of both a reduction pool (as NADPH) and a supply of high-energy phosphate compounds, both necessary for completion of the thermochemical (dark) reactions in photosynthesis.

It is evident that most of the chlorophyll molecules in the photosynthetic units serve only to harvest light and that the energy conversion takes place in only one part of the unit called the *reaction center*. In a photosynthetic unit of the purple bacteria all the chlorophyll molecules except one at the reaction center can be deprived of their magnesium (which in chlorophyll lies in a strategic position like that of iron in hemoglobin), and yet the photosynthetic unit continues to collect light for use at the reaction center. On the other hand, even if all the other molecules are intact, when the chlorophyll at the reaction center is altered, photosynthesis ceases.

In photosystem I of green plants the reaction center also consists of a single chlorophyll *a* molecule with absorption in the very long red: for this reason it is called P700 (pigment 700 absorbing maximally at wavelength 700 nm). Only one molecule in the 300 molecules constituting the photosynthetic unit is P700. Its peculiar properties and its characteristic absorption spectrum apparently result from its specific binding with protein within the photosynthetic unit (a type of allosterism?), since in extracts it cannot, on the basis of its properties, be separated from other chlorophyll present in the unit. It alone of all the light-absorbing chlorophyll molecules in the unit is reversibly bleached (oxidized) by long wavelength red light. Chlorophyll *a* predominates in photosystem I but chlorophyll *b* is also present.

In photosystem II of green plants both chlorophyll *a* and *b* are present. Chlorophyll *a*, absorbing strongly at wavelength 680 nm, referred to as P680, is found at the reaction center. The antennas (light collectors) of both photosystems I and II are complex in composition, containing carotenoid-protein complexes, chlorophyll *a*-protein complexes, and a chlorophyll *b* complex in addition to the reaction centers. The same chlorophyll complexed to a different protein shows a different absorption maximum (Govindjee and Govindjee, 1974). In algae the phycobilins (phycocyanin and phycoerythrin) are present in photosystem II, and in diatoms the carotenoid fucoxanthin is also present in addition to some chlorophyll *a* (Fork, 1977). Immunological evidence suggests localization of photosystem I on the outside face and photosystem II on the inside face of the chloroplast lamellae (Briantais and Picaud, 1972).

The photosynthetic purple bacteria use a variety of hydrogen donors in place of water (e.g., hydrogen sulfide or various organic compounds);* therefore, it appeared that they lack photosystem II, which, among other things, in green plants is involved in oxygen production (from OH^-) (Cheniae, 1970). The photosynthetic bacteria show no enhancement (Blinks and van Niel, 1963).

The scheme of Figure 14.14 does not suggest direct photolysis of water by light. Water is slightly ionized. The OH^- ion serves in photosystem II as electron donor (combining therewith to form water and oxygen), and the H^+ ion is used in photosystem I to combine with $NADP^+$ on acceptance of electrons from ferredoxin. Light in photosystem II excites the electrons made available by OH^-, only indirectly causing separation of water into OH^- and H^+ by continued withdrawal of electrons from OH^-.

Despite the many important discoveries in photosynthesis, the nature of the light reactions and their sequence during photophos-

*In some anaerobic photosynthetic bacteria using hydrogen donors other than hydrogen or water (e.g., succinate) not only is carbon dioxide reduced but also atmospheric nitrogen is reduced to ammonia (fixed). Such nitrogen fixation occurs at the expense of photic energy. This fact is of considerable importance in view of the great importance of nitrogen fixation in the economy of nature (see Fig. 2.9).

phorylation and photoreduction in the chloroplast remain uncertain. However, many new techniques are being applied to study photosynthesis, some of them with high temporal resolving power. These will enable the investigator to follow even reactions that are completed in a very brief period. Therefore, we have reason to expect that the sequence of reactions by which the photic reactions occur will be elucidated in as clear a manner as the dark reactions.

COUPLING BETWEEN ELECTRON TRANSPORT AND PHOTOPHOSPHORYLATION

As noted, quantitative analyses early demonstrated that during photosynthesis ATP is produced from ADP and inorganic phosphate. The source of energy for this synthesis is the energy of electrons in chlorophyll excited by absorption of light. The energy is given off by the electrons as they pass through a series of electron carriers coupled to the enzymes and ADP and inorganic phosphate required for synthesis of ATP.

Biochemists have sought a chemical intermediate coupling electron flow to ATP production, but the substance has resisted identification or isolation. However, an antibody produced against the coupling factor stops photophosphorylation. Light-induced proton uptake then continues until a pH gradient of four units is established (Avron, 1977).

Swelling and shrinking of chloroplasts has been observed during photosynthesis, accompanying electron flow and formation of ATP. Translocation of ions across the illuminated thylakoid membranes accompanied by water uptake resulted in swelling of the membranes along with a change in potential across the membranes (Packer and Deamer, 1968). Interference with the ion gradients or membrane potentials by inhibitors (e.g., valinomycin) or other factors resulted in cessation of ATP synthesis. Presumably, then, energy-yielding electron transport and storage of this energy in ATP are coupled with the ionic gradients and potentials developed (Fig. 14.15).

Mitchell (1966) suggested that coupling be-

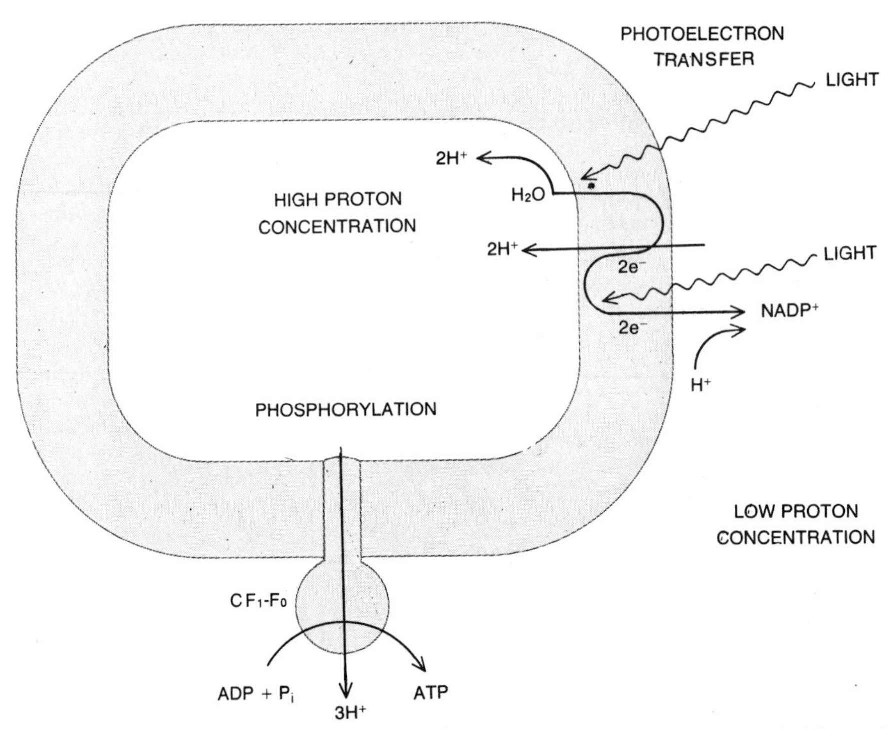

Figure 14.15. *Photosynthetic phosphorylation in chloroplasts derives the energy needed for making ATP from light. As in oxidative phosphorylation, hydrogen ions are transported across the membrane to create a proton gradient, and ATP is synthesized as the protons flow back across the membrane down the gradient. In chloroplasts, however, the direction of proton flow is reversed: the light-driven movement of electrons pumps protons inward, making the interior acidic, and phosphorylation is driven by an outward flow. Moreover, the stoichiometry, or ratio of reactant molecules and ions, is different from that in mitochondria. Each two electrons cross the membrane only twice, translocating only four protons, and for each molecule of ATP formed three protons must pass through the enzyme complex, which is designated CF_1-F_0. (From Hinkle and McCarty, 1978: Sci. Am. (Mar.) 238: 104–123. Copyright © 1978 by Scientific American, Inc. All rights reserved.)*

tween electron flow and ATP synthesis may be electrochemical, resulting from a difference in concentration of protons (H^+) on two sides of a chloroplast membrane and the resultant electrical potential across the membrane (see Chapters 12 and 21). To produce a difference in concentration of H^+ on two sides of a membrane requires input of energy which, in some way still undetermined, presumably comes from the flow of electrons excited by absorption of light (Hind and McCarty, 1973).

That chemiosmotic coupling (transduction of osmotic potential energy to chemical energy) can produce ATP from ADP and inorganic phosphate was demonstrated by soaking a suspension of chloroplasts at pH 4.0 and then transferring them directly to pH 8.0. A temporary gradient of hydrogen ions thus existed across the thylakoid membranes with a greater concentration of hydrogen ions inside resembling the results of light-induced pumping of H^+ into the thylakoids. If ADP and inorganic phosphate were added at the time of transfer from pH 4 to pH 8, ATP was produced in amounts equal to 100 cycles of the photochemically active material (Jagendorf and Uribe, 1965). Whether such soaking damages the coupling factor has not been determined.

While this experiment is more readily explained by chemiosmotic coupling than by chemical coupling (with equilibrium constants possibly shifted by the change in hydrogen ion concentration) too little information is available to resolve the issue of which of the two possibilities is correct (Fork, 1977).

THE DARK REACTIONS IN PHOTOSYNTHESIS

In the dark reactions that follow the photochemical reactions, ATP, produced by photophosphorylation, and NADPH, produced by photoreduction, are put to use to synthesize glucose and other products with a high chemical potential. In this way the light energy transduced into chemical potential in formation of ATP is conserved in compounds that can be stored in quantity, compounds that serve the plant cells in glycolysis and the tricarboxylic acid cycle necessary to maintain life, much as in animal cells. The NADPH is also used in other synthetic reactions.

Several techniques have been responsible for the successful analysis of the thermochemical "dark" reactions in photosynthesis. One of these is the use of radioactive carbon, making it possible to tag the carbon dioxide ($^{14}CO_2$) that is incorporated into organic compounds during photosynthesis. Another technique is paper chromatography (see Appendix 5.4), by which the exceedingly small amounts of substances that appear even after a second of illumination can be identified. Each compound containing ^{14}C can then be determined quantitatively if sensitized photographic films are exposed to the radioactive spots on the paper chromatogram.* From the relative density of spots on the developed films the relative concentrations of the various compounds containing ^{14}C can be determined, and if the film is calibrated by a simultaneous exposure to a compound with a known amount of ^{14}C, the actual concentration of each compound can be determined quantitatively. By increasing the length of exposure of cells to light, one can chart the spread of radioactive carbon into a variety of organic compounds. It is also possible to compare the pathway by which plants incorporate ^{14}C into compounds in the dark with the pathway taken during illumination. Still another technique that has proved very fruitful in analyzing thermochemical reactions is the isolation of the enzymes involved in each of the thermochemical steps in photosynthesis (Zelitch, 1975).

In most of the experiments tracing the incorporation of ^{14}C into photosynthetic products, the radioactive carbon in a solution of bicarbonate is provided to cells that have been brought to a steady state of photosynthesis in the presence of adequate carbon dioxide. At a determined time the cells are killed and analyzed quantitatively by chromatography and sensitized photographic film for various compounds containing ^{14}C. In cells killed immediately after a 1-second exposure to light the ^{14}C is found mainly in phosphoglyceric acid. In cells killed after 5 seconds of illumination, traces of phosphoglyceric acid, alanine, aspartic acid, and malic acid containing ^{14}C appear. In cells killed after 15 seconds of illumination, in addition to the other compounds listed, faint traces of sucrose containing ^{14}C may be detected, and also some pentose, glucose phosphate, and fructose phosphate, all containing ^{14}C.

After a minute of illumination has elapsed, before the cells are killed, abundant radioactive glucose phosphate and fructose phosphate are found. At the end of 5 minutes of illumination, radioactivity appears even in the lipids of the cell, and by way of the amino acids such radioactivity may be able to reach the proteins. It is quite evident that the radioactive carbon obtained from carbon dioxide is rapidly spread throughout the various classes of compounds present in the plant cell. In view of the operation of the metabolic cycles and their interlocking this is not surprising (see Chapter 12).

Since in most plants the 3-carbon acid, phosphoglyceric acid, is the first compound so far

*An alternate method is to cut out the paper containing the radioactive compound and measure its radioactivity directly with a Geiger counter.

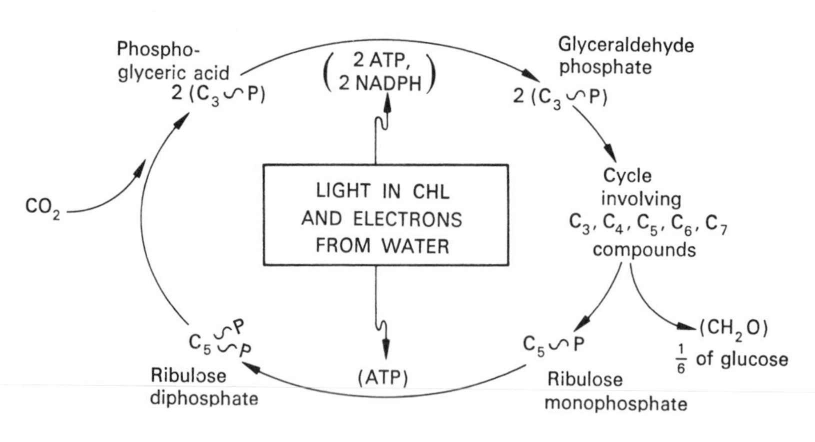

Figure 14.16. The Calvin-Benson cycle for photosynthetic carbon dioxide assimilation. The light reactions furnish the necessary energy and reducing power in the form of ATP and NADPH. (From Clayton, 1971: Light and Living Matter. Vol. 2. McGraw-Hill Book Co., New York.)

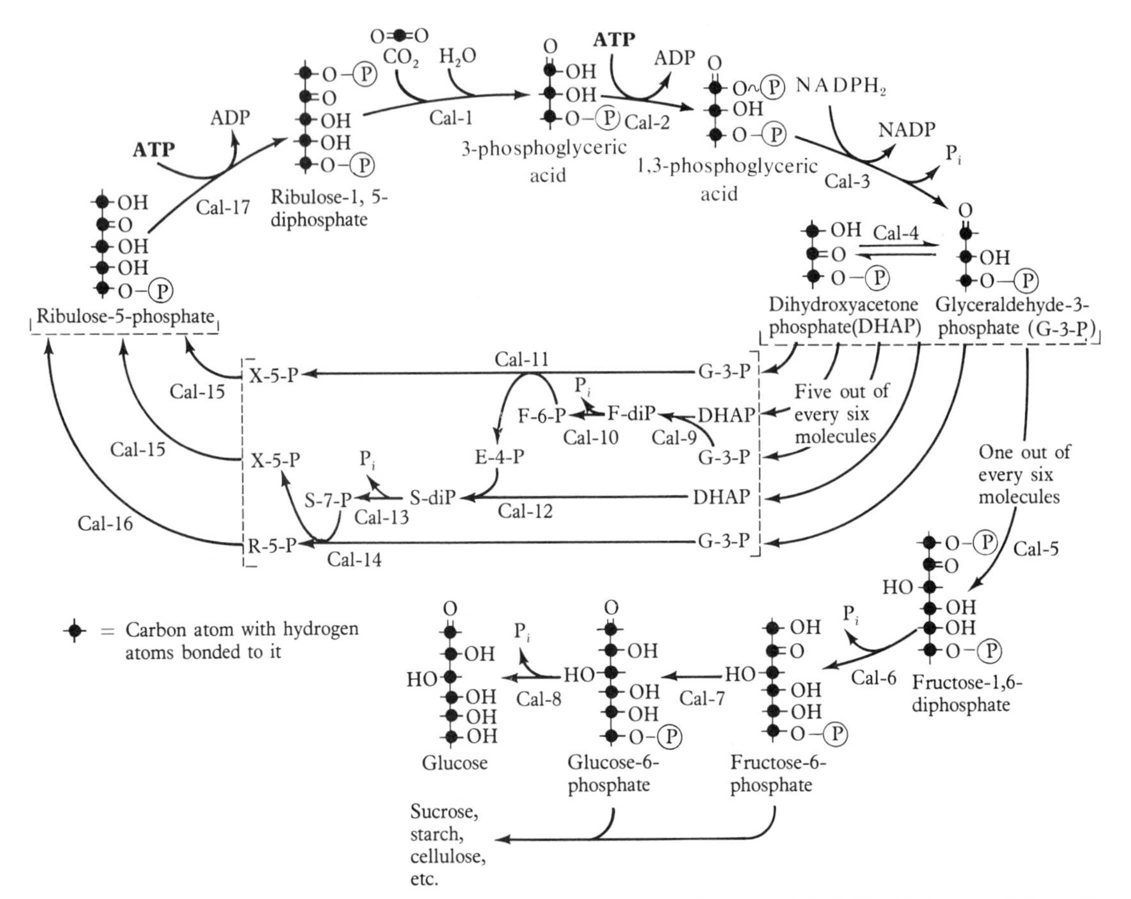

Figure 14.17. *The Calvin cycle for photosynthetic carbon metabolism. Carbon as CO_2 is fixed into organic form, then reduced to the oxidation level characteristic of carbon in sugar molecules. The black dots with crosses represent carbon atoms with attached hydrogen atoms where no other attachment is shown. For net synthesis of one glucose molecule, six molecules of CO_2 must be fixed, resulting in the formation of 12 molecules of glyceraldehyde-3-phosphate (G-3-P). Only one out of every six molecules of G-3-P (two per glucose synthesized) is actually used for sugar synthesis, since the other five out of every six molecules (ten per glucose) are reutilized within the cycle to regenerate the initial five-carbon acceptor molecule, ribulose-1, 5-diphosphate. Abbreviations used in the regeneration sequence shown in brackets are as follows: G, glyceraldehyde; DHAP, dihydroxyacetone phosphate; F, fructose; E, erythrose; S, sedoheptulose; X, xylulose; and R, ribose. (From Becker, 1977: Energy and The Living Cell. J. B. Lippincott Co., Philadelphia, p. 155.)*

found (even after very brief illumination) to incorporate radioactive carbon, its genesis becomes of prime importance. About half of its radioactivity resides in the carboxyl group, suggesting that the $^{14}CO_2$ had been added to a larger molecule from which the phosphoglyceric acid has been split off. Since radioactivity accumulates rapidly in pentose sugar, the pentose sugar is the molecule which most likely incorporates the carbon dioxide. The enzymes involved in this transfer are now known.

The series of dark reactions in photosynthesis elucidated by Calvin has been named the Calvin cycle. The overall scheme of dark reactions is shown in Figure 14.16 and detailed in Figure 14.17 and Table 14.4. In the following discussion, the reactions illustrated in Figure 14.17 are abbreviated Cal-1, etc.

Since researchers in photosynthesis use alternate names and abbreviations for the same substances, these names must be equated here to serve as a guide to figures and tables taken from the literature. Of interest in this respect are the sugars: trioses (3-carbon), tetroses (4-carbon), pentoses (5-carbon), hexoses (6-carbon), and heptoses (7-carbon). When only the category of sugar is specified the terms triose phosphate, triose diphosphate, tetrose phosphate, pentose phosphate, hexose phosphate, hexose diphosphate and so on are used as in Table 14.4. Specific sugars in these categories are named in accordance with their

TABLE 14.4. PHOTOSYNTHETIC DARK REACTIONS*

Number	Reactants	Products	Enzymes
	Reactions Fixing Carbon Dioxide into Pentose Forming Hexose		
1.	Pentose phosphate + ATP	Ribulose diphosphate + ADP	Phosphopentokinase
2.	Ribulose diphosphate + CO_2 + H_2O	2 Phosphoglyceric acid	Pentose diphosphate carboxylase
3.	2 Phosphoglyceric acid + 2ATP	2 Diphosphoglyceric acid + 2ADP	Phosphoglyceric acid kinase
4.	2 Diphosphoglyceric acid + 2NADPH + $2H^+$	2 Triose phosphate + $2NADP^+$ + 2 phosphate (inorganic)	Triose phosphate dehydrogenase
5.	2 Triose phosphate	Hexose diphosphate	Aldolase, triose phosphate isomerase
	Reactions Regenerating Pentose		
6.	Hexose diphosphate + H_2O	Hexose monophosphate + phosphate (inorganic)	Hexose diphosphatase (magnesium dependent)
7.	Hexose monophosphate + triose phosphate	Pentose phosphate + erythrose (tetrose) phosphate	Transketolase
8.	Hexose monophosphate + erythrose phosphate	Sedoheptulose phosphate + triose phosphate	Transaldolase
9.	Sedoheptulose phosphate + triose phosphate	2 Pentose phosphate	Transketolase
	Sum of (1 to 9) × 3: $3CO_2$ + 9ATP + $5H_2O$ + 6NADPH + $6H^+$	1 triose phosphate + 9ADP + $6NADP^+$ + 8 inorganic phosphate	
	Reactions Regenerating NADPH and ATP and Providing H^+		
10.	H_2O + $NADP^+$	NADPH + H^+ + $\frac{1}{2}O_2$ + 2e	Splitting of water
11.	NADPH + 3ADP + 3 inorganic phosphate + H^+ + $\frac{1}{2}O_2$ + 2e	$NADP^+$ + 3ATP + $4H_2O$	
	Sum of (1 to 11) × 3: $3CO_2$ + $2H_2O$ + inorganic phosphate	1 Triose phosphate + $3O_2$	

*ATP = adenosine triphosphate; ADP = adenosine diphosphate; $NADP^+$ = nicotinamide adenine dinucleotide phosphate oxidized, NADPH + H^+ (or $NADPH_2$), reduced; e = electron. (Data from Racker, 1955.) Note that photosynthetic bacteria use NAD and NADH in place of NADP and $NADPH_2$, otherwise the reactions are similar. This table does not take into account C_4 plants—see Figure 14.18.

structures: the trioses, dihydroxyacetone phosphate (DHAP), glyceraldehyde-3-phosphate (G-3-P), and glyceraldehyde-1,3-diphosphate (G-diP); the pentoses: ribulose-5-phosphate (R-5-P), ribulose-1,5-diphosphate (R-diP), and xylulose-5-phosphate (X-5-P); the hexoses: glucose, glucose-6-phosphate, fructose-6-phosphate (F-6-P), and fructose-1,6-diphosphate (F-1,6diP); the heptoses: sedoheptulose-7-phosphate (S-7-P), and sedoheptulose-1,7-diphosphate (S-diP). A synonym for phosphoglyceric acid is glycerate-3-phosphate, and for 1,3-diphosphoglyceric acid, glycerate-1,3-diphosphate.

After the pentose ribulose-1,5-diphosphate combines with carbon dioxide (Cal-1), it almost immediately splits to form two molecules of 3-phosphoglyceric acid. The molecules of phosphoglyceric acid are further phosphorylated (Cal-2) to form 1,3-phosphoglyceric acid and in the presence of NADPH they are reduced (Cal-3) to the trioses, dihydroxyacetone phosphate and glyceraldehyde-3 phosphate and inorganic phosphate (P_i) is released. Two glyceraldehyde-3-phosphate molecules unite to form the hexose sugar, fructose-1,6-diphosphate, as indicated by the fact that a higher percentage of carbon labeling in this molecule appears in the centrally located carbons 3 and 4.

From fructose-1,6-diphosphate, fructose-6-phosphate is formed (Cal-6) by loss of a phosphate, and from it glucose-6-phosphate is formed (Cal-7) in the presence of an isomerase that converts fructose to glucose. From these molecules various disaccharides and polysaccharides may then be synthesized (lowest arrow). The acids of the Krebs cycle containing ^{14}C are formed from the glucose used in respiration (primarily in the dark, since illumination inhibits these reactions). Carbon-14 appears in amino acids when some of the keto acids are aminated, and it appears in lipids and proteins when these compounds are built from fragments in the Krebs cycle (see Chapter 12).

The pentose ribulose-1,5-diphosphate, which appears to serve as the carbon dioxide acceptor in the thermochemical reactions of photosynthesis, is produced by a scheme similar to the pentose shunt already discussed for heterotrophic cells (Chapter 12). Fructose-1,6-diphosphate formed by combination of dihydroxyacetone phosphate and glyceraldehyde-3-phosphate (Cal 9), upon losing one of its high energy phosphates (Cal-10), combines with glyceraldehyde-3-phosphate, in the presence of the enzyme transketolase, to form (Cal-11) the pentose xylulose-5-phosphate and the 4-carbon sugar erythrose-4-phosphate. In the presence of the same enzyme erythrose-4-phosphate combines with dihydroxyacetone (Cal-12) to form the 7-carbon sugar, sedoheptulose diphosphate from which one phosphate is split (Cal-13), forming sedoheptulose-7-phosphate. The latter combines with glyceraldehyde-3-phosphate (Cal-14) to form xylulose-5-phosphate and ribulose-5-phosphate. It is not clear why such an indirect pathway is followed, but tracer carbon indicates that such is the case. The xylulose-5-phosphate is converted to ribulose-5-phosphate by an isomerase (Cal-16) and this molecule acts as the recipient of carbon dioxide after phosphorylation (Cal-17).

In this manner continuous supplies of pentose are maintained for the reception of CO_2. These supplies are adequate to fix carbon dioxide in the thermochemical reactions following the photochemical reactions of photosynthesis. This scheme is supported by the fact that the sugar pentose is always isolated in illuminated cells along with the 7-carbon sugar, sedoheptulose. The latter contains about half of its radioactivity in the central carbons, indicating that it is formed by the combination of smaller fragments. For his share in elucidating the dark reactions in photosynthesis, Melvin Calvin received the Nobel Prize in 1961.

Some facts not easily explained by the Calvin cycle require further study. Thus, with low CO_2 concentration, considerable glycolic acid is produced. This may be a secondary product of the reduction of CO_2 in the Calvin cycle, but a good explanation for its formation is not at hand. Calvin used CO_2 at a concentration of 1 percent. The enzyme binding CO_2 has a low affinity for CO_2, making the higher concentration of CO_2 desirable, but plants must normally photosynthesize at the low CO_2 concentration in air. Furthermore, the data are based on short-term kinetic studies, which favor rapid reactions at the expense of slow ones (Devlin and Barker, 1971).

In summary, the photochemical reactions of photosynthesis provide the ATP and reductant (NADPH), while the dark reactions of photosynthesis use these to reduce CO_2 indirectly. The CO_2 is first accepted by a pentose sugar from which phosphoglyceric acid (PGA) is split off. PGA is then reduced to glyceraldehyde phosphate, which condenses to form a hexose phosphate. The pentose acceptor must be regenerated and this occurs in a number of steps, which make the sequence of events appear very complicated.

THE CHLOROPLAST AS A COMPLETE PHOTOSYNTHETIC ENTITY

Intact chloroplasts are capable of complete photosynthesis outside the cell, provided all the co-factors are supplied (Arnon, 1960). This is true only if the outer membrane is not damaged during preparation.

The chloroplasts in the cells of green plants contain highly organized units of chlorophyll called *grana* (see Fig. 6.12). The grana are organized into lamellae, as shown in Figures 6.12 and 6.13. By chemical analysis grana have been shown to consist of lipids, proteins, and other compounds which are thought to be arranged in a characteristic manner as diagrammed in Figure 6.15. The organization of a granum is destroyed when chloroplasts are disrupted by grinding or other methods.

Grana comparable to those in cells of green plants are lacking in algal cells. *Chromatophores* analogous to free grana are found in some photosynthetic bacteria, although some have lamellae (ingrowth of the cell membrane; see Figs. 1.25 and 1.26) (Clayton, 1965).

Three sets of reactions are observed in intact chloroplasts: (1) the Hill reaction (sustained photoproduction of oxygen in isolated chloroplasts) following light absorption; (2) photosynthetic phosphorylation, during which adenosine diphosphate forms adenosine triphosphate—a reaction occurring in the absence of oxygen and distinct from metabolic phosphorylation; and (3) carbon dioxide fixation. That they are three separate sets of reactions can be demonstrated as follows: carbon dioxide fixation ceases when iodoacetamide is added, but photosynthetic phosphorylation and the Hill reaction continue; methylene blue inhibits both carbon dioxide fixation and photosynthetic phosphorylation but not the Hill reaction; *o*-phenanthroline inhibits all three reactions (Arnon, 1963). Presumably these reactions all occur in the chloroplasts inside the living cell. In the cytoplasm of the cell occur many other reactions and during the course of these reactions the photosynthetic products are built into various derivatives.

C_4-ACID PHOTOSYNTHESIS

In a scattered taxonomic array of plants, mainly plants in arid habitats and tropical grasses (which include the crop plants maize, sugar cane, and sorghum) the first detectable products of photosynthesis are the C_4-dicarboxylic acids, malic and oxaloacetic (OAA). Labeled carbon dioxide ($^{14}CO_2$) is incorporated in the carboxyl groups of these acids at the C_4 position (Fig. 14.18). Aspartic acid is also detectable as an early product of photosynthesis in such plants, but it is probably a secondary product resulting from transamination of oxaloacetic acid. Similar products are formed in darkness in such plants, but the rate of formation in light is about 250 times more rapid. The primary acceptor of carbon dioxide in both light and darkness is the 3-carbon acid phosphoenolpyruvic acid (PEP), which unites with carbon dioxide at a rapid rate, probably because of the high affinity of the enzyme phosphoenolpyruvate carboxylase for carbon dioxide. A supply of phosphoenolpyruvic acid is made available in C_4-photosynthetic plants by an enzyme unique to them, phosphoenolpyruvate synthetase. Conversely, in cells in which C_4 type of CO_2 fixation occurs, the enzyme that catalyzes fixation of carbon dioxide in C_3-photosynthetic plant tissues is only weakly developed (Bjorkman, 1973).

In C_4-photosynthetic plants the leaf structure and chloroplast organization are unique. Around the vascular bundles are two concentric layers of cells containing chloroplasts, an inner bundle sheath, and an outer mesophyll layer (Fig. 14.18). C_4-dicarboxylic acid synthesis occurs in the mesophyll cells, from which the acids are transported to the bundle sheath cells, where decarboxylation occurs, supplying the carbon dioxide to the Calvin cycle enzymes located there (Downton, 1971; Bishop and Reed, 1976). Thereafter, the reactions proceed as in the Calvin cycle already discussed. The C_4 scheme is therefore only a variant of the C_3 scheme. The mesophyll cells apparently act primarily as a pump to absorb carbon dioxide even at very low partial pressure, as, for example, when the stomata are slightly open in plants in arid environments. They do this at the cost of ATP (Fig. 14.18), but judging from the success of C_4 plants in nature, the cost must be justified.

C_4-photosynthetic cells differ in one other important respect from C_3-photosynthetic plants —photorespiration appears to be much reduced or absent.* By photorespiration is meant the increased output of carbon dioxide by plants upon illumination, as compared to such emission in darkness. That such respiration is a result of action of light on chlorophyll is suggested by its absence on illumination of leaves on the same plant lacking chlorophyll (chlorotic leaves). The action spectrum for photorespiration is almost identical to the absorption spectrum of the plants (Sörensen and Halldal, 1977). Photorespiration has been shown to result from oxidation of glycolate produced during photosynthesis (Fig. 14.19). Inhibitors of glycolate metabolism reduced the labeled carbon dioxide resulting from glycolate decomposition. Suggested reactions are the following:

$$CH_2OH{-}^{14}COOH + O_2 \rightarrow CHO{-}^{14}COOH + H_2O_2$$

$$CHO{-}^{14}COOH + H_2O_2 \rightarrow HCOOH + {}^{14}CO_2 + H_2O$$

*Possibly photorespiration occurs, but the CO_2 produced is used too rapidly to be detected (Halldal, personal communication).

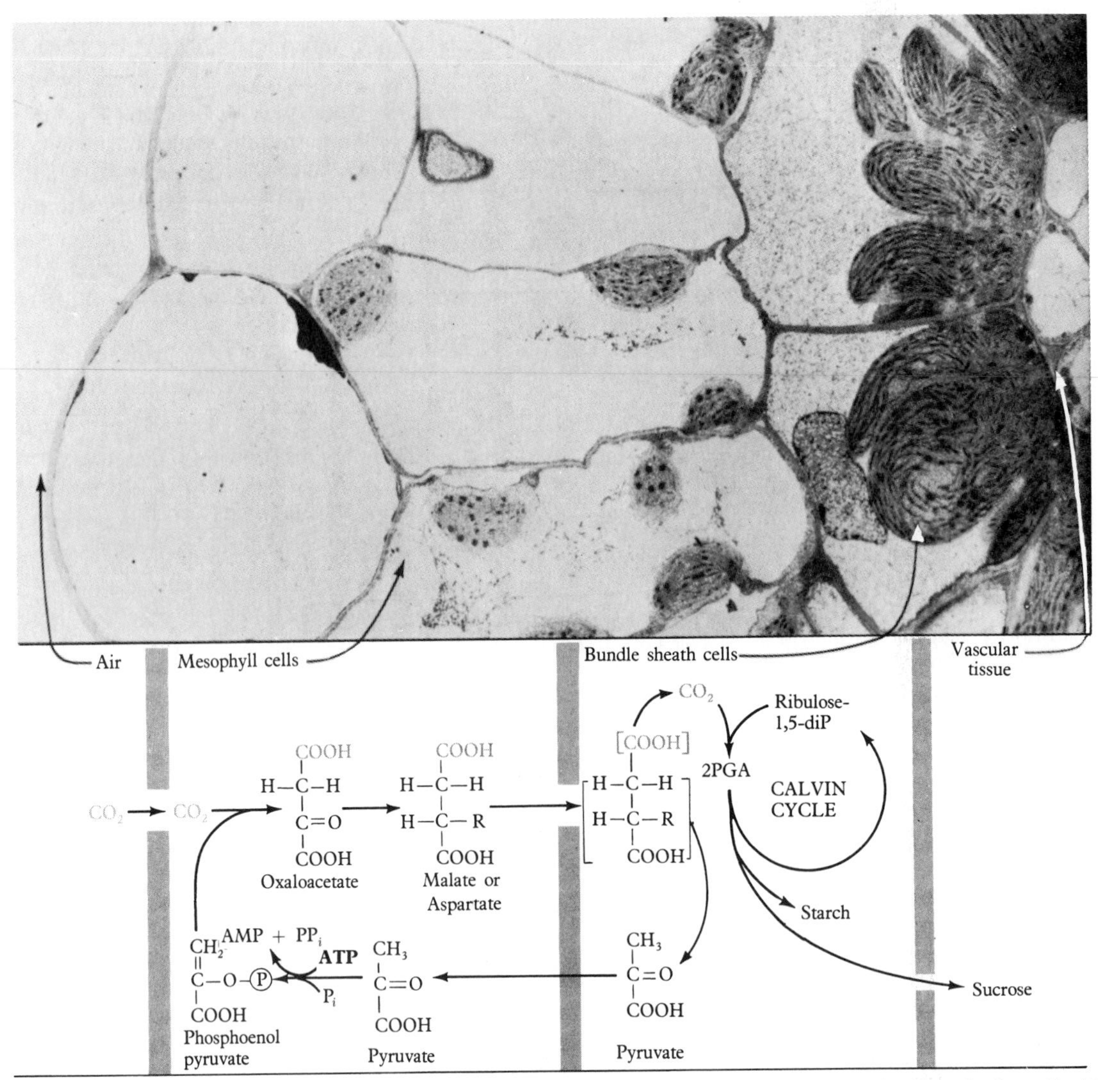

Figure 14.18. *Spatial localization of the Hatch-Slack and Calvin cycles within the leaf of a C_4 plant. Carbon dioxide entering the leaf through a stomate (left) is fixed into organic form by the Hatch-Slack pathway localized within the mesophyll cells. The immediate product of phosphoenol pyruvate carboxylation is oxaloacetate, but this is quickly converted to malate (by reduction; R = OH) or aspartate (by transamination; R = NH_2), depending upon the particular species of plant. Carbon moves into the bundle sheath cells in the leaf interior as the four-carbon acid, which is then decarboxylated. The CO_2 thus released is refixed by the Calvin cycle, yielding ultimately starch and sucrose. The sucrose passes into the adjacent vascular tissue for translocation to other parts of the plant. Meanwhile the pyruvate remaining from the decarboxylation reaction returns to the mesophyll cells for regeneration of phosphoenol pyruvate. The Hatch-Slack pathway operating in the mesophyll cells can be viewed as a pumping mechanism designed to maintain a locally high concentration of CO_2 for the carboxylating enzyme of the Calvin cycle operating in the bundle sheath cells. The particular C_4 leaf shown here is that of the grass* Panicum mileaceum. *(Courtesy of M. D. Hatch.) (From Becker, 1977: Energy and The Living Cell. J. B. Lippincott Co., Philadelphia, pp. 166–167.)*

Photorespiration increases with increase in oxygen concentration (Devlin and Barker, 1971). The enzyme glycolate oxidase required for the reactions is present in the peroxisomes (see Chapter 5) of photorespiratory plants and is lacking or low in concentration in peroxisomes of nonphotorespiratory plants.

The net photosynthetic rate is higher in C_4-photosynthetic plants than in their C_3 counterparts; consequently they are capable of more rapid growth. It has been found that inhibition of photorespiration in C_3-photosynthetic plants results in a higher net rate of photosynthesis, making them more comparable to the C_4 plants. It thus seems likely that the greater net photosynthetic rate of C_4 plants is also in part a consequence of their lack of photorespiration, resulting in a more effective channeling of carbon dioxide into carbohydrate or cell substance. Because of the "cost" of photorespiration to the

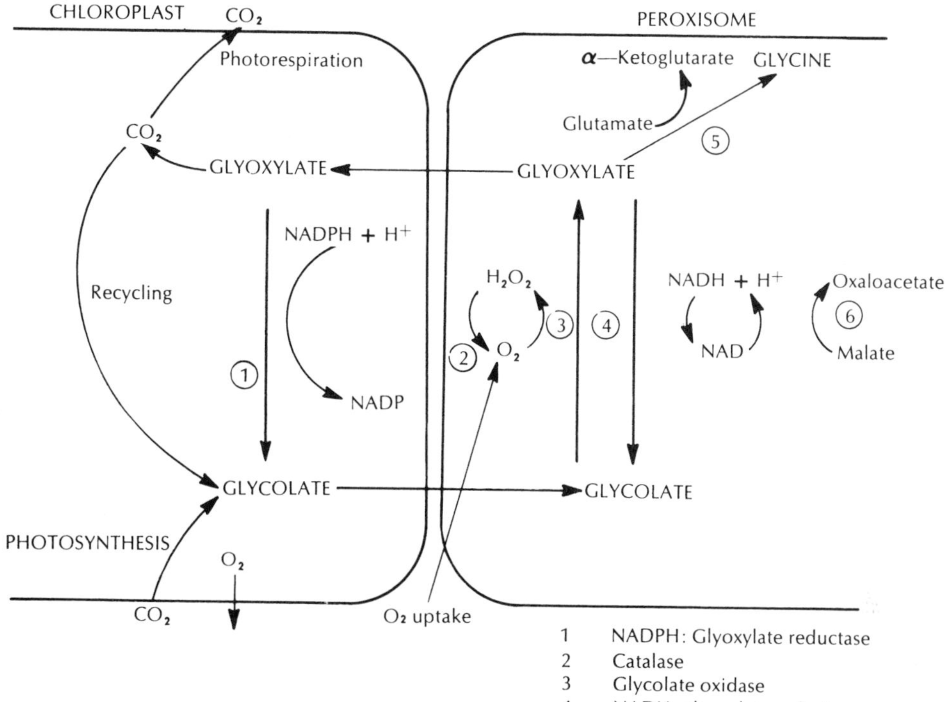

Figure 14.19. *Distribution of glycolate metabolism between peroxisomes and chloroplasts. Carbon dioxide release is from the chloroplast, whereas the oxygen-uptake component of photorespiration is in the peroxisomes. Numbers designate the enzymes for the steps indicated. (From Devlin and Barker, 1971: Photosynthesis. Van Nostrand Reinhold, New York.)*

plant, reduction of the process in C_3 crop plants is a goal of some agronomists.

In conclusion, it appears that C_4-photosynthetic plants, which are chiefly of tropical and arid origin, grow well at higher temperatures and higher light intensities than their C_3-photosynthetic counterparts and have a higher net photosynthetic rate. They also produce a unit weight of dry substance with half the water requirement of the C_3 plants. They are thus well adapted to the conditions in which they evolved (Hatch *et al.*, 1971).

CRASSULACEAN PLANT PHOTOSYNTHESIS

These plants, mainly succulents (with thick fleshy leaves or expanded stems only), live under semiarid conditions and conserve water by opening their stomata only at night, during which time the plants become highly acidified, take in oxygen, but give off no carbon dioxide. The carbon dioxide is fixed by the enzyme phosphoenolpyruvic acid (PEP) carboxylase, present in the cytoplasm of the chlorophyll-bearing cells, to form oxaloacetate. The oxaloacetate is reduced by malic dehydrogenase to malic acid, which accumulates in the vacuoles of the leaf cells. By day, the stomata close and malate is transported out of the vacuoles to the cytoplasm where it is decarboxylated by a NADP-linked malic enzyme to yield pyruvate and CO_2. The reductive pentose phosphate system is photoactivated and the CO_2 released from malate enters the chloroplasts where it is reduced in the Calvin cycle.

Succulents are generally slow-growing, probably because photosynthesis is limited by the potential size of the storage pool for organic acids and of stored carbohydrate for synthesis of the PEP substrate. Perhaps for this reason, succulents, although widely distributed, are never a dominant flora (Osmund, 1978).

Interestingly, a succulent *Agave americana*, if well watered, will switch to a typical C_3 metabolism and open its stomata by day. It then grows more rapidly, as well (Woolhouse, 1978).

LITERATURE CITED AND GENERAL REFERENCES

Allen, J. F., 1977: Superoxide and photosynthetic reduction of oxygen. *In* Superoxide and Superoxide Dismutase. Michelson, McCord and Fridovich, eds. Academic Press, New York, pp. 419–436.

Arnon, D. I., 1960: The role of light in photosynthesis. Sci. Am. (Nov.) *203*: 105–118.

Arnon, D. I., 1963: Photosynthetic photophosphorylation and a unified concept of photosynthesis. Proceedings of the Fifth International Congress of Biochemistry, Pergamon Press, Oxford, pp. 201–232.

Avron, M., 1977: Energy transduction in chloroplasts. Ann. Rev. Biochem. *46*: 143–155.

Bassham, J. A., 1977: Increasing crop production through more controlled photosynthesis. Science *197*: 630–638.

Bassham, J. A. and Calvin, M., 1962: The Photosynthesis of Carbon Compounds. W. A. Benjamin, Menlo Park, Calif.

Becker, W. M., 1977: Energy and the Living Cell. J. B. Lippincott Co., Philadelphia.

Beevers, L. and Hageman, R. H., 1972: The role of light in nitrate metabolism in higher plants. *In* Photophysiology. Vol. 6. Giese, ed. Academic Press, New York, pp. 85–113.

Bishop, D. G. and Reed, M. L., 1976: The C_4 pathway of photosynthesis: Ein Kranz-Tpy Wirtschaftswunder. *In* Photochemical and Photobiological Reviews. Vol. 1. Smith, ed. Plenum Press, New York. pp. 1–69.

Bishop, N., 1973. Analysis of photosynthesis through mutation studies. *In* Photophysiology. Vol. 8. Giese, ed. Academic Press, New York, pp. 65–96.

Bjorkman, O., 1973: Comparative studies on photosynthesis in higher plants. *In* Photophysiology. Vol. 8. Giese, ed. Academic Press, New York, pp. 1–63.

Bjorkman, O. and Berry, J., 1973: High efficiency photosynthesis. Sci. Am. *29* (Oct.) 80–98.

Blinks, L. R. and van Niel, C. B., 1963: The absence of enhancement (Emerson effect) in the photosynthesis of *Rhodospirillum rubrum*. *In* Microalgae and Photosynthetic Bacteria. Special issue of Plant and Cell Physiology, pp. 297–307.

Briantais, J. M. and Picaud, M., 1972: Immunological evidence for a localization of System I on the outside face and System II on the inside face of the chloroplast lamella. Fed. Eur. Biol. Sci. Let. 1971: 100–104.

Brown, J. S., 1973: Separation of photosynthetic systems I and II. *In* Photophysiology. Vol. 8. Giese, ed. Academic Press, New York, pp. 97–108.

Brown, J. S., 1977: Spectroscopy of chlorophyll in biological and synthetic systems (Review). Photochem. Photobiol. *26*: 319–326.

Calvin, M., 1962: The path of carbon in photosynthesis. Science *135*: 879–889 (Nobel lecture).

Calvin, M., 1976: Photosynthesis as a resource for energy and materials. Am. Sci. *64*: 266–278.

Cheniae, G. M., 1970: Photosystem II and oxygen evolution. Ann. Rev. Plant Physiol. *21*: 467–490.

Clayton, R. K., 1965: Modern Physics in Photosynthesis. Blaisdell Publishing Co., Waltham, Mass.

Clayton, R. K., 1971: Light and Living Matter: a Guide to the Study of Photobiology. The Biological Part. Vol 2. McGraw-Hill Book Co., New York, pp. 1–66.

Cramer, W. A. and Whitmarsh, J., 1977: Photosynthetic cytochromes. Ann. Rev. Plant Physiol. *28*: 133–172.

Devlin, R. M. and Barker, A. V., 1971: Photosynthesis. Van Nostrand Reinhold, New York.

Downton, W. J. S., 1971: Adaptive and evolutionary aspects of C_4 photosynthesis. *In* Photosynthesis and Photorespiration. Hatch, Osmond, and Slayter, eds. Wiley-Interscience, New York, pp. 3–17.

Emerson, R. and Arnold, W., 1932: A separation of the reactions in photosynthesis by means of intermittent light. J. Gen. Physiol. *15*: 391–420.

Emerson, R., Chalmers, R. and Cederstrand, C., 1957: Some factors influencing the long wave limit of photosynthesis. Proc. Natl. Acad. Sci. U.S.A. *43*: 133–143.

Fork, D. C., 1977: Photosynthesis. *In* The Science of Photobiology. K. C. Smith, ed. Plenum Press, New York, pp. 329–369.

Fork, D. C. and Amesz, J., 1970: Spectrophotometric studies of the mechanism of photosynthesis. *In* Photophysiology. Vol. 5. Giese, ed. Academic Press, New York, pp. 97–126.

Gantt, E., 1975: Phycobilisomes: light harvesting pigment complexes. Bioscience *25*: 781–788.

Gibbs, M., 1970: Inhibition of photosynthesis by oxygen. Am. Sci. *58*: 634–640.

Govindjee, ed., 1975: Bioenergetics of Photosynthesis. Academic Press, New York.

Govindjee and Govindjee, R., 1974: The absorption of light in photosynthesis. Sci. Am. (Dec.) *231*: 68–82.

Glazer, A. N., 1976: The phycocyanins: structure and function. *In* Photochemical and Photobiological Reviews. Vol. 1. Smith, ed. Plenum Press, New York, pp. 71–115.

Halldal, P., 1978: Personal Communication.

Hatch, M. D., Osmond, C. B. and Slayter, R. O., eds., 1971: Photosynthesis and Photorespiration. Wiley-Interscience, New York.

Hatch, M. D., 1971: Mechanism and function of the C_4 pathway in photosynthesis. *In* Photosynthesis and Photorespiration. Hatch, Osmond and Slayter, eds. Wiley-Interscience, New York, pp. 139–152.

Haxo, F. T. and Blinks, L. R., 1950: Photosynthetic action spectra of marine algae. J. Gen. Physiol. *33*: 389–422.

Hill, R., 1939: Oxygen produced by isolated chloroplasts. Proc. Roy. Soc. London *B127*: 192–210.

Hind, G. and McCarty, R. E., 1973: The role of ion fluxes in chloroplast activity. *In* Photophysiology. Giese, ed. Vol. 8. Academic Press, New York, pp. 113–156.

Jagendorf, A. T. and Uribe, E., 1965: ATP formation caused by acid-base transition of spinach chloroplasts. Proc. Natl. Acad. Sci. U.S.A. *55*: 170–177.

Jensen, S. L., 1963: Carotenoids of photosynthetic bacteria. *In* Bacterial Photosynthesis. Gest, San Pietro, Vernon, eds. Antioch Press, Yellow Springs, Ohio, pp. 19–34.

Junge, W., 1977: Membrane potentials in photosynthesis. Ann. Rev. Plant Physiol. *28*: 503–536.

Kelly, G. J., Latzko, E. and Geliks, M., 1976: Regula-

tory aspects of photosynthetic carbon metabolism. Ann. Rev. Plant Physiol. *27*: 181–205.
Krinsky, N. I., 1968: The protective function of carotenoid pigments. *In* Photophysiology. Current Topics. Vol. 3. Giese, ed. Academic Press, New York, pp. 123–195.
Levine, P., 1969: The mechanism of photosynthesis. Sci. Am. (Dec.) *221*: 58–70.
Loach, P. A., 1977: Primary photochemistry in photosynthesis. Photochem. Photobiol. *26*: 87–94 (Review).
Lumsden, J., Henry, L. and Hall, P. O., 1977: Superoxide dismutase in photosynthetic organisms. *In* Superoxide and Superoxide Dismutase. Michelson, McCord and Fridovich, eds. Academic Press, New York, pp. 437–450.
Miller, K. R., Müller, G. J. and McIntyre, K. R., 1976: The light-harvesting chlorophyll-protein complex, photosystem II. J. Cell Biol. *71*: 624–638.
Mitchell, P., 1966: Chemiosmotic coupling in oxidative and photosynthetic phosphorylation. Biol. Rev. *41*: 445–502.
Mohr, H., 1978: Phytochrome and chloroplast development. Endeavour NS *1*: 107–114.
Myers, J., 1971: Enhancement studies in photosynthesis. Ann. Rev. Plant Physiol. *22*: 289–312.
Oliver, D. J. and Zelitch, I., 1977: Increasing photosynthesis by inhibiting photorespiration with glyoxylate. Science *196*: 1450–1451.
Osmund, C. B., 1978: Crassulacean acid metabolism: a curiosity in context. Ann. Rev. Plant Physiol. *29*: 379–414.
Packer, L. and Deamer, D. W., 1968: Studies on the effect of light on chloroplast structure. *In* Photophysiology. Current Topics. Vol. 3. Giese, ed. Academic Press, New York, pp. 91–122.
Pfenig, N., 1977: Phototrophic green and purple bacteria: a comparative systematic review. Am. Rev. Microbiol. *31*: 275–290.
Rabinowitch, E. I. and Govindjee, 1965: The role of chlorophyll in photosynthesis. Sci. Am. (Jul.) *213*: 74–83.
Rabinowitch, E. I. and Govindjee, 1969: Photosynthesis. John Wiley & Sons, New York.
Racker, E., 1955: Synthesis of carbohydrate from CO_2 and H_2 in a cell-free system. Nature *175*: 249–251.
Radmer, R. and Kok, B., 1975: Energy capture in photosynthesis: photosystem II. Ann. Rev. Biochem. *44*: 409–433.
Radmer, R. and Kok, B., 1977: Photosynthesis: limited yields, unlimited dreams. Bioscience *27*: 599–605.
Ruben, S., Randall, M., Kamer, M. and Hyde, J. L., 1941: Heavy oxygen (^{18}O) as a tracer in the study of photosynthesis. J. Am. Chem. Soc. *63*: 877–879.
Sörensen, L. and Halldal, P., 1977: Comparative analyses of action spectra of glycolate excretion ("photorespiration") and photosynthesis in the blue-green alga *Anacystis nidulans*. Photochem. Photobiol. *26*: 511–518.
Stanier, R. Y., Adelberg, E. A. and Ingraham, J., 1976. The Microbial World. 4th Ed. Prentice-Hall, Englewood Cliffs, N.J.
Trebst, A., 1974: Energy conservation in photosynthetic electron transport of chloroplasts. Ann. Rev. Plant Physiol. *25*: 423–458.
van Niel, C. B., 1949: The comparative biochemistry of photosynthesis. *In* Photosynthesis in Plants. Frank and Loomis, eds. Iowa State College Press, Ames, Iowa, pp. 437–495.
Wang, R. T., Stevens, C. L. R. and Myers, J., 1977: Action spectra for photoreactions I and II of photosynthesis in the blue green alga *Anacystis nidulans*. Photochem. Photobiol. *25*: 103–108.
Weaver, E. C. and Weaver, H., 1972: Electron resonance studies of photosynthetic systems. *In* Photophysiology. Giese, ed. Academic Press, New York, *7*: 1–32.
Woolhouse, H. W., 1978: Light-gathering and carbon assimilation processes in photosynthesis: their adaptive modifications and significance for agriculture. Endeavour *2*NS: 35–46.
Zelitch, I., 1971: Photosynthesis, photorespiration and plant productivity. Academic Press, New York.
Zelitch, I., 1975: Pathways of carbon fixation in green plants. Ann. Rev. Biochem. *44*: 123–145.
Zelitch, I., 1975: Improving the efficiency of photosynthesis. Science *188*: 626–633.

CHAPTER 15

BIOSYNTHESIS

NUTRITIONAL PATTERNS

Photosynthesis represents only one aspect of the synthetic activity of a chlorophyll-bearing cell. The products of photosynthesis may be used to synthesize a complex spectrum of small molecules (monosaccharides, oligosaccharides, fatty acids, amino acids, vitamins) and macromolecules (lipids, polysaccharides, proteins, nucleic acids) that the cell requires for maintenance, growth, and reproduction.

Colorless cells depend upon the products of the photosynthetic cell for their organic supplies: organic carbon, often organic nitrogen, and some vitamins. With these, and with inorganic salts, they too can often synthesize the small molecules and macromolecules that are required for maintaining life and for growth and cell division.

Photosynthetic cells (*photoautotrophic* cells) represent the pattern of nutrition with minimal requirements, inasmuch as they synthesize all their organic requirements from low-energy compounds (carbon dioxide, water, and salts) at the expense of light energy. In all other types of cells (*heterotrophic* cells) some of the synthetic activities unique to photoautotrophic cells were never acquired because the cells lived in a medium containing these compounds (see Chapter 1). If these activities were acquired, they were progressively lost, the loss being greatest in heterotrophic cells requiring the most complex media for growth.

Heterotrophic cells depend to different degrees upon organic nutrients. Some prokaryotic organisms (e.g., wild-type *E. coli*) require only organic carbon (usually glucose) and various salts to synthesize all their other organic requirements. Others must obtain even small organic molecules such as amino acids or vitamins from their nutrients; in this respect, animal cells generally are especially dependent. This does not mean that animal cells are incapable of synthetic activity—witness the genome of such cells (see Chapter 7), which codes for a tremendous number of proteins. Rather, it means that they begin synthesis of macromolecules from the level of supplied organic molecules, and from these they synthesize their complex structures.

In some cases, however, complex nutrients are required by cells without complex structures, as, for example, the mycoplasmas (also called pleuropneumonia-like organisms, PPLO). Here the general synthetic activity has declined as the genome has become correspondingly simplified. Phages represent replicating systems with even more simplified genomes, replication of their own kind being dependent upon the synthetic activity of the host cell. Some viruses have only two genes coding for two proteins (see last section of Chapter 7).

Among the organic requirements of many cells are vitamins needed in small amounts as components of various enzyme systems (see Chapters 3 and 10). Vitamins, like many other common organic compounds, are interchangeable in cells of different organisms with similar enzyme systems. A discussion of the nutritional requirements of different kinds of cells and a classification of modes of nutrition can be found in textbooks of microbiology and biochemistry.

Chemoautotrophic bacteria synthesize all their organic cell requirements from carbon dioxide, water and salts by oxidation of inorganic compounds, as already discussed in Chapter 1. They represent a group of organisms limited to highly specific environments.

BIOSYNTHESIS OF SMALL ORGANIC MOLECULES

The study of biosynthesis of small molecules such as amino acids and purines was pioneered by Beadle and Tatum (1941), largely with the bread mold *Neurospora*. For this work they received the Nobel Prize in 1958. Wild-type *Neurospora* supplied with a medium containing sucrose, certain salts, and the B vitamin biotin produces all the other growth factors and

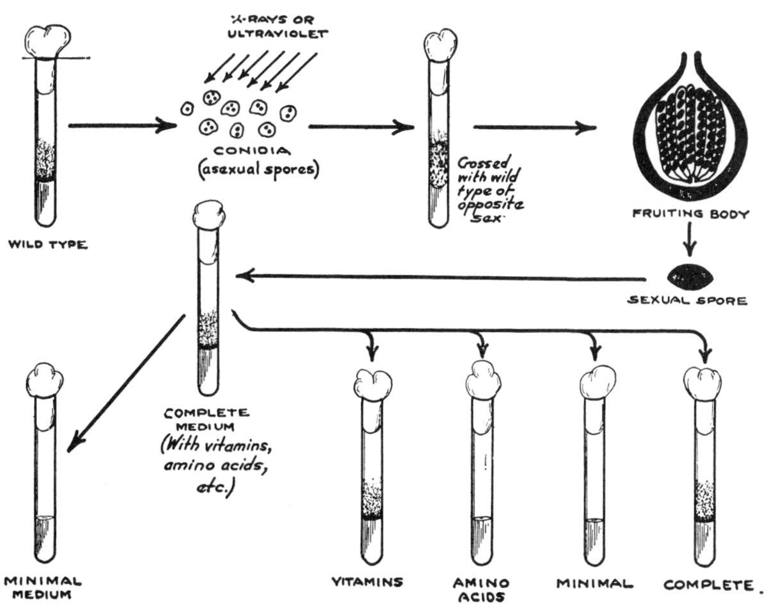

Figure 15.1. *Experimental method for testing biochemical mutants in* Neurospora. *In practice the irradiated asexual spores must be sprouted to form a mycelium, which is fused with another mycelium developed from a nonirradiated asexual spore. The fused mycelia form a fruiting body, in which are developed the ascospores. It is the ascospores that are isolated for the tests shown. If the ascospores fail to grow in minimal medium, a mutation resulting in the loss of the capacity to synthesize a growth requirement is indicated. Growth in minimal medium plus vitamins indicates deficiency in synthesizing a vitamin; growth in minimal medium supplemented with amino acids indicates a deficiency in synthesizing these. If the spores grow in none of these media but grow in the complete medium, the mutational loss is elsewhere and further testing of media is required. In any case further tests are required to determine the precise deficiency. (From Beadle, 1947:* Science in Progress *5: 176.)*

all the amino acids and other compounds it needs. However, when wild-type *Neurospora* spores are exposed to ultraviolet or ionizing radiation, gene "blocks" may occur among the resulting clones; that is, mutants may appear that are unable to synthesize some factors required for growth synthesized by the wild type. These mutants *(auxotrophs)* grow only if the particular compound required is added to the medium (Fig. 15.1). This finding provided a tool for studying the synthesis of many factors required by the cell and has yielded results that would have been difficult by any other methods.

To illustrate, a number of mutants have been obtained by irradiation that fail to produce tryptophan owing to a "block" at some point in tryptophan synthesis. If tryptophan is added to the basal medium (glucose, biotin, and salts) growth occurs. By determining which of a number of precursors of tryptophan also promotes growth, the course of tryptophan synthesis has been determined. As illustrated in Figure 15.2, after mutation of gene 1, a culture of the mutant *Neurospora* synthesizes tyrosine or phenylalanine, which accumulate in the medium, but is unable to make anthranilic acid from these. If it is supplied with an external source of anthranilic acid, the metabolic pathway after the block may proceed and the culture forms indole, which is then coupled with serine to form tryptophan. Mutation of gene 2 "blocks" synthesis of indole from anthranilic acid, although all the preceding syntheses occur. Provided with indole, *Neurospora* with mutated gene 2 readily synthesizes tryptophan. If *Neurospora* with a mutation at gene 2 is not supplied with indole, anthranilic acid accumulates (Beadle, 1959).

Many naturally occurring metabolic defects resulting from spontaneous mutations have been found that lead to such accumulation of intermediates. By identifying these intermediates it has often been possible to map the pathways of biosynthesis and degradation of many compounds. Discussion of biosynthesis of various amino acids, purines, pyrimidines, nucleotides, and vitamins is found in textbooks of biochemistry and in brief form in a book on synthesis of small molecules (Cohen, 1967). The agreement between data gathered on animal cells, *Neurospora,* and *E. coli* is impressive and leads one to appreciate the unity in the biochemistry of the cell, be it in the plant, animal, or microorganism. Such unity for degradation of nutrients was realized long ago (Kluyver and Donker, 1926), but it is now apparent for the synthetic processes as well. The oldest homologies in life are therefore at the cellular level (Szent-Györgyi, 1940). This view is reinforced by the current findings in DNA recom-

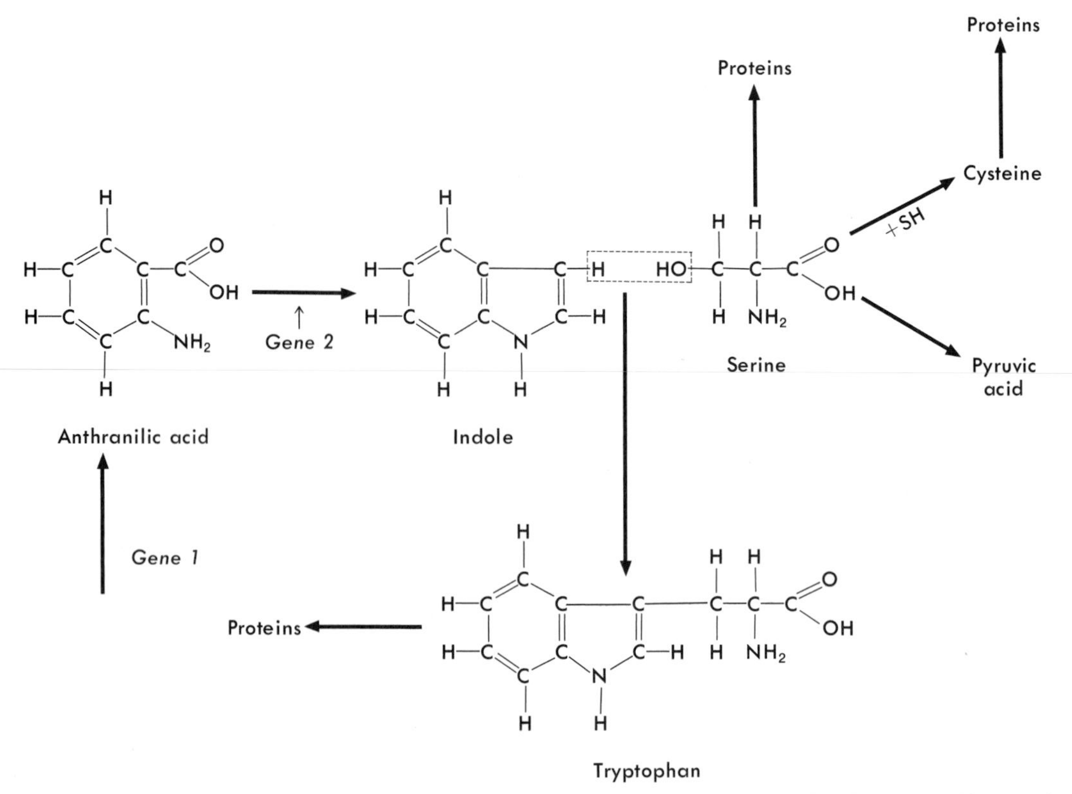

Figure 15.2. The action of various genes on tryptophan synthesis in Neurospora. *(After Tatum and Bonner, from Beadle, 1946: Am. Scientist* 34: *37.)*

binant research. It is possible to introduce animal genes into bacteria to produce hormones such as insulin, using the synthetic machinery of the bacteria.

Study of auxotrophic mutants has been augmented by the use of isotopic tracers for determination of biosynthetic pathways. Labeled intermediate can be isolated by chromatography and identified by comparison with known substances. However, the data gathered by these techniques are not always clear-cut. It is therefore necessary to verify in cell-free extracts the presence or absence of enzymes involved in the reactions by determination of the specific activity of each purified enzyme. By this means, for example, it could be shown that an auxotrophic mutant of *E. coli* requiring homoserine for its growth (or a mixture of threonine and methionine) lacks the enzyme homoserine dehydrogenase. This enzyme is responsible for the synthesis of homoserine from aspartic semialdehyde. The results confirm the data from growth studies with the mutants and indicate clearly that homoserine is an intermediate in the synthesis of threonine and methionine (Cohen, 1967).

The biosynthetic method used by Beadle and Tatum on the mold *Neurospora* has been applied especially to bacteria (e.g., *E. coli*). By this method much has been learned about cellular biosynthesis of vitamins, purines, pyrimidines, amino acids, lipids, proteins, and carbohydrates, as well as about the enzymes concerned.

The biochemical studies demonstrate that each of the biosynthetic steps is catalyzed by an enzyme that is encoded in a gene composed of DNA. Therefore, it is loss of a functional enzyme as a result of gene mutation that causes blocking of a step in the synthetic pathway.

BIOSYNTHESIS OF LIPIDS

Lipids (see Figs. 1.13 and 3.11) generally are molecules of small molecular weight, seldom over 1000. Since they are often assembled from several types of molecules, the synthesis of lipids appears somewhat more complex than that of carbohydrates. However, in essence, synthesis consists of activation of each of the reacting molecules through reaction with ATP and CTP. As an example, formation of a phospholipid, phosphatidyl ethanolamine, is shown in Table 15.1.

TABLE 15.1. ENZYMATIC STEPS IN THE BIOSYNTHESIS OF PHOSPHATIDYL ETHANOLAMINE*

Activation of Fatty Acids (RCOOH and R′COOH)

RCOOH + ATP + CoA—SH → R—CO—SCoA + AMP + pyrophosphate
R′COOH + ATP + CoA—SH → R′CO—SCoA + AMP + pyrophosphate
2AMP + 2ATP → 4ADP
2 pyrophosphate + $2H_2O$ → 4 phosphate

Activation of Glycerol

glycerol + ATP → glycerol 3-phosphate + ADP

Formation of CDP-Diacylglycerol

RCO—SCoA + glycerol 3-phosphate → monoacylglycerol 3-phosphate + CoA—SH
R′CO—SCoA + monoacylglycerol 3-phosphate → diacylglycerol 3-phosphate + CoASH
diacylglycerol 3-phosphate + CTP → CDP—diacylglycerol + pyrophosphate
pyrophosphate + H_2O → 2 phosphate

Attachment of Ethanolamine

CDP—diacylglycerol + serine → phosphatidyl serine + CMP
phosphatidyl serine → phosphatidyl ethanolamine + CO_2
ATP + CMP → ADP + CDP
ATP + CDP → ADP + CTP

Sum: RCOOH + R′COOH + glycerol + serine + 7ATP → phosphatidyl ethanolamine + 7ADP + 6 phosphate + CO_2

In phosphorylated compounds, C stands for cytidine.
*From Lehninger, 1971: Bioenergetics. 2nd Ed. W. A. Benjamin, Menlo Park, Calif. p. 143.

MACROMOLECULAR SYNTHESES

DNA, RNA, protein, and polysaccharides are the major macromolecules synthesized in the cell. Because the mechanism of biosyntheses, especially that of protein biosynthesis, is complex, most of this chapter is devoted to this topic and related subjects. Regulation of biosynthesis is considered in Chapter 16.

Biosynthesis of DNA: Replication

The biosynthesis of deoxyribonucleic acid (DNA) *in vitro,* through polymerization of nucleoside triphosphates in the presence of the appropriate enzyme, was first successfully completed by Kornberg, and for this he received the Nobel Prize in 1959. The nucleic acid DNA consists of four kinds of nucleotides: units of purine or pyrimidine attached to a phosphoric acid residue through a pentose sugar (Freifelder, 1978). In DNA the nucleotides attach to one another in strands, the phosphoric acid and pentose forming the backbone of the spiraling structure of the Watson-Crick DNA model (see Fig. 1.18). The two complementary DNA strands are held together by hydrogen bonds between their respective complementary purine and pyrimidine bases—thymine (T) to adenine (A), and guanine (G) to cytosine (C).

The first step in the synthesis of DNA is the formation of the nucleoside triphosphates (activation) by the union of an ATP molecule with each of the four nucleoside monophosphates (phosphorylation) (Kornberg, 1960). This mechanism is not unique—it is the means whereby amino acids are activated before protein synthesis and fatty acids and monosaccharides are activated before any of their synthetic reactions. It is also the means whereby coenzymes are phosphorylated.

For the four deoxyribonucleoside phosphates (Fig. 15.3), the reactions are:

$$\text{dAMP} + \text{ATP} \rightarrow \text{dATP} + \text{AMP} \qquad (15.1)$$

$$\text{dGMP} + \text{ATP} \rightarrow \text{dGTP} + \text{AMP} \qquad (15.2)$$

$$\text{dCMP} + \text{ATP} \rightarrow \text{dCTP} + \text{AMP} \qquad (15.3)$$

$$\text{dTMP} + \text{ATP} \rightarrow \text{dTTP} + \text{AMP} \qquad (15.4)$$

where d refers to the pentose deoxyribose; ATP refers to adenosine triphosphate; dAMP is deoxyadenosine monophosphate; dGMP, deoxyguanosine monophosphate; dCMP, deoxycytidine monophosphate; dTMP, deoxythymidine monophosphate; dATP, deoxyadenosine triphosphate; dGTP, deoxyguanosine triphosphate; dCTP, deoxycytidine triphosphate; dTTP,

Ribonucleoside 5′-monophosphates	2′-Deoxyribonucleoside 5′-monophosphates
General structure	General structure
Names	Names
Adenosine 5′-phosphoric acid (adenylic acid; adenosine monophosphate; AMP)	Deoxyadenosine 5′-phosphoric acid (deoxyadenylic acid; deoxyadenosine monophosphate; dAMP)
Guanosine 5′-phosphoric acid (guanylic acid; guanosine monophosphate; GMP)	Deoxyguanosine 5′-phosphoric acid (deoxyguanylic acid; deoxyguanosine monophosphate; dGMP)
Cytidine 5′-phosphoric acid (cytidylic acid; cytidine monophosphate; CMP)	Deoxycytidine 5′-phosphoric acid (deoxycytidylic acid; deoxycytosine monophosphate; dCMP)
Uridine 5′-phosphoric acid (uridylic acid; uridine monophosphate; UMP)	Deoxythymidine 5′-phosphoric acid (deoxythymidylic acid; deoxythimidine monophosphate; dTMP)

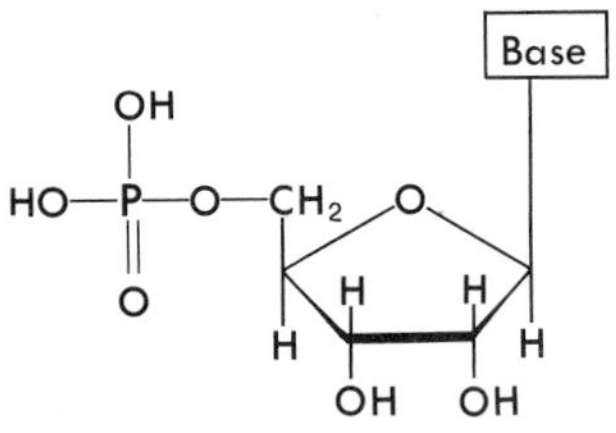

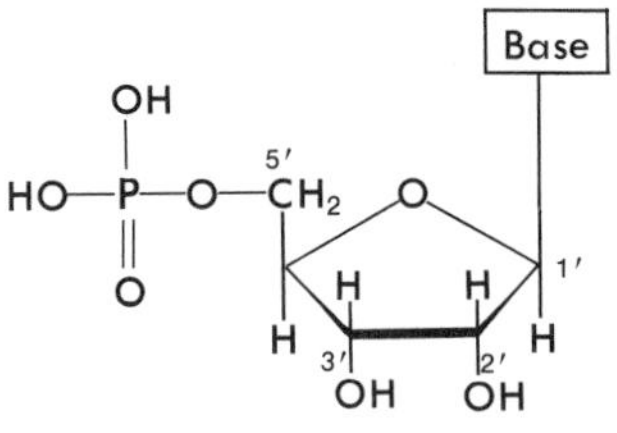

Figure 15.3. *The major ribonucleotides and deoxyribonucleotides. (From Lehninger, 1970: Biochemistry. 2nd Ed. Worth Publishers, New York, p. 310.)*

deoxythymidine triphosphate; and PP, pyrophosphate. An enzyme *(phosphorylase)* must be present to catalyze the reaction in each case. The nature of the reaction of the ATP with each nucleoside phosphate is shown in Figure 15.4*A*.

The next step in the synthesis of DNA is union of the activated nucleoside triphosphates by 3′ to 5′ phosphodiester bonds from one pentose to the other. They are added one at a time, using a DNA molecule supplied to serve as a primer or *template* in the synthesis. If one of the triphosphates is omitted, synthesis is reduced by about 100-fold. Addition of a nucleoside triphosphate is accompanied by splitting off a pyrophosphate (PP). These reactions are shown in Figure 15.4*B*. The pyrophosphate is, in turn, decomposed by the always-present enzyme pyrophosphatase, with the liberation of two molecules of inorganic phosphate. The high phosphate transfer potential (about 7300 cal/mole) of the nucleoside triphosphates supplies the energy required for the synthesis. This occurs only in the presence of an enzyme, *DNA polymerase.* It also proceeds only if some DNA is supplied as a primer. The overall reaction is shown in Figure 15.4*C*.

DNA of high molecular weight serves best as primer. Denaturing the DNA by heat increases its efficiency, perhaps because breaking the hydrogen bonds holding the two antiparallel chains of nucleotides characterizing the DNA molecule makes the template information more accessible. During replication each strand of the double helix of DNA acting as a template is presumably acted upon by the polymerase. The DNA enzymatically synthesized *in vitro* has many physical and chemical properties indistinguishable from the natural primer; for example, it shows hyperchromicity on heating and melting (denaturation). It does renature more rapidly than the primer after heating, but it requires the same temperature for 50 percent melting (Tm). The sedimentation coefficient and intrinsic viscosity of the product also approach that of the primer (Ingram, 1972).

IN PROKARYOTES

In *E. coli,* three DNA polymerases have been isolated: one (I) primarily concerned with repair of DNA, another (II) of uncertain function, and a third (III) for normal replication. DNA polymerase I, by its endonuclease activity, removes

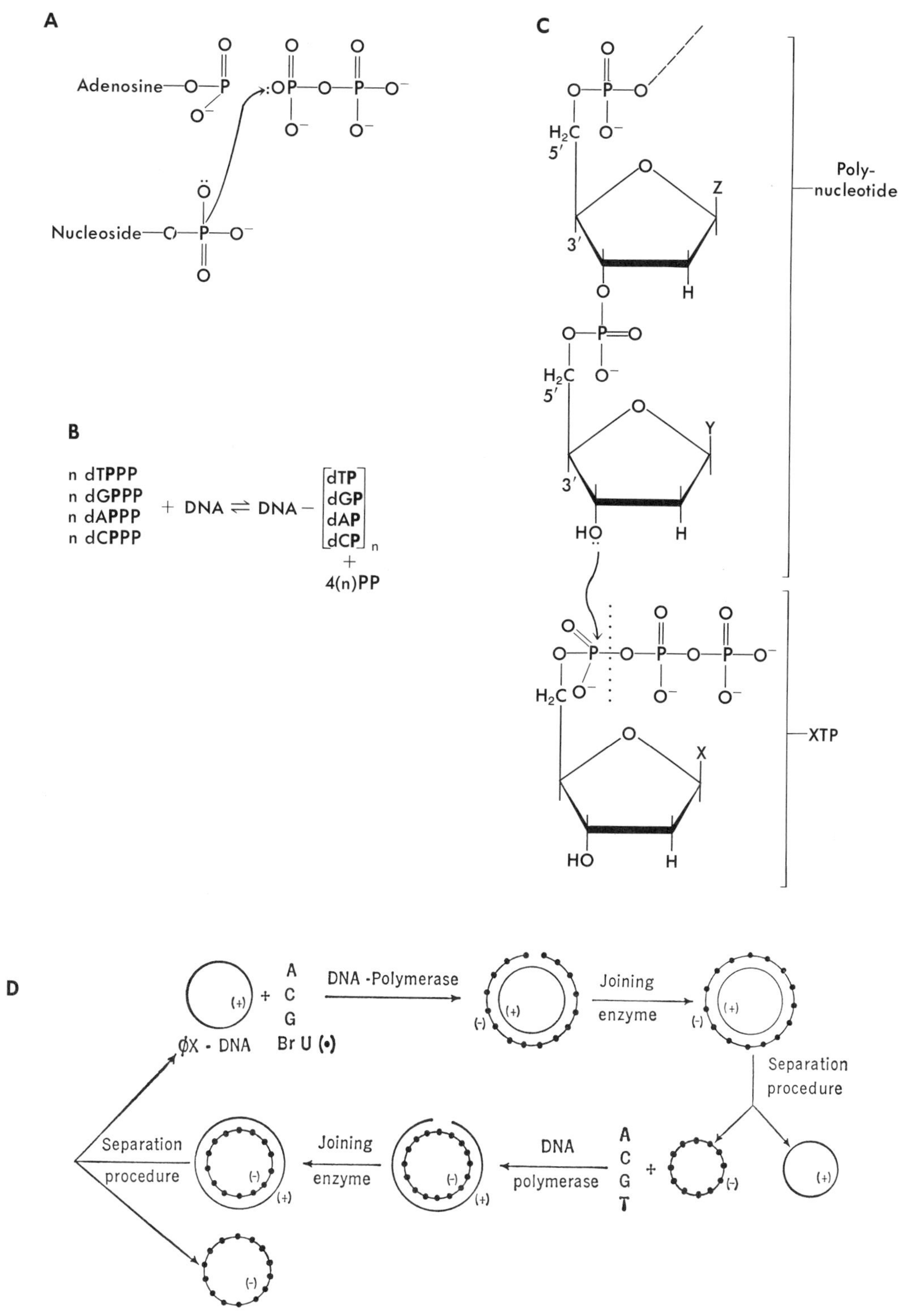

Figure 15.4. *Biosynthesis of DNA.* A, *Reaction between adenosine triphosphate and a nucleoside phosphate, yielding a nucleoside triphosphate and adenosine monophosphate (AMP). The nucleoside triphosphate reacts as in C and a pyrophosphate is released. This is then decomposed by a pyrophosphatase to inorganic phosphate.* B, *Overall equation for polymerization of the nucleoside triphosphates formed as a result of the reaction shown at* A. C, *Mechanism of attachment of a nucleoside to the polynucleotide chain. X, Y, and Z represent purine or pyrimidine bases. (*A, B *and* C *from Kornberg, 1960: Science* 131: *1505.)* D, *Synthesis of infective viral DNA. Bromouracil (BrU) acts as an analog of thymine, allowing identification of strands into which it is incorporated. (After Goulian et al., 1967.)*

all types of damage in DNA that it recognizes (see section on repair in this chapter). Polymerase III is less stable than the others and was more difficult to isolate, but it is 15 times more active than polymerase I in replicative activity. The active form of polymerase III is a DNA dimer complexed with a copolymerase protein (Lehninger, 1975).

Before the polymerase can act in replication the DNA must be opened up by an unwinding protein that attaches to the double-stranded DNA already nicked on one strand by the action of a "swivelase," or untwisting protein.

DNA replication is preceded by the formation of a short strand of RNA complementary to a section of double-stranded DNA by a DNA-directed RNA polymerase. Only then can the DNA polymerase begin to add deoxyribonucleotide units to the 3′ end of the short RNA priming strand from a mixture of the 5′ phosphorylated deoxyribonucleotides. When a considerable length of DNA has been formed, the RNA primer is removed by the enzyme, nuclease. Priming is necessary for each of the Okazaki fragments (pieces of DNA) (see Chapter 7) in which replication occurs. Any nucleotides necessary for completion of a DNA strand between the replicated fragments are filled in with nucleotides by DNA polymerase I and the end of the pieces are then linked together by a ligase.

The function of the DNA primer in the synthesis of DNA is to supply a template. The exactness of the duplication is perhaps best illustrated by the nature of the DNA produced when different template models are supplied. Each DNA, it will be recalled, always has an equal number of adenine and thymine residues (A = T) and an equal number of guanine and cytosine (G = C) residues, but the ratio of A + T to G + C is a characteristic property of each type of DNA. In Table 15.2 are listed the primers used and the base analysis of the primer and product. The agreement is remarkably close. The DNA polymerase, in some unknown and perhaps unique way, takes directions from the template and adds the nucleotides in the order and in the numbers appropriate to the primer supplied. It is always the double-stranded DNA that is produced by the enzyme, even when a single-stranded DNA, such as that from the tiny ΦX174 virus (Sinsheimer, 1959), is provided as template. When DNA, which is normally double-stranded but which had uncoiled and become single-stranded in heating, was used as a template, the formation of only the normal double-stranded DNA was obtained. Evidently enough information is available in a single strand of DNA to enable the enzyme to synthesize the appropriate double-stranded DNA. With highly purified enzymes the single-stranded DNA (obtained by heating the double-stranded one) alone is effective (Kornberg, 1974).

Kornberg used an ingenious method consisting of labeling the nucleosides with ^{32}P, incorporating them at one bond of the pentose moiety, and then selectively splitting them at another bond, by which it proved possible to determine the nucleoside sequence (nearest neighbor analysis) in a DNA molecule. Then, by using as a primer a sample of synthesized DNA in which the sequence of nucleotide bases was known and again analyzing the product for nucleotide sequence, it was demonstrated that the sequence of nucleotides in the primer was duplicated in the newly synthesized DNA. Furthermore, it was possible to distinguish between two possible general models of DNA, one in which the strands were parallel (with the 5′

TABLE 15.2. CHEMICAL COMPOSITION OF DNA ENZYMATICALLY SYNTHESIZED WITH DIFFERENT PRIMERS*

	A	*T*	*G*	*C*	$\frac{A+G}{T+C}$	$\frac{A+T}{G+C}$
Mycobacterium phlei						
Primer	0.65	0.66	1.35	1.34	1.01	0.49
Product	0.66	0.65	1.34	1.37	0.99	0.48
Escherichia coli						
Primer	1.00	0.97	0.98	1.05	0.98	0.97
Product	1.04	1.00	0.97	0.98	1.01	1.02
Calf thymus						
Primer	1.14	1.05	0.90	0.85	1.05	1.25
Product	1.12	1.08	0.85	0.85	1.02	1.29
Bacteriophage T-2						
Primer	1.31	1.32	0.67	0.70	0.98	1.92
Product	1.33	1.29	0.69	0.70	1.02	1.90
A-T copolymer	1.99	1.93	<0.05	<0.05	1.03	40

A, adenine; *T*, thymine; *G*, guanine; *C*, cytosine.
*From Kornberg, 1960: Science *131*.

to 3′ OH's of the phosphoric acid backbone the same in both) with respect to their linkages and one in which they were antiparallel (with the 5′ to 3′ OH's of the phosphoric acid backbone opposite to one another). Apparently, the two strands in a DNA molecule are antiparallel (see review in Ingram, 1972).

When a copolymer made up only of two nucleotides—deoxyadenylate and deoxythymidilate—is supplied as a template to the polymerase and activated nucleotide triphosphates, only the AT copolymer is duplicated even though all four nucleoside triphosphates are present in the reaction mixture. This polymer (Table 15.2, last line) indicates the extent to which the enzyme copies what is presented as a template and also it shows that the hydrogen bonding between pyrimidine and purine must be of considerable importance in the structure of the polymer produced.

It has proved possible, with the highly purified DNA polymerase in an *in vitro* system containing the four nucleoside triphosphates, to synthesize ΦX174 virus DNA that is infective to *E. coli* (Goulian *et al.*, 1967). This is a major advance over the previous attempts in which the artificially synthesized DNA (e.g., the transforming principle) was biologically inactive even though biochemically similar to the template provided. It will be remembered that the ΦX174 virus is a single-stranded circular DNA that is infective only in the circular form. When the DNA strand (arbitrarily designated the + strand) enters the bacterium, it induces by complementation the formation of the − strand. In the *in vitro* experiment it is necessary to include a polynucleotide-joining enzyme (ligase) in addition to the template DNA, the DNA polymerase, and the four nucleoside triphosphates. The polynucleotide-joining enzyme unites the ends of the DNA molecule. By supplying the *in vitro* system with a deoxythymidylic acid analogue (5-bromodeoxyuridylic acid) the DNA made in complementation to the +DNA template becomes heavier than the DNA template, thus making it possible to separate the two by density gradient centrifugation. Possible contamination of one DNA strand with the other was checked by labeling the +DNA with ^{32}P and the components of the −DNA with ^{3}H. The synthetic DNA was obtained free of template DNA and, in circular form, was infective to *E. coli* protoplasts and induced the formation of +DNA. The +DNA was then separated from the −DNA and also was found to be infective (Fig. 15.4*D*). These experiments have many important implications.

It was noted earlier that the ΦX174 virus consists of a single + strand of DNA that induces in the host bacterium a complementary − strand of DNA. The − strand of DNA in turn induces the synthesis of the infective + strands. In *in vitro* experiments with the ΦX174 virus, when the polynucleotide ligase (which has the capacity to join the free 3′ and 5′ ends of nucleotides) was added, the intact circular double-stranded DNA molecules formed had no template activity with the DNA polymerase alone. When, however, a nuclease which nicks the DNA (opens the circle) was subsequently added, another period of synthesis ensued, this time consisting of appearance of intact + strands of DNA. Thus, either strand of DNA can serve as a template in the *in vitro* system. In contrast, however, the assembly of infective particles produced in a cell consists of + strands only, although initially about 20 paired strands appear. The − strand serves for transcription of an RNA complementary to the original DNA; this RNA serves for translation into viral proteins. Then only + strands are replicated until about 200 appear. These are encapsulated by the proteins to complete the viral particles (Fiddes, 1977). The way in which a choice is made between the + and − strand for replication in the formation of the infectious virus particles is not known. However, it is known that the viral strand of DNA remains open, while the complementary strand is closed by the ligase and coated with protein ("packaged"). As a result, it cannot serve as a template for replication of DNA. Upon infecting a cell, therefore, the released virus DNA cannot multiply unless an enzyme nicks it. This involves selecting a site for the nicking.

In the bacterial cell (e.g., *E. coli*), the circular DNA molecule cannot serve as a template until it is nicked by an endonuclease, which presumably occurs at a fixed point (near the arginine H gene). At this point, replication then begins, and growth proceeds in both directions around the circle. About 20 endonucleases that nick the DNA at specific and different points have been isolated and were used to identify the nucleotide sequences in the DNA of ΦX174 (Fiddes, 1977).

While the mechanism of replication of DNA is still unresolved, we now have evidence that initially intermediates much smaller than the DNA molecule, called *Okazaki fragments,* are formed. The size of the intermediates detected depends upon the time at which they are removed from the cell and tested after replication has begun. In 2 seconds the pieces of DNA are single-stranded and very small, with free ends. After a few more seconds the pieces of DNA are double-stranded and contain about 1000 to 2000 nucleotides. For this work a ligase-deficient mutant of *E. coli* was used to prolong the life of the small pieces. Another problem in the experiments is the rapid degradation of the short pieces by an exonuclease. Double-stranded DNA is free from attack (Ingram, 1972).

It is interesting to note that a DNA

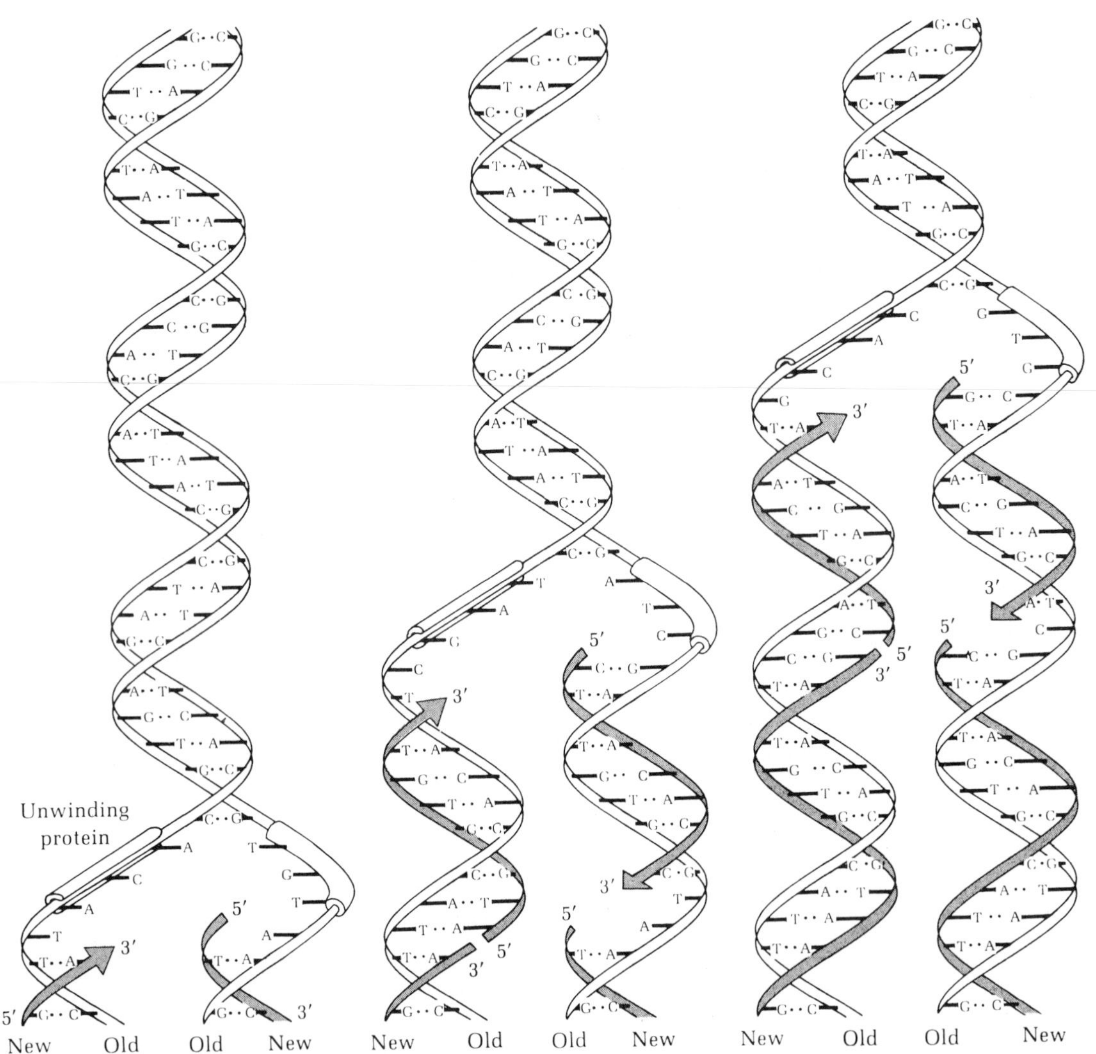

Figure 15.5. *Highly schematic representation of the replication of DNA. As strands of the parent molecule unwind—most likely with the aid of unwinding proteins, which are more numerous and cover more bases than shown here—complementary nucleoside-5′-triphosphates are bound to the exposed bases. The triphosphates react with 3′-hydroxyl groups of the preceding nucleotide in the growing strands with the formation of a new 3′,5′-phosphodiester linkage and the loss of inorganic pyrophosphate. Thus, the new chains are formed in the 5′ → 3′ direction along both parental strands. As the new chains grow, sections of new double helices (each with many more bases than shown here) are formed on each of the parent strands. The daughter strands in each new double helix are stiched together at their ends with the enzyme DNA ligase. (From Volk, 1978: Essentials of Medical Microbiology. J. B. Lippincott Co., Philadelphia.)*

polymerase-deficient mutant of *E. coli* (containing only 5 to 10 molecules of polymerase per cell instead of perhaps 100) seemingly replicates normally. Evidently this small number of molecules serves for normal polymerization. However, such mutants are very sensitive to ultraviolet radiation; evidently for repair of the numerous damaged loci on the chromosome other polymerases are required to avoid the abnormal replication that otherwise occurs (Ganesan *et al.*, 1973).

The mechanism of replication has been the subject of a number of hypotheses, none of which is accepted by all workers in the field. One is shown in Figure 15.5. All models of replication must be regarded as tentative (Gefter, 1975).

IN EUKARYOTES

In the cells that are eukaryotic the mechanism of DNA replication is now thought to be very similar to that in prokaryotic cells and viruses. For replication the long eukaryotic chromosome is thought to be subdivided into many shorter pieces known as replication units or *replicons*. On each replicon, replication begins at some point (the *origin*) and proceeds in both directions from a fork-like region to a *terminus*. A bubble forms with the two daughter strands connected at the two replication forks (see Fig. 7.21). The fact that cloning of animal genes in bacteria is feasible reemphasizes the basic similarity of the mechanism of replication in prokaryotes and eukaryotes.

In eukaryotic cells the movement of the replication fork is of the order of 0.2 to 2 μm/min and the replicon size is of the order of 10 to 250 μm. Bidirectional replication has been observed by autoradiography in mammalian, chicken, and amphibian cells. Replication begins at different times in different replicons, some preceding others. As a result, replication of an entire chromosome takes several hours. Three DNA polymerases have been found in mammalian cells in addition to one in the mitochondria. In plant cells one might expect to find another DNA polymerase in the chloroplast as well. The details of replication in eukaryotic cells are discussed in a review (Edenberg and Huberman, 1975).

BIOSYNTHESIS OF RNA: TRANSCRIPTION

Three types of RNA are found in cells: *ribosomal RNA* (rRNA), the major RNA component, comprises 75 to 80 percent of the total; *messenger RNA* (mRNA) makes up 5 to 10 percent of the total; and *transfer RNA* (tRNA) makes up the remainder (Table 15.3). It will be recalled (see Chapter 7) that upon heating DNA and RNA to disrupt helices, then gradually cooling the mixture of DNA and RNA, all three RNA's conjugate with DNA of the same species, indicating complementation and sharing of information between DNA and RNA. Messenger RNA, when so conjugated, occupies a larger portion of the DNA molecule than rRNA and tRNA. Therefore, it is no surprise that the synthesis of all three types of RNA is DNA-dependent. The interrelationships of the three types of RNA are indicated in Figure 15.6 and the overall reactions in transcription (synthesis of RNA) in Figure 15.7.

TABLE 15.3. TYPES OF CELLULAR RNA (*E. coli*)*

	Amount, % of Total RNA	***Molecular Weight***
Ribosomal RNA (rRNA)	75–80	1.2×10^6, 0.55×10^6, and 30,000
Messenger RNA (mRNA)	5–10	Up to 2×10^6 (heterogeneous)
Transfer RNA (tRNA)	10–15	25,000

*From Ingram, 1972: Biosynthesis of Macromolecules. 2nd Ed. W. A. Benjamin, Menlo Park, Calif.

About 60 to 65 percent of a ribosome consists of rRNA, while 35 to 40 percent consists of protein. The prokaryotic ribosome consists of two subunits, 50S and 30S. The 50S unit consists of 5S and 23S rRNA and 30 polypeptide chains. The 30S unit consists of a 16S rRNA and 20 polypeptide chains (Fig. 15.8). Ribosomal rRNA possesses a degree of helical structure (Ingram, 1972). The ribosomes of mitochondria and chloroplasts resemble those of prokaryote cells.

The cytoplasmic ribosomes of eukaryotes consist of two particles: 60S and 40S. The 60S particle consists of 28S, 7S, and 5S rRNA and more than 50 polypeptides. The 40S particle consists of 18S rRNA and more than 30 polypeptides. The ribosomes of plants are not exactly like those of animals, and those of ani-

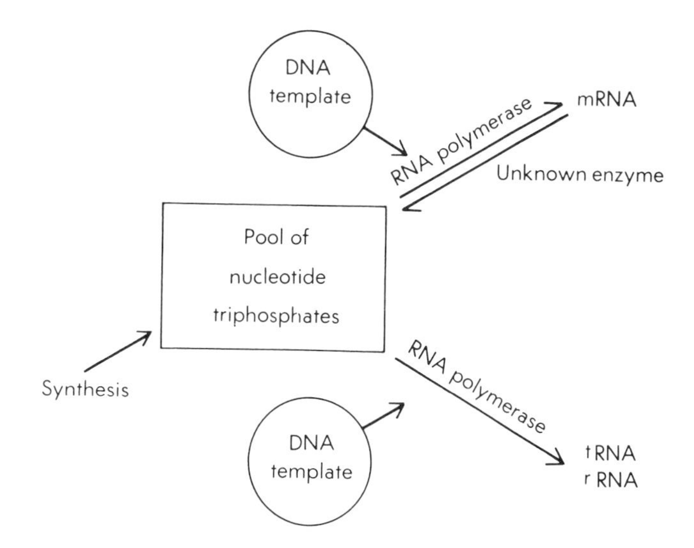

Figure 15.6. Model for the flow of RNA precursors into various forms of RNA in the bacterial cell. (From Ingram, 1972: Biosynthesis of Macromolecules. 2nd Ed. W. A. Benjamin, Menlo Park, Calif.)

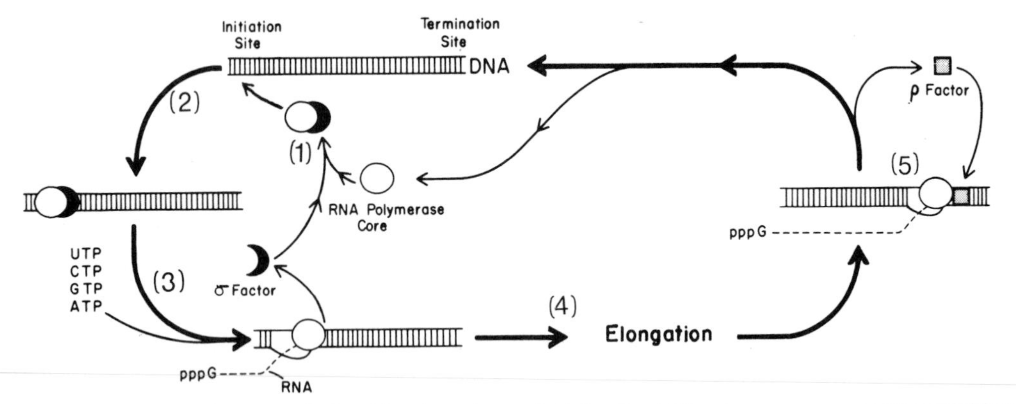

Figure 15.7. Scheme for transcription. The sigma factor (σ) joins the RNA polymerase core at (1) and initiates RNA transcription on DNA (2); in the presence of the high-energy phosphate phosphates UTP, CTP, GTP, and ATP synthesis occurs (3) with elongation of the transcribed message (4) until the rho factor (ρ) terminates the transcription and the DNA and RNA polymerase can return to begin another cycle of transcription.

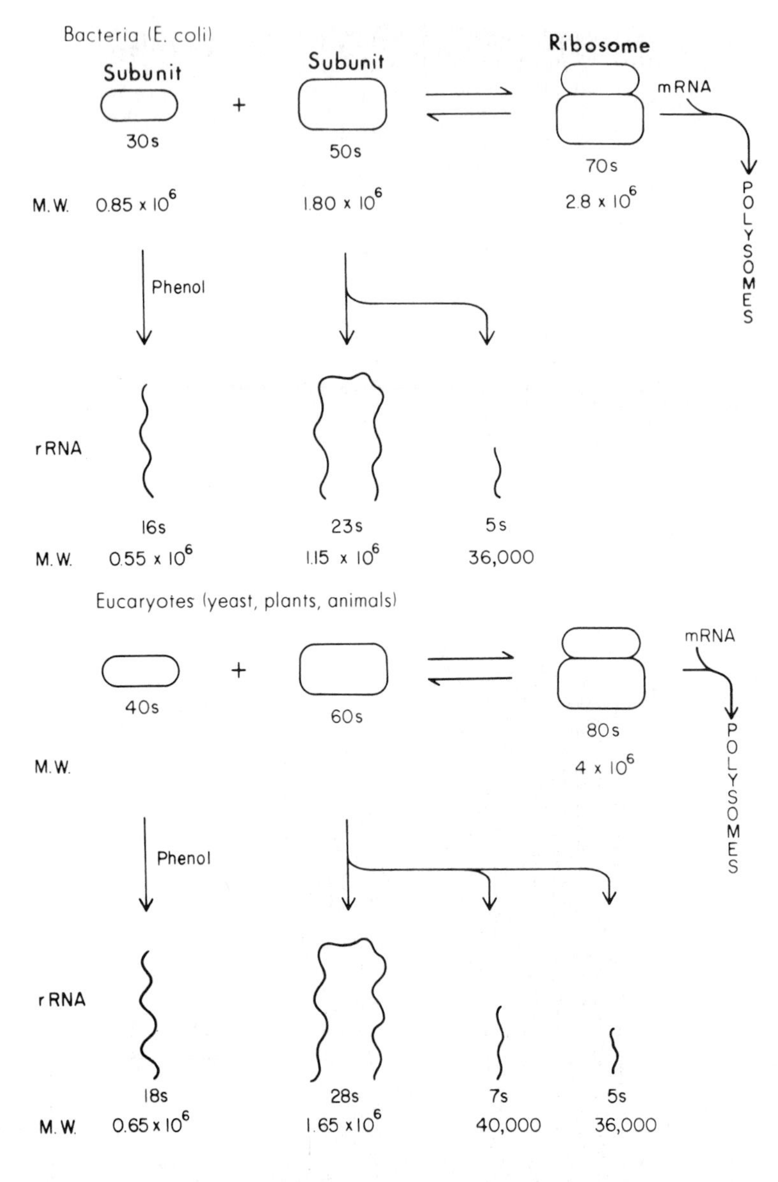

Figure 15.8. The in vitro dissociation and association reactions of ribosomes. Note also the types of rRNA that can be extracted with phenol. Protein synthesis occurs only on ribosomes and even though the ribosomes of prokaryotes and eukaryotes differ, synthesis of proteins occurs in almost identical manner in both. (From Ingram, 1972: Biosynthesis of Macromolecules. 2nd Ed. W. A. Benjamin, Menlo Park, Calif.)

mals vary somewhat with species and phylogenetic position (Lehninger, 1975).

Messenger RNA (mRNA) is heterogeneous, with molecular weights up to two million and with a correspondingly wide sucrose zone centrifugation band, the size variation being a measure of the amount of message represented on the messenger DNA. The mRNA for rabbit hemoglobin peptide chains has been isolated, and in a cell-free system containing ribosomes, tRNA, and enzymes derived from *E. coli* it was able to direct the synthesis of complete peptide chains (Ingram, 1972). This clearly indicates that mRNA codes for a protein.

In contrast to mRNA, transfer RNA (tRNA) from a wide variety of cells is similar in size, with a molecular weight of about 25,000 and a sedimentation constant of 4S. It consists of about 75 nucleotides. All tRNA molecules appear to have at the 3′ end of the nucleotide sequence the termination pCpCpA and at the 5′ end generally the nucleotide guanylic acid, to which an extra 5′ phosphate is attached.

It is known that tRNA acts as an *acceptor* for an activated amino acid and as an *adaptor* for transporting the amino acid to the site of protein synthesis on the mRNA template. Because of the ease of its extraction and its small size (73 to 93 nucleotide residues), considerable work has been done on the configuration of the tRNA molecule, which is known to be of cloverleaf pattern (see Figure 1.20). In the case of one with a longer chain, a fifth arm may appear (Lehninger, 1975). X-ray scatter data indicate that all tRNAs are similar in general. The three-dimensional structure of one tRNA has been determined (Rich and Kim, 1978). Because tRNA has considerable helical structure (up to 60 or 70 percent in some cases), a melting point can be determined for it, although with less precision than with DNA (Ingram, 1972).

The enzyme RNA polymerase exists in three forms. It is responsible for incorporating the four ribonucleoside triphosphates, ATP, GTP, CTP, and UTP, which contain the bases adenine, guanine, cytosine, and uracil, respectively, in the highly energetic triphosphate form. DNA is necessary for RNA synthesis and the synthesized RNA is complementary to the DNA (template) used for its induction, but no primer is required as is the case for DNA replication (Lehninger, 1975). Thus when the DNA is obtained from *E. coli,* the ratio of adenine plus uracil to guanine plus cytosine $\left(\text{i.e., } \frac{A + U}{G + C}\right)$ is 1.0 in the RNA synthesized *in vitro,* which is characteristic of corresponding *E. coli* RNA *in vivo.* When the DNA used in cell-free synthesis is obtained from calf thymus, the $\left(\frac{A + U}{G + C}\right)$ ratio in the RNA synthesized is found to be 1.3, which is characteristic of calf thymus RNA corresponding to its DNA.

Additional evidence of the specificity of DNA as an inducer of RNA synthesis was obtained by using a synthetic DNA-like polydeoxyribonucleotide, e.g., one containing adenine and thymine deoxyribonucleotides (A-T copolymer). In this case the RNA-like polyribonucleotide, synthesized under DNA influence, consisted of polymers of uracil and adenine polyribonucleotides in alternating sequence. When polythymine deoxyribonucleotide was used as the template, only adenine polyribonucleotide was synthesized.

The *DNA-dependent RNA polymerase* appears to be distributed in bacteria and in the nuclei, mitochondria, chloroplasts, and cytoplasm of eukaryotic cells. Either magnesium or manganese ion is required for functioning of this polymerase. Since magnesium is found in higher concentration in the cells, it is probably the ion that generally performs this catalytic function. The DNA-dependent RNA polymerase also requires an active sulfhydryl group for functioning and is therefore readily poisoned by substances such as chloracetophenone that react with this group. During its purification, a sulfhydryl reagent is generally added to insure activity of this group in the enzyme.

For the activity of the RNA polymerase all four of the nucleoside triphosphates (ATP, CTP, GTP, and UTP) are required, absence of any one of these reducing the activity to almost zero. Transcription is initiated with a triphosphate, which remains as part of the chain throughout its elongation (Ingram, 1972). The RNA product of the polymerase is subject to phosphorolysis, catalyzed by the same enzyme in the presence of pyrophosphate (a product of the polymerization), but the reaction is slow.

The DNA-dependent RNA polymerase therefore shows several properties resembling those of the DNA polymerase, and the polymerization is somewhat similar. In both, four nucleoside triphosphates are required; in both, the base analogues are incorporated in a similar manner; in both, the base compositions of the DNA template and the product are complementary. Both are inhibited by actinomycin D and proflavin, although the resemblance in effects is qualitative, not quantitative; in both, pyrophosphate exchange reactions occur dependent on the presence of DNA but not on the presence of all four nucleoside triphosphates; for both, magnesium ion is required.

As polymerization proceeds in the presence of all four ribonucleoside triphosphates, pyrophosphate is formed. The pyrophosphate is split into inorganic phosphate by the enzyme pyrophosphatase, which is always present in cells.

The DNA template used in the RNA syntheses is not used up in the reaction and separates from the RNA synthesized. For example, a sample of transforming DNA has the same transforming activity as it had before it was used in synthesizing six to ten times its weight of RNA (Ingram, 1972).

It appears that both strands of the DNA serve as templates during polymerization *in vitro,* but only one strand is used *in vivo,* as indicated by the base ratios of the RNA produced, which correspond to only one of the DNA strands. However, the DNA from the single-stranded DNA of phage ΦX174 serves as a primer for RNA polymerase *in vitro,* although it is less active than double-stranded DNA. *In vivo* one of the DNA strands is intact, while the other is nicked by an enzyme and is the one transcribed. The RNA strands produced by polymerization have been shown to be antiparallel to DNA by the same procedures used to demonstrate the antiparallel nature of the two strands in a DNA molecule. The new RNA chain separates from the DNA as it forms at the transcription fork.

Transcription is regulated by the *sigma factor* (σ), a protein attached to the RNA polymerase core. The two proteins act together but are separable by appropriate analytical procedures (e.g., on phosphocellulose columns). When the sigma factor is removed from the enzyme-sigma unit, the RNA polymerizing activity of the enzyme is almost completely lost in both *E. coli* and vertebrate liver cell extracts. The polymerase is itself composed of a β, a β', and an α_2 polypeptide chain, the active unit having the composition $\alpha_2\beta\beta'\sigma$. It is thought that the function of the sigma factor is to initiate transcription, after which it is released from the complex with the polymerase and from the DNA (Travers and Burgess, 1969) (see Fig. 15.7). Actually, not one but several sigma factors are present, some highly specific as to the DNA template to which they will become attached with the polymerase. In *Bacillus subtilis* during sporulation a special sigma factor displaces the factor found during the vegetative portion of the life cycle (Ingram, 1972).

A "terminal adding" enzyme completes the synthesis of tRNA begun by the polymerase by adding to its amino acid–acceptor end the characteristic three nucleotides pCpCpA (p refers to phosphate). The precursors of these nucleotides are the corresponding nucleoside triphosphates.

Termination of transcription is apparently carried out by another protein, the *rho factor* (ρ), with a molecular weight of about 60,000. (Possibly another protein (κ) is also involved.) While its function is uncertain, the rho factor may recognize a particular base sequence in the DNA being copied. The RNA transcribed in the manner just described is not in its final chemical state and may be modified in one of several ways: in chain length (prominent in ribosomal RNA) or in alteration of some of the bases by methylation, alkylation, reduction, and isomerization.

The synthesis of all three types of RNA (DNA-dependent) is inhibited by the antibiotic actinomycin D, which probably binds the guanine-containing regions of the RNA polymerase system. In partial inhibition, there is preferential inhibition of the G- and C-copying sequences in DNA.

RNA is also synthesized from nucleoside diphosphates in the presence of *polynucleotide phosphorylase,* an enzyme discovered in extracts of the microbe *Azotobacter vinelandii.* The enzyme has since been extracted and purified from a variety of microorganisms, plant cells, and pig liver cell nuclei. The enzyme catalyzes the synthesis of high-molecular-weight polyribonucleotides with the release of inorganic phosphate:

$$nBRPP \rightarrow (BRP)_n + nP_i \qquad (15.5)$$

where B stands for base (adenine, hypoxanthine, uracil, or cytosine), R for ribose, and P_i for inorganic phosphate.

Single polymers containing only one of the nucleotides—AMP (adenosine monophosphate), UMP (uridine monophosphate), CMP (cytidine monophosphate), or IMP (inosine monophosphate)—have been obtained by incubating the enzyme with the corresponding nucleoside diphosphate in each case, and even nucleosides not normally present in RNA can be polymerized by polynucleotide phosphorylase. Mixed polymers of adenine and uridine ribonucleoside phosphates or of all four nucleoside phosphates of RNA can be obtained if two or four of the nucleoside diphosphates, respectively, are supplied.

The polymers obtained using the DNA-independent enzyme have molecular weights of the order of 30,000 to 2,000,000; therefore, they are in the range of molecular weights for RNA, as obtained from different sources. They display many RNA-like properties. For example, the polynucleotides even react with tobacco mosaic virus protein to form noninfective virus-like particles. It is interesting that no template DNA molecule is required for the polymerization of RNA in the presence of polynucleotide phosphorylase. However, without a DNA template no specific information-containing nucleotide sequence is found in the synthesized RNA, the sequence appearing to be a matter of chance. Such synthetic polynucleotides formed from various mixtures of nucleoside diphosphates have played an important part in deciphering the genetic code for protein synthesis but probably have no specific function in the cell.

The function of polynucleotide phosphorylase* in the cell is not known but it may function to decompose mRNA, which in bacterial cells generally has a short life span. Neither rRNA nor tRNA is decomposed at an appreciable rate by the enzyme (Ingram, 1972).

RNA-DIRECTED DNA SYNTHESIS: REVERSE TRANSCRIPTION

The central dogma of molecular biology states that within a cell information is transferred from DNA to RNA, and from RNA to protein, but not in reverse order—that is, information does not flow from protein to RNA and from RNA to DNA (Crick, 1970). However, it was shown that infection of host cells (chick or rat) with Roux's sarcoma RNA virus transforms them into cancer cells that may (in chicken) or may not (in rat) produce new virus particles. The virus has been shown to induce a new DNA in the host cell complementary to its own RNA genome. The induced DNA, in turn, acts as a template for transcription of mRNA, and the RNA is then translated into new proteins in the host cell, now transformed into a cancerous cell (Hooks *et al.*, 1972). Roux's sarcoma virus thus induces in the host cell an RNA-directed DNA synthesis. The DNA so induced is only about one tenth the size of the viral RNA genome but by hybridization tests is found to be complementary to the virus RNA. It is double-stranded. The DNA-polymerase active in DNA synthesis (reverse transcriptase) is apparently present in the virus and in purified form is capable of directing DNA synthesis on a variety of templates, including natural and synthetic DNA, RNA and DNA-RNA hybrids. A ligase required for completion of DNA synthesis is also present in the virus (Temin, 1972).

Lest the presence of RNA-directed DNA synthesis appear to be a property of only the genome of carcinogenic RNA viruses, it should be pointed out that a RNA-directed DNA polymerase has been found in uninfected rat embryo cells and in cases of gene amplification (Ficq and Brachet, 1971). It is therefore possible that some flow of information from RNA to DNA occurs generally in cells. Pending further discoveries, the central dogma of molecular biology may be restated tentatively in more general terms to be that information in cells flows from nucleic acids to proteins but not from protein to nucleic acids. At least no case of information flow from protein to nucleic acid has yet been disclosed.

CODING FOR mRNA, rRNA AND tRNA ON DNA

Since essentially all the information for synthesis in the cell resides in its DNA, with the exception noted earlier of the infrequent possibilities that some may be in RNA, some portions of the chromosome must code for the various types of cellular RNA. Anticodes to the DNA must therefore exist for each type of RNA present in the cell. (In tRNA the anticodon arm is at one end of the cloverleaf and the codon (3-letter code "word") at the other (see Fig. 1.20). It has been possible to show this by hybridization between DNA and various types of cellular RNA. Hybridization is the pairing of strands of RNA and DNA (see Chapter 7). For hybridization to occur, the hydrogen bonds between the two strands of DNA and within the RNA strand must first be broken by heat. Then the mixture is slowly cooled. In locations in which RNA and DNA are complementary they will hybridize upon cooling (Spiegelman, 1964).

Hybridization reveals that approximately 0.32 percent of the chromosome of *E. coli* codes for the ribosomal RNA. An even smaller portion, approximately 0.025 percent, codes for tRNA. However, there are at least 20 tRNA molecules encoded. The amount of DNA coding for one mRNA is also a relatively small part of the entire chromosome. However, there are at least a thousand kinds of mRNA coding for the thousand (or more) types of proteins in the cell.

It is of interest to determine whether both strands of DNA, or only one, serve for coding the various types of RNA. Hybridization techniques with the single-stranded DNA of ΦX174 bacteriophage have provided information (Sinsheimer, 1962). A virus inhibits syntheses normally occurring in the host cell but initiates and directs syntheses in the cell to produce more of its own DNA and the proteins required for its replication and packaging (see Chapter 1). The virus therefore induces mRNA formation in the bacterium. However, the mRNA isolated from the ΦX174 virus-infected *E. coli* does not hybridize with the DNA isolated from the mature ΦX174 phages. When the mRNA induced in the host cell was mixed with the active virus before packaging had occurred in the bacteria, hybridization was obtained. It has been shown that the replicating form of the virus in the bacterial cell is double-stranded, the second strand being formed by complementation from the single strand initially intro-

*It will be recalled (Chapter 10) that enzymes catalyze the reaction in both directions to equilibrium *(in vitro)*. Therefore, the polynucleotide phosphorylase in presence of RNA alone will decompose it until an equilibrium between the RNA and the products is reached.

duced by a mature phage particle. Therefore, hybridization appears to involve only the replicated second strand. From this and other experiments it can be inferred that only one of the two strands of DNA carries the codes for the various types of RNA.

THE GENETIC CODE

Since usually only four purine and pyrimidine bases are present in RNA and DNA, it would, at first glance, seem difficult to make a code from just four units sufficient to locate 20 amino acids on a ribosome, as is done in the synthesis of proteins. However, it was found possible to develop codes that are specific to each one of the 20 amino acids by using only four nucleotides. Thus, designating the nucleotides 1, 2, 3, and 4 for purposes of argument, one can make 64 unique sequences of three numbers, each designating one and only one amino acid. (Several sequences might designate the same amino acid, but any one sequence would never designate several amino acids.) If transfer RNA molecules with one of these three-unit sequences were to pick up one amino acid each and unite with complementary sequences on messenger RNA, it would be possible to line up the amino acids to build any protein, even with this relatively simple code.

Experimental support for DNA coding of *in vitro* protein synthesis was first provided by the brilliant discovery that protein-like molecules are produced in response to artificial RNA-like polyribonucleotides of known composition (for example, a polyuridine made up only of uridine [U]). Thus, when an RNA-like polyribonucleotide composed of three repeating units of uridine (the U-U-U sequence of the polyuridine being complementary to the adenine [A] sequence, A-A-A, of the DNA) is supplied, the synthesized protein-like polypeptide molecule consists of repeating phenylalanine residues (polyphenylalanine) (Nirenberg and Matthei, 1961). When other synthetic RNA-like compounds are used (e.g., polyribonucleotides made up of uridine [U], cytosine [C], guanine [G], and adenine [A], in each case consisting of repeating units of three—for example, UCAUCAUCAUCA, and so forth), the protein-like polypeptides synthesized are characteristic for each type of RNA-like polyribonucleotide supplied. In the example the polypeptide would be characteristic of repeating sequences of UCA, CAU, or AUC, depending upon which letter was first at the start of synthesis. Experimental data suggesting the RNA code (and so also the complementary DNA code) for all of the amino acids have now been gathered (Watson, 1976).

When it proved possible to synthesize regular polymers (e.g., CUCUCU), additional code words were discovered. For example, CUCUCU, etc., could only provide polypeptides made up of the amino acid leucine, if the reading frame was CUC, or the amino acid serine if the reading frame was UCU. Others were determined by the specific tRNA binding to ribosome-mRNA complexes. The RNA code words for amino acids determined in these various ways are summarized in Table 15.4.

It is evident from Table 15.4 that several code words exist for many individual amino acids; to this extent the code is said to be degenerate (to lack specificity). For example, six *codons* exist for arginine and leucine, four for alanine, threonine, proline, and glycine, and several for a number of other amino acids. Only for tryptophan and methionine do single codons exist (Lehninger, 1975). However, it is possible that lack of specificity may in itself give a certain amount of flexibility to synthesis of proteins in the cell. This probably derives from physical flexibility in the relation of the anticodon in the tRNA carrying an amino acid to the codon. In the tRNA the base at the 5′ end of the anticodon is not as spatially confined as the other two bases, enabling it to form hydrogen bonds with either of the two bases at the 3′ end of the codon. For example, U at the more flexible (or "wobble") position can pair with adenine or guanine, while inosine (an unusual base, derived from adenine by deamination) can pair with uridine, cytosine, or adenine. The *wobble hypothesis* (Crick, 1966a) permits predictions on the number of tRNAs for amino acids having multiple mRNA codons. Thus, instead of 61 anticodons, one for each of the 61 codons for the 20 amino acids (3 of the 64 possible sequences are nonsense codons for termination of protein synthesis), a much smaller number exists because some of the tRNAs recognize several codons each (Watson, 1976).

Although a given gene may code for one polypeptide, one mRNA molecule may have resulted from sequential transcription of several genes and thus may code for several polypeptide chains. In addition to coding for amino acids, code signals must be given to start and stop polypeptide chain growth. When such information is lacking—for example, in a UUU polynucleotide serving as mRNA mixed with ribosomes and phenylalanine—the polypeptide produced (polyphenylalanine) does not become detached from the polynucleotide. The code words for beginning (AUG, GUG) and stopping (UAA, UAG, UGA) polypeptide synthesis are similar to the code words for amino acids, e.g., triplets of bases (Watson, 1976).

When a codon for one amino acid is mutated to that for another amino acid, the mutation is called a *missense mutation*. For example, ab-

TABLE 15.4. THE GENETIC CODE*

1st ↓ 2nd →	U	C	A	G	↓ 3rd
U	Phe	Ser	Tyr	Cys	U
	Phe	Ser	Tyr	Cys	C
	Leu	Ser	Terminate	Terminate	A
	Leu	Ser	Terminate	Trp	G
C	Leu	Pro	His	Arg	U
	Leu	Pro	His	Arg	C
	Leu	Pro	Gln	Arg	A
	Leu	Pro	Gln	Arg	G
A	Ile	Thr	Asn	Ser	U
	Ile	Thr	Asn	Ser	C
	Ile	Thr	Lys	Arg	A
	Met†	Thr	Lys	Arg	G
G	Val	Ala	Asp	Gly	U
	Val	Ala	Asp	Gly	C
	Val	Ala	Glu	Gly	A
	Val†	Ala	Glu	Gly	G

The abbreviations for the amino acids are explained in Table 1.1.
*After Crick, 1966: J. Mol. Biol. *19*: 548.
†Initiation codons.

normal hemoglobins are produced by substitutions of a single wrong amino acid in one location (Ingram, 1963). When a mutation occurs in which one codon for a particular amino acid is changed in such a way that it no longer corresponds to any amino acid but rather to a termination sequence, the mutation is called a *nonsense mutation,* and it too results in abnormal protein synthesis. Such mutations induce codons for terminating protein synthesis prematurely. It is important to point out that in the case of *in vitro* protein synthesis, small changes in conditions (e.g., concentration of magnesium ions) may result in misreading of the message and in mistakes in the protein synthesized. In the cell, the latter type of mutations would presumably be minimized by maintenance of conditions favorable for normal protein synthesis (Watson, 1976). A shift in the frame of reference (reading frame) by deletion or addition of a nucleotide results in a mutation because the entire message from that point on is misread as to the amino acids to be inserted into position on the polypeptide produced (Watson, 1976).

The genetic code appears to be universal. Thus poly-U promotes synthesis of polyphenylalanine in cell-free extracts of many organisms, and poly-C the production of polyproline, and so forth, provided that all the components of the system are included in each case. Ribosomes from one organism serve for mRNA originating in another organism to produce proteins in conformity with the mRNA message. This is another instance that at the level of "biochemical universals" there is "little difference between cabbages and kings" (Szent-Györgyi, 1940).

The analysis of the genetic code was incomplete until the colinearity between amino acids and nucleotides in messenger mRNA, and again between the messenger RNA and DNA, had been shown (Watson, 1976). Now that the nucleotide sequence has been determined in the DNA of the virus ΦX174, in which 5375 nucleotides are present, such correlation is possible (Fiddes, 1977). It had previously been established for small RNA viruses only. It is expected that nucleotide sequences will be determined in the DNA of many organisms, now that suitable methods have been developed (Fiddes, 1977).

The kind of protein produced from a pool of amino acids depends upon the mRNA presented. The mRNA depends upon the gene that produces it, since presumably many types of genes are present on the chromosomes of a cell, each capable of producing a corresponding messenger RNA. On this basis, a cell can have only as many types of polypeptides as it has types of genes (Watson, 1976). This statement may have to be modified as data accumulate, in view of the finding of overlapping genes in DNA phage G4, a close relative of ΦX174. In this case, as an illustration, a piece of the DNA chromosome sequence T-T-C-T-G-A-T-G-A-A-A- is read by Gene B beginning at the first T to synthesize phenylalanine (and terminate); by Gene A, to begin at the second T to synthesize serine, asparagine, and glutamic acid, and by Gene K to begin at ATG to synthesize methionine and lysine (Shaw *et al.,* 1978; Szekely, 1978).

BIOSYNTHESIS OF PROTEIN: TRANSLATION

Proteins have been of great interest to researchers for many years. A great deal is known of protein composition and much has been learned about protein synthesis. An important early observation that dinitrophenol, which is known to uncouple phosphorylations, also inhibits protein synthesis led to the realization that not the amino acids but rather their "activated" phosphorylated forms are probably used in synthesis of proteins in the cell. Use of amino acids labeled with radioactive substances has made it possible to measure their rate of incorporation into proteins, and thus to follow the steps of protein synthesis in the living cell.

Although the pathway by which the present knowledge of protein synthesis has been achieved is somewhat devious, it is now clear that the synthesis of protein in the cell includes a number of steps and the participation of mRNA, rRNA, and tRNA (Figs. 15.9 to 15.15).

The first step in protein synthesis involves activation of the amino acid by reaction with ATP in the presence of a specific activating enzyme, whose function it is to activate the carboxyl group of that amino acid, forming an amino acyl-adenosine monophosphate complex (aa-AMP) (Fig. 15.10, top). The aa-AMP remains bound to the activating enzyme. The next step is the preparation of an acceptor site for the amino acid on transfer RNA (tRNA) by reaction with cytidine triphosphate (CTP) and adenosine triphosphate (ATP). The aa-AMP-enzyme complex is next attached by its carboxyl end to the phosphorylated tRNA specific to that amino acid, forming an ester linkage to either the 2′ or 3′ hydroxyl group of the ribose in the terminal adenosine group of the tRNA molecule; then AMP and the enzyme are released (Fig. 15.10).

Several times 20 kinds of tRNA molecules occur in *E. coli.* They are dissolved in the ground substance of the supernatant of disintegrated

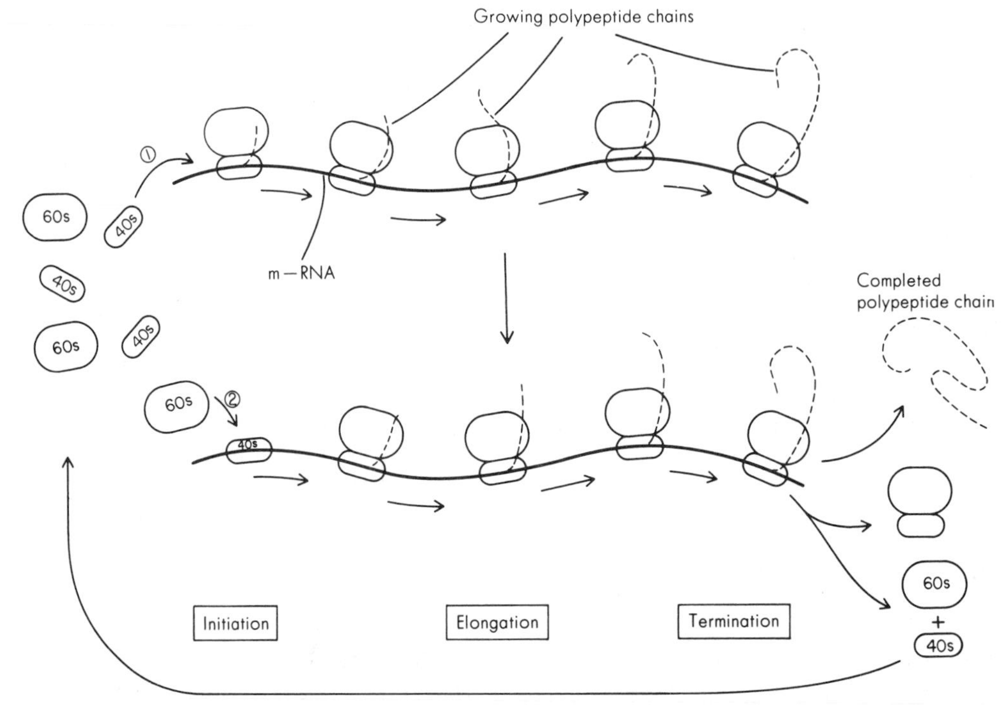

Figure 15.9. *A model for mammalian polysome function. The scheme depicts the attachment of a 40S subunit to mRNA on the first step, followed by the attachment of a 60S subunit to that 40S particle. (From Ingram, 1972: Biosynthesis of Macromolecules. 2nd Ed. W. A. Benjamin, Menlo Park, Calif.)*

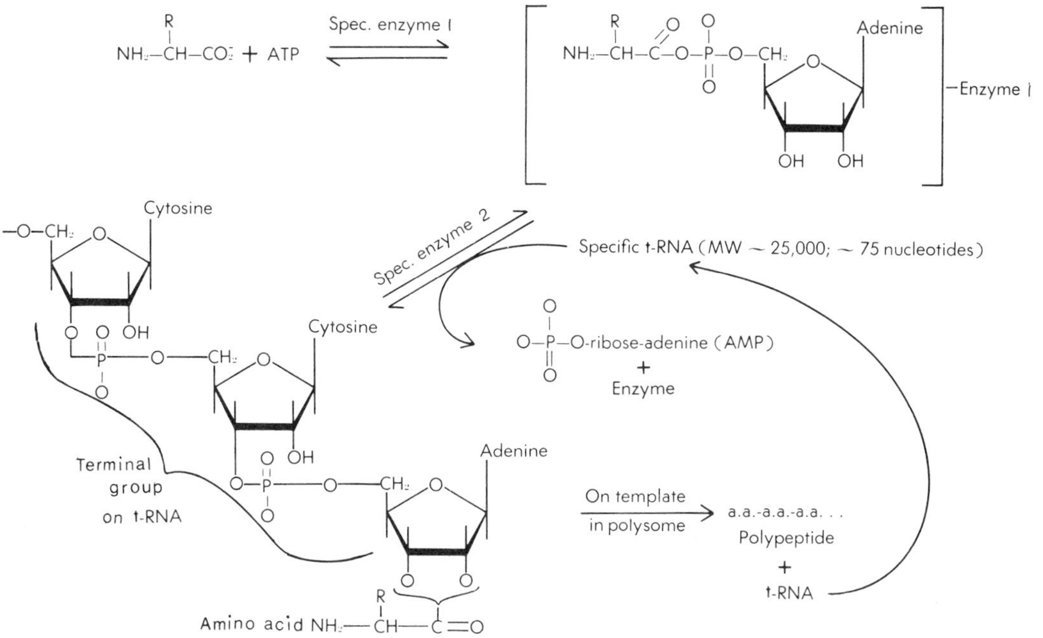

Figure 15.10. *The activation of an amino acid and its attachment to tRNA. (From Ingram, 1972: Biosynthesis of Macromolecules. 2nd Ed. W. A. Benjamin, Menlo Park, Calif.)*

cells, and precipitated at pH 5. When a particular amino acid (labeled) is added until the saturation level is reached, addition of any other labeled amino acid at this time leads to further labeling of the protein, indicating a lack of competition between the two amino acids. This could best be explained by postulating for the second amino acid a separate tRNA molecule different from that involved with the first amino acid. A specific enzyme is probably involved in the attachment of each of the 20 amino acids to the respective molecule of tRNA (Fig. 15.10). No reason has been supplied for the multiplicity of tRNA's. Perhaps some are active under one set of circumstances in the cell, others under other conditions.

It has been found that a given amino acid may attach to any one of several different tRNA molecules. This might be expected in view of multiple coding for a specific amino acid ("degeneracy"). Transfer RNA recognizes its location by base pairing on mRNA, and if several different base triplets (codons) code for the same amino acid, one might expect that the corresponding triplets on tRNAs (anticodons) with complementation to each codon might also do so.

On heating, tRNA shows hyperchromicity (increased absorption at 260 nm), indicating internal base pairing that lowers absorption at 260 nm in its native state. It is suggested that the single strand of which tRNA is composed loops back upon itself (see Fig. 1.20), attachment between adjacent portions of the loop occurring where base complementarity occurs. Four sites of activity are expected on tRNA: (1) the site recognizing the amino acid (the terminal pCpCpA) (see Fig. 15.10), which is alike in all tRNA molecules; (2) the site recognizing the specific amino acid–activating enzyme; (3) the coding recognition site; and (4) the ribosome recognition site. A hypothetical alignment for tRNA is shown in Figure 15.11.

Not only transfer RNA but also ribosomal RNA is needed for amino acid incorporation into a protein. Presumably, tRNA carries the activated amino acid to the mRNA on the ribosome, where, in the presence of guanosine triphosphate (GTP) as a source of energy, the amino acid is transferred to the growing peptide and the tRNA is liberated (Fig. 15.12). Ribosomes are inactive until combined with messenger RNA, for which magnesium ions are required. Magnesium also stabilizes the ribosomes themselves. It is the message on mRNA that is translated by the orderly arrangement of the acyl-amino molecules, one by one by complementation of the base sequences on tRNA (Fig. 15.11).

The requirements for protein synthesis are summarized in Table 15.5. The ribosomes attach to the mRNA, forming a polysome (Fig. 15.9). The function of GTP is still uncertain, but it may supply the energy for movement of the ribosome along the mRNA. Some information is now available on the initiation factors, the elongation factors, and the termination factors in *E. coli* and to a lesser extent on similar factors in the mammalian cell. In cell-free systems it is necessary to add an SH-compound to keep

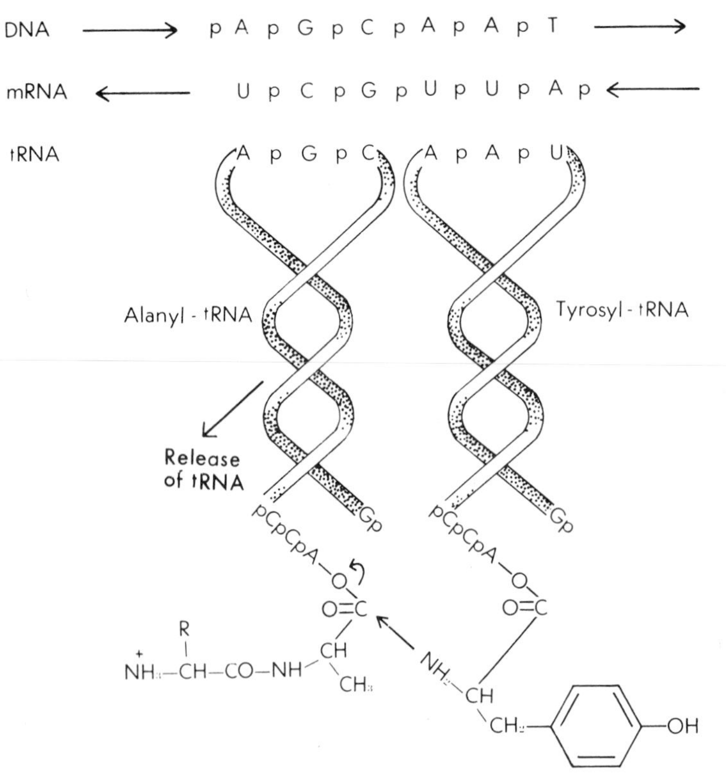

Figure 15.11. The alignment of amino acyl-tRNA on the template and the condensing of two amino acyl groups to form a peptide bond. Note the release of one of the tRNA molecules. The arrows indicate the polarities of the DNA and the mRNA chains. (From Ingram, 1972: Biosynthesis of Macromolecules. 2nd Ed. W. A. Benjamin, Menlo Park, Calif.)

the SH-requiring enzymes in full activity. These various factors are discussed separately later. Returning to Table 15.5 after studying the factors will serve to coordinate the information.

The idea of a messenger RNA and its functioning in this manner was first developed by Jacob and Monod in 1961. The strongest early evidence for the concept was obtained by Volkin and Astrachan in 1956, who infected *E. coli* with T-2 phage and found that synthesis of protein and nucleic acids of the bacterium ceased and that, during the lag before the beginning of synthesis of the protein characteristic of the T-2 coat, a rapid RNA synthesis took place. The synthesized RNA had a short life span and was

TABLE 15.5. FACTORS IN PROTEIN SYNTHESIS*

Factors	*Bacterial*	*Mammalian*	*Function*
Initiation	$IF_1 = A, F_I$ $IF_2 = C, F_{III}$ $IF_3 = B, F_{II}$		
Elongation	EF-T	Rat liver amino acyl transferase I, rabbit reticulocyte binding enzyme	Attachment of amino acyl-tRNA to ribosome in amino acyl (A) site
	G (EF-G)	Rat liver amino acyl transferase II, rabbit reticulocyte factor TFII	Permits reaction with GTP to provide energy for conformational change to move ribosome to next codon
Termination (release)	R_1 R_2 R_3		Translocation of peptidyl-tRNA from A to P (peptidyl) site permitting hydrolysis of polypeptide from tRNA

*Data from Ingram, 1972: Biosynthesis of Macromolecules. 2nd Ed. W. A. Benjamin, Menlo Park, Calif., and Lehninger, 1975: Biochemistry. 2nd Ed. Worth Publishers, New York.

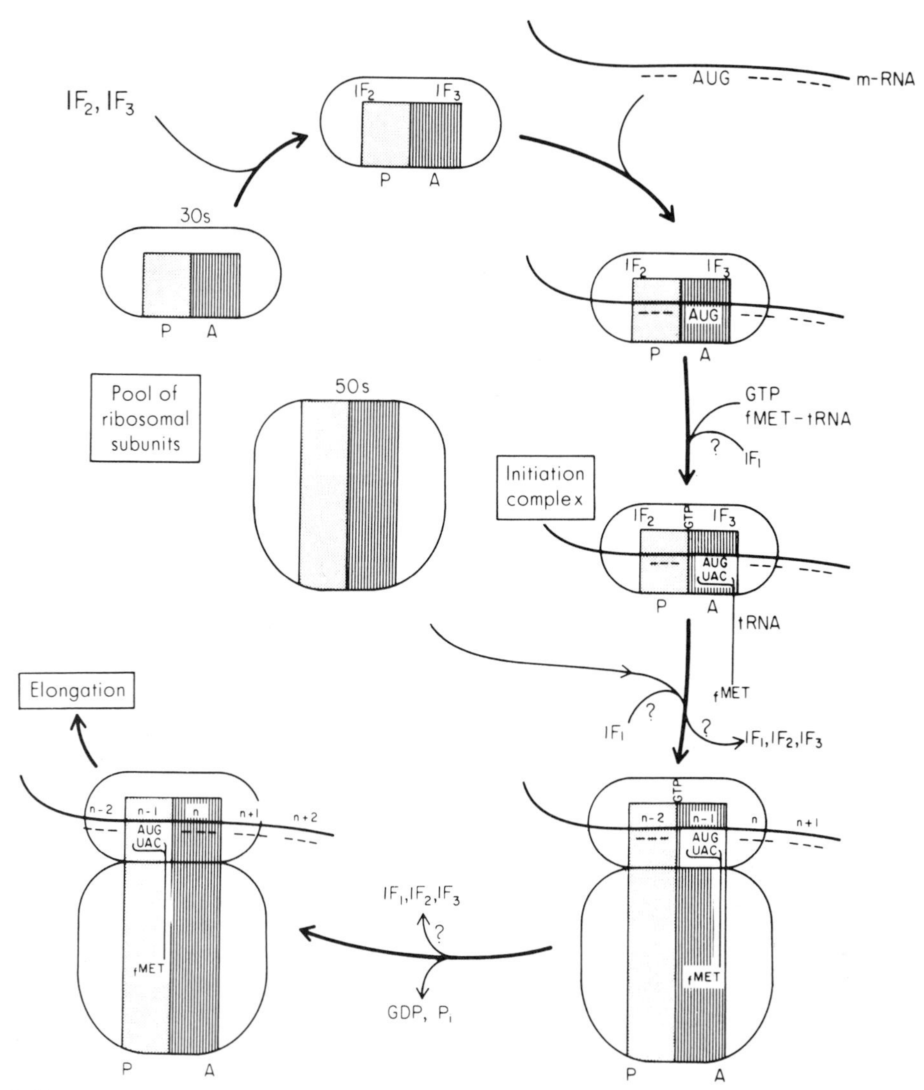

Figure 15.12. *Initiation of protein synthesis in* E. coli. *P, the peptidyl site, is the peptidyl-binding site; A is the hypothetical amino acid-tRNA binding site; AUG is the codon on mRNA, UAC the anticodon on the tRNA; F_1, F_2, and F_3 are the protein initiation factors; 50S and 30S are the major subunits of the ribosome. The steps are described in the text. (From Ingram, 1972: Biosynthesis of Macromolecules. 2nd Ed. W. A. Benjamin, Menlo Park, Calif.)*

complementary in its base composition to the phage DNA but not to that of the bacterial cell. Since then, more conclusive evidence from tracer experiments has accumulated so that at present the concept is well documented and accepted. Messenger RNA is short-lived in bacteria (e.g., half-life of two minutes in *B. subtilis,* enough to permit 15 rounds of transcription of protein). However, it may have a long life in some cells, as in reticulocytes, which produce hemoglobin and give rise to red blood cells (Lipmann, 1963), and perhaps in all plant and animal cells. The unfertilized sea urchin egg has much inactive mRNA, awaiting activation with fertilization (Gross, 1968).

Ribosomes act in groups in cells or extracts that carry on active protein synthesis. Such groups are called polyribosomes or *polysomes* (see Fig. 5.7) (Rich, 1963). Such aggregation was discovered when the ribosomes from actively synthesizing extracts were separated by centrifugation. The sedimentation constants of the "ribosomes" were found to be much larger than those of preparations in which no protein synthesis was occurring and in which a low concentration of magnesium ions was present. Electron micrographs of the sediment confirmed the presence of aggregates rather than single ribosomes. This is thought to indicate that several ribosomes transcribe the message at the same time; as one drops off after completing its protein molecule, another slips on (see Fig. 15.9).

How the ribosome moves along the mRNA to

attach the appropriate amino acid residue to the growing polypeptide chain is unknown. It seems likely that the proteins clustered to form the 30S and 50S subunits of the 70S ribosomes effect conformational changes by which they move along the mRNA. They may therefore be mechanical-chemical systems somewhat like the actin and myosin of muscle or the microtubules of eukaryotic flagella (Lehninger, 1975).

Protein synthesis can be readily followed by using ascites tumor cells at 20°C. Incubation of tumor cells with a ^{14}C-labeled amino acid is followed by the appearance of radioactivity, first in the transfer RNA fraction, more slowly in the ribosomal RNA, then in the ribosome protein, and finally in the soluble protein of the cell. For these reactions guanosine triphosphate is likely the source of high-energy phosphate bonds. GTP probably serves to attach the amino acid–tRNA complex to the ribosome.

Initiation of protein synthesis (Fig. 15.12) occurs at the nucleotide triplet AUG (coding for methionine). Once this "frame of reference" is chosen, the remainder of the message is read three nucleotides at a time in accordance with the code. Three initiation factors, IF_1, IF_2, and IF_3 are required along with GTP and Mg^{2+}, the two major ribosomal subunits (30 S and 50 S in prokaryotes) and mRNA, as well as a special initiator tRNA with a specific methylated methionyl-tRNA (possessing a formyl group on its amino nitrogen). A complex is formed between the 30S ribosome subunit, mRNA, and the formyl-Met-tRNA. Next the 50S ribosomal subunit is attached to the complex, forming a complete 70S ribosome. As the formyl-Met-tRNA is moved from the *acceptor (A) site* on the ribosome to the *protein* or *peptidyl (P) site,* the GTP is cleaved to GDP and inorganic phosphate (P_i). Another amino acid can now be added, the choice depending upon the codon next in line on the mRNA (the codon now at the A site), and the procedure is repeated. The initiation is complete with the acceptance of a second amino acid-tRNA and elongation of the polypeptide now begins. Initiation requiring formyl-Met-tRNA is found in chloroplasts and mitochondria of eukaryotic cells, but only Met-tRNA functions in the cytoplasm of plant and animal cells (Ingram, 1972). Initiation factor IF_2 (molecular weight, 75,000) serves to prepare the peptidyl (P) site on the ribosome, to which the nascent polypeptide moves after attachment of an amino acid residue. IF_3 prepares the tRNA acceptor site (A). The function of factor IF_1 is somewhat uncertain, but it is thought to stimulate the functions of the other two initiation factors (Last, 1971).

Elongation proceeds in a similar manner in 70S and 80S ribosomes (of prokaryotes and eukaryotes, respectively) from the amino terminal end of the peptide chain toward the carboxyl end, the ribosome moving along the mRNA in the 5′ to the 3′ direction. The message on mRNA is read three nucleotides, or one codon, for each step. For this to occur it is necessary for elongation factors Tᴜ, Ts, and G to be present. All these factors are proteins, of molecular weights 67,000, 42,000, and 84,000, respectively. Keeping the amino acyl-tRNA bound to the receptor site A on the ribosome requires the factors Ts and Tᴜ, as well as GTP. As GTP is split to GDP and inorganic phosphate, another peptide bond is formed on the growing peptide chain. The chain is transferred from its tRNA at the P site to the amino group of the newly arrived amino acyl-tRNA at the A site (Fig. 15.13). Now the tRNA molecule that has yielded its amino acid to the peptide is detached from the ribosome and released along with the G factor (Ts and Tᴜ were released earlier) (see Fig. 15.13), and GTP is split to GDP and inorganic phosphate (Ingram, 1972).

In *E. coli* at 37°C new amino acid molecules are added to the growing polypeptide chain at the rate of about 20 per second, and a protein molecule of 150 amino acids is finished in 8 seconds. In eukaryotic cells at 37°C it takes a little more than a second to add an amino acid residue to a nascent protein. For elongation of the growing peptide chain in eukaryotes, two elongation factors, EF-T and EF-G, are also required (Lehninger, 1975).

That the peptide chain is formed in linear fashion from one end is apparent from the studies of Dintzis (1961), who used a cell-free system derived from reticulocytes, which carry on an active synthesis of primarily one protein (hemoglobin). He incubated the cells with ^{14}C-labeled amino acids for a long time to mark the peptide chains evenly. He then exposed the preparations to ^{3}H-labeled amino acids for graded pulse periods. It is argued that if synthesis is sequential there should be a progressively greater labeling with ^{3}H relative to ^{14}C with increased pulse periods. Brief labeling with ^{3}H was found to mark only the portion of the polypeptide chain nearest the carboxyl (terminal) end, whereas longer periods labeled proportionally larger portions away from the terminus until the entire chain was labeled.

Peptide chain termination is attended by enzymatic hydrolytic splitting of the last polypeptide-tRNA ester linkage, releasing the polypeptide chain and permitting the ribosome-mRNA complex to dissociate. For this a codon release signal in mRNA (UAA, UAG, or UGA) is required along with one termination factor, R_1, R_2, or R_3, all of which are proteins. R_1 responds to UAA and UAG, while R_2 responds to UAA and UGA; R_3 accelerates the action of R_1 and R_2. The releasing factors, by binding to the ribosome, cause the shift of the polypeptidyl-tRNA from the A site to the P site.

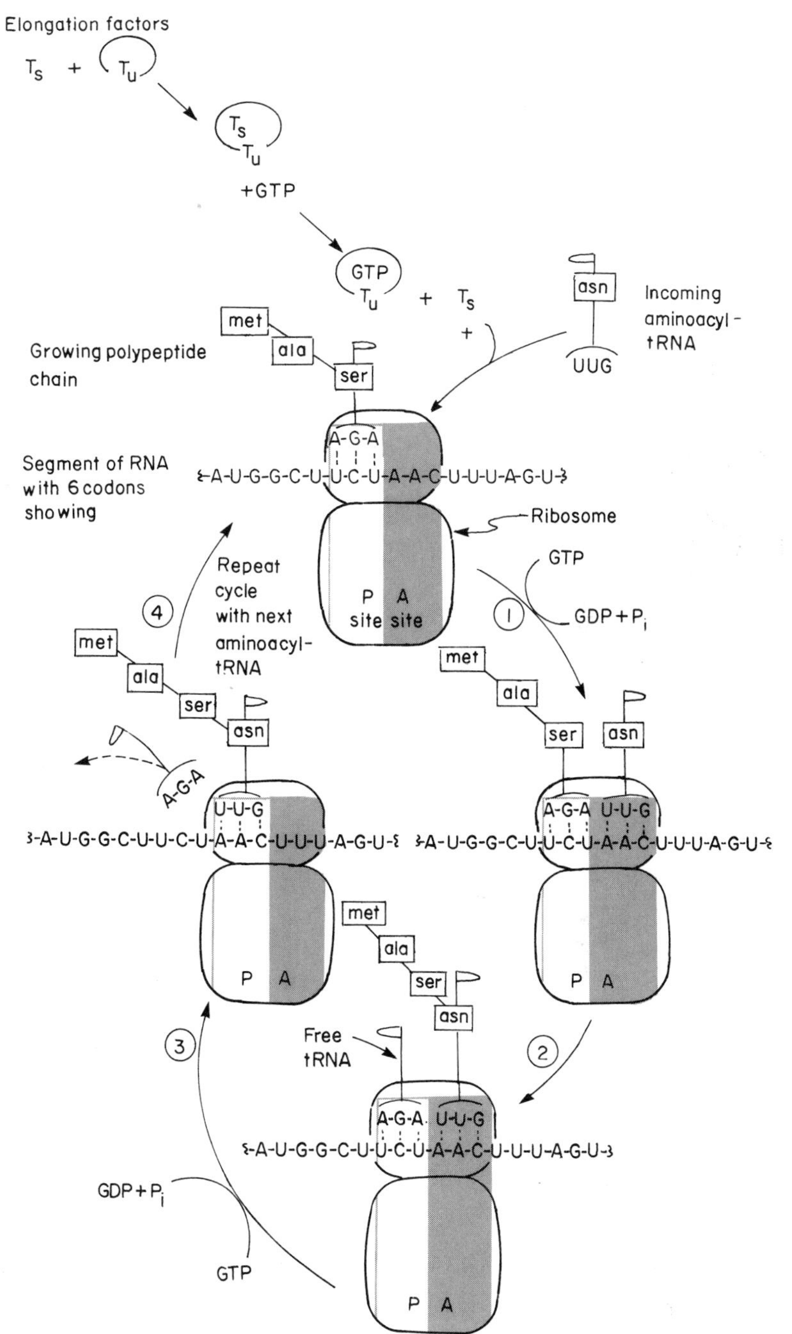

Figure 15.13. Elongation of the peptide chain by addition of one amino acid to a growing polypeptide chain. At the top is shown the series of events that includes the addition of the elongation factors Ts and Tu, their combination with GTP providing energy for combination with tRNA and its attached amino acid, asn. Below is shown the entry of the tRNA (on the right) and its attachment to the ribosome.

The particular step shown is the addition of the fourth amino acid to the polypeptide chain for a specific protein, the coat protein, of the bacteriophage R17 (a protein for which both the amino acid sequence of the polypeptide and the nucleotide sequence of the mRNA are known). The A (aminoacyl) site is shaded in each case. Amino acids, their abbreviations, and codon assignments are as follows: methionine (met): AUG; alanine (ala): GCU; serine (ser): UCU; and asparagine (asn): AAC. The incoming amino acid (asparagine in the step shown) arrives in activated form as the aminoacyl tRNA, having been previously esterified to its proper tRNA carrier by the tRNA synthetase reaction driven by the splitting of ATP to AMP and PP_i. The cyclic process illustrated here is repeated for each successive amino acid until one of the chain-terminating codons (UAG, UAA, or UGA; see Table 15.4) is encountered, at which point the final peptidyl-tRNA bond is hydrolyzed to release the completed polypeptide. (After Becker, 1977. Energy and The Living Cell, J. B. Lippincott Co., Philadelphia, p. 225.)

IF_3, the initiation factor already discussed, probably speeds up the process of termination, readying the ribosome for another occupancy elsewhere (Lehninger, 1975) (Fig. 15.14).

Two experiments confirm the fact that these schemes work. Polynucleotides of known nucleotide sequences have been synthesized and used to direct polypeptide synthesis giving known amino acid sequences corresponding to the codons in the polynucleotides. The synthesized nucleotide sequences used even included initiation, elongation, and termination codons (Khorana, 1968). In a different experiment, purified mRNA for rabbit hemoglobin protein (omitting the heme) isolated from blood-forming cells (reticulocytes) was added to a cell-free extract from *E. coli;* it was found to induce the synthesis of rabbit hemoglobin in the *E. coli*–derived material (Laycock and Hunt, 1969).

Some posttranslational modifications of the polypeptide chain occur after its completion. Microsomes probably catalyze the oxidation of —SH groups to disulfide linkages. Among other

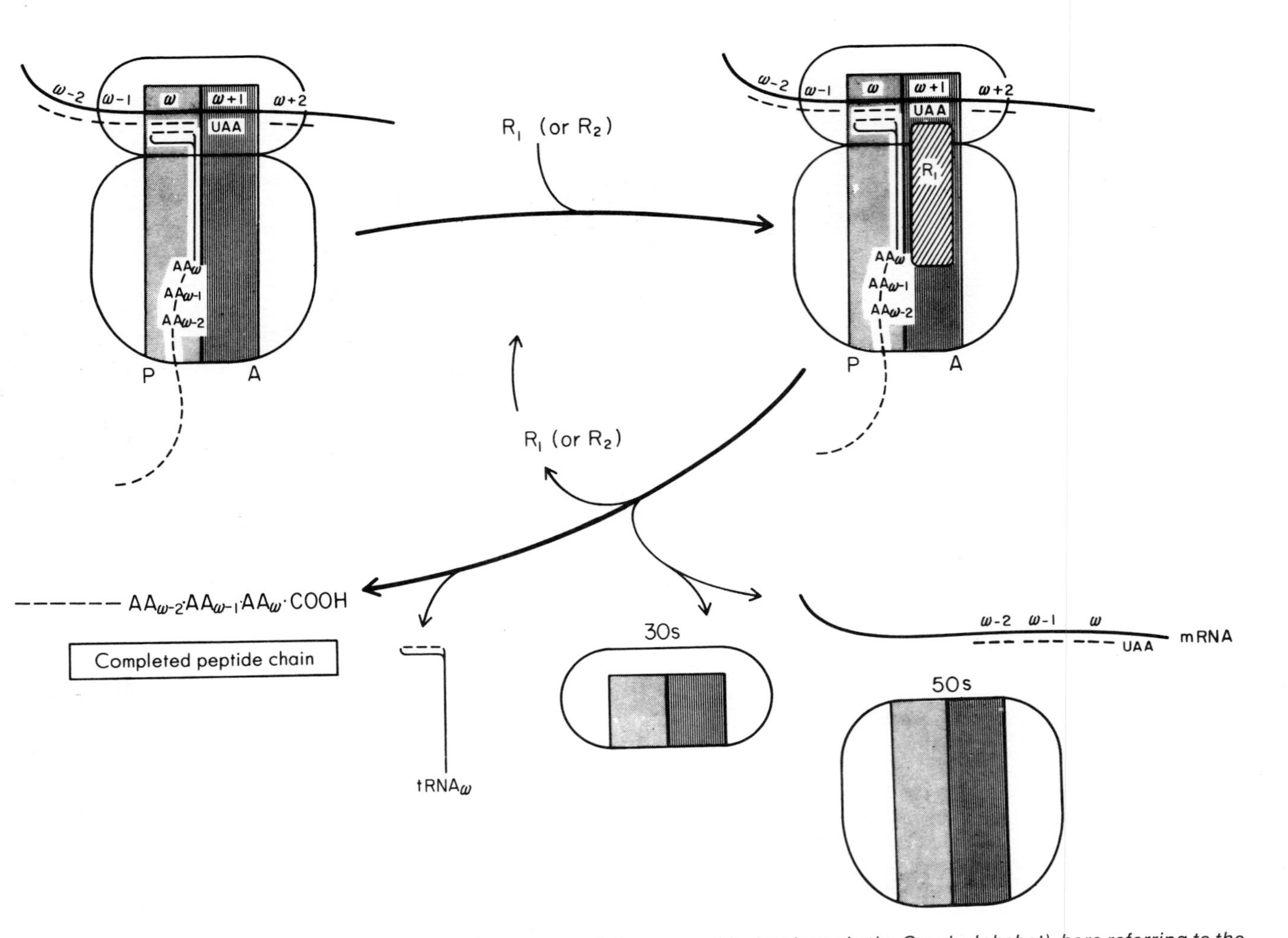

Figure 15.14. *Termination of peptide chain synthesis, ω stands for omega (the last letter in the Greek alphabet), here referring to the last site on the mRNA; when used as a subscript to AA (amino acid) omega refers to the terminal amino acid residue. The solid line running through the ribosome signifies to the messenger RNA and the dashed line attached to the tRNA (with the flag) is the peptidyl chain. R_1 and R_2 are the termination factors that facilitate the termination of the peptidyl chain. (From Ingram, 1972: Biosynthesis of Macromolecules. 2nd Ed. W. A. Benjamin, Menlo Park, Calif.)*

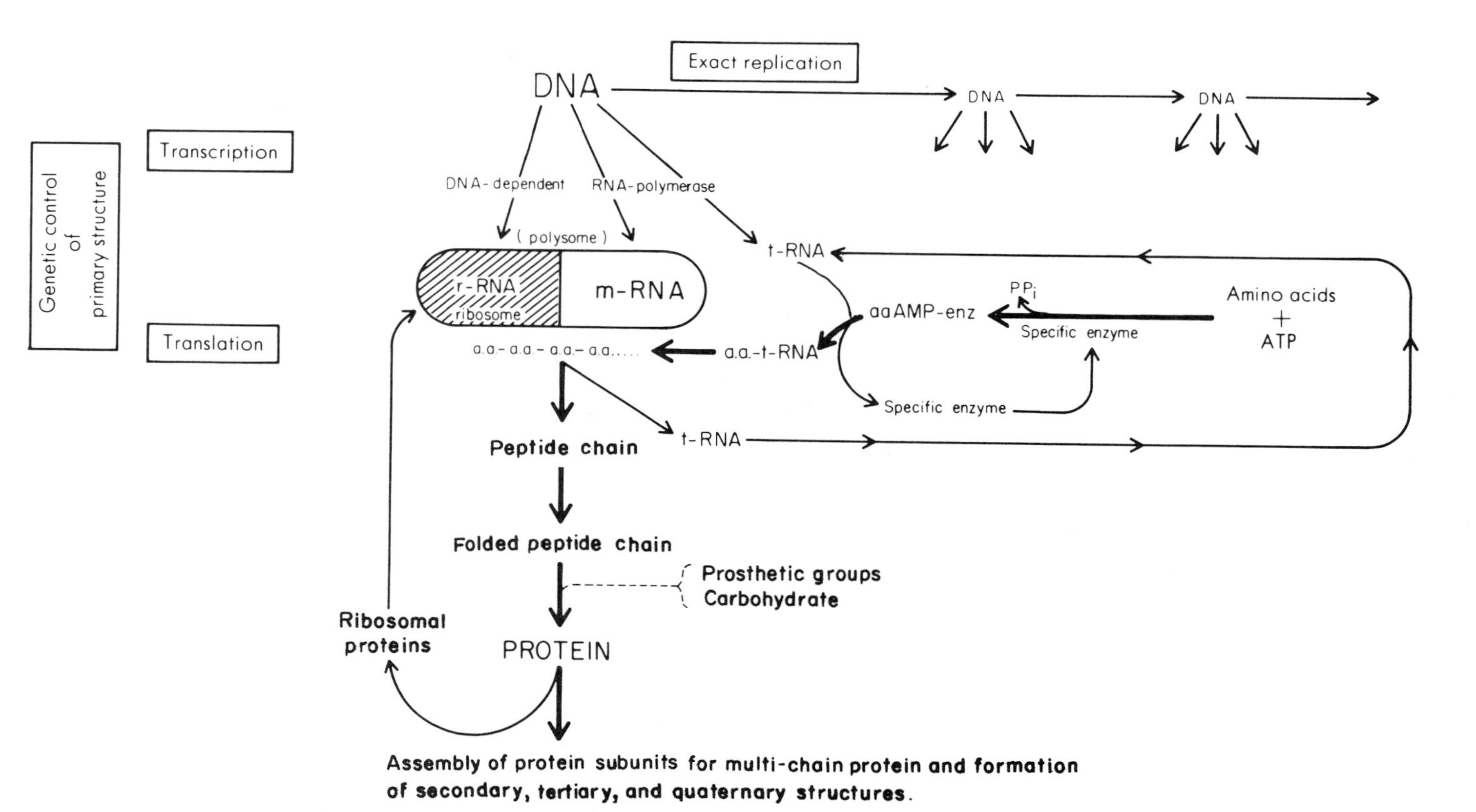

Figure 15.15. *The overall scheme for DNA, RNA, and protein synthesis. (From Ingram, 1972: Biosynthesis of Macromolecules. 2nd Ed. W. A. Benjamin, Menlo Park, Calif.)*

covalent modifications that can occur are methylation, phosphorylation, and hydroxylation (Lehninger, 1975).

It is of great importance to note that specific protein syntheses may be induced in cells. For example, it will be recalled that enzymes can be induced in microorganisms in response to the presence of a specific nutrient for which the organism does not have a constitutive enzyme. Thus, *E. coli* develops a β-galactosidase in response to the presence of lactose in the medium (see Chapter 18). Another case of induced protein synthesis is the development of immunity, that is, the production of antibodies in response to antigens. The active group (hapten) in the antigen is generally a carbohydrate (polysaccharide) or a protein, and the antibody protein developed in response to challenge by a particular antigen is highly discriminatory in its action, indicating its "custom-built" synthesis.

Protein synthesis in prokaryotic cells and in mitochondria of eukaryotic cells is readily inhibited by the antibiotic chloramphenicol, which has a structure resembling phenylalanine. Presumably it takes the place of phenylalanine on the tRNA and is transferred to the polypeptide in process, terminating its growth. However, RNA and DNA synthesis may continue under these conditions. Radiation can inhibit DNA synthesis without immediately affecting other syntheses, indicating partial separation of the three processes. Stopping both protein and RNA synthesis, however, also stops DNA synthesis.

Protein synthesis in eukaryotes is inhibited by cyclohexamide, which has no effect on protein synthesis in prokaryotes or mitochondria. Cyclohexamide blocks peptide bond formation (Lehninger, 1975).

The antibiotic puromycin has a structure resembling that of amino acyl-tRNA. Experiments indicate that in its presence small peptide chains are released from the ribosomes, protein synthesis failing to go to completion because puromycin does not have the precise structure for it to be recognized by the translocation apparatus and to be moved from the A site to the P site on the ribosome (Lehninger, 1975). One puromycin molecule was found to be bound to each N-terminal end, presumably forming a premature C-terminal end.

The newly synthesized polypeptides assume secondary, tertiary, and quaternary structure spontaneously in response to weak attractive forces forming bonds such as already discussed in Chapters 1 and 3 (see Table 1.2). These structures are thermodynamically most stable and require no additional information or enzyme action.

Figure 15.15 summarizes the macromolecular syntheses. The seeming simplicity belies the true state of affairs, especially when one realizes that each of these reactions must be kept in proportion to the cell's needs. Regulation of these precursors is discussed in Chapter 16.

BIOSYNTHESIS OF POLYSACCHARIDES

In the "informational" molecules such as DNA, RNA, and proteins, the units forming the information-carrying sequences that compose the polymer are arranged in a highly specific manner. In polysaccharides the molecules are generally not ordered into information-carrying sequences, nor are they of a definite chain length. This is perhaps understandable since polysaccharides serve either as storage molecules or as structural components. Polysaccharides, as the name implies, are composed of many subunits, and synthesis of polysaccharides requires that individual subunits be in an activated state. For example, synthesis of sucrose from fructose and glucose in organisms is a result of three enzymatic reactions, the source of energy being compounds of high phosphate transfer potential, as shown in Figure 15.16. However, it will be recalled that sucrose is hydrolyzed by invertase with a small change in free energy (see Chapter 10). Theoretically, by increasing the concentration of the products, one should be able to reverse the hydrolytic reaction. This does not occur in the case of action of enzymes hydrolyzing carbohydrates.

In the synthesis of *glycogen,* the enzyme *glycogen synthetase* serves to catalyze the transfer of glucose residues in activated state from UDP-glucose to the nonreducing ends of a glycogen primer (Table 15.6). As a result, the chain is lengthened and UDP (uridine diphosphate) is released in stoichiometric ratio. The reaction proceeds with a decrease in free energy

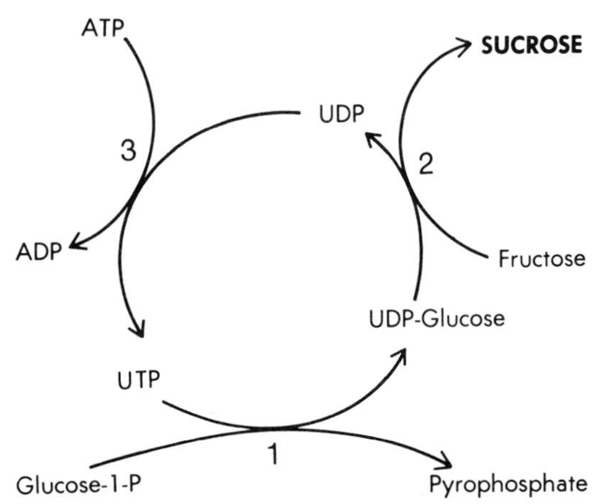

Figure 15.16. Scheme of sucrose synthesis involving three enzymatic reactions (1, 2, 3), UDP is uridine diphosphate; UTP is uridine triphosphate.

TABLE 15.6. THE ENZYMATIC STEPS IN THE BIOSYNTHESIS OF GLYCOGEN*

1. $\text{Glucose} + \text{ATP} \xrightarrow{\text{hexokinase}} \text{glucose 6-phosphate} + \text{ADP}$

2. $\text{Glucose 6-phosphate} \xrightarrow{\text{phosphoglucomutase}} \text{glucose 1-phosphate}$

3. $\text{Glucose 1-phosphate} + \text{UTP} \xrightarrow[\text{pyrophosphorylase}]{\text{UDP-glucose}} \text{UDP-glucose} + \text{pyrophosphate}$

4. $\text{UDP-glucose} + \text{glycogen}_n \xrightarrow[\text{synthetase}]{\text{glycogen}} \text{glycogen}_{n+1} + \text{UDP}$

5. $\text{ATP} + \text{UDP} \xrightarrow[\text{diphosphokinase}]{\text{nucleoside}} \text{ADP} + \text{UTP}$

6. $\text{Pyrophosphate} + H_2O \xrightarrow{\text{pyrophosphatase}} 2\ \text{phosphate}$

Sum: $\text{glucose} + 2\ \text{ATP} + \text{glycogen}_n \longrightarrow 2\ \text{ADP} + 2\ P_i + \text{glycogen}_{n+1}$

*From Lehninger, 1971: Bioenergetics. 2nd Ed. W. A. Benjamin, Menlo Park, Calif.

of approximately 3000 cal/mole. Glycogen consists of a branching system of linearly arranged glucose units. For the formation of branch points, enzymes other than the synthetase are necessary. Thus, the *liver-branching enzyme* forms branch points on the straight polyglucose chain by transferring to it segments of straight chains (Lehninger, 1975).

Starch is a polysaccharide whose main constituent is the unbranched polysaccharide amylose. Synthesis of starch also requires activated glucose units with ATP and UTP serving as energy sources. *Cellulose* is a polysaccharide in which the glucose units are linked by β-D-(1,4)glucosidic bonds. The chains of cellulose are of indefinite length, and resistance of cellulose to reagents may stem in part from the close packing in parallel alignment of such linear units. Synthesis of cellulose requires the presence of GTP, the guanine analog of UTP, but the details of the mechanism by which this occurs are still under investigation (Lehninger, 1975).

RATE OF BIOSYNTHESIS

Lehninger (1971) has made some interesting calculations of the rate of biosyntheses in the *E. coli* cell. The chemical constitution of *E. coli* is well known and the rate of duplication under most optimal conditions is about once every 20 minutes. On the basis of these data it can be shown that 12,500 molecules of lipid and about 32 molecules of polysaccharide are synthesized per second by a single *E. coli* cell. The *E. coli* cell produces about 1400 molecules of protein per second, each one containing some 500 amino acids of 20 different kinds, arranged precisely in order as dictated by the genetic code. About 12 RNA molecules are synthesized per second, but only one duplication of the DNA molecule occurs during the division interval (time between divisions) in *E. coli*.

The rate of production of molecules occurs more slowly in cells with a long division interval (such as 24 hours for some mammalian cells in tissue culture).

Synthesis of lipids and polypeptides in the laboratory is slow. A true protein of relatively large molecular weight (insulin) was synthesized in the laboratory after considerable effort and time (Katsoyannis, 1966). This emphasizes the efficiency and speed of the methods used by the cell.

The rate of ATP turnover is remarkable in an active *E. coli* cell. Lehninger (1971) has calculated that about two and a half million molecules of ATP are used per second in an *E. coli* cell growing at its maximal rate. About 88 percent of the energy released for synthesis by the ATP is used in the synthesis of protein, 3.7 percent in lipid synthesis, 3.1 percent in RNA synthesis, 2.5 percent in DNA synthesis, and 2.7 percent in carbohydrate synthesis.

As discussed previously, the terminal phosphate of the ATP is split off with a decrease of free energy of about 7300 cal/mole *(orthophosphate cleavage)*. However, in many of the biosynthetic reactions considered here, pyrophosphate is split from ATP, forming AMP and pyrophosphate (POP) *(pyrophosphate cleavage)*. The change in free energy is about the same in both pyrophosphate and orthophosphate cleavage. The pyrophosphate still retains one high-energy phosphate bond, but this energy cannot be used to recharge AMP or ADP. The pyrophosphate is split by the

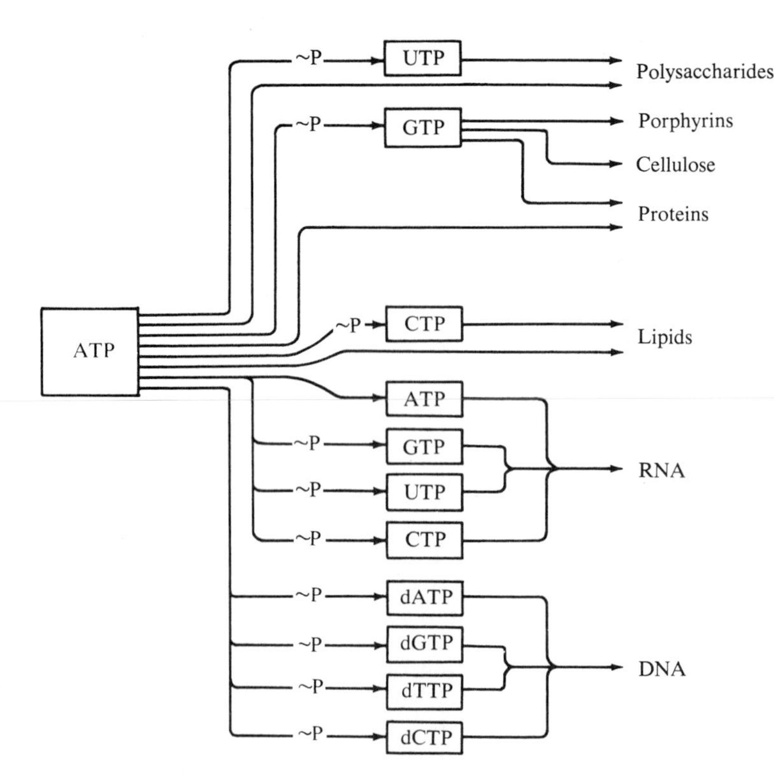

Figure 15.17. Channeling of phosphate bond energy of ATP into specific biosynthetic routes. (From Lehninger, 1971. Bioenergetics. 2nd Ed. W. A. Benjamin, Menlo Park, Calif.)

enzyme pyrophosphatase with a decrease in free energy of about 7300 cal/mole; the total cost in loss of free energy for biosynthetic reactions in which POP is formed is about 2 × 7300 cal/mole. There is, therefore, what Lehninger calls a greater thermodynamic "pull" in such coupled reactions than when only one high-energy phosphate is removed.

It is interesting to note that ATP energy is channeled through many other triphosphates during the synthesis of various compounds. These relationships are summarized in Figure 15.17. No satisfactory explanation has been suggested for such specific channeling.

It is evident that the synthesis of the highly ordered macromolecules proceeds with a decrease in entropy. The system as a whole, however, shows an increase in entropy, the ordered energy in various triphosphates having been expended in increasing the orderliness of the macromolecules.

CHANGES IN BIOSYNTHESES—MUTATIONS

A mutation is a sudden but relatively permanent change in heredity. It is to be expected, on the basis of probabilities, that in the course of a long series of replications of genic DNA a mistake will occur. Such a mistake might well produce changes in synthesis of the specific protein coded for by the section of DNA in which the change occurs, with profound effects on the organism; the frequency of mistakes is greatly accelerated by some mutagenic (and carcinogenic) chemicals and by ultraviolet and ionizing radiations. Considerable evidence has been documented to support the concept that one gene codes for one polypeptide chain. The gene expresses itself by the formation of a polypeptide chain which, if enzymatic, carries out the syntheses designated by the code in the DNA. A slight change in this polypeptide chain—even the replacement of a single amino acid at a critical point on the molecule—results in a change of its properties such that the reaction it catalyzes is carried out more slowly or not at all. A deficiency in the genic DNA template is thus reflected in a deficiency of the polypeptide chain produced under direction of the particular gene. The resultant deficiency shows up as a mutation in the cell or the organism.

The chromatographic method of "fingerprinting" proteins has made it possible to document some mutations and to demonstrate that a mutation with gross effects on the cell or organism may rest upon a simple alteration in synthesis of a protein molecule, perhaps changes in only one amino acid at a given location in the protein molecule. For example, the hereditary disease of man known as sickle cell anemia results from the substitution of a molecule of the amino acid valine for glutamic acid, the fourth amino acid of the β polypeptide chain of the hemoglobin molecule. (A molecule of hemoglobin has two α and two β polypeptides.) As a result, the hemoglobin of the mutant (individual with the

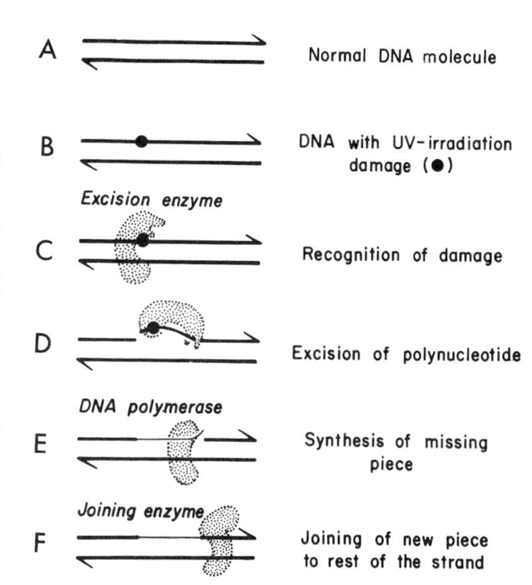

Figure 15.18. *Possible mechanism of excision repair by "cut and patch" of damage induced by ultraviolet radiation in a DNA molecule. DNA, shown in a simplified model with antiparallel strands, (A) is injured by UV-B or UV-C radiation as shown by the black dot (B). When the injury is recognized (C), the excision enzyme snips it out as a polynucleotide (D). It is then replaced by a piece of DNA synthesized under action of the DNA-polymerase, using the information on the complementary strand (E). The new piece is then patched in (F) by the joining enzyme (ligase). The differing shapes of the enzymes are for diagrammatic purposes only. (From Giese, 1976: Living with Our Sun's Ultraviolet Rays. Plenum Press, New York; after Hanawalt, 1972, Endeavour 31: 84.)*

disease) is so altered in its physical properties that in the presence of low oxygen partial pressure the entire erythrocyte takes on a new form —that of a sickle instead of a disc—which is much less effective in transporting oxygen. In another type of mutation, replacement of the glutamic acid by the amino acid lysine at the same position on the hemoglobin molecule gives rise to still another pathological condition. Several other amino acid substitutions on hemoglobin have been found to have corresponding effects on the protein molecule, inducing abnormalities in the red blood cell of the mutants (Lehninger, 1975).

Many other mutations have also been analyzed. In *Neurospora crassa* several mutations affect the enzymes tyrosinase and tryptophan synthetase, with over 100 available mutants in the latter case. A single gene mutation involves a single amino acid alteration in the enzymic proteins produced and this may cause drastic effects on the properties of the enzyme. In *E. coli,* tryptophan synthetase, alkaline phosphatase, and β-galactosidase have all been similarly studied after mutation and in all cases the fully analyzed "fingerprint" shows substitution of an amino acid at one locus on the enzyme protein molecule (Yanofsky, 1976).

EXTRACHROMOSOMAL INHERITANCE

Extrachromosomal inheritance has been known for a long time under the name of maternal inheritance (Sager, 1972). It is now clear that DNA is present in both mitochondria and chloroplasts and might serve as the carrier of extrachromosomal inheritance. Such DNA is self-replicating and apparently codes for protein synthesis, which occurs in these organelles. The chromosomes of both mitochondria and chloroplasts are circular, resembling those of prokaryotes. Chloramphenicol, which inhibits protein synthesis in prokaryotes, also inhibits it in mitochondria and chloroplasts. The ribosomes of mitochondria and chloroplasts resemble in size those of prokaryotes (Lehninger, 1975).

Extrachromosomal mutants are not produced by the mutagens that cause chromosomal mutations; therefore, study of such inheritance has been slow. Recently some chemicals have been found suitable for the purpose (e.g., sublethal concentrations of streptomycin) and a fairly large number of mutations have been accumulated in the single-celled alga *Chlamydomonas reinhardii* (Sager, 1977). The mutations have been observed to occur in linkage groups, with crossing over between them; therefore, the behavior of the extrachromosomal DNA appears to be much like that of the chromosome (Gillham, 1974). Two such linkage groups have been found, one presumably corresponding to that of the mitochondrial DNA, the other to chloroplast DNA (see Chapter 3) (Kung, 1977).

CORRECTION OF ERRORS IN REPLICATION AND REPAIR OF DAMAGE TO DNA

Innate to the living cell is the correction of errors in replication and the repair of damage to DNA. It is unlikely that replication is so error-free as to make natural mutation a relatively rare event. Furthermore, damage to DNA may occur from natural radiation and chemicals in the environment. Studies inves-

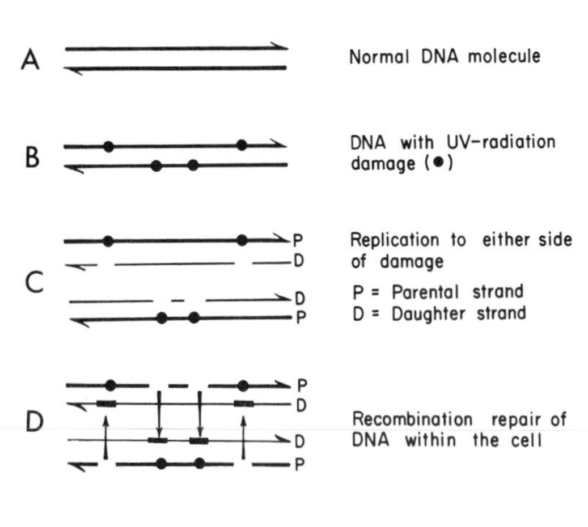

Figure 15.19. *A simplified model for postreplication repair of ultraviolet-damaged DNA molecules (shown as antiparallel lines with arrows): (A) normal DNA molecule; (B) dots indicate damage in DNA; (C) DNA synthesis (replication under direction of DNA polymerase (not shown) proceeds on either side of the damage in the parental strands, leaving gaps in the daughter strands. After replication (C) the missing pieces are added to the incomplete DNA daughter strands (D) by recombination with corresponding uninjured pieces of the parental DNA strands (P), within the cell. The remaining pieces of the parental DNA strands (D) may be repaired by enzymatic excision by the DNA polymerase of damaged regions in the DNA and insertion of polynucleotides in these gaps and in those resulting from recombination repairs. It is possible that in some cases the gaps are repaired without recombination. (From Giese, 1976: Living with Our Sun's Ultraviolet Rays. Plenum Press, New York; After Smith, 1974: in* Sunlight and Man. *Fitzpatrick* et al., *eds., University of Tokyo Press, p. 72.)*

Gaps in replicated DNA opposite damaged spots in the mother strand are repaired in several ways in the daughter strand other than by post-replication repair, although the mechanisms are not yet understood (P. Hanawalt, lecture Jan. 15, 1979).

tigating the repair from radiation damage revealed that the cell has an excision enzyme that uses information from the complementary DNA strand to scan the DNA. Upon detecting such variations as an altered nucleotide, mismatched regions, or thymine dimers, the enzyme excises that portion of the DNA along with several other nucleotides. The excised portion of the strand is then replaced through replication by DNA polymerase I using information on the complementary strand. Finally, the loose end of the newly synthesized section of nucleotides is bound to the DNA molecule by a ligase (joining enzyme) (Fig. 15.18). The DNA is thus restored to its native state. The excision mechanism is sometimes called the *cut and patch mechanism* (Hanawalt and Setlow, 1975).

Excision repair is probably universal in cells, occurring even in the mycoplasmas (PPLO) (see Chapter 1), but it has been most extensively studied in bacteria and in mammalian cells in tissue culture. Evidence for excision repair also exists in plant cells. When repair fails to occur, mutation may result. Nucleotide turnover after irradiation suggests "idling" as the cause of mutation. It appears that excision and insertion of the removed nucleotides is followed by repeated subsequent excision and reinsertion without achieving repair. The cause of idling has not been determined (Villani *et al.*, 1978).

DNA may also be repaired after replication (postreplication repair). During replication of damaged DNA, gaps appear around the erroneous or damaged regions of the parental strands. The gaps in the daughter strands are presumably filled by uninjured sections of the parental strands. Repair of the parental strands after replication consists essentially of synthesis of the missing nucleotides in the DNA gaps, the DNA polymerase using the information on the complementary undamaged DNA strand for the repair. In both parental and daughter strands binding of the newly formed nuclectide chains to the loose ends of the DNA is done by a ligase (Fig. 15.19). Although the facts are clear, the exact manner in which postreplication repair is accomplished is still unexplained. Postreplication repair also appears to be rather general, although it has been studied mostly in bacteria and mammalian tissue culture cells. Little is known about the mechanism of postreplication repair in mammalian cells (Smith, 1977; Lavin, 1978).

Photoreversal of one type of ultraviolet radiation damage, the formation of thymine dimers, that occurs only in the light in the presence of a photoenzyme is considered in Chapter 27.

RECOMBINANT DNA

DNA of different genotypes of a species may be interchanged in a variety of ways. In bacteria that conjugate, the male member of the pair may transmit pieces of DNA to the female member of another genotype. Pieces of DNA of one bacterial genotype may also be transmitted and combined to the DNA of another genotype by a phage vector (transduction). Pieces of dissolved DNA of one genotype may also enter a bacterium of another genotype through its surface, transforming its genotype (transformation). In eukaryotic germ cells DNA recombination occurs during crossing over between chromosomes contributed by the two parents in prophase I of meiosis. Pieces of chromosomes shattered by ionizing radiation may recombine. Recombination of two different pieces of DNA is therefore a common occurrence in nature. Recombinant DNA is composed of pieces of DNA from different sources. It now has come to mean a combination of DNA from different species.

Recently, exciting exchanges of DNA between cells of different species even as distantly related as bacteria and animal cells or bacteria and plant cells have been accomplished *in vitro*. This occurs by way of a vector, either a phage or a *plasmid* (small circular piece of DNA; one or more may be found in all kinds of cells). The possibility of such exchanges rests largely upon the presence in bacteria of enzymes known as *restriction endonucleases*. These enzymes are active on all DNA, which is now thought to be identical in general chemical composition and structure wherever it is found, inasmuch as the genetic code seems to be universal to all life, cellular and viral.

Restriction endonucleases recognize and nick a strand of DNA at unique, rotationally symmetrical $\left(\text{e.g., } \begin{matrix}\text{AAG} & \text{CTT} \\ \text{TTC} & \text{GAA}\end{matrix}\right)$ sequences of nucleotides on the two strands of DNA, usually four to six nucleotides long, thereby chopping the DNA into large discrete fragments. The size of the fragment is dependent upon the length of DNA between the unique sequences recognized by the endonuclease. Fragments of 3000 to 4000 nucleotide pairs could include a structural gene (assuming 3000 nucleotides for a sequence of 1000 amino acids and a polypeptide of 30,000 MW). Some fragments may be larger and include several genes; some may be smaller. Fragments may be sorted to size by sucrose gradient band centrifugation (Murray, 1976).

Restriction endonucleases of special interest for the present discussion nick the DNA phosphodiester bonds a few base pairs apart in opposite strands of the DNA duplex, producing staggered breaks. The loose ends of the staggered breaks are said to be "sticky," inasmuch as they are capable of attachment to other pieces of DNA having complementary nucleotide base sequences (Fig. 15.20). They are therefore especially useful for recombination studies. Rejoining of the matched loose ends of the pieces of DNA is catalyzed by action of a polynucleotide ligase. The DNA of a bacterial species is protected from its own restriction endonuclease by methylation of nucleotides after replication so that the endonuclease cannot sever the phosphodiester bonds.

DNA can be isolated from an organism (even single genes may in some instances be isolated) (Brown, 1973) and subjected to the restriction enzymes, resulting in the formation of fragments of DNA consisting of thousands of nucleotide pairs. Bacterial plasmids chosen as vectors may be cleaved by the same enzyme. In a mixture of the organismal and plasmid DNA some of the plasmids combine with pieces of the foreign DNA (Fig. 15.20). When such plasmids bearing organismal DNA are inserted into *E. coli*, for example, they can be "cloned," that is, multiplied at will with each bacterial division until large populations are produced. In this way it is possible to obtain innumerable copies of some sections of animal DNA that would be much more difficult to obtain by extraction from animal cells.

Theoretically, it should be possible to insert a gene for almost any part of an animal or a plant into a bacterium for study of its function or for analysis of its nucleotide sequences. DNA-containing ribosomal genes from the clawed toad *Xenopus laevis*, histone genes from sea urchin eggs, and fragments of mouse mitochondrial DNA have all been inserted into *E. coli* and cloned successfully (Murray, 1976).

The primary technique for cloning DNA involves reverse transcription from messenger RNA to produce the complementary DNA by use of reverse transcriptases. In cells that carry on limited types of protein synthesis this is especially easy. For example, the mRNA for syntheses of hemoglobin can be purified from the developing erythrocyte (reticulocyte). *Reverse transcriptase* makes possible synthesis of DNA complementary to this mRNA from a pool of the appropriate nucleoside triphosphates. A similar experiment has been performed with insulin mRNA; the mRNA is much easier to isolate than a particular gene. The DNA produced from mRNA in each of these cases is subjected to restriction endonucleases and mixed with nicked plasmids, then inserted into *E. coli* for cloning. The enormous possibilities of these methods as an aid to research should be evident. They could be extremely effective in producing such substances as hormones, antibodies, and nutritive polypeptides and could replace the laborious methods of extracting these substances from animal tissues.

It would seem that there should be little to fear from research on recombinant DNA, provided that the NIH guidelines are followed. *E. coli* does not pick up plasmids from the medium at an appreciable rate. If their surface is modified by heating cells in the presence of calcium chloride, the cells take up plasmids at a detectable frequency. Less than one plasmid in a million is incorporated into a cell, and only a few of these will insert sequences of DNA.

Davis (1977) also pointed out that the K12 strain of *E. coli* used in all recombinant studies was isolated from a patient over 50 years ago and has been transferred over and over on artificial media. Selection of mutants that showed improved growth on the media has altered the strain; therefore, it is doubtful that one of the mutants would now even grow in the human gut. Furthermore, the EK2 strain of K12 has been purposely impaired in its ability to grow except under very special conditions, including use of thymine and other constituents that must be supplied in the medium.

Research on recombinant DNA could prove

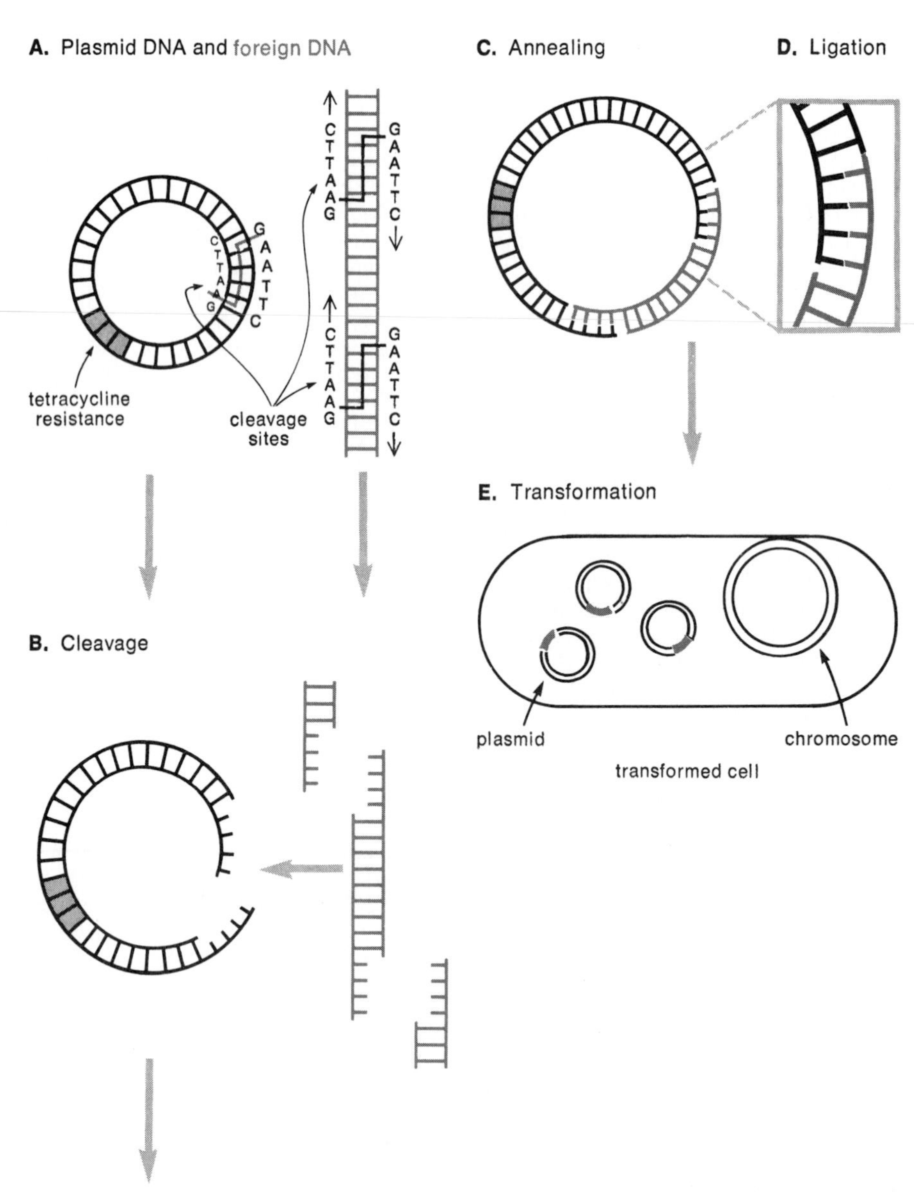

Figure 15.20. *In molecular recombination, foreign DNA is joined to plasmid DNA and introduced with the plasmid into a bacterial cell. The steps in the recombination process are shown in this simplified diagram.* A, *The plasmid has one sequence, and the foreign DNA has many, that are cleaved by a particular restriction enzyme.* B, *In these sequences the enzyme cleaves the two strands of DNA at staggered sites, creating short "sticky" ends that are all complementary to each other.* C, *Under annealing conditions an opened plasmid will sometimes add a fragment of foreign DNA and then close the circle.* D, *The "nicks" in the annealed ends are replaced by covalent bonds through the action of an enzyme, polynucleotide ligase, which forms a stable, covalently closed circle.* E, *The expanded plasmid is occasionally taken up by a cell, in this case a bacterium, in a process called transformation. The plasmid is replicated by the cell, and thus when the cell divides each daughter cell receives not only a copy of the bacterial chromosome but a copy of the plasmid as well. (The size of the plasmid corresponds to about 1 percent and the foreign DNA to about 0.1 percent of the bacterial chromosome.) (From Davis, 1977: Am. Sci.* 65: *548.)*

For their part in the identification, isolation and use of restriction enzymes in recombinant DNA research Werner Arber, Hamilton O. Smith and Daniel Nathans were awarded the 1978 Nobel Prize in physiology or medicine (Science 1978: 202: *1069).*

useful in solving many problems in cell physiology, especially those in which research depends upon enzymes or their products, by the cloning in bacteria of the DNA coding for these enzymes. Ample supplies of enzymes could be obtained from heat- or cold-resistant organisms, isozymes from heat- or cold-adapted organisms, enzymes involved in synthesis of cell membrane constituents, and digestive and oxidoreductive enzymes; these are just a few examples. Proteins regulating cell metabolism and muscular function are other examples. Enzyme products that inhibit or stimulate cell division, interferons that attack animal viruses, neurotransmitters, and hormones are present in such small quantities in cells as to make study of their properties and effects laborious. These products might be obtained in quantities permitting resolution of controversial findings.

One example of biological engineering is given in Figure 15.21, illustrating the transfer of a nitrogen-fixing gene (nif^+) from the nitrogen fixer *Klebsiella pneumoniae* to a deficient strain of the same species, to *E. coli* C, to *Klebsiella aerogenes,* to *Salmonella typhimurium,* to *Aerobacter tumefacians* (which lacks a factor permitting expression of nitrogen fixation) on to *E. coli,* and to *Azotobacter vinelandii* unable to fix nitrogen. The value of nitrogen fixation in species other than the limited number present in nature is evident (Postgate, 1978).

Research on recombinant DNA has caused considerable public opposition. However, scientists themselves, recognizing the possibility of danger, held a conference at Asilomar, California, in August 1975 to formulate rules to prevent damage to people and ecosystems from their researches. A danger is the possible development of new recombinants that would have characteristics harmful to people, other animals, plants, and essential prokaryotes—the "Andromeda strain" danger (Davis, 1977). For example, should a gene for cancer induction be incorporated from foreign DNA into *E. coli* and escape confinement, it could cause cancer on a vast scale, considering the ubiquity of *E. coli* in the human gut, in soil, and in water. There is always the lurking danger of use in biological warfare of such strains or ones that give antibiotic resistance or have other toxic effects.

Controversy on recombinant DNA continues, with numerous articles appearing both in scientific and popular literature. Competent scientists support both sides of the arguments. It is not fruitful to review these arguments here; reference is made to a defense for recombinant DNA research with appropriate guidelines (Grobstein, 1977) and a criticism of the research (Sinsheimer, 1977). The United States National Institutes of Health (NIH) have drawn up guidelines, with both graded physical and graded biological restrictions, depending upon the possible inherent dangers of the planned research being considered for their sponsorship. For example, a "shotgun" approach, in which the entire DNA of an animal cell is subjected to restriction endonucleases and then mixed with plasmids of *E. coli* for cloning, is considered dangerous. Undesirable genes, which are repressed in the original environment, might be released from repression in such cloned cells and confer highly undesirable characteristics upon *E. coli.* Experiments with small DNA fragments, especially if purified in some manner, are considered much safer.

Research effort on recombinant DNA has

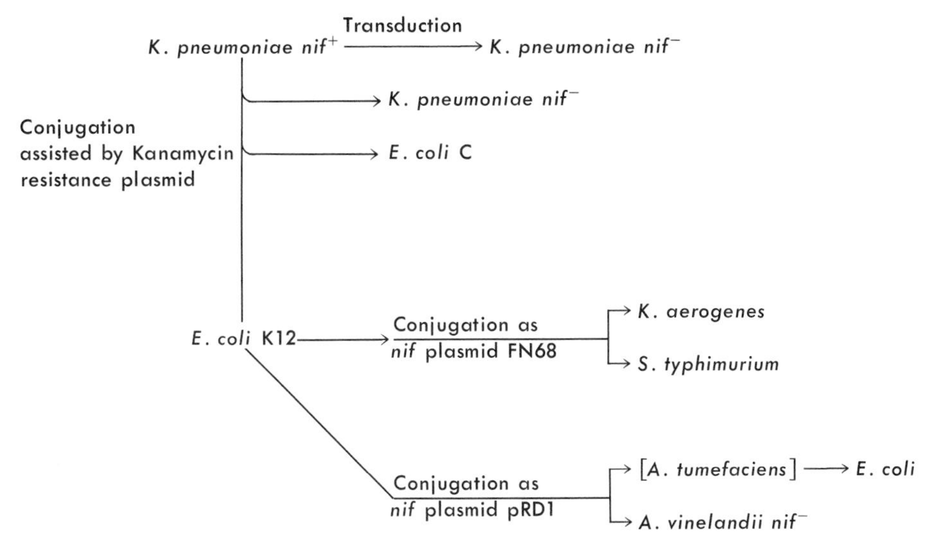

Figure 15.21. *Some transfers of* Klebsiella pneumoniae *nitrogen fixation genes to new bacterial hosts. (From Postgate, J., 1978: Nitrogen Fixation. Edward Arnold, London, p. 62.)*

now become worldwide, requiring relatively little equipment, and it will go on regardless of the arguments. Therefore, use of safeguards agreed upon by all is more important than attempts to restrict such experimentation. The possibility always remains of use of such research by maniacs bent on harming mankind. However, it is not that scientists produce evil things, rather that a small fraction of mankind has a tendency to make evil use of scientific creations.

DEVELOPMENT OF CELL ORGANELLES

Little is known about the molecular biology of the origin and development of organelles and other structures in the cell. It is likely that to a considerable extent such structures owe their origin to operation of weak cohesive forces attracting molecules into the most stable conformation. Proteins, for example, show, in addition to the primary peptide bond structure, secondary, tertiary, and quaternary structure mostly as a result of hydrogen bonds and other types of weak forces, upon which depend some of their specific properties (see Chapter 1, proteins). Denaturation by heat (and other agents) ruptures the bonds that form such structures. For example, an enzyme loses its specific catalytic properties when it is heated, but on cooling the catalytic activity reappears, essentially at the original level of activity. This indicates that on cooling the higher order of structure destroyed by heat has re-formed spontaneously, since it occurs without input of energy.

Chemical disaggregation of protein also is sometimes reversible; for example, hemoglobin is split by urea into two half-molecules of molecular weight 34,000. When urea is removed, the active hemoglobin molecule is reconstituted.

This principle of "self-assembly" (Watson, 1976) may be extended to more complex structures such as cell membranes. Thus, when a cell membrane is ruptured a new membrane is quickly re-formed (e.g., Fig. 8.2). It is thought that lipid molecules, held in reserve or newly synthesized, are quickly attracted to the remaining membrane in existence, forming cohesive bonds with it more readily than with the polar cytoplasm and thus extending it.

Some attempts have been made to gain insight into the processes involved in the formation of various cell organelles by studying what appear to be analogous reactions in colloidal systems. The interaction of colloids of opposite charge results in the formation of tactoids and coacervates (formed by various weak forces). Some globular inclusions in the cell have been likened to *coacervates* (produced by partial discharge or dehydration of a colloid leading to the formation of a larger, less hydrated unit), and spindle-shaped fibers, and even structures such as the division spindle, have been likened to *tactoids* (coacervates of fibrous proteins).

Alternatively, it is thought that many of the organelles in the living cell may arise *de novo* as outgrowths of the cell membrane. It is also postulated that they result from replication of existing organelles that act as templates, guiding the enzymes in depositing the new material (DeRobertis *et al.*, 1975).*

The determination of structures in cells and their heredity falls within the realm of molecular embryology; the reader is referred to monographs on the subject (Wessells, 1977).

*It is amusing at this point to indicate that although the crude constituents of the human body were once considered worth about 97¢, if one were to assemble the purified biological chemicals constituting a person, the cost would be many millions of dollars. If these chemicals were to be assembled into organelles, even had we the knowledge to do so, the cost would be in the trillions. This economics lesson indicates the value of the information in DNA as well as how far we are from knowing how to assemble chemicals into biological structures (Morowitz, 1976).

LITERATURE CITED AND GENERAL REFERENCES

Abdel-Monem, M. and Hoffmann-Berling, H., 1977: Enzymatic unwinding of DNA: III. Mode of activity of *E. coli* DNA-unwinding enzyme. J. Mol. Biol. *110*: 667–686.

Beadle, G. W., 1959: Genes and chemical reactions in *Neurospora*. Science *129*: 1715–1719.

Beadle, G. W. and Tatum, E. L., 1941: Genetic control of biochemical reactions in *Neurospora*. Proc. Natl. Acad. Sci. U.S.A. *27*: 499–506.

Becker, W. M., 1977: Energy and the Living Cell. J. B. Lippincott Co., Philadelphia.

Benzer, S., 1962: The fine structure of the gene. Sci. Am. (Jan.) *206*: 70–84.

Brown, D. D., 1973: The isolation of genes. Sci. Am. (Aug.) *229*: 20–29.

Burdon, R. H., 1976: RNA Synthesis. Halsted Press (Wiley), New York.

Campbell, A. M., 1976: How viruses insert their DNA into the DNA of the host cell. Sci. Am. (Dec.) *235*: 103–122.

Clark, B. F. C. and Marcker, K.A., 1968: How proteins start. Sci. Am. (Jan.) *218*: 36–42.

Cohen, G. N., 1967: Biosynthesis of Small Molecules. Harper & Row, New York.
Crick, F. H. C., 1966a: Codon-anticodon pairing: the wobble hypothesis. J. Mol. Biol. *19*: 548–555.
Crick, F. H. C., 1966b: The genetic code: II. Sci. Am. (Oct.) *215*: 55–60. (Also Oct. 1962, *207*: 66–74.)
Crick, F. H. C., 1970: Central dogma of molecular biology. Nature *227*: 561–563.
Curtis, R., III, 1976: Genetic manipulation of microorganisms—potential benefits and hazards. Ann. Rev. Microbiol. *30*:587–633.
Davis, B. D., 1977: The recombinant DNA scenarios: Andromeda strain, chimera and golem. Am. Sci. *65*: 547–555.
DeRobertis, E. D. P., Saez, F. A. and DeRobertis, E. M. F., 1975: Cell Biology. 6th Ed. W. B. Saunders Co., Philadelphia.
Dintzis, H. M., 1961: Assembly of the peptide chains of hemoglobin. Proc. Natl. Acad. Sci. U.S.A. *47*: 247–261.
Fenster, B. S., 1974: Eukaryotic DNA polymerases: their association with the nucleus and relationship to DNA replication. Int. Rev. Cytol. Suppl. *4*: 363–415.
Fersht, A. R., 1977: Enzyme structure and mechanism. W. H. Freeman, San Francisco.
Ficq, A. and Brachet, J., 1971: RNA-dependent DNA polymerase: possible role in amplification of ribosomal DNA in *Xenopus* oocytes. Proc. Natl. Acad. Sci. U.S.A. *68*: 2774–2776.
Fiddes, J. C., 1977: The nucleotide sequence of a viral DNA. Sci. Am. (Dec.) *237*: 55–67.
Freifelder, D., 1978: The DNA Molecule. W. H. Freeman, San Francisco.
Ganesan, A. T., Yehle, C. O. and Yu, C. C., 1973: DNA replication in a polymerase I deficient mutant and the identification of polymerases II and III in *B. subtilis*. Biochem. Biophys. Res. Commun. *56*: 155–163.
Gefter, M. L., 1975: DNA replication. Ann. Rev. Biochem. *44*: 45–78.
Genetic Code, 1967: Cold Spring Harbor Symp. Quant. Biol. Vol. 31, 750 pp.
Giese, A. C., 1976: Living With Our Sun's Ultraviolet Rays. Plenum Press, New York.
Gillham, N. W., 1974: Genetic analysis of the chloroplast and mitochondrial genomes. Ann. Rev. Genet. *8*: 347–391.
Goulian, M., 1971: Biosynthesis of DNA. Ann. Rev. Biochem. *40*: 855–898.
Goulian, M., Kornberg, A. and Sinsheimer, R., 1967: Enzymatic synthesis of DNA. XXIV. Synthesis of infectious phage ΦX174 DNA. Proc. Natl. Acad. Sci. U.S.A. *58*: 2321–2328.
Grobstein, C., 1977: The recombinant-DNA debate. Sci. Am. (Jul.) *237*: 22–33.
Gross, P. R., 1968: Biochemistry of differentiation. Ann. Rev. Biochem. *37*: 645–660.
Grossman, L., Brown, A., Feldberg, R. and Mailler, I., 1975: Enzymatic repair of DNA. Ann. Rev. Biochem. *44*: 19–43.
Grunberg-Manago, M., 1963: Polynucleotide phosphorylase. Prog. Nucleic Acid Res. Mol. Biol. *1*: 93–123.
Hanawalt, P. C. and Setlow, R. B., 1975: Molecular Mechanisms for Repair of DNA. 2 vols. Plenum Press, New York.
Hollaender, A., ed., 1971: Chemical Mutagens—Principles and Methods for Their Detection. 2 Vols. Plenum Press, New York.
Holley, R. W., 1966: The nucleotide sequence of a nucleic acid. Sci. Am. (Feb.) *214*: 30–39.
Hood, L. E., Wilson, J. H. and Wood, W. B., 1975: Molecular Biology of Eucaryotic Cells: A Problems Approach. W. A. Benjamin, Menlo Park, Calif.
Hooks, J., Gibbs, C. J., Jr., Chopra, H., Lewis, M. and Gaidusek, D. C., 1972: Spontaneous transformation of human brain cells grown *in vitro* and description of associated virus particles. Science *176*: 1420–1422.
Hurwitz, J. and Furth, J. S., 1962: Messenger RNA. Sci. Am. (Feb.) *206*: 41–49.
Ingram, V. M., 1958: How do genes act? Sci. Am. (Jan.) *198*: 68–74.
Ingram, V. M., 1963: The Hemoglobins in Genetics and Evolution. Columbia University Press, New York.
Ingram, V. M., 1972: Biosynthesis of Macromolecules. 2nd Ed. W. A. Benjamin, Menlo Park, Calif.
Jacob, F. and Monod, J., 1961: Genetic regulatory mechanisms in the synthesis of proteins. J. Mol. Biol. *3*: 318–356.
Jukes, T. H., 1963, 1965: The genetic code. Am. Sci. *51*: 227–244 and *53*: 477–494.
Katsoyannis, P. G., 1966: Synthesis of insulin. Science *154*: 1509–1514.
Khorana, H. G., 1968: Nucleic acid synthesis. Pure Appl. Chem. *17*: 349–381.
Kluyver, A. J. and Donker, H. J. L., 1926: Die Einheit in der Biochemie. Chemie der Zelle und Gewebe *13*: 134–190.
Kondo, S., 1975: DNA repair and evolutionary considerations. Adv. Biophys. *7*: 91–162.
Kornberg, A., 1960: Biologic synthesis of deoxyribonucleic acid. Science *131*: 1503–1508. (Nobel lecture.)
Kornberg, A., 1974: DNA Synthesis. W. H. Freeman, San Francisco.
Kung, S., 1977: Expression of chloroplast genomes in higher plants. Ann. Rev. Plant Physiol. *28*: 401–437.
Lande, A. and Ross, W., 1977: Viral integration and excision: structure of the lambda *att* sites. Science *197*: 1147–1160.
Lappe, M. and Morrison, R. S., 1976: Ethical and scientific issues posed by human use of molecular genetics. Ann. New York Acad. Sci. *265*: 1–208.
Last, J. A., comp., 1971: Protein Biosynthesis in Bacterial Systems. Marcel Dekker, New York.
Lavin, M. F., 1978: Postreplication repair in mammalian cells after ultraviolet radiation. Biophys. J. *23*: 247–256.
Laycock, D. G. and Hunt, J. A., 1969: Synthesis of rabbit globin by a bacterial cell-free system. Nature *221*: 1118–1122.
Lehninger, A. L., 1971: Bioenergetics. 2nd Ed. W. A. Benjamin, Menlo Park, Calif.
Lehninger, A. L., 1975: Biochemistry. 2nd Ed. Worth Publishers, Inc., New York.
Lipmann, F., 1963: Messenger RNA. Prog. Nucleic Acid. Res. Mol. Biol. *1*: 135–161.
Maclean, N. and Hilder, V. A., 1977: Mechanism of chromatin activation and repression. Int. Rev. Cytol. *48*: 1–49.
McKinnell, R. G., 1978: Cloning: nuclear transplantation in amphibia. University of Minnesota Press, Minneapolis.
Mechanism of Protein Synthesis, 1970: Cold Spring Harbor Symp. Quant. Biol., Vol. 34.

Merrifield, R. B., 1968: The automatic synthesis of protein. Sci. Am. (Mar.) *218*: 56–74.
Meselson, M. and Stahl, F. W., 1958: The replication of DNA in *E. coli*. Proc. Natl. Acad. Sci. U.S.A. *44*: 671–682.
Morowitz, H. J., 1976: The six million-dollar man. Bioscience *26*: 451–452.
Murray, K., 1976: Biochemical manipulation of genes. Endeavour *35*: 129–133.
Nirenberg, M. W. and Matthei, J. H., 1961: The dependence of cell-free protein synthesis in *E. coli* upon naturally occurring or synthetic polynucleotides. Proc. Natl. Acad. Sci. U.S.A. *47*: 1588–1602.
Nomura, M., 1969: Ribosomes. Sci. Am. (Oct.) *221*: 28–35.
Pestka, S., 1971: Inhibitors of ribosome functions. Ann. Rev. Biochem. *40*: 697–710.
Postgate, J., 1978: Nitrogen Fixation. Edward Arnold, London.
Reanney, D., 1977: Gene transfer as a mechanism of microbial evolution. Bioscience *27*: 340–344.
Replication of DNA in Microorganisms, 1969. Cold Spring Harbor Symp. Quant. Biol., Vol. 33.
Rich, A., 1963: Polyribosomes. Sci. Am. (Dec.) *209*: 44–53.
Rich, A. and Kim, S. H., 1978: The three-dimensional structure of transfer RNA. Sci. Am. (Jan.) *238*: 52–62.
Riley, D. E. and Keller, J. M., 1978: The ultrastructure of non-membranous nuclear ghosts. J. Cell Sci. *32*: 249–268.
Sager, R., 1972: Cytoplasmic Genes and Organelles. Academic Press, New York.
Sager, R., 1975: Patterns of inheritance of organelle genomes: molecular basis and evolutionary significance. *In* Genetics and Biogenesis of Mitochondria and Chloroplasts. Birky, Perlman and Byers, eds. Ohio State University Press, Columbus, pp. 225–267.
Sager, R., 1977: Cytological inheritance. *In* Cell Biology: a Treatise. Goldstein and Prescott, eds. Academic Press, New York, pp. 279–317.
Shaw, D. C., Walker, J. E., Northroup, F. D., Barrell, B. G., Godson, G. N. and Fiddes, J. C., 1978: Gene K, a new overlapping gene in bacteriophage G4. Nature *272*: 510–515.
Sinsheimer, R., 1959: A single-stranded DNA from bacteriophage ΦX174. J. Mol. Biol. *1*: 43–53.
Sinsheimer, R., 1977: Recombinant DNA. Ann. Rev. Biochem. *46*: 415–438.
Smith, A. E., 1976: Protein Biosynthesis. Halsted Press (Wiley), New York.
Smith, K., 1977: Ultraviolet radiation effects on molecules and cells. *In* The Science of Photobiology. Smith, ed. Plenum Press, New York, pp. 113–142.
Spencer, J. H., 1972: The Physics and Chemistry of DNA and RNA. W. B. Saunders Co., Philadelphia.
Spiegelman, S., 1964: Hybrid nucleic acids. Sci. Am. (May) *210*: 48–56.
Stein, G. and Stein, J., 1976: Chromosomal proteins: their role in the regulation of gene expression. Bioscience *26*: 488–498.
Stroun, M., Anker, P., Maurice, P. and Gahan, P. B., 1977: Circulation of nucleic acids in higher organisms. Int. Rev. Cytol. *51*: 1–48.
Szekely, M., 1978: Triple overlapping genes. Nature *272*: 492.
Szent-Györgyi, A., 1940: Vitamins. *In* The Cell and Protoplasm. Taylor, ed. Science Press, New York, pp. 159–165.
Temin, H. M., 1972: RNA-directed DNA synthesis. Sci. Am. (Jan.) *226*: 25–53.
Tomasz, A., 1969: Cellular factors in genetic transformation. Sci. Am. (Jan.) *220*: 38–44.
Travers, A. A. and Burgess, R. R., 1969: Cyclic reuse of the RNA polymerase sigma factor. Nature *222*: 537–540.
Villani, G., Boiteux, S. and Radman, M., 1978: Mechanism of ultraviolet-induced mutagenesis: extent and fidelity of *in vitro* DNA synthesis on irradiated templates. Proc. Nat. Acad. Sci. U.S.A. *75*: 3037–3041.
Watson, J. D., 1976: Molecular Biology of the Gene. 3rd Ed. W. A. Benjamin, Menlo Park, Calif.
Wessells, N., 1977: Tissue Interactions in Development. W. A. Benjamin, Menlo Park, Calif.
Witkin, E., 1969: UV-induced mutation and DNA repair. Ann. Rev. Genetics *3*: 525–552. Ann. Rev. Microbiol. *23*: 487–514.
Woese, C. R., 1970: Genetic code in prokaryotes and eukaryotes. Symp. Soc. Gen. Microbiol. *20*: 39–54.
Wood, W. B. and Edgar, R. S., 1967: Building a bacterial virus. Sci. Am. (Jul.) *217*: 61–74.
Yanofsky, C., 1967: Gene structure and protein structure. Sci. Am. (May) *216*: 80–93.
Yanofsky, C., 1976: The search for the structural relationship between gene and enzyme. *In* Reflections on Biochemistry. Pergamon Press, New York, pp. 263–271.
Zamecnik, P. C., 1960: Historical and current aspects of the problem of protein synthesis. Harvey Lect. *54*: 256–281.

CHAPTER 16

REGULATION OF CELLULAR METABOLISM

In normal cells few soluble intermediate metabolic products accumulate. Only in cells in which mutation has blocked some step in a synthetic or degradative pathway or in cells that have been exposed to metabolic inhibitors is there accumulation of intermediates that can be identified. Therefore, studies on such cells have been of great value in elucidation of intermediary metabolism and synthetic pathways. The fleeting nature of the intermediates of metabolic reactions in normal cells suggests that cellular metabolism is regulated in such a manner that various substances are present in the cell only in concentrations optimal for cell functions.

Regulation of biosynthetic reactions is independent of regulation of corresponding catabolic (energy-liberating) reactions. This results from the fact that *anabolic (synthetic) reactions* are not a reversal of catabolic reactions and occur by different pathways. One of the ways in which anabolic and catabolic reactions are separated is by the use of NADPH as the reductant in anabolic reactions and NADH as the reductant in catabolic reactions (see Chapters 12, 14, and 15). The rate of anabolic and catabolic reactions is thus regulated by different mechanisms. "Dual regulation is a general principle characteristic of all corresponding or parallel anabolic and catabolic pathways," Lehninger (1975).

Regulation of both anabolic and catabolic reactions in the cell is achieved in a number of ways. To illustrate mechanisms, several examples of intracellular regulation and two examples of intercellular regulation are given, with no attempt at complete coverage.

TYPES OF REGULATION

Some metabolic regulation may result from operation of the law of mass action and some may be due to competition of enzymes for a common substrate. However, most metabolic regulation is of a more complex nature, and occurs by affecting either enzyme activity or enzyme synthesis.

Enzyme activity is dependent upon many factors (Fig. 16.1). The manner in which the polypeptide chain folds (secondary and tertiary structures), the attachment of prosthetic groups to the protein and, in enzymes made up of several polypeptide chains, the way the subunits complex (quaternary structure) affect the enzyme activity. Also, metabolic end products may cause feedback activation or feedback inhibition of the enzyme.

The rate of enzyme action depends in part upon the concentration of the enzyme. The enzyme concentration is determined by the balance between the rate of synthesis and the rate of degradation of the enzyme.

Enzyme synthesis can be induced or repressed (see Chapter 15). The rate of polypeptide synthesis is determined in part by the concentration of the mRNA specific to the polypeptide. The concentration of the mRNA is determined by the balance between its synthesis and degradation. The rate of mRNA synthesis is determined by the rate of assembly of the nucleoside triphosphates on the DNA template and the rate of release of the finished product from the template.

Finally, the rate of translation of the mRNA message is affected by the rate of polypeptide

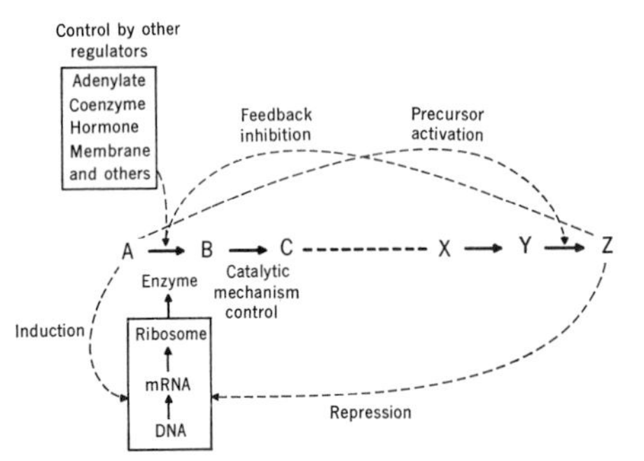

Figure 16.1. *Schematic representation of mechanisms for the regulation of enzyme activity. (From Hammes, G. G., and Wu, C.-W., 1971: Science* 172: *1205–1211. Copyright 1971 by the American Association for the Advancement of Science.)*

initiation, ribosomal movement, and release of the finished polypeptide chain from the ribosomes. Conceivably regulation could occur at any one, several, or all of these levels. Some of these regulatory mechanisms are considered in more detail in the following paragraphs.

MASS ACTION IN REVERSIBLE REACTIONS

Two substances, A and B, react to form C and D and the process is reversible. When C and D begin to accumulate, the rate of formation of A and B from C and D increases:

$$A + B \rightleftharpoons C + D \qquad (16.1)$$

Thus, unless C and D are used up or diffuse away, the reaction reaches an equilibrium in which as much C and D are formed as A and B. The *Lohmann reaction,* in which the high-energy phosphate bond of ATP is transferred to creatine, forming creatine phosphate (CP), is an example of such a simple regulation:

$$ATP + C \rightleftharpoons CP + ADP \qquad (16.2)$$

Here CP will accumulate as long as an excess of ATP is present. When ATP is in short supply, high-energy phosphate will pass from the CP to ADP.

COMPETITION FOR SHARED SUBSTRATE

A "bottleneck" occurs when each of two metabolic pathways compete for a common metabolite or cofactor. An example is the coenzyme NADH produced during the oxidative step in muscle glycolysis. NADH can be oxidized either by lactic dehydrogenase or by oxidative reactions in the electron transport chain. When aerobic conditions prevail, the oxidative reactions predominate.

Such competition is commonplace in intermediary metabolism. For example, α-ketoglutarate is a participant in the Krebs cycle (see Fig. 12.8) as well as in the glutamate dehydrogenase cycle. Glutamate is formed in the presence of ammonia and α-ketoglutarate. In other words, in the presence of NH_3, the Krebs cycle may be suppressed.

It is possible that the *Pasteur effect* is of this nature also. Many years ago Pasteur noticed that fermentation (glycolysis) in yeast is suspended when oxygen is available. It is well known that both oxidative and glycolytic pathways require inorganic phosphate and ADP. Since phosphorylation proceeds at a much higher rate in mitochondrial oxidative metabolism than in glycolysis, the shift from glycolysis to oxidative metabolism may rest upon this fact.

It is also possible that the suppression of aerobic oxidative processes by addition of glucose to some tissues is explainable by competition for a common substrate. Tissues such as muscle in which this effect is noticed are ones with an exaggerated glycolytic pathway. Aerobic glycolysis depletes the inorganic phosphate and adenine nucleotides, thus interfering with oxidative metabolism.

ALLOSTERIC INHIBITION AND ACTIVATION

One of the most prevalent controls for a series of reactions is by *negative feedback inhibition* (Gerhart and Pardee, 1964; Whitehead, 1970). Suppose that in the series of reactions:

$$A \xrightarrow{E_1} B$$

$$B \xrightarrow{E_2} C$$

$$C \xrightarrow{E_3} D$$

$$D \xrightarrow{E_4} F$$

it is found that the end product (F) inhibits the first of the series of reactions catalyzed by enzyme 1 (E_1). Inhibition of the first reaction causes the whole reaction series to be slowed down. For example, it has been found that the end product sometimes reacts with the enzyme, reducing its activity. The presence of F on enzyme 1, in addition to A, causes a change called the *allosteric effect,* a configurational alteration of the enzyme, making it less catalytically ac-

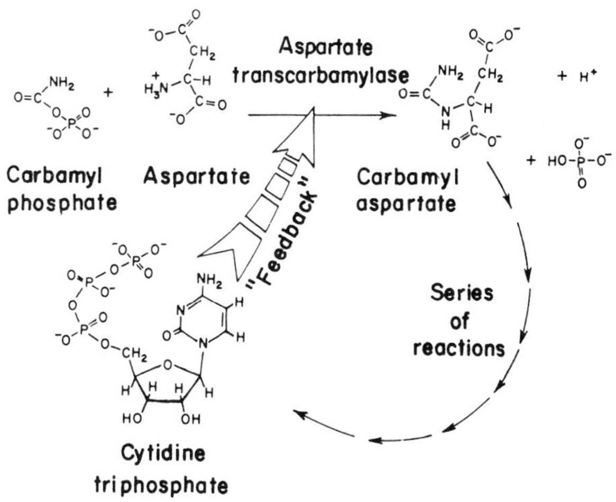

Figure 16.2. *Schematic diagram of feedback inhibition in the pyrimidine pathway of* E. coli *(the end product, cytidine triphosphate, inhibits the activity of the enzyme aspartate transcarbamylase). (From Gerhart and Pardee, 1964: Fed. Proc. 23: 728.)*

tive. As soon as the excess of the end product F is used up in subsequent reactions, the reaction series accelerates again, because F will have then become detached from E_1. Allosteric literally means "another space," that is, the enzyme not only binds the substrate at one site but also the end product at another place. The end product binding causes a non-covalent change, probably a rotation of one part of the enzyme molecule with respect to another.

For example, consider the inhibition of aspartate transcarbamylase by CTP (cytidine triphosphate). The normal substrate, aspartate, has little in common in its structure with CTP, the inhibitor (Fig. 16.2). The probability that this enzyme might accept the two substrates at two different points, e.g., at two different *sites,* was indicated by heat treatment of the enzyme: it was found that heat treatment could destroy the inhibitor site without affecting the catalytic site. However, heat treatment caused dissociation of the enzyme into four subunits, each subunit still possessing catalytic activity. It is thought that aspartate reduces the reactions between the subunits, causing looser association, whereas CTP increases them, bringing them closer together (Gerhart and Pardee, 1964). Another example of feedback inhibition is found in bacteria, in which nitrogen conservation is important; amino acid synthesis is closely regulated by feedback inhibition. In larger eukaryotic cells, such synthesis is less closely regulated (Lehninger, 1975).

An example of feedback inhibition in a more complex series of reactions is shown in Figure

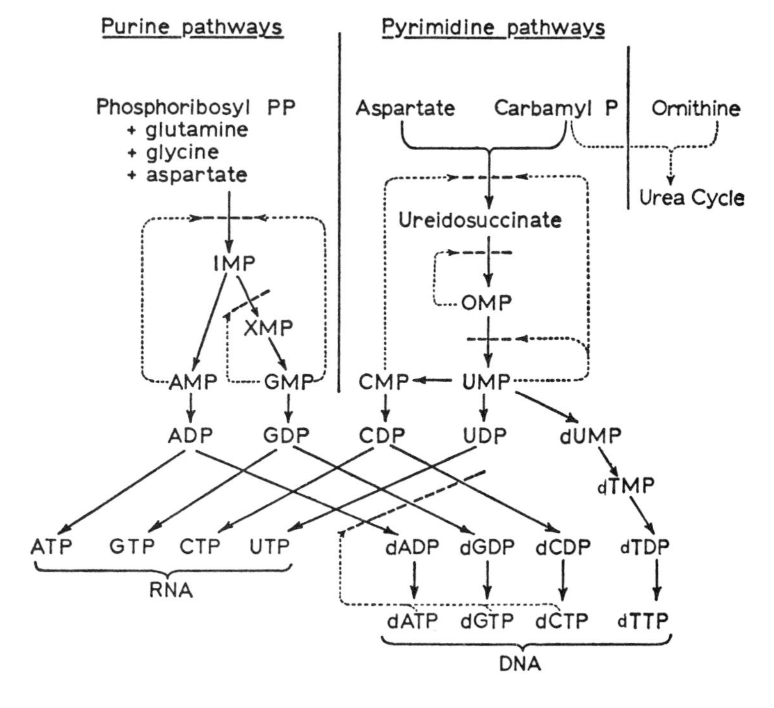

Figure 16.3. *Some feedback relationships (dotted arrows) involved in the regulation of RNA and DNA synthesis. I, inosine; X, xanthosine; A, adenosine; G, guanosine; O, orotidine; U, uridine; C, cytidine; T, thymidine. (From Paul, 1964: Cell Biology. Stanford University Press, Stanford, Calif.)*

16.3 for some of the reactions involved in RNA and DNA synthesis. In this series of reactions, end products affect enzymes at ten different places (dotted arrows). Some similar examples are found in pyrimidine synthesis and purine synthesis (Henderson, 1962). Such regulation is also found in many reactions during development and differentiation of embryos (Walker, 1965).

The allosteric effect of metabolite interaction with the enzyme of the primary reaction need not always be inhibitory. A configurational change might well increase the activity of the enzyme. Thus, it has been found in many cases that a small molecule that does not take part in a reaction is necessary for the reaction; in other words, it acts as an enzyme activator (Koshland, 1973; Kvamme and Pihl, 1968). Similarly, if a metabolite had such an activating effect upon an enzyme in the sequence, the net effect would be to accelerate the series of reactions. An example is observed in the reactions involved in glycogen control. Glycogen is made from glucose-6-phosphate in three enzymatic steps. First glucose-6-phosphate is isomerized to glucose-1-phosphate, and then glucose-1-phosphate reacts with uridine triphosphate (UTP) to form uridine diphosphate-D-glucose (UDPG). The glucose moiety from UDPG is incorporated into glycogen. A good supply of glucose-6-phosphate stimulates the synthesis of glycogen in a cell by activating the enzyme that catalyzes the conversion of UDPG to glycogen.

The advantage to the cell of end product control is the speed with which changes in rate of reaction may be instituted. In this way an equilibrium state may readily be maintained by constant adjustment and readjustment of rates of reaction.

The allosteric effect is also observed in the action of *regulator enzymes*. The curve for the effect of substrate concentration on the rate of reaction for a regulator enzyme is different from that for most enzymes; for most enzymes the curve is hyperbolic, and the rate of reaction falls off as the substrate concentration increases, presumably because the active sites on the enzyme have been saturated (Fig. 10.13). For a regulator enzyme like aspartate transcarbamylase, however, the curve for the rate of reaction with rising substrate concentration is sigmoid (Fig. 16.4). This is thought to indicate that binding one substrate molecule to the enzyme makes binding the next one easier, and the catalytic effect of the enzyme is greatest when four substrate molecules are bound, the allosteric effect presumably reaching its optimal state. This resembles the uptake of oxygen by hemoglobin, which also is made up of four subunits (two α and two β), each with a heme group (see Fig. 2.15). Uptake of oxygen by hemoglobin causes the α units to move closer to one another and the β units to separate more. This allosteric change also affects the properties of hemoglobin such that its oxygen binding is sensitive to pH. When the oxygenated hemoglobin comes into contact with tissue fluid, which by virtue of accumulated carbon dioxide has a lower pH, the reduced oxygen binding by hemoglobin permits oxygen to be given off at even relatively low oxygen partial pressures, a valuable adaptation for supplying cells in need of oxygen. In both cases this represents a regulation of the function of the enzyme activity by an allosteric effect induced by the substrate (Changeoux, 1965).

REGULATION BY COVALENT CHANGES IN ENZYMES

An example of an enzyme in animal tissues in which regulation is achieved by a *covalent*

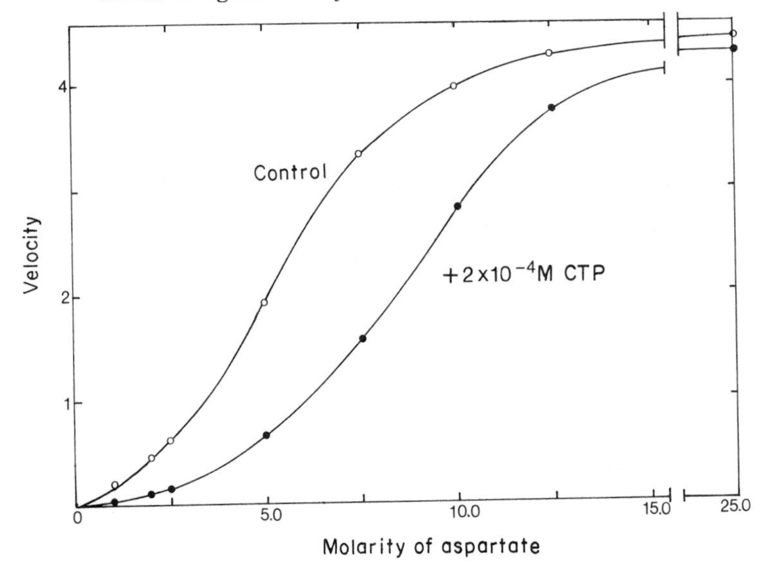

Figure 16.4. Effect of increase in concentration of substrate on rate of reaction of a regulator enzyme (control). The aspartic transcarbamylase is a regulator enzyme and its rate shows a characteristic sigmoid curve. Shown also is the negative feedback inhibition between the end product cytidine triphosphate and the substrate with increasing substrate (aspartate) concentration. (Velocity = units of activity per milligram of protein $\times 10^{-3}$. The reaction mixture contained 3.6×10^{-3}M carbamyl phosphate, with aspartate varied as indicated; 2×10^{-4}M CTP when used; 0.04M potassium phosphate buffer, pH 7.0: and $9.0 \times 10^{-2}\mu g$ of enzyme protein per ml.) (From Gerhart and Pardee, 1964: Fed. Proc. 23: 731.)

change in the enzyme is glycogen phosphorylase. This enzyme catalyzes the breakdown of glycogen to yield glucose-1-phosphate. Phosphorylase exists in two forms, phosphorylase *a*, the more active form and phosphorylase *b*, the less active form. Phosphorylase *a* is made up of four subunits, each subunit containing a phosphorylated hydroxyl end group. When these phosphates are removed by the enzyme phosphorylase phosphatase, the phosphorylase *a* dissociates into two half-units that are much less active in cleaving glycogen than phosphorylase *a*.

Relatively inactive phosphorylase *b* can be reactivated to phosphorylase *a* by the enzyme phosphorylase kinase, which phosphorylates the serine residues at the expense of ATP. Regulation in this case is achieved by the action of two enzymes, which shift the balance between breakdown and synthesis in accordance with the needs of the cells.

Other examples of regulation by covalent changes in enzymes are the proenzymes in the vertebrate gut: pepsinogen and trypsinogen. Each proenzyme is secreted in inactive form and only after hydrolysis of a polypeptide from the proenzyme do the active enzymes, pepsin and trypsin, appear. This occurs as need arises with the appearance of food in the gut (Lehninger, 1975).

REGULATION OF NUCLEIC ACID SYNTHESIS

In bacteria, DNA synthesis goes on almost continuously (for about 80 to 95 percent of the generation time) (see Chapter 7). In eukaryotic cells, DNA synthesis occurs for about half the division interval during the synthetic (or S) period. The synthetic period lasts 7.5 hours in the cells of the broad bean *Vicia faba,* 10.5 hours in the cells of *Tradescantia paludosa,* 9.9 hours in fibroblast cells of the mouse and 5.8 hours in fibroblasts of the Chinese hamster (Kihlman, 1966). During the remainder of the division interval time no DNA appears to be synthesized (see Fig. 26.1). The nature of the signal given for beginning and ending DNA replication has not been determined. Presumably replication stops when the genome has been duplicated (Lark, 1969).

The nature of regulation of DNA synthesis in salivary gland cells of the flies *Drosophila* and *Chironomus,* in which the chromosomes multiply to form many parallel units, is not known. The duplicated polytene chromosomes become easily visible under the light microscope, and the location of various genes can easily be spotted because all the strands are in register with one another (see Fig. 7.11). Gene amplification to facilitate transcription when RNA is needed in quantity (e.g., rRNA) has already been discussed (see Chapter 7).

Transcription of genes to various types of RNA poses certain problems. Presumably rRNA and tRNA are formed continuously in the bacterial cell and they remain in the cell, being much more resistant to nucleases than mRNA. Messenger RNA for production of constitutive enzymes is formed continuously, but in bacteria it is rapidly broken down. The life span of an mRNA molecule is, apparently, related to the needs for its translational activity. The short life span of mRNA in *E. coli* and the long life span of mRNA in the reticulocytes of a mammal probably represent two extremes. However, precise data do not seem to be available about the life span of mRNA in other types of eukaryotic cells, although generally it appears to be long. Some messages may even be stored in eggs awaiting activation through fertilization. Messenger RNA for synthesis of induced enzymes is formed only in the presence of the inducer. When the inducer is absent repression occurs, as already discussed (see Chapters 7 and 15).

Although ribosomes are normally rather stable in active cells, they disappear rapidly when the cell is starved, perhaps because active ribosomes with attached mRNA appear to be resistant to nuclease, whereas inactive ones are not and consequently are degraded. A starved cell, therefore, has fewer ribosomes. Addition of nutrient leads to rapid synthesis of rRNA and the formation of ribosomes. Thus the quantity of rRNA is quickly adjusted to the needs of the cell (Koningsberger and Bosch, 1967).

REGULATION OF PROTEIN SYNTHESIS

In some cells a given protein is made in large quantity and for a prolonged time, e.g., the production of hemoglobin in reticulocytes (progenitors of red blood cells). In such cells a long-lived mRNA has survival value, that is, it conserves energy that would be wasted in producing supplies of mRNA quickly degraded. Therefore, it is not surprising to find that in reticulocytes the mRNA may last for a number of days (Pestka, 1971).

In bacteria undergoing rapid growth and division it is likely that only a few protein molecules of each kind are needed per cell. For example, some data indicate that only about 15 copies of some of the enzymatic proteins are made per cell. Usually, when the required number of proteins has been synthesized the mRNA is hydrolyzed. Such a short-lived mRNA suits the bacteria, which have no need for continued synthesis of one protein. A bacterium that continued synthesis would soon be eliminated in competition with well-regulated cells. It has been shown that the *E. coli* cell has only about 1000 mRNA molecules, which at any one instant must meet all the varying synthetic and catabolic needs of the cell (Yanagisawa, 1963).

ENZYME INDUCTION, REPRESSION, AND ACTIVATION

When *E. coli* cells are placed in a medium containing a galactoside sugar (e.g., lactose) as the sole source of carbon, they are at first incapable of utilizing the sugar because they lack the enzyme β-galactosidase. After a few minutes, however, lactose has induced the *E. coli* to form β-galactosidase, which is capable of digesting lactose into the usable products glucose and galactose. After induction of β-galactosidase the enzyme can be extracted from the lysed cells. The induced enzyme disappears again if the cells are supplied with glucose without a β-galactoside for a while. The enzyme must again be induced if such cells are again placed in a β-galactoside medium.

This experiment indicates that the bacteria may possess enzymes of two kinds: constitutive and induced. *Constitutive enzymes* are always present, at least in small quantities. *Induced enzymes* are absent from the cell and appear only on presentation of an inducing substrate (Vogel, 1971), but they fail to appear if such protein synthesis inhibitors as puromycin or chloramphenicol are added to the substrate. This indicates that protein synthesis is necessary for induction of the enzymes. However, mutants of *E. coli* are known that are incapable of developing the enzyme under conditions favorable for induction in inducible strains.

The experiments on enzyme induction led to far-reaching findings in the hands of Jacob and Monod (both of whom, with Lwoff, received the Nobel Prize in 1966), because they opened the way for an analysis of the genetic basis of induction and repression (Ptashne and Gilbert, 1970).

From what has been considered in Chapters 7 and 15 it is evident that a polypeptide, one or more of which form an enzyme, is synthesized from a message transcribed from a gene by mRNA. This means that a gene must encode the information for each inducible enzyme, just as another gene encodes the information for synthesis of a constitutive enzyme. Since no inducible enzyme is present in the absence of the inducing substance, the gene for this enzyme, if present, must in some manner be repressed, because it is being neither transcribed into mRNA nor translated into protein. A possible mechanism for explaining the phenomenon is given here.

Adjacent genes concerned with some common activity may be grouped into units of the chromosome called *operons* (Stent, 1963). In *E. coli* the gene for the permease concerned with the active transport of the β-galactoside into the cells and the gene for the β-galactosidase are in the same operon, next to the β-galactosidase operator. The genes of the operon, known as the *lac operon,* are under the control of the *operator,* a short segment of the chromosome that turns the β-galactosidase gene on and off, depending on the presence or absence of an inducer. Present also is a regulatory gene, which codes for a relatively low-molecular-weight protein that can combine with the operator to turn off the production of β-galactosidase mRNA. In the presence of an inducer (a β-galactoside), the repressor substance is inactivated by reaction with the inducer. The inactivated repressor cannot combine with the operator, and the production of the β-galactosidase is allowed. All of these ideas are summarized in Figure 16.5.

By the use of a special technique of combining a repressor with a labeled inducer (in this case an analogue for lactose, isopropyl thiogalactoside, which is not metabolized), it proved possible to isolate the repressor for the lac operon but in quantities too small for an effective study of its nature. However, mutants of *E. coli* were obtained in which not only was the inducer bound more firmly to the repressor, making assay easier, but also more repressor was produced, providing sufficient quantity for purification and study. The repressor for the lac operon proved to be a protein with a site that recognizes the DNA bases of the operator (probably 12 in number) and a site that recognizes the inducer. It is a tetramer, each unit of molecular weight 37,000, composed of 347 amino acids. The protein binds to native but not to denatured *E. coli* DNA, presumably at the lac operon. Thus the target of the repressor proves to be DNA. Presumably repression results from interference by the protein with the activity of the RNA polymerase, which cannot initiate transcription of mRNA for transcription of galactosidase in the presence of the repressor. Several other repressors have been isolated, all with properties similar to those of the lac repressor. Sometimes, however, a protein product *(activator)* of a gene turns a system on rather than off, as in the case of the enzymes for utilization of the sugar arabinose in *E. coli* (Ptashne and Gilbert, 1970).

It is seen in Figure 16.5 that all the genes in the operon are affected by the inducer and the repressor; in this case both the permease and the β-galactosidase are affected at the same time. However, as discussed in Chapter 18, the permease and the β-galactosidase systems may be uncoupled by mutations. Thus some mu-

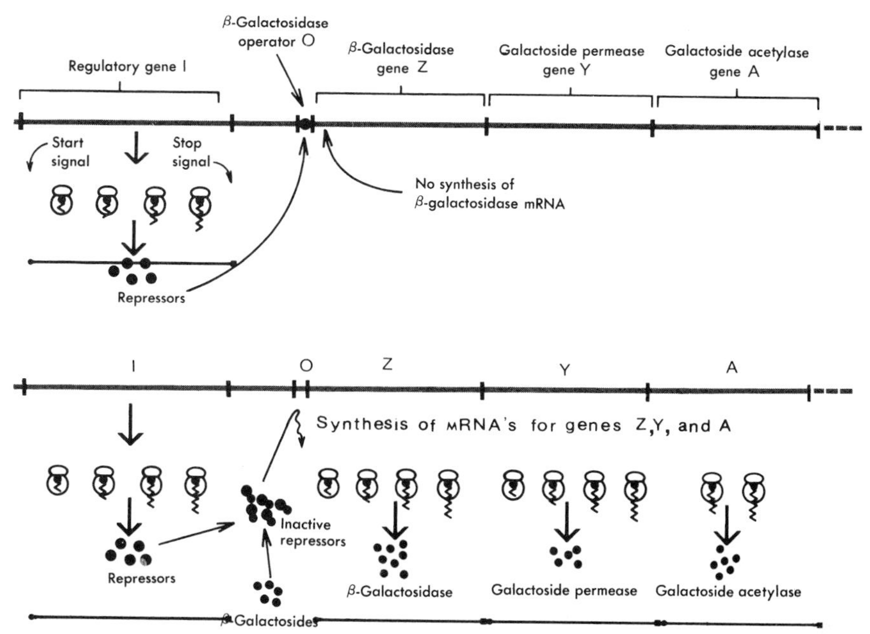

Figure 16.5. The operon and its functioning. The hypothesis currently favored for how the interaction of repressor, corepressor, and operator controls the synthesis of the E. coli *proteins β-galactosidase and β-galactoside permease. Repressors are shown combining with DNA at the operator. This point has not yet been proved, and the alternative hypothesis exists that repressors act by combining with mRNA, thereby preventing its attachment to ribosomes and so allowing its rapid enzymatic breakdown to nucleotides. It is important to note that, under both sets of hypotheses, a specific repressor decreases the amount of a specific mRNA molecule. The genes X, Y, and A are in one operon called the lactose operon. Galactoside acetylase is required in the metabolism of galactoside, but the details are of no consequence here. (From Watson, 1976: The Molecular Biology of the Gene. 3rd Ed. W. A. Benjamin, Menlo Park, Calif.)*

tants are found that can accumulate a β-galactoside but cannot digest it, whereas other mutants possess the capacity to synthesize β-galactosidase, as can be demonstrated by disrupting the cells and liberating the enzyme, but they cannot obtain the sugar because they lack the permease system.

An interplay between repressors and inducers enables the cell to synthesize enzymes in relation to needs and availability of nutrients in the medium. This is important for survival because it would be difficult to maintain at all times in the confines of a small cell the variety of enzymes that might possibly be useful.

Many cases of induction have now been found in animal and plant cells, as well as in many bacteria, and the phenomenon of induction and repression appears to be general, although differences may well exist in details. For example, animal and plant cells are diploid, in contrast to most bacteria, which are haploid, so that recessive mutants will not be expressed unless homozygosity is present at the locus for the mutated gene. This makes their detection more difficult in eukaryotic than in prokaryotic cells.

Although most inborn errors of metabolism in man have in the past been considered mutations in structural genes, it is now necessary to reassess them in view of inducer and repressor regulations found in bacteria. Many metabolic disorders of man have been shown to be hereditary and probably variations of regulation (McKusick, 1975). Many of the articles in the series *Advances in Human Genetics* deal with such abnormalities. A hopeful approach to metabolic studies and controls in mutant cells of man is the use of tissue culture (especially skin) derived from individuals with metabolic disorders and fusion of cells of different genotypes (see Chapter 7).

If a product from a reaction is normal in structure and biological activity but is variable quantitatively, it is likely to be under control of an operator. When the operator fails in its function, probably by mutation, intermediary metabolic products may accumulate. Excessive production of porphyrin followed by its leakage into the urine and elsewhere in the body may be an example of this type of mutation in humans, being expressed by a failure to repress the production of the porphyrin.

It has sometimes been pointed out that genetic controls are slow to act, so that regulation of metabolism by induction and repression presents a rather slow mechanism for metabolic control. Enzyme induction, for example, requires several minutes or more, whereas feedback inhibition of a reaction may be immediate. Repression is also slow inasmuch as the enzymes in the cell, which are degraded only slowly, enable the reaction to continue as long as a supply of enzymes and substrates is available and conditions are otherwise favorable. Although repression is rather slow, it is often accompanied by end product inhibition of activity of the first enzyme in the series of reactions. This negative feedback mechanism makes possible very quick reduction of enzymatic activity, in addition to the slower repression. It is advantageous to the cell to have both rapid and slow regulatory mechanisms, the slow response being like a coarse adjustment, the rapid response like a fine one.

CYCLIC ADENOSINE MONOPHOSPHATE AS A REGULATOR

An important regulatory substance to come to attention in recent years is 3′,5′-cyclic adenosine monophosphate *(cyclic AMP)* (Fig. 16.6). It has been shown to regulate the activity of enzymes, hormones, and genes and serves as the intermediary for the action of hormones on cells in target organs. Sutherland started his studies on this substance a quarter of a century ago and received the Nobel Prize in medicine and physiology in 1971 for elucidation of its function (Robison *et al.,* 1971; Sutherland, 1972).

Cyclic AMP is present in cells to only a thousandth the concentration of ATP. However, its level rises considerably on stimulation with epinephrine (adrenalin). Cyclic AMP is formed from ATP by action of adenylate cyclase, located in the cell plasma membrane. Apparently the binding of the epinephrine to the receptor site in the membrane increases the activity of the adenylate cyclase located there, resulting in the production of cyclic AMP on the inner surface of the membrane where ATP is present in abundance. The cyclic AMP then diffuses throughout the cell. One of its actions is the conversion of the enzyme phosphorylase from the inactive to the active state. Phosphorylase then catalyzes the breakdown of glycogen stored in the cell to glucose-1-phosphate. The activation of phosphorylase is not direct. Cyclic AMP activates protein kinase; it in turn activates phosphorylase kinase, which then activates phosphorylase. The synthesis of glycogen is suspended when the need for glucose arises, since cyclic AMP inactivates glycogen synthetase during this interval (Pastan, 1972).

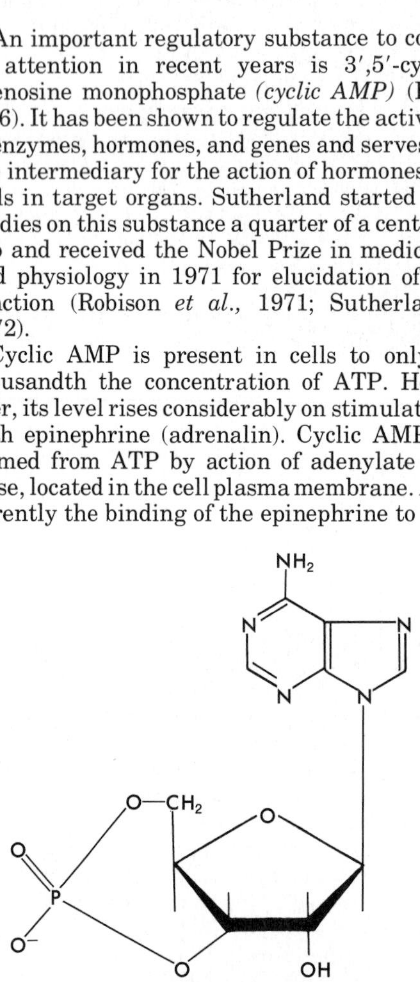

NH2
N N
N N
O—CH2 O
O
P
O⁻ O OH

Figure 16.6. *Cyclic AMP, formed when ATP reacts with itself in the presence of the enzyme adenyl cyclase. The "cycle" involves the attachment of both the 3′ and 5′ carbons of the ribose to the phosphate as shown. Cyclic AMP speeds or slows many different enzymes.*

Soon after its stimulatory action, the cell decomposes cyclic AMP to AMP by action of the enzyme phosphodiesterase, permitting the cell to return to its steady state condition. The level of phosphodiesterase thus depends upon the concentration of cyclic AMP in the cell (Fig. 16.7) (Rosen, 1968).

Cyclic AMP modulates not only epinephrine action but also that of other hormones—for example, the thyroid hormone, the parathyroid hormone, luteinizing hormone (which prepares the uterus to receive the embryo), adrenocorticotropic hormones, and vasopressin (concerned with blood pressure in the kidneys). Cyclic AMP regulates the sugar levels in muscle and is involved in visual excitation and in promoting aggregation of social amoebae of slime molds (Jost and Rickenberg, 1971).

Absence of cyclic AMP is favorable for transformation of cells in a tissue culture to cancerous types (Green, 1970). The details of these regulations are not of concern here and are considered in textbooks of biochemistry and physiology (e.g., Lehninger, 1975; Guyton, 1976) and review volumes (Horecker and Stadtman, 1971). Drugs may also produce some effects by affecting cyclic AMP levels (Rabin, 1972).

It has also been discovered that cyclic AMP regulates genic expression. Thus, when cyclic AMP was added to a cell-free *E. coli* preparation containing DNA from many genes for lactose metabolism, synthesis of the galactosidase was stimulated. In purified preparations it was also essential to add a special protein (catabolite gene activation protein) to make possible galactosidase synthesis (Pastan, 1972). Cyclic AMP is present in *E. coli* and evidence indicates that it stimulates the transcription of many genes, while repressors at the same time prevent unnecessary transcriptions. Thus *E. coli* achieves considerable flexibility in needs for utilization of available food substrates (Pastan, 1972).

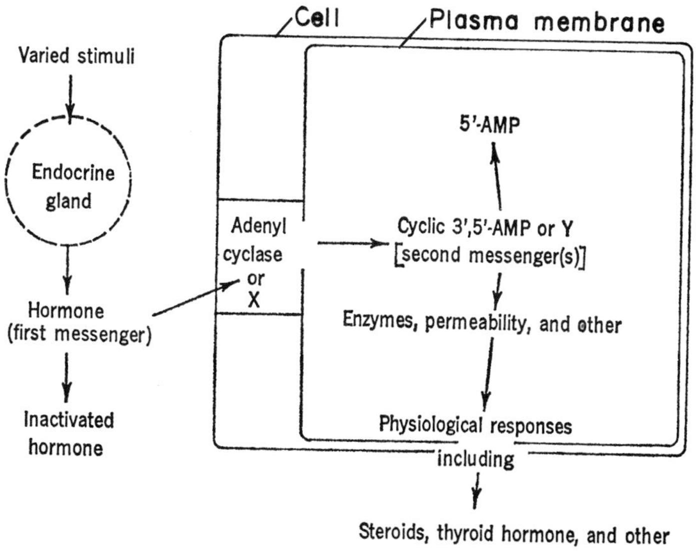

Figure 16.7. *Schematic representation of the second messengers concept. Thus the first messenger does not penetrate the plasma membrane, but rather is represented by a second messenger (cyclic AMP) inside the cell. (From Sutherland, 1972: Science 177: 403. © The Nobel Foundation 1972.)*

REGULATION BY ISOZYME FORMATION

Isozymes (isoenzymes) are enzymes in multiple forms, all performing the same general function at different rates. They are different to some extent in chemical composition so that they are separable by electrophoresis. Lactate dehydrogenase, for example, is found in five electrophoretically distinct fractions. Each of these electrophoretic species of lactic dehydrogenase is a tetramer consisting of two polypeptide chain units, H and M, present in different proportions: H_4, H_3M, H_2M_2, HM_3, and M_4. The five lactate dehydrogenase isozymes differ in many properties: catalytic activity (affinity for the substrate pyruvate as measured by the Michaelis constant K_M; see Chapter 10); amino acid composition; sensitivity to heat; and immunological responses. The two peptides H and M are coded by different genes; therefore, the type of enzyme present is under genetic control and depends upon evocation by the conditions imposed upon the organism (Hammes and Wu, 1971).

During development of an organism different proportions of the various isozymes are produced in cells of differentiating tissues, presumably in relation to the specific requirements of the cells of each tissue. For example, species of lactate dehydrogenase with high content of H are most active at low pyruvate concentrations and are inhibited by high concentrations of pyruvate. Such species are found in the muscle cells of the heart. The properties of the isozyme high in H would tend to channel the lactate into the aerobic reactions of the Krebs cycle rather than to the formation of lactic acid, a regulation obviously of adaptive significance for heart cells. Conversely, lactate dehydrogenases with a high M content remain active at high pyruvate concentrations; therefore, they channel the lactate metabolism toward formation of lactic acid. Such lactate dehydrogenase isozymes are found in skeletal muscle cells. They also represent regulations of adaptive value because skeletal muscle cells undergo periods of intense activity during which they go into oxygen debt. At such times lactic acid formation permits rapid release of energy. The lactic acid is rapidly removed from the muscle cells by the blood (Guyton, 1976).

Isozymes may also form in response to stresses upon cells. Fish exposed to low temperatures acclimate in such a manner that the respiratory rate increases over the initial rate at such temperatures. Isozymes have been found in the muscle cells of such fish, and it has also been noticed that the energy of activation for some of the reactions involved in metabolic activity is decreased. The change in respiratory rate presumably occurs as a result of development of isozymes with decreased energy of activation (Hochachka and Somero, 1973).

An interesting case of isozyme formation coupled with negative feedback control is that of aspartokinase. This enzyme catalyzes the reaction between aspartic acid and ATP to form aspartyl phosphate. In a series of subsequent reactions aspartyl phosphate forms the amino acids lysine, threonine, and methionine. In a number of species of bacteria the aspartokinase exists in two isozymes, one of them sensitive to feedback inhibition by lysine and the other by threonine. In this way if either of the amino acids accumulates a reduction occurs in the formation of aspartyl phosphate, but the reaction is not stopped (Fig. 16.8).

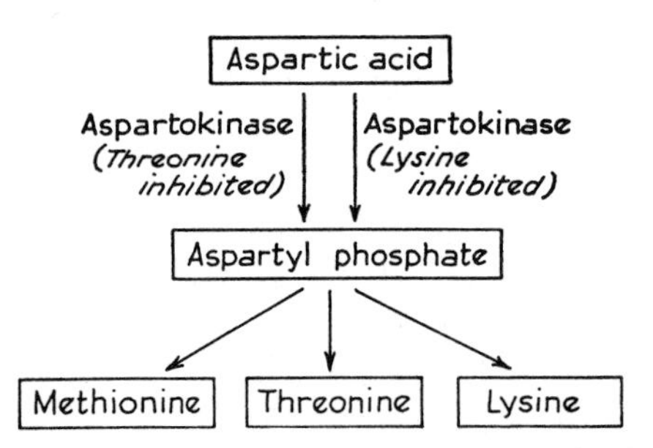

Figure 16.8. The way in which the synthesis of aspartyl phosphate is regulated by different end products through two isozymes of aspartokinase. (From Paul, 1964: Cell Biology. Stanford University Press, Stanford, Calif.)

REGULATION OF RATE OF PROTEIN DEGRADATION

Degradation rates of enzymes may be an important aspect of regulation of metabolic reactions. In the starved rat the content of arginase in liver cells increases. It was shown that this is not the result of increased synthesis, which continues, nor of activation of preexisting inactive enzyme, but rather of a slower rate of degradation of enzyme present. This is a highly specific, sparing regulatory effect because most proteins are degraded actively during starvation.

Enzymes are degraded less readily *in vitro* when combined with substrate than when free. That they are also degraded less readily *in vivo*, if combined with substrate, was demonstrated by injection of tryptophan into the rat. The several-fold increase in concentration of the enzyme tryptophan pyrolase in the cells of the liver in injected animals as compared to controls was presumably attributable to the slower rate of breakdown of the enzyme in the presence of its substrate. Different enzymes are known to be degraded at characteristic rates; that is, they have characteristic *turnover times*. For example, tryptophan pyrrolase is degraded more rapidly than arginase in liver cells of the rat. Therefore, the turnover time determines the steady state concentration of enzymes in cells. This is a means of regulation of cell metabolism although the mechanism is not known (Kleinsmith, 1972).

COMPARTMENTATION

Certain enzymes involved in particular reactions are segregated in the cell, often within membrane-bound units. Such compartmentation may serve as regulatory control in the cell. For example, the reactions of the Krebs cycle can occur only in the mitochondria, within which are the required enzymes. Enzymes involved in lipid synthesis appear to be confined to the membranes of the endoplasmic reticulum. Protein synthesis is shown to occur on the surface of ribosomes. On the other hand, enzymes involved in glycolysis seem to be in the particle-free cytoplasm. This, perhaps, helps to explain the Pasteur effect, since one of the competing enzyme systems is in the mitochondria, and the other is in the particle-free cytoplasm.

The reaction rate of an enzyme free in solution may be different from that of an enzyme bound to a membrane, as has been shown for a number of mitochondrial enzymes. In this manner cell membranes may regulate the rate of enzymatic reactions. Although the way in which this occurs is not understood, it is thought to be akin to the allosteric effect.

It is known that the plasma membrane and other membranes in the cell are sites of active transport of substances across the membrane involved, which depends upon metabolic energy. The transport of various metal ions by the membranes may play a part in regulating activity of the enzymes contained within the membrane-bound structures. For example, in muscle cells the calcium ion concentration is dependent upon the activity of the endoplasmic (sarcoplasmic) reticulum. The calcium ion is required for the functioning of actomyosin as an ATPase.

Membranes act in one other form of regulation—as the targets of hormones, which may alter their permeability properties as discussed in the following paragraphs.

REGULATION BY HORMONES—INTERCELLULAR REGULATION

In the multicellular organism, metabolism and growth may be regulated in a complex manner from some coordinating center. Since these are topics concerned with the entire organism they are more appropriately considered in textbooks of plant and animal physiology. Only two examples will be given to illustrate the nature of the cellular reactions to hormones.

The giant chromosomes of insects, studied especially in tissues of *Drosophila* and *Chironomus* (particularly the salivary glands), are shown to be the result of multiple and progressive replication of the chromatids without mitotic separation. Such chromosomes may measure ten times the length and 10,000 times the cross-sectional area of the original chromatids. The bands seen on such chromosomes (the *chromomeres*) are probably each composed of the tightly folded-up portion of the constituent double DNA-histone chromatid strand. On such giant chromosomes, puffs (*Balbiani rings*) or enlargements are often seen; the puffs may be different in different tissues. It has been found that injection into the larva of *ecdysone,* the molting-metamorphosis hormone of insects, results in development of puffs where none were seen before. Such puffs form within 30 minutes of injection in one of the chromosomes (No. 1) and 30 minutes later in another (No. 4). The hormone is effective in a concentration as low as 10^{-7} μg per mg of larva for the first puff. Once puffing has been induced, the condition persists until after molting is completed, although the details depend upon the stage at which the injection is made.

As discussed in Chapter 7, the puff consists of extruded sections of DNA-histone filament upon which the mRNA develops. The mRNA then translates the message into protein. The hormone apparently activates or derepresses the genes at a particular locus on the chromosome. Different genes may be activated in different tissues by the same hormone, and different hormones may activate different genes in the same chromosome. In this manner hormones serve to coordinate the activities concerned with a particular function (e.g., in the case of ecdysone, molting and metamorphosis) (Beermann, 1965).

The cells of the mammalian liver are known to store a considerable amount of glycogen, which serves to regulate the blood glucose level, largely through the interplay of various hormones upon the enzymes in the liver cells. After a rich meal an abundance of glucose is present in the blood. Such a high glucose level leads to secretion of insulin into the blood from the islets of Langerhans in the pancreas. Insulin not only increases the permeability of cells to glucose but also stimulates the formation of glycogen (glycogenesis) in the liver cells, possibly by action on glucokinase, the enzyme that transfers high-energy phosphate from ATP to glucose. As soon as the level of blood glucose

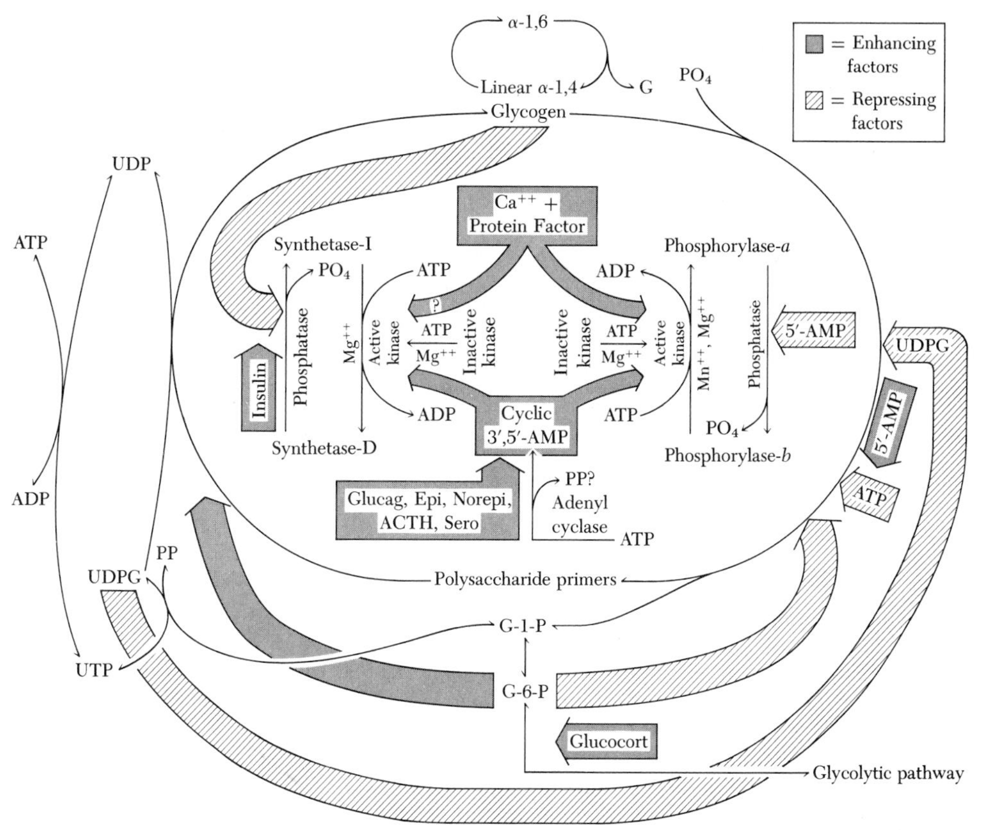

Figure 16.9. Pathways and regulatory factors in glycogen metabolism. ACTH, Adrenocorticotropic hormone; Epi, epinephrine; G, glucose; G-1-P and G-6-P, glucose-1- and glucose-6-phosphate; Glucag, glucagon; Glucocort, glucocorticoid hormones; Norepi, norepinephrine; PO_4, phosphate; PP, pyrophosphate; Sero, serotonin; UDP and UTP, uridine di- and triphosphate. The hormones epinephrine and norepinephrine are produced by the adrenal medulla and stimulate glycogenolysis. Glucagon, produced by certain cells (alpha cells) of the islets of Langerhans in the pancreas, activates liver phosphorylase, thus stimulating glycogenolysis. Glucocorticoid hormones are produced by the adrenal cortex and regulate carbohydrate metabolism. The phosphorylases exist in two forms in all tissues studied, the a form being physiologically more active. (From Cantarow and Schepartz, 1967: Biochemistry. 4th Ed. W. B. Saunders Co., Philadelphia.)

has fallen, secretion of insulin essentially stops, as does the synthesis of glycogen. On the other hand, thyroxine, the secretion of the thyroid gland, activates a phosphorylase to form glucose phosphate from glycogen. The glucose phosphate is decomposed to glucose and inorganic phosphate in the presence of a phosphatase. Secretion of thyroxine may be induced by exposure of the individual to cold. Similarly, epinephrine, appearing in the blood as a result of excitement, leads to greater glycogen breakdown by phosphorolysis (glycogenolysis), as does muscular exercise, perhaps by the same mechanism.

When the supply of carbohydrate in the diet is low, glycogen is still formed, but now from amino acids (see the metabolic mill, Fig. 12.8). The enzymes involved in *glyconeogenesis,* as the process is called, are presumably under control of some of the ketosteroid hormones of the adrenal cortex. Some of these interrelationships are shown in Figure 16.9.

LITERATURE CITED AND GENERAL REFERENCES

Atkinson, D. E., 1977: Cellular Metabolism and its Regulation. Academic Press, New York.

Beermann, W., 1965: Cytological aspects of information transfer in cellular differentiation. *In* Molecular and Cellular Aspects of Development. Bell, ed. Harper & Row, New York, pp. 204–212.

Britten, R. J. and Davidson, E. H., 1969: Gene regulation for higher cells: a theory. Science *165*: 349–357.

Canovas, J. L., Ornston, L. N. and Stanier, R. Y., 1967: Evolutionary significance of metabolic control systems. Science *156*: 1695–1699.

Changeoux, J., 1965: The control of biochemical reactions. Sci. Am. (Apr.) *212*: 36–45.

Cohen, G. N., 1968: The Regulation of Cell Metabolism. Holt, Rinehart & Winston, New York.

Cramer, H. and Schultz, J., 1977: Cyclic 3′,5′-Nucleotides: Mechanism of Action. John Wiley & Sons, New York.

Denkewalter, R. G. and Hirschmann, R., 1969: The synthesis of an enzyme. Am. Sci. *57*: 389–409.

Eagle, H., 1965: Metabolic controls in cultured mammalian cells. Science *148*: 42–51.

Finter, N. B., 1967: Interferons. John Wiley & Sons, New York.

Galston, A. W. and Davies, P. J., 1970: Control Mechanisms in Plant Development. Prentice-Hall, Englewood Cliffs, N.J.

Gerhart, J. C. and Pardee, A. B., 1964: Aspartate transcarbamylase, an enzyme designed for feedback inhibition. Fed. Proc. *23*: 727–735.

Green, M., 1970: Effect of oncogenic DNA viruses on regulatory mechanisms of cells. Fed. Proc. *29*: 1265–1275.

Guyton, A. C., 1976: Textbook of Medical Physiology. 5th Ed. W. B. Saunders Co., Philadelphia.

Hammes, G. G. and Wu, C. W., 1971: Regulation of enzyme activity. Science *172*: 1205–1214.

Hansen, J. N., Spiegelman, G. and Halvorson, H. O., 1970: Bacterial spore outgrowth: its regulation. Science *168*: 1291–1298.

Henderson, J. F., 1962: Feedback inhibition of purine biosynthesis in ascites tumor cells. J. Biol. Chem. *237*: 2631–2635.

Hochachka, P. W. and Somero, G., 1973: Strategies of Biochemical Adaptation, W. B. Saunders, Philadelphia.

Hommes, F. A. and van den Berg, C. J., eds., 1975: Normal and Pathological Development of Energy Metabolism. Academic Press, New York.

Horecker, B. L. and Stadtman, E. R., eds., 1971: Current Topics in Cellular Regulation. Vols. 3 and 4. Academic Press, New York.

Hsia, D. Y., 1966: Inborn Errors of Metabolism. 2nd Ed. Part I. Clinical Aspects. Year Book Medical Publishers, Chicago.

Ingram, V. M., 1963: The Hemoglobins in Genetics and Evolution. Columbia University Press, New York.

Jacob, F. and Monod, J., 1961a: Genetic regulatory mechanisms in the synthesis of proteins. J. Mol. Biol. *3*: 318–356.

Jacob, F. and Monod, J., 1961b: On the regulation of gene activity. Cold Spring Harbor Symp. Quant. Biol. *26*: 193–211.

Jacob, F. and Monod, J., 1963: Genetic repression, allosteric inhibition and cellular differentiation. *In* Cytodifferentiation and Macromolecular Syntheses. Locke, ed. Academic Press, New York, pp. 30–64.

Jacob, F., Perrin, D., Sanchez, C. and Monod, J., 1960: L'opéron: groupe de gènes à expression coordonnée par un opérateur. Comp. Rend. Acad. Sci. Paris *250*: 1727–1729.

Jost, J. P. and Rickenberg, H. V., 1971: Cyclic AMP. Ann. Rev. Biochem. *40*: 741–774.

Karlson, P., 1965: Biochemical studies of ecdysone control of chromosomal activity. J. Cell. Comp. Physiol. *66*(Suppl. 1): 69–75.

Kihlman, B. A., 1966: Action of Chemicals on Dividing Cells. Prentice-Hall. Englewood Cliffs, N.J.

Kleinsmith, L. J., 1972: Molecular mechanisms for the regulation of cell function. Bioscience *22*: 343–348.

Koningsberger, V. V. and Bosch, L., ed., 1967: Regulation of Nucleic Acid and Protein Biosynthesis. Biochimica Biophysica Acta Library, Vol. 10. American Elsevier, New York.

Koshland, D. E., Jr., 1973: Protein shape and biological control. Sci. Am. (Oct.) *229*: 52–64.

Kvamme, E. and Pihl, A., eds., 1968: The Regulation of Enzyme Activity and Allosteric Interactions. Academic Press, New York.

Lark, K., 1969: Initiation and control of DNA synthesis. Ann. Rev. Biochem. *38*: 569–604.

Lehninger, A. L., 1975: Biochemistry. 2nd Ed. Worth Publishers, Inc., New York.

Lodish, H. F., 1976: Translational control of protein synthesis. Ann. Rev. Biochem. *45*: 39–72.

Luria, S. F., 1970: Phage, colicins and macroregulatory phenomena. Science *168*: 1166–1170.

Maaløe, O. and Kjeldgaard, N. O., 1966: Control of

Macromolecular Synthesis. W. A. Benjamin, Menlo Park, Calif.

McKusick, V. A., 1975: Mendelian Inheritance in Man: Catalogs of Autosomal Dominants, Autosomal Recessive and X-linked Phenotypes. 4th Ed. Johns Hopkins University Press, Baltimore.

Markert, C. L., ed., 1975: Isozymes, 4 vols. Academic Press, New York.

Masters, C. J. and Holmes, R. S., 1974: Isozymes, multiple enzyme forms and phylogeny. Adv. Comp. Physiol. Biochem. *5*: 110–195.

Mayr, O., 1970: The origins of feedback control. Sci. Am. (Oct.) *223*: 111–118.

Newell, P. C., 1977: How cells communicate; the system used by slime moulds. Endeavour N.S. *1*: 63–68.

Paik, W. K. and Kim, S., 1971: Protein methylation. Science *174*: 114–119.

Pastan, I. and Perlman, R., 1970: Cyclic AMP in bacteria. Science *169*: 339–344.

Pastan, R., 1972: Cyclic AMP. Sci. Am. (Aug.) *227*: 97–105.

Paul, J., 1964: Cell Biology. Stanford University Press, Stanford, Calif.

Perutz, M. F., 1967: Some molecular controls in biology. Endeavour *26*: 3–8.

Pestka, S., 1971: Protein biosynthesis: mechanism, regulation and potassium dependency. *In* Membranes and Ion Transport. Vol. 2. Bittar, ed. Wiley-Interscience, New York, pp. 279–296.

Ptashne, M. and Gilbert, W., 1970. Genetic repressors. Sci. Am. (Jun.) *222*: 36–44.

Rabin, B. E., 1972: Effects of Drugs on Cellular Control Mechanisms. University Park Press, Baltimore.

Rasmussen, H., and Pechet, M. M., 1970: Calcitonin. Sci. Am. (Oct.) *223*: 42–50.

Rasmussen, H., 1970: Cell communication, calcium ion, and adenosine monophosphate. Science *170*: 404–412.

Robison, G. A., Butcher, R. W., and Sutherland, E. W., 1971: Cyclic AMP. Academic Press, New York.

Rosen, R., 1968: Recent developments in the theory of control and regulation of cellular processes. Int. Rev. Cytol. *23*: 25–88.

Runge, W., 1972: Photosensitivity in porphyria. *In* Photophysiology. Vol. 6. Giese, ed. Academic Press, New York, pp. 149–162.

Stanbury, J. B., Wyngaarden, J. B. and Frederickson, D. C., eds., 1966: The Metabolic Basis of Inherited Disease. Blakiston Division, McGraw-Hill Book Co., New York.

Stein, G. and Stein, J., 1976: Chromosomal proteins: their role in the regulation of gene expression. Bioscience *26*: 488–498.

Stent, G. D., 1963: The operon: on its third anniversary. Science *144*: 816–820.

Sutherland, E. W., 1972: Studies on the mechanism of hormone action. Science *177*: 401–408 (Nobel lecture).

Tomkins, G. M., Gelherter, T. D., Granner, D., Martin, D., Samuels, H. H. and Thompson, E. B., 1969: Control of specific gene expression in higher animals. Science *166*: 1474–1480.

Tomkins, G. M. and Martin, D. W., Jr., 1970: Hormones and gene expression. Ann. Rev. Genet. *4*: 91–106.

Tomkins, G. M., 1972: Specific enzyme production in eukaryotic cells. *In* Advances in Cell Biology. Vol. 2. Prescott, Goldstein and McConkey, eds. Appleton-Century-Crofts, New York, pp. 299–322.

Tomkins, G. M., 1975: The metabolic code. Science *189*: 760–763.

Umbarger, H. E., 1964: Intracellular regulatory mechanisms. Science *145*: 674–679.

Vogel, H. J., ed., 1971: Metabolic Regulation. Vol. 5. Metabolic Pathways. Greenberg, ed. Academic Press, New York.

Walker, J. B., 1965: End product repression in the creative pathway of the developing chick embryo. *In* Molecular and Cellular Aspects of Development. Bell, ed. Harper & Row, New York, pp. 285–296.

Watson, J. D., 1976: The Molecular Biology of the Gene. 3rd Ed. W. A. Benjamin, Menlo Park, Calif.

Watson, J. D., 1972: The regulation of DNA synthesis in eukaryotes. *In* Advances in Cell Biology. Vol. 2. Prescott, Goldstein and McConkey, eds. Appleton-Century-Crofts, New York, pp. 1–46.

Weber, G., ed., 1977: Advances in Enzyme Regulation. Pergamon Press, Oxford, Vol. 15.

Weber, G., 1970: Enzyme regulation in mammalian tissues. Science *167*: 1018–1020.

Whitehead, E., 1970: The regulation of enzyme activity and allosteric transition. Prog. Biophys. Mol. Biol. *21*: 321–397.

Willingham, M. C., 1976: Cyclic AMP and cell behavior in cultured cells. Int. Rev. Cytol. *44*: 319–363.

Yanagisawa, K., 1963: Genetic regulation of protein biosynthesis at the level of the ribosome. Biophys. Biochem. Res. Commun. *10*: 226–231.

Section IV

TRANSPORT ACROSS CELL MEMBRANES

The concentration of molecules and ions inside a healthy cell is quite different from that of the bathing medium. Therefore, regulation of entry and exit of substances through cell membranes is of prime importance for maintenance of life in the cell. Movement of water and solutes through cell membranes is dynamic and during the life of the cell is never in equilibrium with the environment; equilibrium is achieved only on death.

In prokaryotic cells the cell plasma membrane alone performs this regulatory function. In eukaryotic cells, in addition, membranes over each of the cell organelles perform a comparable regulatory function, showing selective permeability to substances. Consequently, many of the general principles of permeability that apply to the plasma membrane apply also to organellar membranes. Organellar membranes insure selective permeability to substances necessary for an organelle's individual function yet permit each organelle to accomplish its specific function.

Currently membrane studies are in great ferment, partly because biophysicists and molecular biologists have provided new techniques for investigating cell membranes. For example, chemically detectable alteration of membrane constituents by induced mutations is being correlated with changes in membrane permeability and composition. The properties of membranes reconstituted from mixtures of compounds isolated from natural cell membranes are being investigated and the properties of artificial membranes simulating natural cell membranes are under study, as well. Further impetus to cell membrane studies is provided by the finding that many enzymes for metabolic reactions are bound to membranes and are only active, or at least most active, when on membranes. Even prokaryotes lacking organelles have their metabolic enzymes on the plasma membrane (and mesosomes when present), indicating the importance of membranes in organizing enzymatic reactions.

No attempt is made to cover the vast literature on permeability; rather, reference is made to reviews summarizing trends in membrane research (e.g., Quinn, 1976). Permeability to molecules that move along a concentration gradient and often require only kinetic energy supplied by temperature for entry through cell membranes is discussed in Chapter 17. Transport that requires metabolic energy, for example, against a concentration gradient, is considered in Chapter 18. Bulk transport of samples of fluid medium or particles and fluid medium is covered in Chapter 19.

A review of Chapter 8 on the structure of cell membranes would facilitate understanding of the transport of molecules across cell membranes.

CHAPTER 17

PERMEATION AT THE EXPENSE OF KINETIC ENERGY

Kinetic energy accounts for random molecular movements (see Chapters 9 and 10) and the higher chemical potential energy of a more concentrated solution of a substance outside than inside the cell gives effective direction to movement into the cell. The kinetic energy is dissipated as molecules of the substance move into the cell and reduce its concentration gradient. For transport of some substances, metabolic energy may be used to provide even more rapid intake along a concentration gradient, for example, in the gut. These cases are discussed in Chapter 18.

OBSTACLES TO PERMEATION

A molecule entering a cell must pass from an external aqueous medium into the cell membrane, through the cell membrane, and then from the membrane into the aqueous medium of the cytoplasm within the cell. Since much of the cell membrane consists of a lipid bilayer (see Chapter 8), passage of molecules from aqueous to lipid phase and lipid to aqueous phase presents barriers to entry and exit. If the molecule entering a cell has high lipid solubility, neither its entry into the membrane nor its passage through the membrane presents an obstacle. However, entry of the molecule from the membrane into the aqueous cytoplasm does present an obstacle because the cohesive bonds it has formed with the lipid molecules of the membrane must be broken before it can leave the membrane (Fig. 17.1). This means expenditure of energy by the permeating molecules.

If, on the other hand, a molecule entering a cell membrane is highly soluble in water, its strong hydrogen bonds with the water in which it is dissolved outside the cell must be broken before the molecule can enter the lipid membrane. Its movement through the lipid membrane also presents an obstacle. However, such a molecule passes readily from the membrane into the aqueous cytoplasm inside the cell where it forms new hydrogen bonds. Thus, for a water-soluble molecule, passage into and through the cell membrane is the most difficult hurdle and requires expenditure of energy except where special aqueous channels penetrate the membrane (see Chapter 8).

KINETIC ENERGY FOR PERMEATION

If the energy required to get across any of these hurdles is large, only molecules with the required energy of activation are successful in moving through the membrane (Rogers and McElroy, 1958b). Accordingly, the rate of entry of many compounds into a cell is increased several-fold by a 10°C rise in temperature (Table 17.1). This indicates that the molecules need to be activated to cross some barrier in the plasma membrane.

While water molecules and ions of small diameter probably move directly through *pores* in the cell membrane (Table 17.2) (Solomon, 1960), evidence exists that larger molecules become attached to a *carrier* with which they form a complex in the cell membrane. The complex shuttles the substance across the membrane, liberating it on the other side. Since the carrier facilitates diffusion, such movement is spoken of as *facilitated diffusion*. Facilitated diffusion can be distinguished from simple diffusion by

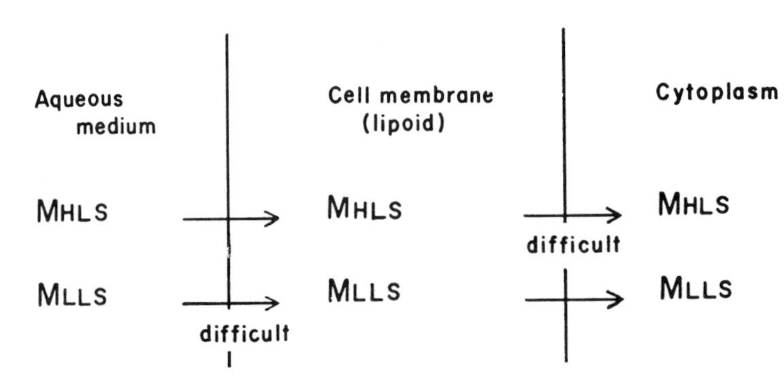

Figure 17.1. Hurdles encountered by a molecule with high lipoid solubility (M_{HLS}) and a molecule with low lipoid solubility (M_{LLS}) when passing through the cell membrane to the cytoplasm. (After Davson and Danielli, 1952: The Permeability of Natural Membranes. 2nd Ed. Cambridge University Press, Cambridge.)

the kinetics of its entry relative to concentration. The *flux* (rate of movement, or quantity of substance passing through a unit area of membrane per unit time) through the membrane by *simple diffusion* increases linearly with increase in concentration of substance (Fig. 17.2, Nos. 1 and 2), while flux by facilitated diffusion levels off with increasing concentration. This is thought to mean that carrier sites in the membrane become more and more completely occupied until, when no additional sites can be occupied, no further increase in flux occurs, a phenomenon called *saturation* (Fig. 17.2, No. 3). Although the carrier complex requires energy of activation to cross the membrane barriers (Fig. 17.2, No. 3), less energy is required than for activated diffusion (Fig. 17.2, No. 2) because of the affinity of the carrier for components (especially lipid) in the membrane. Therefore, the flux for a given substance is greater in the presence of a carrier than in its absence (Davson, 1970).

Because water and solute pass through cell membranes at the same time, the water flux influences the solute flux and the solute flux in turn influences the water flux, so that estab-

TABLE 17.1. ΔH VALUES FOR PERMEATION

Permeant	Q_{10}	ΔH† (Kcal Mole^{-1})	No. of H Bonds	ΔH† Per H Bond (Kcal Mole^{-1})
Penetration into bovine erythrocytes				
Glycerol	3.9	24	6	4
Ethylene glycol	2.8	18½	4	4½
Diethylene glycol	2.8	18½	4	4½
Triethylene glycol	3.2	20½	4	5
1,2-Propandiol	3.0	19½	4	5
1,3-Propandiol	2.9	19	4	5
Propanol	1.3	4½	2	2
Thiourea	2.2	13½	4	3½
Urea	1.4	6	5	1
Penetration into eggs of *Arbacia*				
Ethylene glycol		23.6	4	6
Propionamide		21.6	3 (or 4?)	7 (or 5½?)
Butyramide		22.8	3 (or 4?)	7½ (or 5½?)
Penetration into ascites tumor cells				
Urea		16.5	5	3½
Ethylene glycol		15.0	4	4
Diethylene glycol		15.6	4	4
Triethylene glycol		18.5	4	4½

*From Stein, 1967: The Movement of Molecules Across Cell Membranes. Academic Press, New York. For definition of Q_{10} see Table 10.1.

†ΔH is the heat of reaction (see Chapter 9).

TABLE 17.2. RELATIONSHIP OF EFFECTIVE DIAMETERS OF DIFFERENT SUBSTANCES TO PORE DIAMETER*

Substance	*Diameter (nm)*	*Ratio to Pore Diameter*	*Approximate Relative Diffusion Rate*
Water molecule	3	0.38	5×10^7
Urea molecule	3.6	0.45	4×10^7
Hydrated chloride ion	3.86	0.48	3.6×10^7
Hydrated potassium ion	3.96	0.49	1×10^2
Hydrated sodium ion	5.12	0.64	1
Lactate ion	5.2	0.65	?
Glycerol molecule	6.2	0.77	?
Ribose molecule	7.4	0.93	?
Pore size	8 (Ave.)	1.00	–
Galactose	8.4	1.03	?
Glucose	8.6	1.04	0.4
Mannitol	8.6	1.04	?
Sucrose	10.4	1.30	?
Lactose	10.8	1.35	?

*From Guyton, 1976: Textbook of Medical Physiology. 5th Ed. W. B. Saunders Co., Philadelphia.

lishment of equilibrium is unattainable and the movements are irreversible. It is therefore necessary to take both fluxes into consideration at the same time, requiring the application of principles of the thermodynamics of irreversible systems. Since a first approximation is attained by ignoring such corrections, in this chapter the simplified treatment is presented. (For corrections see Appendices 17.1 and 17.2.)

EXCHANGE DIFFUSION

What is usually measured in experiments is the net flux for a substance, that is, the net gain by, or net loss from, cells. However, much movement occurring across cell membranes is not observed because no net flux exists, the influx being equal to the outflow (efflux). By use of tracers it has been shown that many sub-

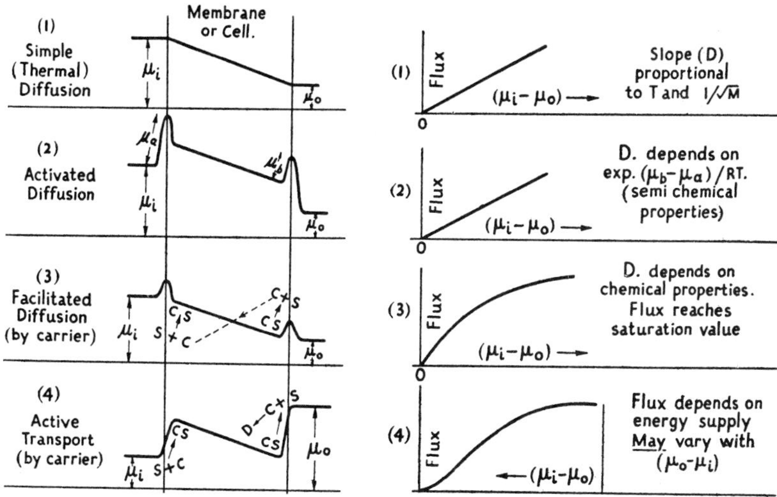

Figure 17.2. Diagrams illustrating various processes by which substances may penetrate membranes. On the left, in each case, are plotted diagrammatically the changes in electrochemical potential (μ) of the substance as it penetrates the membrane or living cell. On the right is plotted the relation between the flux and the difference in electrochemical potential across the membrane or cells. In (1) and (2), the ratio between the flux and the difference in electrochemical potential measures the diffusion, or permeability, constant D. In (3) and (4) this ratio is not constant, and may be regarded as defining the apparent permeability. μ is the initial electrochemical potential of a substance, μ_a the electrochemical potential after activation, μ_o the electrochemical potential after the substance has passed through the cell membrane, T is the absolute temperature, M the molarity of the diffusing substance, R the gas constant, C the carrier, and S the substance passing through the membrane. (From Bayliss, 1956: Modern Views on the Secretion of Urine. Winton, ed. J. and A. Churchill, London.)

stances (e.g., water and Na^+) move in and out of the cell at all times.

For study of exchange diffusion, cells are kept in a solution of ^{24}Na for a period of time to enable them to acquire radioactive sodium. If the cells are then placed in "cold" sodium solution, the radioactive sodium leaves the cells in proportion to its relative concentration inside and outside the cells. The loss of ^{24}Na presumably occurs by exchange diffusion, a radioactive sodium ion being exchanged at random for a cold sodium ion. This is substantiated by the observation that if the bathing solution containing cold sodium is replaced by one of isotonic mannitol, the loss of ^{24}Na is at the expense of kinetic energy, not metabolic energy, as shown by its independence of metabolic poisons such as cyanide (Davson, 1970).

Exchanges of one molecular (or ionic) species for another are also known to occur when the two species are transported by a common carrier. For example, the amino acids histidine and tryptophan are readily exchanged for each other by Ehrlich ascites tumor cells. These exchanges do not require metabolic energy, nor are they subject to the inhibitors that affect metabolically coupled transport (Quastel, 1965).

Hydrogen ions resulting from metabolic activity are always available for exchange with cations *(cation exchange diffusion)*, and bicarbonate ions of similar origin are available for exchange with anions *(anion exchange diffusion)*. Equal numbers of anions and cations need not be taken up by the cell for maintenance of the electric potential across the plasma membrane, but the exchange of an ion for an ion of the same charge must be equal (Ussing, 1952). Steady state exchange (equal numbers of ions moving in both directions) is thought to occur largely by diffusion rather than by carriers, which are likely to be polarized (Davson, 1970). Some cation-for-cation and anion-for-anion exchanges require metabolic energy. Such exchanges are discussed in Chapter 18.

WATER FLUX

While for the conceptual simplicity of the process it is instructive to discuss water flux through membranes separately from solute flux, such a separation is arbitrary. Water, like any other substance, moves in response to a concentration gradient and its cellular influx and efflux are governed by the same principles. However, water moves so much more rapidly than any other substance that many cells act as if they were permeable to it alone; that is, some cells act as almost perfect osmometers. However, many solutes that move more slowly than water pass through cell membranes.

The diffusion of water or any solvent, in more general terms, through any membrane in response to a concentration gradient is called *osmosis*. The principles governing osmotic transfers have been studied with nonliving (both synthetic and natural) and living cell membranes.

Osmosis Across Artificial Membranes

Many artificial membranes have been devised as analogues to the plasma membrane to facilitate study of osmotic transfers, a membrane of copper ferrocyanide being a favorite. In 1877 Pfeffer placed a ferrocyanide solution inside an unglazed pottery jar, surrounded it by a solution of copper sulfate and passed an electric current between the two solutions. The membrane of copper ferrocyanide precipitated in the small pores of the unglazed jar allows only water to pass through it. If the jar is filled with 1 molal NaCl and is suspended in a beaker of distilled water, water molecules pass through the membrane but salts cannot, and since the water molecules cannot get out as readily as they enter because of the interference of sodium and chloride ions inside the jar, water accumulates inside the jar. Water rises in a tube fitted to the mouth of the jar until the pressure developed by its rise equals the *osmotic pressure* (the tendency of the water molecules to enter the jar in response to the concentration gradient between the inside and the outside of the jar) (deLancey *et al.*, 1967).

The osmotic pressure can be measured by a mercury *osmometer*, as illustrated in Figure

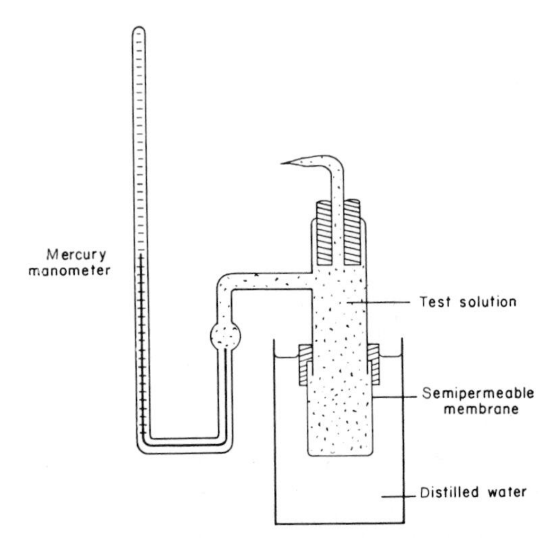

Figure 17.3. *Pfeffer's osmometer, with mercury manometer to measure the osmotic pressure. The test solution is kept inside the jar, in the walls of which has been precipitated a semipermeable copper ferrocyanide membrane. Entry of water occurs until the osmotic pressure is counterbalanced by the pressure of the column of mercury.*

17.3. The physical chemist measures the pressure that just prevents the entry of water (solvent) through a *semipermeable membrane* (a membrane that permits the passage through it of solvent but not solute). In the Hepp osmometer, very small quantities of material can be used and a direct reading of the osmotic pressure may be made. The osmotic pressure registered on a diaphragm may be transduced to an electrical change and readily measured and recorded (Mauro, 1965).

If cellophane dialysis tubing is fastened over a thistle tube and the tube is filled with concentrated sodium chloride solution and immersed in distilled water, the volume of solution in the tube will increase, indicating that cellophane is permeable to water. Tests, however, show that sodium chloride is also passing through the cellphane into the water. But the rise in the level of the tube solution proves that inward movement of water is greater than outward movement of salt. Inward movement stops when the hydrostatic pressure equals the osmotic pressure. A membrane that lets one material (water) through more readily than another (salt or sugar) is *selectively permeable.*

Natural membranes, such as goldbeater's skin (outside membrane of the ox intestine), can be used to study the movement of water in the same manner. The results for natural membranes are similar to those for cellophane.

Permeability of the Plasma Membrane to Water

Experiments indicate that cells are generally highly permeable to water. For example, a sea urchin egg, placed in sea water diluted with distilled water, swells. Swelling increases with increasing dilution of the sea water until the membrane bursts. Diluted sea water is said to be *hypotonic* to the sea urchin egg. When the egg is placed in sea water concentrated by evaporation, it shrinks, and the shrinkage increases as the medium becomes more concentrated. Such a medium is said to be *hypertonic* to the sea urchin egg. In unaltered (isotonic) sea water the egg neither swells nor shrinks.

Red blood cells (erythrocytes) placed in a hypertonic solution shrink, becoming crenated (Fig. 17.4). Placed in a hypotonic salt solution, they swell and become spherical instead of discoidal. Erythrocytes are hemolyzed and lose their hemoglobin in excessively dilute salt solutions. Water permeability in animal cells is discussed by Dick (1959, 1966).

Plant cells, too, are quite permeable to water (Dainty, 1963). The cells of the fresh water plant *Elodea* are normally turgid; that is, the protoplast is held firmly against the cell wall. If placed in the concentrated salt solution (over 0.6 percent), they lose water and become plasmolyzed, and the protoplast of each cell contracts about the shrunken vacuole (Fig. 17.5). Replaced in fresh water, each cell regains water and the protoplast again fills the space within the cell wall. The rigid cellulose cell wall, limiting the volume, prevents excessive uptake of water. If the cell wall is removed by the enzyme cellulase, the naked protoplast swells with uptake of water even to the bursting point (Ruesink and Thimann, 1966). Similar results have been obtained with yeast spheroplasts (cell walls removed by enzymatic action) (Diamond and Rose, 1970), and bacteria (cell

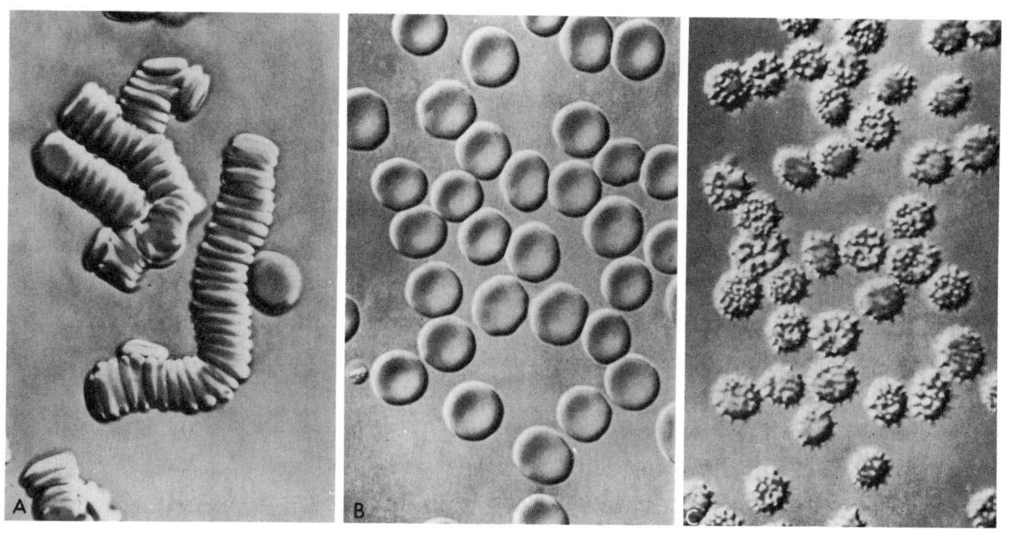

Figure 17.4. Photomicrographs of fresh mammalian red blood corpuscles. A, In vitro the corpuscles stack up like coins, forming "rouleaux." B, Normal biconcave shape of dispersed corpuscles. C, Spiny appearance of corpuscles resulting from crenation on loss of water in hypertonic solutions. Nomarski interference optics. (By courtesy of Dr. Marcel Bessis: From Bloom and Fawcett, 1975: A Textbook of Histology. 10th Ed. W. B. Saunders Co., Philadelphia, p. 138.)

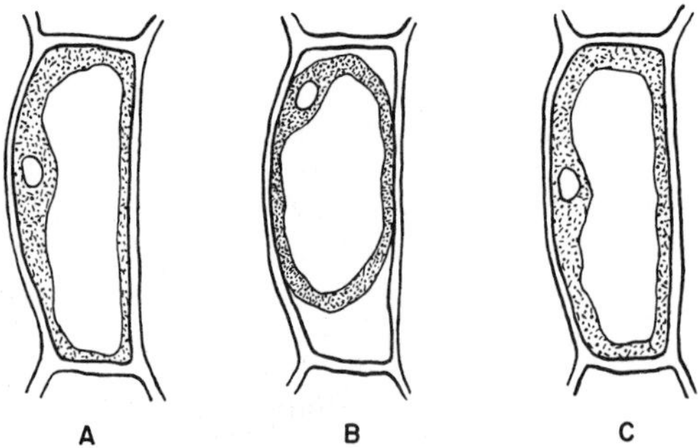

Figure 17.5. *A plant cell (A) undergoes plasmolysis (B) when it is placed in a hypertonic solution. Deplasmolysis (C) occurs after a period of time, but only if the solute can penetrate. Time for deplasmolysis depends upon the rate of entry of the solute.*

walls removed by lysozyme) (Acker and Hartsell, 1960).

Cells plasmolyzed by hypertonic solutions deplasmolyze as solute molecules enter the cells and water, in turn, enters until the cell volume returns to normal. Deplasmolysis occurs rapidly if solute enters rapidly (Fig. 17.6).

Cell organelles with unit membranes similar to the plasma membrane shrink or swell in response to the concentration of solute in the bathing media (see Fig. 6.7).

At all times an exchange of water molecules and solute molecules takes place through the plasma membrane between the inside and the outside of a cell. One cannot dissociate the passage of water from the passage of solutes. Therefore, the membrane of a cell is selectively permeable, selectively allowing movement of various solutes to lesser or greater degrees but always allowing movement of water most readily.

Osmotic Pressure of Nonelectrolytes

The plant physiologist Pfeffer (1877) found that the osmotic pressure of various solutes divided by the concentration of solute is constant (Table 17.3). The physical chemist van't Hoff took this as evidence of the applicability of the gas laws to osmotic pressure. For example, a gram molecular weight of a gas at atmospheric pressure occupies 22.4 liters of space. If it is compressed into the confines of a liter it exerts 22.4 atmospheres pressure. Pfeffer's data showed that a gram molecular weight of a nonelectrolyte dissolved in a liter of water exerts an osmotic pressure of approximately 22.4 atmospheres.

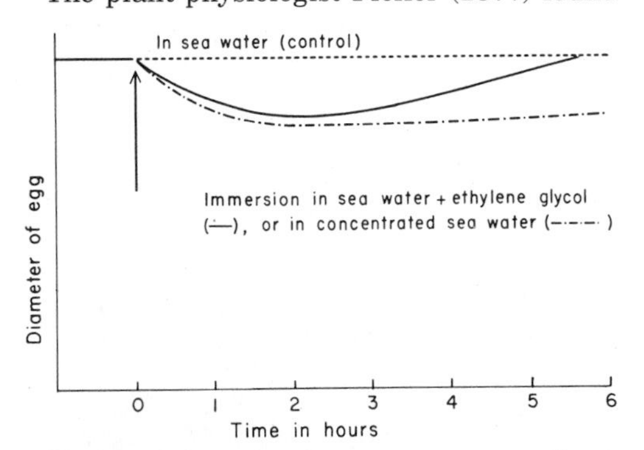

Figure 17.6. *Return to normal volume as a measure of permeability. Sea urchin eggs placed in sea water rendered hypertonic by addition of ethylene glycol in one case and by evaporation of water in the other case. Ethylene glycol enters more rapidly than salts; therefore, the eggs return sooner to their initial volume in ethylene glycol than in sea water.*

TABLE 17.3. PFEFFER'S DATA FOR OSMOTIC PRESSURE AT DIFFERENT CONCENTRATIONS OF SUCROSE*

Concentration (C) in Percent	*Osmotic Pressure (π) in Atmospheres*	$\frac{\pi}{C}$
1	0.70	0.7
2	1.34	0.67
4	2.74	0.68
6	4.10	0.68

*To determine whether the rise in osmotic pressure is proportional to the concentration of the solution, Pfeffer divided the osmotic pressure (π) by the concentration (C) of the solution being tested. As seen in the third column, π/C is fairly constant.

The gas law may be written:

$$PV = nRT \quad (17.1)$$

where P is the pressure in atmospheres, V the volume in liters, n the number of moles, R the gas constant (equal to 0.082 liter atmospheres per degree per mole) and T the absolute temperature. By rearrangement:

$$P = \frac{n}{V} \times RT \quad (17.2)$$

Since moles of solute divided by the liters of solvent in which it is dissolved is equal to the molar concentration (C), C may be substituted for n/V,* and π substituted for P.

Then the equation for osmotic pressure becomes:

$$\pi = CRT \quad (17.3)$$

*The physical chemist uses molal concentrations in dealing with solutions. A *molal* (m) solution contains 1 gram molecular weight of a solute in 1000 grams of solvent, whereas a *molar* (M) solution is a liter of solution containing 1 gram molecular weight of a solute. Use of molal solutions has the advantage that molal solutions of different substances contain the same ratio of solute to solvent molecules. At very low concentrations, molar becomes equivalent to molal, since the effect of molecular volume of the solute becomes negligible.

in which π denotes the *osmotic pressure*. This relationship can also be derived from basic thermodynamic principles (Dick, 1966).

For a gram molecular weight of osmotically active ideal nonelectrolyte, C is 1, R is 0.082 liter atmospheres per degree per mole, and the absolute temperature at 0°C is 273°. Then,

$$\pi = (1 \text{ mole/liter}) \times (0.082 \text{ liter atm/degree/mole}) \times (273^\circ)$$

$$\text{or } \pi = 22.4 \text{ atmospheres}$$

Since the gas laws were developed for perfect gases whose molecules have no volume and do not attract one another, their application to solutions is even more of an approximation than with gases. They are applicable only to weak solutions at best, deviations becoming marked at higher concentrations (Dick, 1966).

Osmotic Pressure of Electrolytes

Since electrolytes dissociate, each molecule gives rise to two or more ions. Because the osmotic pressure is determined by the number of particles actually present in solution, an electrolyte will exert a greater osmotic pressure than a nonelectrolyte for a given molal concentration. The magnitude of the difference depends on the degree of dissociation and the number of ions produced. Therefore, to obtain the osmotic pressure of an electrolyte, a factor correcting for dissociation must be added. The correction factor is less than the total number of ions formed from a molecule of an electrolyte because the activity of the ions, rather than their concentration, determines the osmotic pressure. Ionic interference reduces the effective concentration of the ions, since unlike ions attract one another, thereby reducing the number of ions free to act. The activity of a given electrolyte varies at different concentrations, becoming greater at lower concentrations, at which ionic interference is least.

The activity for each concentration of electrolyte is most easily determined by freezing point determination (cryoscopy), which measures the number of particles active, not just their concentration. Then, by comparing the freezing point depression by an electrolyte with that by an ideal nonelectrolyte of the same molal concentration, a factor called the *cryoscopic coefficient*, G, may be obtained:

$$G = \frac{\overline{\Delta T}_{fp}}{\Delta T_{fp}} \qquad (17.4)$$

where $\overline{\Delta T}_{fp}$ is the freezing point depression produced by a given molal concentration of the electrolyte in question and ΔT_{fp} is the lowering of the freezing point by the same molal concentration of an ideal nonelectrolyte. A few cryoscopic coefficients are given in Table 17.4. The osmotic pressure of an electrolyte can be calculated from the gas law equation (17.3), but the concentration term, C, must be multiplied by the cryoscopic coefficient, G, for each concentration of electrolyte,

$$\pi = (C \times G)RT \qquad (17.5)$$

since G measures the relative increase of number of active particles by ionization.

Colligative Properties of Solutions

To measure the osmotic pressure of cell contents that are present only in small quantities, use is made of other properties of solutions that are functions of the number of active particles per unit volume of solution, not the *kind* of particles. These four properties *(colligative properties)* of a solution are the osmotic pressure, the freezing point depression, the boiling point elevation, and the vapor pressure. They represent a manifestation of a phenomenon common to them all: solute particles decrease the ability of water molecules to move.

TABLE 17.4. CRYOSCOPIC COEFFICIENTS (G)* AT VARIOUS MOLAL CONCENTRATIONS

Electrolyte	*0.02*	*0.05*	*0.1*	*0.2*	*0.5*
$MgCl_2$	2.708†	2.677	2.658	2.679	2.896
$MgSO_4$	1.393†	1.302	1.212	1.125	—
$CaCl_2$	2.673†	2.630	2.601	2.573	2.680
LiCl	1.928	1.912	1.895	1.884	1.927
NaCl	1.921	—	1.872	1.843	—
KCl	1.919	1.885	1.857	1.827	1.784
KNO_3	1.904	1.847	1.784	1.698	1.551

Data *from* Heilbrunn, 1952: An Outline of General Physiology. 3rd Ed. W. B. Saunders Co., Philadelphia.

*$G = \frac{\overline{T}_{fp}}{T_{fp}}$, as defined in equation 17.4.

†0.025 molal

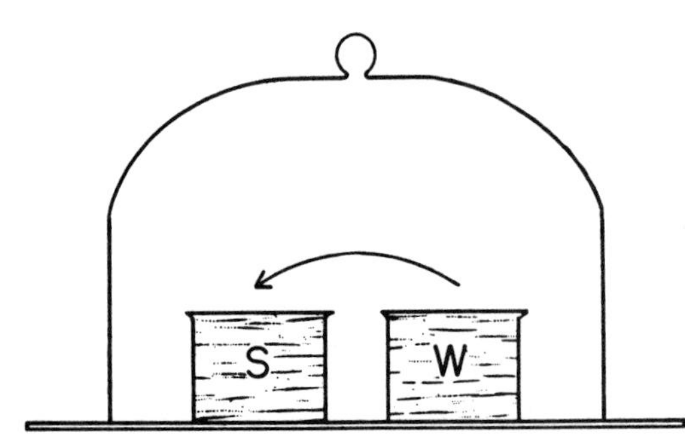

Figure 17.7. *Water distills over from a solution of lower concentration of solute (W) to one of higher concentration of the solute (S). The solute molecules reduce the escaping tendency of the water molecules in each solution, more so where they are more concentrated. The temperature of the solution of lower concentration is lowered by evaporation of water from its surface. The lowering of the temperature can be measured by a thermocouple (Fig. 17.8).*

The greater the number of active particles the greater will be their interference with movement of water molecules from a solution through a membrane. In like manner, the freezing point is lowered because the active particles of a solute interfere with the movement of water molecules in formation of ice crystals. The boiling point is raised because the active solute particles in solution interfere with the movement of water molecules in establishment of an equilibrium, through evaporation, between the water molecules and the atmospheric pressure. The vapor pressure, which depends upon the evaporation of water from the solution, is decreased by the interference of the active solute particles with the escape of water molecules from the solution. Therefore, it is evident that if one of the colligative properties is known, the others can be calculated from it, since they are all related through the same variable.

Vapor pressure determinations are perhaps simplest. Two beakers containing aqueous solutions of different concentrations of solute are placed in an air-tight enclosure, such as the bell jar in Figure 17.7. Because exit of water molecules is less impeded in the dilute solution, more water will leave it than leaves the more concentrated solution. Thus the dilute solution loses more water than it receives and the concentrated solution receives more water molecules than it loses until the two approach equality in concentration. Since the temperature falls in the beaker losing water, and since the change in temperature is proportional to the difference in concentrations of the solutions in the two beakers, the vapor pressure can be calculated from the change in temperature, as measured by a thermocouple (Fig. 17.8). Using this principle, osmotic pressures of very small samples of fluid such as found in a cell can be measured.

In biological work, determination of the freezing point of a solution has been used most

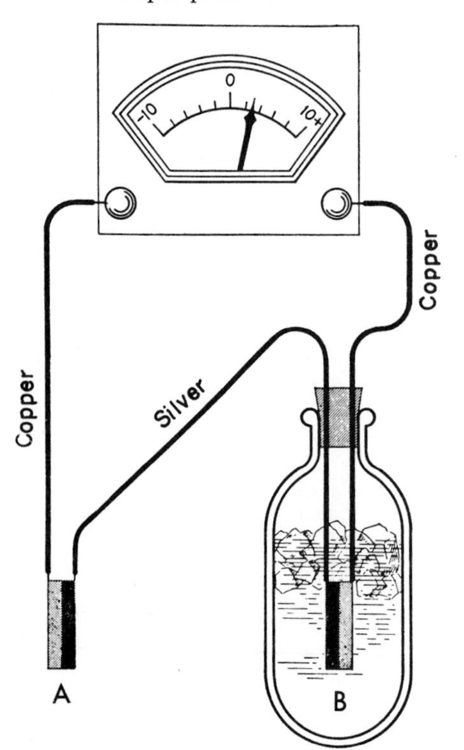

Figure 17.8. *An elementary form of thermocouple for measuring change in temperature. Junction B is always kept at some convenient known temperature, e.g., on ice. When junction A is placed on a substance of different temperature, the thermoelectric current produced deflects the galvanometer. Galvanometric deflection is calibrated with A at several known temperatures, and a calibration curve is constructed to relate the deflection to the temperature at A. Then junction A can be used to measure unknown temperatures within the range of the scale. (From Amberson and Smith, 1948: Outline of Physiology. Appleton-Century-Crofts, New York.)*

often for measuring osmotic pressure because it can be made without altering the solution. The freezing point of an aqueous solution containing one gram molecular weight per kilogram of water (1 molal) of ideal nonelectrolyte is depressed by 1.86°C. From the freezing point depression, the osmotic pressure is readily calculated:

$$\frac{\Delta_x}{\Delta_m} = \frac{x}{22.4 \text{ atm}} \quad (17.6)$$

$$x = \frac{\Delta_x}{\Delta_m} \times 22.4 \text{ atm} \quad (17.7)$$

or

$$x = \frac{\Delta_x \cdot 22.4}{1.86} = 12.06\ \Delta_x \quad (17.8)$$

Where Δ_x = the depression of the freezing point produced by the "unknown" solution, Δ_m = the depression of the freezing point produced by a 1 molal solution of ideal nonelectrolyte and x = the osmotic pressure of the "unknown" solution. Variations of this method are available for freezing point determinations on microscopic quantities of fluid (Prager and Bowman, 1963).

Osmotic pressure can also be determined by the rise in boiling point of a solution, since it is well known that the rise in boiling point is 0.52°C per molal concentration of dissolved nonelectrolyte. However, this method is unsatisfactory for biological solutions because proteins coagulate when boiled.

Cells as Osmometers

A cell with a selectively permeable membrane could hardly be expected to behave like a perfect osmometer. It contains much protein and lipid, which constitute nonsolvent volume, excluding a corresponding volume of osmotically active solution. A term for the *nonsolvent volume* must be subtracted from the term for cell volume in the equation $\pi V = nRT$, which becomes:

$$\pi(V - b) = nRT \quad (17.9)$$

where b represents the nonsolvent volume and other terms have the meanings previously assigned. When correction for the nonsolvent volume (0.125) is made, the sea urchin egg in diluted sea water is observed to function reasonably well as an osmometer, the values in the fourth column of Table 17.5 being more nearly alike than those in the third column, where no correction was made. Some cells into which solutes pass more rapidly than in the sea urchin egg do not even begin to approximate osmometers.

To compare net flux in different cells, it is necessary to determine the volume of water

TABLE 17.5. SEA URCHIN *(ARBACIA)* EGG AS AN OSMOMETER*

Relative Pressure	Volume Observed†	πV	$\pi(V-b)$
1.0	2121	2121	1881
0.9	2316	2084	1868
0.8	2570	2056	1864
0.7	2922	2045	1878
0.6	3420	2053	1909
0.5	4002	2002	1881

*After Lucké and McCutcheon, 1932. The value of b was found to be 12.5 percent (or 0.125) of the total volume of the egg.

†Volume in arbitrary units.

entering per unit area of cell surface per unit time per unit difference in osmotic pressure between the inside and the outside of different kinds of cells. A permeability constant, k, for water is given in these terms for each cell by the following equation (Lucké and McCutcheon, 1932):

$$\frac{dV}{dt} = k\ A(\pi_{in} - \pi_{out}) \quad (17.10)$$

where V is the volume of the cell and dV is a small change in volume assumed to be due to the uptake of water during dt, the brief interval of time; A is the area of the cell surface; π_{in} is the osmotic pressure of the cell; and π_{out} is the osmotic pressure of the external medium.

Equation 17.10 is only an approximation because it takes into account the movement of water only, whereas both water and solute move through the membrane at the same time. The equations taking this into account are given in Appendices 17.1 and 17.2. Some comparative values (Table 17.6), as determined primarily by swelling of cells in hypotonic media, indicate almost a 200-fold variation. The permeability constant is high for cells with

TABLE 17.6. COMPARISON OF PERMEABILITY CONSTANTS FOR WATER OF SOME CELLS*

Species	Permeability Constant†
Amoeba proteus	0.026–0.031
Pelomyxa carolinensis	0.023
Fresh water peritrichs	0.125–0.25
Arbacia egg	0.4
Human erythrocyte	3.0

*From Prosser, 1973: Comparative Animal Physiology. 3rd Ed. W. B. Saunders Co., Philadelphia.

†In cubic micrometers of water, per square micrometer of surface area of cell, per atmosphere difference in pressure between the inside and outside of the cell, per minute.

a higher surface-to-volume ratio than for those with a lower ratio (Dick, 1966). It is perhaps significant that fresh-water species with the greatest osmotic gradient between inside and outside the cell have the lowest permeability constants for water.

It is curious that some cells secrete water; for example, the gland cells in the carnivorous bladderwort *Utricularia* close a trap by quick secretion of water, enabling them to catch insect larvae and small crustaceans (Lewin, 1970).

SOLUTE FLUX

Volume Change

The rate of deplasmolysis measured by the change in volume as cells in solutions of various solutes reabsorb water indicates the entry rate of the solute. For example, in equimolal hypertonic solutions of ethylene glycol, glycerol, or glucose, deplasmolysis occurs first in ethylene glycol and last in glucose. This means that penetration of ethylene glycol is fastest and glucose slowest. If no deplasmolysis occurs, the solute is not entering the cell.

For minute cells the volume of a mass of cells, packed together by centrifugation, is determined before and after immersion in known concentrations of hypertonic solutions. This is a favorite method for permeability studies with red blood corpuscles. A special centrifuge tube called the *hematocrit,* with a calibrated cylindrical portion 1 or 2 mm in diameter, is used for this purpose. If the series of experiments with different solutes is carried out under similar conditions and with the same centrifugal force applied for the same duration, the rate of entry of the various substances can be obtained by the volume changes of the packed cells. However, it is virtually impossible to pack the cells without intercellular fluid; therefore, such experiments compare packing relative to a control.

Hemolysis

Hemolysis, indicated by the sudden appearance of a clear red solution of hemoglobin in a previously murky solution, has been used to measure the rate of penetration of substances into erythrocytes. For this purpose red blood cells are placed in a hypertonic solution of a substance such as ethylene glycol. As ethylene glycol molecules enter, the osmotically active molecules inside the cells increase in concentration, which results in a corresponding uptake of water. When the cell membrane has stretched to its limit, the cells undergo *hemolysis*—that is, the membrane becomes porous enough to permit escape of hemoglobin molecules. Erythrocytes are hemolyzed very slowly in hypertonic solutions of glucose. The relative permeation rate of various kinds of molecules may be measured in this manner. However, comparisons of red blood cells in different species may depend in part on the strength of their membranes.

Quantitative Analyses

Perhaps the most direct method for measuring solute permeability is quantitative analysis of the internal contents of a cell exposed to a solute of a known concentration for known periods of time. For this purpose large plant cells, such as those of *Valonia, Chara,* and *Halicystis,* are ideal because the uptake of the solute can be measured by analysis of the contents of the large central sap vacuole. The vacuole of *Valonia* or *Halicystis* often contains more than 1 ml of sap, making possible microchemical tests on a single cell. The squid giant axon has been used in similar experiments. Permeability constants can be calculated from the data; information obtained for a variety of substances may be compared for a variety of cells.

Tracer Method

Valuable quantitative data on permeability of ions such as sodium and potassium have been obtained by the use of radioactive tracers (see Chapter 21). Some ions (sodium, for example) move in and out of cells so that little net change occurs in concentration for quantitative analysis. Such movement could be detected only by the tracer method (Brooks, 1938).

Use of the tracer method has been confined largely to the study of electrolyte flux for ions that have isotopes with sufficiently long half-life periods to permit measurement. Some of the isotopes available for tracer studies are ^{3}H, ^{14}C, ^{24}Na, ^{32}P, ^{35}S, ^{36}Cl, ^{42}K, ^{45}Ca, ^{59}Fe, ^{64}Cu, ^{65}Zn, ^{82}Br, ^{89}Sr, and ^{131}I. However, the tracer method need not be confined to electrolytes, since organic molecules may be labeled with radioactive carbon, and so forth (Ussing *et al.,* 1974).

CORRELATION OF PERMEABILITY WITH OIL TO WATER SOLUBILITY

A high correlation has been observed between permeability and solubility of compounds in oil (or lipid) as compared to their solubility in water. The less polar a compound the greater its solubility in oil and the lesser its solubility in water. With increasing polarity

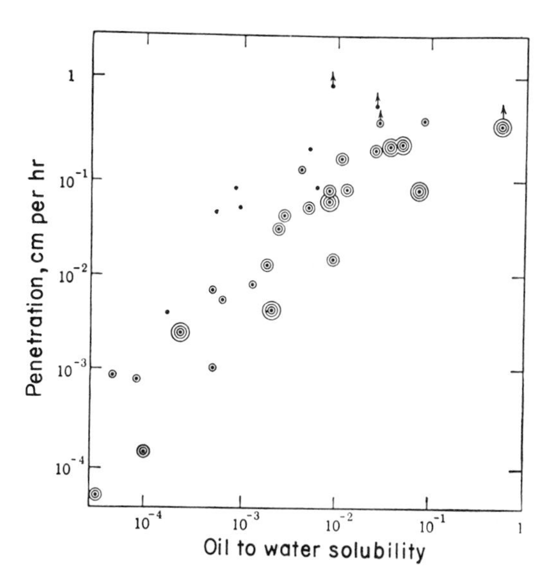

Figure 17.9. *Relation between relative solubility of a substance between oil and water and its penetration* (P) *into the* Chara *cell. The small dots indicate small molecules, and successive circles around these indicate corresponding increases in molecular size. Note that even large molecules of high solubility in oil penetrate the cell readily. (From Collander, 1937: Trans. Faraday Soc. 33: 986.)*

the compounds become correspondingly less soluble in oil and more soluble in water. The *partition coefficient* measures the relative solubility of a substance in oil as compared to water.

Nonpolar compounds, in which electrons are equally shared by the two atoms forming a bond, as in the paraffins, olefins, and cyclic compounds, have high solubility in oils and enter cells rapidly. *Nonionic polar compounds,* derivatives of nonpolar compounds into which electronegative atoms or radicals have been introduced, have lesser oil solubility but greater water solubility than their nonpolar parental compounds and enter cells less readily than nonpolar compounds. *Ionic compounds,* in which the electrons of a valency bond are held by only one of the atoms or radicals so that ions are present even in the crystalline state of such compounds, have little solubility in oil but a great degree of solubility in water. They enter cells slowly.

The entry of many compounds of varying oil-to-water solubility into large cells of the alga *Chara* discloses that molecules with the largest relative solubility of oil to water enter the cell most readily, regardless of their molecular size, although molecular diameter is of consequence when the molecules have the same oil-to-water solubility (Fig. 17.9).

EFFECTS OF MOLECULAR DIAMETER UPON PERMEABILITY

Molecular diameters depend upon steric configuration (especially in organic molecules), symmetry (an asymmetrical molecule having a larger effective diameter than a symmetrical one), and molecular weight. Other things being equal the larger molecule will have a bigger diameter. Molecules of very large diameter have only a slight chance of entering the plasma membrane. Sucrose, for example, penetrates cells in very small amounts, if at all, while larger carbohydrate molecules, such as starch, glycogen, or inulin, have not been shown to penetrate. Perhaps for this reason they serve effectively as food stores. The absorption of proteins (antigens) and other large molecules indicates penetration by another mechanism, probably bulk transport (see Chapter 19).

When the oil-to-water solubility of compounds is the same, the smaller the molecular diameter the greater their rate of penetration. For example, cyanamide, propionamide, succinamide, and diethylmalonamide have about the same oil-to-water solubility, but their molecular weights increase in the order given (42.04, 73.09, 116.12, and 158.20, respectively). Accordingly, their ability to penetrate into the alga *Chara* decreases correspondingly in the order given (Collander, 1959) (Fig. 17.9).

However, molecular size may sometimes be of even greater importance than oil-to-water solubility in determining the penetration of molecules into some cells, as illustrated by the data in Figure 17.10 for *Beggiatoa,* a large sulfur bacterium. Here permeation is correlated with molecular size rather than with oil-to-water solubility (Fig. 17.11). This has also been found for penetration of aldehydes into luminous bacteria (Rogers and McElroy, 1958a, b).

Data relating oil-to-water solubility of molecules and permeability can be put in more dynamic terms indicated by the number of hydrogen bonds broken during movement of a molecule across the cell membrane, as shown in Figure 17.12. The energy required for movement of various molecules, given in Table 17.1, is quite large when several hydrogen bonds

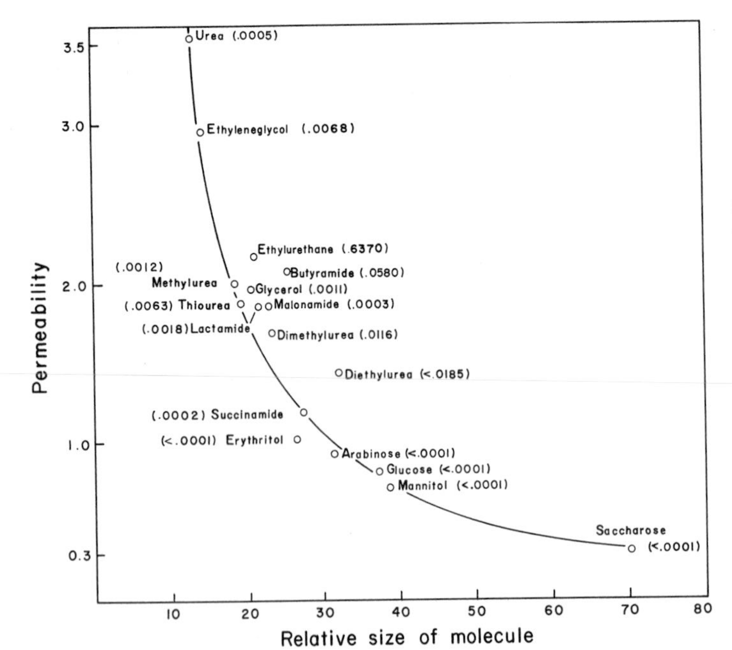

Figure 17.10. *Relation between molecular size and permeability in the sulfur bacterium* Beggiatoa. *Molecular size is measured by molecular refraction in yellow light. Permeability is measured by the* $\log_{10}(10^4 \times$ *the threshold plasmolytic concentration moles per liter). Partition coefficients in ether:water are given in parentheses. Permeability is clearly correlated with molecular size. (Data from Ruhland and Hoffmann, 1925: Planta 1: 1.)*

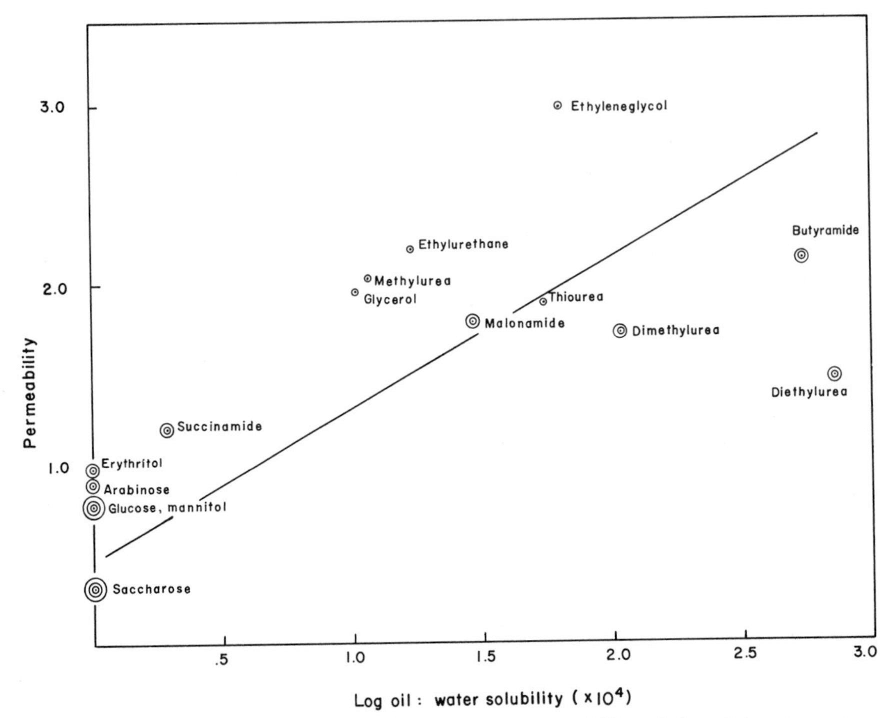

Figure 17.11. *Relation between partition coefficient and permeability of* Beggiatoa *to various organic substances. Although the permeability increases with the partition coefficient, the correlation is not nearly so good as that found in* Chara. *Permeability is measured by* $\log_{10}(10^4 \times$ *threshold plasmolytic concentration in moles per liter). The abscissa is* $\log_{10}(10^4 \times$ *partition coefficient between ether and water). (Data from Ruhland and Hoffmann, 1925: Planta 1: 1.)*

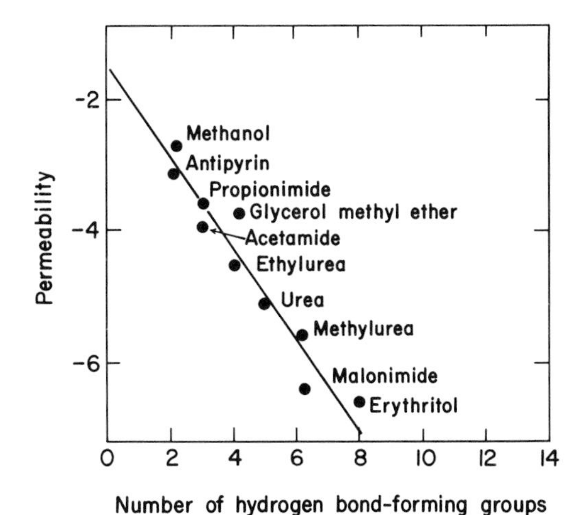

Figure 17.12. *Permeability of the alga* Chara *to various compounds in relation to hydrogen bond-forming groups. (After Stein, 1967: The Movement of Molecules across Cell Membranes. Academic Press, New York.)*

must be broken. For this reason fairly high temperature coefficients are found for penetration of solutes into cells, as seen in Table 17.1 (see also Stein, 1967). However, molecular size may be as important as the number of hydrogen bonds in determining entry of molecules (Quinn, 1976).

THE EFFECT OF IONIZATION ON PERMEABILITY

The relation between charge and penetration is best seen by studying the entry rate of weak electrolytes. The degree to which a weak electrolyte is ionized is determined by the pH. For example, at pH 6.34 carbonic acid is one half dissociated (Fig. 17.13). As the pH rises, dissociation increases, but the entry of carbonic acid decreases. Conversely, with a fall in pH the amount of dissociation decreases, carbon dioxide is formed, and the entry of carbon dioxide into the cell increases (Osterhout and Dorcas, 1926). Similar relations hold for hydrogen sulfide, hydrocyanic acid, auxins (plant growth hormones), dinitrophenol (metabolic stimulant), and other biologically active weak acids. The entry of weak bases such as the amines and alkaloids follows similar patterns except that in their case a rise in pH suppresses dissociation and enhances penetration. In experiments with members of a related series of weak organic acids (Table 17.7), it was found that penetration of the anions increased in the order given in the table. In spite of increase in molecular size, the smaller the degree of dissociation of the acids the more rapid their entry. The rate of entry increases with a rise in lipid-to-water solubility. The data indicate that the presence of a charge on an ion decreases its chance of entry.

The stronger the charge on an ion, the less probable is the ion's entry into a cell. It was found that monovalent cations such as Na^+ and K^+ enter cells more readily than divalent cations such as Ca^{2+} or Mg^{2+}, and divalent cations enter more readily than trivalent cations such as Fe^{3+}. Similarly, monovalent anions such as Cl^- or I^- enter cells more readily than divalent anions such as SO_4^{2-} and divalent more readily than trivalent anions (Ussing *et al.,* 1974).

Not all ions (cations and anions) of the same valency enter cells at the same rate. For example, ammonium ion enters more rapidly than potassium, potassium more rapidly than sodium, and sodium more rapidly than lithium.

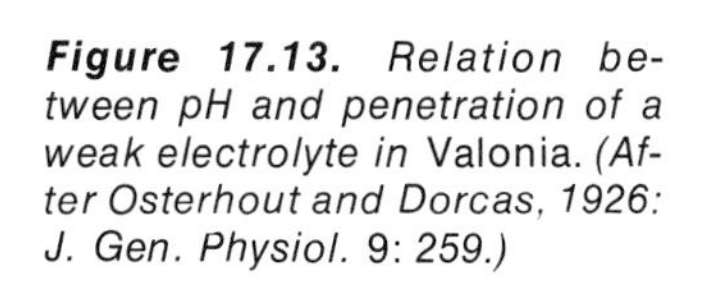
Figure 17.13. *Relation between pH and penetration of a weak electrolyte in* Valonia. *(After Osterhout and Dorcas, 1926: J. Gen. Physiol. 9: 259.)*

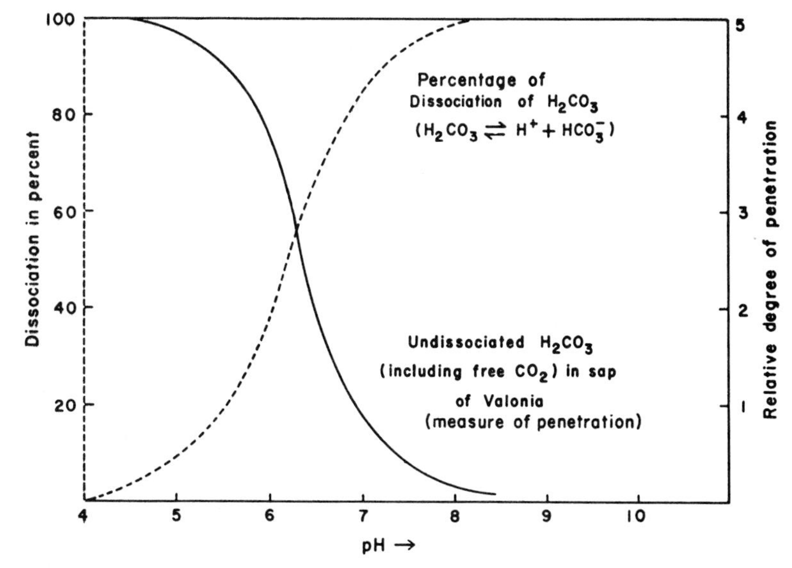

TABLE 17.7. RELATIVE LIPID-TO-WATER SOLUBILITY OF A RELATED SERIES OF WEAK ORGANIC ACIDS*

Acid	Number of Carbons	Molecular Weight	Dissociation Constant	Water Solubility in Grams/100 ml	Relative Solubility, Benzene:Water
Formic	1	46.03	1.77×10^{-4}	Very high	—
Acetic	2	60.05	1.76×10^{-5}	Very high	0.055
Propionic	3	74.08	1.34×10^{-5}	Very high	0.28
Butyric	4	88.10	1.48×10^{-5}	5.62 at −1.1°C	1.08
Caproic	6	116.16	1.43×10^{-5}	0.4	3.73
Caprylic	8	144.22	—	0.25 at 100°C	16.10

*Data from Handbook of Chemistry and Physics, 1968, 49th Ed. Chemical Rubber Co., Cleveland; and Höber, 1945.

This has been explained on the basis of the relative size of the hydrated ions. For instance, potassium is surrounded by several shells of electrons and has a lower net charge density than does lithium, with one shell of electrons. Consequently, less water is attracted to potassium than to lithium. For the entry of anions into cells, a similar explanation has been adduced. Hydration weakens but does not neutralize ionic charge.

The fact that uncharged particles enter a cell more readily than do ions suggests that the plasma membrane itself is charged. Since both anions and cations are affected, the plasma membrane is considered to be a *mosaic* of negatively and positively charged areas. However, the overall charge on the plasma membrane of most cells is positive; therefore, anions enter cells more readily than do cations.

Cells vary in their permeability to electrolytes. Some cells appear to be freely permeable to anions while at the same time they largely exclude cations, while other cells may be freely permeable to cations but may largely exclude anions.

All these considerations on the permeability of cells to ions apply only to situations in which no marked ion accumulation occurs and when no active transport is involved (see Chapter 18).

OSMOTIC RELATIONS IN THE PRESENCE OF PROTEIN

Proteins also form ions. When a potassium proteinate solution, at the appropriate pH, is placed in a cellophane bag (impermeable to protein ions) and immersed in a solution of potassium chloride, the initial distribution of ions is similar to that shown in Figure 17.14*A*. *Inside,* the initial concentration of potassium ions is designated C_i, the concentration of chloride is zero, and the concentration of proteinate is likewise designated C_i. *Outside,* the initial concentration of potassium ions at this time is designated C_o, and the concentration of chloride ions is also designated C_o.

Chloride ions diffuse inward in response to the concentration gradient. To maintain electric neutrality, potassium ions accompany them. Therefore, at equilibrium, a certain concentration, X, of potassium and chloride ions will have entered through the membrane (Fig. 17.14*B*). At that time the concentration of potassium ions inside the cellophane bag will be (C_i + X); that of the chloride ions inside will be X. The concentration of potassium ions outside the cellophane bag will be (C_o − X) and that of the chloride ions outside the bag will be (C_o − X).

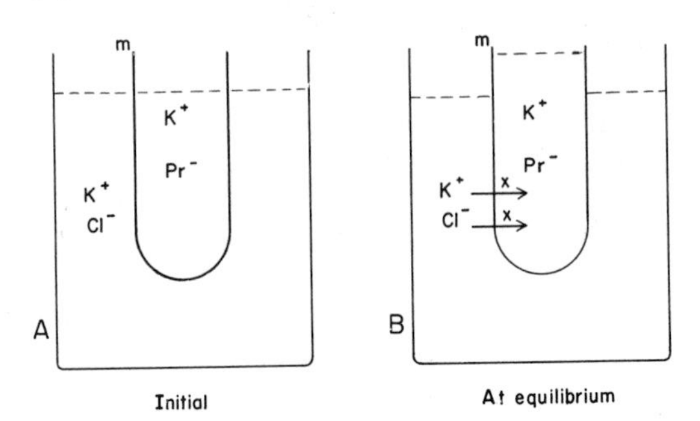

Initial concentration	Final concentration
Inside	Inside
$K^+ = C_i$	$K^+ = C_i + X$
$Pr^- = C_i$	$Pr^- = C_i$
$Cl^- = 0$	$Cl^- = X$
Outside	Outside
$K^+ = C_o$	$K^+ = C_o - X$
$Cl^- = C_o$	$Cl^- = C_o - X$

Figure 17.14. *Donnan equilibrium. The membrane (m) within which the protein solution is contained is impermeable to protein ions but permits all the other ions to pass. Note: Because salts are taken up into the tube containing the protein, the osmotic pressure will rise and water will enter.*

In 1927 Donnan found that at equilibrium the product of the concentrations of the diffusible ions inside the bag equals the product of the concentrations of the diffusible ions outside the bag. He derived the following equation for the electrical potential, ΔE, resulting from the unequal distribution of cations, which must be identical to that due to the unequal distribution of the anions, for univalent ions, where R is the gas constant, T the absolute temperature and $\mathscr{F}$ the Faraday (96,500 volt-coulombs).

$$\Delta E = \frac{RT}{\mathscr{F}} ln \frac{[K]_{in}}{[K]_{out}} = -\frac{RT}{\mathscr{F}} ln \frac{[Cl]_{in}}{[Cl]_{out}} \quad (17.11)$$

Whence:

$$[K^+]_{in}[Cl^-]_{in} = [K^+]_{out}[Cl^-]_{out} \quad (17.12)$$

Substituting for these symbols the concentration terms obtained (for the simplest case, in which the compartments are equal in volume), one can write the following equation:

$$(C_i + X)X = (C_o - X)(C_o - X) \quad (17.13)$$

Solving for X:

$$X^2 + C_iX = C_o^2 - 2C_oX + X^2 \quad (17.14)$$

$$X = \frac{C_o^2}{C_i + 2C_o} \quad (17.15)$$

Knowing the initial concentrations of ions inside and outside a membrane, one can use this equation to solve for X, the concentration (at equilibrium) of the diffusible ions that have entered the membrane (see Overbeek, 1956).

If a mixture of uni-univalent, diffusible, and membrane-penetrating electrolytes such as NaCl, KNO_3, LiBr, and HI is used outside the tube containing the protein, which is present as a salt of potassium (Fig. 17.13), at equilibrium the following relations will hold:

$$\frac{[Na^+]_i}{[Na^+]_o} = \frac{[K^+]_i}{[K^+]_o} = \frac{[Li^+]_i}{[Li^+]_o} = \frac{[H^+]_i}{[H^+]_o} = \frac{[Cl^-]_o}{[Cl^-]_i} = \frac{[NO_3^-]_o}{[NO_3^-]_i} = \frac{[Br^-]_o}{[Br^-]_i} = \frac{[I^-]_o}{[I^-]_i}$$

SALT ANTAGONISM

The salt environment affects the permeability of the plasma membrane to other substances. In studies with cells in a piece of the alga *Laminaria,* the electrical conductivity served as an index of ion entry. When immersed in sodium chloride solutions isotonic with sea water (0.52 molar), the electrical resistance of *Laminaria* cells decreases, indicating an increase in permeability to sodium chloride and water. If not exposed to sodium chloride too long, *Laminaria* cells reimmersed in sea water recover and their electrical resistance increases (Fig. 17.15).

When this experiment is repeated with calcium chloride isotonic to sea water (0.278 molar solution), the electrical resistance of the cells first increases, indicating decreased permeability to calcium chloride and water (Fig. 17.16), and then falls as the cells are injured and become more permeable to calcium and water. Isotonic potassium chloride produces effects similar to those of sodium chloride, and isotonic magnesium chloride to those of calcium chloride.

When *Laminaria* cells are placed in an isotonic mixture of sodium chloride and calcium chloride in sea water proportions, the electrical resistance remains about the same as the control in sea water (17 hours in one test), indicating that cellular permeability is practically normal. Since calcium and sodium chlorides have opposite initial effects on the permeability of the plasma membrane, these two ions are said to be *antagonistic* to one another. The permeability of *Laminaria* cells is normal if potassium and magnesium chlorides are also added to the solution in sea water proportions. *Salt antagonism,* then, is the antagonistic action of different salts in maintain-

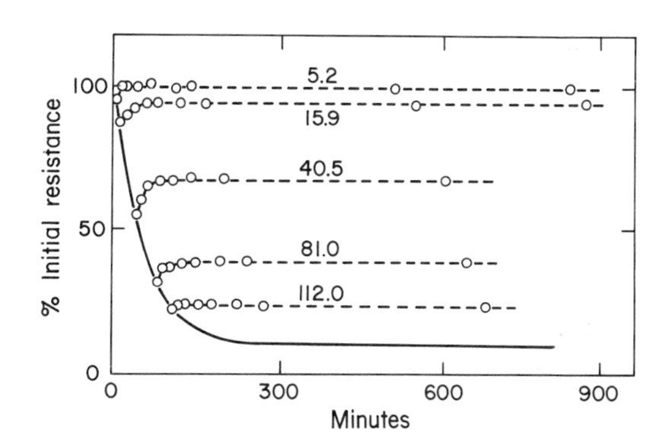

Figure 17.15. *The effects of isotonic NaCl on the permeability of disks cut out of the kelp* Laminaria. *The curves show the net electric resistance (ordinate) in percent of the initial resistance of* Laminaria agardhii *in 0.52 molar NaCl (unbroken line) and recovery in sea water (dotted lines). The figure attached to each recovery curve denotes the time of exposure (in minutes) to the solution of NaCl. (After Osterhout, 1922: Injury. Recovery and Death in Relation to Conductivity and Permeability. J. B. Lippincott Co., Philadelphia.)*

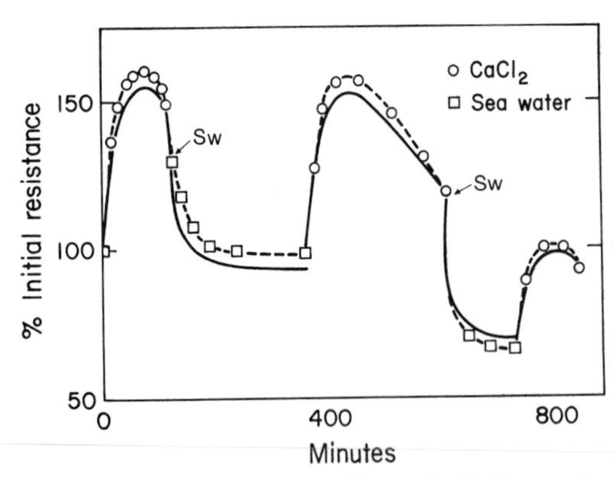

Figure 17.16. *The effect of isotonic $CaCl_2$ on the permeability of disks cut out of the kelp* Laminaria. *The curves show the net electric resistance (ordinate) in percent of the initial resistance of* Laminaria agardhii *in 0.278 molar $CaCl_2$ and in sea water. The unbroken line represents calculated values, and the broken line represents observed values. The curves are based on an average of 10 or more experiments. (After Osterhout, 1922: Injury, Recovery and Death in Relation to Conductivity and Permeability. J. B. Lippincott Co., Philadelphia.)*

ing normal permeability of the plasma membrane.

Salt antagonism is also observed on animal cells. For instance, a frog heart placed in pure sodium chloride solution stops beating but revives if calcium chloride is added in the proportion found in blood or sea water. However, the beat is not normal unless potassium chloride and a buffer are added (Fig. 17.17). Ordinarily glucose is also added to the solution to supply nutrient to the heart muscle cells. Such a balanced salt solution is called *Ringer's solution* in honor of its discoverer, the English physician Sidney Ringer, in 1880. However, it is not known whether the salts exert their action by affecting the permeability of heart muscle cells or for other reasons (e.g., calcium may be required for inciting muscle contraction) (see Chapter 23). Similar solutions have been developed for protozoans, marine eggs, blood cells, tissues, organs, and tissue cultures (Prosser, 1973).

A solution is *isosmotic* if it has the same osmotic pressure as the fluid inside the cells, as measured by the freezing point method. If the solution outside the cell has a higher osmotic pressure or a lower one than inside the cells, it is said to be *hyperosmotic* or *hypoosmotic*, respectively.

Salt antagonism explains why the terms isosmotic and isotonic are not synonymous. A solution may be isosmotic with cells without being isotonic if it does not provide salt balance. For example, isosmotic $CaCl_2$ is not isotonic and causes changes in cell volume. Furthermore, a solution that is isotonic for one type of cell may not be isotonic for another.

Osmotic pressure is measured in *osmoles*. An osmole is a mole of particles—molecules or ions—which can act independently of one another. Therefore, an osmole measures solute activity rather than solute concentration. Accordingly, a molar solution of monovalent electrolyte may be almost 2 osmolar. On the other hand, if association occurs between particles, molar and osmolar nonelectrolyte are not equivalent, osmolar being less than the molar concentration. Since we often do not know the chemical composition of biological fluids, osmolar measurements are most useful. The same is true of an environmental solution.

The antagonistic relations between various cations are more complex than would be suggested by this discussion. A satisfactory interpretation in terms of molecular interactions in the cell membrane has not been found (Lockwood, 1961).

THE EFFECT OF NARCOTICS AND ANESTHETICS UPON PERMEABILITY

Narcotics and anesthetics generally affect cells in proportion to their oil solubility, indicating that they need to enter the cell. Some values for the narcotic action of a number of compounds, correlated with their solubility in olive oil, are given in Tables 17.8 and 17.9.

Effects of anesthetics on cellular permeability are not always the same and the results are

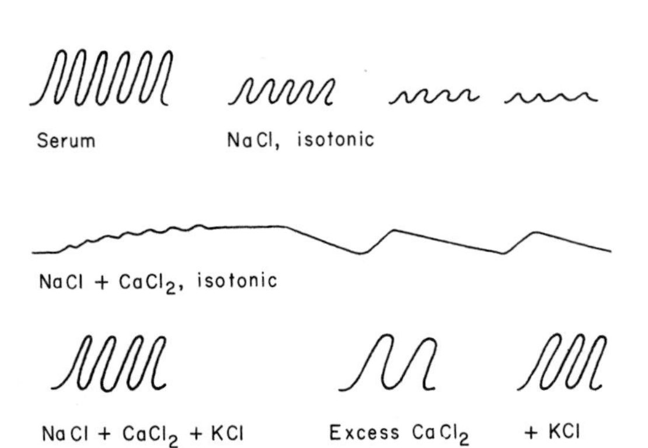

Figure 17.17. *The effects of salts on a frog's heart. The heart stops in isotonic NaCl* (top) *but beats slowly when $CaCl_2$ is added in the proportions found in blood* (center). *The heart must be stimulated once. The beat is almost normal when KCl is added in the proportion found in blood and when the mixture contains a buffer* (bottom). *Excess $CaCl_2$ prolongs the beat of the heart and is antagonized by KCl. (After Ringer, 1882–83: J. Physiol. 4: 29, 422.)*

TABLE 17.8. RELATION BETWEEN NARCOTIC ACTIVITY OF RELATED ALCOHOLS AND THEIR OIL SOLUBILITY*

Alcohol	*Molecular Weight*	*Solubility in gm per 100 ml Water*	*Relative Solubility, Oil:Water*	*"Narcotic Concentration"*
Methyl	31.06	∞	0.0097	5.0
Ethyl	46.07	∞	0.0357	1.6
Propyl	60.09	∞	0.156	0.8
n-Butyl	74.12	7.9	0.588	0.15
i-Amyl	88.15	slight	2.13	0.045

*Data from Handbook of Chemistry and Physics, 1968. 49th Ed.; and Höber, 1945. The narcotic concentration is in moles needed to give 50 percent inhibition of oxygen consumption by goose erythrocytes.

difficult to explain. Perhaps the anesthetic molecules, being highly lipid soluble, accumulate at the cell surface and either increase or decrease entry of other molecules.

THE EFFECT OF PHYSIOLOGICAL STATE UPON PERMEABILITY

Permeability of cells is dependent on their physiological state. An active muscle cell, for example, is permeable to amino acids, glucose, and other materials, whereas a resting one is not (Steinbach, 1954).

When a muscle cell or nerve cell is stimulated, its permeability to ions is increased. Electric stimulation of various plant and animal cells is accompanied by an increase in their permeability to solutes at the negative electrode, where the electrons flow into the cell. The leakage of potassium ions at this electrode in stimulated *Nitella* and the squid giant axon may be sufficient even for measurement by microchemical methods. Sodium ions also pass through the membrane of a stimulated cell more freely than they do through the membrane of a resting cell (Hodgkin and Keynes, 1953).

TABLE 17.9. RELATION BETWEEN NARCOTIC ACTIVITY OF ANAESTHETICS AND THEIR OIL SOLUBILITY*

Compound	*Vol % Required to Just Immobilize Mice*	*Solubility in Olive Oil at 37°C*
Ethane	80	1.3
Acetylene	65	1.8
Dimethyl ether	12	11.6
Methyl-chloride	6.5	14.0
Dimethyl acetal	1.9	100
Chloroform	0.5	265

*Data from Höber, 1945: Physical Chemistry of Cells and Tissues. Blakiston, Philadelphia.

Injury of *Laminaria* cells by such means as heat, radiations, overstimulation with electricity, anesthetics, pH change, or salt unbalance results in an increased permeability of the cells. Slight injury produces a slight increase in conductivity. If exposure to injurious factors has not been excessive, complete recovery often follows on return to sea water, although only partial recovery may occur (Figs. 17.15 and 17.16). All gradations of permeability change are possible, depending upon the extent of damage to the plasma membrane. Death, as measured by permeability, is a quantitative phenomenon.

COMPARISON OF THE PERMEABILITY OF DIFFERENT KINDS OF CELLS

To compare the permeability of various kinds of cells to a substance, it is best to determine the permeability constant that measures the amount of the substance entering a unit area of the cell per unit time, per unit concentration difference of the substance between the inside and outside for each cell.

Calculations of a permeability constant incorporating these quantitative data are based on Fick's law of diffusion, which states that a substance, S, will diffuse through an area, A, at a rate, dS/dt (amount, dS, per unit time, dt), that is dependent upon the difference in concentration between the substances at a certain distance apart:

$$\frac{dS}{dt} = DA\frac{dC}{dx} \qquad (17.16)$$

where D is the diffusion coefficient and is given in moles per unit area per unit concentration gradient, and dC is the difference in concentrations ($C_1 - C_2$) over the distance, dx.

For passage through a membrane such as that of a cell, equation 17.16 becomes

$$\frac{dS}{dt} = \frac{DA}{M}(C_{out} - C_{in}) \qquad (17.17)$$

Here C_{out} represents the concentration of the substance, S, outside, and C_{in} is its concentration inside the cell. A is the area of the membrane, and M, which is substituted for dx of equation 17.16, represents the thickness of the membrane.

Since the amount of substance (S) per volume (V) of the cell is C_{in}, $C_{in}V$ may be substituted for S. When k, the permeability constant, is substituted for D/M, the equation is written as follows:

$$\frac{dC_{in}}{dt} = k\frac{A}{V}(C_{out} - C_{in}) \qquad (17.18)$$

If the volume of fluid outside the cell is large, integration of equation 17.18 and solution for k gives:

$$k = \frac{V}{At} ln \frac{C_{out} - C_{in}}{C_{out} - C^{1}_{in}} \qquad (17.19)$$

Here C_{in} and C^{1}_{in} are the concentrations inside the cell before and after time, t. The permeability constant, k, is therefore given in moles of the substance entering per square micrometer of cell surface per second per mole difference in concentration of substance S outside and inside the cell (Davson, 1970). Equation 17.19 is only an approximation because it takes into account the movement of solutes only, whereas both water and solute move through the cell membranes at the same time. The equations developed in Appendix 17.1 and 17.2 at the end of the chapter take both into account.

Permeability constants were determined for 16 species of plant cells for entry of a variety of substances, some of which enter readily, others slowly. Although for any one substance tested the permeability constant is not the same for cells of different species, the cells of each of these various species often show the same relative order of differences in permeability constants for different substances tested on each of them, as illustrated in Figure 17.18. This suggests that the principles of permeability, developed from experiments on a variety of plants and animal cells, are likely to apply to all cells at least in a qualitative way. No method yet exists for predicting permeability quantitatively.

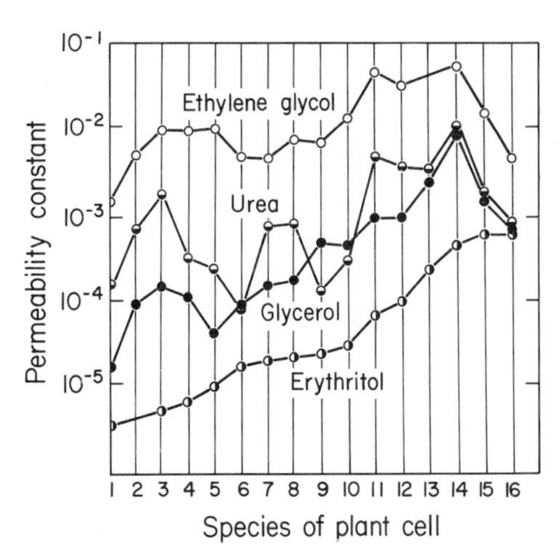

Figure 17.18. *Permeability constants of plant cells to nonelectrolytes. Cells from 16 species are represented: 1, Leaf cells of* Plagiothecum denticulatum*; 2,* Oedogonium *sp.; 3, root cells of* Lemna minor*; 4,* Pylaiella litoralis*; 5,* Zygnema cyanosporum*; 6, subepidermal cells of* Curcuma rubricaulis*; 7,* Spirogyra *sp.; 8, leaf cells of* Elodea densa*; 9, epidermal cells of* Rhoeo discolor*; 10, epidermal cells of* Taraxacum pectinatiforme*; 11, "leaf cells" of* Chara ceratophylla*; 12, internodal cells of* Ceramium diaphanum*; 13,* Bacterium paracoli*; 14,* Oscillatoria princeps*; 15,* Melosira *sp.; 16,* Licmophora *sp. (After Collander, 1937: Trans. Faraday Soc.* 33: *989.)*

COUNTERTRANSPORT

Countertransport, which occurs via a carrier, is the interchange of related molecules across the cell membrane. For example, perfused rat heart can be loaded with the pentose sugar L-arabinose to a concentration in equilibrium with the concentration outside the cell. Since arabinose is not metabolized, it remains available for diffusion outward should circumstances permit.

If the heart so loaded with arabinose is immersed in a solution containing both glucose and L-arabinose, the glucose binds to the cell membrane carrier for sugars and is transported inward where, upon entry into the cytoplasm, it is phosphorylated to form glucose-6-phosphate. Glucose-6-phosphate does not diffuse through the cell membrane, and some of it is metabolized. Glucose competes favorably with arabinose at the external cell surface and by occupying the membrane affinity sites cuts off arabinose entry. However, arabinose meets with no competition at the inner surface of the cell membrane and attaches to the sugar affinity site of a carrier that has just discharged glucose. Arabinose is thereby carried out of the cell membrane to the surface, where it diffuses into the medium (Fig. 17.19). Consequently, a gradient of arabinose-loaded carrier is set up from the inside of the cell to the outside, and the net flow of arabinose is outward. A gradient of glucose-loaded carrier is set up across the membrane from the outside to the inside of the cell, and the net flow of glucose is inward.

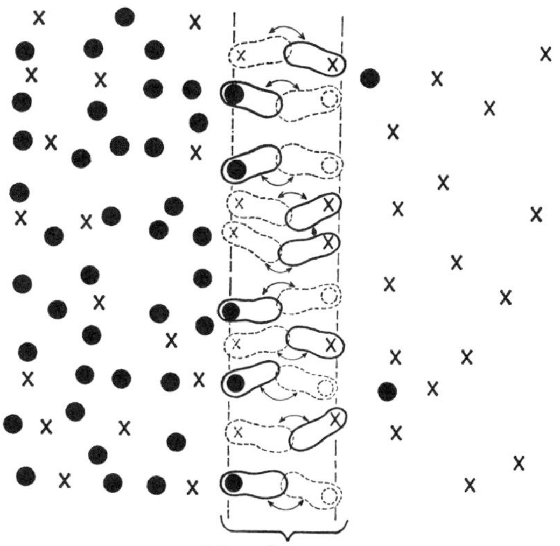

Figure 17.19. *Schematic representation of sugar-carrier complexes in a cell membrane during counterflow of* L*-arabinose* (×) *induced by* D*-glucose* (●). *The figure is explained in the text. (From Park* et al., *1968: J. Gen. Physiol. 52: 296s–318s.)*

The ultimate source of energy for the movements of both arabinose and glucose is the chemical potential of the concentration gradient for each sugar across the cell membrane. What therefore appears at first sign to be active transport is actually movement down concentration gradients. Such movement is not susceptible to metabolic poisons but it is sensitive to poisons that bind to the carrier at the affinity site (or at the sulfhydryl groups in the carrier protein). In the presence of phlorizin the carrier transport of sugars is reduced or stopped, depending on how fully the carrier affinity sites for glucose have been occupied by this poison (Park *et al.*, 1968).

MOVEMENT OF FLUID INTO CELLS BY HYDROSTATIC PRESSURE

In an osmometer the rise of fluid in the tube stops when the hydrostatic pressure of the column, acting on the membrane, just balances the osmotic pressure. Pressure is force per unit area. Therefore, a difference in hydrostatic pressure between the fluids on two sides of a membrane is in effect much like the chemical potential of a difference in concentration of solutions on two sides of a membrane.

A higher hydrostatic pressure outside than inside of the cell moves water through the cell membrane even if the cell is hypotonic to the medium, provided the hydrostatic pressure is more than enough to counter the osmotic pressure (Davson, 1970). It is probable that the flow of water through blood capillaries depends on hydrostatic pressure generated by the heart muscle. From this relationship, hydrostatic pressure has been used to determine experimentally the size of pores in membranes, such as artificial model membranes and erythrocyte cell membranes (Solomon, 1960). Hydrostatic pressure is also important in ultrafiltration in kidney tubules (nephrons).

Sea water can be desalinated by hydrostatic pressure: water passes through the pores of the filter but the solute does not, the reverse of uptake of water in a semipermeable membrane separating sea water from fresh water at atmospheric pressure (Scholander, 1972). The separation of solute from solvent is accomplished at the expense of the force of pressure acting over the distance the water is moved.

APPENDIX

17.1 THE STAVERMAN REFLECTION COEFFICIENT

The reflection coefficient is a measure of the selectivity that a membrane demonstrates between solute and solvent. The reflection coefficient is the ratio of the actual measured osmotic pressure to the value of the osmotic pressure calculated on the basis of the van't Hoff equation (17.3). It varies from unity when the solute does not penetrate the membrane to zero when both solute and solvent have the same degree of restraint on their movement (Staverman, 1948; Davson, 1970).

It can be shown on thermodynamic principles that the reflection coefficient, σ, is related to the membrane pore size available to solvent and solute, as shown in the following equation:

$$\sigma = 1 - A_{sf}/A_{wf} \qquad (17.20)$$

where A_{sf} is the effective pore area available to the solute molecules and A_{wf} is the effective pore area available to the solvent molecules. By determining the radius of the pores in artificial membranes, it could be shown that the reflection coefficient comes closer to unity as the radius of the penetrating molecule approaches in size the radius of the effective pore, preventing its entry. However, it cannot be assumed that pore radius alone determines passage of molecules through a membrane pore, since a hindrance factor must be considered even though its origin is not entirely clear (Beck and Schultz, 1970). The reflection coefficient can also be determined by studying swelling or shrinking of cells placed in solutions at various concentrations of solute and extrapolating to the point of zero the change in cell volume (Davson, 1970).

17.2 THE KEDEM-KATCHALSKY EQUATION FOR MOVEMENT OF A SOLVENT AND SOLUTE THROUGH A CELL MEMBRANE

The following equations apply to any solvent and any solute passing through a membrane. Since in a cell the solvent is almost always water, the discussion here is cast in terms of aqueous solutions.

When two aqueous solutions differing in solute concentration are separated by a membrane permeable to both water and solute (although not equally permeable to both), the equation for water flux from the higher to the lower concentration of water assumes the following form (after Davson, 1970):

$$dV/dt, \text{ or } J_v = L_p\Delta P + L_{pd}RT\Delta C_s \qquad (17.21)$$

where J_v, the flow of water (given in $cm^3cm^{-2}sec^{-1}$), is equal to the volume of water, dV, crossing unit area of cell membrane in unit time dt, under the influence of a one atmosphere osmotic pressure difference. The coefficient L_p describes the effect of a difference in hydrostatic pressure (equivalent to the osmotic pressure), and the coefficient L_{pd} describes the effect of the difference in concentration of solute ΔC_s on movement of water; R is the gas constant and T the absolute temperature.

The movement of solute in this solution is given by equation:

$$J_d = L_{dp}\Delta P + L_dRT\Delta C_s \qquad (17.22)$$

where J_d is the ratio of solute flux to water flux of both solute and solvent through the membrane, and L_{dp} is the equivalent of L_{pd} in equation 17.21. Thus both water flux and solute flux are affected by solute concentration difference, ΔC_s, and pressure difference, ΔP.

When the membrane is completely impermeable to solute, the driving force on the water is the osmotic pressure, equal to $RT\Delta C_s$ and the difference in hydrostatic pressure ΔP, and as the same constant applies to both:

$$J_v = L_p\Delta P + L_pRT\Delta C_s \qquad (17.23)$$

Here L_p is like the classical permeability coefficient for water (see equation 17.10).

When, on the other hand, the membrane is permeable to solute, but not to water, L_p is less than L_{pd} and if there is no net flux, that is, when J_v is zero,

$$\frac{-L_{pd}}{L_p} + \frac{\Delta P}{RT\Delta C_s} = \sigma \qquad (17.24)$$

Here σ is the reflection coefficient which Staverman (1948) defined as the ratio of the osmotic pressure difference and the theoretical osmotic pressure for solute impermeability.

Then $L_{pd} = -L_p$ and equation 17.21 becomes:

$$J_v = L_p\Delta P - \sigma L_pRT\Delta C_s \qquad (17.25)$$

Once the reflection coefficient is determined experimentally it is possible to measure the flow of water under set conditions and then to calculate L_p.

From equation 17.22 it may be shown that when solute, S, moves through the membrane in the absence of water flow, that is, when $\Delta P = 0$, the equation becomes:

$$J_d = L_dRT\Delta C_s \qquad (17.26)$$

where L_d is the ratio of velocities of solute to solvent movement in the membrane; movement of solute may then be stated in terms of solute flux, in moles per second:

$$dS/dt = \omega RT\Delta C_s + J_v(1 - \sigma)C_s \qquad (17.27)$$

where dS/dt is the solute flux, or the quantity of solute, dS, passing through unit membrane area per unit time, dt, and C_s is the mean concentration of solute in the cell membrane and is approximately equal to $C_s/ln\ (C_1/C_2)$ where C_1 is the concentration of solute on one side of the membrane, C_2 its concentration on the other side; and ω is defined in equation 17.30.

When the water flow is zero, that is, when $J_v = 0$, equation 17.27 becomes:

$$dS/dt = \omega RT\Delta C_s \quad (17.28)$$

Since dS/dt also equals $k_sA\Delta C_s$, where k_s is the permeability constant for solute, A is the area of the membrane (C_s has the same meaning as before).

Then,

$$Ak_s = \omega RT \quad (17.29)$$

where ω is related to the coefficients of equations 17.23 and 17.24 in the following complex manner:

$$\omega = \frac{L_pL_d - L_{pd}^2}{L_p} C_s \quad (17.30)$$

It is evident that when both solute and solvent penetrate the membrane, the mobility of the solute through the membrane cannot be given by a single constant, but is a function of ω, J_v and σ (Davson, 1970).

LITERATURE CITED AND GENERAL REFERENCES

Acker, R. F. and Hartsell, S. E., 1960: Flemming's lysozyme. Sci. Am. (Jun.) *202*: 132–142.

Beck, R. E., and Schultz, J. S., 1970: Hindered diffusion in microporous membranes with known pore geometry. Science *170*: 1302–1305.

Bresler, E. A., Wendt, R. P. and Mason F. A., 1971: Steady-state sieving across membranes. Science *172*: 858–859.

Brooks, S. C., 1938: The penetration of radioactive KCl into living cells. J. Cell. Comp. Physiol. *11*: 247–252.

Brooks, S. C., 1951: Penetration of radioactive isotopes P^{32}, Na^{24} and K^{42} into *Nitella*. J. Cell. Comp. Physiol. *38*: 83–93.

Brooks, S. C. and Brooks, M.,M., 1941: The Permeability of Living Cells. Protoplasma Monographien. Borntraeger, Berlin, Vol. 19.

Collander, R., 1959: Cell membranes: their resistance to penetration and their capacity for transport. *In* Plant Physiology, a Treatise. Vol. 2. Steward, ed. Academic Press, New York, pp. 3–102.

Collander, R. and Bärlund, H., 1933: Permeabilitatsstudien in *Chara ceratophylla*. Acta Bot. Fennica *11*: 1–114.

Dainty, J., 1963: Water relations of plant cells. Adv. Bot. Res. *1*: 279–326.

Davson, H., 1970: Textbook of General Physiology. 4th Ed. Vol. 2. Williams & Wilkins, Baltimore.

Davson, H. and Danielli, J. F., 1952: The Permeability of Natural Membranes. 2nd Ed. Macmillan Co., New York.

deLancey, G. B., Stana, R. R. and Chiang, S. H., 1967: A technique for producing an artificial membrane suitable for diffusion studies. *In* Chemical Engineering in Medicine and Biology. Hershey, ed. Plenum Press, New York, pp. 365–389.

Diamond, J. M., 1966: Non-linear osmosis. J. Physiol. *183*: 52–82.

Diamond, R. J. and Rose, A. H., 1970: Osmotic properties of sphaeroplasts from *Saccharomyces cerevisiae* grown at different temperatures. J. Bacteriol. *102*: 311–319.

Dick, D. A. T., 1959: Osmotic properties of living cells. Int. Rev. Cyt. *8*: 388–448.

Dick, D. A. T., 1966: Cell Water. Butterworth, Washington, D.C.

Donnan, F. G., 1927: Concerning the applicability of thermodynamics to the phenomena of life. J. Gen. Physiol. *8*: 685–688.

Fenichel, I. R., 1969: Intracellular transport. *In* Biological Membranes. Dowben, ed. Little, Brown & Co., Boston, pp. 177–221.

Fiske Osmometer, Fiske Associates, Inc., Bethel, Conn.

Giese, A. C., 1968: Effect of ultraviolet radiation on some activities of animal cells. *In* Ultraviolet Radiation (Conference on Biological Effects of UV-Radiation). Urbach, ed. Pergamon Press, Oxford, pp. 61–76.

Green, P. and Stanton, F. W., 1967: Turgor pressure: direct measurement in single cells of *Nitella*. Science *155*: 1675–1676.

Guyton, A. C., 1976: Textbook of Medical Physiology. 5th Ed. W. B. Saunders Co., Philadelphia.

Hammel, H. T., 1976: Colligative properties of a solution. Science *192*: 748–756.

Höber, R., ed., 1945: Physical Chemistry of Cells and Tissues. Blakiston Co., Philadelphia.

Hodgkin, A. L. and Keynes, R. D., 1953: The mobility and diffusion coefficient of potassium in giant axons from *Sepia*. J. Physiol. *119*: 513–528.

House, C. R., 1974: Water Transport in Cells and Tissues. Edward Arnold, London.

Katchalsky, A., 1969: Membrane thermodynamics. *In* Membranes Permeabilité Selec. Colloq., 1967, pp. 19–28.

Kedem, O. and Katchalsky, A., 1958: Thermodynamic analysis of the permeability of biological membranes to nonelectrolytes. Biochim. Biophys. Acta *27*: 229–246.

Krogh, A., 1939: Osmotic Regulation in Aquatic Animals. Cambridge University Press, London.

Kwant, W. O. and Seeman, P., 1970: The erythrocyte ghost is a perfect osmometer. J. Gen. Physiol. *55*: 209–219.

Lambert, C. C. and Lambert, G., 1978: Tunicate eggs utilize ammonium ions for flotation. Science *200*: 64–65.

Lewin, S., 1970: Water extrusion in biological reactions. J. Theoret. Biol. *26*: 481–495.

Lockwood, A. P. M., 1961: Ringer solutions and some notes on the physiological basis of their ionic composition. Comp. Biochem. Physiol. *2*: 241–289.

Lucké, B. and McCutcheon, M., 1932: The living cell in osmotic systems and its permeability to water. Physiol. Rev. *12*: 68–139.

Mauro, A., 1965: Osmotic flow in a rigid porous membrane. Science *149*: 867–869.

Northcote, D. H., 1977: Plant Biochemistry II. University Park Press, Baltimore. (Section on osmoregulation and membrane transport.)

Ochsman, J. Z., Wall, B. J. and Gupta, B. L., 1977: Cellular basis for water transport. Soc. Exp. Biol. Symp. *28*: 305–350.

Ospina, B. and Hunter, F. R., 1966: Facilitated diffusion in mouse and rat erythrocytes. Nature *211*: 851.

Osterhout, W. J. V., 1922: Injury, Recovery, and Death in Relation to Conductivity and Permeability. J. B. Lippincott Co., Philadelphia.

Osterhout, W. J. V. and Dorcas, M. J., 1926: The penetration of carbon dioxide into living protoplasm. J. Gen. Physiol. *9*: 255–267.

Overbeek, J. T., 1956: The Donnan equilibrium. Prog. Biophys. *6*: 57–84.

Papahadjopoulos, D., 1972: Studies on the mechanism of action of local anaesthetics with phosphorylated model membranes. Biochim. Biophys. Acta *265*: 169–186.

Park, C. R., Crofford, O. B. and Kono, T., 1968: Mediated (nonactive) transport of glucose in mammalian cells and its regulation. J. Gen. Physiol. *52* (Part 2): 265s–318s.

Peters, E. and Saslow, G., 1939–40: Performance of the Hepp micro-osmometer. J. Gen. Physiol. *23*: 177–184.

Pfeffer, W., 1877: Osmotische Untersuchungen. Engelmann, Leipzig.

Prager, D. J. and Bowman, R. L., 1963: Freezing point depression: new method for measuring ultra-micro quantities of fluids. Science *142*: 237–239.

Prescott, D. M. and Mazia, D., 1954: Permeability of nucleated and enucleated fragments of *Amoeba proteus* to D_2O. Exp. Cell Res. *6*: 117–126.

Prosser, C. L., ed., 1973: Comparative Animal Physiology. 3rd Ed. W. B. Saunders Co., Philadelphia.

Quastel, J. H., 1965: Molecular transport at cell membranes. Proc. Roy. Soc. London Ser. B. *163*: 169–196.

Quinn, P., 1976: The Molecular Biology of Cell Membranes. University Park Press, Baltimore.

Rehm, W. S., 1944: An easily constructed Hepp osmometer. Science *100*: 364.

Ringer, S., 1880: Concerning the influence exerted by each of the constituents of the blood on the contraction of the ventricle. J. Physiol. *3*: 380–392.

Robinson, J. R., 1960: Metabolism of intracellular water. Physiol. Rev. *40*: 112–149.

Rogers, P. and McElroy, W. D., 1958a: Enzymatic determination of aldehyde permeability in luminous bacteria. I. Effect of chain length on light emission and penetration. Arch. Biochem. Biophys. *75*: 87–105.

Rogers, P. and McElroy, W. D., 1958b: Enzymatic determination of aldehyde permeability in luminous bacteria. II. Effect of temperature on enzymatic activity and penetration. Arch. Biochem. Biophys. *75*: 106–116.

Rose, A. T., 1976: Osmotic stress and microbial survival. Soc. Gen. Microbiol. Symp. *26*: 155–182.

Ruesink, A. W. and Thimann, K. V., 1966: Protoplasts: preparation from higher plants. Science *154*: 280–281.

Saier, M., Jr. and Stiles, C. D., 1975: Molecular Dynamics in Biological Membranes. Springer Verlag, New York.

Solomon, A. K., 1960: Pores in the cell membrane. Sci. Am. (Dec.) *203*: 146–156.

Sourirajan, S., 1970: Reverse Osmosis. Academic Press, New York.

Staverman, A. J., 1948: Nonequilibrium thermodynamics of membrane processes. Trans. Faraday Soc. *48*: 176–185.

Stein, W. D., 1967: The Movement of Molecules Across Cell Membranes. Academic Press, New York.

Stein, W. D., 1968: The transport of sugars. Br. Med. Bull. *24*: 146–149.

Steinbach, H. B., 1954: The regulation of sodium and potassium in muscle fibers. Symp. Soc. Exp. Biol. *8*: 438–452.

Terepkin, A. R., Coleman, J. R., Armbrecht, H. J. and Gunter, T. F., 1976: Transcellular transport of calcium. Soc. Exp. Biol. Symp. *30*: 117–140.

Tosteson, D. C., Cook, P., Andreoli, T. and Tiffenberg, M., 1967: The effect of valinomycin on K and Na permeability of HK and LK sheep red cells. J. Gen. Physiol. *50*: 2513–2525.

Ussing, H. H., 1952: Some aspects of the application of tracers in permeability studies. Advances Enzymol. *13*: 21–65.

Ussing, H. H., Erlij, D. and Lassen, V., 1974: Transport pathways in biological membranes. Ann. Rev. Physiol. *36*: 17–49.

Warren, K. B., 1966: Intracellular Transport. Symp. Soc. Cell Biol. *5*: 1–325.

Zimmerman, U., 1978: Physics of turgor and osmoregulation. Ann. Rev. Plant Physiol. *29*: 121–148.

CHAPTER 18

METABOLICALLY COUPLED ACTIVE TRANSPORT

It requires work to move a substance through a cell membrane "uphill" against a concentration gradient; this work is usually done at the expense of metabolic energy. Such movement is called metabolically coupled *active transport.**

Active transport is visibly demonstrated in the action of vertebrate kidney tubule cells. Pieces of kidney tubule, separated with a micromanipulator from a chick embryo, often round up into spheres one cell layer thick (Fig. 18.1). When such spheres are immersed in dilute solutions of phenol red dissolved in embryonic tissue fluid, each sphere accumulates the dye in its internal cavity so that the cavity, which is slightly alkaline, becomes deep pink in color. The dye may be seen passing through the cells of the tubule, in which it lingers in transit and becomes orange in color, a change to be expected of phenol red at the pH (about 6.7 to 6.8) inside the cells. If phenol red were diffusing through the cells in response to the difference between the concentrations of the dye in the bathing solution and in the vesicle, it would move inward only until equilibrium was established. The appearance of the dye in much higher concentration inside the cavity of the vesicle than elsewhere means that the dye passing through the cells is being secreted into the cavity against a concentration gradient (Chambers *et al.*, 1935).

*Active transport may occur even when a downhill gradient favorable for diffusion is present, as, for example, glucose uptake in the vertebrate gut. In this case the cell expends energy in active transport primarily to accelerate the rate of glucose uptake (by as much as 10,000-fold) as compared to the rate of uptake by diffusion alone (see section on active transport in gut). However, diffusion transport accompanies active transport.

If the kidney cell preparation is placed on ice at the time phenol red is added, it fails to accumulate the dye, although a small amount will diffuse inward. If it is placed on ice after accumulation of dye inside the vesicle, the dye diffuses out and distributes itself equally between the medium and the vesicle lumen. Warmed to chick body temperature (39°C), the vesicle again accumulates the dye. The experiment demonstrates that although the dye continues to diffuse through tubule cells even at low temperature, it is no longer accumulated against a concentration gradient. Some metabolic process, essential to accumulation, is inhibited by the low temperature and measurements show that respiration of the kidney tubule cells comes virtually to a stop when the tubule is placed on ice. Since respiration is the main source of energy for cellular work, the inference is drawn that the movement of the dye against a concentration gradient requires cellular work.

If the kidney tubule cells at 39°C are inhibited by a metabolic poison so that respiration virtually ceases, dye is no longer accumulated. Diffusion, however, continues in response to the concentration gradient and a preparation poisoned after it has accumulated dye is still able to lose it until an equilibrium has been reached with the medium.

The aglomerular kidney tubule (nephron) (Fig. 18.2) found in a few marine fishes like the toadfish and the pipefish further illustrates the ability of kidney cells to transport substances against a concentration gradient. The only way in which the nephron of such a kidney can form urine is by transport of water and solutes across the cells of the tubule from the blood into the lumen. If the concentration of a compound in the tubule is less than that in the blood, entry

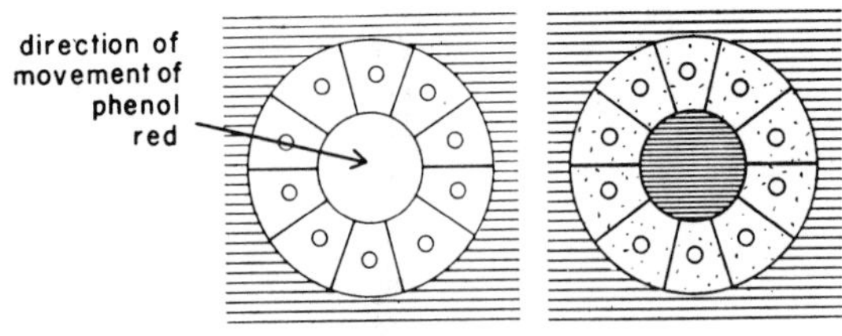

Figure 18.1. Diagrammatic illustration of accumulation of phenol red in sections of proximal kidney tubules of chick embryo. Phenol red moves inward and becomes more concentrated in the inside of the vesicle than on the outside.

of that compound into the tubule can occur by simple diffusion through the cell. On the other hand, the urine of aglomerular fishes is found to contain creatine, creatinine, uric acid, magnesium, potassium, chloride, and sulfate, each often in a concentration considerably greater than in the blood. The aglomerular kidney also is shown to concentrate in the urine a variety of foreign test substances that have been injected into the blood, such as nitrate, thiosulfate, sulfocyanides, indigo carmine, neutral red, and phenol red. This activity ceases when oxygen is withheld or when the metabolism of the kidney cells is inhibited by appropriate metabolic poisons. Therefore, all evidence indicates that active transport against a concentration gradient requires metabolic energy.

The glomerular kidney (Fig. 18.2), which is characteristic of all vertebrates except the fishes cited, presents special problems of organization and is dealt with in textbooks of vertebrate physiology (e.g., Guyton, 1976).

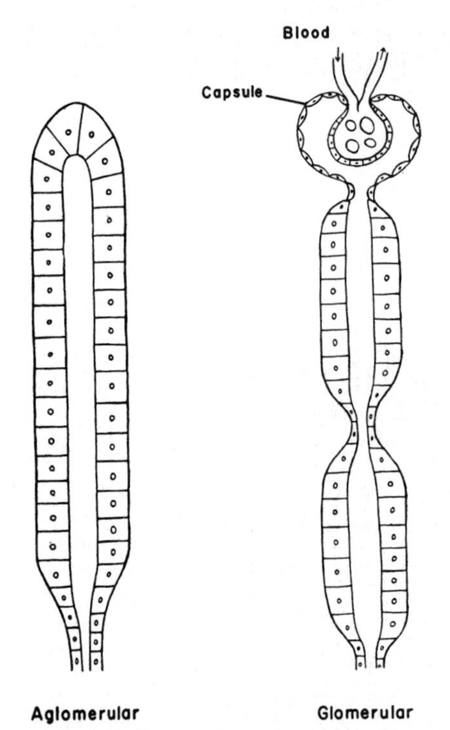

Figure 18.2. Diagram of aglomerular and glomerular kidney tubules (nephrons). Both are copiously supplied with blood capillaries along the outside. Only the glomerular tubule has a capillary tuft in a capsule, through which filtration under physical blood pressure occurs in the spaces between the cells rather than through the cells.

ENERGY INVOLVED IN ACTIVE TRANSPORT

The experiments described demonstrate that kidney tubule cells can do work required for accumulation of solute against a concentration gradient only if they are able to liberate energy. Energy, then, is needed for the cell to carry dye molecules against a concentration gradient, just as energy is needed to climb a hill (Fig. 18.3).

The work, W, required to carry a given amount of a substance against the chemical potential (concentration) gradient, is related to the concentration difference between the medium and the inside of the cell (of the substance in question) in the following manner:

$$W = RT \ln \frac{C_1}{C_2} \qquad (18.1)$$

where R is the gas constant
T the absolute temperature
C_1 the higher concentration of the substance and
C_2 the lower concentration of the substance

A charge on an ionized substance being transported, such as a salt ion, also poses a hindrance to entry because the cell surface is charged. For example, potassium is positively charged, as is much of the cell surface. In addition to the work against the concentration gradient, work must be done that is proportional to the difference in charge on the surface of the cell and the charge on the ion, multiplied by the number of ions transported for potassium to enter. The equation takes the form (Cirillo, 1966):

$$W = RT \ln \frac{C_1}{C_2} + n\mathscr{F}E \qquad (18.2)$$

where n is the valency of the ion
$\mathscr{F}$ is the Faraday (96,500 coulombs) and

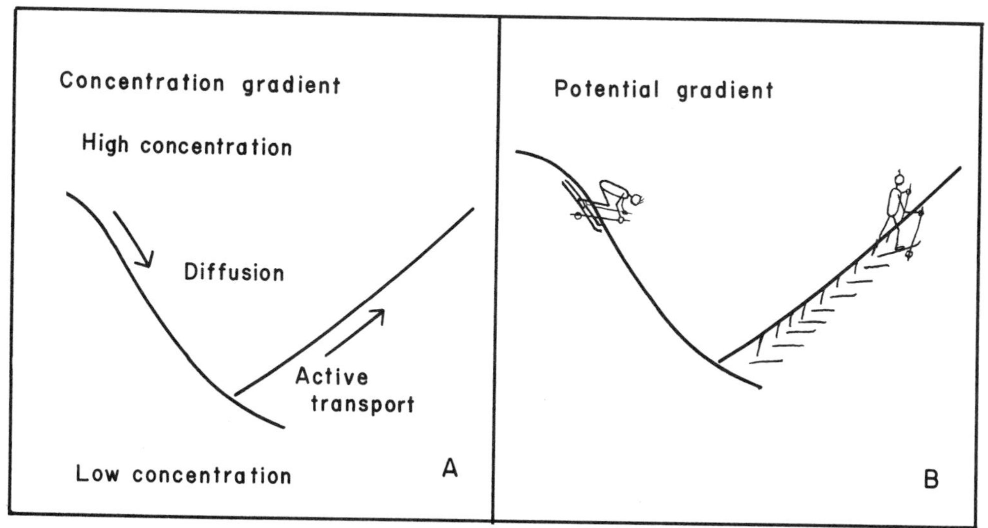

Figure 18.3. Analogy between moving a quantity of a substance down a concentration gradient (left side of A) or downhill (left side of B) and against a concentration gradient (right side of A) or uphill (right side of B). Moving downhill in both cases occurs as a result of the potential resident in the system (downhill concentration gradient in A, height in B). Moving uphill in both cases requires expenditure of energy equivalent to the product of the concentration gradient (in A) or height (in B) and the quantity of substance moved (work = force times distance). Equations 18.1 and 18.2 give the quantitative relation of the work required to move a substance against a concentration gradient and against an electrical potential gradient.

E is the potential difference in volts between the medium and the surface of the cell.

Because both solvent and solute move across the cell membrane, the corrections considered in Appendices 17.1 and 17.2 must be taken into account. This is especially desirable for solutes passing through water-filled channels (Quinn, 1976). Equations 18.1 and 18.2, as first approximations, serve the present purpose.

To visualize the energy requirement in transporting an ion against a concentration and potential gradient, consider transport of K^+ from a medium in which its concentration is 20 mM into a cell (e.g., squid giant axon) where it is 400 mM against a membrane charge of 70 mv at room temperature (300°K).

$$W = RT\, ln \frac{C_1}{C_2} + n\mathscr{F}E \qquad (18.3)$$

The first term is in calories, the second in volt-coulombs or joules. To convert joules to calories, divide by the number of joules in a calorie, 4.183. To convert ln to $\log_{10}$ multiply $\log_{10}$ by 2.3. Then,

$$\begin{aligned} W &= (2 \times 300 \times 2.3 \times 1.30103) + \\ &= (1 \times 96{,}500 \times 0.07)/4.183 \qquad (18.4) \\ &= 3410 \text{ cal/mole} \end{aligned}$$

A determination of the kidney respiratory rate indicates that the energy released by respiration is more than adequate to transport the amount of solutes known to move against concentration gradients.

Therefore, the decisive criterion for active transport is its dependence upon metabolic energy. If accumulation ceases when respiration is blocked by a respiratory poison (e.g., cyanide), lack of oxygen, or low temperature, active transport is indicated. In cells in which the energy source for active transport is anaerobic metabolism (glycolysis), blocking anaerobic metabolism (e.g., by iodoacetic acid) will stop active transport, as will low temperature, but lowering the oxygen partial pressure will have no effect. Active transport always occurs at the expense of either aerobic or anaerobic metabolism except in the special case of the halophilic bacteria *(Halobacterium halobium),* in which light energy absorbed by a pigment serves this purpose. As much as 30 percent of the metabolism of a mammalian red blood cell and 15 percent of the metabolism of an ascites (tumor) cell are thought to be expended in active transport processes (Stein, 1967).

COMPARISON OF ACTIVE TRANSPORT AND FACILITATED DIFFUSION

It is of interest to note similarities and differences in active transport and facilitated diffusion. In both processes a carrier mechanism is involved, that is, a protein in the membrane attaches to the solute and "carries" it through the membrane. Therefore, both processes may show saturation; in other words, the permeability increases with increasing concentration of test substance until the carrier is translocating all it can (much like an enzyme saturated with substrate) (see Fig. 10.13). In both active transport and facilitated diffusion related substrates compete for the carrier and stereospecificity (e.g., of D- and L-compounds) is of consequence in complexing of the substrate with the carrier. In both, the Q_{10} for transport may be 2 or more.

Permeation (penetration through the cell membrane) by a carrier is also greatly accelerated over what could occur by simple diffusion without a carrier, in both cases. However, a temperature low enough to stop metabolism stops active transport but only decreases the rate of facilitated diffusion. Metabolic poisons stop active transport but not facilitated diffusion. Finally, active transport can occur against an electrochemical gradient, whereas facilitated diffusion cannot.

It will become evident that the cell is versatile and capable of simultaneously using multiple pathways for uptake of a substance as necessity dictates. It can use active transport when the substance is less concentrated outside than inside the cell, but it can take it up by diffusion, either simple or facilitated, when the concentration of the substance is higher outside than inside the cell. Active transport may occur simultaneously with diffusion, as in monosaccharide transport in the gut after a meal, because it is a more rapid pathway. In fact, all three types of transport may occur at the same time.

DIRECT MEASUREMENT OF ACTIVE TRANSPORT

From the standpoint of the present chapter, perhaps the most interesting direct measurement of active transport was obtained with frog skin cells by the sodium current registered when the voltage between the two sides of the skin had been shorted out. The procedure for this purpose is shown in Figure 18.4. A piece of frog skin, S, is spread between the chambers C and C′, both containing identical Ringer's solution bathing the skin. The potential (skin potential) between the two chambers is measured and shorted out by the following method. Two agar-Ringer bridges, A and A′, on either side of the skin connect through calomel-KCl electrodes to a potentiometer, P, by which the potential between the two chambers is measured. Another pair of agar-Ringer bridges, B and B′, dipping into chambers C and C′, connect, by way of a KCl-AgCl solution into which dip silver electrodes, to batteries and a microammeter, M. Through the silver electrodes a voltage is applied to the frog skin preparation until the voltage on the potentiometer, P, reads zero, thus short-circuiting the skin potential. The current, which then flows through the circuit, as measured by the microammeter M, is considered to be a measure of the quantity of

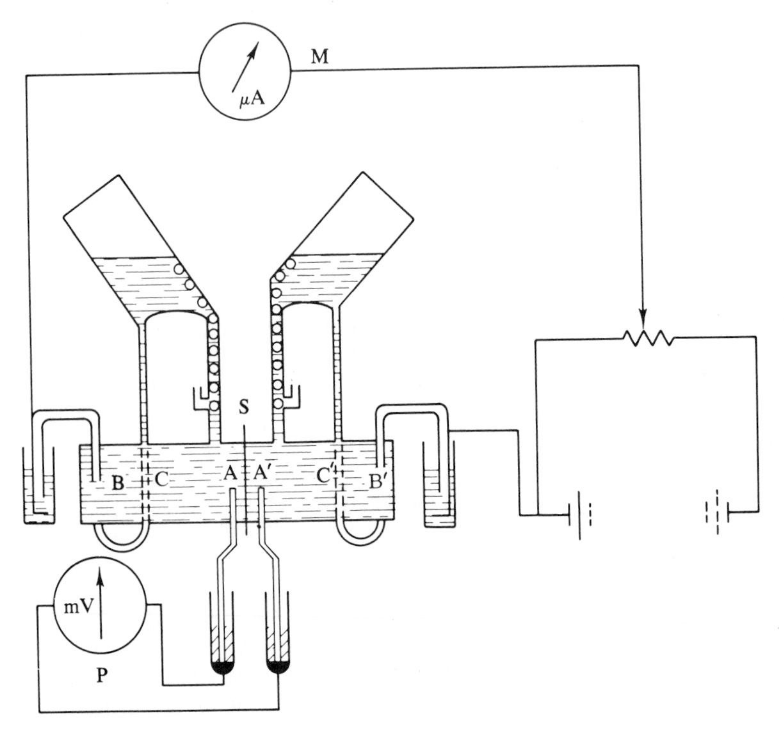

Figure 18.4. *Diagram of the short-circuit method. See text for explanation. (After Ussing and Zerrahn, 1951: Acta Physiol. Scand. 23: 111.)*

TABLE 18.1. MEASUREMENT OF SHORT CIRCUIT CURRENT USING ^{22}Na (INFLUX) AND ^{24}Na (OUTFLUX) IN FROG SKIN*

	Microamperes			
Addition	***In***	***Out***	***ΔNa^+***	***Current***
Control	233	19.4	213.6	195
Epinephrine	208	110	98	156

*From H. H. Ussing, 1952: Advan. Enzymol. *13*: 21. Note that epinephrine, the hormone of the adrenal glands, increases outflow of sodium and decreases inflow.

sodium ions (from Ringer's solution) passing through the membranes of the skin cells. It is a direct measure of the movement of sodium ions passing through the cell membranes by active transport. Theoretically, this current (Table 18.1) could be used to light a lamp or run a motor.

That sodium ions pass in and out through the cell membrane of the cells in the skin can be demonstrated by the use of isotopes. For example, ^{22}Na is used on the inner surface of the skin and ^{24}Na on the outer surface. After a brief period of time, by measuring the current and the amount of each tracer passing through, it is possible to calculate the fractions of the sodium current attributable to inward and outward movements of sodium (see Table 18.1) (Ussing *et al.,* 1974).

EXAMPLES OF ACTIVE TRANSPORT

Active transport seems to be quite general in plant cells, animal cells, and bacteria. A few interesting cases are cited here.

Accumulation of Potassium in Cells

A great accumulation of potassium is found in many plant and animal cells and in bacteria. For example, *Nitella,* a fresh-water plant with large cells, has 1190 times as much potassium as the water in which it grows (Fig. 18.5); *Valonia,* another large plant cell, has 41.6 times as much; *Chara* has 63 times as much. A somewhat similar difference in concentration of potassium between the inside of the cell and the bathing medium is found in muscle and nerve cells (see Fig. 3.5) and in microorganisms such as yeast and bacteria (Luria, 1975).

That potassium is present in cells as an osmotically active ion is demonstrated by the level of electrical conductivity of cells; otherwise the conductivity would be much lower because potassium is the main cation in the cell. Consequently the potassium gradient between cell interior and medium can be maintained only by active transport. Therefore, when metabolism is inhibited, potassium should leak out of cells. This is what happens when a squid giant axon or a *Halicystis* (large algal) cell is placed under anaerobic conditions. If oxygen is then bubbled through the water, potassium is again accumulated. In these cells anaerobic metabolism obviously does not provide the energy required for active transport of potassium.

In human erythrocytes absence of oxygen does not prevent maintenance of the potassium gradient between cell interior and medium, provided glucose is present, because energy for active transport is provided by glycolysis. Glycolytic poisons such as fluoride or iodoacetic acid stop active transport, as does lowering the temperature until glycolysis stops. Both yeast and *E. coli* also accumulate potassium anaerobically in the presence of glucose.

Potassium leakage following stimulation of cells has been the subject of a number of studies on the squid giant axon and *Nitella.* However, it has been found that potassium is not the only ion that moves through the membrane of a stimulated cell. By carefully timing the movements of radioactive sodium and potassium in separate experiments, it has been shown that in an electrically stimulated squid giant axon cell sodium diffuses in, then potassium diffuses out. On recovery, sodium is subsequently actively extruded and potassium is again accumulated (see Chapter 21).

One mechanism handles both sodium and potassium, and changing the concentration of one of the ions outside the cell also affects the transport of the other ion (Stein, 1967). Evidence from use of tracers indicates that potassium transport in muscle and nerve cells is dependent upon metabolism, even when the electrochemical gradient is favorable for diffusion entry (Baker, 1966). Somewhat similar data are found for the cells of the red alga *Porphyra perforata* (Eppley, 1959). In yeast, potassium is taken up in exchange for hydrogen ions, but only if nutrient is available to sustain metabolism (Ussing *et al.,* 1974).

On the other hand, the erythrocytes of some carnivorous species of mammals, such as the dog and the cat, accumulate sodium instead of potassium. Red cells of some sheep accumulate potassium, while those of other breeds accumulate sodium (Garrahan, 1970). Nor is ion transport the same in cells of different tissues in the same species. For example, the epithelial cells of the small intestines actively transport both sodium and chloride ions, whereas cells of the kidney tubules transport sodium actively while chloride moves by diffusion in response to concentration gradients (Ussing *et al.,* 1974).

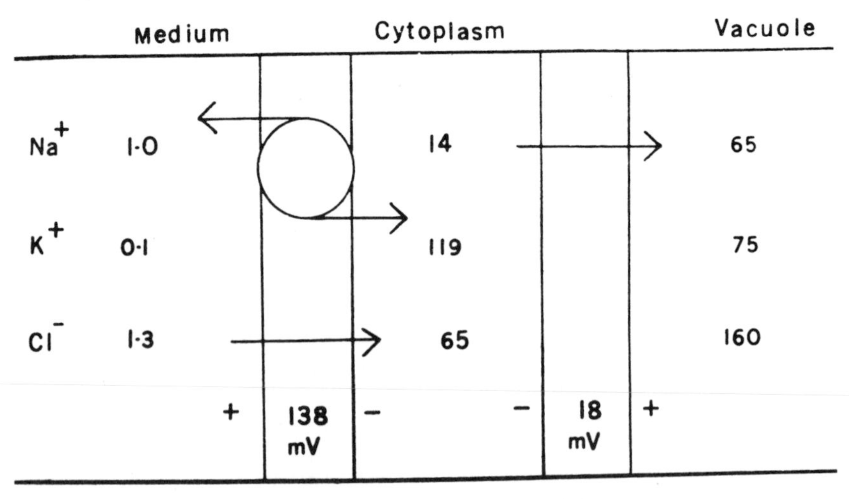

Figure 18.5. *Ionic and electrical gradients across the tonoplast (lining the central sap vacuole) and plasmalemma (on the surface of the cell) of* Nitella translucens. *The medium is 138 mV positive to the cytoplasm; the vacuole, on the other hand, is 18 mV positive to the cytoplasm. Active pumping of ions is indicated by arrows. The differences between E_m (the potential) and equilibrium potentials of sodium potassium, and chloride at the tonoplast were +56, −5, and +4 mV, respectively, and at the plasmalemma, −72, +40, and −236 mV, respectively. Such active pumping of ions leads to the large difference in potential between the inside and the outside of the cell. (From Spanswick and Williams, 1964: J. Exp. Bot. 15: 193.)*

Active transport of salts also occurs in the tear glands of birds and reptiles. This interesting topic is considered in detail elsewhere (Peaker and Linzeli, 1975).

Vacuolar Function

Freshwater protozoans presumably make use of active transport in voiding water. The medium surrounding them contains a higher concentration of water than the protozoan protoplasm does; therefore, water enters, in response to the concentration gradient, but it must again be voided into the medium against a concentration gradient. The magnitude of water elimination becomes evident when it is realized that every 15 or 20 minutes a *Paramecium* empties a quantity of water equal to its entire volume through its contractile vacuole (Kitching, 1967).

On the other hand, a brackish-water protozoan in slightly diluted sea water may take in little water, and the activity of its contractile vacuoles may be very low. However, if the medium is greatly diluted, as after a heavy rain, the rate of vacuolar contraction is greatly increased.

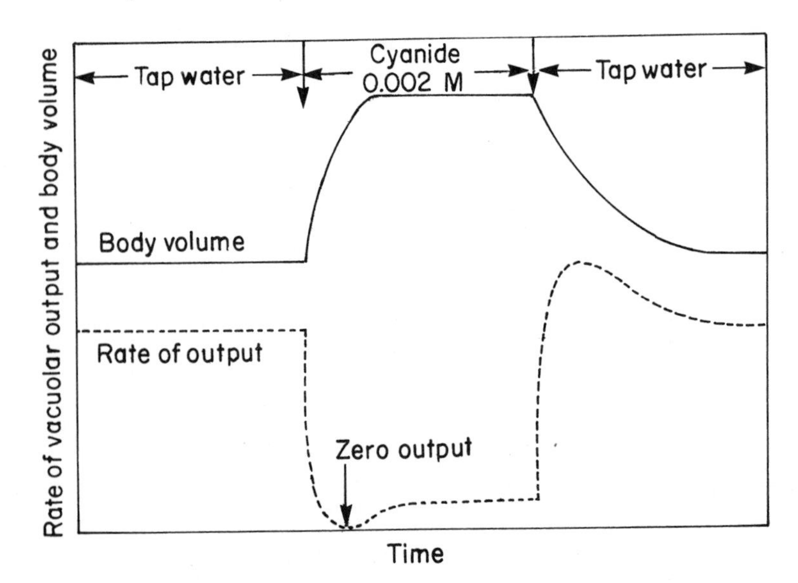

Figure 18.6. *The effects of cyanide on the swelling and vacuolar output of the fresh water protozoan* Zoothamnion. *The time of a complete experiment is about 2 hours. (After Kitching, 1938: Biol. Rev. 13: 423.)*

When a freshwater protozoan is poisoned with cyanide and its respiration is thus suspended, the contractile vacuoles fail to function and the cell swells (Fig. 18.6). If it is not washed free of cyanide, it bursts.

Evidence from micropuncture and analysis of the fluid withdrawn from the contractile vacuole of *Amoeba* at various times in the vacuolar cycle suggests that the fluid secreted into the contractile vacuole is isosmotic with the cytoplasm. Subsequently some salt is actively absorbed against a concentration gradient, and a fluid hypoosmotic to the cytoplasm is voided (Schmidt-Nielsen and Schrauger, 1963; Riddick, 1968).

Active Transport in Mitochondria and Endoplasmic Reticulum

Active transport at the expense of metabolic energy occurs not only through the plasma membrane but also through the membranes of the mitochondria, endoplasmic reticulum, and perhaps other organelles. Thus, active accumulation of calcium phosphate against a concentration gradient has been observed many times in the endoplasmic reticulum (Quinn, 1976). As much as a 50-fold increase in calcium of rat kidney mitochondria has been recorded, leading to practically complete exhaustion of calcium in the medium. Such uptake occurs only if an oxidizable substrate is present and stops if the respiration is blocked by a metabolic inhibitor (e.g., cyanide). Also required is the high-energy phosphate compound adenosine triphosphate (see Chapter 12) and inorganic phosphate as well as magnesium ion. Uptake does not occur at 0°C. Inorganic phosphate absorption accompanies calcium ion and phosphate is also accumulated (Quinn, 1976).

The endoplasmic (sarcoplasmic) reticulum of striated muscle cells has also been found to be an active site of calcium accumulation that shows all the characteristics of active transport. The uptake has been traced autoradiographically to those elements of the muscle cells that are portions of the endoplasmic reticulum. The rapid and active uptake of calcium here is of great value to the cell, since calcium is involved in muscular contraction and relaxation (Quinn, 1976).

The various examples cited demonstrate the great variety of biological processes that are dependent upon active uptake of compounds or ions against concentration gradients by the plasma membranes and membranes of cell organelles. Such active uptake occurs in plant cells, animal cells, and microorganisms. It may appear surprising that cells bring in some substances by active transport that might enter by their own kinetic energy, i.e., when present outside the cell in higher concentration than inside. However, active transport makes it possible for the cell to take in a particular substance at a *rate* determined solely by the needs of the cell at any time, regardless of how favorable or unfavorable the concentration gradient may be for its uptake at that time. Thus, to a cell, active transport of a substance at its instant command, although metabolically expensive, has sufficient advantage over the wait for a favorable concentration gradient to warrant its cost.

CARRIERS

Although active transport of sugars, amino acids, and inorganic ions is quite different in bacteria and animal cells and will be treated separately, it depends upon carriers in both. A carrier is a molecule that combines with a substance (substrate) and transports it from one side of the cell membrane to the other. By facilitating passage across the cell membrane a carrier increases permeation. Carriers seem to be proteins located in the cell membrane, extending through it in some spots. It is thought that carriers participate in both facilitated diffusion along a concentration gradient and in active transport uphill against a concentration gradient, but that they function in the second case only if the cell can provide the necessary metabolic energy. Carriers are discussed here in relation to active transport, but what is said about their general properties pertains to carriers participating in facilitated diffusion as well. Carriers are known for monosaccharides, amino acids, other organic substances, and inorganic ions (Quinn, 1976). Some carriers in bacteria have been purified and characterized, and the carrier proteins have many of the characteristics of enzymes. For example, they are highly specific for substrate. One carrier will transport a number of substances with similar chemical structures (e.g., glucose and the related synthetic sugars 3-*o*-methylglucose and 2-deoxyglucose) but not some of the other sugars tested. Stereospecificity also exists, as previously mentioned. The influx of a substance transported by a carrier depends upon the concentration of the substance up to a maximum flux that appears to be limited by the number of independent adsorption sites available for the substrate on the carriers. The flux of one substance may affect the flux of another, as if the two substances shared a common membrane carrier that alternates between inward- and outward-oriented states. The activities of carriers for permeation in one direction are subject to inhibition by compounds with structural similarity to the substrate; if these preferentially combine with the carrier site for a particular substrate, they prevent combination of carrier with substrate. As expected, carrier transport dependent on metabolic energy is subject to metabolic inhibitors. Carriers are also inhibited by phenylisothiocyanate, which acts on proteins. Some carriers in bacteria are *constitutive,* that is, they are present in wild-type cells without previous exposure to the substrate. However, carriers are often induced in bacteria only after exposure to the substrate, thereby reducing the need for a large stock of proteins. This is generally taken to imply *induction* by synthesis of a specific protein. Inhibitors of protein synthesis (such as chloramphenicol) prevent synthesis of induced carriers in bacteria.

Mutant bacteria in which a carrier for a particular substrate cannot be induced (synthesized) make possible identification of the carrier protein. In the method, cells of wild type that synthesize the carrier are labeled with one isotope (for example, ^{14}C-arginine); the other cells, the mutants, which are not genetically capable of making the carrier, are labeled with another isotope (e.g., ^{3}H-arginine). The carrier protein, then, will be labeled with ^{14}C but not with ^{3}H, whereas all the other proteins should have both labels. When the cells labeled in these two ways are mixed, disrupted, and fractionated, the fraction with the higest ratio of ^{14}C to ^{3}H should contain the carrier protein (Pardee, 1968). This approach has been used to isolate the protein carrier for β-galactosides. By modification and improvement of this technique, the pure membrane-bound protein (M-protein) specific for transport of β-galactosides of *E. coli* has been isolated. Its molecular weight is 31,000, and about 10^4 molecules of the protein are present per bacterium. The protein is absent in wild-type bacteria but its synthesis can be induced by exposing the wild-type cells to galactosides. Following such exposure, carrier-negative mutants of these bacteria are incapable of induced synthesis of the carrier of galactosides.

Carrier has binding specificity for substrates in a variety of ways, for example, by measuring the uptake of substrate through dialysis tubing, or by the desorption (removal) by substrate of proteins absorbed to a resin column. By use of this specificity, proteins specific for sulfate, leucine, galactose, phosphoenolpyruvate, and galactosides have been isolated from bacteria. For animal cells proteins specific for calcium and sodium are obtained from chick duodenum, and for potassium from beef brain. All these proteins are of molecular weight of about 30,000 and have a dissociation constant breakdown into subunits of 10^{-5} to 10^{-6}. They are not identical, however, as indicated by the differing amino acid constitutions that have been determined in some instances.

SUGAR TRANSPORT IN BACTERIA

In active transport energy can be provided for the carrier proteins by coupling with high-energy phosphate through enzymes (Fig. 18.7). The energy donor can then effect a change either in the substrate or in the carrier. The carrier system in bacteria appears to be made up of four protein components. An energy-supplying protein named HPr has been found to relate to transport of nine sugars. This donor protein itself phosphorylated on a histidine residue by phosphoenolpyruvate (PEP), a reaction catalyzed by an enzyme (called Enzyme I):

$$\text{PEP} + \text{HPr} \xrightarrow{\text{Enzyme I, Mg}^{2+}} \text{pyruvate} + \text{P} \sim \text{HPr}$$

$$\text{P} \sim \text{HPr} + \text{factor III} \longrightarrow \text{HPr} + \text{factor III} \sim \text{P}$$

$$\text{factor III} \sim \text{P} + \text{sugar} \xrightarrow{\text{Enzyme II}} \text{sugar-6P} + \text{factor III}$$

Each of the sugars transported across the cell membrane is released inside the cytoplasm of the cell in a phosphorylated form, the phosphorylation being catalyzed by Enzyme II. The phosphorylated sugar cannot return to the exterior through the cell membrane because of its lack of affinity for either the membrane or the carrier. Mutants lacking the ability to produce HPr or Enzyme I are capable of only minimal active transport, and little simple diffusion appears to occur. The HPr protein has been purified and turns out to be a relatively small molecule with a molecular weight of 9400. It has no affinity for sugar substrates. The component of the carrier system that recognizes a specific membrane-bound protein is Enzyme II, and apparently a specific Enzyme II is required for each sugar. Therefore, Enzyme II indicates a class of proteins rather than a single one. Factor III is also a protein. Some of these components are constitutive, some induced; induction is described in the next section.

The complete system of Enzyme I, Enzyme II, factor III, and HPr may not be required for transport in all cases. For example, mutants of *E. coli* that lack Enzyme I will transport lactose, which in this case is not phosphorylated. Facilitated transport of lactose also occurs even when energy flow is inhibited by appropriate metabolic poisons, but only downhill along a concentration gradient. Differences from genus to genus in transport system requirements are also found among bacteria. In some cases ATP serves as the high-energy source; in other cases other phosphorylated compounds are used.

Although much of the evidence linking these proteins to active transport is indirect, regions

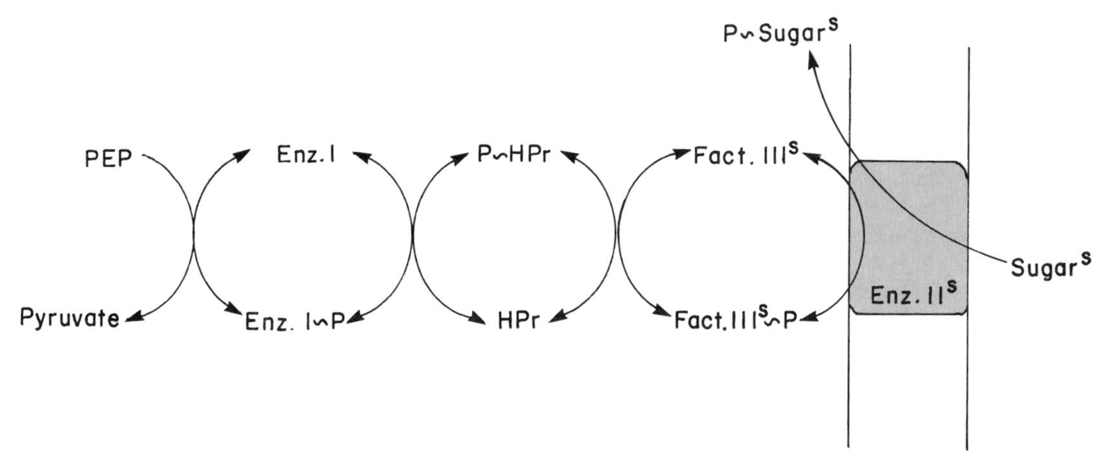

Figure 18.7. *The reaction sequence of sugar uptake into* Staphylococcus aureus *(gram-positive). Superscript s refers to sugar-specific and inducible components. (From Quinn, 1976: The Molecular Biology of Cell Membranes. University Park Press, Baltimore, p. 159.)*

of the chromosome have been identified with components of the transport system, and a defect in any one of them may interfere with active transport. When the transport system is modified by peculiar growth conditions, a parallel change in transport of the substrate in question is observed.

Restoration of the transport function by addition of parts of a system that was removed by chemical or other means is thought to be the most convincing evidence of the function of these parts as carriers. Success in restoring galactose transport in *E. coli* cells, which were osmotically shocked to remove the galactose transport system, has been achieved largely with crude extracts of the bacteria. The HPr system for transport of sugars in *E. coli* has also been reconstituted from the separated parts of extracts of the bacteria. Some experiments have also been tried with limited success by adding some transport proteins to artificial lipid membranes. For example, cyclic peptide antibiotics, such as valinomycin, specifically increase transport of some substances across artificial membranes.

The fact that about 5 percent of the proteins of a cell are released by subjecting cells to gentle osmotic shock suggests that among the released proteins are the carriers, which are presumably intrinsic proteins spanning the cell membranes. However, among the proteins so released are nucleases and phosphatases. Others may be present that appear to be involved because they are easily released from the cell surface, but these do not necessarily participate in carrier transport. The HPr is released by gentle osmotic shock, but the ATPase, required for active transport of sodium and potassium in bacteria and red blood cells, is released only after the use of detergents.

ENZYME AND CARRIER INDUCTION IN BACTERIA

Wild-type *E. coli,* placed in a solution of lactose (a disaccharide) uses this sugar only after a lapse of time. The delay could be interpreted as the time required either for the entry of the sugar into the cell or for the induction of the enzyme (β-galactosidase) (see Chapter 16) necessary to digest the disaccharide into monosaccharides before its use in metabolism. It can readily be demonstrated by disruption and testing for the disaccharide that the digestive enzyme is absent in wild-type *E. coli,* but that it is readily induced in the presence of lactose. However, another protein, the permease that permits passage of lactose through the cell membrane, is also induced in the presence of lactose. The fact that the two proteins are separate has been analyzed by the use of a number of mutant strains of *E. coli.*

A mutant of *E. coli* was isolated in which the digestive enzyme could not be induced. Yet after an induction period the mutant is able to remove lactose from the medium, as shown by quantitative determinations before and after the induction period. This indicates that the *permease* or carrier system has now been induced and lactose enters the cell. It was also shown, however, that the digestive enzyme could not be induced in the mutant, since lactose began to accumulate inside the cell. If the cell walls are removed (by lysozyme), the mutants accumulate lactose extensively and even burst, taking up water in response to the increased osmotic pressure. As much as 22 percent of the dry weight of the bacteria may consist of lactose at this time. Entry of lactose in this case is obviously not by diffusion but rather by enzymatically controlled transport.

The separateness of permease and digestive enzyme can be shown by the use of a metabolically inactive analogue of lactose such as thiomethyl-D-galactoside, which is structurally similar to lactose and induces production of both galactosidase and permease. However, this lactose analogue cannot be metabolized by wild-type *E. coli* and when supplied to wild type protoplasts, it is accumulated to the point of bursting the cells. The cells cannot digest the lactose analogue, yet they can accumulate it; the permease system fails to distinguish between it and lactose (Cohen and Monod, 1957). As might be expected, inhibitors of protein synthesis such as chloramphenicol and ultraviolet radiation retard or stop induction of permease and galactosidase (see Chapter 15).

It is evident that uptake of lactose occurs against a concentration gradient and requires work. One might, therefore, expect it to be blocked by metabolic poisons such as azide and dinitrophenol, which are known to interfere with the formation of high-energy phosphate bonds. These poisons are indeed found to stop accumulation of sugars in permease systems, indicating that such bonds are perhaps involved in transport of materials by permease systems (Cohen and Monod, 1957).

Permeases have been postulated for a variety of substrates, each one specific to a substrate. *E. coli* is thought to have 30 to 60 such systems. It is conceivable that some, if not all, are coupled to energy donors. Permeases have been described for a number of species of bacteria and for yeast. Carriers with a similar function occur in plant and animal cells but are much more difficult to demonstrate there. The name permease has not been generally accepted. Instead, the proteins involved are called *carriers* or *transport proteins* (Pardee, 1968), and apparently they are not enzymes, as was first suggested for permease. However, enzymes are associated with carriers when active transport occurs, as in the monosaccharide systems described in the preceding paragraphs.

SUGAR TRANSPORT IN ANIMAL CELLS

Sugar transport in animal cells is different from that in bacteria, being dependent upon the presence of sodium and potassium ions. Sugar moves from the gut lumen through the lining epithelium into the blood. It moves by diffusion along a concentration gradient when the concentration of glucose, for example, is higher in the gut than in the bloodstream. To learn whether transport occurs only by diffusion, or whether active transport also occurs, the metabolism of gut cells was inhibited by appropriate poisons in isolated loops of dog intestine. It was found that although the pentoses moved as freely as before poisoning, the hexoses glucose and galactose were transported much more slowly than before poisoning, suggesting that the movement of glucose and of galactose depends in part upon active transport. To test this hypothesis, the bloodstream was loaded with glucose (or galactose) so that any movement of glucose from unpoisoned gut cells to the blood would have to occur against a concentration gradient. It was found that glucose (or galactose) still entered the blood at a relatively rapid rate, indicating active transport. In the reciprocal experiment, in which a concentration of glucose (or galactose) less than that in the blood was placed in the gut so that any movement from gut to blood again would have to occur against a concentration gradient, uptake of sugars into the blood still continued.

If the glucose concentration in the gut is increased, the rate of its transport into the blood rises until a concentration limit is reached, beyond which further increase in its concentration has no effect on the rate of its transport. The rate of transport is seemingly limited by what the glucose carrier mechanism can transport per unit time (saturation). Nonetheless, this maximum rate is 10,000 times the diffusion rate (Elbrink and Bihler, 1975).

The mechanism of active transport of sugars was thought to be linked to sodium transport in the following manner *(cotransport hypothesis)*. The sodium pump* extrudes sodium ion through the plasma membrane into the medium (lumen of the gut), keeping the sodium concentration inside the cell lower than outside. Therefore, the sodium ion tends to diffuse into the cell. Sodium ion entering the membrane of the cell binds to a carrier, thereby increasing the carrier's affinity for glucose and for some other monosaccharides. Glucose at the surface of the cell then binds to the carrier at the affinity site, and the ternary complex (sodium, carrier, and glucose) undergoes conformational change across the cell membrane, powered by the *concentration difference* (chemical potential) between the sodium outside and inside the membrane (Fig. 18.8). When the ternary complex meets the cytoplasm where potassium concentration is high, the sodium on the complex is displaced by potassium. This lowers the affinity of the carrier for glucose.

Glucose is discharged into the cytoplasm, where it is phosphorylated. Because the cell membrane is essentially impermeable to it, phosphorylated glucose does not leave the cell. The carrier again becomes available for transport and meets a sodium ion diffusing inward, picks up another sugar molecule, and re-

*Deguchi *et al.* (1977: J. Cell Biol. *75*: 619–634) have studied the ultrastructure of the sodium pump.

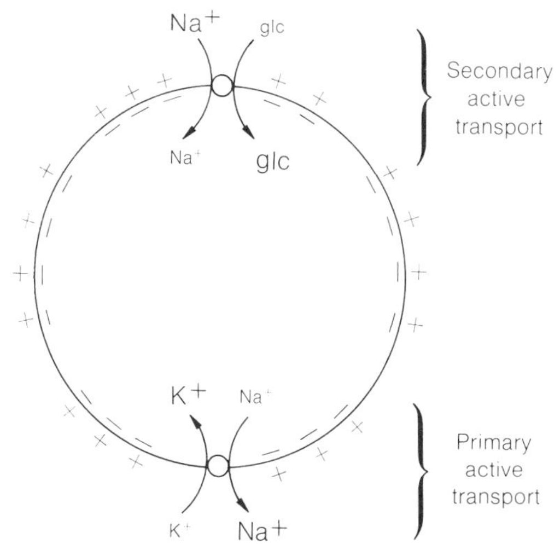

Figure 18.8. *Schematic diagram of active transport processes in an intestinal epithelial cell. The process at the bottom shows the primary active transport of* Na^+ *and* K^+ *by the* Na^+, K^+*-translocating ATPase. The process at the top illustrates the glucose—*Na^+ *cotransport hypothesis catalyzed by the glucose carrier. Primary active transport is considered to be that of the* K^+ *and* Na^+ *ions that provide the concentration difference enabling* Na^+ *to aid in the transport of glucose. (From Saier and Stiles, 1975: Molecular Dynamics of Biological Membranes. Springer Verlag, New York.)*

peats the cycle (Crane, 1967a and b; Newey *et al.,* 1970; Schultz and Curran, 1970). This hypothesis, which is appealing because it includes a unitary mechanism for active transport of many organic materials, suggests that active transport is powered by the sodium pump. However, in the toadfish gut the sodium gradient between the lumen of the gut and the cell interior can be lowered tenfold, abolished, or even reversed without changing the quantitative aspects of sugar transport, a finding incompatible with the cotransport hypothesis (Farmanfarmaian *et al.,* 1971). Other difficulties also exist with this attractive unifying concept.

However, we do know that an ATPase that is dependent for its activation upon the presence of sodium and potassium ions is involved. Many of its properties have been determined. Some pentoses probably move through the plasma membrane from gut cells to blood by simple diffusion and only when the concentration gradient is downhill. Glucose and galactose also move similarly to a small extent when the concentration gradient is downhill. However, in addition, they are actively transported, even against a concentration gradient; much of the transport of glucose apparently occurs in this manner. Fructose is not transported in this manner. Glucose and fructose have hexokinase specificity and are used in metabolism, making energy available for active transport; galactose does not have such specificity. Both glucose and fructose stimulate net transport of water across the gut wall; galactose does not (Davson, 1970). The detailed mechanism of transport of monosaccharides across the gut wall remains largely an open question.

It is interesting to note that when two sugars (e.g., D-glucose and D-galactose), both known to be actively transported, are tested at the same time, they compete with one another, presumably for the carrier to which either one sugar or the other can be bound at any one time. The binding is stereospecific, since neither L-glucose nor L-galactose can be actively transported. Of some 49 sugars tested, only 14 were found to be actively transported in the intestine (Crane, 1967a and b).

APPLICABILITY OF THE MICHAELIS-MENTEN EQUATION TO CARRIER TRANSPORT KINETICS

In facilitated diffusion and active transport the carrier (C) has an affinity site to which the substrate (S) binds, forming a complex (CS). At the cytoplasmic side of the membrane, the substrate is released. The higher the concentration of complex CS, the more rapid the substrate transport rate. The substrate transport rate therefore increases with increase in concentration of substrate to a maximum at which substrate binds all the available carrier. Increasing the substrate concentration beyond this point *(saturation)* has no further effect on rate of substrate transport because no more carrier is available to transport it. At saturation, the carrier has become the limiting factor (excepting, of course, for a slight amount of simple diffusion of the substrate, when such occurs).

The binding of substrate to carrier and subsequent release of the carrier resembles an enzymatic reaction. Consequently, the Michaelis-Menten equation for enzymatic kinetics (see Chapter 10) has been adapted for carrier transport in the following form:

$$J_s = J_s^{max} \cdot [S]_m/K_t + [S]_m \qquad (18.5)$$

where J_s is the solute flux at the time tested, J_s^{max} is the maximum solute flux, K_t is a constant showing the concentration of substrate when half the carrier is bound to substrate and is an indication of the affinity of the carrier for the substrate (if the affinity is high the concentration of substrate at which half the carrier is bound to substrate will be low), and $[S]_m$ is the concentration of substrate.

Plotting J_s against substrate concentration

gives a curve from which K_t can be determined, but it is easier to determine it from a plot of the reciprocal of J_s against the reciprocal of the substrate concentration, which gives a straight line, as is illustrated in Figure 10.21*A* and *B*.* Some of the problems in applying the plots effectively to permeability relations are considered elsewhere (Dowd and Riggs, 1965).

THE EFFECT OF ENVIRONMENT UPON ACTIVE TRANSPORT

Because temperature affects metabolic processes, it is expected to influence the rate of active transport. If a substance is carried through a cell membrane by active transport, the rate of liberation of energy by respiration is likely to be the limiting factor, and a temperature coefficient characteristic of thermochemical reactions is to be expected. In other words, the entry rate of such a compound should be doubled or tripled by a rise of 10°C in the environment (see Chapter 10). However, a high temperature coefficient obtained for transport of any substance would not by itself prove that the substance is taken in by an active process. It must be corroborated by other means as well.

Other environmental factors that affect the metabolic system are likely to affect active transport also. Some of these are narcotics and anesthetics, heavy metals, radiations, salts, and pH, but analysis of their effects on active transport has not yet been extensive.

Hormones alter the surface properties of cells, affecting diffusion down a concentration gradient as well as active transport against a concentration gradient. Thus, neurohypophyseal extract increases the entry of sodium ions into cells of target organs. Epinephrine increases sodium influx and efflux of cells in target organs and it increases glucose uptake in muscle cells. Acetylcholine increases ionic fluxes in nerve cells (see Table 21.2). Thyrotropin activates a Na^+–K^+-dependent ATPase located in the cell membranes of the thyroid gland. Aldosterone, a steroid from the adrenal cortex, causes significant retention of sodium in the kidney cells and increases sodium transport. This is thought to occur by induction of carrier protein synthesis via an effect on DNA-dependent RNA synthesis; aldosterone has no effect on ATPase in the cell membranes. Parathyroid hormone increases the permeability of mitochondria to both potassium and magnesium, resulting in their swelling. Insulin increases glucose uptake in muscle cells but not in erythrocytes (Bentley, 1971). A hormone called muscle activity factor released by contracting muscle has been suggested to explain the increased uptake of glucose by active muscle. Anoxia (lack of oxygen) also has this effect (Elbrink and Bihler, 1975; Quinn, 1976). It is likely that hormones attach to receptors in the membrane and open channels to the molecules in question, much as do certain antibiotic ionophores (see Chapter 8 and Quinn, 1976).

Drugs also affect the permeability of cells in certain target organs. For example, cardiac glycosides (e.g., strophanthin) that increase the contraction of heart muscle inhibit the Na^+–K^+-dependent ATPase and active transport of cations. Evidence exists that the drug quinidine, used to abort or prevent certain cardiac arrhythmias, may act by an effect on the sodium carrier in the sodium pump. Some colicins (El and K), compounds produced by *E. coli* that kill cells of related bacterial strains, are known to block active transport of lactose, various amino acids, and potassium ions in bacteria (Luria, 1975). Many other drugs are suspected to affect transport of various substances across the cell membrane, but the mechanism by which they produce their effects is still unknown.

MECHANISM OF ACTIVE TRANSPORT

Many suggestions for the possible mechanism of active transport have appeared in the literature. Summaries of these hypotheses, including figures and references to other literature, are available elsewhere (Quinn, 1976). It appears likely that active transport occurs by way of carriers, which are protein molecules that probably change conformation in the process of active transport, thereby passing the molecule or ion from one side of the membrane to the other.* The conformational change requires energy, and ATP usually serves this purpose, involving an ATPase. The concept is illustrated by the influx of potassium and efflux of sodium in the erythrocyte. For this purpose erythrocyte ghosts are prepared.

In this preparation red cells are hemolyzed in a hypotonic solution in which the membrane pores enlarge, permitting hemoglobin and other solutes to escape from the cell. Washed

*In Figure 10.21*A* and *B*, J_s of equation 18.3 is given as v, J_s^{max} as V, and K_t as K_m (the Michaelis constant), as is usually done for enzyme kinetics.

*Although carriers have been postulated by some to make flip-flop movements in cell membranes, by which they carry the attached molecule from one side of the membrane to the other, no evidence exists from spin resonance studies that such movements can occur. Carrier molecules are probably capable only of rotational and lateral movements and conformational changes.

ghosts are then placed in a solution of similar hypotonicity containing ATP, which enters the enlarged pores of the cell. When such cells are subsequently placed in solution isotonic with blood, the pores return to their former size, trapping within the cell the ATP, which cannot pass through the cell membrane. Such cells extrude sodium and convert ATP to ADP. Ghosts without internal ATP but to which ATP is applied externally are unable to extrude sodium. It has now been demonstrated that the ATPase reacts with sodium only at the cytoplasmic sites and potassium ion reacts with it only externally. The enzyme (ATPase) is quite large, with a molecular weight of 190,000 to 500,000. Assuming a molecular weight of 250,000 and a spherical molecule, its diameter would be about 8.5 nm, enough to extend completely across the cell membrane (Quinn, 1976).

When the enzyme transport system is activated by ATP, the outer side has a greater affinity for potassium ion than for sodium ion, whereas the inner site has a greater affinity for sodium ion than for potassium ion. Splitting of a high-energy phosphate bond from the ATP (with the formation of adenosine diphosphate and inorganic phosphate) reverses these affinities, as a consequence of which sodium ion is expelled from the inside of the cell while potassium ion is taken in, completing the cycle. The enzyme system is sensitive to low concentrations of the cardiac glycoside G-strophanthin (or ouabain, as it is more generally called). It requires Mg^{2+} and both Na^+ and K^+ for its operation; in absence of K^+, Na^+ is not "pumped out." It is probable that in the case of Na^+ and K^+, and perhaps other cations, the enzyme system is also the transport system. In ouabain-treated corneal cells sodium was localized in the cell by electron microscopic studies as the precipitate, sodium pyroantimonate (Kaye and Cole, 1965). Evidence for such a system has been found in the cells of 29 of 36 tissues tested in the cat and in 10 other species. Although the quantitative details vary, the essential qualitative features remain the same (Skou, 1969; Jorgensen *et al.*, 1971; Quinn, 1976). When organic molecules are transported, it is likely that other carriers are present, but the energy is probably supplied by ATP in the presence of an ATPase.

Apparently in some bacteria oxidative energy for active transport of potassium ions, sugars, and amino acids is supplied without requiring transfer of phosphate groups. One model of the possible mechanism suggests an accumulation of protons on one side of the membrane and hydroxyl ions on the other, leading to a potential difference of about 200 mv across the cell membrane, the energy being stored much as in an electrical condenser. Discharge of the energy in some way energizes active transport. How the energy is applied in these bacteria is one of the blank spots in our knowledge of the energetics of active transport.

Even more intriguing is the source of energy for pumping protons out of halophilic bacteria (*Halobacterium halobium*). Absorption of light (bleaching the pigment rhodopsin present in high concentration in the cell membrane) provides the energy for extrusion of protons. As a result, a potential gradient of several hundred mv is produced between the inside and the outside of the cell membrane, like the condenser effect mentioned. However, whether this activates the active transport directly or by way of the production of ATP that occurs in the cells as extruded protons fall through the potential gradient on returning to the cell, is not known (Stoeckenius, 1976). Spectroscopic evidence for a conformational change in bacteriorhodopsin on absorption of light has been suggested as the possible mechanism by which protons are pumped across the cell membrane (Ippen *et al.*, 1978). Light has also been used to "pump" carbon monoxide against a concentration gradient in an artificial membrane containing hemoglobin (Schultz, 1977).

One big void in our knowledge of the mechanism of active transport is the way in which the energy from any of the mentioned sources performs its tasks. Flip-flop movements of protein molecules have been largely disqualified and diffusion of proteins across the cell membrane is difficult to conceive; this leaves conformational changes in carrier molecules as the most likely mechanism. But the manner in which the change in conformation of a carrier performs its task remains unknown.

At present scientists are far from accepting a single hypothesis concerning the nature of active transport, but certain issues have been settled. General consensus seems to be that a carrier is involved in both facilitated diffusion and active transport and that the carrier is a protein. Some of the proteins have been isolated and characterized. It seems likely that movement of the substrate occurs by a conformational change in the carrier protein. The carrier protein itself may be an enzyme or be linked to an enzyme or enzymes and thus energy from ATP is made available for active transport, although alternate sources occur in some bacteria. In some cases, sugar uptake in bacteria, for example, the enzymes have been isolated and characterized. The ATPase has also been well characterized. Although for animal cells considerable evidence has accumulated that points to the sodium pump as a possible prime mover, suggesting that movement of sugars and amino acids was powered by cotransport, contrary evidence has been reported. The problem remains open.

LITERATURE CITED AND GENERAL REFERENCES

Alvarado, F., 1966: Transport of sugars and amino acids in the intestine: evidence for a common carrier. Science *151*: 1010–1013.

Ashton, R. and Steinrauf, L. K., 1970: Thermodynamic considerations of ion transporting antibiotics. J. Mol. Biol. *49*: 547–556.

Baker, P. F., 1966: The sodium pump. Endeavour *25*: 166–172.

Bentley, P. J., 1971: Endocrines and Osmoregulation. Springer Verlag, New York.

Berlin R. D. and Oliver, J. M., 1975: Membrane transport of purines and pyrimidines, bases and nucleoside in animal cells. Int. Rev. Cytol. *42*: 287–326.

Bittar, E. E., ed., 1970–1971: Membranes and Ion Transport. 3 Vols. Wiley-Interscience, New York.

Blinks, L. R., 1949: The source of the bioelectric potentials in large plant cells. Proc. Natl. Acad. Sci. U.S.A. *35*: 566–575.

Blobber, G. and Dobberstein, B., 1975: Transfer of proteins across membranes. J. Cell Biol. *67*: 852–862.

Bonting, S. L., 1970: Na-K activated ATPase and cation transport. *In* Membranes and Ion Transport. Vol. 1. Bittar, ed. Wiley-Interscience, New York, pp. 257–303.

Brooks, S. C., 1929: The accumulation of ions in cells—a non-equilibrium condition. Protoplasm *8*: 389–412.

Chambers, R., Beck, L. V. and Belkin, M., 1935: Secretion in tissue cultures. I. Inhibition of phenol red accumulation in the chick kidney. J. Cell. Comp. Physiol. *6*: 425–439; 441–455.

Christensen, H. N., 1970: Linked-ion and amino acid transport. *In* Membranes and Ion Transport. Vol. 1, Bittar, ed. Wiley-Interscience, New York, pp. 365–394.

Cirillo, V. R., 1966: Membrane potentials and permeability. Bacteriol. Rev. *30*: 68–79.

Cohen, G. and Monod, J., 1957: Bacterial permeases. Bacteriol. Rev. *21*: 169–194.

Crane, R. C., 1967a: Na^+-dependent transport of carbohydrates through intestinal epithelium. Protoplasma *63*: 36–40.

Crane, R. C., 1967b: Gradient coupling and the membrane transport of water-soluble compounds: a general biological mechanism? *In* Peptides of the Biological Fluids. Vol. 15. Elsevier, Amsterdam, pp. 227–235.

Davson, H., 1970: Textbook of General Physiology. 2 Vols. 4th Ed. Williams & Wilkins, Baltimore.

Deguchi, H., Jørgensen, P. L. and Maunsbach, A. B., 1977: Ultrastructure of the sodium pump. J. Cell. Biol. *75*: 619–634.

Dowd, J. E. and Riggs, D. S., 1965: A comparison of estimates of M-M constituents from various linear transformations. J. Biol. Chem. *240*: 863–869.

Eisenbach, M. *et al.,* 1977: Light driven sodium transport in sub-bacterial particles of *Halobacterium halobium*. Biochem. Bioph. Acta *465:* 599–613.

Elbrink, J. and Bihler, L., 1975: Membrane transport: its relation to cellular metabolic rates. Science *188*: 1177–1184.

Ellory, J. C., 1977: Membrane Transport in Red Cells. Academic Press, New York.

Eppley, R. W., 1959: Potassium accumulation and sodium efflux by *Porphyra perforata* tissues in lithium and magnesium sea water. J. Gen. Physiol. *43*: 29–38.

Erlanger, B. F., 1976: Photoregulation of biologically active macromolecules. Ann. Rev. Biochem. *45*: 267–283.

Farmanfarmaian, A., Ross, A. and Mazal, D., 1971: Coupled transport of sodium and sugar—in vivo evaluation in the intestine of the toadfish *Opsanus tau*. Biol. Bull. (Abstr.) *141*: 385.

Garrahan, P. T., 1970: Ion movements in red blood cells. *In* Membranes and Ion Transport. Vol. 2. Bittar, ed. Wiley-Interscience. New York, pp. 155–215.

Glynn, I. M., and Karlion, S. J. D., 1975: The sodium pump. Ann. Rev. Physiol. *37*: 13–55.

Greenawalt, J. W., Rossi, C. S. and Lehninger, A. L., 1964: Effect of active accumulation of calcium and phosphate ions on the structure of rat liver mitochondria. J. Cell Biol. *23*: 21–38.

Guyton, A. C., 1976: Textbook of Medical Physiology. 5th Ed. W. B. Saunders Co., Philadelphia.

Hajdu, S. and Leonard, E. J., 1976: A calcium transport system for mammalian cells. Life Sci. *17*: 1527–1533.

Harris, J. W. and Kellermeyer, R. W., 1970: The Red Cell. Revised Ed. Harvard University Press, Cambridge, Mass.

Hasselbach, W., 1964: ATP-driven active transport of calcium in the membranes of the sarcoplasmic reticulum. Proc. Roy. Soc. (London) *B160*: 501–504.

Hokin, L. E. and Hokin, M. R., 1965: The chemistry of cell membranes. Sci. Am. (Oct.) *213*: 78–86.

Hope, A. B., 1971: Ion Transport and Membranes: A Biophysical Outline. University Park Press, Baltimore.

Hurlbut, W. P., 1970: Ion movements in nerve. *In* Membranes and Ion Transport. Vol. 2. Bittar, ed. Wiley-Interscience, New York, pp. 95–143.

Ippen, E. P., Shank, C. V., Lewis, A. and Marcus, M. A., 1978: Subpicosecond spectroscopy of bacteriorhodopsin. Science *200*: 1279–1281.

Jørgensen, P. L., Skou, I. C. and Solomon, L. P., 1971: Purification and characterization of (Na^+ and K^+) ATPase. Biochim. Biophys. Acta *233*: 381–394.

Kaback, H. R., 1974: Transport studies in bacterial membrane vesicles. Science *186*: 882–892.

Katchalsky, A. and Oster, G., 1969: Chemicodiffusional coupling in biomembranes. *In* Molecular Basis of Membrane Functions. Tosteson, ed. Prentice-Hall, Englewood Cliffs, N.J.

Katchalsky, A. and Spangler, R., 1968: Dynamics of membrane processes. Quart. Rev. Biophys. *1*: 127–175.

Kaye, G. I. and Cole, J. D., 1965: Electron microscopy: sodium localization in normal and ouabain-treated transporting cells. Science *150*: 1167–1168.

Keynes, R. D., 1961: The energy source for active transport in nerve and muscle. *In* Membrane Transport and Metabolism. Kleinzeller and Kotyk, eds. Academic Press, New York, pp. 131–139.

Kirschner, L. B., 1970: The study of NaCl transport in aquatic animals. Am. Zool. *10*: 365–376.
Kitching, J. A., 1967: Contractile vacuoles, ionic regulation and excretion. *In* Research in Protozoology. Vol. 1. Chen, ed. Pergamon Press, New York, pp. 307–336.
Koch, A. R., 1970: Transport equations and criteria for active transport. Am. Zool. *10*: 331–346.
Korenbrot, J. I., 1977: Ion transport in membranes: incorporation of biological ion transporting proteins in model membrane systems. Ann. Rev. Physiol. *39*: 19–49.
Kotyk, A. and Janacek, K., 1975: Membrane Transport: Principles and Techniques. Plenum Press, New York.
Lehninger, A. L., 1964: The Mitochondrion. W. A. Benjamin, Menlo Park, Calif.
Lieb, W. R. and Stein, W. B., 1972: Carrier and noncarrier models for sugar transport in the human red cell. Biochim. Biophys. Acta *265*: 187–207.
Lindley, B. D., 1970: Fluxes across epithelia. Am. Zool. *10*: 355–364.
Luria, S. F., 1975: Colicins and the energetics of cell membranes. Sci. Am. (Dec.) *233*: 30–37.
MacRobbie, F. A., 1970: The active transport of ions in plant cells. Quart. Rev. Biophys. *3*: 251–294.
Meech, R. W., 1976: Intracellular calcium and the control of membrane permeability. Soc. Exp. Biol. Symp. *30*: 161–192.
Newey, H., Rampone, A. J. and Smyth, D. H., 1970: The relation between L-methionine uptake and sodium in rat small intestine *in vitro*. J. Physiol. *211*: 539–549.
Onsager, L., 1970: Possible mechanisms of ion transport. *In* Physical Principles of Biological Membranes. Snell, ed. Gordon and Breach, London, pp. 137–141.
Pardee, A. B., 1968: Membrane transport proteins. Science *162*: 632–637.
Peaker, M. and Linzeli, J. L., 1975: Salt Glands in Birds and Reptiles. Physiological Soc. Monograph.
Poole, R. J., 1978: Energy coupling for membrane transport. Ann. Rev. Plant Physiol. *29*: 437–460.
Potts, W. T. W. and Parry, G., 1964: Osmotic and Ionic Regulation in Animals. Macmillan Co., New York.
Quinn, P., 1976: The Molecular Biology of Cell Membranes. University Park Press, Baltimore.
Rappaport, S. I., 1970: Sodium-potassium exchange pump. Relation of metabolism to electrical properties of the cell. I. Theory. Biophys. J. *10*: 246–259.
Repke, D., Spratt, J. and Katz, A., 1976: Reconstitution of an active calcium pump in the sarcoplasmic reticulum. J. Biol. Chem. *251*: 3169–3175.
Riddick, D. H., 1968: Contractile vacuole in the ameba *Pelomyxa carolinensis*. Am. J. Physiol. *215*: 736–740.
Robinson, J. R., 1960: Metabolism of intracellular water. Physiol. Rev. *40*: 112–149.
Sachs, J. R., 1977: Kinetics of the inhibition of the sodium-potassium pump by external sodium. J. Physiol. (London) *264*: 449–470.
Saier, M. H., Jr. and Stiles, C. D., 1975: Molecular Dynamics in Biological Membranes. Springer Verlag, New York.
Schmidt-Nielsen, B. and Schrauger, C. R., 1963: *Amoeba proteus:* studying the contractile vacuole by micropuncture. Science *139*: 606–607.
Schmidt-Nielsen, K., 1959: Salt glands (of birds). Sci. Am. (Jan.) *200*: 109–116.
Schultz, J. S., 1977: Carrier-mediated photodiffusion membranes. Science *197*: 1177–1179.
Schultz, S. G. and Curran, P., 1970: Stimulation of intestinal sodium absorption by sugars. Am. J. Clin. Nutrition *23*: 437–440.
Schultz, S. G. and Zalusky, R., 1964: Ion transport in isolated rabbit ileum. II. The interaction between sodium and active sugar transport. J. Gen. Physiol. *47*: 1043–1059.
Simoni, R. D. and Postma, P. W., 1975: The energetics of bacterial active transport. Ann. Rev. Biochem. *44*: 523–544.
Skou, J. C., 1969: Role of membrane ATPase in the active transport of ions. *In* Molecular Basis of Membrane Function. Tosteson, ed. Prentice-Hall, Englewood Cliffs, N. J., pp. 455–482.
Slayman, C. L., 1970: Movement of ions and electrogenesis in microorganisms. Am. Zool. *10*: 377–392.
Smith, D. C., 1974: Transport from symbiotic algae and symbiotic chloroplasts to host cells. Soc. Exp. Biol. Symp. *28*: 485–520.
Solomon, A. K., 1962: Pumps in the living cell. Sci. Am. (Aug.) *207*: 100–108.
Stein, W. D., 1967: The Movement of Molecules Across Cell Membranes. Academic Press, New York.
Stoeckenius, W., 1976: The purple membrane of salt-loving bacteria. Sci. Am. (Jun.) *234*: 38–46.
Terepkin, A. R., Coleman, J. R., Armbrecht, H. J. and Gunter, T. E., 1976: Transcellular transport of calcium. Soc. Exp. Biol. Symp. *30*: 117–140.
Ussing, H. H., 1969: The interpretation of tracer fluxes in terms of membrane structure. Quart. Rev. Biophys. *1*: 365–376.
Ussing, H. H., Erlij, D. and Lassen, U., 1974: Transport pathways in biological membranes. Ann. Rev. Physiol. *36*: 17–47.
Voute, C. L. and Ussing, H. H., 1970: Quantitative relation between hydrostatic pressure, extracellular volume and active sodium transport in the epithelium of the frog skin *(Rana temporaria)*. Exp. Cell Res. *62*: 375–383.
Warren, K. B., 1966: Intracellular Transport. Symp. Soc. Cell Biol. *5*: 1–325.
Weed, R. I., Jaffe, E. R. and Miescher, P. A., 1970: The Red Cell. Grune and Stratton, Inc., New York.
Whittam, R. and Wheeler, K. P., 1970: Transport across cell membranes. Ann. Rev. Physiol. *32*: 21–60.
Wiley, W. R. and Schneider, R. P., 1970: Transport of metabolites in microorganisms. Am. Zool. *10*: 405–411.

CHAPTER 19

BULK TRANSPORT

The cell has developed elaborate mechanisms for bulk transport of fluid and of particles too large to gain entry through the channels of the cell membrane. The particles in bulk transport range in size in a continuously graded series from the macromolecular to whole cells.

ENDOCYTOSIS

Endocytosis is the bulk uptake of particles or fluid through the cell membrane by formation of a vacuole. The size of the vacuoles varies with the size of the entering material (Fig. 19.1). In *pinocytosis* (cell drinking) the minute vacuoles, at the limit of resolution with the light microscope, are called *pinosomes*. Uptake of particles as large as whole eukaryotic cells by amoeboid cells is termed *phagocytosis* (cell eating) and the vacuole is called a *phagosome*. Uptake of particles by ciliary movement into a funnel and a mouth is spoken of as *phagotrophy* and the vacuole is termed a *food vacuole,* regardless of whether it contains only fluid or both fluid and particles (Allison and Davies, 1974).

The terms designating various types of endocytosis are arbitrary and do not represent fundamental subdivisions of bulk transport. They are retained because they are useful in summarizing the information available. Endocytosis is considered a possibility in all eukaryotic cells, although it is more prominent in some cell types than others (Silverstein *et al.*, 1977).

PHAGOCYTOSIS

Phagocytosis occurs most readily in amoeboid cells wherever found; the cells are called *phagocytes*. The engulfed particle (phagosome), which is surrounded by a unit membrane, usually contains some medium in which the particles are suspended (Fig. 19.2).

Phagocytosis is an important aspect of digestion in all invertebrates in which intracellular digestion occurs. However, in most animals phagocytosis serves chiefly as a defense against invading organisms, as first emphasized by Metchnikoff (1883), who studied phagocytosis in a wide range of vertebrate and invertebrate animals. Phagocytes also remove cellular debris in injured or diseased tissues. Phagocytes have been shown to possess proteinases and lipases, which enable them to digest protein and lipids. Also present is lysozyme, which enables them to digest the cell walls of bacteria, thereby gaining entry to the bacterial cell contents. Phagocytes also serve to transport macromolecules and hormones and participate in immunological reactions (Silverstein *et al.,* 1977).

Phagocytes may be free or fixed. Lining the channels or cavities in the liver, spleen, lymphatic tissue, and bone marrow of vertebrates are fixed amoeboid cells known as fixed macrophages. The fixed cells are known collectively as the reticuloendothelial system, although it is unlikely that either cells of the reticulum or endothelial cells join the macrophages in phagocytosis. These fixed phagocytes remove bacteria, cellular debris, and other foreign matter from passing fluid. Fixed phagocytic cells also occur in sponges, coelenterates, free-living turbellarian flatworms, and in the lining of the digestive epithelia or canals of bivalve mollusks. Phagocytic cells complete digestion intracellularly (Prosser, 1973).

Freely moving amoeboid cells in blood or perivisceral fluid, called *leukocytes* or white blood cells, occur in the blood of all animals investigated. Mammalian leukocytes are classified into two main categories, namely

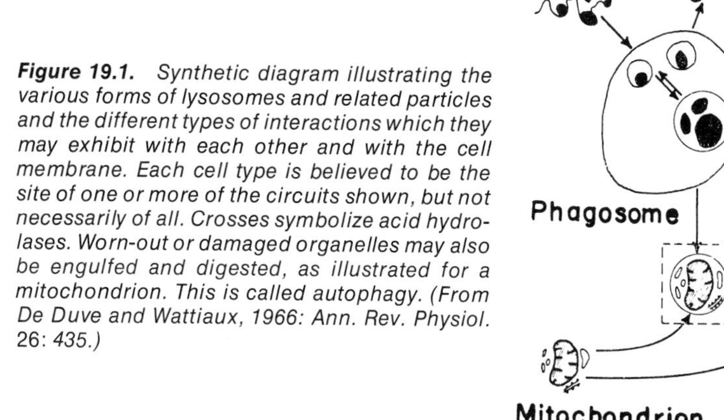

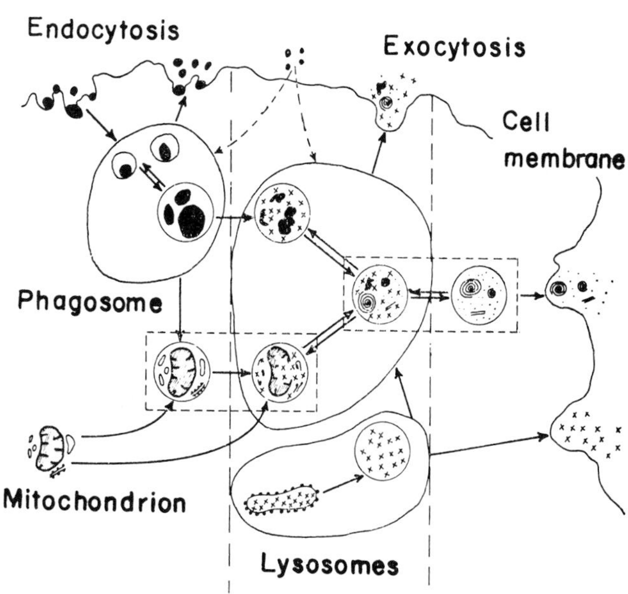

Figure 19.1. Synthetic diagram illustrating the various forms of lysosomes and related particles and the different types of interactions which they may exhibit with each other and with the cell membrane. Each cell type is believed to be the site of one or more of the circuits shown, but not necessarily of all. Crosses symbolize acid hydrolases. Worn-out or damaged organelles may also be engulfed and digested, as illustrated for a mitochondrion. This is called autophagy. (From De Duve and Wattiaux, 1966: Ann. Rev. Physiol. 26: 435.)

polymorphonuclear (irregular, multilobed nuclei) and *mononuclear* with a single oval nucleus. The polymorphonuclear types are (1) neutrophils (62.0 percent), (2) eosinophils (2.3 percent), and (3) basophils (0.4 percent), so named because they are stained by neutral, acid, and basic dyes, respectively. The mononuclear types are (1) monocytes (5.3 percent), and (2) lymphocytes (30 percent), which have less cytoplasm surrounding them than monocytes. All leukocytes are phagocytic at some stage. Most prominent in the blood in this respect are the *neutrophils* (polymorphonuclear neutrophils, or PMN's) and *monocytes,* which develop into macrophages, both fixed and wandering. The two together are the so-called "professional" phagocytes. Lymphocytes and monocytes squeeze out of the blood capillaries and turn into active phagocytes when wandering in tissue spaces. In invertebrates, granular and agranular leukocytes are found fixed in blood and coelomic channels, or wandering in tissue spaces. Many interesting and peculiar types of leukocytes are found, but they have not been studied sufficiently to make possible a meaningful classification.

Monocytes, as noted, develop into both motile and fixed *macrophages.* Macrophages can engulf large particles and may even engulf whole eukaryotic cells, such as erythrocytes and malarial parasites.

Phagocytes are selective of materials engulfed; otherwise,.they might well digest themselves or tissue cells of the same animal. It has been shown that roughness of surface and coating with opsonin, a globulin antibody protein, determine suitability for phagocytosis. Some pathogenic bacteria are encapsulated and resist uptake, but after they are coated with antibodies specific to them, they may be phagocytosed by PMN's or macrophages with specific binding sites for the particular antibodies involved. Phagocytes show marked chemotaxis that causes them to migrate to infected or damaged cells (McCutcheon, 1946).

A receptor site is necessary for phagocytosis although the nature of the receptor is not known. Some receptors are trypsin-sensitive and some are not (Silverstein *et al.,* 1977). Erythrocytes from another species, apparently lacking such recognizable sites, are phagocytized slowly or not at all. However, treatment of the erythrocytes with aldehyde or alteration by a variety of other agents that damage them appears to uncover receptor sites in the cells, vastly changing their acceptability to phagocytes. This has been found true for phagocytes from such diverse groups of organisms as the rat (LS cells) and the wax moth *Galleria,* and *Acanthamoeba,* all fed with red cells. Uptake is favored by the appropriate sodium and calcium ion concentrations in the medium (Fig. 19.2), and by appropriate osmolarity and ionic strength (Rabinowitch and de Stefano, 1970, 1971).

Lysosomes fuse with the phagosome membrane, discharging their hydrolytic enzymes into it (Fig. 19.2*A*, 6, *B* and *C*). A leukocyte will engulf undigestible polystyrene (synthetic plastic) beads of appropriate sizes, and the phagosome formed will fuse with lysosomes. The phagosome thus becomes a secondary lysosome (Rodesch *et al.*, 1970).

Ingested material is digested rapidly in white blood cells. For example, within three hours after ingestion of bacteria labeled with tracers, 80 percent of the label appears in small molecules. The total number of bacteria that a leukocyte can ingest in its life span is limited—5 to 25 for a small polymorphonuclear

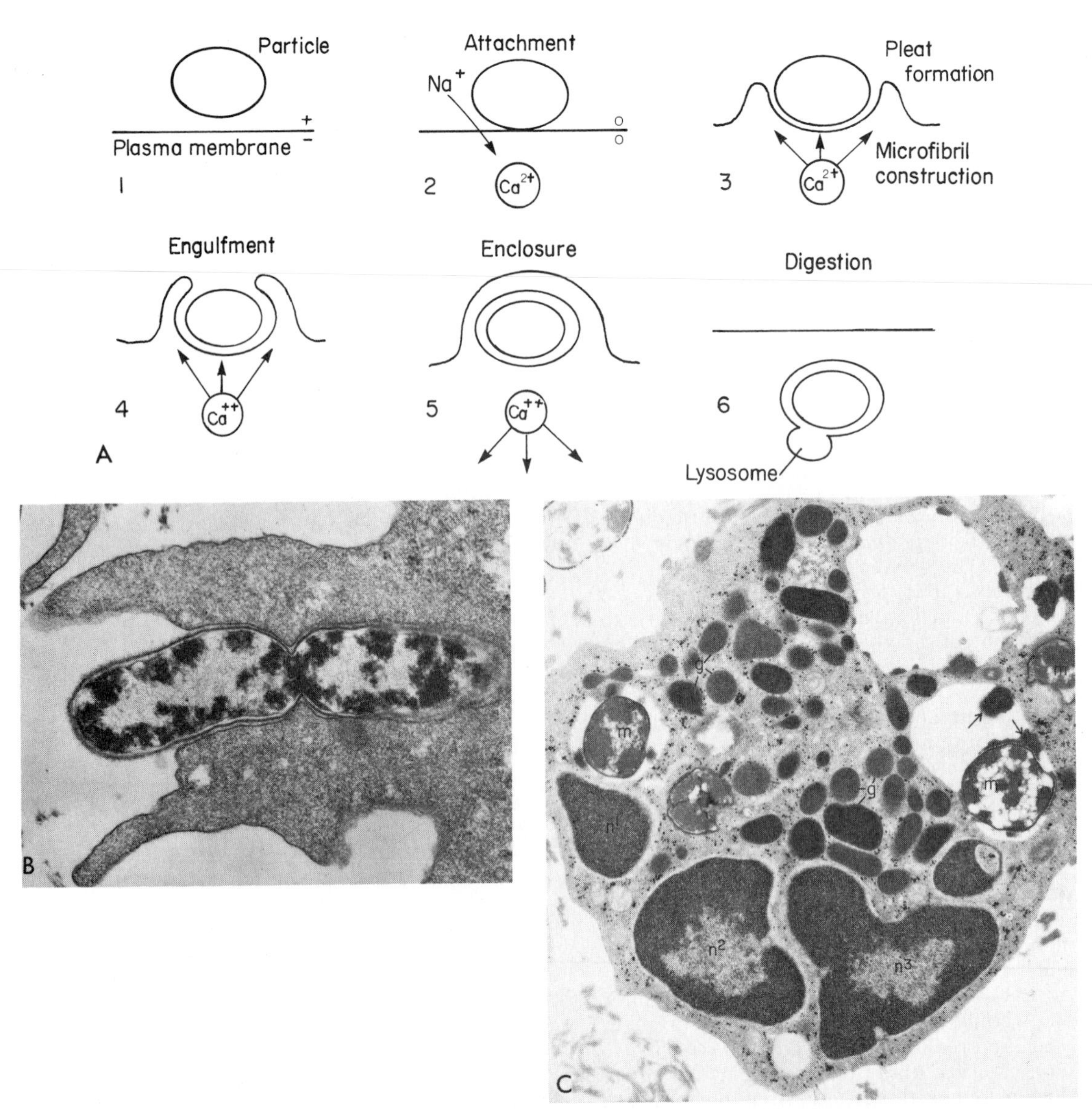

Figure 19.2. A, *Diagram of a possible mechanism of phagocytosis. (1) Attachment of a particle to the plasma membrane, (2) influx of sodium ions depolarizing the membrane and releasing calcium ions from an endoplasmic reticulum store, (3) beginning of contraction of microfilaments induced by calcium forming a pleat around the particle, (4) extension of the pleat by a rolling movement of the membrane, (5) enclosure of the particle and removal of calcium by the endoplasmic reticulum, (6) engulfment of the particle and fusion with a lysosome. (After Allison and Davies, 1974: Symp. Soc. Exp. Biol.* 28: *419–446.)* B, *Electronmicrograph of a polymorphonuclear leukocyte (neutrophil) engulfing a bacterium. Note the plasma membrane surrounding the pseudopodia and the enclosed bacterium. (By courtesy of Dorothy F. Bainton, University of California Medical School, San Francisco.)* C, *Lower magnification of a polymorphonuclear leukocyte showing sections through the nucleus (n^1, n^2, and n^3), a bacterium (m) in a phagosome both on the left and the right, and granules (g) staining with the dye azure. Granules (lysosomes, products of the Golgi complex) are seen combining with the bacteria (arrows) in the phagosome. (By courtesy of Dorothy F. Bainton, University of California Medical School, San Francisco.)*

neutrophil, about 100 for a large macrophage. Whether this limit is set by inability to produce more surface membrane or other factors is not known. The life span of a leukocyte is brief, only a few days at most, although macrophages are thought to live longer. Some leukocytes continue to phagocytize indefinitely through their life spans (Silverstein *et al.*, 1977).

Oxidative metabolism is apparently not necessary for phagocytosis in neutrophil leukocytes, since neither lack of oxygen nor poisoning with cyanide to inhibit the cytochrome oxidase pathway was found to interfere with uptake. Poisoning with glycolytic inhibitors such as iodoacetic acid and fluorite, however, decreases or stops phagocytosis (Karnovsky, 1962). These results are not definitive because, in one study, iodoacetic acid was ineffective (Bodel and Malawista, 1969). As expected, lactic acid accumulates as a result of glycolysis and glycogen is depleted in the cells. Therefore, glucose is usually supplied to leukocytes in such studies to permit replacement of the depleted glycogen. It is estimated that about 10^9 molecules of ATP are needed to ingest one polystyrene sphere 1.17 μm in diameter; the ATP is presumably replaced by glycolysis. The molecular mechanism by which a vacuole is formed is still unknown. Direct oxidation of added glucose in presence of oxygen is favored during phagocytosis and hydrogen peroxide is formed. This occurs by the pentose phosphate shunt, which increases twofold to tenfold during phagocytosis (Silverstein *et al.*, 1977). It is thought, therefore, that since this produces raw products for lipid synthesis, its activity may be correlated with need for new lipid to replace that used in formation of phagosomes, in addition to replacing glycogen (Karnovsky, 1962). Oxidative phosphorylation is required for phagocytosis in lung macrophages (Oren *et al.*, 1963).

Phagocytosis may lead to tissue damage in some cases. For example, crystals of sodium urate deposited in the joints of individuals with gout are engulfed by phagocytes. Phagocytes kill bacteria that they engulf by producing hydrogen peroxide or possibly superoxide. They also normally combine with lysosomes that liberate digestive enzymes to lyse the cells engulfed. Neither of these efforts affects the crystals, which are then voided into the tissue along with the hydrolytic enzymes and oxidants. The tissue is thereby damaged, leading to inflammation (Johnson and Lehmeyer, 1977; Willoughby, 1978).

Phagocytosis is temperature-sensitive, suggesting its dependence upon energy-liberating processes, an increase of about threefold occurring with a 10-degree rise in temperature (Silverstein *et al.*, 1977).

No essential difference exists between phagocytosis by leukocytes and food vacuole formation in amoeboid protozoans. In an amoeba the *food vacuole* is enclosed in a unit membrane just as in a phagocyte. The number of food vacuoles formed may be very large: the amoeba *Chaos chaos* feeding on *Paramecium aurelia* may form as many as 100 food vacuoles in a 24-hour period. Since each vacuole requires ingestion of about 10 percent of the plasma membrane, the amoeba will have ingested ten times as much plasma membrane in a day as it has at any one time. Feeding thus requires a very rapid turnover of membrane, or reutilization of the membrane ingested, by reassembly of subunits. In this respect an amoeba appears to better the performance of a phagocyte (Christiansen and Marshall, 1965). A suggestion presented is reutilization of membranes of vesicles in both *Acanthamoeba* and phagocytes. *Acanthamoeba* was provided with latex beads that make possible easy isolation of the vesicles for testing. Amoebas provided with labeled lipid precursors showed no appreciable increase in labeling when they had phagocytized the beads, over the label that was incorporated into controls. Of several labeled compounds tested only a small increase occurred in incorporation of phosphatidyl-1,2-^{14}C. Similar results were obtained with macrophages (Ulsamer *et al.*, 1971).

Several defects in phagocytosis may make humans particularly susceptible to infection. In chronic granulomatous disease phagocytes (polymorphonuclear neutrophils) are deficient in an enzyme that generates hydrogen peroxide. Bacteria are engulfed normally but not killed inside the phagosome, suggesting that H_2O_2 is necessary for killing. In Chediak-Higashi syndrome phagocytes have abnormal giant lysosomes (among various defects), and these do not fuse readily with pathogen-containing phagosomes. The details of the immune system and immune diseases may be found in textbooks and reviews on immunology.

Food vacuole formation in an amoeba (*Amoeba proteus*) has been studied with phase and electron microscopy and the steps in the process analyzed. Once a *Tetrahymena* comes into contact with an amoeba its movement becomes restricted because of entanglement of its cilia in the polysaccharide coating the amoeba's surface (see Fig. 8.4). In turn, the surface of the amoeba becomes attached to its cortex below the point of contact with the *Tetrahymena* by what appear to be microfibrils; the cytoplasm of the amoeba then flows around these attachments, forming the cup of the food vacuole much as in the formation of a pseudopod. The food cup thus appears to be built up from a

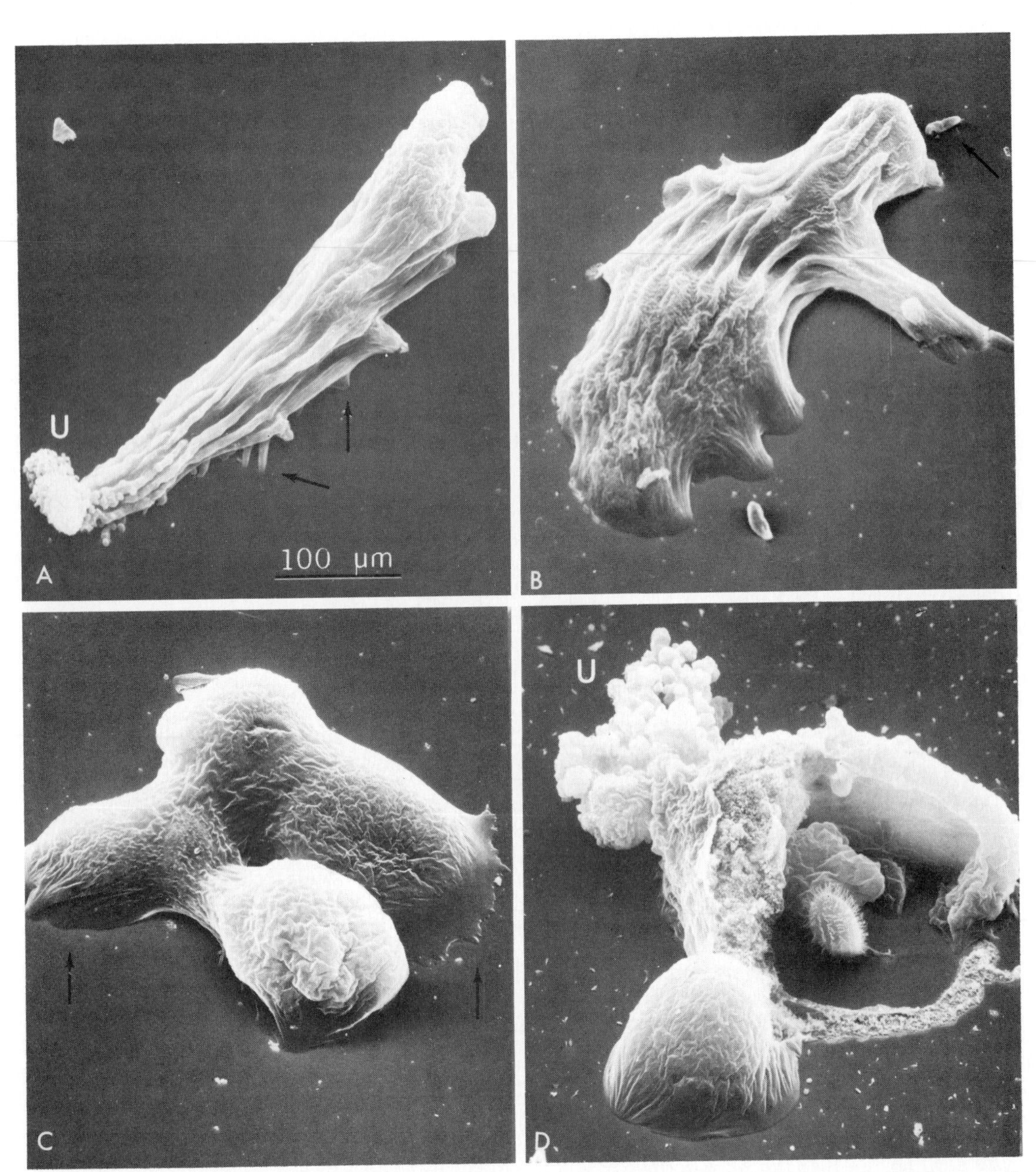

Figure 19.3. *Phagocytosis by* Amoeba proteus *feeding on the ciliate* Tetrahymena. A *Shows a nonfeeding amoeba with a convoluted posterior portion called the uroid (U), ridges over the posterior half of the cell, and small adhering pseudopods (arrows).* B, *One minute after addition of* Tetrahymena *suspension, note the smoothed uroid in response to the ciliate and the broad pseudopodia forming.* C, *An amoeba with four food cups covering ciliates.* D, *An amoeba with two food cups, one of which is broken, showing the* Tetrahymena *within the cup. The uroid (U) has retained its convolutions. (By courtesy of K. W. Jeon and M. S. Jeon, 1976: J. Protozool.* 23: *83–86.)*

Legend continued on the following page

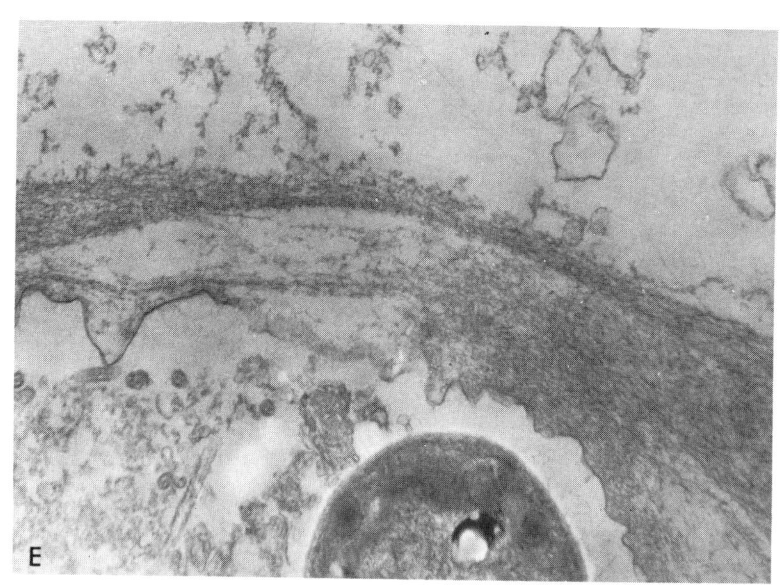

Figure 19.3. *Continued.* E, *Transmission electron micrograph of a food vacuole in an amoeba* (Chaos chaos) *three and a half hours after it had ingested a ciliate* (Stentor). *At the periphery of the food vacuole note the ring of microfilaments (near the top) extending from 0.1 to 2 μm into the cytoplasm. Projecting into the vacuole are indentations of the vacuolar membrane. The large body (bottom) is an unchanged symbiotic alga found in the digested* Stentor *(×8000). (From C. Chapman-Andresen and J. R. Nilsson, 1967: Comp. Rend. Lab. Carlsberg 30: 189.)*

single pseudopod around a fixed point or central cavity. Presumably, the attachment of an amoeba to its prey leads to a local change in properties of its plasma membrane, and contraction occurs by action of the actin and myosin fibrils beneath the membrane. It now seems likely that all eukaryotic cells possess actin and myosin, although in proportions different from those in muscle (Silverstein *et al.*, 1977). An increase in calcium ions is the trigger for contraction, just as in muscle contraction (Allison and Davies, 1974). Scanning electron microscope studies of phagocytosis in *Acanthina castellanii* and in *Amoeba proteus* (Fig. 19.3) show three-dimensionally some of the movements involved in phagocytosis (Goodale and Thompson, 1971; Jeon and Jeon, 1976).

PHAGOTROPHY

The intake of food by ciliates has sometimes been spoken of as phagocytosis, or even pinocytosis, since the invagination of the membrane in formation of a food vacuole is similar in general nature in all three processes. However, the term *phagotrophy* is considered preferable for ciliates because ingestion differs in a number of respects from that in an amoeba or phagocyte in that particulate or liquid matter passes into a cytostome (cell mouth), resulting in the formation of a food vacuole.

In nature the ciliate *Tetrahymena* ingests particulate food, and even undigestible particles are ingested, provided that they are coated with digestible matter. Inasmuch as protozoans lose metabolites into the medium, including some enzymes (probably from lysosomes), undigestible particles added to the medium are quickly coated with digestible organic matter. Yet *Tetrahymena* placed in a mixture of bacteria and colloidal gold selects the bacteria. The basis for such selection is unknown. *Paramecium* when placed in a medium containing particles of various sizes preferentially selects the smaller ones, possibly because of the greater ease of entry of such particles. Interestingly enough, in an axenic (free of organisms other than the one in culture) medium, such as protease peptone, *Tetrahymena* still forms membrane-bound food vacuoles (Conner, 1967).

Many ciliates have developed adaptations for preying on other protozoans. Some are selective; for example, *Didinium* attacks and ingests the contents of a *Paramecium* (various species) much larger than itself, while various similar organisms (except *Colpidium*) are rejected even when the didinia are starving to death. On the other hand, some other protozoans are not as selective, e.g., the ciliate *Blepharisma*, usually a bacterial filter-feeder, may also eat small protozoans and sometimes even large ones. Protozoans such as *Blepharisma* may even become cannibalistic if the only edible matter available consists of small individuals of their own kind. Cells in multicellular organisms (invertebrates) also may become cannibalistic on occa-

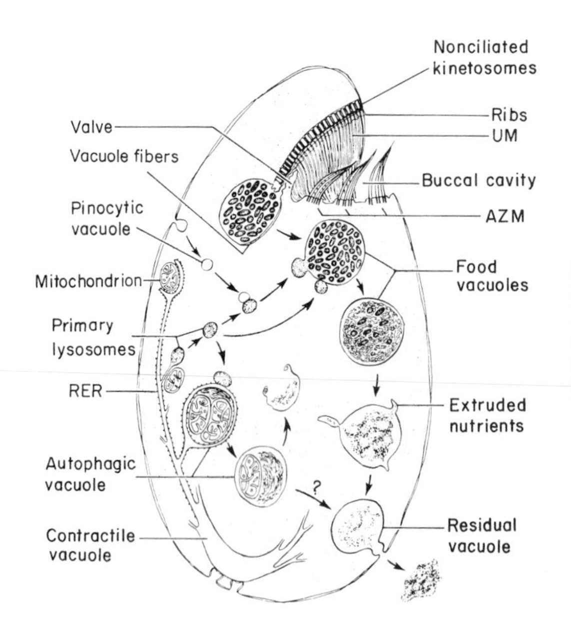

Figure 19.4. Schematic representation of the possible structural pathways of intracellular digestion in Tetrahymena pyriformis. *RER, rough endoplasmic reticulum; JM, undulating membrane; AZM, adoral zone of membranelles. (From Elliott and Clemmons, 1966: J. Protozool.* 13: *311.)*

sion (e.g., in ovules or ovaries some cells may engulf others, producing large gametes containing much stored material). Similarly, amoeboid scavenger cells ingest relict gametes from gonads and form a mass of tissue containing nutrients for the next generation of gametes (Giese, 1973).

Lysosomes pour their contents into food vacuoles formed by phagotrophy much as in phagosomes, and digestion in a ciliate is completed within a few hours (Elliott and Clemmons, 1966) (Fig. 19.4). The food vacuole shows changes in acidity at different phases in the digestive cycle, presumably providing the appropriate pH for action of the hydrolytic enzymes (Muller, 1967). Large prey (e.g., in cannibals) may swim inside the food vacuole for several minutes to several hours. Food vacuoles circulate in the cytoplasm of the cell and the length of the route depends on the size and shape of the ciliate (Fig. 19.4). The soluble digested nutrients pass into the cytoplasm and the food vacuole shrinks, ultimately containing only the undigestible residues that are voided to the medium via an anal pore (exocytosis, discussed later in this chapter).

AUTOPHAGY

Under some conditions, for example, starvation or injury, autophagy (self-eating) may occur by engulfment and digestion of an individual's own organelles. Electron micrographs have demonstrated in vacuoles the presence of mitochondria in various stages of digestion. The enzymes for digestion are provided by lysosomes, which fuse with the membrane-bound vacuole in the same manner as they fuse with a food vacuole (Fig. 19.1). It is suggested that autophagy may be a means of providing nutrients during starvation for the remainder of the cell by removal of less essential organelles. It is also possible that autophagy is a normal activity, in which defective organelles are removed that can be replaced by the cell as needed (Ericsson, 1969).

PINOCYTOSIS

Edwards, in 1925, working with *Amoeba,* noticed the formation of small channels in the cell membrane followed by the separation of small vacuoles, or *pinosomes,* in the cytoplasm

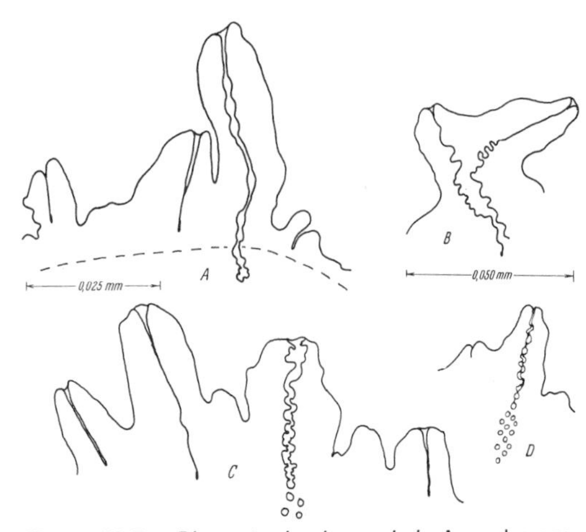

Figure 19.5. Pinocytosis channels in Amoeba proteus. *A, Channels in small pseudopods in various stages of formation. B, Convoluted channels. C, Channel beginning to disintegrate at the inner end. D, Further disintegration, showing that many drops of fluid are ingested in the formation of each channel. (After Mast and Doyle, 1934: Protoplasma* 20: *555.)*

(Fig. 19.5). Little study was done on this observation until fairly recently, when workers in a number of laboratories developed an intense interest in the process. Some workers distinguish between macropinocytosis and micropinocytosis. An example of the former is the uptake of colloid 0.2 to 0.3 μm in diameter by thyroid epithelial cells (Allison and Davies, 1974).

Pinocytosis is treated in somewhat more detail here than is warranted by its relative importance, largely because of the insight it gives into the nature of the cell membrane and its multiple functions.

Occurrence

The study of pinocytosis has been facilitated by modern technological methods. For example, the development of the electron microscope made possible resolution of much smaller channels and pinosomes. Also, proteins labeled with dyes that fluoresce in the ultraviolet may be used to induce pinocytosis experimentally and can be identified readily inside the cell by fluorescence microscopy (Holter and Holtzer, 1959). They can be traced, following pinocytosis, in the large number of minute vacuoles scattered in the cytoplasm. Organic molecules labeled with ^{14}C tracers may also be used. In this case, autoradiographs are made of the cells; if pinocytosis has occurred, small vacuoles containing radioactive material are evident. Ferritin, a protein containing much iron, has also been used. The iron, having a high density, is opaque to electrons. Following pinocytosis, electron micrographs reveal it in small vacuoles in the cytoplasm.

Pinocytosis occurs in amoeba and such amoeboid cells as slime molds, leukocytes, brush border cells of the kidney (active in fluid exchanges), intestinal epithelial cells, fixed macrophages, liver Kupffer (phagocytic) cells, cells lining capillary walls, fibroblasts, some malignant cells in tissue culture, and some nonamoeboid cells. It has also been observed in endothelial cells lining capillaries, mucosal cells in the intestine, kidney tubule cells, mammalian neurons, adipose cells, and bladder epithelial cells (Silverstein *et al.*, 1977). Pinocytosis was not observed (using labeled protein) in cardiac, smooth, or skeletal muscle cells or in cells of a number of other tissue cultures, including human amnion cells (Holter and Holtzer, 1959).

Little convincing evidence has been found for the existence of pinocytosis in cells of plant roots, although labeled proteins and enzymes were found to penetrate into these cells from the outside and to appear in the cytoplasm around the nucleus. Electron micrographs of plant cells in some instances suggest canals leading from the surface into the cytoplasm that might be interpreted as pinocytosis channels, but reevaluation of these results has led to the conclusion that pinocytosis is questionable in plant cells (Bradfute *et al.*, 1964). However, claims that it occurs in some plant cells continue to appear in the literature (Findlay and Mercer, 1971), and pinocytosis has been clearly demonstrated in plant cells from which the cell walls have been removed (Gunning and Steer, 1975; Cocking, 1977).

Indirect evidence also indicates that pinocytosis probably occurs on a wide scale in cells of embryos, since many proteins are known to pass through the placenta and immunologically specific proteins have been found to enter a wide variety of tissue cells, even into the cell nuclei (Schechtman, 1956). The entry of the enzyme ribonuclease into many types of cells has also been demonstrated indirectly by its hydrolysis of the cellular RNA, as revealed by the subsequent decreased staining by basic dyes and the decreased absorption of ultraviolet radiation of the cells. Pinocytosis has also been observed in developing eggs of the snail *Lymnaea stagnalis*.

It is thought by some that pinocytosis is perhaps a general phenomenon and negative evidence is discounted on the basis that appropriate conditions may not yet have been supplied to the cells. On the other hand, the phenomenon is perhaps largely limited to cells capable of forming pseudopodia or with highly mobile membranes. No evidence has yet been obtained for action of actin and myosin fibers in the formation of pinosomes, as is considered likely for phagosomes (Silverstein *et al.*, 1977). However, microfilaments underlie the plasma membrane and are present between the cell organelles. Since cytochalasin B both depolymerizes microfilaments and interferes with pinocytosis, it is possible that microfilaments are involved in formation of pinosomes. Because colchicine, which depolymerizes microtubules, has no effect on pinocytosis in some cases, microtubules are probably not involved in pinocytosis (Allison and Davies, 1974).

Induction

Mast and Doyle (1934) showed that in appropriately starved cells of *Amoeba proteus*, pinocytosis is induced by salts and proteins. It is now known that amino acids (Fig. 19.6) and viruses are also effective, but that neither carbohydrates nor nucleic acids induce pinocytosis. In some instances basic dyes have also induced pinocytosis (Rustad, 1961). In general it appears that to induce pinocytosis the ion must be positively charged.

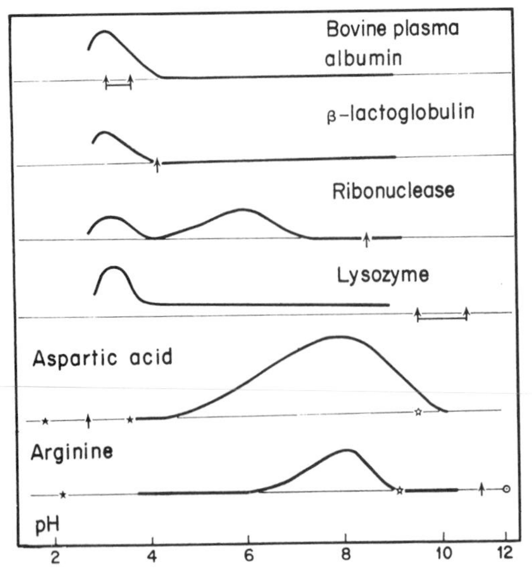

Figure 19.6. *Effectiveness of induction of pinocytosis, as judged by the mean number of channels induced in 10 amoebas in 20 minutes, of solutions of various proteins (above) and amino acids (below). Cysteine, asparagine, triglycine, glutamine, methionine, glycine, and histidine were ineffective except for a very slight action at very high pH values. Pepsin, hemocyanin, β-lactoglobulin, and insulin were ineffective except for slight activity at very low pH. The isoelectric pH values (where the molecule is equally charged positively and negatively) are indicated by arrows. (From Chapman-Andresen, 1962: Comp. Rend. Lab. Carlsberg 33: 172, 186.)*

Chapman-Andresen (1962), in an extensive study on *Amoeba proteus,* made quantitative comparisons between various conditions for inducing pinocytosis by counting the number of channels induced in 10 amoebas (starved 2 days) in 20 minutes. Some data gathered in this manner are cited from her studies in Figures 19.6 to 19.9.

Salts induce pinocytosis also, cations being more effective than anions, as shown in Figure 19.7. Potassium, sodium, and magnesium salts cause induction more readily than does calcium. Sodium and magnesium induce narrower channels than potassium. A pH of 6.5 to 8.0 (Fig. 19.8) and a temperature between 12 and 22°C are most favorable (DeTerra and Rustad, 1959) (Fig. 19.9). The optimal concentration is different for each salt, although 0.1 M. solutions are adequate in most cases. Pinocytosis has been seen in the amoeboid cells (eleocytes) of the sea urchin (Holter, 1959), which are already accommodated to a salt concentration of about 0.5 M.

It was found that more pinocytotic channels were induced in enucleated than in nucleated pieces of *Amoeba proteus* (in 0.1 M NaCl) 90 minutes after cutting. However, if the tests are made 5 days after cutting (during which time the enucleated pieces have presumably deteriorated), many more channels appear in nucleated than in enucleated pieces.

Visible light has no effect on pinocytosis, but ultraviolet induces it (Rinaldi, 1959). Hydrostatic pressure, which induces cytoplasmic solation, inhibits pinocytosis at 3000 pounds per square inch (Zimmerman, 1965).

Once pinocytosis has been induced in amoebas, it goes on actively for 15 to 45 minutes and then decreases. To induce another period of activity it was found necessary to wash the amoebas free of ionic medium and reimmerse them in salt or protein solutions (Holter, 1959).

Leukocytes continue pinocytosis for longer periods than *Amoeba* do, and only a portion of the surface of the leukocyte is active at any one time. However, treatment of a macrophage with proteolytic enzymes stops uptake of substances from the medium. Recovery occurs after a lapse of time but requires protein synthesis, inasmuch as pinocytosis is not resumed in the presence of cyclohexamide, an inhibitor of protein synthesis (Lagunoff, 1971).

Inducers of pinocytosis may cause a 50-fold decrease in electrical resistance of the plasma membrane prior to formation of typical channels and vacuoles. At this time structural changes in the unit membrane can be observed; the middle electron-transparent layer increases about twofold. Such changes in the *Amoeba* membrane are presumably dependent upon the calcium concentration in the external medium (Brandt and Freeman, 1966).

Pinocytosis in Sarcoma 180 II tumor cells resembles that observed in *Amoeba* (Ryser, 1968). Also, basic polyamino acids of molecular weight greater than 1000 are effective; the higher their molecular weight the greater their effectiveness. Steric factors play a role since poly-D-lysine is more effective than poly-L-lysine. Acidic polyamino acids may be inhibitory (*e.g.,* polyglutamic acid). Pinosomes combine with lysosomes, resulting in digestion of foreign proteins.

Pinosomes

When an amoeba is placed in a solution favorable for induction of pinocytosis, it becomes detached from the substrate, the contractile vacuole temporarily ceases operation and it forms a rosette. Most of its surface is covered by small active pseudopodia, each made up almost entirely of clear ectoplasm. In the pseudopodia may be seen the clear channels that are the forerunners of the pinosomes (Fig. 19.10). By the use of proteins labeled with fluorescent dyes, the pinosomes, formed by pinched-off vacuoles, may be readily seen inside the amoeba.

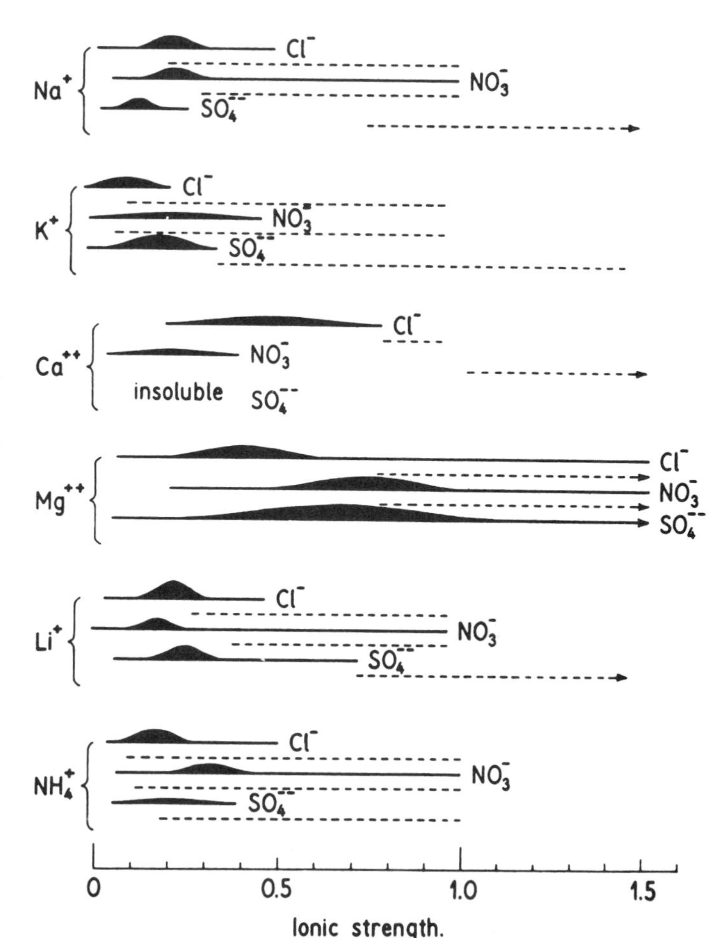

Figure 19.7. *Pinocytosis and toxicity in relation to concentrations of salts expressed in terms of ionic strength (the ionic strength is half of the sum of the terms obtained by multiplying the molality of each ion in the solution by the square of its valence). Dotted line indicates toxicity, solid line indicates pinocytosis, the width of the lines giving an indication of the intensity of the reaction. (From Chapman-Andresen, 1958: Comp. Rend. Lab. Carlsberg 31: 77.)*

The pinocytotic vacuoles contain within them the same solutes as the external medium. Holter (1959) has shown that the vacuoles in an amoeba do not centrifuge down soon after formation but rather stay at the centripetal end with the lighter materials of the cell. If the amoeba is centrifuged after a lapse of time, the vacuoles are carried down to the centrifugal end with the heavier materials in the cell. Presumably in the meantime the material in the vacuoles has been concentrated by the withdrawal of water. This can be shown by the use of fluorescent proteins, which gradually become concentrated into a dense, brilliantly fluorescent globule. Finally, only a fluorescent granule remains when most of the water is withdrawn.

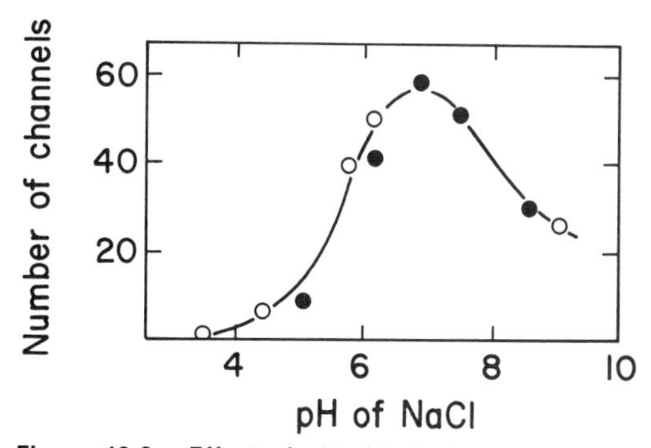

Figure 19.8. *Effect of pH of inducing solution on response of* Amoeba proteus *to 0.125M NaCl in distilled water. The light and dark circle represent two series of experiments. (From Chapman-Andresen, 1962: Comp. Rend. Lab. Carlsberg 33: 147.)*

It is of interest to know whether the membrane of the pinocytotic vacuole is like the plasma membrane and whether other materials besides water can also enter the cytoplasm from the vacuole. As a test case, glucose marked with ^{14}C was included in the inducing solution containing protein. Autoradiographs following pinocytosis indicated that the glucose had entered into pinosomes, and later autoradiographs showed that the glucose became dispersed in the cell. Ultimately, about 15 percent of the radioactive carbon became incorporated into the cytoskeleton of the cell, and about 75 percent of the radioactive carbon was given off by the amoeba as carbon dioxide. Similar experiments carried out with methionine labeled with ^{35}S showed it to enter into the cytoplasm also.

Less has been learned about the fate of pro-

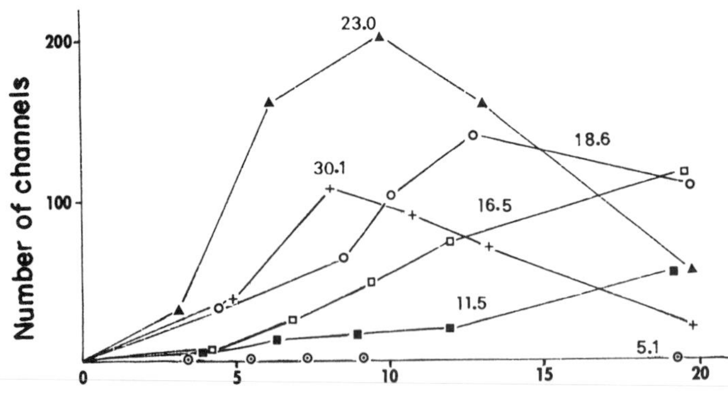

Figure 19.9. Effect of temperature on pinocytosis in Amoeba proteus in 0.125M NaCl at pH 6.5. The temperature in °C is indicated to the side of each of the curves. No pinocytosis occurred at 5.1°C. (From Chapman-Andresen, 1962: Comp. Rend. Lab. Carlsberg 33: 137.)

teins taken into the cell. The fluorescein labeling on a fluorescent protein is removed from the cell much more rapidly than the ^{14}C label on glucose, but whether the protein is digested or the labeling material is merely voided is not known (Holter, 1959).

Pinocytotic vacuoles have been seen to shrink, coalesce, and even divide, but never to disappear. Reutilization of membrane subunits appears likely, as in the case of phagosomes already discussed.

Mechanism

Pinocytosis in the amoeba *Chaos chaos* has been described as a result of movement of the elastic plasma membrane (by microfilaments?). The plasma membrane wrinkles when a flow of a pseudopod begins and the membrane is not yet freed from the gel adjoining it. This leads to invagination of parts of the plasma membrane, which are consequently separated from the larger advancing portion of the membrane. It is thought that the charged solute particles used to induce pinocytosis may weaken the structural rigidity of the membrane, allowing parts of it to be drawn into the cytoplasm during movement by its remaining attachments to the gel sheet. These vacuoles or channels are then separated from the membrane and released into the interior of the cell as pinosomes (Brandt, 1958).

Bennett (1956), on the other hand, postulated

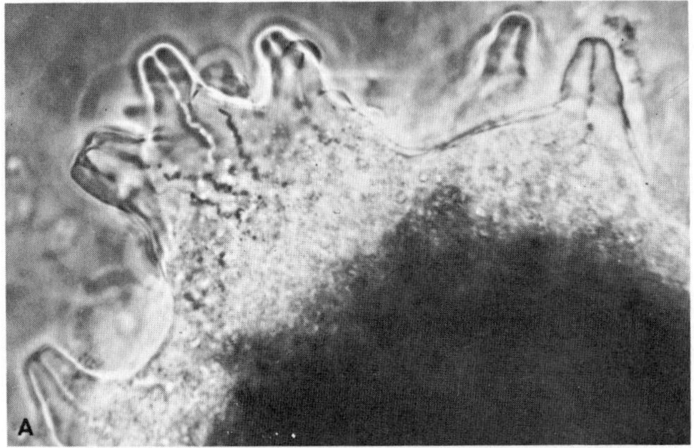

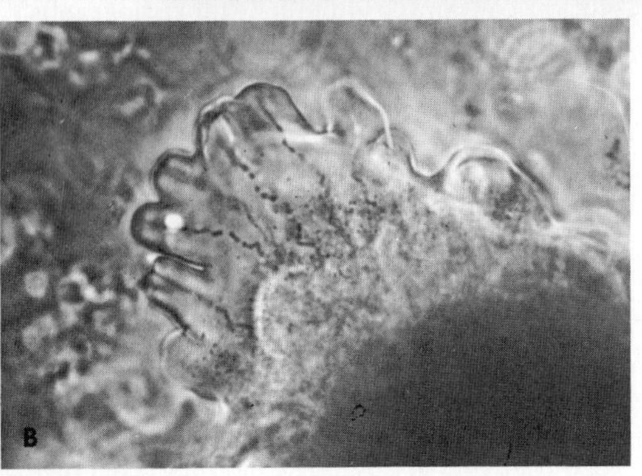

Figure 19.10. A, Pinocytosis in Amoeba proteus as seen with phase contrast microscopy. The Amoeba was starved for 24 hours, then immersed into 1 percent egg albumin in balanced salt solution. The pinocytotic channels are clearly visible in the pseudopodia. In some of the channels deep inside the cell small pinosomes may be seen forming from the engulfed fluid. B, 35 seconds later, during which time the amoeba has moved; therefore, the pseudopodia do not match. The most prominent channels in A have become the most prominent chains of vacuoles (pinosomes) in B. (From Holter, 1960: Int. Rev. Cytol. 8: 484–485. Photographs by courtesy of David Prescott, Oak Ridge National Laboratory.)

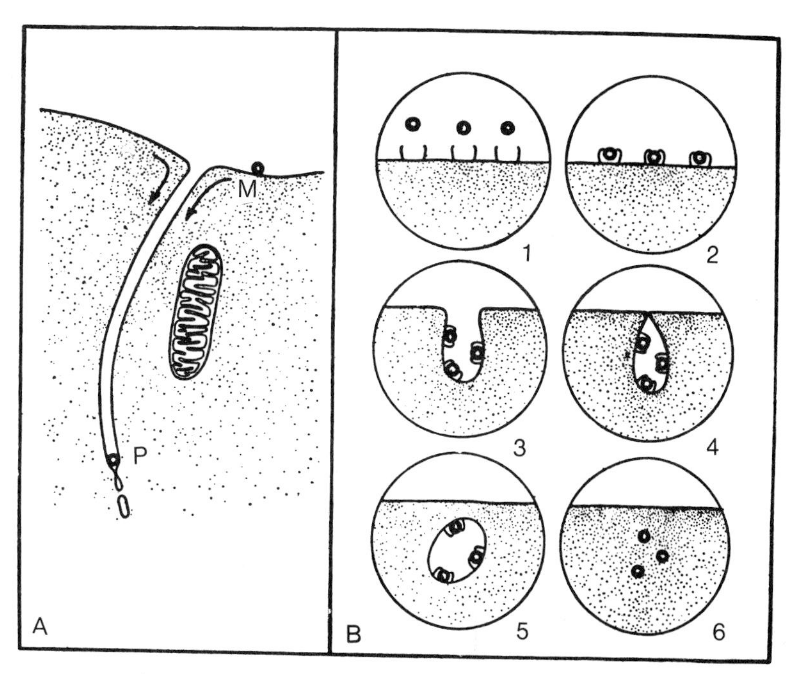

Figure 19.11. *Induction of pinocytosis by attachment of particles, as envisaged by Bennett.* A, *A particle (P) induces membrane flow (M¹, M) forming a channel containing the particle (P). This is taken inward into a pinosome such as those illustrated forming.* B, *The series of figures illustrates the approach (1) and attachment (2) of a particle, invagination of the cell membrane (3), formation of a pinosome or pinocytotic vacuole (4, 5), and, finally, incorporation of the particulate material into the cytoplasm (6). (From Bennett, 1956: J. Biophys. Biochem. Cytol.* 2: *Part* 2 *(Suppl.), 99.)*

that pinocytosis does not involve just the engulfment of fluid but that materials must attach to the surface of the cell before the vacuole can form, as shown in Figure 19.11. Decisive evidence for this mechanism has been obtained, using protein as the inducer of pinocytosis. The amoebas were immersed in an antigen protein for varying periods. They were then freeze-dried and sectioned, and the sections were exposed to an antibody protein labeled with fluorescein. The sections so treated showed a brilliant fluorescence under ultraviolet radiation, indicating that the surface of the amoeba was heavily coated with the antigen to which the antibody containing fluorescein had become attached. Further proof of attachment of protein came from studies using protein labeled with radioactive material. In 5 minutes an amoeba binds 50 times the amount of protein contained in its own volume (of solution). It would therefore appear that such binding may play an important role in pinocytosis and may be a necessary first step, so that pinocytosis is not just "cell drinking" but is a process consisting of several steps (Rustad, 1961).

Basic dyes (e.g., alcian blue, toluidine blue) are found to induce pinocytosis, also, and are bound to the surface of an amoeba, most likely by electrostatic forces. Proteins compete for the surface of an amoeba in this reaction. Some photomicrographs indicate that the surface of *Amoeba proteus* has hairlike projections of extracellular material 100 to 200 nm long and 6 nm wide, which greatly increase the area of the surface available for binding. The extraneous membranes, composed of mucopolysaccharides, carry a negative charge, reacting with the cations as inducers. Experiments show that particles become attached to these structures as a first step in pinocytosis (Rustad, 1961).

Binding, or surface adsorption, of protein on the surface of the cell continues even in the presence of carbon monoxide or cyanide, either of which is inhibitory to the oxidative metabolism of the cell. It also continues at low temperatures, which are inhibitory to completion of pinocytosis. This indicates that the first step in the process of pinocytosis is a physical reaction.

However, the reactions that follow, including the formation of channels and vesicles, are stopped by metabolic inhibitors in *Amoeba* but not in sarcoma cells; these reactions are also stopped by low temperatures in both kinds of cells. This indicates that they are dependent upon the oxidative metabolism of the cell in *Amoeba* but may depend upon glycolysis in sarcoma cells, although some workers have considered the possibility of no metabolic energy requirement (Allison and Davies, 1974). Thus, it seems that at least two processes are involved in pinocytosis.

Significance

Evidence indicates that such macromolecules as proteins (enzymes, hormones, antibodies) and polypeptides probably enter embryonic and often mature cells by pinocytosis because the small pores in the cell membrane would not permit entry of such large entities. Neither would these molecules dissolve into the membrane for carrier transport. Pinocytosis might also be the route of entry for transforming DNA (see Chapter 7) and may be of special importance for transfer of the iron compound ferritin in reticulocytes (young red blood cells). Transport of lipids, plasma protein, glycoproteins, and large polysaccharides (dextrans) may also occur by pinocytosis. The blood "dust" (chylomicrons), consisting of fine fat droplets, may well be liberated into the blood by exocytosis, which is the reverse of pinocytosis.

In secretory cells pinocytosis may be a way of returning cell membrane added in exocytosis to its source, namely the Golgi body or endoplasmic reticulum, or both (Silverstein *et al.*, 1977).

Is pinocytosis active transport? If active transport is defined as the selective uptake of a solute by a carrier at the expense of metabolic energy, then pinocytosis does not qualify as active transport. In pinocytosis the active step requiring energy is the formation of a vacuole or channel and its separation from the cell membrane, a nonselective random sampling of the environmental fluid. The binding of proteins and other molecules to the surface of the cell is selective but it is a physical event not requiring metabolic energy. When a pinosome is formed, particles and solution are included; but they may move passively along or against a concentration gradient.

Solutes included in the contents of a pinosome are a random sample of the environmental fluid and are subsequently concentrated. Whether these solutes pass out of the pinosome by diffusion or by active transport through the vacuolar wall and whether the solutes are concentrated by removal of the water from the vacuole by osmosis or active transport is not known. Some solutes from the pinosome are readily incorporated into the cytoskeleton or used in metabolism, e.g., ^{14}C-glucose, mentioned before. Some cells also void secretory vacuoles or granules—presumably by the reverse of pinocytosis. It must be recognized that pinocytosis appears to be even more complex than active uptake.

EXOCYTOSIS

Extrusion of cell material has been called *exocytosis* (see Fig. 19.1). Secretion of enzymes and other secretion granules from cells is a general form of exocytosis found in a wide variety of organisms. Perhaps the most discussed example of a secretory process is to be found in the exocrine pancreatic cell. In one study the organelles of the pancreatic cell were fractionated and enzyme and other protein syntheses were determined by tracer studies on electron micrographs of the organelles and on fractions. Enzyme activity was tested on the fractions. It was observed that leucine-1-^{14}C was incorporated into mixed proteins as well as into a well-characterized secretory protein, alpha chymotrypsinogen. After a meal, the ribosomes of the rough endoplasmic reticulum showed marked activity in protein synthesis, and large tracer-labeled granules appeared in the cavities of the reticulum. The tracer next appeared in the zymogen granules first in the Golgi assembly, then at the apex of the cells. The granules exhibiting enzymatic activity were then extruded from the apex of the cells into the lumen of the pancreatic acini (tubules) (Figs. 19.12 and 19.13). Exocytosis is calcium-dependent, as shown by ^{45}Ca influx during the process. Exocytosis is also increased by

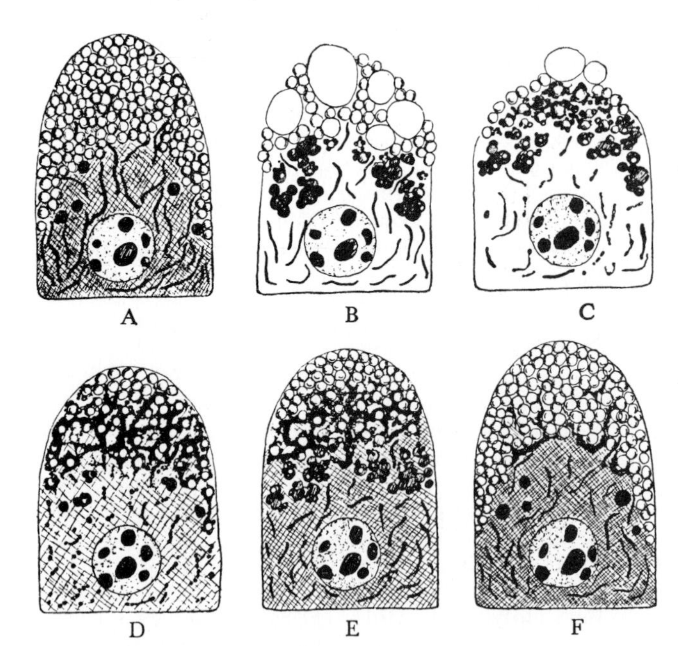

Figure 19.12. *Secretory cycle of the pancreatic cell of the white mouse.* A, *Cell from a fasting animal. Zymogen granules (circles) and mitochondria (elongated bodies).* B, *Same, a half hour after injection of pilocarpine, a drug that induces secretion. Vacuolization and excretion of the zymogen, Golgi apparatus (black aggregations) increased in size, disappearance of the basophilic (dark) substance.* C, *One hour later, excretion almost complete.* D, *After four hours. Typical Golgi net with newly formed granules.* E, *After seven hours, the process of recovery continuing.* F, *After 14 hours. Recovery completed. (From De Robertis* et al., *1970: Cell Biology. 5th Ed. W. B. Saunders Co., Philadelphia.)*

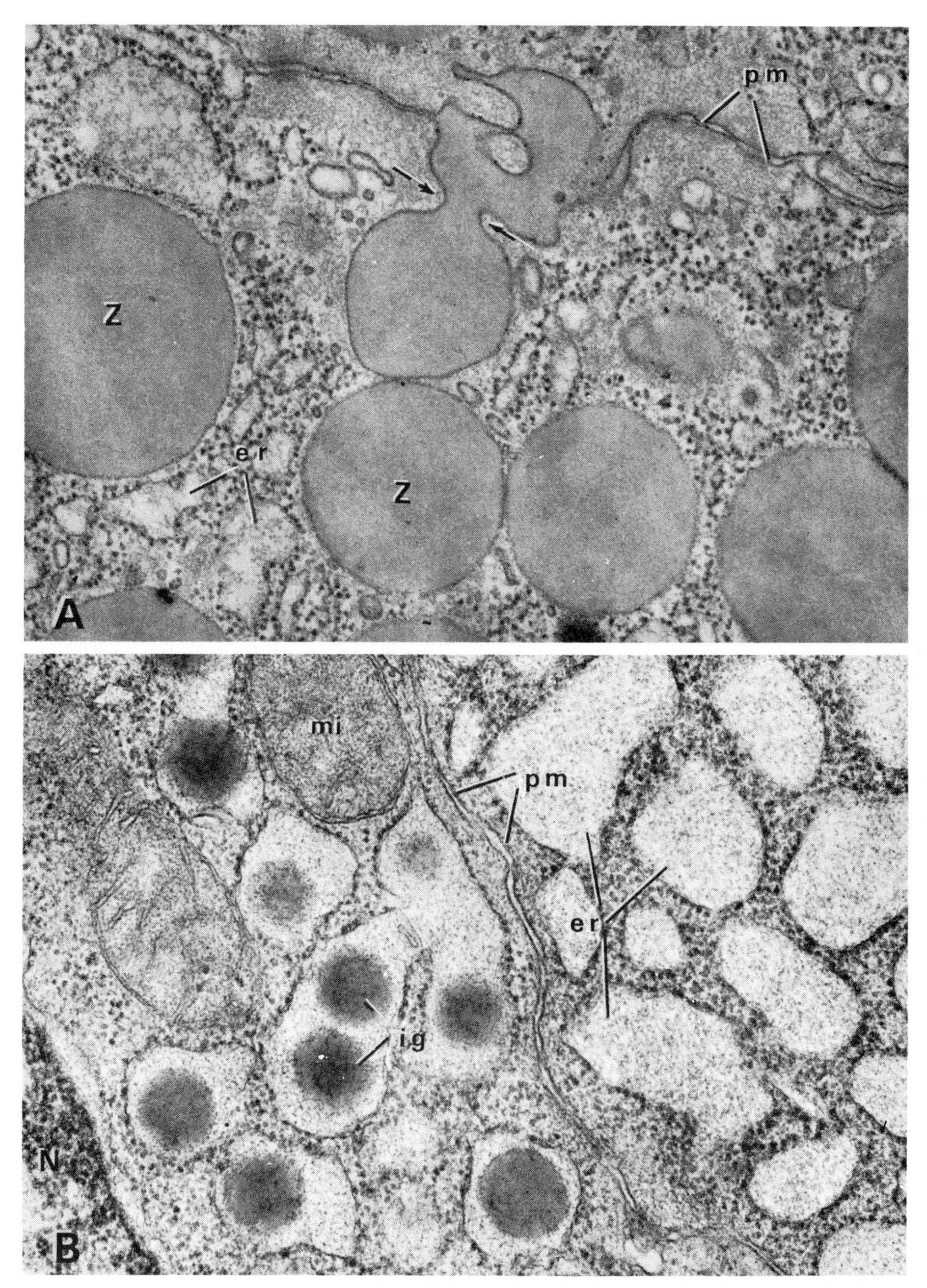

Figure 19.13. A, *Apical region of an acinar cell from the pancreas of a guinea pig, showing zymogen granules (Z), one of which is being expelled into the lumen by exocytosis followed by membrane fusion;* er, *granular endoplasmic reticulum;* pm, *plasma membrane (× 30,000). (Courtesy of G. E. Palade.)* B, *The same as* A, *but from the basal portion, showing the enlarged cisternae of the endoplasmic reticulum* (er), *some of which contain intracisternal granules* (ig); mi, *mitochondria;* N, *nucleus;* pm, *plasma membrane (×30,000). (Courtesy of D. Zambrano.)*

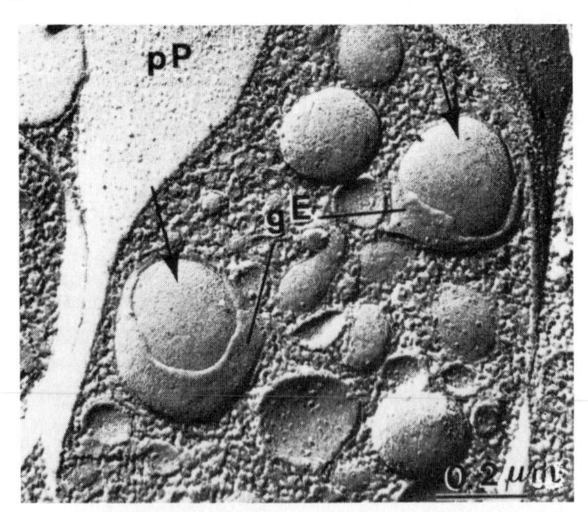

Figure 19.14. Replica of a freeze-fractured axon from the neurohypophysis (pituitary) showing: secretory vesicles (arrows), cross-fractured cytoplasm, the inner leaflet of the plasma membrane (pP), and the inner granule leaflet membrane (gE). Note granular interior of the secretory vesicles (arrows). ×61,000. (From Theodosis, Dreifuss, and Orci, 1978: J. Cell Biol. 78: 542–553.)

ionophores that increase calcium influx (Foreman *et al.*, 1976). No evidence was found for direct participation of the nucleus or the mitochondria in synthesis of the enzymes (Caro and Palade, 1964). A three dimensional view of the secretory vesicles is seen in a freeze-fracture electronmicrograph of a neurosecretory cell in the neurohypophysis in Figure 19.14. Each vesicle is enclosed in a unit membrane that splits into outer and inner (cytoplasmic) faces (Theodosis *et al.*, 1978).

It has been shown that wastes are voided from some cells by exocytosis in much the same manner as enzymes. When solid wastes cannot be voided from the cells, they accumulate and fill the cells. For example, in vertebrates wastes accumulate in some cells in what are known as lipofuscin globules. In other cells various types of crystals accumulate. Whether, as has sometimes been argued, accumulation of nonvoided wastes is a cause of aging, or merely accompanies aging, cannot be answered (DeDuve and Wattiaux, 1966).

In both vertebrates and invertebrates, cells that accumulate wastes may be sacrificed when full. It is well known, for example, that the vertebrate pus cells are leukocytes that have become choked with bacteria and cellular debris from a wound or injured area. They are either drained off, washed off, or when dry, scale off as the wound heals. Cells may eliminate small amounts of particulate or fluid waste into blood or tissue fluid by exocytosis. The wastes eliminated in a given discharge range from relatively small particles or volumes to very large ones. Exocytosis may also be a way of replenishing the plasma membrane depleted by endocytosis.

All free-living protozoans are able to void solid indigestible wastes through the cell anus (cytopyge). Usually such residues accumulate at the posterior end of the cell. Extrusion (cell defecation) is slow except when the cell is stimulated, as for example by gentle centrifuging, when the extrusion may be quite violent following strong contraction. The rate of extrusion of wastes depends on the rate of feeding and the content of undigestible residue in ingested food. Little is known about the relation of extrusion of wastes to the various parameters of the protozoan's existence. In some cases exocytosis may be an expensive process. For example, it is estimated that extrusion of one trichocyst in *Paramecium* requires 14×10^6 ATP molecules (Mott *et al.*, 1978).

Protozoans living in fresh water have contractile vacuoles by which they periodically void the excess water continuously entering the cell. The contractile vacuole represents a specialization of exocytosis for maintaining osmotic balance in the cell. Protozoans living in the sea generally lack contractile vacuoles, being in osmotic equilibrium with sea water. However, a contractile vacuole may sometimes be induced by dilution of sea water medium with distilled water (Connor, 1967). The contractile vacuole is discussed in Chapter 18. Contractile vacuoles are also found in sponges and zoospores of some algae (Riddick, 1968).

SUMMARY

Bulk transport inward as well as outward occurs across the plasma membrane (and possibly across the membranes of cell organelles) by invagination and evagination of the membrane. Bulk transport is useful in carrying large molecules which would have otherwise have difficulty passing through the cell membrane. Bulk transport occurs over a wide range of particle and vacuole sizes. Distinguishing certain size ranges is arbitrary although useful for discussion. Endocytosis may even serve to concentrate substances in the volume of fluid brought in. Exocytosis permits cell products such as enzymes to be extruded and also permits voiding of cell wastes that would accumulate and would probably injure the cell.

Unknown at present is the cause of leakage of various nutrients and enzymes from cells into the surrounding medium. Leakage is thought to occur by exocytosis; molecules may also pass outward across the membrane. In general, however, the differential permeability of the cell membrane retains nutrients and other characteristic molecules and ions within the cell, permitting minimal leakage.

LITERATURE CITED AND GENERAL REFERENCES

Allison, A. C. and Davies, P., 1974: Mechanism of endocytosis and exocytosis. Soc. Exp. Biol. Symp. *28*: 419–446.

Bainton, D. F., 1972: Origin, content and fate of PMN (polymorphonuclear) granules. *In* Phagocytic Mechanisms in Health and Disease. Stratton-Intercontinental Medical Book Corp., New York, pp. 123–136.

Bellanti, J. A. and Dayton, D. H., eds. 1975: The Phagocytic Cell in Host Resistance. Raven Press, New York.

Bennett, H. S., 1956: The concepts of membrane flow and membrane vesiculation as mechanisms for active transport and ion pumping. J. Biophys. Biochem. Cytol. Suppl. *2*: 99–103.

Bodel, P., and Malawista, S. E., 1969: Phagocytosis by human blood leucocytes during suppression of glycolysis. Exp. Cell Res. *56:* 15–23.

Bradfute, O. E., Chapman-Andresen, C. and Jensen, W. A., 1964: Concerning morphological evidence for pinocytosis in higher plants. Exp. Cell Res. *36*: 207–210.

Brandt, P. W., 1958: A study of the mechanism of pinocytosis. Exp. Cell Res. *15*: 300–313.

Brandt, P. W. and Freeman, A. R., 1966: Plasma membrane substructure changes correlated with electrical resistance and pinocytosis. Science *155*: 582–585.

Caro, L. G. and Palade, G. E., 1964: Protein synthesis, storage and discharge in the pancreatic exocrine cell, an autoradiographic study. J. Cell Biol. *20*: 473–495.

Carr, I., 1973: The Macrophage: A Review of Ultrastructure and Function. Academic Press, New York.

Chapman-Andresen, C., 1962: Studies on pinocytosis in amoebae. Compt. Rend. Lab. Carlsberg *33*: 73–264.

Chapman-Andresen, C. and Christensen, S., 1970: Pinocytic uptake of ferretin by the ameba *Chaos chaos,* measured by atomic uptake of iron. Compt. Rend. Trav. Lab. Carlsberg *38*: 19–57.

Christiansen, R. G. and Marshall, J. M., 1965: A study of pinocytosis in the ameba *Chaos chaos*. J. Cell Biol. *25*: 443–457.

Cline, M. J., 1975: The White Cell. Harvard University Press, Cambridge, Mass.

Cocking, F. C., 1977: Uptake of foreign genetic material by plant protoplasts. Int. Rev. Cytol. *48*: 323–345.

Connor, R. L., 1967: Transport phenomena in protozoa. *In* Chemical Zoology. Vol. 1. Kidder, ed. Academic Press, New York, pp. 309–350.

Cushman, S. W., 1970: Phagocytosis and factors influencing its activity in the isolated adipose cell. J. Cell Biol. *46*: 342–353.

Damon, D., 1970: Recognition by macrophages of alteration in the membrane of old red cells and expelled nuclei. *In* Permeability and Function of Biological Membranes. Bolis, *et al.,* eds. North Holland Publishing Co., Amsterdam, pp. 57–73.

DeDuve, C. and Wattiaux, R., 1966: Functions of lysosomes. Ann. Rev. Physiol. *26*: 435–492.

DeTerra, N. and Rustad, R. C., 1959: The dependence of pinocytosis on temperature and aerobic respiration. Exp. Cell Res. *17*: 191–195.

Edwards, J. G., 1925: Formation of food cups in Amoeba induced by chemicals. Biol. Bull. *48*: 236–239.

Elliott, A. M. and Clemmons, G. L., 1966: An ultrastructual study of ingestion and digestion in *Tetrahymena pyriformis.* J. Protozool. *13*: 311–323.

Ericsson, J. L. E., 1969: Studies on induced cellular autophagy. II. Characterization of the membrane bordering autophagosomes in parenchymal liver cells. Exp. Cell Res. *56*: 393–405.

Findlay, N., and Mercer, F. V., 1971: Nectar production in *Abutilon.* II. Submicroscopic study of the nectary. Aust. J. Biol. Sci. *24*: 657–664.

Foreman, J. C., Garland, L. G. and Mongar, J. L., 1976: The role of calcium in secretory processes. Model studies in mast cells. Soc. Exp Biol. Symp. *30*: 193–218.

Giese, A. C., 1973: Blepharisma—The Biology of a Light Sensitive Protozoan. Stanford University Press, Stanford, Calif.

Goodale, R. J. and Thompson, J. E., 1971: A scanning electron microscope study of phagocytosis. Exp. Cell Res. *64*: 1–8.

Gropp, A., 1963: Phagocytosis and pinocytosis. *In* Cinematography in Cell Biology. Rose, ed. Academic Press, New York.

Gunning, B. F. S. and Steer, M. W., 1975: Ultrastructure and the Biology of Plant Cells. Crane, Russak & Co., New York.

Guyton, A. C., 1976: Textbook of Medical Physiology. 5th Ed. W. B. Saunders Co., Philadelphia.

Hausmann, K., 1978: Extrusive organelles in protists. Int. Rev. Cytol. *52*: 197–276.

Hill, D. L., 1972: The Biochemistry and Physiology of *Tetrahymena.* Academic Press, New York.

Holter, H., 1959: Pinocytosis. Int. Rev. Cytol. *8*: 481–504.

Holter, H., 1961: The induction of pinocytosis. *In* Biological Approaches to Cancer Chemotherapy. Harris, ed. Academic Press, New York, pp. 77–88.

Holter, H. and Holtzer, H., 1959: Pinocytotic uptake of fluorescein-labelled proteins by various tissue cells. Exp. Cell Res. *18*: 421–423.

Holtzer, H. and Holtzer, S., 1960: The *in vitro* uptake of fluorescein labelled plasma proteins. I. Mature cells. Compt. Rend. Lab. Carlsberg *31*: 373–408.

Holzman, E. and Peterson, E., 1969: Uptake of protein by mammalian neurons. J. Cell Biol. *40*: 863–869.

Jeon, K. W. and Jeon, M. S., 1976: Scanning electron micrographs of *Amoeba proteus* during phagocytosis. J. Protozool. *23*: 83–86.

Johnson, R. B., Jr. and Lehmeyer, J. E., 1977: The involvement of oxygen metabolites from phagocytic cells in bactericidal activity and inflammation. *In* Superoxide and Superoxide Dismutase. Michelson, McCord and Fridovich, eds. Academic Press, New York, pp. 291–305.

Karnovsky, M. L., 1962: Metabolic basis of phagocytic activity. Physiol. Rev. *42*: 143–168.

Lagunoff, D., 1971: Macrophage pinocytosis: the removal and resynthesis of a cell surface factor. Proc. Soc. Exp. Biol. Med. *138*: 118–123.

Lampen, J. D., 1974: Movement of extracellular enzymes across cell membranes. Soc. Exp. Biol. Symp. *28*: 351–374.

Ledoux, L., 1965: Uptake of DNA by living cells. Prog. Nucleic Acid Res. Molec. Biol. *4*: 231–267.

Lewis, W. H., 1931: Pinocytosis. Bull. Johns Hopkins Hospital *49*: 17–28.

McCutcheon, M., 1946: Chemotaxis in leucocytes. Physiol. Rev. *20*: 319–336.

Marshall, J. M. and Nachmias, V. T., 1965: Cell surface and pinocytosis. J. Histochem. Cytochem. *13*: 92–104.

Mast, S. O. and Doyle, L., 1934: Ingestion of fluid by *Amoeba*. Protoplasma *20*: 555–560.

Metchnikoff, E., 1883: Untersuchungen uber die mesodermalen Phagocyten einige Wirbeltiere. Biol. Zentr. *3*: 500–505.

Mott, H., Belinski, M. and Plattner, H., 1978: Exocytosis in *Paramecium*. J. Cell Sci. *32*: 67–86.

Muller, M., 1967: Digestion. *In* Chemical Zoology. Kidder, ed. Academic Press, New York, pp. 351–380.

Normann, T. C., 1976: Neurosecretion by exocytosis. Int. Rev. Cytol. *46*: 2–78.

Oren, R., Farnham, A. F., Saito, K., Milofsky, E. and Karnovsky, M., 1963: Metabolic patterns in three types of phagocytizing cells. J. Cell Biol. *17*: 487–501.

Palade, G. E. and Siekevitz, P., 1956: Pancreatic microsomes. J. Biophys. Biochem. Cytol. *2*: 671–690

Poste, G. and Nicholson, G. L., eds., 1977: The Synthesis, Assembly, and Turnover of Cell Surface Components. North Holland Publishing Co., Amsterdam.

Prosser, C. L., ed., 1973: Comparative Animal Physiology. 3rd ed. W. B. Saunders Co., Philadelphia.

Quinn, P., 1976: The Molecular Biology of Cell Membranes. University Park Press, Baltimore.

Rabinowitch, M., 1969: Uptake of aldehyde-treated erythrocytes by L2 cells. Exp. Cell Res. *54*: 210–216.

Rabinowitch, M. and de Stephano, M. J., 1970: Interaction of red cells with phagocytes of the wax moth *(Galleria mellonifera)* and mouse. Exp. Cell Res. *59*: 272–289.

Rabinowitch, M. and de Stephano, M. J., 1971: Phagocytosis of erythrocytes by *Acanthameba* sp. Exp. Cell Res. *64*: 275–289.

Ricketts, T. R., 1970: Effects of endocytosis upon phosphatase activity of *Tetrahymena pyriformis*. Protoplasma *71*: 127–137.

Ricketts, T. R., 1971: Endocytosis in *Tetrahymena pyriformis*. Exp. Cell Res. *66*: 49–58.

Riddick, D. H., 1968: Contractile vacuole in the amoeba *Pelomyxa carolinensis*. Am. J. Physiol. *215*: 736–740.

Rinaldi, A. R., 1959: The induction of pinocytosis in *Amoeba proteus* by ultraviolet radiation. Exp. Cell Res. *18*: 70–75.

Rodesch, F. R., Neve, P. and Dumont, J. E., 1970: Phagocytosis of latex beads by isolated thyroid cells. Exp. Cell Res. *60*: 354–360.

Romanenko, A. S. and Salyaev, R. K., 1975: Pinocytosis in the myxomycete *Physarum polycephalum:* light and electron microscopical observations. Tsitologia *17*: 1124–1127.

Rustad, R. C., 1961: Pinocytosis. Sci. Am. (Apr.) *204*: 120–130.

Saier, M. J. and Stiles, C. D., 1975: Molecular Dynamics in Biological Membranes. Springer Verlag, New York.

Satir, B., 1974: Membrane events during the secretory process. Soc. Exp. Biol. Symp. *28*: 399–418.

Schechtman, A. M., 1956: Uptake and transfer of macromolecules by cells with special reference to growth and development. Int. Rev. Cytol. *5*: 303–322.

Siekevitz, P. and Palade, G. E., 1958: A cytochemical study of the pancreas of the guinea pig. J. Biophys. Biochem. Cytol. *4*: 309–318.

Silverstein, S. C., Steinman, R. M. and Cohn, Z. A., 1977: Endocytosis. Ann. Rev. Biochem. *46*: 669–732.

Stimmersling, R. and Stockton, W., 1975: Cytological studies on endo and exocytosis in acellular slime molds. Protoplasma *85*: 243–260.

Straus, W., 1964: Occurrence of phagosomes and phagolysosomes in different segments of the nephron in relation to reabsorption, transport, digestion and extrusion of intravenously injected horseradish peroxidase. J. Cell Biol. *21*: 295–308.

Theodosis, D., Dreifuss, J. and Orci, L., 1978: A freeze fracture study of membrane events during neurophysiological secretion. J. Cell Biol. *78*: 542–553.

Ulsamer, A. G., Wright, P. L., Wetzel, M. G. and Korn, E. D., 1971: Plasma and phagosome membranes of *Acanthamoeba castellanii*. J. Cell Biol. *51*: 193–215.

Willoughby, D. A., 1978: Inflammation. Endeavour N. S. *2*: 57–65.

Zimmerman, A. M., 1965: Effects of high pressure on pinocytosis in *Amoeba proteus*. J. Cell Biol. *25*: 397–400.

Section V

EXCITABILITY AND CONTRACTILITY

The response of an organism to its environment depends upon *cellular excitability,* one of the fundamental properties of life. If an organism is to survive, it must obtain nutrients, resist unfavorable changes in environment, and, if its species is to continue, reproduce. Chance plays a part in the struggle for existence, but those organisms that survive must respond in an adaptive manner. Excitability is the capacity to react to changes in the environment, external and internal, and energy is continuously expended by cells to maintain a condition of readiness to react.

Organisms respond to the environment as a whole, but for experimental purposes it is more convenient to vary each factor singly. An environmental change large enough to be detected constitutes a *stimulus.* Response to the stimulus occurs by way of *effectors,* such as muscles, glands, and cilia in animals. In plants, response occurs by growth or turgor movements.

A stimulus may be detected by an entire organism, as happens with an amoeba; most often, however, a specialized structure, the *receptor,* is especially sensitive to an environmental factor, for example, the eye to light. Receptors may occur even in a unicellular organism, one example being the colored light spot of the flagellate *Chlamydomonas.*

In a metazoan animal the stimulus sets off a disturbance (the *nerve impulse*) that is carried through cells *(neurons)* in the nervous system to the effectors. In multicellular plants a similar disturbance, slower in propagation, may also be passed from receptor to effector.

Most of the chapters in this section are devoted to analysis of excitability and response in animal cells. These cells have been investigated most extensively and illustrate the properties of excitability and contractility more effectively than the responses of plants. However, some plant cells are considered for comparison.

The nerve impulse is discussed in Chapter 20, its origin and propagation in Chapter 21, contractility in Chapter 22, and the chemistry of contractile processes in Chapter 23.

CHAPTER 20

ACTION POTENTIALS OF CELLS

The concentration of ions inside all cells is different from that in the medium bathing them; potassium is often much more concentrated and sodium much less concentrated inside cells than in the bathing medium. Since the resting (unexcited) cell is generally permeable to potassium but essentially impermeable to sodium, potassium ions leak out from the cell until a potential difference of about 60 to 70 mv develops across the plasma membrane, with the inside of the cell negative to the outside owing to loss of potassium ions unaccompanied by loss of anions. Conversely, to the same degree the outside of the cell becomes positive to the inside. In highly excitable cells such as neurons a stimulus, for instance, the application of electrical current, alters the permeability of the plasma membrane, permitting sodium ions to enter the cell. As sodium ions enter, the inside of the cell becomes momentarily positive to the outside. The change in permeability at the point of stimulation induces a like change in the oppositely charged adjacent part of the membrane. That in turn excites the section beyond it so that the excitation is ultimately propagated the full length of the neuron. Action potentials are especially prominent in nerve and muscle cells but they also occur in other excitable animal and plant cells.

THE NATURE OF THE NERVE IMPULSE

Neurons (nerve cells) are fundamentally similar to other cells in that they have a cell body with cytoplasm and its characteristic organelles, a nucleus, and a cell membrane. Therefore, they are subject to the same principles that govern the behavior of all living cells. But by their specialized structures they are admirably fitted for conduction of nerve impulses. *Dendrites,* short extensions from the cell body, carry impulses to the cell body, and an *axon,* a long outgrowth, carries the impulse to the next neuron, or muscle cell, or other target cell. An axon is often referred to as a nerve fiber, but, in a loose sense, nerve fiber may also refer to the whole neuron.

As seen in Figure 20.1, the *axon* of the vertebrate myelinated neuron is wrapped in a *myelin sheath.* This sheet consists of multiple membranes of the *Schwann cells,* the nuclei of which occur occasionally along the sheath (see also Figs. 8.5 and 8.6). Schwann cells in nerve tissue culture have actually been observed to wrap around the neuron, thus laying the myelin sheath (as had been postulated for many years) (Ernyei and Young, 1966). The myelin sheath is thinner on invertebrate neurons, where it may be only a single layer. It is practically absent in the finer fibers, called *nonmyelinated* fibers, of the vertebrate autonomic nervous system. The myelin sheath does not enclose the dendrites, the cell body, or the terminations of the axon. *Myelin,* containing a considerable amount of lipid, acts as an insulator and myelinated neurons transmit the nerve impulse more rapidly than do nonmyelinated ones of the same diameter (Table 20.1).

Neurons vary in size and form (Bodian, 1962). The largest mammalian nerve fibers are designated A fibers, those of intermediate size, B fibers, and small ones, C fibers (Erlanger and Gasser, 1937). Fibers of different sizes serve different purposes.

A *nerve* consists of a bundle of many neurons. However, single fibers can be dissected out of a

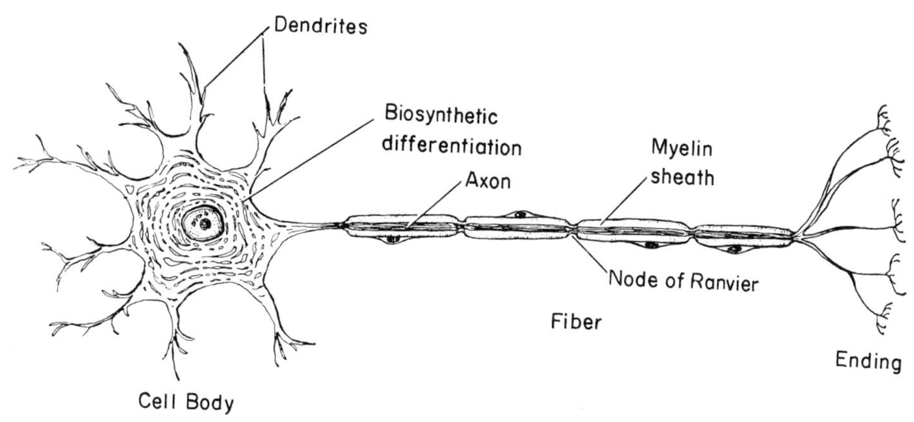

Figure 20.1. *Diagram of a myelinated neuron. The nuclei along the axon are in Schwann cells which form the myelin sheath. Such cells cover the axon except at the nodes of Ranvier and at the terminal branching portion. The terminal endings make synaptic connections with effector cells (muscle or gland cells). (From Schmitt, 1959: Biophysical Science. A Study Program. Oncley, ed. Copyright © 1959. Reprinted by permission of John Wiley & Sons, New York.)*

bundle and studied individually, and much of the information on action potentials has been obtained in this manner (Tasaki, 1953).

Glial cells are the cells with long fibrous processes found in the central nervous system. They show resting potentials like most cells, but not action potentials. Glial cells connect to one another by low-resistance bridges but they are separated from nerve cells by intercellular clefts (Kuffler and Nicholls, 1977).

Measurement and Conduction

The muscles in the leg of a frog twitch when the sciatic nerve is stimulated. Pioneer investigators used the muscle twitch as evidence of a nerve impulse from the point of stimulation to the muscle, and the nerve-muscle preparation became a classic tool in the study of nerve transmission. Direct measurements of nerve activity, however, were initiated in 1786 by Galvani, who obtained evidence for the presence of electric currents in nerve as well as in muscle, even when stimuli other than electricity (e.g., pinching) were used.

The introduction of the cathode ray oscillograph by Gasser and Erlanger in 1922 provided an instrument now used to measure action potentials and other electrical manifestations of living things. The electron beam, having very slight inertia, responds to an imposed potential with practically no delay. Activated by an oscillating circuit, the beam sweeps at any desired frequency across the screen from side to side between two vertical plates in the tube. An action potential from a nerve fiber can be passed through an amplifier and imposed upon two horizontal plates, causing the beam to move upward or downward according to the potential (Fig. 20.2). Because the image has afterglow (phosphorescence), it can be readily photographed for a permanent record.

Recording electrical changes at the surface of an excited cell is used in limited areas of research. However, analysis of the molecular nature of the nerve action potential demands quantitative information on the potential that exists between the inside and the outside of the resting cell and the changes in the potential that accompany its passage along the excitable cell. Therefore, only recordings between an electrode in the cell and an indifferent electrode in the bathing medium are considered here (see Appendix 20.1).

Excitation may also be evoked by an electrode pair, one inserted into the cell, the other placed in the medium. Nonpolarizable electrodes (see Chapter 11) with silver chloride wires leadings into microelectrodes containing

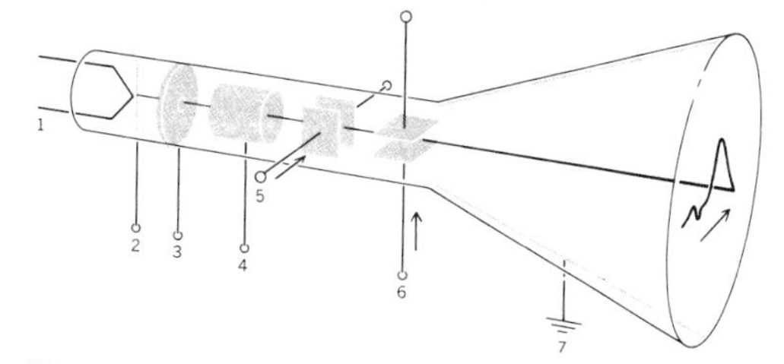

Figure 20.2. *Cathode ray oscilloscope. The cathode (1) emits electrons. Control grid or "brightness control" (2) governs strength of electron current. Main anode, or "gun" (3), accelerates electrons. Focussing anode (4) forms sharp electron beam. Pair of vertical deflection plates (5) produce "time base" deflection. Pair of horizontal deflection plates (6) receive nerve signal from amplifier. Metallized screen (7) discharges electron beam after it has produced bright spot by collision with fluorescent screen on front of tube. (From Katz, 1966. Nerve, Muscle and Synapse. Copyright 1966 by McGraw-Hill, Inc. Used by permission of McGraw-Hill Book Company, New York.)*

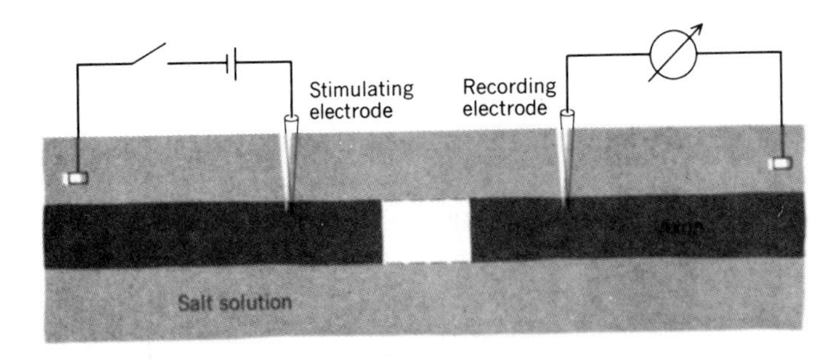

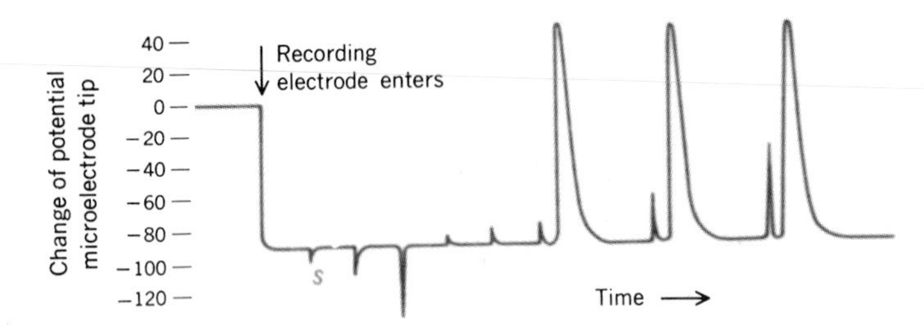

Figure 20.3. *The response of a single axon. Stimulation and recording are done with the help of intracellular microelectrodes. The points of stimulation and recording are kept far apart. After insertion of the recording electrode, eight stimuli (very brief shocks, s) of varying polarity and intensity are applied. Provided the shock is of the right polarity and exceeds a critical threshold strength, an action potential of fixed magnitude (all-or-none response) results. (From Katz, 1966: Nerves, Muscle and Synapse. Copyright 1966 by McGraw-Hill, Inc. Used by permission of McGraw-Hill Book Co., New York.)*

KCl solution are generally used for this purpose.

Once an impulse has been initiated, regardless of the intensity of stimulation, its potential on a single neuron under the same conditions is always the same, whether measured near the stimulus or far from it. The impulse passes to the end of the neuron. These properties have been termed the *all-or-none* principle (Fig. 20.3). The nerve impulse has been likened to a trail of gunpowder, along which each bit is set off by the fire next to it, regardless of whether a match or a blowtorch was first used to ignite it. The neuron appears to be ready to discharge, needing only to be triggered.

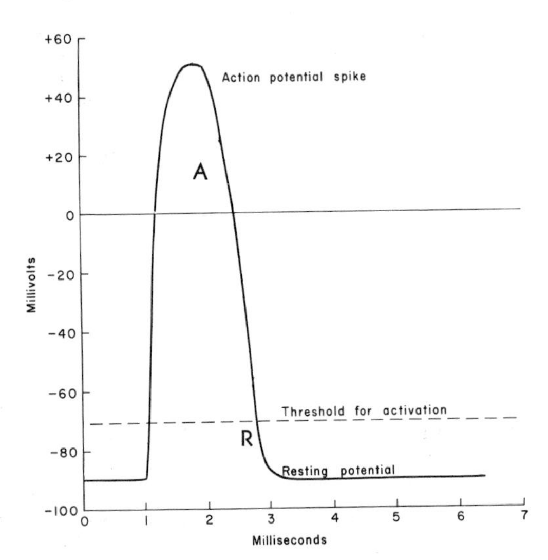

Figure 20.4. *Change in electrical potential upon stimulation of a single isolated electroplax of the electric fish* Electrophorus electricus. *Note complete depolarization and overshoot, the inside of the cell changing from negative to positive in the overshoot (the portion of the spike over the zero line). A stands for absolute refractory period, R for the relative refractory period. (After Gerebzoff and Schoffeniels, 1960: Comp. Biochem. 2: 525.)*

Cold and anesthetics decrease a neuron's excitability locally, but the magnitude of the action potential measured in front of and behind an anesthetized (or chilled) portion of a neuron remains the same provided the impulse has passed the affected area, i.e., the excitation field is able to bridge the gap. Therefore, transmission occurs *without decrement*. Once a nerve impulse has passed over a neuron, a *refractory period* ensues, a sort of "busy signal" period, during which the neuron recovers its capacity to react. If the neuron has been repeatedly stimulated to carry a train of impulses, its recovery is slower than after a single impulse. The *spike* is the maximal action potential. For any neuron the spike is of constant magnitude (Fig. 20.3) and passes unaltered the full length of the neuron. No additional activity can be induced by a stimulus of any strength until the spike at the point of stimulation has been completed; this is the *absolute refractory period* (Figs. 20.4*A*, 20.20*D*). During the tail of the spike, a stimulus much stronger than normal will excite the neuron. This period is called the *relative refractory period*. Finally, the neuron resumes normal sensitivity. The voltages and the time for the spike and the refractory periods are different for different types of neurons. In large mammalian fibers (A fibers) the spike reaches a magnitude of a few hundredths of a volt, and lasts about 1 millisecond. The spikes might be considered the message carriers of the neuron, whereas the refractory period might be considered the indicator of the readiness with which messages will be accepted. The ionic basis for the spike and the refractory period are discussed in Chapter 21.

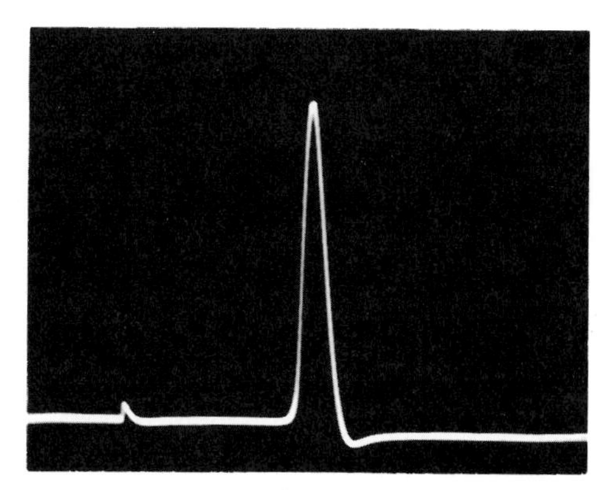

Figure 20.5. *Single fiber (central interneuron) isolated from the crayfish ventral nerve cord and recorded in oil. Preceding the spike is seen the stimulus artifact. (By courtesy of D. Kennedy.)*

Figure 20.5 shows an oscilloscopic record of a nerve action spike. The small pip which precedes the spike is a *stimulus artifact* and is a physical detection of the electric field set up by the stimulus. A piece of string soaked in balanced salt solution will serve as well as the nerve fiber for producing this artifact.

THE RATE OF CONDUCTION OF THE NERVE IMPULSE

Although the nerve impulse travels so fast that at least one prominent physiologist (J. Müller, 1801–1858) despaired of its ever being measured, its speed was recorded later by one of his students, Hermann von Helmholtz 1821–1894), who used a smoked drum, a coil and a pendulum. This crude method nonetheless gave values close to those accepted today. The speed of the nerve impulses on large mammalian nerve fibers has since been shown to be seldom over 100 meters per second (Brazier, 1968). This shows that the impulse is not a simple electrical wave (which would travel at an extremely high velocity) but rather some reaction in the nerve cell accompanied by an electrical wave.

Other factors being equal, the impulse is conducted more rapidly the greater the cross-sectional diameter of the neuron. A large diameter allows easier axial current flow, thus allowing more rapid discharge of the membrane capacitance, also called condenser property (ability to store electric charge), ahead of the advancing spike. The speed of impulse conduction is also increased with increasing thickness of the myelin sheath because of decreased transmembrane current loss between nodes that results from improved insulation. Furthermore, a thickened myelin sheath describes the overall membrane-sheath capacitance of the nerve axon. The thickness of the myelin sheath may be measured by its birefringence (double refraction). When the product of the birefringence and the diameter is nearly constant, the rate of propagation is about the same even for neurons of diverse origins, as shown in the last column of Table 20.1. For myelinated nerve fibers of mammals, the rate of conduction of the nerve impulse varies with the cross-sectional diameter, as illustrated in Figure 20.6. The rate of transmission for a given neuron is constant; it may be slowed at a partial block, but beyond the block it continues at its characteristic rate. The rate of nerve impulse conduction at a specified temperature in nerve cells of a wide variety of animals is shown in Table 20.2.

TABLE 20.1. VELOCITY OF PROPAGATION OF THE NERVE IMPULSE AS AFFECTED BY AXON DIAMETER AND SHEATH THICKNESS*

Nerve Cell	*Fiber Diam. (FD) in μm*	*Axon Diam. (AD) in μm*	$\frac{AD}{FD}$	*Birefringence (B)*	*Velocity (CA) (m/sec)*	*B × FD*
Squid, giant	650	637	0.98	−0.0001	25	—
Earthworm, giant	100	90	0.90	0.0010	25	0.10
Shrimp, giant	50	43	0.87	0.0024	25	0.12
Frog, sciatic	10	7.5	0.75	0.0105	25	0.105
Cat, saphenous (20°C.)	8.7	6.6	0.76	0.014	25	0.12
Catfish	8.8	5.8	0.58	0.012	25	0.105

*Sheath thickness is determined by birefringence. Birefringence is measured in terms of the degree of rotation of the analyzer prism in a polarizing microscope required to extinguish the polarized light from the nerve. The thicker the layer of myelin the greater the birefringence, as seen in the vertebrate nerves compared with the invertebrate nerves. For use of a polarizing microscope, see Chapter 3. (Modified from Taylor, 1942: J. Cell. Comp. Physiol. *20*.)

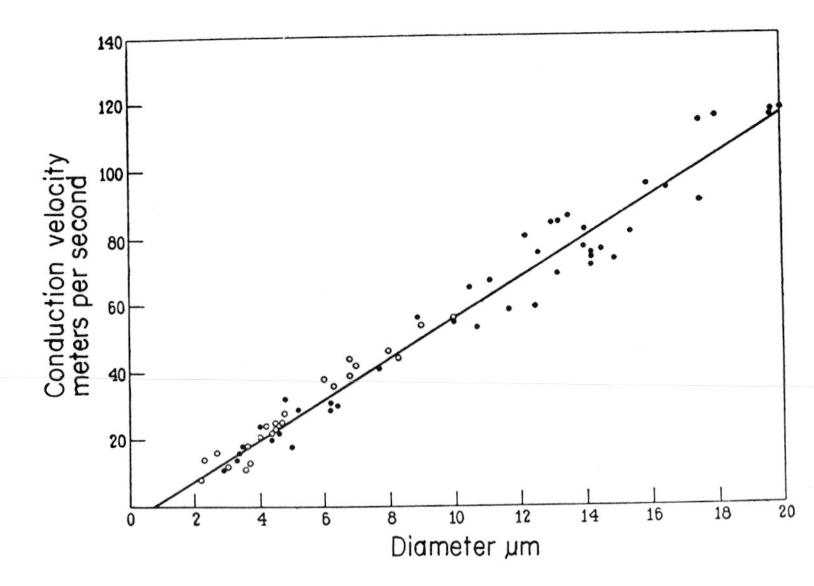

Figure 20.6. *The linear relation between diameter and conduction velocity in meters per second of mammalian myelinated nerve fibers. Each point represents a determination of the maximum conduction velocity of a given nerve, plotted against the diameter of the largest fiber in that nerve. Dots represent adult nerves; circles represent immature nerves. (After Hursh, from Gasser, 1941: Ohio J. Sci. 41: 145.)*

TABLE 20.2. COMPARISON OF CONDUCTION RATES OF NERVE IMPULSES IN VERTEBRATE AND INVERTEBRATE NERVES*

Type of Nervous Structure	*Group of Animals*	*Genus and Type of Nerve*	*Temperature in °C*	*Velocity in Meters/Second*
Nerve net	Coelenterate	Anemone (*Metridium*)	21	0.121–0.146
		Calliactis, net	—	0.04–0.15
		Calliactis, through tracts	—	1.2
Ganglionic cord	Annelid	Earthworm (*Lumbricus*)	—	0.6
	Arthropod	Myriapod (*Scolopendra*)	—	2.5
		Crayfish (*Cambarus*)	—	1.2
Nerve fibers	Mollusk	Snail (*Helix*), <1 μm diam.	15–18	0.4–0.05
		Mytilus, pedal nerve	—	0.64
		Sepia, <50 μm diam.	—	3.5, 2.26
	Arthropod	*Limulus*, leg nerve	—	4.6, 1.3
		Lobster, leg nerve	—	9.2, 1.8
	Vertebrate	Turtle, vagus, nonmyelinated	—	0.8, 0.3
		Frog, nonmyelinated	21.5	0.5–0.4
		Frog, dorsal root	21.5	42.0
		Rabbit, depressor, 2–4 μm	38	5.0
		Dog, saphenous, 17 μm	38	83.3
Giant fibers	Mollusk	*Loligo*, 718 μm diam.	—	22.3
	Annelid	Earthworm (*Lumbricus*),		
		lateral	10–12	7–12
		median	10–12	17–25
	Arthropod	*Leander*, 35 μm diam.	17	18–23
	Vertebrate	*Ameirus*, 22–43 μm	10–15	50–60

*After Prosser, 1946: Physiol. Rev. *26*.

THE STRENGTH-DURATION RELATIONSHIP

Although changes in pressure, heat, chemicals, and other stimuli may also evoke a nerve impulse, these stimuli cannot be conveniently applied experimentally without damaging the nerve cell, nor can they be readily quantified. Ever since the pioneering experimental studies of Galvani, electrical stimulation has been used almost exclusively for experimental studies of the nerve impulse. For this purpose an electronic stimulator, which develops the desired voltage almost instantly and in which the duration of current flow can be precisely controlled, is now used.

To stimulate a neuron the cathode of the stimulating electrode is placed toward the pickup electrodes of the recording instrument, since at the anode the stimulation threshold rises even to the point of blocking the nerve impulse arising at the cathode. In the absence of an anodal block, a nerve cell transmits impulses in both directions from an electrically stimulated spot (Brazier, 1968).

A minimal *threshold current* is necessary to stimulate neurons. The threshold current that, after operating for a long period, is just able to stimulate the neuron is called the *rheobase*. Below the rheobase no response occurs because the neuron recovers as rapidly as it is affected. The time required to stimulate a neuron with a current of rheobasic strength is called the *utilization time*.

The stronger the stimulating current the shorter the time needed to stimulate the neuron. This is most clearly illustrated in a curve relating the strength of the current and the time it must act to initiate a nerve impulse. Such a curve is called the *strength-duration curve* (Fig. 20.7). The strength-duration relationship is usually described in terms of the current rather than in terms of the voltage, and this convention is followed here.* On the basis of this curve, provided the product of the current and the square root of the time is maintained constant, the neuron is always stimulated. This product is referred to as *threshold stimulation*. Any product less than this constitutes a *subthreshold stimulation*. Although the quantitative values may be quite different from diverse excitable systems, the general shape of the strength-duration curve is usually similar (Lapique, 1926).

A current above rheobase but applied to a neuron for less time than "required" by the strength-duration curve fails to evoke a response. However, the same stimulus repeated

*Since by Ohm's law the current in a circuit is directly proportional to the electromotive force and inversely proportional to the resistance of the circuit, the current in amperes is directly proportional to the voltage in volts.

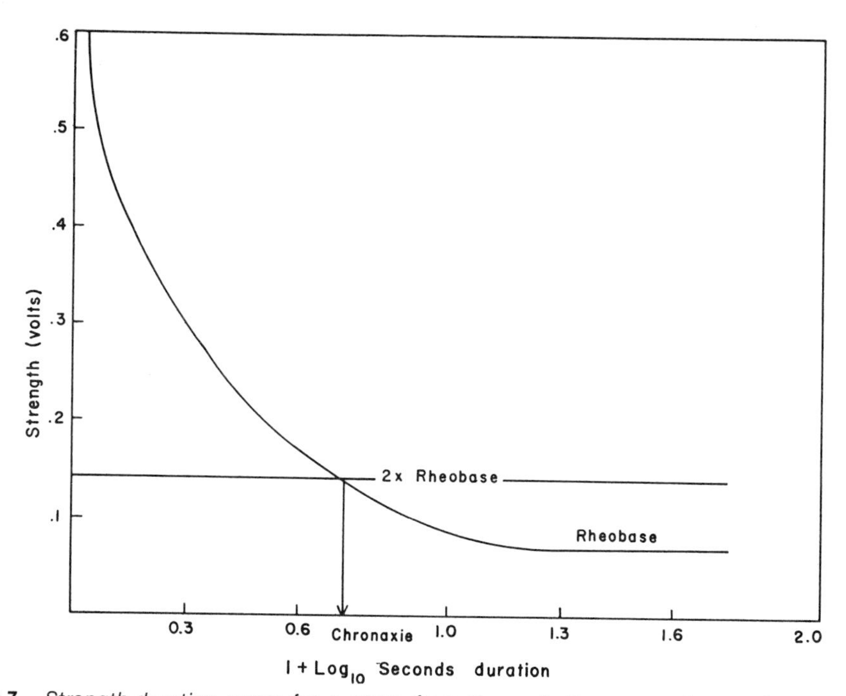

Figure 20.7. *Strength-duration curve for a nerve from the crab* Cancer productus. *Similar curves are obtained for any excitable system although absolute values vary. The chronaxie is the time required to stimulate when the current is twice the rheobase. It is marked at the base of the arrow on the time scale. The threshold current is the rheobase—the lowest current that acting for a long time will stimulate the nerve cell.*

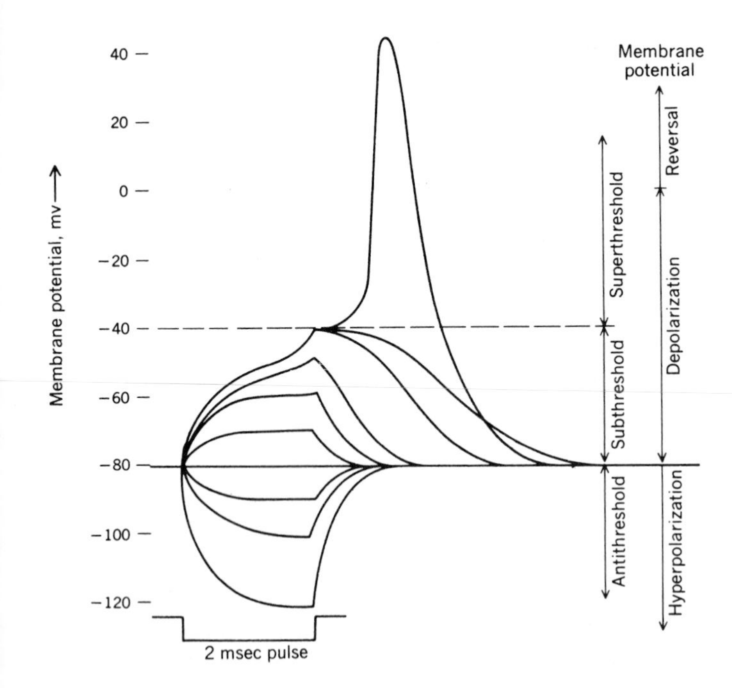

Figure 20.8. A stimulating current must be greater than threshold to initiate a spike discharge in a neuron. Below threshold local depolarization (electrotonus) occurs but within milliseconds returns to normal. Once threshold is reached, not only is the neuron depolarized but it becomes positively charged inside relative to the outside of the membrane. This reversal of potential is called the overshoot. Although the neuron is refractory to stimulation during the spike, sensitivity to stimuli returns when the neuron recovers its normal potential within a matter of milliseconds. The resting potential is here given as −80mV inside. (From Katz, 1966: Nerves, Muscle and Synapse. Copyright 1966 by McGraw-Hill, Inc. Used by permission of McGraw-Hill Book Co., New York.)

rapidly enough (e.g., every 0.15 milliseconds) for a sufficient number of times stimulates by temporal *summation,* provided the sum of all the stimuli is greater than the minimum demanded by the strength-duration curve.* Stimulation under such conditions therefore does not violate the strength-duration principle. Summation, even when adequate, will still fail to stimulate if the time lapse between stimuli is too great because the system recovers from each stimulus. To trigger excitatory processes both the strength of the current and its application time must be sufficient to overcome the tendency of the neuron to recover.

In a special case stimulating voltage is gradually increased up to, and kept at, a value that normally stimulates the nerve fiber. However, stimulation does not occur because the neuron has become *accommodated.* During the gradual change in voltage a new (rising) base level is established at each moment from which the added current is too small to cause stimulation. Therefore, the rate of voltage change is still another factor necessary to evoke a response, even when the strength-duration relationships are presumably satisfied.

Defining excitation of a nerve cell has been likened to "defining quantitatively the effort needed to bring a leaky vessel to overflowing" (Johnson *et al.,* 1974). Although a subthreshold stimulus is incapable of evoking a propagated spike potential, it alters the neuron and a measurable, local, nonpropagated potential appears, indicating a mild state of excitation that declines exponentially with time since a recovery process always counters excitation. Only when stimulation causes the local potential to decline from −80mv to less than −40mv is excitation successful in provoking a propagated spike potential (Fig. 20.8). Both excitation and recovery processes have been formulated mathematically, but the subject is too complex for discussion here (see Junge, 1976).

Excitation, as exemplified by neurons, is initiated electrically in experimental studies, but may also be initiated by such factors as chemicals, pressure, and light. Because the action potential is self-propagated, like a burning fuse, it is of constant magnitude. The action potential so initiated passes without decrement the full length of the neuron. A threshold current is necessary to excite a neuron and the greater the current, the less the time of its application needed to excite the neuron. Excitation consists essentially of a permeability change to ions, a topic to be considered in detail in the next chapter.

**Spatial summation* occurs in synapses. Impulses impinging on a synapse from many neurons, each impulse alone inadequate to stimulate the postsynaptic neuron, may, if simultaneous, together stimulate it.

CORE CONDUCTOR THEORY OF THE NERVE IMPULSE

An electrotonic potential develops on the nerve fiber and in the vicinity of an electrode inserted into the cell. The potential indicates that current is flowing along specific lines as shown in Figure 20.9 top right. The potential change is greatest immediately underneath the electrodes and falls off exponentially at points progressively farther away (bottom right). It is proposed that the development of these potentials and their associated currents can best be accounted for if the nerve fiber is considered to act as a *core conductor,* that is, as a cylinder of conducting fluid material (axoplasm, the cytoplasm of the axon) with a sheath (cell membrane and myelin sheath) of high electrical resistance, surrounded by a layer of conducting medium (balanced salt solution). The cell membrane, which becomes polarized, is thus presumed to impose a capacity on the system such that the nerve fiber may be viewed as a condenser capable of being charged on application of a potential difference across it. Such an electrical model of a nerve fiber has been constructed and studied. However, under steady-state conditions (continued maintenance of the potential) during application of a difference in potential to the nerve fiber, a slow flow (leak) of current is found to occur. To allow for this leak in the electrical model, it is necessary to couple a resistance that permits such leakage with a condenser.

A measurable time is usually required for the electrotonic potential to develop both at the electrodes applied to a nerve fiber and in their immediate neighborhood. To account for the time, it is necessary to consider the nerve fiber as being made up of a series of condenser-resistance units (Fig. 20.9, middle). When a potential is applied through an electrode in the nerve fiber, its structures, acting like a condenser, are charged, with the result that the area around the electrode becomes hyperpolarized and opposes further flow of current.

The resistance and capacitance of the nerve fiber membrane and the core (axoplasm) can be calculated from experimentally determinable properties of nerve fibers. From such data it is possible to predict quantitatively, with time as a factor, the development of local charge *(electrotonus)* in a nerve fiber if the nerve fiber is considered to act like the electrical model described here. When the steady-state condition has been attained, the distribution of electrotonic potentials around the electrodes can also be computed. The predictions made on the basis of such a model of a nerve

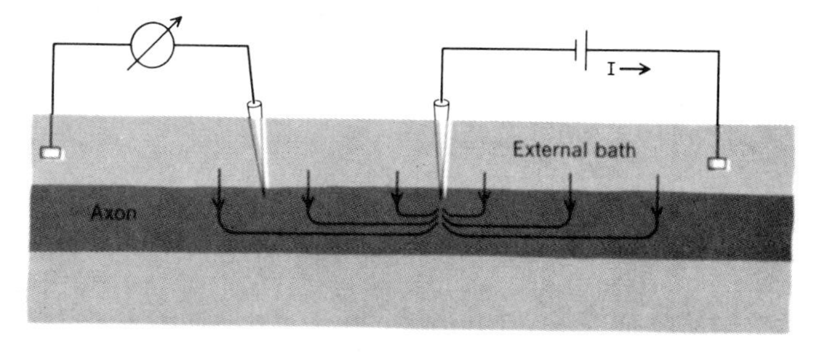

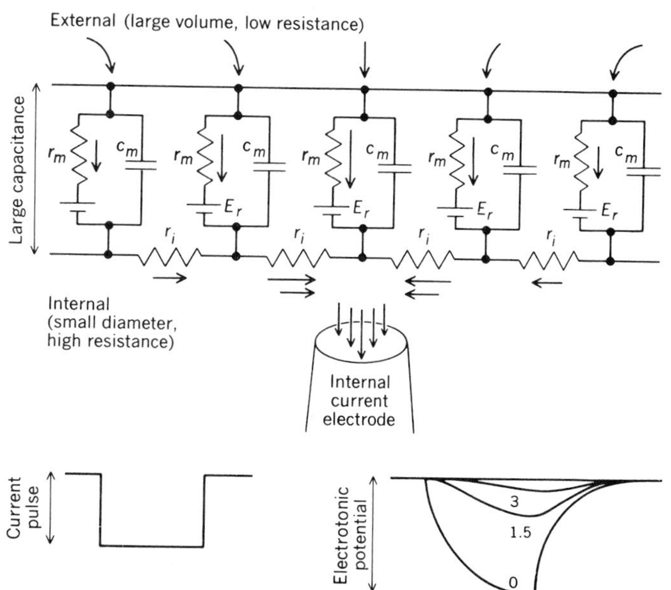

Figure 20.9. *Localized (electrotonic) changes of membrane potential shown below (right), produced by a current pulse. Diagram at top shows direction of current (I) flow and the positions of current-passing and potential-recording microelectrodes, which are inserted close together in a single axon. The electric circuit (middle) represents distributed resistance (r_m, r_i) and capacitance (c_m) of axon core and axon membrane, and the battery E_r, the source of the current. Diagram at bottom shows square current pulse and resulting potential change across the membrane, recorded at 0, 1.5, and 3 lengths away from the current-passing electrode. (From Katz, 1966: Nerves, Muscle and Synapse. Copyright 1966 by McGraw-Hill, Inc. Used by permission of McGraw-Hill Book Co., New York.)*

fiber are well in agreement with the facts, and the core conductor formulation has been useful in planning experimental studies, especially in defining more clearly the properties of parts of the nerve fiber.

For passive conduction, such as occurs in an insulated electric wire (or perhaps as a better comparison, a submarine cable), a cable made like a nerve fiber would be of little use because its losses are great: the surface leakage and the resistivity of its core are 10^8 greater and its sheath capacity 10^6 greater than commercial cables. In it a weak signal (just sufficient to produce a spike) fades out within a few millimeters. The "cable" model would fail to conduct an impulse were it not for the regenerative processes on a nerve fiber that reamplify a signal above threshold along each point of the line (Katz, 1966; Clark and Plonsey, 1966).

The peculiar cable properties of a nerve cell are, however, of importance in permitting subthreshold stimulation to remain sufficiently long for summation. This is important, especially in the dendrites of a neuron, since their excitability is local, graded, and decremental. Therefore, succeeding increments of depolarization from the dendrites may summate to the threshold for a propagated all-or-none spike on the axon.

SALTATORY CONDUCTION OF THE NERVE IMPULSE

Conduction in myelinated as compared to nonmyelinated nerve fibers of similar size is exceedingly rapid. The myelin sheath is interrupted periodically by the *nodes of Ranvier* (see Fig. 20.1), where the membrane of the neuron makes contact with the bathing fluid. The nodes are farther apart in neurons of large diameter than in neurons of small diameter. For example, the distance between nodes is about 0.2 mm for a bullfrog neuron of 4 μm diameter, about 1.5 mm for a neuron of 12 μm diameter and 2.5 mm for one of 15 μm diameter (Tasaki, 1953). It has been suggested that perhaps conduction in myelinated neurons is so rapid because the action potential skips from node to node rather than traveling the full length of the membrane. Such skipping from node to node is called *saltatory conduction* (see Fig. 21.8*D*).

The nodal threshold for stimulation is much lower than the internodal threshold (Fig. 20.10). Upon stimulation of a nerve cell an electric field is formed, which has been shown to operate from the point of excitation over a span of about two nodes. Therefore, this electric field is much more likely to evoke a change in potential in a region of low threshold (node) than in a region of high threshold (internode). Thus when an unanesthetized node is cathodally stimulated and both the node to the left and the node to the right of it are anesthetized, conduction of the impulse is still achieved across these anesthetized nodes to the next unanesthetized neighboring nodes. The same stimulus fails if applied to the unanesthetized internode. However, if two nodes to the left and two to the right of a node are anesthetized, the same stimulus applied to the unanesthetized node fails to excite the fiber because the third node away from the stimulated node is beyond operation of the electric field. A squid giant axon fails to conduct the impulse across a dead section unless a metal wire is laid across this section. The electric field is not strong enough in water but is adequate in the better conductor furnished by the wire bridge (Tasaki, 1953).

One objection to the theory of *saltatory conduction* is that nodes have not generally been seen in the central nervous system, yet rate of conduction in the brain and cord equals that in

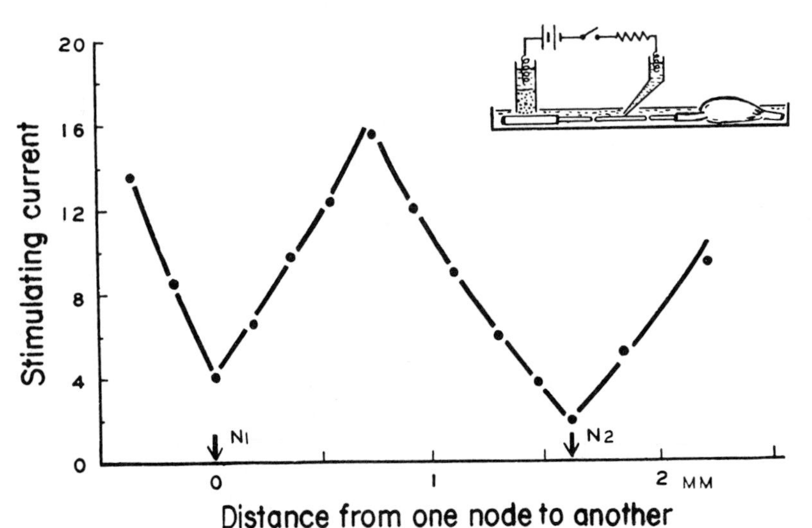

Figure 20.10. A graph demonstrating that nodal stimulation of a myelinated neuron is lower than internodal stimulation. N_1 and N_2 refer to nodes. (After Tasaki, 1953: Nervous Transmission, Charles C Thomas, Springfield, Ill.)

the peripheral nerve fibers. However, improved techniques have revealed nodes in preparations of the central nervous system, and at present most workers appear to be convinced of the generality of saltatory conduction.

The theory of saltatory conduction of the nerve impulse is incomplete. Conduction in demyelinated fibers, in which the myelin sheath has been removed by application of diphtheria toxin, is still saltatory (Rasminsky and Sears, 1972).

RESTING METABOLISM OF NERVE CELLS

Resting nerve cells respire at about the same rate per unit of mass as resting muscle cells. Improvements in respirometry and especially the development of microrespirometers have made possible measurement of gaseous exchange, and a respiratory quotient of approximately 0.8 has been found for the neuron, indicating that several nutrients are being oxidized (Connelly, 1959; Arshavski, 1960). Although glucose is believed to be the main source of energy because it is lost rapidly in the respiring nerve cells, the presence of creatine, ammonia, and phosphate as metabolic wastes produced in respiration suggests that protein and lipid are also being consumed. When nerve cells are subjected to anaerobiosis, lactic acid accumulates, indicating that during anaerobic metabolism the source of energy for maintaining excitability is mainly glycogen. The tolerance of a nerve cell to anaerobic conditions, in fact, depends partly upon its glycogen reserve. Crab nerve fibers, with much glycogen, conduct without oxygen for 5 hours, whereas mammalian nerve fibers, with little glycogen, stop in 20 minutes (Brazier, 1968).

Resting metabolism of nerve cells resembles endogenous metabolism of microorganisms in that it is not susceptible to the metabolic poison azide, even in concentrations that block the activity metabolism in stimulated nerve cells. Many of the enzymes participating in aerobic and anaerobic metabolism in other cells and also enzymes required for phosphorylations have been identified in nerve cells, but, partly because the material is limited and so much supporting connective tissue surrounds a nerve, enzymatic studies on nerve cells have not been extensive (Abood and Gerard, 1954).

ACTIVITY METABOLISM OF NERVE CELLS

The metabolic rate of stimulated myelinated nerve cells may increase from 1.3 to 4.0 times the resting rate. The respiratory quotient of stimulated nerve cells is 1.0, suggesting carbohydrate utilization (Gerard, 1932). Biochemical evidence indicates that glucose is oxidized in central nervous system neurons and in nonmyelinated nerve fibers. Decreases in ribonucleic acid content have been measured after extensive nerve activity (Hyden, 1960) (see Chapter 7). The RNA may, however, act as a mediator in the manufacture of other materials (Katz, 1959).

It has been claimed that RNA is involved in learning and that such learning can be transferred from one individual to another by transfer of RNA (Babich *et al.*, 1965). Synthesis of protein appears to be involved in memory (Agranoff, 1967). That learning has a physical basis seems reasonable, but the topic is beyond the realm of this discussion.

It is thought that the immediate source of energy for conduction of the nerve impulse in vertebrate nerve cells is probably creatine phosphate. Creatine and inorganic phosphate, in addition to ammonia, are among the metabolic products that increase in a perfusate of tetanized nerve fibers. Creatine phosphate breakdown is more rapid during nerve activity than at rest, and presumably the high energy of the phosphate bond is involved in the transmission of the nerve impulse. In invertebrate nerve cells other phosphates may be utilized (Brink *et al.*, 1952).

Another metabolic change observed in active nerve cells is the breakdown of acetylcholine:

$$\underset{\text{acetylcholine}}{[(CH_3)_3\overset{+}{N}CH_2CH_2OCOCH_3]OH^-} \xrightarrow[\text{acetylcholine esterase}]{HOH} \underset{\text{choline}}{[(CH_3)_3\overset{+}{N}CH_2CH_2OH]OH^-} + \underset{\text{acetic acid}}{[CH_3COO^-]H^+} \quad (20.1)$$

Acetylcholine is widespread in the nervous system and is synthesized in the presence of the enzyme choline acetylase, which is also widespread in nerve cells. Acetylcholine is spontaneously hydrolyzed into the inactive compounds choline and acetic acid, a reaction greatly speeded by the enzyme acetylcholine esterase, found in nerve cells. Acetylcholine enclosed in vesicles occurs in high concentration on the terminal axon surface of most neurons. The action of acetylcholine is inhibited by the drug eserine. Acetylcholine is probably concerned with passage of the nerve impulse across synapses and the neuromuscular junction (see Chapter 21).

A clear picture of the rapid sequence of metabolic events in the nerve fiber is not yet available. Although instruments exist for measuring the evolution of heat and electric charge, resistance, and capacitance (function as a condenser) during and after passage of the nervous impulse, no microchemical methods of comparable speed have been developed to measure the chemical changes preceding or concomitant with the nerve impulse.

HEAT PRODUCTION DURING NERVE ACTIVITY

It is difficult to measure directly the heat evolved during the passage of a single nerve impulse. However, when the preparation is treated with veratrine, which exaggerates the relative refractory period (Brink *et al.*, 1952), the train of nerve impulses resulting from tetanic stimulation produces measurable heat. From thermopile measurements of the total heat produced by a train of nerve impulses, the heat evolved in a single impulse is estimated to be 10^{-7} calories per gram of nerve fiber. Only a small fraction (about 1/30) of the heat is apparently given off during the initial action potential; the remainder of the heat develops during the recovery process. It is thought that the initial spike of the action potential may correspond to the period of breakdown of high-energy phosphate bonds. The recovery period may thus be the time during which glucose and other compounds are decomposed and the high-energy phosphate bonds are rebuilt (Hill, 1933; Abbott *et al.*, 1958).

The more frequent the stimulation, the smaller is the increment of heat produced and oxygen consumed per nerve impulse. The adaptive efficiency of a frequently stimulated neuron is thought to result from the changed physical-chemical state of the excited cell.

In an atmosphere of pure nitrogen the heat production rate of resting nerve cells declines over a period of 2 hours, but even after this time heat continues to be evolved, although at about one fifth the rate in air. Metabolism continues anaerobically, accumulating lactic acid. When on return to air the nerve cells recover from the oxygen debt, they evolve extra heat before resuming normal heat production (Fig. 20.11) (Gerard, 1946; Schallek, 1949).

EFFECTS OF VARIOUS FACTORS ON ACTION POTENTIALS OF NERVE CELLS

Fatigue

Because the refractory period permits neurons to recover their normal excitability, it is practically impossible to fatigue nerve cells in the presence of adequate oxygen and nutrients. Experimentally it is possible to obtain 1000 full-sized responses per second from a nerve fiber (A fibers from a frog), and with stronger shock up to 2000 have been obtained briefly (Adrian, 1950). Normal stimulation of receptor cells *in vivo* produces trains up to 160 impulses per second. For example, a cat foot pad sends in 9 to 100 impulses per second in response to tactile stimulation. Such a relatively low frequency of impulses allows for ready recovery. If stimulated to discharge at abnormally rapid rates, a nerve fiber conducts more slowly, its refractory period becomes longer, the initial potential declines and the threshold rises—all signs of a failure to return completely to normal. To this extent a nerve fiber may be fatigued (Adrian, 1950).

Anoxia

Lack of oxygen ultimately blocks transmission of impulses in a nerve cell. For example,

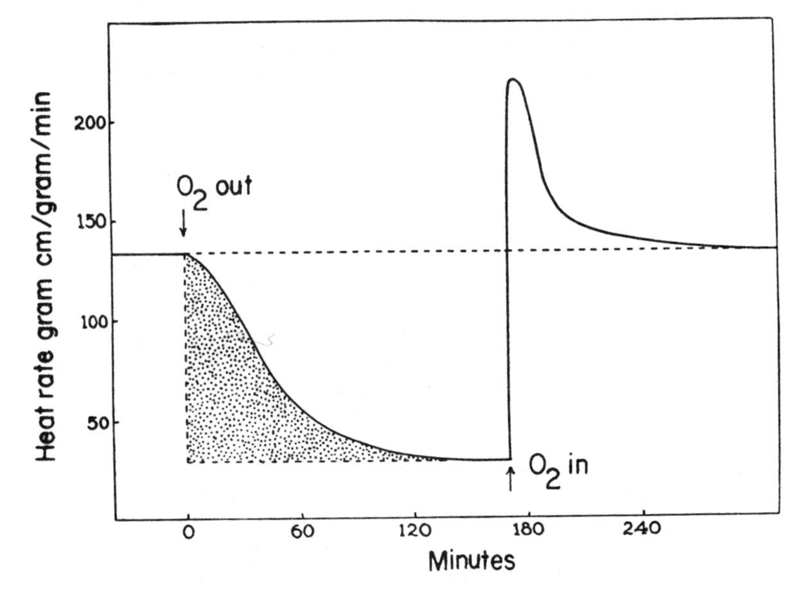

Figure 20.11. *Resting heat production of frog nerve. At the first arrow the nerve is placed in nitrogen, with resulting slow fall in heat rate. At the second arrow the nerve is replaced in oxygen. (Modified from Feng, 1936: Ergbn. Physiol. 38: 73.)*

the action potential of a nerve fiber so stimulated that the nerve impulse has to travel through a section kept in pure nitrogen shows progressively smaller spikes in that section for about an hour (i.e., for a frog nerve fiber in nitrogen, a 3 percent loss in spike size occurred in 15 minutes, 13 percent in 30 minutes, and 49 percent in 40 minutes). At the same time the rate of transmission of the nerve impulse declines (Schmitt and Cori, 1937; Fenn and Gerschman, 1950). Beyond the section in nitrogen, the spikes are of normal size, as expected on the basis of the all-or-none principle. After an hour, transmission of the nerve impulse in the section kept in nitrogen ceases, indicating that a block has been established. The higher the temperature, within the limits of viability of the nerve fiber, the more rapidly the block appears.

A nerve fiber kept anoxic accumulates an oxygen debt. When returned to air the nerve fiber takes up oxygen rapidly for a while until it has paid its debt. Presumably, anaerobic processes are unable to effect complete recovery (Lundberg and Oscarsson, 1953). If a nerve in a state of anoxia is poisoned by iodoacetate so that glycolysis also ceases, conduction ceases much sooner. This experiment indicates that in the absence of oxygen, glycolysis, at least in part, supplies the energy required to synthesize the compounds that maintain the nerve cell in a condition of readiness to respond (Gerard, 1946). This is suggested also by the heat changes during anaerobic metabolism, as seen previously.

Respiration in the presence of oxygen is necessary for continued maintenance of the nerve cell's readiness to respond to stimulation, indicating a continuous expenditure of energy to maintain excitability.

Anesthetics

Many anesthetics and narcotics block conduction of the nerve impulse. The time required to block conduction is dependent upon the concentration of anesthetic; the higher the concentration, the shorter the time required. The span or area of the nerve fiber to which anesthetic is applied also affects the speed with which the anesthetic acts, conduction being blocked more rapidly the larger the area treated. Some anesthetics (e.g., cocaine) produce a definite but reversible effect; the others, like chloral hydrate, produce an effect that becomes progressively more injurious and less reversible the longer its time of action. It is not possible to review here the complexities of the pharmacology of nerve cells (Brazier, 1968).

Application of an anesthetic raises the stimulation threshold for a nerve fiber, and larger stimuli are then required to evoke a spike. Other properties of the nerve fibers change as anesthesia progresses; the sensitivity to a chemical stimulus such as citrate is reduced, the velocity of conduction of an action potential decreases, and the size of the spike evoked by a stimulating shock successively declines, while the refractory period increases in length. All these changes indicate a progressive decrease in neuron excitability (Tasaki, 1953; Stevens, 1966).

ACTION POTENTIALS IN MUSCLE FIBERS

Action potentials occur also in various muscle fibers. Thus the excited portion of the muscle fiber also becomes negative relative to the resting portion, and an action potential as high as 100 millivolts may be recorded on a muscle fiber. Many years ago Galvani demonstrated the electric nature and similarity of muscle and nerve excitation by connecting two sciatic nerve–gastrocnemius muscle preparations in series (Fig. 20.12). When the first muscle was excited to contraction, for example, by pinching the nerve connected to it, it excited the sciatic nerve of the second muscle, following which the second muscle contracted soon after the first.

Action potentials are demonstrable in muscle cells of gut, heart, oviduct, uterus, and other organs. The tracing of the action potential of the heart is called an *electrocardiogram* (Fig. 20.13). The remarkable feature of an electrocardiogram of a large mammal is that in the contracting heart the field set up by simultaneous activity in the heart muscle fibers is large enough to be picked up on the surface of the body in spite of the activities of an enormous amount of intervening tissue (Brazier, 1968).

The most unique action potentials appear in electric fishes (Bennett, 1971). In some electric fishes a considerable number of muscle fibers

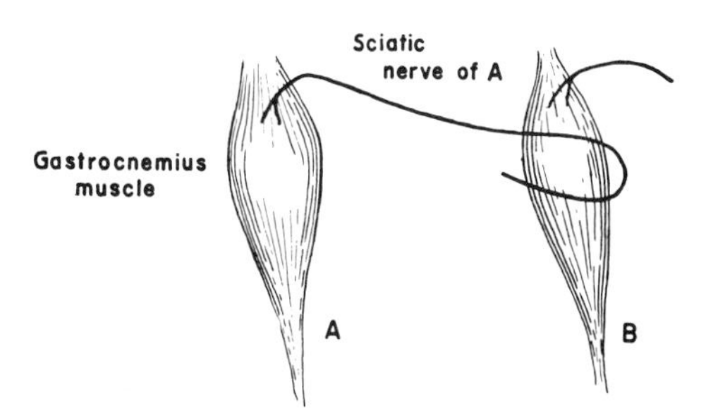

Figure 20.12. *Excitation of a nerve by a muscle. When the sciatic nerve that makes its ending on muscle B is pinched, muscle B is excited. Action currents on the surface of muscle B then excite the sciatic nerve lying on its surface leading to muscle A causing its contraction.*

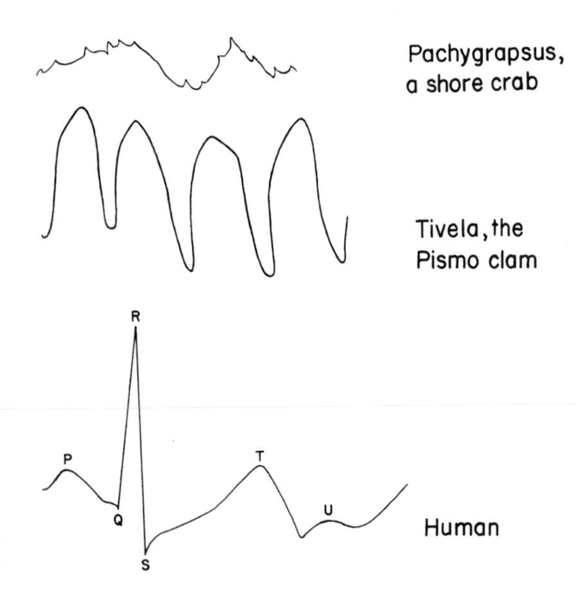

Figure 20.13. *Electrocardiograms of hearts of several animals. Note that the electrocardiogram for the crab,* Pachygrapsus, *shows a number of sharp peaks indicating discharge of ganglion cells in this* neurogenic heart *(that is, a heart in which the beat originates in discharges of nerve cells), whereas the electrocardiograms of the clam* Tivela *and the human are smooth, as expected in a* myogenic *heart (that is, one in which the beat is initiated in the heart muscle itself). Both kinds of hearts may be regulated by nerves and hormones, however.*

In the human heart the P wave precedes the contraction of the auricle and is probably the wave of excitation of this structure. The R wave precedes the contraction of the ventricle, T, representing the end of the ventricular systole (contraction). The events correlating with Q, S and U are not determined.

have become modified to form the electric organs, which are composed of a series of electric units. Each such electric unit, known as an *electroplax,* begins during early embryonic development as a normal skeletal muscle fiber but later develops into what looks like a flat plate attached to a myoneural junction (Fig. 20.14). Like a myoneural junction, the electroplax fatigues rapidly and is susceptible to the drug curare, which readily blocks synaptic transmission to the electric organ.* A single electroplax gives off a peak voltage of 50 to 150 millivolts. The large voltage measured in electric fishes is made possible by the arrangement of the electroplaxes in an organ as a series of batteries, summing the voltages of the individual electroplaxes. By parallel connections between series of electroplaxes, large currents can be obtained. The electric eel *Electrophorus electricus,* for example, has been found to produce electric shocks of several hundred volts (Cox, 1943).

It is interesting to point out that Alessandro Volta (1745–1827), who designed the first voltaic cell, used the electric organ of the electric fish as a model. Just as many electroplaxes in series give off a high voltage, so did his "pile" of copper and zinc discs in which the voltage was summated as in a series of batteries. The example illustrates a case in which physics received a great impetus from investigation of a biological phenomenon (Keynes, 1957; Bennett, 1970).

*Curare is an alkaloid arrow poison used by some South American Indians in hunting. It is extracted from plants of a number of families, in one case from the bark and in another from a gourd.

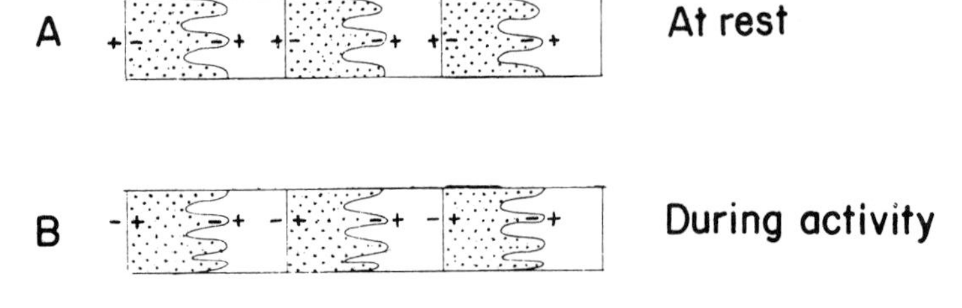

Figure 20.14. *Diagrammatic representation of the changes in the electroplaxes (single units) of the electric organ of a fish,* Electrophorus electricus. *The nerve cell is stippled, the non-nerve cell is not. The unit shown is a single series of electroplaxes. Many series "hooked up" in parallel make possible a large current as well as a high voltage. At rest* (A) *both the nerve cell membrane and the nonnervous cell membrane have a positive charge, the inside of the cell being negative. When excited the nerve cell membrane is not only depolarized, but becomes negatively charged* (B). *Under such conditions, because of the structure of the electric organ, the voltages of the series of electroplaxes add up, like the voltages in a series of little batteries. (After Gerebtzoff and Schoffeniels, 1960: Comp. Biochem.* 2: *519.)*

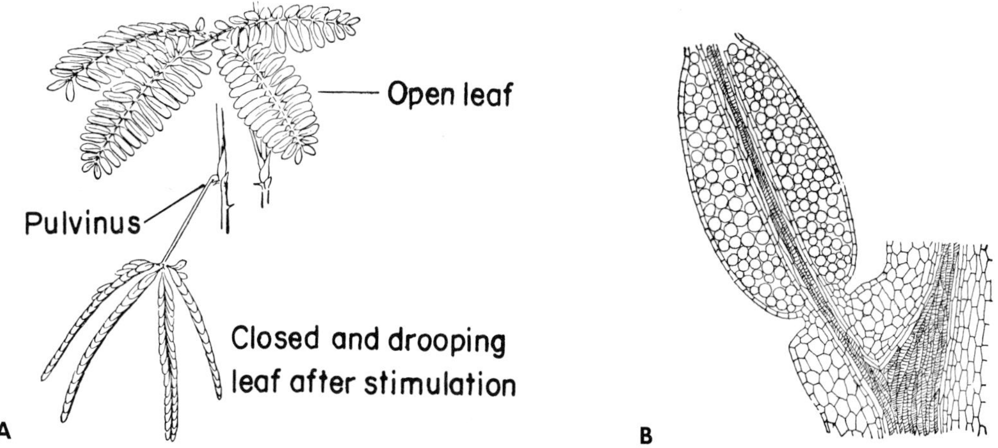

Figure 20.15. A, *The sensitive plant* Mimosa, *showing the closing of the pinnate leaflets and the drooping the entire leaf as a result of mechanical stimulation, (After Coulter,* et al., *1911: A Textbook of Botany. American Book Company, New York.)* B, *Diagram of a lengthwise section of a pulvinus of* Mimosa. *Note the loose cortical cells and the vascular bundle coursing through them. When the cells on the lower (left) side lose turgor the stem droops. In the leaflets the opposite occurs; turgor is lost in the cells on the upper side, causing the leaflets to close. (From Smith,* et al., *1929: General Botany. Macmillan, New York, p. 161. Copyright 1929 by the Macmillan Co.)*

ACTION POTENTIALS IN EXCITABLE PLANT CELLS

Excitable plant cells also develop action potentials when stimulated. For example, in the sensitive plant *Mimosa,* action potentials develop immediately after stimulation of the sensitive cells (Figs. 20.15 and 20.16) and may be recorded by a galvanometer (Sibaoka, 1966; Higinbotham, 1973; Hope and Walker, 1975). When the action potentials reach the pulvini, turgor changes lead to movement. (A pulvinus is an enlargement at the base of a leafstalk, as seen in Figure 20.15*B*.)

Even more remarkable are the action potentials in *Nitella,* a fresh water alga with very large multinucleate cells (Osterhout, 1936; Blinks, 1955) (Fig. 20.17). When a *Nitella* cell is stimulated by pressure, heat, light, chemicals, or electricity, the action potential (about a tenth of a volt) is propagated at the rate of several centimeters per second in air, and about ten times as fast in water with electrolyte (Sibaoka, 1966). The action potential usually passes to the ends of the cell; however, it may die out after having traversed only a part of it, a behavior contrary to the all-or-none principle characteristic of the nerve impulse. Like nerve cells, however, *Nitella* cells show temporal summation of stimuli. After excitation a *Nitella* cell becomes refractory or incapable of being stimulated for a brief time. The refractory period may last 30 seconds to several minutes, or, after rough handling as during isolation of an internode cell for an experiment,

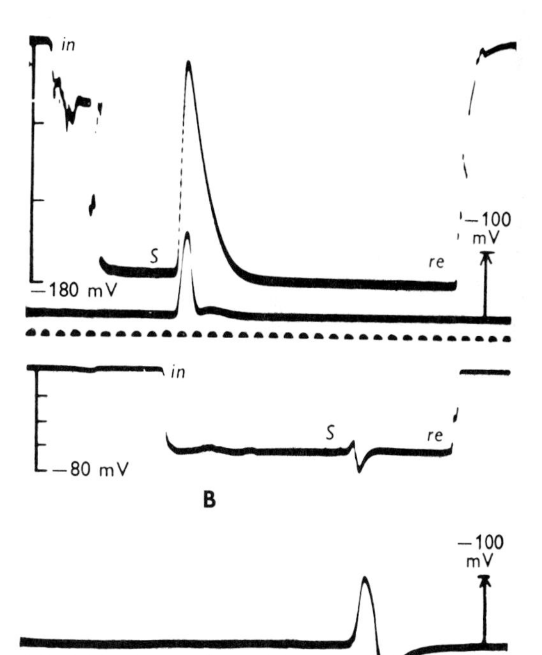

Figure 20.16. Mimosa pudica *(the sensitive plant). Membrane potentials (upper tracing) in an excitable cell of protoxylem* (A) *and an inexcitable cell of pith* (B) *in the petiole. Microelectrodes inserted into cells at* in *and removed from cells at* re. *Petiole is stimulated electrically at* s. *In an inexcitable cell* (B) *only a slight change in potential is observed. This is probably an electrotonic change due to the action current of the surrounding excitable cells. Diphasic action potentials (lower tracing) led from the cell surface are simultaneously recorded. Time marks, 1 sec. (After Sibaoka, 1962: Symp. Soc. Exp. Biol.* 20: *62.)*

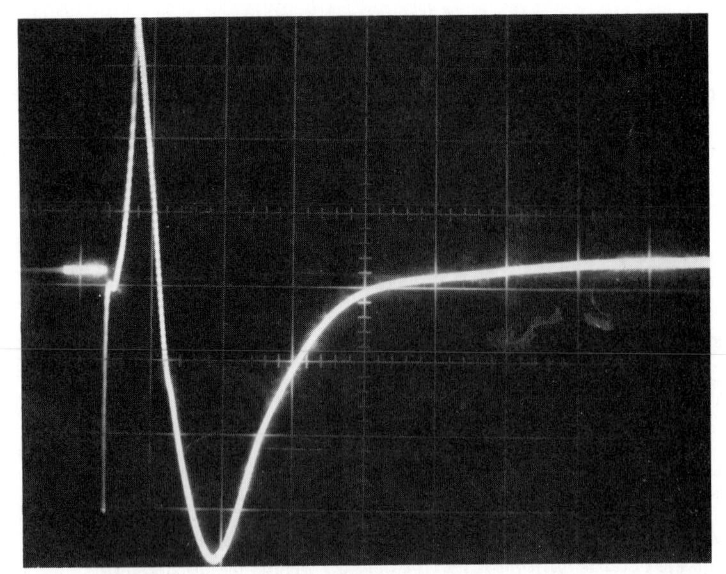

Figure 20.17. *Action potential in* Nitella *in air. Stimulation by 10 v, 0.1 msec; surface recording electrodes 3 mm from surface stimulating electrodes. Ordinate 10 mv per square, abscissa 2 sec per square. The first downward deflection is the stimulus artifact. The spike is biphasic as is characteristic of surface recording (see Appendix 20.1). (By courtesy of Nicholas Holland.)*

Nitella may be refractory for hours (Blinks, 1955).

The action potential of a *Nitella* cell is not just an electrical wave, since it travels more slowly than electronic current flow in a conducting medium, but it is apparently a change in the cell membrane accompanied by an electrical disturbance (Fig. 20.17). That the electrical wave is important for propagation of the membrane change is demonstrated experimentally. Thus when a portion of a *Nitella* cell is blocked by cold or chloroform, the action potential does not pass the block. Yet if a U-shaped piece of filter paper, or string, moistened with balanced salt solution bridges the block, the action potential is conducted through the bridge (Fig. 20.18).

Passage of an action potential in *Nitella* kept in the dark is accompanied by increased oxygen consumption and carbon dioxide production. Heat is also given off. These metabolic accompaniments of the action potential are thought to be the result of reactions involved in regenerating the cellular membrane altered by excitation (Blinks, 1955; Hope and Walker, 1975). Similar potentials have been recorded in the insectivorous sundew *Drosera* (Williams and Pickard, 1972) and in the carnivorous plant Venus's flytrap (Benolken and Jacobson, 1970).

Action potentials may be more widespread in plant cells than previously supposed. For example, action potentials have been recorded in the mycelium of the mold *Neurospora crassa* (Slayman *et al.*, 1976) and in *Acetabularia* upon cutting off the cap. In the case of *Acetabularia*, the action potential is thought to signal certain morphogenetic reactions (Melkimyan *et al.*, 1976).

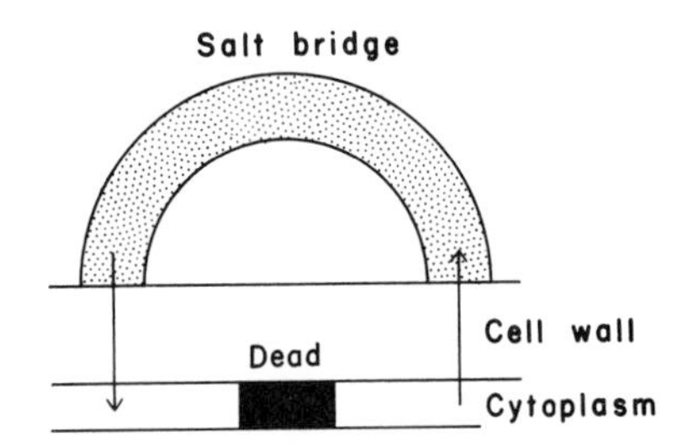

Figure 20.18. *Diagram of a bypass for a block in a* Nitella *cell. Part of the cytoplasm has been killed with chloroform. The salt bridge, consisting of filter paper or string soaked in salt solution, completes the circuit and makes transmission possible. (After Osterhout and Hill, 1929–30: J. Gen. Physiol.* 13: *548.)*

APPENDIX

20.1 SURFACE RECORDING OF ACTION POTENTIALS

A nerve can be stimulated to discharge action potentials in any one of a number of ways—by pinching (Fig. 20.19), by heat, by intense light in the presence of a photosensitizer, by electricity, and naturally by receptors to environmental stimuli. It is easiest to measure electrical stimuli quantitatively; therefore, electrical stimulation is generally the method of choice.

For some research purposes, and especially for class use, it is convenient to record action potentials from the surface of a nerve lying in air in a moist chamber on two sets of electrodes, one to stimulate, another to record the response

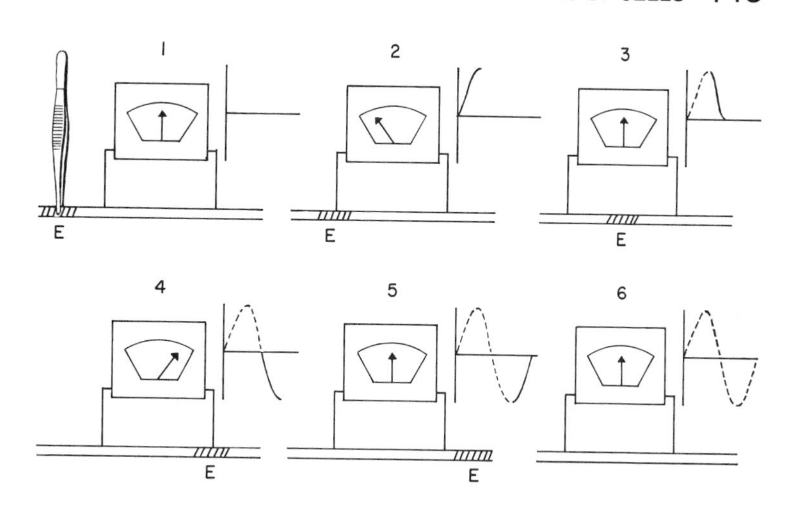

Figure 20.19. *The diphasic nature of the action potential recorded on the surface of a nerve. A pinch of the forceps stimulates the nerve, and a wave of excitation (E) passes the left pickup electrode (that is, the one nearest the point of stimulation), making it temporarily negative to the right one. Later, when the impulse reaches it, the right pickup electrode becomes temporarily negative to the left one. Finally, after the wave of excitation has passed both pickup electrodes, the hand on the dial returns to zero. If the movement of the hand on the dial is plotted, a graph similar to those at the upper right of these diagrams is obtained.*

with a recording device, usually an oscilloscope. The record obtained in this manner is a *diphasic action potential* because when the excitation reaches the first recording electrode, that electrode becomes negative relative to the second recording electrode. When the excitation reaches the second recording electrode, provided the two wires are separated by an appropriate gap, the nerve will have recovered at the first recording electrode and the second recording electrode negative to the first one (Figs. 20.19 and 20.20*B* and *D*).

In oscilloscopic recording the second half of a diphasic action potential is not a mirror image of the first half (Fig. 20.20), because the nerve at the first electrode will not have recovered fully when the excitation reaches the second one.

When a nerve is stimulated a second time very soon after the first stimulation while the

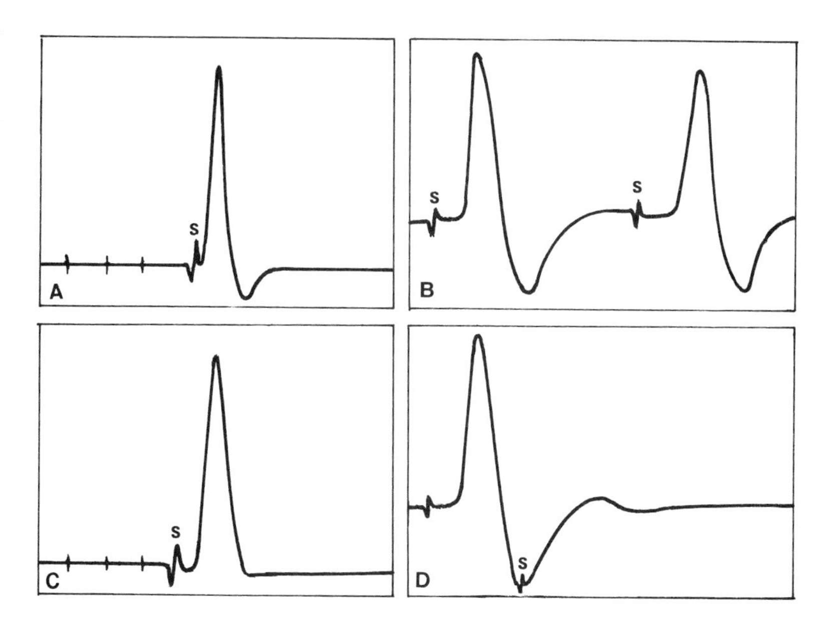

Figure 20.20. *Action potentials (spikes) on a frog sciatic nerve. A, Compound diphasic action potential from the fastest (alpha) fibers of the frog sciatic nerve, recorded by a pair of electrodes 1 cm. apart, located 1 cm. from the stimulating electrode. B, Two stimuli applied to the frog nerve in rapid succession 0.007 second apart. Two diphasic spikes appear. C, Compound monophasic action potential recorded from the nerve in A, at a pair of electrodes 1 cm. apart located 2 cm. from the stimulating electrodes. The nerve has been crushed between the recording electrodes. Note that the first phase of the action potential is now broader than in A, because of the elimination of the second electrode response, which in A hastens the falling phase. Markers for A and C indicate 1 millisecond. D, Two stimuli given 0.003 second apart. The nerve is unable to respond to the second stimulus because it is still in the refractory phase. In all the records the small break in the record is the stimulus artifact. (A and C by courtesy of D. Kennedy, B and D redrawn from Katz, 1952: Sci. Am. (Nov.) 187: 55.)*

nerve is still in a refractory state, no response occurs (Fig. 20.20*D*). However, if the nerve is stimulated again after it has recovered, a second spike similar to the first is obtained (Fig. 20.20*B*). If the nerve is crushed between the recording electrodes the response is *monophasic* (Fig. 20.20*C*) because the action potential does not reach the second stimulating electrode.

LITERATURE CITED AND GENERAL REFERENCES

Abbott, B. C., Hill, A. V. and Howarth, J. F., 1958: The positive and negative heat production associated with a nerve impulse. Proc. Roy. Soc. London *B148*: 149–183.

Abood, L. G. and Gerard, R. W., 1954: Enzyme distribution in isolated particulates of rat peripheral nerve. J. Cell. Comp. Physiol. *43*: 379–392.

Adrian, E. D., 1950: The control of nerve-cell activity. Symp. Soc. Exp. Biol. *4*: 85–91.

Agranoff, B. W., 1967: Memory and protein synthesis. Sci. Am. (Jun.) *216*: 115–122.

Aidley, D. J., 1971: The Physiology of Excitable Cells. Cambridge University Press, Cambridge.

Arshavski, Y. I., 1960: The role of metabolism in the production of bioelectric potentials. Russian Rev. Biol. *50*: 55–68.

Babich, F. R., Jacobson, A. J., Bubash, S. and Jacobson, A., 1965: Transfer of a response to naive rats by injection of RNA attracted from trained rats. Science *149*: 656–657.

Bennett, M. V. L., 1970: Comparative physiology: electric organs.Ann. Rev. Physiol. *32*: 471–528.

Bennett, M. V. L., 1971: Electric organs. *In* Fish Physiology. Vol. 5. Hoar and Randall, eds. Academic Press, New York, pp. 347–491.

Benolken, R. M. and Jacobson, S. L., 1970: Response properties of a sensory hair excised from Venus's flytrap. J. Gen. Physiol. *56*: 64–82.

Berg, H. C., 1975: Chemotaxis in bacteria. Ann. Rev. Biophys. Bioengineer. *4*: 19–136.

Bergstrom, S. R., 1969: Protein chemistry and learning. Mod. Kemi. *10*: 54–55.

Blinks, L. R., 1955: Some electrical properties of large plant cells. *In* Electrochemistry in Biology and Medicine. Shedlovsky, ed. John Wiley & Sons, New York, pp. 187–212.

Bodian, D., 1952: Introductory survey of neurons. Cold Spring Harbor Symp. Quant. Biol. *17*: 1–13.

Brazier, M. A. B., 1968: The Electrical Activity of the Nervous System. 3rd Ed. Macmillan Co., New York.

Brink, F., Jr., 1951: Anesthetizing action. *In* Second Conference on Nerve Impulse Transactions. Nachmansohn, ed. Josiah Macy, Jr., Foundation, New York, pp. 124–175.

Brink, F., Jr., Bronk, D. W., Carlson, F. D. and Connelly, C., 1952: The oxygen uptake of active axons. Cold Spring Harbor Symp. Quant. Biol. *17*: 53–67.

Bullock, T. H., Orkand, R. and Grinell, A., 1977: Introduction to Nervous Systems, W. H. Freeman Co., San Francisco.

Clark, J. and Plonsey, R., 1966: A mathematical evaluation of the core conductor model. Biophys. J. *6*: 95–112.

Connelly, C. M., 1959: Recovery processes and metabolism of nerve. *In* Biophysical Science—A Study Program. Oncley, ed. John Wiley & Sons, New York, pp. 504–514.

Cooke, I. M. and Lipkin, M., Jr., eds., 1972: Cellular Neurophysiology, a Source Book. Holt, Rinehart & Winston, New York.

Cox, R. T., 1943: Electric fish. Am. J. Physiol. *11*: 13–22.

DiCara, L. V., 1970: Learning in the autonomic nervous system. Sci. Am. (Jan.) *222*: 31–39.

Drachman, D. B., ed., 1974: Trophic functions of the neuron. Ann. N.Y. Acad. Sci. *228*: 1–413.

Erlanger, J. and Gasser, H. S., 1937: Electrical Signs of Nervous Activity. University of Pennsylvania Press, Philadelphia.

Ernyei, S. and Young, M. R., 1966: Pulsatile and myelin-forming activities of Schwann cells in vitro. J. Physiol. *183*: 469–480.

Fenn, W. O. and Gerschman, R., 1950: The loss of potassium from frog nerve in anoxia and other conditions. J. Gen. Physiol. *33*: 105–204.

Gerard, R. W., 1932: Nerve metabolism. Physiol. Rev. *12*: 469–592.

Gerard, R. W., 1946: Nerve metabolism and function. A critique of the role of acetylcholine. Ann. N.Y. Acad. Sci. *47*: 575–600.

Greulach, V. A., 1955: Plant movements. Sci. Am. (Feb.) *192*: 100–106.

Gutknecht, J., 1970: The origin of bioelectrical potentials in plant and animal cells. Am. Zool. *10*: 347–354.

Guyton, A. C., 1976: Textbook of Medical Physiology. 5th Ed. W. B. Saunders Co., Philadelphia.

Higinbotham, N., 1970: Movement of ions and electrogenesis in plant cells. Am. Zool. *10*: 393–403.

Higinbotham, N., 1973: Electropotentials of plant cells. Ann. Rev. Plant Physiol. *24*: 25–46.

Hill, A. V., 1933: The three phases of nerve heat production. Proc. Roy. Soc. London *B 113*: 345–365.

Hope, A. B. and Walker, N. A., 1975: Physiology of Giant Algal Cells. Cambridge University Press, New York.

Hyden, H., 1960: The neuron. *In* The Cell. Vol. 4. Brachet and Mirsky, eds. Academic Press, New York, pp. 215–323.

Jack, J. J. B., Noble, D. and Tsien, R. W., 1975: Electric Current Flow in Excitable Cells. Oxford University Press, New York.

Johnson, F. H., Eyring, H. and Storer, D. S., 1974: The Theory of Rate Processes in Biology and Medicine. Wiley-Interscience, New York.

Junge, D., 1976: Nerve and Muscle Excitation. Sinauer Associates, Sunderland, Mass.

Katz, B., 1952: The nerve impulse. Sci. Am. (Nov.) *187*: 55–64.

Katz, B., 1959: Nature of the nerve impulse. *In* Biophysical Sciences—A Study Program. Oncley, ed. John Wiley & Sons, New York, pp. 466–474.

Katz, B., 1966: Nerve, Muscle, and Synapse. McGraw-Hill Book Co., New York.

Keynes, R. D., 1956: The generation of electricity by fishes. Endeavour *15*: 215–222.

Kinosita, H., Dryl, S. and Naitoh, Y., 1964: Changes

in membrane potential and the responses to stimuli in *Paramecium*. J. Fac. Sci. Univ. Tokyo *10*: 291–301.

Kishimoto, U., 1957: Studies on the electrical properties of a single plant cell. Internode cell of *Nitella*. J. Gen. Physiol. *40*: 663–682.

Kishimoto, U., 1972: Characteristics of the excitable *Chara* membrane. Adv. Biophys. *3*: 149–226.

Kobataki, Y., Inoue, I. and Ueda, T., 1975: Physical chemistry of excitable membranes. Adv. Biophys. *7*: 43–89.

Kuffler, S. W. and Nicholls, J. G., 1977: From Neuron to Brain. Sinauer Associates, Sunderland, Mass.

Lapique, L., 1926: L'excitabilite en function du temps. Université de France, Paris.

Lundberg, A. and Oscarsson, O., 1953: Anoxic depolarization of mammalian nerve fibers. Acta Physiol. Scand. *30*(Suppl. *111*): 99–110.

McConnell, J. V., 1966: Comparative physiology: learning in invertebrates. Ann. Rev. Physiol. *28*: 107–136.

Maio, J., 1958: Predatory fungi. Sci. Am. (Jul.) *199*: 67–72.

Melkimyan, V. G., Rogatykh, N. P. and Zubner, T. N., 1976: Informational role of action potentials in cell of *Acetabularia*. Dokl. Akad. Nauk SSSR, Ser. Biol. *224*: 1223–1225.

Nobel, P. S., 1974: Introduction to Biophysical Plant Physiology. W. H. Freeman Co., San Francisco.

Oschs, S., 1972: Fast transport of materials in mammalian nerve fibers. Science *176*: 252–260.

Osterhout, W. J. V., 1936: Electrical phenomena in large plant cells. Physiol. Rev. *16*: 216–237.

Pickard, B. G., 1971: Action potentials resulting from mechanical stimulation of pea epicotyls. Planta *97*: 106–115.

Rasminsky, M. and Sears, T. A., 1972: Internodal conduction in undissected demyelinated nerve fibers. J. Physiol. *227*: 323–350.

Ruch, T. C. and Patton, H. D., eds., 1965: Physiology and Biophysics. 19th Ed. W. B. Saunders Co., Philadelphia.

Salmoiraghi, G. S. and Bloom, F. E., 1964: Pharmacology of individual neurons. Science *144*: 493–499.

Schallek, W., 1949: The glycogen content of some invertebrate nerves. Biol. Bull. *97*: 252–253.

Schmitt, F. O., 1965: The physical basis of life and learning. Science *149*: 931–943.

Schmitt, F. O. and Cori, C. F., 1937: Lactic acid formation in medullated nerve. Am. J. Physiol. *106*: 339–349.

Scott, B. I. H., 1962: Electricity in plants. Sci. Am. (Oct.) *207*: 107–117.

Sibaoka, T., 1962: Excitable cells in *Mimosa*. Science *137*: 226.

Sibaoka, T., 1966: Action potentials in plant organs. Symp. Soc. Exp. Biol. *20*: 49–73.

Slayman, C. L., Long, W. S. and Gradman, D., 1976: Action potentials in *Neurospora crossa*. Biochina. Biophys. Acta *426*: 732–744.

Sleigh, M. A., 1966: The coordination and control of cilia. Symp. Soc. Exp. Biol. *20*: 11–31.

Stevens, C. F., 1966: Neurophysiology: A Primer. John Wiley & Sons, New York.

Suckling, E. E., 1961: Bioelectricity. McGraw-Hill Book Co., New York.

Tasaki, I., 1953: Nervous Transmission. Charles C Thomas, Springfield, Ill.

Tyler, A., Monroy, A., Kao, C. Y. and Grundfest, H., 1955: Electrical potential changes upon fertilization of the starfish egg. Biol. Bull. *109*: 352–353 (Abstract).

Williams, S. E. and Pickard, B. G., 1972: Receptor potentials and action potentials in *Drosera* (sundew) tentacles. Planta *103*: 193–221.

Williams, S. E. and Pickard, B., 1972: Properties of action potentials in *Drosera* tentacles. Planta *103*: 222–240.

Young, J. Z., 1965: Two memory stores in one brain. Endeavour *24*: 13–26.

CHAPTER 21

DEVELOPMENT, PROPAGATION AND TRANSMISSION OF THE ACTION POTENTIAL

Since the action potential is the "language" by which excitable cells communicate and evoke response, its nature, origin, propagation along a cell, and cell-to-cell transmission are of profound interest. A potential already exists between the inside and the outside of a resting cell. How the process of stimulating the cell alters the resting potential and results in an action potential is the subject of this chapter.

A clear and well-illustrated account of the subject for neurons and muscle fibers with references to the original literature is to be found in Kuffler and Nicholls (1977), Chapters 4 to 12. Some studies on plant cells are included here to illustrate the generality of electrical activity in cells (Hope and Walker, 1975).

RESTING POTENTIALS ACROSS THE CELL MEMBRANE

The resting potential is measured by inserting one electrode inside a cell and placing another in the medium (Fig. 21.1). Cells such as the squid giant axon, large muscle fibers, and large plant cells like *Valonia, Halicystis,* or *Nitella* are best for this purpose because of their size. For small cells minute transverse electrodes are being used with success. A resting potential of approximately 10 to 100 millivolts (mv) is registered in most cells. In the squid giant axon and frog muscle fiber the potential is about 70 mv; in *Halicystis,* 70 to 80 mv; in a *Nitella* cell, 100 to 200 mv. In all cases the inside of the cell is negative to the outside, except in one species of *Valonia,* on which the medium is 10 to 20 mv negative to the inside. However, this is not a contradiction and has been satisfactorily explained (Blinks, 1940).

It is probable that the true intracellular potential of cells is somewhat larger than measured by these methods, because cells are injured during electrode penetration and because chances for electrical shorting are considerable. For example, the action potential calculated for the squid giant axon is 75 mv, that of an axon measured intact in the animal is 70 mv, while that for one removed from the body is 60 to 65 mv.

To explain the resting potential of a cell it is useful to consider an inanimate analogy. A potential difference arises when a membrane separating two solutions is selectively permeable to either anions (negatively charged ions) or cations (positively charged ions), and a difference in concentration of some nondiffusible ion exists between the inside and the outside of the membrane (Fig. 21.2). Such a difference in potential is called a *concentration chain potential.*

When an artificial membrane separates solutions containing different concentrations of the same electrolyte, the nature of the potential depends upon the permeability of the membrane to the cations and anions of this electrolyte. Four different cases are described in the following paragraphs and in Figure 21.2.

1. If the membrane is equally permeable to both the cation and anion, e.g., K^+ and Cl^-, the potential is transitory, since the ions ulti-

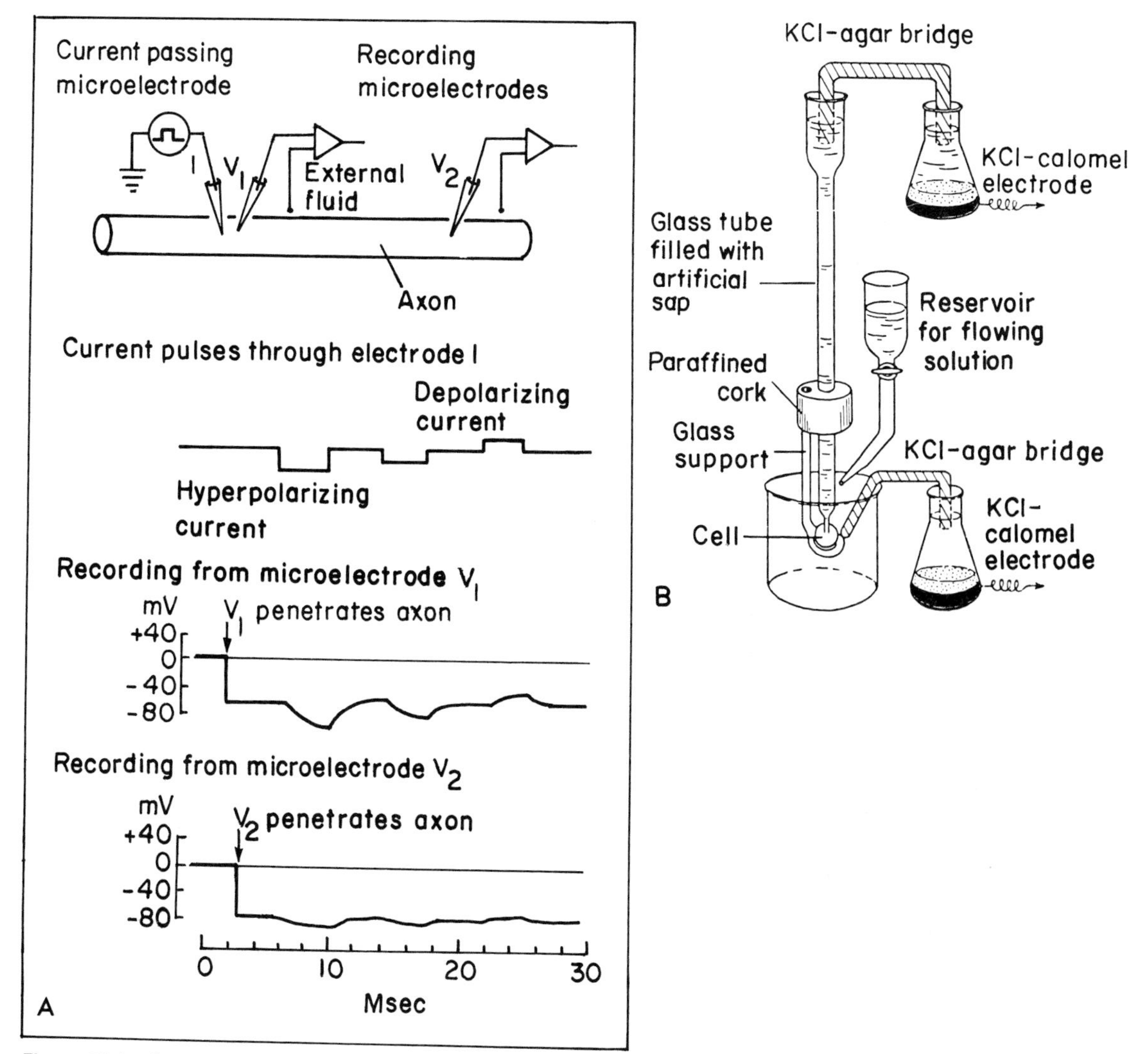

Figure 21.1. *Intracellular recording.* A, *Squid giant axon;* B, *Halicystis, a large algal cell. In* A, *one microelectrode (V_1) is inserted into the axon and records a resting potential of −70 mV (inside negative) with respect to the outside. A second electrode (I), next to V_1, is used to pass pulses of current that produce localized graded potentials. The first two are hyperpolarizations and the third is a depolarization. Electrode V_2, about 1 mm away from V_1, also measures a resting potential of −70 mV when it penetrates the axon. But the localized graded potentials are smaller and slower than at V_1, owing to the passive electrical properties of the axon. (From Kuffler and Nicolls, 1977: From Neuron to Brain, © 1977 by Sinauer Associates, Inc., Sunderland, Mass., p. 80.) In* B, *one electrode is placed inside a "giant"* Halicystis *cell and another electrode is placed on its surface. (From Blinks, 1933: J. Gen. Physiol. 17: 111.)*

mately distribute themselves equally between the two solutions. The mobilities of the various ions determine how long the potential will last. K^+ and Cl^- have almost equal mobilities; therefore, the potential would be slight and would fall much sooner than if NaCl were used on each side of the membrane, since Na^+ is less mobile than Cl^-. Therefore, the side of the membrane with a higher concentration of NaCl would remain positive to the other for a brief but finite time. The greater the difference in mobilities, the longer the temporary potential would last.

2. If the membrane is permeable to cations but not to anions (e.g., to K^+ but not to Cl^-) then a potential develops between the two sides, and the solution of higher initial cation concentration becomes negative to the one of lower concentration. The magnitude of the potential can be calculated from the *Nernst equation:*

$$E = \frac{RT}{n\mathcal{F}} \ln \frac{C_1}{C_2} \qquad (21.1)$$

In this equation E is the potential in volts, R the gas constant (taken as 8.312 joules per degree

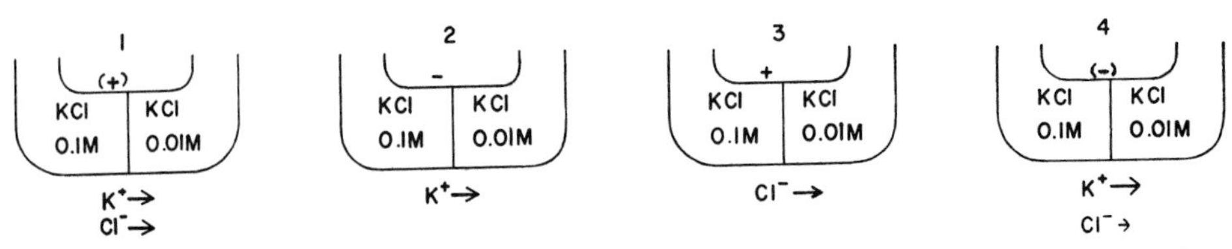

Figure 21.2. *Concentration chain potentials. Potentials develop when solutions of the same electrolyte are present in different concentrations on opposite sides of a membrane (the vertical line in each drawing). The arrows show only the overall movement, since the movement of a given ion is not restricted entirely to passage through the membrane in one direction. The charge on the membrane may be temporary (as indicated by parentheses) or continued, depending upon the conditions. When the membrane is permeable to both ions, as in case 1, the potential is fleeting. When it is permeable only to cations, the solution of higher concentration becomes negative (case 2). When it is permeable only to anions, the solution of higher concentration becomes positive to the other solution (case 3). When the membrane is a mosaic of areas, some permeable to cations and some permeable to anions, the charge depends upon the ratio of the two kinds of areas. If the areas permeable to cations predominate (case 4), the solution of higher concentration becomes negative to the other solution, but only briefly since the concentrations of the two solutions eventually become equalized by the movement of cations and anions from the solution of higher concentration to the solution of lower concentration.*

per mole), T the absolute temperature in degrees, n the valency change, $\mathscr{F}$ the Faraday (96,500 coulombs per gram equivalent), *ln* the natural logarithm (2.3 × $\log_{10}$), C_1 the higher concentration and C_2 the lower concentration (molal) of the electrolyte.

The equation can be simplified by combining the constants R, n and $\mathscr{F}$ and converting *ln* to $\log_{10}$. Then setting up the equation for room temperature, 27°C (300° absolute) and converting to millivolts:

$$E = 59.5 \log_{10} \frac{C_1}{C_2} \qquad (21.2)$$

For a tenfold difference in concentration, $\log_{10} C_1/C_2$ equals 1; therefore, E is 59.5 mv.

An example of such a membrane is one made of celloidin (cellulose nitrate), which bears a negative charge. A potential is established across such a membrane separating two solutions of KCl because K^+ penetrates to some extent, whereas Cl^- does not (Teorell, 1962). The tendency of K^+ to penetrate is expressed by the magnitude of the potential developed.

In general, whenever an indiffusible ion is present on one side of the membrane only, and diffusible ions are present on both sides, the situation is much like the Donnan equilibrium previously formulated for proteins (see Chapter 18).

3. If the membrane is permeable to anions but not to cations (e.g., to Cl^- but not to K^+), a potential develops, but the solution of higher initial anion concentration in this case becomes positive to the one of lower concentration because of the movement of the anions. The tendency for the anions to diffuse is expressed in the potential developed. An example is celloidin membrane made up with the dye rhodamine B, or coated with protamine, which is positively charged and permeable to anions only.

4. If a mosaic membrane permeable to anions in certain portions and to cations in other portions separates solutions of different concentrations, both anions and cations will enter. For example, a membrane composed partly of rhodamine B-celloidin and partly of celloidin will permit anions (e.g., Cl^-) to penetrate the rhodamine B-containing portion and cations (e.g., K^+) to penetrate the celloidin portion. The magnitude and sign of the potential developed depend upon the ratio of the cation-permeable portion of the membrane to the anion-permeable portion of the membrane, upon the rate of penetration of the two ions, and also upon their relative mobility. The greater the mobility of the ions, the greater their penetration through the membrane, and the more nearly equal the areas permeable to them, the more transitory will be the potential. The solution of high concentration will be positive to the one of low concentration if the total membrane area permeable to anions is large and negative if the area permeable to cations is large. The potential will be less than the calculated concentration chain potential between the two solutions because both ions pass from the higher to the lower concentration, reducing the concentration difference. Furthermore, the potential will eventually fall to zero when the two solutions become equalized in concentration.

Origin of Resting Potential Across the Cell Membrane

The resting potential is presumably maintained so long as the cell is alive and active, as was found when the potential was tested periodically for over a week on an impaled living *Halicystis* cell (see Fig. 21.1*B*). This indicates that the source of the potential is also persistent. It is postulated that such a potential

TABLE 21.1. CONCENTRATION OF SOME IONS IN CELLS AND BATHING MEDIA*

Cell	K^+	Na^+	Cl^-	Organic Anions
Nitella[1]	54.3	10.0	90.7	—
Pond water[1]	0.05	0.2	0.9	—
Squid giant axon[2]	400	50	40	381.4
Squid blood[2]	20	440	560	—
Sea water[2]	10	460	540	traces
Crab nerve fiber[3]	112	54	52	—
Crab blood[3]	12.1	469	524	—
Frog muscle fiber[4]	124	10.4	1.5	70
Frog plasma[4]	2.25	109	77.5	12.9†

*Data are in mM per liter (1) and mM per kg water (2, 3, 4).
†Accounting incomplete.
[1]Hoagland and Davis, 1922–23: J. Gen. Physiol. *5*:629.
[2]Hodgkin, 1958: Proc. Roy. Soc. *B148*:1. Slicks of sea water may contain considerable organic matter.
[3]Shaw, 1955: J. Exp. Biol. *32*:383.
[4]Conway, 1957: Metabolic Aspects of Transport Across Membranes. Murphy, ed. University of Wisconsin Press, Madison.

could result from a sustained difference in either cation or anion concentration between inside and outside of cells, as discussed earlier (cases 2 and 3), but not when both ions penetrate the membrane, because then the potential would decline as equilibrium is reached. If the potential across the plasma membrane is due to such a concentration chain, some persistent concentration difference in either anions or cations should be demonstrable. Furthermore, if the concentration of the ion responsible for the resting potential is equalized on the two sides of the plasma membrane, either by washing the cell free of it or by matching the concentration of the ion inside the cell with a solution applied to the outside of the cell, the resting potential should fall to zero.

The concentration of various ions inside cells and in their bathing medium is given in Table 21.1. It will be noted that K^+ is present in a much higher concentration inside the cells than outside (e.g., a *Nitella* cell contains K^+ in the central sap vacuole in a concentration 1065 times as great as in the pond water in which it grows). Although it has been claimed that the potassium in cells may be bound and not exchangeable with that in the medium, the high conductivity of the cells (in which potassium is the main cation) and the ready diffusion of labeled potassium from one end of the cell to the other indicate that, in cells tested, most of the potasssium is in ionic form (Hodgkin, 1951; Hill, 1970).

In *Nitella* Na^+ and Cl^- are also present in much higher concentration in the central sap vacuole than in pond water, but in the frog nerve fiber, the squid giant axon, and in most cells, both these ions are present in much lower concentration inside the cell than in the bathing medium.

Whenever indiffusible anions (or indiffusible cations) are present inside the cell and diffusible anions (or cations) are in the medium, a steady state develops. The Nernst equation can be used to calculate the potential. Since several ions are present in different concentrations on the two sides of the cell membrane, in theory, all the diffusible ions present in the cell and in the bathing fluid to which the cell is exposed should be considered in any equation purporting to relate the concentration differences of ions outside and inside the cell to its membrane potential. However, equation 21.3 (derived by Hodgkin and Katz, 1949) is a version of the Nernst equation (21.1) modified to take into consideration the development of the resting potential of the excitable cell in terms of only the three ions present in highest concentration in the cell and the bathing medium (Hodgkin, 1951; Chandler and Hodgkin, 1965).

$$E = \frac{RT}{\mathscr{F}} \ln \left(\frac{P_{K^+}(K^+)\ in + P_{Na^+}(Na^+)\ in + P_{Cl^-}(Cl^-)\ out}{P_{K^+}(K^+)\ out + P_{Na^+}(Na^+)\ out + P_{Cl^-}(Cl^-)\ in} \right) \qquad (21.3)$$

where E is the potential across the cell membrane, P_K, P_{Na}, and P_{Cl} refer to the permeability coefficients of K^+, Na^+, and Cl^- respectively, and "in" and "out" refer to the inside and outside of the cell.

This formulation assumes that the structure of the membrane is stable and that changes in the ionic environment in no way affect the permeability characteristics of the membrane. It is therefore an approximation, since it is known

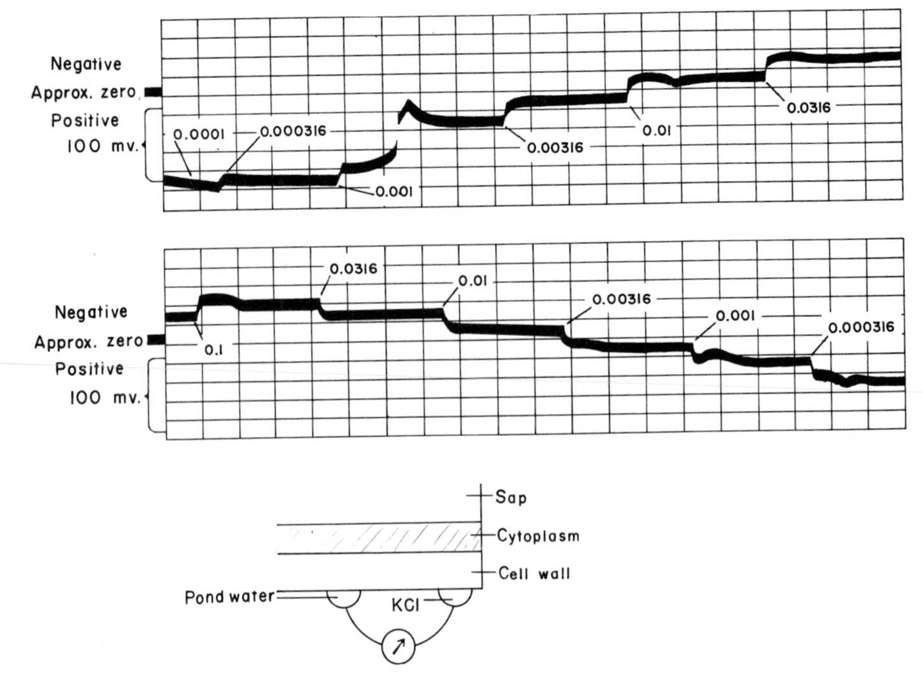

Figure 21.3. *Changes in potential produced in a* Nitella *cell when pond water and different concentrations of KCl are applied to the cell wall in the manner shown in the diagram beneath the graphs. The figures on the graphs indicate the molal concentrations of KCl applied. In the upper graph the potential is progressively decreased as the concentration of KCl is progressively increased, until at 0.0316M the potential is reversed. The lower graph shows the reverse effect as the concentration of KCl on the outside is lowered. The figure below shows the arrangement of the recording instrument (with arrow) and the two electrodes, one in KCl, one in pond water. (After Hill and Osterhout, 1937–38: J. Gen. Physiol. 21: 541.)*

that certain ions actually do affect the plasma membrane. Increase in outside K^+ concentration results in a small decrease in the resistance of the cell membrane; increase in Ca^+ has the reverse effect (Gerebtzoff and Schoffeniels, 1960).

Equation 21.3 may be simplified by dropping out the term for Na^+, since permeability for Na^+ in the resting state is equal only to about 1 percent of the permeability for K^+. Furthermore, when the Na^+ concentration of the bathing medium is varied over a wide range, the resting potential of the cell is not affected, indicating that the Na^+ gradient across the cell membrane plays little, if any, role in establishing and maintaining the resting potential (Katz, 1959a). If the permeability of the cell membrane to K^+ and Cl^- is assumed to be about equal in both directions, the permeability constants in the equation cancel out, leaving only the terms for concentrations of K^+ and Cl^- inside and outside the cell, the ones for K^+ being of greater importance in most cells (Baker, 1966).

Increasing the concentration of K^+ in the medium of *Nitella* causes a progressive decline in its resting potential, which is abolished when the K^+ concentration in the medium equals that inside the cell. It is even possible to reverse the sign of the resting potential if the concentration of K^+ becomes greater in the medium than inside the cell (Blinks, 1955). The effects are completely reversible, indicating that the cells remain healthy during the tests (Fig. 21.3).

Since K^+ concentration inside most cells is higher than in the medium bathing them, it is tempting to postulate that it acts in other cells as in *Nitella*. In fact, experiments have shown that increasing concentrations of K^+ applied to the outside of the squid giant axon or frog sartorius muscle fiber or frog myelinated nerve fiber (Hodgkin *et al.*, 1949) reduce the resting potentials in these cells as in *Nitella*. However, the experimental results, especially for muscle fibers, are in best agreement with theory if the Cl^- concentrations are also taken into account, indicating that the Cl^- contributes to the net potential, even though its share in most cells is small.

In one experiment the axoplasm of a squid giant axon was replaced by a solution containing K^+ and anions of varying concentration. When the gradient of K^+ between the inside and outside of the axon approximated that between axoplasm and blood, the resting potential was found to be only slightly less than that in an isolated but unaltered axon. A reduction

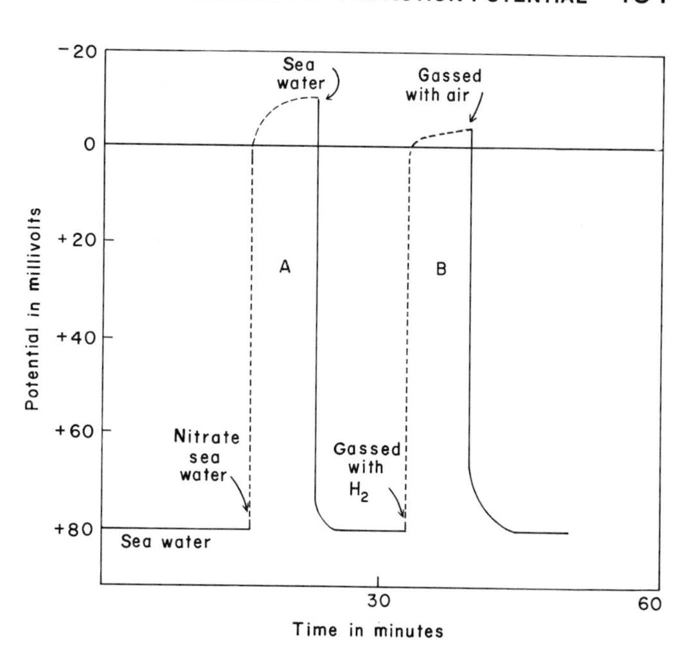

Figure 21.4. *The potential of* Halicystis *as affected by the replacement of the chloride of sea water with nitrate.* B, *The potential of* Halicystis *as affected by the removal of oxygen. The dotted line in* A *shows the fall in potential when* Halicystis *is placed in nitrate sea water. The dotted line in* B *shows the fall in potential when* Halicystis *is freed of oxygen by being gassed with hydrogen. The solid vertical line in* A *shows the return of the potential to the original resting value when the cell is returned to sea water. The solid vertical line in* B *shows the return of the potential when the cell is supplied with oxygen by being gassed with air. In each case the cell was set up as illustrated in Figure 23.1. (After Blinks, 1940: Cold Spring Harbor Symposia* 8: *204.)*

of the potassium gradient, however it was achieved, caused a proportional decrease in the "resting potential" in accordance with the Nernst equation. By making the K^+ concentration outside greater than inside the axon, the potential could even be reversed. Changing the kind of anion (but not its concentration) had little effect on the potential (Baker *et al.*, 1961; Baker and Connelly, 1966).

In most cells K^+ and Cl^- appear to be very nearly in electrochemical equilibrium, the large concentration difference between the inside and the outside of the cell for each of these ions being very nearly balanced by the charge on the cell membrane.

Both the concentration gradient and the electrochemical gradient tend to bring Na^+ into the cell. Sodium ion has been shown to enter the cell slowly, yet in the resting cell its concentration remains low. Therefore, it is postulated that Na^+ is extruded by cells (see Chapter 18). Na^+ extrusion from cells (by the sodium pump) appears to be linked with active uptake of K^+ from the medium by what appears to be a type of cation-exchange process. It has been shown, for example, that extrusion of Na^+ becomes reduced greatly when the K^+ content of the medium is reduced, and in general, the rates of both processes change simultaneously and in a parallel manner and appear to be linked by a common carrier (Katz, 1966).

It is of interest to note that in one species of *Halicystis* the concentration of potassium in the sap is about the same as that in sea water, yet the outside of the cell is 70 mv positive to the inside. The potential persists even if the vacuolar content is removed and sea water is substituted. It is evident that resting potentials in such cells must be explained on some basis other than by analogy with the potassium concentration.

To determine whether anion gradients could explain the resting potential in this case, tests were made with diverse anions. The anions inside a living cell are largely organic ions resulting from metabolism—such as acetate, pyruvate, lactate, and amino acids—plus some inorganic ions such as sulfate and phosphate, all of which pass through the cell membranes very slowly. These ions, collectively called X^-, are absent from the natural bathing medium, which is rich in chloride ion. Chloride ion penetrates through cell membranes relatively readily but is always present inside cells in low concentration. A greater concentration of chloride ion outside than inside the cell favors its inward penetration from the medium, leaving the outside of the cell positively charged. If the reasoning is correct, substitution of chloride ion in the bathing medium with a very slowly diffusing or nondiffusing anion (nitrate) should reduce or abolish the potential. This was verified in experiments on *Halicystis* (Blinks, 1955) (Fig. 21.4).

Relation of the Resting Potential to Metabolism

Maintenance of the resting potential of the cell has been found to depend upon the presence of oxygen. Thus, when free oxygen is lacking in *Halicystis,* the resting potential falls (Fig. 21.4*B*), but it rises again when oxygen is readmitted or the plant is illuminated (Blinks, 1955; Gaffey and Mullins, 1959). Lack of oxygen reduces the resting potential of the squid giant axon and other neurons, and the reduction is accompanied by a loss of potassium (Hodgkin and Keynes, 1953). Since glycolytic metabolism occurs in the absence of oxygen (see Chapter 12), the nerve cell continues to liberate energy, but does so at a reduced rate, as indicated by a decrease in liberation of heat. Inhibition of glycolysis by poisons specific to glycolytic processes (e.g., iodoacetic acid, fluoride, or phlorhizin*) should reduce the resting potential even more. Experiments verify this prediction, although in no case is the resting potential abolished completely so long as the nerve cell remains alive. The residual potential probably depends upon physical phenomena such as the Donnan equilibrium of the proteins. On the other hand, in an anoxic muscle fiber poisoned with iodoacetic acid, the potential falls to zero in 2 to 3 hours (Hodgkin, 1951).

The very gradual decline in the membrane potential of cells in which metabolic poisons have eliminated energy sources for the sodium pump indicates how slow transmembrane leakage really is. The decrease in potential is attributable to the gradual diminution of the ionic gradients (Connelly, 1959).

To sum up, some cells maintain across the cell membrane a resting potential between the inside and the outside of the cell on the basis of differences in concentration of some cations, whereas others do so on the basis of differences in concentration of some anions. Since metabolism appears necessary to maintain the resting potential of a cell, it is inferred that metabolic energy is needed to maintain the difference in concentration of ions between the outside and the inside of the cell. Inasmuch as the polarized state is necessary for excitability, the cell must pay for maintaining its state of excitability by continuously expending energy (Fig. 21.5).

*These poisons act in very different ways. Iodoacetic acid inactivates sulfhydryl groups—specifically, in this case, triosephosphate dehydrogenase (see Table 12.2). Fluoride inhibits enolase. Phlorhizin inhibits glucose uptake.

The cost of maintaining an excitable state is suggested by the following calculation, based on the Nernst equation (21.1). The energy (W) required to move 1 mole of Na^+ from inside (in) to outside (out) of a two compartment system, and 1 mole of K^+ in the opposite direction, is:

$$W = \frac{RT}{\mathscr{F}}\left(ln\,\frac{Na^+_{out}}{Na^+_{in}} + ln\,\frac{K^+_{in}}{K^+_{out}}\right) \qquad (21.4)$$

where Na^+_{out}, Na^+_{in}, K^+_{out}, and K^+_{in} are ion concentrations (or activities, strictly speaking). It is found that for the concentration gradient of Na^+ and K^+ found in the nerve fiber, about 3000 cal/mole are required to move a mole of Na^+ and K^+. If about 7300 cal/mole are available from the hydrolysis of 1 mole of ATP or ADP, it should be possible to move several moles of Na^+ for each mole of ADP formed. For frog nerve fibers at rest, the oxygen uptake is 1.5 μmole/gram (wet)/hour, and the Na^+ flux is about 6.6 μmole/gram (wet)/hour. Thus, the ratio of Na : O (oxygen atom) is 2.2. The ratio of ADP converted to ATP per oxygen atom reduced (P : O) in mitochondria is about 3. It would thus appear that only about 30 percent of the resting metabolism is available for work other than ion transport (Connelly, 1959).

A similar calculation for nerve fibers repeatedly stimulated to produce action potentials for an hour at 50 impulses per second shows that even with the severalfold increase in respiratory metabolism, more than two thirds of the energy made available (as high-energy phosphate) from oxygen consumption is probably devoted to ion transport (Connelly, 1959).

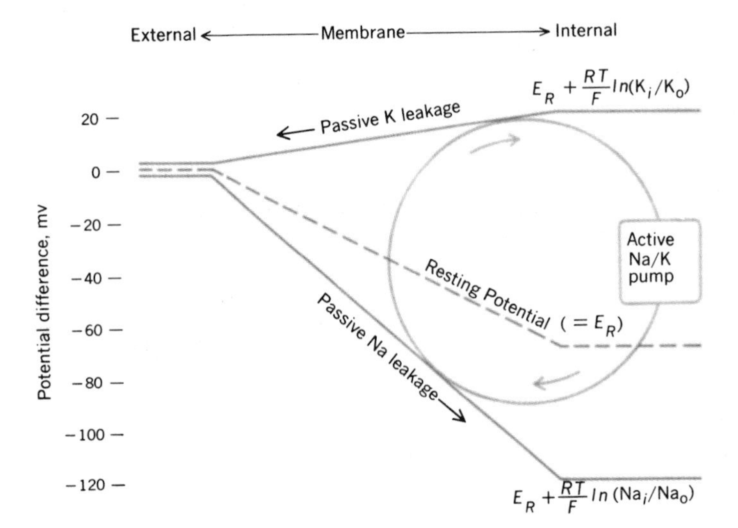

Figure 21.5. Electrochemical (downhill) gradients and ionic (uphill) pumps. Electrochemical potential differences cause K to leak outward and Na to leak inward through the cell membrane. Ionic distribution is maintained by an active secretory process, requiring continuous supply of energy. Note: The electric potential gradient across the membrane is indicated by the broken line. (From Katz, 1966: Nerve, Muscle and Synapse. McGraw-Hill Book Co., New York.)

ORIGIN OF THE PROPAGATED ACTION POTENTIAL

How is the resting potential of an excitable cell altered during the development of an action potential? The positive charge on the outer surface of the cell membrane, because of the presence of positively charged ions, is delicately balanced by the negatively charged anions of salts, organic acids, and so on, inside the cell. Application of a stimulus through an electrode disturbs the resting arrangement and distribution of these ions. If the membrane is slightly disturbed, only a local potential develops, the disturbed area becoming negative to the surrounding, more positive region on the cell membrane. It is thought that at this time sodium ions enter the cell in small numbers. This local potential remains near the point of stimulation only briefly and the ions soon redistribute themselves to the steady-state pattern of a resting cell. The stronger the stimulus, the larger the local potential developed and the greater the area of disturbance on the cell membrane. If a local potential on a neuron reaches one third to one fifth the height of a spike, it suddenly develops into a spike which passes the full length of the axon in the all-or-none manner of a propagated action potential (see Figs. 20.4 and 21.6). It is estimated that about 6×10^9 electrons are needed to stimulate a nerve cell.

That stimulation of an excitable cell produces a profound change is seen by the altered electrical properties of the cell's membrane. Thus, upon stimulation the electrical resistance in a *Nitella* cell falls from 100,000 to 500 ohms per square centimeter, and in the squid giant axon it falls from 1000 to 25 ohms per square centimeter (Hodgkin, 1951; Katz, 1966). Furthermore, potassium is lost in measurable quantities from squid giant axons and *Nitella*. Both lines of evidence indicate that on stimulation the cell membranes of these excitable cells have become much more freely permeable to certain ions. A free interchange of ions between the outside and the inside of a cell at the point of stimulation would depolarize and abolish the resting potential in that portion of the cell. Recording with electrodes inside some excitable cells has demonstrated that the point of stimulation on the membrane of the excited cell not only loses its charge (depolarizes) but reverses charge and becomes negative to the inside (Fig. 21.7). It must be remembered throughout, particularly when studying figures such as 21.7, that the resting potential may be stated either as positive when one considers the outside relative to the inside (see preceding sentence) or negative when one considers the inside relative to the outside (see Fig. 21.7). Either point of view is correct, with the magnitude of the change in potential being the real concern. Measurements on the squid giant axon indicate that the stimulated portion of the surface of the nerve fiber becomes almost as negative as it was previously positive. This has been verified on many animal celis but does not occur in cells of *Nitella,* in which only depolarization occurs upon stimulation (Hodgkin, 1951).

By use of radioactive tracers the movements of sodium and potassium ions across the membrane of the nerve fiber have been followed during stimulation and recovery of the squid giant axon (Hodgkin and Huxley, 1952). At the point of stimulation the permeability to Na^+ is suddenly increased several hundred times by the opening of some 20 to 500 sodium gates per square μm of membrane and reaches its peak in 100 microseconds. The potential gradient across the cell membrane is thus temporarily abolished at the point of stimulation, the outside surface becoming first depolarized and then developing a negative charge by entry of

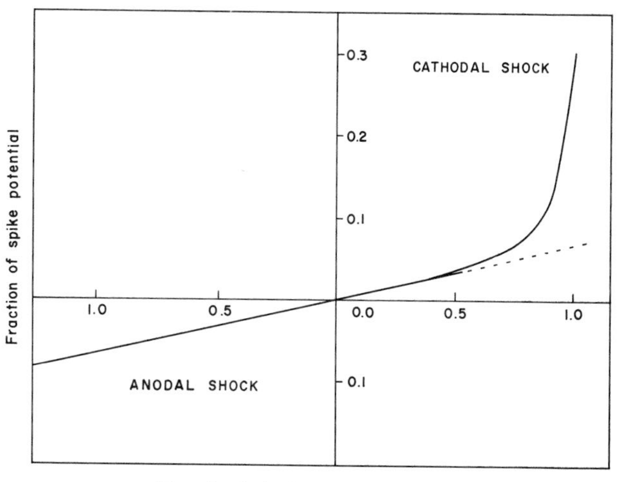

Figure 21.6. *Relations between electrotonic potentials and spike potentials on a nerve cell. The spike does not arise from anodal shock even when the shock is very large, but it does arise from cathodal shock when a given fraction (about 0.3) of the spike potential has been reached. Each potential was measured 0.29 millisecond after the shock. The dotted line in the cathodal shock quadrant is an extrapolation showing the potential that would be developed if a spike did not arise. Threshold is about one fifth to one third depolarization (see Fig. 20.4). (After Hodgkin, 1938: Proc. R. Soc. B126: 96.)*

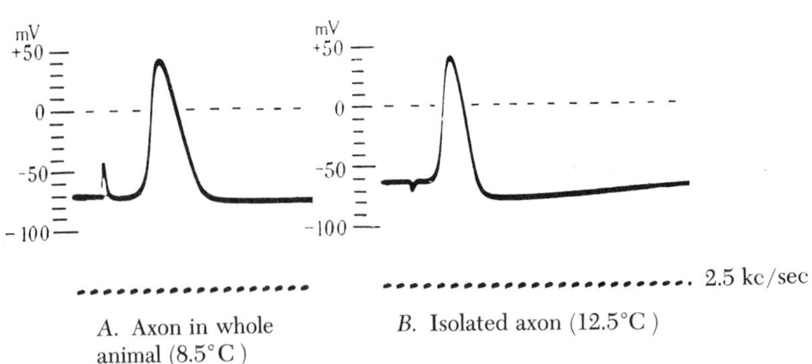

Figure 21.7. *Potential of the inside relative to the outside of an excitable cell during spike formation. A is from an intact giant axon of the squid* (Lologo forbesi), B *is from an isolated squid giant axon. (From Hodgkin, 1958: Proc. R. Soc.* B148: *1.)*

Na^+ into the cell. At the close of this period (the peak of the spike) the membrane again becomes essentially impermeable to sodium, but its K^+ permeability increases and K^+ leaks out from the point of stimulation. The leakage of K^+ repolarizes the cell and the spike potential falls. Since K^+ leaks out at a slower rate than Na^+ moves inward, the repolarization of the membrane to resting potential is slower than its depolarization. Presumably, at the end of the spike, the sodium pump once more comes into operation, as does the coupled active transport of K^+ into the cell, and the membrane begins to recover its normal permeability to ions at this point (Cross *et al.*, 1965).

The spike cannot be explained simply by temporary suspension of the sodium pump because, when the latter is suspended by metabolic poisons, the action potential of the cell changes only very slowly and gradually. The action potential is completely independent of energy-liberating processes (Kuffler and Nicholls, 1977). The events that transpire during spike development involve large numbers of ions and are extremely rapid, but neither the nature of the changes in the cell membrane resulting in increased permeability nor the nature of the restorative processes is fully understood.

However, it should not be inferred that sodium ions, which have entered the cell during depolarization, will at once be expelled or that potassium ions that have left the cell will have to be regained by the nerve fiber before it can recover. The quantity of ions lost (about 3 to 4 $\times$ 10^{-12} M per square cm of membrane) during a single impulse is but a small fraction (about 0.0001 percent in the squid giant axon) of that present in the nerve fiber. Even if the sodium pump and the associated uptake of potassium ions did not become active again, the squid giant axon could be stimulated thousands of times before the ion stored would become exhausted. But to recover its sensitivity, the membrane of the nerve fiber must first shut off the sodium influx as the potassium efflux repolarizes it. If the nerve fiber is not stimulated again, the sodium pump presumably resumes operation and the accumulated Na^+ is voided, while the lost K^+ is simultaneously regained. For these and other fundamental studies on nerve action potentials Hodgkin and Huxley, jointly with Eccles, received the Nobel Prize in 1963.

Nerve cell dendrites generally show local, nonpropagating, decremental potentials. In the dendrites both Na^+ entry and K^+ loss occur simultaneously, in constrast to the ionic movements that occur at the nerve axon. The degree of response is determined by the number of channels open for Na^+ and K^+ and the duration and extent of each channel opening. Such graded, nonpropagating potentials are of importance in receptor-sensory neuron systems as well as in synapses, making possible the effective passage of information as well as integration of the information.

That sodium is the conductor of current during the rise of the action potential is demonstrated by substituting for it, in the medium bathing the neuron, an organic cation, choline chloride, that does not penetrate the cell as readily as Na^+ does. The decrease in height of the spike is then proportional to the decrease of Na^+ in the bathing fluid. The sodium ion is thus as vital to the height of the spike in most cells as is the potassium ion to the resting potential of cells. However, Na^+ is apparently not always the ion that carries the current. In some algae (e.g., *Chara*) it appears that leakage of Cl^- from the cell serves the same purpose. In some crustacean muscle fibers (e.g., barnacles), mammalian cardiac muscle fibers, and some invertebrate neurons, evidence exists for influx of calcium ions instead of sodium ions (Kuffler and Nicholls, 1977).

Tetrodotoxin, a neurotoxin from the Japanese puffer fish, in low concentration blocks Na^+ permeability in nerve and muscle cells but does not affect K^+ movement. Consequently, it blocks spike formation in muscle and nerve cells, which depend upon Na^+ for carrying the inward current during excitation. It is without effect in those cells in which calcium plays this role. The receptors for tetrodotoxin appear to be on the outer surface of the excitatory cells, since injection of the poison inside such cells does not produce a block. It has been shown that tetraethylammonium salts selectively block the potassium channel without affecting the sodium channel. The separateness of the Na^+ and K^+ gates, observed in studies already discussed, is thus corroborated (Kuffler and Nicholls, 1977).

THEORY OF THE PROPAGATION OF THE ACTION POTENTIAL

It is to be expected that local currents will flow from the unstimulated, positively charged areas of the cell membrane to the stimulated, depolarized, or negative portions, and as each new area becomes depolarized or negative, it in turn acts as the sink toward which the current flows from the adjacent area of a neuron and serves to give rise to a new regenerative impulse (Fig. 21.8). Progressive depolarization, or reversal of charge, of the entire length of the neuron therefore follows from the point of stimulation outwardly, as if an electrode were traveling along the nerve fiber at the rate of conduction characteristic of the fiber (Brazier, 1968). It has been demonstrated that the source of the current is the flow of Na^+ into the cell. This, briefly, is the so-called *local circuit theory* of propagation of the action potential.

Repolarization of the surface of an excited fiber with a positive charge sets in as soon as the absolute refractory period is over, and it progresses so rapidly that never more than a fraction of a nerve or muscle fiber is depolarized at a time. However, this may be a region of a few millimeters to a few centimeters, depending upon the duration of the spike and its speed of propagation. Most likely recovery occurs at the expense of high-energy phosphate bonds, since the concentration of creatine phosphate declines after stimulation. Because a neuron can continue to discharge for some time in the absence of free oxygen, even when glycolysis is inhibited by a poison, a store of high-energy phosphate bonds must be available for the purpose. These are presumably rebuilt at the expense of the metabolic activities of the neuron—slowly in the complete absence of oxygen and much more rapidly in the presence of oxygen. Experiments with tracers indicate

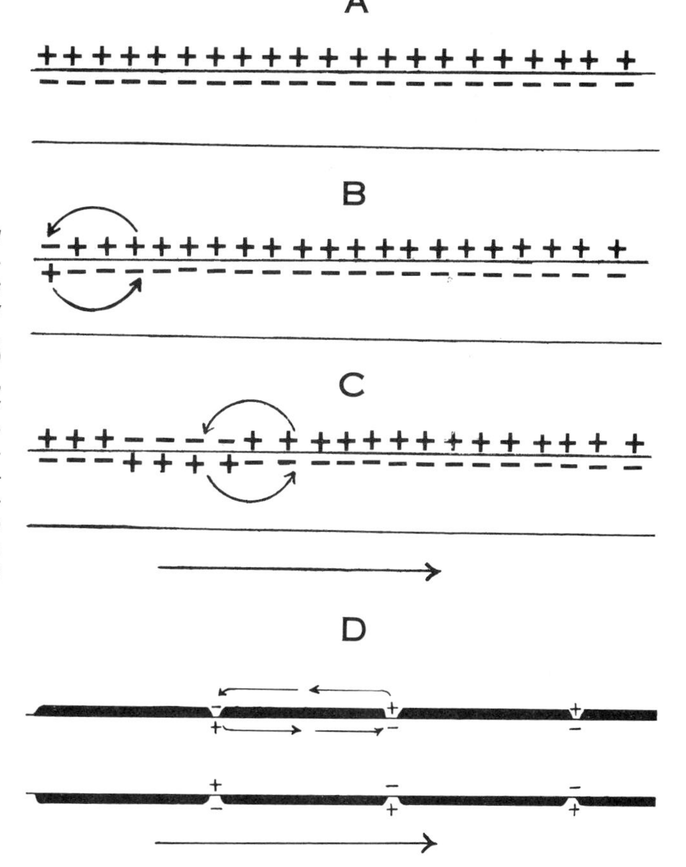

Figure 21.8. *Diagram to illustrate the local circuit theory of propagation of the action potential* (A, B *and* C) *in unmyelinated neurons and muscle fibers as compared to the saltatory conduction in myelinated neurons* (D). A, *Membrane of an unexcited nerve (or muscle) fiber;* B, *cell membrane excited at one end;* C, *movement of the action potential, followed by recovery;* D, *node-to-node saltatory conduction. In large nerve fibers less than one-hundredth as much ionic exchange occurs during an impulse in saltatory conduction as compared to conduction in an unmyelinated nerve fiber. The arrows in* C *and* D *show the direction of impulse propagation. (After Hodgkin, 1957: Proc. R. Soc.* B148: *1.)*

that, during recovery, Na^+ leaves the cell and K^+ reenters (Keynes, 1949). Work that requires metabolic energy must be done during recovery of a stimulated cell, as is indicated by the evolution of heat.

It is quite evident that various properties of the propagated action potential (nerve impulse) may be interpreted on the basis of the local circuit theory of propagation. Thus, when either a stimulus below rheobase is used or a stimulus above rheobase is applied for an inadequate length of time, the membrane is not altered enough to produce a sink, and the ions soon redistribute themselves into the normal pattern for the resting potential of the unstimulated cell. All-or-none response and conduction without a decrement are to be expected of a nerve fiber if the membrane of the entire fiber is initially equally polarized (charged). The reaction is self-generating if the same amount of electrical charge occurs along the entire length of the nerve fiber. The absolute refractory period is a consequence of the closing of the sodium gate and the opening of the potassium gate. During this period the membrane potential is driven toward the potassium potential below the resting level (more negative to the outside than at rest) and the recovery mechanism is temporarily inactive (Katz, 1966). Consequently, the neuron is inexcitable. The initial state of polarization is presumably regained during the relative refractory period. The brief states of increased and, later, of decreased excitability (as seen in vertebrate A fibers) occur

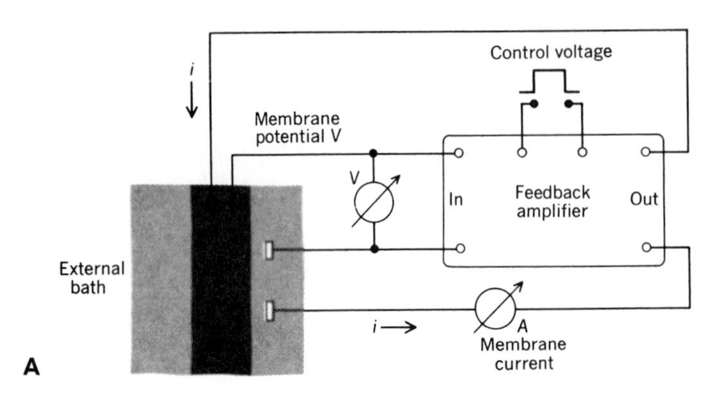

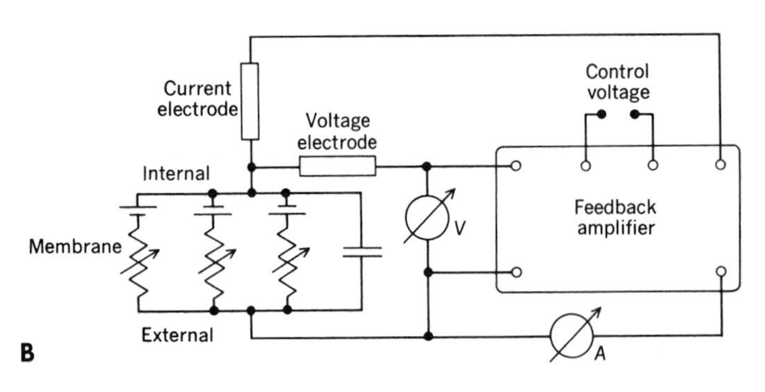

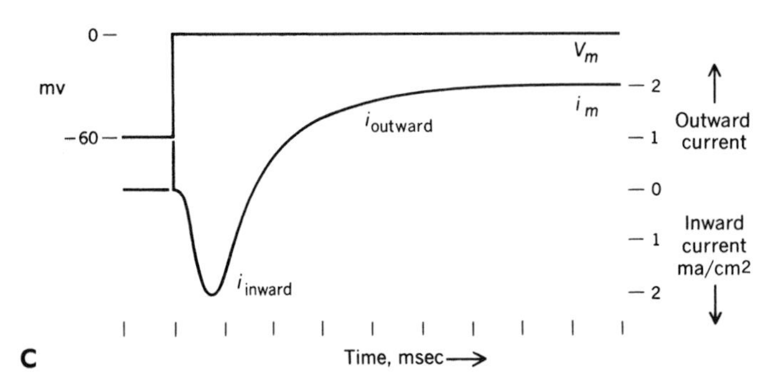

Figure 21.9. *Diagram of "voltage clamp" (A and B) and of the measurement of membrane current during voltage displacement (C). V_m is the membrane voltage, $i_{outward}$ is the outward current, i_m is the membrane current, i_{inward} is the inward current, ma is milliamperes. (From Katz, 1966: Nerve, Muscle and Synapse. McGraw-Hill Book Co., New York.)*

probably because the membrane has not fully regained normal permeability. Therefore, a response may be elicited, but the threshold for stimulation during the period of decreased excitability is higher.

The strongest support for the sodium theory of the action potential is provided by controlling the membrane potential and measuring the current flow (voltage clamp experiment). For this a feedback amplifier is connected by two electrodes on either side of the membrane to automatically supply the current needed to shift the membrane potential and maintain it at any desired level (Fig. 21.9). The membrane can be completely depolarized. The accompanying current flow then shows three components: a transient outward current, a transient inward current, followed by a persistent outward current that continues as long as the membrane remains depolarized. So long as the membrane is depolarized, any current flow observed must be attributable to ions entering or leaving the axoplasm through the membrane. Therefore, the inward current flow must be due to the cations entering or anions emerging from the axoplasm. This phase is converted into the persistent outward current, which was shown by tracers to be attributable to the efflux of potassium ions from the axoplasm. Calculations using the Nernst equation showed that the inward current was very close to that expected on the basis of sodium entry. The voltage clamp was then set at various levels and the conductances were determined. From these measurements of the increase in sodium permeability, its decline, and the increase in potassium permeability, empirical equations for the dependence of these processes on the membrane potential were developed. These equations were then used to predict the changes in membrane potential in response to stimulation under normal conditions (Hodgkin and Huxley, 1952; Katz, 1966). The correspondence between the two was close (Fig. 21.10).

The neuron cell membrane registers many events, and its exact state and response are largely dependent upon its history. For example, in A fibers after-potentials become prominent during tetanic stimulation. Should the stimuli arrive before the neuron has recovered entirely, when the neuron is still somewhat depolarized, the spike appears to be higher by the amount of the negative after-potential. The subsequent hyperpolarization is also gradually augmented by frequent stimulation, and the neuron becomes less excitable because a greater stimulus is required for depolarization (Erlanger and Gasser, 1937). Furthermore, subthreshold stimulation by a steady cathodal current lowers the excitability because it leads to accommodation when the rate of displacement of ions is equaled by that of recovery. When a neuron is depolarized by asphyxia, it is rendered incapable of response, yet it can be repolarized by a battery current and induced to respond at least once (Bishop, 1951). A state of polarization seems to be necessary for maintaining excitability.

Saltatory conduction, as discussed previously for myelinated nerve fibers (see Fig. 21.8*D*), is thought to be the result of local cir-

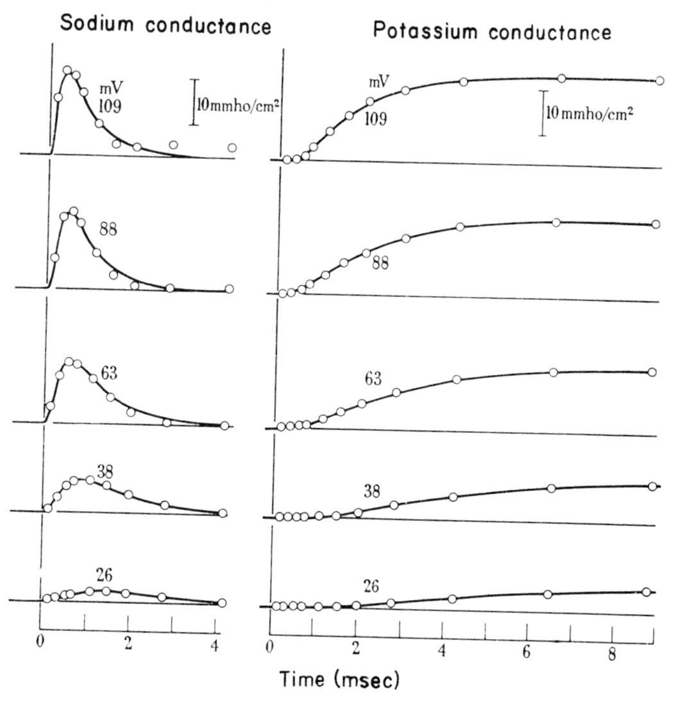

Figure 21.10. *Changes in sodium and potassium conductance associated with different depolarizations at 6°C. The numbers attached to the curves give the depolarizations used. The circles are experimental estimates and the smooth curves are solutions of the equations used to describe the changes in conductance. The conductance is given in millireciprocal ohms (mmho/cm²) because in Ohm's law the current is the reciprocal of the resistance in ohms. (From Hodgkin, 1958: Proc. R. Soc. London* B148: *1.)*

cuits occurring only from node to node. The amount of naked axon is very slight even at the nodes, perhaps a few tens of nanometers at most, because, according to electron micrograph studies, the Schwann cells of the myelin sheath send finger-like growths along the axon. The amount of Na^+ and K^+ exchange is thus greatly reduced and the net work required of the cell for recovery is thereby much lowered. However, the fact that in a demyelinated nerve fiber propagation of an impulse still occurs in saltatory manner presents a puzzling problem (Rasminsky and Sears, 1972).

SENSORY FIBER ACTION POTENTIALS

Receptor cells have much lower thresholds to stimuli than do other cells; for example, a pressure receptor is much more sensitive to pressure than neurons in a nerve trunk. Sometimes the distal branches of the afferent neuron are modified to serve as receptors (e.g., muscle sense organs and pain receptors). Sometimes the entire cell serves this purpose, as in the olfactory epithelium; sometimes epithelial cells innervated by afferent neurons serve as receptors, as in taste buds. Stimulated receptor cells excite adjacent afferent (sensory) neurons, which inform the central nervous system about the strength of the stimulus (Fig. 21.11). Since Adrian's classic experiments many decades ago, it has been recognized that varying the strength of a stimulus (e.g., pressure on a pressure receptor) affects only the frequency of its discharge, not the size of the impulses sent into the central nervous system. The neuron code, then, appears to be frequency modulation of a single type of message for a given neuron. For his many contributions to nerve physiology Adrian shared the 1932 Nobel Prize with Sherrington.

In 1950 Katz demonstrated that stretching a muscle mechanoreceptor, which is a modified part of the fibers of an afferent neuron, generates in the sensitive fibers a local, graded, decremental potential, the magnitude and duration of which depend upon the intensity and duration of stimulation. This local *generator potential,* when above threshold, sets off a train of propagated action potentials on the axon of the afferent neuron. Discharges continue on the afferent axon as long as the depolarization of the receptor fibers remains above threshold. It was soon demonstrated that upon stimulation similar local generator potentials occur in other types of receptors as well.

A pressure receptor consists of a structure enclosing the distal fibers of an afferent neuron. By removing parts of the receptor, an attempt has been made to determine its sensitive region. It was found that when only a core of the receptor remains around the nerve fiber, stretching or deforming the core induces trains of impulses along the attached fiber but that when the enclosed nerve fiber ending itself is damaged, discharge stops. It would appear that the *biological transducer,* which transforms the mechanical energy into electrical energy and nerve impulse, is the terminal portion of the neuron. Evidence exists that following stimulation of such a receptor neuron, the electrical membrane resistance of its distal portion is reduced, permitting ions to enter, partially depolarizing its membrane. When depolarization has reached threshold, propagated action potentials are initiated on the afferent nerve fiber (axon). The local depolarization characteristic of a receptor is spoken of as a generator potential, because it generates the propagated action potential on the axon.

The excitation of an afferent axon does not seem to differ significantly from excitation of any nerve fiber or excitable cell. However, a fundamental difference apparently exists between electrical stimulation of a nerve fiber and stimulation through receptor cells. A single electrical stimulation above threshold applied to an afferent nerve fiber results in a single spike. However, even brief stimulation of a receptor almost invariably results in a

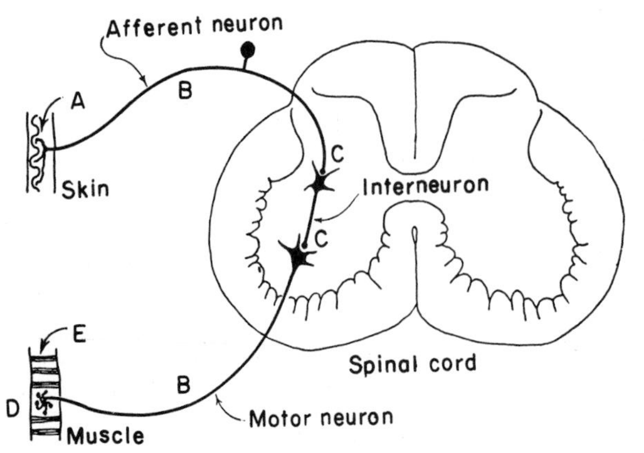

Figure 21.11. *Diagram of a simple reflex arc.* A *shows the receptor at which excitation of afferent neurons occurs. The impulses generated at* A *are conducted to the central nervous system by the afferent neuron, in the cord by interneurons and out to the effector muscle* (E) *by the motor neuron (efferent neuron). Synaptic transmission occurs at* C *and neuromuscular transmission at* D. *(From Ruch and Patton, 1965: Physiology and Biophysics. 19th Ed. W. B. Saunders Co., Philadelphia.)*

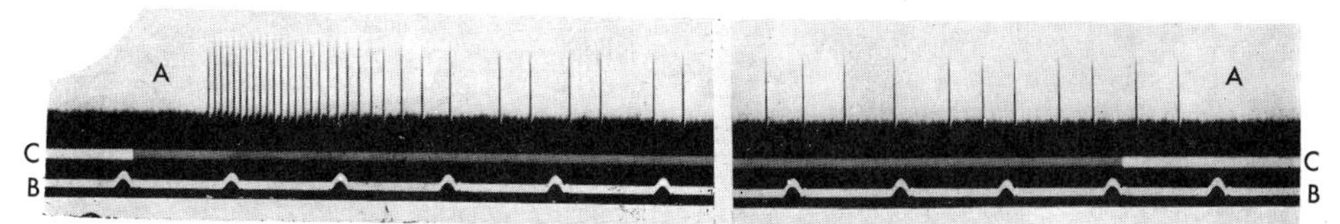

Figure 21.12. *Train of impulses* (A) *on an afferent neuron. Single nerve fiber from an ommatidium (compound eye unit) of the horseshoe crab* Limulus. *The lower notched markings* (B) *are fifths of a second. The line above the time signal* (C) *shows when the light went on and off. The vertical break represents a time lapse of 2.4 seconds. (From Hartline and Graham, 1932: J. Cell. Comp. Physiol.* 1*: 285.)*

train of impulses along the afferent nerve fiber, indicating that a generator potential continues to excite the neuron (Fig. 21.12). For a given kind of receptor, the frequency and duration of the train of impulses depend upon the duration and intensity of stimulation of the receptor. These factors largely determine how long the generator potential is maintained at a sufficient level to continue to excite the afferent fiber. If the stimulation has been weak and brief, the subsequent generator potential on the fine terminations of the afferent fiber will be small and will decay rapidly. Consequently, only a few spikes will be induced in the train. If stimulation is intense, the corresponding generator potential will be greater both in electrical charge and in the area of the neuron altered, and decay will be slower. As a consequence of slower decay, the resultant train of impulses will be more prolonged.

It is everyday experience that our sense receptors "get accustomed" to a given stimulus continuously applied, unless it is increased in intensity. It can readily be demonstrated, by electrical stimulation, that the afferent neuron is not fatigued, but rather that the stimulus fails to maintain a generator potential sufficient to produce discharge. Such *sensory adaptation* may have several causes. One is that the stimulating energy may fail to reach the sensory terminal. Another is accommodation of the spike-generating membrane. Sensory adaptation occurs at different rates in diverse receptors (Fig. 21.13). For example, touch receptors adapt fairly readily, but pressure receptors in the carotid artery show almost no adaptation and continue to discharge as long as pressure is applied. In a receptor adapted to a given intensity of stimulation, impulses can again be induced by sufficiently increasing the strength of the stimulus. In this case a generator potential has again been induced, but the adaptation will occur again at a rate characteristic for the receptor (Guyton, 1976).

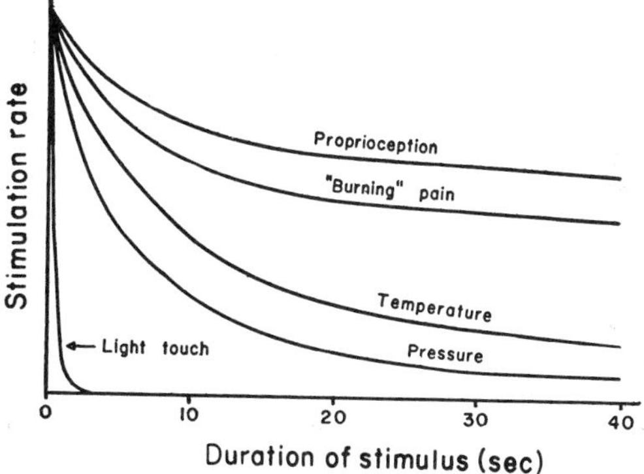

Figure 21.13. *Sensory adaptation in various receptors, as measured by frequency of impulse discharge after continuous application of a stimulus of the same intensity. (From Guyton, 1971: Textbook of Medical Physiology. 4th Ed. W. B. Saunders Co., Philadelphia.)*

Ultimately a receptor cell may react feebly or fail to respond when it becomes fatigued. This suggests that some material necessary for action of receptors is temporarily exhausted and the receptor is less responsive or that the change in responsiveness may represent a change in the state of the receptor membrane.

SYNAPTIC TRANSMISSION OF ACTION POTENTIALS

The *synapse* is defined as the functional connection between two neurons. It was so named in 1898 by Sherrington, who considered the synapse to be a region of contact, not confluence, of two excitable cells. Electron microscopy has since demonstrated that the cells at a synapse are indeed separate—not only are presynaptic and postsynaptic cell membranes present, but between them generally exists a synaptic cleft several tens of nanometers wide (Fig. 21.14) (Kuffler and Nichols, 1977). It is interesting that neurons and muscle cells in tissue culture form unions resembling synapses (Fischbach, 1970).

Synaptic contacts vary in structure. For example, a crossing of cells with similar fiber membranes lying alongside one another *(en passant synapses)* is found in coelenterates. A more prolonged contact of fiber membranes in the *loop synapses* (neuropile connections) is found in many invertebrates (Bullock and Horridge, 1965). The most highly differentiated contacts between neurons occur in vertebrates, in which the modified ends of the nerve fibers are applied to dendrites or to the nerve cell body. Many of the sensory nerve connections

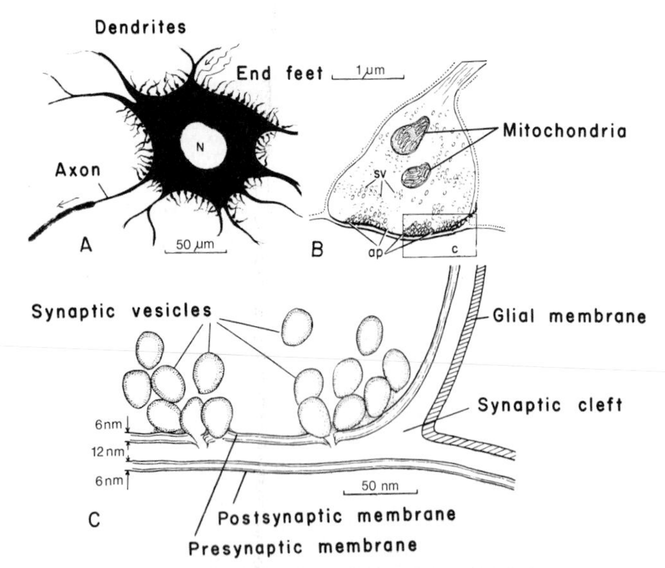

Figure 21.14. *Diagram showing a synaptic junction at various magnifications. A, A motor neuron under medium power of a light microscope. Note the numerous end feet—the terminations of axons of interneurons. N represents the nucleus of the motor neuron. B, One end foot seen with an electron microscope, magnified 60 times over A. In addition to the mitochondria, note the synaptic vesicles (sv) and the active points (ap) where the synaptic vesicles appear to aggregate and open onto the synapse. The dotted line represents the glial membrane. C, One sector of B seen using high resolution electron microscopy. Note that some of the synaptic vesicles are rupturing into the synaptic cleft through the presynaptic membrane. Observe that the membranes bordering the cleft are much like any other cell membranes. The synaptic cleft is a real separation between the cells enclosed by the membrane of a supporting or glial cell. (Adapted from DeRobertis* et al., *1965: Cell Biology. 4th Ed. W. B. Saunders Co., Philadelphia.)*

are of this type, and the nerve fiber endings may be cuplike, clublike, or basket-like. *End feet synapses* (boutons termineaux) are endings found on the motor neurons (Fig. 21.14). It is interesting to note that in the vertebrate central nervous system, however, the synapses of special cells, such as the Purkinje cells, are of a rather simple type, nerve fiber abutting against nerve fiber *(en passant)*. The *myoneural junction* (muscle end plate) is the synapse between a motor neuron and a muscle fiber and is an area of considerable contact between nerve fiber and muscle cell, with synaptic boutons lying in depressions of the muscle fiber and separated by synaptic clefts (Kuffler and Nicholls, 1977) (Fig. 21.15). These varied types of synapses have functional significance and are discussed in *The Synapse* (1976), the subject of a symposium at Cold Spring Harbor, New York.

While we think of synapses between neurons as being from the axon of one to the cell body of the other, in the animal they may occur from one axon to another axon (axoaxonic) or from axon to dendrite of another neuron (axodendritic). In diagrams few synapses are shown between neurons. In preparations from tissues they may be quite numerous. For example, some 6000 axoaxonic and axodendritic synapses may be found on a single motoneuron. Some of these come from axons of afferent fibers of stretch receptors and from muscle spindles, for example; most of them come from the central nervous system (Schmidt, 1975).

When the surface of contact between two neurons is similar (e.g., in the *en passant* synapses of coelenterates), the synapse may transmit in both directions. When the surface of contact between the two neurons (or neuron and muscle cell) is more complex, for example, consisting of a number of synaptic boutons, the synapse transmits in one direction only, that is, it is *polarized*. Polarization probably results from differentiation in neurosecretory activity at presynaptic and postsynaptic membranes, and it may occur even where the contacts between these membranes are apparently structurally undifferentiated, e.g., in the central nervous system of the vertebrate.

Impulses are transmitted across synapses in some cases electrically, in others chemically. The type of transmission is largely determined by the structure of the synapse.

Electrical Synaptic Transmission

Synaptic transmission by electrotonic spread from the presynaptic to the postsynaptic membrane was first demonstrated in the crayfish giant motor synapse. Here the synaptic delay is very short—0.1 millisecond from beginning of prespike (local potential) to beginning of the excitatory postsynaptic potential (Fig. 21.16*A*). Presynaptic depolarization actually spreads across the synapse and is measurable as an appreciable potential difference in the postsynaptic membrane. While the postsynaptic depolarization in turn causes presynaptic electrotonic effects, it does not cause presynaptic depolarization. Thus, rectification (i.e., passage of current in one direction only) of the synaptic junctional membrane is indicated. It is proposed that the synaptic membrane acts as a rectifier of positive current from the presynaptic to the postsynaptic fiber, with high resistance to flow in the opposite direction (Furshpan and Potter, 1959a, b). The synaptic rectifier is oriented in the right direction to allow the presynaptic action current to stimulate the post-fiber. The electrotonic current resulting from the spike on the presynaptic fiber is sufficient to account for normal transmission. The contact between the neurons showing electrical synaptic transmission consists of gap junctions in which the intercellular space is only two nm instead of tens of nm as in chemical synaptic transmitting synapses. The two neurons interconnect by channels in small structures forming polygonal arrays that span the gap between them. These structures contain a protein ("connexin") that may play a role in electrical link-

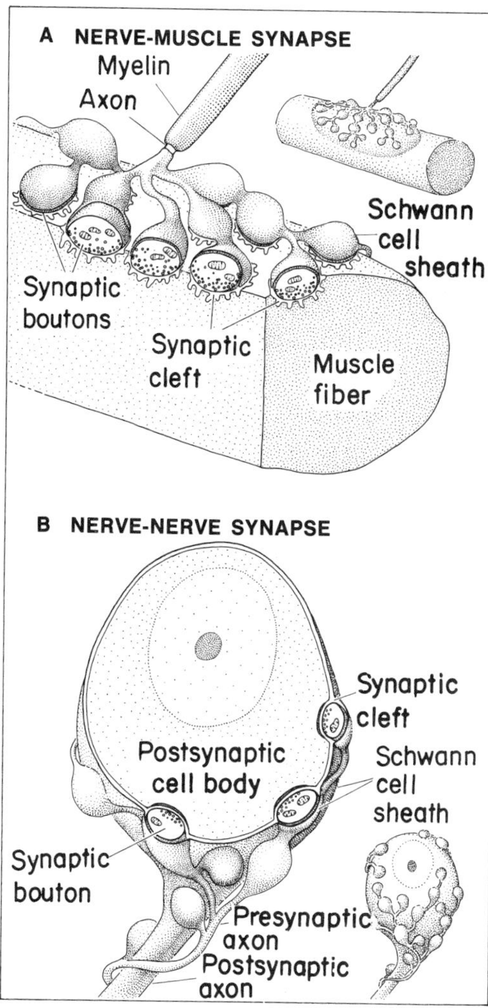

Figure 21.15. *Structural features of chemical synapses. A, The end plate formed by a motor nerve on a skeletal muscle fiber of the snake consists of a collection of synaptic boutons. Boutons contain synaptic vesicles and mitochondria and are always separated from the postsynaptic cell by a cleft about 50 nm wide. They are covered by fine lamellae of Schwann cells. B, Similar synaptic boutons are distributed on a ganglion cell in the heart of the frog. (After McMahan and Kuffler, 1971: from Kuffler and Nicholls, 1977: From Neuron to Brain. © 1977 by Sinauer Associates, Inc. Sunderland, Mass., p. 80.)*

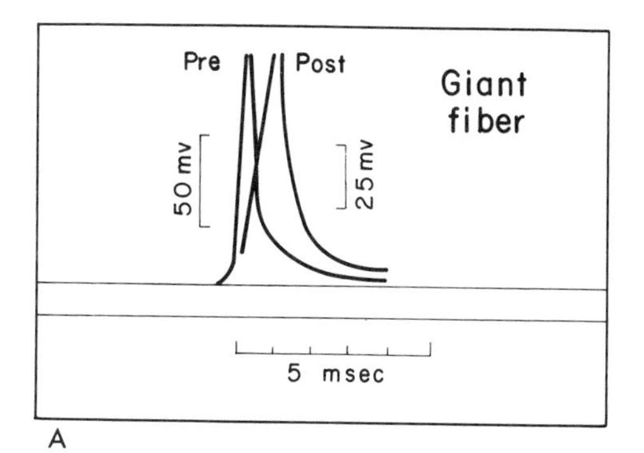

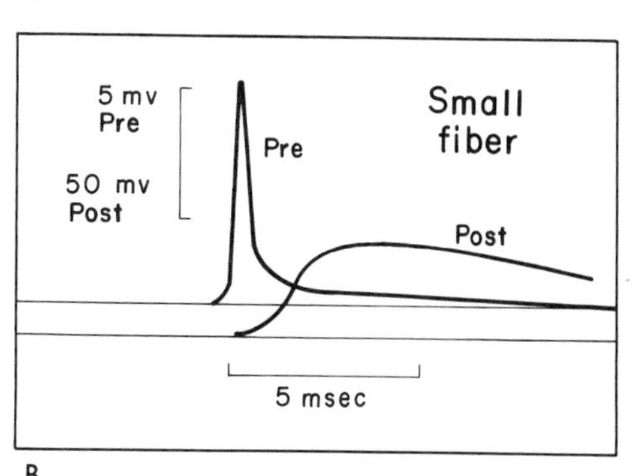

Figure 21.16. *Pre- and postsynaptic electrical events in crayfish nerve cord neurons. A, Development of a spike potential in the postsynaptic fiber within 1 msec of the presynaptic spike in a giant fiber with tight junction. B, Slow postsynaptic electrotonic spread (note difference in mv scales for pre- and postsynaptic events) following a presynaptic spike. The giant fiber illustrates electrical synaptic transmission and the small fiber the typical events in a synapse with chemical transmission. (After Furshpan and Potter, 1959a: J. Physiol. 145: 289.)*

age between the two neurons (Kuffler and Nicholls, 1977) (see Fig. 8.13).

A similar situation obtains in the goldfish Mauthner cells (in the medulla, having very numerous synaptic connections; Furukawa and Furshpan, 1963). The synaptic delay in some electric organs is so short as to essentially rule out the possibility of chemical transmission (Bennett, 1970).

Examples of electrical transmission across gap junctions have now been found in the nervous systems of some annelids, mollusks, and arthropods, in the medulla of the puffer fish, the anterior horn ganglion of the frog, and in the ciliary ganglion of the chick (Kuffler and Nicholls, 1977).

Chemical Synaptic Transmission

In most cases studied the synaptic delay is much greater than in the crayfish giant motor synapse, and varies between about 0.3 and 10 msec (Fig. 21.16*B*). Consequently, it is quite unlikely that electrotonic spread occurs from presynaptic to postsynaptic fibers because the field on the presynaptic fiber is disappearing or gone before the postsynaptic local potential develops. Internal electrodes fail to indicate induction of electrotonus in the postsynaptic fiber at the time of maximal electric field on the presynaptic fiber (Eccles, 1965). It is suggested that on reaching the synapse the presynaptic potential activates a chain of events that results in the secretion of a *transmitter substance,* which, when released, attaches itself to special receptor molecules on the surface of the contacting postsynaptic excitable cell. This chemical combination leads to a membrane change that depolarizes the postsynaptic cell locally. When depolarization exceeds the threshold level, a propagated action potential is initiated (Table 21.2). A rise in temperature decreases synaptic delay, a finding in keeping with the theory of chemical transmission across the synapse; a rise in temperature should have little effect on electrical transmission (Katz and Miledi, 1965).

The neuromuscular junction is an example of a synapse at which chemical transmission is well supported by experimental data (Dale *et al.,* 1936). It has been possible to demonstrate that a transmitter substance (acetylcholine) is produced by the presynaptic nerve fibers. When it is applied to the end plate of the muscle fiber, this substance causes a local depolarization of the cell membrane and sets up action potentials, followed by contraction of the muscle fiber. The chemical effect is localized to the synaptic area of the muscle fiber. In this junction there is an irreducible synaptic delay of 0.5 to 1 msec between the arrival of the nerve impulse and the start of the local (end plate) potential, part of this time being necessary for secretion of the acetylcholine. Histochemical studies have demonstrated a localization at this junction of the enzyme *acetylcholine esterase,* the function of which is decomposition of the acetylcholine to choline and acetic acid (Chapter 20), thereby preventing a prolonged depolarized state. When this enzyme is inactivated (e.g., by the drug *eserine*), stimulation of the motor nerve is followed by multiple contractions.

Stimulation of the motor neuron produces no measurable potentials at the end plate at the time the spike reaches the terminal portion of the neuron. When potential changes of either polarity are imposed on the terminal portion of the motor nerve fiber, they do not spread beyond the nerve terminal, but they do increase the rate of acetylcholine discharge. No evidence has been found of cable transfer of electric current from motor fiber to muscle fiber, but we have adequate evidence of chemical changes leading to action potentials and contractions. It has been possible to apply acetylcholine solutions directly to the motor end plate (to reach the surface of the receptor areas), and under these conditions as little as 10^{-16} gram equivalents of acetylcholine gives rise to an effective depolarization of the muscle fiber followed by a spike and contraction. Application of a similar concentration of acetylcholine inside the muscle fiber at the postsynaptic membrane has no effect. Therefore, the muscle end organ acts much like a chemoreceptor, a small quantity of the substance sufficing to depolarize the cell. Acetylcholine receptors have been shown to be glycoproteins of about 300,000 molecular weight, consisting of 5 to 6 subunits (Kuffler and Nicholls, 1977).

It is interesting to note that calcium ion, which is known to be the activator of muscle contraction, is apparently also the activator of neurotransmitter secretion. When the impulse reaches the synapse the excitation causes a permeability change at the presynaptic membrane such that calcium ion enters. Entry of calcium ions is thought to induce coalescence of the membranes of the synaptic vesicles with the presynaptic membrane, leading to extrusion of the neurotransmitter substance into the space between the two excitable cells (Katz and Miledi, 1967; Gilula and Epstein, 1976).

The way in which the transmitter substance causes depolarization is not entirely understood, but the most likely mechanism is the opening of some ionic channel in the cell membrane. Evidence exists that acetylcholine simultaneously increases the membrane permeability of the motor end plate to several monovalent ions (e.g., sodium, potassium, ammonium) and possibly opens up an ionic channel to all small ions on both sides of the membrane. The result is a partial depolarization to a level of about 10 to 20 mv (negative inside), which corresponds to a free diffusion (liquid junction) potential between the cytoplasm and the external fluid. This local nonpropagated depolarization sets off the propagated spike potential (Katz, 1966).

When resting potentials are recorded on the end plate (postsynaptic), minute potential changes of about 0.5 mv are observed on oscillographic records, with a rapid rise (1 msec) and a slower decay (about 20 msec) indicating momentary depolarizations of the end plate (Dudel and Kuffler, 1961). This "miniature" end plate potential appears to indicate spontaneous release of "packets" of acetylcholine by the motor neuron, since gradual application of a very low concentration of acetylcholine gives a smooth change, not the sudden changes seen in the records. The miniature end plate potentials are reduced in size by curare, and their amplitude and duration increase when the action of the acetylcholine esterase at the end plate is blocked by eserine, preventing local hydrolysis of the acetylcholine. All these experiments suggest that acetylcholine is continually manufactured and expelled from the nerve cell endings, but the spontaneous release inducing the small potentials occurs not continually but in packets containing about 10,000 acetylcholine molecules and opening about 1000 ionic channels for about 1 msec. Presumably an impulse reaching the synapse causes a much larger number of such packets (200 to 300) to be released at once and is accompanied by a much larger potential change that serves to depolarize the cell (Kuffler and Nicholls, 1977). It is interesting that in the crustacean myoneural junction, where another chemical transmitter is active, miniature potentials are also recorded.

Electron microscopy has shown that the synaptic vesicles are seen at synapses, groups of them seemingly moving to active spots (see Fig. 21.14*B* and *C*) where they appear to rupture into the synaptic cleft. However, such vesicles are lacking at synapses in which electrical transmission has been demonstrated, as shown in Figure 21.17 (Whittaker, 1968).

Acetylcholine probably plays a role in other synapses besides myoneural. One example is in autonomic ganglia, where nerve fibers (preganglionic) from the central nervous system synapse with other (postganglionic) nerve fibers passing directly to effectors in the viscera. Experiments show that when the preganglionic nerve fibers are stimulated and their synapses are perfused with balanced salt solution during the excitation, acetylcholine appears in the perfusate. This perfusate has been found to produce electrical changes leading to excitation and an action potential when applied at other autonomic synapses. However, experimental application of acetylcholine is usually ineffective in exciting synapses of the central nervous system (Prosser, 1973).

It is of interest that the same transmitter substance may have diametrically opposed effects in different kinds of cells. Thus acetylcholine acts as a neuromuscular transmitter, inducing contraction in skeletal muscles, and it is a cardiac inhibitor, slowing or stopping the heart.

Other transmitter substances also play a part in synaptic transmission. For example, norepinephrine (formerly called noradrenalin) is important in the postganglionic fibers of the sympathetic nervous system. Some evidence exists that packets of norepinephrine are secreted in such synapses, much as in the case of acetylcholine in most synapses. A "sensory" substance of unknown chemical nature has been postulated for the synapses of afferent nerve fibers (Florey, 1966). Transmitter substances are found in the central nervous system and have been isolated and tested but not iden-

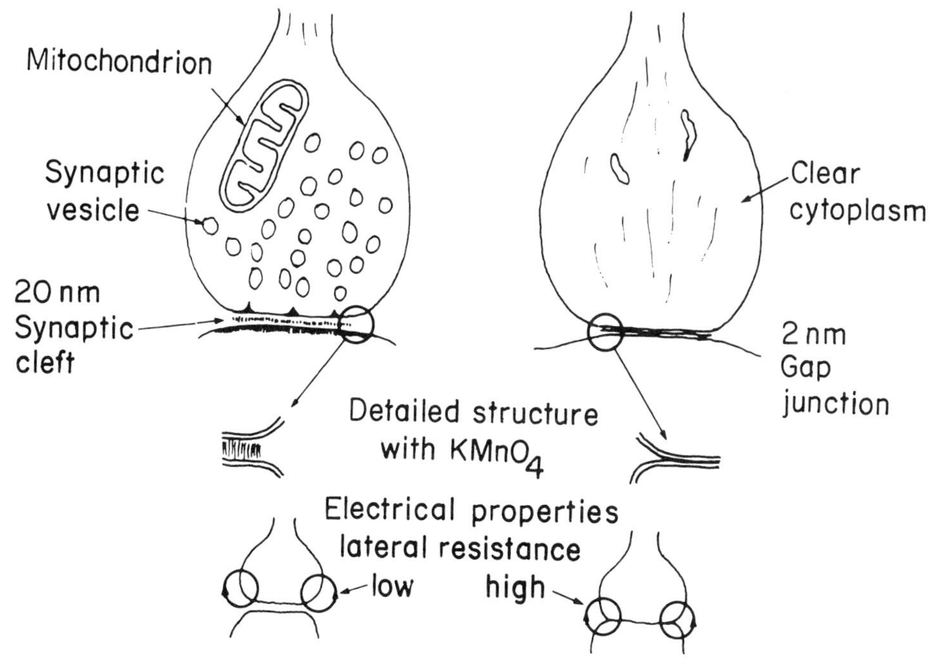

Figure 21.17. *Characteristics of open, chemical synapse* (left) *and gap function electrical synapse* (right). *Note in the latter synapse the absence of synaptic vesicles and cleft and the invasion of the postsynaptic elements by action currents which in the chemical synapse are short-circuited by the low-resistance cleft. (After Whittaker, 1968: Proc. Nat. Acad. Sci. U.S.A.* 60: *1081.)*

tified chemically with certainty. The potential changes that they produce, however, have been measured and correlated with the existing ionic gradients. At the synapses between interneurons and motor neurons, for example, the application of transmitter substance increases the permeability of the postsynaptic region to all ions, resulting in depolarization (Katz, 1966).

In spite of an enormous effort to identify neurotransmitters, relatively few are known with certainty: acetylcholine, norepinephrine, epinephrine in vertebrates, and γ-aminobutyric acid (GABA) in crustacean inhibitory fibers. Among other substances suggested as neurotransmitters are 5-hydroxytryptamine (serotonin), dopamine, octopamine, substance P, substance I, glycine, and glutamic acid, but evidence is incomplete (Kuffler and Nicholls, 1977). Substance I in Crustacea is GABA. Substance P is a peptide, supposedly an excitatory transmitter to motor neurons in the spinal cord of vertebrates.

Although biochemical identity of the transmitter substances remains doubtful in many instances, the fact that a transmitter substance is involved in transmission at synapses seems to be rather generally accepted. Thus it is considered that in most neurons a chemical change produced by a transmitter substance results in an electrical change (local generator potential). This sets off a conductile process (propagated action potential) ending in an effector action—the secretion of another transmitter substance. The action of the stimulus upon receptor cells probably also produces secretion of a transmitter substance that induces the depolarization of the afferent neuron. From this point of view the molecular mechanism for activation of various synapses may be quite similar in principle, despite diversity of transmitter substances.

Inhibition at Synapses

Inhibition of synaptic transmission is part of the mechanism of integration of responses to stimuli. Attempts to explain synaptic inhibition on the basis of electrotonic spread from presynaptic to postsynaptic membranes have been as unsuccessful in most cases as those purporting to explain synaptic transmission of excitation by electrotonic spread, and for the same reasons. However, inhibitory effects of some secretions and drugs upon synapses have been known for a long time.

An example of inhibition is the action of the vagus nerve (tenth cranial nerve) upon the heart beat. It is known that stimulation of the vagus releases acetylcholine, which causes hyperpolarization or stabilization of the resting potential of the heart muscle fibers and results in a slowing or stopping of the heart beat. The mechanism, seemingly, is a change in ionic conductance, but the channel opened in this case is restricted to K^+ and does not include Na^+. As a consequence, the membrane of the muscle cells affected tends to be held at the K^+ equilibrium potential, which is somewhat greater than the existing resting potential (Katz, 1966; Kuffler and Nicholls, 1977).

In the synapses of the vertebrate central nervous system some evidence of an inhibitory substance ("I" substance) has been gathered on the basis of the action of perfusates from inhibited synapses. Such substances generally act as hyperpolarizers of the postsynaptic junctions by increasing permeability to both K^+ and Cl^- (Table 21.2). The inhibitory substance found at crayfish muscle fiber end organs has been shown to be γ-aminobutyric acid (GABA) (Kuffler and Nicholls, 1977). It is likely that different inhibitory substances occur in different groups of animals. It is also possible that different inhibitory substances act in different parts of the nervous system of the same species. These problems are the basis of much current research.

Drugs may also inhibit the passage of the nerve impulse across a synapse. Perhaps the best known action of this type is curare, the active ingredient in South American arrow poison, which paralyzes animals by preventing transmission of the nerve impulse across the myoneural junction. Curare competes with acetylcholine for the receptor sites on the postsynaptic membrane and acts as a competitive inhibitor on the motor end-plate membranes. Curare itself has no effect on the properties of the postsynaptic membrane or on the permeability of the cells.

Facilitation at Synapses

At most synapses more than one incoming impulse is required to excite a postsynaptic neuron to a propagated action potential. When each successive impulse induces an increment of local postsynaptic potential greater than the preceding one, rather than merely adding to the preceding one (as in summation), *facilitation* is said to have occurred. On the basis of the chemical transmitter theory of synaptic transmission, facilitation could be explained as the release of successively larger amounts of the transmitter substance with each successive impulse. When the transmitter substance has induced sufficient depolarization of the post-

TABLE 21.2. CHANGES INDUCED IN EXCITABLE CELLS BY STIMULATION

Excitable Cells	*Excitation Induced*	*Excitor*	*Ionic Changes at Membrane*
Dendrites	Local (nonpropagated), graded, decremental depolarization	Acetylcholine in some, X* in others	Simultaneous Na^+ influx and K^+ efflux
Axon of neuron	Propagated all-or-none action potential (spike), with overshoot	See text	Na^+ influx followed by K^+ efflux
Vertebrate skeletal muscle fiber	Spike with overshoot	Acetylcholine	Na^+ influx followed by K^+ efflux
Invertebrate muscle fiber	Local, graded, decremental depolarization (slow fiber) or	X	Ionic flux† increased
	spike with overshoot (fast fiber)	X	Ionic flux increased sequentially for different ions
Receptor cell and afferent neuron junction‡	Local, graded, decremental, depolarizing generator potential (at transducer)	Sensory substance?	Ionic flux increased
Synapse: excitatory	Local, graded, decremental depolarization	In some, acetylcholine; in others, norepinephrine; in still others, X	Simultaneous Na^+ influx and K^+ efflux; possibly other ions
Synapse: inhibitory	Local, graded, decremental hyperpolarization; or stabilization of resting potential	Substance "I"#	Change in ionic flux of some ions§
Synapse of vagus on vertebrate heart	Local, graded, decremental hyperpolarization	Acetylcholine	K^+ efflux
Nitella cell	Depolarization and spike which may be all-or-none or decremental, no overshoot	?	Ionic flux changed

Data mainly from Katz, 1959, 1961; Florey, 1961; Frank and Fuortes, 1961; and Kuffler and Nicholls, 1977.
*X—an unknown substance present in extracts; electrical transmission in some crayfish synapses (Furshpan and Potter, 1959a,b).
†May involve entry of divalent cations in some, sodium and other ions in others, see Katz (1959b).
‡The axon of the afferent neuron acts like any other axon.
§Hyperpolarization could occur by entry of Cl^-.
#GABA, or γ-aminobutyric acid in crayfish myoneural junctions.

synaptic junction, a propagated action potential is initiated. Facilitation thus prevents impulses generated by minor stimuli from spreading immediately through the nervous system and serves as a device to keep responses proportional to stimulation. In this way, facilitation, like inhibition, serves as a mechanism for coordination of responses of an organism. In some giant fiber systems, cases have been recorded in which one presynaptic impulse induces one postsynaptic response, but they are unusual (Bullock and Horridge, 1965).

LITERATURE CITED AND GENERAL REFERENCES

Many classics in cell neurophysiology have been collected in a single volume by Cooke and Lipkin (1972). This makes it easy to obtain most of the references in this chapter and in Chapter 22.

Adrian, R. H., 1968: Ionic basis of excitability. *In* Molecular Basis of Membrane Function (Symposium). Tosteson, ed. Prentice-Hall, Inc., Englewood Cliffs, N.J.

Aidley, D., 1974: The Physiology of Excitable Cells. Cambridge University Press, Cambridge.

Axelrod, J., 1971: Noradrenalin: fate and control of its synthesis. Science *173*: 598–606.

Baker, P. F., 1966: The nerve axon. Sci. Am. (Mar.) *214*: 74–82.

Baker, P. F. and Connelly, C. M., 1966: Some properties of the external activation site of the sodium pump. J. Physiol. *185*: 270–297.

Baker, P. F., Hodgkin, A. L. and Shaw, T. I., 1961: Replacement of the protoplasm of a giant nerve fiber with artificial solutions. Nature *190*: 885–886.

Baker, P. F. and Reuter, H., 1975: Calcium Movement in Excitable Cells. Pergamon Press, New York.

Bennett, M. V. L., 1970: Comparative physiology: Electric organs. Ann. Rev. Physiol. *32*: 471–528.

Bishop, G. H., 1951: The nerve impulse. *In* Transactions of the Second Macy Foundation Conference, New York.

Blinks, L. R., 1940: The relation of the bioelectric phenomena to ionic permeability and to metabolism in large plant cells. Cold Spring Harbor Symp. Quant. Biol. *8*: 204–215.

Blinks, L. R., 1955: Some electrical properties of large plant cells. *In* Electrochemistry in Biology and Medicine. Shedlovsky, ed. John Wiley & Sons, New York, pp. 187–212.

Brazier, M. A. B., 1968: The Electrical Activity of the Nervous System. 3rd Ed. Macmillan Co., New York.

Bullock, T. H. and Horridge, G. A., 1965: Structure and Function in the Nervous System of Invertebrates. W. H. Freeman Co., San Francisco.

Chandler, W. K., and Hodgkin, A. L., 1965: The effect of internal sodium on the action potential in the presence of different internal anions. J. Physiol. *181*: 595–611.

Cold Spring Harbor Symp. Quant. Biol., 1966: Sensory Receptors, Vol. 30.

Connelly, C. M., 1959: Recovery processes and metabolism of nerve. *In* Biophysical Science—A Study Program. Oncley, ed. John Wiley & Sons, New York, pp. 475–484.

Cooke, I. M. and Lipkin, M., Jr., 1972: Comparative Cellular Neurophysiology: A Source Book. Holt, Rinehart & Winston, New York.

Cross, S. B., Keynes, R. D. and Rybora, R., 1965: The coupling of sodium efflux and potassium influx in frog muscle. J. Physiol. *181*: 865–880.

Dahlstrom, A., Haggendal, J., Heiwell, P. O., Larsson, P. A. and Saunders, N. R., 1974: Interaxonal transfer of neurotransmitters in mammalian neurons. Soc. Exp.Biol. Symp. *28*: 229–248.

Dale, H. H., Feldberg, W. and Vogt, M., 1936: The release of acetylcholine at voluntary nerve endings. J. Physiol. *86*: 353–380.

del Castillo, J. and Katz, B., 1967: Biophysical aspects of neuromuscular transmission. Prog. Biophys. Mol. Biol. *6*: 121–170.

De Robertis, E., 1971: Molecular biology of synaptic receptors. Science *171*: 963–971.

Dudel, J. and Kuffler, S. W., 1961: The quantal nature of transmission and spontaneous miniature potentials at the crayfish neuromuscular junction. J. Physiol. *155*: 514–529.

Eccles, J. C., 1964: Ionic mechanism of postsynaptic inhibition. Science *145*: 1140–1147.

Eccles, J. C., 1965: The synapse. Sci. Am. (Jan.) *212*: 56–66.

Erlanger, J. and Gasser, H. S., 1937: Electrical Signs of Nervous Activity. University of Pennsylvania Press, Philadelphia.

Fischbach, G. D., 1970: Synaptic potentials recorded in cell cultures of nerve and muscle. Science *169*: 1331–1333.

Florey, E., 1961: Transmitter substances. Ann. Rev. Physiol. *23*: 501–528.

Florey, E., 1966: An Introduction to General and Comparative Animal Physiology. W. B. Saunders Co., Philadelphia.

Frank, K. and Fuortes, M. G. F., 1961: Excitation and conduction. Ann. Rev. Physiol. *23*: 357–387.

Furshpan, E. J. and Potter, D. D., 1959a: Transmission at the giant motor synapses of the crayfish. J. Physiol. *145*: 289–325.

Furshpan, E. J. and Potter, D. D., 1959b: Slow postsynaptic potentials recorded from the giant motor fiber of the crayfish. J. Physiol. *145*: 326–335.

Furukawa, T. and Furshpan, E. J., 1963: Two inhibitory mechanisms in the Mauthner neurons in goldfish. J. Neurophysiol. *26*: 140–176.

Gaffey, C. T. and Mullins, L. J., 1959: Ion fluxes during the action potential in *Chara*. J. Physiol. *144*: 505–520.

Gerebtzoff, M. A. and Schoffeniels, E., 1960: Nerve conduction and electrical discharge. Comp. Biochem. *2*: 519–544.

Gilula, N. B. and Epstein, N. L., 1976: Cell to cell communicating gap junctions and calcium. Soc. Exp. Biol. Symp. *30*: 259–272.

Griffin, R. B. and Livett, E. G., 1971: Synaptic vesicles in sympathetic neurons. Physiol. Rev. *51*: 98–157.

Grundfest, H., 1965: Electrophysiology and pharmacology of different components of bioelectric transducers. Cold Spring Harbor Symp. Quant. Biol. *30*: 1–14.

Guyton, A., 1976: Textbook of Medical Physiology. 5th Ed. W. B. Saunders Co., Philadelphia.

Heslop, J. P., 1974: Fast transport along nerves. Soc. Exp. Biol. Symp. *28*: 209–228.

Hill, B., 1970: Ion channels in nerve membranes. Prog. Biophys. Mol. Biol. *21*: 1–32.

Hodgkin, A. L., 1951: The ionic basis of electrical action in nerve and muscle. Biol. Rev. *26*: 339–409.

Hodgkin, A. L., 1958: The Croonian Lecture: Ionic movements and electrical activity in giant nerve fibers. Proc. R. Soc. B. *148*: 1–37.

Hodgkin, A. L., 1964a: The ionic basis of nervous conduction. Science *145*: 1148–1153.
Hodgkin, A. L., 1964b: The Conduction of the Nervous Impulse. Charles C Thomas, Springfield, Ill.
Hodgkin, A. L. and Huxley, A. F., 1945a: Resting and action potentials in single nerve fibres. J. Physiol. *104*: 176–195.
Hodgkin, A. L. and Huxley, A. F., 1945b: Action potentials recorded from inside a nerve fibre. Nature *144*: 710–711.
Hodgkin, A. L. and Huxley, A. F., 1952: Movement of sodium and potassium ions during nervous activity. Cold Spring Harbor Symp. Quant. Biol. *17*: 43–52.
Hodgkin, A. L., Huxley, A. F. and Katz, B., 1949: Ionic currents underlying activity in the giant axon of the squid. Arch. Sci. Physiol. *3*: 129–150.
Hodgkin, A. L. and Katz, B., 1949: The effect of sodium ions on the electrical activity of the giant axon of the squid. J. Physiol. *108*: 37–77.
Hodgkin, A. L. and Keynes, R. D., 1953: The mobility and diffusion coefficient of potassium in giant axons from *Sepia*. J. Physiol. *119*: 513–528.
Hope, A. B. and Walker, N. A., 1975: Physiology of Giant Algal Cells. Cambridge University Press, Cambridge.
Horridge, G. A., 1968: Interneurons: Their Origin, Action, Specificity, Growth and Plasticity. W. H. Freeman Co., San Francisco.
Hoyle, G., 1962: Comparative physiology of conduction in nerve and muscle. Am. Zool. *2*: 5–25.
Huxley, A. F., 1964: Excitation and conduction in nerve: quantitative analysis. Science *145*: 1154–1159.
Junge, D., 1976: Nerve and Muscle Excitation. Sinauer Associates, Sunderland, Mass.
Katz, B., 1959a: Nature of the nerve impulse. *In* Biophysical Science—A Study Outline. Oncley, ed. John Wiley & Sons, New York, pp. 466–474.
Katz, B., 1959b: Mechanisms of synaptic transmission. *In* Biophysical Science—A Study Outline. Oncley, ed. John Wiley & Sons, New York, pp. 524–531.
Katz, B., 1961: How cells communicate. Sci. Am. (Sept.) *205*: 209–238.
Katz, B., 1966: Nerve, Muscle and Synapse. Macmillan Co., New York.
Katz, B., 1969: The Release of Neural Transmitter Substances (Sherrington Lectures, No. 10). Charles C Thomas, Springfield, Ill.
Katz, B. and Miledi, R., 1965: The effect of temperature on the synaptic delay at the neuromuscular junction. J. Physiol. *181*: 656–670.
Katz, B. and Miledi, R., 1967: The release of acetylcholine from nerve endings by graded electric pulses. Proc. R. Soc. B *167*: 23–28.
Katz, B. and Miledi, R., 1971: The effect of prolonged depolarization on synaptic transfer in the stellate ganglion of the squid. J. Physiol. *216*: 503–512.
Keynes, R. D., 1949: The movements of radioactive ions in resting and stimulated nerve. Arch. Sci. Physiol. *3*: 165–175.
Keynes, R. D., 1958: The nerve impulse and the squid. Sci. Am. (Dec.) *199*: 83–90.
Keynes, R. D. and Lewis, P. R., 1951: The resting exchange of radioactive potassium in crab nerve. J. Physiol. *113*: 99–114.
Korn, H. and Faber, D. S., 1976: Vertebrate central nervous system; same neurons mediate both electrical and chemical inhibitions. Science *194*: 1166–1169.
Krnjevic, K., 1966: Chemical transmission in the central nervous system. Endeavour *25*: 8–12.
Kuffler, S. W. and Nicholls, J. G., 1977: From Neuron to Brain, Sinauer Associates, Sunderland, Mass. (Chapters 4–12, cellular aspects.)
Kuno, M., 1971: Quantal aspects of central and ganglionic synaptic transmission in vertebrates. Physiol. Rev. *51*: 647–678.
Lester, H. A., 1977: The response to acetylcholine. Sci. Am. (Feb.) *236*: 107–116.
Lissmann, H. W., 1963: Electric location by fishes. Sci. Am. (Mar.) *208*: 50–59.
Loewenstein, W. R., 1960: Biological transducers. Sci. Am. (Aug.) *203*: 99–108.
McLennan, H., 1970: Synaptic Transmission. 2nd Ed. W. B. Saunders Co., Philadelphia.
Nathanson, J. A. and Greengard, P., 1977: "Second messengers" in the brain. Sci. Am. (Aug.) *237*: 108–119.
Neumann, E. and Bernhardt, J., 1977: Physical chemistry of excitable membranes. Ann. Rev. Biochem. *46*: 117–141.
Neuroid and neuronal mechanisms of coordination in simple systems, 1974. Am. Zool. *14*: 883–1080.
Nobel, P. S., 1974: Introduction to Biophysical Plant Physiology. W. H. Freeman Co., San Francisco.
Parsons, R. L., 1969: Mechanism of neuromuscular blockade by tetra ethylammonium. Am. J. Physiol. *216*: 925–931.
Patterson, P. H., Potter, D. D. and Furshpan, E. J., 1978: The chemical differentiation of nerve cells. Sci. Am. (Jul.) *239*: 50–59.
Prosser, C. L., ed., 1973: Comparative Animal Physiology. 3rd Ed. W. B. Saunders Co., Philadelphia.
Rasminsky, M. and Sears, T. A., 1972: Internodal conduction in undissected demyelinated nerve fibers. J. Physiol. *227*: 323–350.
Schmidt, R. F., ed., 1975: Fundamentals of Neurophysiology. Springer-Verlag, Heidelberg.
Shepherd, G. M., 1978: Microcircuits in the nervous system. Sci. Am. (Feb.) *238*: 93–103.
Slayman, C. L., Long, W. S. and Gradmann, D., 1976: Action potentials in *Neurospora crassa,* a mycelial fungus. Biochim. Biophys. Acta *426*: 732–744.
Suckling, E. E., 1961: Bioelectricity. McGraw-Hill Book Co., New York.
Takeuchi, A. and Takeuchi, N., 1966: A study of the inhibitory action of aminobutyric acid on neuromuscular transmission in the crayfish. J. Physiol. *183*: 418–432.
Tasaki, I. and Singer, I., 1965: A macromolecular approach to the excitable membrane. J. Cell. Comp. Physiol. *66*(Suppl. 2): 137–145.
Teorell, T., 1962: Excitability phenomena in artificial membranes. Biophys. J. *2*: 27–52.
The Synapse. 1976: Cold Spring Harbor Symp. Quant. Biol. Vol. 15.
Twiggle, D. J., 1971: Neurotransmitter-receptor Interactions. Academic Press, New York.
von Euler, U. S., 1971: Adrenergic neurotransmitter functions. Science *173*: 202–206.
Whittaker, V. P., 1968: Synaptic transmission. Proc. Nat. Acad. Sci. U.S. *60*: 1081–1091.
Zacks, S. I., 1964: The Motor Endplate. W. B. Saunders Co., Philadelphia.

CHAPTER 22

CONTRACTILITY

Contractility, the capacity for movement, is one of the fundamental properties of life. Although some cells lack the power of independent motion, their cellular structures move actively in some stage of their development (e.g., mitosis) and probably slowly at all times. Two kinds of cell contractility are manifested: cytoplasmic streaming inside the cell and movement of the cell relative to its surroundings. Ameboid and ciliary movements are the primary means of locomotion in protozoans; larval invertebrates and zoospores and gametes of some plants move by cilia or flagella. Sliding movements of uncertain origin occur in diatoms and some blue-green algae. In mature plants, movement relative to the surroundings is primarily by growth or by turgor changes. In mature animals only gametes, ciliated epithelia, wandering ameboid cells, and muscle cells perform easily visible movements. Embryonic cells in animals move from places of formation to sites often distantly located, where the cells become parts of organs. Cells in tissue culture and healing wounds and some cancer cells also show free movement.

The general characteristics of various types of movement occurring in cells of organisms are considered in this chapter. The chemical basis of contractility, primarily as represented in striated muscle but with some comparisons to movement in less differentiated systems, forms the subject matter of Chapter 23.

NONMUSCULAR MOVEMENT

Cytoplasmic Streaming

Cytoplasmic streaming (cyclosis) is readily seen in many cells, e.g., the alga *Nitella,* mycelia of molds *(Neurospora),* slime molds, and protozoans *(Amoeba, Paramecium).* Streaming is also readily seen in epidermal cells, hair cells, and phloem cells of the seed plant.

Cytoplasmic streaming has long been a subject of study, and the effects of various environmental conditions upon streaming have been extensively investigated. Streaming can be initiated in some cells by visible light and also by certain chemicals. Streaming slows when the cell is chilled and accelerates when the temperature is raised. It is stopped by various stimuli, such as electric shock and ultraviolet radiation. The rate of streaming is affected by various physiological and chemical factors in much the same manner as the viscosity of cytoplasm (Jahn and Bovee, 1969).

Although it is inherently a fascinating topic, relatively little is known about the mechanism of cytoplasmic streaming, despite considerable research (reviewed by Hope and Walker, 1975). Microfilaments are present and cytochalasin B inhibits streaming without affecting the microfilaments. Colchicine has no effect. Streaming has been likened to ameboid movement, for which the mechanism is still somewhat obscure.

Ameboid Movement

Ameboid movement is the characteristic motion of amoebas, including wandering ameboid cells present in the blood and body fluid of diverse animals. Ameboid wandering cells serve as scavengers and may serve to regenerate injured regions. Phagocytosis of foreign particles by ameboid white blood cells is important as a defense mechanism in multicellular animals. In organogenesis during embryonic development, some cells disperse to their definitive lo-

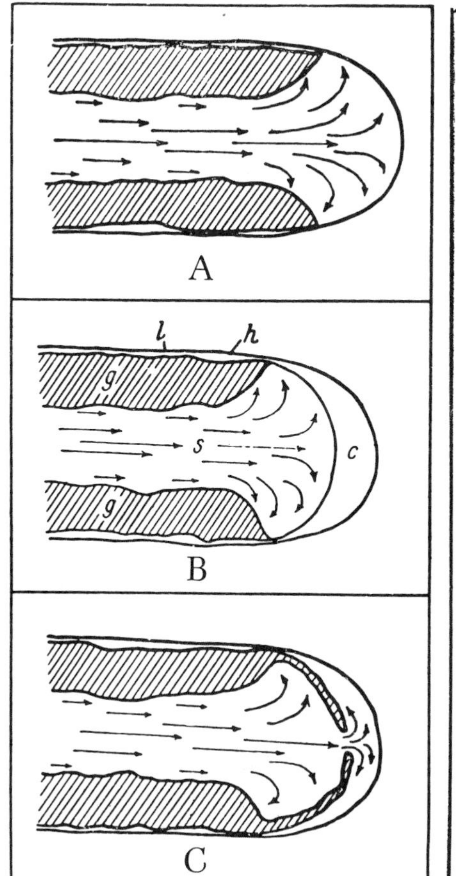

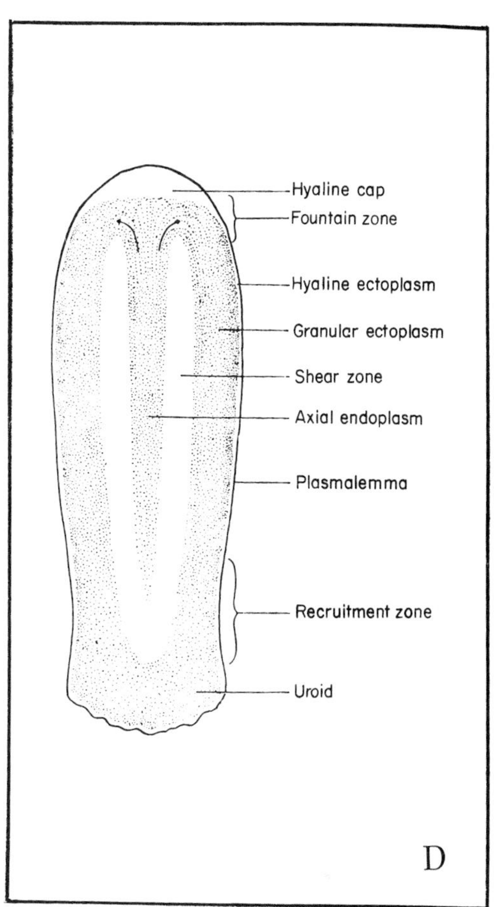

Figure 22.1. A, B *and* C, *Ameboid movement illustrating the growth of a pseudopod. The shaded region represents gelled cytoplasm. In* A *the flowing endoplasm reaches the plasmalemma (membrane) at the tip of the advancing pseudopod. In* B *there is a well-defined hyaline cap (c). In* C *the hyaline cap has broken. S, plasmasol; g, plasmagel; h, hyaline layer; l, plasmalemma. (After Mast, 1931: Protoplasma* 14*: 323.)* D, *A new scheme for ameboid structure and movement based on studies of cytoplasmic flow. A new terminology is proposed for regions of the cell which are differentiated with respect to consistency.* Uroid *refers to the tail process. (From Allen, 1960: J. Biophys. Biochem. Cytol.* 8*: 395.)*

cations in organs by ameboid movement. In rare instances, as in *Ascaris,* spermatozoa are ameboid. Ameboid motion is found rarely in plant cells; an example is the plasmodium of the slime mold, a huge multicellular ameboid mass formed by the aggregation and coalescence of vast numbers of small ameboid cells. The plasmodium, as a unit, then moves about like a giant amoeba.

Despite a considerable amount of research, a satisfactory analysis of ameboid movement is still not available (Allen and Taylor, 1975). It has been shown that in an amoeba, contraction is preceded by a spike potential. The potentials are variable but in some preparations they are equivalent in magnitude to those shown in muscle cells (Kamiya, 1964). The most generally accepted explanation of the mechanism of ameboid movement is the conception of a reversible gel-sol transformation. According to Mast's theory solation of the anterior end of the amoeba accompanied by the contraction of the cortical gel at the posterior end ("tail") results in propulsion of a pseudopod (Fig. 22.1*A, B, C*).

Certain experimental observations on ameboid movement could be interpreted by assuming that instead of propelling itself forward by a contraction at the rear end, an amoeba, instead, literally pulls itself forward by contraction in the endoplasm at the anterior end. When amoebas are subjected to precisely controlled centrifugal exposures, the particulates in the axial endoplasm move only after a certain yield value has been reached, indicating structure such as is found in some colloids. The only regions that show a low viscosity and where the particles move with high velocity are the "shear zone," surrounding axial endoplasm, and the posterior endoplasm or "recruitment zone" (Fig. 22.1*D*). The latter is so named because endoplasm is recruited from the walls of the ectoplasm in the posterior third of the cell.

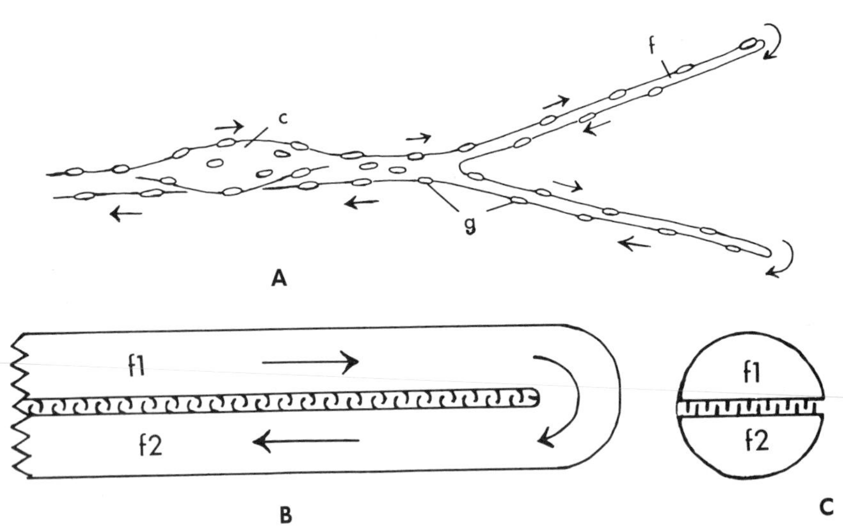

Figure 22.2. *Movement in a foraminiferan protozoan.* A, *The general shape and structure of the tip of one of the finer pseudopodia with a single bifurcation into branches about 1 μm in diameter. Arrows show movement of the granules (g), and of a small cytoplasmic mass (c), all of which are attached to the actively moving filament (f).* B, *Arrangement of actively moving filaments (f1, f2) of pseudopod as seen in side view. The direction of movement is shown by arrows. The moving material is assumed to be in the form of a semicylindrical filament turned back upon itself at the tip with the flat surfaces opposed. The shearing force is assumed to be between the adjacent surfaces and is designated by the short curved lines.* C, *Cross section of a filament. (From Jahn and Rinaldi, 1959: Biol. Bull.* 117: *103, 111.)*

The endoplasm appears to contract actively as it enters the "fountain zone" (Fig. 22.1*D*) just posterior to the hyaline cap, where the axial endoplasm becomes everted to form part of the ectoplasmic tube. The tension developed between the anchored advancing rim of the ectoplasmic tube and the movable endoplasm displaces the latter anteriorly. Continuous streaming is thought to be maintained by propagation of the contraction posteriorly along the ectoplasm at the same velocity as that at which the endoplasm advances relative to the pseudopodial tip. At the anterior end of the cell the contraction phase of the contractility cycle expresses water and solutes to form the hyaline cap (Allen and Kamiya, 1964). This view has been criticized on the basis of cinematographic studies of particle movements (Rinaldi and Jahn, 1963). The experimental observations adduced in support of the new hypothesis are not considered by others to be incompatible with the old hypothesis, namely, that it is the isometric contraction of the posterior end of an amoeba that provides the propulsive force for ameboid movement. Thus, while the new observations provide interesting details on the state of the internal cytoplasm and other aspects of ameboid movement and the formation of pseudopodia, they are not considered convincing support for the new hypothesis (Jahn and Bovee, 1969; Rinaldi *et al.*, 1975). Most telling is the finding that no contraction was localized in the anterior portion of glycerinated models of *Amoeba proteus* and *Chaos chaos* (Rinaldi *et al.*, 1975). This topic is further discussed in Chapter 23.

A type of "ameboid" movement has been described for foraminiferan protozoans, e.g., *Allogromia,* which possess *reticulopodia,* forming a network instead of simple lobose or thread-like pseudopodia. In each of these relatively stiff filamentous reticulopodia, which may extend for many millimeters around the body of the shelled protozoan, the streaming is always in two directions simultaneously—toward and away from the body—as can be seen by watching the ever-moving particles adhering to the moving surfaces on the two sides of the reticulopod (Fig. 22.2). The reticulopod is unlike a pseudopodium inasmuch as no gel-sol differences between its outer and inner cytoplasm can be detected. When the cell is injured the reticulopod separates into two filaments. It is postulated that this type of movement consists of shearing forces in opposite directions on the part of the two opposing faces of the filament, except at the tip where the filaments appear to be continuous as if one filament turned back upon itself (Fig. 22.2*B*). This concept has been extended to explain cytoplasmic movement in *Nitella* and other cells as well. The filaments in a reticulopodium suggest the two types of protein filaments present in muscle cells that have been demonstrated by electron microscopy to move along one another in the course of contraction (see Chapter 23) but on a different scale (Jahn and Rinaldi, 1959). Unfortunately, electron microscopic studies do not disclose any structure that supports such a mechanism (Wohlfarth-Bottermann, 1964a, b). However, membrane-bound protoplasmic fibrils containing filaments are present, several of them making up a reticulopodium. These may play a role in the movement observed. In contraction, tension may be applied to one side of the fiber loop while compression is applied to the other, as a result of which the two sides may be forced in opposite directions. Such a model permits directive forces without requiring structures not present in the electron micrographs. Any model must be considered tentative pending accumulation of decisive data.

Another type of ameboid movement by *filopodia* is observed in tissue-cultured fibroblasts and epithelium cells, in amebocytes, blood platelets, and at the growth cone of nerve cells. Filopodia consist of a single bundle of microfilaments enclosed in the plasma membrane. They extend from the cell surface and seem to be exploring the environment touched; often they attach to the substrate. They may pick up particles from the surface of the substrate and take them into the cell; for example, mouse cell fibroblasts (3T3 cells) remove gold particles from a surface in half an hour. Whether movement occurs by sliding filaments as in muscle has not yet been determined (Albrecht-Buehler, 1978).

Perhaps akin to ameboid movement is the

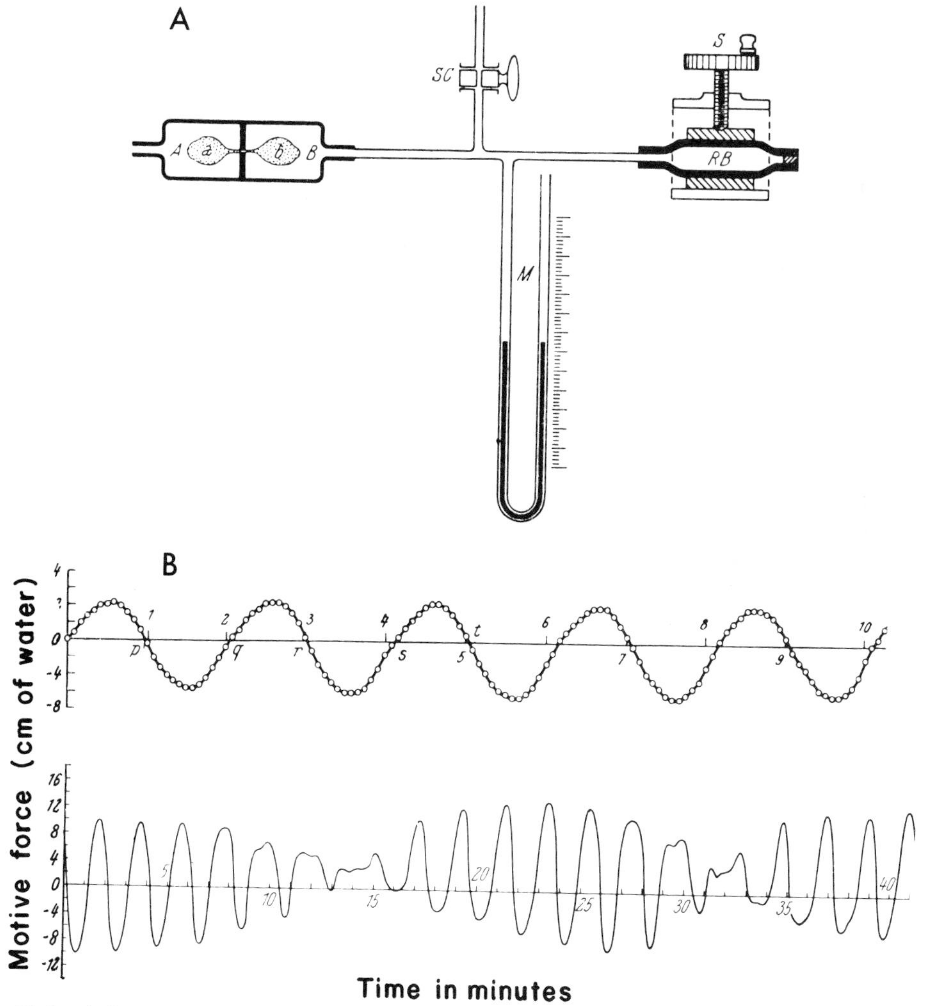

Figure 22.3. A, *Diagram showing the general arrangement for measuring the motive force of protoplasmic streaming in a slime mold plasmodium. The motive force of the plasmodium is measured in terms of the water pressure (height) required to just prevent movement. The movements in a plasmodium appear to be spontaneous and go on periodically as long as it is in the vegetative state. Note double chamber (A, B) with myxomycete (a, b) connected through a small hole between the chambers. M is a water-filled manometer, SC is a stopcock and S is a screw to control the pressure applied to the bulb RB.* B, *A recording of the motive force in centimeters of water. Note the beatlike wave pattern under normal conditions. The upper line is a recording showing all the points recorded several times a minute. The lower line shows a record in which the points are left out and the abscissa is compressed to make possible the showing of data for a longer period of time. It is possible to study the effect of temperature, anesthetics, and various injurious agents on the wave pattern using this equipment. (From Kamiya, 1959: Protoplasmatologia* 8: *Sect. 3a, pp. 41, 43.)*

movement of slime molds, e.g., *Physarum polycephalum,* the "many-headed" one, which produces many pseudopodia, sometimes moving in several directions (Kamiya, 1959, 1960; Steward, 1964). An interesting aspect of the movement is its pulsating nature, the pseudopodium advancing for a while, then retreating. Movement is usually greater in one direction than in the other; therefore, the slime mold as a whole advances or retreats. The nature of the pulsations and the motive force has been recorded by measuring the pressure just necessary to prevent movement (Fig. 22.3). A contractile protein has been isolated from *Physarum,* and adenosine triphosphate plays a role in movement here as in muscle cells (Pollard, 1973). Bursts of heat evolution synchronized with the rhythmic contractions of the plasmodium have also been recorded (Allen *et al.,* 1963).

Movement of Pigment Granules in Chromatophores

Pigment cells *(chromatophores)* are of interest in the present discussion because of the remarkable movements that may be seen in them. These cells are found in the skin of many types of animals and in many cases enable the animal to simulate the background, as in the chameleon. Although chromatophores are classified in a variety of ways with regard to the pigment they contain, there are essentially three main structural types: (1) individual reticulate cells such as those found in vertebrates (Fig. 22.4*A, B*); (2) larger reticulate units made up of several fused cells (syncytia), such as those found in Crustacea; and (3) multicellular units consisting of a central pigment-containing cell surrounded by radially arranged muscle cells, found in some cephalopods (Fig. 22.4*C, D*) (Fingerman, 1970).

The pigments in chromatophores are of diverse colors and chemical nature, and in most cases the pigments move inside the cell. The black and brown pigments are melanins, the red and yellow ones are carotenoids, the blue ones are protein-carotenoid complexes, and the reflecting white ones are guanine crystals. The guanine crystals are usually immobile, and the reflecting chromatophores are deeper-seated than the other kinds. A single chromatophore may contain several types of pigment granules

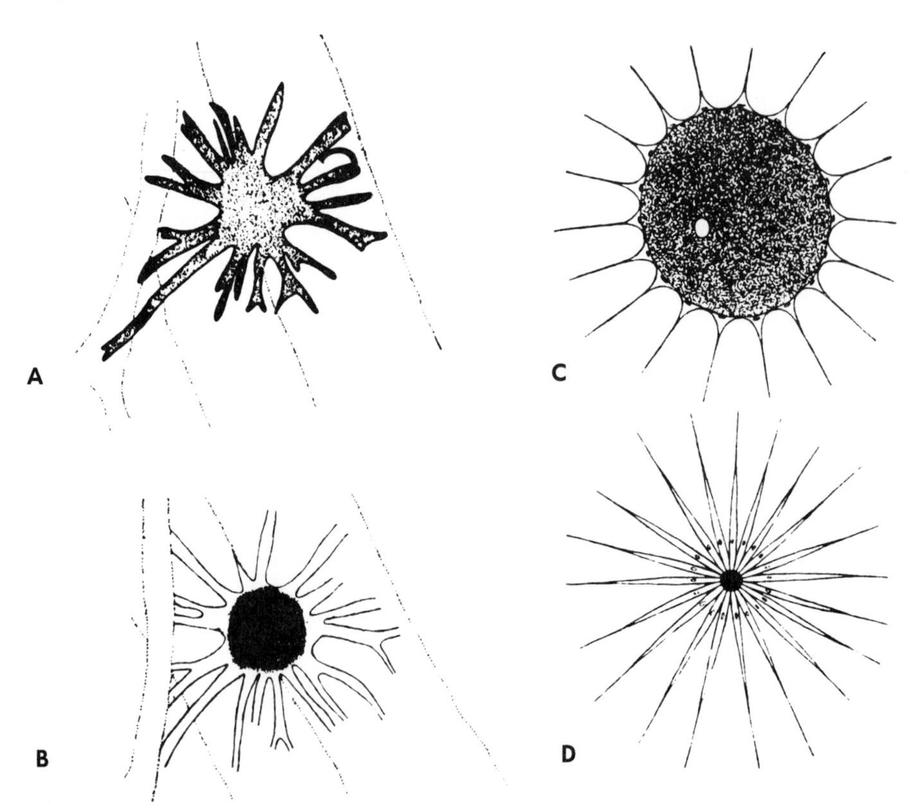

Figure 22.4. *Chromatophores.* A *and* B *represent camera lucida drawings of a melanophore from a scale of the killifish* Fundulus *in the expanded and contracted state.* C *and* D *represent the expanded and contracted states of the pigment in the chromatophore of the squid* Loligo. *Expansion of the pigment in this case results from contraction of the circle of muscle cells surrounding the pigment cell. (*A *and* B *from Matthews, 1931: J. Exp. Zool.* 58: *471;* C *and* D *from Bozler, 1928: Z. Verg. Physiol.* 7: *379.)*

that move independently of one another, each pigment granule individually moving from the central mass into the branching ramifications of the cell or flowing back into the central zone. Three stages of pigment spread are recognized: *punctate* or contracted, *stellate* or intermediate and *reticulate* or fully expanded. Some chromatophores respond directly to light, electricity, and hormones. For example, the squid chromatophore shows a response much like that obtained for the twitches and summation of a muscle fiber (Hill and Solandt, 1934).

In the guppy fish, epinephrine induces the punctate state in chromatophores. Cyclic AMP introduced iontophorically (by micropipette) inside the pigment cell also induces the punctate state but not when applied outside the cell. It is likely that hormones and possibly other agents act on pigment cells by the release of cyclic AMP in the cell (Fujii and Miyashita, 1976). Melatonin also induces the punctate state in catfish melanophores, suggesting it as a possible hormone activating receptors on the chromatophores (Fujii and Miyashita, 1978).

The mechanism by which the pigment moves in the reticulate chromatophore cell is not fully understood. The pigment granules appear to flow into the arborizations or channels of the cells. The channels seem to be permanent structures clearly visible, at least in tissue cultures of such cells, even when the cell is in punctate form and the pigment is in the fully contracted stage (Bagnara *et al.,* 1968). However, it has been stated that in some cases the channels collapse when the punctate state is assumed (Fingerman, 1970). The channels are lined with microtubules. It is possible that in the punctate state the microtubules disappear because the proteins of which they are composed are known to disassociate under a variety of conditions, e.g., low temperature (Jahn and Bovee, 1969).

Experiments show that application of pressure to the chromatophores of the killifish *Fundulus* brings about pigment dispersion in the chromatophores. Since the pressure required to do this (7000 to 8000 pounds per square inch) usually induces a sol-like condition in cytoplasm, it was thought that perhaps dispersion of pigment is caused by solation, whereas, conversely, the contraction of pigment is caused by gelation. To test this hypothesis Marsland (1944) centrifuged chromatophores in the punctate state and was unable to stratify the pigment granules, even with a force 70,000 times gravity. However, he was easily able to stratify the pigment granules in the reticulate state with less centrifugal force. If solation and gelation do accompany pigment, the exact method by which this achieves the movement of the pigment granules is not clear.

The microtubules in fish melanophores seem to be permanent structures, and the pigment granules appear to glide along the microtubules during aggregation and dispersion. No evidence has been obtained for either contraction or dissolution of microtubules as the motile force for pigment granule migration under normal circumstances. However, colchicine and vinblastine cause depolymerization of the microtubules (Murphy and Tilney, 1974). With immunofluorescence techniques, Schliwa (1978) was able to get striking pictures of microtubules in fish melanophores. However, he could not detect changes in microtubules during pigment aggregation, though he points out that the number of microtubules is too great for detection of a small change in number.

Kinosita (1963) found that the pigment granules in chromatophores of the medaka fish *Oryzias* are negatively charged. Therefore, application of electric fields causes displacement of the granules (electrophoresis). He postulated that when the agent that produces the contracted state (punctate state) of the chromatophore pigment is applied, the central part of the chromatophore becomes more positively charged than the periphery of the cell, attracting the negatively charged particles to the center. On removal of the stimulus and loss of charge difference (or possibly, the development of negativity in the central portion of the chromatophore), the pigment particles once again disperse and reach a reticulate state. Other theories are considered by Fingerman (1970). The motile force for pigment granule migration has not been identified.

In cephalopods the pigment cell is in an expanded state when the radial muscle cells are contracted and in a punctate state when these muscle cells are relaxed (Fig. 22.4*C, D*). The muscle cells are under the control of the nervous system. The chromatophores of crustaceans are coordinated by hormones, whereas those of fishes are controlled by both hormones and nerve impulses (see Fingerman, 1970).

Ciliary and Flagellar Movements

Cilia, the numerous hairlike motile processes on a cell, occur in all animal phyla except in nematode worms and arthropods, but they are especially prominent in members of the protozoan class Ciliata and in free-swimming larvae of many marine animals (see Fig. 22.5 and Fig. 3.21). A *flagellum* is a whiplike motile organelle, present singly or in small numbers at one end of a cell but in basic structure indistinguishable from a cilium. It is present in all flagellates, in the spermatozoa of most animals and plants, in the flagellated cells of sponges and coelenterates, and in the zoospores of algae.

Cilia and flagella have many characteristics

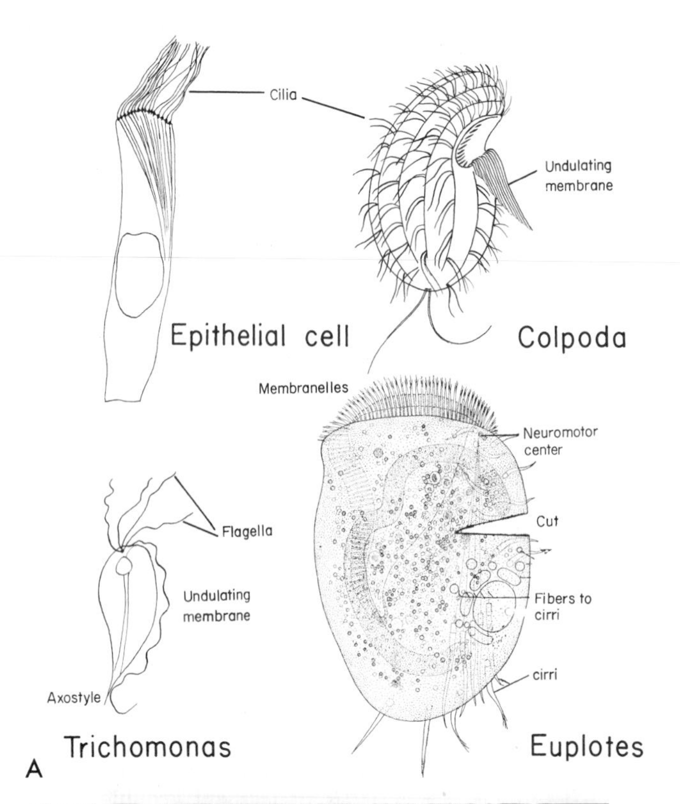

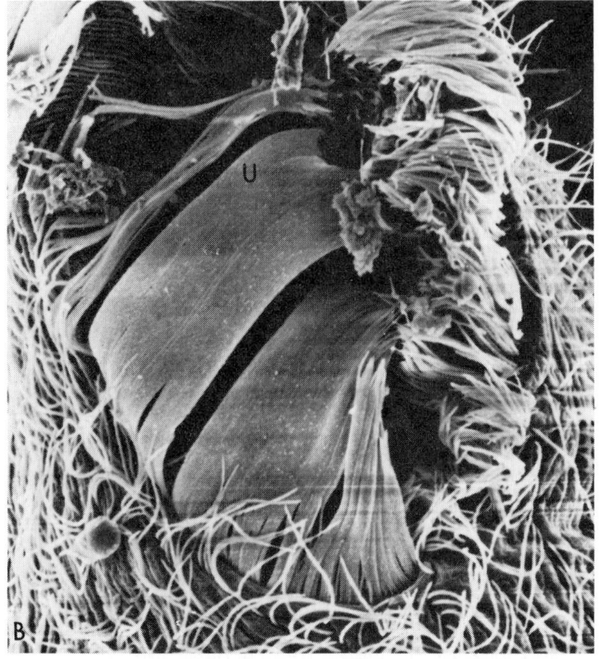

Figure 22.5. A, *Cilia and flagella of various types of cells. The epithelial cell is from the hepatic duct of a snail. (After Gray, 1928: Ciliary Movement. Macmillan Co., New York.) In* Colpoda duodenaria, *a ciliate, note the fusion of cilia to form an undulating membrane and the orderly arrangement of cilia in rows. In* Trichomonas, *from the intestine of a termite, note especially the undulating membrane made by fusion of one flagellum to the cell membrane and the supporting skeletal rod (the endostyle).* Euplotes *is a ciliate with cilia fused to form membranelles at the top and along the groove leading to the mouth and into cirri, or walking appendages, on the ventral surface of the animal. Each* cirrus *is innervated by a fiber from the* neuromotor center *near the anterior end of the animal. A cut across this set of fibers causes uncoordinated movement of the cirri to which the fibers connect. A similar cut elsewhere in the body has no effect on coordination. The large sickle-shaped body is the macronucleus. Note also the numerous vacuoles. In this species the contractile vacuole is formed by fusion of many small vacuoles. (After Taylor, 1923: J. Exp. Zool.* 37: *259; and 1920: Univ. Calif. Publ. Zool.* 19: *403.)* B, *Mouth parts of* Blepharisma americanum *showing the undulating membrane (U) made up of fused cilia, and the bordering cilia. The undulating membrane has broken into several pieces during the drying process (scanning electron micrograph ×25,000). (By courtesy of D. Shamlian and H. Wessenberg.)*

in common, but flagellar motion is often much more complicated than ciliary motion. The flagellum may be used like an oar, or it may move only at the tip like a propeller. The cilium is always stroked like an oar, although not always in the same manner (Sleigh, 1971; Blake and Sleigh, 1974).

Cilia and flagella serve many purposes and their movements either propel the organism or move the medium past a fixed cell. They usually beat continuously and act as independent effectors. However, in a few invertebrate embryos they are probably under nervous control, since their movement may be stopped upon stimulation of the embryo. In ciliates they were thought to be coordinated by a neuromotor cen-

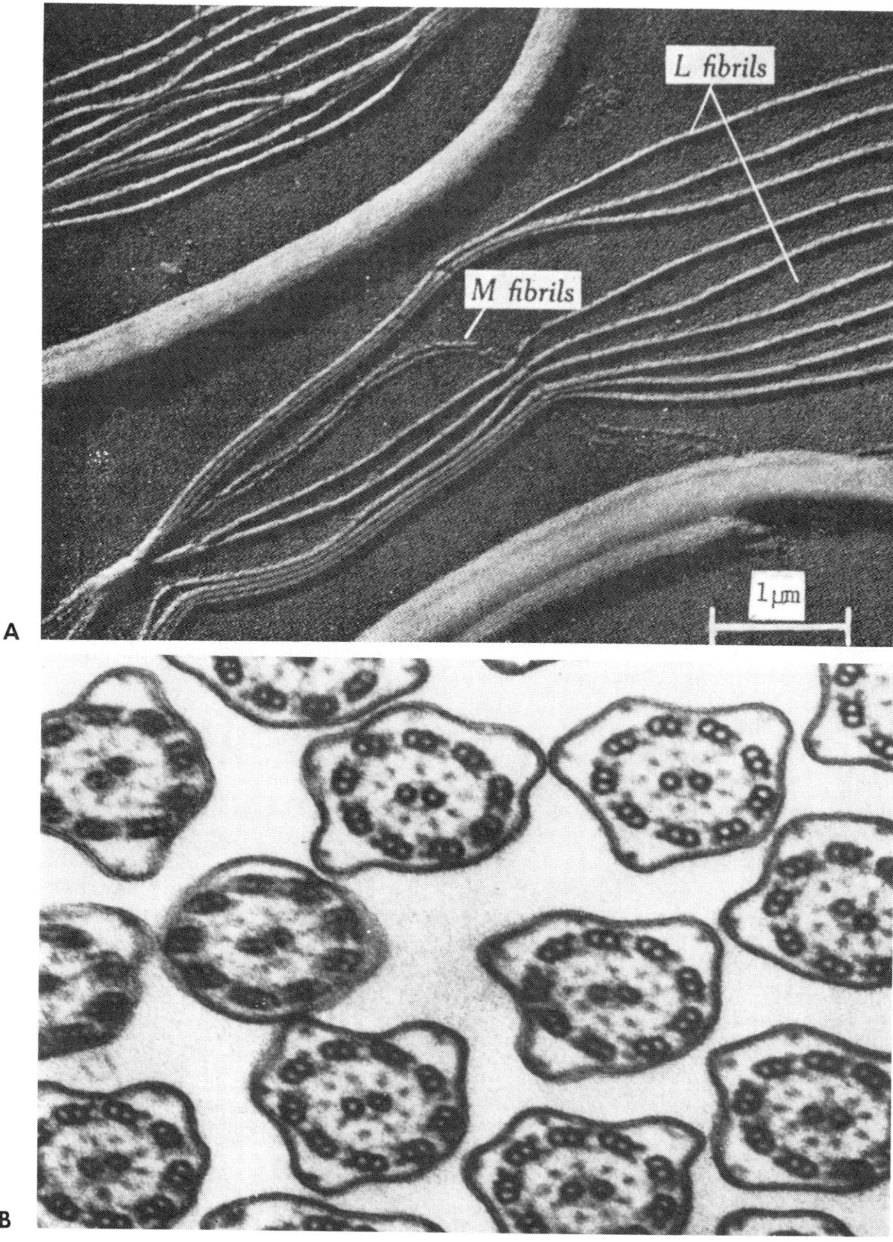

Figure 22.6. Electron micrographs of flagella. A, intact (large cable) and dismembered tails of cock spermatozoons, showing the eleven fibrils of which the flagellum is composed. M fibrils refers to the two median, more delicate fibrils; L fibrils refers to the nine larger, lateral fibrils. (From Grigg and Hodge, 1949: Austral. J. Sci. Res. B2: 271.) B, Cross sections of flagella of Pseudotrichonympha *(a flagellate from a termite). Note that the median fibrils are single while the lateral fibrils are double. (From Gibbons and Grimstone, 1960: J. Biophys. Biochem. Cytol. 7: 697.)*

ter near the mouth since destruction of the fibers connecting the center to the cilia results in uncoordinated movement (Taylor, 1920). However, Okajima (1966) reported coordinated movement in *Euplotes* even after complete dissection of the neuromotor fibers. He attributed coordination to a quick spread of electrical "events" on the surface of the cell.

Numerous studies of ciliary and flagellar movements have been made (Sleigh, 1971; Blake and Sleigh, 1974) and a great deal of information is available on the effect of various environmental conditions on the motion of cilia and flagella. It is interesting to note that in electron micrographs of all cells observed each cilium and flagellum has a characteristic pattern of nine circumferential and two central paired microtubules, the latter microtubules being somewhat more delicate than the former (Fig. 22.6).

Another contractile element, the *spasmoneme,* makes possible the extremely rapid contraction of vorticellid ciliates, for example, in the stalk of *Vorticella,* but especially observable in the large basal stalks of colonies of *Carchesium* and *Zoothamnium.* This enables the organisms to remove themselves from the path of a predator. Analysis of high-speed films of the movements indicates a duration of contraction between 2 msec and 10 msec, or a maximum contraction rate of 200 lengths a second compared to 22 lengths a second for the very rapidly contracting mouse finger muscle. *Myonemes* that permit rapid contraction of the entire body of the ciliates *Stentor* and *Spriostomum* may be similar to spasmonemes but much smaller (Routledge *et al.*, 1976).

One of the oddest nonmuscular movements described is a "rotary" motor of the axostyle of some flagellates (devescovinids) that inhabit the gut of some termites. The axostyle, made up of about 400 microtubules, passes through the posterior end of the body like a drive shaft and rotates 1.5 times per second. The torque-generating mechanism appears to be distributed along the entire length of the axostyle since microlaser beam section of the rod alters but does not stop its rotation. It is presumed that the actin microfilaments of the sheath surrounding the axostyle interact with myosin microfilaments in the cytoplasm to provide the motile force, although conclusive evidence is still lacking (Tamm, 1976).

MUSCULAR MOVEMENT

Contractility has reached its highest development in muscle cells, in which, perhaps, it is most exaggerated and diagrammatic. Being adapted to their special function, the muscle cells are modified in many ways. In general, the units of structure are elongated and spindle-shaped, and they consist either of single cells (e.g., smooth muscle cells) or of "composite cells" called *muscle fibers* (skeletal muscle fibers). In each case the muscle cells, or fibers, possess *contractile myofibrils* (fine elongated structures) and one or more nuclei, and are enclosed in a membrane called the *sarcolemma.* Although some studies have been made upon single muscle fibers (Ramsey and Street, 1941; Street and Ramsey, 1965), most investigations of muscle contraction deal with the entire muscle or with muscle-nerve preparations, despite the fact that the major concern is the nature of cellular contractility. It will become clear that the behavior of muscle is explainable only in terms of the cells or fibers of which it is composed.

Three kinds of vertebrate muscle cells are recognized: *smooth, cardiac* (heart), and *skeletal* (striated) muscles. Only the high points of the contractile properties of these various types of muscle cells can be discussed here.

Contraction of Smooth Muscle Cells

The contents of the gut and various other tubular internal organs are propelled by the contraction of the smooth muscle cells lining their walls.

Contraction of smooth muscle cells shows some interesting and unique features. In structure, smooth muscle consists of tiny, elongated spindle-shaped cells, each with a single nucleus and with interconnecting bridges (Rice and Brady, 1972) (Figs. 22.7 and 22.8). Minute longitudinal myofibrils (units of contraction) have been described in these cells and clearly seen in electron micrographs (Hanson and Lowy, 1957). The cells are arranged in the wall of the tubular organ in opposing circular and longitudinal layers in such a way that contraction of the longitudinal layer shortens an organ and contraction of the circular layer lengthens it. The single cells are difficult to study individually; therefore, for experiments on contraction, organs (e.g., stomach, intestine) or parts of them containing smooth musculature have generally been used.

A striking property of the contraction of smooth muscle cells of the digestive tract is the rhythmic "beat," which occurs even when all nervous connections are severed if the preparation is aerated and bathed in a balanced salt solution (Figs. 22.9 and 22.10). Therefore, the tendency to contract is inherent or *myogenic,* that is, the contractions are initiated in the muscle cells themselves. This rhythmic contraction leads to *peristalsis,* or waves of contraction, at intervals of about 1 minute. A peristaltic wave is highly coordinated; co-

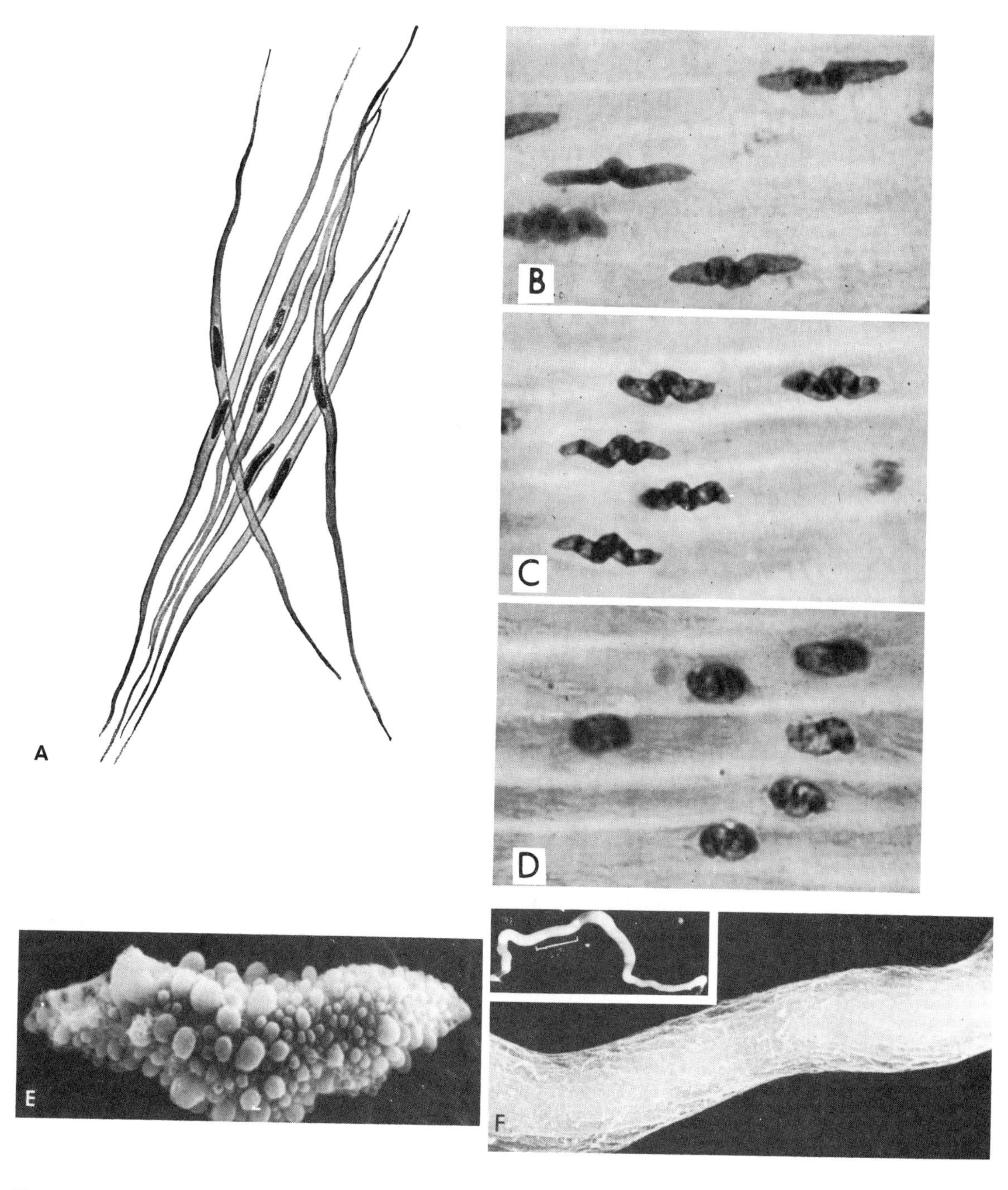

Figure 22.7. A, *Isolated smooth muscle cells from the wall of the stomach of a cat (×220).* B, C, D. *High-powered photomicrographs of smooth muscle of the intestine, showing how the elongated nuclei of the fibers are thrown into folds (pleats) when the fibers, on contraction, become shorter and thicker. In* D *they are folded so tightly that they appear, on superficial examination, to be ovoid.* E, F. *Scanning electron microscope studies of single smooth muscle cells isolated from toad stomach by digestion. In the contracted state* (E) *blebs of cytoplasm are extruded; these disappear as the cell relaxes to its extended state* (F). *Box in* F *shows entire relaxed muscle cell, below is a detailed view of the smooth surface at higher magnification. (*A *from Bloom and Fawcett, 1975: A Textbook of Histology. 10th Ed. W. B. Saunders Co., Philadelphia;* B, C and D *from Hamm, 1974: Histology. 7th Ed. J. B. Lippincott, Philadelphia;* E and F *from Fay, Cooke, and Canaday, 1976:* In *Physiology of Smooth Muscle. Bulbring and Shuba, eds. pp. 249–264. Raven Press, New York.)*

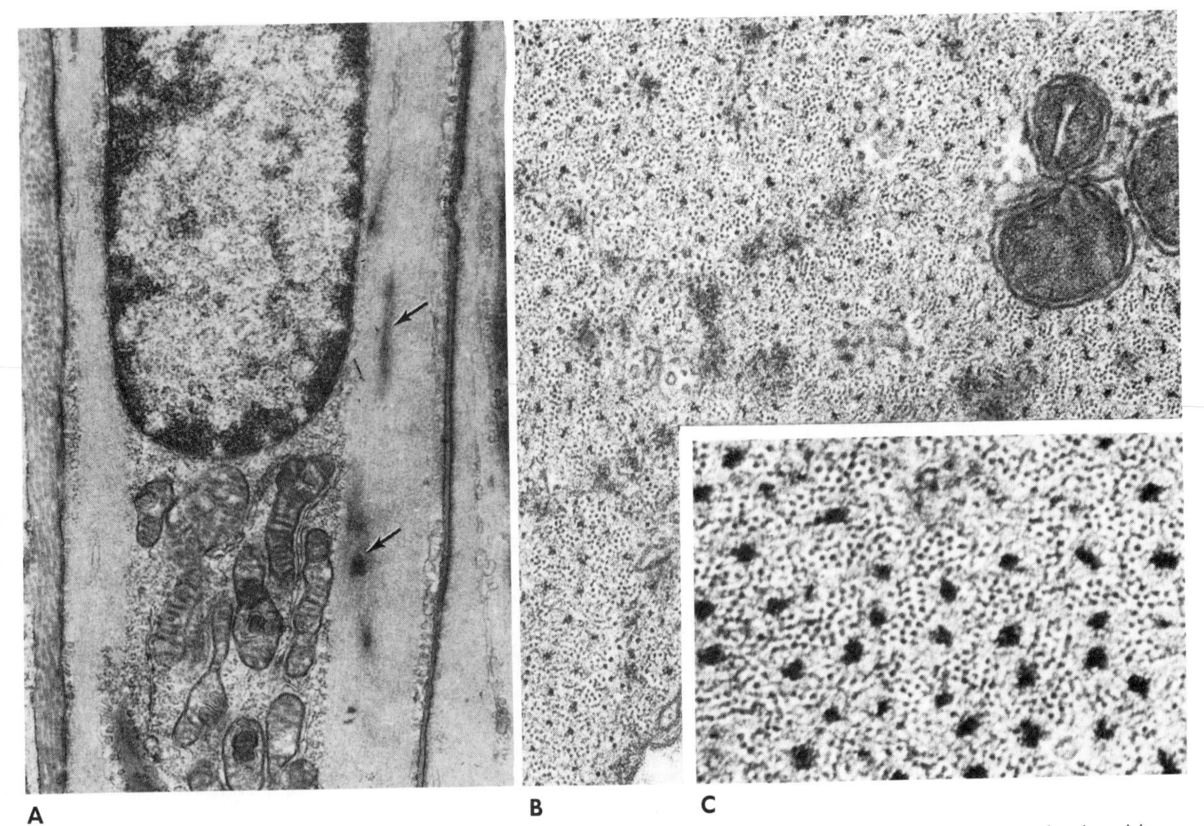

Figure 22.8 A, *Central portion of a smooth muscle in longitudinal section. Below the nucleus note the mitochondria, some endoplasmic reticulum, and numerous ribosomes. Microfilaments in the uniform grey material do not show at this magnification (×20,000).* B, *Vascular smooth muscle cell in cross section showing thick and thin filaments (×86,000).* C, *Higher magnification of a section of B (×186,000). (*A *from Bloom and Fawcett, 1975: A Textbook of Histology. W. B. Saunders Co., Philadelphia;* B *and* C *from Somlyo, Devine and Somlyo, 1972:* In *Vascular Smooth Muscle. Springer-Verlag, New York.)*

ordination is achieved by a slow spread of excitation (action potential). (This may be initiated or potentiated by discharges from the parasympathetic nervous system. On the other hand, such activity may be reduced or suppressed by discharges from the sympathetic nervous system.) Preceding a contraction, an action potential can be measured at a "segment" (a small portion between two regions of contracted circular muscle) of the intestine, and after it has contracted, the electrical wave excites another segment, causing it to contract, until the contraction wave has passed the full length of the intestine. Long before this contraction wave has completed its course, another series of contraction waves has been initiated.

The resting potentials of smooth muscle cells are somewhat unstable, and every once in a while a cell depolarizes and "fires," that is, sets off an action potential. This serves as a pacemaker and starts depolarization in other cells.

Although contraction can be induced and maintained in smooth muscle cells without nervous connections, it is generally controlled by the nervous system through double innervation, that is, by excitatory and inhibitory (autonomic) nerve fibers. When denervated smooth muscle cells are stimulated with electricity, they remain contracted for many seconds before they relax. The time for a complete

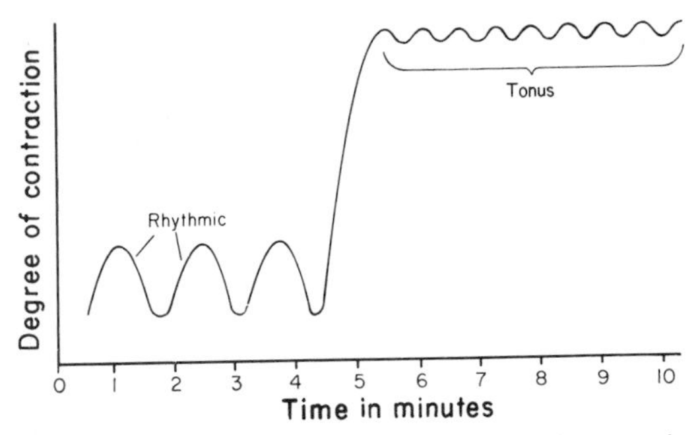

Figure 22.9. *Record of rhythmic and tonic smooth muscle contraction. In Figure 22.10C, one rhythmic beat is magnified to show the accompanying set of action potentials. (From Guyton, 1976: Textbook of Medical Physiology. 5th Ed. W. B. Saunders Co., Philadelphia.)*

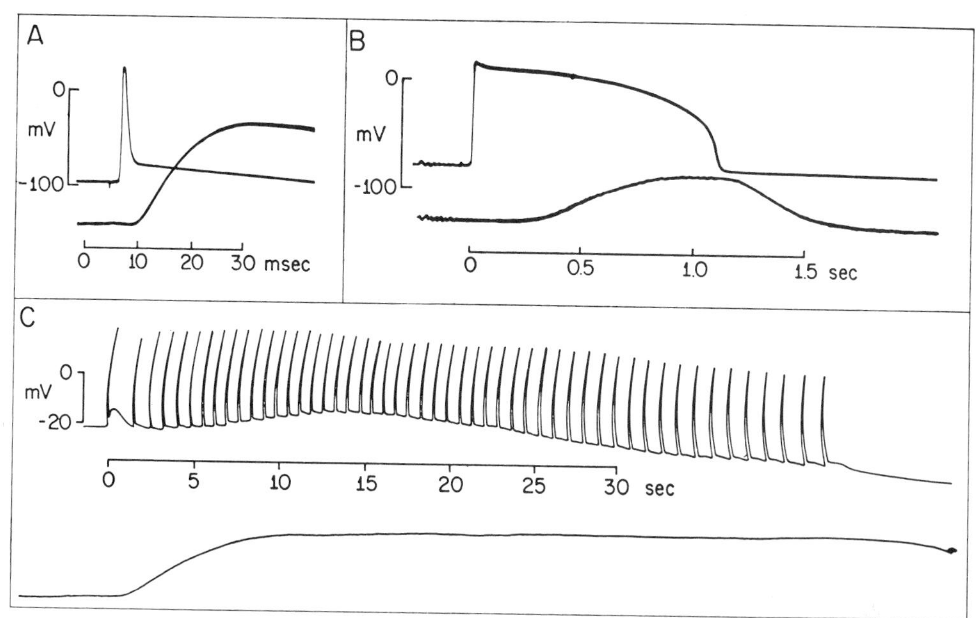

Figure 22.10. *Simultaneously recorded transmembrane potentials (upper curves) and contraction (lower curves) in three types of muscle.* A, *Isolated frog skeletal muscle fiber.* B, *Whole frog ventricle; action potential recorded from one "cell."* C, *Strip of pregnant rat uterus (smooth muscle); action potential recorded from one "cell, spontaneous activity." This record represents one rhythmic beat of the sort illustrated in Figure 22.9.* Abscissae, *Time in milliseconds* (A) *or seconds* (B *and* C). Ordinates, *Upper tracing, millivolts; lower tracing, arbitrary units of contractile tension. (Part* A *after Hodgkin and Horowicz, J. Physiol.* 136: *17P–18P.) (From Ruch and Patton, 1965: Physiology and Biophysics. 19th Ed. W. B. Saunders Co., Philadelphia.)*

cycle of contraction and relaxation is called a *muscle twitch*. The twitch time for smooth muscle is generally long, and the *refractory period* (that is, the period during which the muscle cells cannot be induced to contract) is relatively short. This means that a sustained state of contraction can be maintained even by a small number of periodic impulses from the nervous system (Bulbring *et al.*, 1970).

Smooth muscle cells maintain *tonus* (the mild state of sustained contraction of a muscle brought about by contraction of a few of its units) over a wide range of levels of contraction. For example, the cells of the stomach distend as the organ receives food, yet the same tonus is exerted for each distention. As food is used up, the cells of the stomach contract and maintain a fairly constant tension on the food that remains. Smooth muscle cells are capable of contraction down to within a few percent of the lengths of the maximal stretch, a contractility far greater than that of any other muscle cells (Bulbring *et al.*, 1970; Ruegg, 1971).

Smooth muscle cells have relatively low sensitivity to stimuli for contraction. For example, for bladder musculature of a rat the time required to stimulate with twice the threshold current (the chronaxie) is about 50 to 100 msec and for uterine muscle of a guinea pig it is 1500 to 2000 msec as compared with 0.15 to 0.5 msec for human skeletal muscle.

Scanning electron microscopic studies on single smooth muscle cells (fibers), isolated by digestion of stomach muscles of a toad *(Bufo marinus)*, demonstrate that on contraction the fibers become covered with blebs. Transmission electronmicroscopic studies show that the contractile fibrils in the fiber are oblique (diagrammed in Fig. 3.2*D*) and that contraction of a fiber results in an increase in diameter and extrusion of cytoplasm in the form of blebs that disappear again on relaxation of the fiber (see Fig. 22.7*E* and *F*; Fay, 1976).

The contraction of smooth muscle cells is influenced by the action of hormones. This field of research is very active, but it is beyond the realm of this discussion. It will suffice to mention that in the vertebrate, epinephrine inhibits the contraction of smooth muscle cells and acetylcholine enhances it. Female sex hormones and the pituitary gland secretion make possible powerful contractions of the uterine musculature (Turner and Bagnara, 1976).

Contraction of Cardiac Muscle Cells

An invertebrate heart may be a very simple organ—merely a contractile tube resembling a more active visceral muscle, as in the annelid worms. In fishes the heart is still tubular, but in all other vertebrates it is a fairly complex struc-

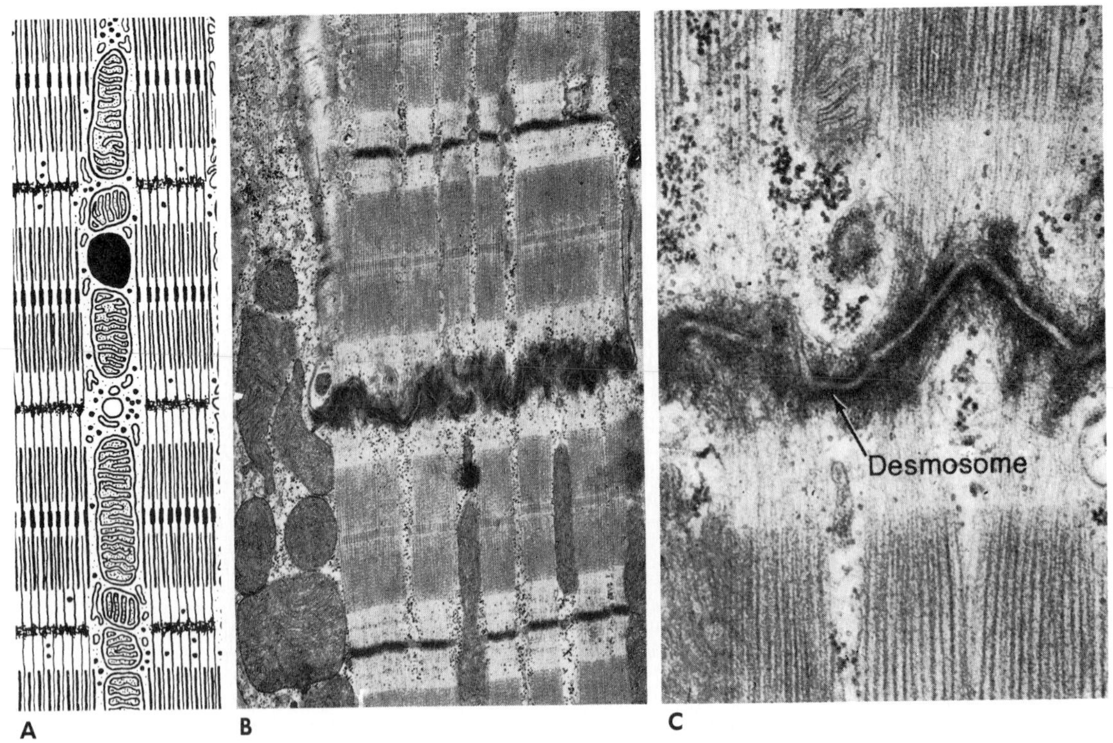

Figure 22.11. Vertebrate cardiac muscle. A, Drawing showing thick and thin filaments, mitochondria, sections of tubules of the sarcoplasmic reticulum (T-system), and intercalated discs (dark lines) separating individual cells. B, Electron micrograph of vertebrate cardiac muscle showing one intercalated disc between two cells and mitochondria that are especially prominent on the left side (×12,000). C, Electronmicrograph showing detail of one intercalated disc (×70,000). In addition to the desmosome some other junctional units already featured in Figure 8.13 are found. (A from Lentz, 1971: Cell Fine Structure. W. B. Saunders Co., Philadelphia; B from Bloom and Fawcett, 1975: A Textbook of Histology. 10th Ed., W. B. Saunders Co., Philadelphia; C from Fawcett and McNutt, 1969: J. Cell Biol. 42: 1.)

ture. The vertebrate heart muscle, viewed in sections under the light microscope, appears to be composed of an interlacing network of striated muscle fibers and was thought to be a syncytium, that is, to form a single multinucleate unit. However, electron micrographs show that the network is composed of separate cells (fibers) that are in intimate contact rather than in a continuous syncytium (Sjöstrand *et al.,* 1958). The separate fibers may be demonstrated also by treatment with strong alkali that causes them to detach from one another. Each cell is indicated by a nucleus, and each mass of cytoplasm is separated from the next by membranes called intercalated discs (Fig. 22.11). The discs are complex structures including desmosomes and tight junctions (see Chapter 8).

The vertebrate heart, even when removed from the body, still continues to beat. The beat is, therefore, said to be myogenic, although a particular balance of ions is necessary for maintenance of this activity (see Chapters 4 and 17).

Even chick cardiac muscle cells in tissue cultures (dispersed by action of proteolytic enzymes) are capable of contraction. They are found to be of two kinds, "leaders" and "followers." The leaders initiate the beat, and the followers contract only if they are in contact with one another, so that a single rhythmic beat may follow (Harary, 1962; Smith and Berndt, 1964). Even cells isolated from adult hearts will beat in appropriate media (Bloom, 1970). The transmembrane potentials of such cells are like those of muscle cells generally, the resting potential being 59 mv and the action potential 71 mv (Lehmkuhl and Sperelakis, 1963).

Although the entire heart is capable of contracting, a gradient of excitability is found in its parts, and the most excitable portion is called the *pacemaker.* In the frog heart the sinus venosus is the pacemaker (Fig. 22.12*A*), in the mammalian heart the *sinoatrial node,* a vestige of the sinus venosus, is the pacemaker. From the pacemaker the wave of excitation (depolarization) sweeps over the heart, initiating the beat in the atria, from which it spreads by way of the atrioventricular node and the *bundle of His* to the ventricle (see Fig. 22.12*B*). Modified muscular fibers (Purkinje fibers in the bundle of His) in the ventricle carry the action potential at the rate of about 5 meters per second to various portions of the

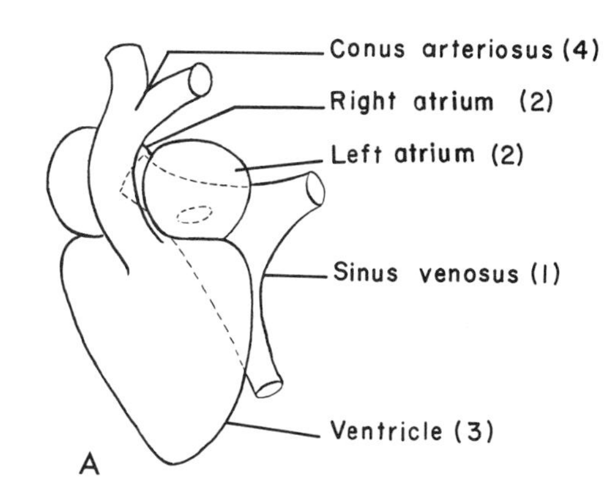

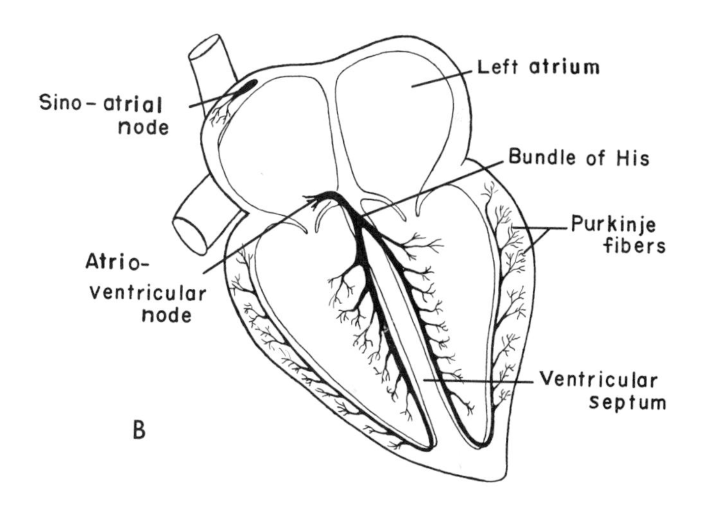

Figure 22.12. A, *Sequence (shown by numbers) in contraction of the parts of a frog's heart.* B, *Diagram of a mammalian heart, showing the bundle of His and the Purkinje fibers.*

heart, making possible a coordinated contraction (Guyton, 1976).

Initiation of the heart beat, as seen in the frog, is not the prerogative of the sinus venosus alone, since if the sinus is cut from the atria, or the atria from the ventricle, each piece, after it has recovered from the shock of the operation, will continue to beat. However, only the beat of the portion containing the sinus venosus equals that of the intact heart; the isolated atria and ventricles beat more slowly. The severed tip of the ventricle shows the least ability to initiate a contraction. Apparently the pacemaker is the most excitable portion of the heart; therefore, it is the first to contract. If it is affected locally by the application of a cold or hot rod, the heart beat is lowered or increased, respectively.

If a ligature is tied between the pacemaker and the rest of the heart, the ventricle no longer beats at its normal rate (Fig. 22.13). If the ventricle, physiologically isolated in this manner, is stimulated by electrical means, a threshold stimulus evokes contraction to the full extent characteristic of the normal tonus level of the heart. Increasing the stimulus does not increase the degree of contraction. Therefore, the cardiac muscle contraction obeys the all-or-none rule under the conditions of these experiments.

The refractory period for cardiac contraction is fairly long, about 0.3 second in man (0.25 second absolute, 0.05 second relative—Guyton, 1976). It depends upon the size of the animal, being shorter for the rapidly beating heart of a small animal. It is shorter also in a warm-blooded animal than in a cold-blooded one of the same size. Because of the long refractory period, the heart muscle fibers cannot be tetanized on continued stimulation. Electrical stimulation is effective only if the heart muscle fibers are in the relative refractory phase (Fig. 22.14), when an extra contraction may be induced. The extra contraction is followed by a compensatory phase (refractory time) during which contraction cannot be induced by further stimulation.

The long refractory period of the heart probably results from permeability changes and

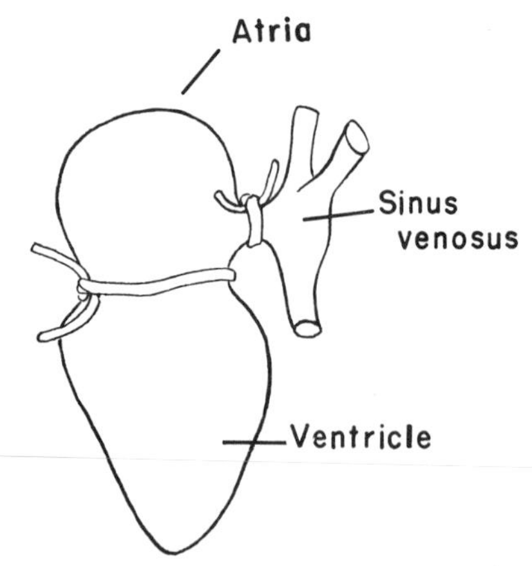

Figure 22.13. Isolating the sinus venosus from the atria and the atria from the ventricle by means of ligatures.

ionic fluxes at the cell membranes. The Na^+ influx is not cut off as rapidly as in nerve, nor does the K^+ efflux rise above the steady-state condition. Repolarization is therefore slower (Katz, 1966).

The force of the isolated heart beat depends upon the stretch to which the fibers are subjected, i.e., the degree to which the heart is filled with blood. This relationship is known as *Starling's law of the heart.* The heart muscle fibers maintain tonus at various levels of contraction; they can contract from any one of these levels. Such a response is adaptive and of importance in maintaining a good blood supply, since the heart may respond to varied demands upon it by a change in the strength or frequency of contraction. It is questionable whether such control operates in the animal in which neural controls are probably of primary importance (Guyton, 1976).

The vertebrate heart beat, although self-initiated, is regulated by inhibitory and excitatory nerve fibers from the autonomic nervous system. Stimulation of the inhibitory nerve fibers, which come by way of the tenth cranial nerve (the vagus), slows or stops the heart (Fig. 22.15). Otto Loewi showed that something diffusible appears in the heart after stimulation of the vagus and that the perfusate from such a heart, when circulated to another heart, will inhibit the second heart also (Fig. 22.16). For studies in this field he received the Nobel Prize in 1936. The diffusible substance, or "vagusstoff," was later identified as acetylcholine, application of which slows or stops the heart beat. Acetylcholine causes hyperpolarization of the heart by potassium ion efflux, thus preventing or delaying depolarization of the muscle when myogenic impulses impinge.

A heart slowed by acetylcholine or by brief stimulation of the vagus nerve, usually beats more forcibly for a while and then resumes its normal rate. The beat is resumed because the acetylcholine is decomposed to choline and acetic acid by a highly active enzyme, choline esterase, which is widely distributed, especially in nerve and muscle cells (see Chapters 20 and 21). Sometimes, even if vagal stimulation is continued, the heart may occasionally "escape" from the inhibitory influence of the vagus and beat more forcibly than normally, thus indicating the strong tendency of the heart muscle cells to continue rhythmic contraction.

Stimulation of the excitatory nerve fibers, which come by way of the sympathetic nerve, accelerates the heart (Fig. 22.17). It is found that the perfusate of a heart excited in this

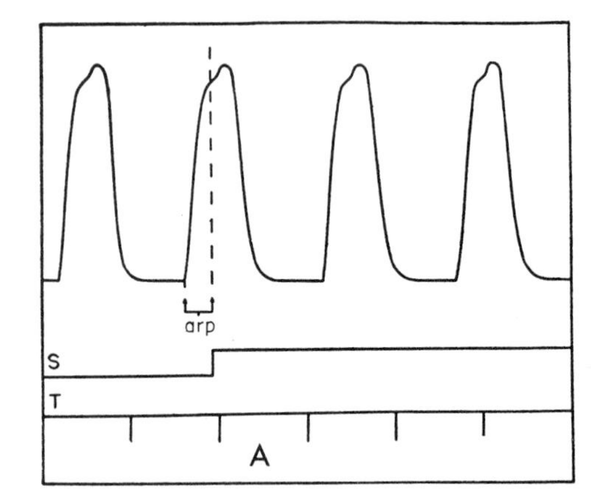

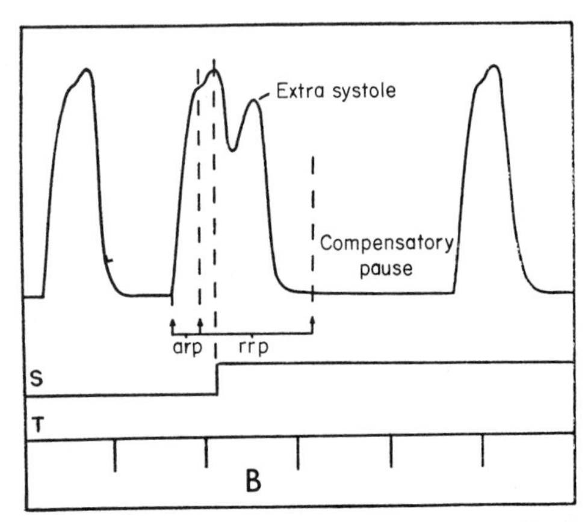

Figure 22.14. A, *Demonstration of the absolute refractory period (arp) of the heart. Each spike is a kymograph record of a heart beat, the peak being the height of ventricular systole, after which the heart begins to relax. When the heart is contracting, a single electric shock (S) has no effect. T is the time in seconds.* B, *Demonstration of the relative refractory period (rrp). When a single electric shock is sent into the ventricle at about the peak of ventricular contraction or soon afterwards, an extra systole appears, subnormal in strength and followed by a compensatory pause. T is the time in seconds. (From Amberson and Smith, 1948: Outline of Physiology. 2nd Ed. Copyright © 1948. The Williams & Wilkins Company, Baltimore.)*

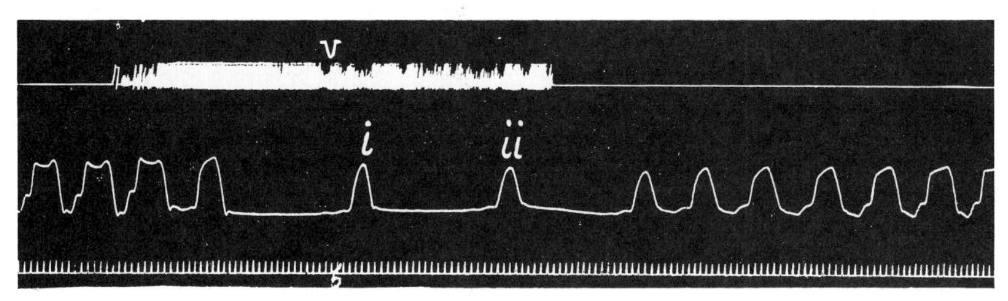

Figure 22.15. A kymograph record of the inhibition of heart beat by stimulation of the vagus. The signal magnet record at v shows the tetanic stimulation of the vagus. At i and ii the heart "escapes" from the inhibitory influence, indicating the strength of the tendency to beat. The lower markings are 0.2 second intervals.

manner contains norepinephrine (differing from epinephrine by a CH_3 radical). Application of norepinephrine to the surface of the heart and its injection into the heart are familiar methods for stimulating heart muscle contractions. Norepinephrine secreted by the terminations of the sympathetic nerves leads to a more rapid depolarization of the pacemaker of the heart so that the threshold for firing the spike is reached sooner; the spike height is also increased (in the overshoot [the spike beyond the point of depolarization]), as is the rate of rise of the spike height (Guyton, 1976).

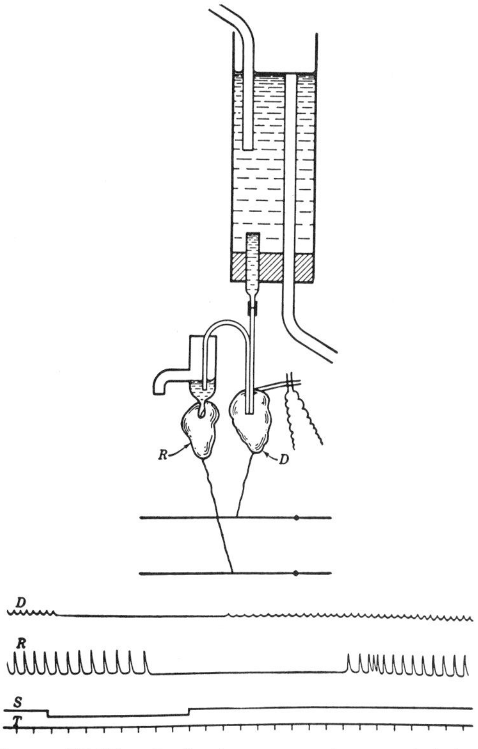

D R S T

Figure 22.16. Perfusion apparatus containing Ringer's solution and two hearts with fluid intercommunication to demonstrate Loewi's classic experiment on regulation of the heart by a hormone. Heart D is stimulated electrically by way of the vagus nerve, and it stops beating, as shown in kymograph record D below. Fluid pumped from donor heart (D) into recipient heart (R) stops heart R somewhat later, as shown in record R below. The effective substance in the perfusate has been shown to be acetylcholine. S (below) indicates the time at which the electric stimulus is applied to the vagus. T represents the time in seconds for the 3 records above it. (From Scheer, 1953: General Physiology. John Wiley & Sons, New York.)

Contractions of some invertebrate hearts share many of the properties of contraction of the vertebrate heart. However, in other invertebrate hearts like that of the adult king crab *Limulus*, the heart is *neurogenic;* that is, it will not beat when severed from nervous connections. It is also interesting to note that in the larval stage of this animal, the heart is myogenic, suggesting that the neurogenic heart beat is the result of a more specialized structure than a heart in which the contractions are myogenic. Studies on vertebrate and invertebrate hearts and their control are discussed in books on human and comparative physiology (Guyton, 1976 and Prosser, 1973).

Whereas recording heart beat and other muscle contractions by kymograph is illustrated here because its directness and simplicity enable the reader to visualize the process, usually an instrument made up of a number of components (black boxes in a figure) collectively called a *physiograph* is used in current experimentation on muscle contraction. In this instrument mechanical movement is transduced to an electrical signal that can be amplified and automatically recorded in a number of ways. One example is recording directly as an oscillographic signal that can be photographed, or transduced again to be mechanically recorded by levers that trace the signal on moving graph paper, or even transduced to an audio signal heard simultaneously with the other record. Audio signals add a dramatic effect in classroom laboratories, especially in studies on the heart, but are also used by researchers to monitor prolonged experiments. Physiographs are also effective in recording blood pressure and electrocardiograms from electrodes on the surface of the body (as in medical practice). A number of physiological records—heart beat, blood pressure, electrocardiograms, and time

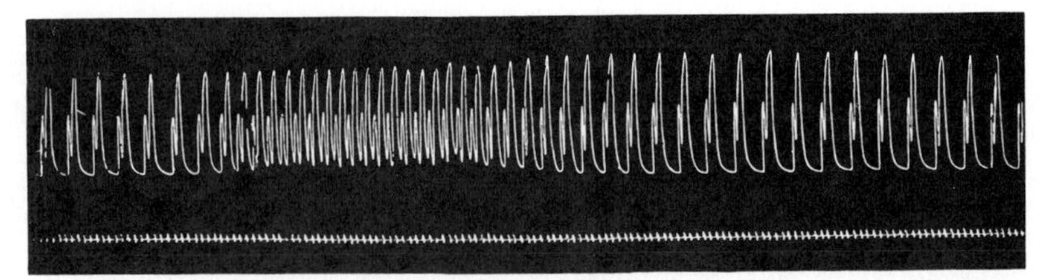

Figure 22.17. Kymograph record showing acceleration of the heart of a frog by stimulation of the sympathetic nerve. (From Zoethout and Tuttle, 1955: Textbook of Physiology. 12th Ed. C. V. Mosby Co., St. Louis.)

lapse—can be recorded simultaneously on a multichannel recorder. The electronic circuitry and details of instrumentation are not relevant to our discussion; the reader is referred to handbooks on this subject (Rudin *et al.*, 1971).

Contraction of Skeletal Muscle Fibers

Locomotion is made possible by skeletal (body) muscle fibers, and the development of these highly differentiated structures indicates the premium placed upon speed in the animal world.

Individual muscle cells of the body musculature of lower invertebrates are relatively simple cells, much like visceral muscle cells of vertebrates and show similar rhythmic pulsations (Prosser, 1973). It is only in some higher invertebrates (cephalopod mollusks and arthropods) and in the vertebrates that skeletal muscle cells become much elongated and show striking transverse striations. Such muscle cells probably represent highly specialized structures in which certain properties of contractility have been exaggerated while other properties, such as myogenicity, have been lost (Bendall, 1969). Striated muscle fibers of intermediate complexity are found in some other invertebrates, even some coelenterates.

The skeletal muscles are made up of multinucleate muscle fibers containing contractile myofibrils, across which run transverse *striations,* enclosed in a *sarcolemma* (Fig. 22.18). The contractile fibrils of a muscle fiber practically fill it, but a certain amount of cytoplasm *(sarcoplasm)* surrounds the fibrils, and in the cytoplasm are found the spindle-shaped nuclei, the entire fiber being a multinucleate unit. It is interesting to note that such multinucleate fibers have arisen not only in the vertebrates but also in several invertebrate groups (Prosser, 1973). This suggests an evolutionary advantage of a unit of this type, in which the contractile fibrils may become much longer than would be possible in a single cell. A long muscle fiber has an advantage in the propagation of excitation for contraction and in the mechanical work performed.

Some vertebrate muscle fibers are red, some

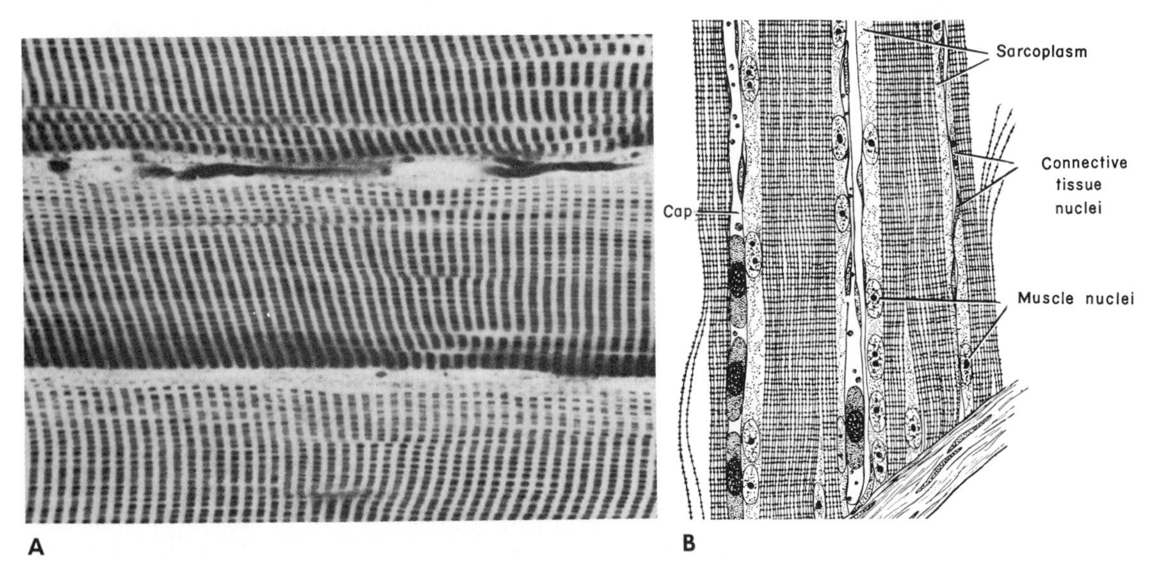

Figure 22.18. Longitudinal sections of striated body muscle. A, Three muscle fibers from the dog. The dark transverse A bands and light I bands are clearly differentiated; in the A bands the dark longitudinal striations correspond to myofibrils. B, Striated muscle from the sucker Castostomus. (Cap refers to capillary.) (A from Bloom and Fawcett, 1975: A Textbook of Histology. W. B. Saunders Co., Philadelphia. B from Dahlgren and Kepner, 1925: Principles of Animal Histology. Macmillan, New York.)

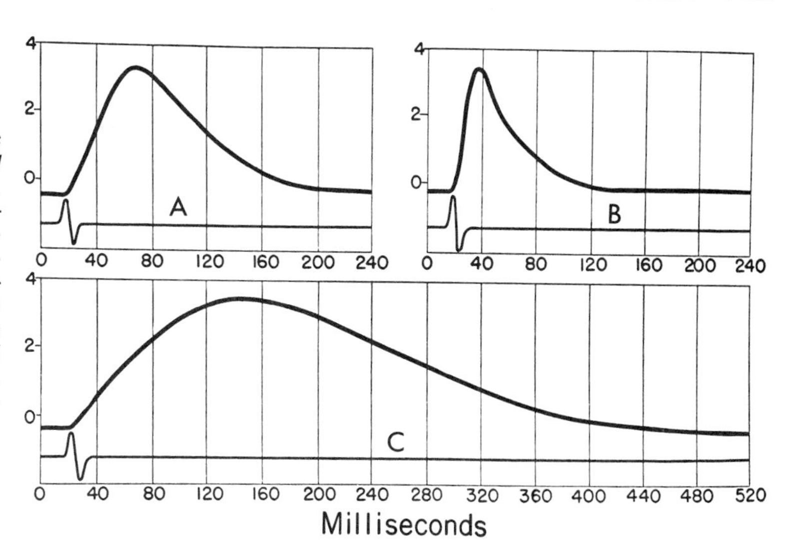

Figure 22.19. *Action potentials (lower record in each example) and muscle twitches in three muscles. A, The mainly white-fibered gastrocnemius. B, The wholly white-fibered internal rectus. C, The red-fibered soleus. The action potentials are diphasic because they were recorded from the surface of the muscles. (From Amberson and Smith, 1948: Outline of Physiology. 2nd Ed. © 1948. The Williams & Wilkins Company, Baltimore.)*

white, and some muscles are made up predominantly of white fibers, some predominantly of red, and some of an equal mixture of the two. Red fibers contain myoglobin, one of the four units that make up a hemoglobin molecule; this gives them a characteristic color and serves as a reserve of oxygen. Because they unload their oxygen only at lower oxygen partial pressure than hemoglobin, they are slower and have less ATPase activity than white fibers. They are low in glycogen stores but have many mitochondria and a copious capillary blood supply, and red fibers carry on active aerobic metabolism. For this reason they are not readily fatigued because they can regenerate ATP supplies as rapidly as they are used up. Such muscles function where long, sustained, slow contractions are required. White muscle fibers, on the other hand, have little myoglobin and few mitochondria, but they have large glycogen reserves and are primarily powered by glycolysis. Since glycolytic ATP synthesis is not efficient, the white fibers fatigue readily because they cannot regenerate ATP rapidly enough. Nevertheless, white muscle fibers have high ATPase activity, which permits rapid contractions for brief periods of time. Such fibers constitute muscles that make rapid movements for short periods of time. (Examples of the characteristic properties of red and white fibers are shown in Figure 22.19.)

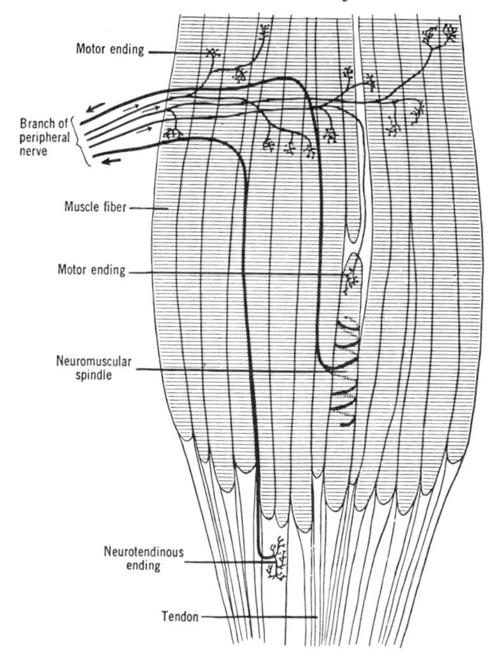

Figure 22.20. *Relations between neurons and muscle fibers. Efferent neurons end in myoneural junctions (motor endings). Two afferent neurons are also shown, one from a muscle spindle and the other from a tendon receptor. (From Gardner, 1968: Fundamentals of Neurology. 5th Ed. W. B. Saunders Co., Philadelphia.)*

The skeletal muscle fibers of vertebrates are innervated only by excitatory nerve fibers (single innervation). An efferent neuron innervates a small number of muscle fibers, each branch of the distal portion of the axon attaching to a muscle fiber by a muscle end plate or *myoneural junction* (see Figs. 21.15 and 22.20). The motor neuron and the muscle fibers innervated by it respond as a unit and constitute the *motor unit.* A large number of motor units make possible a fine gradation of force—from delicate movements to powerful sustained ones—depending upon the number of motor units called into action. When the nerve impulse traveling on the efferent neuron reaches the myoneural junction, it first induces a local excitatory state (local potential) on the muscle fiber near the myoneural junction. If this change is of sufficient magnitude (above threshold), an action potential (see Chapter 20) sweeps the muscle fiber, following which the muscle fiber contracts (Fig. 22.19).

The muscle fibers may also be directly stimulated by heat, light, chemicals, pressure, or electricity, and the excitation is always accompanied by action potentials. When a muscle fiber is stimulated electrically with a current below threshold, a local excitatory state develops around the cathode. This state of excitation grows with continued stimulation, but only when the current is above threshold and the time of application sufficiently long will the excitation become propagated and induce contraction of the muscle fiber. Muscle fibers, like nerve fibers, maintain a resting potential of about 70 mv, the outside being positive to the inside. The sign of the potential reverses during excitation. Evidence indicates that during the development of the propagated action potentials ionic changes occur in muscle fibers much as in nerve fibers. The action potential travels over the muscle fibers at the rate of 1 or 2 meters per second, which is much slower than the rate of the action potential in nerve fibers. Electrical excitation induces muscular contraction by liberation of calcium ions from the *sarcoplasmic* (endoplasmic) *reticulum* in excitation-contraction coupling (Peachey, 1968). This is discussed in Chapter 23.

The chronaxie (the minimum time at which a current twice the rheobase will excite contraction) for muscle fibers is comparable to that for the motor neurons innervating it. The strength-duration curves for muscle resemble those for nerve fibers, only the absolute values being different. Muscle fibers are thus highly excitable systems and have many properties in common with nerve fibers.

If a muscle fiber is stimulated with a current above rheobase but of a duration insufficient to excite, contraction does not occur. If the stimuli are frequently repeated, they may summate, each subliminal excitation producing a local potential that may be added to others until the local potential sets off a propagated action potential (spike) that, in turn, is followed by a contraction (Fig. 22.19). When single muscle fibers are isolated by removal or destruction of surrounding muscle fibers and then stimulated with a graded series of shocks, the individual muscle fiber always contracts as a unit and to the same extent. When individual muscle fibers in an intact muscle are labeled by spraying with mercury and then excited with electrodes small enough to contact only single fibers, individual fibers are seen to contract completely or not at all. These experiments show that single muscle fibers in the intact muscle, like single neurons, obey the all-or-none law under the conditions of the experiment.

An interesting but little understood exception to the all-or-none principle occurs in some frog muscle fibers of the back, which are innervated by small nerves. Stimulation of these nerves causes individual muscle fibers to contract, sometimes to a greater degree than at other times. Analysis has shown that a local excitatory state in such muscle fibers is accompanied by a local contraction, the degree of local contraction depending upon the magnitude of the local excitatory potential. The purpose served by local contractions in these muscle fibers is unknown. On the other hand, in some invertebrates, such as the Crustacea, local contractions are the main method for obtaining graded responses, since the number of motor units is very small (Wiersma, 1952).

When a muscle is stimulated with electricity, by varying the voltage, it is possible to find a minimal stimulus threshold below which no response is observed. As the voltage is increased above the threshold, the force of contraction of the muscle increases in degree, until a maximal contraction is observed (Fig. 22.21; Guyton, 1976). The same graded response is found upon stimulation of an intact muscle in the animal by way of the nerve and is the consequence of the response of successively larger numbers of motor units, as stimulation increases in strength. A muscle has many motor units, each of which has a characteristic threshold. In a population of fibers some have low, some high, and some intermediate thresholds, the sensitivity being distributed about a mean. When a muscle is stimulated, gradation is obtained by calling into action all those fibers for which the stimulus is above threshold (Prosser, 1973).

A single electric shock induces a single *twitch* of a muscle fiber. The twitch is a laboratory phenomenon primarily useful for analysis of the properties of muscular contraction. The record of a muscle twitch may be spread out by being recorded on a rapidly moving drum of a kymograph, or a physiograph recording sheet, thereby making possible a study of the time relations in the contraction and relaxation of the muscle fiber. The entire cycle for the "form" curve of the internal rectus muscle of a frog occupies about 0.1 second (Fig. 22.19). Immediately after stimulation, a latent period of about 0.0075 second elapses before the contraction begins. During this period the action po-

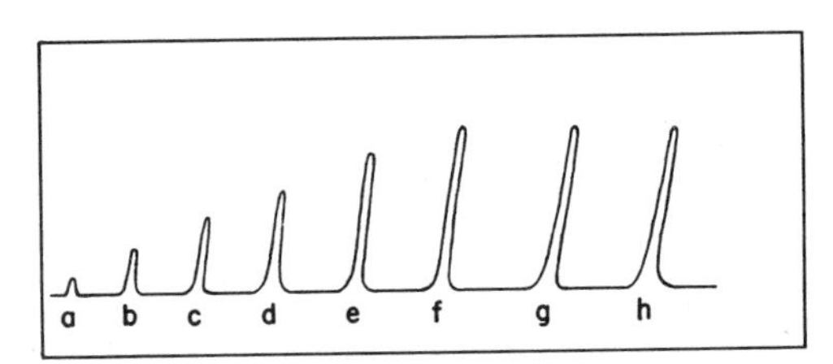

Figure 22.21. *Graded response of a skeletal muscle to increasing stimulation. A maximal response is reached at f.*

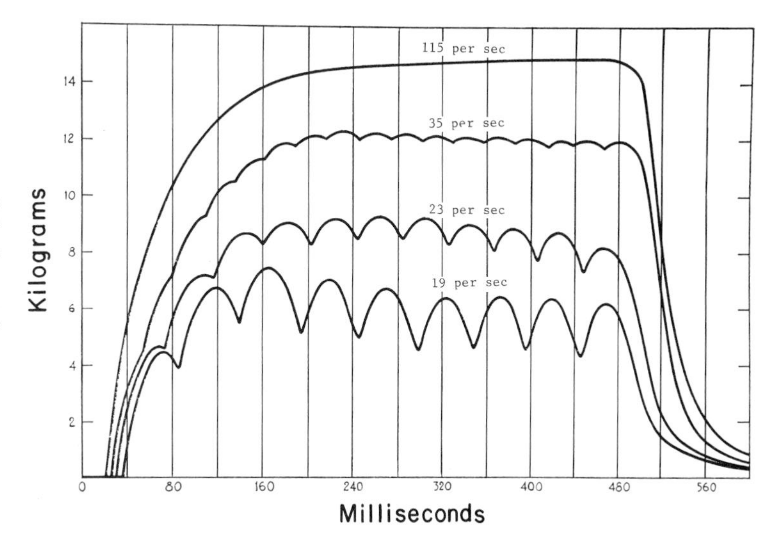

Figure 22.22. *Partial and complete tetanus caused by varying rhythms of stimuli applied to the gastrocnemius muscle. (From Amberson and Smith, 1948: Outline of Physiology. 2nd Ed. © 1948. The Williams & Wilkins Company, Baltimore.)*

tential, beginning at the point of stimulation, sweeps the muscle. Contraction occupies about 0.04 second, relaxation about 0.05 second. The exact times vary with the type of muscle fiber and the temperature. After stimulation muscle fibers become *refractory* for a period lasting a fraction of a millisecond to several milliseconds, during which they cannot be excited again. This is the period when the potassium gate of the cell membrane is open and the membrane is not capable of a spike potential (see Chapter 21). Since the twitch of a frog internal rectus muscle reaches its full contraction 0.04 second after excitation and relaxes only after another 0.05 second, it is evidently excitable long before it has relaxed. Consequently, it is possible for a skeletal muscle fiber to maintain continuous tension if stimulated frequently enough within a given time. A sustained contraction of this type is called *tetanus* (Fig. 22.22).

If the stimuli are spaced so that they reach the muscle after it has begun to relax, incomplete tetanus is obtained (Fig. 22.22). The record shows that a relaxing muscle may be induced to contract from any degree of relaxation, as might be expected from the shortness of the refractory period.

Contractions of the muscle fibers in the living animal are normally initiated by continuous trains of discharge along motor neurons. The impulses are spaced about a hundredth of a second apart, leading to tetanic contractions of some muscle fibers in various muscles of the body. By this means, muscular *tonus* is maintained. This tonus is responsible for maintenance of posture and for keeping the musculature in readiness for action. In a muscle stimulated to tetanic contraction, the degree of contraction increases during the course of the train of impulses, and the final state of contraction achieved is greater than the initial state. This indicates that a muscle kept contracted does more work per stimulus than one stimulated only from a state of rest, as shown by the force it can exert or the weight it can lift (indicated in kilograms in Fig. 22.22), a behavior in apparent contradiction to the all-or-none principle. The fact is, individual skeletal muscle fibers will also contract to a greater extent if stimulated when already contracted than if stimulated from a state of rest, thus showing

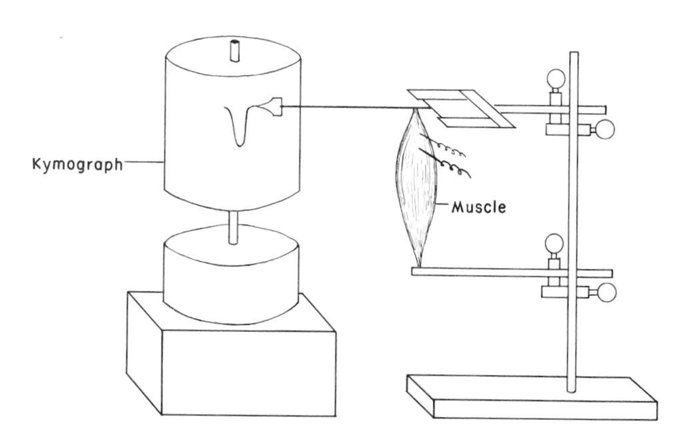

Figure 22.23. *Isometric muscle contraction. The muscle is attached to a lever which is soldered to a steel spring held tightly in the clamp. It therefore exerts tension with only a minimum of contraction which is magnified by the lever. The kymograph has been largely replaced by physiographs that record several physiologic parameters at one time. The figures with the kymography are retained because they illustrate that most of the information needed for our discussion can be obtained with simple equipment.*

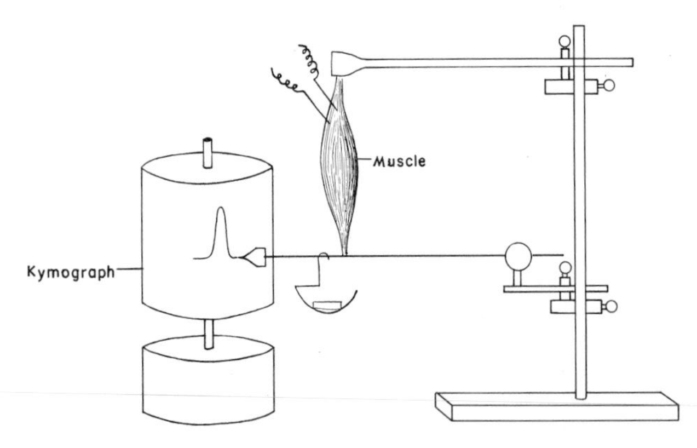

Figure 22.24. Isotonic contraction. The muscle pulls an easily movable lever and shortens as it does so.

increased capacity for work under these conditions. The all-or-none law, then, holds under a given set of circumstances.

When a stimulated muscle is unable to move its load and does not shorten, the muscular contraction is said to be *isometric* ("same length"). Although tension (force) is developed and the muscle tonus is increased, no mechanical work is done and all the expended energy appears as heat. For recording the maximum tension exerted by a particular muscle, use may be made of a lever and an attached spring (see Fig. 22.23). Muscles maintain body posture by isometric contractions.

On the other hand, if a muscle is able to move a load, that is, when the resistance offered by the load is less than the tension developed, the muscle shortens and performs mechanical work. The same tonus is maintained, irrespective of the degree of contraction; hence, this contraction is said to be *isotonic* ("same tonus") (Fig. 22.24). The muscles used in lifting a load are contracting isotonically. In lifting a load, an average of 20 to 25 percent of the total chemical potential energy expenditure appears as mechanical work, the remainder being dissipated as heat. The muscle is therefore comparable in efficiency to the gasoline engine (Guyton, 1976). For quantitative studies a more refined lever and auxiliary recording equipment are required. An example of the lever is shown in Figure 22.25.

FATIGUE

A muscle fiber is capable of maintaining tetanus for a considerable period of time, the limit being set by development of *fatigue* and resultant failure of contraction (Fig. 22.26).

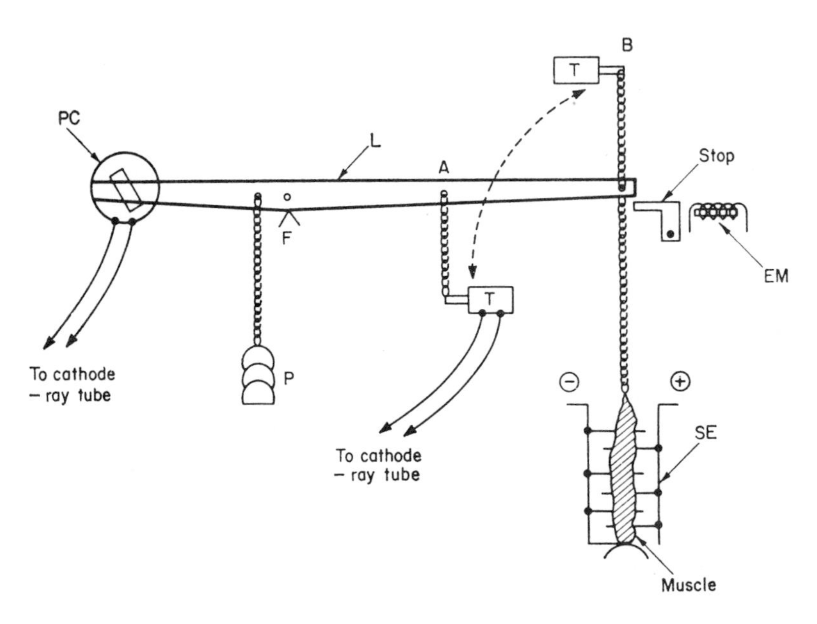

Figure 22.25. Experimental setup for twitch and tetanus experiments. Note the various stops for controlling the length of the muscle, and also how the tension transducer can be moved so as to measure tension either in an isotonic (A) or an isometric contraction (B). The lever, L, is made of light metal and is pivoted at F. The muscle is attached at the end of L by a chain tied firmly to the tendon of the muscle. The other end of the muscle, still attached to the bone, is attached to a rigid support in a bath. The muscle rests on platinum stimulating electrodes. To study isotonic twitches, a load is placed at P. T is a tension transducer, EM an electromagnet used to stop the contraction at any desired muscle length, and PC is a photocell to record movements of the lever. (From Bendall, 1969: Muscles, Molecules and Movement. American Elsevier, New York.)

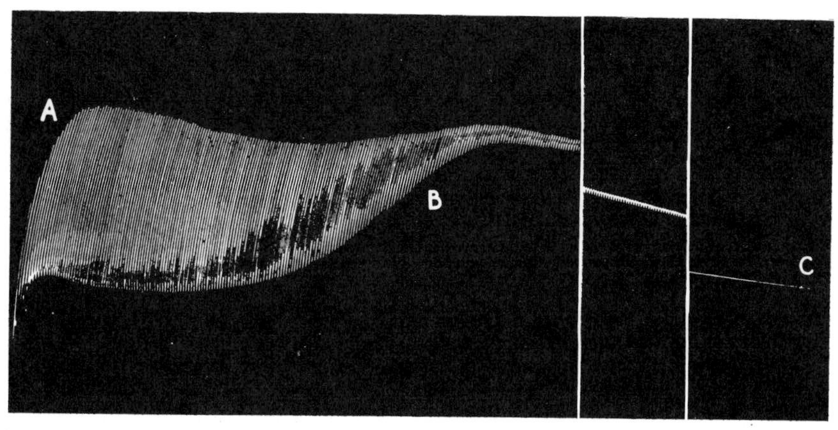

Figure 22.26. *Kymograph records of fatigue curve, showing* A, *staircase,* B, *contracture, and* C, *complete fatigue. (From Zoethout and Tuttle: 1955: Textbook of Physiology. 12th Ed. C. V. Mosby Co., St. Louis, p. 109.)*

Fatigue is also observed if individual twitches are induced continuously—by electricity, for example—although a frog leg muscle contracts about 12,000 times in 3⅓ hours before fatiguing (Labhardt, 1927). As a muscle fatigues, its degree of contraction decreases gradually because some of the component fibers fail to contract; consequently, the muscle's capacity for work decreases. The cause of muscular fatigue is not known with certainty. It is thought to be due to several changes occurring concurrently: first, exhaustion of the high-energy phosphate substrates (adenosine triphosphate, ATP, and creatine phosphate, CP) serving as the immediate source of energy and, second, accumulation of waste products. The molecular mechanism and the role of ATP and CP in muscular contraction are considered in Chapter 23. Experiments have demonstrated that removal of wastes by continually washing a thin sheet of muscle in balanced salt solution delays fatigue. In any event, if stimuli are applied at such intervals that the muscle fibers have a chance to recover, they may be able to contract for hours without fatigue. Apparently, recovery from a contraction is complete, or nearly so, in about 1 second.

Intact muscle fibers, stimulated *in vivo* through a nerve, fatigue much more rapidly; e.g., frogs are fatigued after 29 to 82 jumps in 1 to 2 minutes (Galperin *et al.,* 1934). In this case the impulse passes by way of the myoneural junction and thence to the muscle fiber. Thus, the myoneural junction fatigues quite readily, even though the muscle fibers of a fatigued animal will continue to contract on direct stimulation. The nerve fiber fatigues even more slowly than the muscle fiber it innervates. The myoneural junction apparently acts as a safety valve, preventing injury to the muscle fiber by too prolonged contraction.

Relaxation of a skeletal muscle fiber depends upon cessation of activities leading to contraction and could conceivably be a passive process, since skeletal muscles occur in pairs that are opposite in action, e.g., an extensor and a flexor. Studies of an isolated muscle show that without any pull or traction upon it, a contracted muscle relaxes and extends itself but is not capable of doing work in the direction of extension. Thus, if the muscle is placed in a bath of mercury, it cannot extend because of the weight of the mercury. Nonetheless, evidence indicates that relaxation of the muscle fiber is a spontaneous process and is largely a consequence of the elasticity of the membrane surrounding each fiber as well as of the elasticity of the various other connective tissue components binding the fibers into units of the muscle and not merely the absence of the active state (see Chapter 23). At the molecular level contraction is stopped by the removal of calcium ions by the sarcoplasmic (endoplasmic) reticulum in muscle fibers. Calcium ions are required for the ATPase activity of the muscle proteins, and ATP is the immediate source of energy for contraction. Once contraction has been completed and the calcium reabsorbed into the sarcoplasmic reticulum, the reserve ATP reacts with the muscle proteins, causing the actin and myosin myofilaments to dissociate from each other. This constitutes relaxation (see Chapter 23).

Contracture is the name given to a particular muscular contraction, which is sustained, reversible, occurs with the liberation of heat, and shows no propagated spike potentials, but in which the muscle fiber remains depolarized. It is brought on by stimuli such as strong electric currents, heat, acids, narcotics, potassium, citrate, caffeine, acetylcholine, nicotine, veratrine, and many other drugs. When veratrine, for example, is dropped on a single muscle fiber, it gives rise to several propagated spikes, after which a prolonged negativity (after-potential) develops accompanied by a contracture. Spontaneous normal repolarization of the membrane of the muscle fiber, or anodal stimulation, brings the contracture to an end. If the muscle fiber is injured, contracture becomes irreversible.

If a skeletal muscle fiber is stretched by weights, it may be considerably extended. After removal of the weights it returns to its original length, indicating that it is highly elastic. The elasticity is accounted for in part by the contractile fibrils and in part by the sarcolemma membrane that encloses each muscle fiber. The membrane remains elastic, even when the contents of the muscle fiber have been expressed, although it is not as elastic as the whole fiber (Ramsey and Street, 1941; Street and Ramsey, 1965).

DEVELOPMENT OF MUSCLE CELLS

Skeletal muscle fibers grow by the fusion of single embryonic myoblasts. In tissue culture these cells are highly motile, as are the epithelial cells and fibroblasts often explanted with them, but only myoblasts stop motion on making contact with one another. Myoblasts fuse only with other myoblasts, never with epithelial cells or fibroblasts. Furthermore, fusion occurs only between myoblasts in postmitotic stages of the cell cycle. In cultures containing myoblasts from two species, for instance, chick and rat, chick and rat cells in postmitotic stages fuse with one another. In the multinucleate myotubules formed, chick and rat components occur in proportion to the population density of cells of each species in the mixed culture. The myofibrils formed contain both chick and rat myoproteins, also in proportion to the number of nuclei of each of the species in the forming fiber. An appropriate calcium ion concentration is required for cell fusion (Ringertz and Savage, 1976).

In the developing fibers of the embryonic chick muscle, thick and thin filaments appear at the same time. Microtubules are also present below the sarcolemma, but they appear to have little relation to the muscle fibers. The sarcoplasmic reticulum appears after the formation of the fibrils (Fischman, 1967). The number of fibers in a muscle is constant after birth, although the fibers may grow in length, but the number of myofibrils in a fiber may increase as much as fourfold by transverse splitting and elongation (Goldspink, 1970; Allen, 1978). The ATPase activity of the myosin and the actomyosin in the filaments shows progressive development in embryos and in some species even after birth. The ATPase activity of these proteins correlates with the speed of the muscles in which they are found (Needham, 1971). The details of myogenesis with relation to function of the muscles are only now being studied.

LITERATURE CITED AND GENERAL REFERENCES

Albrecht-Buehler, G., 1978: The tracks of moving cells. Sci. Am. (Apr.) *238*: 69–76.

Allen, E. R., 1978: Development of vertebrate skeletal muscle. Am. Zool. *18*: 101–111.

Allen, R. D., 1964: Cytoplasmic streaming and locomotion in marine Foraminifera. *In* Primitive Motile Systems in Cell Biology. Allen and Kamiya, eds. Academic Press, New York, pp. 407–432.

Allen, R. D. and Kamiya, N., eds., 1964: Primitive Motile Systems in Cell Biology. Academic Press, New York.

Allen, R. D., Pitts, W. R., Jr. and Brault, J., 1963: Shuttle-streaming: synchronization with heat production in slime mold. Science *142*: 1485–1487.

Allen, R. D. and Taylor, D. L., 1975: The molecular basis of ameboid movement. *In* Molecules and Cell Movement. Inoue and Stephens, eds. Raven Press, New York, pp. 239–258.

Ambrose, E. J., 1965: Cell movements. Endeavour *24*: 27–32.

Bagnara, J. T., Taylor, J. D. and Hadley, M. E., 1968: The dermal chromatophore. J. Cell Biol. *38*: 67–79.

Bendall, J. R., 1969: Muscles, Molecules and Movement. An Essay in the Contraction of Muscles. American Elsevier, New York.

Blake, J. R. and Sleigh, M. A., 1974: Mechanism of ciliary locomotion. Biol. Rev. *49*: 85–125.

Bloom, S., 1970: Spontaneous rhythmic contraction of separated heart cells. Science *167*: 1727–1729.

Bovee, E. C. and Jahn, T. L., 1973: Locomotion and behavior. *In* The Biology of Amoeba. Jeon, ed. Academic Press, New York, pp. 249–290.

Bulbring, E., Brading, A. F., Jones, A. W. and Tomita, T., 1970: Smooth Muscle. Edward Arnold, London.

Carlson, F. D. and Wilkie, D. R., 1974: Muscle Physiology. Prentice-Hall, Englewood Cliffs, N.J.

Child, F. M., 1967: The chemistry of cilia and flagella. *In* Chemical Zoology. Florkin and Scheer, eds. Vol. I. Protozoa. Kidder, ed. Academic Press, New York, pp. 381–393.

Condulis, J. S. and Taylor, D. L., 1977: The contractile basis of amoeboid movement. V. The role of gelation, solation and contraction in extracts from *Dictyostelium discoideum*. J. Cell Biol. *74*: 901–927.

Craig, R. and Megerman, J., 1977: Assembly of smooth muscle myosin into side polar filaments. J. Cell Biol. *75*: 970–996.

Fabczak, S., Korohoda, W. and Walczak, T., 1973: Studies on the electrical stimulation of contraction in *Spirostomum*. Cytobiology *7*: 152–163.

Fay, F. S., 1976: Structural and functional aspects of isolated single smooth muscle fibers. *In* Cell Motility. A: Motility, Muscle and Non-Muscle Cells. Goldman, Pollard and Rosenbaum, eds. Cold Spring Harbor Conf. on Cell Proliferation *3*: 185–201.

Fingerman, M., 1970: Comparative physiology: chromatophores. Ann. Rev. Physiol. *32*: 345–372.

Fischman, D. A., 1967: An electron microscope study of myofibril formation in embryonic chick skeletal muscle. J. Cell Biol. *48*: 557–575.

Fuchs, F., 1975: Striated muscle. Ann. Rev. Physiol. *37*: 461–502.

Fujii, R. and Miyashita, Y., 1976: Beta adenoceptors, cyclic AMP and melasome dispersion in guppy melanophores. *In* Pigment Cell. Vol. 3. Riley, ed. Barger, Switzerland, pp. 336–344.

Fujii, R. and Miyashita, Y., 1978: Receptor mechanisms in fish chromatophores-IV. Effects of melatonin and related substances on dermal and epidermal melanophores of the siluroid, *Parasilurus asotus*. Comp. Biochem. Physiol. *59C*: 59–63.

Galperin, L., Okun, M., Simonson, E. and Sirkina, G., 1934: Beiträge zur Physiologie der Ermüdung. Arbeitsphysiologie *8*: 407–423.

Geschwind, T. I., Horowitz, J. M., Mikukis, G. M. and Dervey, R. D., 1977: Iontophoric release of cAMP and dispersal of melanosomes within a single melanophore. J. Cell Biol. *74*: 928–939.

Goldspink, G., 1970: The proliferation of myofibrils during muscle fiber growth. J. Cell Sci. *6*: 593–603.

Grinell, F., 1978: Cellular adhesiveness and extracellular substrata. Int. Rev. Cytol. *33*: 65–144.

Guyton, A. C., 1976: Textbook of Medical Physiology. 5th Ed. W. B. Saunders Co., Philadelphia.

Hanson, J. and Lowy, J., 1957: Structure of smooth muscles. Nature *180*: 906–909.

Harary, I., 1962: Heart cells *in vitro*. Sci. Am. (May) *206*: 141–152.

Hayashi, T., 1961: How cells move. Sci. Am. (Sept.) *205*: 184–204.

Hill, A. V. and Solandt, D. Y., 1934: Myograms from the chromatophores of *Sepia*. J. Physiol. *83*: 13–14.

Hitchcock, S. E., 1977: Regulation of motility in nonmuscle cells. J. Cell Biol. *74*: 1–15 (review).

Hoffmann-Berling, H., 1964: Relaxation of fibroblast cells. *In* Primitive Motile Systems in Cell Biology. Allen and Kamiya, eds. Academic Press, New York, pp. 365–375.

Hope, A. B. and Walker, N. A., 1975: The Physiology of Giant Algal Cells. Cambridge University Press, London. (Chapter 12, Protoplasmic Streaming, pp. 162–175.)

Huxley, A. F., 1964: Muscle. Ann. Rev. Physiol. *27*: 131–152.

Huxley, H. E., 1965: The mechanism of muscular contraction. Sci. Am. (Dec.) *213*: 18–27.

Jahn, T. L. and Bovee, E. C., 1969: Protoplasmic movements within cells. Physiol. Rev. *49*: 793–862.

Jahn, T. L. and Rinaldi, R., 1959: Protoplasmic movement in the foraminiferan *Allogromia laticollaris;* and a theory of its mechanism. Biol. Bull. *117*: 100–118.

Jeon, K., ed., 1973: The Biology of Amoeba. Academic Press, New York.

Kamiya, N., 1959: Protoplasmic Streaming. Protoplasmatologia VIII.3a. Heilbrunn and Weber, eds. Springer-Verlag, Vienna.

Kamiya, N., 1960: Physics and chemistry of protoplasmic streaming. Ann. Rev. Plant Physiol. *11*: 323–340.

Kamiya, N., 1964: The motive force of endoplasmic streaming in the amoeba. *In* Primitive Motile Systems in Cell Biology. Allen and Kamiya, eds. Academic Press, New York, pp. 257–277.

Karnovsky, M. L., 1962: Metabolic basis of phagocytic activity. Physiol. Rev. *42*: 143–168.

Katz, A. M., 1977: Physiology and Biophysics of the Heart. Raven Press, New York.

Katz, B., 1966: Nerve, Muscle and Synapse. McGraw-Hill Book Co., New York.

Kimes, B. W. and Brandt, B. L., 1976: Properties of a clonal muscle cell line from the rat heart. Exp. Cell Res. *98*: 367–381.

Kinosita, H., 1963: Electrophoretic theory of pigment migration within the fish melanophore. Ann. N.Y. Acad. Sci. *100*: 992–1003.

Kinosita, H. and Murakami, A., 1967: Control of ciliary motion. Physiol. Rev. *47*: 53–82.

Kuroda, K., 1964: Behavior of naked cytoplasmic drops isolated from plant cells. *In* Primitive Motile Systems in Cell Biology. Allen and Kamiya, eds. Academic Press, New York, pp. 31–41.

Labhardt, E., 1927: Untersuchunger der Muskelermüdung am ausgeschnitten Froschmuskel unter physiologischen Bodingungen. Zeits. Biol. *86*: 27–38.

Lehman, W., 1976: Phylogenetic diversity of the proteins regulating muscular contraction. Int. Rev. Cytol. *44*: 55–92.

Lehmkuhl, D. and Sperelakis, N., 1963: Transmembrane potentials of trypsin-dispersed chick heart cell cultures *in vitro*. Am. J. Physiol. *205*: 1213–1220.

Lester, H. A., 1977: The response to acetylcholine. Sci. Am. (Feb.) *236*: 107–118.

Marsland, D., 1944: Mechanism of pigment displacement in unicellular chromatophores. Biol. Bull. *87*: 252–261.

Matthews, S. A., 1931: Observations on pigment migration within the fish melanophores. J. Exp. Zool. *58*: 471–486.

Mooseker, M. S., 1976: Brush border motility. J. Cell Biol. *71*: 417–433.

Murphy, D. B. and Tilney, L. G., 1974: The role of microtubules in the movement of pigment granules in teleost melanophores. J. Cell Biol. *61*: 757–779.

Mustacich, R. V. and Ware, B. R., 1976: A study of protoplasmic streaming in *Nitella* by laser Doppler spectroscopy. Biophys. J. *16*: 373–388.

Nakajima, H., 1964: The mechanochemical system behind streaming in *Physarum*. *In* Primitive Motile Systems in Cell Biology. Allen and Kamiya, eds. Academic Press, New York, pp. 111–123.

Needham, D. M., 1971: Machina Carnis: The Biochemistry of Muscular Contraction and its Historical Development. Cambridge University Press, Cambridge.

Okajima, A., 1966: Ciliary activity and coordination in *Eupolotes eurystomus*. I. The effect of microdissection of neuromotor fibers. Comp. Biochem. Physiol. *19*: 115–131.

Peachey, L. D., 1965: Transverse tubules in excitation-contraction coupling. Fed. Proc. *24*: 1124–1134.

Peachey, L. D., 1968: Muscular contraction. Ann. Rev. Physiol. *30*: 401–440.

Pollard, T. D., 1973: Progress in understanding amoeboid movement at the molecular level. *In* The Biology of Amoeba. Jeon, ed. Academic Press, New York, pp. 291–317.

Prosser, C. L., ed., 1973: Comparative Animal Physiology. 3rd Ed. W. B. Saunders Co., Philadelphia.

Prosser, C. L., 1974: Smooth muscle. Ann. Rev. Physiol. *36*: 503–535.

Ramsey, R. W. and Street, S. F., 1941: Muscle function as studied in single muscle fibers. Biol. Symp. *3*: 9–34.

Rebhun, L. I., 1972: Polarized intracellular transport, saltatory movements and cytoplasmic streaming. Int. Rev. Cyt. *32*: 93–137.

Rice, R. V. and Brady, A. C., 1972: Biochemical and ultrastructural studies on vertebrate smooth muscle. Cold Spring Harbor Symp. Quant. Biol. *37*: 429–436.

Rinaldi, R. A. and Jahn, T. L., 1963: On the mechanism of ameboid movement. J. Protozool. *10*: 344–357.

Rinaldi, R. A. and Jahn, T. L., 1964: Shadowgraphs of protoplasmic movement in *Allogromia laticollaris* and a correlation of this movement to striated muscle contraction. Protoplasma *58*: 369–390.

Rinaldi, R., Opas, M. and Hrebenda, B., 1975: Contractility of glycerinated *Amoeba proteus* and *Chaos chaos*. J. Protozool. *22*: 286–292.

Ringertz, N. R. and Savage, R. E., 1976: Cell Hybrids. Academic Press, New York.

Routledge, L. M., Amos, W. B., Yew, F. F. and Weis-Fogh, T., 1976: New calcium-binding contractile proteins. *In* Cell Motility. A. Motility, Muscle and Non-Muscle Cells. Goldman, Pollard and Rosenbaum, eds. Cold Spring Harbor Conf. Cell Proliferation *3*: 93–113.

Rudin, S. G., Foldvari, T. L. and Levy, C. K., 1971: Bioinstrumentation, Experiments in Physiology. Harvard Apparatus Foundation, Millis, Mass.

Ruegg, J. C., 1971: Smooth muscle tone. Physiol. Rev. *51*: 201–248.

Sale, W. S. and Satir, P., 1976: Splayed Tetrahymena cilia. J. Cell Biol. *71*: 589–605.

Satir, P., 1974: How cilia move. Sci. Am. (Oct.) *231*: 45–52.

Schliwa, M., 1978: Microtubule apparatus of melanophores, three dimensional organization. J. Cell Biol. *76*: 605–614.

Schliwa, M., Osborn, M. and Weber, K., 1978: Microtubule system of isolated fish melanophores as revealed by immunofluorescence microscopy. J. Cell Biol. *76*: 229–236.

Sjöstrand, F. S., Andersson-Cedergran, E. and Dewey, M. M., 1958: Ultrastructure of intercalated discs of frog, mouse and guinea pig cardiac muscle. J. Ultrastruct. Res. *1*: 271–287.

Sleigh, M. A., 1962: The Biology of Cilia and Flagella. Pergamon Press, New York.

Sleigh, M. A., 1971: Cilia. Endeavour (Jan.) *30*: 11–17.

Smith, T. E., Jr. and Berndt, W. O., 1964: The establishment of beating myocardial cells in long-term culture in fluid medium. Exp. Cell Res. *36*: 179–191.

Stephens, N. H., 1977: The Biochemistry of Smooth Muscle. University Park Press, Baltimore.

Steward, P. A., 1964: The organization of movement in slime mold plasmodia. *In* Primitive Motile Systems in Cell Biology. Allen and Kamiya, eds. Academic Press, New York, pp. 69–78.

Street, S. F. and Ramsey, R. W., 1965: Sarcolemma: transmitter of active tension in frog skeletal muscle. Science *149*: 1379–1380.

Tamm, S. L., 1976: Properties of a rotary motor in eukaryotic cells. *In* Cell Motility, Book C: Microtubules and Related Proteins. Goldman, Pollard and Rosenbaum, eds. Cold Spring Harbor Conf. Cell Proliferation *3*: 949–967.

Taylor, C. V., 1920: Demonstration of the function of the neuromotor apparatus of *Euplotes* by the method of microdissection. Univ. Calif. Publ. Zool. *19*: 403–471.

Thaenert, J. C., 1957: Intercellular bridges as protoplasmic anastomoses between smooth muscle cells. J. Biophys. Biochem. Cytol. *6*: 67–70.

Turner, G. D. and Bagnara, J. T., 1976: General Endocrinology. 6th Ed. W. B. Saunders Co., Philadelphia.

Vahouny, G. V., Wei, R., Starkweather, R. and Davis, C., 1970: Preparation of beating heart cells from adult rats. Science *167*: 1616–1618.

Wiersma, C. A. G., 1952: Comparative physiology of invertebrate muscle. Ann. Rev. Physiol. *14*: 159–176.

Wilkie, R. D., 1968: Muscle. Edward Arnold, London.

Witman, G. B., Plummer, J. and Sander, G., 1978: *Chlamydomonas* flagellar mutants lacking radial spikes and central tubules. J. Cell Biol. *76*: 729–747.

Wohlfarth-Bottermann, K. E., 1964a: Differentiations of the ground cytoplasm and their significance for the generation of the motive force of ameboid movement. *In* Primitive Motile Systems in Cell Biology. Allen and Kamiya, eds. Academic Press, New York, pp. 79–109.

Wohlfarth-Bottermann, K. E., 1964b: Cell structures and their significance in ameboid movement. Int. Rev. Cytol. *16*: 61–131.

Yoneda, M., 1962: Force excited by a single cilium of *Mytilus edulis*. II. Free motion. J. Exp. Biol. *39*: 307–317.

CHAPTER 23

THE CHEMICAL BASIS OF CONTRACTILITY

Cytoplasmic streaming, ameboid and ciliary movement, and muscular contraction are possible only because of molecular and ionic movements taking place inside living cells. The nature of these molecular and ionic movements and the source of energy by which they occur are fundamental to an understanding of the mechanism of contractility.

Because energy conversions occur on such a large scale in muscle fibers and because the quantities of enzymes and metabolic intermediates are so large, vertebrate skeletal muscle fibers have been extensively used in metabolic studies. Many of the enzymes have been isolated and their reactions have been studied *in vitro* (see Chapter 12).

Because muscle fibers are so highly differentiated, they show most clearly the basic features of a contractile system. Therefore, they serve most effectively for an analysis of the molecular mechanism of contraction. The major objective of this chapter is to describe the mechanism of contraction at the molecular level and to consider how metabolic energy is transduced to mechanical energy in the muscle fiber.

Studies on vertebrate cardiac and visceral muscle fibers and on invertebrate muscle fibers are also available in the literature and they indicate essential similarity in the chemical systems involved in all muscular contraction.

Contraction in nonmuscle cells has been studied intensively only recently; at the molecular level, some similarity to muscular contraction has been observed. These studies are briefly reviewed at the end of the chapter.

EVIDENCE OF MOLECULAR ORIENTATION IN MUSCLE FIBERS

To perform work it is necessary that a muscle fiber exert force in a system with a favorable spatial relation. Muscle cells and individual muscle fibers are always elongated, and they may even run the full length of a muscle and join with the tendon. The membrane *(sarcolemma)* around each muscle fiber is electrically polarized, the inside being about 70 mv negative to the outside (see Chapters 21 and 22). A muscle fiber contains elongated *myofibrils* (Fig. 23.1), which lie in the cytoplasm *(sarcoplasm)*, and around or between the myofibrils are found the nuclei. Each myofibril is about 1 micrometer (μm) in diameter and is presumably the seat of muscle contraction.

The striated appearance of a skeletal muscle fiber seen under the microscope results from the striations on the myofibrils, the striations being similarly oriented ("in register") across all the myofibrils of a muscle fiber. Under the polarizing microscope the dark bands are found to be doubly refracting, and the light bands singly refracting. This suggests molecular orientation.

Under high power of the light microscope the striation pattern on an isolated myofibril is seen as a regular alternation of isotropic I-bands (the light bands), through which light passes equally in all directions, and anisotropic A-bands (the dark bands), with different refractive indices in different directions (Fig. 23.1, center, 23.2 and 23.4). When a vertebrate striated muscle fiber is almost fully relaxed, the length of one A-band of a myofibril is commonly about 1.5 μm and that of one I-band is about 0.8 μm. In the middle of the A-band is a portion called the H-zone, which is lighter than the rest, and in the middle of the I-band is a darker Z-line, which bisects the band. From one Z-line to the next is one repeating unit *(sarcomere)* of the myofibril (Figs. 23.1, 23.2, and 23.4) (H. E. Huxley, 1957). This microscopic structure also indicates a regular arrangement of molecular units.

Evidence of molecular orientation in muscle fibers comes also from x-ray diffraction studies.

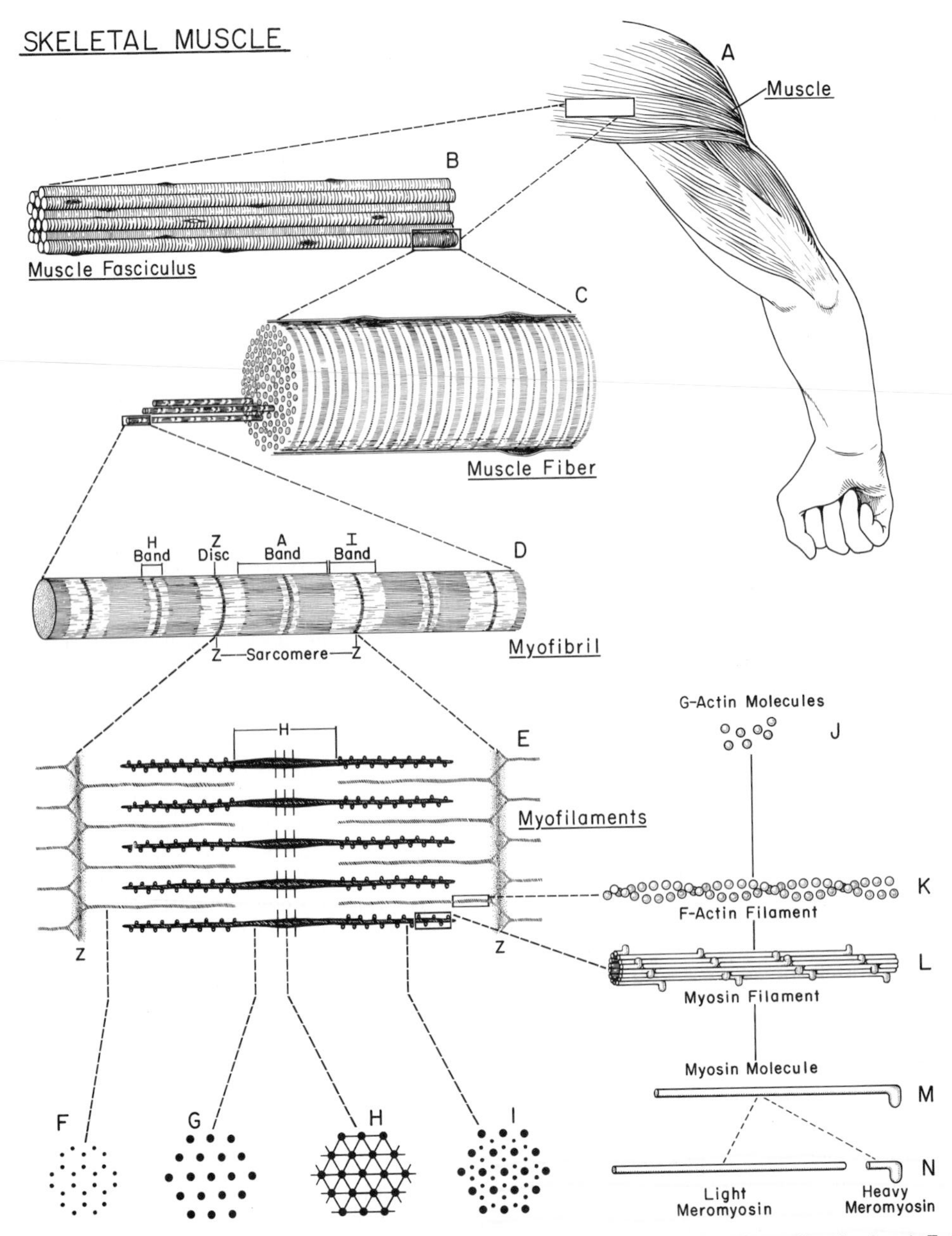

Figure 23.1. *Diagram of the organization of skeletal muscle from the gross to the molecular level.* F, G, H, and I *are cross sections at the levels indicated. The H-band is usually called the H-zone and the Z-disc is called the Z-line. (Drawing by Sylvia Colard Keene.) (From Bloom and Fawcett, 1975: A Textbook of Histology. 10th Ed. W. B. Saunders Co., Philadelphia.)*

Briefly, such studies are based on the fact that x-ray wavelengths are diffracted by layers of molecules (or ions). If a beam of x-rays is passed through a crystal, the diffracted beam forms a circle of discontinuous images about the central beam. If a beam of x-rays is passed through a powdered salt crystal, the diffracted beam does not form discontinuous images but rather forms a series of continuous rings about the central beam, indicating less perfect orientation because of fragmentation than that found in true crystals (Fig. 23.3).

When x-rays are passed through muscle fibers, the diffracted beam forms a series of continuous rings surrounding the central beam, indicating imperfect orientation. Each ring is the result of diffraction of the x-rays by a different set of oriented molecules. By measuring the angles between the central beam of x-rays and the rings it is possible to deduce the distance between the molecular units in the muscle fiber and the dimensions and arrangements of the structural units (Astbury, 1950). The x-ray diffraction method has an advantage

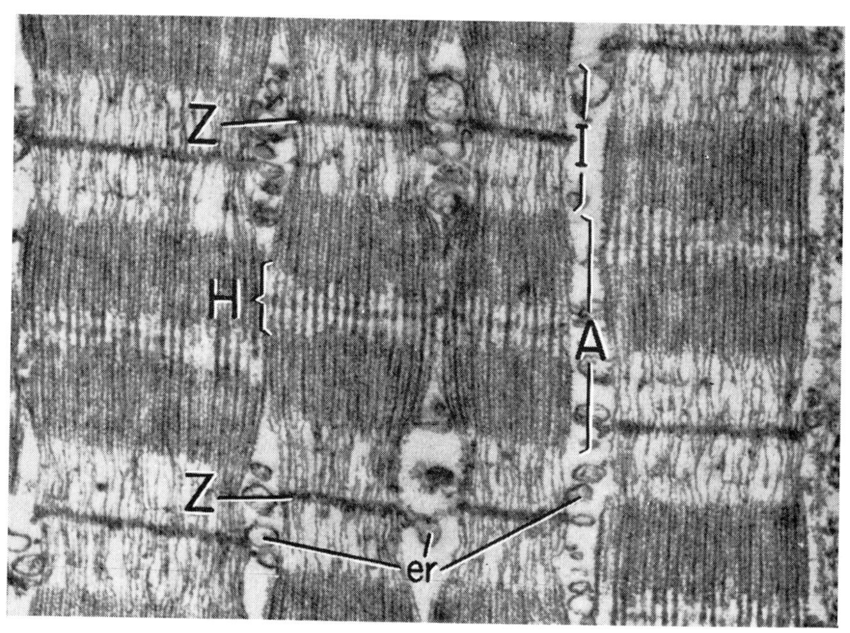

Figure 23.2. *Electron micrograph of four myofibrils, showing the alternating sarcomeres with the Z-lines and the H-, A-, and I-bands. The sarcoplasmic reticulum (er) is situated between the myofibrils. The finer structure of the myofibril represented by the thin (actin) and the thick (myosin) filaments is also clear (×60,000). (By courtesy of H. Huxley.)*

in studies of the molecular structure of muscle fibers in that it may be used on living muscle fibers when at rest or contracted. The data so gathered give the true distances between the ordered molecular units, unmodified by fixatives and other agents (Fig. 23.3).

Without going into details, it is possible to state that the contractile units in the fibrils are oriented along the length of the fibrils with a regularity suggestive of a crystal. Since the indicated size of the contractile unit is that of a large protein molecule in every case, the x-ray studies make it seem likely that the regularly arranged constituents are protein molecules.

When the myofibril is examined under the electron microscope, it is found to be made up of two kinds of smaller *filaments,* one about twice as thick as the other. The thicker filaments are about 10 nm in diameter and 1.5 μm long, and the thinner ones are about 6 nm in diameter and 2 μm long. The thick filaments have been shown to consist mainly of the protein *myosin,* while the thin filaments consist mainly of the protein *actin.* Each filament is seen arrayed with the other filaments of the same kind, and the two arrays of filaments overlap for part of their length. It is apparently this overlap that gives rise to the cross bands of the myofibril: the dense *A-band* consists of overlapping thick and thin filaments; the light *I-band* consists of thin filaments alone; the *H-zone* consists of thick filaments alone. Halfway along their length, the thin filaments pass through a narrow zone of dense material; this constitutes the *Z-line* (A. F. Huxley, 1957; H. E. Huxley, 1957, 1965).

In the A-band the two kinds of filaments are arranged in such a manner that each thin filament lies in a spatial symmetry among three thick ones (Figs. 23.1 and 23.4). The two kinds of filaments are linked together by myosin *cross-bridges,* which project outward from the thick filaments at a fairly regular interval of about 6 to 7 nm, each bridge being 60 degrees further around the axis of the filament, so that the bridges form a helical pattern that is repeated every six bridges, or about every 40 nm, along the filament (Fig. 23.5). In the M-line, which is in the middle of the A band, is a pair of proteins called the M-line proteins, one of molecular weight 155,000, the other 88,000. They are present in small amounts and their function is not known.

The structures seen with the electron microscope are, of course, being observed on fixed and dried material. However, the regularity of filaments and their cross linkages is also indicated in x-ray diffraction studies, although the evidence is indirect and difficult to present.

Thick and thin filaments have been found in the myofibrils of many types of skeletal muscle fibers of both vertebrates and some invertebrates. In vertebrate smooth muscle fibers myosin is present in long ribbon-like structures. Actin is found in filaments, as in striated muscle, but the filaments are arranged in small, regular arrays between and around the myosin ribbons (Lowy and Small, 1970). The two major structural proteins of skeletal muscle fibers, myosin and actin, can be extracted from such smooth muscles, but it would seem

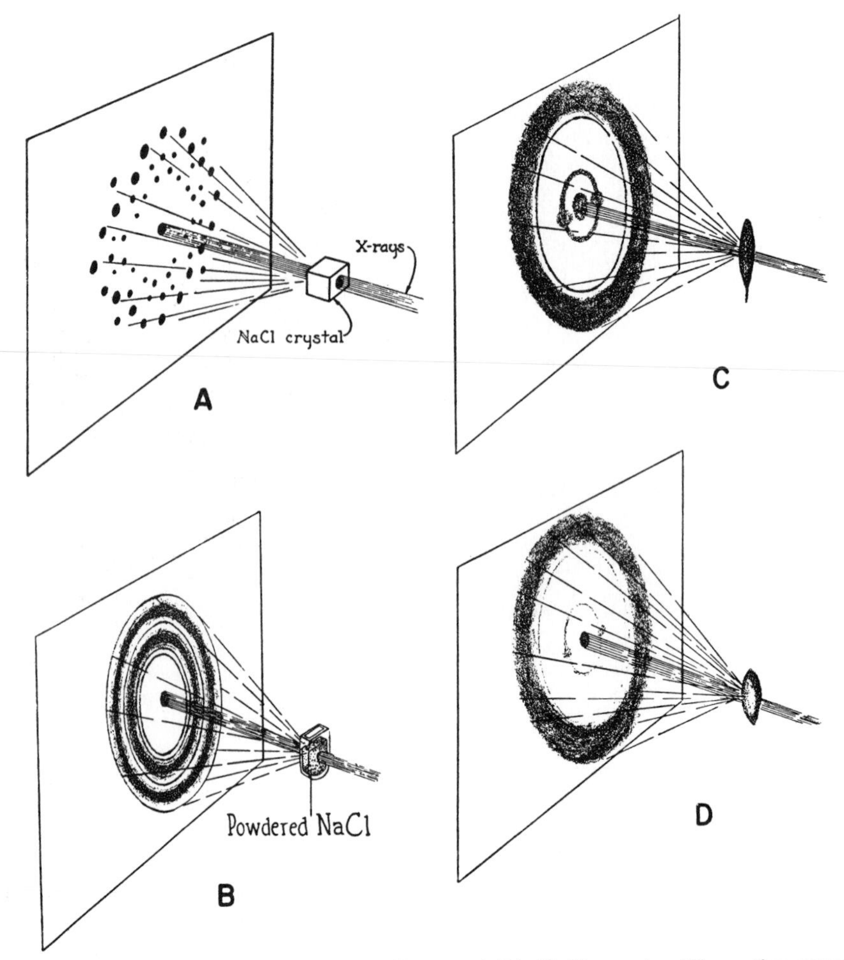

Figure 23.3. X-ray diffraction diagrams for NaCl crystal (A), *NaCl powder* (B), *resting muscle* (C) *and contracted muscle* (D). *Modified from Amberson and Smith, 1948: Outline of Physiology. Appleton-Century-Crofts, New York.)*

that the molecular organization of these proteins in the smooth muscle cells differs in some fundamental manner from their organization in the skeletal muscle fibers, perhaps running obliquely across the individual fibers (see Fig. 3.2) (Huxley, 1963, 1965, 1977).

Perhaps the most convincing evidence of the reality of the filaments and their cross-bridges in myofibrils is obtained when these are separated intact by treatment of a live muscle in a Waring Blendor. Surprisingly, this drastic treatment separates a muscle fiber into its filaments and no further. In such preparations the cross-bridges often bind the thick and thin filaments together strongly enough to resist separation, and some electron micrographs actually show the two types of filaments connected by cross-bridges. Also, thin filaments are often seen attached to the Z-line. If the fresh muscle tissue or glycerol-extracted muscle tissue is treated with the appropriate salt solutions, the cross-bridges are broken. The separated thick and thin filaments can then be made visible in electron micrographs with negative staining and measured. They are found to have the same dimensions as measured in thin sections (Huxley, 1963, 1965).

THE STRUCTURAL PROTEINS OF MUSCLE CELLS

About 20 per cent of the wet weight of the muscle fibers is represented by proteins. If a mass of muscle fibers is minced, wrapped in cheesecloth and squeezed in a press under many atmospheres of pressure, the fluid that exudes is called the *press juice*. Press juice contains the fluid from inside the cells as well as some intercellular fluid. If the muscle residue left after squeezing out the press juice is washed in very dilute salt solution, water-soluble albumins and readily soluble globulins are removed. The remaining material that did not dissolve contains, among other substances, myosin and actin, two major proteins already

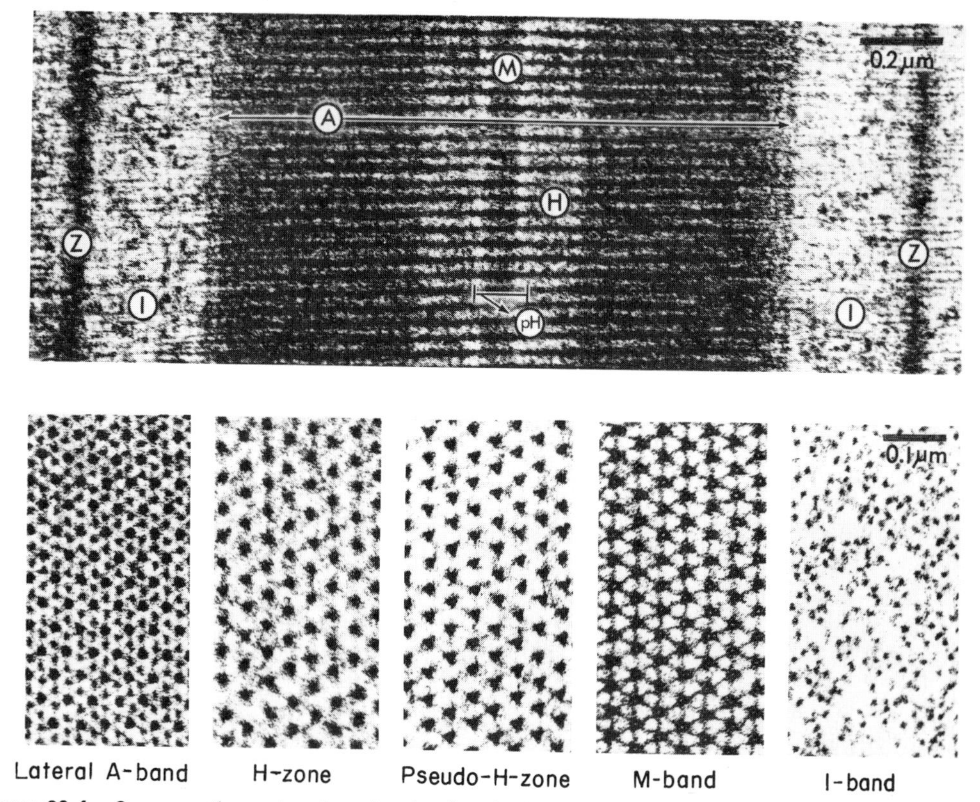

Figure 23.4. *Cross sections at various levels of a single muscle sarcomere (from lateral muscles of the freshwater killifish* Fundulus diaphanus*) shown above in longitudinal section, extending from Z-band (Z) to Z-band:* Lateral A-band, *cross section made in the region of overlap of thick and thin filaments; note that the thick filaments are solid and have circular or triangular profiles.* H-zone, *cross section of a region that shows only thick filaments.* Pseudo-H-zone, *cross section through the pseudo-H-zone where the thick filaments have no cross bridges; note that the thick filaments are solid or triangular in cross section.* M-band, *cross section through a region where each thick filament is connected to each of its six neighboring thick filaments by bridges made of M-proteins; the latter probably bind myosin molecules of the thick filaments together.* I-band, *cross section through the region where only thin filaments are present. (By courtesy of F. A. Pepe, University of Pennsylvania Medical School. From Pepe, F. A., 1971: Structural components of the striated muscle fiber.* In *Subunits in Biological Systems. Timasheff and Fasman, eds. Marcel Dekker, Inc. New York, Vol. 5: 323–353.)*

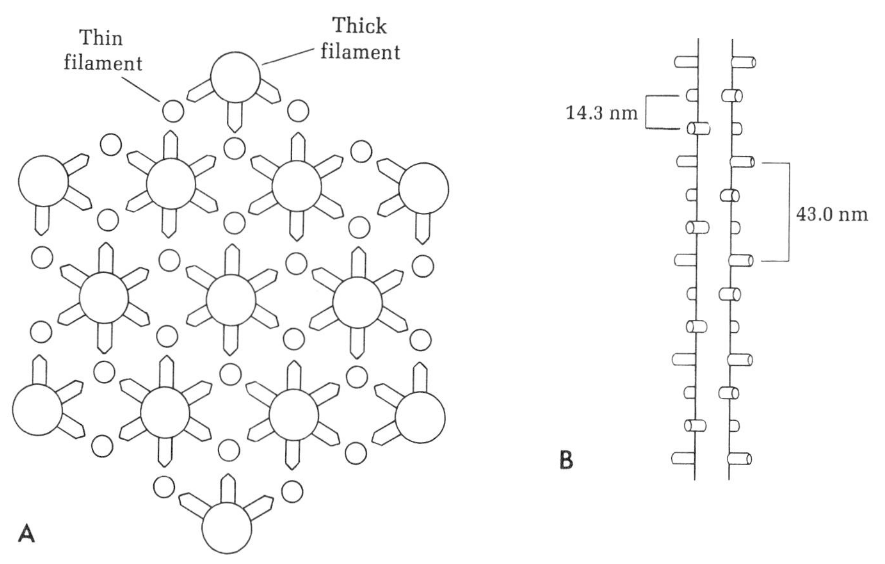

Figure 23.5. *The arrangement of cross bridges on the thick filaments in frog sartorius muscle, deduced from low-angle X-ray diagrams.* A, *as seen in cross section (note the 6:2 arrangement);* B, *as seen in longitudinal view. (From Huxley and Brown, 1967: J. Mol. Biol.* 30: *383.)*

introduced that make up 80 percent of the mass, and several minor proteins making up the remainder. If this initial mixture is extracted with a solution of 0.3 M potassium chloride, buffered with 0.15 M potassium phosphate to give the solution a pH of 6.5, and if the extraction is not too prolonged, myosin will be more concentrated than actin in the solution. It can be purified by diluting the solution slightly, whereupon its complex with actin, namely *actomyosin,* precipitates out (Lehninger, 1975).

Molecular Characteristics of Thick Filaments

Myosin is found to constitute 50 to 55 percent of the myofibrillar protein in mammalian skeletal muscle fibers. Myosin is a double helix type of protein of high molecular weight (about 460,000). The molecule consists of two polypeptide chains, each in helical configuration, twined together in a supercoil (Fig. 23.6). At the end of each of the polypeptides is a globular

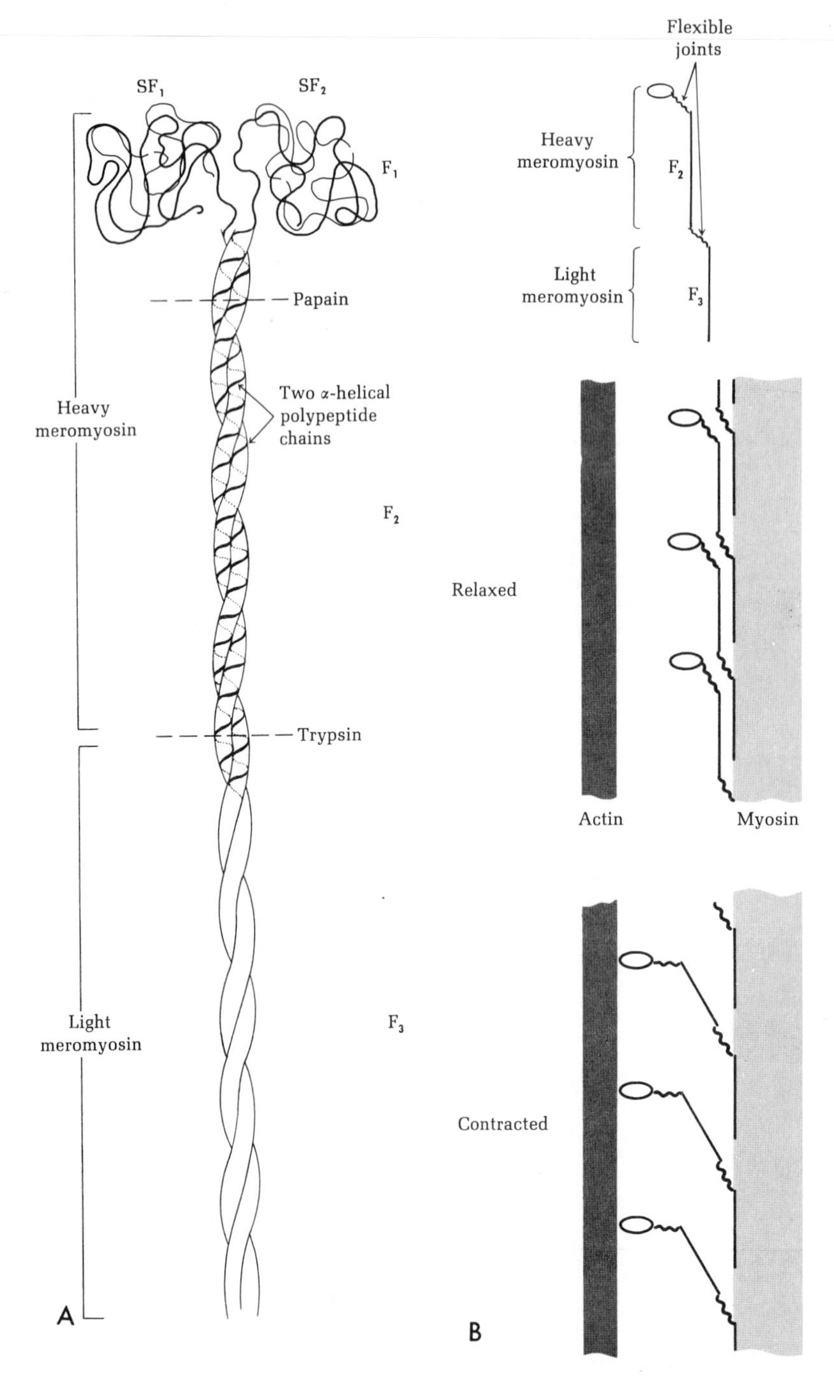

Figure 23.6. A, *structure of the myosin molecule showing its component parts and the points of cleavage by the enzymes papain and trypsin.* B, *Top, diagrammatic representation of the molecule of myosin shown in A; middle, relation between myosin and actin in the relaxed state; bottom, relation between myosin and actin in the contracted state. (From Lehninger, 1975. Biochemistry. 2nd Ed. Worth Publishers, New York.)*

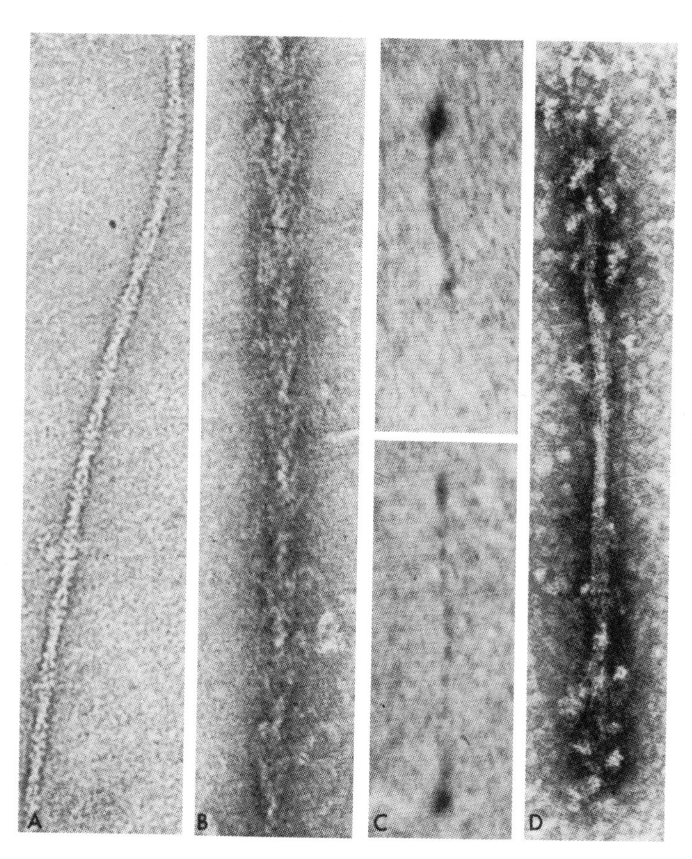

Figure 23.7. *Actin and myosin.* A, *Actin showing characteristic structure visible in an electron micrograph at enlargement of 100,000 diameters. The filament has the appearance of two coils of globular units wound in a double helix.* B, *Arrowheads pointing in one direction along each filament of actin labeled with heavy meromyosin, implying that actin has a polarity of its own.* C, *Myosin molecules as seen in electron micrographs prepared by the shadow casting method. The wide "head" has the enzymatic property and combines with actin; the straight "tail" can aggregate with other myosin molecules (×300,000).* D, *Aggregation of several molecules of myosin, negatively stained and magnified 175,000 diameters. Note that tails join in the center. (By courtesy of H. E. Huxley.)*

head or enlargement. The myosin head contains four small polypeptide chains, two of identical molecular weight (18,000) and one each of molecular weight 16,000 and 21,000. Therefore, the myosin molecule consists of six polypeptide chains, two large and four small.

The large polypeptides of myosin are among the longest that are known. The two large chains of myosin can be separated experimentally from each other by various reagents (e.g., urea). Each head section is also separable into two subunits by appropriate reagents (Lehninger, 1975). As seen in electron micrographs myosin molecules appear to be long and thin (160 nm by 5 nm) (Fig. 23.7*C, D*). Myosin is composed of many dicarboxylic acids; therefore, it has many ionizable residues, and, near neutrality, it has one charge for almost every 0.1 nm of its length. This very high charge density explains why the molecule maintains its fibrous form, since one part of the molecule repels another part of the same molecule as well as another molecule of its own kind. Because of its needle-like, linear form, myosin shows beautiful streaming birefringence.

Myosin has been fragmented by limited hydrolysis with trypsin or chymotrypsin into two fractions *(meromyosins),* a heavy one with the head and a light one. These fractions are of particular interest because electron micrographs indicate them to be of the right order of magnitude to account for some of the angular deviations of the x-ray diffraction rings. From measurements on electron micrographs of the two fragments of myosin (light and heavy meromyosin) it appears that one kind of fragment is linked end to end with the other in the myosin molecule.

When the meromyosins are shadow cast, heavy meromyosin appears to have a double globular head and a short tail, and light meromyosin is just a straight rod, the three units having flexible connections with one another. Thus, the myosin molecule is asymmetrical, with the head, 4 nm in diameter, accounting for one-sixth of the length, and the remainder a tail 2 nm in diameter.

Treatment of myosin with papain separates the head from the tail. Apparently the ATPase activity of intact myosin and the affinity of myosin for actin resides only in its heavy meromyosin head, whereas the affinity for other myosin molecules resides in its light meromyosin tail. However, the character of the ATPase depends upon whether the myosin is free to attach to actin. When myosin is free, it requires calcium ion for ATPase activity and is inhibited by magnesium. When myosin is attached to actin not only is the ATPase activity of the actomyosin higher than that of myosin but it is also stimulated by magnesium. The head has been used for studies on the contractile mechanism because it is more convenient to handle than heavy meromyosin (Cohen, 1975).

It is of interest that no matter whether the filaments from the myofibrils are obtained by homogenizing a muscle in a Waring Blendor or reconstituted from solution (precipitated in dilute potassium chloride solution), they have the same linear form and head shape. The myosin molecules thus display a marked tendency to form long threads, like the filaments in muscle, by attaching to one another spontaneously.

About 400 myosin molecules are present in each thick filament. The myosin molecules are oriented with their heads away from the midline of the thick filaments. The myosin heads, resembling barbs, project in a regular helical manner from the filament for the shortest distance in relaxed muscle, when they are detached from the adjacent thin filaments.

Thick filaments contain other proteins in addition to myosin; C-protein and M-line protein. C-protein (molecular weight 140,000) constitutes 3.5 percent of the total protein in a thick filament. It binds strongly to the myosin tail and is wound around the thick filament at regular intervals. It may serve to clamp the bundle of heavy filaments together. The M-line protein is thought to bind the myosin molecules of a thick filament at the M-line.

Molecular Characteristics of Thin Filaments

Actin, the other major structural protein of muscle fibers, is less readily extracted in salt solutions than is myosin and dissolves only when the concentration of KCl is raised and the solution is alkalinized. Actin constitutes 20 to 25 percent of the protein in skeletal myofibrillar material of mammals. In the absence of salts, actin exists in globular form (G-actin), with a molecular weight of about 47,000, but in the presence of 0.1 M potassium chloride, it polymerizes to form strands of F-actin. Polymerization is accompanied by splitting of ATP; the ADP formed becomes firmly bound to the actin and inorganic phosphate is given off. Each molecule of actin appears to bind one ADP. The F-actin appears to be asymmetrical and to be made up of two coils of globular units wound in a double helix with crossover points 360 degrees apart (see Figs. 23.7*A* and 23.8). This means that when myosin combines with actin *in vitro* it will amplify the coiled arrangement (Huxley, 1963). The actin helix has a period of 43 nm according to both x-ray diffraction and electron microscope studies.

In polymerized form, F-actin shows streaming birefringence, like myosin, another sign of the linear entities of which it is composed. Like logs in a stream of water, the long molecules are lined up in solution and differentially affect light passed at right angles to the long axes as compared with light passed parallel to the long axes of the molecules.

In addition to actin, several other proteins are present in thin filaments, chiefly tropomyosin and troponin, as well as small amounts of α and β actinin of unknown function. Tropomyosin consists of two subunits and troponin of three subunits. The two strands of F-actin, each composed of G-actin monomers, are coiled around each other forming the thin filaments. *Tropomyosin,* making up about 10 percent of the total contractile protein in muscle, consists of two different kinds of α helical polypeptide strands of molecular weight 33,000 and 37,000. These polymers form a twisted coil 40 nm long. The tropomyosin coils are arranged end to end, fitting into the shallow grooves between the coiled F-actin strands in such a way that each of the two tropomyosin continuities present contacts a different strand (Fig. 23.8).

The second accessory protein of the thin filament, which serves a regulatory function, is *troponin,* a large globular protein (Ebashi, 1977). Troponin consists of three subunits: TNC, TNI, and TNT (Fig. 23.8). Subunit TNC, also called troponin-A and of molecular weight 18,000, binds two calcium ions causing a conformational change in troponin. The inhibitory subunit TNI of molecular weight 23,000 has a site for binding actin but not for binding cal-

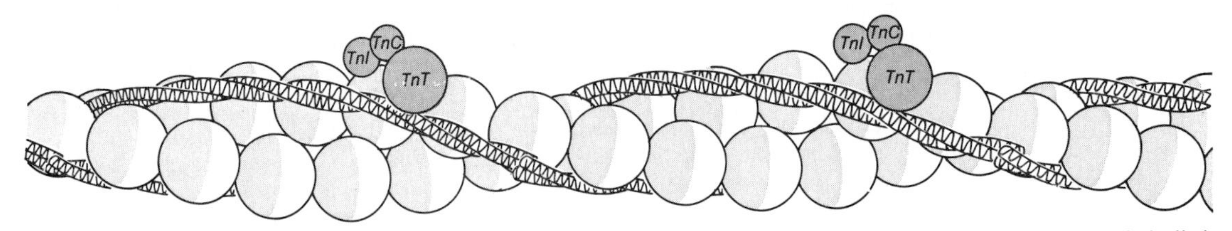

Figure 23.8. A, *filament of tropomyosin molecules (coiled coils) wound in each of the two grooves of the actin helix in a thin filament of muscle. The actin molecules (grey and white circles) are polar; they all point in the same direction in a double helical array. The tropomyosin filaments (smaller, rope-like helix) consist of polar tropomyosin molecules, bonded head to tail, that lie near the grooves, each molecule spanning seven actin monomers. A troponin complex (dark grey smaller circles) is about a third of the way from one end of each tropomyosin molecule. In this schematic diagram only the troponin complexes on one side of the actin helix are shown. (From Cohen, 1975: Sci. Am. (Nov.)* 233: *139. Copyright © 1975 by Scientific American, Inc. All rights reserved.)*

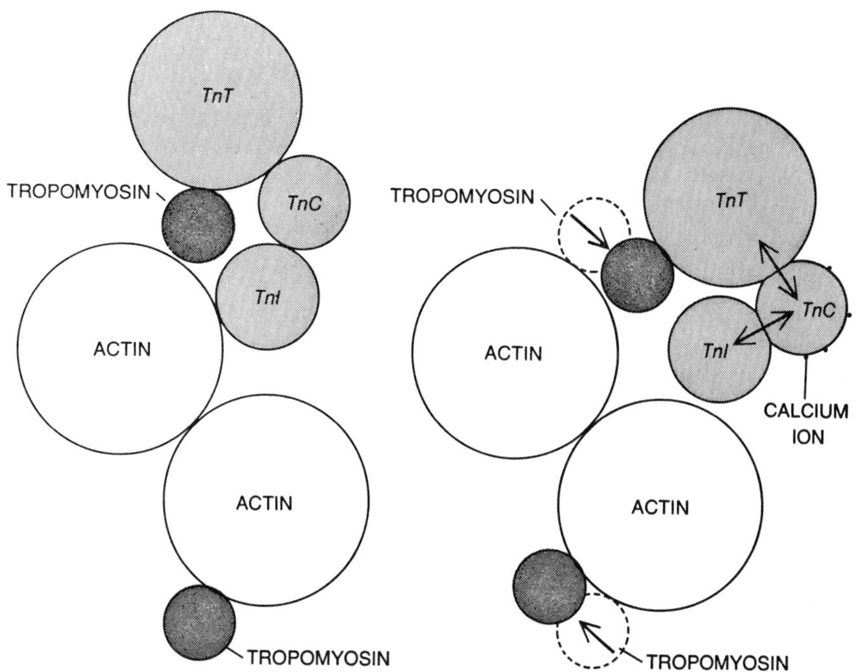

Figure 23.9. *Subunits of troponin may function as suggested in this tentative model of the regulation of muscle contraction. The thin filament is seen schematically, end on. In the resting state* (left) *the* TnT *subunit binds to tropomyosin and the inhibitory subunit* TnI *binds to actin; the calcium-ion concentration is low and the links between troponin subunits are relatively loose. In the active state* (right), *above a critical calcium level, linkages between troponin subunits are tightened and the link between* TnI *and actin is weakened; tropomyosin moves deeper into the actin groove, exposing the site at which myosin can bind. (From Cohen, C., 1975: Sci. Am. (Nov.)* 233: *45. Copyright © by Scientific American, Inc. All rights reserved.)*

cium ions; it inhibits contact of actin and myosin-head cross-bridges. The third component of troponin, TnT, is the tropomyosin-binding subunit. The complete troponin molecule contains one each of the three subunits and has a globular shape (Ebashi *et al.,* 1976; Collins, 1976).

Each troponin molecule attaches to the thin filament at two binding sites, one specific to the actin strand and the other to the tropomyosin strand. The binding site of troponin to tropomyosin is fixed but the binding site to actin is not; depending upon the binding of calcium ions, it either binds to or separates from actin (Fig. 23.9). One troponin is found every 40 nm of the thin filament. For every seven globular actin molecules there is one troponin and one tropomyosin molecule (Fig. 23.8).

Regulation of contractions by proteins is not the same in all animal phyla. In animals of some phyla the regulatory proteins bind to both actin and myosin, in some to actin alone, and in still others to myosin alone. The interesting implications of these findings are discussed by Szent-Györgyi (1976).

ACTOMYOSIN

When pure actin and pure myosin are mixed in the presence of salts normally found in the muscle fibers, a striking change occurs: the viscosity of the mixture becomes much greater than the sum of the viscosities of the two constituent solutions. Electron micrographs show that a network is formed by electrostatic association between the two proteins. This complex is called *actomyosin.* Actin and myosin have been found to combine in stoichiometric ratios.

The actomyosin complexes undergo dissociation in the presence of ATP and Mg^{2+}, accompanied by a large decrease in viscosity. After dissociation, ATP is hydrolyzed. When ATP is completely hydrolyzed to ADP and P_i, reassociation of actin and myosin occurs. It will later be seen that these are steps in the breaking and making of cross linkages between actin and myosin of the muscle fibers.

It has been possible, through studies with the electron microscope, to localize some of the proteins in the filaments of the myofibrils. Thus, when myosin of a muscle fiber is extracted with an appropriate potassium chloride solution, the thicker filaments of the myofibrils are much less dense, which indicates that they had contained the myosin. When muscle fibers are treated to remove actin, the electron micrographs of the thin filaments in the fibrils are much less dense, indicating that these consist mostly of actin. Corroborating data come from analyses of the filaments using immunochemical techniques; fluorescent antibodies may be

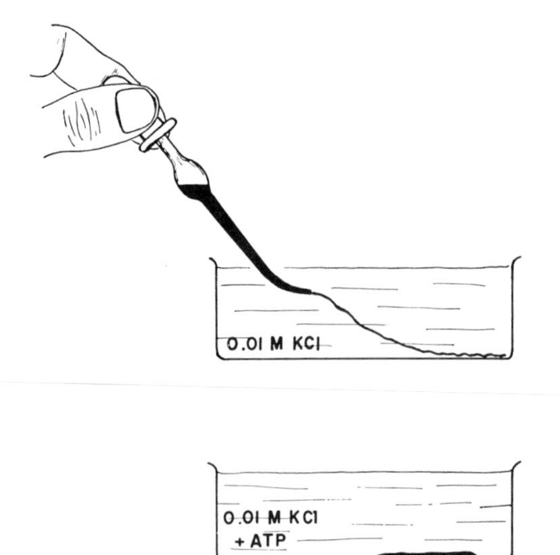

Figure 23.10. Formation of an actomyosin thread by adding a jet of actomyosin to a dilute solution of KCl. Contraction of the thread is shown after addition of ATP.

produced against either myosin or actin (Marshall *et al.*, 1959) and used to visually identify the protein in tissue preparations.

Actomyosin dissolved in 0.6 M potassium chloride solution shows brilliant streaming birefringence. The electron microscope discloses actomyosin particles to be long threads of about 15 nm diameter and ten or more times the length of a myosin molecule (Fig. 23.7*B*).

When actomyosin dissolved in 0.6 M potassium chloride is squirted into 0.01 M potassium chloride (Fig. 23.10), a gelled thread of actomyosin is obtained. When such a thread is placed in ATP solution, it contracts (Weber and Porzehl, 1954). Actomyosin is also found to possess an enzymatic function (adenosine triphosphatase), splitting ATP to ADP and inorganic phosphate. These observations created a sensation when they were first announced, because actomyosin thus appeared to be a model, or analogue, of muscular contraction. However, important differences were soon found between a muscle fibril and an actomyosin thread. The latter loses about 40 percent of its water on contraction, whereas a muscle fibril does not. Furthermore, study of the actomyosin thread under the electron microscope shows that the molecules are in disarray, resembling the brush network in a protein gel rather than the linearly arranged and regularly spaced units disclosed by x-ray analysis of a muscle fiber. Moreover, if the thread represents a model of a muscle fiber, it should perform work like a muscle fiber. Yet when the actomyosin thread is loaded and ATP is added, the thread stretches rather than contracts.

In muscle, myosin and actin exist as filaments, with connections by cross-bridges. They are not only aligned but also polarized. A model

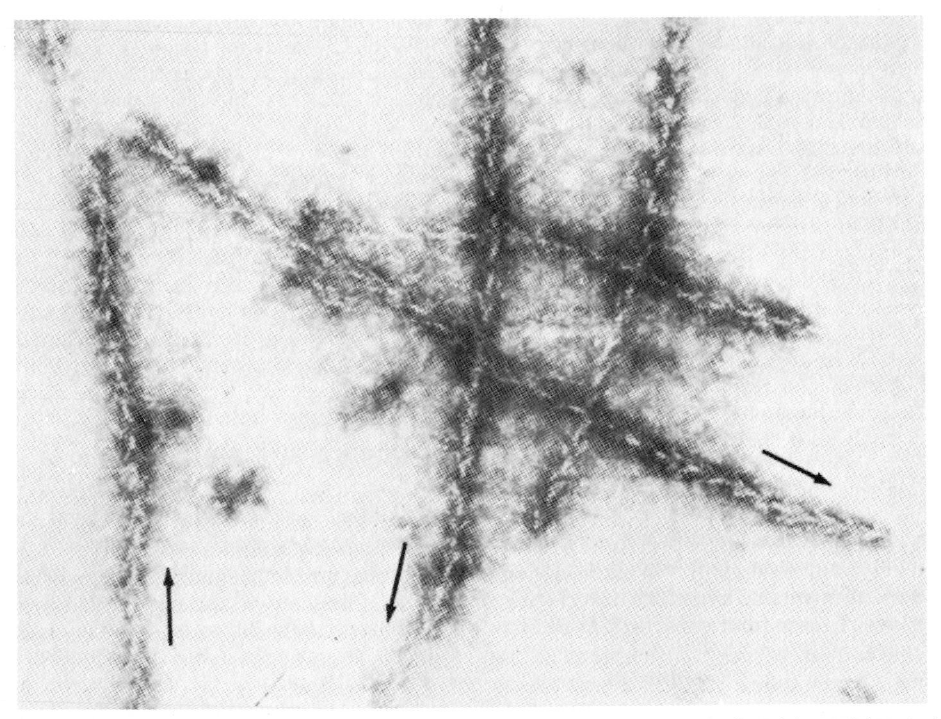

Figure 23.11. When actin is added to heavy meromyosin, which has a head and tail end, it combines to form a strand with distinct polarity (as shown by the arrows) resulting in an arrowhead appearance. Seen by negative staining the arrows all point in the same direction (×155,000). (By courtesy of H. E. Huxley.)

must therefore duplicate these arrangements. Although no model has yet been made that shows all the properties of muscular contraction, the elements of the filament systems have been reassembled. Assembly of actin and myosin filaments occurs spontaneously from solutions of actin and myosin, indicating that the filaments are the result of properties of the respective actin and myosin molecules.

It will be recalled that actin has the form of a double helix, being made up of a double coil of globular units with a crossover distance of 360 degrees (Fig. 23.8). When actin is labeled with heavy meromyosin, which has a head and tail end, it combines to form a strand with distinct polarity and with an arrowhead appearance. Seen by negative staining, the arrows all point in the same direction (Fig. 23.7*B* and Fig. 23.11). All the actin molecules of a given strand react in the same way with the meromyosin molecule.

To be effective in its function of contraction, the filament of the myofibril must have polarity and the polarity should be opposite on the two sides of the Z-line. To determine whether such polarity existed Huxley (1963) made use of the occasional Z-lines with attached thick filaments that appear in muscle homogenates. To a homogenate with such filaments attached to a Z-line he added heavy meromyosin and then studied the resulting structure with the electron microscope. He found that the arrow-like structures made by the attachment of the heavy meromyosin to the actin were in opposite directions on the two sides of the Z-line, pointing away from it (Fig. 23.7*D*). The polarity of the filaments is less readily seen when entire myosin molecules are used instead of the heavy meromyosin fraction, because the light meromyosin fraction attracts other myosin molecules to form the large myosin filament. Presumably the same orientation exists. In any case, the myosin molecules are arranged in such a way that all the globular portions are at the two ends of the filaments, the middle portion being free of "heads." Such a strand is interpreted as consisting of myosin molecules attached to one another by the light meromyosin portions and attached to the actin filaments in opposite polarities at the two ends, as shown in Figure 23.11.

SOURCE OF ENERGY FOR CONTRACTION

Whereas the fibrils of the muscle fiber are the seat of muscular contraction, the sarcoplasm is the seat of the biochemical reactions from which high-energy compounds are made available to the muscle fibrils. Mitochondria, with all the essential oxidation-reduction and phosphorylation enzyme systems of the Krebs cycle, are found in muscle fibers. They are arranged in close proximity to the region requiring energy liberation (Fig. 23.12). They are also more complex internally than mitochondria of less active cells, containing more cristae, presumably in connection with their greater activity in oxidation-reduction than in less active cells. These mitochondria are especially prominent in size and number in the flight muscle fibers of birds and insects. The number of mitochondria in a muscle increases with regular sustained exercise.

The press juice of minced muscle fibers is largely extracted from the sarcoplasm and contains most of the high-energy phosphate compounds, as well as salts and the globular proteins that constitute almost all the enzymes. Many of the glycolytic enzymes characteristic of the Krebs cycle may be identified in the press juice (Mommaerts, 1969).

Glycogen constitutes about 1 percent of the wet weight of whole muscle fiber, creatine phosphate (CP) about 0.5 percent, and adenosine triphosphate (ATP) about 0.025 percent, indicating a considerable store of ready nutrient. Experiments show that during muscular contraction, glycogen disappears and a corresponding quantity of lactic acid appears when the rate of activity is too rapid to permit complete oxidation of the pyruvic acid formed during glycolysis. In vertebrates the excess lactic acid is carried by the blood to the liver, where about four-fifths of it is resynthesized into glycogen, while one-fifth of the lactic acid is completely oxidized to carbon dioxide and water in the process. New supplies of glucose are continually delivered to muscle cells by the blood.

The diagram of the anaerobic breakdown of glucose (see Fig. 12.5) reveals the mechanism whereby pyruvic acid produced in glycolysis is converted into lactic acid. In the absence of oxygen, lactic acid accumulates, since the reactions cannot go any further. When oxygen is available, lactic acid is oxidized in the reactions of the Krebs cycle, as shown in the diagram of the metabolic mill (see Fig. 12.8). The glycolytic reactions occur whether oxygen is present or absent. If oxygen is present the glycolytic reactions are then followed by the reactions of the Krebs cycle.*

According to measurements of the heat produced during muscular contraction, energy is

*Muscle cells make use of fat as well as glycogen, red muscles more so than white. During migration, butterflies, fishes, and birds make extensive use of fat for muscular contraction and show a corresponding low respiratory quotient (Drummond and Black, 1960).

Figure 23.12. *Mitochondria (M) between sarcomeres of frog muscle. (From Franzini-Armstrong, 1970: J. Cell Biol. 47: 488.)*

liberated in two stages. A small amount of heat is given off upon contraction, during breakdown of high-energy phosphates, and a much greater amount of heat is liberated for many seconds, or even for several minutes, after contraction. During this time, in the absence of oxygen, glycolysis occurs. In the presence of oxygen some of the lactic acid is decomposed completely to carbon dioxide and water, accompanied by restitution of the high-energy phosphate pool.

If a muscle fiber is stimulated in the presence of adequate oxygen, it fatigues slowly. In the absence of oxygen, the muscle fatigues much more rapidly because it must then depend upon glycolysis for its energy, and anaerobic processes always yield considerably less energy than aerobic processes. Since glycolysis precedes aerobic oxidations and continues in their absence, is it, then, the immediate source of energy for contraction of muscle fibers? To test this, Lundsgaard (1930) used iodoacetic acid to poison the enzymes catalyzing the glycolytic reactions. He found that contraction still occurred but that the muscle fatigued even more rapidly than an unpoisoned anaerobic control. Clearly, some immediate source of energy other than glycolysis for muscular contraction had to be sought.

Analyses showed that the compounds of high-phosphate transfer potential decreased in amount as contractions continued in the poisoned anaerobic preparations. Presumably, these compounds were being used in muscular contractions, but in the absence of both glycolysis and aerobic reactions they could not be replenished, resulting in rapid fatigue of the muscle fibers.

If ATP is the immediate source of energy for contractility it should provoke contraction when applied directly to a muscle fiber (Bozler, 1953). As expected, it has been demonstrated that when a solution of ATP is injected into an artery, tetanus is induced in a muscle fiber supplied with blood by this artery. If a muscle is cut up into small bundles and placed at low temperature in glycerin, all the soluble constituents leach into the glycerin, but the structural proteins presumably remain intact and arranged as they were in the live muscle. Muscle strands leached in this manner are called *muscle models* and have been used extensively in research. Contraction with a tension equal to that exerted by living muscle fibers may then be induced by placing the dead leached muscle fibers in either fresh muscle press juice or a balanced salt solution containing ATP (Filo *et al.,* 1965). Press juice in which the ATP has been decomposed does not induce contraction.

That ATP is involved in some way in muscular contraction is incontestable. The decomposition of ATP to ADP, liberating the energy of one high-energy phosphate bond, appears as the likely immediate source of energy for muscular contraction. ADP is "recharged" to ATP either by glycolysis or by the reactions of the Krebs cycle and the electron transport chain as described in Chapter 12. ATP can be immediately replenished by the reaction between ADP and creatine phosphate, so CP may be considered a reserve of high-energy phosphate bonds in the muscle cells.

It is inviting to assume that in the living muscle fibers ATP must play the same role as in contraction of muscle fiber models and actomyosin threads. That ATP is present in muscle fibers in abundance is shown by biochemical extraction. Furthermore, a muscle fiber that has exhausted its ATP goes into a state of rigor (stiffness), which is reversed by addition of ATP. In fact, one of the functions of ATP is believed to be the maintenance of suppleness and plasticity of muscle fibers (Weber, 1959).

An attempt has been made to localize ATP in the bands of the sarcomeres in muscle fibers by autoradiography, using tritiated adenine in the ATP. The resolution of the photographs was not quite adequate to make localization certain, but concentration in the boundary between the A-bands and the I-bands is suggested (Huxley and Hanson, 1959; Hill, 1959).

THE SLIDING FILAMENT MECHANISM OF MUSCULAR CONTRACTION

Another approach to the study of muscular contraction has been used by H. E. Huxley (1957, 1971), J. Hanson (1959) and A. F. Huxley (1957), who investigated the structural changes occurring during a cycle of contraction, as seen in a series of electron micrographs of ultrathin sections (about 15 nm) of muscle fibers. Such an electron micrograph is shown in Figure 23.2 and an interpretation of what is seen is given diagrammatically in Figures 23.1 and 23.5.

During muscular contraction, as seen with electron micrographs, the length of the A-band remains constant, but the length of the I-band varies with the state of contraction. Since the length of the A-band is apparently equal to that of the thick filaments, it can be assumed that the length of these filaments is constant. The length of the H-zone in the middle of the A-band increases and decreases proportionally with that of the I-band, so that the distance from the end of one H-zone to the beginning of the next H-zone remains approximately the same except when the H-zone is eliminated during maximal contraction as the actin filaments slide past one another. Since this distance is equal to the length of the thin filaments extending to either side of a Z-line, they also appear to remain approximately constant in length. Consequently, it is concluded that when a muscle fiber changes length, as during contraction, the two sets of filaments slide past one another (Huxley and Hanson, 1959).

When a muscle fiber is contracted to its fullest extent, the thin filaments slide past one another (Fig. 23.13). Under such conditions, new bands appear on the electron micrographs, which are explained as a result of the shortening process (H. E. Huxley, 1970, 1971).

As was pointed out earlier, actin and myosin must be combined to form actomyosin before contraction will occur in the presence of ATP. However, since the actin and myosin filaments in the myofibrils are too far apart for any "action at a distance," the motive power for muscular contraction is postulated to be the formation of chemical couplings between the myosin and actin filaments through cross-bridges between the two (see Fig. 23.6). The bridges are permanent parts (heads) of the myosin filaments jutting outward toward the actin filaments, and the number of such bridges appears to be the same as the number of molecules of myosin. A chemical reaction is postulated to occur between the myosin head and actin filaments, as a result of which the actin filament slides along the myosin filaments for perhaps 10 nm, and the head then returns to its original configuration, ready for another pull, completing a cycle

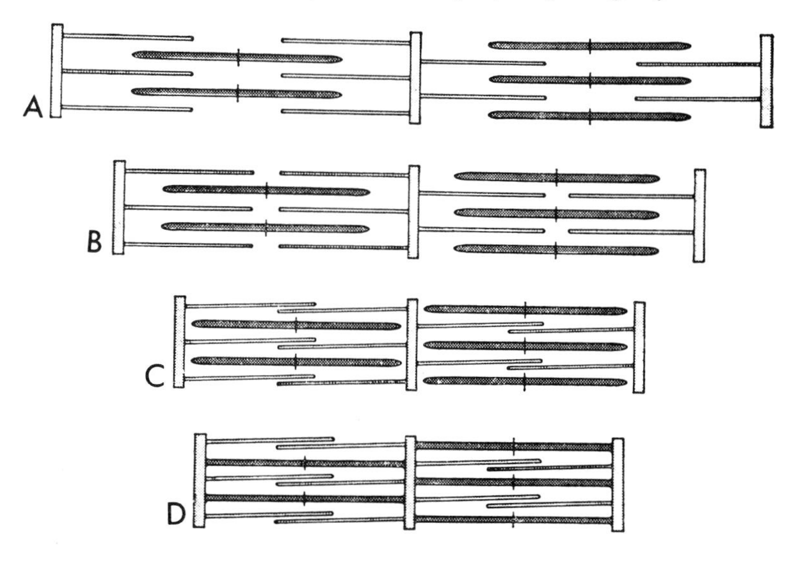

Figure 23.13. Diagram showing the sliding model of contraction. A, relaxed condition; the H-bands are wide and contain only thick filaments; B, beginning of contraction; the thin filaments slide toward the center of the sarcomere, and the H-bands become thinner; C, further contraction; the thin filaments penetrate the entire H-bands (inversion of the banding); and D, maximal contraction; thick filaments are pushed against the Z-line. (From Huxley, H. E., Proc. Roy. Inst. Gr. Br. 44: 274, 1970.)

(Huxley, 1965) (Fig. 23.14). To account for the rate of contraction and of energy liberation in the rabbit psoas muscle fiber, for example, it has been calculated that it is necessary for each bridge to go through 50 to 100 cycles per second. Each such reaction is thought to be at the expense of one ATP molecule. When hydrolysis of phosphate from ATP stops (presumably as a result of inactivation of ATPase activity upon removal of calcium ions), the myosin and actin dissociate from each other and the fiber returns to its relaxed state.

In muscle fibers binding sites on actin can be occupied either by tropomyosin or myosin. When the calcium ion concentration is low (10^{-7} to 10^{-8}), tropomyosin binds to actin and formation of cross-bridges with myosin is blocked; the muscle fiber is then relaxed. When the concentration of calcium ions rises (10^{-6}), as happens on excitation, calcium binds to troponin; this causes conformational changes in tropomyosin such that it dissociates from its binding site on actin. The actin is then free to react with myosin. This results in a ratchet-like movement, cumulatively leading to muscle contraction.

The sliding filament theory effectively explains a number of the features of muscular contraction. For example, it is possible to envisage why a rapidly contracting muscle fiber exerts less tension than one contracting slowly, i.e., because it takes a certain amount of time to form the bridges between actin and myosin. Hence, if the contraction is too fast, fewer

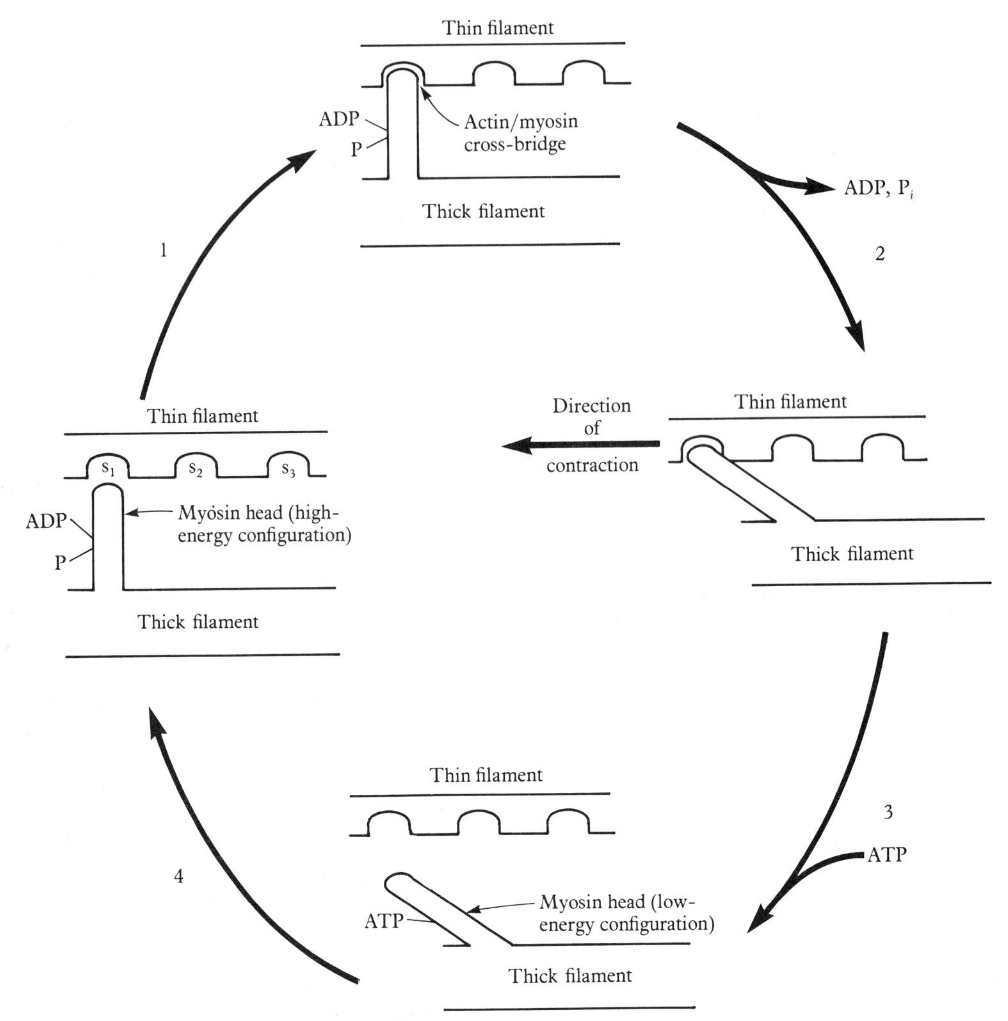

Figure 23.14. *Possible mechanism by which a muscle transduces the chemical energy of ATP into mechanical energy of movement. In step 1 the myosin head, in a high energy configuration (as a result of reaction with ATP) combines with the thin filament by the mechanism described in the text and moves the thin filament, as shown in step 2. It now is recharged with ATP and disconnects from the thin filament, which has been moved one notch. After reacting with ATP it returns to its original orientation opposite site 2 on the thin filament and repeats the cycle, each repetition moving the units one notch with respect to one another (contraction). (From Becker, 1977: Energy and the Living Cell. J. B. Lippincott Co., Philadelphia, p. 283.)*

bridges can form, and fewer bridges means less tension exerted. Again, it is possible to see why a muscle fiber carrying a larger load uses more energy (heat plus work) than one under a lesser load. When the load is larger, more bridges form and more ATP molecules are used up. It is also possible to conceive how a muscle fiber can continue to exert tension, since at any site the reaction can occur more than once, the bonds between actin and myosin breaking and re-forming, permitting a muscle fiber to continue to do work.

Many hypotheses have been proposed to explain the molecular reactions by which the energy liberated in the hydrolysis of ATP to ADP serves to propel the myosin myofilament relative to the actin myofilament. Each hypothesis accounts for some of the experimental findings, some more than others, but none accounts for all the facts of muscular contraction.

The currently favored hypothesis for explaining transduction of chemical to mechanical energy in the muscle fiber is an angular change in the protruding head of a myosin molecule attached to an actin molecule and energized by ATP. The actin resembles a ratchet in which movement of the myosin filament in relation to the actin filament occurs one G-actin unit at a time (Fig. 23.14). The myosin head makes contact with an actin molecule when troponin is activated by calcium ions. The calcium ions probably bring about a conformational change between the three subunits of troponin (TnT, TnC, and TnI) resulting in their closer binding to one another, thereby exposing the active site on actin to combine with the myosin head (see Fig. 23.9) (Cohen, 1975). This is followed by a conformational change in myosin such that the myosin head assumes a 45-degree angle with its tail (instead of the previous 90-degree angle). Hydrolysis of ATP to ADP and P_i causes the head to revert to the 90-degree angle, preparing it to react with the next G-actin monomer in F-actin for another round of movement (Fig. 23.14). A series of such rounds occurring rapidly in succession constitutes muscular contraction.

WORK AND HEAT RELATIONS IN MUSCLE CONTRACTION

A muscle can contract with shortening *(isotonic contraction)* or without shortening *(isometric contraction)*. The muscle fibers can exert maximal tension (force) when they do not shorten; for example, a striated muscle can exert a tension of 3 kg per square centimeter of cross section. Even though a muscle fiber in this state does no visible work in the physical sense of moving an object through a given distance, it needs energy to maintain the state of tension, and this energy appears as heat.

In review, muscular contraction is powered immediately by adenosine triphosphate (ATP). ATP is hydrolyzed to adenosine diphosphate (ADP) and inorganic phosphate (P_i) in the presence of actomyosin, which acts as an adenosine triphosphatase (ATPase). ADP can be rephosphorylated by transphosphorylation from creatine phosphate (CP) in the presence of creatine kinase. The creatine is rephosphorylated by metabolic reactions (either glycolysis or aerobic reactions; see Chapter 12). The chemical energy of ATP used in muscular contraction is converted partly to potential energy and partly to heat. When no useful work is done by the muscle, for example, when a weight lifted by a muscle is permitted to fall on relaxation of the muscle, all the energy is converted to heat. The ratio of useful work to heat is the efficiency of the muscle, and a value of 35 percent is cited as the highest efficiency achieved for frog and toad muscle when the muscle is

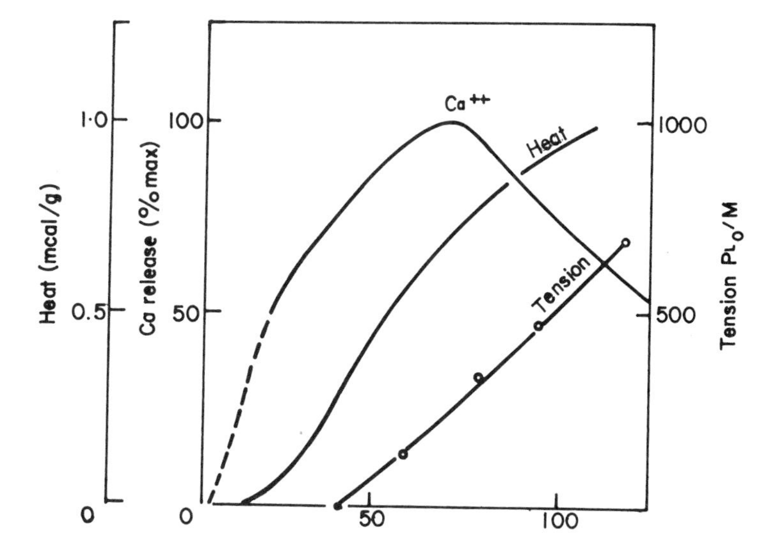

Figure 23.15. *Composite diagram illustrating the release of Ca ions, the heat output and the development of tension during the very early stages of a twitch in a toad sartorius muscle at 0°C. Max. tension = 1200 g/cm²; heat at height of twitch = 3.0 mcal/g. (From Bendall, 1969: Muscles, Molecules and Movement. American Elsevier, New York.)*

doing work at constant speed (in a specially designed instrument). However, somewhat higher values have been claimed for tortoise muscle (Bendall, 1969).

The heat given off is thus determined by a variety of conditions, some of which are considered here. As might be expected, the greater the number of bondings between actin and myosin the greater will be the tension exerted by the muscle and the greater the amount of heat liberated. Maximum tension is exerted at a sarcomere length of 2.2 μm in frog muscle because at this length the actin filaments are almost touching in the A-band and are therefore in contact with the maximum possible number of myosin heads. As the number of points of contact decreases during relaxation, tension decreases until at 3.65 μm the two are pulled completely out of contact with one another, the tension falls to zero, and the heat produced falls off precipitously.

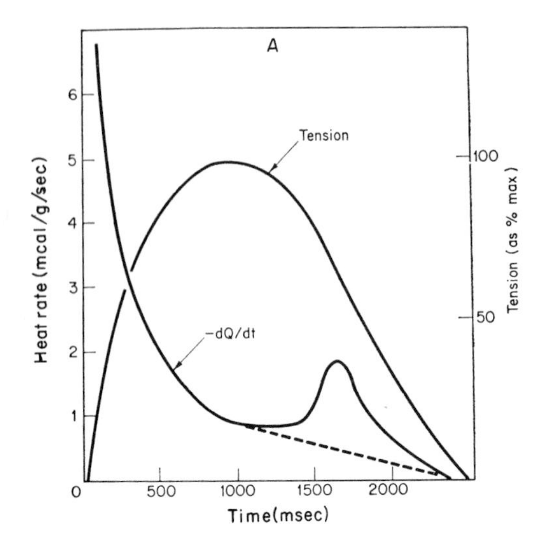

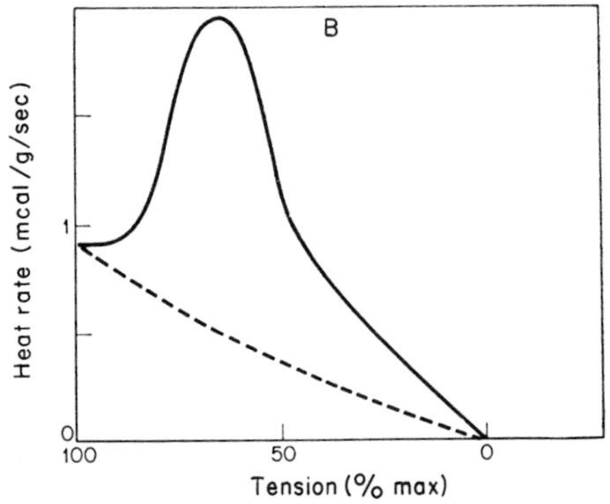

Figure 23.16. *A, The rate of heat production (−dQ/dt) during an isometric twitch of a toad semi-membranosus muscle at 0°C. Tension development shown for comparison. B, Plot of the heat rate against the tension during relaxation. Dotted line in both figures represents the likely heat rate in an ideal isometric twitch, in which there would be no stretching of the series elastic components (the "give" in sarcomeres, the connection of terminal sarcomeres to connective tissue, and the tendons). (From Bendall, 1969: Muscles, Molecules and Movement. American Elsevier, New York.)*

The heat given off by a muscle is divided into *activation heat* and *latent heat,* as already mentioned. Activation heat appears, after calcium release but before the tension develops, and rises rapidly (Fig. 23.15); the latent heat then continues, although the rate of heat emission (dQ/dt, where Q is the heat and t is the time) declines until the tension also declines, at which time it rises briefly before it accompanies the fall of the tension to zero (Fig. 23.16). The first quick heat liberation (activation heat) is attributed to the activation process, during which calcium ions are released from the triads and subsequently attached to the active sites on troponin (on actin). The latent heat is a result of the breakdown of ATP as the filaments glide over one another during shortening of the sarcomeres, as well as in maintaining any level of continued contraction and tension.

The amount of heat liberated depends partly on whether the contraction of the muscle is isometric or isotonic.

In isometric contraction the muscle does not change length, but energy is used to stretch the series of elastic components (sarcolemma and connective tissue) and connections of the muscle and is stored as potential energy in these components. The remainder is given off as heat. When, after stimulation ceases, the muscle relaxes, the potential energy so stored in the elastic components is released as heat (Bendall, 1969). One can demonstrate the effect of shortening of a muscle on heat given off by first allowing a loaded muscle to contract isometrically and then suddenly permitting it to shorten isotonically to a predetermined length. Shortening is always accompanied by an extra heat component, which is related to the degree of shortening and to the load. When shortening comes to an end, the rate of heat given off returns to the level of the curve for an otherwise identical isometric muscle contraction (Fig. 23.17) (Bendall, 1969).

In isotonic contraction the heat and work are dependent on the degree of shortening and on the load. The potential energy gained in these cases by the work done in carrying a load a certain distance is released as heat if the muscle is allowed to relax with the load. Thus the two components can be measured by the heat given off before and after relaxation. Inasmuch as more work is done in isotonic contraction, more heat is given off than in isometric contraction (Bendall, 1969).

Forced lengthening of a muscle fiber during contraction decreases the energy liberated. Hill (1960) has gathered evidence to indicate that

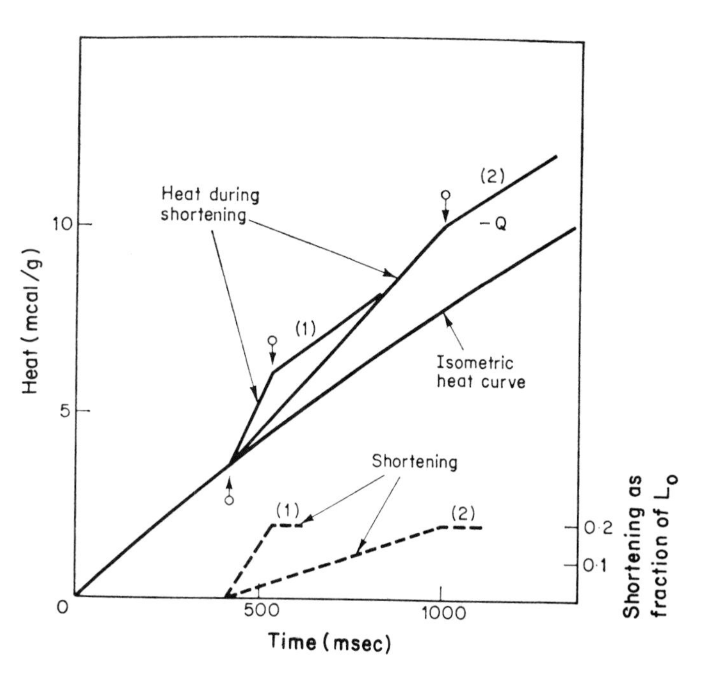

Figure 23.17. *The effect on the heat output of release of a muscle from isometric restraint during a 0.4 sec tetanus. The release = 0.2 L_0 (where L_0 is the original length of the muscle, here released from 1.1 L_0 to 0.9 L_0); the tension (pull), P_0 = 2000 g/cm²; velocity, V_m = 2 muscle lengths per sec. Frog sartorius at 0°C. Release at 0.4 sec after shock at upward arrow ♂ : (1) under load of 0.05 P_0 = 100 gm/cm², (2) under load of 0.50 P_0 = 1000 gm/cm². Upper curves show heat evolved, lower curves (dotted) show the shortening, which comes to an end at downward arrow ♀ . (From Bendall, 1969: Muscle, Molecules and Movement. American Elsevier, New York.)*

the chemical processes induced in a muscle fiber by applying a stimulus are reversed by the application of external mechanical work (force acting through a distance), although it is not possible to store energy in the muscle fiber in this way.

Many studies have been made on muscle tension in isotonic and isometric contraction and on the effect of stretch and other factors on the tension developed. These studies are mainly pertinent to muscle as it operates in a metazoan and are treated in textbooks of comparative and mammalian physiology (Prosser, 1973; Guyton, 1976). Attention here has centered on contractility as a cellular problem, primarily on the chemical nature of the contractile mechanisms, the liberation of energy, and the way in which this provides for the relative movements of the myofilaments with respect to one another, and the molecular basis of movement.

EXCITATION OF MUSCULAR CONTRACTION

Although evidence for activation of the muscle fiber by acetylcholine secreted by the terminations of efferent neurons at the myoneural junction is convincing (see Chapter 21), the problem still remains of how the filaments are activated. It has been shown that if minute microelectrodes (about 2 μm in diameter) are brought alongside the isotropic and anisotropic bands of a single isolated frog muscle fibril and a flow of current is allowed to depolarize it, contraction occurs when the stimulating electrode is against the Z-line but not if it is placed midway between the two Z-lines. This holds true even though the current is increased several fold (Huxley and Taylor, 1958). The Z-lines, as was pointed out previously, appear to line up and perhaps connect transversely through a whole series of muscle fibers. It is tempting to suggest that the Z-lines are involved in coupling the fibers during excitation, but the mechanism by which this is accomplished has not as yet been defined.

Electron micrographs of the myoneural junction show a tubular system extending from the membrane of the muscle fiber inward as a sort of microsynaptic system (subdivision of a synaptic connection) to the fibrils. This system has been called the *triad* or *T-system* (Figs. 23.18 and 23.19). It consists of two types of extensions. One is a central tubular extension inward of the sarcolemma (plasma membrane of the muscle cell), which contacts the sarcomeres (the myofibril units) at the Z-line (or at the boundary of the A- and I-bands in some muscles). The second is two lateral tubular extensions of the endoplasmic (sarcoplasmic) reticulum. That the tubules are truly extensions of the cell membrane has been demonstrated by placing muscle fibers in a suspension of labeled ferritin (an electron-dense, iron-containing compound), in thorium dioxide (Philpott and Goldstein, 1967), or in a solution of fluorescent protein (Endo, 1966). These tracers are shown to appear in sarcolemmal tubules of the triad next to the sarcomeres (Peachey, 1965). The lateral vesicles of the triad (extensions of the sarcoplasmic reticulum) accompany the sarcolemmal tubules to the sarcomere. Such a T-system is quite prominent in fibers of active

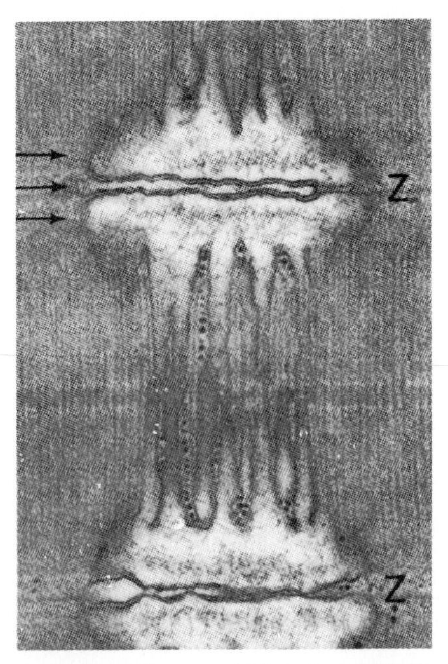

Figure 23.18. *Diagram of triads by which the impulse is spread in excitation of muscular contraction. A triad consists of the incursion of the sarcolemma (follow middle arrow) and the tubules of the sarcoplasmic reticulum (upper and lower arrows), which impinge upon the incursion from the sarcolemma. The impulse comes to the incursive tubule of the sarcolemma and a current is incited. This leads to the release of calcium ions from the tubules of the sarcoplasmic reticulum at the Z-line, leading to muscular contraction. (By courtesy of K. R. Porter.)*

muscle, but less so in cardiac and slow muscles (Franzini-Armstrong, 1973).

Because the T-system functions in both excitation and relaxation of the muscle (discussed later), it is more than a telegraphic connection signaling contraction. In invertebrate muscles studied the triad is replaced by a *dyad* of sarcolemmal and sarcoplasmic tubules. In the dyad the invagination of the cell membrane into a single element of the sarcoplasmic reticulum forms a two-part unit (Hoyle, 1970).

Details of the method by which the triad functions are not fully understood. Excitation of the sarcolemma induces a spike potential with depolarization of the T-system (Needham, 1971). This in turn results in liberation of calcium ions from the outer sarcoplasmic reticulum vesicles of the triad (Ebashi, 1965). That calcium is liberated from the triads has been demonstrated by treating the muscle preparation during stimulation with oxalate and then studying it by electron microscopy. Vesicles so treated show deposits of electron-dense calcium oxalate that are not present in unstimulated controls (Constantin and Podolsky, 1965). Calcium release from the triads is also shown by autoradiography experiments with ^{45}Ca. Muscle stimulated to contract in the presence of ^{45}Ca shows the presence of silver grains, which appear to be in the filaments over the triad; muscles treated with relaxant (calcium chelators) show much less (Winegrad, 1965a).

The release of calcium as the excitant to contraction has also been demonstrated using aequorin, the luminous protein from the jellyfish *Aequoria.* Aequorin requires only calcium to induce luminescence (see Chapter 13). Stimulation of muscle fibers (giant fibers from a barnacle) in the presence of aequorin resulted in a flash of light. Since aequorin was present in excess, the muscle could be excited time and again, each time with emission of a flash of light as calcium was released from the muscle (Hoyle, 1970; Ashley, 1971). Calcium thus appears to be the excitant in all muscle systems tested (Szent-Györgyi, 1976), even in as primitive a contractile system as that in the ciliate protozoan *Spirostomum.* This elongated organism, which has contractile longitudinal myonemes composed of microfilaments, shortens markedly on stimulation. Injected aequorin emitted light each time contraction occurred (Ettienne, 1970).

It is believed that the T-system would help explain spread of excitation, since no ordinary diffusion of a substance across a membrane would be fast enough to account for the rapidity of a contractile response. However, calcium ion may be released in small amounts causing a membrane change that then induces release of larger amounts—a two-stage activation (Taylor and Godt, 1976).

RELAXATION OF MUSCLE CELLS

Since ATP is used in muscular contraction, either it is present in an inactive form to be released for action only when the impulse reaches the muscle fiber or the enzymatic breakdown of ATP is normally inhibited until the action potential reaches the muscle fibers. Then the inhibitory influence vanishes and the ATP in the region of the active enzyme is decomposed.

In a number of studies the attempt has been made to separate a soluble inhibitory substance from a particular fraction of muscle homogenate, but this has failed. Meanwhile, it has been demonstrated that the equivalent of relaxation in muscle fibers can be brought about in muscle homogenates or model systems by chelating agents such as ethylene-diamine tetraacetic acid (EDTA), which removes calcium ions. It has also been demonstrated that the microsomes in muscle homogenates and the sarcoplasmic reticulum in muscle fibers per-

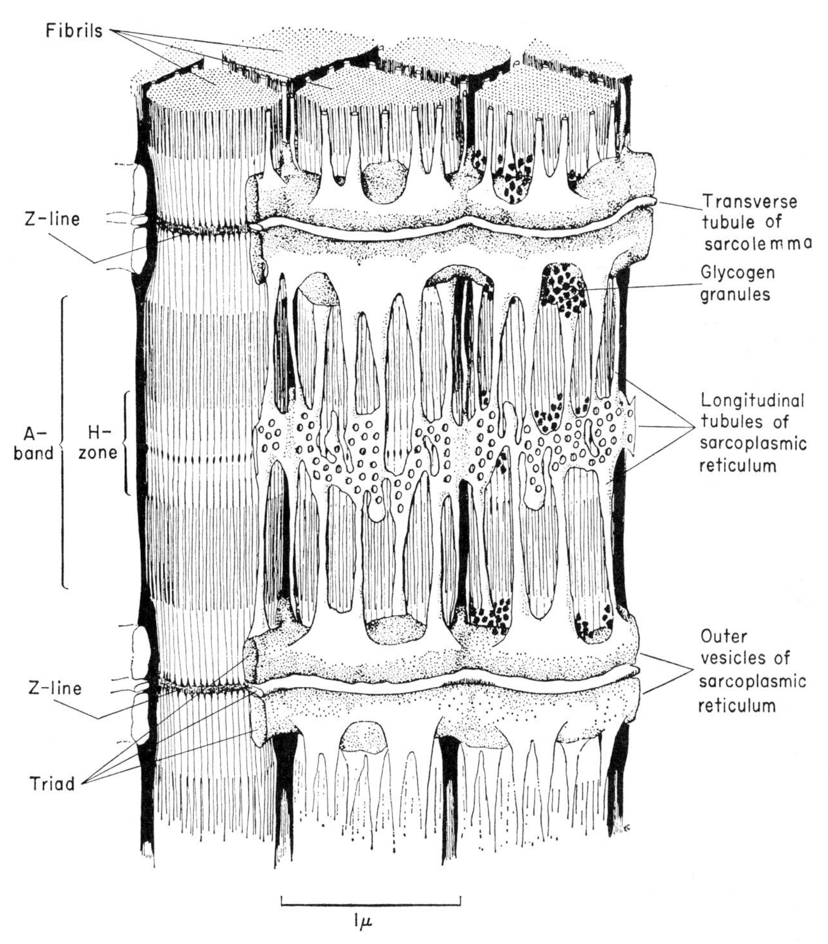

Figure 23.19. *The sarcoplasmic reticulum and transverse tubules of the frog sartorius muscle. (After Peachey, 1965: J. Cell Biol.* 25*: 222.)*

form similar functions in removing the calcium ions (Hasselbach, 1965).

The manner in which microsomes in muscle analogues and sarcoplasmic reticulum in muscle fibers remove calcium is now known to occur through the operation of a calcium pump. The mechanism of active transport of calcium across the membranes of the sarcoplasmic reticulum against a concentration gradient is similar to that for other substances already discussed in Chapter 18. Presumably a carrier, X, attaches to the calcium ion at the outer boundary of the membrane of the sarcoplasmic vesicle and, at the expense of ATP, moves it to the other side and into the cavity of the vesicle (Philpott and Goldstein, 1967).

Quantitative estimates indicate that at highest efficiency two calcium ions are carried per ATP molecule decomposed (Hasselbach, 1964). Such ionic movements must occur rapidly, indeed, since liberation of calcium-inciting contraction followed by reabsorption of the ions permitting relaxation must be accomplished within the short period of a muscle twitch.

While the consensus is that ATP is closely linked with muscular contraction, the evidence for its action in the live muscle fiber can only be inferred. In live muscle fibers it is difficult to demonstrate any changes in ATP concentration of sufficient magnitude, even by spectrophotometric means. Changes in concentration of creatine phosphate, however, can be demonstrated, and the decrease in creatine phosphate is proportional to the number of contractions of the muscle. Creatine phosphate, it will be remembered, is always ready to donate its high-energy phosphate bond to ADP, resulting in the formation of ATP and creatine (see Chapter 12). The equilibrium of CP in the reaction with ADP lies in the direction of ATP production (in the presence of the enzyme creatine phosphokinase). When creatine phosphokinase is inactivated by a specific poison, fluorodinitrobenzene, however, ATP is decomposed to ADP and inorganic phosphate in proportion to the number of contractions induced. This considerably strengthens the hypothesis that ATP furnishes the energy for muscular contraction (Wilkie, 1966).

From the foregoing, one finds that the muscle fiber is a machine that elegantly converts chemical energy to mechanical energy. Mechanical work is done directly at the expense of chemical energy in ATP, which is converted to ADP and inorganic phosphate. In man-made machines the chemical energy must first be converted to electrical or heat energy. Engineers have yet to find a way to imitate the direct conversion of chemical to mechanical energy that occurs in the muscle cell. The biochemical mechanism for liberation of the chemical potential energy in the living cell is well known. However, one large and critical portion of the puzzle yet remains to be found: how is the final chemical to mechanical energy conversion completed inside the living muscle fiber?

MECHANISM OF CONTRACTION IN NONMUSCULAR CELLS

Ameboid Movement

The molecular basis of ameboid movement has been difficult to determine because of the unstable location of the contractile elements (see Chapter 22). Both actin and myosin have been identified biochemically in amoebas and thick and thin filaments have been seen in electron micrographs. It thus seems reasonable to suggest that ameboid movement is a result of the interaction of these microfilaments in a manner similar to what happens in muscle (Huxley, 1973). One filament, against which the other filament may glide, is attached stably in a rigid structure such as a gel in an amoeba. However, we do not yet know with certainty which filament this is. If the contractile elements have polarity, movement can be achieved by reactions between them.

That glycerinated amoebas will contract on addition of ATP was demonstrated by Simard-Duquesne and Couillard in 1962. In a more recent study glycerinated amoebas were observed to contract over the entire surface, including the uroid (tail). Such models contract only if calcium ions are present in a favorable concentration, indicating that, as in muscle, the calcium ion is an excitant to contraction. Magnesium ion is also required for contraction. Whether regulator proteins are present in amoebas as in muscle is not known.

Contraction in glycerinated models occurs to the extent of about 50 percent of the original dimensions, but no pseudopods are formed, although those present retain their shape. Contraction is limited to the ectoplasm; no contraction is seen in the endoplasm. Thick and thin filaments are found in the ectoplasmic tube, sometimes near the plasma membrane in glycerinated amoebas. The data favor the ectoplasmic contraction mechanism of ameboid movement (see Chapter 22) (Rinaldi *et al.*, 1975), although disagreement in the literature on this point continues (Allen and Taylor, 1975; Taylor and Godt, 1976).

Studies on ameboid movement in the slime mold *Physarum polycephalum* have yielded more biochemical data than those on *Amoeba* because larger amounts of actin and myosin are present. Actin comprises about 2 to 5 percent of the protein in *Physarum;* about one fifth as much myosin is present. *Physarum* actin resembles its muscle analogue in sedimentation rate, molecular weight, diffusion coefficient, filament formation, reaction with myosin, and even in amino acid composition. However, it differs from muscle actin; for example, it is more readily extracted (at lower ionic strength of solution) than muscle actin. *Physarum* myosin is similar to muscle myosin in size, shape, and function but it differs in several respects. As an example, it is more soluble and has a more limited capacity for aggregation than muscle myosin. There seems little reason to question that an actin-myosin interaction is the basis of movement in *Physarum* both in the small ameboid stage (from which actin and myosin have also been extracted) and in the large multinucleate plasmodial stage. Tubulin is absent in these stages but is likely to be present in the flagellated stage of the life cycle (Jacobson *et al.*, 1976).

Ciliary Movement

The mechanism of ciliary movement has been analyzed much more effectively than has ameboid movement because of the distinctive and permanent structure of the organelles in cilia in contrast to the continuous flux of whatever structure underlies ameboid movement. Cilia and flagella show an identical and characteristic structure, each being composed of two central single microtubules and nine peripheral double microtubules, as shown in Chapter 22 and Figure 23.20. Henceforth only cilia are mentioned as a rule, but the information is relevant to flagella as well.

The microtubules of a cilium are composed of a globular protein, tubulin (molecular weight 120,000). Tubulin consists of two nonidentical polypeptide chains, called α and β chains, and contains tightly bound GTP. A single cylindrical tubule is built of alternating α and β subunits, 13 such subunits being present at the circumference of the tubule. As seen in Figure 23.21, from microtubule (or subfiber) A toward microtubule B of another microtubule pair protrude two arms that are composed of another type of protein called *dynein.* There are two

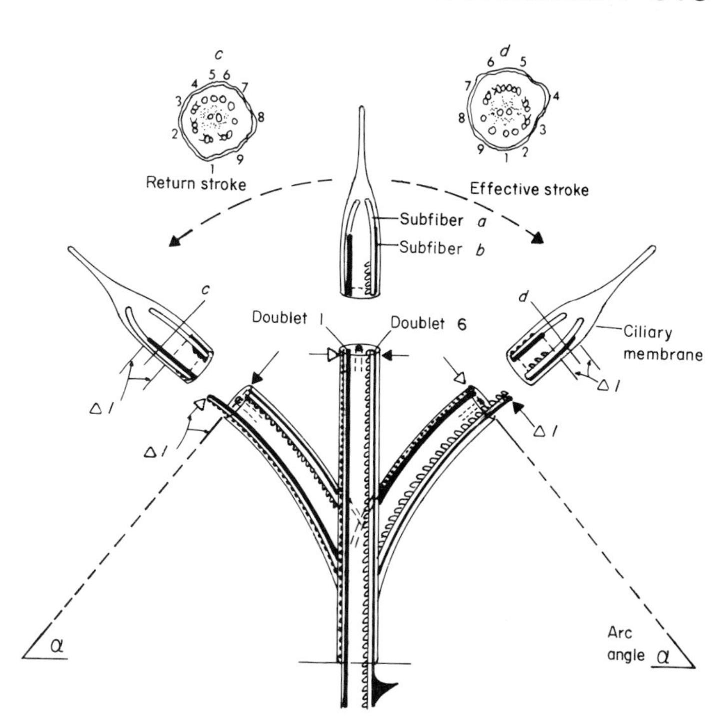

Figure 23.20. *Sliding microtubule-subfiber hypothesis for ciliary action based on electron micrographs of differences in the microtubular subfibers near the tip of the cilium (diagrammed at* c *and* d*). The relations are shown for microtubule doublets 1 and 6 in straight up (as shown in Fig. 23.21), stroke position (right at* d*), and return position (shown at* c*, left). The displacement of the microtubule subfiber on the outside of the cilium with respect to the one on the inside (△ 1) is shown for the stroke and return positions. The cross sections of the cilia at* c *and* d *above are taken between the levels of △ 1 on the left and right, respectively. When the ATPase (dynein arms) causes microtubule subfibers to slide past one another, shear resistance in the cilium changes sliding to bending since the subfibers do not contract or stretch. (After Satir, 1968: J. Cell Biol.* 39*: 77.)*

dyneins, a 14S dynein (molecular weight 600,000) and a 30S dynein, both of them with ATPase activity.

The microtubules (subfibers A and B), of which a cilium is composed, could either contract individually or they could slide along one another in the manner of muscle filaments. A decision as to which mechanism was the one actually used depended upon scientists getting detailed information on the lengths of the paired microtubules during various phases of ciliary movement, and it was possible to get such information from cilia on the cells of a mussel gill. The cilia on these cells move in a metachronal wave like the waves produced by the wind blowing on a grain field. Some cilia are straight, others are bent in contraction, others are returning to the straight position. Electron micrographs of a series of cilia in a metachronal beat with some straight and others bent showed no change in the length of individual microtubules but clearly showed a sliding of individual microtubules A and B with respect to one another (Satir, 1974; see Figs. 23.20 and 23.21 for details).

In addition to the two single microtubules in the center and the nine paired outer microtubules, spokes (and a nexin band) tie the microtubules together (Figs. 23.20 and 23.21). It is thought that the anchored spokes provide the shear resistance that converts the sliding movement of the microtubules to the bending of the cilia (Satir, 1974). In electron micrographs the spokes appear to be stretched in the regions in which cilia are bent, but not in the areas in which the cilia are straight.

That ATP provides the energy for ciliary action was first demonstrated on glycerinated preparations of cilia by Hoffmann-Berling in 1960. Glycerinated preparations lack the cell membrane over the cilia as well as the other substances that surround the axoneme (the 9 + 2 microtubular portion), resembling in this respect glycerinated muscle preparations. Glycerinated preparations are inactive, but on addition of ATP ciliary motion is resumed. Gibbons and Rowe (1965) showed that when the dynein bridges were removed with a chelating agent, the glycerinated cilia no longer responded to ATP. When extracted dyneins were added to the treated cilia, the arms were reformed, protruding again from subfiber A of one microtubule doublet to subfiber B on the neighboring doublet; on addition of ATP ciliary motion was resumed. In all these experiments the appropriate concentrations of calcium and magnesium ions were supplied. It was later demonstrated that ATP is expended the full length of the cilium. Ciliary movement, although dependent upon a different set of structures, in many respects resembles muscular contraction—in cilia the microtubules glide along one another, reminiscent of the actin and myosin microfilaments in muscle.

Flagella may be removed from sperm by a laser beam; they then swim by whipping movements in a characteristic manner although they have been separated from the cell. This continues for as long as they have a supply of ATP. When this is exhausted and movement stops, addition of more ATP enables them to resume movement. Cilia and flagella are thus

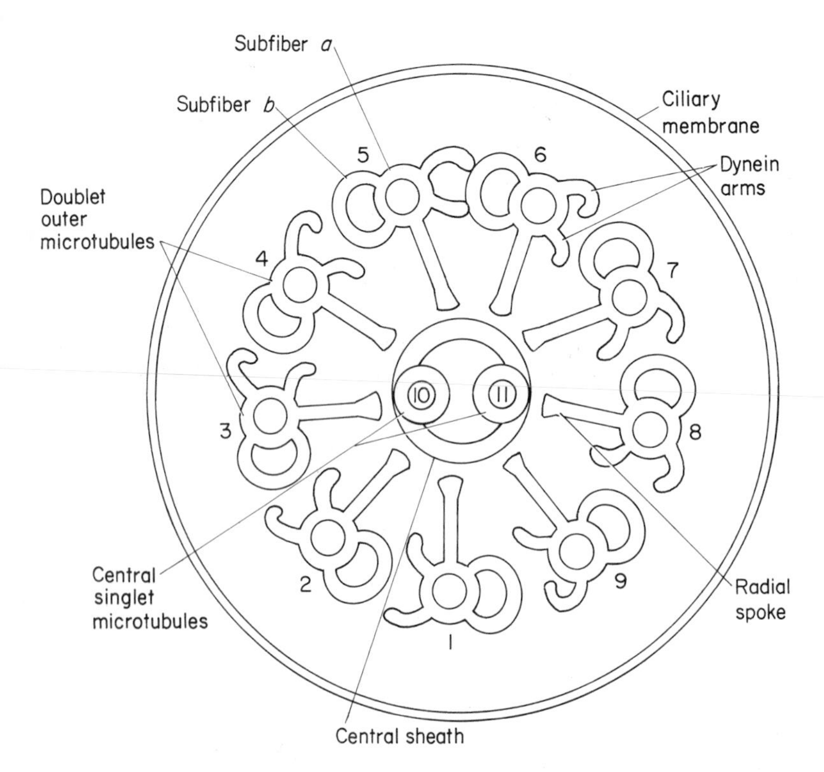

Figure 23.21. *Diagrammatic cross section of an individual cilium: note the 9 + 2 microtubule pattern. Each microtubule doublet has two subfibers,* a *and* b; *subfiber* a *consists of 11 or 12 tubulin subunits in circumference and subfiber* b *of 13. Arms composed of the ATP-ase dynein, project from subfiber* a *towards subfiber* b *of the next doublet. An imaginary axis could be drawn through the middle of the dynein arm forming a bridge between doublets 5 and 6 and the middle of subfiber* a *of doublet 1. (After Satir, 1968: J. Cell Biol.* 39: *77 and Sci. Am. (Oct.)* 23: *45–52.)*

independent of the cell except for the required supply of ATP.

When cilia are glycerinated and then treated with trypsin, the spokes that hold the microtubules together are dissolved. When such preparations are supplied with ATP the A and B microtubules literally walk away from one another. Because there is nothing to hold the axoneme together to permit the characteristic bending of the cilia, the reaction between the dynein bridges and the two microtubules proceeds until the ratchet mechanism moves one microtubule past the other to the point of separation.

In conclusion, it seems that a reasonable hypothesis to explain ciliary movement is the sliding of members of a doublet of microtubules in relationship to one another by a ratchet-like mechanism. The dynein arms act as ATPases, analogous perhaps to the ratchet-like mechanism between the myosin head and the actin filament. The sliding of outer doublets in cilia is resisted by the radial connection leading to bending.

Regulation of Ciliary Movement

It has become evident that calcium ions are important in ciliary movement. Although a number of organisms have now been studied, the discussion here is mainly based on work done with *Paramecium*.

In *Paramecium* the magnitude of the action potential (regenerative depolarization potential) depends upon the extracellular calcium ion concentration, much as in the giant muscle fibers of the barnacle. In most cells the magnitude of the action potential depends instead upon the presence of sodium in the external medium. In *Paramecium* calcium conductance is responsible for depolarization of the cell membrane and the overshoot (positive potential of the spike) that develops during an action potential. Under experimental conditions other ions may be substituted for calcium, but they are not likely to play such a role in nature. The following discussion is taken from Eckert (1972; see also Eckert *et al.*, 1976).

The membrane of *Paramecium* responds to stimuli by graded depolarizations that depend upon the intensity of the stimulus, the region of the cell stimulated, and the genotype of the cell. Thus, in response to a series of graded stimuli the membrane responds by graded depolarizations to zero and then progressive positive potential (inside) as a function of the extracellular calcium ion concentration (Fig. 23.22).

The stimulus—for example, a tap from a microstylus—evokes different membrane responses depending on whether it is delivered to the anterior or the posterior end of the cell. At the anterior end such stimulation produces depolarization, resulting in an action potential. At the posterior end, on the other hand, it produces hyperpolarization only. Stimuli evoke the responses mentioned in most strains of *Paramecium*, but in a mutant (Pawn) of *P. au-*

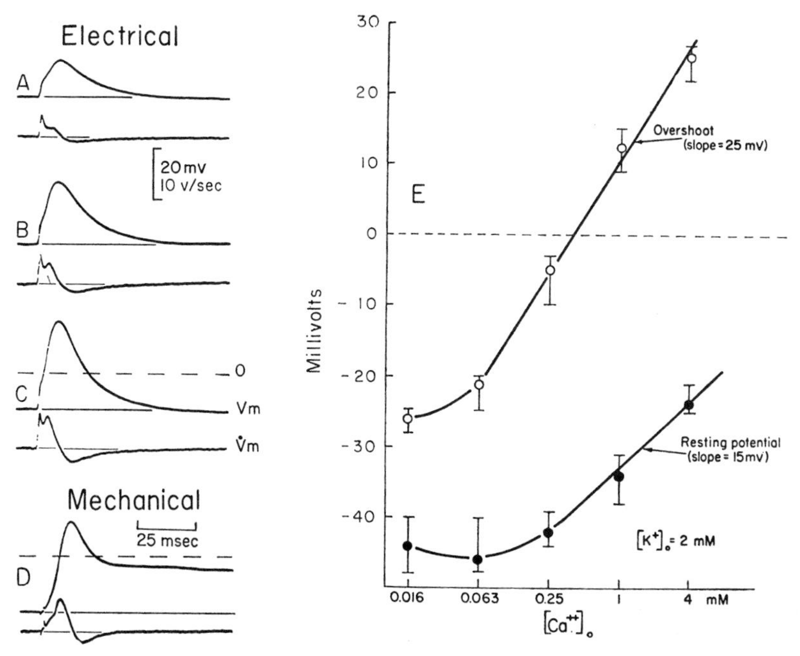

Figure 23.22. *The calcium response in* P. caudatum. A–C, *Regenerative excitation (calcium responses) graded in amplitude* (V_m) *and rate of rise* ($\dot{V}_m$) *with increasing intensity of 2-msec current pulse delivered with intracellular electrode.* D, *Electrical response to mechanical stimulation of anterior surface. Results in* A *through* D *were recorded from the same cell in 1 mM* $CaCl_2$ *and 2 mM KCl.* E, *Resting potential and peak amplitude (overshoot: the positive inside potential beyond depolarization) of maximal regenerative responses to electrical stimulation plotted against extracellular* Ca^{2+} *concentration. The potential of the cell depends upon the calcium level: with increasing calcium concentration not only is the cell depolarized but it becomes positive inside (overshoot). The concentration of KCl was held constant at 2 mM throughout. (From Eckert, R., 1972: Science, 176: 473–481, Copyright 1972 by the American Association for the Advancement of Science.)*

relia stimulation produces electrotonic changes only, without depolarization and action potentials. Apparently it lacks the requisite membrane properties to give the characteristic response (Eckert, 1972).

Reversal of ciliary beat in ciliates as well as in epithelia has long intrigued investigators of ciliary movement. Ciliary reversal occurs in response to changes in environmental stimuli. These are thought to operate by way of changes in internal calcium concentration affecting the ciliary apparatus. Some of the relationships between extracellular calcium concentration, ciliary frequency, and degree of ciliary reversal are shown in Figure 23.23. Evidently a stimulus that hyperpolarizes the cell membrane (using an inserted microelectrode) incites little ciliary reversal, while one that depolarizes the membrane causes definite ciliary reversal (Fig. 23.23*C*). Since depolarization of the membrane increases the calcium conductance (Fig. 23.23*A*), the internal calcium concentration is increased. The increase in internal calcium concentration increases ciliary reversal, and the quantitative ratios suggest a close relationship between the two. The fact that such induced ciliary reversal continues for only a limited period of time suggests that the membrane gate permitting entry of calcium is shut soon after stimulation. A calcium pump then eliminates the excess internal calcium against a concentration gradient, inasmuch as the concentration of calcium ions is normally lower inside the cell than outside. If this is so, lowering the temperature should prolong the period of ciliary reversal resulting from a given stimulus inasmuch as it would reduce the amount of metabolic energy available for the active transport of calcium to the exterior. Experiments confirm this hypothesis. It thus appears that ciliary reversal is primarily dependent upon an effect of an increase in the internal calcium concentration on the ciliary apparatus.

Spasmonemes

Spasmonemes, the bands of contractile microfilaments especially prominent in the basal stalk of colonial flagellates (*Carchesium*, *Zoothamnium* discussed in Chapter 22) and in *Vorticella* have properties similar to muscle fibers but are more rapid in action. Gel electrophoresis of isolated spasmonemes reveals a pair of new contractile proteins called *spasmins*

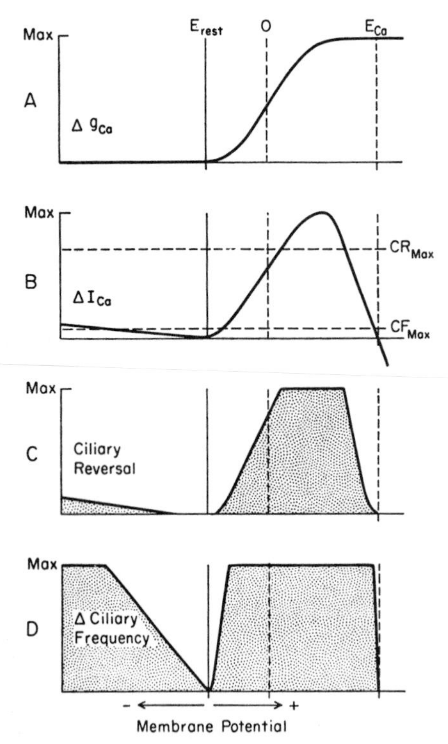

Figure 23.23. *Proposed relations between membrane potential, calcium conductance, calcium current, ciliary reversal, and ciliary frequency.* A, *Peak transient increases in calcium conductance, Δg_{Ca}, produced by sudden changes in V_m from resting potential (E_{rest}) to various hyperpolarized (toward left) or depolarized (toward right) membrane potentials. Sigmoid curve is characteristic of electrically excitable membranes.* B, *Corresponding transient change in inward calcium current, ΔI_{Ca}. When ΔI_{Ca} is above the level labeled CF_{max}, ciliary frequency is maximal, and when ΔI_{Ca} is above the level labeled CR_{max}, ciliary reversal is maximal.* C, *Reversed beating increases with the intracellular concentration of Ca^{2+}, which is closely related to ΔI_{Ca} of plot B.* D, *Frequency of beating is more sensitive to I_{Ca} than is reversal. Plots are hypothetical and are not intended to depict precise quantitative relations. (From Eckert, R., 1972: Science,* 176: *473–481. Copyright 1972 by the American Association for the Advancement of Science.)*

A and B, both of molecular weight 20,000, but separable by gel electrophoresis. Spasmonemes also contain higher molecular weight proteins whose function is unknown; possibly they act as regulators. Proteins with the gel electrophoretic characteristics of actins and tubulins are absent from the extracts of spasmonemes (Routledge *et al.*, 1976).

Spasmonemes are composed of microfilaments 2 to 4 nm in diameter and include membranous tubules. Contraction of spasmonemes differs from muscular contraction in that it does not require ATP and magnesium ions nor is it stopped by poisons that inhibit metabolic release of energy. Calcium ions (10^{-6}M) not only initiate contraction but also appear to drive it, although how this occurs is still a matter of conjecture. Relaxation occurs when the calcium ion concentration has fallen to 10^{-8}M. Calcium ions are thought to be taken up and stored in the tubules; for this ATP and metabolism are required (Routledge *et al.*, 1976).

Prokaryotic Locomotion

Locomotion in flagellated bacteria is quite different from flagellar movement in eukaryotic cells. A "flagellum" is a totally different structure in the prokaryote; it does not resemble a eukaryotic flagellum in either structure or mechanism of action. The prokaryotic flagellum is made up of a single protein, *flagellin,* of molecular weight about 40,000. Flagellin forms chains, three of which intertwine to form a prokaryotic flagellum. No cell membrane is present over a bacterial flagellum as in a eukaryotic flagellum. The prokaryotic flagellum passes through the cell wall taking origin in a flattened M ring or rotor (Fig. 23.24). Above the M ring the flagellum passes through an S ring attached to the cell wall (Fig. 23.24). Presumably reaction between the two rings motivates the propeller-like action of the rigid flagellum. In gram-negative bacteria L and P rings are attached to the outer portion of the wall.

Flagellin has no ATP-ase activity and evidence has accumulated that ATP is not the source of energy for movement of the flagellum. The source of energy for movement of the flagellum has not been ascertained. The flagellum may change its direction of rotation and sometimes it does so periodically. Rotation in one direction causes straight-line locomotion whereas rotation in the other causes tumbling. Much remains to be learned about the mechanism of bacterial flagellar movement (De Pamphilis and Adler, 1971; Berg, 1975).

The mechanism of gliding movement of some blue-green algae and related bacteria is even more a mystery than bacterial flagellar movement. A similar movement has been described for eukaryotic diatoms.

Gliding movements of some bacteria (e.g., *Myxococcus*) may be as rapid as 15 to 70 μm per min. Gliding movements, often going backward and forward many times, produce slime trails and slime trails are preferentially invaded. The bacteria may be in groups associated by pili. Such movement may be important in fruiting myxobacteria as a means of aggregating a large number of cells that secrete a common polysaccharide stalk on the top of which sporulation occurs (Kaiser, 1978). Although microfibers have been observed in some myxobacteria *(Chondrococcus,* Pate and Ordal, 1972), there is no evidence that they function

A

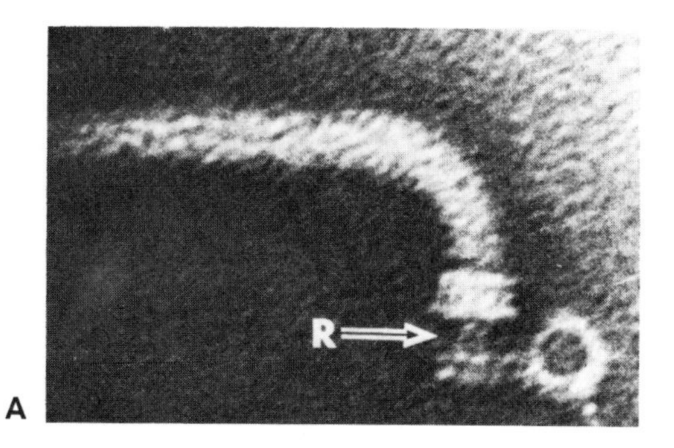

B

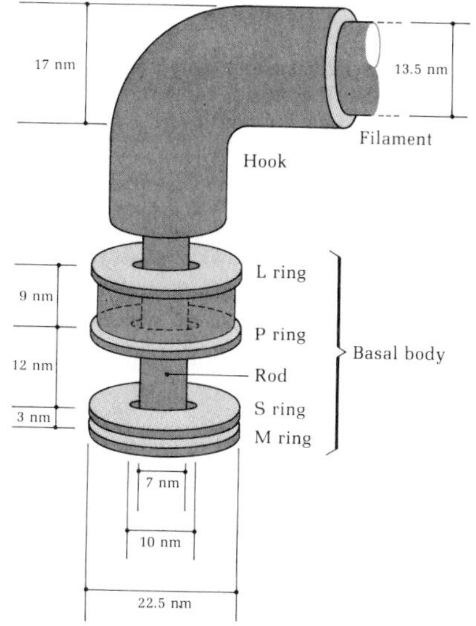

Figure 23.24. A, *Electronmicrograph of the flagellum of* E. coli. *The bent portion is called the hook and the terminal portion the filament. R points towards the ring system that is best interpreted from the diagram in* B *(×395,000). (From DePamphilis, M. L. and Adler, K., 1971: J. Bacteriol.* 105: *384–395.)* B, *Diagram of the structure of the flagellum of a bacterium. (From Volk, W. A., 1978: Essentials of Medical Microbiology, J. B. Lippincott Co., Philadelphia, p. 31.)*

in movement. Gliding occurs without visible body distortions or motile organs. One suggestion of old vintage is slime secretion, another a proton pump motivating a series of surface rotary assemblies each resembling the rotor of a flagellum. Poisons that selectively inhibit bacterial flagella also inhibit gliding motility. Latex spheres added to a suspension of gliding bacteria move along the cell surface as if propelled by rotary assemblies in the cell envelope (Pate and Chang, 1978).

Jerky movements called twitching occur in some bacteria without detectable motile organs and the mechanism is unknown. Sheet movement and sliding of colonies have also been described (Henrichsen, 1972).

SUMMARY OF NONMUSCULAR CONTRACTION

It now appears that actin is ubiquitous in nonmuscular eukaryotic cells and that myosin is present in most of them, although lacking in some sperm. It also appears likely that ATP (or another nucleotide) acts as the energy source for contraction and that calcium ions serve to regulate contraction.

It is interesting that in cells as diverse as brain, macrophages, fibroblasts, leukocytes, and liver, as well as in platelets, the actin is very similar in amino acid constitution to that in muscle. Even more suprising is that in cells of organisms as diverse as protozoans, plants, fungi, and animals the actin also shows great similarity in amino acid sequences. Differences in properties are found; for example, nonmuscle actins are much more labile than those from muscle. Although their localization has not been determined, it appears likely that in many cases actins are close to, or attached to, the plasma membrane. Actins are present in much lower concentration in nonmuscular systems than in muscle. For example, 17 percent of the protein in chicken breast muscle is actin whereas only 0.25 percent of the chick embryo fibroblast protein is actin (Anderson, 1976).

Myosin is generally found where actin is present in nonmuscular cells, with the exception noted. Myosin is present in much lower concentration than actin, about 1/100 as much. It consists of heavy and light meromyosins and attaches to actin under appropriate conditions that do not always resemble those in muscle. Regulatory proteins have been found in addition to actin and myosin in some of the nonmuscle contractile cells, although they have not been fully characterized. Because of the low concentration of myosin, contractile processes are much less readily studied in nonmuscle systems than in muscle; superprecipitation rather than contraction must be used as a sign of interaction between actin and myosin. The details of contraction and its regulation in such systems await more extensive analysis.

In conclusion, in eukaryotic cells two general types of movement appear to be present. One takes place between two types of filaments, actin and myosin, as, for example, in muscle. A second takes place between two types of microtubules (composed of tubulin A and B respectively), with a ratchet-like mechanism by way of dynein bridges from subtubule A to subtubule B, as in cilia and flagella. In both eukaryotic cases sliding between the elongated structures brings about contraction. In both cases, ATP is the source of energy for contraction. In both cases regulatory proteins are present and excitation occurs by way of increased calcium concentration while relaxation is achieved by resorption of calcium ions into the endoplasmic reticulum, lowering their concentration around the contractile structures.

In nonmuscle eukaryotic cells actin and probably myosin are almost always present; tubulin is ubiquitous. Therefore, when nonmuscle cells show movement, it is tempting to suggest that this is the result of either the actin-myosin system or the tubulin-dynein system, although direct corroborative evidence is difficult to obtain for either. The difficulty of analysis of movement in many nonmuscle nonciliated cells is compounded by lack of fixed structures to study by electron microscopy or with which to experiment. Nor is chemical isolation of the responsible labile contractile proteins easy. Furthermore, when fixed structures are lacking, the proteins involved in regulation of contraction are difficult to separate from the many other proteins in the cells. The results of experiments on nonmuscle systems are therefore more tenuous and hypotheses are more common than decisive results. Even more tenuous are results on movement in prokaryotic cells. However, much progress has been made in the last decade and intensive study of nonmuscular contraction continues (Pollard, 1975; Clark and Spudich, 1977).

LITERATURE CITED AND GENERAL REFERENCES

Adrian, R. H., Costantin, L. L. and Peachey, L. D., 1969: Radial spread of contraction in frog muscle fibers. J. Physiol. *204*: 231–257.

Allen, R. D. and Taylor, D. L., 1975: The molecular basis of ameboid movement. *In* Molecules and Cell Movement. Inoué and Stephens, eds. Raven Press, New York, pp. 239–258.

Amos, W. B., Rutledge, L. M., Weis-Fogh, T. and Yew, F. F., 1976: The spasmoneme and calcium dependent contraction in connection with specific calcium-binding proteins. Soc. Exp. Biol. Symp. *30*: 273–301.

Anderson, P., 1976: The actin of muscle and fibroblast. Biochem. J. *155*: 297–301.

Ashley, C. C., 1971: Calcium and the activation of skeletal muscle. Endeavour *30*: 18–25.

Ashley, C. C., Caldwell, P. C., Campbell, A. K., Lea, T. J. and Moiseseu, D. G., 1976: Calcium movements in muscle. Soc. Exp. Biol. Symp. *30*: 397–472.

Astbury, W. T., 1950: Muscular contraction and relaxation: X-ray studies of muscle. Proc. Roy. Soc. (London) *B 137*: 58–63.

Bendall, J. R., 1969: Muscles, Molecules and Movement. American Elsevier, New York.

Berg, H. C., 1975: How bacteria swim. Sci. Am. (Aug.) *233*: 36–44.

Blake, J. R. and Sleigh, M. A., 1974: Mechanism of ciliary locomotion. Biol. Rev. *49*: 85–125.

Bozler, E., 1953: The role of phosphocreatine and ATP in muscular contraction. J. Gen. Physiol. *37*: 63–70.

Brokaw, C. J., 1975: Cross bridge behavior in a sliding filament model for flagella. *In* Molecules and Cell Movement. Inoué and Stephens, eds. Raven Press, New York, pp. 165–179.

Bulbring, E. and Shuba, M. F., eds., 1975: Physiology of Smooth Muscles. Raven Press, New York.

Child, F. M., 1967: The chemistry of cilia and flagella. *In* Chemical Zoology. Florkin and Scheer, eds. Vol. I. Protozoa. Kidder, ed. Academic Press, New York, pp. 381–393.

Clarke, M. and Spudich, J. A., 1977: Nonmuscle contractile proteins: the role of actin and myosin in cell motility and shape determination. Ann. Rev. Biochem. *46*: 797–822.

Cohen, C., 1975: The protein switch of muscle contraction. Sci. Am. (Nov.) *233*: 36–45.

Collins, J. H., 1976: Structure and evolution of troponin C and related proteins. Soc. Exp. Biol. Symp. *30*: 303–334.

Condeelis, J. S., 1977: The isolation of microquantities of myosin from *Amoeba proteus* and *Chaos carolinensis*. Anal. Biochem. *78*: 374–394.

Constantin, L. L. and Podolsky, R. J., 1965: Calcium localization of the activation of striated muscle fibers. Fed. Proc. *29*: 1141–1145.

Costantin, L. L., Franzini-Armstrong, C. and Podolsky, R. J., 1965: Localization of calcium accumulating structures in striated muscle fibers. Science *147*: 158–160.

De Pamphilis, M. L. and Adler, J., 1971: Fine structure and isolation of the hook-basal body complex of flagella from *E. coli* and *B. subtilis*. J. Bacteriol. *105*: 384–395.

Drummond, G. I. and Black, E. C., 1960: Comparative physiology: fuel of muscle metabolism. Ann. Rev. Physiol. *22*: 169–190.

Ebashi, S., 1977: Troponin and its function. *In* Search and Discovery. Kaminer, ed. Academic Press, New York, pp. 77–89.

Ebashi, S., Nonomura, Y., Yoyo-oku, T. and Katayama, E., 1976: Regulation of muscle contraction by the calcium-troponin-tropomyosin system. Soc. Exp. Biol. Symp. *30*: 349–360.

Ebashi, S., Endo, M. and Ohtsuki, I., 1969: Control of muscle contraction. Quart. Rev. Biophys. *2*: 351–384.

Eckert, R., 1972: Bioelectric control of ciliary activity. Science *176*: 473–481.

Eckert, R., Naitoh, Y. and Machemer, H., 1976: Calcium in the bioelectric and motor functions of *Paramecium*. Soc. Exp. Biol. Symp. *20*: 233–255.

Endo, M., 1966: Entry of fluorescent dye into the sarcotubular system of the frog muscle. J. Physiol. *185*: 224–238.

Ettienne, E. M., 1970: Control of contractility in *Spirostomum* by dissociated calcium ions. J. Gen. Physiol. *56*: 168–179.

Excitation-Contraction Coupling in Striated Muscle. 1966, Fed. Proc., Vol. 24.

Fawcett, D. W., 1961: The sarcoplasmic reticulum of

skeletal and cardiac muscle. Circulation *24*: 336–348.

Filo, R. S., Bohr, D. F. and Ruegg, J. C., 1965: Glycerinated skeletal and smooth muscle: calcium and magnesium dependence. Science *147*: 1582–1583.

Fleischer, M. and Wohlfarth-Bottermann, K. L., 1975: Correlation between tensile force generation, fibrillogenesis, and ultrastructure of cytoplasmic actomyosin during isometric and isotonic contraction of protoplasmic strands. Cytobiology *10*: 339–365.

Franzini-Armstrong, C., 1973: Membranous systems in muscle fibers. *In* Structure and Function of Muscle. Bourne, ed. Academic Press, New York, pp. 531–619.

Gergely, J., 1977: Molecular movements and conformational changes in muscle contraction and regulation. *In* Search and Discovery. Kaminer, ed. Academic Press, New York, pp. 91–98.

Gibbons, I. R., 1968: Biochemistry of motility. Ann. Rev. Biochem. *37*: 521–546.

Gibbons, I. R., 1975: The molecular basis of flagellar motility in sea urchin spermatozoa. *In* Molecules and Cell Movement. Inoué and Stephens, eds., Raven Press, New York, pp. 207–232.

Gibbons, I. R. and Rowe, A. J., 1965: Dynein: a protein with adenosine triphosphatase activity from cilia. Science *149*: 424–425.

Goldman, R., Pollard, T. and Rosenbaum, J., eds., 1976: Cell Motility, Muscle and Non-muscle Cells. Cold Spring Harbor Conference on Cell Proliferation, Cold Spring Harbor, New York. (Many excellent articles.)

Guyton, A. C., 1976: Textbook of Medical Physiology. 5th ed. W. B. Saunders Co., Philadelphia.

Hasselbach, W., 1964: ATP-driven active transport of calcium in the membranes of the sarcoplasmic reticulum. Proc. R. Soc. (London) *B 160*: 501–504.

Hasselbach, W., 1965: Relaxing factor and the relaxation of muscle. Prog. Biophys. Mol. Biol. *14*: 167–222.

Hayashi, T., 1961: How cells move. Sci. Am. (Sept.) *205*: 184–204.

Henrichsen, J., 1972: Bacterial surface translocation: a survey and a classification. Bacteriol. Rev. *36*: 478–503.

Hepler, P. K. and Palevitz, B. A., 1974: Microtubules and microfilaments. Ann. Rev. Plant Physiol. *25*: 309–362.

Hill, A. V., 1960: Production and absorption of work by muscle. Science *131*: 897–903.

Hill, D. K., 1959: Autoradiographic localization of adenine nucleotide in frog's striated muscle. J. Physiol. *145*: 132–174.

Hitchcock, S. E., 1977: Regulation of motility in non-muscle cells. J. Cell Biol. *74*: 1–15.

Hoffmann-Berling, H., 1960: Other mechanisms producing movements. Comp. Biochem. *2*: 341–370.

Hoyle, G., 1969: Comparative aspects of muscle. Ann. Rev. Physiol. *31*: 43–84.

Hoyle, G., 1970: How is muscle turned on and off. Sci. Am. (Apr.) *222*: 85–93.

Huang, B. and Mazia, D., 1975: Microtubules and filaments in ciliate contractility. *In* Molecules and Movement. Inoué and Stephens, ed. Raven Press, New York, pp. 389–409.

Huxley, A. F., 1957: Muscle structure and theories of contraction. Prog. Biophys. Mol. Biol. *7*: 255–319.

Huxley, A. F. and Taylor, R. E., 1958: Local activation of muscle fibers. J. Physiol. *144*: 426–441.

Huxley, H. E., 1957: The double array of filaments in cross-striated muscle. J. Biophys. Biochem. Cytol. *3*: 631–647.

Huxley, H. E., 1963: Electron microscope studies of natural and synthetic protein filaments from striated muscle. J. Mol. Biol. *7*: 281–308.

Huxley, H. E., 1965: The mechanism of muscular contraction. Sci. Am. (Dec.) *213*: 18–27.

Huxley, H. E., 1969: Mechanism of muscular contraction. Science *164*: 1356–1366.

Huxley, H. E., 1971: The structural basis of muscular contraction. Proc. R. Soc. *178*: 131–149 (the Croonian lecture).

Huxley, H. E., 1973: Muscular contraction and cell motility. Nature *243*: 445–449.

Huxley, H. E., 1977: Past and present studies on the interaction of actin and myosin. *In* Research and Discovery. Kaminer, ed. Academic Press, New York, pp. 63–75.

Huxley, H. E., and Hanson, J., 1959: The structural basis of the contraction mechanism in striated muscle. Ann. N.Y. Acad. Sci. *81*: 403–408.

Inoué, S. and Stephens, R. E., eds., 1975: Molecules and Cell Movement. Raven Press, New York. (The articles are mainly addressed to the specialist rather than the general reader.)

Jacobson, D. N., Johnke, R. M. and Adelman, M. R., 1976: Studies on motility in *Physarum polycephalum*. *In* Cell Motility, B. Actin, Myosin and Associated Proteins. Goldman, Pollard and Rosenbaum, eds. Cold Spring Harbor Conf. on Cell Proliferation. *3*: 749–770.

Jahn, T. L. and Bovee, E. C., 1969: Protoplasmic movements within cells. Physiol. Rev. *49*: 793–862.

Kaiser, D., 1978: Personal communication.

Kamin, B. and Bell, A. L., 1966: Synthetic myosin filaments. Science *151*: 323–324.

Katsura, I. and Noda, H., 1971: Studies on the formation and physical chemical properties of synthetic myosin filaments. J. Biochem. (Tokyo) *69*: 219–229.

Katz, A. M., 1976: The Biochemical and Biophysical Basis of Cardiac Function. Raven Press, New York.

Keller, J. B. and Rubinow, S. I., 1976: Swimming of flagellated microorganisms. Biophys. J. *16*: 151–170.

Kersey, Y. M., Hepler, P. K., Palevitz, B. A., and Wessels, N. K., 1976: Polarity of actin filaments in Characean algae. Proc. Nat. Acad. Sci. U.S.A. *73*: 165–167.

Konigsberg, I. R., 1964: The embryological origin of muscle. Sci. Am. (Aug.) *211*: 61–66.

Laki, K., ed., 1971: Contractile Proteins and Muscle. Marcel Dekker, New York.

Lehman, W., 1976: Phylogenetic diversity in the proteins regulating muscular contraction. Int. Rev. Cytol. *44*: 35–92.

Lehninger, A., 1975: Biochemistry. 2nd Ed. Worth Publishers, New York.

Lieberman, M. and Sano, T., eds., 1975: Developmental and Physiological Correlates of Cardiac Muscle. Raven Press, New York.

Localzo, J., Reed, G. H. and Wever, A., 1977: How

active is actin? Propagated conformational changes in the actin filament. *In* Search and Discovery. Kaminer, ed. Academic Press, New York, pp. 99–106.

Lowy, J. and Small, J. V., 1970: The organization of myosin and actin in vertebrate smooth muscle. Nature (London) *227*: 46–51.

Lundsgaard, E., 1930: Studies on muscle contraction without heat production. Biochem. Z. *217*: 162–177.

MacLennon, D. H. and Holland, P. C., 1975: Calcium transport in sarcoplasmic reticulum. Ann. Rev. Biophys. Bioengineer. *4*: 377–404.

Mannherz, H. G. and Goody, R. S., 1976: Proteins of contractile systems. Ann. Rev. Biochem. *45*: 426–465.

Marshall, J. M., Jr., Holtzer, H., Finck, H. and Pepe, F., 1959: The distribution of protein antigens in striated myofibrils. Exp. Cell Res. 7(Suppl.): 219–233.

Mechanism of Muscle Contraction, 1972: Cold Spring Harbor Symp. Quant. Biol., Vol. 30.

Merton, P. A., 1972: How we control the contraction of our muscles. Sci. Am. (May) *226*: 30–37.

Mommaerts, W. F. H. M., 1969: Energetics of muscle contraction. Physiol. Rev. *49*: 427–508.

Morkin, E., 1970: Postnatal muscle fiber assembly: localization of newly synthesized myofibrillar proteins. Science *167*: 1499–1511.

Murray, J. M. and Weber, A., 1974: Cooperative action of muscle proteins. Sci. Am. (Feb.) *233*: 59–71.

Needham, D. M., 1971: Machina carnis: The Biochemistry of Muscular Contraction in its Historical Development. Cambridge University Press, Cambridge.

Niedergerke, R., Ogden, D. C., and Page, S., 1976: Contractile activation and calcium movements in heart cells. Soc. Exp. Biol. Symp. *30*: 381–395.

Pate, J. L., 1978: Personal communication.

Pate, J. L. and Ordal, E. J., 1967: The fine structure of *Chondrococcus columnaris*. III. The surface. J. Cell Biol. *35*: 37–51.

Pate, J. L. and Chang, L. E., 1978: Evidence that gliding motility in procaryotic cells is driven by rotary assemblies in the cell envelope. Personal communication of unpublished manuscript.

Peachey, L. D., 1965: Transverse tubules in excitation-contraction coupling. Fed. Proc. *24*: 1124–1134.

Peachey, L. D., 1967: Muscle. Ann. Rev. Physiol. *30*: 401–440.

Pease, D. C., Jenden, D. J. and Howell, J. N., 1965: Calcium uptake in glycerol-extracted rabbit psoas muscle fibers. II. Electron microscope localization of uptake sites. J. Cell Comp. Physiol. *65*: 141–154.

Pepe, F. M., 1971: Structure of the myosin filament of striated muscle. Prog. Biophys. Mol. Biol. *22*: 75–96.

Perry, S. V., Margreth, A. and Adelstein, R. S., 1977: Contractile Systems in Non-Muscle Tissues. Elsevier-North Holland, Amsterdam.

Philpott, C. W. and Goldstein, M. A., 1967: Sarcoplasmic reticulum of striated muscle: localization of potential calcium binding sites. Science *155*: 1019–1021.

Pollard, T. D., 1973: Progress in understanding ameboid movement at the molecular level. *In* The Biology of Amoeba. Jeon, ed. Academic Press, New York, pp. 291–317.

Pollard, T. D., 1975: Functional implications of the biochemical and structural properties of cytoplasmic proteins. *In* Molecules and Cell Movement. Inoué and Stephens, eds. Raven Press, New York, pp. 259–286.

Pollard, T. D. and Korn, E. D., 1971: Filaments of *Amoeba proteus*. II. Binding of heavy meromyosin by thin filaments in motile cytoplasmic extracts. J. Cell Biol. *48*: 216–219.

Porter, K. R. and Franzini-Armstrong, C., 1965: The sarcoplasmic reticulum. Sci. Am. (Mar.) *212*: 73–80.

Prosser, C. L., ed., 1973: Comparative Animal Physiology. 3rd ed. W. B. Saunders Co., Philadelphia.

Rinaldi, R. A. and Baker, W. R., 1969: A sliding filament model of amoeboid motion. J. Theor. Biol. *23*: 463–474.

Rinaldi, R., Opas, M. and Hrebenda, B., 1975: Contractility of glycerinated *Amoeba proteus* and *Chaos chaos*. J. Protozool. *22*: 286–292.

Routledge, L. M., Amos, W. B., Yew, F. F. and Weis-Fogh, T., 1976: New calcium-binding contractile proteins. *In* Cell Motility. A. Motility, Muscle and Non-Muscle Cells. Goldman, Pollard and Rosenbaum, eds. Cold Spring Harbor Conf. on Cell Proliferation. *3*: 93–113.

Ruegg, J. C., 1964: Tropomyosin-paramyosin system and "prolonged contraction" in a molluscan smooth muscle. Proc. R. Soc. (London) *B 160*: 536–542.

Satir, P., 1974: How cilia move. Sci. Am. (Oct.) *231*: 45–52.

Simard-Duquesne, N. and Couillard, P., 1962: Ameboid movement. I. Reactivation of glycerinated models of *Amoeba proteus* with ATP. Exp. Cell Res. *28*: 85–91.

Smith, D. S., 1966: The organization and function of the sarcoplasmic reticulum and T-system of muscle cells. Prog. Biophys. Mol. Biol. *16*: 107–142.

Sofer, D., ed., 1975: The biology of microtubules. Ann. New York Acad. Sci. *253*: 1–848. (Summary pages 797–802.)

Squire, J. M., 1975: Muscle filament structure and muscle contraction. Ann. Rev. Biophys. Bioengin. *4*: 137–163.

Stephens, R. E., 1970: Thermal fractionation of outer fiber doublet microtubules into A- and B-subfiber components: A- and B-tubulin. J. Mol. Biol. *47*: 353.

Szent-Györgyi, A. G., 1960: Proteins of the myofibril. *In* The Structure and Function of Muscle. Vol. 2. Bourne, ed. Academic Press, New York, pp. 1–43.

Szent-Györgyi, A. G., 1976: Comparative survey of the regulatory role of calcium in muscle. Soc. Exp. Biol. Symp. *30*: 335–347.

Szent-Györgyi, A. G., 1977: Myosin-linked regulation of muscle contraction. *In* Search and Discovery. Kaminer, ed. Acadmic Press, New York, pp. 107–116.

Taylor, S. R. and Godt, R. E., 1976: Calcium release and contraction in vertebrate skeletal muscle. Soc. Exp. Biol. Symp. *30*: 361–380.

Tolney, L. G., 1975: The role of actin in nonmuscular cell motility. *In* Molecules and Cell Motility. Inoué and Stephens, eds. Raven Press, New York, pp. 339–388.

Weber, H. H., 1959: The relaxation of the contracted actomyosin system. Ann. N.Y. Acad. Sci. *81*: 409–421.

Weber, H. H. and Porzehl, H., 1954: The transference of the muscle energy in the contraction cycle. Prog. Biophys. Mol. Biol. *4*: 60–111.

Weihing, R. R. and Korn, E. D., 1971: *Acanthamoeba*

actin. Isolation and properties. Biochemistry *10*: 590–600.
Wilkie, D. R., 1966: Muscle. Ann. Rev. Physiol. *38*: 17–38.
Wilkie, D. R., 1968: Muscle. Edward Arnold, London.
Winegrad, S., 1965a: Role of intracellular calcium movements in excitation-contraction coupling in striated muscle. Fed. Proc. *24*: 1146–1152.
Winegrad, S., 1965b: Autoradiographic studies of intracellular calcium in frog skeletal muscle. J. Gen. Physiol. *48*: 455–479.
Wilson, D. M., 1968: The catch property of ordinary muscle. Proc. Nat. Acad. Sci. U.S.A. *61*: 909–916.
Wooledge, R. C., 1971: Heat production and chemical change in muscle. Prog. Biophys. Mol. Biol. *22*: 39–74.

Section VI

BIOLOGICAL RHYTHMS

Not only is the cell highly organized in structure, with its many membrane-bound organelles, but in addition its activities are organized into sequences of enzymatically catalyzed chemical reactions that make possible life itself. All these reactions are probably integrated by interactions between the organelles. And all reactions in the organelles are in turn integrated into sequences that form a *temporal pattern* for the cell. This pattern of rhythmic repetition of reactions, some with short periods, some with long ones, and the mechanisms by which such periodicity is maintained form the subject matter of this section.

Daily and seasonal rhythms in activities of organisms, such as feeding, sleeping, running, and mating, have long been observed as natural phenomena but recognition of the nature of these rhythms, especially their internal origins, is mostly of recent date. In fact, intensive analysis of the biology of the rhythms has proceeded actively only in the last four decades. At present, extensive research continues, including its application to the biology of humans in health and disease.

However, a few early observations should be mentioned. Pliny (A.D. 23–79) observed that leaves of some plants take night positions different from day positions, the movement being called sleep movements by Linnaeus in 1751. The astronomer De Mairan in 1727 found that such movements occur also in plants kept in continuous darkness and de Candolle in 1835 demonstrated that the interval between movements in constant conditions (continuous dim light) was less than 24 hours.

The literature on rhythms, both *diurnal* (daily) and *annual,* especially the descriptive, is overwhelming and continues to increase in volume. Citations are therefore made primarily to reviews and emphasis is placed especially on the cellular aspects of rhythms and rhythms in unicellular organisms.

CHAPTER 24

TEMPORAL ORGANIZATION OF THE CELL

Any continually recurring event may be described graphically as *cyclic,* and the interval of time it takes to complete a cycle is its *period.* The period is measured from one phase in the cycle (such as its peak) to the next. The *frequency* of the rhythm per unit of time (day, year) is the reciprocal of the period, or the number of times it will recur during that unit of time. The degree of change during a given periodic function is called its *amplitude.*

In any population individuals may show periodicity of a specific function, but any two individuals may or may not be *in phase.* If the function in the two individuals peaks at the same time, they are said to be in phase with one another. The *phase* is the time location of a function in the cycle. A convenient model (Fig. 24.1) by which to visualize a rhythm is a helical spring attached rigidly at one end and free but loaded at the other. If friction is ignored, such a spring once set in motion will continue to oscillate up and down with a period that is a function of the physical makeup of the spring, moving in either direction until it reaches zero velocity, then reversing itself until it again reaches its original displacement, after which it repeats the cycle. The recurrent oscillation constitutes a rhythm, the repeating unit of which is a cycle, and the time for a complete cycle is its period. If, in plotting the cycle as a circle, the minimal displacement is arbitrarily set at 0°, a maximum displacement occurs at 180°, and the return portion of the cycle is completed at 360°. A specific point on the circle, which may be given in degrees, is considered the phase of the cycle, and the angle corresponding to the point connected to the center of the circle and the horizontal is the *phase angle.* The peak phase of the cycle for any activity being measured is called its *acrophase* (Halberg, 1977 Fig. 24.6). The rope that a child jumps seen from the side describes such a cycle, the phase angle being the relation at any time of the tip of the rope to the pivot (hand). Some acrophases for human rhythms are shown in Figure 24.10.

CIRCADIAN RHYTHMS

Much attention has been focused on daily rhythms that occur in the life activities of unicellular and multicellular organisms exposed to the solar day. Day-active (diurnal) animals, for example, squirrels, run and feed only at certain times of the day and become relatively inactive during the night. Conversely, nocturnal animals, for example, mice, are most active at night and quiescent during the day. Similarly, leaf and flower petal movements in plants in daylight have been observed to occur only during certain hours of the day.

It had long been supposed that the natural alternating rhythm of day and night on earth imposes diurnal (daily) rhythms on all living things. But if the rhythm is related in this simple causative way, it should disappear when the organism is placed in continuous light or darkness. It was easily demonstrated that isolation of diurnal organisms in continuous darkness (or continuous dim light) did not disturb the cycle of activity, which continues much as in nature. However, an important difference for such isolated organisms is that the period of the activity cycle is never exactly 24 hours, being slightly more or slightly less. For this reason these rhythms have been called *circadian* (*circa dies,* meaning about a day) by Halberg and the term has been generally accepted (Halberg and Howard, 1958; Aschoff, 1965b). The rhythms of

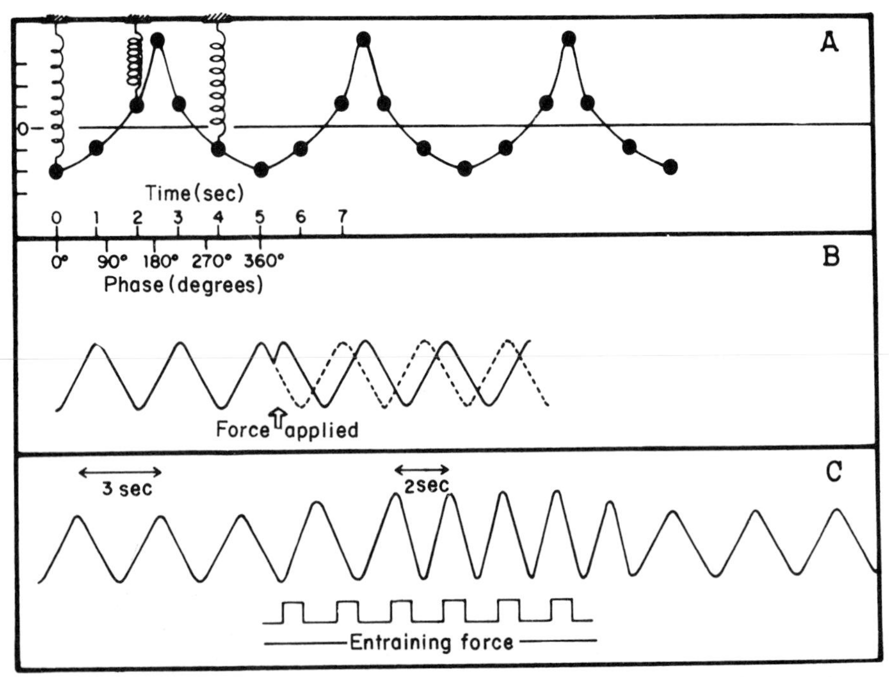

Figure 24.1. Diagram illustrating the behavior of three different oscillating nonlinear springs when: A, Free-running; B, phase shifted with a single stimulus; C, entrained by a periodic force. In A the position of the mass on the end of the spring is plotted as a function of time. The period of the oscillation for this system is 5 seconds. In B a curve is drawn representing the variation of position as a function of time for another oscillating system. An instantaneous force, applied during the third cycle, causes a phase delay of 90°. This can be seen as the difference between the dashed line, which represents the system before application of the force, and solid lines. In C the curve represents an oscillator with a free-running period of 3 seconds. Before the first three cycles, a force with a period of 2 seconds is imposed on the system, as represented by the square wave in the lower part of the figure. After two transient cycles, the oscillator becomes entrained and its period is now the same as that of the force. After the periodic force is discontinued, the oscillator begins to run free once again with its initial free-running period. From Menaker, 1969: Am. Inst. Biol. Sci. Bull. 19: 681–689.

nocturnal organisms are also circadian. Green plants cannot be kept in darkness for long, but in dim light, which permits sufficient photosynthesis to maintain them in good condition, the circadian rhythms continue.

A circadian rhythm is apparently endogenous and indicates the presence of a temporal organization or a *biological clock* in an organism. It is evident that when a circadian rhythm, which is slightly less or slightly more than a solar day, is free-running under constant environmental conditions (continuous light or darkness), its rhythm will get out of phase with the solar day, gaining or losing the same increment of time each day until a solar day has been either gained or lost; at this time it once again more or less coincides in phase with the solar day. The fact that it gets out of phase with the solar day only when free-running shows that the biological clock must normally be reset or synchronized with the solar day from cues in nature.

Circadian rhythms have been observed in all major groups of organisms: in protozoans, coelenterates, flatworms, annelids, mollusks, and various arthropods (especially in insects), in birds and small mammals, several fungi (including *Neurospora*), in some algal cells (such as *Acetabularia* and diatoms), and in higher plants. Linnaeus (1707–1778), capitalizing on his knowledge of petal movements, was able to design a "flower clock" in which the petals of different species in a bed opened at different hours of the day. (Bees, in turn, know from which plants to gather nectar at each hour of the day.) Activity rhythms of organisms have been studied mostly because of the ease with which they can be recorded, but circadian rhythms occur at all levels of organization of the organism, including the cellular, among which are biochemical rhythms. Many of the rhythms in a given organism have been found to occur in sequences; that is, not all the rhythms are in phase at a given time but the phase relationships among the various rhythms are the same from day to day. Some rhythms persist even in isolated organs and in cells of a tissue culture (Andrews, 1968).

Circadian rhythms are thus prevalent in eukaryotic unicellular and multicellular or-

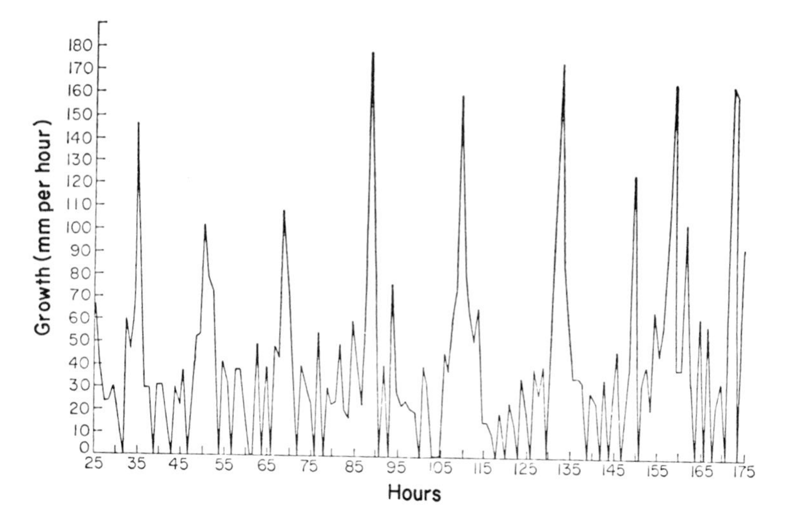

Figure 24.2. *The rate of advance of a culture of* E. coli *down a race tube (a long tube containing culture medium on the bottom). Note the distinct peaks, which occur about 20 hours apart. (From Rogers and Greenbank, 1930: J. Bacteriol. 19: 181–190.)*

ganisms (Still, 1972; Palmer, 1976). No convincing evidence for circadian rhythms has been presented for prokaryotic cells, although an experiment in which a culture of *E. coli* advanced down a race tube at a variable rate with peaks about 20 hours apart has sometimes been considered evidence for a circadian rhythm (Rogers and Greenbank, 1930) (Fig. 24.2). However, the data can be explained in other ways, and more recent studies with prokaryotes are also subject to criticism.

CIRCADIAN RHYTHMS IN UNICELLED ORGANISMS

Circadian rhythms have been demonstrated in the phototactic responses and in the division rate of the flagellate *Euglena* (Bruce and Pittendrigh, 1955, 1960); in bioluminescence, steady glow, in photosynthesis, and in division rate of the dinoflagellate *Gonyaulax*. They have also been discovered in the division rate and mating reactions of *Paramecium* (Volm, 1964; Cohen, 1965; Barnett, 1966), in movements of some diatoms (Eppley *et al.*, 1967; Jørgensen, 1966), and in the photosynthetic and other biochemical activities of the alga *Acetabularia* (Hastings and Keynan, 1965). A rhythm known since ancient times is the rhythm of reproduction of the genus *Plasmodium,* the infecting organism of malaria. The fever accompanying malaria in birds comes every 24 hours, the *P. vivax* infection in humans comes every 48 hours, while that accompanying *P. malariae* infection in humans, every 72 hours. This is the time required for the invading organisms to multiply to the point at which they rupture the red blood cells of the host and toxins are liberated, resulting in fever and chills. These rhythms have recently been shown to be circadian (Hawkins, 1970).

The circadian rhythms in *Gonyaulax* (Fig. 24.3) are here used to illustrate many of the features of a cellular circadian rhythm because they have been investigated rather extensively (Sweeney, 1963, 1969b; Hastings, 1964; Hastings and Keynan, 1965; McMurry and Hastings, 1972). *Gonyaulax* is a photosynthetic cell that emits a brief flash of light (90 msec) when stimulated by agitation. Cultures grown in a daily cycle of day and night display rhythmically a greater luminescence upon agitation during the night than during the day, as shown in Figure 24.3. When such cultures are kept in a dark chamber, it is found that the rhythm continues, but its amplitude decreases progressively because photosynthesis, the source of nutrients for the cell, ceases in the dark. Bright light permits photosynthesis but inhibits the rhythm. Dim light, which would permit just enough photosynthesis to supply necessary nutrients to the cell, does not inhibit the rhythmic bioluminescent response, which persists without decrease in amplitude. Even cultures grown in bright light for a year, during which time the rhythm disappears, resume the rhythm when again grown in dim light. The *innate* (endogenous) *period* of the rhythm is always almost 24 hours, being somewhat less than 24 hours under some conditions, somewhat more under others (Hastings, 1964; Njus *et al.*, 1977).

The rhythm of luminescence in *Gonyaulax* is relatively independent of temperature. Thus, the period is about 22.3 hours at 16°C, which is approximately the temperature of its normal environment in the ocean, and it is essentially the same at 26.8°, 23.6°, 22°, and 19°C. At higher temperatures the period is longer, becoming 26.5 hours at 26.7°C. At 11.5°C and 32°C the rhythm practically ceases (Fig. 24.3). Relative independence of temperature has been observed for circadian rhythms in many types of organisms whose internal temperature ap-

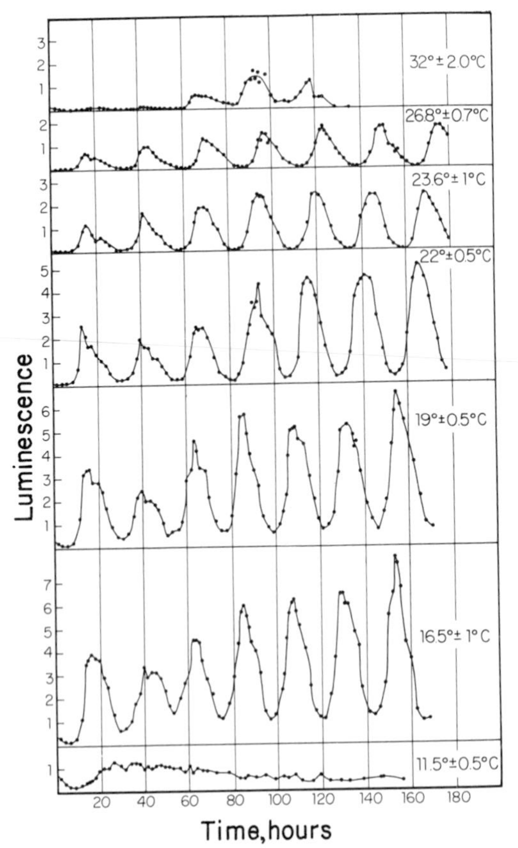

Figure 24.3. *Characteristics of the persistent rhythm of luminescence in the dinoflagellate* Gonyaulax *at several different temperatures. The cells were kept at 22°C in LD (i.e., alternate 12 hours light and 12 hours dark) conditions prior to the beginning of the experiment shown on the graph. At that time the cells were transferred from the dark at 22°C to the various temperatures, all in constant dim light, and left in these conditions throughout the experiment. The luminescence of the cells was determined approximately every 2 hours. At the dim light intensity used, there is little cell division, even at the optimal temperature for growth. The number of cells therefore remained essentially the same in all tubes at all temperatures. (From Hastings and Sweeney, 1957: Proc. Nat. Acad. Sci. U.S.A. 43:804.)*

proximates the environmental temperature (Sweeney and Hastings, 1960; Sweeney, 1977). However, at critical temperatures and light intensities the two factors may amplify one another (synergism) (Njus *et al.,* 1977).

It is possible to shift the phase of the rhythm in *Gonyaulax* by appropriate manipulation of the environment. For example, if a culture of *Gonyaulax,* which has been grown under natural periods of night and day, is exposed to bright light for several hours during the night, and then placed in continuous dim light, it will once again show an endogenous rhythm but with a *phase shift* of several hours, the shift span depending especially upon sufficiently bright light and its duration. The rhythm will now operate from this new base line and without reference to the solar day, since the organisms are in continuous dim light (Fig. 24.4). Clearly, the biological clock of *Gonyaulax* is influenced by bright light, but its innate period is set by endogenous factors. So long as conditions are constant it will continue with its innate rhythm (Sweeney, 1969b).

Gonyaulax taken from a rhythmic culture with an innate period of about 24 hours and exposed to alternate periods of 7 hours of darkness and 7 hours of light for a while become entrained to a new rhythm (Fig. 24.5). However, soon after they are again placed under continuous dim illumination they revert to the innate period of almost 24 hours. Similar entraining periods of 6, 8, and 16 hours give like results. It appears that the light, as a timer, has a powerful effect but not a persistent one (Sweeney and Hastings, 1960; Hastings, 1964).

Luminescence, steady glow, photosynthesis, and division rate all show circadian rhythms in *Gonyaulax,* but they are not in phase with one another, each reaching the acrophase at a different time of day. More technically stated, at any one time the phase angles of the four rhythms differ from one another (Fig. 24.6). Interestingly, the different phase angles persist. That this relationship is not fortuitous is demonstrated when a phase shift is induced. All four rhythms are shifted, but they maintain the temporal relationships of their phase angles to one another. In the free-running condition in dim light, an increase or a decrease in light intensity for a few hours will induce a phase shift. For example, when the light is reduced to zero (e.g., darkness for a few hours), all four rhythms are shifted to an equal degree. The direction of a phase shift depends upon the time at which the interruption occurs (Fig. 24.7) (McMurry and Hastings, 1972). Most often the phase shift of a rhythm is a delay if the interruption is applied just after the onset of the particular cellular activity, and an advance if it is applied some hours afterward (Menaker, 1969). Ultraviolet light also induces a phase shift (advance) in all four rhythms in *Gonyaulax* and white light undoes the effect of the ultraviolet light (Sweeney, 1963).

In unicellular organisms circadian rhythms have been most extensively studied in the flagellate *Euglena gracilis* (Edmunds and Halberg, in press). The organism offers special advantages inasmuch as it can be grown autotrophically in the light or heterotrophically in darkness and can be maintained in the stationary phase of the cell cycle, divorced from the forces involved in cell division, and in mutant strains deficient in photosynthesis (with little or no chlorophyll). The circadian rhythms are summarized in Figure 24.8 as acrophases (peaks in the rhythms where 360° represents

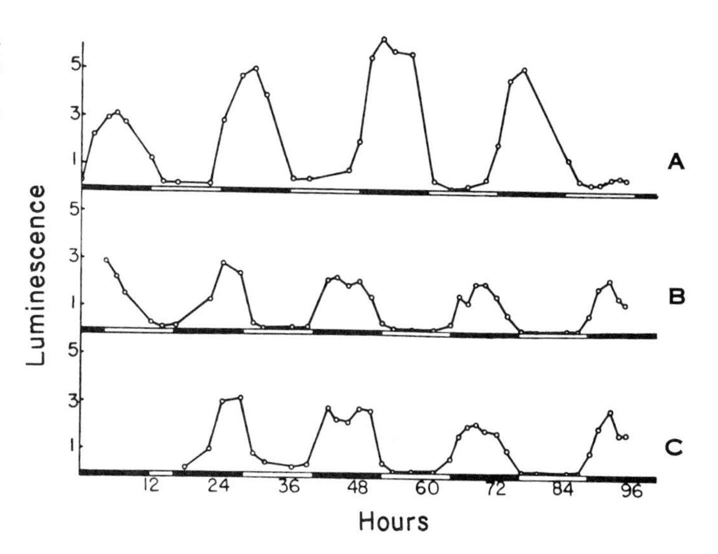

Figure 24.4. Phase shift. This experiment illustrates the effect of changing the solar time at which the light and dark periods occur. Curve A shows the pattern of luminescence changes in the dinoflagellate Gonyaulax in an LD culture—one exposed to a diurnal rhythm of 12 hours light (L) and 12 hours darkness (D)—which had been on the schedule indicated for some time. The black bars on the time axis indicate dark periods. The lower two graphs, B and C, illustrate the effect of imposing upon cultures (which were previously on the schedule shown in the top graph) an LD schedule in which the light and dark periods were at a different time of day. The new schedules were started at zero hours on the graph. Temperature, about 26°C. Light intensities used, about 250 foot-candles. (From Hastings and Sweeney, 1958: Biol. Bull. 115: 443.)

the period) with the 95 percent confidence limits. A large number of strains have been studied (first set of acrophases). In all cases the cultures were first subjected to light regimens to synchronize them and then left free-running in continuous light or darkness. In some of the biochemical analyses, studies were made only with cells on a daily light-dark regimen.

CIRCADIAN RHYTHMS IN MULTICELLULAR ORGANISMS

Many cellular activities in man and other organisms are found to be rhythmic. Although normally set to diurnal rhythms by the day-night cycle in nature, when under constant conditions they persist as circadian rhythms, being nearly—but not exactly—of 24-hour periodicity (Fig. 24.9). For example, a man in a cave or in an underground isolated bunker, without knowledge of time of the solar day or night, continues to show a circadian body temperature cycle and peaks in excretion of sodium, calcium, and ketosteroids in the urine (Aschoff, 1965a). Evidence exists that the numbers of white blood cells, the concentration of various hormones and certain nutrients in

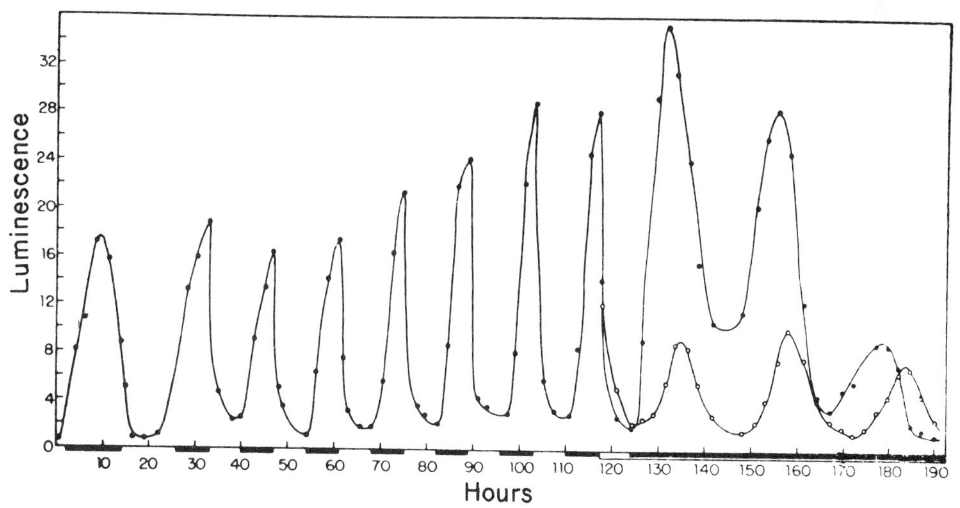

Figure 24.5. This chart illustrates the entrainment of the luminescence rhythm in the dinoflagellate Gonyaulax to a 14-hour cycle and the manifestation of an endogenous diurnal rhythm when the cells are placed in constant conditions subsequent to the treatment. Dark periods are indicated by black bars on the time axis. The cells were on an LD—12 hours light (L), 12 hours darkness (D)—schedule previous to the time when the 14-hour cycle was started (at 26 hours). Light intensity throughout the 14-hour cycling was 800 foot-candles. At 117 hours some aliquots were removed from the dark and placed in constant dim light at 230 foot-candles. The luminescence changes in these cultures are shown by the open circles. From 124 hours on, the other aliquots were left in the dark. Luminescence changes are plotted with solid circles. Note rapid decline of peak luminescence in darkness. Temperature, 21°C. (From Hastings and Sweeney, 1958: Biol. Bull. 115: 444.)

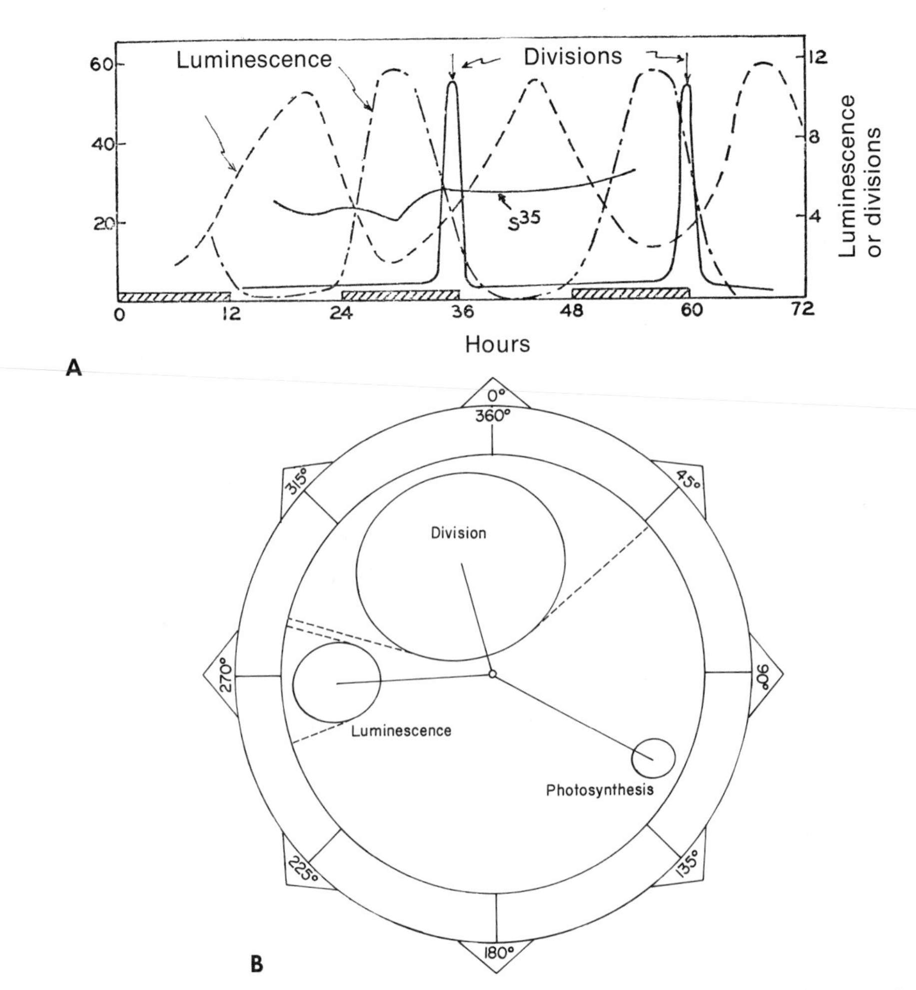

Figure 24.6. A, *This figure illustrates diurnal rhythms of luminescence, photosynthesis, and cell division in cultures of* Gonyaulax polyhedra *kept under conditions of alternating light and dark periods of 12 hours each at 25°C. The dark periods are indicated on the abscissa by cross-hatched rectangles. Ordinates for luminescence and cell division are in relative units. Each of the rhythms shown persists when the cells are transferred to constant conditions. Rhythms of a spontaneous luminescent flashing and of a steady glow have also been observed but are not shown here. The curve labeled* S^{35} *illustrates an experiment in which it was found that the rate of inorganic sulfate incorporation, measured in counts per second on left ordinate, does not show a diurnal rhythm. (From Hastings, 1959: Ann. Rev. Microbiol.* 13: *297–312.)* B, *The data of* A *put on the basis of 360° = period of a rhythm. The ovals are the 95 percent confidence limits. In this case 360° = 24 hours, because the cells were on a 12L–12D schedule.*

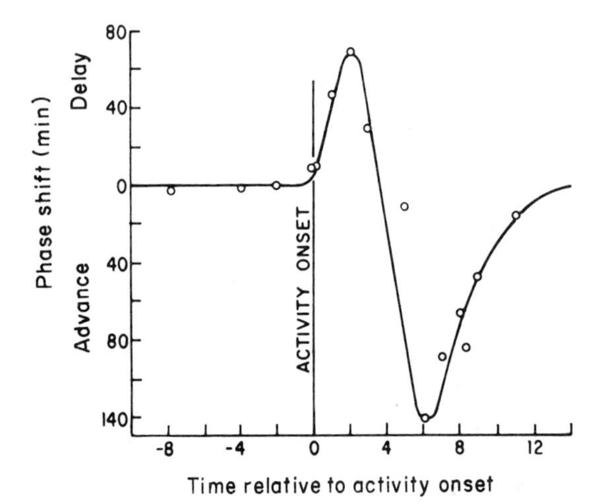

Figure 24.7. *Direction of phase shift in a circadian rhythm, depending upon time relative to activity onset. Phase response of individual hamster to 10 minute light signals. Light pulses were administered to the same animal at different time points relative to his activity onset. The advance (or delay) of the phase shifts induced has been plotted on the ordinate as a function of the time at which each signal was administered (abscissa). Each point on such a curve is quite reproducible for a single individual, but phase-response curves vary considerably among individuals. (From Menaker, 1969: Am. Inst. Biol. Sci. Bull.* 19: *681–689; redrawn from De Coursey, J. Cell. Comp. Physiol.* 63: *189.)*

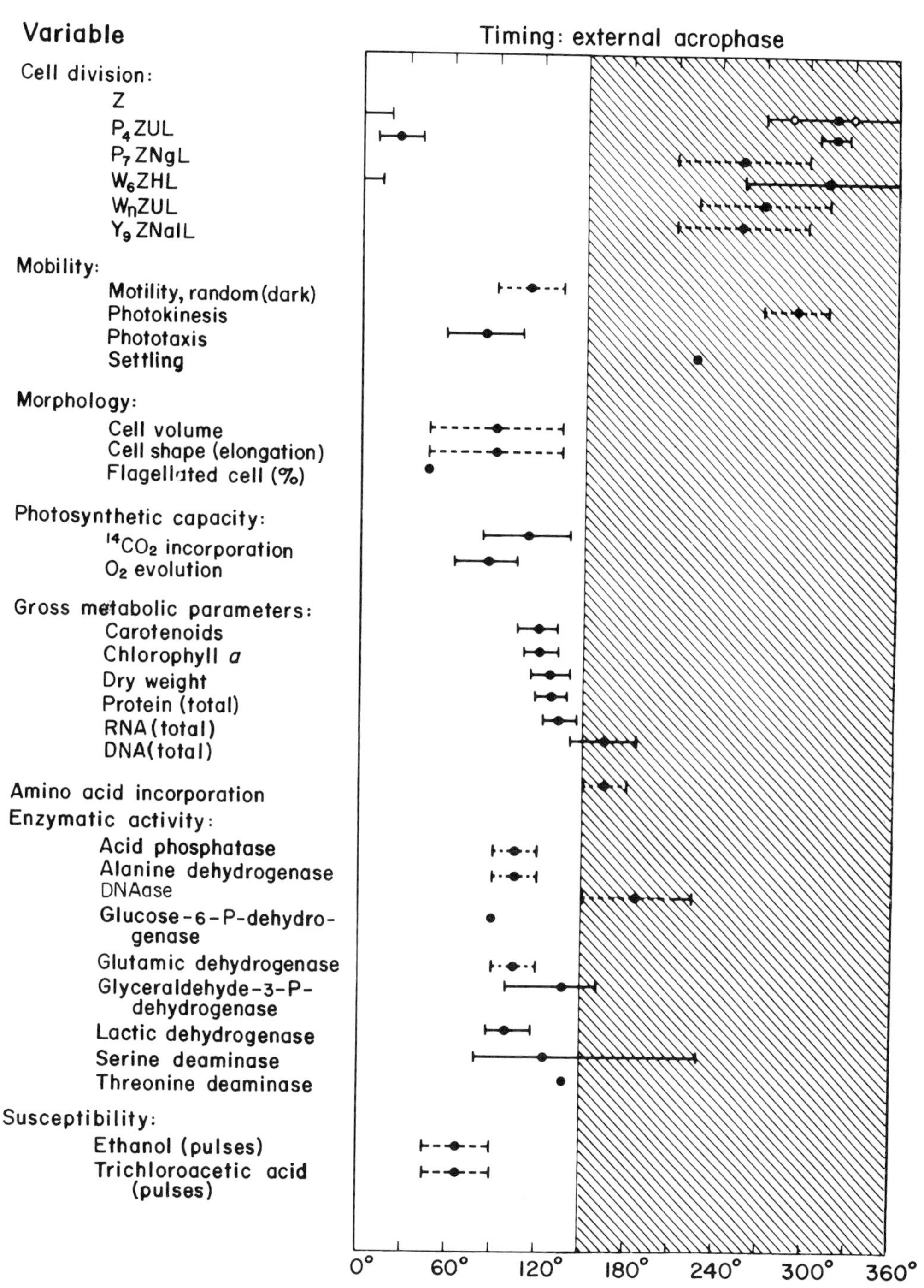

Figure 24.8. Circadian rhythms in Euglena gracilis. The 360° approximates 24 hours of time and the light and dark areas are the natural day and night periods. The lines through the acrophases (peaks of the rhythms) are the 95 percent confidence limits. (From Edmunds and Halberg [in press]: Comparative Pathology of Abnormal Growth. Kaiser, ed., Raven Press, New York.)

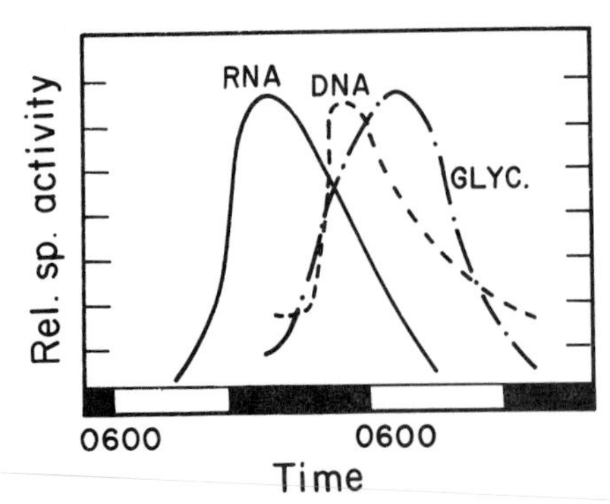

Figure 24.9. Rhythms in a liver cell of a mouse. These are shown here for a day but they repeat day after day when free-running. The amplitude (relative specific activity) of the biochemical synthesis (ordinate) has been plotted against the time in a 24-hour cycle of 12 hours light and 12 hours darkness. RNA synthesis (solid line), DNA synthesis (dashed line), glycogen synthesis (dash dot line). (After Halberg, 1960: Cold Spring Harbor Symp. Quant. Biol. 25: 289.)

the blood, the synthesis of nucleic acids in some cells of the body, and the susceptibility of cells to drugs and toxins also show circadian rhythms (Halberg and Howard, 1958). Action of drugs at different phases of the circadian cycle is therefore of great interest in medical practice (Halberg, 1977).

The period of circadian rhythms in mammals is sensitive to light intensity; it is shortened by high intensity and lengthened by low intensity in diurnal species and is inversely affected by light in nocturnal species. The phase of mammalian circadian rhythms has, in some instances, been shifted by appropriate lighting regimens (e.g., light by night, darkness by day) (Aschoff, 1965b; Aschoff *et al.,* 1975). The malaise experienced for several days by jet travelers on arrival at a distant point in east-west or west-east travel is partly the result of displacement of the phase of their circadian rhythms, since north and south flights of equal distance without shift in latitude (although day length does change) do not produce comparable effects (Hauty and Adams, 1965; McFarland, 1975; Halberg *et al.*, 1977). Although circadian rhythms persist under constant conditions, they are disrupted by random lighting (Holmquist *et al.*, 1966).

In experiments with the mouse it has been shown that the susceptibility to various poisons, such as endotoxins produced by bacteria, and to various drugs follows definite daily rhythms. Doses that kill during one part of the 24-hour cycle are harmful but not lethal during other parts of the cycle. Since drugs can be much more effective when taken at one time of the day than at another, it is important to discover whether treatment with a particular drug would be most effective at the time when the organism is most susceptible to the drug. The results could have significant and important bearing on medical treatment (Luce, 1970; Halberg, 1977). Timing of medication has long been known to be of extreme importance in treatment of malaria.

Of special interest are the circadian rhythms of biochemical synthesis in the liver cells of mammals. It has been found that RNA synthesis reaches its peak before DNA synthesis does, and DNA before glycogen (Fig. 24.9). Significantly, a phase shift induced by appropriate lighting regimens shifts all three syntheses comparably, the three syntheses maintaining their characteristic temporal phase relationships to one another. Mitosis is also circadian in mouse epidermal cells and adrenal cells, although it occurs at a different time of day in each of them (Halberg *et al.,* 1959; Jerusalem, 1967). Significantly, also, the changes in activity of various enzymes in cells of mammals show a circadian rhythm (Klevecz and Ruddle, 1968). Such temporal sequences in the cell indicate a high degree of regulation of each of the reactions in relation to each other (see Chapter 16).

A method has been described that makes possible detection of many circadian rhythms using statistical computer analysis of data in which the "noise" level from many interfering variables is too high to permit detection by ordinary graphic methods of presentation. These methods are applicable to other organisms, although they have been used primarily in data from mammals, including humans (Halberg *et al.,* 1958, 1959; Halberg, 1969, 1977). A general chart of some human circadian rhythms is given in Figure 24.10. The method has also been applied to rhythms in *Gonyaulax* (see Fig. 24.6*B*) and to *Euglena* (Fig. 24.8).

SYNCHRONIZATION OF CIRCADIAN RHYTHMS

It has been shown that circadian rhythms are synchronized with certain factors of the environment. The single most important of these factors is light, which has been amply illustrated in preceding examples. However, factors other than light play a part in synchronizing endogenous rhythms in some organisms. For example, in some birds entrainment of circadian rhythms has been reported with sound (Menaker and Eskin, 1966) and with social interaction (Menaker, 1969; 1977). Atmospheric pressure changes are reported to be effective in humans (Hayden and Lindberg, 1969). Experiments with suspensions of *Gonyaulax,* however, have shown that no relationship apparently exists between environmental mechanical disturbance of any kind and the

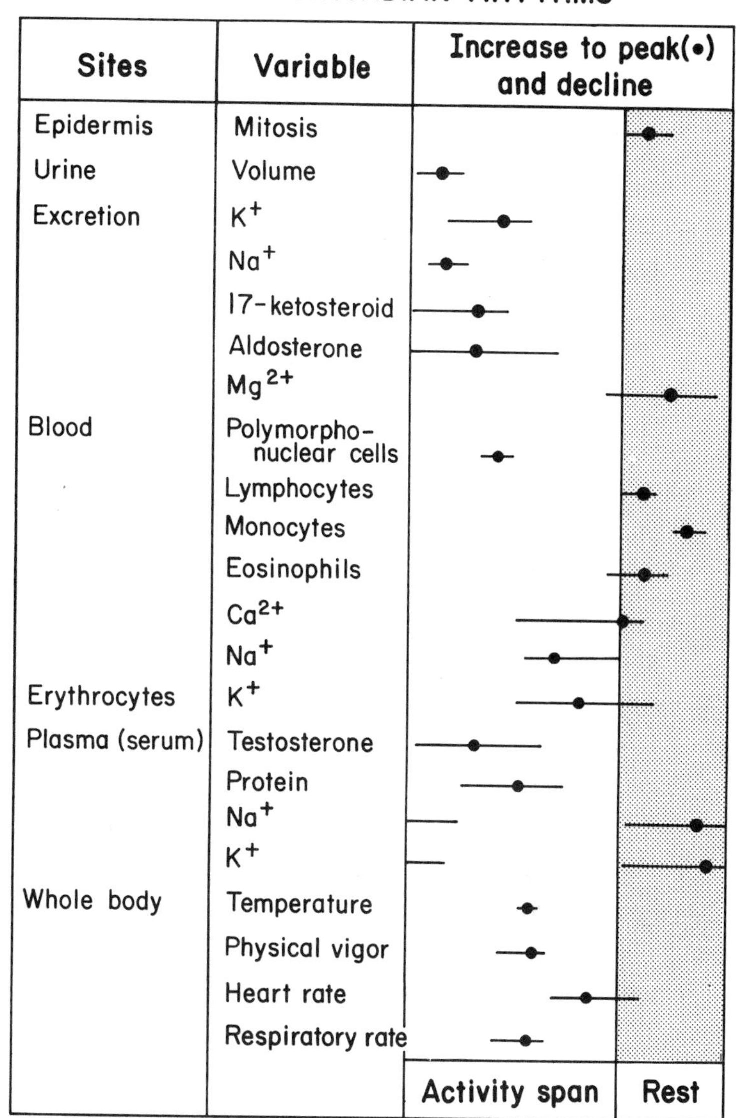

Figure 24.10. *Human circadian rhythms. (Data courtesy of Franz Halberg, University of Minnesota.) The activity span and rest span together equal 24 hours. The acrophase is shown as a dot; it represents the peak for the rhythm in question. The lines are the 95 percent confidence limits. (After Luce, 1970: Biological Rhythms in Psychiatry and Medicine. U.S. National Institutes of Mental Health, Chevy Chase, Md.)*

innate rhythms in this organism (Hastings, 1964).

Temperature is another important factor in synchronization of circadian rhythms. For example, let us again consider the case of the unicellular parasite (*Plasmodium*) that causes malaria. Light is unlikely to penetrate the bloodstream and would be useless as a cue for synchronizing the division rate of the organism with its environment. While the details of the life cycle of the parasitic cells are not of moment here, experiments have shown that the small cyclic change in body temperature (known to be circadian in birds and mammals, including humans) apparently serves as the cue for the periodicity of the reproductive cycle in the malarial parasite, making available the gamete stage that is infective to the mosquito at the time most appropriate for continued transmission of the disease. This cyclic body temperature change also serves as a cue for the gathering of filarial parasites (causing elephantiasis) in the bloodstream at a time most appropriate for their transmission by mosquitoes to the host (Hawkins, 1970). Temperature changes in the environment also serve as a cue for circadian activity rhythms in a number of cold-blooded animals (Menaker, 1969). Under certain conditions (transitional light intensities and temperatures) temperature and light may even reinforce each other (synergism). This produces a greater effect than would be achieved by a simple combination of temperature and light or by either of the two alone (Njus *et al.*, 1977).

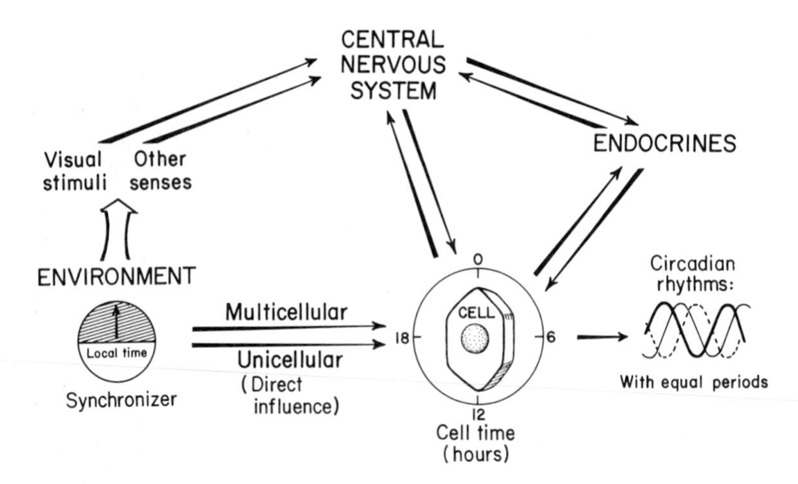

Figure 24.11. Factors and pathways known or postulated to participate in frequency synchronization among mammalian circadian rhythms as well as in synchronization between rhythms and environmental timers. Such relations may possibly obtain in other multicellular animals but have not been as well documented as in mammals. (After Halberg, 1960: Perspectives Biol. Med. 3: 505 and Cold Spring Harbor Symp. Quant. Biol. 25: 209.)

It is argued that circadian rhythms in multicellular organisms may originate in particular cells that then serve as pacesetters, and this function has been claimed for certain cells in the nervous system of the cockroach (Harker, 1964) and of the mollusk *Aplysia* (Strumwasser, 1965; Lickey, 1967; Jacklett, 1969). Such pacesetter cells may possibly also exist in other organisms and may evoke circadian secretion of hormones, which in turn may affect distant cells. Endogenous circadian rhythms of secretion of corticosteroid hormones from the adrenal cortex cells of mammals have been observed, and such cells continue to show a circadian rhythm of secretion in tissue culture as well. However, even though the adrenocortical steroids might in turn pace other biochemical activities, some endogenous rhythms (blood serum level of certain hormones) in mammals are not abolished by adrenalectomy (Halberg *et al.,* 1959, 1969).

The pineal gland (in the brain), which has a circadian pattern of activity, is thought by some to be a major pacesetter in some circadian rhythms of vertebrates, but it is under control of sympathetic nerves. However, some of the activities of the pineal gland have been shown to be controlled by light (Wurtman and Axelrod, 1965; Wurtman *et al.,* 1968; Binkley *et al.,* 1977), which makes possible synchronization of circadian rhythms with nature. Removal of the pineal abolishes the endogenous rhythms but exposure of the animal to light-dark cycles still induces a rhythm corresponding to the environmental regimen. Removal of the pituitary gland, which regulates and coordinates output of many other glands and has control of metabolic and sexual events, abolishes some circadian rhythms such as the color changes in lizards (Wurtman *et al.,* 1968).

It is noteworthy that when a phase shift of a specific rhythm is induced by a different light-dark regimen (for man or mouse), the various other rhythms do not maintain the same temporal phase relationships to one another. Thus the rhythm of body temperature is immediately entrained to either a 21- or 27-hour cycle, while the rhythm of potassium excretion remains on the 24-hour period of the original routine. The activity rhythm also is capable of entrainment to a wider range of light-dark cycles than the temperature rhythm. Furthermore, in some individuals body temperature rhythms and activity rhythms have different periods in absence of environmental cues (see McMurry and Hastings, 1972, for references). Such desynchronization during entrainment to light-dark cycles different from the solar day suggests the presence of several clocks or pacesetters in the mammal. In view of the complexity of the mammalian economy (Fig. 24.11), with integration by both nervous and endocrine elements, this is perhaps not surprising.

It is not possible at present to do more than make a consistent verbal framework for the description and interpretation of existing data on rhythms, since specific cellular components have not yet been identified as parts of the mechanism of the clock. In general, the circadian rhythm is considered to be largely unaffected by the input of information coming to the cell from the environment in the form of pressure (including osmotic pressure), mechanical disturbance, ionizing radiation (e.g., cosmic rays), and to a considerable extent temperature, although a contrary view has also been expressed (see Brown *et al.,* 1970; Brown, 1973). Pressure change has been considered effective in man (Hayden and Lindberg, 1969). The endogenous period depends upon a metabolic supply of energy, as already discussed for *Gonyaulax,* but it is unlikely that its period is determined by metabolic reactions in a simple manner. The rate of most metabolic reactions in the cell is doubled by a 10° rise in temperature within the viable range (see Chapter 10), whereas such a change in temperature has little effect on the endogenous period of various circadian rhythms, as we have seen illustrated for *Gonyaulax* (see Fig. 24.3). It is postulated by some that two coupled systems are involved, one of them light-sensitive and temperature-independent that serves as the pacesetter and a second one temperature-sensitive but light-insensitive, phased by the first. Circadian rhythms in the cell probably originate in concatenations of molecular reactions, some of them temperature-independent but with sustained circadian oscillation in absence of external synchronizers.

The importance of the nucleus (and thus presumably of DNA) for circadian rhythms is suggested by the fact that circadian rhythms are typical of eukaryotic cells. This is also suggested by nuclear control of the phase of the rhythm. For example, if two cells of *Acetabularia* are entrained by differing light-dark cycles to be opposite in phase, an interchange of the nuclei between the two cells by microsurgical technique results in a change in phase in each of the cells.

To determine whether circadian rhythms depend upon macromolecular syntheses, poisons specifically inhibiting various syntheses have been tested on cells. Actinomycin D, which inhibits DNA-dependent RNA synthesis, was found to inhibit circadian rhythms in *Gonyaulax* (Karakasian and Hastings, 1962), and cyclohexamide, which inhibits protein synthesis, lengthened the period of circadian rhythms in *Euglena* (Feldman, 1967), as did heavy water (Bruce and Pittendrigh, 1960). Neither actinomycin D, nor puromycin, nor chloramphenicol (the latter two affecting protein synthesis) affected the phase or circadian rhythms (photosynthesis, and so forth) in the algal cell *Acetabularia,* although the amplitude was damped (Sweeney *et al.,* 1967). The damping effect of actinomycin and of ribonuclease was observed in nucleate but not in enucleate cells (van den Driessche, 1966). Transplanted nuclei were found to impose their rhythm upon the host cell (Schweiger *et al.,* 1964), although cells maintain their circadian rhythms for many days after enucleation in absence of other influences. The results suggest the importance of the nucleus in circadian rhythms but do not permit correlation with any specific reaction. It is perhaps significant in this respect that true circadian rhythms have not been clearly demonstrated in prokaryotic cells.

On the basis of known facts, a possible model for a circadian clock in *Gonyaulax* has been constructed (Fig. 24.12).

Circadian rhythms persist in some cases when the nucleus is removed and continued translation in the cytoplasm (not in the organelles) is necessary for the cell clock to function (Schweiger, 1977). Unless one postulates a very long-lived mRNA, the model in Figure 24.12 is probably not accurate.

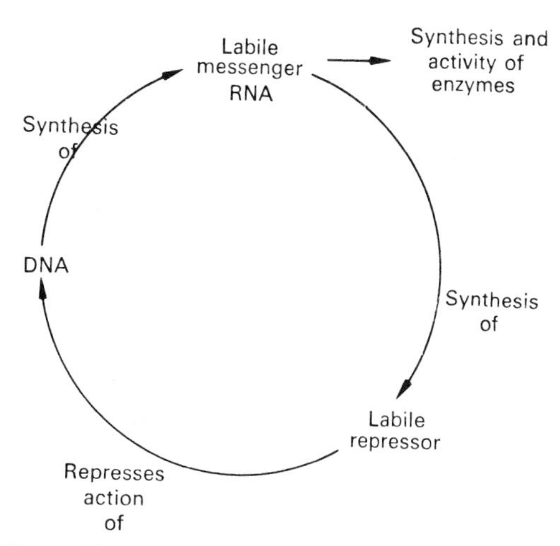

Figure 24.12. *A possible model for an oscillating circadian clock involving DNA-directed synthesis of RNA. (From Clayton, 1971:* Light and Living Matter *2:169, based on data for* Gonyaulax *from Hastings.)*

Another model attempts to localize the biological clock in the cell membrane. Supposedly the clock consists of two components with a feedback loop: an ion pump in the cell membrane (see Chapter 18) and the ion being pumped. The pump functions until the ion level on the two sides of the membrane reaches a critical level, at which time it shuts off. By passive diffusion the ion concentration tends to become equalized; then the pump is again activated. Illumination of *Acetabularia,* for example, changes the resting potential between the inside and the outside of the cell. Such a change would alter the clock feedback loop. A phase advance would be expected when the pump is working and a phase delay when it stops. However, little evidence is at present available to support the accuracy of this model (Sweeney, 1977). We now know many of the things that the clock is *not,* but detailed explanation of its nature continues to be elusive (Schweiger, 1977).

Some speculation is offered that the circadian oscillation is probably an evolutionary adaptation to a highly rhythmic planet in which the predominant periodicity is the 24-hour solar day. Rhythmic conditions in the solar day in nature regularly impose changes in the factors that act on organisms, and natural selection would have preserved those individuals that were best adapted to such changes. To survive, the cell (or organism) must essentially be ready for the next change in conditions in nature and must be synchronized with its natural environment. If life exists on a planet with a different length of solar day, one would expect that the circadian rhythm on that planet would approximate its solar day much as life on earth does the 24-hour solar day.

TIDAL RHYTHMS

Many marine invertebrates display rhythms correlated with the tidal day, which is 24.8 hours rather than 24 hours long. The rhythms are shown by activity and by the state of chromatophores in fiddler crabs, for example. But even some diatoms *(Hantzchia virgata)* that are present about a millimeter below the surface of the sand at high tide, migrate to the surface at low tide, presumably by secretion of slime. The tidal rhythms may be retained for many days after the organisms are placed under constant conditions.

Tidal rhythms overlay circadian rhythms, making a complex behavior pattern that is beyond the scope of the present discussion (see DeCoursey, 1976 and Palmer, 1975, 1976). *Lunar rhythms* are also found among marine invertebrates (Still, 1972).

SEASONAL CIRCANNUAL RHYTHMS

The passage of the seasons in northern and southern hemispheres and the changes in activities of organisms correlated with the seasons have always been recognized as the rhythm of life. Away from the equator the temperature varies noticeably with the seasons, so it was assumed that the temperature change controlled seasonal rhythms in organisms. Consequently, when Garner and Alard (1920) reported that the seasonal blooming of plants was determined primarily by daylength, not by temperature, their findings were regarded as questionable. However, many experiments soon verified these findings and demonstrated that most plants can be classified into either of two categories—those that bloom as daylength increases and those that bloom when daylength decreases—yet others are neutral, and change in daylength has no effect on their blooming.

A few years later it was shown that in some insects and birds breeding was controlled by daylength also. For example, the testes of the Canadian junco could be brought into breeding condition, even when the temperature outside was 4.4°C, by exposing the birds to winter days lengthened after sunset by relatively weak artificial light (Rowan, 1938). Thus, it is clear that light is an important factor in seasonal rhythms, as it is in circadian rhythms.

However, temperature proves to be a modifier of *circannual* (*circa annum,* meaning about a year) rhythms even when they are primarily controlled by light, and in some cases it may even act as a primary synchronizer, e.g., in some poikilothermal (cold-blooded) organisms whose temperature depends upon that of the environment.

Circannual Rhythms in Multicellular Plants: Photoperiodism

Photoperiodism for plants is the response of cells in plants to alternate periods of light and darkness. It is correlated with the length of the light and dark periods. Experiments have demonstrated that upon exposure to light the cells in the young leaves of a plant produce a diffusible hormone. This hormone is known to stimulate the cells of the floral organs, resulting in their growth. The hormone can pass by direct contact from illuminated cells to others not

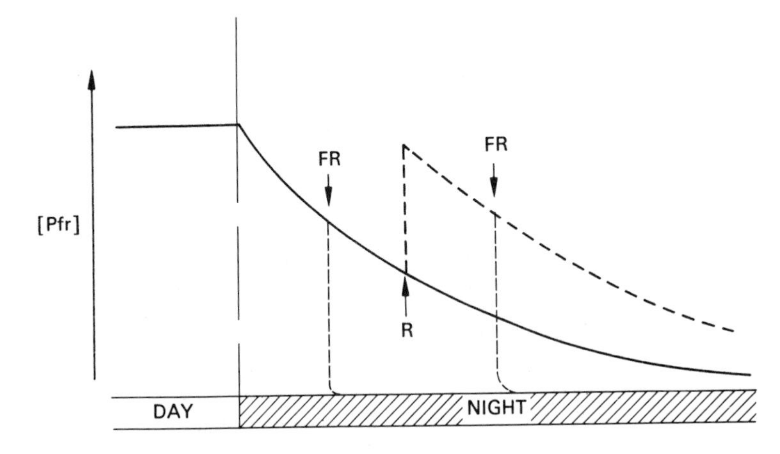

Figure 24.13. *The predicted level of the concentration of the far-red-absorbing form of phytochrome; P_{fr} is plotted against the time. By day the balance between P_r and P_{fr} is indicated by a photostationary state ($P_r \rightleftharpoons P_{fr}$). At night, as a result of the reaction $P_r \rightleftharpoons P_r$, P_{fr} decays exponentially (solid curve). A red flash (R, upward dashed curve) retards the decay by converting some of the P_r back to P_{fr}. A far-red flash (FR, downward dashed curves) lowers the concentration by rapid conversion of P_{fr} to P_r. (From Clayton, 1971:* Light and Living Matter 2: *153.)*

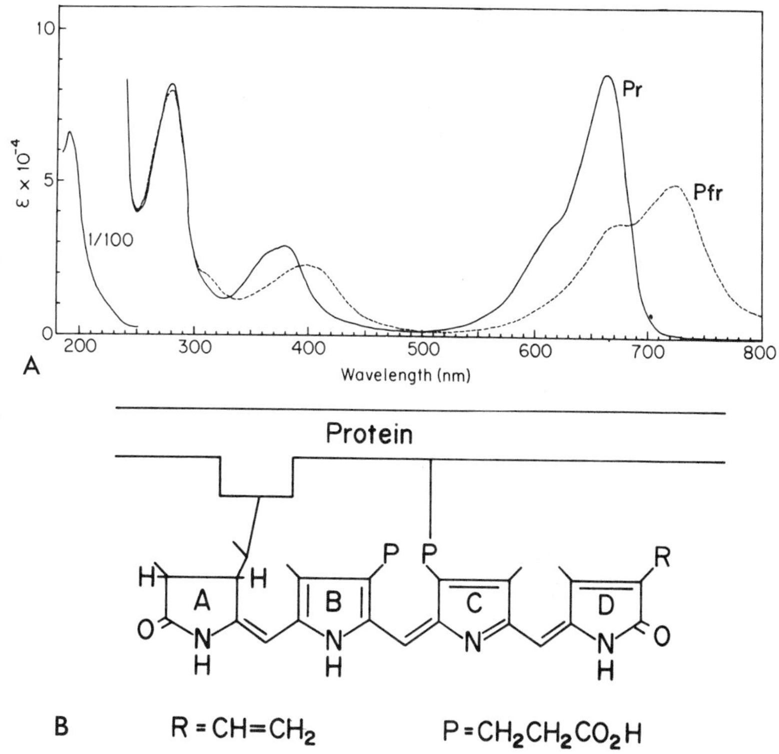

Figure 24.14. A, *Absorption spectra of highly purified phytochrome extracted from dark-grown oat seedlings.* P_r, *red-absorbing form of phytochrome,* P_{fr}, *far-red absorbing form of phytochrome.* B, *Postulated molecular structure of phytochrome chromophore (light-absorbing portion). The chromophores for both* P_r *and* P_{fr} *have identical ring structures but the* P_{fr} *lacks the double bond between rings A and B.* (A *from Anderson, Jenner and Mumford, 1970: Biochem. Biophys. Acta* 221:*69–73.* B *from Grombein, Rudiger and Zimmermann, 1975: Z. Physiol. Chem.* 356: *1709–1714.)*

treated, but will not pass through agar or cellophane. Variations in the intensity of the light (above a certain threshold), such as are caused by clouding of the skies, do not affect the flowering response, which is regulated only by the number of hours of minimal light exposure. The alternate period of darkness is also of primary importance and its interruption, even with a brief flash of light, alters the response, in some cases having the effect of continuous light between flashes. In other words, alternate periods of light and darkness of definite duration are required.

Many other growth responses, including germination of seeds at the soil surface, are regulated by the same rhythms of day and night (Shropshire, 1972, 1977). In some cases, by manipulation of the light periods, several flowering cycles have been induced in one calendar year, although under natural conditions only one cycle occurs.

The red end of the spectrum, particularly at wavelength 660 nm, is the region of maximal stimulation for photoperiodic response in plant cells. Blue light is relatively ineffective. The photoperiodic response induced by red light (660 nm) can be reversed by illumination with far red (730 nm); it can be reactivated by another exposure to red (660 nm) and again reversed by far red (730 nm). These sequences may be repeated many times. As a result of these experiments, the hormone is considered to be a pigment called *phytochrome*.

The pigment was conceived to exist in cells in two forms, one absorbing strongly in the red, and another in the far red (Fig. 24.13):

$$P_r \underset{730\text{ nm}}{\overset{660\text{ nm}}{\rightleftharpoons}} P_{fr} \qquad (24.1)$$

P_{fr} is considered to be the active form that induces the photoperiodic responses. Supposedly P_{fr} gradually and spontaneously reverses to form P_r, quickly in cells for long-day (short-night) plants, and slowly in cells of short-day (long-night) plants. Inasmuch as P_{fr} reverts spontaneously to P_r, its concentration is continuously falling. It is likely that P_{fr} must be present above a critical concentration to be active in photomorphogenesis (regulation of form by light) in plants. Once its concentration has fallen below the critical level it ceases to produce its effect. However, a single flash of red light at the right time might conceivably raise the concentration of P_{fr} over the critical level and thus act much like a long light period (Fig. 24.14). The dark conversion of P_{fr} to P_r is somewhat affected by temperature. Phytochrome has been called a "biological switch," turning cellular processes on and off (Shropshire, 1972, 1977).

Although phytochrome is present in exceedingly small amounts (1 part in 10 million), it has been extracted from the cells of the tips of young albino plants and was shown to be a bluish compound consisting of a protein with a smaller prosthetic molecule attached to it. The extract has all the properties predicted for it from photoperiodic studies. Thus, it shows maximal absorption in one form at 660 nm, being converted to the other form with maximal absorption at 730 nm (Fig. 24.14*A*). The 730 nm wavelength reconverts it to the form with maximal absorption at 660 nm. The conversions are entirely photochemical and occur as readily at very low temperatures as at high temperatures. Phytochrome is as widely distributed as chlorophyll and is to photoperiodism what chlorophyll is to photosynthesis.

The protein moiety of phytochrome is a dimer with a molecular weight of about 240,000. It is made up of two subunits of molecular weight about 120,000. The amino acid composition of the protein has been determined and the relative proportion of the various amino acid residues in the molecule is known, but the steric structure of phytochrome is still undetermined (Shropshire, 1977).

The prosthetic group of phytochrome is an open tetrapyrrole of the general nature of the phycobilins of red and blue-green algae (see Fig. 24.14*B*). The molecular change involved in the transformation of phytochrome from the P_r to the P_{fr} form is a change in one bond in the tetrapyrrole, and the changes can be followed in cells of live tissues by rapid spectrophotometry (Shropshire, 1977).

While much information is being accumulated on the molecular biology of phytochrome, it is considered premature to accept any of the schemes proposed for its action (Shropshire, 1977). The evidence, however, indicates that phytochrome perhaps affects specific enzyme systems, since some biochemical reactions in plant cells are unaffected by the change from the P_r to the P_{fr} form, while other reactions are either depressed or accelerated. Phytochrome affects transcription (RNA synthesis) and translation (protein synthesis) in cells. Its action on enzyme systems appears to be indirect via the genes (Shropshire, 1977).

Some plants, like the tomato, are photoperiodically neutral. Subjection of the plant to an appropriate temperature is more important than a change in daylength for induction of flowering and the setting of fruit in such cases. Temperature change is also the cue that releases underground plant bulbs from dormancy, inducing growth and development. The utility of a cue other than light for underground bulbs is evident (Withrow, 1959). The importance of phytochrome in plant growth responses should not exclude the possibility of action of other environmental factors as useful cues in special cases.

The interrelationship between circadian and circannual rhythms is shown in the effects of phytochrome on circadian plant leaf movements. For example, the sensitive plant *Mimosa* responds to a transition from light to darkness by folding its leaves (see Fig. 20.15). Loss of turgor in the lower half of the base of the leaflet results in its folding. A flash of red light interposed between the end of the light period and the beginning of the dark period inhibits this response. The effect of far red light is reversed by a pulse of red light; this in turn is erased by another flash of far red light, and so on. Changes in permeability induced by phytochrome might underlie these responses (Shropshire, 1972).

Circannual Rhythms in Multicellular Animals: Photoperiodism

Considerable information is now available on seasonal rhythms in the photoperiodic responses of birds, mammals, and some invertebrates (mainly insects), but the analysis of either the cellular or the molecular photobiological mechanisms involved has only begun. Some of the quantitative relationships, however, between exposure to light as a synchronizer and the responses have been determined (Sollberger, 1965).

Many birds breed in the spring when days are getting longer. Precocious breeding can be induced in winter by artificially lengthening the days, even when the temperature is below freezing. By appropriate lighting regimens several cycles of breeding have been induced in each of several species of birds although under normal conditions only one cycle would have occurred in one year. However, in some species of birds it is the male that is induced into breeding condition by the light regimen; the female is stimulated to maturity by the sexually mature male. Thus, light affects the female only indirectly. The converse pattern is found in some mammals—they breed in the fall when the days are short and produce their young in the spring. Here the stimulus of the light apparently acts directly upon cells in the anterior lobe of the hypophysis. When the hypophysis is activated, it stimulates growth of gonadal cells. Some evidence exists that photoperiodic control regulates breeding in many higher and lower forms of life, as well as migration and other behavior patterns and even fat deposition (Farner and Lewis, 1971). In some cases (e.g., some insects) reproduction can be controlled practically at will by manipulation of the photoperiod, although nutrition and temperature play a role as modifiers (Lees, 1968).

It had been long suspected that the seasonal breeding of birds and mammals might be circannual and dependent upon the operation of an endogenous rhythm. That breeding in nor-

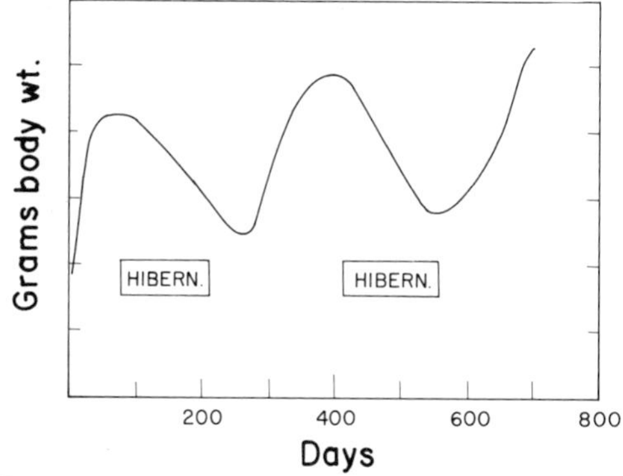

Figure 24.15. Smoothed curves for the circannual rhythm of body weight in a ground squirrel kept in a room at 22°C illuminated 12 hours daily. Hibernation occurred during the periods indicated. (Data from Pengelley and Asmundson, 1971.)

mal ferrets could occur in darkness clearly indicates that light was not necessary for breeding (Thompson, 1954); in other words, the cycle was endogenous and circannual, although light normally synchronized the circannual rhythm with the solar year. Similarly, ground squirrels, kept at 0°C in a constant schedule of 12 hours light and 12 hours darkness, maintained their body temperature at 37°C until they stopped eating in October, at which time they went into hibernation until April, just as those in nature were doing (Pengelley and Fisher, 1963). When they were kept at 22°C the same sequence was repeated. The periodicity measured over a four-year period was between 324 and 329 days, or circannual (Fig. 24.15). When kept at 35°C, which is close to their body temperature, the squirrels could not hibernate but showed a decrease in feeding and a loss of weight corresponding to that in the hibernating animals. A test was also made with laboratory-born and -bred squirrels on which a natural cycle could not have been imprinted; they were kept in a constant environment at 3°C and in darkness. Again, a circannual rhythm was established and hibernation ensued at approximately the programmed time. In nature this circannual year would have been synchronized with the solar year by cues from the environment, primarily light (Pengelley and Asmundson, 1971; Pengelley, 1974).

A circannual rhythm (of molting and migrating restlessness) has been demonstrated in willow warblers that under natural conditions molt and show restlessness once a year just before migrating. Kept under constant conditions of temperature and light (12 hours light, 12 hours darkness) laboratory warblers molted and became restless at about the same time as those in nature (Gwinner, 1968). A circannual rhythm of molting and reproduction of somewhat less than a year has also been shown in the translucent cave crayfish *Orconectes,* even when kept at an average temperature of 13°C. The crayfish were handled or disturbed for only 15 minutes several times a month when examined by the experimenters. This circannual rhythm is synchronized with the solar year in nature by environmental cues, possibly temperature (Jegla and Poulson, 1970).

Some evidence (especially the appearance of ketosteroids in the urine) exists that humans also display long-term rhythms, possibly circannual and synchronized with the solar year by daylength (Halberg *et al.,* 1965). As already seen, the circannual rhythms of reproduction in many organisms are synchronized with nature by daylength. This is easily understood, since change in daylength is one of the natural rhythms that is repeatable from year to year

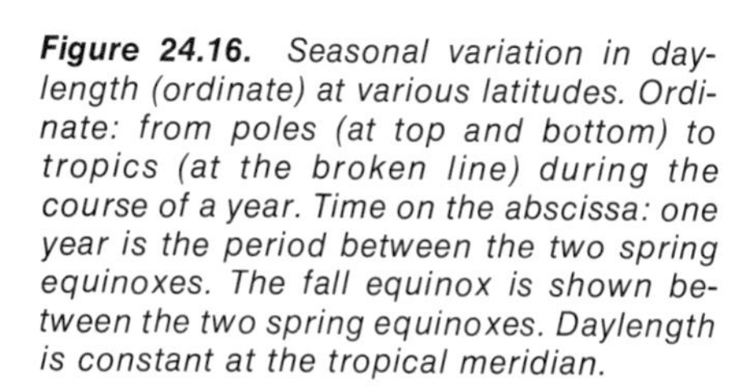
Figure 24.16. Seasonal variation in daylength (ordinate) at various latitudes. Ordinate: from poles (at top and bottom) to tropics (at the broken line) during the course of a year. Time on the abscissa: one year is the period between the two spring equinoxes. The fall equinox is shown between the two spring equinoxes. Daylength is constant at the tropical meridian.

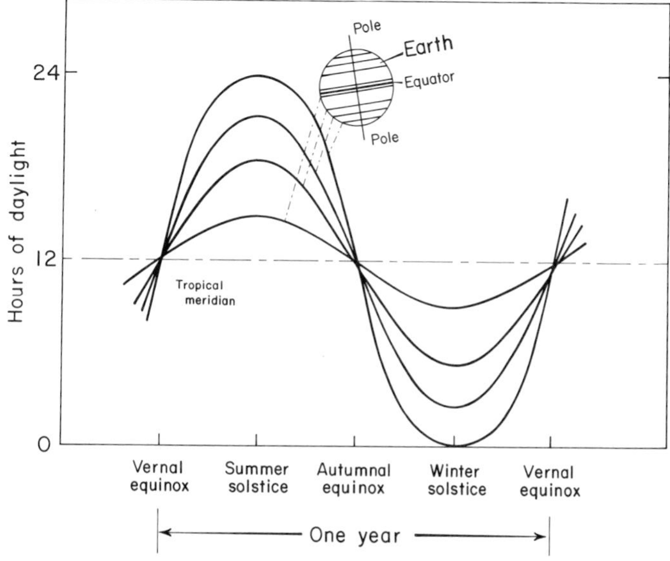

without change for a given latitude above or below the equator (Fig. 24.16). On land or bodies of fresh water it is much more useful as a cue of the season than is temperature change because temperature varies fitfully with atmospheric currents.

The question of how an organism detects the length of day remains unanswered. However, Bünning many years ago suggested that the circadian clock may subserve the function. For organisms with circadian rhythms time is divided into subjective day (illuminated period) and subjective night (darkness) that together make up the "almost" 24-hour circadian period; normally these are synchronized with the 24-hour cycle of the solar day by light. In spring, daylight comes progressively earlier and earlier, and the light accordingly interrupts the subjective night sooner and sooner, thus conceivably cuing the organism to the change in the environment. In fall the converse is true, the night later and later delaying the onset of subjective day (Bünning, 1973). The circadian clock might also explain the effectiveness of a flash of light interrupting the dark period and having essentially the effect of a long light period; if a flash comes at the time of subjective day, it has the same signalling effect to appropriate receptors as continuous illumination between two light signals—that is, between the end of the light period and the flash of light.

Factors Other than Light Regulating Circannual Rhythms and Conclusion

The temperature of the sea changes only slowly with changes in temperature of the air because of its vast size and the large heat capacity of water compared to that of air. Because of this, the seasonal change in the temperature of sea water is much more repeatable from year to year than is the change in temperature on land or in small bodies of fresh water. Therefore, it is not surprising that the reproductive cycle of many marine invertebrates is more often correlated with seasonal temperature change of sea water than with daylength (Fig. 24.17). Some species, however, do show correlation with daylength (Giese and Pearse, 1974). None of these cycles has been studied under free-running conditions in a constant environment for more than brief periods, although an endogenous circannual rhythm is suggested for some—for example, for the ochre sea star (Halberg *et al.*, 1969).

Circannual cycles are probably the result of the operation of a complex sequence of events of shorter cycles of activity and the interplay between hormones and their target cells. In vertebrates, gonads are stimulated to growth and gametogenesis by hormones (gonadotrophic hormones) from the hypophysis. The gonads develop and the gametes reach maturity. These events are accompanied by secretion from cells in gonadal accessory tissues of hormones that have a number of effects, some of which ready the organism for reproduction. Others, in a feedback upon the hypophysis, stop production of any more gonadotrophic hormone. The gonadal tissues then become refractory, and it is only after the passage of time composed of short days and really long nights that they again become capable of stimulation. Under natural conditions this approximates an annual cycle. As already pointed out, some environmental cue induces the secretion of gonadotrophic hormones, once again synchronizing the circannual cycle with the annual solar cycle in nature. While the mechanism in invertebrates may vary considerably, hormones have been demonstrated to play an important role in reproduction, and their secre-

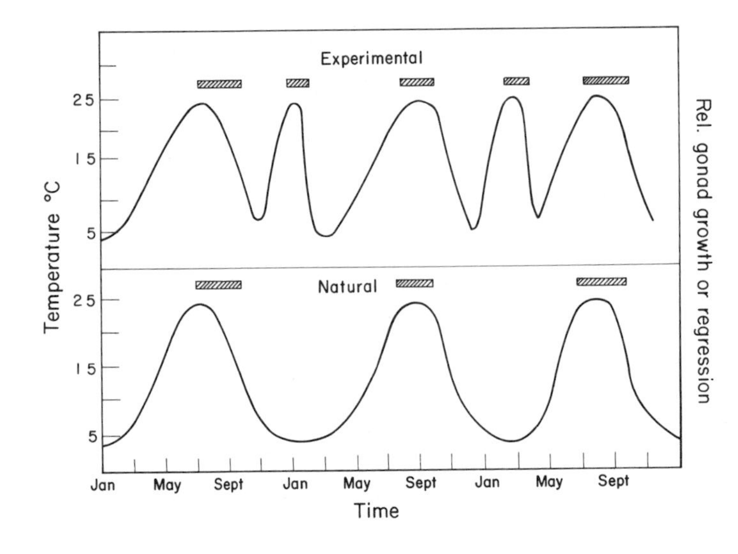

Figure 24.17. *Breeding (shaded bars) in the oyster* (Crassostrea virginica) *correlated with temperature (curves); in nature (bottom) and under experimental conditions (top). During winter the experimental animals were held in the laboratory at temperatures equivalent to those in summer in nature. Note that extra breeding cycles were induced by the higher temperatures in mid-winter. (Data from Loosanoff and Davis, 1952: Science* 115: *675–676.)*

tion is cued by environmental factors, often by daylength (Durchon, 1967).

It is clear that a circannual rhythm has survival value and that it is probably a product of natural selection. It is important that nutrients (sources of energy) be stored in advance for a forthcoming change in conditions, for reproduction, for migration, for hibernation, and against starvation. This storage can be insured by a circannual clock, which is synchronized with nature by environmental cues. Circannual rhythms seem to be characteristic of multicellular organisms; they have not been described in unicellular forms. Therefore, they are probably a result of multicellularity and apparently a means of coordinating the multiple events that result from multicellularity.

It must also be mentioned here that evidence has been advanced for the presence of rhythms other than circadian and circannual, some based on a week, some on the lunar month and tides (Fingerman, 1960), and still others. Almost nothing is known of the cellular aspects of these rhythms (Luce, 1970; Brown, 1973, DeCoursey, 1976).

LITERATURE CITED AND GENERAL REFERENCES

Andrews, R. V., 1968: Temporal secretory responses of cultured hamster adrenals. Comp. Biochem. Physiol. *26*: 179–193.

Aschoff, J., 1965a: Circadian rhythms in man (a summary). Science *148*: 1427–1432.

Aschoff, J., ed., 1965b: Circadian Clocks. North Holland Publishing Co., Amsterdam.

Aschoff, J., Hoffmann, K., Pohl, H. and Wever, R., 1975: Reentrainment of circadian rhythms after phase shifts of the zeitgeber. Chronobiologia *2*: 23–78.

Barnett, A., 1966: A circadian rhythm reversal in *Paramecium multimicronucleatum,* syngen 2, and its genetic control. J. Cell Physiol. *67*: 239–270.

Beck, S., 1968: Insect Photoperiodism. Academic Press, New York.

Binkley, S. A., Riebman, J. B. and Reilly, K. B., 1978: The pineal gland: a biological clock in vitro. Science *202*: 1198–1201.

Binkley, S., Riebman, J. B. and Reilly, K. B., 1977: Timekeeping by the pineal gland. Science *197*: 1181–1183.

Biological Clocks, 1960: Cold Spring Harbor Symp. Quant. Biol. *25*: 87, 131, 149 and 217.

Brown, F. A., Hastings, J. W. and Palmer, J. D., 1970: The Biological Clock: Two Views. Academic Press, New York.

Bruce, V. G. and Minis, D. H., 1969: Circadian clock action spectrum in a photoperiodic moth. Science *163*: 583–585.

Bruce, V. G. and Pittendrigh, C. S., 1955: Temperature independence in a unicellular clock. Proc. Nat. Acad. Sci. U.S.A. *42*: 671–682.

Bruce, V. G. and Pittendrigh, C. S., 1960: An effect of heavy water on the phase and period of the circadian rhythm in *Euglena*. J. Cell Comp. Physiol. *56*: 25–31.

Bünning, E., 1973: The Physiological Clock: Circadian Rhythms and Biological Chronometry. 3rd Ed. Springer-Verlag, New York.

Clayton, R. K., 1971: Light and Living Matter: A Guide to the Study of Photobiology. Vol. 2. The Biological Part. McGraw-Hill Book Co., New York.

Cohen, L. W., 1965: The basis for the circadian rhythm of mating in *Paramecium bursaria.* Exp. Cell Res. *37*: 360–367.

Colquhoun, W. R., 1971: Biological Rhythms and Human Performance. Academic Press, New York.

Conroy, L., 1970: Human Circadian Rhythms. Churchill, London.

Danielewskii, A. S., 1965: Photoperiodism and Seasonal Development of Insects. Oliver and Boyd, London.

DeCoursey, B. J., ed., 1976: Biological Rhythms in the Marine Environment. University of South Carolina Press, Columbia, S. C.

Driesche, T. V., 1966: The role of the nucleus in circadian rhythms of *Acetabularia mediterranea.* Biochim. Biophys. Acta *12*: 456–470.

Durchon, M., 1967: L'Endocrinologie de Vers et des Molluscues. Masson et Cie, Paris.

Eakin, R. M., 1974: The Third Eye. University of California Press, Berkeley. (The pineal eye and glands.)

Edmunds, L. N., Jr., 1969a: Circadian rhythm of cell division in *Euglena:* effects of a random illumination regimen. Science *165*: 500–503.

Edmunds, L. N., Jr., 1969b: Effect of "skeleton" photoperiods and high frequency of light cycles on the rhythm of cell division in synchronized cultures of *Euglena*. Planta *87*: 134–163.

Edmunds, L. N., Jr., 1977: Clocked cell clocks. Waking and Sleeping *1*: 227–252.

Edmunds, L. N., Jr. and Halberg, F., 1978: Circadian time structure of *Euglena:* a model system amenable to quantification. *In* Comparative Pathology of Abnormal Growth. Kaiser, ed. Raven Press, New York (in press).

Ehret, C. F. and Trucco, E., 1967: Molecular aspects of the circadian clock. J. Theor. Biol. *15*: 240–262.

Eling, W., 1967: The circadian rhythm of nucleic acids. *In* The Cellular Aspects of Biorhythms. von Mayersbach, ed. Springer-Verlag, New York, pp. 105–114.

Eppley, R. W., Holmes, R. W. and Paasche, E., 1967: Periodicity in cell division and physiological behavior of *Ditylum brightwellii,* a marine planktonic diatom, during growth in light-dark cycles. Arch. Mikrobiol. *56*: 305–323.

Farner, D. S. and Lewis, R. A., 1971: Photoperiodism and reproductive cycle in birds. *In* Photophysiology. Vol. 6. Giese, ed. Academic Press, New York, pp. 325–370.

Feldman, J. F., 1967: Lengthening the period of a biological clock in *Euglena* by cyclohexamide, an inhibitor of protein synthesis. Proc. Nat. Acad. Sci. U.S.A. *57*: 1080–1087.

Fingerman, M., 1960: Tidal rhythmicity in marine organisms. *In* Biological Clocks. Cold Spring Harbor Symp. Quant. Biol. *25*: 481–489.

Garner, W. W. and Alard, H. A., 1920: Effect of relative length of day and night and other factors of the environment on growth and reproduction in plants. J. Agric. Res. *18*: 553–606.

Giese, A. C., and Pearse, J. S., 1974: Introduction. *In* Reproduction of Marine Invertebrates. Vol. 1. Academic Press, New York, pp. 1–47.

Gwinner, E., 1968: Circannuale Periodik als Grundlage des Jahreszeitlichen Funktionswandels bein Zugvögeln. J. Ornithologie *109*: 70–95.

Halberg, F., 1969: Chronobiology. Ann. Rev. Physiol. *31*: 675–725.

Halberg, F., 1977: Implications of biological rhythms for clinical practice. Hospital Practice *12*: 139–149.

Halberg, F. and Howard, R. B., 1958: 24 hour periodicity and experimental medicine, examples and interpretations. Postgrad. Med. *24*: 349–358.

Halberg, F., Engeli, M., Hamburger, C. and Hillman, D., 1965: Spectral resolution of low frequency, small amplitude rhythms in excreted 17-ketosteroids; probably androgen-induced circaseptan desynchronization. Acta Endocrinol. *50*(Suppl. 103): 5–54.

Halberg, F. E., Halberg, F. and Giese, A. C., 1969: Estimation of objective parameters for circannual rhythms in marine invertebrates. Rass. Neurol. Veg. *23*: 173–186.

Halberg, F., Halberg, E., Barnum, C. P. and Bittner, J. J., 1959: Physiologic 24-hour periodicity in human beings and mice, the lighting regime and daily routine. *In* Photoperiodism and Related Phenomena in Plants and Animals. Withrow, ed. American Association for the Advancement of Science, Washington, D.C.

Halberg, F., Johnson, E. A., Nelson, W., Runge, W. and Sothern, R., 1972: Autorhythmicity procedures for physiologic self-measurements and their analysis. Physiology Teacher *1*: 1–11.

Halberg, F., Reinberg, A. and Reinberg, A., 1977: Chronobiologic serial sections gauge circadian rhythm adjustments following transmeridian flight in novel environment. Waking and Sleeping *1*: 259–279.

Hamner, K. C. and Hoshizaki, T., 1974: Photoperiodism and circadian rhythms: an hypothesis. Bioscience *24*: 407–414.

Harker, J., 1964: The Physiology of Diurnal Rhythms. Cambridge University Press, Cambridge.

Hastings, J. W., 1959: Unicellular clocks. Ann. Rev. Microbiol. *13*: 292–312.

Hastings, J. W., 1964: The role of light in persistent daily rhythms. *In* Photophysiology. Vol. 1. Giese, ed. Academic Press, New York, pp. 333–361.

Hastings, J. W., and Keynan, A., 1965: Molecular aspects of circadian rhythms. *In* Circadian Clocks. Aschoff, ed. North Holland Publishing Co., Amsterdam, pp. 167–182.

Hastings, J. W. and Schweiger, H-G., eds., 1976: The molecular basis of circadian rhythms. Life Science Research Report No. 1. Dahlem Workshop on the Molecular Basis of Circadian Rhythms, Berlin.

Hauty, G. T. and Adams, T., 1965: Phase shifting of the human circadian system. *In* Circadian Clocks. Aschoff, ed. North Holland Publishing Co., Amsterdam, pp. 413–425.

Hayden, P. and Lindberg, R. G., 1969: Circadian rhythm in mammalian body temperature entrained by pressure changes. Science *164*: 1288–1289.

Hawkins, F., 1970: The clock of the malarial parasite. Sci. Am. (Jun.) *222*: 123–131.

Hendricks, S. B., 1968: How light interacts with living matter. Sci. Am. (Sept.) *219*: 175–186.

Hendricks, S. B. and Siegelman, H. W., 1967: Phytochrome and photoperiodism in plants. *In* Comprehensive Biochemistry. Vol. 27. Florkin and Statz, eds. Elsevier, Amsterdam.

Holmquist, D. L., Retienne, K. and Lipscomb, H. S., 1966: Circadian rhythms in rats: effects of random lighting. Science *152*: 662–664.

Jacklett, J. W., 1969: Circadian rhythm of optic nerve impulses recorded in darkness from the isolated eye of *Aplysia*. Science *164*: 502–503.

Jegla, T. C. and Poulson, T. L., 1970: Circadian rhythms: 1. Reproduction in the cave crayfish *Orconectes pellucidus intermis*. Comp. Biochem. Physiol. *33*: 347–355.

Jerusalem, C., 1967: Circadian changes in the DNA content in rat liver cells as revealed by histophotometric methods. *In* The Cellular Aspects of Biorhythms. von Mayersbach, ed. Springer-Verlag, New York, pp. 115–123.

Jørgensen, E. G., 1966: Photosynthetic activity during the life cycle of synchronous *Skletona* cells. Physiol. Plantarum *19*: 789–799.

Karakasian, M. W. and Hastings, J. W., 1962: The inhibition of a biological clock by Actinomycin D. Proc. Nat. Acad. Sci. U.S.A. *48*: 2130–2137.

Klevecz, P. R., and Ruddle, F. H., 1968: Cyclic changes in enzyme activity in synchronized mammalian cell cultures. Science *159*: 634–636.

Lees, A. D., 1968: Photoperiodism in insects. *In* Photophysiology. Vol. 4. Giese, ed. Academic Press, New York, pp. 47–137.

Lickey, M. E., 1967: Effect of various photoperiods on a circadian rhythm in a single neuron of *Aplysia*. *In* Conference on Invertebrate Nervous Systems. University of Chicago Press, Chicago, pp. 321–328.

Lofts, B., 1970: Animal Photoperiodism. Edward Arnold, London.

Luce, G. C., 1970: Biological Rhythms in Psychiatry and Medicine. National Institutes of Health, Bethesda, Md.

McFarland, R. A., 1975: Air travel across time zones. Am. Sci. *63*: 23–30.

McMurry, L. and Hastings, J. W., 1972: No desynchronization among four circadian rhythms in the unicellular alga, *Gonyaulax polyhedra*. Science *175*: 1137–1139.

Menaker, M., 1969: Biological clocks. Am. Inst. Biol. Sci. Bull. *19*: 681–689.

Menaker, M., ed., 1971: Biochronometry. National Academy of Science, Washington, D.C.

Menaker, M., 1972: Nonvisual light perception. Sci. Am. (Mar.) *226*: 22–29.

Menaker, M., 1977: Extraretinal photoreception. *In* The Science of Photobiology. Smith, ed. Plenum Press, New York, pp. 227–240.

Menaker, M. and Eskin, A., 1966: Entrainment of circadian rhythms by sound in *Passer domesticus*. Science *154*: 1579–1581.

Miles, L. E. M., Raynal, D. M. and Wilson, M. A., 1977: Blind man living in normal society has circadian rhythm of 24.9 hours. Science *198*: 421–423.

Mills, J. N., 1966: Human circadian rhythms. Physiol. Rev. *46*: 128–171.
Mohr, H., 1969: Photomorphogenesis. *In* An Introduction to Photobiology. Swanson, ed. Prentice-Hall, Englewood Cliffs, New Jersey, pp. 99–141.
Njus, D., Gooch, V. D., Mergenhagen, D., Sulzman, F. and Hastings, J. W., 1976: Membranes and molecules in circadian systems. Fed. Proc. *35*: 2353–2357.
Njus, D., McMurry, L. J. and Hastings, J. W., 1977: Conditionality of circadian rhythmicity: synergistic action of light and temperature. J. Comp. Physiol. *117*: 335–344.
Palmer, J. D., 1975: Biological clocks of the tidal zone. Sci. Am. (Feb.) *232*: 70–79.
Palmer, J. D., 1976: An Introduction to Biological Rhythms. Academic Press, New York.
Palmer, J. D., 1977: Human rhythms. Bioscience *27*: 93–99.
Pengelley, E. T., ed., 1974: Circannual Clocks. Academic Press, New York.
Pengelley, E. T. and Asmundson, S. J., 1971: Annual biological clocks. Sci. Am. (Apr.) *224*: 72–79.
Pengelley, E. T. and Fisher, K. C., 1963: The effect of temperature and photoperiod on the yearly hibernating behavior of captive golden mantled ground squirrels *(Citellus lateralis tescorum)*. Can. J. Zool. *41*: 1103–1120.
Pittendrigh, C. S., 1965: On the mechanism of the entrainment of a circadian rhythm by light cycles. *In* Circadian Clocks. Aschoff, ed. North Holland Publishing Co., Amsterdam, pp. 277–297.
Rogers, L. A. and Greenbank, G. R., 1930: The intermittent growth of bacterial cultures. J. Bacteriol. *19*: 181–190.
Rowan, W., 1938: Light and seasonal reproduction in animals. Biol. Rev. *13*: 374–402.
Salisbury, F. B., 1971: The Biology of Flowering. Natural History Press, Garden City, New York.
Sargent, M. L. and Briggs, W. R., 1967: The effects of light on circadian rhythm of conidiation in *Neurospora*. Plant Physiol. *42*: 1504–1510.
Saunders, D. S., 1976: The biological clock of insects. Sci. Am. (Feb.) *234*: 114–121.
Schering, L. E., 1976: The dimension of time in biology and medicine—chronobiology. Endeavour *35*: 66–72.
Schweiger, E., Wallraff, H. G. and Schweiger, H. G., 1964: Endogenous circadian rhythm of the cytoplasm of *Acetabularia:* Influence of the nucleus. Science *146*: 658–659.
Schweiger, H. G., 1977: Circadian rhythms in unicellular organisms. An endeavour to explain molecular mechanisms. Int. Rev. Cytol. *51*: 315–342.
Shropshire, W., Jr., 1972: Phytochrome-photochromic sensor. *In* Photophysiology. Vol. 7. Giese, ed. Academic Press, New York, pp. 33–72.
Shropshire, W., Jr., 1977: Photomorphogenesis. *In* The Science of Photobiology. Smith, ed. Plenum Press, New York, pp. 281–312.
Smolensky, M., Halberg, F. and Sargent, F., II, 1972: Chronobiology of the life sequence. *In* Advances in Climatic Physiology. Ito, Ogata and Yoshimura, eds. Igaku Shoin, Ltd., Tokyo, pp. 281–318.
Sollberger, A., 1965: Biological Rhythm Research. American Elsevier Publishing Co., New York.
Still, H., 1972: Of Time, Tides and Inner Clocks. Stackpole Books, Harrisburg, Pa.
Strumwasser, F., 1965: The demonstration and manipulation of a circadian rhythm in a single neuron. *In* Circadian Clocks. Aschoff, ed. North Holland Publishing Co., Amsterdam.
Sweeney, B. M., 1963: Resetting the biological clock in *Gonyaulax* with ultraviolet light. Plant Physiol. *38*: 704–708.
Sweeney, B. M., 1969a: Transducing mechanisms between circadian clock and overt rhythms in *Gonyaulax*. Can. J. Bot. *47*: 299–308.
Sweeney, B. M., 1969b: Rhythmic Phenomena in Plants. Academic Press, New York.
Sweeney, B. M., 1977: Chronobiology (Circadian rhythms). *In* The Science of Photobiology. Smith, ed. Plenum Press, New York, pp. 209–226.
Sweeney, B. M. and Hastings, J. W., 1958: Rhythmic cell division in populations of *Gonyaulax polyhedra*. J. Protozool. *5*: 217–224.
Sweeney, B. M. and Hastings, J. W., 1960: Effects of temperature upon diurnal rhythms. Cold Spring Harbor Symp. Quant. Biol. *25*: 87–104.
Sweeney, B. M., Tuffli, C. F., Jr. and Rubin, R. H., 1967: The circadian rhythm in photosynthesis in *Acetabularia* in the presence of actinomycin D, puromycin and chloramphenicol. J. Gen. Physiol. *50*: 647–659.
Thompson, A. P. D., 1954: The onset of oestrus in normal and blinded ferrets. Proc. Roy. Soc. London B *142*: 126–135.
van den Driessche, J., 1966: Circadian rhythms in *Acetabularia:* photosynthetic capacity and chloroplast shape. Exp. Cell Res. *42*: 18–30.
Volm, M., 1964: The daily rhythm of cell division in *Paramecium bursaria*. Z. Vergl. Physiol. *48*: 157–180.
Wilkins, M., 1965: The influence of temperature and temperature changes in biological clocks. *In* Circadian Clocks. Aschoff, ed. North Holland Publishing Co., Amsterdam, pp. 146–163.
Withrow, R. B., 1959: Photoperiodism and Related Phenomena in Plants and Animals. American Association for the Advancement of Science, Washington, D.C.
Wurtman, R. J. and Axelrod, J., 1965: The pineal gland. Sci. Am. (Jul.) *213*: 50–60.
Wurtman, R. J., Axelrod, J. and Kelly, D. E., 1968: The Pineal. Academic Press, New York, pp. 107–144.

Section VII

THE CELL CYCLE

The events of a eukaryotic cell cycle were described by Howard and Pelc in 1953 as interphase, mitosis, and cell division. Interphase was subdivided into G_1 (Gap 1) from telophase to the beginning of DNA synthesis, S (DNA synthesis), and G_2 (Gap 2) from completion of DNA synthesis to mitosis *(karyokinesis),* which usually ends in cell division (cytokinesis). Mitosis generally occupies a small fraction of a cell cycle span, S somewhat more time, G_1 generally the largest and most variable fraction, and G_2 a smaller fraction than G_1 (Fig. 25.1). Much of the increase in cell size occurs in

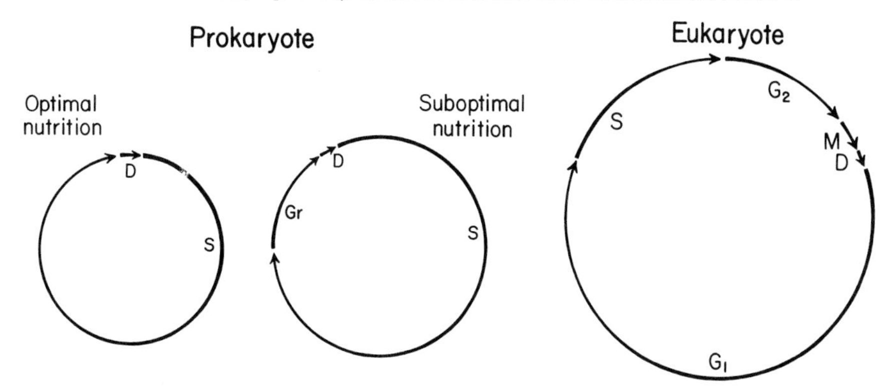

Figure 25.1. *Cell cycle in prokaryotic and eukaryotic cells compared. DNA synthesis (S) is continuous in prokaryotic cells up to the time of division (D) in optimal nutritive conditions. It is as long but followed by a period of growth (G) before division under suboptimal conditions. In the eukaryotic cell, DNA synthesis occupies only a part of the cell cycle. DNA synthesis is followed by G_2 (Gap 2) that ends in mitosis (chromosome separation) ending in cell division (cytokinesis). Division is followed by G_1 (Gap 1), during which synthesis of proteins and other cell requirements occurs and the cell grows in size.*

interphase, during most of which cell syntheses continue, although generally a pause occurs in synthesis and growth before cell division.

In prokaryotes the cell cycle depends upon nutritional conditions; if those are optimal, DNA synthesis (S) *(replication)* occupies the entire period between cell divisions. Under suboptimal conditions S occupies a time span equal to that under optimal conditions, but growth of the cell then continues until the volume has doubled, at which time the cell divides. The replicated DNA molecules *(chromosomes)* separate without a mitotic spindle.

Either a cell doubles in volume and divides or it continues to function until it dies. In dividing it reproduces its kind, for all cells arise only from preexisting cells. In cells that divide it is almost impossible to study growth separately from cell division, since growth precedes cell division. To the degree that separation of the two subjects is possible, cell growth is considered in Chapter 25, cell division in Chapter 26 and acceleration, retardation and blocking of cell division by various agents in Chapter 27.

CHAPTER 25

NORMAL AND ABNORMAL CELL GROWTH

Because a cell generally doubles in size before it divides, cell division can be used as an index of cell growth. Exceptions are eggs of marine invertebrates in which nutrients required for cell division have been accumulated by the egg in the ovary. In such eggs the reproductive phase of the cycle can run freely, not interlocking with the growth phase that is necessary when size doubling must precede cell division. However, the cells developed from divisions of a marine invertebrate egg become smaller and smaller. The same is true of yeast deprived of nitrogen. Cell divisions continue at the normal rate, but the resulting cells are smaller and smaller, converting some of the protein already present in the cell into proteins specifically needed for cell division (Mazia, 1974). This chapter is concerned with the growth aspects of the normal cell cycle.

The cell grows from within. Growth occurs only if temperature, oxygen, and other environmental conditions are favorable, and if the required nutrients and adequate water are available. In the absence of water, the cell may pass into dormant resting stages such as those of cysts or spores, perhaps remaining dormant for many years (see Chapter 4). Many cell lines can be preserved by freezing and revived by raising the temperature. If some nutrient is lacking, growth may stop or become abnormal. Growth is also easily inhibited by poisons, radiations, and, in fact, by all agents that injure the cell or interfere with normal cell functions.

GROWTH STUDIES ON SINGLE CELLS

To determine the rate of growth during a cell cycle it is desirable to gather periodic information on the size of the cell at various phases of the cycle. This is done by making a series of measurements on single living cells. From linear measurements it is possible to determine the cell volume and through use of an interference microscope or diver balance the dry mass (the total constituents of the cell minus water), and so study the same cell's growth during all phases of the cell cycle.

In both prokaryotes and eukaryotes the most common pattern of growth during the cycle indicates continuous increase in cell volume and dry mass. For example, the cell volume and dry mass increase linearly in *E. coli*. However, both cell volume and dry mass appear to increase exponentially in mammalian cells in tissue culture, but it is difficult to be certain that the deviations of the data points from a straight line do not result from imprecise measurements. Yet other deviations from a pattern of linear increase in cell volume and dry mass are found; thus, in *Schizosaccharomyces pombe* (a yeast) the linear growth pattern in cell volume ends in a plateau before cleavage (Fig. 25.2), while in the bacterium *Streptococcus faecalis* growth in volume may start later than growth in dry mass and both then continue at a declining rate until division (Fig. 25.3). In *Amoeba proteus*, on the other hand, the rate of growth in cytoplasmic volume declines, but it ends in a plateau before division (Fig. 25.4) (Mitchison, 1971).

From the total dry mass and volume measurements it is possible to determine the percent concentration (density) of solute. It is evident from Figures 25.2 and 25.3 that a cell may undergo changes in concentration of solute throughout the cell cycle, and that marked changes sometimes occur. Some of these changes in solute concentration are correlated

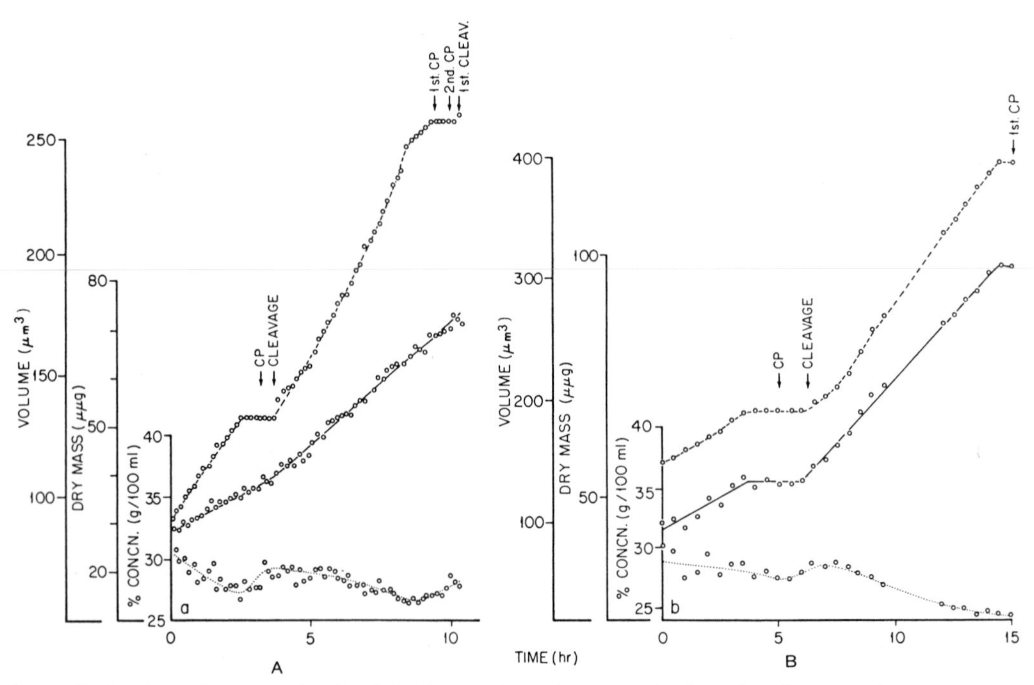

Figure 25.2. Growth curves of cells of Schizosaccharomyces pombe from interference microscope measurements. Top curve, volume; center curve, total dry mass; lower curve, concentration of cells. A, Growth at 23°C. One cell is followed through division to two daughter cells and then to the division of one of these daughters. This pattern of growth is shown by all cells in the range 23°C to 32°C. CP, first appearance of cell plate. B, Growth at 17°C. One cell is followed through division to two daughter cells and then to the appearance of a cell plate in one of these daughters. (From Mitchison, 1963: J. Cell Comp. Physiol. 62, Suppl. 1: 1–13.)

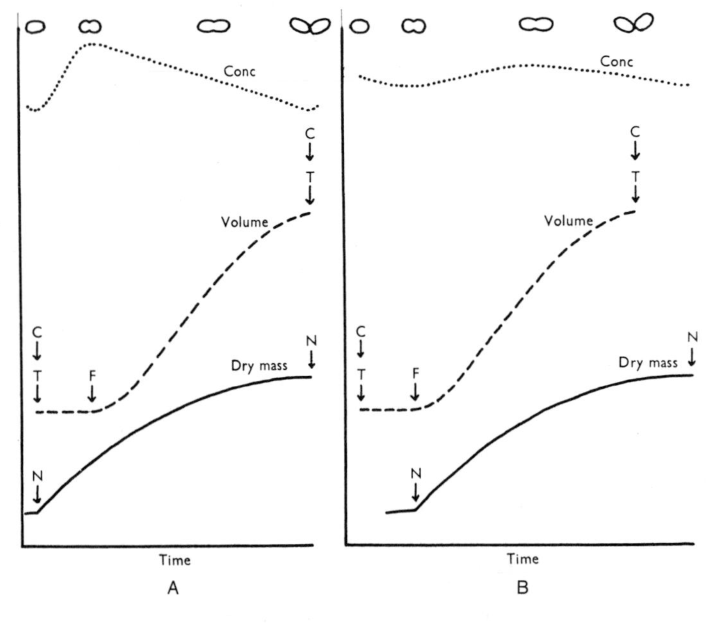

Figure 25.3. Diagrammatic growth curves of individual cells of Streptococcus faecalis from interference microscope measurements. A, The majority pattern at 17°C. B, The majority pattern at 40°C. Top curve, concentration of cells (number per unit volume); middle curve, cell volume; lower curve, total dry mass. T is the "turning" point when the two daughter cells finally separate and usually turn at an angle to each other. F is when a furrow first appears at the surface. C may mark the start of the cell cycle and N of the chromosome cycle. (replication) (From Mitchison, 1961: Exp. Cell Res. 22: 208.)

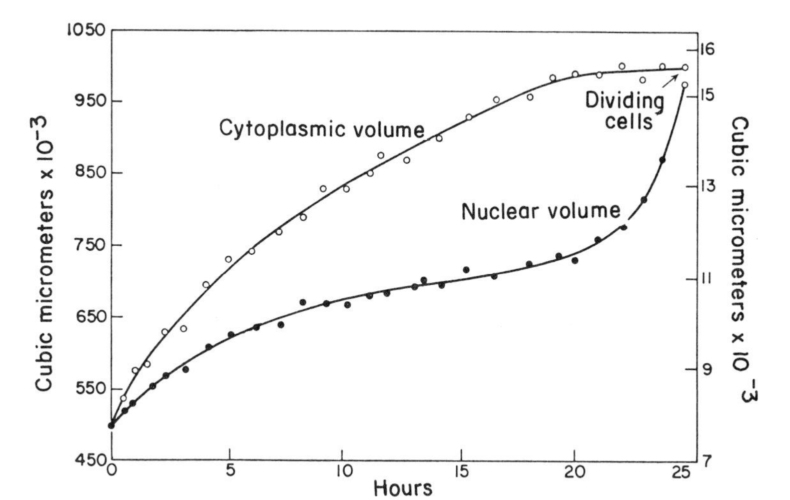

Figure 25.4. *Cytoplasmic volume (open circles) and nuclear volume (closed circles) during the cell cycle of* Amoeba proteus. *(From Prescott, 1955: Exp. Cell Res. 9: 328.)*

with stages in the cell cycle considered in Chapter 26. Changes in concentration of solute are evident during the cell cycle when the cells are compared at different temperatures (Figs. 25.2 and 25.3).

Other useful measurements that can be made on single cells during the cell cycle include the concentration of macromolecules and the concentration of compounds of low molecular weight (Fig. 25.5). For this purpose cells are treated with acid, after which the molecules of low molecular weight leak from the cells. Such compounds constitute as much as a quarter of the total dry mass of a cell, the remainder consisting of macromolecules. Inasmuch as by far the greatest proportion of macromolecular mass in an animal cell is protein, a dry mass interferometer measurement made on fixed cells is an approximate measure of the protein content of the cell. To make such measurements on a progression of stages it is necessary to have many cells at the same phase in the cell cycle, because for each measurement a cell is killed (Mitchison, 1971).

The volume of organelles may also vary during the course of the cell cycle. Most data available at present are on the nucleus and nuclear growth does not parallel that of the cell as a whole (Fig. 25.4). Generally the nucleus shows a large increase in volume near the end of the cell cycle just before cell division. The increase in nuclear mass differs in character from that of the entire cell in that it is a net rather than a gross increase owing to proteins moving across the nuclear envelope from nucleus to cytoplasm and from cytoplasm to nucleus (Mitchison, 1971).

Further refinement calls for measurements to be taken during the cell cycle of synthesis of specific proteins such as enzymes, synthesis of the various types of RNA, synthesis and replication of DNA, and correlation of all these syntheses with changes in the conformation of the chromosomes and other organelles, some in preparation for karyokinesis and others for cytokinesis. As these events are closely allied to cell division they are considered in Chapter 26.

Cells do not always exactly double in volume during the growth phase in the cell cycle; for example, deviations of about 12 percent are found in mouse connective tissue cells in culture. Smaller cells may reattain the average double volume during the next round of growth but take more time to do so than do larger cells. Also, division is not always equal; one daughter cell may be larger than the other. Nonetheless, the average volume for a population of cells is about half the average mother cell volume (Mazia, 1974).

MEASURING GROWTH RATES OF CELLS

Inasmuch as under identical conditions cells of a given kind reach approximately the same

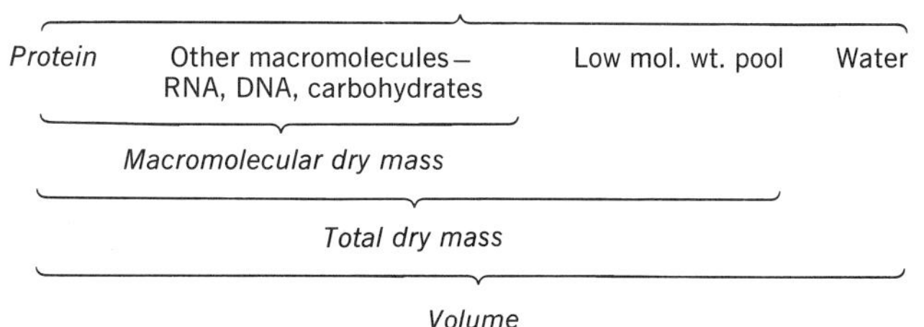

Figure 25.5. *Criteria (in italics) which can be used for measuring cell growth. (From Mitchison, 1971: The Biology of the Cell Cycle. Cambridge University Press, Cambridge.)*

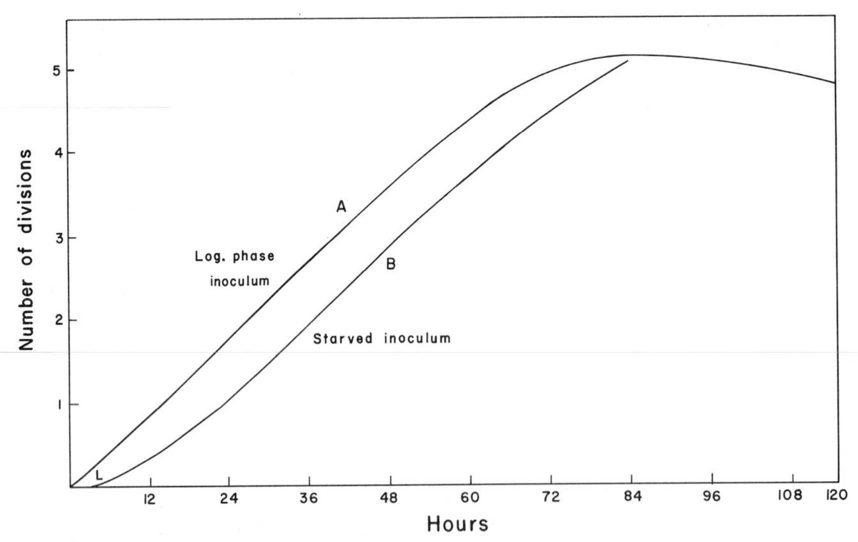

Figure 25.6. Population curve for a protozoan culture (Paramecium caudatum). *If paramecia dividing at the maximal rate (logarithmic phase) are used in inoculum, division continues at a maximal rate. If starved animals are used as inoculum, a lag (L) precedes division at the maximal rate.*

size when ready to divide, the number of cells may be used as a measure of growth. Single-celled eukaryotes can be counted directly in samples of the medium, even under low magnification. Bacteria must be counted under greater magnification in a chamber containing a grid and so arranged as to hold a fixed volume; such a device is the Petroff-Hauser counting chamber. Such test materials as blood cells, tissue culture cells, and protozoans may also be counted electronically, for example, in a Coulter counter. In such electronic devices the change in resistance as a cell passes between two oppositely stationed electrodes is recorded as a pulse. It is even possible to count cells of different sizes and to record the number of each size separately. Measurement of the optical density of a suspension of cells has been widely used for study of growth of bacterial populations. The optical density of a population of bacteria is measured at the original density and at various known dilutions. The number of cells is then determined by counts of samples at several optical densities, and a calibration

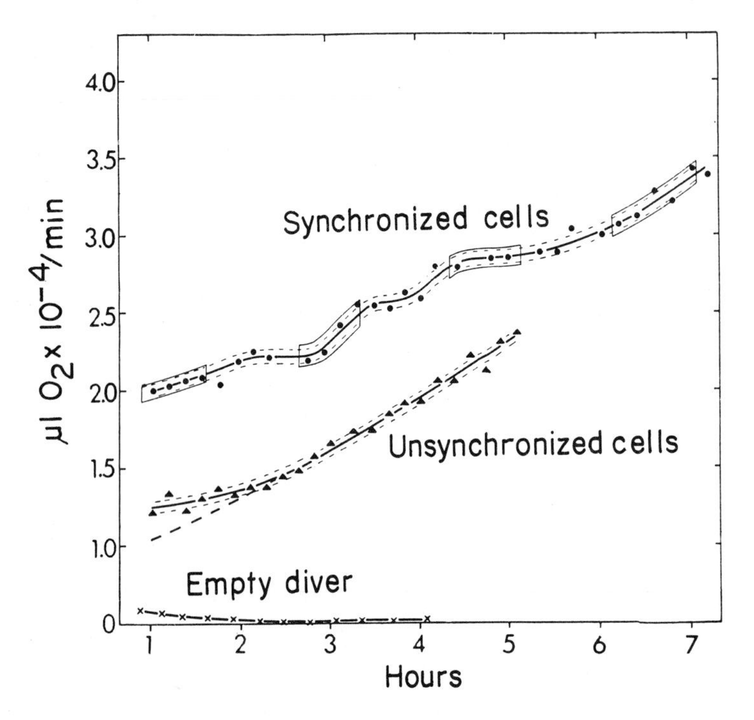

Figure 25.7. Respiration in populations of Tetrahymena *cells plotted on arithmetical scales against time in hours.* Top, *11 synchronized cells were placed in the diver, 70 were present at the end.* Middle, *20 unsynchronized cells were placed in the diver.* Bottom, *(x) is a control run with an empty diver. (Experiments by H. Thormar; from Zeuthen and Scherbaum, 1954: Colston Papers 7: 151.)*

curve may then be constructed relating optical density and number of cells. When, instead of dividing, the bacterial cells enlarge or form filaments, the optical density may increase linearly even though the genome may not have replicated.

Neither cell counts nor optical density measurements tell how many of the cells recorded are capable of division. To determine this a sample of a bacterial culture is seeded on a nutrient agar plate and incubated. Since a colony is visible only when a relatively large number of cells is present, the existence of a colony at a given spot in the plate indicates many rounds of replication from an initial viable, healthy cell. With protozoans it is possible to determine periodically the number of cells in a small test tube (or other container) and so to follow the division rate of a single cell lineage (Fig. 25.6) (Giese, 1946).

Under strictly controlled environmental conditions oxygen consumption is proportional to the number of cells, provided that an adequate oxygen supply is available. Therefore, it is possible to determine the growth rate of some cells by the increase in oxygen consumption (Fig. 25.7). The change in total concentration of certain respiratory enzymes in growing cells under an unlimited oxygen supply may also be used as a measure of the rate of growth of cells.

Growth of certain cell constituents may be followed in living cells by the rate of tracer incorporation (e.g., ^{35}S incorporated into protein). Since protein increases continuously in a healthy, growing culture, the label concentration in the culture may be a satisfactory measure of the overall growth rate. Tritiated thymidine and ^{14}C-thymidine have been used in a similar manner to label DNA.

LIFE CYCLE OF CELLS

One bold theory holds that while unicellular organisms are potentially immortal, only the cells of the "germ plasm" transmitted from generation to generation in multicellular organisms are immortal; the specialized cells of the body (soma) are destined to die (Weismann, 1904). Some protozoans have been grown for many generations in the absence of other microorganisms (in axenic culture) on a well-defined medium, for example, the ciliate *Tetrahymena pyriformis* (Table 25.1). Growth continues without loss of vigor; in fact, *Tetrahymena* in cultures multiply more rapidly than newly isolated clones in a natural culture medium with bacteria (Hall, 1967). Conjugation, once thought to rejuvenate senescent protozoan cultures, has not been observed in the axenic cultures, although it can be induced in stocks isolated from nature when the appropriate mating types are mixed (Elliott and Hayes, 1953). A variety of bacteria, fungi, and unicellular algae kept axenically in defined media and in absence of sexual processes have been grown at a constant rate for thousands of generations in repeated transfers or in steady-state devices (Brock, 1966). To that extent, it may be stated that microorganisms are potentially immortal and that cell growth and division occur without abatement if nutrient is available and other conditions are favorable.

Yet a growth cycle ending in death will occur in a culture of microorganisms if the environment is unfavorable. For example, if a single vigorous protozoan is inoculated into a limited volume of culture medium, it divides regularly until lack of food limits its growth. A plot of the number of divisions against time gives a

TABLE 25.1. NUTRIENT REQUIREMENTS OF *Tetrahymena pyriformis**

L-*Amino Acids*	*Vitamins, etc.*	*Salts*
Arginine	Thiamine†	$MgSO_4$
Histidine	Riboflavin	$Fe(NH_4)_2(SO_4)_2$
Isoleucine	Nicotinic acid	$MnCl_2$
Leucine	Pyridoxin†	$ZnCl_2$
Lysine	Folic acid	$CaCl_2$
Methionine	Pantothenic acid	$CuCl_2$
Phenylalanine	—	$FeCl_3$
Serine†	Purine (guanine)	KH_2PO_4
Threonine	Pyrimidine (uracil, cytidine)	K_2HPO_4
Tryptophan		
Valine	α Lipoic acid	

*Data from Kidder *et al.*, 1951: Physiol. Zool. *24*: 70; and Holz, G. G., 1966: J. Protozool. *13*: 2.

†The dagger indicates that mutants exist that do not require these nutrients. Culture media usually contain, in addition to the nutrients listed, some carbohydrate or acetate, which supposedly have sparing action for amino acids, but neither these nor sterols nor unsaturated fatty acids are required. For other species of *Tetrahymena* the requirements may be different.

characteristic curve (Fig. 25.6*A*). After the initial rise the curve flattens out and forms a plateau (the stationary phase) as food becomes less available. The curve declines as the individuals begin to die for lack of nourishment. If a starving individual is inoculated into identical medium, a lag in division precedes the rising straight line; the lag represents the time the cell needs to recover from the effects of starvation (Fig. 25.6*B*). If individuals are removed from a culture in which they are about to die for lack of food and are inoculated into fresh medium, after a lag phase they resume growth and continue to multiply at the same rate as a control from a vigorous culture.

Even if food is available, microorganisms may die when other conditions become unfavorable. For example, yeast grown anaerobically produces alcohol and carbon dioxide; alcohol, as it accumulates, eventually prevents growth and division of the yeast. Similarly, lactic acid bacteria produce lactic acid in a concentration that first kills off all other microorganisms and finally the lactic acid bacteria themselves (Brock, 1966). Humans have made use of some of these special byproducts of microorganisms, for example, yeast in baking and manufacture of alcoholic beverages and lactic acid bacteria to preserve ensilage and make sauerkraut.

If microorganisms did not die for lack of food or because of other unfavorable conditions such as extremes of pH, temperature, radiations, and effects of poisons, or because of natural enemies, they would soon exhaust the nutrients of the earth through exponential growth:

$$N = 2^n \qquad (25.1)$$

In this equation N is the number of individuals, and n is the number of divisions.

Such a growth curve resembles a compound interest curve, because each increment resulting from a cell division doubles the population, which will double again at the next division. Such an exponential increase in population is said to be explosive.

Equation 25.1 is more convenient in its logarithmic form:

$$\text{Log } N = n \log 2 \qquad (25.2)$$

$$\text{or } n = \frac{\log N}{\log 2} \qquad (25.3)$$

Equation 25.3 was used in plotting the data in Figure 25.8. The equation may be put into a more generalized form:

$$N = 2^{t/\tau} \qquad (25.4)$$

where t is the time elapsed and τ the generation time (time from one cell division to the next) in the same units.

In multicellular organisms cells may be divided into two categories: *cycling* and *noncycling*. Embryonic cells divide rapidly but the division rate gradually declines. Upon reaching maturity only limited numbers of cells continue to cycle in the organism, in tissues such as the gametogenic cells in the germinal epithelium, the blood-forming cells, and the basal cells of the epidermis. In most body cells cycling

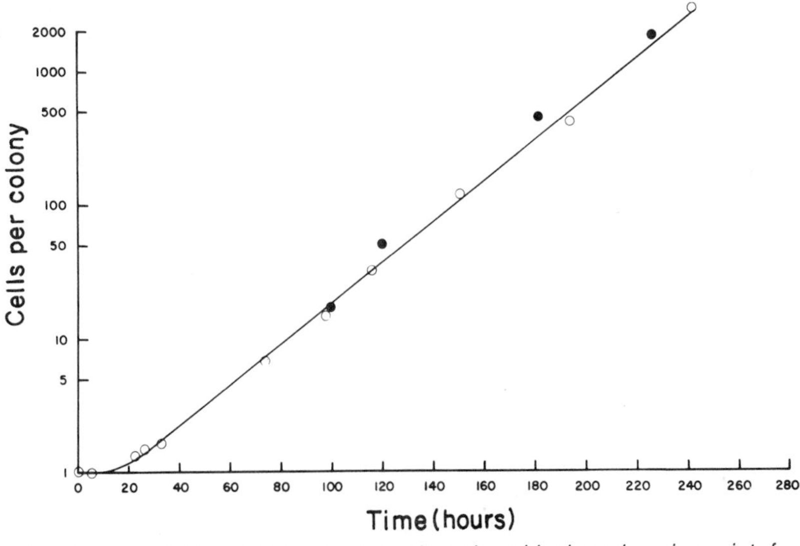

Figure 25.8. *Growth curve of HeLa cell colonies plotted on a logarithmic scale, using points from two sets of experiments (closed and open circles). The cells require about 24 hours to become established (lag phase). Then they multiply at a steady rate (logarithmic phase) that can be maintained almost indefinitely if food supply of the cells is continually renewed. Note the resemblance to Figure 25.6. The HeLa cell is from a tumor. (From Puck, 1956: J. Exp. Med.* 103: *273.)*

stops; the cells become noncycling and DNA synthesis stops. Cycling may be induced in cells that are otherwise noncycling, for example, after injury, when fibroblasts multiply to heal a wound in the skin. Or when immune reactions are induced, noncycling lymphocytes begin cycling, although only after they have passed through the G_1 phase. Cycling cells continue to go through S phase, noncycling cells do not. Noncycling lymphocytes may also be transformed into cycling cells by action of plant proteins known as lectins that induce the G_1 phase, after which the cells go into the S phase.

Although most mature and aging cells no longer grow and cycle, it is held by some on the basis of tracer experiments that biochemical replacement of macromolecules apparently continues at all times (Schoenheimer, 1942). Even such inert cell products as collagen in tendons were thought to be continually replaced. This conclusion has been questioned because tracers could also be incorporated into new cells replacing those that have died. Apparently such cell replacement takes place at a more rapid rate than previously supposed (Leblond and Walker, 1956). Nevertheless, some evidence indicates that animal cells in tissue culture kept from dividing for a long time do indeed incorporate some new material (Mazia, 1963).

Cells of aging multicellular individuals usually become somewhat dehydrated, respire more slowly (Zeuthen, 1947, 1953), show decreased enzymatic activity, and accumulate calcium, various pigments (e.g., lipofuscin granules), and other inert matter (Shock, 1962; Strehler, 1977). The inert matter may be insoluble excretory residues of metabolism incapable of passing through a cell membrane. Dehydration of the cytoplasm, resulting perhaps from decrease in solubility of proteins and loss of their water-binding capacity, is noted in aging cells of many, but not all, species. These changes could be the result of accumulation of random deleterious mutations or the consequence of failure of life-maintenance systems (Sacher, 1978).

It would be of interest in this respect to follow the molecular changes of various cellular constituents with aging of the cell. Such a comparative study has been made with collagen, which forms perhaps one third of the protein in the human body and is the major constituent in tendon and connective tissue. Experiments indicate that on heating, the tension exerted by collagen fibers from old individuals is greater than that of young individuals, and the change is progressive with age. Although the molecular basis of the change in collagen has not been fully established, it appears that cross-linking by ester linkages increases with age, conferring on collagen this particular property.

Why do cells of multicellular organisms age? It could be because the individual cells making up the body have completed their life cycle. Or it could be that the body, like an exhausted culture medium for unicellular organisms, is no longer able to provide adequately for the constituent cells. Testing with clonal tissue cultures to determine whether cells do or do not age is considered in the next section.

GROWTH OF CELLS IN TISSUE CULTURE

Cells of multicellular organisms in tissue culture, transferred periodically to fresh, controlled medium, offer a unique opportunity to study cell growth and aging processes.

Cells from various higher plant and animal tissues have been successfully cultured in the absence of other living things. In a typical culture medium for animal tissue cells, plasma serves as a substrate for cell attachment, and an extract from an embryo usually serves as nutrient, although synthetic media are now widely used, as shown for HeLa cells in Table 25.2 (Waymouth, 1965; Paul, 1970). The cells must be periodically washed free of metabolic wastes. If these conditions are met, the tissue cells are capable of living, growing, and dividing. It was first believed that they could continue dividing year after year and be alive and fully vigorous many years after the animal or plant from which the tissue cells were taken would normally have died. Thus Carrel isolated fibroblasts from a piece of chick heart and kept them alive in culture from 1913 to 1946, when the experiment was terminated by choice. However, it has been shown that new fibroblasts were added with the nourishing embryo extracts, and it was these cells that made possible seemingly indefinite multiplication of cells in tissue culture (Hayflick, 1975). Plant cells, too, have been grown in tissue culture and in completely synthetic media, and the requirements for different types of plant tissues have been determined (White, 1963; Steward, 1969).

When cells of human tissues kept in tissue culture are dissociated, clones can be grown from single cells in colonies, much like microorganisms (Puck, 1961; 1972). Such cultures are invaluable for studying many problems (Telling and Badlett, 1970). They show that human epithelial cells and fibroblasts, like those of the chick, grow rapidly and multiply if supplied with the proper nutrients and washed free of wastes (Paul, 1970; Cameron and Thrasher, 1971). Protein-free culture media have been developed for this purpose (Eagle, 1960; Katsuta and Takaoka, 1973). However, only a few cell types have been adapted to grow in defined media. In some cases a medium conditioned by previous growth of cells may serve

TABLE 25.2. A SERUM-FREE MEDIUM THAT SUPPORTS QUANTITATIVE GROWTH OF SINGLE S3-HeLa CELLS*

Components	*Grams/Liter*	*Components*	*Grams/Liter*
1. Amino Acids			
L-Arginine HCl	0.015	L-Valine	0.010
L-Histidine HCl	0.015	L-Glutamic acid	0.030
L-Lysine HCl	0.032	L-Aspartic acid	0.012
L-Tryptophan	0.008	L-Proline	0.010
β-Phenyl-L-alanine	0.010	Glycine	0.040
L-Methionine	0.010	Glutamine	0.080
L-Threonine	0.015	L-Tyrosine	0.016
L-Leucine	0.010	L-Cystine	0.003
L-Isoleucine	0.005		
2. Vitamins and Growth Factors			
Hypoxanthine	0.010	Choline	0.0012
Thiamine HCl	0.0020	Calcium pantothenate	0.0012
Riboflavin	0.00020	Niacinamide	0.0012
Pyridoxine HCl	0.00020	Inositol	0.0004
Folic acid	0.00004	Vitamin B_{12}	0.0010
Biotin	0.00004		
3. Salts and Other Small Molecules			
NaCl	7.40	$CaCl_2 \cdot 2H_2O$	0.016
KCl	0.285	$NaHCO_3$	1.20
$Na_2HPO_4 \cdot 7H_2O$	0.29	Glucose	1.10
KH_2PO_4	0.083	Phenol red	0.0012
$MgSO_4 \cdot 7H_2O$	0.154		
4. Proteins			
Normal serum albumin	2.0	Fetuin	2.0

*From Puck, T. T., 1959–1961: Quantitation of growth of mammalian cells. Harvey Lect. *55*: 1–12. A completely synthetic medium for mammalian cells has been developed for growth of one of the HeLa strains designated NCTC strain 2071, for example. The medium contains 69 crystalloidal compounds. Growth is somewhat slower than when protein, especially blood serum, is added.

for other cells not previously adapted. But for most tissue cultures serum remains a necessary constituent. Attempts have been made to isolate a growth factor that would stimulate DNA synthesis in cultured cells. Such a factor has been isolated from fibroblasts, and a nerve growth factor has been found to stimulate DNA replication in fibroblasts, chick epidermal cells (embryonic), and L 3T3 cells. Much remains to be learned about a synthetic medium for growth of mammalian tissue culture cells (Gospodarowicz and Moran, 1976; Gospodarowicz *et al.*, 1977).

It is claimed that *in vivo* the growth rate of mouse mammary epithelium declines during serial propagation, the decline being proportional to the number of cell divisions occurring (Daniel and Young, 1971). It has been stated that *in vitro* human cells in tissue culture also have a limited ability to multiply. In serial transfer of embryonic cells in tissue culture, the division rate was found to decline. A maximum of about 50 (40 to 60) population doublings was found to be the rule before the cells stopped dividing. If samples of cells are frozen at various stages after isolation from an embryo and then reconstituted in culture medium, they complete only what is left of their approximately 50 cycles of doubling (Hayflick, 1975).

Fibroblasts from animals with different life spans show a maximal number of doublings proportional to life span: 90 to 125 for a Galapagos tortoise, 40 to 60 for a human, 15 to 35 for a chicken, and 14 to 28 for a mouse. The supposed immortality of cells in tissue culture is therefore apparently based on a technical error in Carrel's chick fibroblast cultures. Yet, on the contrary, it seems that cells in adult tissues of multicellular organisms are activated to divide upon injury to the organism. Cell proliferation so induced makes possible repair of the injury and occurs to some extent in all multicellular organisms (McMinn, 1969). Perhaps such cells capable of proliferation still have a reserve of doublings left at the time they become dormant. It has been found that the number of remaining cell doublings for cells in tissue culture is inversely proportional to the age of adults from whom cells are taken (Hayflick, 1975). Most cells in the human body probably stop doubling before their reserve is exhausted, some perhaps very early. The doubling probably is held in check by division-retarding substances secreted by the tissue cells themselves or by immune mechanisms (see next section).

CELL AGING

A number of hypotheses have been suggested to account for aging of cells in multicellular organisms. One is that aging is inherent in the genome of the species, a result of the presence of "aging" genes. Another is that the cell genome runs out of genetic information for division. More appealing is a third hypothesis that the DNA code gradually accumulates inaccurate information, causing proteins to be synthesized that lack the required specificity for continued divisions. Repair systems for correcting errors in the code do exist and are of great importance in preservation of the genetic code (see Chapter 15). However, it may be that even these systems fail as damage to the genome accumulates during the lifetime of an organism, and perhaps it is the miscoding for the very repair enzymes themselves that presages aging and death of cell lines (Strehler, 1977).

The fact that the ciliate *Tetrahymena,* a eukaryotic cell, as well as many prokaryotic microorganisms have grown with unabated vigor for countless generations on fully defined media raises a doubt as to the validity of senescence (aging) in mammalian fibroblasts. Inasmuch as the culture medium for mammalian cells is still incompletely defined and adaptation of a freshly isolated line of cells is necessary for growth, it seems possible that aging of mammalian cells in culture is an artifact of the culture conditions.

Another possible explanation for supposed senescence in mammalian cell lines in culture has recently been proposed, namely, that "the finite lifespan of human fibroblasts may be due to the decline and loss of a subpopulation of immortal cells" (Holliday *et al.,* 1977). Potentially immortal cells may generate two types of cells: those with a commitment to senescence after a limited number of generations and other cells potentially immortal. Some evidence has been presented that in the process of transfer of small numbers of cells from culture to culture the cells with potential immortality may have been diluted out. Larger samples give populations with longer life spans. Senescence then, as observed in the usual culture procedures, may be an artifact of the method.

It is significant that cells in tissue culture taken from mammals with a longer life span have a more effective repair system for damage to DNA by ultraviolet radiation than those with a short life span. Cells of seven species ranging in size from shrew and mouse to man and elephant were tested. Improved repair with life span would have survival value (Hart and Setlow, 1974).

Tissue and organ cultures of plants have been maintained in periodic subcultures for years. It seems likely that plant cells in this respect resemble microorganisms rather than mammalian cells and that they are potentially immortal. In many cases whole plants can be reconstituted from a tissue culture callus. Species of plants that propagate by shoots, as exemplified by buffalo grass of the great plains of North America, are natural clones that may have persisted since the last glacial period, reemphasizing their seeming "immortality" (Leopold, 1975).

Fishes and many invertebrates (not insects) lack a definitive maximal size. They continue to grow all their lives and die as a consequence of accident or inability to cope with enemies. This facility may only mean that their cells are capable of many more doublings than mammalian, bird, or reptilian cells. Even more remarkable are trees like the ancient sequoias that may reach an age of 3000 years and bristle cone pines estimated to be 4000 years old (Kozlowski, 1971). The problem remains open.

GROWTH OF CANCER CELLS

Possibly the only way a cell of a mammal or bird can acquire immortality—in the sense of indefinitely continued doubling—is for it to become malignant or cancerous. HeLa cells, so named for the patient from whose cervical cancer they were obtained, have been grown in suspension for many years and are extensively used in research on animal cells. They quickly overrun any other cell line when allowed to contaminate a culture. Cancer cells generally grow at an exponential rate comparable to that of microorganisms, as shown in Figures 25.8 and 25.9. The cancer may remain localized to form a single tumor or it may spread by way of lymph or blood *(metastasize)* to all parts of the body. The undifferentiated cells grow and disorganize the tissues among which they become localized. Cancer cells do not appear to be

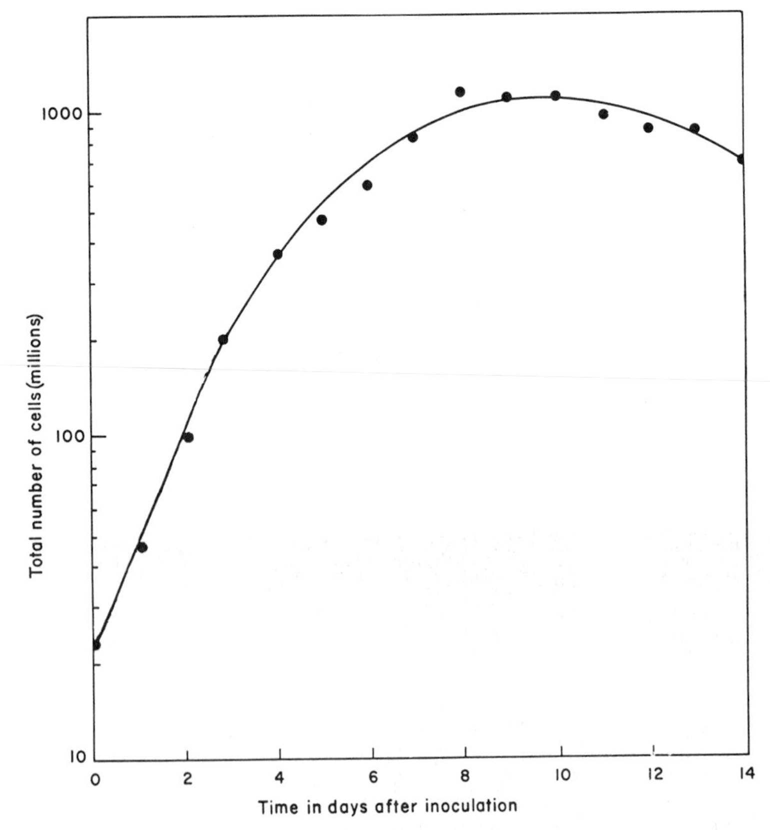

Figure 25.9. Population growth curve for cells of the Ehrlich ascites tumor grown in the mouse. This tumor is made up of individual cells growing in the body (ascitic) fluid. The initial injection on "0" days consisted of 22 × 10^6 cells. Note the initial logarithmic rise for almost 4 days, the later decline in rapid growth, and then the decrease in number of cells as some of them die, the whole curve resembling that in Figure 25.6. The mean survival time for mice with the tumor so inoculated is 14 days. Each point is the average of six determinations. (By courtesy of A. Furst and M. Dollinger, Stanford University.)

markedly different in nucleolar (Busch *et al.*, 1963) or nuclear fine structure (Kollner, 1963), or in macromolecular syntheses and biochemistry (Busch, 1962).

However, cancer cells are different from normal cells in several respects. Normal cells in tissue culture generally stop moving on making contact with one another *(contact inhibition)* and arrange themselves with respect to one another. (In mixed cell cultures they also differentiate.) Cancerous cells, on the other hand, move right on, crawling over one another, and they do not differentiate (Fig. 25.10). In fact, they may dedifferentiate to a embryonic-like state. Contact inhibition is not always observable in normal cells nor is lack of it in malignant cells, but it is rather general. Contact-inhibited cells are noncycling; those not so inhibited are cycling (Mazia, 1974). To prevent or cure cancer it will be necessary to gain a fundamental understanding of the nature of cell growth and the mechanisms by which the body normally controls growth and cell division.

In normal noncycling tissues of a full-grown mammal an equilibrium state is reached in which mitosis is suspended except in the few types of cycling cells already mentioned. These cycling cells replace only cells that are spent; therefore, they are also regulated by some feedback mechanism. When this fails in skin epidermal cells and cell divisions occur too rapidly, disease known as psoriasis results. Normally, dead epidermal cells are shed as rapidly as new ones form. In psoriasis the excess cells pile up in layers to form epidermal scales that may cause almost intolerable itching. Psoriasis is not true cancer, although it resembles this disease.

Inhibition of cell division in the normal body may be the result of tissue-specific glycoproteins called *chalones*. Chalones are not cytotoxins—their effects are presumably reversible upon removal. It has been claimed that when tissue culture cells are freed of chalones by washing, they resume mitosis and proliferation, provided they have not yet doubled about 50 times. Addition of appropriate chalones again stops mitosis. Chalones have been isolated from 20 tissue types (Bullough and Mitrani, 1976).

Two chalones have been extracted from epidermis, one that affects phase G_1 in the cell cycle and another that affects G_2. The first determines whether the cell goes into mitosis or stays in interphase. The action of the second chalone is less clear (Bullough and Mitrani, 1976). Chalones are opposed in action by the so-called mesenchyme factor that elicits mitoses in epidermis. Detailed discussion of these an-

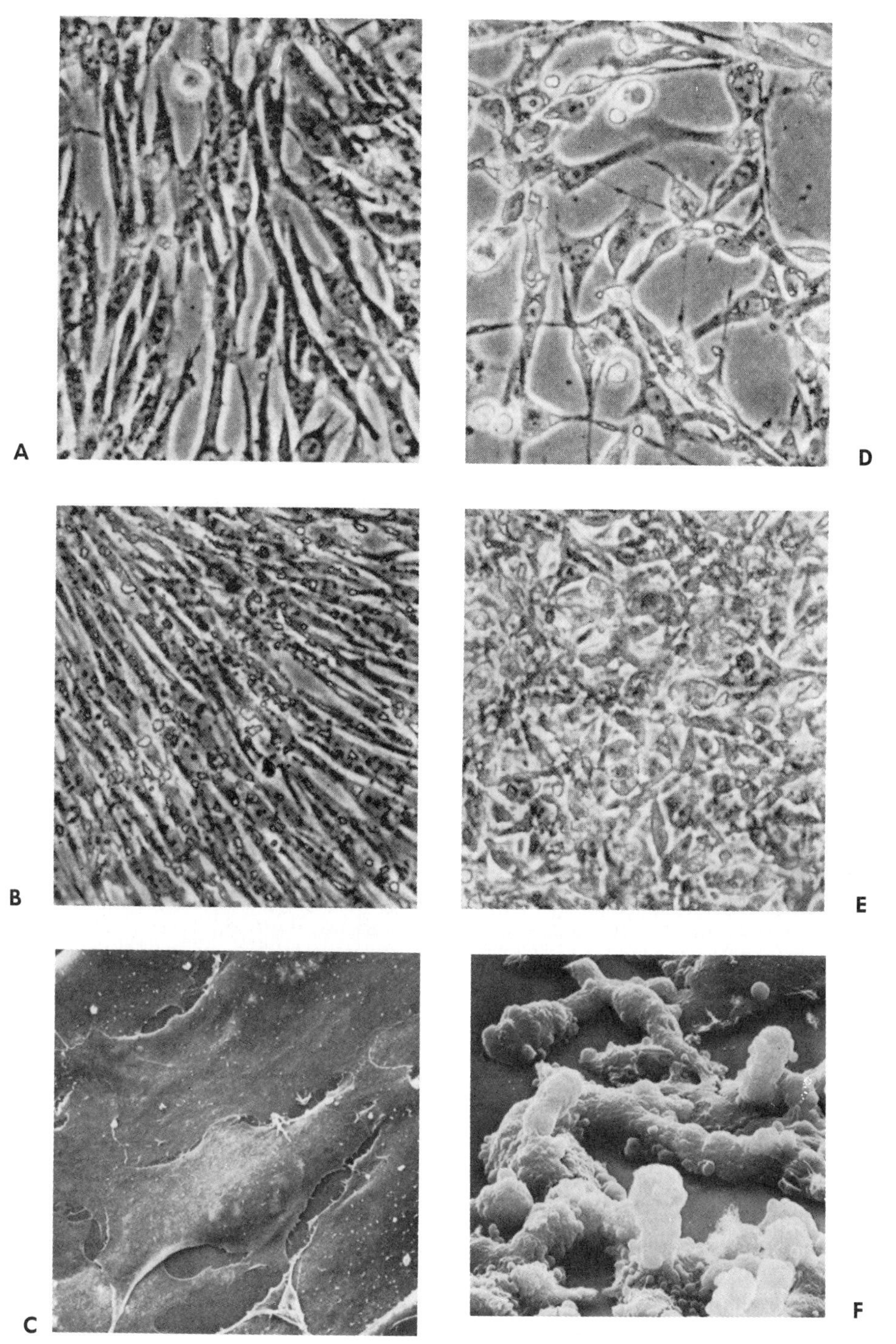

Figure 25.10. A, B, C, *normal and* D, E, F, *virus-transformed hamster cells in tissue culture.* A, B, D, E *are kidney cells (BKH strain) taken by light microscope;* C, F *are embryonic cells (LK-9 strain) taken with scanning electron microscope,* C ×*1620,* F ×*3145. Note lack of contact inhibition and overgrowth of virus-transformed (cancerous) cells. This is especially evident in* F, *where the cells pile up in contrast to the normal cells in* C *flattened against the substrate. Normal cells round up only during mitosis. Microfilaments decrease in number in virus-transformed cells, conceivably accounting for their inability to flatten. (*A, B, D, E *from Dulbecco, 1967: Induction of Cancer by Viruses. Sci. Am.* 216 *(Apr.). Copyright © 1967 by Scientific American, Inc. All rights reserved.* C *and* F *from Goldman* et al., *1974: Cold Spring Harbor Symp.* 39: *601–614.)*

tagonisms is more in the realm of physiology of organisms.

The epidermal chalones have been purified to a considerable extent. The one acting at stage G_1 has a molecular weight of 10^5 daltons, the one acting at stage G_2 has a molecular weight of 3-4 $\times 10^4$ daltons. The most highly purified preparations do not inhibit DNA synthesis (Marks, 1976). It has been suggested, without evidence to back the suggestion, that chalones release cyclic AMP. This in turn might lead to exhaustion of DNA precursors. The problem of chalone action is open.

Chalones presumably attach to specific receptor sites on cell membranes, but until chalones are purified their mechanism of action will be difficult to ascertain. Purification is difficult because chalones are present in such minute quantities. In malignant tissue less chalone is produced than in normal tissue, and that small amount is released into the blood, since it is not attached to receptor sites. Paradoxically, the blood that drains malignant tissues thus contains more chalones than the blood draining their normal counterparts (Bullough and Lawrence, 1970; Maugh, 1972). Much information on chalones and their possible action has appeared in the literature, as summarized by Houck (1976). Inhibition of growth may occur by means of immunological reactions rather than chalone action. Precise information is lacking.

At present cancer is treated by various techniques that aim only to suppress growth and division rate of the component cells. Examples are use of ionizing radiations that suppress division of cells in certain types of cancer (Hall, 1976) and selective incorporation of radioactive salts into a tissue containing cancerous cells to suppress cell division, for example, radioactive iodine (^{131}I) for thyroid tumors (see Chapter 27). Another treatment method is use of poisons that block the synthesis of metabolites necessary for growth and division of cancerous cells (Pullman and Pullman, 1955). Still other treatments are exposure to ultrasound (Li *et al.*, 1977) and to microwaves that selectively heat solid tumors with poor circulation (Hahn, 1978). Use of natural mitotic inhibitors such as chalones awaits more information on their chemical nature and availability in pure form.

Interpretation of cancer induction as it affects the cell is still lacking, in spite of many attempts. It is known that the cancerous state can be induced in mammalian cells by polyoma virus. To transform the normal cells, the circular virus DNA genome must be integrated into the genome of the cell that forms covalent linkages with the cellular DNA (Dulbecco, 1976). The surface of such cells is visibly affected (as seen in scanning electron micrographs) and lacks receptors for inhibitors that presumably stop mitosis in normal cells.* It is likely that a transforming protein produced by the altered DNA induces the cancerous state. The genome of the altered cells resembles the fetal more than the adult genome.

Quantitative studies reveal that the enzyme pattern in malignant cells resembles that in the fetus more than that in the adult, as expected from the genome change. The resultant metabolism of tumor cells thus resembles that taking place in fetal cells rather than that in cells from which they arose. This is presumably a result of dedifferentiation in tumor cells of the normal enzyme pattern, suggesting that the genes that code for the normal enzyme pattern in the adult have been affected by repressors produced by the transformed host cells (Knox, 1972). It would be interesting to know whether the enzyme pattern returns to normal (by derepression) in those plant and animal cells that revert from cancerous to normal (Braun, 1970).

The fact that glycolysis becomes emphasized in tumors is interesting. This is probably a secondary change, although it was formerly thought to be a cause for the transformation of normal cells into tumor cells. Glycolysis results from inadequacy of the vascular supply when the tumor gets large.

It now seems probable that a mutation that stabilizes the transformation (preventing reversion) is induced when a normal cell becomes cancerous. This removes the cancer cell from any controls, if such there be, which would normally repress cell multiplication in the body. As an example of a possible normal type of repression, bone marrow cells grown in tissue culture from a normal individual and from a leukemic individual are inhibited by antagonists to folinic acid, one of the B vitamins required for cell division (see Table 25.2). Within 15 minutes, division stops dramatically. Normal cells remain inhibited from dividing so long as they are exposed to the antagonist. However, leukemic cells begin to divide actively after 24 hours, although still exposed to the antagonist. During this time they appear to have inactivated the antagonist because the filtrate from this culture no longer inhibits either normal cells or leukemic cells, whereas the filtrate from a normal cell culture is still inhibitory. (The inactivated material has been separated, crystallized, and reactivated.) Apparently, leukemic cells differ from normal cells in being able to develop an enzyme system

*The plant proteins known as lectins transform noncycling lymphocytes into cycling lymphocytes. Analysis of this mechanism may throw light on the nature of carcinogenesis. Lectins react with the surface of cells showing contact inhibition causing clumping, but not with cells that do not show contact inhibition (Sharon, 1977).

capable of inactivating the folinic acid antagonist. Such enzymes are considered to be produced by genes, so it would seem likely that leukemic cells differ genetically from normal ones (Jacobson, 1958). Therefore, it is probable that transformed cells lack the receptor proteins present on the surface of normal cells; thereby they are removed from control by regulatory substances in the body. Even the microvilli observable with the scanning microscope on the surface of normal cells are altered by the development of malignancy (Spring-Mills and Elias, 1975).

Still another interpretation of carcinogenesis is release of "hidden" viruses. It will be recalled that lysogenic viruses remain inactive in nonpermissive bacterial cells unless the cell is damaged. But the hidden viruses do produce a change in the host cell because such bacteria are not infected by the active or lytic variety of the same type of virus. It is conceivable that cells become leukemic when invaded by viruses that make them immune to the various controls that the body normally exerts on noninfected cells. Other agents, for example, carcinogens, may activate a transformation to malignancy by action upon the hidden virus, perhaps by mutation in the virus, since most carcinogens can be converted to known mutagens by a metabolic change in the host cell.

Chromosomal abnormalities also accompany the change in cells from normal to malignant in some cases. The situation is complex and is reviewed elsewhere (German, 1974). As an example, in chronic myeloid leukemia one of the somatic chromosomes (No. 22) is abnormal and multiplies with progression of the disease. Chromosome abnormality implies genetic imbalance. Whether a virus is also involved is not stated.

Two striking changes have been noticed in virus-infected cells. Like cancerous cells they lack the contact inhibition that is characteristic of normal cells in tissue culture, and they do not stop dividing as do normal cells. This indicates not only a change in surface properties observable under scanning electron micrographs but also a signal for replication (Dulbecco, 1976).

Virus infection provides a favorable opportunity to study the genetic changes that convert a normal cell into a cancer cell because the number of genes present in most such viruses is small, perhaps nine or ten in all. An attempt has been made to identify all the viral genes and then to learn which of these genes direct cancer induction. In the polyoma virus enough DNA (about 5000 base pairs) is present to code for perhaps eight or nine proteins. Probably one or two code for the protein in the viral coat. Six functions, each arising in the host cell after infection and each coded by one gene in the virus, have now been studied in the infected cells. These functions, which involve changes in biochemical or immunological properties, probably depend upon proteins produced by these genes. None of these functions has yet been specifically related to the two cancerous properties that were observed in virus-infected tissue culture cells: loss of contact inhibition and uncontrolled multiplication (Dulbecco, 1967, 1976).

It is recognized that possibly mutation without observable virus infection may occur in a cell of an animal or a plant. Such a mutation could result in a change to a cancerous condition by releasing the cell from the regulation that normally controls cell division. However, it is difficult to identify such a genetic change because the numerous genes in the animal cell genome make analysis immensely more complicated than in the virus.

In some cases, reversal of the malignant condition of cells in plants and animals has raised the question of whether the original carcinogenesis in these cases might not have been an altered expression of a gene rather than a change in genome. Thus, the crown gall tumor, a stem enlargement presumably caused by *Agrobacterium tumefaciens,* can be reversed by a series of transfers of tumor cells to healthy plants. Reversion of tumors has also been observed in both newts and mice (Braun, 1965; 1970).

A variety of evidence implicating viral and genetic origins of cancer has accumulated, but none has yet been found that would permit a successful analysis of the basic change involved. Cancer remains an important problem to the cell physiologist (Knox, 1972; Dulbecco, 1976).

VASCULARIZATION OF TUMORS

A tumor that does not become vascularized does not become malignant and cannot grow beyond a relatively small size. For example, tumorous tissue placed in the anterior chamber of a rabbit eye fails to develop because it is too far removed from capillaries that might vascularize it. Placed nearer blood vessels, it quickly induces profuse vascularization and grows to a large size by exponential cell growth and multiplication. The tumor's vasculogenic substance induces development of capillaries even through a Millipore filter with holes too small to permit passage of cells but big enough to allow passage of molecules. In this manner a tumor induces vascularization. This occurs even through a tissue basement membrane of tough connective tissue fibers that separates, for example, the bloodless epidermis from the vascularized dermis of the skin and the bloodless lining of various glands from the dermis in which the glandular epithelium is suspended (Folkman, 1976).

Because tumor cells acquire greater malignancy with the passage of cell generations and vascularization promotes cell division, vascularization hastens the development of malignancy and the spread of tumors through blood and lymph.

REGULATION OF FORM AND SIZE IN CELLS

The ability of a cell to maintain its organization and characteristic size at a given temperature and in the presence of adequate nutrients is perhaps best shown by the regeneration of protozoans after removal of parts of the cell (Fig. 25.11). In fact, if the entire hypostome (mouth area) is cut off, it is regenerated, enabling a *Blepharisma,* for example, to feed within 6 hours of the original operation. If large portions of the cell are removed, but macronuclear fragments remain, regeneration will result in a miniature edition of the species. Through transcription the macronucleus supplies messenger RNA to direct protein synthesis required to reconstitute the structures removed by cutting. Agents that stop DNA-dependent RNA synthesis stop regeneration (Giese, 1973). After feeding, the regenerated organism rapidly regains its original size. Redifferentiation is usually restricted to portions of the cell adjoining those removed by microsurgery; however, even then, substances required for regeneration must be obtained by dissolution of existent structures. In some ciliates, amazingly enough, even a small injury results in dedifferentiation of the entire cell and redifferentiation of all the organelles (such as membranelles, motile organs, and mouth parts). It is also an indication of the regulation of form that when *Stentor* is minced into about 40 adherent pieces, the pieces reaggregate to form a normal individual (Tartar, 1960). As a result of such morphogenetic activity, organelles that have been removed or damaged are regenerated, and the characteristic form and size of the species are reconstituted (Tartar, 1967). Since only the macronucleus is required for regeneration, all the information is present in the highly redundant genome of the macronucleus. Regeneration offers an excellent opportunity for study of cell growth uncomplicated by division (Suzuki, 1973).

When pieces are removed from eggs of some chordates and echinoderms, the eggs produce complete embryos of reduced size. This shows that the egg, like the unicellular organism, maintains a certain organization. When pieces are removed from the eggs of annelids or mollusks, the eggs develop into partial embryos only. In such eggs the pattern is supposedly already laid down, and each part is predestined to form only a particular organ.

It is probable that most vertebrate cells can recover from minor injuries, although experi-

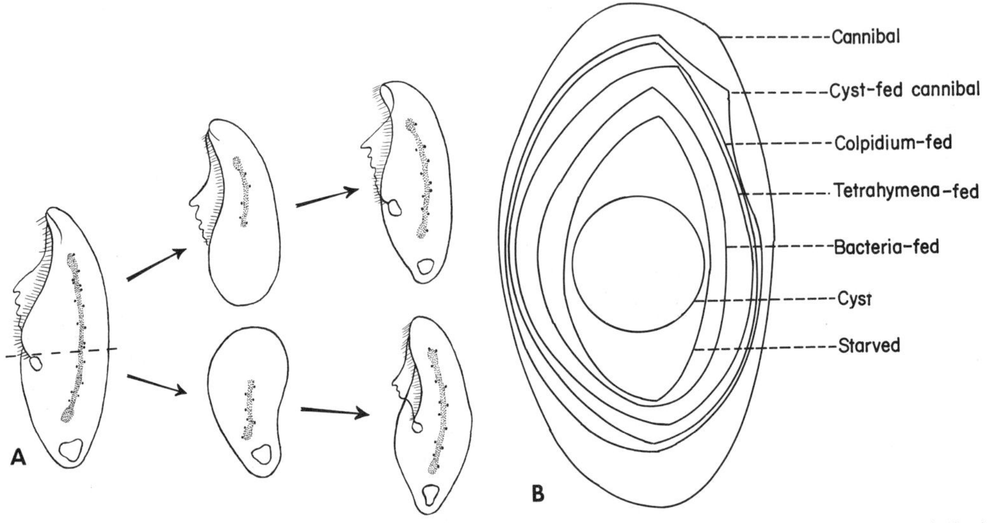

Figure 25.11. *A, Regeneration of missing parts in two pieces of a bisected* Blepharisma. *Note especially the regeneration of the peristome (mouth region) in the posterior fragment. Only immediate and end results are shown here. A nuclear cycle also occurs after cutting, the macronucleus condensing before it elongates again and grows in size. The micronuclei (of which there are many) multiply also. Regeneration takes about 5 to 6 hours at room temperature. (From a classroom experiment. For details see Moore, 1924: J. Exp. Zool. 39: 249.) B, Change in size of the ciliate* Stylonychia curvata *with diet.* Tetrahymena *and* Colpidium *are ciliates; the bacteria used were* Pseudomonasovalis. *All at 20°C in balanced salt medium. Cannibals divided more rapidly than controls on bacteria.*

ments on regeneration in vertebrate cells have been limited. The neuron of a vertebrate, for example, regenerates damaged axons or dendrites, provided that the cell body containing the nucleus is intact.

The underlying nature of cell regeneration, redifferentiation, and also cellular interactions leading to differentiation in development and morphogenesis is not known. In the slime mold *Dictyostelium* it has been shown that the chemical attractant (acrasin) causing the processes of cellular interaction, aggregation of the ameboid cells to form the plasmodium, and differentiation of fruiting stalks is cyclic AMP (Bonner, 1971).

Some ciliates, such as *Stylonychia* and *Blepharisma,* both ordinarily feeders on bacteria, may be fed protozoans of successively increasing size, ultimately including their own kind (cannibalism). As a result, they grow to many times the bulk of bacteria-fed individuals. The mouth parts, membranes, motile organs, macronuclei, and other structures become proportionally enlarged, and the large size is maintained so long as the abnormal diet is continued. In all cases, however, when such giants are returned to a bacterial diet, division is more rapid than in controls given bacteria, so that a return to the characteristic species size is soon accomplished (Fig. 25.11*B*) (Giese, 1938; Giese and Alden, 1938).

Conversely, ciliates may be starved for long periods, during which they become much reduced in size. *Tetrahymena,* for example, will become a minute particle with little resemblance to its former self. Yet on addition of nutrients it quickly enlarges to its normal size. When it has doubled its volume it divides, thus maintaining the normal species size.

That cell size is regulated by division is most clearly demonstrated with *Amoeba proteus* (Prescott, 1956a and b). During growth of an amoeba the mass increases until it just doubles. Then no further increase occurs, and after a pause it divides, each daughter cell receiving half the mass of the mother cell. If a part of an amoeba is removed during the period of growth, division occurs only after the mass of protoplasm removed has been replaced by growth. An amoeba may be prevented from dividing, presumably indefinitely, by the periodic removal of a piece of protoplasm, although some evidence exists to the contrary (Hirshfield *et al.,* 1960).

Heredity plays an important role in the maintenance of size and shape of cells, since clones of a species of microorganisms under a given set of environmental conditions tend to be quite constant in size and shape for numerous generations and a constant ratio of nucleus to cytoplasm is maintained. However, overall size in a clone varies with changes in environmental factors (Adolph, 1931).

Experiments have shown that increasing the number of chromosomes by polyploidy is attended by an increase in cell size (Fankhauser, 1952). If the number of chromosomes is increased without decrease in their size, the amount of chromatin in the polyploid nucleus is correspondingly increased. The nucleus enlarges proportionally and the cytoplasm compensates. A polyploid salamander, for example, has giant cells. The same is true of cells in mosses that are usually haploid but may be experimentally induced to become diploid, triploid, or tetraploid. Wherever found, polyploidy is not always correlated with increase in cell size (Brodsky and Uryvaeva, 1977). Vertebrate tissue culture cells become giant cells when they are exposed to an x-ray dose that stops division but permits replication and growth to continue (Hall, 1976).

In plant cells, growth might appear to be limited by the rigidity of the cell walls. However, the cell wall is rendered soft and plastic by the plant growth hormone auxin (indole acetic acid), as occurs to undifferentiated cells in the region of growth. For example, when illuminated on one side only, plants turn toward the light. Auxin accumulates on the side away from the light, inducing extra elongation of young cells in the region of growth; therefore, curvature in response to light is a result of increase in cell size and not cell multiplication. In darkness application of auxin to one side of a plant also results in curvature. Whenever plant cells increase in volume after cell division, the cell walls soften but the action of auxin is less evident than in tropistic responses to light and gravity. During embryogeny plant cells are small and lack a central sap vacuole. They enlarge and differentiate as the vacuole develops. It is conceivable that plant cell growth regulation may be controlled in part by the auxin supply.

UNBALANCED GROWTH

When an *E. coli* mutant that requires the pyrimidine base thymine for replication is grown in a medium deficient only in thymine, growth (as measured by turbidity) apparently occurs at essentially the normal rate. However, when this culture is grown in a thymine-deficient medium for a period equivalent to one normal generation and is then seeded on a complete medium including thymine, about 90 percent of the individuals fail to produce colonies. Injury therefore appears to have occurred when replication was attempted in the absence of thymine. If grown on a thymine-deficient medium for two generations, about 99 percent fail to form colonies upon transplantation to a complete medium. Tests show that protein and RNA increase in the cells, but only a slight increase occurs in DNA (Cohen and Barner, 1954). Such disproportionate growth of different cell constituents is known as *unbalanced growth.* A similar phenomenon has been observed in yeast (Ridgeway and Douglas, 1958) and *Neurospora* (Strauss, 1958). However, it does not appear to be general, since some *Neurospora* cells recover when returned to a complete medium even after becoming much enlarged.

Studies on repair mechanisms in cells (see Chapters 15 and 27) have indicated that during the course of replication single-strand breaks occur in the double-stranded DNA as it opens before the growing point. Such breaks are readily repaired, provided that the nucleoside triphosphates and the necessary enzymes are present. But if thymine is withdrawn from the medium of a thymineless mutant of *E. coli* and DNA replication is attempted, the DNA molecule breaks and cannot be repaired because of the lack of the nucleoside containing thymine. Consequently the cell cannot replicate, the chromosome becomes fragmented, and the cell dies (Hanawalt, 1966). A similar course of events may occur in other cases as well. When recovery occurs after growth in deficient media it is likely that postreplication repair has occurred (see Chapter 15).

However, when a mutant requiring some nutrient common to synthesis of both RNA and DNA is kept on a medium deficient in the nutrient, very little growth occurs; normal growth resumes when the cells are placed in a complete medium containing the nutrient common to synthesis of RNA and DNA (Strauss, 1958). Presumably injury occurs only when growth is unbalanced, that is, when most constituents can increase in quantity but one does not.

It might be expected that any factor other than limitation of nutrients that leads to unbalanced growth might injure the cell. DNA synthesis is one of the processes of the cell that is most susceptible to ultraviolet insult. Experiments with ultraviolet radiations show that the irradiated cells in which only DNA synthesis is inhibited may continue to grow (as measured by optical density) and individual cells accumulate RNA and proteins. Plated out, however, they fail to produce colonies; that is, they are unable to divide (Cohen and Barner, 1954). It is argued that unbalanced growth in absence of DNA synthesis has so deranged the cell that it cannot reproduce. If, however, the cells are irradiated sufficiently to stop DNA synthesis and then kept for several normal cell generations in nonnutrient solution so that they cannot grow, addition of required nutrients restores balanced growth. These are the very conditions under which DNA dark repair is most effective. The cells thus recover the ability to synthesize DNA and growth is normal (Setlow, 1964; Hanawalt, 1972).

REGULATION OF GROWTH

Balanced growth is a steady-state condition in which every cell component doubles each cell generation. The generation time depends upon the species and the environment. Thus, *E. coli* on a medium containing water, glucose, ammonium chloride, potassium basic phosphate, sodium sulfate, and trace metals, kept at 37°C, divides about every 40 minutes. If supplied with amino acids and some other organic substances at 37°C, it divides every 20 minutes. The cells have thus reached a new steady state in which duplication of cell constituents occurs more rapidly. Division is considerably retarded at 15°C at which temperature a cell gives rise to only about 128 cells after 24 hours, whereas at 37°C the number would be about 2×10^9.

Cell growth remains a challenging problem for experimental study. The question at hand is to determine how a cell achieves balanced growth regulation. Since growth is a result of metabolic activity along various fronts, the questions really asked are: How is metabolic activity regulated? How are the various types of metabolic activity coordinated?

The induction of enzyme synthesis and the repression of enzyme synthesis have already been discussed (see Chapter 16). End-product inhibition is a means of regulating many of the steps in synthesis, producing immediate changes in rate of reactions and thus preventing accumulation of cell products. In Figure 16.2 the interplay between the synthesis and end-product inhibition of various substances needed for the synthesis of DNA and RNA is considered. The net effect is a proportionality between synthesis of the two important nucleic acids. A similar interplay in other synthetic pathways perhaps leads to production of useful amounts of proteins and other substances needed by the cell. Thus, a balanced synthesis may be achieved in the amounts of the various requirements for cell growth, thereby achieving the steady state necessary for balanced growth.

LITERATURE CITED AND GENERAL REFERENCES

Adelman, R. C. and Britton, G. W., 1975: The impaired capability for biochemical adaptation during aging. Bioscience *25:* 639–643.

Adler, W. A., 1975: Aging and immune function. Bioscience *25*: 652–657.

Adolph, E. F., 1931: The Regulation of Size as Illustrated in Unicellular Organisms. Charles C Thomas, Springfield, Ill.

Attallah, A. M., 1976: Regulation of cell growth *in vitro* and *in vivo;* point/counterpoint. *In* Chalones. Houck, ed. American Elsevier, New York, pp. 141–172.

Bell, G. I., 1976: Models of carcinogenesis as an escape from mitotic inhibitors. Science *192:* 569–572.

Berenblum, I., 1975: Carcinogenesis as a Biological Problem. American Elsevier, New York.

Bonner, J. T., 1971: Aggregation and differentiation in the cellular slime molds. Ann. Rev. Microbiol. *25*: 75–92.

Braun, A. C., 1965: The reversal of tumor growth. Sci. Am. (Nov.) *213:* 75–83.

Braun, A. C., 1970: On the origin of the cancer cells. Am. Sci. *58:* 307–320.

Braun, A. C., 1974: Biology of Cancer. Addison-Wesley Publishing Co., Reading, Mass.

Brock, T. D., 1966: Principles of Microbial Ecology. Prentice-Hall, Englewood Cliffs, N.J.

Brodsky, W. A. and Uryvaeva, I. V., 1977: Cell polyploidy: its relation to tissue growth and function. Int. Rev. Cytol. *50:* 273–332.

Bullough, W. S. and Laurence, B., 1970: Chalone control of mitotic activity in sebaceous glands. Cell Tissue Kinet. *3:* 291–300.

Bullough, W. S. and Mitrani, F., 1976: An analysis of the epidermal chalone control mechanism. *In* Chalones. Houck, ed. American Elsevier, New York.

Busch, H., 1962: Biochemistry of the Cancer Cell. Academic Press, New York.

Busch, H., Byvoet, P. and Smetana, K., 1963: The nucleolus of the cancer cell: a review. Cancer Res. *23:* 313–338 (401 references).

Cameron, I. L. and Thrasher, J. D., eds., 1971: Cellular and Molecular Renewal in the Mammalian Body. Academic Press, New York.

Cohen, S. S. and Barner, H. D., 1954: Studies on unbalanced growth in *Escherichia coli.* Proc. Nat. Acad. Sci. U.S.A. *40:* 885–893.

Croce, C. M. and Koprowski, H., 1978: The genetics of human cancer. Sci. Am. (Feb.) *238:* 117–125.

Cutler, E. G., ed., 1976: Cellular Aging: Concepts and Mechanisms. Karger, Basel.

Daniel, C. W. and Young, L. J. T., 1971: Influence of cell division on an aging process: Life span of mouse mammary epithelium during serial propagation *in vivo.* Exp. Cell Res. *65:* 27–32.

Daudel, P. and Daudel, R., 1966: Chemical Carcinogenesis and Molecular Biology. Wiley-Interscience, New York.

Donache, W. D., Jones, N. C. and Teather, R., 1973: The bacterial cell cycle. Symp. Soc. Gen. Microbiol. *23:* 9–44.

Dulbecco, R., 1967: The induction of cancer by viruses. Sci. Am. (Apr.) *216:* 28–37.

Dulbecco, R., 1976: From the molecular biology of oncogenic DNA viruses to cancer. Science *192:* 437–440.

Eagle, H., 1960: The sustained growth of human and animal cells in a protein-free environment. Proc. Nat. Acad. Sci. U.S.A. *46:* 427–432.

Eagle, H., 1963: Population density and nutrition of cultured mammalian cells. *In* The General Physiology of Cell Specialization. Mazia and Tyler, eds. McGraw-Hill Book Co., New York, pp. 151–170.

Edelman, G. M., 1976: Surface modulation in cell recognition and cell growth. Science *192:* 218–226.

Elliott, A. M. and Hayes, R. E., 1953: Mating types in *Tetrahymena.* Biol. Bull. *105:* 269–284.

Evans, V. J., Bryant, J. C., Kerr, H. A. and Schilling, E. L., 1964: Chemically defined media for cultivation of longer-term cell strains from four mammalian species. Exp. Cell Res. *36:* 439–474.

Fankhauser, G., 1952: Nucleo-cytoplasmic relations in amphibian development. Int. Rev. Cytol. *1:* 165–193.

Finch, C. E. and Hayflick, L., 1977: Handbook of the Biology of Aging. Van Nostrand Reinhold Co., New York.

Folkman, J., 1976: The vascularization of tumors. Sci. Am. (May) *234:* 59–73.

Frei, E., III and Freireich, E. J., 1964: Leukemia. Sci. Am. (May) *210:* 88–96.

Gamborg, D. L. and Wetter, L. R., 1975: Plant Tissue Culture Methods. Nat. Res. Council. Saskatoon, Canada.

German, O. J., ed., 1974: Chromosomes and Cancer. John Wiley & Sons, New York.

Giese, A. C., 1938: Cannibalism and gigantism in *Blepharisma.* Trans. Am. Microsc. Soc. *57*: 245–255.

Giese, A. C., 1946: A simple method for division rate determination in *Paramecium.* Physiol. Zool. *18:* 158–161.

Giese, A. C., 1973: Blepharisma—the Biology of a Light-Sensitive Protozoan. Stanford University Press, Stanford, Calif.

Giese, A. C., and Alden, R. H., 1938: Cannibalism and giant formation in *Stylonychia.* J. Exp. Zool. *78:* 17–134.

Goldman, R. D., Chang, C. and Williams, J. F., 1974: Properties and behavior of hamster embryo cells transformed by human adenovirus Type 5. Cold Spring Harbor Symp. *39*: 601–614.

Gospodarowicz, D. and Moran, J. S., 1976: Growth factors in mammalian cell cultures. Ann. Rev. Biochem. *45:* 531–558.

Gospodarowicz, D., Weseman, J., Moran, J. S. and Lindstrom, J., 1977: Effect of fibroblast growth factor on the division and fusion of bovine myeloblasts. J. Cell Biol. *70*: 395–405.

Grinell, F., 1978: Cellular adhesiveness and extracellular substrate. Int. Rev. Cytol. *53*: 65–144.

Hahn, G. M., 1978: The use of microwaves for the hypothermic treatment of cancer: advantages and disadvantages. *In* Photochem. Photobiol. Rev. Smith, ed. Vol. 3, pp. 277–301.

Hall, E. J., 1976: Radiations and Life. Pergamon Press, New York.

Hall, R. P., 1967: Nutrition and growth of protozoa. *In* Research in Protozoology. Vol. 1. Chen, ed. Pergamon Press, New York.

Hanawalt, P. C., 1966: The ultraviolet radiation sensitivity of bacteria: its relation to the DNA replication cycle. Photochem. Photobiol. *5:* 1–12.

Hanawalt, P. C., 1972: Repair of genetic material in living cells. Endeavour *113:* 83–87.

Hart, R. W. and Setlow, R. B., 1974: Correlation between DNA excision-repair and life-span in a number of mammalian species. Proc. Nat. Acad. Sci. U.S.A. *71*: 2169–2173.

Hayflick, L., 1975: Cell biology of aging. Bioscience *25:* 629–637.

Heidelberger, C., 1975: Chemical carcinogenesis. Ann. Rev. Biochem. *44:* 79–126.

Hillman, W. S., 1962: The Physiology of Flowering. Holt, Rinehart & Winston, New York.

Hirshfield, H. I., Tulchin, H. and Fong, B. A., 1960: Regeneration and cell division in two protozoan species. Ann. N. Y. Acad. Sci. *90:* 523–528.

Holliday, R., Huschtscha, L. I., Tarrant, G. M. and Kirkwood, T. B. L., 1977: Testing the commitment theory of cellular aging. Science *198:* 366–372.

Houck, J., ed., 1976: Chalones. American Elsevier, New York.

Howard, A. and Pelc, S. C., 1953: Synthesis of DNA in normal and irradiated cells and its relation to chromosome breakage. Heredity *6*: Suppl. 261–273.

Jacobson, W., 1958: The toxic action of drugs on the bone marrow. *In* A Symposium on the Evaluation of Drug Toxicity. Walpole and Spinks, eds. Little, Brown & Co., Boston, pp. 76–101.

Katsuta, H. and Takaoka, T., 1973: Cultivation of cells in protein-free and lipid-free synthetic media. *In* Methods in Cell Biology. Vol. 6, Prescott, ed. Academic Press, New York, pp. 1–42.

Kelner, A., 1953: Growth, respiration and nucleic acid synthesis in ultraviolet-irradiated and in photoreactivated *Escherichia coli*. J. Bacteriol. *65:* 252–262.

Knox, W. E., 1972: The protoplasmic pattern of tissues and tumors. Am. Sci. *60:* 480–488.

Kollner, P. C., 1963: The nucleus of the cancer cell. Exp. Cell Res. *9*(Suppl. 90): 3–14.

Kozlowski, T. T., 1971: Growth and Development of Trees. Vol. 1. Academic Press, New York.

Leblond, C. P. and Walker, B. E., 1956: Renewal of cell populations. Physiol. Rev. *36:* 255–276.

Leopold, A. C., 1975: Aging, senescence, and turnover in plants. Bioscience *25*: 659–662.

Levi-Montalcini, R., 1968: Nerve growth factor. Physiol. Rev. *48:* 535–569.

Li, G. C., Hahn, G. M. and Tolmach, L. J., 1977: Cellular inactivation by ultrasound. Nature *235*: 2188–2200.

Maramarosch, K., 1976: Invertebrate Tissue Culture. Academic Press, New York.

Marks, F., 1976: The epidermal hormones. *In* Chalones. Houck, ed. American Elsevier, New York, pp. 173–227.

Maugh, T. H., II, 1972: Chalones: chemical regulators of cell division. Science *176:* 1907–1908.

Maugh, T. H. II, and Murphy, D. G., eds., 1978: Chemical carcinogens: the scientific basis for regulation. Science *201*: 1200–1205.

Mazia, D., 1963: Synthetic activities leading to mitosis. J. Cell. Comp. Physiol. *62:* 123–140.

Mazia, D., 1974: The cell cycle. Sci. Am. (Jan.) *230:* 55–64.

McMinn, R. M. H., 1969: Tissue Repair. Academic Press, New York.

Miller, O. J., 1974: Cell hybridization in the study of the malignant process including cytogenetic effects. *In* Chromosomes and Cancer. German, ed. John Wiley & Sons, New York, pp. 521–563.

Mitchison, J. M., 1961: The growth of single cells. III. *Streptococcus faecalis*. Exp. Cell Res. *22*: 208–225.

Mitchison, J. M., 1963: Pattern of synthesis of RNA and other cell components during the cell cycle of *Schizosaccharomyces pombe*. J. Cell. Comp. Physiol. *62*(Suppl. 1): 1–13.

Mitchison, J. M., 1971: The Biology of the Cell Cycle. Cambridge University Press, Cambridge.

Mitchison, J. M., 1973: The cell cycle of a eukaryote. Symp. Soc. Gen. Microbiol. *23:* 189–208.

Nichols, W. W., ed., 1977: Senescence. Dominant or Recessive in Somatic Cell Crosses? Plenum Press, New York.

Nicolson, G. L., 1978a: Cell and tissue interactions leading to malignant tumor spread (metastasis). Am. Zool. *18:* 71–80.

Nicolson, G. L., 1978b: Experimental tumor metastasis: characteristic organ specificity. Bioscience *28:* 441–447.

Padilla, G. M., Whitson, G. L. and Cameron, I. L., 1969: Cell Cycles: Gene-Enzyme Interaction. Academic Press, New York.

Paul, J., 1970: Cell and Tissue Culture. 4th Ed. Williams & Wilkins, Baltimore.

Pollack, R., ed., 1975: Readings in Mammalian Cell Culture. Cold Spring Harbor Laboratory, New York.

Prescott, D. M., 1955: Relations between cell growth and cell division. I. Reduced weight, cell volume, protein content and nuclear volume of *Amoeba proteus* from division to division. Exp. Cell Res. *9:* 328–337.

Prescott, D. M., 1956a: Relation between cell growth and cell division. II. The effect of cell size on cell growth rate and generation time in *Amoeba proteus*. Exp. Cell Res. *11:* 86–94.

Prescott, D. M., 1956b: Relation between cell growth and cell division. III. Changes in nuclear volume and growth rate and prevention of cell division in *Amoeba proteus* resulting from cytoplasmic amputations. Exp. Cell Res. *11:* 94–98.

Prescott, D. M., 1976a: The cell cycle and control of cellular reproduction. Adv. Genet. *18:* 99–177.

Prescott, D. M., 1976b: Reproduction of Eukaryotic Cells. Academic Press, New York.

Puck, T. T., 1961: Quantitation of growth of mammalian cells. Harvey Lect. *55:* 1–12.

Puck, T. T., 1972: The Mammalian Cell as Microorganism. Holden-Day, Inc., San Francisco.

Pullman, A. and Pullman, B., 1955: Electronic structure and carcinogenic activity of aromatic molecules. Advances Cancer Res. *3:* 117–169.

Racker, E., 1972: Bioenergetics and the problem of tumor growth. Am. Sci. *60:* 56–63.

Ridgeway, G. J. and Douglas, H. C., 1958: Unbalanced growth of yeast due to inositol deficiency. J. Bacteriol. *76:* 163–166.

Sacher, G. A., 1978: Longevity and aging in vertebrate evolution. Biosci. *28*: 497–501.

Sanford, K. N., 1965: Malignant transformation of cells *in vitro*. Int. Rev. Cytol. *18:* 249–311.

Schoenheimer, R., 1942: The Dynamic State of Body Constituents. Harvard University Press, Cambridge, Mass.

Searle, C. E., 1976: Chemical Carcinogens. Am. Chem. Soc. Monogr. 173.

Setlow, R. E., 1964: Physical change and mutagenesis. J. Cell. Comp. Physiol. *64*(Suppl. 1): 51–68.

Sharon, N., 1977: Lectins. Sci. Am. (Jun.) *236:* 108–119.

Sheldrake, A. R., 1974: The aging, growth and death of cells. Nature *250:* 381–385.

Shock, N. W., 1962: The physiology of aging. Sci. Am. (Jan.) *206:* 100–110.

Spring-Mills, E. and Elias, T. J., 1975: Cell surface differences in ducts from cancerous and noncancerous human breasts. Science *188:* 947–948.

Steward, F. C., ed., 1969: Plant Physiology. Vol. 5B. Analysis of Growth: The Responses of Cells and Tissues in Culture. Academic Press, New York.

Strauss, B. S., 1958: Cell death and "unbalanced growth" in *Neurospora.* J. Gen. Microbiol. *18:* 658–669.

Strehler, B. L., 1977: Time, Cells and Aging. Academic Press, New York.

Suzuki, S., 1973: Morphogenesis. *In* Blepharisma, The Biology of a Light Sensitive Protozoan. Giese, ed. Stanford University Press, Stanford, Calif.

Tartar, V., 1960: Reconstitution of minced *Stentor coeruleus.* J. Exp. Zool. *144:* 187–207.

Telling, R. C., and Badlett, P. J., 1970: Large scale cultivation of mammalian cells. Adv. Appl. Microbiol. *13:* 91–119.

Timiras, P. S., 1978: Biological perspectives on aging. Am. Sci. *66*: 605–613.

Verzár, A., 1963: The aging of collagen. Sci. Am. (Apr.) *208:* 104–114.

Walford, R. L., 1969: The Immunologic Theory of Aging. Munksgaard, Copenhagen.

Warburg, O., 1956: Origin of cancer cells. Science *123:* 309–314.

Waymouth, C., 1965: Construction and use of synthetic media. *In* Cells and Tissues in Culture. Willmer, ed. Academic Press, New York, pp. 99–142.

Weismann, A., 1904: Vortrage über Deszendenztheorie. G. Fischer, Jena.

White, P. R., 1963: The Cultivation of Animal and Plant Cells. 2nd Ed. Ronald Press Co., New York.

Willmer, E. N., ed., 1965: Cells and Tissues in Culture. Academic Press, New York.

Woolhouse, H. W., 1978: Senescence processes in the life cycle of flowering plants. Bioscience *28:* 25–31.

Zeuthen, E., 1947: Body size and metabolic rate in the animal kingdom. Compt. Rend. Lab. Carlsberg Series Chim. *26:* 17–161.

Zeuthen, E., 1953: Oxygen uptake as related to body size in organisms. Quart. Rev. Biol. *28:* 1–12.

Zimmerman, A. M., Padilla, G. M. and Cameron, I. L., eds., 1973: Drugs and the Cell Cycle. Academic Press, New York.

CHAPTER 26

CELL DIVISION

It is always exciting to watch cell division, especially when the division is synchronous, as in a batch of freshly fertilized sea urchin eggs under a microscope. The elongation of spindles in the eggs and the development of cleavage furrows are dramatic events beautiful to observe; the separation of the daughter cells climaxes a long series of cytological and biochemical events that form the substance of this chapter.

Much has been learned about cell division but it still is one of the less developed areas of cell physiology. Many of the remaining problems are evident in this chapter.

TIME SPAN OF THE CELL CYCLE

The cell cycle or generation time from one cell division to the next varies with species and environmental conditions. *E. coli* at 37°C and supplied with water, glucose, ammonium chloride, potassium basic phosphate, sodium sulfate, and trace metals divides every 40 to 60 minutes; but if, in addition, amino acids and other useful organic substances are added, it divides every 20 minutes. Some other bacteria living in concentrated brines (halophiles) (see Chapter 4) may divide only twice a day.

Eggs of marine animals from warm climates may divide every 30 minutes, and those from cooler climates may divide every hour. However, such cells have undergone a period of growth in the ovary, and they have supplies contained within themselves that will serve for many cell divisions before additional nutrients are needed. Protozoan cells divide somewhat more slowly; *Tetrahymena pyriformis,* under favorable conditions, divides every 3 to 4 hours and *Paramecium aurelia* every 6 hours. Cells of higher plants and animals divide once or twice a day. For example, the generation time for cells of the broad bean *Vicia faba* is 14.3 hours; for the cells of *Tradescantia paludosa,* 17 hours; for Chinese hamster fibroblasts, 11 hours; and for mouse fibroblasts, 22 hours.

The cell cycle includes many events in both nucleus and cytoplasm, although attention in the past has centered on the nucleus. In prokaryotes the cell grows in volume as protein, RNA, and DNA are synthesized during replication, and the cycle ends in the formation of a cell plate separating the new cells. In eukaryotes the cell grows in volume mainly as a result of protein synthesis; RNA and DNA are synthesized, mitosis occurs, and finally *cytokinesis* (cell division) takes place, with formation of two daughter cells *(fission)*. The term *mitosis* is often used as a synonym for cell division in eukaryotes, but it represents only one series of events in the cell cycle, during which the duplicated chromosomes are separated from each other. Mitosis often occupies a relatively short time span (Fig. 26.1), most of the cell cycle being *interphase.* In mitosis, the replicated chromosomes condense during *prophase,* become aligned on the equatorial plane during *metaphase,* and members of each pair of chromosomes separate during *anaphase.* The chromosomes aggregate at the two poles of the dividing cell during *telophase,* and the nuclear envelope once again covers the nucleus (Fig. 26.2). *Interphase* consists of three main stages in relation to DNA synthesis: G_1 (Gap 1) preceding DNA synthesis from telophase to the beginning of S (synthesis); S, during which DNA replication takes place; and G_2 (Gap 2), the period following DNA synthesis to the beginning of mitosis (Howard and Pelc, 1953). The subdivisions of the cell cycle are not as precisely demarcated as Figure 26.1 might lead one to believe, the time span for each stage shown in the figure being an average. The variation in the time spans for individual cells in a population is shown in Figure 26.3.

The time required for DNA replication

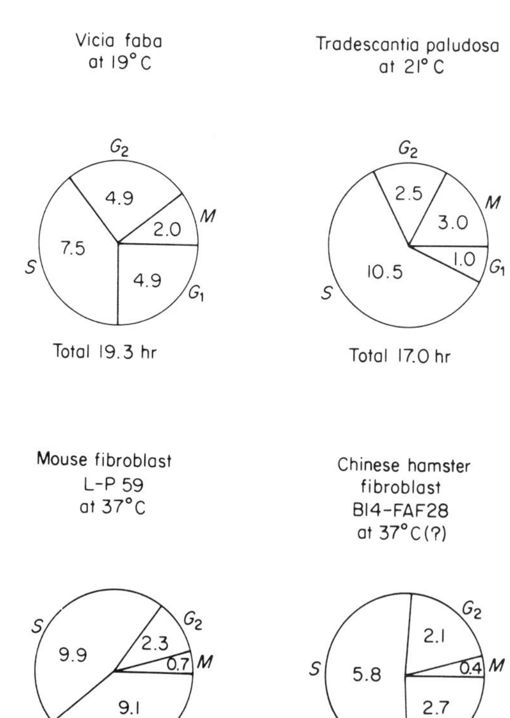

Figure 26.1. *Duration in hours of phases in the division cycle of a cell: G_1 (Gap 1), S (synthetic phase), G_2 (Gap 2), and M (mitosis). G_1 follows mitosis and lasts until the beginning of DNA synthesis (S); S lasts until replication of DNA is over, when G_2 begins and lasts until the beginning of mitosis. Data are for root tip cells of* Vicia faba *and* Tradescantia paludosa *and tissue culture cells of mouse fibroblast and Chinese hamster fibroblast. (From Kihlman, 1966: Actions of Chemicals on Dividing Cells. Prentice-Hall, Englewood Cliffs, N.J.)*

varies, as seen in Figure 26.1. Division of the cell cycle into stages in terms of DNA synthesis indicates the extreme importance attached to this event. However, it obscures the importance of other events also of interest for an analysis of the cell cycle, for example, the period when rRNA and important enzymes are synthesized.

In multicellular plants and multicellular animals cell division continues until adulthood. Most cells in the adult differentiate from embryonic progenitor cells but no longer divide. Division continues only in germinal tissues like the cambium in the stem and root of a plant, in gametogenic cells in gonads,* in the basal germinal cells of the epidermis, and in blood cell-forming tissues of an animal. Cell division repairs wounds in both plants and animals.

At present investigators are attempting to ascertain the exact sequence of events occurring throughout the cell cycle, such as where in the sequence various types of RNA are synthesized and where specific enzymatic or control proteins appear. The object is to learn what triggers each event in the cell cycle, including mitosis and cyokinesis. Inasmuch as methods are not yet refined enough to make biochemical determinations on single cells, it is necessary to use cell populations. In a routine culture, cells are in various stages of development, from those that have just divided to those just getting ready to divide. When small numbers will suffice, dividing cells can be picked by micropipette and used for chemical determinations at various times after cell separation. When larger numbers of cells are required in biochemical determinations, cell division must be synchronized so that many cells in the same stage of the cell cycle are available.

SYNCHRONIZED CELL DIVISION

Eggs of marine animals have been favorite objects for studies on synchronized cell division because these eggs reach a similar stage of development in the ovary and then began to "idle" biochemically. When activated by fertilization they begin from the same base line and so reach the division stage almost simultaneously. When the egg is damaged by temperature, pH, or radiation, not only is cleavage delayed but it is desynchronized. Desynchronization probably occurs because susceptibility to damage varies about a mean for the cell population; a characteristic distribution curve is obtained for onset of division in the population (Fig. 26.4).

The high degree of synchrony of bacteria and fungi germinated from spores can be explained on similar grounds. The spores are dormant, growth has been suspended, and metabolism is idling. When a suitable culture medium is supplied, growth begins from a common starting point. Therefore, division occurs at almost the same time for most germinated bacteria and fungal spores. Divisions are synchronized in decreasing degree for several generations thereafter.

In unsynchronized cultures of bacteria, protozoa, or tissue culture cells, on the other hand, at any one time the ratio of cells in mitosis to those not in mitosis (called the *mitotic index*) may average 5 to 10 percent. The mitotic index is an inverse function of the generation time, being larger for a short generation time and smaller for a long generation time, even if the

*After proliferation of gametogenic cells, meiosis with disjunction of members of a pair of chromosomes occurs during the formation of gametes. While the control mechanisms that trigger meiosis instead of further mitosis are of interest in cell physiology, little is known about it. The complex subject is discussed in textbooks of cytology (De Robertis *et al.*, 1975).

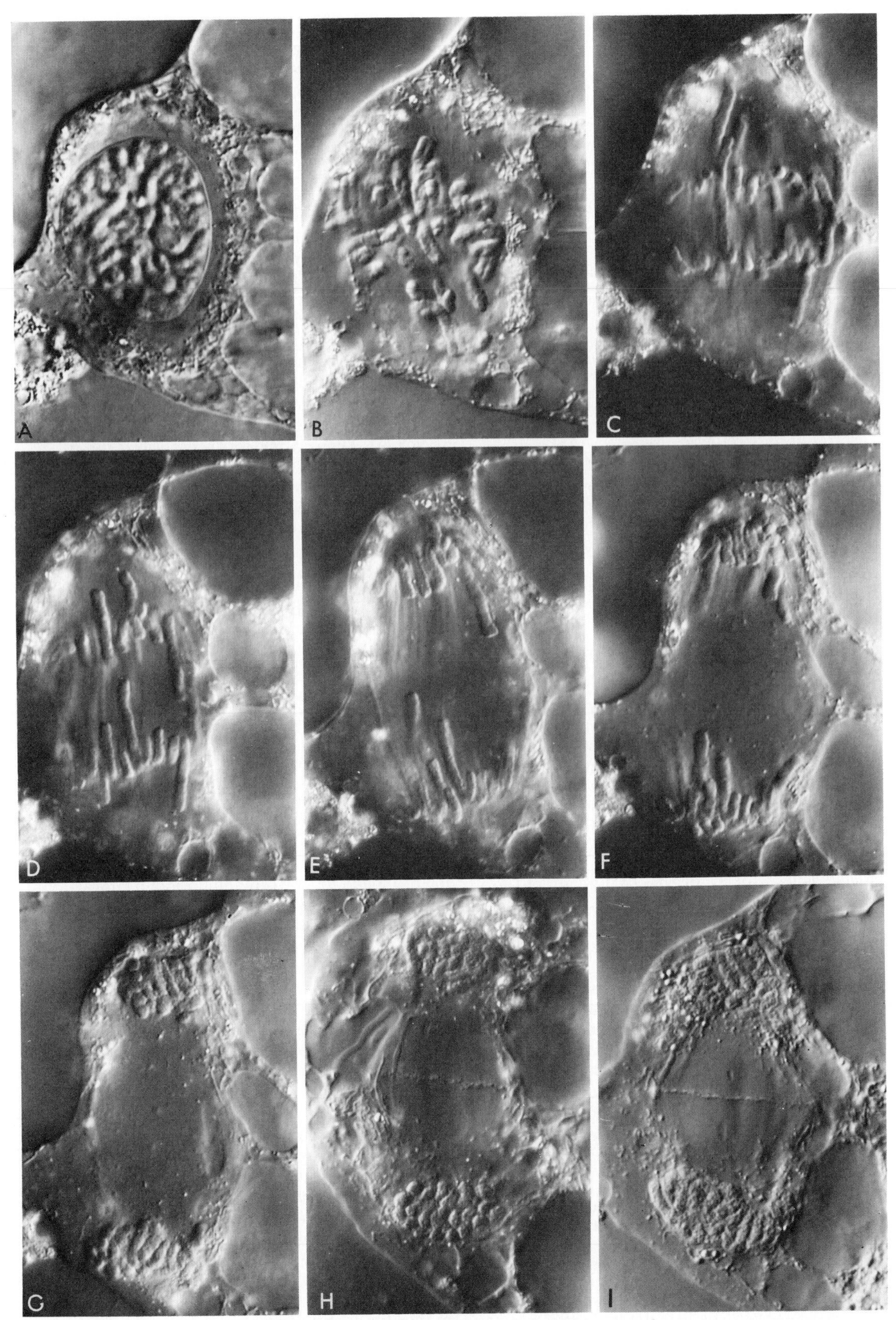

Figure 26.2. See legend on the opposite page.

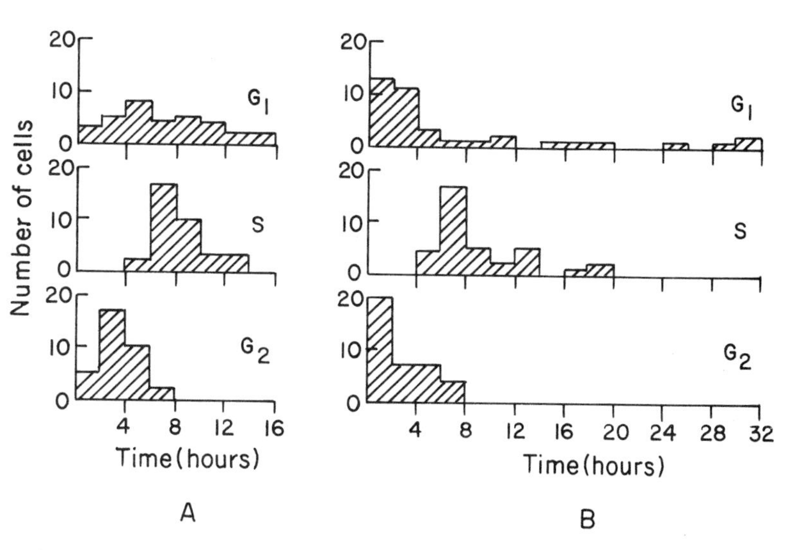

Figure 26.3. *Frequency histograms of the lengths of G_1, S and G_2 in mammalian cells.* A, In vitro *results from 15 cases.* B, In vivo *results from 38 cases, excluding spermatogonia. There is an uncertainty of about half a mitotic time (c. 0.5 h) in the lengths of G_1 and G_2, since in some cases this has been included and in others excluded. (From Mitchison, 1971: The Biology of the Cell Cycle. Cambridge University Press, Cambridge.)*

number of cells entering mitosis is the same in both (Zeuthen and Scherbaum, 1954).

It is of interest to consider the respective roles of nucleus, cytoplasm, and environment in synchronizing cell division. In multinucleated cells, whether giant amoebae, insect embryos, or plant endosperm, the internal environment is similar and nuclear division is usually perfectly synchronized. Grafts are made between two multinucleated amoebas *(Chaos chaos)*, each individual with nuclei dividing synchronously but with the cells out of phase with one another. After one cell cycle all the nuclei from both amoebas (now in one cell)

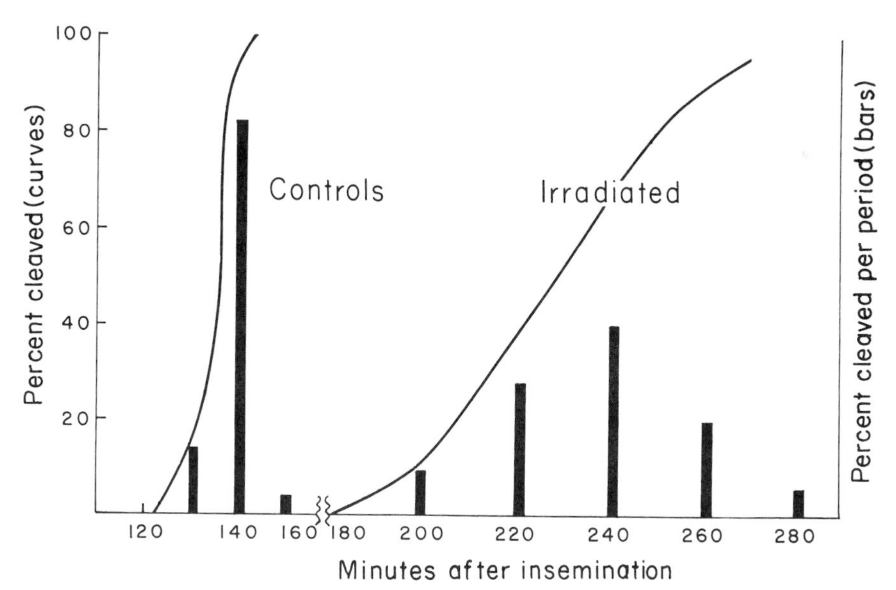

Figure 26.4. *High degree of synchrony in division of normal sea urchin eggs (inseminated simultaneously) as shown in the curve for "controls." Note that the histogram plotting the number of eggs cleaving in a 10-minute period indicates that over 80 percent of the eggs cleave simultaneously. The percentage cleaving at any one time is the mitotic index. The same sample of eggs fertilized with ultraviolet radiation-treated sperm (50 ergs/nm² of wavelength 265 nm) shows much less synchrony in division. The division figures were recorded photographically over a period of about 300 minutes. (Data from P. Wells, 1952: Doctoral dissertation, Stanford University.)*

Figure 26.2. *Stages in mitosis of endosperm cells of the blood lily,* Haemanthus katherinae. *Mitosis takes about 2.3 to 3 hours. A, Prophase; B, metaphase; C and D, anaphase; E and F, telophase; G, condensation of chromosomes at the poles; H and I, beginning and later stages of phragmoplast formation. The phragmoplast is an accumulation of microtubules first a ring then a plate, forming a precursor to the cell wall as it accumulates vesicles (from the Golgi body or dictyosome) that secrete the cell wall fibers. Nomarski interference-contrast microscopy, all ×700. (By courtesy of A. Bajer. For details see Bajer and Mole-Bajer, 1973.)*

divide synchronously, the larger original cell imposing its division timing (Erikson, 1964; Agrell, 1964). Spermatogenic cells lining the lumen of a germinative tubule and connected to one another by cytoplasmic bridges also divide synchronously (Fawcett *et al.*, 1959). The common lack of synchrony in mass cultures of independent cells might therefore result from failure of nuclear secretions from dividing cells to reach the nuclei of other cells in order to effect synchronous division.

On the other hand, the cytoplasm may affect the nucleus and stimulate nuclear division, for example, when the nucleus of a differentiated, nondividing cell (young intestinal cell) is inserted into an enucleated frog egg (see Chapter 7).

METHODS FOR SYNCHRONIZING CELL DIVISION

Selection

If only small numbers of cells are needed, cells at the same stage in the cell cycle, for example, those just dividing, can be picked by pipette from a mass culture under low-power magnification. A Coulter counter (electronic device) permits mechanized, rapid selection of cells in one stage in a cycle (about a thousand per second). To obtain much larger masses of cells for biochemical determinations, filtration, sucrose gradient separation, and membrane elution have been successfully used.

When a growing culture of microorganisms is passed through several layers of filter paper, small cells pass through first and may be used to start a culture in which the cells will divide with an appreciable degree of synchrony (Murayama, 1964). Small cells that are present at the top of a sucrose gradient after centrifugation of a mass culture may be used in a similar manner to start synchronous cultures of budding and fission yeasts (Fig. 26.5), *E. coli*, and mammalian cells in tissue culture (Mitchison, 1971).

Greater separation than by either of these methods is achieved by the membrane elution method, in which the growing culture is collected on a membrane filter (or ion exchange paper). Fresh warm culture fluid is then passed slowly but continuously through the inverted filter. This washes off the excess adherent cells but leaves those that at a particular stage in the cell cycle are adsorbed most readily to the available sites on the surface of the filter. These cells divide and so produce a population that is in the same stage in the cell cycle, which in subsequent culture show a high degree of synchrony. If the filter has a large area, a very large population of synchronized cells can be obtained with each division of the adhering cells.

Mitotic selection is useful for mammalian cells in tissue culture. During mitosis such cells round up and are more readily washed off from a culture on a glass plate than cells in other phases of the cell cycle (Prescott, 1976a).

Chemical Shock

One type of chemical shock consists of limiting the supply of some essential nutrient for a period and then supplying it to the culture. For example, DNA synthesis and cell division are blocked in *E. coli* T-15 (a thymine-requiring mutant, "thymineless") by suspension in a thymine-free nutrient medium. However, RNA synthesis and protein synthesis continue, apparently at the usual rate. When thymine is added 30 minutes later, DNA synthesis is resumed and after a lag of about 35 to 40 minutes nearly all the cells divide simultaneously (Barner and Cohen, 1956). A *Lactobacillus*

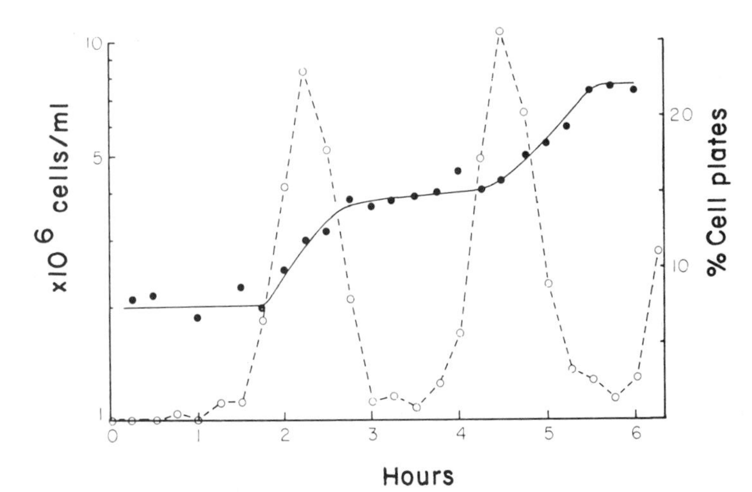

Figure 26.5. *Synchronous culture of the fission yeast* Schizosaccharomyces pombe *prepared by gradient separation. A fission yeast divides equally rather than budding, as do many other yeasts. Closed circles are cell numbers and open circles are cell plate indexes (equivalent to the mitotic index). (From Mitchison, 1970:* In Methods in Cell Physiology. *Prescott, ed. Academic Press, New York, Vol. 4.)*

acidophilus "thymidineless" mutant reacts in the same way to thymidine deficiency and resupply (Burns, 1964). Temporarily withholding a variety of nutrients followed by resupplying them is also effective for synchronizing division in a number of other microorganisms. Examples are silicon depletion in the diatom *Navicula,* amino acid starvation in *E. coli, Streptococcus faecalis,* and in hamster tissue cells (isoleucine and glutamine being critical here), and even general starvation in some cases (yeast) (Mitchison, 1971). In an example of an arthropod, synchronous division in epidermal cells of the blood-sucking bug *(Rhodnius)* occurs after its periodic ingestion of blood.

Another type of chemical shock is temporary inhibition of some phase of mitosis. For example, colcemid, nitrous oxide at high pressure, thymidine block, and vinblastine inhibit metaphase. Cells accumulate at this phase of mitosis, and when the block is removed, a higher degree of synchronized cell division occurs (Mitchison, 1971; Prescott, 1976a).

Physical Shock

Temperature shock has been found effective as a physical synchronizer, presumably because some reactions in the predivision period are more sensitive to heat or cold than others. A high temperature stops division without stopping biosyntheses. Cells lagging in preparations for division then catch up with others

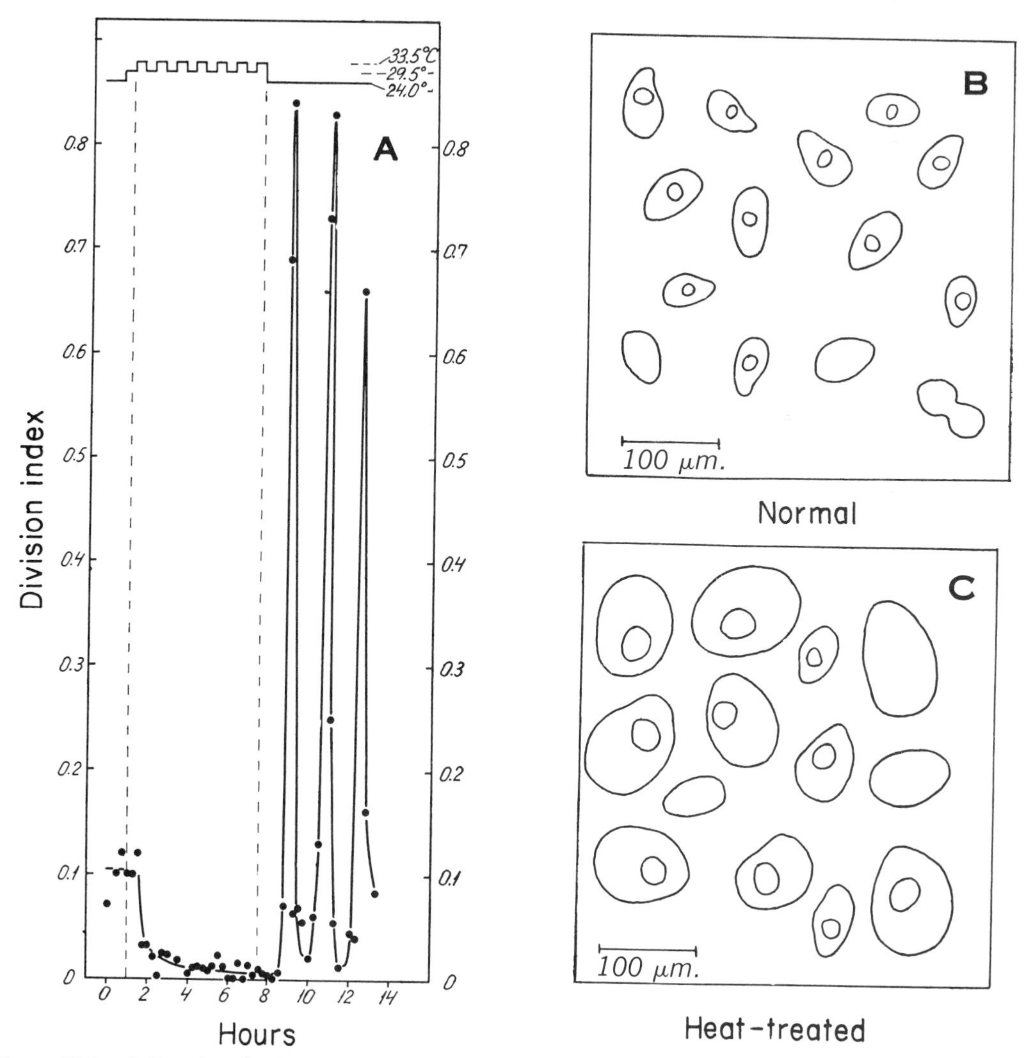

Figure 26.6. *A, Synchronization of division by heat-shocking* Tetrahymena. *Seven shocks at 33.5°C are given in succession, alternating with 29.5°C (top line). At the conclusion of the heat shocks, the temperature is dropped to 24.0°C. The three post-heat treatment divisions shown are synchronized to the extent indicated by the division index, the percent of cells in cytokinesis.* B, *Normal* Tetrahymena *drawn with camera lucida.* C, Tetrahymena *after heat treatment. The circles inside the cells are the macronuclei. (From Zeuthen and Scherbaum, 1954: Colston Papers 7: 141.)*

after temperature shock. A single temperature shock synchronizes only a small proportion of the cells because it allows only a small proportion of the cells to accumulate in predivision stages. On the other hand, a series of high temperatures alternating with exposure to near-optimal temperature for a brief period synchronizes most of the cell population by gradual accumulation of a large proportion of the cells in predivision stages.

Zeuthen and Scherbaum (1954) found that alternate exposure of suspensions of the ciliate *Tetrahymena* to half-hour periods at 29.5°C (optimal) and 34°C (inhibitory) for seven cycles synchronized division in 85 percent of the cells when the cultures were subsequently kept at 24°C. Synchrony persisted for several cell generations before division became random (Fig. 26.6). Continuous synchrony can be achieved by shocking every generation (Zeuthen, 1974). Thermal shocks have also been effective for synchronizing division of many other kinds of cells, including bacteria (Fig. 26.7) (Zeuthen, 1958).

Cold shocks induce synchronization of cell division in a number of protozoans and bacteria (Fig. 26.7). Nutritional deficiency, coupled with temperature shocks, has also been very effective (Scherbaum, 1960a and b). X-rays have also been shown to induce synchrony in division, presumably by holding back the cells in

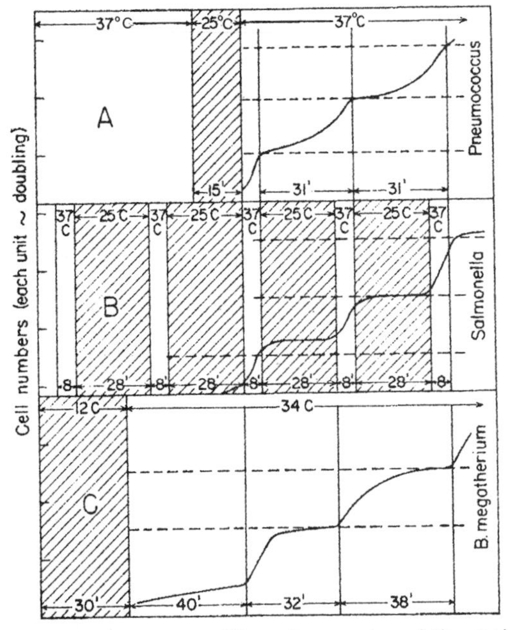

Figure 26.7. *Three different examples of thermal synchronization in bacteria.* A, Pneumococcus *(now* Streptococcus pneumoniae) *(Hotchkiss, 1954).* B, Salmonella typhimurium *(Maaløe and Lark, 1954).* C, Bacillus megatherium *(Szybalski and Hunter-Szybalska, 1955). Shaded areas are for periods of lower temperature exposure.*

the radiation-sensitive predivision stages and causing accumulation of cells that will ultimately divide at nearly the same time (Spoerl and Looney, 1959).

Synchronized cultures of various cells in suspensions are now being widely used in biochemical and cytochemical research (Zeuthen, 1971a and b). It is important to note, however, that changes in size and composition of the cells after synchronization must be taken into consideration (Prescott, 1976b). For example, resistance of such cells to ultraviolet radiation is markedly altered (Iverson and Giese, 1957).

Bruce (1965) lists the generation time of cells from a large number of species of plants and animals. In many cases the division of cells appears to be circadian, almost every 24 hours; but it has not yet been determined whether the day-night cycle will synchronize many of these to a 24-hour rhythm, although it does so in *Gonyaulax* (Fig. 26.8) and in *Chlorella* (Tamiya, 1964).

DNA SYNTHESIS IN THE EUKARYOTIC CELL CYCLE

Much of the material summarized here is from Mitchison's excellent review (1971) and updated from Prescott's reviews (1976a and b). Macromolecular syntheses during the cell cycle have also been reviewed by Monesi (1969).

DNA synthesis in eukaryotes is generally periodic, occupying only a fraction of the cell cycle (see Fig. 26.1), except possibly in amphibian embryos. In mammalian cells S lasts for about 6 to 8 hours, G_2 about half as long, regardless of type of tissue and species; M usually lasts 1 to 2 hours. Stage G_1, on the other hand, is quite variable; differences in duration of the cell cycle are mainly related to the length of the G_1 period.

Interestingly, duration of plant cell cycles and of their G_1, S, G_2, and M is comparable to that in animal cells, despite the fact that they are recorded at temperatures about 15°C lower. Data are inadequate to state whether G_1 in plant cells is the most variable portion of the cell cycle.

Information on the duration of stages in the cell cycle in lower eukaryotes is also inadequate for making comparisons with higher eukaryotes. However, when trophic and genetic DNA are separated, as in the ciliate macronucleus and micronucleus, respectively, the S period occurs at a different time in each component. Also, in lower eukaryotes G_1 may be absent (e.g., in *Amoeba proteus*, the slime mold *Physarum*, the fission yeasts, and the micronuclei of the ciliates *Tetrahymena* and *Euplotes*). Mitochondrial and chloroplast DNA

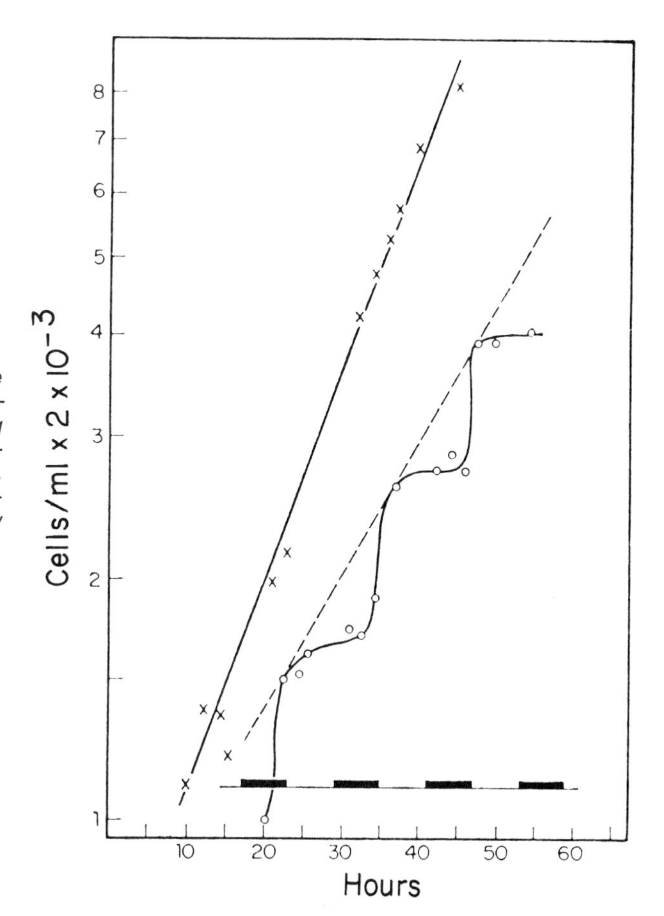

Figure 26.8. *Increase in cell population of the photosynthetic dinoflagellate* Gonyaulax polyhedra *in continuous light (X) and in alternating light and dark periods (open circles). Light intensity, 900 foot-candles; temperature, 21.5°C. Note the synchronization achieved by the alternating light-darkness regimen. (From Sweeney and Hastings, 1958: J. Protozool.* 5*: 217.)*

have a different time pattern of synthesis than nuclear DNA.

DNA replication occurs at many points on a single eukaryotic chromosome, although not simultaneously; about 25 banks or families of replicons (units of replication in a chromosome) exist in mammalian cells. Within a chromosome each of the sections of DNA being replicated is short, and replication is completed in about 17 minutes. Initial evocation in the first bank(s) is probably genetically controlled, but the initial bank(s) may then send signals to the next bank(s); successive evocations of DNA synthesis thus lead to completion of replication in the whole chromosome in about 7 hours. However, the exact mechanism by which this occurs is not known (Prescott, 1976a). The rate of DNA synthesis in eukaryotic chromosomes is about an order of magnitude slower than in bacteria (*E. coli;* 30 μm per minute; in eukaryotic chromosomes about 3 μm per minute). Replication in eukaryotic cells is bidirectional from an initiation point and occurs in a timed sequence that appears to be genetically determined (see Chapter 15).

Initiation of DNA synthesis requires some stimulus that presumably comes from the cytoplasm. The cytoplasm of a cell in the S stage in the cycle will induce DNA synthesis in newly transplanted nuclei from cells out of phase with its cycle. The cytoplasmic stimulatory substance is apparently needed in a critical concentration to be effective, since protein synthesis is required to initiate DNA synthesis; possibly this substance may be either an enzyme of DNA synthesis, a histone, or some other chromosomal protein. Inhibition of DNA synthesis arrests further progress in the cell cycle (Prescott, 1976a). Not unexpected is the finding that histone synthesis is confined to the S phase of the cell cycle, presumably correlated with DNA synthesis (see section on protein synthesis).

DNA SYNTHESIS IN THE PROKARYOTIC CELL CYCLE

The prokaryotic genome consists of a single circular DNA molecule. In slow growth a single replicating fork moves around the chromosome from its origin; in rapid growth two such forks may be present, the second beginning before the first has completed its round of replication. The chromosome is attached to the cell membrane in the region of the replicating fork and in some species evidence exists for attachment

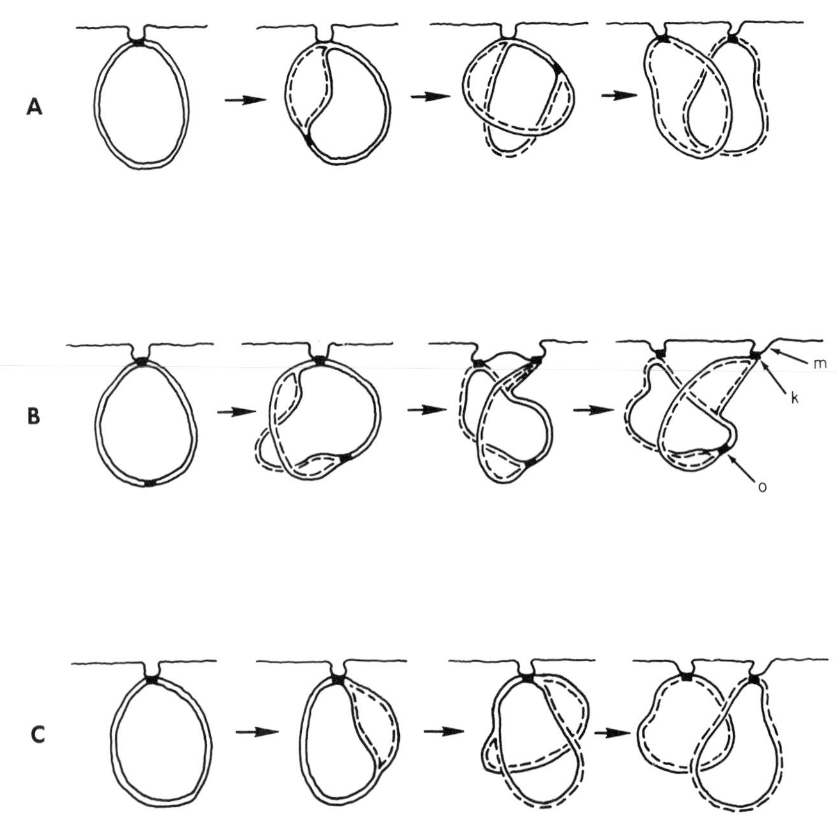

Figure 26.9. *Three models of the replication of a circular bacterial chromosome and separation of the daughter chromosomes.* A, *Site of replication associated with cell surface (mesosome) throughout replication cycle.* B, *Specialized site (kinetochore), 180° from the origin, attached to mesosome; active separation of daughter chromosomes can begin when replication is half completed.* C, *Origin, acting as kinetochore, attached to mesosome throughout replication cycle; as in* A, *active separation of daughter chromosomes cannot begin until replication has been completed. m, Mesosome; k, kinetochore; o, origin of replication; dashed lines represent newly synthesized polynucleotide chains. (From Luykx, 1970: Cellular Mechanisms of Chromosome Distribution. Academic Press, New York.)*

at the point of origin of replication as well (Fig. 26.9*B*). The chromosomes may be aided in separation by the growth of a membrane between these two points of attachment (Fig. 26.9). Most information has come from studies on *E. coli* and *Bacillus subtilis*.

In a slow-growing prokaryote DNA synthesis is periodic, with a pause between successive replications, but in a fast-growing one it is continuous. *Initiator protein* must first be synthesized to initiate a new round of DNA synthesis, but it is not necessary for completion of a round already begun. The notion of a critical mass for a cell before replication and division may be explainable in terms of the constant accumulation of an initiator protein that must reach a critical level in the cell, perhaps followed by the formation of a structure composed of several proteins. However, the increase in cell volume may dilute an inhibitor that prevents replication and division, although no evidence has been reported for it.

RNA SYNTHESIS DURING THE CELL CYCLE

A major fraction of cellular RNA is rRNA. Most of the remaining RNA is tRNA. Like all the other RNA, messenger RNA, present in small quantity, comes from the nucleus but functions primarily in the cytoplasm, where its message is translated into protein. However, most of the rapidly labeled RNA of the nucleus is not mRNA and is rapidly degraded within the nucleus. Its function is debatable but it might serve for gene regulation.

Except during the various phases of mitosis, RNA appears to be synthesized continuously. The rate of RNA synthesis increases through the cell cycle, although the rate of increase is not the same in all types of cells (Mitchison, 1971). Inhibition of RNA synthesis arrests further progress of the cell cycle (Prescott, 1976a).

Little is known of the relative rates of synthesis of the various classes of RNA. From the available evidence it would appear that all kinds of RNA are synthesized continuously except during mitosis. When division occurs by amitosis (without formation of a spindle and separation of chromosome pairs), as in the macronucleus of ciliates, RNA synthesis continues even during cell division. In prokaryotes, RNA synthesis is continuous (Mitchison, 1971).

PROTEIN SYNTHESIS DURING THE CELL CYCLE

The overall increase in cell volume discussed earlier in this chapter is largely the result of the total protein synthesis in the cell. Histone synthesis occurs only during the S phase of the cell cycle, but nonhistone synthesis on chromosomes (and in the cytoplasm) occurs to varying degrees throughout interphase. Of greater interest at this point is consideration of the synthesis of specific proteins, such as enzymes. Many enzymes are synthesized during the cell cycle at times specific to each enzyme. Enzymes are of two major types, stable and unstable; unstable enzymes are degraded after a period of function. Stable enzymes are synthesized in a step pattern (Fig. 26.10*A*), while the unstable ones show a peak pattern (Fig. 26.10*B*). Some enzymes are synthesized continuously during the cell cycle, more commonly in mammals than in lower eukaryotes and prokaryotes. In some cases such continuous synthesis results in a pattern of increase in which, at a particular time in the cell cycle, the rate of synthesis doubles (Fig. 26.10*D*); one example is sucrase in the fission yeast *Schizosaccharomyces*. Details on protein synthesis during the cell cycle are reviewed elsewhere (Mitchison, 1971). Inhibition of protein synthesis arrests the progress of the cell cycle (Prescott, 1976b). Timing of enzyme synthesis is complicated by the possible synthesis of proenzymes (e.g., trypsinogen). What is measured is perhaps the time of activation of an enzyme rather than its synthesis.

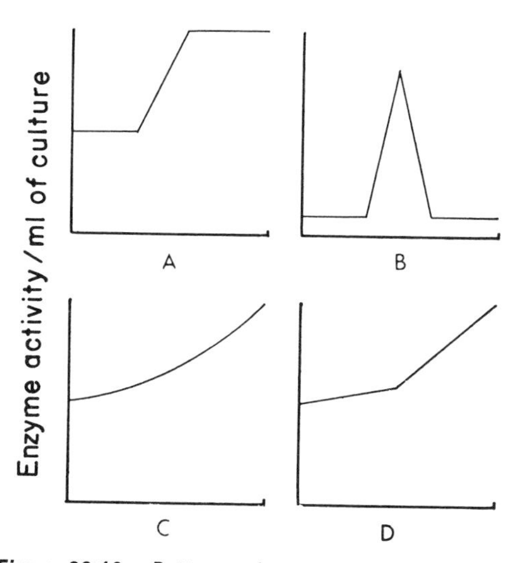

Figure 26.10. Patterns of enzyme synthesis in synchronous cultures during one cell cycle. A, Step. B, Peak. C, Continuous exponential. D, Continuous linear with an increase in rate at a point about halfway through the cycle. (From Mitchison, 1969: Science 165: 657. Copyright 1969 by the American Association for the Advancement of Science.)

In most prokaryotes and eukaryotes it is possible to induce some enzymes at any time during the cell cycle. The amount of enzyme induced doubles at a particular time in the cell cycle that presumably corresponds to replication of the gene coding for the enzyme. Doubling in prokaryotes occurs simultaneously with DNA replication; doubling in the eukaryotic fission yeast, which divides equally rather than by budding, occurs later.

Periodic synthesis of enzymes might result from feedback repression and end-product inhibition (see Chapter 16). When the enzyme concentration reaches a particular level, synthesis may be turned off; then the enzyme concentration will fall to a low level before its synthesis again becomes evident. If enzyme synthesis were entrained to the cell cycle, the kind of oscillation observed would occur. Linear synthesis of enzymes might be explained on the basis of linear reading of the message on the chromosome during transcription, although information is meager.

The cell cycle has been analyzed by the use of temperature-sensitive mutants in the yeast *Saccharomyces cereviseae;* such mutants have also been obtained in Chinese hamster cells in tissue cultures and in *Tetrahymena*. The temperature-sensitive mutants show blocks in the cell cycle when grown at nonpermissive temperatures. This allows demonstration of the sequence of events in the cell cycle much the way biochemical mutants permit determination of the steps in the synthesis of a molecule such as tryptophane, as discussed in Chapter 15. In *Saccharomyces* a divergence of pathways occurs from the "start" point of the cell cycle either into bud emergence or into initiation of another round of DNA replication and nuclear division (Fig. 26.11). The two paths converge for cytokinesis and bud separation (see Prescott, 1976a for review).

Assessment of much data on the cell cycle has led Prescott (1976b) to consider that a checkpoint in G_1 probably exists at which arrest of the cell cycle generally occurs. Many factors may cause such an arrest. Therefore, G_1, when present, is the most variable period of the cell cycle. For example, the cycles of mammalian cells in tissue culture may be arrested by

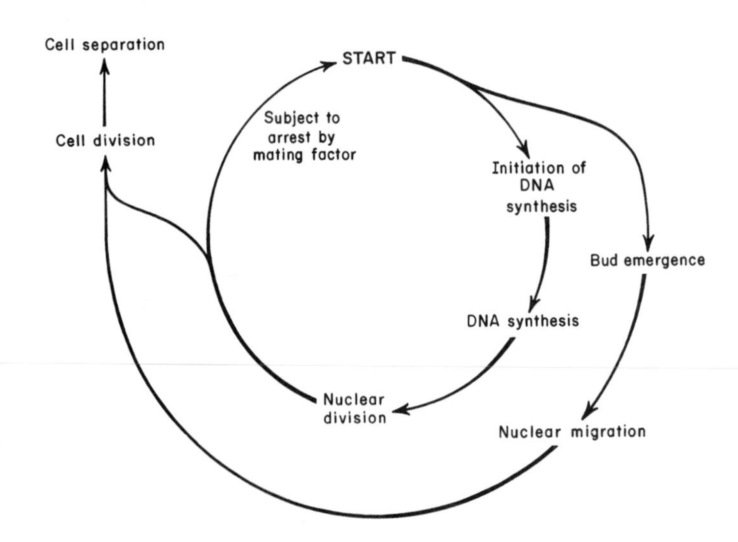

Figure 26.11. *The cell cycle in a budding yeast, as determined by temperature-sensitive mutants. (After Hartwell* et al., *1974. Science* 183: *46–51.)*

either lack of amino acids or adequate serum, by the presence of cAMP in too high a concentration, or by chalones. The longer the arrest the more difficult it is for the cell to come out of G_1. In fact, cells may enter a stage called G_0, from which it is very difficult for them to reenter the cell cycle. Sometimes, as in some epidermal cells, arrest occurs in G_2. Such cells are the first to reenter the cell cycle upon injury to the epidermis, appearing to be reserves for this function.

The role of cAMP in control of the normal cell cycle remains controversial because the evidence for action of cAMP on the cycle is so conflicting. Conclusions depend on whether the author takes an optimistic or a pessimistic view (Rebhun, 1977). Addition of cAMP prolongs G_1, and the mechanism for release of cAMP from action on the cell membrane is known. But whether this mechanism performs a controlling role in the cell cycle is unclear. Biochemical studies on the cell cycle are complicated by the imperfect synchrony of cell division achieved, even that achieved by the best methods (Prescott, 1976a, b). Resolution of this issue awaits accumulation of decisive data.

SOURCE OF ENERGY FOR CELL DIVISION

During cell division the cell does work at the expense of energy derived from nutrients, as in synchronized cell division induced when starved yeast cells are supplied with nutrients such as glucose.

Even if glucose is supplied to starved epidermal cells, division fails if oxygen is not available (Rasmussen, 1963). Glucose may be replaced by lactate, glutamate, fumarate, or citrate, suggesting that operation of the Krebs TCA cycle supplies the energy for cell division. Mitosis is inhibited by poisons of the Krebs TCA cycle such as malonate, cyanide, and carbon monoxide, and by phlorhizin, a phosphorylation inhibitor (Bullough, 1952). Inhibition by phlorhizin suggests that high-energy phosphate bonds are direct or indirect energy sources. Marine eggs and protozoa also require oxygen for division. Frog eggs and many embryonic tissue cells do not require oxygen for division, presumably supplying energy for division by glycolysis. Cell division in cells that can supply their energy needs for division by glycolysis is very sensitive to glycolytic inhibitors such as iodoacetic acid (Bullough, 1952).

To have an effect on cell division, energy sources must be supplied early in the cell cycle, during what Bullough calls the *antephase*. Once a critical concentration of the energy-rich substances has been accumulated and prophase begins, nothing short of killing the cell by oxygen deprivation, metabolic poisons, or radiations will stop it from dividing. For example, cells subjected to massive doses of radiation divide and immediately cytolyze. Initiation of mitosis appears to begin a series of concatenated and irreversible reactions that stop only when the cells have divided (Bullough, 1952).

The rate of cleavage of eggs of the purple sea urchin *Strongylocentrotus purpuratus* is decreased when the concentration of carbon monoxide is at a level that inhibits respiration. At this time the ATP level is also decreased. When the ATP level drops to 50 percent of normal, mitosis is completely inhibited. By varying the carbon monoxide concentration to inhibit ATP production to varying degrees, it can be demonstrated that the mitotic rate parallels the ATP level in cells. Division can be blocked at any stage of mitosis if the inhibitor is applied at the apporpriate time. When phosphorylation is uncoupled by dinitrophenol

(DNP), ATP fails to accumulate and cell division is also blocked. A concentration of DNP that uncouples phosphorylation increases respiration, but the respiration is useless and idling (Epel, 1963). This indicates that ATP is the likely energy source for cell division.

Heat shock does not stop growth, although it prevents division. Consequently, *Tetrahymena* so treated may attain a volume four times that of the controls. On return to the optimal temperature after one or more heat shocks, fission occurs in less than the normal generation time. Presumably the mitogenic and growth channels are separate at their definitive ends, although both use the same sources of energy and materials. When the mitogenic channel is blocked, the energy and substances pass into synthetic reactions leading to extra cell growth. Consequently, after removal of the block, some time must elapse before the specific molecules or structures necessary for cell division are synthesized to the necessary level (Zeuthen, 1964). When heat shocks to *Tetrahymena* are continued well beyond seven cycles, the cells may still ultimately divide, even during a heat shock. This is interpreted as accumulation of a critical concentration of a substance or completion of a structure needed for cell division. Division can then no longer be prevented by the type of thermal shocks that previously blocked it.

Tetrahymena that are heat-shocked in nutrient medium and transferred to balanced salt medium with no nutrients subsequently divide at least twice in absence of nutrients (Hamburger and Zeuthen, 1957). This indicates a clear-cut separation of growth from cell division.

Experiments with dinitrophenol distinguish two phases of metabolism with respect to cell division in *Tetrahymena.* One, constituting about 30 percent of total metabolism, relies on endogenous reserves and can support cell division; the other phase (about 70 percent of total metabolism) depends upon exogenous supplies and is coupled with growth (Scherbaum, 1960a and b). It now appears likely that a division protein, which participates in forming structures necessary for both the oral apparatus and division, is critical for induction of cell division (Zeuthen, 1971a). It has also been shown that ATP, GTP, and RNA accumulate in the period just preceding cytoplasmic division and that the changes in concentration of these substances are reversed at mitosis (Zeuthen, 1961). However, such an accumulation of free energy sources is interpretable on the basis of protein synthesis, because protein synthesis decreases for a period beginning just about two-thirds of the time to the next cell division. At this time the protein formed on the ribosomes is less readily released than previously (Plesner, 1961). The call for ATP is then less than its concentration in the cell, along with that of GTP, and RNA increases until it is controlled by feedback inhibition. The change in concentration of ATP, GTP, and RNA thus appears to be incidental to the slowing or cessation of protein synthesis rather than to accumulation of an energy reservoir. The high-energy phosphates then play a part in reversing the binding of the protein to the ribosomes (Plesner, 1964).

INITIATION OF CELL DIVISION

The cell grows to a certain stage and then divides. In some unknown way each unit (organelle?) in the cell presents a template upon which duplication occurs. When cell division is completed, each daughter cell contains in itself a duplicate of all the varied structures present in the mother cell.

Mitosis and cell division (cytokinesis) have been studied in a multitude of eukaryotic cells. During mitosis, which precedes cell division, chromosomes condense during prophase and appear on an equatorial metaphase plane (see Fig. 26.2). The daughter chromosomes, that are replicated during interphase, separate from one another beginning at each centromere (spindle attachment to the chromosome), and a duplicate set of chromosomes is moved toward each pole during anaphase. The chromosomes condense during late telophase and pass into the resting state of interphase.

The trigger that initiates cell division has long been of interest and numerous suggestions as to its nature have been made. Complete doubling in cell mass, the last of a series of growth requirements that each result in doubling of cell organelles, has been suggested as the trigger mechanism because in it the ratio of cytoplasm to nucleus is upset. Evidence for this can be seen in an amoeba shaken when about to divide. It divides unequally but each daughter amoeba grows until it equals the mass of the mother amoeba before it in turn divides (Fig. 26.12). Amputation of a portion of a growing amoeba is followed by replacement of the part removed before division occurs. However, in all cases when an amoeba has doubled its mass it enters a rest period of about one-sixth the generation time (4 hours out of 24) before it divides. During this time lapse something occurs that triggers division (Mazia, 1961). A similar rest period is observed in *Tetrahymena* (Zeuthen and Scherbaum, 1954), but in *E. coli B* synthesis continues in exponential fashion (Scherbaum, 1960a).

Another possible trigger for cell division is an upset in the surface-to-volume ratio. However, if a piece of cytoplasm is removed at a "critical" time, just before an amoeba has doubled its

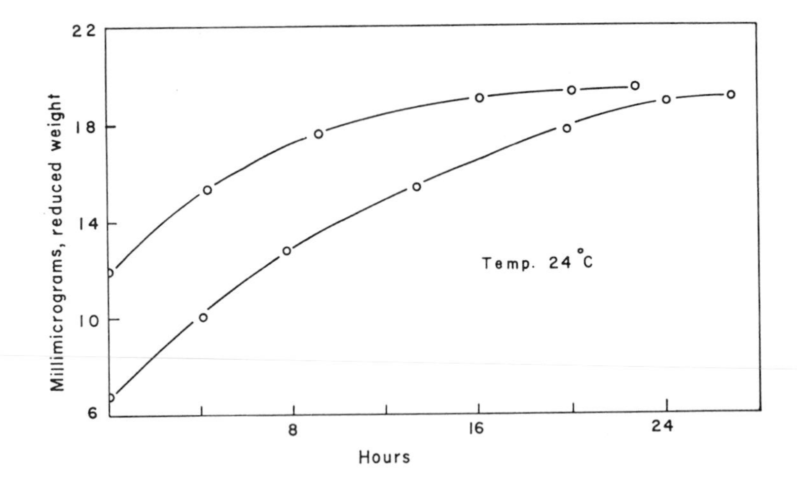

Figure 26.12. Relation between the weight of Amoeba proteus *at "birth" and the weight at maturity. Curves represent the growth of sister cells following unequal division of their mother cell. The last point on each curve gives the weight at the time of division. (Data from Prescott, after Mazia, 1956: Am. Scientist 44: 11.)*

mass, division will still occur, even though offspring that are smaller than normal are produced. This experiment might be taken to indicate that by this time division has been determined and follows even if the amoeba is not as large as a typical mother cell about to divide. In such an interpretation neither the ratio of nucleus to cytoplasm nor the ratio of surface to volume can in itself be the probable trigger for cell division. When cells are starved but undergo division, they produce smaller daughters. However, such alterations in cell size are also produced by a number of other changes in environment—heat, for example (Prescott, 1964).

Doubling the DNA content has been suggested as a possible trigger. By use of labeled precursors it has been shown that in most cells DNA is doubled in the interphase during the S period (see Fig. 26.1) even before visible chromosomal doubling. As the nucleolus re-forms in the "resting" nucleus, intense protein synthesis appears to occur just outside the nuclear envelope. Interphase is therefore not a period of cell rest but rather a period during which necessary cell constituents are synthesized.

The fact that damage to the nucleolus of a grasshopper neuroblast by microbeam irradiation at certain critical times (telophase to middle prophase) stops cell division, while irradiation of surrounding cell areas does not, suggests that the nucleolus might have a trigger function (Gaulden and Perry, 1958). Decisive evidence of the nucleolar role as the trigger is lacking.

It was also thought possible that the chromosomes or other organelles produce a specific substance that induces cell division. Some evidence has been reported that kinetin (6-furfurylaminopurine) may serve this function in some plant cells in which division is otherwise infrequent (Miller *et al.*, 1955; Stiles and Cocking, 1969). No convincing evidence has accumulated to show this to be general in plant cells, and kinetin does not affect animal cells. In fact, nuclear structures need not be present for the sea urchin egg cytoplasm to divide; parthenogenetic agents induce non-nucleated egg fragments to divide, but they develop abnormally and eventually die (see Fig. 7.3).

It has been suggested that cell division is initiated by a coagulative process similar to the formation of a clot in blood: any stimulus that induces clot formation would therefore be likely to initiate formation of spindles and cell division. Agents that inhibit clot formation (e.g., heparin) interfere with division of fertilized eggs. Agents that release calcium, which is also required for blood clotting, facilitate cell division, whereas agents (oxalate or citrate) that bind calcium inhibit cell division (Heilbrunn, 1955). These facts are probably interpretable on the basis that calcium is required for contracting the band of microfilaments in the cleavage furrow.

The two most critical events in the cell cycle at which signals must be given for action are the transition from G_1 to S, when replication of DNA occurs, and from G_2 to M, when mitosis begins. By fusing cells in different phases of the cell cycle with Sendai virus, the triggers can be described more precisely than by inferences drawn from the experiments just described.

When cells in the S phase are fused to cells in the G_1 phase of the cell cycle, the nuclei in the G_1 phase begin to synthesize DNA sooner than they would have done if allowed to proceed in the normal manner. Therefore, an activator must have developed during S phase to trigger DNA synthesis in the G_1 nuclei. The same happens even if mouse-hamster cell hybrids are used; hamster cells begin S phase sooner than do mouse cells, but the nuclei of the mouse cells are triggered to begin DNA synthesis in the fused cell at the same time as hamster cells.

However, each nucleus—whether from hamster or mouse—retains the length of its own period for DNA synthesis. Therefore, the signal for beginning DNA synthesis has no control over the length of time required to complete it (Mazia, 1974).

Initiation of M phase has also been studied by cell fusion. If an M phase cell is fused to a G_2 phase cell, the G_2 nucleus develops condensed chromosomes. Since replication of DNA occurred in the cells while they were in the S phase, the chromosomes are double. However, when an M phase cell is fused to a G_1 cell, chromosome condensation occurs in the G_1 nucleus, but the chromosomes are single because replication has not yet occurred in the G_1 cell. Again the result is not strictly species-specific and can be performed by fusing hamster and mouse cells in different cell cycle phases.

It remains an enigma why the chromosomes condense when M phase cells are fused to S phase cells. They condense in fragments, a phenomenon called pulverization. Again the result is not species-specific. One possible explanation for pulverization is that in eukaryotic cells replication occurs in many places on chromosome subdivisions (replicons). It is possible that the chromosomes fragmented into replicons condense into smaller units.

Whatever provides the trigger for cell division, many factors contribute toward a favorable state for cell division and preparations must be completed along many parallel lines before cell division can occur. The question of what provides the trigger for cell division, if indeed there is one, is open.

Formation and Function of the Mitotic Apparatus

That the division spindles are gels has long been known. This was demonstrated by micromanipulative studies and by experiments in which intact mitotic figures (spindle and chromosomes) were isolated from dividing eggs by the use of mild detergents that dissolve the rest of the cell (Fig. 26.13) (Mazia, 1961). The mitotic apparatus has also been isolated from dividing eggs by changes in pH (Kane, 1962) and from dividing eggs from which the membrane and the hyaline outer cytoplasm has been removed by immersion in ethylenediamine tetraacetic acid (EDTA), dextrose, and dithioglycol solutions (Mazia *et al.*, 1961). The mitotic apparatus of other cells has also been isolated, for example, from the giant amoeba (Goode and Roth, 1969) and from mammalian cells (Cande *et al.*, 1977). The isolated apparatus of the sea urchin egg is made up largely of one type of protein, namely tubulin, the protein characteristic of microtubules (Inoué and Ritter, 1975). In addition, about 3 to 5 percent RNA is present. Electrophoretic diagrams indicate two peaks, one of which is tubulin and the other its conjugate with RNA; both act as antigens (Zimmerman, 1963). Electron microscopic studies of fixed animal cells demonstrate that the spindle consists of definitely organized and oriented protein microtubules, the centers of orientation being the centrioles (see Chapter 3) (Mazia, 1961). When three pairs of centrioles are present, as in polyspermic eggs (eggs into which two or more sperm have entered), the orientation of the microtubules is toward three poles. Presumably, something in or from or around the centrioles orients microtubules, although evidence of its nature or its origin and transport is still lacking.

Plants generally lack centrioles, yet in their division spindle microtubules are oriented in parallel. Microtubule-associated proteins or other localized proteins are considered to exert lateral interaction on the microtubules leading to their orientation (Bajer, 1978). The nature of the orienting stimulus for microtubules in the division spindle therefore has not been explained.

That a spindle is necessary for cell division is demonstrated by experiments with agents that depolymerize microtubules. When a cell approaching division is poisoned with colchicine or colcemid, new microtubules do not appear, and some of those already present depolymerize; not all microtubules are sensitive to these drugs. The remaining microtubules are disoriented, and separation of the doubled chromosomes fails to occur. Because of this, polyploid cells may be produced by colchicine. Cells washed free of colchicine later divide but remain polyploid. The division spindle therefore provides at least a skeleton within which some agent or process separates the chromosomes.

Some investigators postulate that chromosomes move as a result of dissolution of spindle microtubules at the poles of the dividing cell as the mechanism by which chromosomes converge at the poles (Nicklas, 1975; Inoué and Ritter, 1975). While this is possible, an active contraction remains an attractive alternative.

Glycerol removes most of the cell constituents in muscle fibers, leaving the contractile fibers intact. Subsequent addition of ATP to such muscle "models" leads to contraction (see Chapter 23). Similar experiments were tried with dividing cells in which it was considered possible that contraction of the spindle fibers separated the paired chromosomes on the metaphase plate and moved them to the poles of the daughter cells. When cells in incipient division (showing spindles) were placed in

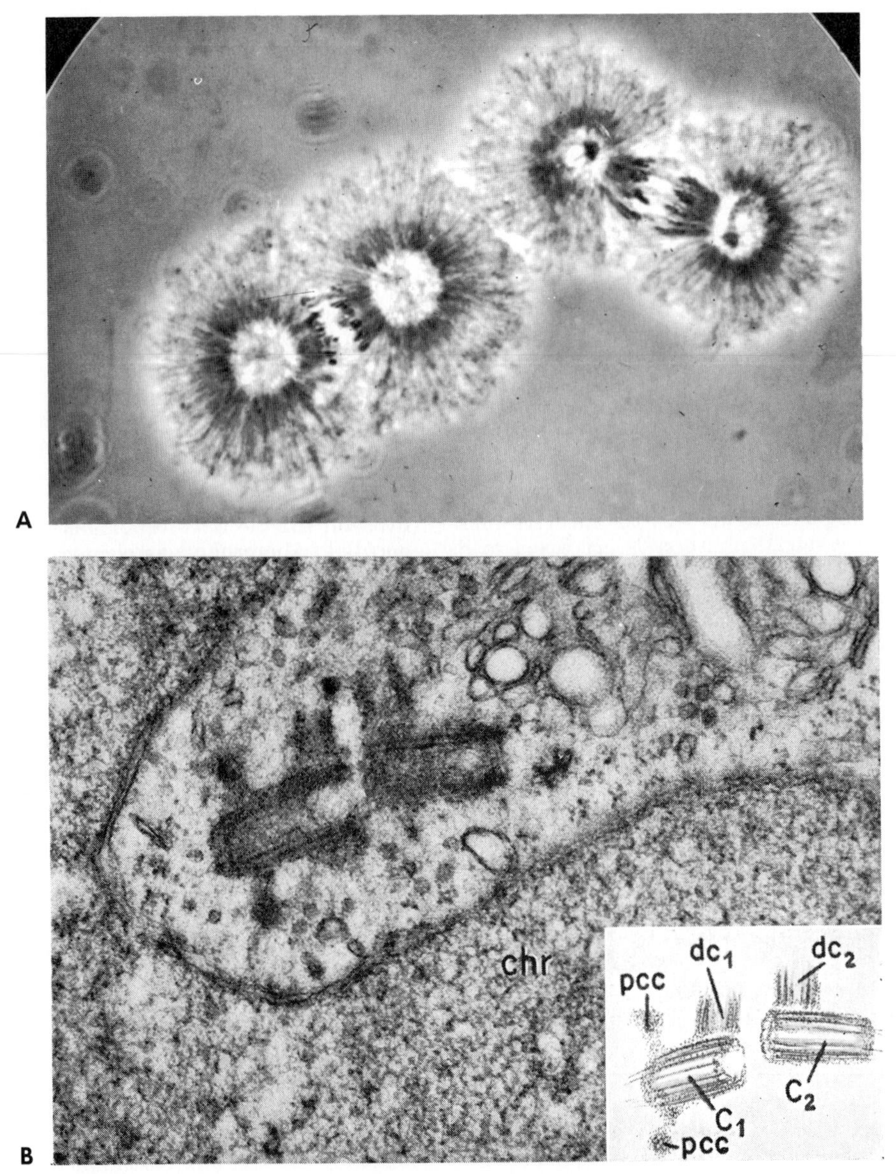

Figure 26.13. A, *Two isolated mitotic apparatuses at anaphase (developing eggs of the sea urchin* Strongylocentrotus purpuratus*). Note mitotic centers, astral rays, chromosomal fibers connecting chromosomes to mitotic centers, and interzonal regions of spindles between separated chromosomes. (From Mazia, 1956: Am. Scientist 44: 21.)* B, *Electron micrograph (and diagram in inset) of two centrioles of an embryonic cell in chicken spleen. The two centrioles, C_1 and C_2, are in the invaginated portion of the nuclear membrane near the Golgi body (at the top, right). Chr, chromatin; pcc, pericentriolar bodies; dc_1 and dc_2, daughter centrioles of C_1 and C_2, respectively (×60,000). (By courtesy of E. DeHaven and W. Barnard.)*

glycerol solutions, division ceased. (Endosperm cells of the blood lily *Haemanthus katherinae* divide in glycerol solutions up to a concentration of about 25 percent [Bajer, 1978].) When ATP was subsequently applied to the "models," the chromosomes moved toward the poles (Hoffmann-Berling, 1960). Interpretation of these results is controversial. It now seems likely that not spindle fiber (microtubule) contraction (which is not observed) but contraction of the cytoplasm squeezed the chromosomes toward the poles. The nature of the contractile element in the cytoplasm has not been determined with certainty.

Because there is now evidence for both actin (Cande *et al.*, 1977) and myosin (Fujiwara and Pollard, 1976) in the spindle of the intact dividing cell, in addition to tubulin of the microtubules, it is attractive to postulate a sliding filament model for chromosome movement, similar to that in muscle. The evidence is admittedly still tenuous. Remaining is the possibility of contraction in the matrix between the microtubules (Bajer, 1978). There is optimism that isolates and *in vitro* models may shed light on the process *in vivo,* although such work is only beginning (McIntosh, 1977).

CYTOKINESIS IN ANIMAL CELLS

Furrow Formation and Cleavage

Furrow formation (the indentation, in some cases circumferential) in a dividing animal cell is accompanied by increased viscosity in the furrow area. High pressures, which liquefy protein gels, also prevent cleavage furrow formation and cause dividing cells to fuse. The furrow disappears when even slight pressure is applied to both sides with a microneedle, but cell division resumes when the pressure is released (Guttes and Guttes, 1960). An egg ruptured at the time of furrow formation disintegrates, but a part of the cortex (the outer portion of the cytoplasm) and the furrow remain intact, indicating that the furrow is rigidly structured. Furthermore, it has been noticed that as the furrow forms in cleaving eggs, a slight separation between the cortex of each of the two blastomeres (the two-celled stage) is maintained for a time, indicating occurrence of some change that prevents coalescence of the two layers (Marsland, 1957).

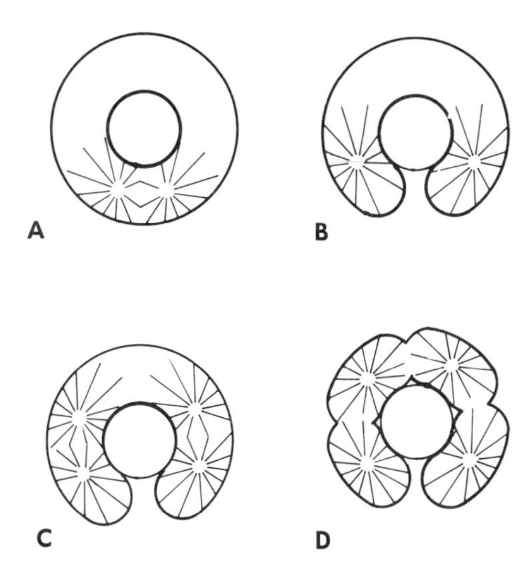

Figure 26.14. *Cleavage of a torus-shaped echinoderm egg.* A, B, *Immediately before and after the first cleavage.* C, *Immediately before second cleavage. Each pair of asters is joined by a spindle.* D, *Beginning third cleavage; furrows form adjacent to the spindles and between the two asters that were not joined by a spindle, resulting in uninucleate cells. The central circle represents a glass sphere. (From Rappaport, 1971: Int. Rev. Cytol.* 31*: 173.)*

It was early postulated that the trigger for cell division was a nuclear secretion that altered the cell surface, but enucleated eggs form spindles and asters and cleave without chromosomes (Harvey, 1936). Cytokinesis requires the presence of the mitotic apparatus in some form, with or without chromosomes, and furrowing will occur even where astral rays alone are present (Fig. 26.14*D*). If the mitotic apparatus is removed before anaphase, furrows do not form; if it is removed after anaphase, cleavage continues to completion (Hiramoto, 1956; 1965). Cleavage furrows are formed in the same place as when the mitotic apparatus is intact. *Tetrahymena* also divides even if the macronucleus is removed from the amicronucleate strain (Nachtwey, 1965). Presumably the mitotic apparatus has already stimulated formation of the furrow by anaphase, and its presence is no longer essential for completion of cleavage. Much of the present discussion is from the excellent summary of Rappaport (1971).

If the stimulus for furrow formation passes from the mitotic apparatus to the cell poles (Fig. 26.15*A*), it should be possible to detect changes in the properties of the polar cell surface before and during cleavage—but no changes have been reported. If the stimulus for furrow formation passes from the mitotic apparatus to the equator of the cell (Fig. 26.15*B*), it should be possible to detect changes in equatorial properties related to cytokinesis. The equatorial cortex indeed thickens and becomes more rigid and less elastic than the remainder of the egg cortex. Its chemical properties and staining also change. Perhaps most interesting of all, at this time microfilaments 4 to 7 nm in diameter, arranged circumferentially, may be seen at the furrow base in electron micrographs of the cortex.

The region of the prospective furrow is altered in preparation for cleavage. It becomes thickened and increases in mechanical strength. Surface stiffness increases at metaphase, rising to a maximum at cleavage, after which it declines rapidly. The alterations are such as would be expected if cytokinesis is to be accomplished by contraction. In some cases (e.g., frog egg), furrow formation continues even in an isolated piece of the egg equatorial surface, as if it were programmed for furrowing. However, not all cells continue to cleave when the equatorial ring is cut, especially from the inside. If the membrane of an egg about to cleave is torn, it disintegrates except for the furrow.

Microfilaments appear in the equatorial ring as it forms (see Fig. 26.19*B*). Agents that interfere with microfilaments also interfere with furrowing and cleavage. For example, surface active chemicals (detergents) disrupt cleavage immediately. Hydrostatic pressure at a

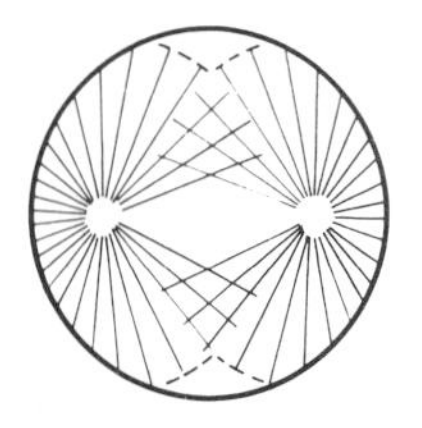
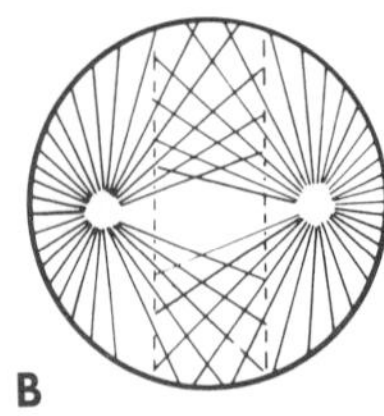

Figure 26.15. Diagrammatic representation of cleavage stimulus patterns. A, *Polar stimulation model. The influence of the asters reaches the polar surfaces but fails to reach the equatorial surface peripheral to the dotted lines. The equatorial surface is not altered by stimulation.* B, *Equatorial stimulation model. The entire cell surface can be reached by influence from the asters or achromatic apparatus, but the equatorial surface between the dotted lines is subjected to greater stimulatory activity. The equatorial surface is altered by stimulation. (From Rappaport, 1971: Int. Rev. Cytol.* 31: *175.)*

threshold value known to cause dissolution of microfilaments reverses cleavage. The effect is reversible; lowering the pressure permits the cleavage furrow to re-form, a cycle that can be repeated a number of times. Addition of ATP to eggs enables them to withstand higher pressures before cleavage reversal; ATP is known to cause contraction of fibrils of various types (see Chapter 23). Conversely, mersalyl acid, which decreases critical gel strength, presumably by disorganizing microfibrils, decreases the threshold required to reverse a cleavage furrow (Rappaport, 1971).

We have good evidence that actin is present in dividing mammalian cells in tissue culture, as demonstrated by the decoration of the extracted actin by heavy meromyosin, much in the manner that characterizes muscle actin (Cande *et al.*, 1977). Evidence gained from immunofluorescence studies indicates the presence of myosin in similar cells (Fujiwara and Pollard, 1976). Therefore, it is most appealing to suggest the hypothesis that an actomyosin type of contraction in a contractile band in the furrow of animal cells serves as the mechanism of division. More evidence must be accumulated; especially convincing would be purified extracts of myosin from such cells and demonstration of its reaction with actin in much the same manner as in muscle extracts. However, as pointed out in Chapter 23, myosin is always present in much smaller quantities than actin in nonmuscular primitive contractile systems.

Information is needed on many aspects of cytokinesis before definite conclusions can be drawn. Thus the nature of the stimulating influence that passes from the mitotic apparatus to the cell surface and the means by which it reaches its destination are unknown. The molecular biology of the contractile mechanism producing division and the way it operates are intriguing problems, as yet scarcely more than outlined.

Work with eggs placed under pressure and subjected to various inhibitory reagents suggests that ATP supplies the energy for whatever movements occur during cleavage, since addition of ATP often relieves the effect of an inhibitor (Zimmerman *et al.*, 1957). The energy required for egg cell division is about three times that normally available from respiration. Since the respiratory rate rises only a few percent during cleavage, the energy for cleavage must come from a supply of high-energy phosphate bonds, accumulated in preparation for each successive cleavage (Zeuthen, 1971a and b). Theories of cytokinesis were developed primarily from data on marine eggs. They do not apply to such phenomena as multiple fission, which occurs in some plant and animal cells, nor do they take into consideration the division of plant cells and bacteria. In this process, development of a new cell wall is of prime importance for cytokinesis.

CYTOKINESIS IN PLANT AND BACTERIAL CELLS

In plant cells and bacteria enclosed in cell walls, cytokinesis differs from that in animal cells by the formation of a *cell plate* (the first visible sign of a wall in a dividing cell) between the daughter cells resulting from a cell division. The cell plate forms the basis of the future cell wall. Centrioles are lacking in most plant cells, but a typical spindle with poles is present, as indicated by the polar convergence of the microtubules of the spindle. When daughter nuclei are re-forming, the cell plate appears in the center of the spindle midway between the poles (Buvat, 1969).

Electron microscopic observations have shown that the formation of a phragmoplast, a modified region of the equatorial plane of the spindle that appears after the chromosomes have moved to the poles, precedes the formation of the equatorial cell plate. The phragmoplast consists of aligned and interdigitated microtubules. Some of the microtubules are probably re-formed from microtubules originally present elsewhere in the cell. Among the oriented microtubules appear vesicles that produce the cell wall materials. The products of the vesicles advance beyond the region of the phragmoplast, ultimately reaching the cell walls to either side of the cell, thus completing the cell plate (see Fig. 26.2) (Lambert and Bajer, 1972). While it is generally stated that the vesicles are Golgi bodies (called dictyosomes in plants), in the blood lily *Haemanthus katherinae* the vesicular bodies show a close resemblance to the endoplasmic reticulum. In

Figure 26.16. *Differentiation in a semicell after cell division in the green alga* Micrasterias denticulata. A, *Early bulge stage (×370).* B, *Initiation of marginal differentiation into five lobes (×340).* C, *Further expansion of the five lobes and beginning of movement of the chloroplast into the forming semicell (×370).* D, *The five lobes have undergone two more series of marginal bifurcations to give the 17-lobe stage. Note that the polar lobe does not bifurcate. (From Pickett-Heaps, 1975: The Green Algae. Copyright © 1975 Sinauer Associates, Sunderland, Mass.)*

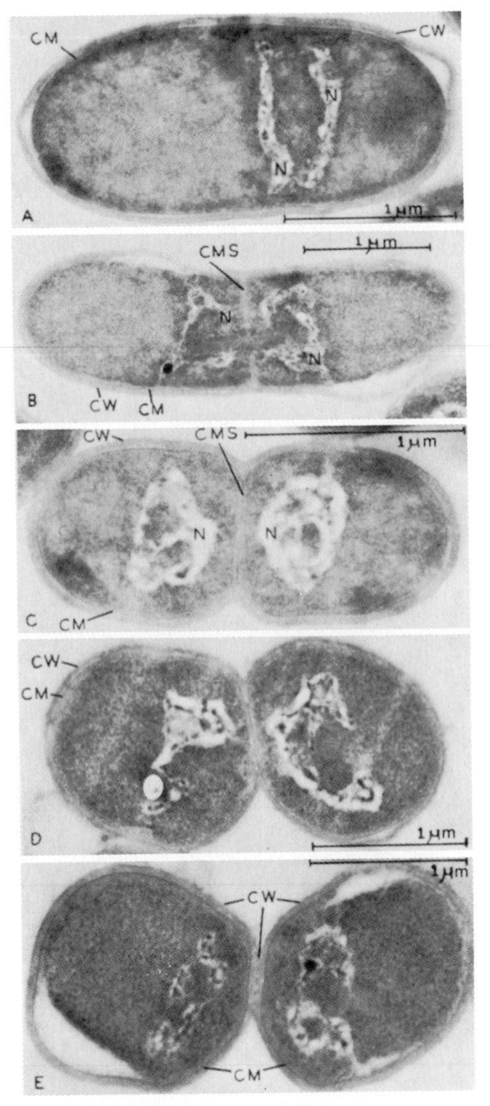

Figure 26.17. Electron photomicrographs of ultrathin sections of an unidentified bacterium showing various stages in cell division. A, Before cell division; the nuclear material (N) is in the form of two bars through which a threadlike component extends. CM, cell membrane; CW, cell wall. B, The nuclear material has divided and the cytoplasmic membrane septum (CMS) nearly separates the cytoplasm into two portions. C, The cytoplasmic membrane septum is complete. D, Two layers of the cytoplasmic membrane septum can be distinguished: the cells are becoming rounded. E, The cell wall is complete between the daughter cells, which have nearly separated. (From Carpenter, P. L., 1977: Fundamentals of Microbiology. 4th Ed. W. B. Saunders Co., Philadelphia.)

this species Golgi bodies are sparse, so perhaps the formation of the cell plate differs from that in the other species of plants studied (Hepler and Jackson, 1968). The vesicles are covered by unit membranes, and it is thought that as they unite they may form the plasma membranes separating daughter cells.

A striking case of semicell morphogenesis and cell wall formation following cell division is seen in the green algae *Micrasterias denticulata.* A bleb of cytoplasm containing the daughter nucleus is extruded from the original semicell. Marginal differentiation into five lobes is then observed. The chloroplast moves into the semicell and the lobes further bifurcate. A cell wall gradually forms around it (Fig. 26.16) (Pickett-Heaps, 1975).

The details of cytokinesis in plants continue to be investigated. However, the mechanisms by which the phragmoplast appears and the cell wall subsequently forms are virtually unknown, although hypotheses abound (Bajer and Molé-Bajer, 1973).

In bacterial and other prokaryotic cells a cell plate is formed about midway between the two ends of a cell in which the "nuclear" material (the chromosome, a double-stranded DNA molecule; see Chapter 15) has divided. However, the cell plate in this case appears as a furrow or an invagination of the cell membrane; cell walls are formed between the apposed cell membranes. When the ends of the invagination meet at the center of the cell, cell division is essentially complete, the cell walls are finished, and the two cells can separate (Fig. 26.17). A possible derivation of the more complex mechanism of cell division in plant eukaryotic cells from simple furrowing characteristic of prokaryotic cells is shown in Figure 26.18.

Perhaps the apparent marked difference in cytokinesis in animal, plant, and prokaryotic cells is an indication of the distant evolutionary separation of these cell lines. On the other hand, it is the formation of a cell wall and the way that the wall is formed that makes it seem that cytokinesis is so different in these groups.

CELL SURFACE CHANGES DURING CYTOKINESIS

Scanning electron microscope studies reveal that the cell surface is very active during cytokinesis; for example, in the zebra fish egg folds and microvilli appear (Fig. 26.19*A*). The microvilli may change in size but they are present throughout cytokinesis. Only a few shallow folds are present on the surface of the newly fertilized egg but as the constriction of the division furrow progresses the folds radiating from the furrow become more prominent, and in a late stage of furrow formation the radiating surface is covered with many shallow folds.

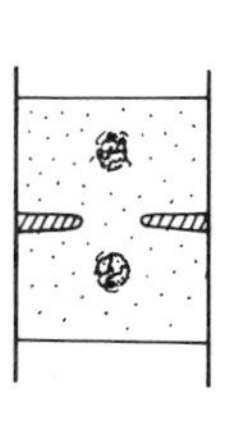

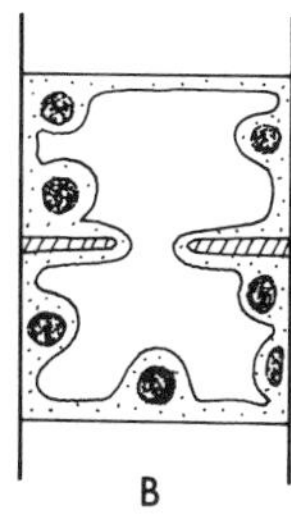

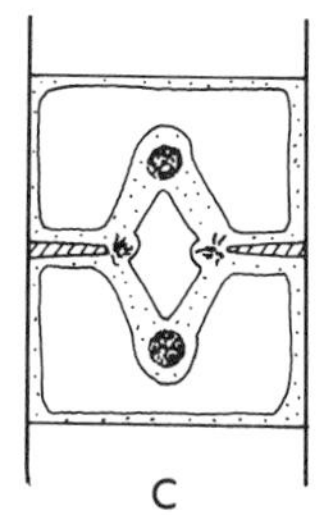

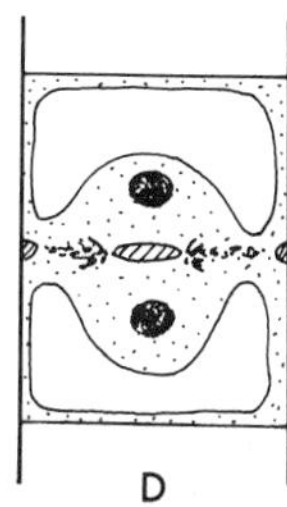

Figure 26.18. *Diagrammatic representation of a possible evolutionary sequence giving rise to the phragmoplast of higher plants, the stages being derived from present-day organisms. A, Blue-green algae, prokaryotic with cytokinesis accomplished by furrowing. B, Eukaryotic algae (multinucleate in this case): cytokinesis involving furrowing, but in no way associated with spindle apparatus (e.g., Cladophoraceae). C, Cytokinesis occurring initially by furrowing, then later involving a phragmoplast-type structure as the membrane impinges on the telophase spindle (e.g.,* Spirogyra*). D, Phragmoplast forming new cell wall, generally from the center outward: some trace of membrane furrowing may possibly exist (e.g., some algae and higher plants generally). (From Pickett-Heaps, 1972:* In *Advances in Cell Biology. Vol. 2. Prescott, Goldstein and McConkey, eds. Appleton-Century-Crofts, New York, p. 225.)*

Transmission electron micrographs of the furrow of a dividing zebra fish egg show a contractile band 100 to 125 μm thick composed of microfilaments 4 to 6 nm in diameter (seen also in the cortex) arranged parallel (as seen in tangential sections) and associated with sections of the endoplasmic reticulum (Fig. 26.19*B*).

In the frog egg microvilli and alternate ridges and furrows in the cleavage furrow may be observed. While the details differ, the surface activity resembles that seen in the zebra fish egg.

Even more striking surface changes than those seen in the zebra fish egg are seen in dividing HeLa and KB mammalian cells in tissue culture: microvilli, filopodia (retractile thin filamentous pseudopodial extensions), blebs, and ruffles (Fig. 26.19*C* and *D*). The cells round up during mitosis, anchored in position by the filopodia, and they remain rounded until the end of mitosis; in G_1 they again flatten over the glass and remain flattened during the G_1, S and G_2 phases of the cell cycle.

The cause and function of blebbing and bubbling during mitosis is not fully understood, nor has an explanation of the waves of ruffling been reported. However, it may be significant that during the period in which the surface changes are prominent lectin-receptive sites are exposed on the cell surface (Beams and Kessel, 1976). Although scanning electron micrographs are from dead dried cells, the quickly dried cells have not been altered by fixatives and probably represent what is present on the living cell.

SUMMARY

In both prokaryotic and eukaryotic cells DNA synthesis is a central event in the cell cycle, but RNA synthesis and protein synthesis are necessary for growth of the cell in bulk. Cells generally double their volume before cell division. Prokaryotic cells generally divide more rapidly than eukaryotic cells. Division of both types of cells can be synchronized by selection of cells in certain stages of the cell cycle, by chemical shock, and by physical shock (usually high or low temperature). Synchronized cells have been extensively used in experiments on the cell cycle.

DNA synthesis is generally continuous in metabolizing prokaryotic cells but confined to the S (synthetic) phase of the cell cycle in eukaryotic cells. Histones are synthesized during the S phase, while non-histone proteins in both nucleus and cytoplasm are largely synthesized during interphase.

Experiments have not yet identified what it is that triggers cell division. However, a DNA-programmed sequence of biochemical and morphogenetic events characterizes the cell cycle, ending in cell division (Fig. 26.20).

Cell division (cytokinesis) in eukaryotic cells generally follows nuclear division (karyokinesis). In this process during mitosis the chromosomes, duplicated at interphase, are separated from one another and aggregated in telophase by the mitotic apparatus at the poles of the prospective daughter cells. In animal cells a spindle forms between the centrioles situated at the poles of the cell; centrioles are lacking in plant cells although the mitotic apparatus is otherwise similar to that in animal cells. Movement of the chromosomes is probably achieved by elements in the mitotic spindle; however, no agreement has been reached as to whether the microtubules or the microfilaments do this. The microtubules do not appear to be contractile and might serve only to shape the mitotic apparatus, depolymerizing late in mitosis. Some workers even postulate that linear depolymerization permits separa-

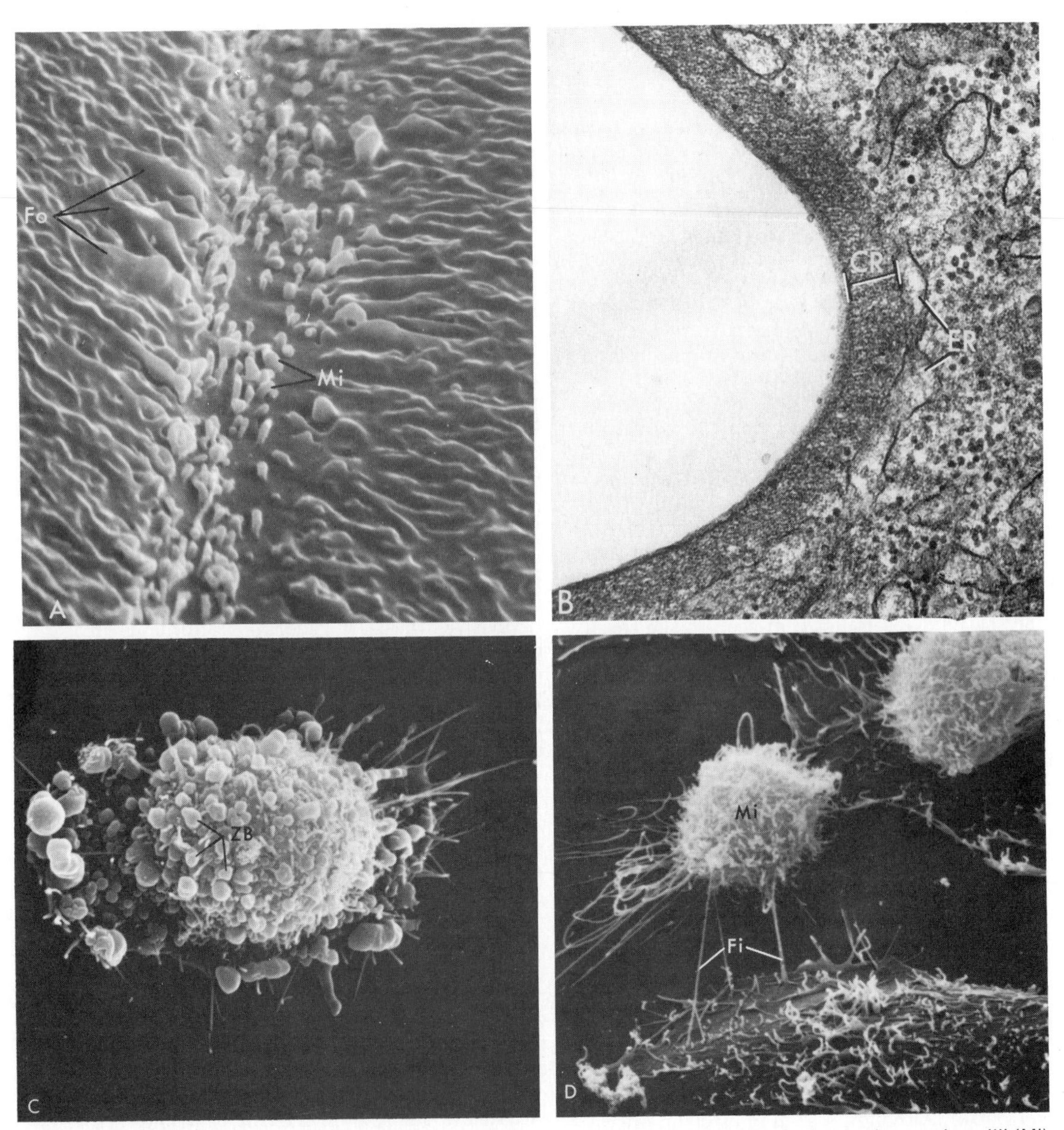

Figure 26.19. A, *Cleavage furrow in the zebra fish egg. Scanning electron micrograph clearly shows microvilli (Mi) and folds (Fo) (×3150).* B, *Transmission electron micrograph of the contractile ring (CR) in the cleavage furrow of the zebra fish egg, four-cell stage (approximately ×78,000).* C, *Blebbing (ZB) on an interphase cell of a mammalian (KB) cell in tissue culture (×1889).* D, *Scanning electron micrograph of a HeLa (mammalian tumor) cell in tissue culture showing numerous filopodia (Fi) and microvilli (Mi) (×1664). (From Beams and Kessel, 1976: Am. Scientist* 64: *279–290.)*

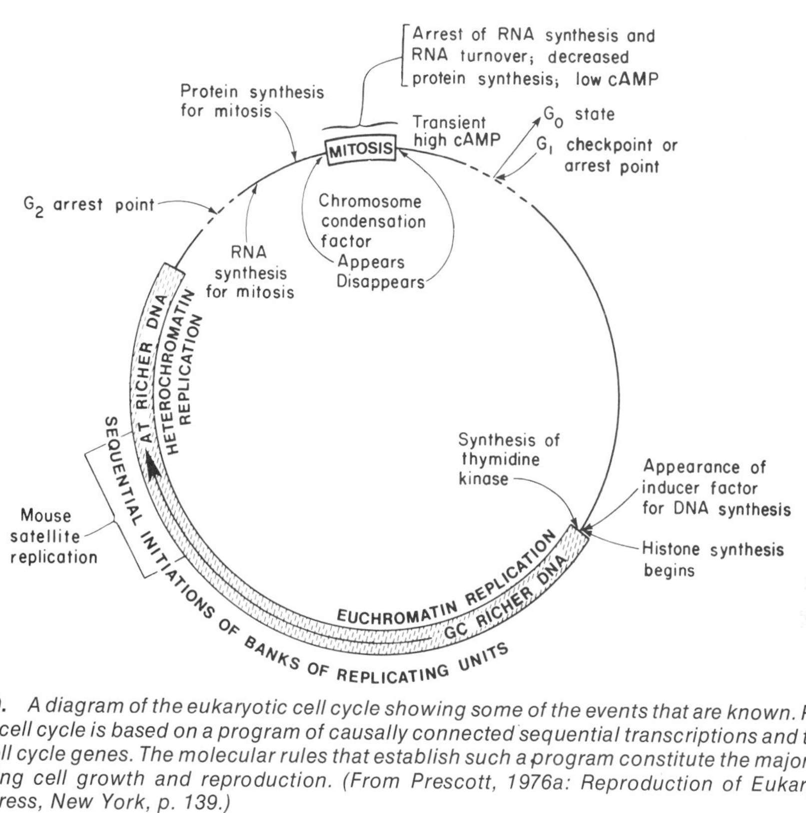

Figure 26.20. *A diagram of the eukaryotic cell cycle showing some of the events that are known. Progression through the cell cycle is based on a program of causally connected sequential transcriptions and translations of a set of cell cycle genes. The molecular rules that establish such a program constitute the major problem in understanding cell growth and reproduction. (From Prescott, 1976a: Reproduction of Eukaryotic Cells. Academic Press, New York, p. 139.)*

tion of the chromosomes. The problem of chromosome movement remains open.

Furrow formation precedes cytokinesis in animal cells. Microfilaments aggregate in the furrow region, forming a contractile band, and the band contracts as the furrow deepens until cleavage occurs. Contraction of the microfilaments in the band may be similar to the sliding filament mechanism of muscular contraction inasmuch as actin has been extracted from the band and indirect evidence for myosin (fluorescence of antibodies to myosin) has been reported. It must be remembered that in nonmuscular primitive contractile systems myosin is always present only in minute amounts. The mechanism of microfilament contraction in cleavage remains to be explained.

In cells with cell walls, cell division is fundamentally different from that in animal cells. In bacteria, division follows cell plate formation between the prospective daughter cells. In plant cells, cell walls are synthesized by phragmoplasts laid down by elements of the Golgi body. Fusion of the membranes of the phragmoplasts gives rise to the daughter cell membranes without furrow formation.

LITERATURE CITED AND GENERAL REFERENCES

Agrell, I., 1964: Natural division synchrony and mitotic gradients in metazoan tissues. *In* Synchrony in Cell Division and Growth. Zeuthen, ed. Interscience Publishers, New York, pp. 39–67.

Albrecht-Buehler, G., 1977: Daughter 3T3 cells: Are they mirror images of each other? J. Cell Biol. *72*: 595–603.

Attallah, A. M., 1976: Regulation of cell growth *in vitro* and *in vivo*. Point/counterpoint. *In* Chalones. Houck, ed. American Elsevier, New York, pp. 141–172.

Bajer, A., 1965: Cinemicrographic analysis of cell plate formation in endosperm. Exp. Cell Res. *37*: 376–398.

Bajer, A. S. and Molé-Bajer, J., 1973: Spindle dynamics and chromosome movements. Int. Rev. Cytol. Suppl. 3, pp. 1–280.

Bajer, A. and Molé-Bajer, J., 1975: Lateral movements in the spindle and the mechanism of mitosis. *In* Molecules and Cell Movement. Inoué and Stephens, eds. Raven Press, New York, pp. 77–96.

Bajer, A., 1978: Personal communication.

Barner, H. D. and Cohen, S. S., 1956: Synchroniza-

tion of division of a thymineless mutant of *Escherichia coli*. J. Bacteriol. *68*: 80–88.
Beams, H. W. and Kessel, R. G., 1976: Cytokinesis: a comparative study of cytoplasmic division in animal cells. Am. Sci. *64*: 279–290.
Berridge, M. J., 1976: Calcium, cyclic nucleotides and cell division. Soc. Exp. Biol. Symp. *30*: 219–231.
Bruce, V. G., 1965: Cell division rhythms and the circadian clock. *In* Circadian Clocks. Aschoff, ed. North Holland Publishing Co., Amsterdam, pp. 123–138.
Bullough, W. S., 1952: Energy relations of mitotic activity. Biol. Rev. *27*: 133–168.
Bullough, W. S. and Mitani, F., 1976: An analysis of the epidermal chalone control mechanisms. *In* Chalones. Houck, ed. American Elsevier, New York, pp. 7–36.
Burns, V. W.. 1964: Synchronization of division in bacteria by nutritional change. *In* Synchrony of Cell Division and Growth. Zeuthen, ed. Interscience Publishers, New York, pp. 433–439.
Buvat, R., 1969: Plant Cells. McGraw-Hill Book Co., New York.
Cameron, J. L. and Thrasher, J. D., eds., 1971: Cellular and Molecular Renewal in the Mammalian Body. Academic Press, New York.
Cande, W. Z., Lazarides, E. and McIntosh, J. R., 1977: A comparison of the distribution of actin and tubulin in the mammalian spindle as seen in direct immunofluorescence. J. Cell Biol. *72*: 552–567.
Caplan, A. I. and Ordahl, C. P., 1978: Irreversible gene repression model for control of development. Science *201*: 120–130.
Cone, D., Jr., 1970: Variation of the transmembrane potential level as a basic mechanism of mitosis control. Oncology *24*: 438–470.
Dan, K., 1966: Behavior of sulfhydryl groups in synchronous division. *In* Cell Synchrony: Studies in Biological Regulation. Cameron and Padilla, eds. Academic Press, New York, pp. 307–327.
Deysson, G., 1968: Antimitotic substances. Int. Rev. Cytol. *24*: 99–148.
De Robertis, E. D. P., Saez, F. A. and De Robertis, E. M. F.: Cell Biology. 6th Ed. W. B. Saunders Co., 1975.
De Terra, N., 1969: Cytoplasmic control over the nuclear events of cell reproduction. Int. Rev. Cytol. *25*: 1–29.
Edmunds, L. N., Jr., 1977: Clocked cell cycle clocks. Waking and Sleeping *1*: 227–252.
Epel, D., 1963: The effects of carbon monoxide inhibition on ATP level and the rate of mitosis in the sea urchin egg. J. Cell Biol. *17*: 315–319.
Erikson, R. O., 1964: Synchronous cell and nuclear division in tissues of the higher plants. *In* Synchrony in Cell Division and Growth. Zeuthen, ed. Interscience Publishers, New York, pp. 11–37.
Fawcett, D. W., Ito, I. and Slautterback, D., 1959: The occurrence of intercellular bridges in groups of cells exhibiting synchronous division. J. Biochem. Biophys. Cytol. *5*: 453–460.
Forer, A. and Jackson, W. T., 1976: Actin filaments in the endosperm of a higher plant *Haemanthus katherinae* Baker. Cytobiologie *12*: 197–214.
Fuge, H., 1978: Ultrastructure of the mitotic spindle. Int. Rev. Cytol. Suppl. *6*: 1–58.
Fujiwara, K. and Pollard, T. D., 1976: Fluorescent antibody localization of myosin in the cytoplasm, cleavage furrow and mitotic spindle of human cells. J. Cell Biol. *71*: 848–875.
Gaulden, M. E. and Perry, R. P., 1958: Influence of nucleolus on mitosis as revealed by ultraviolet microbeam irradiation. Proc. Nat. Acad. Sci. U.S.A. *44*: 554–559.
Goode, D. and Roth, L. E., 1969: The mitotic apparatus of a giant ameba. Solubility properties and induction of elongation. Exp. Cell Res. *58*: 343–352.
Greenspan, H. P., 1977: On the dynamics of cell cleavage. J. Theor. Biol. *65*: 79–99.
Guttes, E. and Guttes, S., 1960: Suppression of cytoplasmic division in *Amoeba proteus*. Nature *187*: 520–521.
Hamburger, K. and Zeuthen, E., 1957: Synchronous divisions in *Tetrahymena pyriformis* as studied in an inorganic medium. The effect of 2,4-dinitrophenol. Exp. Cell. Res. *13*: 443–453.
Harding, C. V., Reddon, J. R., Unakar, N. J. and Bagchi, M., 1971: The control of cell division in the ocular lens. Int. Rev. Cytol. *31*: 216–300.
Hartwell, L. H., 1978: Cell division from a genetic perspective. J. Cell Biol. *77*: 627–637.
Hartwell, L. H. and Unger, M. W., 1977: Unequal division in *S. cereviseae* and its implication for the control of cell division. J. Cell Biol. *75*: 422–435.
Harvey, E. B., 1936: Parthenogenic merogony or cleavage without nuclei in *Arbacia punctulata*. Biol. Bull. *71*: 101–121.
Hastings, J. W. and Sweeney, B. M., 1964: Phased cell division in marine dinoflagellates. *In* Synchrony in Cell Division and Growth. Zeuthen, ed. Interscience Publishers, New York, pp. 307–349.
Heilbrunn, L. V., 1955: Dynamics of Living Protoplasm. Academic Press, New York.
Hepler, P. K., 1976: The blepharoplast of *Marsilia*: its *de novo* formation and spindle association. J. Cell Sci. *21*: 361–390.
Hepler, P. K. and Jackson, W. T., 1968: Microtubules and early stages of cell-plate formation in the endosperm of *Haemanthus katherinae* Baker. J. Cell Biol. *38*: 437–446.
Hepler, P. K. and Newcomb, E. H., 1967: Fine structure of cell plate formation in the apical meristem of *Phaseolus* roots. J. Ultrastruct. Res. *19*: 498–513.
Hiramoto, Y., 1956: Cell division without mitotic apparatus in sea urchin eggs. Exp. Cell Res. *11*: 630–656.
Hiramoto, Y., 1965: Further studies on cell division without mitotic apparatus in sea urchin eggs. J. Cell Biol. *25*: 161–167.
Hoffmann-Berling, H., 1960: Other mechanisms producing movements. Comp. Biochem. *2*: 341–370.
Houck, J., ed., 1976: Chalones. American Elsevier, New York.
Howard, A. and Pelc, S. R., 1953: Synthesis of DNA in normal and irradiated cells and its relation to chromosome breakage. Heredity *6*: 261–273.
Inoué, S., 1976: Chromosome movement by reversible assembly of microtubules. *In* Cell Motility. Goldman, Pollard and Rosenbaum, eds. Cold Spring Harbor Conferences on Cell Proliferation *3*: 1317–1328.
Inoué, S. and Ritter, H., Jr., 1975: Dynamics of mitotic spindle organization and function. *In* Molecules and Cell Movement. Inoué and Stephens, eds. Raven Press, New York, pp. 3–30.
Iverson, R. M. and Giese, A. C., 1957: Nucleic acid

content and ultraviolet susceptibility of *Tetrahymena pyriformis*. Exp. Cell Res. *13*: 213–223.

Jeter, J. R., Jr., Cameron, J. L., Padilla, G. M. and Zimmerman, A. M., 1978: Cell Cycle Regulation. Academic Press, New York.

Kane, R. E., 1962: The mitotic apparatus—isolation by controlled pH. J. Cell Biol. *12*: 47–55.

Kubai, D. R., 1975: The evolution of the mitotic spindle. Int. Rev. Cytol. *43*: 167–227.

Lambert, A. and Bajer, A. S., 1972: Dynamics of spindle fibers and microtubules during anaphase and phragmoplast formation. Chromosoma *39*: 101–144.

Luykx, P., 1970: Cellular Mechanisms of Chromosome Distribution. Academic Press, New York.

McIntosh, J. R., 1977: Mitosis in vitro: isolates and models of the mitotic apparatus. *In* Mitosis: Facts and Questions. Little, Palewitz, Petzelt, Ronstingl, Schroeter and Zimmerman, eds. Springer-Verlag, New York, pp. 167–195.

McIntosh, J. R., Hepler, P. K. and Van Wie, D. G., 1969: Model for mitosis. Nature *224*: 659–663.

McIntosh, R., Cande, W. Z. and Snyder, J. A., 1975: Structure and physiology of the mammalian mitotic spindle. *In* Molecules and Cell Movement. Inoué and Stephens, eds. Raven Press, New York, pp. 31–76.

Marks, F., 1976: The epidermal hormones. *In* Chalones. Houck, ed. American Elsevier, New York, pp. 173–227.

Marsland, D., 1957: Temperature-pressure studies on the role of sol-gel reactions in cell division. *In* Influence of Temperature on Biological Systems. Johnson, ed. American Physiological Society, Washington, D.C., pp. 111–126.

Mazia, D., 1961: How cells divide. Sci. Am. (Sept.) *205*: 100–120.

Mazia, D., 1961: Mitosis and the physiology of cell division. *In* The Cell. Vol. 3. Brachet and Mirsky, eds. Academic Press, New York, pp. 77–412.

Mazia, D., 1963: Synthetic activities leading to mitosis. J. Cell. Comp. Physiol. *62*: 123–140.

Mazia, D., 1974: The cell cycle. Sci. Am. (Jan.) *230*: 54–64.

Mazia, D., Chaffee, R. R. and Iverson, R. M., 1961: Adenosine triphosphatase in the mitotic apparatus. Proc. Nat. Acad. Sci. U.S.A. *47*: 788–790.

Mazia, D., 1977: Future research on mitosis. *In* Mitosis: Facts and Questions. Little, Palewitz, Petzelt, Ronstingl, Schroeter, and Zimmerman, eds. Springer-Verlag, New York, pp. 196–219.

Meeker, G. L. and Iverson, R. M., 1971: Tubulin synthesis in fertilized sea urchin eggs. Exp. Cell Res. *64*: 129–132.

Miki-Noumura, T., 1968: Purification of the mitotic apparatus protein of sea urchin eggs. Exp. Cell Res. *50*: 54–64.

Miki-Noumura, T. and Kondo, H., 1970: Polymerization of actin from sea urchin eggs. Exp. Cell Res. *61*: 31–41.

Miller, C. O., Skoog, F., Von Saltza, M. H. and Strong, F. M., 1955: Kinetin, a cell division factor from deoxyribonucleic acid. J. Am. Chem. Soc. *77*: 1392.

Mitchison, J. M., 1952: Cell membranes and cell division. Symp. Soc. Exp. Biol. *6*: 105–127.

Mitchison, J. M., 1969: Enzyme synthesis in synchronous culture. Science *165*: 657–663.

Mitchison, J. M., 1971: The Biology of the Cell Cycle. Cambridge University Press, Cambridge.

Mitchison, J. M., 1974: Sequences, pathways and timers in the cell cycle. *In* Cell Cycle Control. Padilla, Cameron and Zimmerman, eds. Academic Press, New York, pp. 125–142.

Monesi, V., 1969: DNA, RNA, and protein synthesis during the mitotic cell cycle. Handbook of Molecular Cytology. Lima-de-Faria, ed. North Holland Publishing Co., Amsterdam.

Murayama, Y., 1964: Synchrony of bacterial populations as established by filtration techniques. *In* Synchrony of Cell Division and Growth. Zeuthen, ed. Interscience Publishers, New York, pp. 421–432.

Nachtwey, D. S., 1965: Division of synchronized *Tetrahymena pyriformis* after emacronucleation. Compt. Rend. Lab. Carlsberg *35*: 23–25.

Nicklas, R. B., 1972: Mitosis. *In* Advances in Cell Biology. Vol. 2. Prescott, Goldstein and McConkey, eds. Appleton-Century-Crofts, New York, pp. 225–297.

Nicklas, R. B., 1975: Chromosome movement: current models and experiments on living cells. *In* Molecules and Cell Movement. Inoué and Stephens, eds. Raven Press, New York, pp. 97–117.

Pickett-Heaps, J. D., 1975: The evolution of the mitotic apparatus: An attempt at comparative ultrastructural cytology in dividing plant cells. Cytobios *1*: 257–280.

Pickett-Heaps, J. D., 1975: The Green Algae. Sinauer Associates, Sunderland, Mass.

Plesner, P., 1961: Changes in ribosome structure and function during synchronized cell division. Cold Spring Harbor Symp. Quant. Biol. *26*: 159–162.

Plesner, P., 1964: Nucleotide metabolism during synchronized cell division in *Tetrahymena pyriformis*. Compt. Rend. Lab. Carlsberg *34*: 1–76.

Prescott, D. M., 1964: The normal cell cycle. *In* Synchrony in Cell Division and Growth. Zeuthen, ed. Interscience Publishers, New York, pp. 71–97.

Prescott, D. M., 1976a: Reproduction of Eukaryotic Cells. Academic Press, New York.

Prescott, D. M., 1976b: The cell cycle and control of cellular reproduction. Adv. Genet. *18*: 99–177.

Puck, T. T., 1957: Single human cells in vitro. Sci. Am. (Aug.) *197*: 91–100.

Puck, T. T., 1960: *In vitro* studies on the radiation biology of mammalian cells. Prog. Biophys. Mol. Biol. *10*: 237–258.

Rappaport, R., 1971: Cytokinesis in animal cells. Int. Rev. Cytol. *31*: 169–213.

Rappaport, R., 1975: Establishment and organization of the cleavage mechanism. *In* Molecules and Cell Movement. Inoué and Stephens, eds. Raven Press, New York, pp. 287–304.

Rasmussen, L., 1963: Delayed divisions in *Tetrahymena* as induced by short-time exposures to anaerobiosis. Compt. Rend. Lab. Carlsberg *33*: 53–71.

Rebhun, L. I., 1977: Cyclic nucleotides, calcium, and cell division. Int. Rev. Cytol. *47*: 1–54.

Rebhun, L., Math, J. and Renillard, S. P., 1976: Sulfhydryl and regulation of cell division. *In* Cell Motility. Goldman, Pollard and Rosenbaum, eds. Cold Spring Harbor Conferences on Cell Proliferation *3*: 1343–1366.

Reid, B. J. and Hartwell, L. H., 1977: Regulation of mating in the cell cycle of *S. cereviseae*. J. Cell Biol. *75*: 355–365.

Rossmann, H. B., 1977: Cell surface enzymes: effects

on mitotic activity and cell division. Int. Rev. Cytol. *50*: 1–23.

Sakai, H., 1960: Studies on sulfhydryl groups during cell division of sea urchin egg. III. —SH groups of KCl-soluble proteins and their change during cleavage. J. Biophys. Biochem. Cytol. *8*: 609–615.

Scherbaum, O. H., 1960a: Possible sites of metabolic control during the induction of synchronous cell division. *In* Second Conference on Mechanism of Cell Division. Gross and Mazia, eds. Ann. N.Y. Acad. Sci. *90*: 565–579.

Scherbaum, O., 1960b: Synchronous division of microorganisms. Ann. Rev. Microbiol. *14*: 283–310.

Schmidt, R. R., 1974: Transcriptional and post-transcriptional control of enzyme levels in eukaryotic organisms. *In* Cell Cycle Control. Padilla, Cameron, and Zimmerman, eds. Academic Press, New York, pp. 201–233.

Scholander, P. F., Leivestad, H. and Sundnes, G., 1958: Cycling in the oxygen consumption of cleaving eggs. Exp. Cell Res. *15*: 501–511.

Schroeder, R. F., 1975: Dynamics of the contractile ring. *In* Molecules and Cell Movement. Inoué and Stephens, eds. Raven Press, New York, pp. 305–334.

Spoerl, E. and Looney, D., 1959: Synchronized budding of yeast cells following x-irradiation. Exp. Cell Res. *17*: 320–327.

Stiles, W. and Cocking, E. C., 1969: Principles of Plant Physiology. Methuen, London.

Subirana, J. A., 1968: Role of spindle microtubules in mitosis. J. Theor. Biol. *20*: 117–123.

Suhr-Jessen, P. B., Stewart, J. M. and Rasmussen, L., 1977: Timing and regulation of nuclear and cortical events in the life cycle of *Tetrahymena pyriformis*. J. Protozool. *24*: 299–303.

Swann, M. M., 1954: The control of cell division. *In* Seventh Colston Symposium: Recent Developments in Cell Physiolgy. Kitching, ed. Academic Press, New York, pp. 185–196.

Sweeney, B. M. and Hastings, J. W., 1958: Rhythmic cell division in populations of *Gonyaulax polyedra*. J. Protozool. *5*: 217–224.

Sylvan, B., Tobias, C. A., Malmgren, H., Otteson, R. and Thorell, B., 1959: Cyclic variations in the peptidase and catheptic activities of yeast cultures synthesized with respect to cell multiplication. Exp. Cell Res. *16*: 75–87.

Tamiya, H., 1964: Growth and cell division of *Chlorella*. *In* Synchrony in Cell Division and Growth. Zeuthen, ed. Interscience Publishers, New York, pp. 247–305.

Taylor, E. W., 1975: Some comments on the mechanism of mitosis. *In* Molecules and Cell Movement. Inoué and Stephens, eds. Raven Press, New York, pp. 1–2.

Tonnesen, T. and Andersen, H. A., 1977: Timing of tRNA and 5S rRNA gene replication in *Tetrahymena pyriformis*. Exp. Cell Res. *106*: 408–412.

Vago, C., ed., 1972: Invertebrate Tissue Culture. Academic Press, New York.

Watanabe, Y., 1965: Isolation of characterization of the division protein in *Tetrahymena pyriformis*. Exp. Cell Res. *39*: 443–452.

Wolpert, L., 1960: The mechanics and mechanism of cleavage. Int. Rev. Cytol. *10*: 163–216.

Wolpert, L., 1963: Some problems of cleavage in relation to the cell membrane. *In* Cell Growth and Cell Division. Harris, ed. Academic Press, New York, pp. 277–298.

Wood, H. N., Braun, C., Brandes, H. and Kende, H., 1969: Studies on the distribution and properties of a new class of cell division-promoting substances from higher plant species. Proc. Nat. Acad. Sci. U.S.A. *62*: 349–356.

Zeuthen, E., 1949: Oxygen consumption during mitosis; experiments on fertilized eggs of marine animals. Am. Nat. *83*: 303–322.

Zeuthen, E., 1958: Artificial and induced periodicity in living cells. Advances Biol. Med. Phys. *4*: 37–73.

Zeuthen, E., 1961: Cell division and protein synthesis. *In* Biological Structure and Function. Vol. 2. Goodwin and Linberg, eds. Academic Press, New York, pp. 537–548.

Zeuthen, E., 1964: The temperature-induced division synchrony in *Tetrahymena*. *In* Synchrony in Cell Division and Growth. Zeuthen, ed. Interscience Publishers, New York, pp. 99–158.

Zeuthen, E., 1971a: Recent developments in the studies of the *Tetrahymena* life cycle. Adv. Cell Biol. *2*: 111–152.

Zeuthen, E., 1971b: Synchrony in *Tetrahymena* by heat shocks spaced a normal cell generation apart. Exp. Cell Res. *68*: 49–60.

Zeuthen, E., 1974: A cellular model for repetitive and free running synchrony in *Tetrahymena* and *Saccharomyces*. *In* Cell Cycle Controls. Padilla, Cameron and Zimmerman, eds. Academic Press, New York, pp. 1–30.

Zeuthen, E., 1977: Studies on the cell cycle. *In* Regulatory Biology. Copeland and Marzluf, eds. Ohio State University, Columbus, Ohio, pp. 243–284.

Zeuthen, E. and Rasmussen, L., 1972: Synchronized cell division in protozoa. *In* Research in Protozoology. Vol. 4. Chen, ed. Pergamon Press, New York, pp. 9–145.

Zeuthen, E. and Scherbaum, O., 1954: Synchronous division in mass culture of the ciliate protozoan *Tetrahymena pyriformis,* as induced by temperature changes. *In* Recent Developments in Cell Physiology. Kitching, ed. Academic Press, New York, pp. 141–157.

Zimmerman, A. M., 1963: Chemical aspects of the isolated mitotic apparatus. *In* The Cell in Mitosis. Levine, ed. Academic Press, New York, pp. 159–179.

Zimmerman, A. M., Landau, J. V. and Marsland, D., 1957: Cell division: pressure-temperature analysis of effects of sulfhydryl reagents on cortical plasma gel structure and furrowing strength of dividing eggs (*Arbacia* and *Chaetopterus*). J. Cell Comp. Physiol. *49*: 395–435.

CHAPTER 27

ACCELERATING, RETARDING, AND BLOCKING CELL DIVISION

Many factors determine the rate of cell division. Some of these factors, temperature, for example, accelerate cell division to a maximum rate over one part of the viable range but retard and even block it over other parts of the range. Only a few factors are discussed in this chapter, special emphasis being given to poisons because of their use in the analysis of division mechanisms as well as in cancer therapy. Also emphasized is radiation because of its threat to the environment as a result of pollution of the stratosphere and proliferation of nuclear power plants. Because of the damage modern civilized man has caused, McHarg has called him a "planetary disease."*

ACCELERATION OF CELL DIVISION

Nutrients have a marked effect on division rate. *E. coli,* for example, in minimal medium consisting of one organic carbon source and a variety of salts divides about every 40 minutes at 37°C. Supplying amino acids and other organic compounds that the cells can use in growth reduces by half the division time at the same temperature (see Chapters 1 and 26 for details). Nutrition is an important factor in all cells, though the results are not always as clear-cut as in the example cited.

*"Man is an epidemic, multiplying at a superexponential rate, destroying the environment upon which he depends, and threatening his own extinction. . . . He treats the world as a storehouse existing for his delectation; he plunders, rapes, poisons, and kills this living system, the biosphere, in ignorance of its working and its fundamental value" (McHarg, 1971).

The genotype of a cell may determine division rate, inasmuch as clones of the same species may divide at different rates. This is true of bacterial biochemical mutants that synthesize a low concentration of an enzyme in a synthetic reaction series, creating a bottleneck that limits growth rate. Less clearly analyzed cases appear among the ciliate protozoans.

Temperature affects the rate of cell division in all organisms. The range of temperature tolerated varies among organisms; thermophiles live best in a high range, and temperatures normal to mesophiles often do not permit growth of thermophiles. At the other end of the scale are the cryophiles that thrive best at low temperatures (see Chapter 4). For each of these three types of organisms, temperatures at both the lower and higher ends of the viable range retard cell division, presumably by affecting enzyme structure (as discussed in Chapter 10).

The effect of temperature on division rate is illustrated in Figures 27.1 and 27.2 for two species of the ciliate *Blepharisma* (mesophiles). The division rate at low temperature may increase after prolonged acclimation to lower temperatures. For example, blepharismas transferred from cultures at 25°C will not divide at 10°C but after several months of acclimation at 13°C, they will divide at 10°C, as shown in Figure 27.3 (Giese, 1973). Acclimation in many organisms occurs by synthesis of isozymes with favorable properties for the particular temperature range (Hochachka and Somero, 1973).

It was once claimed that slight doses of ultraviolet radiations and x-rays accelerated cell division. No decisive data documenting such acceleration have substantiated this claim.

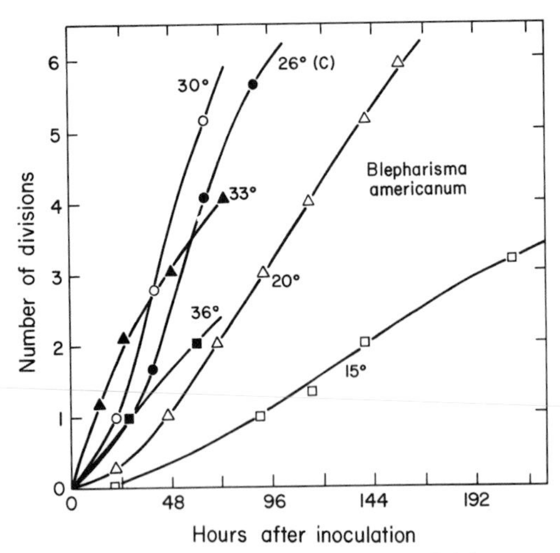

Figure 27.1. *Effect of temperature on division rate of the ciliate* Blepharisma americanum. *(From Giese, 1973: Blepharisma, The Biology of a Light-Sensitive Protozoan. Stanford University Press, Stanford, Calif. p. 100.)*

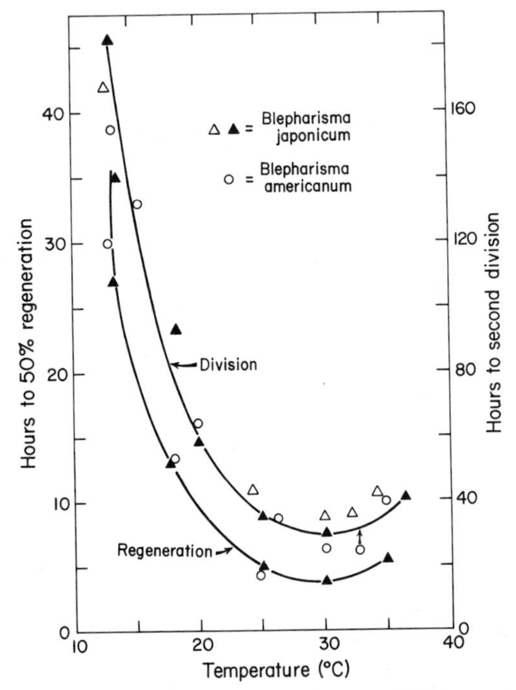

Figure 27.2. *Effect of temperature on division and regeneration after transection in two species of* Blepharisma. *(From Giese, 1973: Blepharisma, The Biology of a Light-Sensitive Protozoan. Stanford University Press, Stanford, Calif. p. 102.)*

Nevertheless, the detection by photomultipliers of short wavelength ultraviolet rays emitted by dividing yeast raises the question of whether experiments using the more critical techniques might not reveal stimulation (Gurwitsch *et al.*, 1974; Quickenden and Que Hee, 1976).

RETARDATION OF CELL DIVISION

From the preceding discussion it is obvious that poor nutritive conditions and low temperatures retard division. Discussion here centers primarily on two other factors, salinity of the medium and radiation damage.

Little is known of how salinity affects cell division, although changes in salinity occur in natural habitats. In freshwater pools salts are diluted after rainstorms and concentrated after dry spells. Most studies on freshwater organisms have documented viability and survival after transfer from one salinity to another (see Chapter 4). Some cells tolerate shifts from freshwater to seawater and vice versa (e.g., *Paramecium calkinsi*). Brackish-water organisms in estuarine habitats, where shifts in salinity are periodic with tidal surges, tolerate wide variations in salinity. However, the effect of salinity change on division rate of cells in these organisms does not appear to have been studied. No cells tolerate distilled water, since for normal membrane function salts are required in the cell environment and salts are readily lost from cells in distilled water. Some protozoa tolerate very low salt concentrations, voiding the water they take in by contractile vacuoles, which become more active with dilution and are suppressed in isosmotic solutions.

Cells probably have an optimum salt concentration for growth and cell division, although this has been little documented. The fresh water ciliate *Blepharisma* grows well in pond water (4.4×10^{-3} osm.) but it will grow even better at 11 to 21 times pond water concentration. Raising the concentration to 41 times pond water concentration kills all the cells, but if the cells are first acclimated to a concentration of 31 times pond water preceding transfer to this high concentration of salts, they grow and divide, but more slowly than do controls (Figure 27.4).

Marine invertebrates tolerate relatively little dilution, being isosmotic with seawater. Unless growing in estuaries, they are normally subjected to relatively small changes in salinity, even after heavy storms. Sea urchin eggs and sperm tolerate about a 10 percent increase or

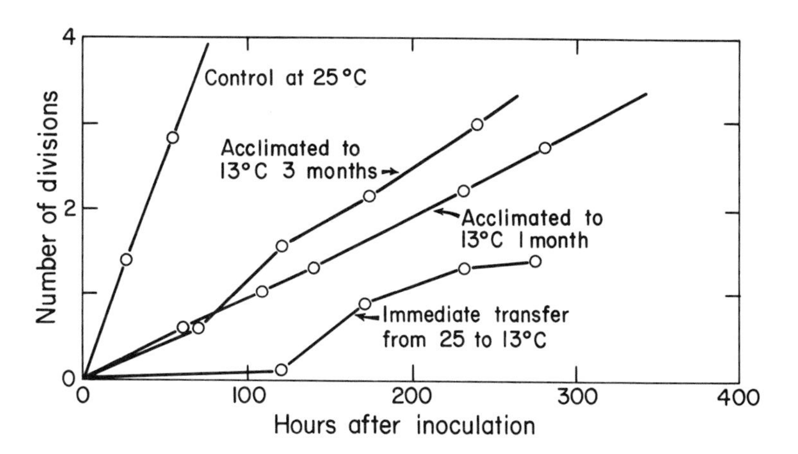

Figure 27.3. *Acclimation of* Blepharisma japonicum *to lower temperatures. Acclimation to higher temperatures was not successful. (From Giese, 1973: Blepharisma, The Biology of a Light-Sensitive Protozoan. Stanford University Press, Stanford, Calif. p. 105.)*

decrease in salt concentration, and cleavage occurs at the same rate as in controls in seawater, but larger changes in salt concentration are damaging and stop cleavage. To a small degree inert organic substances, for example, sucrose, compensate for low salt concentration (Giese and Farmanfarmaian, 1963).

Ultraviolet C and ultraviolet B radiation retard division in a variety of cells (Fig. 27.5). The major locus of action of ultraviolet radiation is DNA replication; it is likely that division is retarded by delayed replication. Since syntheses other than DNA are also affected by these radiations, their slowing may add to the retarding effect of ultraviolet radiation on eukaryotic cells.

It is interesting that individual irradiated ciliate cells have been kept alive for a month or more without dividing, although slowly increasing in volume because the syntheses other than DNA are less susceptible. Once DNA synthesis occurs again, irradiated cells divide at the same rate as controls. Irradiated bacteria may also grow into "spaghetti" forms without dividing for the same reason; on recovery they divide into cells of characteristic size.

X-irradiated cells in tissue culture about to undergo mitosis complete the process and then show a delay before the next division, but irradiated cells in earlier stages become quiescent. Recovery from small doses of x-rays is rapid, and on recovery the cells appear to divide more synchronously than controls because both the crop of cells recovering from radiation injury and those not damaged by the radiation reach the division stage at the same time.

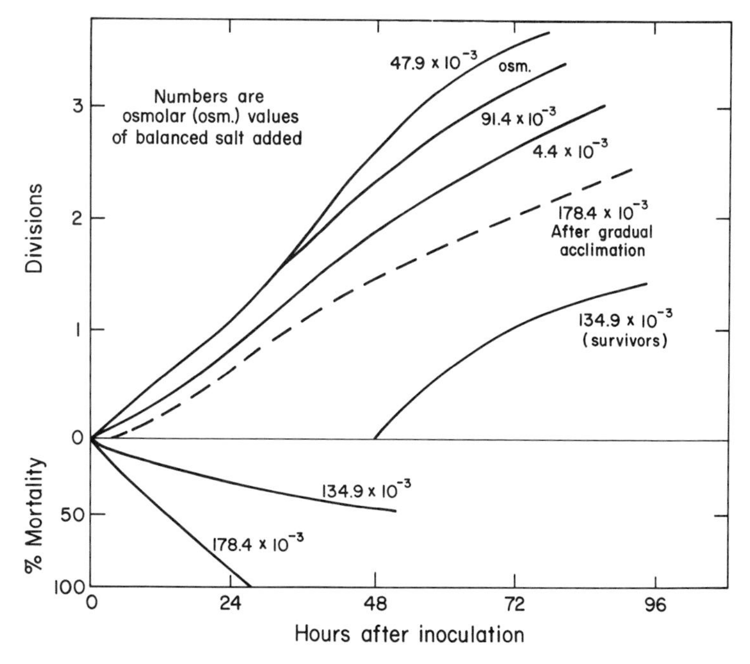

Figure 27.4. *Effect of salinity on division rate and acclimation to changes in salinity in* Blepharisma japonicum. *Pond water concentration is about 4.4×10^{-3} osm. and 11, 21, 32 and 40.5 × pond water concentration are given as 47.9, 91.4, 134.9 and 178.4 times 10^{-3} osm. (From Giese, 1973: Blepharisma, The Biology of a Light-Sensitive Protozoan. Stanford University Press, Stanford, Calif. p. 109.)*

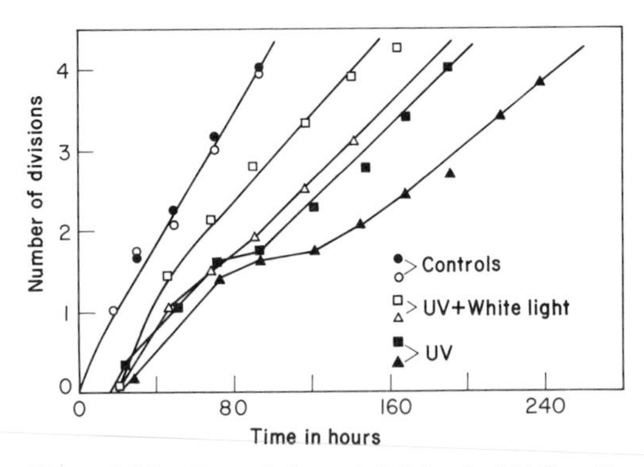

Figure 27.5. *Retardation of division by UV-C radiation (dose: squares 2475 ergs/mm² at wavelength 265 nm; triangles 3,700 ergs/mm² same wavelength). Note the high degree of photoreactivation by post-irradiation exposure to illumination by daylight fluorescent bulb. (From Giese, 1973: Blepharisma, The Biology of a Light-Sensitive Protozoan. Stanford University Press, Stanford, Calif. p. 295.)*

MEANS OF BLOCKING CELL DIVISION

Poisons

A great number of poisons block cell division, but only representative ones that help identify metabolic reactions or events in cell division are mentioned here (Table 27.1). The action of such poisons has been studied extensively because of their possible practical use in control of cancer. Much of the information cited here is from Zimmerman *et al.* (1973). No attempt is made to cover the vast amount of literature on the subject.

Many drugs interfere with interphase growth. For example, metabolic poisons that inhibit glycolysis (iodoacetic acid) or respiration (cyanide, azide, antimycin) reduce energy release and prolong G_1 or prevent division. Reagents that uncouple phosphorylation from energy-yielding reactions (e.g., dinitrophenol) prevent accumulation of high-energy compounds required for synthesis and cell division (Table 27.1).

Drugs that block nucleic acid synthesis stop cell division in different ways. Mitomycin C prevents cross linking of DNA strands. Azaserine blocks synthesis of purine nucleotides. Other drugs compete with some normal metabolite for the surface of an enzyme; for example, 5-fluorouracil and 5-fluorodeoxyuridine at relatively low concentrations interfere with DNA synthesis by inhibiting thymidine synthetase. (At higher concentrations these drugs also inhibit RNA synthesis.)

Actinomycin D at concentrations that have no effect on DNA synthesis blocks DNA-dependent RNA synthesis, probably by preventing mRNA production. Streptomycin inhibits tRNA and ribosome interaction. Tetracyclines prevent binding of tRNA onto the surface of both 70S and 80S ribosomes. Rifampycin blocks RNA polymerase.

The antibiotic chloramphenicol resembles and competes with phenylalanine on the surface of protein-synthesizing enzymes and thereby blocks protein synthesis, prolonging G_1 and even blocking cell division. Puromycin also interferes with protein synthesis by terminating the nascent peptide chain on both 70S and 80S ribosomes.

Such antagonists to folic acid (a B vitamin, see Table 10.5) as amethopterin and aminopterin affect a specific step in the chromosome separation between metaphase and anaphase in chick fibroblasts. Therefore, they have been classed as specific metaphase inhibitors (Jacobson and Webb, 1952). Amino acid analogues (e.g., p-fluorophenylalanine) greatly prolong the duration of metaphase without affecting the duration of the rest of the cell cycle (Sisken and Iwasaki, 1969).

The mitotic inhibitors colchicine, podophyllum, chloracetophenone, vinblastine, and vincristine prevent polymerization of tubulin into microtubules for the mitotic spindle without stopping growth or duplication of various organelles in the cell. Chromosomes are duplicated but the spindle fibers are disoriented, resulting in polyploidy; in some species polyploidy persists even after the poison is removed. They do not interfere with making or breaking disulfide bonds between proteins forming the spindle fibers but seem to interfere with the secondary bonding in the formation of an oriented and symmetrical mitotic spindle. When colchicine is removed tubules re-form, but they remain disoriented (Wunderlich and Speth, 1970).

Alkylating agents (e.g., those that replace an H atom of a compound with an alkyl group, such as the methyl or ethyl) do not affect growth but prevent cell division. Such alkylating agents are the nitrogen mustards used as mutagens and in the treatment of some cancers, and methane sulfonates, e.g., Myleran (methane sulfonic acid, tetramethylene ester) used in treatment of chronic myeloid leukemia. The exact manner in which they inhibit cell division is not known, but some alkylating agents cause chromosome breakage or failure of normal chromosome movement much like x-radiation (Mazia and Gontcharoff, 1964; Hall, 1976). This could presumably result from translocation of parts of two chromosomes, resulting in two kinetochores (chromosome constrictions for spindle fiber attachment) on a single chromosome. Consequently the chromo-

TABLE 27.1. ACTION OF REPRESENTATIVE DRUGS ON THE CELL CYCLE*

Stage in Cycle Affected	*Drug*	*Mechanism of Action*
Interphase growth	Iodoacetate	Inhibits glycolysis
	Antimycin, cyanide, azide	Affect cytochrome oxidase, decrease respiration
	Oligomycin and dinitrophenol	Interfere with phosphorylation
DNA synthesis	Mitomycin C	Interferes with cross-linking DNA strands
	Azaserine	Blocks synthesis of purine nucleotides
	5-Fluorouracil	Interferes with thymine synthetase action
	Various anticancer drugs†	Various actions
RNA synthesis	Actinomycin D	Inhibits DNA-dependent RNA synthesis
	Streptomycin	Inhibits tRNA and ribosome interaction
	Tetracyclines	Inhibits binding of tRNA on ribosomes (70S, 80S)
Protein synthesis	Rifampycin	Blocks RNA polymerase
	Chloramphenicol and erythromycin	Inhibit peptide bond formation and translocation of 70S ribosomes
	Cyclohexamide	Specifically inhibits 80S ribosomes
	Puromycin	Prematurely terminates growing peptide chain on 70S or 80S ribosomes
Metaphase	Aminopterin	Prevents separation of metaphase chromosomes
Mitotic apparatus	Colchicine Podophyllum Chloroacetophenone Vinblastine Vincristine	Prevent polymerization of tubulin into microtubules
Chromosome integrity	Alkylating agents (Myleran-methane sulfonate)	Unknown
Duplication of centrioles	Mercaptoethanol	Unknown
Cytokinesis	Cytochalasin B	Interferes with formation of contractile band of microfilaments
Cell wall formation	Penicillin Cycloserine	Block murein synthesis and cell wall formation
Induction of mitosis	Concanavalin Phytohemagglutinins	Lymphocyte mitosis induced by unknown mechanism (cells removed from G_0 or G_1

*Most data from Zimmerman *et al.*, 1973. For microtubule inhibitors see Wilson, 1975.

†Cyclohexamide, 5-fluorodeoxyuridine, hydroxyurea, cytosine arabinoside, and hydroxycortisone might be added to the list. Cozzarelli (1977) states that naladixic acid, arylhydrazinopyrimidines, arabinosyl adenine, edeines, and neocarbozinostatin are also DNA-synthesis inhibitors. Novobiocin and phosphonoacetic acid inhibit both RNA and DNA synthesis, as does azaguanine. Cantell (1978) states that interferon inhibits cell division but he has not determined where it acts.

somes would be torn in half during anaphase. Abnormal disjunction could also occur from failure of spindle attachment to kinetochores. These agents also appear to affect spindle fiber formation, generally reducing the cytoplasmic viscosity. Esterification of carboxyl groups of proteins and combinations with nucleic acids have also been found.

Another interesting poison the study of which has given considerable insight into mitosis is mercaptoethanol, which prevents the duplication of mitotic centers (centrioles) of the animal cell. In animal cells centrioles always appear in pairs. Each member of a pair must duplicate before it can act as a division center for the cell. The precise way in which mercaptoethanol interferes with cell division depends upon just when it is applied during the division cycle. Exposure of sea urchin eggs to mercaptoethanol affects the mitotic centers, which cannot duplicate until the cells are removed from the poison. Cell division is abnormal because mercaptoethanol specifically affects only duplication centrioles, not their separation or movement in the cell (Mazia *et al.*, 1960). Mercaptoethanol also blocks and delays division in *Tetrahymena* (Mazia and Zeuthen, 1966).

The mold metabolite cytochalasin B, while permitting nuclear division to proceed, inhibits cytokinesis in cells by preventing microfilament assembly (Estensen, 1971). The equatorial contractile band fails to reach full development, preventing cleavage (Carter, 1972).

Penicillin and cycloserine prevent bacterial cell division by interfering with synthesis of murein (peptidoglycan) required for cell walls. The bacteria develop as protoplasts (without cell walls) and cannot divide, perhaps because the chromosome attached to the cell membrane cannot separate. Since animal cells do not have cell walls, their division is unaffected by penicillin.

A search has been made for other naturally occurring substances that either stimulate or retard cell division. Plant concanavalin and hemagglutinins are mentioned in Table 27.1 as mitogens in specific instances. Fibroblast-stimulating substances of proteinaceous nature have been isolated and a nerve growth factor has been known for several years. Mediation of bradykinin-induced stimulation of mitotic activity in rat thymocytes by cyclic AMP through initiation of DNA synthesis has been reported (Whitfield *et al.*, 1970). Chalones are discussed in Chapter 25 as tissue-specific cell division inhibitors. Konyshev (1976) has reviewed the occurrence of mitogens and the evidence for both mitogenic and retarding effects of various tissue extracts. At present the subject is controversial, with conflicting results reported by many investigators.

Excessively High and Low Temperatures

The temperature range over which cell division will occur depends on the normal habitat of the species. Organisms are classified according to how they respond to temperature: *thermophiles* (heat lovers) grow at much higher temperatures than do *mesophiles* (medium-heat lovers) and often do poorly at temperatures favorable for mesophiles; in other words, they require higher temperatures for growth. Mesophiles, the commonest organisms on earth, have lower temperature requirements than thermophiles and variable thermal ranges related to their environmental origins. *Cryophiles* (cold-lovers) grow at lower temperatures than other organisms tolerate during growing; some organisms might tolerate these temperatures only in dormancy. Figure 27.6 relates temperature and division rate in the three types of organisms. Each type grows over a limited range of temperatures; on either side they are dormant or dead. For each species an increase in temperature accelerates cell divi-

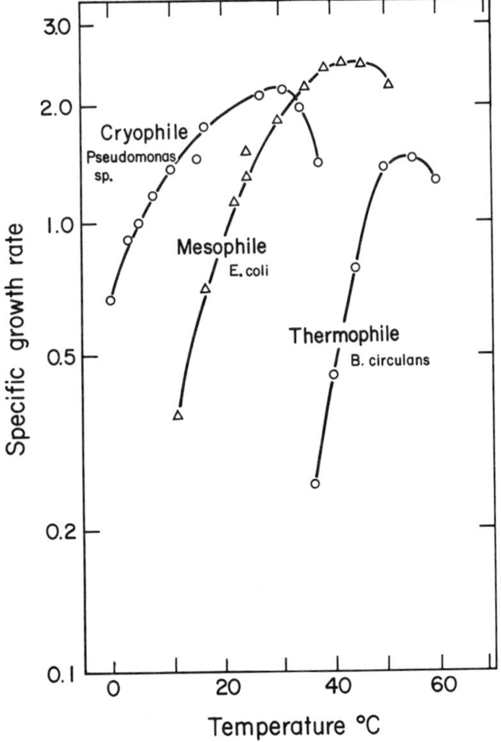

Figure 27.6. *Effect of temperature on specific growth rates (measured by relative optical densities) of a cryophilic pseudomonad bacterium, a mesophilic* E. coli *and a thermophilic* Bacillus circulans. *(After Farrell and Rose, 1967:* In Thermobiology. *Rose, ed. Academic Press, New York.)*

sion up to a point (Fig. 27.6), indicating a high activation energy requirement for some key reaction.

As discussed in Chapter 10, temperature affects the reaction rate in two ways, first by affecting the amount of kinetic energy available for activating molecules and second by affecting the enzyme configuration. Either too high or too low a temperature may induce an enzyme configuration unfavorable for its catalytic action. It is likely that the limit of temperature tolerance by a species is largely determined by the effect of the temperature on enzymic configuration, as considered in Chapters 4 and 10.

Ultraviolet Radiation

Ultraviolet radiation in sunlight beyond our atmosphere covers a span from 15 to 390 nm, but at the surface of the earth the span is from 286 to 390 nm as a result of absorption of the shorter wavelengths by oxygen and ozone in the higher atmosphere (see Figures 2.12 to 2.14). Ultraviolet B (286 to 320 nm) kills cells and causes sunburn and damage to human skin more readily than does ultraviolet A (320–390 nm). Doses of ultraviolet A orders of magnitude larger than ultraviolet B are required to cause such effects. Ultraviolet C radiation (15–280 nm) is present in sunlight beyond our atmosphere and is available as artificial light from various lamp sources. Ultraviolet C radiation is even more damaging than the other two types but is interesting for analytical photochemical and photobiological experiments and is extensively used in various laboratories (Giese, 1976).

Ultraviolet radiation is strongly absorbed by nucleic acids and to a lesser extent by proteins (see Fig. 3.18), but because RNA and proteins (including enzymes) are highly redundant, whereas DNA is usually present in duplicate only, DNA is most vulnerable to ultraviolet radiation. Furthermore, it carries the information by which RNA and proteins are synthesized; therefore, its destruction stops synthesis of other macromolecules.

Ultraviolet radiation induces several types of damage in DNA, as shown in Figure 27.7, the dominant change being pyrimidine dimer formation between neighboring pyrimidines on a DNA strand. The commonest dimer is the *thymine dimer,* but smaller numbers of cytosine-to-cytosine and cytosine-to-thymine dimers are also formed. When the DNA polymerase attempts to synthesize a DNA strand complementary to the one altered by exposure to ultraviolet radiation, it stops at the defect, probably because of steric hindrance in the molecule. If the defect is repaired, DNA

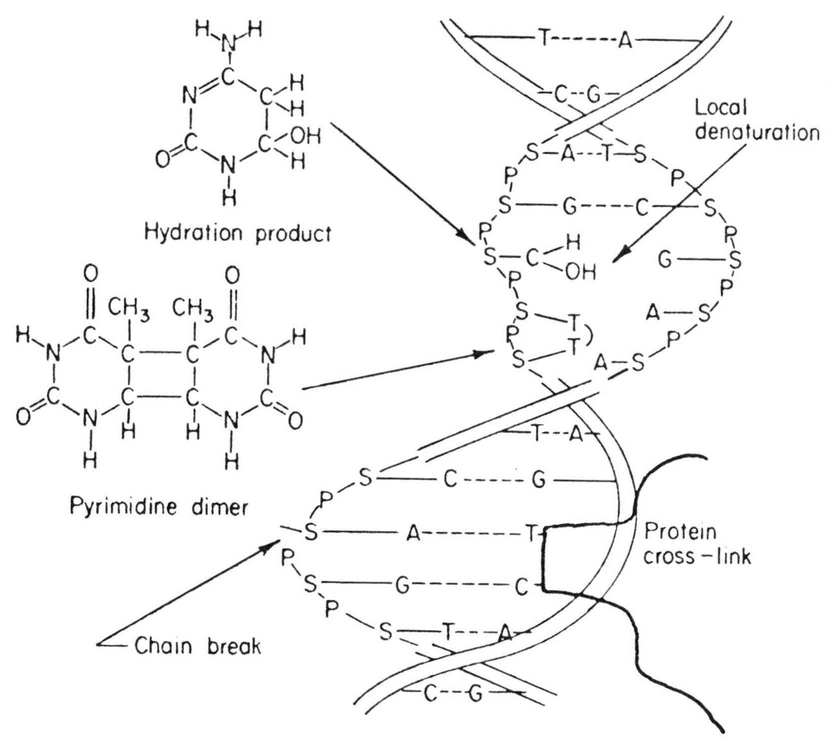

Figure 27.7. *Various possible alterations in DNA extracted from UV-B or UV-C radiation-treated cells. (After Deering, 1962: Sci. Am. (Dec.) 207: 135.)*

synthesis can be resumed. If replication cannot be accomplished the damaged cell is unable to divide; therefore, it is reproductively dead, although metabolic reactions may continue for a while and the cell may grow in size.

Excision repair and postreplication repair of DNA, often called dark repair mechanisms because they can occur in the dark, were discussed in Chapter 15 as mechanisms for correction of mistakes (or damage) in DNA. Both excision and postreplication repair will occur in the dark; they will also occur in light. Another type of repair, the first to be discovered (Kelner, 1949), is *photoreactivation*. To a considerable extent, damage to DNA in cells affected by absorption of ultraviolet B and ultraviolet C radiation can be reversed by simultaneous or subsequent exposure of the cells to blue-violet and ultraviolet A radiation. Photoreactivation occurs in all cells with photoenzymes. In studies with extracted DNA, this enzyme was found to attach to pyrimidine dimers of the ultraviolet radiation-treated DNA (Fig. 27.8). Neither altered DNA nor photoenzyme alone absorbs radiation in the photoreactivating part of the spectrum, but the complex of the two does, resulting in the splitting (monomerization) of the dimer. This returns DNA to its original (native) condition. Because the photoenzyme attaches specifically only to dimers in DNA, it fails to reverse ultraviolet radiation–induced damage to RNA or to proteins. Since photoreactivation repairs up to 90 percent of DNA damage and only pyrimidine dimers are photoreversed, it appears that the major change in DNA induced by ultraviolet radiation is the production of thymine dimers.

Photoreactivation has been demonstrated in many kinds of microorganisms, plant cells, and animal cells. Exceptions occur among mutants of various species, which may show various degrees of deficiency in photoreactivation. The only cells that seem to lack photoreactivation are found in placental mammals. Photoenzymes have been demonstrated in mammalian white blood cells and fibroblasts, using removal of thymine dimers in extracted DNA as test object for the studies, but photoreactivation has not yet been demonstrated in the cells themselves.

Photoreactivation is a less general type of repair mechanism than dark repair because it is specific for repair of pyrimidine dimers. Other types of damage to the DNA molecule (Fig. 27.7) induced by ultraviolet radiation are not subject to photoreversal. Excision repair and postreplication repair correct dimers and ultraviolet-induced defects in DNA other than dimers as well as damage inflicted on DNA by ionizing radiation, chemicals, and natural errors in replication.

Like excision and postreplication repair, photoreactivation is not complete. It produces what has been called a *dose-reduction effect;*

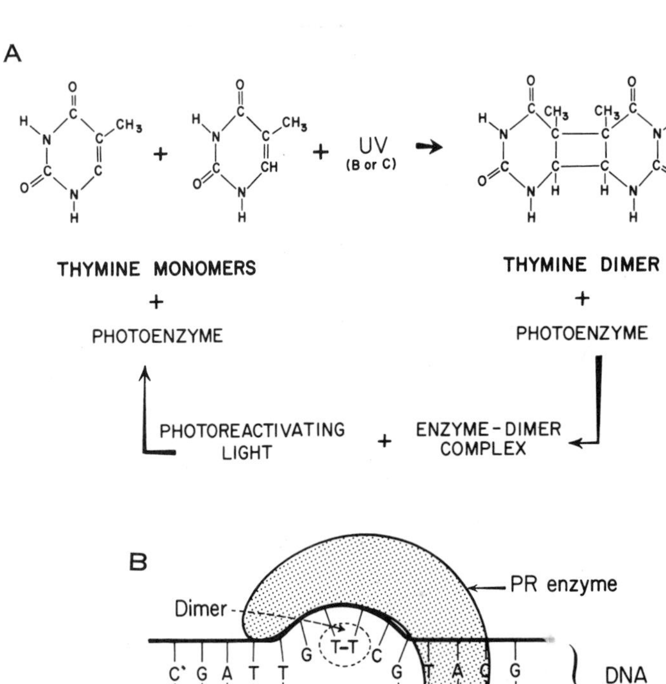

Figure 27.8. A, *Thymine dimer formation between neighboring thymine residues in DNA after exposure to UV-B or UV-C radiation.* B, *Possible structural distortion in DNA by dimer formation providing steric hindrance to DNA-polymerase action. The photoreactivating enzyme is shown attached to the distorted region. Adsorption of light by the enzyme-DNA complex splits the dimer restoring DNA to its native state. (After Hanawalt, 1972: Endeavour 31: 84.)*

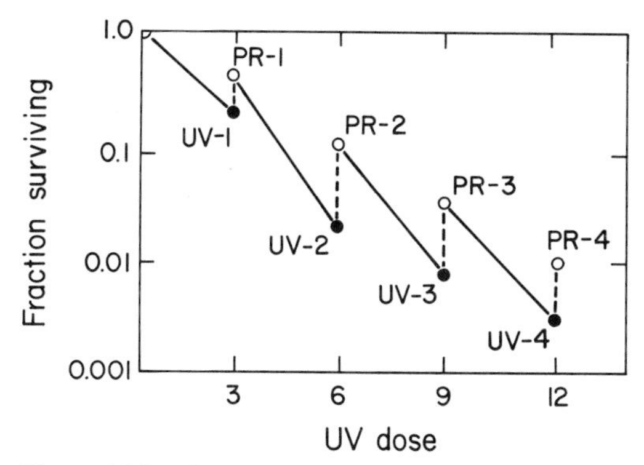

Figure 27.9. *Four cycles of ultraviolet inactivation and photoreactivation in* E. coli. *UV-1 to UV-4 cycles of irradiation, PR-1 to PR-4, cycles of photoreactivation. Note graduated decline in surviving fraction of cells. Dose is in relative units. (After Nishiwaki, 1954: Yokohama Med. Bull. 4: 21.)*

that is, the population of cells photoreactivated after ultraviolet irradiation is one that appears as if it had been exposed to a lesser dose of ultraviolet radiation than actually was the case. If the same population of cells is irradiated, photoreactivated, and again irradiated and photoreactivated in a series of experiments, fewer cells survive in each step of the series until essentially all of them are killed (Fig. 27.9).

Since photoreactivation is so prevalent in the living world, it probably represents a system of repair that entered into the genome of living things at a very early stage of evolution. Perhaps it was essential to survival when cells were bombarded with the entire span of ultraviolet radiation in sunlight before an effective ozone layer had developed in the earth's atmosphere (see Chapter 2).

The usefulness of repair systems becomes quite evident when the resistance to ultraviolet radiation of mutants defective in one or another DNA repair system are compared with the wild type possessing them. In *E. coli,* for example, a mutant deficient in excision repair is very much more sensitive to ultraviolet radiation than the wild type. Almost equally sensitive is a mutant deficient in postreplication repair. A double mutant lacking both excision and postreplication repair is so sensitive that a small dose of radiation, scarcely detectable with the wild-type population, essentially eliminates the population (Fig. 27.10*A*).

Repair systems are of equal importance to mammalian cells. Fibroblasts from humans with the disease xeroderma pigmentosum are also highly sensitive to ultraviolet radiation, although in this case a detailed genetic analysis has not yet been made because of the paucity of cases. Several genotypes, perhaps representing different degrees of deficiency in repair, have been described, some also having defects other than the sensitivity to ultraviolet radiation. Fibroblast cells from such individuals are very much more sensitive to ultraviolet radiation than fibroblasts from normal individuals, as seen in Figure 27.10*B*. It will be

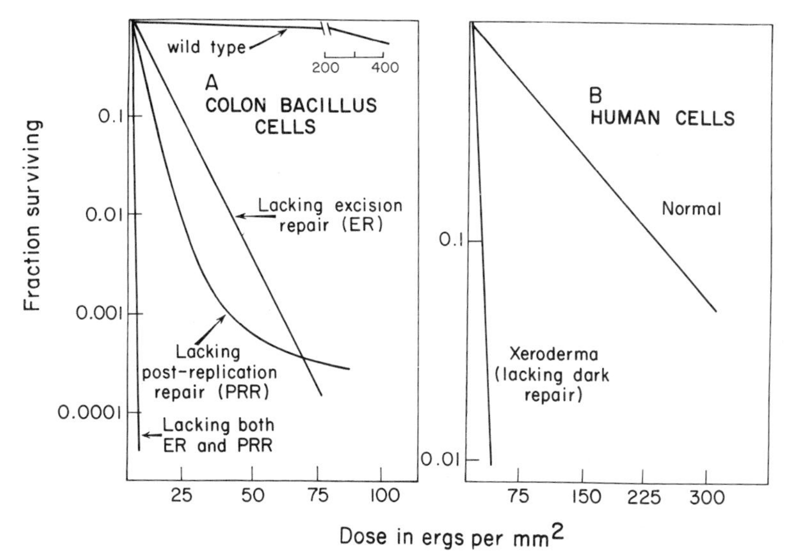

Figure 27.10. *Importance of repair mechanisms to survival of cells, as shown in* A *by the increased sensitivity of* E. coli *with deficient excision repair or with deficient post-replication repair, and the extraordinary sensitivity of a double mutant lacking both repair systems. (After Howard-Flanders and Boyce, 1966: Radiation Res. Suppl. p. 156.)* B, *High sensitivity of human fibroblasts from a patient with the disease xeroderma pigmentosum in which repair systems are deficient as compared to fibroblasts from a normal individual. (After Takebe, 1974, Furuyama, Miki and Kondo, 1974:* In *Sunlight and Man. Fitzpatrick, ed. Tokyo University Press, Tokyo, Japan, p. 109.)*

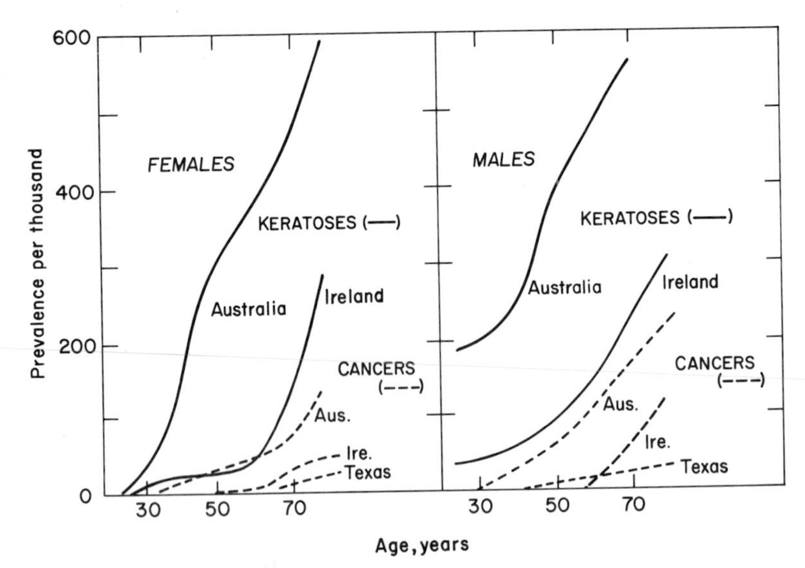

Figure 27.11. Increased prevalence of human skin cancer and keratoses with age and for populations from different latitudes but very similar heritage. Note the much greater prevalence in low-latitude Australia compared to high-latitude Ireland, in both of which Celtic peoples predominate. The low prevalence in Texas is a reflection of the more deeply pigmented skin of the population of different genetic stock. (After Urbach, Rose, and Bonnem, 1972: In Environment and Cancer, Williams and Wilkins, Baltimore, p. 367.)

recalled that fusion of cells from individuals with different types of xeroderma may result in reconstitution of a repair system (Chapter 15). The skin of individuals with this disease is correspondingly sensitive to sunlight. Even short exposures to sunlight damage the skin, resulting in blotchy pigmentation and dry, coarse skin; repeated exposures lead to papular growths of epidermal cells (keratoses) and finally cancers. Most individuals with this affliction die of cancer in their teens.

It is evident that systems for repair of ultraviolet radiation damage must be coded for in the DNA of cells of organisms or they would not survive exposure to sunlight. In cells of wild-type organisms repair systems are extraordinarily effective. Furthermore, photoreactivation and dark repair coexist in many types of cells, with the exceptions noted. However, it should be emphasized that repair is probably never quite complete. It is sufficient to reverse most of the damage from normal exposure to sunlight. But even human skin can be afflicted by overexposure to sunlight, and the skin of normal individuals with little protective pigment exposed in sunny regions of the earth may become badly damaged and aged in appearance. The skin of such individuals at age 25 has been found to resemble that of a person in the late seventies. Furthermore, skin cancers, the commonest cancers of mankind, are much more common on light-skinned individuals living in highly sunny climates than in those in less sunny ones (Figure 27.11) (Ambrose, 1975; Blum, 1976).

Ultraviolet radiation is used extensively to induce mutations in microorganisms. Mutation probably is a consequence of improper repair of ultraviolet radiation damage to DNA. By induction of mutations ultraviolet radiation probably played an important part in the early evolution of life. While most mutations are deleterious, some must have been advantageous in the early history of the earth. Once higher forms of life became dominant ultraviolet radiation probably was no longer important for induction of mutations because it is absorbed superficially by the surface layers of the skin; therefore, it cannot reach the gonadal cells.*

Ionizing Radiation

Ionization on reaction with matter is caused by x-rays, γ-rays (from radioactive elements in the earth's crust and from space), and particulate radiations. Particulate radiations are protons (hydrogen nuclei, mass about 2000 times an electron), neutrons (uncharged particles of the mass of hydrogen nuclei), α-particles (helium nuclei, of 2 protons and 2 neutrons), negative π mesons (mass 273 times an electron), and β particles (speedy electrons). In the cell, both organic molecules and water are ionized; in the presence of oxygen this leads to peroxidation (Fig. 27.12). Peroxidation changes the specific biological activity and

*Even visible light can damage cells and produce mutations when a photosensitizer such as a natural pigment or one acquired from the environment transmits the absorbed light energy to cell constituents (Giese, 1976). In the presence of oxygen, the dye induces photochemical formation of the powerful oxidant, singlet oxygen, which reacts with cell constituents, including DNA in some cases. Singlet oxygen is also formed in the atmosphere by absorption of short wavelength ultraviolet radiation by molecules and accounts for some aspects of atmospheric photochemistry, not discussed here. The subject is complex and the interested reader is referred to the literature (Ogryzlo, 1970; Wilson and Hastings, 1970; Johnston, 1971, 1972).

Biochemical effects of ionizing radiation

$$H_2O \xrightarrow{\text{x-ray}} H_2O^+ + e^- \text{ (ionized molecule)}$$

$$H_2O^+ \longrightarrow H^+ + OH\cdot \text{ (hydroxyl radical)}$$

$$RH \xrightarrow{\text{x-ray}} RH^+ + e^- \text{ (ionized molecule)}$$

$$RH^+ \longrightarrow H^+ + R\cdot \text{ (organic radical)}$$

In the absence of oxygen and in presence of reducing compounds, e.g. —SH

$$R\cdot + \text{—SH} \longrightarrow RH + \text{—S}\cdot \text{ (chemical recovery)}$$

In presence of oxygen, characteristic of tissues in the body:

$$R\cdot + O_2 \longrightarrow RO_2^{\cdot} \longrightarrow ROOH \text{ (peroxide formation)}$$

Peroxidation of the DNA leads to base damage and single strand breaks

Figure 27.12. *Diagrammatic representation of peroxidation of biological systems by x-irradiation. DNA (serving in place of R above) is the most vulnerable cell chemical because it is least redundant, usually present in no more than two copies per cell; furthermore, it contains the information for syntheses of all the other cell compounds. Both base damage and single-strand breaks are repairable by dark repair mechanisms. Double-strand breaks are produced by two single-strand breaks opposite one another or by radiation such as alpha particles that produce a high-density ionization trail (high linear energy transfer), as shown in Figure 27.13. Double-strand breaks are generally not repairable. Hydroxyl radicals produced in water are strong oxidants and may indirectly damage DNA. Tissue generally consists of 75 to 80 percent water. (By courtesy of J. M. Brown.)*

structure of macromolecules. The relative biological effectiveness of various ionizing radiations varies largely with the density of the ionization trail; a dense trail reaching its objective is more damaging than a sparse one (Fig. 27.13) (Alexander, 1965; Sparrow *et al.*, 1970).

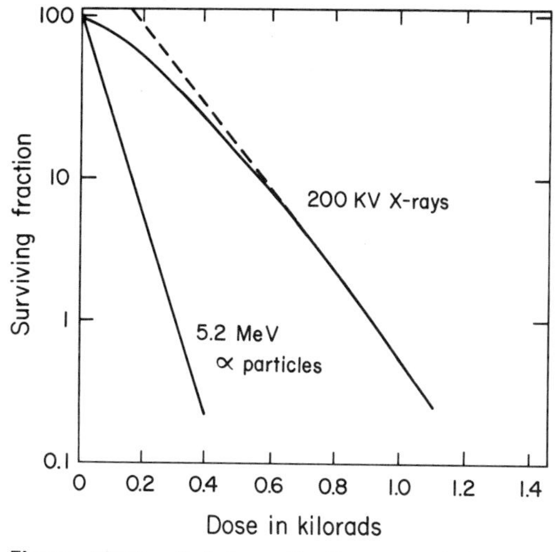

Figure 27.13. *Relative effectiveness of ionizing radiation with high- and low-density ionization trails from alpha particles (5.2 MeV alpha particles) and 200 KV x-rays, respectively. The straight line relation for alpha particles indicates that even one "hit" on the chromosome is effective, whereas for 200 KV x-rays several "hits" are required; the number of hits required can be determined by extrapolating from the straight portion of the curve to the ordinate. (After Berendsen* et al., *1963: Rad. Res.* 18: *106.)*

Ionizing radiation effects are closely related to the amount of DNA present in the cell's genome (Fig. 27.14). When the radiation's ionizing trail crosses a DNA molecule a single- or double-strand break may occur. A single-strand break is easily repaired by the mechanisms discussed in Chapter 15; a double-strand break blocks cell division and results in reproductive death.

Blocking of cell division by x-radiation was observed soon after the discovery of x-rays, and this led to the use of ionizing radiations in the treatment of cancer. Almost without exception dividing cells are most sensitive to x-rays during a short period just prior to the onset of and during mitosis (Fig. 27.15) (Sinclair, 1968).

Synthesis of macromolecules continues in a *radiation-sterilized* cell (a cell that will eventually die), but cell division is often blocked. When this occurs, the result is the production of giant cells. Unbalanced growth of this type is ultimately followed by death, but the undivided cell may live for a considerable time. During this time, various nutrients leak from the radiation-sterilized cells. These nutrients proved effective as a growth medium for normal unirradiated cells isolated on a mat of radiation-sterilized cells in the first demonstration of the growth of single mammalian cells *in vitro* (Puck, 1972). Single cells give rise to a colony of their descendants, called a clone, a population likely to be genetically homogeneous.

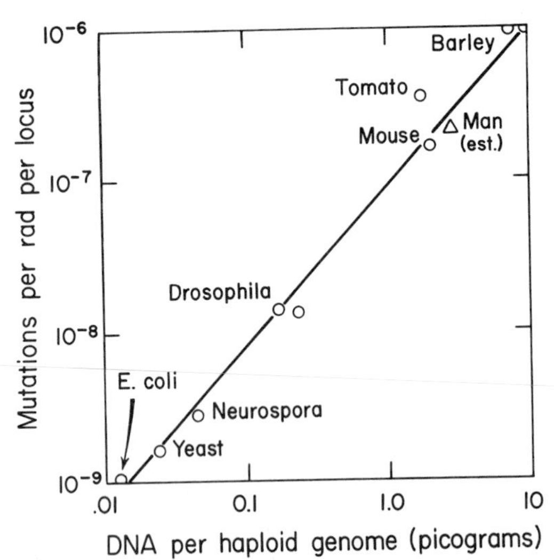

Figure 27.14. *Relation between quantity of DNA in a cell's genome and effects of ionizing radiation. The value for humans is estimated from DNA content. (After Abramson* et al., *1973: Nature* 245: *460.)*

Cells in tissue culture given 10 to 100 times the reproductive death dose of radiation may grow up to a millimeter in diameter. The cytoplasmic increase in a giant cell indicates that many metabolic activities continue in a reproductively dead cell, including uptake of nutrients, liberation of energy, and macromolecular synthesis. Such radiation-sterilized cells even permit an infecting virus to replicate in much the same manner as in normal cells. It is apparent that a cell's reproductive capacity is infinitely more sensitive to radiations than are its other functions (Puck, 1972).

Clonal colonies are currently grown from single human cells in a synthetic medium. Being genetically homogeneous, such cells are useful in the study of radiation effects. Experiments show that, provided they are tested in the same physiological state (actively growing and dividing), mammalian cells of different tissue types have the same degree of resistance to ionizing radiation, as measured by reproductive death. A dose of 100–200 rads is generally sufficient to kill about half the cells in any of the tissue cell types.*

However, radiosensitivity of cells in the same physiological state but in different species varies markedly. For example, the dose inducing reproductive death in half the cell population is 100 to 200 rads for mammalian cells, 10,000 for *E. coli,* 30,000 for yeast, 100,000 for amoeba and 300,000 for *Paramecium.*

THE THREAT TO LIFE FROM AN INCREASED RADIATION LOAD ON EARTH

Life on earth tolerates the *radiation load* coming from radioactive elements in the earth's crust, the cosmic high-energy radiation coming from space and the infrared, visible, and ultraviolet wavelengths in sunlight. Damage is sustained, but cells repair the radiation damage; this facility makes them appear to be resistant to the natural radiation load. Damage from ionizing radiation is repaired by excision and postreplication gap repair, while damage from ultraviolet B radiations is repaired by similar dark repair and by photoreactivation. At current natural dosages of radiation, repair

*The rad, the basic ionizing radiation unit, is an amount of absorbed energy equal to 100 ergs per gram of tissue. Most human beings would be killed by a single whole-body dose of 400 rads. Each year each person in the United States receives approximately one-tenth of a rad from background irradiation (from cosmic rays and from radioactive materials in soil, rocks, and food).

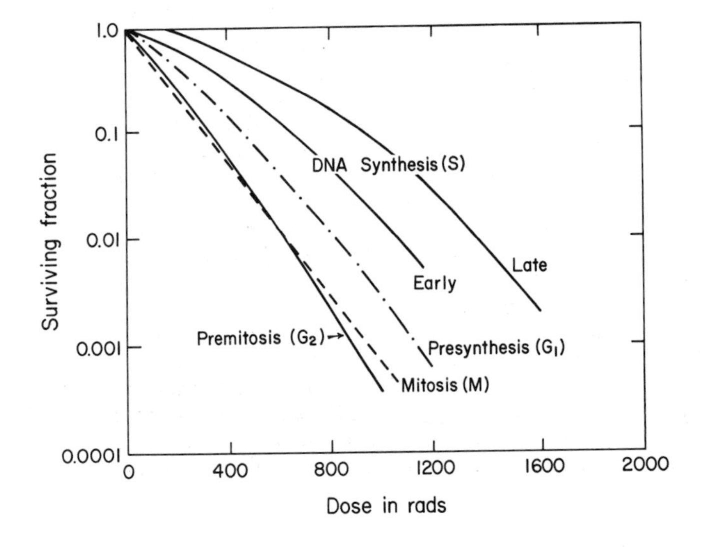

Figure 27.15. *Relative sensitivity of mammalian cells to x-rays in tissue culture at various phases in the cell cycle. (After Sinclair, 1965: Rad. Res.* 33: *33.)*

balances damage. Mutants with deficient repair mechanisms are probably eliminated by natural selection.

Civilized man continues to assail and pollute the natural environment, resulting in an increased radiation load on life. It is probable that both ionizing radiation and ultraviolet B radiation in the environment will increase as a result of these practices (Smith, 1973; Handler, 1975; Giese, 1976). Our assault on nature in the quest for more and more energy has led to development of nuclear power plants, necessary for the long-touted "peaceful" use of atomic energy (Fig. 27.16). At the time of this writing, some 80 nuclear plants are producing electricity in the United States alone, and a plan has been developed to add about 1000 plants by the year 2000 (Reynolds, 1976). The low accident record of these plants currently allays fears that release of their radioactive wastes into the air can adversely affect life. However, the possibilities for accidents are vastly increased as a result of more chances for human error that accompany the escalating number of nuclear power plants. These plants also represent more numerous targets for aberrant individuals or groups in our world's unstable societies.

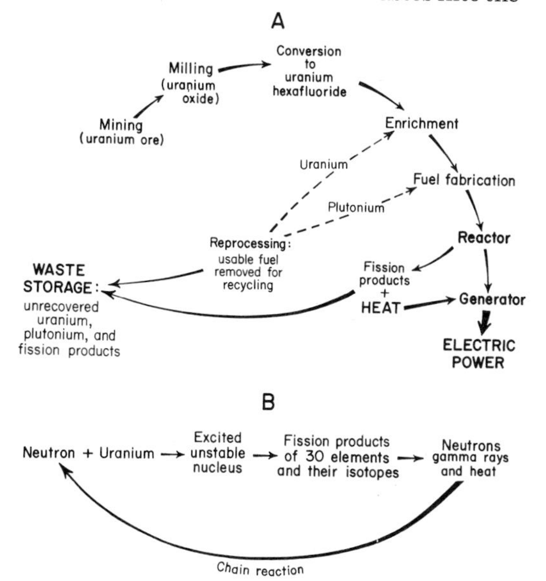

Figure 27.16. *A, Energy from fission of Uranium 235. Many initial pairs of products are possible from U-235 fission (atoms of atomic weights 99 and 133 + neutrons, for example); some 200 fission products have been identified from both the original fissions and the fission products from initial fission pairs. Among them is strontium 90, which is deposited in bone and shells and can be used to measure contamination of the atmosphere with fission products from atomic bomb tests. Concentrated in the bone, it can give rise to leukemia and bone tumors. Cesium 137 resembles potassium in its properties and concentrates in soft tissues. Other radioactive elements are also concentrated by organisms. They accumulate in animals at the top of food chains, including the human consumer of their flesh. (Data from Reynolds, 1976, p. 71.) B, The nuclear fuel cycle. At present the fission waste products are stored in the reactors but fuel reprocessing is being studied with a view to more efficient use of the fuel. The dotted lines in A show the proposed recycling. (Data from Reynolds, 1976.)*

The problem of disposal of radioactive wastes has never been satisfactorily resolved. It is necessary that none of these wastes be directly released into the atmosphere or the hydrosphere, because the lifetime of the most highly toxic of the wastes is about 500 years (about fifteen human generations!) and for others as much as 500,000 years. The wastes must be sunk in the ocean or buried in the geologically stable sands of deserts. That the buried wastes can escape into the environment is shown by the fact that nearly half a million gallons of toxic wastes have leaked from the steel tanks in the deserts in the state of Washington near Hanford (Brown, 1976). Medvedev (1976) has described the tragic consequences to the people in and around Blagoveshensk in the Ural Mountains of the Soviet Union from a nuclear dump explosion of volcanic violence. Strong winds scattered the radioactive waste for hundreds of miles and a large population was affected. Hundreds of people were killed outright and thousands received large doses of radiation bound to be very harmful in time. Details of these latent effects are not likely to be released to the world.

We are assured again and again by nuclear physicists and politicians that nuclear power plants are safe; it is planned that the wastes will be fused in vitreous medium and buried so deeply that they will not reach ground water level (Cohen, 1976). However, who can predict the geological changes that might occur in thousands or hundreds of thousands of years?

Building nuclear plants in underdeveloped countries whose technicians and workers lack proper training increases the chance of accidents. The radioactive products in a nuclear power plant permit any country to develop an atomic bomb as India did from reactors supplied by Canada for peaceful purposes on a foreign aid program (Hall, 1976). Plutonium, an element produced artificially from uranium, is the most toxic of all radioactive elements. It is used as a source of efficient radioactive breeder reactions for multiplying the energy obtained from the original uranium source (Healy, 1976). It is supposed to be the perfect element for making an atomic bomb with relatively little scientific know-how, although not all scientists agree (Cohen, 1976).

In a sense we are bringing radioactive elements to the earth's surface out of its bowels; there they could not harm us because of the rocky shields between them and our cells. At the surface we can not only concentrate them but, by the breeder reaction, produce the radioactive element plutonium, which is more toxic than any in the crust of the earth.*

An atomic war would be the most devastating result of the proliferation of nuclear plants. According to some experts (Drell and von Heppel, 1976; Alfren *et al.*, 1976), no safety in nuclear war can be expected, even in a "limited" nuclear war proposed by military strategists. An atomic war could greatly increase radioactivity over wide areas of the world; the scope of this effect would depend only on the war's extent.

In his search for luxury and gadgets, modern civilized man may have released an uncontrollable atomic demon that may change the environment of our cells for thousands of years to come. The demand for energy could be reduced by an estimated 33 to 45 percent if homes and buildings were better insulated, and were designed by architects to offset the need for air-conditioning and heating fuels, and if industrial operations were redesigned for energy conservation (Brown, 1976). A savings in energy could be realized also from banning such energy-wasting devices as large luxury automobiles and campers. New sources such as geothermal energy and energy from sunlight, wind, and tides could be developed for general use. At much less cost in energy per capita than in the United States, Sweden enjoys a standard of living equal to that in the United States by attention to these possibilities. Such steps could reduce the need for more atomic power plants (Schumacher, 1973; Biró, 1978), although a modest increase in the number of such plants is perhaps unavoidable.

As we have seen, ionizing radiation damages cells and sterilizes them; in larger organisms chronic radiation damage may lead to leukemia and other types of cancer, and other changes. In addition, the radiations we produce will probably come back to us in the form of induced mutations, a genetic load for many future generations. Since mutations are generally harmful and only the very lowest level of radiation may be without mutational effect (Fig. 27.17; National Academy of Sciences report, 1972; see also Alper, 1974; Novitski, 1976), the prospect is not a pretty one.

Another possible change in the radiation environment is the type of ultraviolet radiation reaching the earth's surface. In Chapter 2 it was pointed out that our stratospheric ozone shield developed only after the release of oxygen into the air following the evolution of photosynthesis in plants using water as the hydrogen donor. Presumably, life appeared on land only after the ozone shield made its development possible. In other words, without the ozone layer, intense ultraviolet B and ultraviolet C radiation from sunlight would have killed life on land (Giese, 1976). The ozone layer is therefore an essential part of our environment. Several developments threaten the ozone layer: supersonic transports (SST's), aerosols, and atomic bombs (Fig. 27.18).

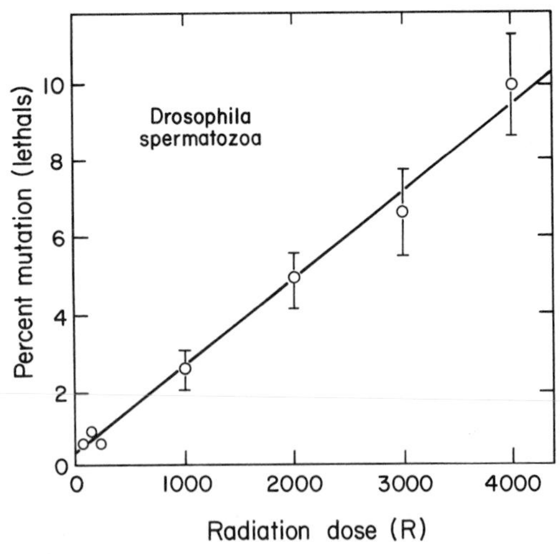

Figure 27.17. *Relationship between ionizing radiation dose (in roentgens, R) administered to* Drosophila *sperm and the number of mutations. (After Spencer and Stern, 1948: Genetics* 33: *43.) For newer data, see* The Effect on Populations of Exposure to Low Levels of Ionizing Radiation, *U.S. National Academy and National Research Council Report, 1972. Washington, D.C.*

All planes emit nitric oxide that results from heat-induced combination of oxygen and nitrogen in the engines. In the troposphere the nitric oxides react with other substances to form some constituents of smog, but eventually they are diluted by winds in the turbulent troposphere and washed out by rains. However, when planes fly in the calmer, rainless stratosphere, removal of nitric oxide is slower because of the inversion block that prevents turbulent exchange between troposphere and stratosphere. The higher in the stratosphere nitric oxide is emitted the slower is its removal. Yet supersonic transports fly in the stratosphere because lower airlanes are not always available and because greater speed can be built up in the stratosphere, which has less air friction. At present, SST's are not a threat. However, should their numbers increase by the hundreds or thousands they may seriously contaminate the stratosphere, unless new motors are designed in which engine combustion temperatures can be lowered. According to aeronautical en-

*Plutonium was probably made in a "natural fission reactor" 2×10^6 years ago in South Africa when a rich deposit of uranium began to operate as a nuclear reactor. Such events are probably rare (Cowan, 1976).

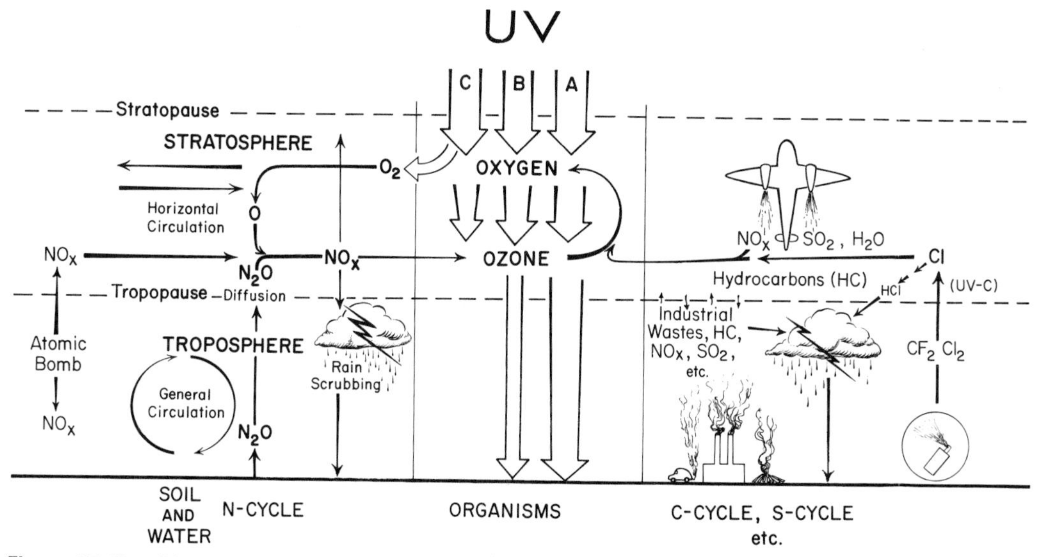

Figure 27.18. Diagram of the probable relationships between the earth's surface, the atmosphere, and sunlight, showing the effects of pollutants on the ozone layer. (From Giese, 1976: Living With Our Sun's Ultraviolet Rays. Plenum Press, New York, p. 173.)

gineers, developing this type of motor is feasible but expensive.

The ozone concentration in the stratospheric shield is controlled by the nitrogen cycle on earth (see Fig. 2.9). Nitric oxide formed by short wavelength ultraviolet rays from nitrous oxide diffusing naturally from the troposphere into the stratosphere decomposes a good proportion of the ozone formed photochemically. This is the equilibrium condition to which life has become adjusted. If we add more nitric oxide to the stratosphere, more ozone is decomposed; therein lies the possible threat from supersonic transports (Booker, 1975; Nachtwey, 1975).

The gases used in aerosols and as refrigerants are halogenated methanes; an example is dichlorodifluoromethane (CF_2Cl_2). It was synthesized and achieved wide use because of its inertness; it affects neither animals nor plants. Microorganisms have not been found to decom-

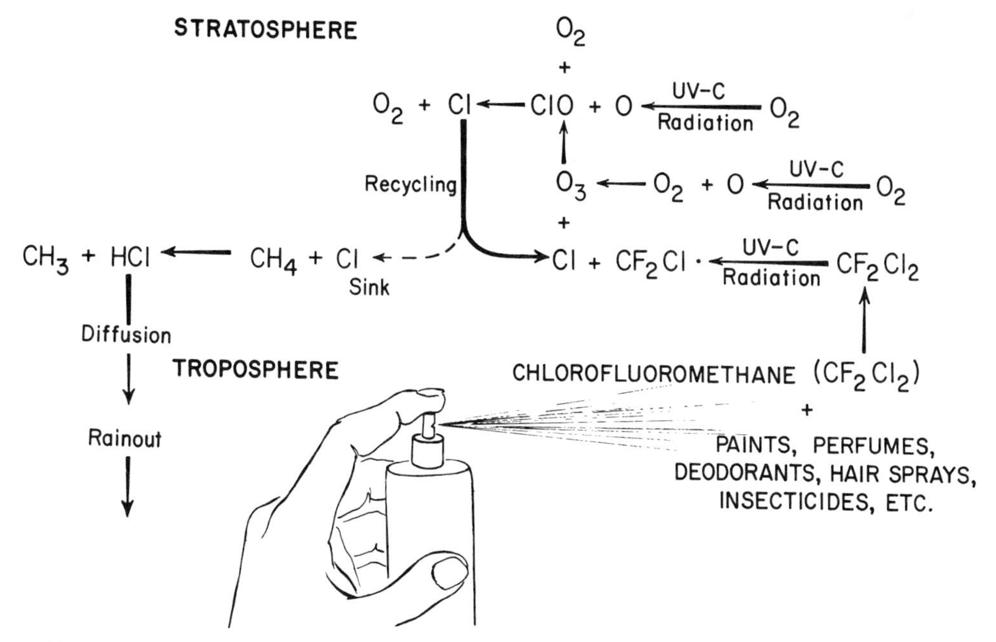

Figure 27.19. Presumed destructive action of chlorofluoromethanes used as propellants in spray cans and refrigerants on the decomposition of ozone in the stratosphere. (From Giese, 1976: Living With Our Sun's Ultraviolet Rays. Plenum Press, New York, p. 171.)

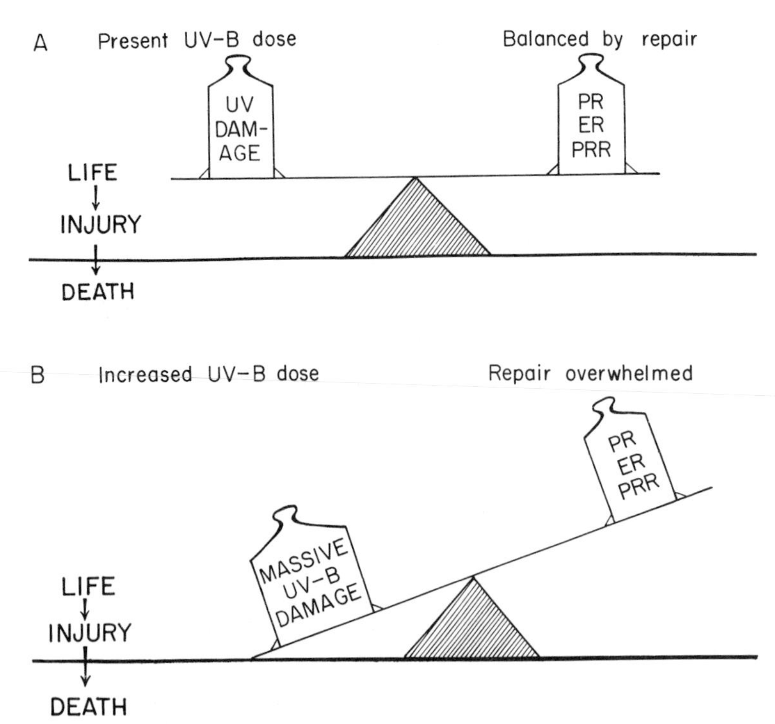

Figure 27.20. Life represented schematically in uneasy balance between damage from ultraviolet radiation and repair. At the present UV-B levels, photoreactivation (PR), excision repair (ER), and post-replication repair (PRR) keep cells viable, counteracting continuous damage from sunlight. Massive UV-B radiation damage resulting from damage to the ozone layer might overwhelm the repair systems because repair is never quite complete (for example, see Fig. 27.9). (From Giese, 1976: Living With Our Sun's Ultraviolet Rays. Plenum Press, New York, p. 174.)

pose it. Consequently all the halogenated methanes produced since their initial synthesis and manufacture still reside in our atmosphere; only a small fraction has risen to the stratosphere. In the stratosphere the chlorofluoromethanes are decomposed by short wavelength ultraviolet radiation, resulting in the release of chlorine free-radicals (Fig. 27.19). In laboratory tests chlorine was found to decompose ozone more actively than nitric oxide and like nitric oxide it recycles, acting as a catalyst. While such action in the stratosphere has not been completely proved to occur, no sinks for chlorofluoromethanes have been found in the atmosphere other than its decomposition in the stratosphere by short wavelength ultraviolet radiation. As already mentioned, the danger exists of great stratospheric ozone destruction by the chlorofluoromethanes when the load currently in the troposphere finally reaches the stratosphere (Stever and Peterson, 1975). Ozone monitoring by satellite has been going on since December 1976. It is now evident that the threat to the ozone layer from chlorofluoromethanes is far greater than the threat from SST's (Hidalgo, 1978) (Fig. 27.19).

Still another threat to the ozone layer is the possibility of an atomic war. To take an extreme case, should the two major atomic powers release half the atomic weapons in their arsenals, it has been estimated that not only would about a hundred million people perish but that the soil of those nations attacked would become incapable of producing crops for some 25 years (Handler, 1975). Possibly even worse, it is estimated that the ozone layer would be reduced by 50 to 70 percent. This reduction, more striking in the northern hemisphere, would occur over the world. We do not know the effect of an increase in damaging ultraviolet radiations on farm crops or the oceanic plankton crops, which serve as the ultimate sources of food for fish, shellfish, and other food sources of humans. The prospect is not pleasant for the millions of people who depend upon seafood for their protein. We must remember that repair of radiation damage is never complete; excessive radiation may overwhelm the cell's repair systems (Fig. 27.20).

We cannot escape the fact that organisms are made of cells. Cells are damaged by ultraviolet B and ultraviolet C radiations. Chronic damage to human skin cells leads to accelerated skin aging and to the development of skin cancers, the commonest cancers in humans. It is estimated that for every percent increase in ultraviolet B rays skin cancers will increase by 2 percent or more, probably mainly by lowering the age at which the cancer appears. Fortunately, a human's reproductive cells are well protected, because ultraviolet rays penetrate the body only superficially. But what the increase in ultraviolet rays does not damage is damaged by penetrating ionizing radiations resulting from radioactive contamination.*

Whatever happens, life on earth need not

*Omitted here is discussion of possible thermal changes in climate resulting from a perturbed ozone layer, because their effects have not been clarified (Bartholic, 1975; Ramanathan, 1975).

cease because of human perturbation of the environment—what is destroyed will ultimately be replaced, if not by the same organisms, then by others. The earth may have suffered irradiation catastrophes in the past when its polarity reversed (Wolfendale, 1978). Nor is atmospheric alteration by any human act permanent. Once the pollution ceases, the atmosphere will eventually return to its original state. The only problem is that time is required for reestablishment of equilibria and that in terms of human life span the time may be long. For reequilibration of the ozone layer, years to a century may be needed; for decline in atomic radiations, perhaps thousands of years. Some type of cell-based life will continue, regardless. Whether *Homo sapiens* will continue depends much upon our awareness and decisions in the coming years.

LITERATURE CITED AND GENERAL REFERENCES

Alexander, P., 1960: Radiation-imitating chemicals. Sci. Am. (Jun.) *202*: 99–108.

Alexander, P., 1965: Atomic Radiation and Life. Revised Ed. Penguin Books, New York.

Alfren, H., Barnaby, F. and York, H., 1976: Big threat from little bombs. New Science *70*: 516–517.

Alper, T., ed., 1974: Cell Survival after Low Doses of Radiation: Theoretical and Clinical Implications. John Wiley & Sons, New York.

Ambrose, E. J. and Roe, F. J., eds., 1975: The Biology of Cancer. 2nd Ed. Halsted Press, New York.

At Issue: Fluorocarbons, 1975: Kaiser Center, Oakland, Calif.

Bartholic, J., ed., 1975: Impacts of Climatic Change on the Biosphere. CIAP Monograph 5. Part 2. Climatic Effects. U.S. Department of Transportation, Washington, D.C.

Bebbington, W. P., 1976: Reprocessing of nuclear fuels. Sci. Am. (Dec.) *235*: 30–41.

Biró, A., ed., 1978: The village: a last resort. Mazingira—The World Forum for Environment. Issue No. 6, 97 pp. Pergamon Press, Oxford.

Blum, H. F., 1976: Ultraviolet radiation and skin cancer in mice and man. Photochem. Photobiol. *124*: 249–254.

Booker, H. C., ed., 1975: Environmental Impact of Stratospheric Flight. Biological and Climatic Effects of Aircraft Emissions in the Stratosphere. National Academy of Sciences, Washington, D.C.

Brown, J. M., 1976: Health, safety and social issues of nuclear power and the nuclear initiative. *In* The California Nuclear Initiative. Reynolds, ed. Institute for Energy Studies, Stanford University, Stanford, Calif., pp. 127-201.

Cantell, K., 1978: Towards the clinical use of interferon. Endeavour *2* NS: 27–30.

Carter, L. J., 1976: Solar and geothermal energy: new competition for the atom. Science *186*: 811–813.

Carter, S. B., 1972: The cytochalasins as research tools in cytology. Endeavour *31*: 77–82.

Cohen, B. L., 1976: Impacts of nuclear energy industry on human health and safety. Am. Sci. *64*: 550–560.

Cowan, G. A., 1976: A natural fission reactor. Sci. Am. (Jul.) *235*: 36–47.

Cozzarelli, N. R., 1977: The mechanism of action of inhibitors of DNA synthesis. Ann. Rev. Biochem. *46*: 641–668.

Deysson, G., 1964: Antimitotic substances. Int. Rev. Cytol. *24*: 99–148.

Drell, S. D. and von Heppel, F., 1976: Limited nuclear war. Sci. Am. (Nov.) *235*: 27–37.

Eggleton, A., Cox, T. and Derwont, D., 1976: Will chlorofluoromethanes really affect the ozone shield? New Scientist *70*: 402–403.

Epstein, J. H., 1970: Ultraviolet carcinogenesis. *In* Photophysiology. Vol. 5. Giese, ed. Academic Press, New York, pp. 235–273.

Estensen, R. D., 1971: Cytochalasin B. I. Effect on cytokinesis of Novikoff hepatoma cells. Proc. Soc. Exp. Biol. Med. *136*: 1256–1260.

Finlayson, B. J. and Pitts, J. N., Jr., 1976: Photochemistry of the polluted atmosphere. Science *192*: 111–119.

Franklin, T. J. and Snow, G. A., 1975: Biochemistry of antimicrobial action. Halsted Press, New York.

Fusion Power—a step in the right direction, 1976: New Scientist *71*: 497.

Giese, A. C., 1973: Blepharisma: The Biology of a Light-Sensitive Protozoan. Stanford University Press, Stanford, Calif.

Giese, A. C., 1976: Living With Our Sun's Ultraviolet Rays. Plenum Press, New York.

Giese, A. C. and Farmanfarmaian, A., 1963: The resistance of the purple sea urchin to osmotic stress. Biol. Bull. *124*: 182–192.

Gnevyshev, M. N. and Ol′, A. I., eds., 1978: Effects of Solar Activity on the Earth's Atmosphere and Biosphere. Halsted Press, Somerset, N.J.

Golding, E. W., 1976: The Generation of Electricity by Wind Power. Halsted Press, New York.

Grobecker, A. J., Coroneti, S. C. and Cannon, R. H. Jr., 1974: The Effects of Stratospheric Pollution. Final Report. U.S. Department of Transportation Climatic Impact Assessment Program. Washington, D.C.

Gurwitsch, A. A., Eremeyev, V. F. and Karabchievsky, Y. A., 1974: Energy Bases of the Mitogenetic Radiation and its Registration on Photoelectron Multiplier. Meditsina, Moscow.

Hall, E. J., 1976: Radiation and Life. Pergamon Press, New York.

Hampson, J., 1974: Photochemical war on the atmosphere. Nature *250*: 189–191.

Hanawalt, P. C. and Setlow, R. B., eds., 1975: IV. Molecular Mechanisms for the Repair of DNA. Plenum Press, New York.

Handler, P., ed., 1975: Long-term Worldwide Effects of Multiple Nuclear Weapon Detonations. National Academy of Sciences, Washington, D.C.

Healy, J. W., ed., 1976: Plutonium—Health Implications for Man. Pergamon Press, New York.

Hidalgo, H., 1978: Understanding anthropogenic effects on ultraviolet radiation and climate. IEEE

transactions on Geoscience Electronics. *GE-16*: 4–22.
Hochachka, P. and Somero, G., 1973: Strategies of Biochemical Adaptation. W. B. Saunders Co., Philadelphia.
Jacobson, W. and Webb, M., 1952: Nucleoproteins and cell division. Endeavour *11*: 200–207.
Johnson, W. H., 1976: Social impact of pollution control legislation. Science *192*: 629–631.
Johnston, H., 1971: Reduction of stratospheric ozone by nitrogen oxide catalysts from supersonic transport exhaust. Science *173*: 517–522.
Johnston, H., 1972: Newly recognized vital nitrogen cycle. Proc. Nat. Acad. Sci. U.S.A. *69*: 2369–2372.
Kaplan, F. M., 1978: Enhanced-radiation weapons. Sci. Am. (May) *238*: 44–51.
Kelner, A., 1949: Effect of visible light on recovery of *Streptomyces griseus* conidia from UV irradiation. Proc. Nat. Acad. Sci. U.S.A. *35*: 73–79.
Kende, H., 1971: The cytokinins. Int. Rev. Cytol. *31*: 301–338.
Kihlman, B., 1966: Actions of Chemicals on Dividing Cells. Prentice-Hall, Englewood Cliffs, N.J.
Konyshev, V. A., 1976: Chemical nature and systematization of substances regulating animal tissue growth. Int. Rev. Cytol. *47*: 195–224.
Leach, J. F., Pingstone, A. R., Han, R. A., Ensell, F. J. and Burton, J. L., 1976: Interrelation of atmospheric ozone and cholecalciferol production in man. Aviation Space and Environmental Medicine *47*: 630–633.
Legault, A. and Lindsey, G., eds., 1976: The Dynamics of the Nuclear Balance. Cornell University Press, Ithaca, N.Y.
Lovelock, J. E., 1975: Natural halogens in the air and in the sea. Nature *256*: 193–194.
Lozzio, B. B., Lozzio, C. B., Bamberger, E. E. and Lair, S. V., 1975: Regulators of cell division: endogenous mitotic inhibition to mammalian cells. Int. Rev. Cytol. *42*: 1–47.
McHarg, I. L., 1971: Man: Planetary Disease. Morrison Memorial Lecture. Agricultural Research Service, U.S. Department of Agriculture, Washington, D.C., p. 1.
Mazia, D., 1970: Regulatory mechanisms of cell division. Fed. Proc. *29*: 1245–1247.
Mazia, D. and Gontcharoff, M., 1964: The mitotic behavior of chromosomes in echinoderm eggs following incorporation of bromodeoxyuridine. Exp. Cell Res. *35*: 14–25.
Mazia, D. and Zeuthen, E., 1966: Blockage and delay of cell division in synchronous populations of *Tetrahymena* by mercaptoethanol (monothioethylene glycol). Compt. Rend. Trav. Lab. Carlsberg *35*: 341–361.
Mazia, D., Harris, P. J. and Bibring, T., 1960: The multiplicity of mitotic centers and the time course of their duplication and separation. J. Biochem. Biophys. Cytol. *7*: 1–20.
Medvedev, Z., 1976: Two decades of dissidence. New Scientist *172*: 264–267.
Molina, M. J. and Rowland, F. S., 1974: Stratospheric sink for chlorofluoromethanes: chlorine atom-catalyzed destruction of ozone. Nature *249*: 810–812.
Nachtwey, D. S., Caldwell, M. M. and Biggs, H. R., eds., 1975. Impacts of Climatic Change on the Biosphere. Climatic Impact Assessment Program (CIAP) Monograph 5. Part 1. UV Radiation Effects. U.S. Department of Transportation, Washington, D.C.
Novitski, E., 1976: The enigma of radiation effects in *Drosophila*. Science *194*: 1387–1390.
Ogryzlo, E. A., 1970: Physical properties of singlet oxygen. *In* Photophysiology. Giese, ed. Vol. 5, pp. 35–47.
Ozone monitoring takes off, 1976: New Scientist *72*: 374.
Parrish, J. A., Anderson, R. R., Urbach, F. and Pitts, D., 1978: UV-A. Human Biologic Responses to UV Radiation with Special Reference to Longwave UV. Plenum Press, New York.
Pizarello, D. J. and Witcofski, R. L., 1975: Basic Radiation Biology. 2nd Ed. Lea and Febiger, Philadelphia.
Pollard, W. G., 1976: The long-range prospects for solar derived fuels. Am. Sci. *64*: 509–513.
Puck, T. T., 1960: Radiation and the human cell. Sci. Am. (Apr.) *202*: 142–153.
Puck, T. T., 1972: The Mammalian Cell as a Microorganism. Holden-Day, New York.
Quickenden, T. I. and Que Hee, S. S., 1976: The spectral distribution of the luminescence emitted during growth of the yeast *Saccharomyces cereviseae* and its relationship to mitogenetic radiation. Photochem. Photobiol. *23*: 201–204.
Ramanathan, V., 1975: Greenhouse effect due to chlorofluorocarbons: climatic implications. Science *190*: 50–51.
Reynolds, W. D., ed., 1976: The California Nuclear Initiative Institute for Energy Studies. Stanford University, Stanford, Calif.
Rickinson, A. B., 1970: The effects of low concentrations of actinomycin D on the progress of cells through the cell cycle. Cell Tissue Kinet. *3*: 335–347.
Rochlin, G. I., 1977: Nuclear waste disposal: two social criteria. Science *195*: 23–31.
Schipper, L. and Lichtenberg, A. L., 1976: Efficient energy use and well-being: the Swedish example. Science *194*: 1001–1013.
Schumacher, E. F., 1973: Small Is Beautiful: Economics As If People Mattered. Harper and Row, New York.
Science, Issue on Energy. 1974: Vol. *184*: 331–386.
Scientific American issue on Ionizing Radiation, Sept. 1959. (Dated but still useful for a general view for a number of topics.)
Sinclair, W. K., 1968: Cyclic X-ray response in mammalian cells *in vitro*. Rad. Res. *33*: 620–643.
Sisken, J. E. and Iwasaki, T., 1969: The effects of some amino acid analogs on mitosis and the cell cycle. Exp. Cell Res. *55*: 161–167.
Smith, K. C., ed., 1973: Biological Impacts of Increased Intensities of Solar Ultraviolet Radiation. Nat. Acad. Sci. National Acad. Engineer., Washington, D.C.
Smith, K. C., 1974: The cellular repair of radiation damage. *In* Sunlight and Man. Fitzpatrick, ed. Tokyo University Press, Tokyo, Japan, pp. 57–66, 67–77.
Sparrow, A. H., Nauman, C. H., Donnelly, G. M., Wells, D. L. and Baker, D. G., 1970: Radioresistivities of selected amphibians in relation to their nuclear and chromosome volumes. Rad. Res. *42*: 353–371.
Stever, H. G. and Peterson, R. W., eds., 1975: Fluorocarbons and the Environment. Report of

Federal Task Force on Inadvertent Modification of the Stratosphere. Council on Environmental Quality. U.S. Govt. Printing Office, Washington, D.C.
Szent-Györgyi, A., 1965: Cell division and cancer. Science *149*: 34–37.
Szent-Györgyi, A., Egyud, L. G. and McLaughlin, J. A., 1967: Keto-aldehydes and cell division. Science *155*: 539–541.
Tanooka, H., 1978: Thermorestoration of mutagenic radiation damage in bacterial spores. Science *200*: 1493–1494.
Vendryes, G. A., 1977: Superphenix: a full-scale breeder reactor. Sci. Am. (Mar.) *236*: 26–35.
Wang, W. C., Yung, Y. L., Lacis, A. A., Mo, T. and Hansen, J. E., 1976: Greenhouse effects due to man-made perturbations of trace gases. Science *194*: 685–690.
Whitfield, J. F., MacManus, J. P. and Gillan, D. J., 1970: Cyclic AMP mediation of bradykinin-induced stimulation of mitotic activity and DNA synthesis in thymocytes. Proc. Soc. Exp. Biol. Med. *133*: 1270–1274.
Wilson, L., 1975: Action of drugs on microtubules. Life Sci. *17*: 303–310.
Wilson, T. and Hastings, J. W., 1970: Chemical and biological aspects of singlet excited molecular oxygen. *In* Photophysiology. Giese, ed. Academic Press, New York, Vol. 5, pp. 49–95.
Wolfendale, A., 1978: Cosmic rays and ancient catastrophes. New Scientist *79*: 634–636.
Wunderlich, F. and Speth, V., 1970: Antimitotic agents and macronuclear division of ciliates. IV. Reassembly of microtubules in macronuclei of *Tetrahymena* adapting to colchicine. Protoplasma *70*: 139–152.
Zimmerman, A. M., 1973: Drugs and the Cell Cycle. Academic Press, New York.

INDEX